HANDBOOK OF MODERN ION BEAM MATERIALS ANALYSIS

HANDBOOK OF MODERN ION BEAM MATERIALS ANALYSIS

Editors:
Joseph R. Tesmer • Michael Nastasi

Contributing Editors:
J. Charles Barbour • Carl J. Maggiore • James W. Mayer

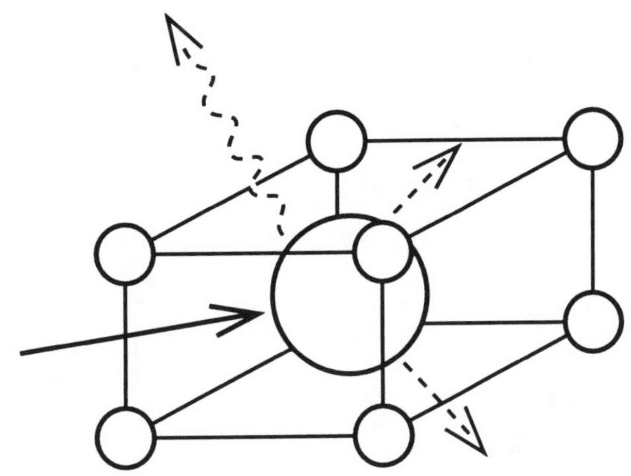

MRS

MATERIALS RESEARCH SOCIETY
Pittsburgh, Pennsylvania

ISBN 1-55899-254-5

Printed in the United States of America

Disclaimer of Liability

Neither the editors, authors, contributors, their affiliated institutions nor the Material Research Society makes any warranty or representation, express or implied, including the warranties of merchantability and fitness for a particular purpose, or assumes any legal liability or responsibility for the accuracy, completeness, or usefulness of any information, apparatus, product, or process disclosed, or represents that its use would not infringe privately owned rights.

CONTENTS

Dedication

This handbook is dedicated to the Böhmische Physical Society. The Society was founded in 1975, and chooses its scientific members based on their contributions to the field of particle-solid interactions through independent, original research. The Society promotes the advancement of knowledge of the interaction of particles with solids and encourages the exchange of information between scientists and engineers engaged in research or development in this field. The scope of this field is particularly broad, and encompasses studies in channeling of energetic particles in crystals, ion-implantation in semiconductors, investigation of near-surface regions and interfaces by backscattering or ion-induced x-rays, radiation damage in solids, photon emission, sputtering effects, and other aspects of particle-solid interactions. Royalties from this handbook will be donated to the Böhmische Physical Society.

Dedication

PREFACE

Producing this handbook was far more time consuming than anyone imagined. Countless hours were devoted to its preparation by the contributing editors and chapter editors, as well as the many contributors and helpers. We applaud their monumental efforts. They have waited a long time to see their efforts in print — so long, in fact, that several have retired. We think their patience will be rewarded.

The development of this handbook was enthusiastically supported by Don M. Parkin, Director of the Los Alamos National Laboratory's Center for Materials Science. It would not exist without his support, nor that of the Department of Energy. Our immediate supervisors, Joe D. Thompson and Frank Gac, were patient and understanding of the time and effort we devoted to this undertaking.

We are grateful to the Materials Research Society for undertaking the formidable task of publishing this handbook. The Director of Publications, Gail Oare, and her staff offered valuable assistance in guiding the book to completion. The book was designed by Cindy Love, MRS Art Director, and the cover was designed by Ed Stiller of MRS, based on an original drawing by Carl Maggiore of Los Alamos National Laboratory. Seth Beckerman of Business and Technical Communications was the Production Editor — his skills and good humor helped transform massive amounts of paper into this volume.

We acknowledge the editors and authors of the *Ion Beam Handbook for Materials Analysis* for providing a superb foundation for this handbook — in fact, some of the authors from that handbook also repeat for this one. We again thank the many contributors to this book who are listed below.

Los Alamos, April 1995

J.R. Tesmer
M. Nastasi

Arizona State University
James W. Mayer

Boston University
Soumendra N. Basu

CNRS-CERI, Orleans, France
Gilbert Blondiaux
Jean-Luc Debrun

IBM Almaden Research Center
John E.E. Baglin
Andrew J. Kellock

Idaho State University
György Vizkelethy

Los Alamos National Laboratory
Michael Bozoian
Aaron Dick
Lynn Foster
Gerald Hale
Michelle Lee
Carl J. Maggiore
Michael Nastasi
Joseph R. Tesmer

Lund Institute of Technology
Harry J. Whitlow

McMaster University
John A. Davies

Middle Tennessee State University
Martha R. Weller

Purdue University
Frank A. Rickey

Sandia National Laboratories
George W. Arnold
J. Charles Barbour
Barney L. Doyle
Peter F. Green
James A. Knapp

State University of New York at Albany
William A. Lanford

Technical Research Centre of Finland
Juha-Pekka Hirvonen

University of Arizona
Richard P. Cox
John A. Leavitt
Laurence C. McIntyre, Jr.

University of Helsinki
Reijo Lappalainen
Eero Rauhala

University of North Carolina
Max L. Swanson

University of Western Ontario
William N. Lennard
Ian V. Mitchell

University of Wisconsin
Paul M. DeLuca, Jr.

Vanderbilt University
Robert A. Weller

CHAPTER 1

INTRODUCTION

J. R. Tesmer and M. Nastasi

Los Alamos National Laboratory, Los Alamos, New Mexico

The seeds for this handbook were sown by J. C. Barbour of Sandia National Laboratories during a 1987 MRS Southwest Regional Conference on the Development of High-Temperature Superconducting Materials. He demonstrated that high-energy alpha backscattering was astoundingly useful for detecting oxygen in high-Tc materials. Concurrently, developments in this area were also underway at many other laboratories. This question arose: "Are there more data and applications that we should be using?" Soon afterwards Jim Mayer suggested writing a new ion beam analysis handbook to update the widely-used (but now out-of-print) *Ion Beam Handbook for Materials Analysis*, edited by J.W. Mayer and E. Rimini, often referred to as the "Green Book." Another strong reason was the unpublished compilation of nuclear reaction data by R. A. Jarjis (1979). This monumental work indicated just how much data were available for possible analyses.

Two conferences, both sponsored by the Center for Materials Science at Los Alamos, were held in New Mexico during the development of this handbook. The first was the High Energy and Heavy Ion Beams in Materials Analysis Workshop in Albuquerque in June 1989 (Tesmer *et al.*, 1990), and later, the Workshop on Ion Beam Analysis in Los Alamos in August 1991. It was at the former workshop that a consensus of those attending enthusiastically endorsed the creation of a new handbook, and many of the attendees became chapter editors.

The Green Book is primarily a compilation of data, and to the novice, not very user friendly. The intent of writing a new handbook was to expand and update the Green Book as well as to produce a different, more organized style of presentation. Emphasis is placed on not only the existing and new data, but also on practical examples of how analysis techniques are applied to common problems. Background materials are also presented where appropriate. The discussions are aimed at an audience with an education equivalent to that of a graduate student in materials science.

The new handbook is divided into two parts — Chapters 2 through 13 present discussions and examples, while Appendices 1 through 18 are compilations of relevant data. A veteran of ion beam analysis might well be able to get along with the appendices while the less experienced reader will find the discussions of interest. The number of techniques discussed has been expanded to include elastic recoil detection and activation analysis. The topic of ion-induced x-rays, or PIXE, was purposely omitted. A recent text by Johansson and Campbell (1988), covers this topic in detail. Material has been added to push the boundaries of ion-beam analysis to higher energies. The detection of light elements has been emphasized. In addition, background material in the areas of energy loss, nuclear theory, instrumentation, counting statistics, analysis pitfalls, and radiation safety is also presented to provide a better understanding of the principles and hazards associated with the techniques. The topics of instrumentation and radiation safety, while of interest to everyone, will be of particular interest to materials scientists who find themselves operating a small ion-beam analysis laboratory with very little assistance.

Clearly, the whole field of ion-beam materials analysis could not be represented in this one-volume handbook. Many lively discussions were centered on which topics should be included. The editors made the final decisions and take sole responsibility for the topics included in this handbook.

We urge that caution be observed when using beam energies exceeding the coulomb barriers of the target elements or of the particles of the beam. Nuclear reactions wanted or unwanted will occur, and there can be associated radiation hazards from both prompt and induced radiation. The extent of these hazards depends on the beam, beam current, and target (sample) material, as well as the facility layout and shielding. Always consult your local health physics professional before undertaking measurements involving nuclear reactions, including those involving elastic resonances.

REFERENCES

Jarjis, R.A. (1979), *Nuclear Cross Section Data for Surface Analysis*, Department of Physics, University of Manchester.

Johansson, S.A.E., and Campbell, J.L. (1988), *PIXE: A Novel Technique for Elemental Analysis*, Wiley, Chichester.

Mayer, J.W., and Rimini, E. (eds.). (1977), *Ion Beam Handbook for Materials Analysis*, Academic Press, New York.

Tesmer, J.R., Maggiore, C.J., Nastasi, M., Barbour, J.C., and Mayer, J.W. (eds.). (1990), *High Energy and Heavy Ion Beams in Materials Analysis*, Materials Research Society, Pittsburgh, Pennsylvania.

CHAPTER

2

ENERGY LOSS

E. Rauhala

University of Helsinki, Helsinki, Finland

CONTENTS

2.1 INTRODUCTION

The knowledge of the slowing down of ions in traversing matter is of fundamental importance in methods of materials analysis using beams of charged atomic particles. Depth perception follows directly from the energy lost by the probing particles and the energy loss affects both quantitative and compositional analysis. The physics of the energy loss phenomena is very complex, involving many kinds of interactions between the projectile ion, target nuclei, and target electrons. Because of their significance in many fields of physics, these phenomena have been subject to intense studies since the beginning of the century.

The theoretical treatment has been reviewed, among many others, by Bohr (1948), Whaling (1958), Fano (1963), Jackson (1962, 1975), Bichsel (1970), Sigmund (1975), Ahlen (1980), Littmark and Ziegler (1980), Ziegler (1977, 1980), Ziegler *et al.* (1985).

The experimental methods have been reviewed and investigated by, e.g., Chu (1979), Bauer (1987), Mertens (1987), Powers (1989).

2.1.1 Stopping and range tables

The number of experimental studies on ion energy loss is overwhelming. This data on ion stopping and ranges are collected in a number of compilations, some of which are listed below.

Stopping tables:

Whaling (1958)
Northcliffe and Schilling (1970)
Bichsel (1972)
Ziegler and Chu (1974)
Andersen and Ziegler (1977)
Hubert *et al.* (1980)
Ziegler (1977, 1980)
Janni (1982)
Ziegler *et al.* (1985)
Hubert *et al.* (1990)

Range tables:

Johnson and Gibbons (1970)
Northcliffe and Schilling (1970)
Brice (1975)
Gibbons *et al.* (1975)
Winterbon (1975)
Hubert *et al.* (1980)
Littmark and Ziegler (1980)
Ziegler *et al.* (1985)
Hubert *et al.* (1990)

2.2 ENERGY LOSS AND STOPPING CROSS SECTION

2.2.1 Basic concepts

The definition of the concepts energy loss, specific energy loss, stopping cross section, and stopping power varies in the literature. Which definition applies in a particular case can be established from dimensional considerations. We adopt the following notation (see Fig. 2.1):

Energy loss dE/dx: $\Delta E/\Delta x \rightarrow dE/dx$, when $\Delta x \rightarrow 0$. ΔE is the amount of energy lost per distance Δx traversed. Units, e.g., MeV/mm, eV/µm. In the literature dE/dx is often called stopping power ($dE/dx \equiv S$) or specific energy loss, and energy loss is denoted by ΔE.

Stopping cross section: $\varepsilon = (1/N)(dE/dx)$ or $\varepsilon = (1/\rho)(dE/dx)$. N is the volume density (e.g., in atoms/cm^3) and ρ is the mass density (e.g. in g/cm^3). Units, e.g., eV/(10^{15}atoms/cm^2) or keV/(mg/cm^2). ε is also sometimes called stopping power.

2.2.2 Models

The relative importance of the various interaction processes between the ion and the target medium depend mostly on the ion velocity and on the charges of the ion and target atoms. At ion velocities, v, significantly lower than the Bohr velocity, v_0, of the atomic electrons,

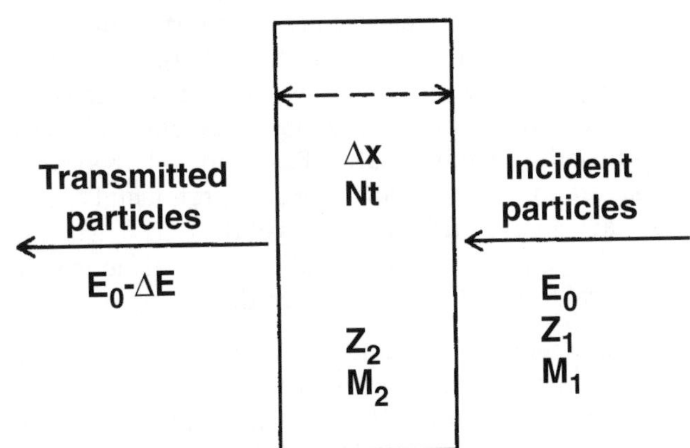

FIG. 2.1. A schematic diagram of particles of atomic number Z_1 and mass M_1 penetrating through a target foil of thickness $\Delta x = t$, composed of atoms with atomic number Z_2 and mass M_2. The energy lost is ΔE. Target thickness Δx corresponds to areal density Nt.

the ion carries its electrons and tends to become neutralized by electron capture. At these velocities elastic collisions with the target nuclei, i.e., the nuclear energy loss, dominates.

As the ion velocity is increased the nuclear energy loss diminishes as 1/E. The electronic energy loss, i.e., inelastic collisions with the atomic electrons, soon becomes the main interaction. The total energy loss is obtained as a sum of the nuclear and electronic contributions. In the ion velocity range from $v \sim 0.1v_0$ to $Z_1^{2/3}v_0$ the electronic energy loss is approximately proportional to velocity or $E^{1/2}$. Ion energy loss in the low ion velocity region was first treated by Bohr (1913, 1915, 1948), Lindhard *et al.* (1953, 1963), and Firsov (1959).

At higher velocities $v >> v_0$ the charge state of the ion increases and finally it becomes fully stripped of its electrons. At a given velocity the energy loss is proportional to the square of ion charge. The theoretical framework of the energy loss of high velocity ions originates from the works of Bohr (1913, 1915, 1948), Bethe (1930, 1932, 1934) and Bloch (1933a, 1933b). The Bethe-Bloch formula has the general form

$$dE/dx = NZ_2 (Z_1 e^2)^2 f(E/M_1) \qquad (2.1)$$

where e is the electron charge and $f(E/M_1)$ is a function depending only on the target, not on the type of the projectile.

To extrapolate the theoretical treatment to the intermediate region where the ion is only partially stripped, a concept of effective charge has been formulated. As a consequence of Eq. (2.1), the effective charge also relates the energy loss of different ions: the ratio of heavy ion energy loss to light ion energy loss is obtained from the square of the effective charge. Based on both theoretical considerations and the analysis of experimental energy loss data, analytical formulations of the effective charge and the energy loss for all kinds of projectile ions in all elemental targets have been derived.

Major theoretical advances accomplished recently include, among others, the works of Sigmund (1982), Oddershede and Sabin (1983), Kürth and Wedell (1981), Brandt and Kitagawa (1982), and Ziegler *et al.* (1985).

Appendix 3 presents Tables A3.1.1 - A 3.1.4 for hydrogen and helium ion electronic and nuclear stopping cross sections in elemental targets for selected energies, as calculated by the semi-empirical model of Ziegler *et al.* (1985).

2.2.3 Conversion methods

2.2.3.1 Units

Atomic mass units are used throughout. $M_1 \equiv M_1$[amu] and $M_2 \equiv M_2$[amu] thus stand for the mass of the ion and target medium in atomic mass units, respectively. For a compound or mixture, M_2 = mean atomic mass, e.g., for quartz $M_{SiO_2} = (M_{Si}+2M_O)/3$. The mass density of the medium is $\rho[g/cm^3]$. The Bohr velocity for atomic electrons is v_0. Unit conversions relevant to energy loss calculations are presented in Table 2.1.

2.2.3.2 Low vs. high energies

The nuclear stopping cross section at ion energy E (in keV) may be calculated from Ziegler *et al.* (1985)

$$\varepsilon_n(E) = \frac{8.462\, Z_1 Z_2\, S_n(E_r)}{(M_1+M_2)\left(Z_1^{0.23}+Z_2^{0.23}\right)}\ \mathrm{eVcm^2/10^{15} atoms} \qquad (2.2)$$

where E_r is the reduced energy and $S_n(E_r)$ the reduced nuclear stopping defined by

$$E_r = \frac{32.53\, M_2\, E}{Z_1 Z_2 (M_1+M_2)\left(Z_1^{0.23}+Z_2^{0.23}\right)} \qquad (2.3a)$$

$S_n(E_r)$ for $E_r \leq 30$ keV:

$$S_n(E_r) = \frac{\ln(1+1.1383\, E_r)}{2\left(E_r + 0.01321 E_r^{0.21226} + 0.19593 E_r^{0.5}\right)} \qquad (2.3b)$$

$$S_n(E_r) \text{ for } E_r > 30 \text{ keV:} \quad S_n(E_r) = \frac{\ln(E_r)}{2\, E_r}$$

To compute the nuclear stopping cross section $\varepsilon_n(E)$ at an energy E, first evaluate E_r from Eq. (2.3a), then select Eq. (2.3b) for $S_n(E_r)$ and use this in Eq. (2.2) for $\varepsilon_n(E)$.

Above 200 keV/amu the nuclear stopping contribution is small, typically below 1% of the electronic stopping. Appendix 3.1 presents nuclear stopping cross sections for ^{1}H and ^{4}He ions in the energy region 10 – 1000 keV.

The electronic energy loss for H and He ions is assumed velocity-proportional in Ziegler *et al.* (1985) below 25 keV/amu and 1 keV/amu, respectively. For heavy ions ($Z_1 > 2$), the definition of the upper limit for the velocity proportional region is more complicated.

Table 2.1. Unit conversions for energy, velocity, thickness, areal density, energy loss and stopping cross section. Masses and densities are found in Appendix 1. Units are shown in brackets [ ].

Multiply units	by	for units	Example
MeV	$\frac{1}{M_1[\text{amu}]}$	MeV/amu	4 MeV ^{4}He ~ 1 MeV/amu
v/v_0	0.1581	$(\text{MeV/amu})^{1/2}$	$v=v_0$ for 0.025 MeV/amu ^{1}H
$(\text{MeV/amu})^{1/2}$	$1.389 \cdot 10^7$	m/s	2 MeV ^{4}He ~ $v_{He} = 9.82 \cdot 10^6$ m/s
10^{15}atoms/cm^2	$\frac{1.661 \cdot 10^{-2} M_2[\text{amu}]}{\rho[\text{g/cm}^3]}$	nm	10^{18}atoms/cm^2 of Au ~ 170 nm
µg/cm^2	$\frac{10}{\rho[\text{g/cm}^3]}$	nm	100 µg/cm^2 of C ~ 285 nm
µg/cm^2	$\frac{10^3}{1.661 M_2[\text{amu}]}$	10^{15}atoms/cm^2	100 µg/cm^2 of Au ~ $305 \cdot 10^{15}$atoms/cm^2
eVcm^2/10^{15}at	$\frac{1}{1.661 M_2[\text{amu}]}$	MeV/(mg/cm^2)	100 eVcm^2/10^{15}atoms in Al_2O_3 ~ 2.95 MeVcm^2/mg. $M_2 = (2M_{Al}+3M_O)/5$. $M_{Al} = 26.98$, $M_O=16.00$
eVcm^2/10^{15}at	$\frac{10^2\ \rho[\text{g/cm}^3]}{1.661 M_2[\text{amu}]}$	keV/µm	30 eVcm^2/10^{15}atoms in Si ~ 150 keV/µm

Energy loss in the relativistic energy region has been reviewed by Ahlen (1980). Relativistic effects must be considered above 10 MeV/amu. These effects have usually been considered in formulations of the Bethe-Bloch equation. The extreme relativistic energy region is, however, often excluded, e.g., > 1 GeV/amu in Ziegler (1980) and > 100 MeV/amu in Ziegler *et al.* (1985).

2.2.3.3 Light to heavy ions

Heavy ion scaling rule

From Eq. (2.1), the stopping cross sections of two different projectiles a and b having the same velocity v in a medium may be derived

$$\left[\frac{\varepsilon}{(\gamma Z_1)^2}\right]_a = \left[\frac{\varepsilon}{(\gamma Z_1)^2}\right]_b \quad . \tag{2.4}$$

In Eq. (2.4) γ is the fractional effective charge, defined as $\gamma = Z_1^*(v, Z_2)/Z_1(v, Z_2)$, where $Z_1^*(v, Z_2)$ is the effective charge of the ion at velocity v in the medium (Z_2). The stopping cross section of a **h**eavy **i**on (HI) is then obtained from the stopping of protons (H) at the **same velocity** (assuming $\gamma_H = 1$)

$$\varepsilon_{HI} = \varepsilon_H Z_{HI}^2 \gamma_{HI}^2 \tag{2.5}$$

where ε_H is the proton stopping cross section, which can be calculated from Table A3.1.1 in Appendix 3 or directly using the formalism presented in Appendix 3.2. Equation 2.5 is the heavy ion scaling rule. At the same velocity we have $E_{HI} = (m_{HI}/m_H) \cdot E_H$. The higher the ion energy, the closer γ approaches unity. When $v/v_0 = 6.325\,(E[\text{MeV/amu}])^{1/2} >> 1$ the ion is assumed fully stripped and $\gamma = 1$.

Quick estimates of heavy ion stopping in elements may be calculated from the proton stopping values obtained using Appendices 3.1 or 3.2, if we assume γ equal to unity. Depending on ion velocities $v >> v_0$, this procedure is usually accurate to better than 20%.

Effective charge formulations

For the heavy ion fractional effective charge γ, many different Z_1 and Z_2 dependent formulations have been proposed in the literature, including Northcliffe and Schilling, (1970), Forster *et al.* (1976), Anthony and Lanford (1982), and Hubert *et al.* (1989, 1990). Semi-empirical parametrizations based on an extensive amount of experimental data were presented by Ziegler (1980), Ziegler *et al.* (1985), and Hubert *et al.* (1990).

For ^{4}He ions, Ziegler *et al.* (1985) and Ziegler and Manoyan (1988) present this simple formula

$$\gamma_{He}^2 = 1 - \exp\left\{ - \sum_0^5 a_i \left[\ln (E / M_1)\right]^i \right\} \qquad (2.6)$$

where E/M_1 is in keV/amu (M_1 = 4.0026) and fitting constants a_i = 0.2865, 0.1266, –0.001429, 0.02402, –0.01135, and 0.001475 when i = 0 to 5, respectively.

Ziegler (1980) gives γ for heavy ion scaling ($Z_1 > 3$) from proton stopping in the energy region 200 keV/amu < E < 22 MeV/amu. Note that E is in keV/amu for Eqs. (2.7) and (2.8).

$$\gamma_{HI} = 1 - \exp(-A) \times \left[1.034 - 0.177 \exp(-0.08114\, Z_{HI})\right]$$
$$A = B + 0.0378 \sin(\pi B / 2) \qquad (2.7)$$
$$B = 0.1772\, E_{HI}^{1/2}\, Z_{HI}^{-2/3} \quad .$$

For Li ions (Z_1 = 3) a separate parametrization is given:

$$\gamma_{Li} = A\left\{1 - \exp\left[-(B + C)\right]\right\}$$
$$A = 1 + (0.007 + 5 \cdot 10^{-5}\, Z_2) \times \exp\left[-(7.6 - \ln E_{Li})^2\right] \qquad (2.8)$$
$$B = 0.7138 + 0.002797\, E_{Li}$$
$$C = 1.348 \times 10^{-6}\, E_{Li}^2 \quad .$$

Recent Z_1 and Z_2 dependent effective charge parametrization formulae for 2.5 – 75 MeV/amu heavy ions, with ^{4}He used as the reference particle, are found in Hubert *et al.* (1990). The parametrization of Ziegler *et al.* (1985), applicable for $Z_1 > 2$, $E > 25$ keV/amu, requires more extensive calculations.

EXAMPLE 2.1. Approximation $\gamma_{He} = 1$.

Assuming $\gamma_{He} = 1$ leads to the mnemonic: 2 MeV ^{4}He ions lose energy 4 times more rapidly than ^{1}H ions at 0.5 MeV. This rule is accurate to 3%, the factor $Z_{He}^2\gamma_{He}^2$ being actually 3.89 instead of 4. At E_{He} = 0.5, 1.0, 1.5, and 3.0 MeV the factors are 2.88, 3.46, 3.75, and 3.97, respectively (Eq. 2.6).

EXAMPLE 2.2. Approximate ^{7}Li ion stopping cross section in carbon at 2, 5, and 10 MeV.

As $M_{Li} \approx 7M_H$, we have $E_H \approx E_{Li}/7 \approx$ 285, 715, and 1430 keV at the same velocities. Taking $\gamma_{Li} = 1$ and $Z_{Li}^2\gamma_{Li}^2 = 9$, we may write Eq. (2.5) as E_{Li}(2, 5, 10 MeV) = 9ε_H(285, 715, and 1430 keV). Table A3.1.1 in Appendix 3 yields by linear interpolation: ε_H(285, 715, and 1430 keV) ≈ 10.2, 5.78, and 3.64 eVcm^2/10^{15}atoms and thus ε_{Li}(2, 5, and 10 MeV) ≈ 92, 52, and 33 eVcm^2/10^{15}atoms.

EXAMPLE 2.3. ^{7}Li ion stopping in carbon at 2, 5, and 10 MeV as calculated from Eqs. (2.5) and (2.8).

E_{Li} = 2, 5, and 10 MeV ~ 285, 715, and 1430 keV/amu (~ 3.4, 5.3, and 7.6 v/v_0). Eq. (2.8) yields γ_{Li} (2, 5, and 10 MeV)= 0.80, 0.97, and 1.00. From Eq. (2.5) and EXAMPLE 2.2 we then find ε_{Li}(2, 5, and 10 MeV) = 10.2·9·(0.80)^2, 5.78·9·(0.97)^2, and 3.64·9·(1.00)^2 eVcm^2/10^{15}atoms = 59, 49, and 33 eVcm^2/10^{15}atoms. These values may be compared to those obtained in EXAMPLE 2.2.

2.2.3.4 Elemental targets vs. compounds and mixtures

Bragg's rule

A simple linear additivity rule of energy loss in compounds may be used on the assumption that the interaction processes between ions and component target elements are independent of the surrounding target atoms. Consider a compound or mixture A_mB_n. The *Bragg's rule* (Bragg and Kleeman, 1905) for the stopping cross section ε^{AB} [eVcm^2/at] for the compound may be written as

$$\varepsilon^{AB} = m\varepsilon^A + n\varepsilon^B \qquad (2.9)$$

where m+n is normalized to unity.

EXAMPLE 2.4. 10 MeV ^{4}He ion stopping in aluminum oxide Al_2O_3.

Table A3.1.2 gives ε^{Al}(10 MeV) = 17.28 eVcm^2/10^{15}atoms and ε^{O}(10 MeV) = 12.43 eVcm^2/10^{15} atoms. For the oxide we then have $\varepsilon^{Al_2O_3} = (2\varepsilon^{Al} + 3\varepsilon^{O})/5 = 0.4\varepsilon^{Al} + 0.6\varepsilon^{O} \approx 14.37$ eVcm^2/10^{15} atoms.

Accuracy, deviations and corrections to Bragg's rule

Bragg's rule assumes the interaction between the ion and the atom to be independent of the environment. The chemical and physical state of the medium is, however, observed to have an effect on the energy loss. The deviations from Bragg's rule predictions are most pronounced, of the order of 10 – 20%, around the stopping maximum for light organic gases and for solid compounds containing heavier constituents, such as oxides, nitrides, etc.

To allow for the chemical state effects Ziegler and Manoyan (1988) have developed the 'cores and bonds' (CAB) - model originally based on the works of Kreutz *et al.* (1988 and earlier refs. therein), Chan *et al.* (1978), Sabin *et al.* (1985), and Oddershede and Sabin (1989). The model assumes the energy loss of ions in a compound to be due to two contributions: the effect of the cores, i.e., the closed electron shells of atoms and the effect of the chemical bonds, such as H–C or C=C bonds. Ziegler and Manoyan converted the available H, He, and Li ion data to H ion data and introduced them in the model calculations. Table 2.2 presents the Ziegler and Manoyan (1988) core-and-bond stopping contributions.

The stopping cross section of any ion with an atomic number Z_1 at 125 keV/amu may thus be written

$$\varepsilon(Z_1) = (\gamma Z_1)^2 \left(\Sigma\varepsilon_{cores} + \Sigma\varepsilon_{bonds} \right) \tag{2.10}$$

where γ is the fractional effective charge of the ion (see Sec. 2.2.3.3). The velocity dependence is given by

$$\varepsilon(v) = (\gamma Z_1)^2\, \varepsilon_{p,Bragg}(v) \left[f(v)\, \frac{\varepsilon_{p,CAB}\,(125\ \text{keV})}{\varepsilon_{p,Bragg}\,(125\ \text{keV})} \right]$$

$$f(v) = \left\{ 1 + \exp 1.48 \left[6.325\,(E/M_1)^{1/2} - 7 \right] \right\}^{-1} , \tag{2.11}$$

where E/M_1 is in MeV/amu and $\varepsilon_{p,CAB}$ and $\varepsilon_{p,Bragg}$ are the proton stopping cross sections calculated by the CAB-model and Bragg's rule, respectively.

To calculate the CAB-correction to Bragg's rule, the bond-structure of the compound needs to be known. For target atoms not included in Table 2.2, use the proton stopping in that element and ignore the bonding.

The model works best for cases where the physical state of the target has little effect and for targets similar to those in the original experiments (tabulated in Ziegler and Manoyan, 1988). The largest differences between the CAB-calculations and simple Bragg's additivity rule predictions are found around the stopping maximum. The differences reduce with increasing energy, disappearing for ion velocities above 10 v_0. The average accuracy of the calculation is better than 2% when compared to data from several hydrocarbon targets.

EXAMPLE 2.5. CAB-correction to ^{4}He stopping in Polycarbonate ($C_{16}H_{14}O_3$) at 600 keV.

There are 14 (H–C) bonds, 10 (C–C) bonds, 6 (C=C) bonds, 4 (C–O) bonds and 1 (C=O) bond in a Polycarbonate molecule. Table 2.2 then gives

$$\sum \varepsilon_{bonds} = \frac{14 \cdot 7.244 + 10 \cdot 3.938 + 6 \cdot 9.790 + 4 \cdot 6.168 + 1 \cdot 13.926}{16 + 14 + 3}$$

$$= 7.216\ \text{eV cm}^2 / 10^{15}\ \text{atoms} \ ,$$

$$\sum \varepsilon_{cores} = \frac{14 \cdot 0 + 16 \cdot 6.145 + 3 \cdot 5.446}{16 + 14 + 3}$$

$$= 3.474\ \text{eV cm}^2 / 10^{15}\ \text{atoms}$$

From Eq. (2.10) we get $\varepsilon_{p,CAB}$(125 keV) = [$\Sigma\varepsilon_{cores} + \Sigma\varepsilon_{bonds}$] = 10.69 eV cm^2/10^{15}atoms. As E_{He} = 600 keV corresponds to 0.15 MeV/amu, Eq. (2.11) gives f(v) = 0.999. By linear interpolation from Table A3.1.1 we obtain ε_p^H (125 keV) = 4.50, ε_p^C (125 keV) = 13.88 and ε_p^O (125 keV) = 15.76 eV cm^2/10^{15}atoms, and for $\varepsilon_{p,Bragg}$ (125 keV)

$$\varepsilon_{p,Bragg}(125\ \text{keV}) = \frac{14 \cdot 4.50 + 16 \cdot 13.88 + 3 \cdot 15.76}{16 + 14 + 3}$$

$$= 10.07\ \text{eV cm}^2 / 10^{15}\ \text{atoms} \ .$$

The CAB-correction $f(v)\varepsilon_{p,CAB}$ (125 keV)/$\varepsilon_{p,Bragg}$ (125 keV) in square brackets in Eq. (2.11) is the factor to multiply the Bragg's rule stopping cross section calculation for 600 keV ^{4}He. We get 1.06 for the correction. More accurate interpolations instead of the linear interpolations from Table A3.1.1 yield $\varepsilon_{p,Bragg}$ (125 keV) = 10.25 eV cm^2/10^{15} atoms and thus 1.04 for the CAB-correction.

Table 2.2. Core-and-bond stopping cross sections per molecule (in $eVcm^2/10^{15}$atoms) for 125 keV protons (from Ziegler and Manoyan, 1988). The symbols –, =, and ≡ stand for single, double and triple bonds, respectively.

Target atom	Cores	Hydrogen bonds		Carbon bonds		Other bonds	
H	0.000	H–H	9.590	C–C	3.938	N≡N	20.380
C	6.145	H–C	7.244	C=C	9.790	N–O	15.796
N	5.859	H–N	8.244	C≡C	15.022	O=O	21.290
O	5.446	H–O	8.758	C–N	5.080	S–H	4.844
F	5.431			C–O	6.168	S–C	1.617
S	32.735			C=O	13.926		
Cl	28.795			C–F	10.998		
				C–Cl	3.713		

Physical state effects, i.e., stopping dependence on the physical state of the medium, have also been observed. In general, most experimental data indicate <20% larger stopping in vapor for light ions than in the solid phase (Thwaites, 1985). For heavy ions, the opposite has been observed, the stopping being < 20% lower in gases than in solids (e.g., Herault *et al.*, 1991 and refs. therein). Experiments on both chemical and physical state effects have been reviewed by Thwaites (1985, 1987) and Bauer (1990).

2.2.3.5 Stopping and ΔE to depth scale

The energy loss dE/dx or the stopping cross section ε may be used to relate the energy ΔE lost by the probing ions to the thickness penetrated in the target material. In principle the calculations involve solving integrals of the form

$$\Delta E = \int_0^x (dE/dx)dx \quad or \quad x = \int_E^{E_0} (dE/dx)^{-1} dE \; . \qquad (2.12)$$

The former gives ΔE when the depth dependence of stopping is assumed, the latter the depth scale x for a given energy interval $\Delta E = E - E_0$.

Depending on the target thickness and the stopping variation in the ion energy region in question, dE/dx or ε may be evaluated either at ion surface energy E_0 or at some mean energy, e.g. $E_{av} = (E+E_0)/2$. For ^{4}He ions around 2 MeV these methods usually differ by a few percent for the first few hundred nm of target. For high accuracy, numerical integration of integrals [Eq. (2.12)] should be used for thick layers and near the stopping maximum.

The same units for thickness should be used as found in stopping, for example, use areal density Nt[10^{15}atoms/cm^2] if stopping is given in $eVcm^2/10^{15}$atoms.

Thin targets

The *surface energy approximation* yields $\Delta E = dE/dx(E_0)\Delta x$ or $\Delta E = \varepsilon(E_0)Nt$. This may also be used for obtaining the mean energy $E_{av} = E_0 - \Delta E/2$ for the *mean energy approximation*, which then gives $\Delta E = dE/dx(E_{av})\Delta x$ or $\Delta E = \varepsilon(E_{av})Nt$.

Thick targets

Numerical integration of Eq. (2.12) proceeds as follows: Estimate ΔE (or Δx or Nt) as above. Divide the target into n slabs. Use surface energy approximation *in each of these slabs*, i.e., calculate the energy lost in *i*th slab ΔE_i and the total ΔE in the target as

$$\Delta E_i = dE/dx(E_{i-1})\Delta x_i \quad or$$
$$\Delta E_i = \varepsilon(E_{i-1})(Nt)_i \qquad (2.13)$$
$$\Delta E = \sum_1^n \Delta E_i$$

where $\varepsilon(E_{i-1})$ and $dE/dx(E_{i-1})$ are the stopping cross section and energy loss evaluated at the energy of the ion at the $(i-1)$th slab.

EXAMPLE 2.6. Proton depth scale in carbon. What is the 2.0 MeV proton energy lost in a carbon target for depths (a) 1000 nm and (b) 20 μm ?

The thicknesses correspond (see Table 2.1 for unit conversions) to areal densities of 17.6 and $353 \cdot 10^{18}$ C atoms/cm^2, respectively.

(a) Table A3.1.1 for ε and the surface energy approximation give ΔE = 2.866 $eVcm^2/10^{15}$atoms·$17.6 \cdot 10^{18}$at/cm^2 ≈ 50 keV

(b) $\Delta E \approx 1000$ keV by surface energy approximation as above. This implies $E_{av} = E_0 - \Delta E/2$ = 1500 keV. Table A3.1.1 gives ε(1500 keV) = 3.506 $eVcm^2/10^{15}$atoms and mean energy approximation thus ΔE = 3.506 $eVcm^2/10^{15}$atoms·$353 \cdot 10^{18}$at/cm^2 ≈ 1235 keV.

Numerical integration over 6 slabs results in ΔE ≈ 1250 keV: Each slab thickness corresponds to $(Nt)_i = (353/6) \cdot 10^{18}$ atoms/cm^2. This gives the energy loss in the first slab as $\Delta E_1 = \varepsilon(E_0)(Nt)_i$ = 2.866 $eVcm^2/10^{15}$ atoms· $(353/6) \cdot 10^{18}$ atoms/cm^2 = 168.5 keV, (The stopping cross sections $\varepsilon(E_i)$ are calculated from Appendix 3.2.) The energy at the end of the first slab is then $E_1 = E_0 - \Delta E_1$ = 2000 − 168.5 = 1832 keV. Energy loss in the second slab is evaluated at this energy: $\Delta E_2 = \varepsilon(E_1)(Nt)_i$ = 3.051 $eVcm^2/10^{15}$atoms·$(353/6) \cdot 10^{18}$ atoms/cm^2 = 179.5 keV. We get $E_2 = E_1 - \Delta E_2$ = 1832 − 179.5 keV = 1652 keV at the end of the second slab. Continuing the procedure, the energy losses in the remaining four slabs are obtained as ΔE_i (i=3–6) = 193.0, 210.3, 233.7, and 268.1 keV. Summing up, the total proton energy loss is $\Delta E = \Sigma \Delta E_i$ (i=1–6) = 1253 keV.

2.2.3.6 Isotopes

The electronic stopping for a given ion X with an atomic number Z_1 at an energy E/A is the same regardless of ion mass number A. The stopping of 2 MeV ^2H ions is thus equal to the stopping of 1 MeV ^1H ions.

For isotopic compositions in the target medium differing from the natural composition with an average mass, M'_{av}, the stopping [$MeVcm^2$/μg] is be approximated by scaling with the average mass ratio: Stopping (medium [M'_{av}, Z_2]) = Stopping (medium [M_{av}, Z_2])·(M_{av}/M'_{av}), where M_{av} is the average mass of the medium with the natural isotopic composition.

2.2.4 Stopping plots and comparisons to experimental data

Figure 2.2 presents nuclear and electronic stopping cross sections for selected ions in silicon. To obtain total stopping from Fig. 2.2, the nuclear and electronic contributions must be added. Figure 2.3 illustrates total semiempirical stopping cross sections and experimental data available in the literature for Mylar ($C_{10}H_8O_4$). The Ziegler *et al.* (1985) model has been used for the stopping calculations, and Bragg's rule and CAB-corrections (Ziegler and Manoyan, 1988) were applied for Mylar. Nuclear stopping was calculated from Eqs. (2.3a) and (2.3b).

In general the calculated stopping cross sections are expected to be accurate to within 5 – 20%. H and He ion stopping predictions are more accurate than those for especially low energy heavy ions. More experimental data on ion energy loss and comparisons with theory are presented in Ziegler (1980). Paul *et al.* (1991) have recently accomplished a survey of all existing hydrogen (10 – 2500 keV/amu) and helium (60 – 7500 keV) ion stopping cross section data in selected target media (Al, Ni, Cu, Ag, and Au). They also propose a new fit of electronic stopping cross sections for these media.

It should be noted that the separate scaling for Li ions (Eq. 2.7) reproduces better the Li stopping data available in the energy range ~ 0.3 – 1.5 MeV/amu than the model in Fig. 2.3. This is also the case for compounds such as Kapton, polycarbonate and Havar (Räisänen and Rauhala, 1990) and for elemental targets C, Al, Ni, Cu, In, Sn, Ag, Ta, and Au (Teplova *et al.*, 1962; Kuronen *et al.*, 1989; Lin *et al.*, 1986).

Plots of semiempirical predictions of stopping cross sections for a number of common elemental target materials and some absorber and stopper foils (Be, C, Al, Si, Cu, Au) and a few compound materials (Kapton, SiO_2, GaAs, stainless steel, $Y_1Ba_2Cu_3O_7$ high-T_C superconductor material) are shown in Appendix 3.3. The stopping cross sections are calculated according to the Ziegler *et al.* (1985) model.

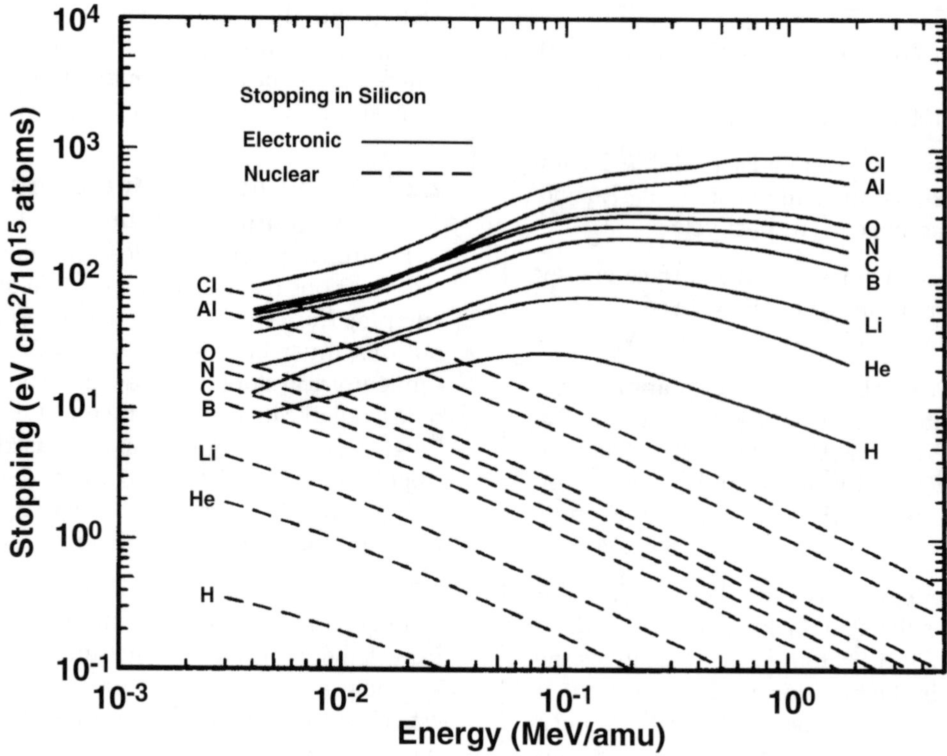

FIG. 2.2. Electronic and nuclear stopping cross sections for ^{1}H, ^{4}He, ^{7}Li, ^{11}B, ^{12}C, ^{14}N, ^{16}O, ^{27}Al and ^{35}Cl ions in silicon. (The curves do not cross.)

2.2.5 Computer calculations

The program package TRIM (Transport of Ions in Matter, version TRIM-91, Ziegler *et al.*, 1991) contains personal computer programs for calculating stopping cross sections of ions in matter. All ions in elemental and compound target media, consisting of up to five different elements may be treated in a wide ion energy range of 1 eV – 2 GeV/amu. The energy loss calculations are based on a large number of experimental data and the semiempirical formalism developed in Ziegler *et al.* (1985). A tabulation of the compositional and structural information of about 100 common nuclear physics and target materials as well as several plastics, polymers, liquids, gases, and metal alloys are included. The energy loss in many of the tabulated compounds is calculated according to the CAB-model of Ziegler and Manoyan (1988).

An earlier version (TRIM-85) program code is presented in Ziegler *et al.* (1985).

2.3 STRAGGLING

When a beam of charged particles penetrates matter, the slowing down is accompanied by a spreading of the beam energy. This phenomenon is called energy straggling. It is due to statistical fluctuations in the number of collision processes. In materials analysis with ion beams straggling broadens the measured energy distributions and resonances and impairs depth and mass resolutions.

2.3.1 Models

2.3.1.1 Bohr's theory

When the energy transfers to target electrons in the individual collisions are small as compared to the width of the energy loss distribution, the distribution is close to a Gaussian. The condition for a Gaussian distribution may be formulated (Besenbacher *et al.*, 1980) as

$$\mathrm{Nt[atoms/cm^2]} \geq 2\cdot 10^{20}\,\frac{1}{Z_2} - \left(\frac{\mathrm{E\,[MeV/amu]}}{Z_1}\right)^2 \quad . \qquad (2.14)$$

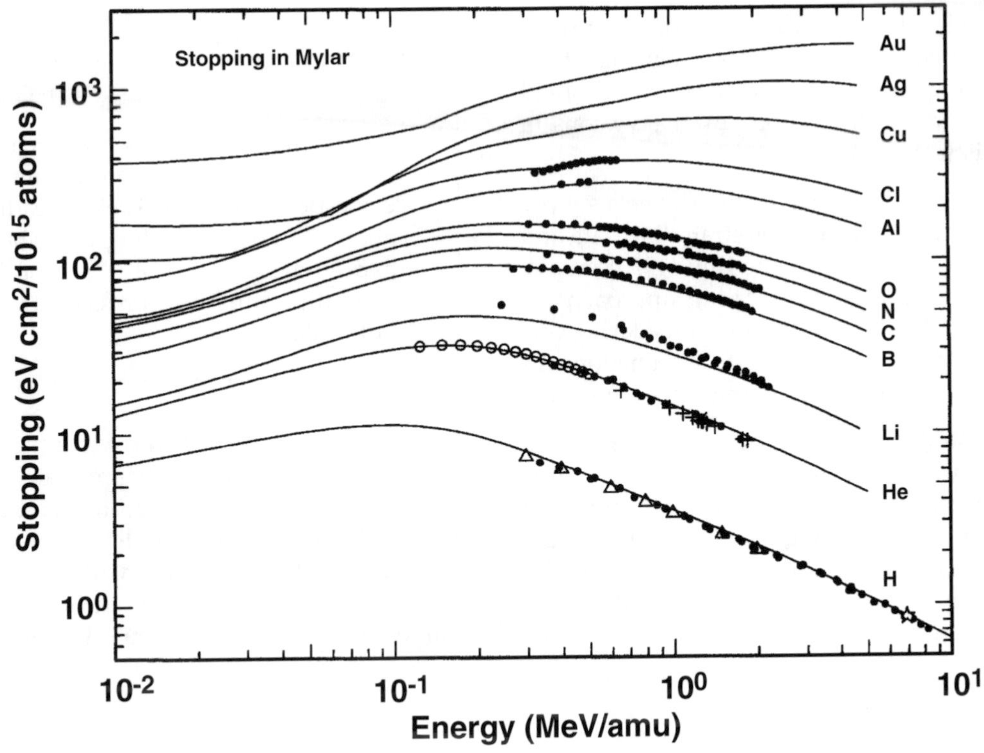

FIG. 2.3. Semiempirical and experimental stopping cross sections for ^{1}H, ^{4}He, ^{7}Li, ^{11}B, ^{12}C, ^{14}N, ^{16}O, ^{27}Al, ^{35}Cl, ^{63}Cu, ^{107}Ag and ^{197}Au ions in Mylar. Experimental data from Refs. L'Hoir and Schmaus (1984)(Δ), Ishiwari *et al.* (1982)(★), Takahiro *et al.* (1990) (o), Santry and Werner (1984)(+), Shepard and Porter (1975)(×) and Räisänen and Rauhala (1990)(•). (The curves do not cross.)

For thick targets, where the energy loss during penetration exceeds 25%, the Gaussian approximation fails.

In the limit of high ion velocity, the energy loss is dominated by electronic excitations. In this region straggling is almost independent of projectile velocity. In the Gaussian distribution regime a simple expression was derived by Bohr (1948)

$$\Omega_B^2[\mathrm{keV}^2] = 0.26\, Z_1^2 Z_2\, Nt[10^{18}\mathrm{at/cm}^2] \qquad (2.15)$$

where Ω_B^2 is the Bohr value for the variance of the average energy loss fluctuation Ω (Full Width at Half the Maximum height is FWHM ≡ 2.355 Ω).

EXAMPLE 2.7. From Eq. 2.15 we get (within 4%) for ^{4}He ions: Ω_B^2 [keV^2] ≈ Z_2 Nt [10^{18}atoms/cm^2]. This relation is helpful for quick estimates of ^{4}He ion Bohr straggling.

2.3.1.2 Corrections to Bohr's theory, other models

Bohr's treatment has been improved by many authors to extend its applicability to lower velocities of light ions ($Z_1 < Z_2$). Lindhard and Scharff (1953) propose a simple correction for ion velocities below E[keV/amu] = 75 Z_2

$$\frac{\Omega^2}{\Omega_B^2} = \begin{cases} 0.5\, L(x), & \text{for } E < 75\, Z_2\ (\mathrm{keV/amu}) \\ 1, & \text{for } E \geq 75\, Z_2\ (\mathrm{keV/amu}) \end{cases}$$

$$L(x) = 1.36\, x^{1/2} - 0.016\, x^{3/2} \qquad (2.16)$$

$$x = E[\mathrm{keV/amu}]/(25\, Z_2)\ .$$

Bonderup and Hvelplund (1971), Chu (1976) and Besenbacher *et al.* (1980) present more refined and extensive calculations. In the non-Gaussian regions and for heavier ions energy, straggling has been described in Livingston and Bethe (1937), Landau (1944), Symon (1948), Vavilov (1957), Schulek *et al.* (1966), Tschalär (1968), and Bichsel and Saxon (1975). For heavier ions, the charge exchange, in addition to the target non-uniformity effects, becomes increasingly important. The charge exchange effects have been treated in Sofield (1990). No appropriate theory for slow heavy ions in light targets exists.

Recently Yang *et al.* (1991) have undertaken a survey of the existing H, He, and heavy ion straggling data and developed a fitting function for the Chu (1976) model. Using the effective charge and scaling approach for energy straggling, the Chu model, and considering correlation effects and charge change effects, functions fitting the data for heavier ion straggling were obtained.

Figure 2.4 illustrates proton and helium ion straggling for the elements as calculated by the Bohr, Lindhard-Scharff, and Chu (fits by Yang *et al.*) models at selected ion energies. Straggling for heavier ions in the regime $Z_1<Z_2$, a condition often met in ion beam analysis, may be approximated by multiplying both Ω^2/Nt and the given energies by M_H/M_{HI} .

EXAMPLE 2.8. Straggling of 1.0 MeV protons in gold.

Figure 2.4 gives $\Omega^2/Nt \approx 10\,keV^2cm^2/10^{18}$ atoms. As 10^{18} atoms/cm^2 (about 170 nm) of gold thus corresponds to about 3 keV spreading in energy, about 20 keV straggling contribution may be expected to the 1.0 MeV proton beam energy resolution after transmission through a 1 µm gold foil.

2.3.1.3 Compounds and mixtures

A similar linear additivity approach for treating compounds (mixtures) for energy loss has been proposed by Chu (1976) for energy straggling. Consider a compound (mixture) A_mB_n (m + n = 1) with an atom density N^{AB} [atoms/cm^3] and the atomic densities N_A and N_B. Assume straggling values Ω^A and Ω^B for the elements in a

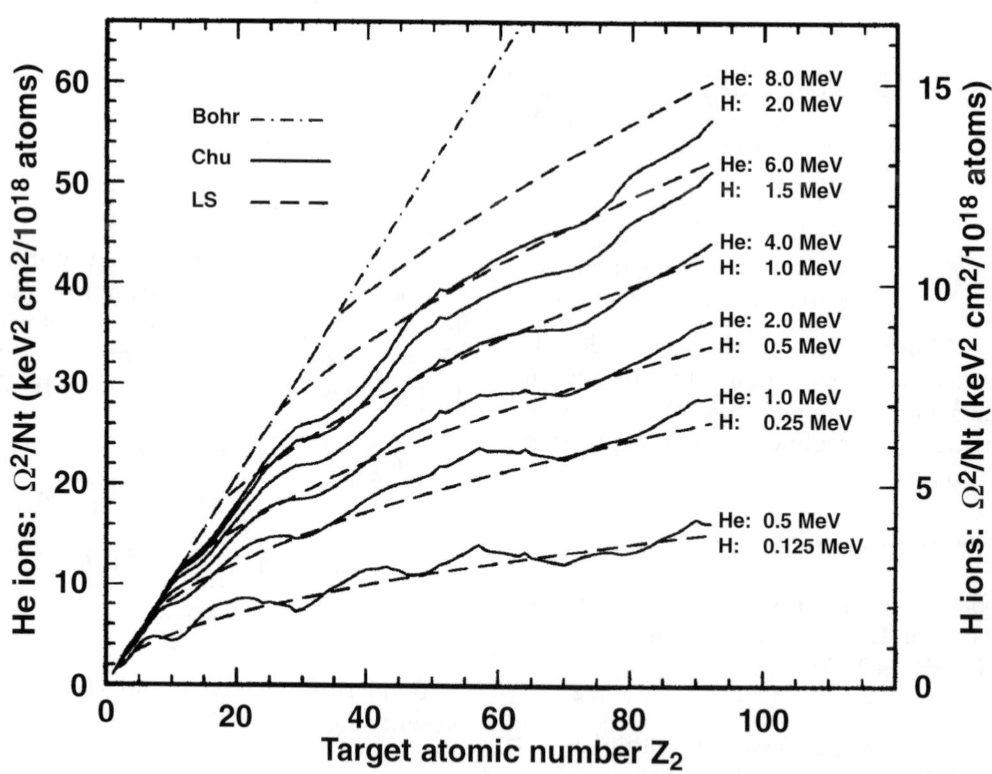

FIG. 2.4. Helium ion straggling per unit target areal density according to Bohr, Chu (fits by Yang *et al.*), and Lindhard and Scharff.

layer of thickness t. Straggling Ω^{AB} in the compound layer A_mB_n may then be obtained from

$$\frac{(\Omega^{AB})^2}{N^{AB}\,t} = \frac{m\,(\Omega^{A})^2}{N_A\,t} + \frac{n\,(\Omega^{B})^2}{N_B\,t} \qquad (2.17)$$

2.3.1.4 Additivity of energy loss fluctuations

The different Gaussian contributions to energy loss fluctuation may be added in quadrature. The energy resolution in a spectrum, for example, arises from several different contributions: the system resolution Ω_{DET}, energy straggling Ω_{STR}, beam energy profile Ω_{BEAM}, etc. These contributions give the total variance of the energy loss fluctuation

$$\Omega^2_{TOT} = \Omega^2_{DET} + \Omega^2_{STR} + \Omega^2_{BEAM} + \ldots \qquad (2.18)$$

The same applies to contributions to energy loss fluctuation originating from multilayered targets or from target thickness variations. In the former case the total variance of energy loss fluctuation is $\Omega^2_{TOT} = \Omega^2_A + \Omega^2_B + \ldots$, where A and B stand for successive layers in a medium penetrated by the ions. A distribution in target thickness with a variance Δt, increases the energy loss fluctuation by $\Omega^2_{\Delta t} = (dE/dx)^2\,(\Delta t)^2$ (Besenbacher *et al.*, 1980).

2.3.2 Comparisons to experimental data

Energy spreading in experimental spectra often results mostly from contributions other than energy straggling. Non-uniformity of targets and correlation effects in energy loss (Besenbacher *et al.*, 1980, Yang *et al.*, 1991) in solid media tend to yield fluctuations in energy loss often much larger than energy straggling. These effects are also expected in routine ion beam analysis.

Qualitative agreement for light ion experimental data with the predictions of Bohr, Lindhard and Scharff, Bonderup and Hvelplund, and Chu is generally observed, although data not corrected for the non-uniformity and correlation effects may in some cases exceed the straggling predictions by hundreds of percent.

2.4 RANGES

Projected range is defined as the mean depth from the target surface at which the ion comes to a halt. Range along the path is the total distance the ion traversed along the trajectory. Since the ion propagation is not rectilinear, the former is smaller than the latter. By range straggling we mean the width of the ion range distribution. The range distribution is further characterized by higher moments, such as skewness and kurtosis.

2.4.1 Calculations and simulations

In range theory the range distribution calculation is regarded as a transport problem describing the motion of the ions during their slowing down to zero energy. The calculations may be broadly classified into the analytic methods and the simulation methods.

The analytic approach has been developed in Lindhard *et al.* (1963), Brice (1970), Sanders (1968), Winterbon *et al.*, (1970), Sigmund (1969), Littmark and Ziegler (1980), and in the PRAL (Projected Range Algorithm) code by Biersack (1981, 1982) and Ziegler *et al.* (1985). The PRAL program, based on transport theory and the stopping formalism developed in Ziegler *et al.* (1985) is included in the TRIM-91 program package (Ziegler *et al.*, 1991). Multilayered targets can not be treated by the transport theoretical calculations.

Simple universal expressions for ion projected ranges, R_p, in the low energy region, where nuclear stopping dominates, have been proposed in Winterbon *et al.* (1970), Kalbitzer and Oetzmann (1980), and later refined in Gupta and Bhattacharya (1984), Izsack *et al.*, (1986), and Kido *et al.* (1986). The expressions are given in terms of reduced variables, energy E_r [see Eqs. (2.3a) and (2.3b)] and length R_r

$$R_r = \frac{0.275\,N[10^{22}\,\text{at/cm}^3]\,M_1M_2}{(M_1+M_2)^2\,(Z_1^{0.23}+Z_2^{0.23})^2}\,R_p\,[\text{nm}]\;. \qquad (2.19)$$

Power law expressions give the projected range in terms of the reduced variables (Kalbitzer and Oetzmann, 1980)

$$\begin{aligned} R_r &= 1.63\,E_r^{2/3} && \text{for } 0.01 < E_r < 0.3 \\ R_r &= 2.2\,E_r && \text{for } 0.3 < E_r < 6. \end{aligned} \qquad (2.20)$$

At the higher reduced energy region a simple relation for ion projected range R_p may be derived

$$R_p\,[\text{nm}] = \frac{72\,(M_1+M_2)\,(Z_1^{0.23}+Z_2^{0.23})}{Z_1Z_2M_1\,N[10^{22}\,\text{at/cm}^3]} \times E\,[\text{keV}] \text{ for } 0.3 < Er < 6 \qquad (2.21)$$

The maximum E_r-value, $E_r = 6$, for example, corresponds to an energy E = 240 keV and range R_p = 510 nm for ^{13}Al ions in ^{14}Si.

The simulation methods are based on molecular dynamic (MD) or binary collision approximation (BCA) calculations. The BCA simulation method involving Monte Carlo calculations has developed into many well-established computer codes such as TRIM (Biersack and Haggmark, 1980; Ziegler *et al.*, 1985, 1991; Biersack *et al.*, 1991), EVOLVE (Rousch *et al.*, 1981), PIBER (Adesida and Karapiperis, 1982), TRIDYN (Möller and Eckstein, 1984), TCIS (Fu-Chai and Heng-De, 1985), and TRIOPS (Chou and Ghoniem, 1987). The simulation approach may yield ranges and range distributions for multilayered targets in a wide ion energy range.

2.4.2 Ranges in compounds

Accurate treatment of ion ranges in compounds requires extensive calculations. Two simple techniques use the corresponding ranges in elemental materials: (a) Average the ranges for the ion in each element of the compound, weighted by the relative composition. (b) Find the range of the ion in a target having an atomic number equal to the average atomic number in the compound. For the reduced variables R_r and E_r method (a) yields for a composite A_mB_n: $E_r^{AB} = mE_r^A + nE_r^B$ and $R_r^{AB} = mR_r^A + nR_r^B$, with A+B normalized to unity.

2.4.3 Isotopic effects

Approximations exist for treating the isotope effects. The range for different isotopes of ions ${}^{A}_{Z_1}X$ at energies E/A may be scaled with the mass number ratio: Range $({}^{A'}_{Z_1}X)$ = Range $({}^{A}_{Z_1}X)\cdot(A'/A)$. The range of 2MeV ^{2}H ions is thus twice the range of 1 MeV ^{1}H ions.

For isotopic compositions in the target medium differing from the natural composition, range may be approximated from: Range (medium $[M'_{av}, Z_2]$) = Range (medium $[M_{av}, Z_2]$)$\cdot(M'_{av}/M_{av})$, where M_{av} is the average mass of the medium with the natural isotopic composition.

2.4.4 Range plots and comparisons to experimental data

Figure 2.5 presents projected ranges of a variety of ions in the energy interval 10 keV – 100 MeV in Mylar and Kapton. Similar data for a number of common elemental

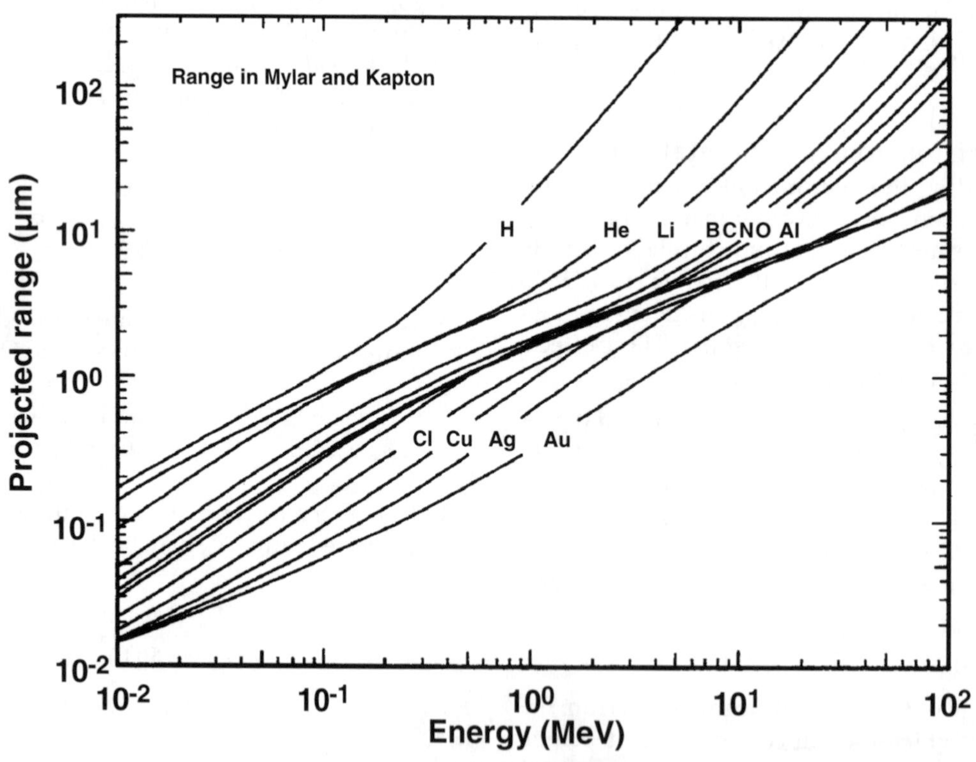

FIG. 2.5. Projected ranges for ^{1}H, ^{4}He, ^{7}Li, ^{11}B, ^{12}C, ^{14}N, ^{16}O, ^{27}Al and ^{35}Cl ions in Mylar and Kapton. The differences in the calculated range values for Mylar and Kapton are less than 2%. (Curves do not cross.)

target materials and some absorber and stopper foils (Be, C, Al, Si, Cu, Au) and a few compound materials (SiO_2, GaAs, stainless steel, $Y_1Ba_2Cu_3O_7$ high-T_C superconductor material) are found in Appendix 3.4. The ranges are calculated by the analytic stopping and range program included in TRIM-91.

In general the calculated projected ranges are expected to be roughly as accurate as the stopping powers, i.e., 5-20%. Some recent experiments on low energy, heavy ion range distributions seem to indicate that the calculations underestimate the data by 20-40%.

ACKNOWLEDGMENT

The author wishes to thank Dr. J.F. Ziegler for valuable suggestions in planning this chapter.

REFERENCES

Adesida, L., and Karapiperis, L. (1982), *Radiat. Eff.* **61**, 223.

Ahlen, S.P. (1980), *Rev. Mod. Phys.* **52**, 121.

Andersen, H.H., and Ziegler, J.F. (1977), *Hydrogen Stopping Powers and Ranges in All Elements*, Pergamon Press, New York.

Anthony, J.M., and Lanford, W.A. (1982), *Phys. Rev.* **A25**, 1868.

Bauer, P. (1987), *Nucl. Instrum. Methods* **B27**, 301.

Bauer, P. (1990), *Nucl. Instrum. Methods* **B45**, 673.

Besenbacher, F., Andersen, J.U., and Bonderup, E. (1980), *Nucl. Instrum. Methods* **168**,1.

Bethe, H.A. (1930), *Ann. Phys.* **5**, 325.

Bethe, H.A. (1932), *Z. Phys.* **76**, 293.

Bethe, H.A., and Heitler, W. (1934), *Proc. R. Soc. London* **A146**, 83.

Bichsel, H. (1970), *Am. Inst. of Phys. Handbook*, 3rd Edition.

Bichsel, H. (1972), *Am. Inst. of Phys. Handbook*, 8-142, McGraw-Hill, New York.

Bichsel, H., and Saxon, R.P. (1975), *Phys. Rev.* **A11**, 1286.

Biersack, J.P. (1981), *Nucl. Instrum. Methods* **182/183**, 199.

Biersack, J.P. (1982), *Z. Phys.* **A305**, 95.

Biersack, J.P., and Haggmark, L.G. (1980), *Nucl. Instrum. Methods* **174**, 257.

Biersack, J.P. and Ziegler, J.F. (1982), In *Ion Implantation Techniques*, Ryssel, H. and Glawischnig, H. (eds.), Springer-Verlag, Berlin, p. 157.

Biersack, J.P., Steinbauer, E., and Bauer, P. (1991), *Nucl. Instrum. Methods* **B61**, 77.

Bloch, F. (1933a), *Ann. Phys.* **16**, 287.

Bloch, F. (1933b), *Z. Phys.* **81**, 363.

Bohr, N. (1913), *Philos. Mag.* **25**, 10.

Bohr, N. (1915), *Philos. Mag.* **30**, 581.

Bohr, N. (1948), *Mat. Fys. Medd. Dan. Vid. Selsk.* **18** (8), **24** (19).

Bonderup, E., and Hvelplund, P. (1971), *Phys. Rev.* **A4**, 562.

Bragg, W.H., and Kleeman, R. (1905), *Philos. Mag.* **10**, S318.

Brandt, W., and Kitagawa, M. (1982), *Phys. Rev.* **B25**, 5631.

Brice, D.K. (1970), *Radiat. Eff.* **6**, 77.

Brice, D.K. (1975), *Ion Implantation Range and Energy Deposition Distributions, 1, High Energies*, Plenum Press, New York.

Chan, E.K.L., Powers, D., Lodhi, A.S. and Brown, R.B. (1978), *Appl. Phys.* **49**, 2346.

Chou, P.S., and Ghoniem, M. (1987), *Nucl. Instrum. Methods* **B28**, 175.

Chu, W.K. (1976), *Phys. Rev.* **A13**, 2057.

Chu, W.K. (1979), In *Accelerators in Atomic Physics*, Richard, P. (ed.), Academic Press, New York.

Fano, U. (1963), *Annu. Rev. Nucl. Sci.* **13**, 1.

Firsov, O.B. (1959), *Zh. Eksp. Teor. Fiz.* **36**, 1517. See also Zh. Eksp. Teor. Fiz. **32**, 1464; Zh. Eksp. Teor. Fiz. **33**, 696; Zh. Eksp. Teor. Fiz. **34**, 447.

Forster, J.S., Ward, D., Andrews, H.R., Ball, G.C., Costa, G.J., Davies, W.G., and Mitchell, I.V, (1976), *Nucl. Instrum. Methods* **136**, 349.

Fu-Zhai, Cui, and Heng-De, Li (1985), *Nucl. Instrum. Methods* **B7/8**, 650.

Gibbons, J.F., Johnson, W.S., and Mylroie, S.W. (1975), *Projected Range Statistics: Semiconductors and Related Materials*, 2nd Edition, Halsted Press Stroudsbury, PA.

Gupta, S.K., and Bhattacharya, P.K. (1984), *Phys. Rev.* **B29**, 2449.

Herault, J., Bimbot, R., Gauvin, H., Kubica, B., Anne, R., Bastin, G. and Hubert, F. (1991), *Nucl. Instrum. Methods* **B61**, 156.

Hubert, F., Fleury, A., Bimbot, R., and Gardes, D. (1980), *Ann. Phys.* (Paris) **5S**, 1.

Hubert, F., Bimbot, R., and Gauvin, H. (1989), *Nucl. Instrum. Methods* **B36**, 357; Atomic Data and Nuclear Data Tables (1990), **46**, 1.

Ishiwari, R., Shiomi, N., and Sakamoto, N. (1982), *Phys. Rev.* **A25**, 2524.

Izsack, K., Berthold, J., and Kalbitzer, S. (1986), *Nucl. Instrum. Methods* **B15**, 34.

Jackson, J.D. (1962), *Classical Electrodynamics*, Wiley, New York Ch. 13.

Jackson, J.D. (1975), *Classical Electrodynamics*, 2nd Edition, Wiley, New York Ch. 13.

Janni, J.F. (1982), *Atomic Data and Nuclear Data Tables* **27**, 147.

Johnson, W.S., and Gibbons, J.F. (1970), *Projected Range Statistics in Semiconductors*, Stanford University Bookstore, Stanford, CA.

Kalbitzer, S., and Oetzmann, H. (1980), *Radiat. Eff.* **47**, 57.

Kido, Y., Konomi, I., Kakeno, M., Yamada, K., Dohmae, K., and Kawamoto, J. (1986), *Nucl. Instrum. Methods* **B15**, 42.

Kreutz, R., Neuwirth, W. and Pietsch, W. (1988), *Phys. Rev.* **A22**, 2598.

Kuronen, A., Räisänen, J., Keinonen, J., Tikkanen, P., and Rauhala, E. (1988), *Nucl. Instrum. Methods* **B35**, 1.

Kürth, E., and Wedell, R. (1981), *Phys. Lett.* **86A**, 54. See also *Phys. Status Solidi* **B116** (1983) 585; with Semrad D., and Bauer, P. (1985), *Phys. Status Solidi* **B127**, 633.

L'Hoir, A., and Schmaus, D. (1984), *Nucl. Instrum. Methods* **B4**, 1.

Landau, L. (1944), *J. Phys.* USSR **8**, 201.

Lin, H.H., Li, L.W., and Norbeck, E. (1986), *Nucl. Instrum. Methods* **B17**, 91.

Lindhard, J., and Scharff, M. (1953), *Mat. Fys. Medd. Dan. Vid. Selsk.* **27** (15).

Lindhard, J., Scharff, M., and Schiøtt, H.E. (1963), *Mat. Fys. Medd. Dan. Vid. Selsk.* **33** (14), See also Lindhard, J. (1954), *Mat. Fys. Medd. Dan. Vid. Selsk.* **28** (8); Lindhard, J., and Winther, A. (1964), *Mat. Fys. Medd. Dan. Vid. Selsk.* **34** (4); Lindhard, J., Nielsen, V., and Scharff, M. (1968), *Mat. Fys. Medd. Dan. Vid. Selsk.* **26** (10); Lindhard, J., Nielsen, V., Scharff, M., and Thomsen, P.V. (1968), *Mat. Fys. Medd. Dan. Vid. Selsk.* **33** (10).

Littmark, U., and Ziegler, J.F. (1980), *Handbook of Range Distributions for Energetic Ions in All Elements*, Pergamon Press, New York Ch. 6.

Livingston, M.S., and Bethe, H.A. (1937), *Rev. Mod. Phys.* **9**, 245.

Mertens, P. (1987), *Nucl. Instrum. Methods* **B27**, 315.

Möller, W., and Eckstein, W. (1984), *Nucl. Instrum. Methods* **B2**, 814.

Northcliffe, L.C., and Schilling, R.F. (1970), *Nuclear Data Tables*, **7**, 233.

Oddershede, J., and Sabin, J.R. (1983), *Phys. Rev.* Lett. **51**, 1332. See also *Atomic Data and Nuclear Data Tables* **31** (1984) 275; *Phys. Rev.* **A35** (1987) 3283; *Phys. Rev.* **26** (1982) 3209; *Phys. Rev.* **A29** (1984) 1757; *Nucl. Instrum. Methods* **B12** (1985) 80; *Nucl. Instrum. Methods* **B24/25** (1987) 339; *Int. J. Quantum. Chem. Symp.* **19** (1985) 733.

Oddershede, J. and Sabin, J.R. (1989), *Nucl. Instrum. Methods* **B42**, 7.

Paul, H., Semrad, D., and Seilinger, A. (1991), *Nucl. Instrum. Methods* **B61**, 261.

Powers, D. (1989), *Nucl. Instrum. Methods* **B40/41**, 324.

Räisänen, J., and Rauhala, E. (1990), *Phys. Rev.* **B41**, 3951; *Phys. Rev.* **B42** (1990) 3877.

Rousch, M.L., Andreadis, T.D., and Goktepe, O.F. (1981), *Radiat. Eff.* **55**, 119.

Sabin, J.R., Oddershede, J. and Sigmund, P. (1985), *Nucl. Instrum. Methods* **B12,** 80.

Sanders, J.B. (1968), *Can. J. Phys.* **46**, 455.

Santry, D.C., and Werner, R.D. (1984), *Nucl. Instrum. Methods* **B1**, 13.

Schulek, P., Golovin, B.M., Kulyukina, L.A., Medved, S.V., and Pavlovich, P. (1966), *Sov. J. Nucl. Phys.* **4**, 996.

Shepard, C.L., and Porter, L.E. (1975), *Phys. Rev.* **B12,** 1649.

Sigmund, P. (1969), *Phys. Rev.* **184**, 383.

Sigmund, P. (1975), in *Radiation Damage Process in Materials*, Du Puy, C.S.H. (Ed.), Noorhoff, Leiden Ch. 1.

Sigmund, P. (1982), *Phys. Rev.* **A26**, 2497. See also *Nucl. Instrum. Methods* **B12** (1986) 1.

Sofield, C.J. (1990), *Nucl. Instrum. Methods* **B45**, 648 and references therein.

Sofield, C.J., Cowern, N.E.B., and Freeman, J.M. (1980), *Nucl. Instrum. Methods* **174**, 611.

Symon, K.R. (1948), Thesis, Harvard University, Cambridge, Mass.

Takahiro, K., Nishiyama, F., Yamasaki, T., Osaka, Y., and Yamaguchi, S. (1990), *Nucl. Instrum. Methods* **B52**, 177.

Teplova, Ya.A., Nikolaev, V.S., Dmitriev, I.S., and Fateeva, L.N. (1962), *Sov. Phys. JETP* **15**, 31.

Thwaites, D.I. (1985), *Nucl. Instrum. Methods* **B12**, 84.

Thwaites, D.I. (1987), *Nucl. Instrum. Methods* **B27**, 293.

Tschalär, C. (1968), *Nucl. Instrum. Methods* **64**, 237. See also Thesis (1967), University of Southern California, Los Angeles; *Nucl. Instrum. Methods* **61** (1968) 141.

Vavilov, P.V. (1957), *Sov. Phys. JETP* **5**, 749.

Whaling, W. (1958), *Handbuch der Physik*, Bd. XXXIV 13, Springer, Berlin.

Winterbon, K.B. (1975), *Ion Implantation Range and Energy Deposition Distributions, 2, Low Energies*, Plenum Press, New York.

Winterbon, K.B., P. Sigmund, P., and Sanders, J.B. (1970), *Mat. Fys. Medd. Dan. Vid. Selsk.* **37** (14).

Yang, Q., O'Connor, D.J., and Wang, Z. (1991), *Nucl. Instrum. Methods* **B61**, 149.

Ziegler, J.F. (1977), *Helium Stopping Powers and Ranges in All Elemental Matter*, Pergamon Press, New York.

Ziegler, J.F. (1980), *Handbook of Stopping Cross-Sections for Energetic Ions in All Elements*, Pergamon Press, New York.

Ziegler, J.F., and Biersack, J.P. (1991), TRIM-91 program package, kindly supplied by Dr. Ziegler.

Ziegler, J.F., and Chu, W.K. (1974), *Atomic Data and Nuclear Data Tables*, **13**, 463.

Ziegler, J.F., and Manoyan, J.M. (1988), *Nucl. Instrum. Methods* **B35**, 215.

Ziegler, J.F., Biersack, J.P., and Littmark, U. (1985), *The Stopping and Range of Ions in Solids*, Vol. 1, Pergamon Press, New York.

CHAPTER 3

NUCLEAR THEORY

F. A. Rickey

Purdue University, West Lafayette, Indiana

Contributors: M. Bozoian and G. M. Hale

CONTENTS

3.1 INTRODUCTION

There is interest in the use of both high energy and heavy ions in ion beam analysis of materials (Tesmer *et al.*, 1990). The potential advantages of both innovations are apparent in the greater depths, enhanced mass resolution, and specificity that can be obtained.

At low enough incident energies, the interaction of a charged particle with an atomic nucleus strictly follows classical laws which depend only on the respective charges. The repulsive Coulomb forces effectively shield the incident ion from the short-ranged nuclear forces. The exact quantum-mechanical solution to the Coulomb scattering problem agrees with the classical one. Thus both nuclear physics and quantum mechanics can be ignored.

As the incident energy increases, the projectile can penetrate to within the range of the attractive nuclear force. Nuclear interactions modify the simple Rutherford scattering results, and nuclear reactions occur. Thus, in order to analyze results, one must have some understanding of both nuclear physics and quantum phenomena.

This chapter is not a course in either. Rather it is intended to convey some of the simple ideas from both fields, and perhaps give hints as to what kind of questions need to be asked. The chapter will begin with basic concepts dealing with reaction energetics and kinematics, followed by a little quantum mechanics that will make some results plausible. In addition, processes which involve nuclear scattering and reactions will be reviewed, and finally, models for dealing with different processes will be introduced.

3.2 BASIC CONCEPTS

Ultimately, we are concerned with the probability of a particular process occurring when a sample of nuclei is bombarded with a beam of particles. In the language of nuclear physics, various measurable quantities related to reaction probabilities are loosely grouped under the heading "cross section." Unfortunately the term cross section is frequently used in reference to very different observables (Krane, 1987), and we had best be specific from the beginning. Roughly speaking, cross section relates to the rate at which particles are removed from the beam. Let the intensity of the incident beam be I_a particles per unit time, and the surface density of the target be Nt nuclei per unit area. If reaction products of type b appear at the rate R_b, then the cross section for this process is

$$\sigma = \frac{R_b}{I_a Nt} \quad . \tag{3.1}$$

This definition shows that σ has the dimensions of area, hence the term cross section. However, R_b refers to the total rate for a particular reaction. In many cases we detect b emitted in a particular direction (θ, ϕ) with a detector subtending a solid angle dΩ. For this, we speak of *differential cross sections*. Since the solid angle dΩ is usually small (solid angles are measured in steradians, with the surface of a sphere subtending 4π steradians at its center), only a small fraction dR_b of R_b is actually counted. If the rate at a particular (θ, ϕ) is r(θ, ϕ), then $dR_b = r(\theta, \phi)\, d\Omega/4\pi$. We then define the differential cross section $\sigma(\theta) = (d\sigma/d\Omega)$ as

$$\sigma(\theta) = \frac{r(\theta,\phi)}{4\pi I_a Nt} \quad . \tag{3.2}$$

The reaction cross section can be found by integrating over all angles. With $d\Omega = \sin\theta d\theta d\phi$, we have

$$\sigma = \int \sigma(\theta) d\Omega = \int_0^{\pi} \sin\theta d\theta \int_0^{2\pi} d\phi \sigma(\theta) \quad . \tag{3.3}$$

3.2.1 Reaction energetics

When a beam of energetic particles is incident on a sample of nuclei, a variety of final products may be formed. Each different "state" is called a reaction channel. In formal nuclear physics courses, each reaction channel is specified by a unique set of quantum numbers. However, here a channel will be taken to simply involve a unique set of products. As an example, consider the case where ^{56}Fe is bombarded by α particles. The incident channel is completely specified if the incident energy is known, but many final channels are possible.

$$\alpha + {}^{56}\mathrm{Fe} \rightarrow \begin{cases} {}^{56}\mathrm{Fe} + \alpha \\ {}^{56}\mathrm{Fe}^* + \alpha \\ {}^{59}\mathrm{Ni} + \mathrm{n} \\ {}^{58}\mathrm{Ni} + 2\mathrm{n} \\ {}^{58}\mathrm{Co} + \mathrm{pn} \\ {}^{57}\mathrm{Co} + \mathrm{p2n} \end{cases} \quad . \tag{3.4}$$

Other final channels are possible; these are just examples. Note that the first two final channels represent elastic scattering, with the target nucleus left in its ground state, and inelastic scattering, where the target nucleus is left in some excited state, E_x, represented by the superscript *. The behavior of any single reaction channel is affected by competition from other channels. Thus it is useful to know which reaction channels are energetically allowed.

Total energy and momentum are always conserved in nuclear reactions. Total energy conservation includes kinetic, mass, and excitation energies. Mass by itself is not conserved because of different nuclear binding energies. Changes in mass in a reaction are described by the reaction Q value. A standard notation for a reaction is (Krane, 1987)

$$a + A \rightarrow B + b \ . \tag{3.5}$$

With the total mass in the incident channel as m_i and the final channel as m_f the Q value is defined as

$$(m_i - m_f)c^2 \equiv Q \ . \tag{3.6}$$

The Q value is the kinetic plus excitation energy in the final channel for zero kinetic energy in the incident channel. Q values for the reaction channels shown in Eq. (3.4) are given in Table 3.1. A Q value can be either positive or negative. If the Q value is positive, the reaction channel is energetically allowed for any projectile energy, although there will be Coulomb barrier effects to deal with. If, however, the Q value is negative, the reaction channel is energetically forbidden until the incident energy exceeds a threshold energy. One can't have negative kinetic energy, even in nuclear physics, and must supply enough kinetic energy in the incident channel to overcome the negative Q value. This threshold energy is (Krane, 1987)

$$(E)_{th} = -Q\left(\frac{m_B + m_b}{m_B + m_b - m_a}\right) \ . \tag{3.7}$$

The negative Q values for several of the reactions of the example given above says that these channels are forbidden by energy until the incident α particle energy exceeds the threshold energy.

One might think that the threshold energy for reaction channels with negative Q values would simply be –Q. This is not true because linear momentum must also be conserved. The kinetic energy of the incident particle cannot be converted completely into mass energy, because some must go into kinetic energy of the reaction products. In a more general way, one might say that not all of the incident energy is available for nuclear reactions or processes.

Table 3.1. Q values for reactions in Eq. (3.4).

Reaction	Q value *(MeV)*
$^{56}Fe + \alpha$	0.0
$^{56}Fe^* + \alpha$	$-E_{ex}$
$^{59}Ni + n$	–5.00
$^{58}Ni + 2n$	–14.05
$^{58}Co + pn$	–13.74
$^{57}Co + p2n$	–22.36

3.2.2 Center of mass system

Nuclear physicists like to use the center of mass system to describe reactions, since the mathematics is simpler. Strictly speaking, one should use the phrase "center of momentum system" in the following discussion, since at high energies relativistic effects are present. However, at the modest energies used in materials analysis, rest masses can be used with great accuracy, and the center of mass and center of momentum systems coincide. "Center of Mass" (CM) is the operational description. In the center of mass reference frame, where the origin moves with the center of mass, the total linear momentum of the reacting particles is zero. (The linear momentum of the center of mass is unchanged by the collision or reaction because no external impulsive forces are present.)

Although many accelerator facilities use colliding beams, where both the "incident" and "target" ions are in motion, the most common case is that in which the target nucleus is at rest in the laboratory system. The details of the transformations between the laboratory and center of mass systems are straightforward, and appear in Appendix 4 as well as standard references (Schiff, 1955). All we need here are a few simple results. The notation used in the following equations is that laboratory quantities are identified with the subscript 0, and masses (m_a and m_A) are identified by the symbol of Eq. (3.5) (a and A).

The total kinetic energy in the CM system is related to the incident energy in the laboratory by

$$E = \frac{A}{a + A} E_0 \quad . \tag{3.8}$$

This energy E is equal to the kinetic energy of a fictitious particle of mass

$$\mu = \frac{aA}{a + A} \tag{3.9}$$

whose speed is the speed of the incident particle in the laboratory system. The quantity μ is called the reduced mass, and will be used in later sections. The relationship between laboratory and CM angles at which a reaction product is emitted is

$$\tan \theta_0 = \frac{\sin \theta}{\gamma + \cos \theta} \quad . \tag{3.10}$$

The quantity γ is equal to the ratio of the speed of the center of mass in the laboratory system to the speed of the observed particle in the CM system. If particle *b* is detected, this is

$$\gamma = \left[\frac{a\,b}{A\,B} \frac{E}{E + Q} \right]^{1/2} \quad . \tag{3.11}$$

If particle B is detected, this is

$$\gamma = \left[\frac{a\,B}{A\,b} \frac{E}{E + Q} \right]^{1/2} \quad . \tag{3.12}$$

For $\gamma < 1$, the laboratory angle θ_0 covers the range from 0 to π. However, if $\gamma \geq 1$, the laboratory angle θ_0 cannot exceed the angle

$$\theta_0 = \sin^{-1}\left(\frac{1}{\gamma}\right) \quad . \tag{3.13}$$

For $\gamma = 1$ this maximum angle is $\pi/2$, and as γ increases the maximum angle decreases toward zero. For elastic scattering ($Q = 0$), γ is always one for the product A, and is ≥ 1 for the product a if $a \geq A$. Notice that if $Q < 0$, γ can also be > 1. This is particularly useful for recoil implantation of the *in situ*-produced nuclear species where the kinematic conditions are such that the implanted species can be easily separated from the beam. (Conlon, 1979).

Theoretical calculations of cross sections are almost always done in the CM system, and in many cases have a simple angular dependence there. Thus it is useful to transform differential cross sections between the two systems. One finds that

$$\sigma(\theta_0) = \sigma(\theta) \frac{d\Omega}{d\Omega_0} \tag{3.14}$$

where

$$\frac{d\Omega}{d\Omega_0} = \frac{\left(1 + 2\gamma \cos \theta + \gamma^2\right)^{3/2}}{\left|1 + \gamma \cos \theta\right|} \quad . \tag{3.15}$$

3.2.3 Basics from quantum mechanics

In order to proceed further, we must consider the quantum-mechanical nature of nuclear scattering and reactions. A few simplified examples can provide a qualitative understanding of the phenomena encountered. (Schiff, 1955)

At the heart of the problem is the wave nature of matter. A particle can be described as a wave whose wavelength, λ, depends on the potential through which it travels. A "free" particle looks like a traveling wave. A traveling wave moving in the $\pm x$ direction may be represented by

$$\Psi \sim e^{-iwt} e^{\pm ikx} \tag{3.16}$$

where the wave number k is

$$k = \frac{2\pi}{\lambda} = \left[\frac{2\mu E}{\hbar^2} \right]^{1/2} \quad . \tag{3.17}$$

A particle confined to some region of space (by some force represented by an attractive potential well) looks like a standing wave. The nuclear reaction problems can then be described (in the CM system) by the interaction of a traveling wave representing a particle of reduced mass μ and CM energy E with a potential fixed at the center of mass. There are many analogies with wave optics, where a change in index of refraction at an interface is analogous to a change in potential. In particular the phenomena of reflection and transmission of waves at an interface and those of diffraction and interference have many similar characteristics.

Many aspects of the general scattering and reaction problem can be understood by breaking down the process into three stages. For incident charged particles, the Coulomb barrier must be penetrated before nuclear effects play a role. Thus the first stage consists of the interaction of the wave with a repulsive potential barrier. Once past the barrier, the particle interacts with the attractive nuclear potential. In this second stage, the particle may be scattered, or may be captured and spend some time in the nuclear interior to form an intermediate state in the process. The third stage deals with the decay of this intermediate state, leading to the final channel of the process.

An accurate solution of the problem must involve the complete three-dimensional case with realistic potentials describing nuclear and Coulomb forces. However, the general behavior in the problem, from a quantum mechanical standpoint, can be seen from the one-dimensional case using simplified potentials. The dominant characteristic of nuclear potentials is that they are strong and short-ranged. Many features can be qualitatively understood by using a simple step function or square potential, which has a constant value $V(x) = V_0$ over some region of space and $V_0 = 0$ elsewhere.

The first stage of the reaction process may be addressed by considering the collision of a particle with a one-dimensional square potential barrier, *V(x)*, shown in Fig. 3.1 (Schiff, 1955). The particle is incident from the left with energy $E > 0$, and may be turned back by the barrier or penetrate through it. This barrier might represent a repulsive force like a Coulomb barrier. This potential has the form

$$
\begin{aligned}
V(x) &= 0 \quad x < 0 \\
V(x) &= V_0 \quad 0 < x < a \\
V(x) &= 0 \quad x > a
\end{aligned}
\tag{3.18}
$$

For $x < 0$ the wave function must represent a particle moving to the right (incident) as well as to the left (reflected). For $x > a$, the wave function must represent a particle moving to the right (transmitted).

Classically, one might expect no transmission if $E < V_0$, and no reflection if $E > V_0$. This is not the case here. The particle can be found within the barrier region with a different wave number,

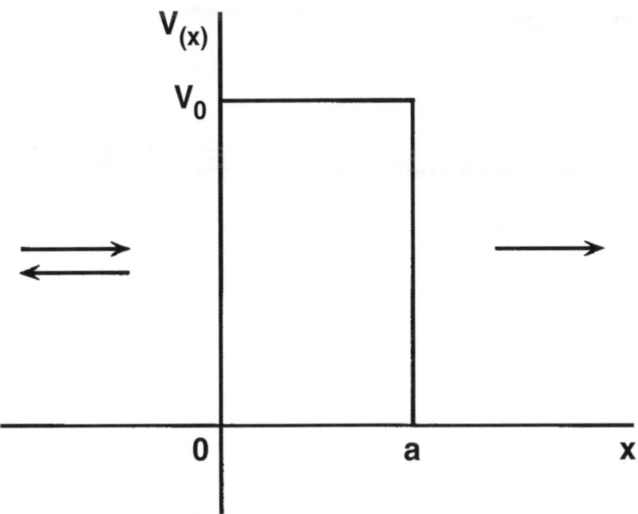

FIG. 3.1. A one-dimensional square potential barrier.

$$\alpha = \left[\frac{2\mu(E - V_0)}{\hbar^2}\right]^{1/2} \tag{3.19}$$

giving rise to both reflection and transmission over a wide range of incident energies. The reflection coefficient $\Re$ and the transmission coefficient $\Im$ (defined as the ratios of the reflected and transmitted fluxes, respectively, to the incident flux) can be evaluated by imposing the boundary conditions that the wave function and its derivative be continuous at $x = 0$ and $x = a$.

For $E > V_0$ the coefficients are

$$\Re = \left[1 + \frac{4E(E - V_0)}{V_0^2 \sin^2 \alpha a}\right]^{-1}$$

$$\Im = \left[1 + \frac{V_0^2 \sin^2 \alpha a}{4E(E - V_0)}\right]^{-1} . \tag{3.20}$$

As $E \to V_0$

$$\Im \to \left[1 + \frac{\mu V_0 a^2}{2\hbar^2}\right]^{-1} . \tag{3.21}$$

The important feature of this is that at energies above the barrier, the wave still "feels" the barrier, and there is partial reflection.

For $E < V_0$ the reflection and transmission coefficients are obtained by replacing α by $i\beta$, where

$$\beta = \left[\frac{2\mu(V_0 - E)}{\hbar^2}\right]^{1/2} . \tag{3.22}$$

The transmission coefficient becomes

$$\Im = \left[1 + \frac{V_0^2 \sinh \beta a}{4E(V_0 - E)}\right]^{1/2} . \tag{3.23}$$

This decreases monotonically below the value for $E = V_0$ as E decreases below V_0. For energies well below the barrier the transmission coefficient drops exponentially.

$$\Im \to \frac{16E(V_0 - E)}{V_0^2} e^{-2\beta a} \tag{3.24}$$

This quantum mechanical effect of barrier penetration, forbidden classically, is called tunneling, and is an important feature of reaction processes. While the form of $\Im$ must be modified for realistic barriers, the exponential dropoff is a significant signature of tunneling.

The second stage of the process, the interaction of the particle with the attractive nuclear forces, may be addressed by considering the problem of scattering by a one-dimensional square well (Schiff, 1955). The potential is as in Eq. (3.18), except that $V(x) = -V_0$ in the region $0 < x < a$. While one would expect no scattering classically, scattering takes place in the quantum picture. Formally one can obtain the scattering coefficients by changing the sign of V_0 in Eq. (3.20) and the expression for α [Eq. (3.19)]. (Since $E > V_0$ always, only that part of the previous discussion applies here.)

$$\alpha = \left[\frac{2\mu(E + V_0)}{\hbar^2}\right]^{1/2} . \tag{3.25}$$

Maximum scattering occurs when $\alpha a = \pi/2, 3\pi/2$, etc., when $\Re$ is maximum, and these maxima are significant only when E is small compared to V_0. These values of αa correspond to the existence of energy levels in a one-dimensional square well. We can think of the potential as containing a virtual energy level slightly above zero binding energy. The incident particle can spend more time in the nuclear interior, which produces a greater distortion of the incident wave function than at neighboring energies. This concept gives rise to resonance phenomena in nuclear reactions.

If a virtual state of an intermediate nucleus is formed, the third stage of the scattering/reaction process must be considered. Such a state may "live" for a time long compared to the normal transit time of the incident particle in the interior. This can be understood by considering the case in which a particle moving with energy E encounters a step potential such as shown in Fig. 3.2, representing a particle trying to leave the nuclear interior (Krane, 1987). Even when $E > 0$, the particle will be reflected at the potential step. The reflection and transmission coefficients are

$$\Re = \left[\frac{\alpha - k}{\alpha + k}\right]^2$$

$$\Im = \frac{4k\alpha}{(\alpha + k)^2} \tag{3.26}$$

where

$$\alpha = \left[\frac{2\mu(E + V_0)}{\hbar^2}\right]^{1/2}$$

$$k = \left[\frac{2\mu E}{\hbar^2}\right]^{1/2} . \tag{3.27}$$

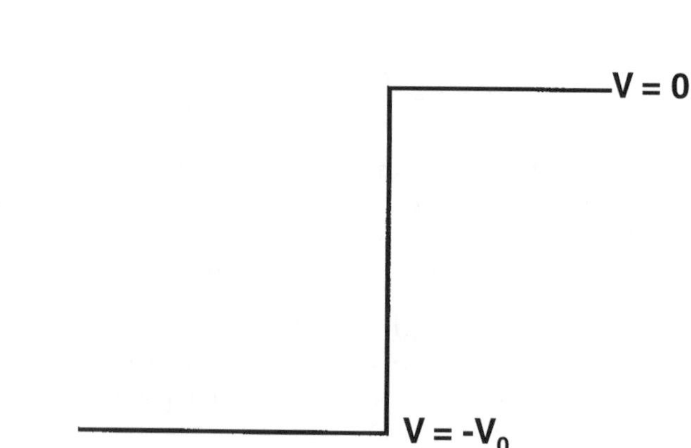

FIG. 3.2. A one-dimensional potential step.

For small E, $\alpha >> k$, and reflection dominates transmission even though the particle is not bound in the normal sense. The particle may undergo several reflections in the nuclear interior before it finally escapes. Only at high energies does $k \to \alpha$ and the transmission coefficient approach unity. This effect also plays an important role in particle emission during nuclear reactions. As the particle spends more time in the nuclear interior, it tends to share its energy, and indeed its identity, with other particles in the intermediate nucleus. If this nucleus lives long enough before it decays so that the formation channel is "forgotten", the intermediate nucleus is called a compound nucleus. This label implies very specific properties of the system, which are not necessarily required for later topics. To avoid possible complications, an intermediate nucleus which lives for any appreciable time will be called a "capture nucleus."

3.2.4 Extension to three dimensions

These one-dimensional examples serve to illustrate important concepts applicable to real systems, because most of the potentials in real life are central potentials with no angular dependence. In spherical coordinates this would say that $V = V(r)$, not $V(r, \theta, \phi)$. The consequence of this is that angular momentum is conserved in a reaction, since the forces are purely radial, and the extension to three dimensions is greatly simplified.

With a central potential, the general wave equation describing the problem can be separated into differential equations involving r, θ, and ϕ alone. The equations in θ and ϕ can be solved easily, leaving the bulk of the work for the solution of the radial equation. Thus, the three-dimensional case reduces to a one-dimensional problem in essence. However, the separation process introduces a term into the radial equation that resembles a new potential involving the angular momentum ℓ

$$\frac{\ell(\ell+1)\hbar^2}{2\mu r^2} \quad . \tag{3.28}$$

This term is called the centrifugal potential, and acts as a barrier to prevent the transmission of particles with $\ell > 0$, just as the Coulomb barrier retards the transmission of charged particles. The term reflects the fact that, for a given energy, particles with higher angular momenta interact with the nuclear potential at greater distances from the center. In these "glancing" collisions, the probability of penetration (interaction) is less than for a head-on collision. The effects of a centrifugal potential for $\ell = 3$ on a square potential is shown in Fig. 3.3. (Cohen, 1971).

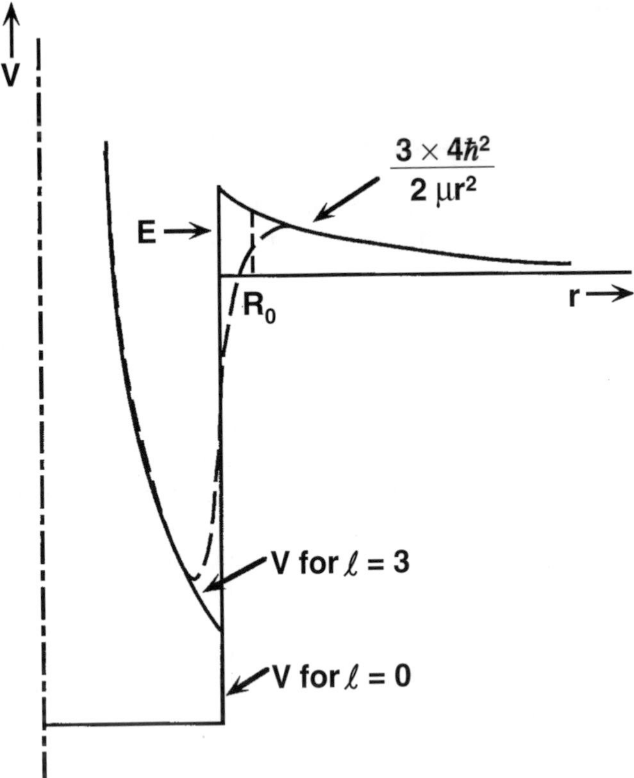

FIG. 3.3. The inclusion of a centrifugal potential with a square potential well. The net potential is shown for $\ell = 3$.

It is customary to expand the incoming and outgoing waves in terms which have a specific angular momentum ℓ. This is called the partial wave expansion. At relatively low incident energies only partial waves with low ℓ contribute. The practical upper limit on ℓ, L, can be estimated by considering that angular momentum is the product of the linear momentum and a radius. The maximum radius expected is the maximum interaction distance, which we can write approximately as $R = R_1 + R_2$, the sum of the radii of the interacting particles. (A satisfactory estimate of a nuclear radius is $R_i = 1.25\,A^{1/3}$, where A is the total number of nucleons.) Thus the upper limit on ℓ is approximately (Cohen, 1971)

$$L \approx Rk \tag{3.29}$$

where $k = (2\mu E/\hbar^2)^{1/2}$ is the wave number of the free particle.

Cross sections for reactions which involve partial waves with $\ell > 0$ exhibit a dependence on the angle of the outgoing particle. The angular component of the *ℓth* partial wave is given in terms of the Legendre polynomial $P_\ell(\cos\theta)$. The angular distribution that results is expressed as a linear combination of the squares of the $P_\ell(\cos\theta)$ that contribute. It is easy to show that

$$|P_\ell|^2 = P_{2\ell} \ . \quad (3.30)$$

Because of the upper limit on ℓ, the maximum complexity of an angular distribution involves terms in

$$(\cos\theta)^{2L} \ . \quad (3.31)$$

3.2.5 Solid angle corrections

Most theoretical calculations ignore the fact that detectors subtend finite solid angles; that is, the calculated cross section is what you would measure with a point detector. Sharp maxima and minima in theoretical angular distributions are always washed out in the real world due to averaging over the solid angle of the detector. In order to make a precise comparison of theory and experiment, one needs to integrate the theoretical cross section $\sigma(\theta, \phi)$ over the θ and ϕ included in the solid angle of the detector. This can be very difficult, since the limits of integration in θ and ϕ depend on the position of the detector. For example, at zero degrees, contours of constant θ are circles, while at ninety degrees they are straight lines. However, if the cross section is given in terms of a partial wave expansion, the problem can be solved in a straightforward way (Krane, 1972).

Consider an angular distribution expanded in a Legendre polynomial series

$$W(\theta) + \sum_\ell A_\ell P_\ell(\cos\theta) \ . \quad (3.32)$$

Since the detector subtends a finite solid angle $d\Omega$, reaction products emitted at a range of angles (θ', ϕ') will be detected with the detector axis at θ. The measured distribution $\overline{W}(\theta)$ is

$$\overline{W}(\theta) = \frac{\int W(\theta')d\Omega}{\int d\Omega} \ . \quad (3.33)$$

This integral is not simple because both θ' and the associated azimuthal angle ϕ' refer to the beam axis. We can, however, transform any $P_\ell(\cos\theta')$ into variables (β, γ) which describe the orientation of the reaction product relative to the detector axis, and (θ, ϕ) which describe the orientation of the detector axis relative to the beam axis. This is accomplished using the spherical-harmonic addition theorem

$$P_\ell(\cos\theta') = \frac{4\pi}{2\ell+1} \sum Y_\ell^{m'*}(\theta, \phi)\, Y_\ell^{m'}(\beta, \gamma) \quad (3.34)$$

$$= P_\ell(\cos\theta)\, P_\ell(\cos\beta) + \text{terms in } m' > 0 \ . \quad (3.35)$$

Because of the general azimuthal symmetry in both ϕ and γ, only the m = 0 terms contribute.

The $W(\theta)$ is a sum, so the expression for $\overline{W}(\theta)$ involves terms:

$$\int P_\ell(\cos\theta')d\Omega = \int P_\ell(\cos\theta)\, P_\ell(\cos\beta)\, d\Omega \quad (3.36)$$

$$= P_\ell(\cos\theta) \int P_\ell(\cos\beta) \sin\beta d\beta$$

$$= P_\ell(\cos\theta)\, J_\ell(\beta) \quad (3.37)$$

where $J_\ell(\beta)$ is a Bessel function. One can then define solid angle correction factors to the various terms in the angular distribution

$$Q_\ell = \frac{J_\ell(\beta)}{J_0(\beta)} \quad (3.38)$$

which are always less than one. The observed angular distribution is then

$$\overline{W}(\theta) = \sum_\ell Q_\ell A_\ell P_\ell(\cos\theta) \ . \quad (3.39)$$

The Q_ℓ depend on the detector geometry only, and are easily calculated. Table 3.2 gives examples of Q_ℓ for a circular detector subtending three different solid angles. It is clear that Q_ℓ decreases as ℓ increases.

Table 3.2. Solid angle corrections.

	Q_ℓ		
ℓ	$d\Omega$ = 1.56%	$d\Omega$ = 0.39%	$d\Omega$ = 0.10%
0	1.00	1.00	1.00
1	0.98	0.99	1.00
2	0.96	0.99	1.00
3	0.91	0.98	0.99
4	0.86	0.96	0.99
5	0.79	0.94	0.98
6	0.72	0.92	0.98
7	0.63	0.90	0.97
8	0.55	0.86	0.96

3.3 PROCESSES

3.3.1 General background

When nuclear interactions take place, cross section calculations must include effects of the additional channels allowed. In addition, interferences between allowed channels must be included. Generally speaking, we can write a cross section as the square of an amplitude.

$$\sigma(\theta) = |f(\theta)|^2 \; . \tag{3.40}$$

The $f(\theta)$ describes the removal of particles from the incident beam, and their emergence in some channel, and may be complex. In general several mechanisms may contribute to a particular channel, so that $f(\theta)$ is the sum of several partial amplitudes. Consider a simple case with only two partial amplitudes

$$f(\theta) = f_1 + f_2 \; . \tag{3.41}$$

Crudely the cross section will be

$$\sigma(\theta) = |f_1|^2 + |f_2|^2 + \text{cross terms} \; . \tag{3.42}$$

The cross terms lead to interference which may be constructive or destructive depending on the relative phases. In most references one will find partial amplitudes expanded in partial waves for each ℓ value, and will encounter phrases like "$\ell = 0$ scattering."

3.3.2 Potential scattering

Let us suppose that the incident energy is high enough for particles to penetrate the Coulomb barrier and sense the nuclear potential, but low enough so that penetration into the nuclear interior is unlikely. We have seen earlier that reflection from the nuclear potential can occur, and this is called *potential scattering*. The scattering amplitude has two components, Rutherford and nuclear.

$$f(\theta) = f_R + f_{nuc} \; . \tag{3.43}$$

The interference term between these components is negative (Krane, 1987). Thus a signature of nuclear potential scattering is the reduction of the scattering cross section relative to Rutherford scattering, $\sigma/\sigma_R < 1$. The magnitude of this effect depends on the minimum separation distance r_{min} between the projectile and the target nucleus, which in turn depends on the incident energy and the observed scattering angle. Figure 3.4 shows a backscattering cross section for ^{4}He on Ge relative to ^{4}He on Au (assumed to be pure Coulomb scattering) measured as a function of CM projectile energy. The relative scattering

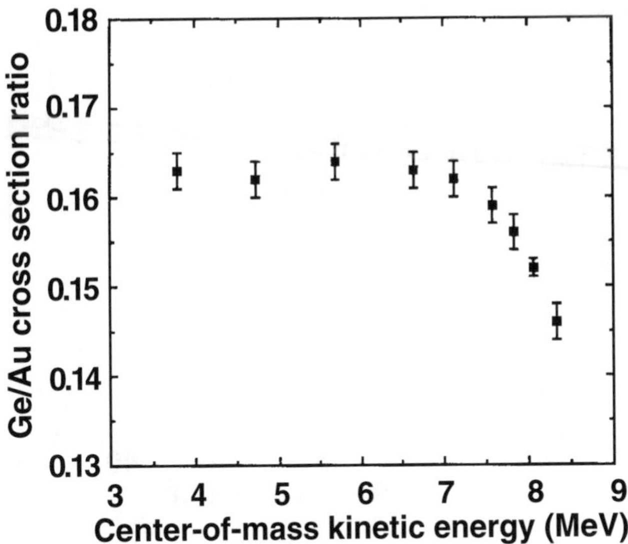

FIG. 3.4. Elastic backscattering cross section for ^{4}He on Ge relative to ^{4}He on Au, from Bozoian (1990).

cross section deviates smoothly from pure Coulomb scattering for projectile energies above approximately 7.2 MeV.

3.3.3 Classical model for deviation from Coulomb scattering

Deviations of the backscattering cross section from the Rutherford value can be calculated using the potential scattering model. Figure 3.5 shows the CM ^{4}He energy, E_{nr}, at which the backscattering cross section deviates by 4% from its Rutherford value versus atomic number for several target elements (including Ge from Fig. 3.4). These data can be accurately reproduced by solving classically the problem of Coulomb backscattering in the presence of a weak Yukawa-like nuclear potential. The resulting formula for E_{nr}, valid for (1) ion projectiles ranging from hydrogen to oxygen and targets ranging from lithium to germanium, (2) deviations from Rutherford cross sections between 2% and 6%, (3) laboratory back angles between 150 and 180 degrees, and (4) projectile to target mass ratios between 0 and 2, is

$$E_{nr} = \frac{\alpha_c}{-R \ln\left[\dfrac{\alpha_c}{2\alpha_n (Z_1^2 A_1)^x}\right]} \text{ MeV} \tag{3.44}$$

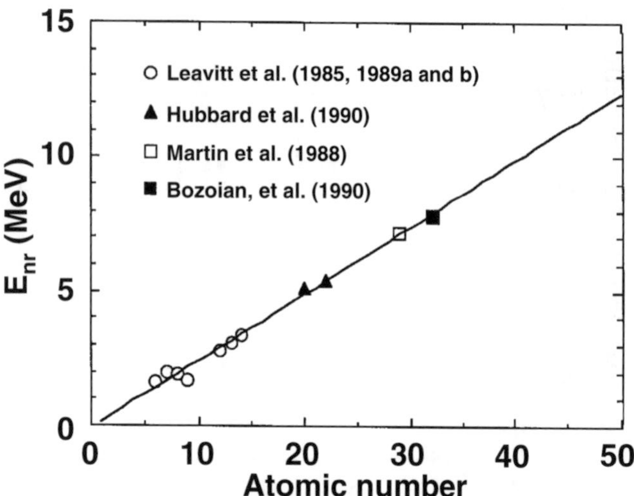

FIG. 3.5. The Center of Mass ^{4}He energy at which the backscattering cross section deviates by 4% from Rutherford, E_{nr}, as a function of atomic number for several elements.

where α_c = 1.44 Z_1 Z_2 MeV-fm, α_n = 390 MeV-fm, $R = 1.3\,A_2^{1/3}$ fm, and the exponent x is $x = Z_1/2$ for $1 \le Z_1 \le 4$ or $x = 1$ for $Z_1 > 4$. The subscripts 1 and 2 refer to projectile and target, respectively.

Using the above formula in conjunction with data and Optical Model calculations (see 3.3.5) yields the following simplified relationships, which underestimate Eq. (3.44) by no more than 10%.

$$E_{nr} = \frac{\alpha_c}{14.4}\ \text{MeV for } Z_1 = 1$$

$$E_{nr} = \frac{\alpha_c}{11.5}\ \text{MeV for } Z_1 > 1 \quad . \tag{3.45}$$

Potential scattering by itself fails to describe deviations at forward angles, or indeed other scattering phenomena, because it does not account for the nuclear interior. The nucleus is a strongly absorbing object, which has serious consequences.

3.3.4 Level densities and widths

When a particle enters a nucleus for anything other than an extremely transient visit, a quantum state of the system must be formed. This state is in an intermediate "capture nucleus," which is at rest in the CM system, with an excitation energy

$$E_{ex} = E_{cm} + Q \tag{3.46}$$

where Q is the Q-value for the a + A → (a + A) reaction.

Although a quantum state is characterized by a definite energy and set of quantum numbers, there is always an uncertainty in its energy related to its lifetime (Cohen, 1971). This is simply seen from the uncertainty principle

$$\Delta E\ \Delta t \ge \hbar \quad . \tag{3.47}$$

This says that if a state has a lifetime Δt, it is impossible to know its energy with any better precision than $\Delta E \simeq \hbar / \Delta t$. All excited states eventually decay with some probability per unit time λ, and thus have an energy width Γ.

$$\Gamma = \hbar\,\lambda \quad . \tag{3.48}$$

When a state can decay in several ways a partial width Γ_i is defined for each type of decay.

$$\Gamma_i = \hbar\lambda_i$$

$$\Gamma = \sum_i \Gamma_i \quad . \tag{3.49}$$

Bound states deep in the potential well have relatively small decay rates since they normally decay only by γ-ray emission, and thus have widths typically less than 10^{-3} eV. However states with positive energies, which can decay by particle emission, have much larger widths. Decay rates are controlled by the potential step at the nuclear radius and penetration through Coulomb or angular momentum barriers. Typical widths might be 0.6 – 100 keV.

Level densities (the average number of levels per unit excitation energy) also play an important role. Level densities increase quite sharply with excitation energy (approximately as the exponential of $\sqrt{E_{ex}}$). It is customary to speak of the average energy spacing between states, D, for states of the same ℓ values.

The nature of nuclear scattering and reactions is quite sensitive to the relative magnitudes of Γ and D. Figure 3.6 shows three hypothetical cases; (a) $\Gamma \ll D$, (b) $\Gamma < D$, and (c) $\Gamma \gg D$. The example of Fig. 3.6(a) would be typical of states deep in the potential well, but is also characteristic of isobaric analogue states which are separated from normal states at high excitation energies because of their isospin. Figure 3.6(b) represents the case in which nuclear states can be formed only at certain narrow ranges of incident energies. This case is characterized by resonances in scattering and reaction processes. In Fig. 3.6(c)

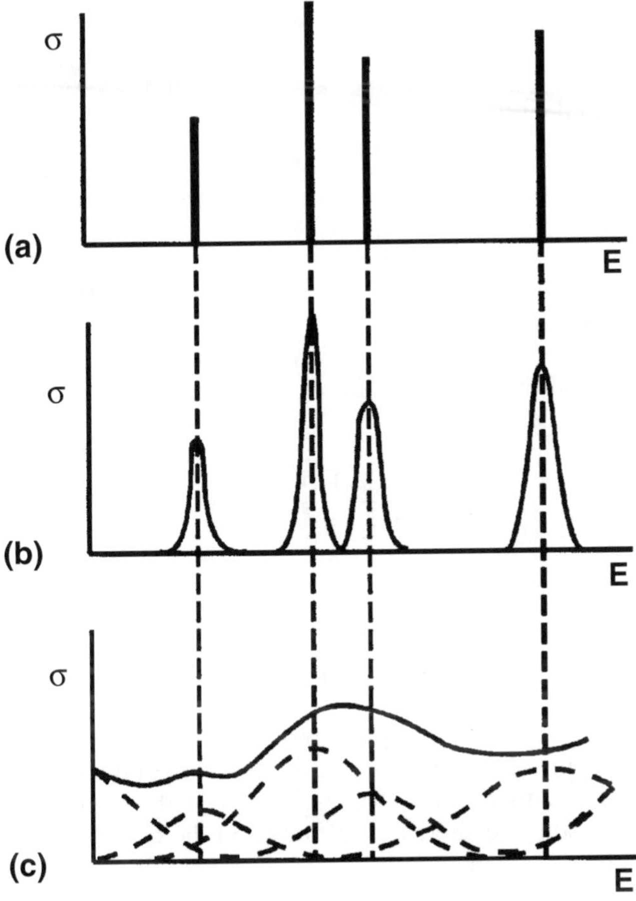

FIG. 3.6. Examples of the relationship of level spacings to widths.

the energy levels begin to form a continuum of overlapping levels, and one expects processes to exhibit a smooth dependence on the incident energy. Models for dealing with processes in these different energy regimes will be discussed below.

3.3.5 Elastic scattering: Optical Model

If the incident particle produces an intermediate capture nucleus in the continuum region of Fig. 3.6(c), particles can be emitted in many channels other than the elastic channel. A microscopic treatment of this many-channel problem would be painful. A simple two-channel model has been developed, which makes use of the assumption that absorbtion in the continuum region is a smooth function of incident energy. This model is called the Optical Model, because the calculation resembles that of light incident on a somewhat opaque glass sphere. In addition to attenuation of the incident light, pronounced interference effects are seen.

The Optical Model assumes that the two channels are elastic scattering and anything else. The absorbtive effects are obtained by introducing an imaginary potential, $iW(r)$, in addition to the real potential $V(r)$. We thus have the complex potential

$$U(r) = V(r) + iW(r) \ . \tag{3.50}$$

Much effort by many research groups has gone into studying the forms of V and W which best describe experimental elastic scattering data, and determining consistent parameters for them. The real potential is usually taken to be (Krane, 1987)

$$V(r) = \frac{-V_0}{1 + e^{(r-R)/a}} \tag{3.51}$$

which is a Wood-Saxon potential used in many Shell Model calculations. The constant, a, is called the diffuser. At reasonably low incident energies, only the "valence" nucleons near the surface contribute to absorbtion. Thus the imaginary potential is often chosen to be proportional to dV/dr, which peaks near the surface. Figure 3.7 shows the radial dependence of V and W for typical parameter choices. A spin orbit term is often added, also peaked near the surface. For charged particles a Coulomb term is necessary.

The success of the Optical Model has been very impressive, and the systematics established for its parameters allow quantitative predictions even when little experimental data are available. It provides a reliable mechanism for calculating background scattering processes when resonance effects are present.

3.3.6 Optical Model and deviations from Coulomb scattering

Optical Model calculations of the departure from Rutherford scattering can apply to any scattering angle, and are not limited to backscattering as are the classical analyses outlined in Section 3.3.3. As an example, Figs. A8.6 and A8.7 in Appendix 8 show the CM incident energies for ^{6}Li and ^{7}Li projectiles at which scattering cross sections deviate by 4% from their Rutherford values versus atomic numbers for elements ranging from Be to Zr. Because of a lack of experimental data, the energies were calculated from 47 optical potentials parameterized from higher energy nuclear studies. The calculated

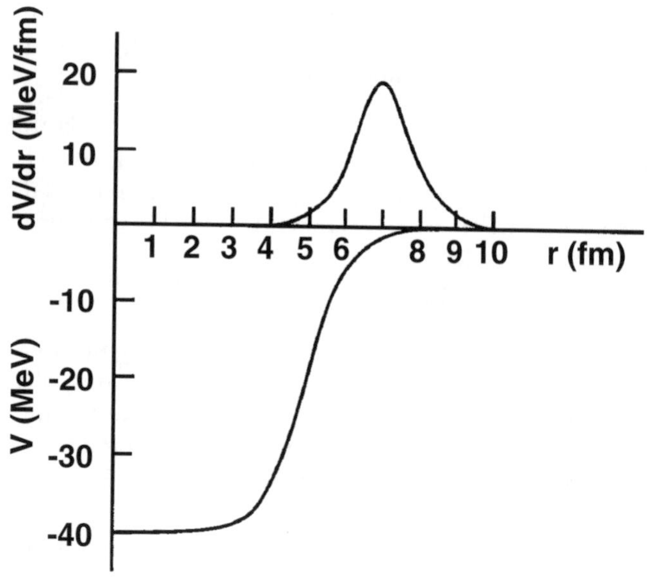

FIG. 3.7. Real and imaginary Optical Model potentials.

E_{nr} exhibits the same kind of linear dependence on target atomic numbers as for ^{4}He projectiles (Fig. 3.5). Moreover, for those targets with more than one optical potential, all the calculated E_{nr}'s are within a few percent of each other. Recent experimental results for ^{6}Li and ^{7}Li ion beams corroborate these calculations. All of this gives confidence that Optical Model calculations of deviations from Rutherford scattering can either augment or be used in the absence of available experimental data. Further Optical Model calculations of deviations from Rutherford scattering are presented in Appendix 8 for projectiles ranging from protons to oxygen.

3.3.7 Resonances in nuclear reactions

If the level structure of the capture nucleus resembles that of Fig. 3.6(b), the absorbtive effects no longer exhibit a smooth dependence on the incident energy, but show pronounced resonances at energies which match the excitation energy of the quasibound nuclear state. If we consider resonance due to an isolated state of the capture nucleus with width Γ, corresponding to a resonant incident energy E_R, we might expect the following behavior.

1. The energy dependence of the cross section should be related to the width Γ of the level.
2. The magnitude of the cross section of the resonance should be governed by the relation between the partial widths for formation and decay of the state and the total width of the state.

The shape of the cross section for an isolated resonance can be derived from reaction theory. The most reliable theory is cast in the "R-Matrix." Section 3.4 will discuss this in more detail and show examples of its use. However, formal R-Matrix theory can obscure the fundamental results expected, so here a simple result is discussed, the *Single Level Breit-Wigner Formula* (Krane, 1987). If the incident channel is denoted by α, and the decay channel by β,

$$\sigma = \frac{\pi}{k^2} g \frac{\Gamma_\alpha \Gamma_\beta}{(E - E_R')^2 + \Gamma^2/4} \tag{3.52}$$

where

$$g = \frac{2I + 1}{(2s_\alpha + 1)(2s_\beta + 1)} \ . \tag{3.53}$$

In this, I is the total angular momentum, and the s's are the spins of the projectile and target. For spinless particles g reduces to $(2\ell + 1)$. The E_R' represents the fact that the resonance energy is shifted from the energy of the state by a small amount.

Clearly a resonance in a particular channel depends not only on the existence of a level, but the partial widths. Partial widths are sensitive to the nuclear structure of the state. Crudely speaking, the partial width for the channel a + A includes a factor which describes how much the state looks like the nucleus A plus particle a. If this overlap is small, then the corresponding partial width is small. In general one finds stronger resonances if the number of decay modes is small.

Resonances in the elastic channel are always affected by resonances in other reaction channels, since particles are removed from the elastic channel. The shape of a resonance in the elastic channel always differs from the simple Breit-Wigner shape because of interference with the background elastic scattering. If the background scattering is potential scattering, this is easy to calculate, and is normally destructive below the resonance energy and constructive above it. One typically gets the shape shown in Fig. 3.8 (Krane, 1987).

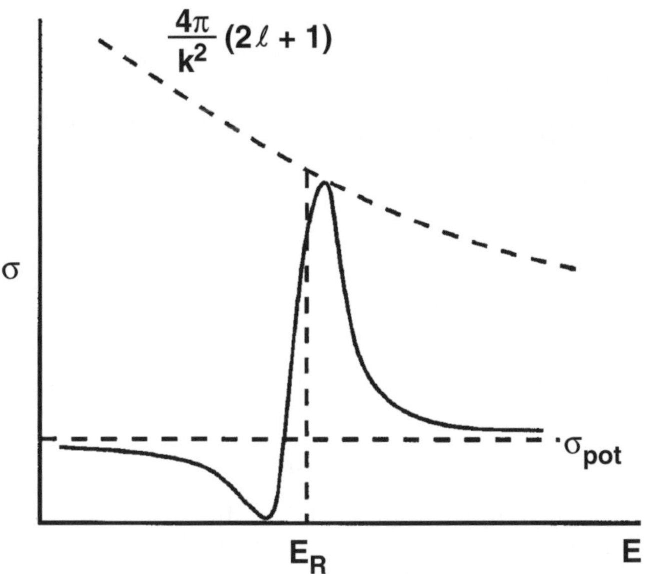

FIG. 3.8. Typical resonance shape.

While this shape qualitatively reproduces resonances seen in real experiments, it fails in quantitative measures because potential scattering ignores the nuclear interior. To get a better description, R-Matrix theory is necessary.

3.4 R-MATRIX ANALYSIS

Many scattering formalisms are based on asymptotic forms of wave functions — wave functions evaluated at large distances from the nuclear interaction region. While resonance behavior may be added to these asymptotic formalisms in an ad-hoc fashion, they cannot accurately describe details of resonance behavior which are sensitive to the nuclear interior.

The R-Matrix is a formalism which deals with quantities evaluated at or near the nuclear surface, defining a region inside of which particles may interact, and outside of which no nuclear interactions take place. Thus the R-Matrix can couple physical pictures of nuclear structure within the interaction region to scattering and reaction cross sections measured outside of the interaction region.

The philosophy of the R-Matrix formalism may be grasped by considering the simple case of potential scattering, in which the possibility of reactions is excluded. The radial wave function $u_\ell(r)$ is defined by the equation

$$\frac{d^2u_\ell}{dr^2}+\frac{2m}{\hbar^2}\left(E-V-\frac{\ell(\ell+1)\hbar^2}{2mr^2}\right)u_\ell=0 \ . \tag{3.54}$$

Quantities R_ℓ may be defined which connect values of u_ℓ with its derivatives evaluated at $r = a$.

$$u_\ell(a)=R_\ell a\left(\frac{du_\ell}{dr}\right)_{r=a} \tag{3.55}$$

This can be extended by expanding u_ℓ in terms of actual states, u_λ with energy eigenvalues E_λ, of the incident particle in the nuclear potential which may be formed at incident energy E. (Recall that several levels may overlap at a given excitation energy.) R_ℓ can be shown to be

$$R_\ell=\frac{\hbar^2}{2ma}\sum_\lambda\frac{u^2_{\ell,\lambda}(a)}{E_\lambda-E}=\sum_\lambda\frac{\gamma^2_{\ell,\lambda}}{E_\lambda-E} \tag{3.56}$$

where

$$\gamma_{\ell,\lambda}=\left(\frac{\hbar^2}{2ma}\right)^{1/2}u_{\ell,\lambda}(a) \tag{3.57}$$

is called the reduced-width amplitude.

This can be generalized to allow for reactions with initial channel c and final channel c'. With Ψ_c analogous to u_ℓ and Δ_c analogous to $(r du_\ell/dr)$, both evaluated at $r = a$, one finds (suppressing ℓ)

$$\Psi_c=\sum_{c'}R_{cc'}\,\Delta_{c'} \ . \tag{3.58}$$

The $R_{cc'}$ are elements of the R-Matrix, and can be cast in the so-called pole expansion

$$R_{cc'}=\sum_{c'}\frac{\gamma_{\lambda c}\gamma_{\lambda c'}}{E_\lambda-E} \tag{3.59}$$

where again the γs are reduced width amplitudes.

The cross section can be written in terms of the R-Matrix, and once the R-Matrix has been successfully parameterized with existing data, all features of the cross section, including resonances, can be accurately reproduced.

In order to parameterize the R-Matrix, all possible reaction channels which have the same intermediate nuclear system in common must be taken into account. An example will clarify this. Consider the case in which ^{4}He is elastically scattered by ^{16}O. The scattering will be Rutherford scattering until a CM incident energy, E_{nr}, of approximately 2 MeV is reached. At higher energies the scattering will deviate from Rutherford and resonances will be encountered due to the p + ^{19}F channel becoming energetically allowed. Both channels, elastic and inelastic, would be incorporated in the R-Matrix analysis. Parameterizing the R-Matrix would involve data from the four reactions, ^{16}O(α, α)^{16}O, ^{19}F(p,p)^{19}F, ^{16}O(α,p)^{19}F, and ^{19}F(p,α)^{16}O, which have the common intermediate nuclear system ^{20}Ne. While this may seem like an unpleasant complication, it has the advantage that the R-Matrix relates all possible reactions and sums them to the total available reaction cross section. Consequently, the R-Matrix calculation generally selects the "correct" datum or data set when there are conflicts.

An example is the R-Matrix analysis of the elastic scattering of ^{4}He from ^{13}C, shown in Fig. 3.9. The intermediate system is ^{17}O, and in addition to the elastic channel, the non-elastic channel n + ^{16}O contributes. The R-Matrix analysis resulted in a new evaluation of the non-elastic channel, for which there were over 2400 very accurate data points. There were only 200 data points for the elastic α + ^{13}C channel, but confidence in the elastic analysis is enhanced because of the very accurate non-elastic channel data. The CM backscattering angle is 169.6°. The cross section is seen to decrease smoothly as the inverse square of the energy, following the well-known Rutherford behavior, until approximately 2.10 MeV laboratory (1.61 MeV CM) energy. This compares with E_{nr} = 1.48 MeV (CM) from Eq. (3.44). From approximately 2 MeV to just less than 2.75 MeV, lab energy, the cross section exhibits deviations from Rutherford with interference effects. The resonance at approximately 2.75 MeV is at least five times greater than background. From 2.75 MeV to 2.80 MeV there is a small but interesting interference effect between the strong resonance and the underlying background. The R-Matrix analysis precisely reproduces all these data, showing that it is capable of describing not only resonances, but complex backgrounds and interferences.

REFERENCES

Barnes, P.K., Belote, T.A., and Risser, J.R. (1965), *Phys. Rev.* **140** B616.

Bozoian, M., Hubbard, K.M., and Nastasi, M. (1990), *Nucl. Instrum. Methods* **B51**, 311.

Cohen, B.L. (1971), *Concepts of Nuclear Physics*, McGraw-Hill, New York.

Conlon, T.W. (1980), *Nucl. Instr. and Meth.* **171**, 297.

Hubbard, K.M., Martin, J.A., Muenchausen, R.E., Tesmer, J.R., and Nastasi, M. (1990), *High Energy and Heavy Ion Beams in Materials Analysis*, Materials Research Society, Pittsburgh, Pennsylvania.

Krane, K. S. (1972), *Nucl. Instr. and Meth.* **98**, 205.

Krane, K. S. (1987), *Introductory Nuclear Physics*, John Wiley and Sons, New York.

Leavitt, J.A., Stoss, P., Cooper, D.B., Seerveld, J.L., McIntyre, Jr., L.C., Davis, R.E., Guiterrez, S. and Reith, T.M. (1985), *Nucl. Instrum. Methods* **B15**, 296.

Leavitt, J.A., McIntyre, Jr., L.C., Stoss, P., Oder, J.G., Ashbaugh, M.D., Dezfouly-Arjomandy, B., Yang, Z.-M. and Lin, Z. (1989), *Nucl. Instrum. Methods* **B40/41**, 776.

Leavitt, J.A. (1989), unpublished data.

Martin, J.A., Nastasi, M. Tesmer, J.R. and Maggiore, C.J. (1988), *Appl. Phys. Lett.* **52**, 2177.

Schiff, L. I. (1955), *Quantum Mechanics*, McGraw-Hill, New York.

Tesmer, J. R., Maggiore, C. J., Nastasi, M., Barbour, J. C., and Mayer, J. W. (1990), *High Energy and Heavy Ion Beams in Materials Analysis*, Materials Research Society, Pittsburgh, Pennsylvania.

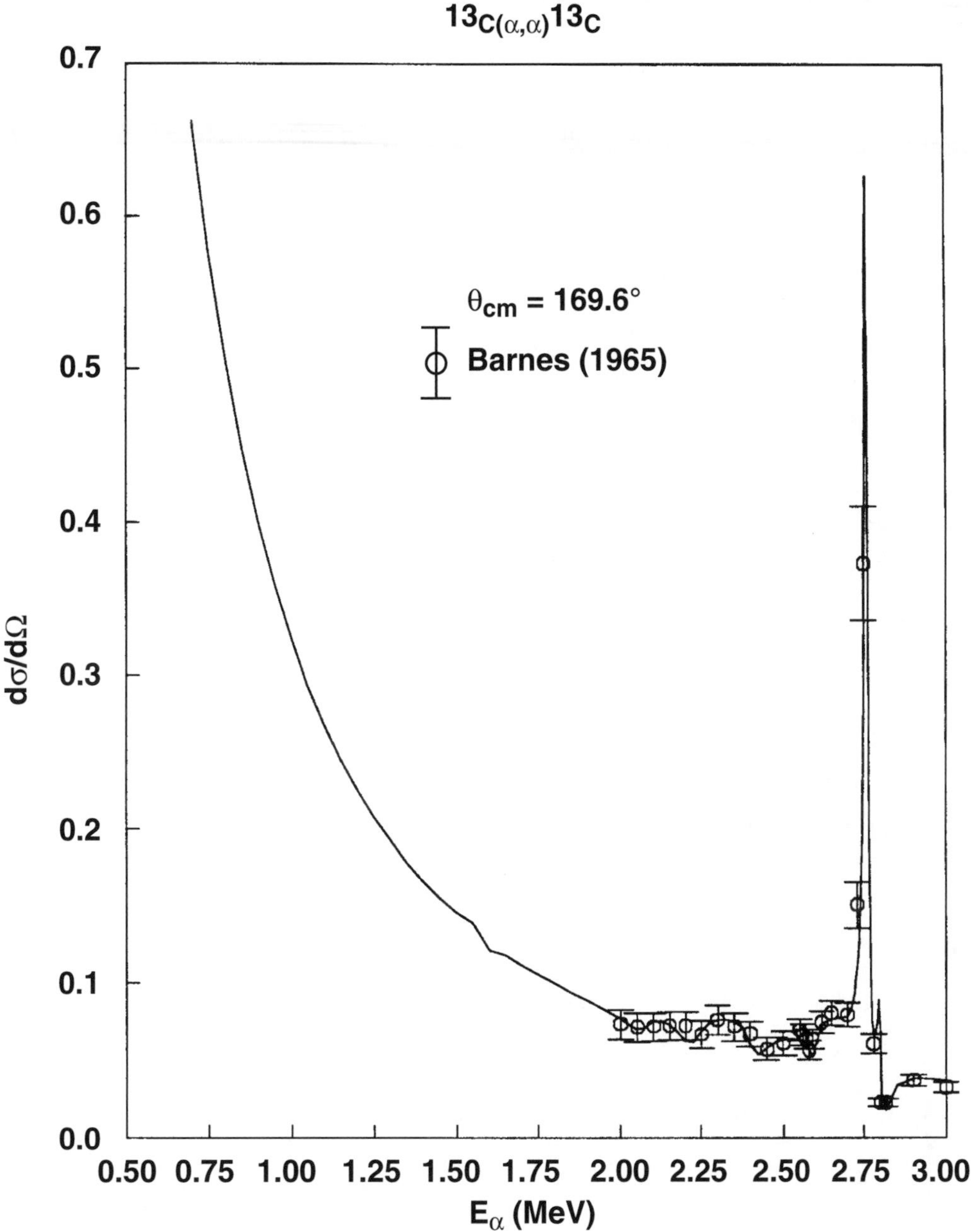

FIG. 3.9. R-Matrix analysis of $^{13}C(\alpha, \alpha)^{13}C$.

BACKSCATTERING SPECTROMETRY

J. A. Leavitt and L. C. McIntyre, Jr.

University of Arizona, Tucson, Arizona

M. R. Weller

Middle Tennessee State University, Murfreesboro, Tennessee

CONTENTS

4.1 INTRODUCTION

Backscattering spectrometry using ion beams with energies in the MeV range has been used extensively for accurate determination of stoichiometry, elemental areal density, and impurity distributions in thin films. Measurement of the number and energy distribution of ions backscattered from atoms in the near-surface region of solid materials allows identification of the atomic masses and determination of the distribution of target elements as a function of depth below the surface.

Application of the technique to thin film analysis is illustrated in Fig. 4.1 for the ideal case of a two-element thin film of uniform composition on a low-mass substrate. Analysis ions scatter elastically from target atoms with energy characteristic of the mass of the struck particle; they also lose energy passing into and out of the film material. Energy analysis of the backscattered ions by the detection system yields the backscattering spectrum displayed in the lower portion of Fig. 4.1 in the form of counts per channel vs. channel number. The channel number is normally linearly related to the backscattered ion energy, E_1. Appearing in the spectrum is a nearly flat-topped "peak" for each element present in the film. The peak widths are caused by the energy loss of the analysis ions in the film material.

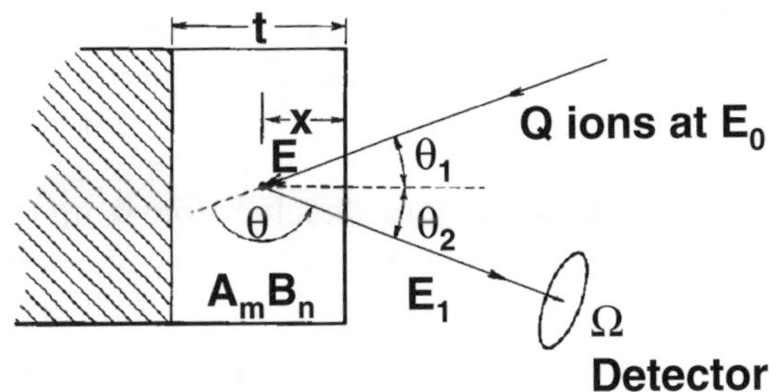

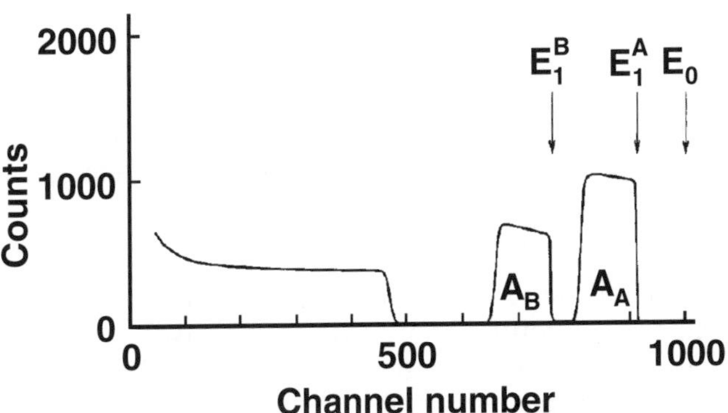

FIG. 4.1. Basic backscattering spectrometry. Experimental geometry (upper figure). Backscattering spectrum (lower figure) for the two-element compound (A_mB_n) film of uniform composition on a low-mass substrate.

The film elements may be identified by insertion of measured energies (E_1^A, E_1^B) of the high-energy sides of the peaks into

$$K_i \equiv E_1^i / E_0 \tag{4.1}$$

to calculate the kinematic factor K for the ith element. E_0 is the incident ion laboratory kinetic energy. In turn, the kinematic factor, K, is given by

$$K = \left[\frac{\left(M_2^2 - M_1^2 \sin^2\theta\right)^{1/2} + M_1 \cos\theta}{M_1 + M_2} \right]^2 \tag{4.2}$$

where θ is the laboratory angle through which the incident ion is scattered, and M_1 and M_2 are the masses of the incident and target particles, respectively. Since the parameters M_1, E_0, and θ are usually known, M_2 is determined and the target element is identified.

The areal density, $(Nt)_i$, in atoms per unit area, may be determined for the ith element from knowledge of the detector solid angle, Ω, the integrated peak count A_i for Q incident ions, and the measured or calculated cross section $\sigma_i(E,\theta)$ using

$$(Nt)_i = \frac{A_i \cos\theta_1}{Q\Omega\sigma_i(E,\theta)} \ . \tag{4.3}$$

Here, N_i is the atomic density (atoms per unit volume) of the ith element and t is the physical film thickness. If the scattering is Rutherford (pure Coulomb scattering), then $\sigma_i(E,\theta)$ may be calculated from

$$\sigma_R(E,\theta) = \left(\frac{Z_1Z_2e^2}{4E}\right)^2 \times \frac{4\left[\left(M_2^2 - M_1^2\sin^2\theta\right)^{1/2} + M_2\cos\theta\right]^2}{M_2 \sin^4\theta \left(M_2^2 - M_1^2 \sin^2\theta\right)^{1/2}} \tag{4.4}$$

where Z_1 and Z_2 are the atomic numbers of the incident and target ions, respectively. This equation is given in cgs units. A useful number in evaluating this equation is $e^2 \simeq 1.44 \times 10^{-13}$ MeV cm. For very thin films, E, the analysis ion energy immediately before scattering, may be taken as E_0. For thicker films, the mean energy of the analysis ions in the film should be used for E (see 4.2.4.6).

The average stoichiometric ratio for the compound film (A_mB_n) may be calculated from Eq. (4.3) to be

$$\frac{n}{m} = \frac{N_B}{N_A} = \frac{A_B}{A_A} \cdot \frac{\sigma_A(E,\theta)}{\sigma_B(E,\theta)} \quad . \tag{4.5}$$

Note that this ratio depends only on the ratio of measured integrated peak counts A_A/A_B and knowledge of the cross section ratio σ_A/σ_B. The hard-to-measure quantities Q and Ω have cancelled.

Conversion of areal densities, $(Nt)_i$, to physical film thickness, t, requires knowledge of the film density, ρ_{AB}. The relevant atomic densities, N_A^{AB} and N_B^{AB}, may be calculated from

$$N_A^{AB} = \frac{m\rho_{AB}N_0}{M_{AB}} \ ; \ N_B^{AB} = \frac{n\rho_{AB}N_0}{M_{AB}} \tag{4.6}$$

and then used in

$$t = \frac{(Nt)_A}{N_A^{AB}} = \frac{(Nt)_B}{N_B^{AB}} \tag{4.7}$$

to calculate t. Here, N_0 is Avogadro's number and $M_{AB} = mM_A + nM_B$ is the molecular weight of compound A_mB_n.

Typical uncertainties in the results are ±3% for areal densities and a few tenths of 1% for average stoichiometric ratios. The actual uncertainty in the physical film thickness is usually unknown since the film densities are usually unknown.

The preceding discussion indicates the power of backscattering spectrometry for analysis of a simple film (see 4.3.1.1 for further experimental details). Historically, the majority of backscattering analyses have been performed using ^{4}He analysis ions with energies in the 1-2 MeV region. The reasons for this are: (1) the available accelerators produced beams with these energies, (2) data for the energy loss of ^{4}He in the elements were better known than for other ions, (3) silicon surface barrier detector energy resolution for ^{4}He is about 15 keV, and most importantly, (4) the backscattering cross sections for ^{4}He incident on all elements more massive than Be are nearly Rutherford in this energy region.

The principal strengths of Rutherford backscattering (RBS) with ^{4}He ions are: (1) it is an absolute method that does not require the use of standards; meaningful uncertainties can usually be assigned to the results; (2) it is quick and easy; typical data acquisition time is about ten minutes; (3) it is frequently non-destructive, and (4) it may be used for depth profiling (with 10-30 nm depth resolution). The technique's principal weakness is that it is not good for trace-element analysis. It has moderate sensitivity ($\sim 10^{-4}$) to heavy elements in or on light matrices but very poor sensitivity ($\sim 10^{-1}$) to light elements in or on heavy matrices.

Recent years have seen increasing use of higher-energy light ions (^{1}H,^{4}He), higher-energy heavy ions (^{16}O,^{35}Cl), and even lower-energy ions (^{4}He,^{12}C) for backscattering analyses. The higher-energy light ions are used to improve the accuracy of measured stoichiometric ratios by reducing backscattering peak overlap, and to improve mass resolution and sensitivity to light elements. The cross sections are non-Rutherford for light target elements. The higher-energy heavy ions are used to improve mass and depth resolution and enhance the sensitivity to medium and heavy elements. The lower-energy ions are used to improve mass and depth resolution and sensitivity for all elements. Recent detector developments, particularly for time-of-flight and electrostatic analyzers, have made the use of the lower-energy ions feasible.

It is the aim of this chapter to present basic equations and tabular and graphical data needed to analyze backscattering spectra. A few examples illustrating how this information is used for the analyses are included. For a more extensive discussion of the principles of backscattering spectrometry, see Chu *et al.* (1978). Although this chapter is not a review of recent advances in the field, it concludes with a succinct discussion of experimental considerations that may be of use to the analyst in choosing the experimental set-up that will produce optimum data.

4.2 FUNDAMENTALS

4.2.1 The kinematic factor and mass resolution

Equation (4.2), the expression for the kinematic factor K, results from application of conservation of energy and momentum to the two-body collision between isolated particles of masses M_1 and M_2. Incident beam particle 1 (at laboratory kinetic energy E_0) is scattered with final laboratory kinetic energy E_1 through laboratory angle θ by target particle 2 (initially at rest in the laboratory). The binding energy of particle 2 in the target is neglected. The kinematic factor is independent of the nature of the force between the particles (as long as energy is conserved). Tables of kinematic factors, K, are included in Appendix 5 for scattering of ^1H and ^4He. The kinematic factors for other ions can be calculated by using Eq. (4.2). Average kinematic factors, $\overline{K}$, are also listed. The average mass $\overline{M}_2$, for natural isotopic abundance, has been used in the calculation of $\overline{K}$ [Eq. (4.2)].

For fixed θ, the energy separation, ΔE_1, for beam particles scattered by target particles of mass difference ΔM_2 is [from Eq. (4.1)]:

$$\Delta E_1 = E_0 \left(\frac{dK}{dM_2} \right) \Delta M_2 \quad . \tag{4.8}$$

If ΔE_1 is set equal to δE, the minimum energy separation that can be experimentally resolved, then δM_2, the mass resolution of the system is

$$\delta M_2 = \frac{\delta E}{E_0 \left(\dfrac{dK}{dM_2} \right)} \quad . \tag{4.9}$$

Figure 4.2 contains information that may be used to estimate δM_2 for a given experimental situation if the overall energy resolution, δE, is known. The quantity δE contains contributions from detector resolution, straggling, beam energy spread, and various geometric effects (O'Connor and Chunyu, 1989).

The mass resolution at the sample surface is usually determined primarily by the detector resolution; straggling dominates for layers deep in the sample. For fixed $\delta E/E_0$, Fig. 4.2 indicates that δM_2 improves with increasing analysis beam mass. This is somewhat deceptive since δE frequently depends on the analysis beam mass. For instance, if a typical surface barrier detector is used with 5 MeV ^4He and ^{12}C analysis beams, then δM_2 at $M_2 = 100$ u is actually somewhat better (2.2 u vs. 3.2 u) for ^4He than for ^{12}C because the detector resolution is ~15 keV for ^4He and ~50 keV for ^{12}C at 5 MeV (Leavitt *et al.*, 1988). There are, of course, instances where the detector resolution improves with increase in beam mass; such is the case for the time-of-flight detector.

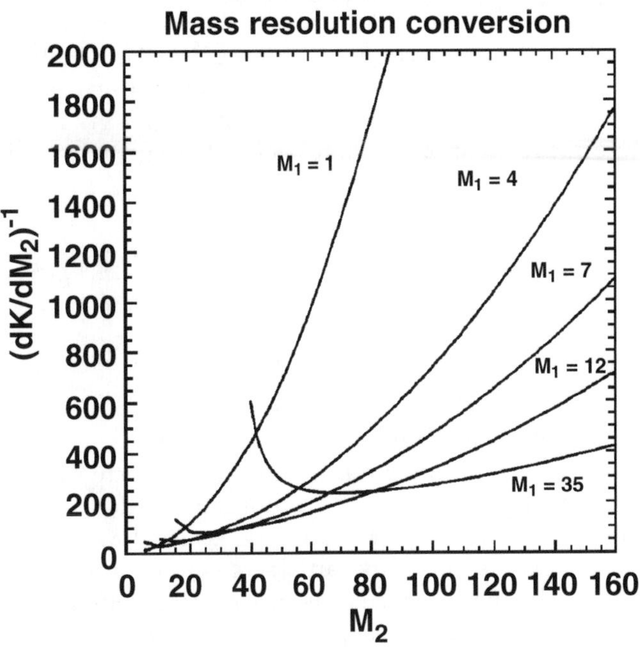

FIG. 4.2. Plots of $(dK/dM_2)^{-1}$ vs. target mass M_2 for several analysis beams (M_1 = ion mass). Units of M_1 and M_2 are u. The laboratory backscattering angle is 180°. The plots may be used to estimate the mass resolution $\delta M_2 = (\delta E/E_0)\,(dK/dM_2)^{-1}$ if $\delta E/E_0$ is specified.

4.2.2 Elastic scattering cross sections

4.2.2.1 Definition of the differential cross section

The average differential cross section, $\sigma(\theta,E)$, for scattering of beam particles of incident energy E by target particles in a thin film is defined by

$$\sigma(\theta, E) \equiv \left(\frac{1}{Nt} \right) \frac{dQ(E)}{Q} \frac{1}{\Omega(\theta)} \tag{4.10}$$

where Nt is the number of target atoms/unit area perpendicular to the beam, and [dQ(E)]/Q is the fraction of incident particles scattered into the small solid angle $\Omega(\theta)$ centered at deflection angle θ. If dQ(E) is replaced by A and subscripts i added, Eq. (4.3) results.

4.2.2.2 Rutherford cross sections

If the force between the incident nucleus (M_1,Z_1e,E) and the target nucleus (M_2,Z_2e, initially at rest) is assumed to be the Coulomb force, $\vec{F}_{12} = (Z_1 Z_2 e^2/r^2)\,\hat{r}$, then use of the above definition results in Eq. (4.4), the expression for the Rutherford cross section in the laboratory system. Numerical values of laboratory Rutherford cross sections for beams of ^{1}H, ^{4}He , ^{7}Li, ^{12}C, and ^{14}Si at 1 MeV incident energies are given in Appendix 6 for several backscattering angles. An accurate approximation for large backscattering angles and $(M_1/M_2) << 1$ is Chu *et al.* (1978)

$$\sigma_R(E,\theta) \simeq 0.02073\left(\frac{Z_1Z_2}{4E}\right)^2\left[\sin^{-4}\left(\frac{\theta}{2}\right)-2\left(\frac{M_1}{M_2}\right)^2\right] \quad (4.11)$$

with E in MeV and σ_R in b/sr [1b (barn) = 10^{-24} cm^2].

4.2.2.3 Non-Rutherford cross sections

Experimental measurements indicate that actual cross sections depart from Rutherford at both high and low energies for all projectile-target pairs. The low-energy departures are caused by partial screening of the nuclear charges by the electron shells surrounding both nuclei. Results of several investigations (L'Ecuyer *et al.*, 1979; Hautala and Luomajärvi, 1980; Andersen *et al.*, 1980; MacDonald *et al.*, 1983; and Wenzel and Whaling, 1952) indicate that these low-energy corrections are given with adequate accuracy by L'Ecuyer *et al.* (1979)

$$\sigma/\sigma_R = 1 - \frac{0.049\, Z_1 Z_2^{4/3}}{E_{CM}} \quad (4.12)$$

or by (Wenzel and Whaling, 1952)

$$\sigma/\sigma_R = 1 - \frac{0.0326\, Z_1 Z_2^{7/2}}{E_{CM}} \quad (4.13)$$

for light-ion analysis beams with MeV energies. Here E_{CM} is the center-of-mass kinetic energy in keV. In practice, replacing E_{CM} by E_{Lab} produces negligible error. A table of low-energy corrections is given in Appendix 6. Algorithms for rapid computations of cross sections for medium energy (50 - 1000 keV) backscattering are discussed by Mendenhall and Weller (1991a).

The high-energy departures of the cross sections from Rutherford behavior are caused by the presence of short-range nuclear forces. Recent measurements and calculations (Bozoian *et al.*, 1990; Bozoian, 1991a and 1991b; and Hubbard *et al.*, 1991) regarding the onset of these high-energy departures are summarized in Fig. 4.3 for ^{1}H, ^{4}He, and ^{7}Li analysis beams. The straight lines on Fig. 4.3 represent rough boundaries separating the region of Rutherford behavior (below the line) from the region of non-Rutherford behavior (above the line). Equations resulting from the least-squares fits to the points in Fig. 4.3 are

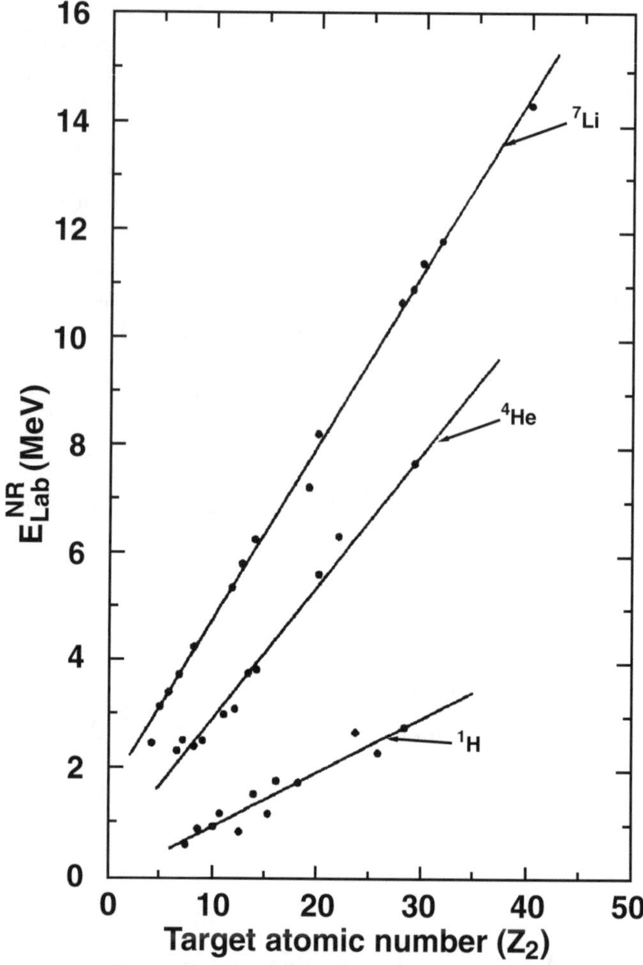

FIG. 4.3. Laboratory projectile energies, E_{Lab}^{NR}, at which backscattering cross sections (for $160° < \Theta_{Lab} < 180°$) deviate from Rutherford by 4% vs. target atomic number Z_2, for ^{1}H, ^{4}He, and ^{7}Li projectiles. The straight lines are least squares fitted to the data points [Eq. (4.14)]. The ^{1}H and ^{4}He points are experimental (see Bozoian, 1991b for references). The ^{7}Li data were obtained from optical model calculations (Bozoian, 1991a).

$$\begin{aligned} &\text{For } ^1\text{H:}\ E_{Lab}^{NR} \simeq (0.12 \pm 0.01)\, Z_2 - (0.5 \pm 0.1) \\ &\text{For } ^4\text{He:}\ E_{Lab}^{NR} \simeq (0.25 \pm 0.01)\, Z_2 + (0.4 \pm 0.2) \quad (4.14) \\ &\text{For } ^7\text{Li:}\ E_{Lab}^{NR} \simeq (0.330 \pm 0.005)\, Z_2 + (1.4 \pm 0.1) \end{aligned}$$

where E_{Lab}^{NR} is the laboratory projectile kinetic energy (in MeV) at which the backscattering cross section (for $160° < \theta_{Lab} < 180°$) deviates from Rutherford by 4% for a target atom of atomic number Z_2. Note that cross sections for backscattering of ^1H at 1 MeV are non-Rutherford for $Z_2 \lesssim 15$, while cross sections for ^4He at 2 MeV are Rutherford for $Z_2 \gtrsim 6$. At present, no practical method exists for rapid accurate calculation of these high-energy non-Rutherford cross sections; they must be measured.

Advantages associated with the use of non-Rutherford cross sections such as improved accuracy in determination of stoichiometric ratios and increased sensitivity for detection of light elements in heavy-element matrices frequently justify the additional work required to make the measurements. According to Eq. (4.5), the average stoichiometric ratio for two film elements depends only on the ratio of integrated peak counts, A_A/A_B, and the cross section ratio, σ_B/σ_A. These stoichiometric ratios may be determined as accurately as a few tenths of a percent by acquisition of sufficient data if the backscattering peaks are well separated and if the cross section ratios are accurately known. However, use of analysis ion energies such that the cross sections are Rutherford (and therefore accurately known) frequently produces backscattering peaks that overlap. The uncertainties in the peak count ratios resulting from deconvolution/ simulation techniques often severely limit the accuracies of the stoichiometric ratios, particularly in cases of non-uniform film composition.

Use of higher analysis ion energies usually results in desired reduction of peak overlap since the energy loss of the analysis ions in matter decreases with increasing energy in the energy range normally used. However, the accuracy of the cross section ratio may be adversely affected if the cross sections of interest are non-Rutherford at the higher ion energy. The analysis ion energy should usually be chosen to be as high as possible to take advantage of the reduction in peak overlap, but with a value such that the relevant cross sections have accurately measured values that do not vary wildly in the region just below the incident ion energy. If the ratio of measured-to-Rutherford cross section varies slowly with energy, the non-Rutherford effect may be easily included in calculations using Eqs. (4.3) and (4.5) by simply dividing the A_i by the non-Rutherford enhancement factors $(\sigma/\sigma_R)_i$ at the mean energy of the projectile in the film and proceeding with the calculation as if the cross sections were Rutherford. Since non-Rutherford cross sections for light target elements are frequently many times Rutherford, while corresponding cross sections for the heavier target elements may remain Rutherford, increased relative sensitivity for light element detection results. For example, the relative sensitivity for detection of C in thin films on Si substrates is enhanced relative to Rutherford by a factor of 7 for 3.8 MeV ^4He and by a factor of more than 100 for 4.27 MeV ^4He. In another instance, ^4He ions with energies 8.1–9.1 MeV have recently been used to enhance the sensitivity for detection of O in superconducting films by a factor of ~25 over Rutherford (Martin *et al.*, 1988; Barbour *et al.*, 1988).

Strong, narrow, isolated resonances in the non-Rutherford cross sections may be used for depth profiling light elements in or on heavy matrices. Examples are the 3.04 MeV resonance in the ^4He-^{16}O cross section (Cameron, 1953), the 2.525 MeV resonance in the ^1H-^9Be cross section (Mozer, 1956; Leavitt *et al.*, 1994), and the 4.26 MeV resonance in the ^4He-^{12}C cross section (Bittner and Moffat, 1954).

There is considerable information in the literature regarding measured non-Rutherford cross sections for ^1H and ^4He projectiles. Most of these data were acquired during the 1950s and 1960s in connection with nuclear level structure studies. The data have usually been presented in graphical form only, as differential cross sections in the center-of-mass system vs. projectile energy in the laboratory system. An unpublished compilation (Jarjis, 1979) gives some tables of numerical values produced by use of a computer digitizer. A few more recent reports contain tabular as well as graphical data on the measured cross sections. Non-Rutherford cross section information for ^1H and ^4He analysis beams is presented in Appendix 7.

4.2.3 Experimental geometry

Two experimental arrangements in common use are referred to as the IBM and Cornell geometries. For both geometries, the incident beam is horizontal and the sample surface vertical. For the IBM geometry, the scattered beam (directed at the detector), the incident beam, and the sample normal are all in the same horizontal plane. In the Cornell geometry the detector is directly below the incident beam; the incident beam and the scattered beam are in a vertical plane. In both geometries the angle between the sample normal and the incident beam is θ_1 and the angle between the sample normal and the scattered beam is θ_2. The "tilt" axis is a vertical axis through the beam spot on the sample surface, so the "tilt" angle is θ_1 (see Fig. 4.1) in both cases. The relation between the scattering angle θ and θ_1 and θ_2 is $\theta = \pi - |\theta_1 \pm \theta_2|$ for the IBM geometry and $\cos\theta_2 = \cos(\pi - \theta)\cos\theta_1$ for the Cornell geometry. In the IBM geometry, for a given tilt angle θ_1, the angle θ_2 depends on the direction of θ_1, i.e., whether the sample normal is rotated toward or away from the direction of the scattered beam. The relation between the inward and outward path lengths and the perpendicular distance x below the sample surface at which a backscattering event took place is given by

$$d_{in} = \frac{x}{\cos\theta_1}\ ;\ d_{out} = \frac{x}{\cos\theta_2}\quad \text{(IBM)}$$

$$d_{in} = \frac{x}{\cos\theta_1}\ ;\ d_{out} = \frac{x}{\cos\theta_1, \cos\theta_2}\quad \text{(Cornell)} \tag{4.15}$$

4.2.4 Effects of energy loss of ions in solids

4.2.4.1 Definitions

Many of the features of a backscattering spectrum are determined by the energy loss of the analysis beam ions as they traverse the sample material. Consequently, a quantitative knowledge of this energy loss is a key element in understanding a backscattering spectrum. This section offers only a brief summary of the relevant quantities and relations relating to energy loss, but a full exposition is given in Chapter 2.

The stopping power of a material for a particular ion is usually defined as the energy loss per distance travelled in the material, denoted as dE/dx. This quantity depends on the ion and the material traversed as well as the energy of the ion. Usual units of the stopping power are eV/Å or eV/nm.

Another quantity, the stopping cross section ε, is defined as the energy loss/atom/cm^2 (areal density) of material traversed. This quantity is independent of the volume density of the material. Usual units are 10^{-15} eV cm^2. The relation between these two quantities is given by

$$\frac{dE}{dx} = N\varepsilon \tag{4.16}$$

where N is the atomic density (atoms/cm^3). The reader should be aware that the literature is not consistent in these definitions, and a common unit of "stopping power" used in nuclear physics is eV/(mg/cm^2) where the material "thickness" is given in units of mg/cm^2 (actually an areal density) (Northcliffe and Schilling, 1970).

Values of ion stopping cross sections in all elements are available from an extensive study based on semi-empirical fitting of experimental data (Ziegler *et al.*, 1985). A parameterization of proton stopping cross sections involving eight parameters per element is given together with scaling rules for extension to any analysis ion (see Appendix 3, Section 3.2). The variation of stopping cross section for ^{4}He ions with energy is similar for all elements, showing a broad maximum below about 1 MeV. The decrease in energy loss of ^{4}He ions with increasing energy above about 1 MeV is responsible for the fact that the elemental peak widths in a backscattering spectrum are narrower at higher incident energies in this energy range.

An approximation, the Bragg rule (Bragg and Kleeman, 1905), is commonly used to calculate stopping cross sections of ions in compounds or mixtures of different elements. This approximation simply assumes that each target atom acts independently in the energy loss process and ignores any effects of chemical bonding in the material. This rule can be expressed for a compound A_mB_n by

$$\varepsilon^{A_mB_n} = m\varepsilon^A + n\varepsilon^B \tag{4.17}$$

which gives the compound stopping cross section in terms of energy loss/molecule/cm^2 traversed. The corresponding stopping power is given by

$$\left(\frac{dE}{dx}\right)^{AB} = N^{AB}\varepsilon^{AB} = N_A^{AB}\varepsilon^A + N_B^{AB}\varepsilon^B \tag{4.18}$$

where N^{AB} is the molecular density (molecules/cm^3) and N_A^{AB} and N_B^{AB} are the atomic densities of A and B in the compound (the m and n subscripts on A and B are suppressed in this notation). (See Example 4.1.) Deviations from the Bragg rule have been reported and a discussion of the effect of chemical binding on stopping powers is given in Ziegler and Manoyan (1988) (also see 2.2.3.4.).

EXAMPLE 4.1. Calculate the stopping cross section and stopping power of 2 MeV $^4He^+$ in Al_2O_3 using the Bragg rule. Use tabulated values of stopping cross sections (see Appendix 3, Section 3.1), $\varepsilon^{Al} = 44 \times 10^{-15}$ eV cm^2 and $\varepsilon^{O} = 35 \times 10^{-15}$ eV cm^2 in Eq. (4.17) to find

$$\varepsilon^{Al_2O_3} = (2 \times 44 + 3 \times 35) \times 10^{-15}$$
$$= 193 \times 10^{-15} \text{ eV cm}^2 \ .$$

To find $(dE/dx)^{Al_2O_3}$ we calculate the molecular density $N^{Al_2O_3}$ as

$$N^{Al_2O_3} = \frac{\rho N_0}{M} = \frac{4 \dfrac{gm}{cm^3} \times 6 \times 10^{23} \dfrac{molecules}{mole}}{102 \dfrac{gm}{mole}}$$

$$= 2.35 \times 10^{22} \frac{Al_2O_3}{cm^3} \ .$$

The stopping power is obtained from Eq. (4.18)

$$\left(\frac{dE}{dx}\right)^{Al_2O_3} = N^{Al_2O_3} \varepsilon^{Al_2O_3}$$
$$= 2.35 \times 10^{22} \times 193 \times 10^{-15}$$
$$= 46 \frac{eV}{\text{Å}} \ .$$

An alternate method, using an atomic basis, involves calculating the atomic densities of each element in the molecule.

$$N_{Al}^{Al_2O_3} = 2 \times 2.35 \times 10^{22} = 4.7 \times 10^{22} \text{ Al}/cm^3$$

and

$$N_{O}^{Al_2O_3} = 3 \times 2.35 \times 10^{22} = 7.1 \times 10^{22} \text{ O}/cm^3 \ .$$

Using the second equality in Eq. (4.18), we find:

$$\left(\frac{dE}{dx}\right)^{Al_2O_3} = N_{Al}^{Al_2O_3} \varepsilon^{Al} + N_{O}^{Al_2O_3} \varepsilon^{O}$$
$$= 4.7 \times 10^{22} \times 44 \times 10^{-15}$$
$$+ 7.1 \times 10^{22} \times 35 \times 10^{-15}$$
$$= 46 \times 10^8 \frac{eV}{cm}$$
$$= 46 \frac{eV}{\text{Å}}$$

The calculation of energy loss, ΔE, of ions traversing a solid involves the integration of the stopping power as in

$$\Delta E = \int \frac{dE}{dx} dx \ . \qquad (4.19)$$

In cases of thin targets the stopping power can often be taken as constant, evaluating the necessary quantities at the incident energy (surface energy approximation) or at the mean energy of the beam in the target (mean energy approximation). In the case of thick targets, a computer is usually used to calculate the energy of ions at a depth below the sample surface by numerical integration of Eq. (4.19), dividing the target material into thin slabs and using a parameterization to calculate the stopping power at successive ion energies. Notice that calculation of ion energy loss requires a knowledge of the composition of the sample, which may be the object of the analysis and therefore unknown. An iterative procedure is often required in such cases, initially using an assumed composition to calculate an initial approximation of energy loss which is in turn used in calculating improved values of the composition (see 4.3.1.1).

The range of ion beams in materials and the phenomena of energy straggling are also of importance in understanding a backscattering spectrum (see Chapter 2). Range information can be used to estimate the maximum analysis depth possible with a given beam at a given incident energy. Recall, however, that the beam must scatter and reemerge from the sample to be useful in backscattering analysis. A rough criteria for accessible depth proposed by Chu *et al.* (1978) is that the energy of the scattered particle at the detector should be greater than 1/4 the incident energy.

4.2.4.2 Depth scale

To utilize the depth profiling capabilities of backscattering it is necessary to relate the energy of the scattered particle to the depth in the sample where the scattering occurred. This depends on the energy loss of the analysis ion traversing the sample, the kinematic factor for the scattering, and the orientation of the sample normal relative to both the incident beam and the detector direction. We denote the difference in energy at the detector of a particle scattered at the surface and a particle scattered at a depth x, measured perpendicular to the sample surface, as ΔE. This quantity is given as a function of x by

$$\Delta E = [S]\, x \qquad (4.20)$$

where $[S]$ is called the energy loss factor and is defined by

$$[S] = \left[K \left(\frac{dE}{dx} \right)_{in} \frac{1}{\cos \theta_1} + \left(\frac{dE}{dx} \right)_{out} \frac{1}{\cos \theta_2} \right] . \qquad (4.21)$$

The kinematic factor K and the angles θ_1 and θ_2 are defined in Sections 4.2.1 and 4.2.3, and the stopping powers are those for the analysis ion on the inward and outward path.

The corresponding relation involving the areal density and stopping cross sections is given by

$$\Delta E = [\varepsilon]\, Nx \qquad (4.22)$$

for a single element sample. The quantity $[\varepsilon]$ is called the stopping cross section factor and is defined by

$$[\varepsilon] = \left[K\, \varepsilon_{in} \frac{1}{\cos \theta_1} + \varepsilon_{out} \frac{1}{\cos \theta_2} \right] . \qquad (4.23)$$

For multi-element samples, we see that the depth-energy relation depends on the struck particle; therefore, separate relations must be calculated for each element in the sample. In the case of a compound A_mB_n, N becomes the molecular density (molecules/cm^3) N^{AB} and corresponding values of K and ε_{out} which apply for scattering from elements A or B are used in the stopping cross section factor. The relations for element *A* are given in Eqs. (4.24) and (4.25).

$$\Delta E_A = [\varepsilon]_A^{AB}\, N^{AB}\, x \qquad (4.24)$$

$$[\varepsilon]_A^{AB} = \left[K_A\, \varepsilon_{in}^{AB} \frac{1}{\cos \theta_1} + \varepsilon_{out,\,A}^{AB} \frac{1}{\cos \theta_2} \right] . \qquad (4.25)$$

Similar relations can be written for element B. The lower index refers to the scattering element and the upper index refers to the stopping material. A surface energy approximation is often used for stopping cross section factors. In this appoximation, ε_{in} is evaluated at the incident energy E_0 and ε_{out} is evaluated at energy KE_0. The resulting stopping cross section factor is usually written $[\varepsilon_0]$. (See Example 4.2.)

Since both the energy loss factor and stopping cross section factor require evaluation of energy dependent parameters, an integral over the relevant energies should be performed. In many cases use of the incident beam energy (surface—energy approximation) or the mean beam energy in the sample (mean—energy approximation) is satisfactory.

4.2.4.3 Depth resolution

The considerations in the previous section lead to an expression for the depth resolution in backscattering analysis. Using Eq. (4.20) we see that the minimum detectable depth difference, δx, is related to the minimum detectable scattered particle energy difference, δE, by

$$\delta x = \frac{\delta E}{[S]} . \qquad (4.26)$$

Sources of energy spread include detector resolution, energy spread in the incident beam, straggling, and kinematic effects. Since energy straggling increases as the ion beam traverses the sample, the depth resolution degrades with depth in the sample. A common practice is to quote depth resolution at the surface which need not include a straggling contribution. A convenient approximation is to assume all sources of energy spread are Gaussian and to add them in quadrature.

The depth resolution can be improved by increasing [S]. This is usually done by tilting the sample normal relative to the incoming beam (i.e., increasing θ_1 and/or

EXAMPLE 4.2. Calculate the depth-scattered ion energy differences for 2 MeV $^4He^+$ in Al_2O_3. We again consider the case $\theta_1 = 0$ and $\theta_2 = 10°$. Here we must calculate separate differences for Al and O. The Al and O stopping cross section factors are calculated using Eq.(4.25). The K factors for ^4He on Al and O are 0.5525 and 0.3625, respectively, resulting in energies after scattering (at the surface) of $0.5525 \times 2.0 = 1.105$ MeV for Al and $0.3625 \times 2.0 = 0.725$ MeV for O. We evaluate the elemental stopping cross sections involved in $\varepsilon_{out,Al}^{Al_2O_3}$ and $\varepsilon_{out,O}^{Al_2O_3}$ at these energies using the surface energy approximation.

$$\varepsilon_{in}^{Al_2O_3} = 2 \times \varepsilon_{in}^{Al} + 3 \times \varepsilon_{in}^{O} = 2 \times 44 \times 10^{-15} + 3 \times 35 \times 10^{-15} = 193 \times 10^{-15} \text{ eV cm}^2$$

$$\varepsilon_{out,Al}^{Al_2O_3} = 2 \times \varepsilon_{out,Al}^{Al} + 3 \times \varepsilon_{out,Al}^{O} = 2 \times 51 \times 10^{-15} + 3 \times 46 \times 10^{-15} = 240 \times 10^{-15} \text{ eV cm}^2$$

$$\varepsilon_{out,O}^{Al_2O_3} = 2 \times \varepsilon_{out,O}^{Al} + 3 \times \varepsilon_{out,O}^{O} = 2 \times 54 \times 10^{-15} + 3 \times 48 \times 10^{-15} = 252 \times 10^{-15} \text{ eV cm}^2 \ .$$

We can now calculate the stopping cross section factors

$$[\varepsilon_0]_{Al}^{Al_2O_3} = K_{Al}\, \varepsilon_{in}^{Al_2O_3} \frac{1}{\cos\theta_1} + \varepsilon_{out,Al}^{Al_2O_3} \frac{1}{\cos\theta_2}$$

$$[\varepsilon_0]_{Al}^{Al_2O_3} = 0.5525 \times 193 \times 10^{-15} + 240 \times 10^{-15} \times 1.015 = 350 \times 10^{-15} \text{ eV cm}^2$$

and

$$[\varepsilon]_{O}^{Al_2O_3} = K_{O}\, \varepsilon_{in}^{Al_2O_3} \frac{1}{\cos\theta_1} + \varepsilon_{out,O}^{Al_2O_3} \frac{1}{\cos\theta_2}$$

$$[\varepsilon_0]_{O}^{Al_2O_3} = 0.3625 \times 193 \times 10^{-15} + 252 \times 10^{-15} \times 1.015 = 326 \times 10^{-15} \text{ eV cm}^2 \ .$$

Using the molecular density $N^{Al_2O_3} = 2.35 \times 10^{22}$ molecules/cm^3, we find

$$\Delta E_{Al} = [\varepsilon_0]_{Al}^{Al_2O_3}\, N^{Al_2O_3}\, x = 82.3 \frac{\text{eV}}{\text{Å}} \times x$$

$$\Delta E_{O} = [\varepsilon_0]_{O}^{Al_2O_3}\, N^{Al_2O_3}\, x = 76.6 \frac{\text{eV}}{\text{Å}} \times x$$

where ΔE_{Al} and ΔE_O are the energies of ions scattered from Al and O relative to ions scattered at the surface and x is the depth in the target where the scattering took place.

θ_2). The effect is to increase the path length required to reach a given depth (measured perpendicular to the surface) in the sample. This increases the scattered particle energy difference for a given depth difference. It should be noted that the use of large tilt angles introduces additional sources of energy broadening and requires that the sample surface be reasonably flat. Several studies concerned with optimizing depth resolution by target tilting can be found in the literature (Williams and Moller, 1978; O'Conner and Chunyu, 1989; and Boerma *et al.*, 1990).

4.2.4.4 Surface spectrum height

An important characteristic of a backscattering spectrum is the height (counts/channel) of the front edge of an elemental peak corresponding to scattering from the top surface of the sample. We consider here cases where the sample produces elemental peaks which are much wider than the energy resolution of the detection system. For a single-element target, the surface height is given by

$$H_0 = \sigma(E_0)\Omega Q \mathcal{E} / ([\varepsilon_0] \cos\theta_1) \tag{4.27}$$

where $\mathcal{E}$ is the energy width per channel and $[\varepsilon_0]$ is the surface energy approximation of the stopping cross section factor [defined in Eq. (4.23)] evaluated at the incident energy. The remaining symbols are those defined in Eq. (4.3). This equation is obtained from Eq. (4.3), where the areal density (Nt) contributing to one channel of the spectrum at the surface is $\mathcal{E}/[\varepsilon_0]$ [see Eq. (4.22)].

The surface heights of the two elemental peaks in a compound A_mB_n are given by

$$H_{A,0} = \sigma_A(E_0)\,\Omega Q m \mathcal{E}/([\varepsilon_0]_A^{AB}\cos\theta_1)$$

$$H_{B,0} = \sigma_B(E_0)\,\Omega Q n \mathcal{E}/([\varepsilon_0]_B^{AB}\cos\theta_1) \quad (4.28)$$

where $[\varepsilon_0]_A^{AB}$ and $[\varepsilon_0]_B^{AB}$ are the compound stopping cross section factors as defined in Eq. (4.25) using the surface energy approximation.

The stoichiometry of a multi-element sample can be calculated by comparing surface heights of the elemental peaks (see 4.3.1.2 and Example 4.3).

4.2.4.5 Peak widths for thin films

The energy width of elemental peaks in single- or multi-elemental samples, where the peaks are at least partially resolved and wider than the system energy resolution, can be calculated from Eqs. (4.22) or (4.24) by replacing x by the film thickness t. Conversely, the elemental areal densities can be calculated from experimentally determined peak widths using these equations and a knowledge of stopping cross sections. This procedure is described in 4.3.1.3.

4.2.4.6 Mean energy in thin films

To calculate the mean energy of the analysis ions in a thin film containing r elements, use Eq. (4.3) with $E = E_0$ to calculate $(Nt)_i^{SEA}$ in the surface energy approximation (SEA). Calculate the energy loss of the ions passing through the film, ΔE_{in}^{SEA}, using

$$\Delta E_{in}^{SEA} = \sum_{i=1}^{r} \varepsilon^i (E_0)(Nt)_i^{SEA} \quad . \quad (4.29)$$

Then calculate the mean energy of the ions in the film, $\overline{E}^{(1)}$, using

$$\overline{E}^{(1)} = E_0 - \frac{\Delta E_{in}^{SEA}}{2} \quad . \quad (4.30)$$

This result represents a first order correction. The procedure should be iterated until $\overline{E}$ changes by less than a specified percentage between successive iterations. For the second iteration, the $(Nt)_i^{(1)}$ should be calculated using Eq. (4.3) with $E = \overline{E}^{(1)}$; then $\Delta E_{in}^{(1)}$ and $\overline{E}^{(2)}$ are calculated using

$$\Delta E_{in}^{(1)} = \sum_{i=1}^{r} \varepsilon^i (\overline{E}^{(1)})(Nt)_i^{(1)} \quad (4.31)$$

and

$$\overline{E}^{(2)} = E_0 - \frac{\Delta E_{in}^{(1)}}{2} \quad . \quad (4.32)$$

EXAMPLE 4.3. Calculate surface heights for 2 MeV $^4He^+$ on Al_2O_3. We assume the following experimental parameters for this calculation:

$\Omega = 10^{-3}$ sr

$\mathcal{E} = 1$ keV/channel

$Q = 6.24 \times 10^{13}$ incident particles (10 μC charge)

$\theta_1 = 0$, $\theta_2 = 10°$ (scattering angle = 170°)

The Rutherford cross sections for Al and O are found in Appendix 6, to be 0.2128×10^{-24} cm^2/sr and 0.0741×10^{-24} cm^2/sr. The compound stopping cross section factors are those found in the depth scale Example 4.2:

$$[\varepsilon_0]_{Al}^{Al_2O_3} = 350 \times 10^{-15} \text{ eV cm}^2$$

$$[\varepsilon_0]_{O}^{Al_2O_3} = 326 \times 10^{-15} \text{ eV cm}^2 \quad .$$

Using Eq. (4.28), we find

$$H_{Al,0} = \frac{\sigma\Omega Q 2\mathcal{E}}{[\varepsilon_0]_{Al}^{Al_2O_3}}$$

$$= \frac{0.2128 \times 10^{-24} \times 10^{-3} \times 6.24 \times 10^{13} \times 2 \times 10^3}{350 \times 10^{15}}$$

$$= 76 \text{ cts} \quad .$$

$$H_{O,0} = \frac{\sigma\Omega Q 3\mathcal{E}}{[\varepsilon_0]_{O}^{Al_2O_3}}$$

$$= \frac{0.0741 \times 10^{-24} \times 10^{-3} \times 6.24 \times 10^{13} \times 3 \times 10^3}{326 \times 10^{15}}$$

$$= 43 \text{ cts (or particles / channel)} \quad .$$

Several iterations may be required if the energy loss in the film is an appreciable fraction of E_0. The final $(Nt)_i^{(f)}$ values are calculated using Eq. (4.3) with $E = \overline{E}^{(f)}$, where $\overline{E}^{(f)}$ is the final mean energy obtained from the iteration process. If the scattering is Rutherford and the film is not too thick

$$(Nt)_i^{(f)} = \left(\frac{\overline{E}^{(f)}}{E_0}\right)^2 (Nt)_i^{SEA} \quad . \qquad (4.33)$$

4.3 SAMPLE ANALYSIS

This section starts with a very brief description of apparatus and operating conditions in a "typical" backscattering laboratory. The specific laboratory which is described was chosen for reasons of familiarity and concreteness. Procedures for determining certain necessary experimental quantities such as analysis beam energy, pulse-height-analyzer energy/channel, etc., are also briefly described. Examples of actual data analysis follow. Several of these analyses are explicitly done "by hand" to clearly illustrate the power and simplicity of the technique. The section concludes with a brief discussion of computer-assisted data analysis, which is, of course, currently used for most analyses.

Most of the data for the sample analyses described in this section were acquired with the standard backscattering setup described in Leavitt (1987). Ion beams from a vertical single-ended 5.5 MV Van de Graaff were deflected through 90° by a bending magnet into a horizontal collimating beam-line that preceded the target chamber. A 25 mm^2 surface barrier detector, placed about 150 mm upstream from the target holder at an angle 10° below the beam (Cornell geometry), subtended solid angle $\Omega = 0.78$ msr at the target. Hence, the backscattering angle, $\theta = 170°$. Detector pulses due to the backscattered analysis ions were preamplified, shaped and amplified, and sorted by a pulse-height-analyzer (PHA). The result, in the form of counts/channel vs. channel number, constituted the backscattering spectrum of the target. These data were sent to a PC for disk storage, integration of peaks, plotting, data analysis, etc. Some typical operating conditions and parameter ranges for this system are given in Table 4.1.

The beam energy (or the bending-magnet field) was calibrated using three (α,γ) resonances in ^{24}Mg (Endt and van der Leun, 1967), and (α,α) resonances in ^{14}N (Herring, 1958) and ^{16}O (Häusser *et al.*, 1972), which covered the ^4He beam-energy range 2437.4 keV to 5058 keV. See Appendix 17 for additional calibration points. The beam energy was known to ± 5 keV and was stable to less than 1 keV; the beam-energy spread was less than 0.5 keV.

Table 4.1. Typical experimental operating conditions and parameter ranges used during acquisition of backscattering spectra described in Section 4.3.

Experimental parameter	Units	Values
Analysis ion energy	MeV	1.0-5.0
Beam cross section	mm × mm	1.5 × 1.5
Beam current	nA	10-200
Integrated charge	μC	5-100
Detector energy resolution for ^4He ions	keV	15
Data acquisition time	min	5-10
Vacuum	Torr	2×10^{-6}
Pump-down time	min	15

The conversion of PHA channel number, n, to backscattered ion energy, E_1, was accomplished by least-squares fitting

$$E_1 = n\,\mathcal{E} + E' \qquad (4.34)$$

to backscattering peak data (n_i, E_1^i) from a very thin film containing Ta, Nb, Al, and O on a C substrate. The value of E_1^i for the *i*th film element was calculated with Eq. (4.1) using known values of K_i and E_0. The value of n_i, the *peak* channel for the *i*th element was read from the spectrum (see Appendix C in Chu *et al.*, 1978). Thus, values of energy/channel, $\mathcal{E}$, and energy intercept, E', were determined for the amplifier gain used.

The number of analysis ions, Q, incident on the target was calculated from the total charge, Q′, deposited in the insulated target chamber during the run; $Q = Q'/e$, where $e = 1.602 \times 10^{-19}$ Coulombs. It was assumed that the incident ions bore charge +e. (See Chapter 12 for possible corrections.) The charge Q′ was divided by the dead-time correction factor, DTR, to account for the fact that the PHA did not accept pulses during a portion of the time while Q′ was being collected. The DTR factor was usually taken as the ratio of the "real" time to "live" time (both supplied by the PHA); this factor was usually $\lesssim 1.02$ for the analyses discussed below. Other methods for determining the DTR factor are discussed in Chapter 12.

Daily measurement of the areal density of a secondary standard Ta film, which had been calibrated with a Bi-implanted-in-Si RBS standard (Eshbach, 1983), provided another correction factor C_{Bi}. This factor was related to the efficiency of charge collection and the value of solid angle used. Its value was that required to give the correct areal density for the Ta standard. Typical C_{Bi} values were near 1.00 with an uncertainty of ±0.03; this uncertainty of ~ 3% included the uncertainties in charge collection and solid angle measurement as well as the uncertainty (2.5%) in the Bi RBS standard itself.

A number of apparatus parameters and measured or calculated quantities were usually considered known prior to acquisition of data for a particular sample; a summary of these is given in Table 4.2. Certain other quantities peculiar to a particular run on a particular sample are listed in Table 4.3.

4.3.1 Thin film analysis

4.3.1.1 The peak integration method

If the backscattering peaks are well separated so the integrated peak counts, A_i, can be accurately determined from the spectrum, then the *peak integration* method may be applied in the simple direct manner discussed in Section 4.1. A slightly modified version of Eq. (4.3)

$$(Nt)_i = \frac{A_i \cos\theta_1 \cdot DTR \cdot C_{Bi} \cdot e}{Q' \cdot \Omega \cdot \sigma_R^i(E,\theta) \cdot \left(\dfrac{\sigma}{\sigma_R}\right)_i} \qquad (4.35)$$

should be used for the calculation of $(Nt)_i$, the areal density of the ith element in the film. The symbols are defined in Tables 4.2 and 4.3. The ratio, $(\sigma/\sigma_R)_i$, is the non-Rutherford correction factor discussed in 4.2.2.3. Note that in instances where the integrated charge and solid angle are not well known, the quantity $C_{Bi}/(Q' \cdot \Omega)$ may be obtained from the substrate (if there is one), assuming the stopping cross section factor and scattering cross section is known for the substrate.

The spectrum shown in Fig. 4.1 is actually that of a Gd Fe film on a Si substrate; the analysis beam ions were 3776 keV ^4He. Example 4.4 uses data taken from this spectrum to provide element identification, elemental areal densities, the stochiometric ratio, and a "thickness" estimate for this film.

Table 4.2. Quantities whose values are usually known prior to acquisition of backscattering data for a particular sample.

Symbol	Quantity
M_1, Z_1	Identity (mass and atomic number) of the analysis ions
E_0	Incident laboratory kinetic energy of the analysis beam ions
θ	Laboratory angle through which the analysis ion is scattered
Ω	Solid angle subtended by the detector at the target
θ_1, θ_2	Angles between the sample normal and the incident and backscattered beams, respectively
$\mathcal{E}$, E′	Energy/channel and energy intercept, respectively; parameters connecting the backscattered energy, E_1, of an analysis ion with the PHA channel number, *n*, by $E_1 = n\mathcal{E} + E'$
C_{Bi}	Correction factor (related to efficiency of charge collection and solid angle measurement) that gives correct $(Nt)_{Bi}$ for Bi RBS standard
$K_i(\theta)$	The kinematic factor for target element *i* and backscattering angle θ
$\sigma_i(E,\theta)$	Cross sections for scattering of analysis ions of laboratory energy, *E*, through angle θ, for the *i*th target element
$\varepsilon^i(E)$	Stopping cross sections for analysis ions of laboratory energy *E* in the *i*th target element

Table 4.3. Quantities whose values are determined by a particular run on a particular sample.

Symbol	Quantity
Q′	Integrated charge deposited on the sample during the run
DTR	Dead-time-ratio for the PHA during the run
n_i	Channel number (at half-maximum) of the high-energy edge of the signal due to scattering from the *i*th target element at the sample surface
n_i'	Channel number (at half-maximum) of the low-energy edge of the backscattering peak due to the *i*th target element
$H_{i,0}$	Spectrum height (counts/channel) of the signal due to scattering from the *i*th target element at the sample surface
A_i	Integrated counts in the peak due to scattering from the *i*th element (in a thin film)
ΔE_i	Energy width of the peak due to scattering from the *i*th target element (in a thin film)

EXAMPLE 4.4. This example is an application of the peak integration method to analysis of the two-element thin film whose spectrum is shown in Fig. 4.1.

• *A priori* acquisition parameters (the symbols are defined in Table 4.2).

$E_0 = 3776$ keV	$\theta = 170°$
$\theta_1 = 0°$	$\theta_2 = 10°$
$\Omega = 0.78$ msr	$C_{Bi} = (0.99 \pm 0.03)$
$\mathcal{E} = (3.742 \pm .005)$ keV/ch	$E' = (8 \pm 3)$ keV
$K_{Fe}\,(170°) = 0.7520$	$K_{Gd}\,(170°) = 0.9039$ (from Appendix 5)

$$\sigma_R^{Fe}\,(E_0, 170°) = \frac{3.521}{(3.776)^2} \times 10^{-24}$$

$$= 0.2469 \times 10^{-24}\ \frac{cm^2}{sr}$$

$$\sigma_R^{Gd}\,(E_0, 170°) = \frac{21.53}{(3.776)^2} \times 10^{-24}$$

$$= 1.510 \times 10^{-24}\ \frac{cm^2}{sr}$$

(from Appendix 6)

$$\left(\frac{\sigma}{\sigma_R}\right)_{Fe} = 1 - \frac{(0.049)(2)(26)^{4/3}}{3776} = 0.998$$

$$\left(\frac{\sigma}{\sigma_R}\right)_{Gd} = 1 - \frac{(0.049)(2)(64)^{4/3}}{3776} = 0.993$$

[from Eq. (4.12)]

ε^{Fe} (3776 keV) $= 51.4 \times 10^{-15}$ eV cm^2

ε^{Fe} (3676 keV) $= 52.2 \times 10^{-15}$ eV cm^2

ε^{Gd} (3776 keV) $= 86.3 \times 10^{-15}$ eV cm^2

ε^{Gd} (3676 keV) $= 87.5 \times 10^{-15}$ eV cm^2

(from Appendix 3)

• Parameter values directly associated with the spectrum shown in Fig. 4.1 (the symbols are defined in Table 4.3).

$$Q' = 20.01\ \mu C\ ;\ DTR = 1.008\ ;$$
$$n_B = (757 \pm 1)\ ;\ n_A = (910 \pm 1)$$
$$n'_B = (660 \pm 1),\ n'_A = (812 \pm 1)$$
$$H_{A,0}^{AB} = (1020 \pm 20)\ \text{cts};\ H_{B,0}^{AB} = (640 \pm 20)\ \text{cts}$$

Integrated counts in spectral regions of interest (initial and final channel numbers are listed):

channels (789 – 918) = 103978 cts ; (920 – 960) = 49 cts

channels (640 – 767) = 64957 cts ; (768 – 788) = 79 cts

• Element identification [using Eqs. (4.34) and (4.1)].

$$E_1^B = n_B\,\mathcal{E} + E' = (757 \pm 1)\,(3.742 \pm 0.005) + (8 \pm 3) = (2841 \pm 6)\ \text{keV}$$

$$E_1^A = n_A\,\mathcal{E} + E' = (910 \pm 1)\,(3.742 \pm 0.005) + (8 \pm 3) = (3413 \pm 7)\ \text{keV}$$

$$K_B = \frac{E_1^B}{E_0} = \frac{(2841 \pm 6)}{(3776 \pm 5)} = 0.752 \pm 0.002$$

$$K_A = \frac{E_1^A}{E_0} = \frac{(3413 \pm 7)}{(3776 \pm 5)} = 0.904 \pm 0.002$$

Therefore, elements A and B are Gd and Fe, respectively. (Note that element A could also be Tb, since $K_{Tb} = 0.9048$.)

• Calculation of elemental areal densities, $(Nt)_i$.

First, the A_i are calculated from the integrated counts in the regions of interest. In this instance, a constant background, determined by the counts/ch just above the particular peak, is subtracted.

$$A_{Fe} = 64957 - \frac{79}{21}(128) = (64475 \pm 261)\ \text{cts}$$

$$A_{Gd} = 103978 - \frac{49}{41}(130) = (103823 \pm 323)\ \text{cts}$$

In this case, the background correction is almost negligible; the calculated uncertainties in the A_i are statistical.

Next, the areal densities in the surface energy approximation, $(Nt)_i^{SEA}$, are calculated using Eq. (4.35) with $E = E_0$

$$(Nt)_{Fe}^{SEA} = \frac{(64475 \pm 261)(1.008)(0.99 \pm 0.03)(1.602 \times 10^{-19})}{(20.01 \times 10^{-6})(0.78 \times 10^{-3})(0.2469 \times 10^{-24})(0.998)} \text{ atoms/cm}^2$$

$$= (2.68 \pm 0.08) \times 10^{18} \text{ atoms/cm}^2$$

$$(Nt)_{Gd}^{SEA} = (0.709 \pm 0.021) \times 10^{18} \text{ atoms/cm}^2 \ .$$

Then the mean energy of the ^{4}He ion in the film, $\overline{E}^{(1)}$ is calculated (in first order) using Eq. (4.29) for the first-order energy loss, ΔE_{in}^{SEA}, of the ions in the film and Eq. (4.30)

$$\Delta E_{in}^{SEA} = \varepsilon^{Fe}(E_0)(Nt)_{Fe}^{SEA} + \varepsilon^{Gd}(E_0)(Nt)_{Gd}^{SEA}$$

$$= (51.4 \times 10^{-15})(2.68 \times 10^{18})$$

$$+ (86.3 \times 10^{-15})(0.709 \times 10^{18}) \text{ eV}$$

$$= 199 \text{ keV}$$

$$\overline{E}^{(1)} = E_0 - \frac{\Delta E_{in}^{SEA}}{2} = 3776 - \frac{199}{2} = 3676 \text{ keV} \ .$$

The areal densities, $(Nt)_i^{(1)}$, including the first-order correction for energy loss in the film are [adjusting the Rutherford cross section in Eq. (4.35) to the mean energy in the sample, using the inverse square dependence on energy]

$$(Nt)_{Fe}^{(1)} = \left(\frac{3676}{3776}\right)^2 (Nt)_{Fe}^{SEA} = 2.54 \times 10^{18} \text{ atoms/cm}^2$$

$$(Nt)_{Gd}^{(1)} = 0.672 \times 10^{18} \text{ atoms/cm}^2 \ .$$

Results of an additional iteration of this procedure using Eqs. (4.31), (4.32), and (4.33) (note that ε^{Fe} and ε^{Gd} are evaluated at $\overline{E}^{(1)}$) are

$$\Delta E_{in}^{(1)} = (52.2 \times 10^{-15})(2.54 \times 10^{18})$$

$$+ (87.5 \times 10^{-15})(0.672 \times 10^{18}) \text{ eV}$$

$$= 191 \text{ keV}$$

$$\overline{E}^{(2)} = 3776 - \frac{191}{2} = 3681 \text{ keV}$$

$$(Nt)_{Fe}^{(2)} = \left(\frac{3681}{3776}\right)^2 (Nt)_{Fe}^{SEA}$$

$$= (2.55 \pm 0.08) \times 10^{18} \text{ atoms/cm}^2$$

$$(Nt)_{Gd}^{(2)} = (0.674 \pm 0.021) \times 10^{18} \text{ atoms/cm}^2 \ .$$

Additional iterations produce no further change in the $(Nt)_i$ values, so these are the final values of energy loss in the film, mean analysis ion energy in the film, as well as the elemental areal densities given by the mean energy approximation.

• The average stoichiometric ratio for this film may be calculated using Eq. (4.5)

$$\frac{N_{Fe}}{N_{Gd}} = \frac{A_{Fe}}{A_{Gd}} \cdot \frac{\sigma_R^{Gd}(E_0, 170°)}{\sigma_R^{Fe}(E_0, 170°)} \cdot \frac{\left(\frac{\sigma}{\sigma_R}\right)_{Gd}}{\left(\frac{\sigma}{\sigma_R}\right)_{Fe}}$$

$$= \frac{(64475 \pm 261)}{(103823 \pm 323)} \cdot \frac{21.53}{3.521} \cdot \frac{0.993}{0.998}$$

$$= 3.78 \pm 0.02 \ .$$

If the "molecular" formula for the film is written $Gd_m Fe_n$, then $m = 0.209 \pm 0.001$, $n = 0.791 \pm 0.001$. The quoted uncertainties are statistical, that is, due to the uncertainty in the A_i ratio. The cross section ratio has been regarded as exact in this calculation. The cross sections for He-Fe and He-Gd, except for the small electron shell corrections, are believed to be Rutherford at the analysis energy used. Of course, an assigned cross section ratio uncertainty may be easily included in the calculation (see Chapter 11).

• The *value* of the physical film thickness, *t*, could be calculated using Eqs. (4.6) and (4.7) if the density of the film were known. It was not, in this instance. However, it is customary to produce an *estimate* of the film thickness in such cases using the following procedure. The actual Gd Fe mixture is replaced by an elemental bilayer with the same areal densities for Gd and Fe. The elemental layers are assumed to have elemental bulk densities. The thickness of this replacement film serves as the *estimate* of the physical thickness of the original film. The values of elemental bulk densities N_{Fe} = 8.44 × 10^{22} atoms/cm^3, N_{Gd} = 3.02 × 10^{22} atoms/cm^3 are obtained from Appendix 1. Hence, from Eq. (4.7)

$$t_{Fe} = \frac{2.55 \times 10^{18}}{8.44 \times 10^{22}} \text{ cm} = 302 \text{ nm}$$

$$t_{Gd} = \frac{0.674 \times 10^{18}}{3.02 \times 10^{22}} \text{ cm} = 223 \text{ nm}$$

$$t_{GdFe} \sim 525 \text{ nm} \ .$$

Note that no uncertainty is assigned to this final thickness estimate. In addition, the judicious analyst should indicate that this *estimate* may not accurately represent the actual physical thickness of the film.

• Summary of the analysis results for the two-element film of Fig. 4.1. The stoichiometric ratio is

$$\frac{N_{Fe}}{N_{Gd}} = 3.78 \pm 0.02$$

so the stoichiometry of the film is $Gd_{(0.209 \pm 0.001)}$ $Fe_{(0.791 \pm 0.001)}$. The areal densities are

$$(Nt)_{Fe} = (2.55 \pm 0.08) \times 10^{18} \text{ atoms/cm}^2$$

$$(Nt)_{Gd} = (0.674 \pm 0.021) \times 10^{18} \text{ atoms/cm}^2 \ .$$

The film thickness is about 0.5 μm. Note that the uncertainty in the ratio N_{Fe}/N_{Gd} is about 0.5%, while the $(Nt)_i$ values are uncertain by about 3%. This is a typical result and reflects the cancellation of the hard-to-measure quantities Q and Ω from the ratio.

The peak integration method may be applied in this simple and direct manner to the analysis of any single-layer film whose backscattering peaks do not overlap. It is not necessary that film composition be uniform as a function of depth. The resulting stoichiometric ratios will, of course, be the *average* values for the film. For very thin films, the backscattering peaks will not have the nearly flat portions on top but will be sharp peaks; in this instance, the channel numbers of the peak *centers* should be used in Eqs. (4.1) and (4.34) for element identification (see Appendix C in Chu *et al.*, 1978).

Multilayer films

The peak integration method may also be applied to the analysis of multilayer films. The spectrum of a trilayer film on a C substrate is shown in Fig. 4.4. The peaks associated with a particular layer may be identified by noting peak locations in spectra taken at different tilt angles (θ_1). The high-energy edges of the peaks associated with the surface layer have the same locations in these spectra. Buried-layer peak locations shift toward lower energies as the tilt angle is increased since the effective overlayer thicknesses increase with tilt angle as $(\cos \theta_1)^{-1}$. The locations of peaks associated with a particular buried layer all shift by about the same number of channels as the tilt angle is varied. Two backscattering spectra, one taken with $\theta_1 \sim 0°$ and the other with θ_1 between 45° and 60°, are usually sufficient for assignment of the peaks to appropriate layers. Once the peaks have been assigned to layers, the peak integration method may be applied to successive layers, starting with the surface layer.

The analysis ion energy incident on a particular layer is, of course, the energy E_0 minus the energy lost by the ions traversing the layers that cover the layer being analyzed. The $(Nt)_i$ for a particular layer should be calculated in the mean energy approximation before proceeding to the next layer. In the case of the spectrum shown in Fig. 4.4, the peak integration method may be only partially applied. The stoichiometry and areal densities of the Al_2O_3(Ar) layers may be completely determined. But the NiFe layer stoichiometry cannot be determined from these data since the Ni and Fe peaks overlap. For this spectrum, the Ni and Fe edges are separated by less than 10 channels. The surface barrier detector energy resolution ($\delta E \doteq 15$ keV) is about 8 channels, and in addition, both Fe and Ni have several isotopes. The result is that this particular NiFe film is probably too thick for accurate determination of its stoichiometry by ^{4}He ion beam analysis with surface barrier detectors. Films of NiFe as thick as 25nm may be analyzed with 5 MeV ^{4}He (Leavitt *et al.*, 1985). In such cases, thinner "witness" films, made by the same deposition procedures, are sometimes used to provide the stoichiometric

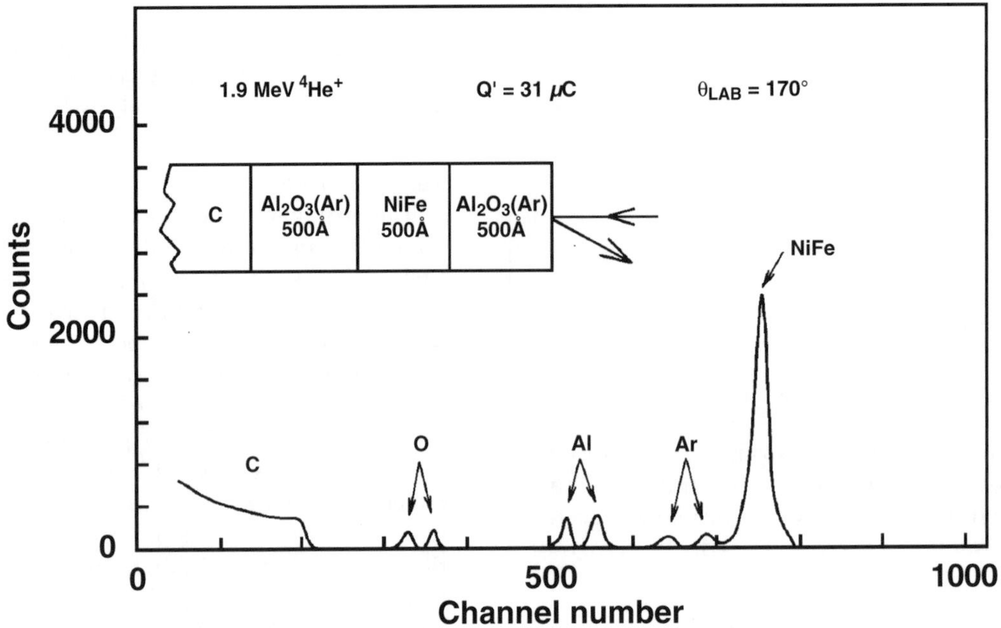

FIG. 4.4. The 1.9 MeV ^{4}He backscattering spectrum of a trilayer film on a carbon substrate. The backscattering signals from the three layers are clearly separated. The Ni and Fe peaks are not resolved.

ratio, N_{Fe}/N_{Ni}. The integrated peak counts, A_{Fe} and A_{Ni}, may then be calculated from [see Eq. (4.3)]

$$A_A = \frac{A_{AB}}{1 + \left(\frac{N_B}{N_A}\right)\left(\frac{\sigma_B}{\sigma_A}\right)} \tag{4.36}$$

and the measured integrated A_{NiFe} peak counts (with A=Ni, B=Fe), thus providing the A_i values for use with the peak integration method. As indicated, the analysis of multilayer films that produce spectra with non-overlapping peaks may be carried out "by hand," but the procedure is tedious and bears a strong risk of numerical error in the calculations. Computer assistance with these calculations is strongly recommended.

Use of elastic scattering resonance

Figure 4.5 shows a case where a strong non-Rutherford resonance (Mozer, 1956; Leavitt *et al.*, 1994; see Appendix 7) in the ^{1}H-^{9}Be cross section causes sufficient enhancement of the Be signal so that the peak integration method may be used to determine the areal density of the Be layer. The net integrated count in the Be peak in this spectrum is A_{Be} = (2353 ± 172) cts; the large uncertainty is caused by subtraction of the substantial signal due to the Al_2O_3 substrate. The backscattering spectrum taken with 1.9 MeV ^{4}He shows no sign of a Be peak. At the 2525 keV proton analysis energy used, the cross sections for Be, O, and Al are all non-Rutherford; only the ^{1}H-Ge scattering is Rutherford. The ^{1}H-^{9}Be cross section enhancement factor for thin targets (Leavitt *et al.*, 1994) is $(\sigma/\sigma_R)_{Be}$ = (60 ± 3). Substitution of these values of A_{Be} and $(\sigma/\sigma_R)_{Be}$ along with the appropriate values of Q', Ω, σ_R^{Be}, θ_1, DTR, and C_{Bi} into Eq. (4.35) gives $(Nt)_{Be}$ = (313 ± 29) × 10^{15} Be atoms/cm^2. If bulk density Be is assumed (Appendix 1), then the estimate of the Be film thickness is t_{Be} = 26 nm. When using a narrow resonance and a thick target, the average cross section over the range of ion energies in the target must be known. This, of course, varies with the target thickness. The principal source of uncertainties in the final results is the uncertainty in the measured value of the cross

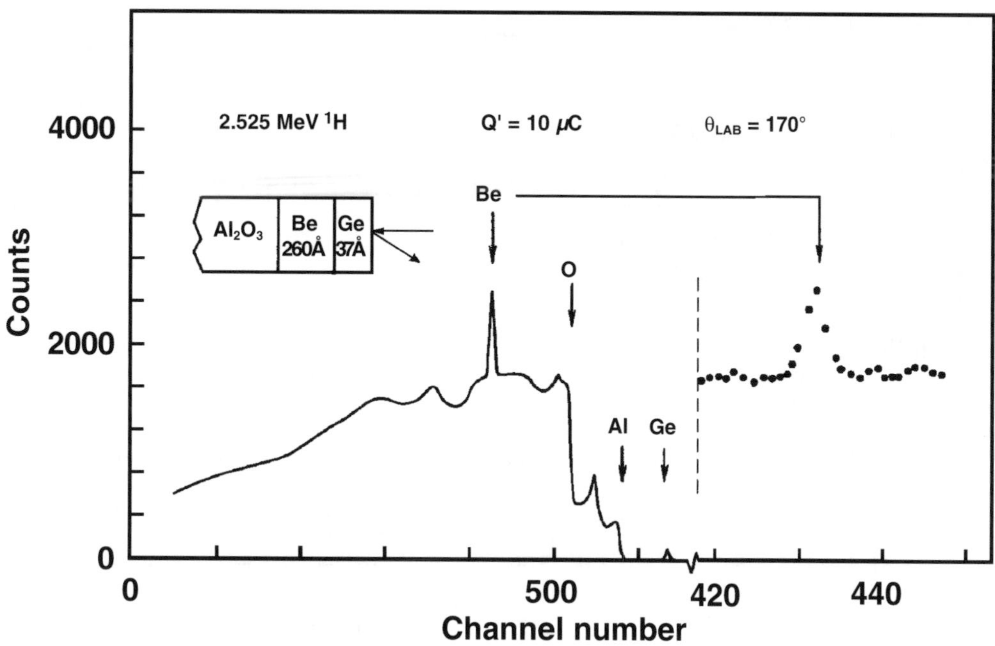

FIG. 4.5. The 2.525 MeV ^{1}H backscattering spectrum of a bilayer film on a sapphire substrate. The fact that the ^{1}H - ^{9}Be cross section is about 60 times Rutherford at this energy allows the Be backscattering signal to be observed.

section. It is clear that considerable work remains to be done concerning measurement of non-Rutherford cross sections. Non-Rutherford resonance cross section data are compiled in Appendix 7.

Overlapping peaks

Figure 4.6 shows the spectrum of a thin film with backscattering peaks that overlap. The film consists of 50 bilayers of Fe/Mo with a total estimated thickness of about 360 nm. The individual layers are so thin that the peaks from individual layers are not resolved by the detection system ($\delta E \doteq 15$ keV, $\Delta E_{in}^{bilayer} \doteq 5$ keV). Therefore, the film may be treated as a compound film of FeMo, and the appearance of the spectrum indicates that this "mixture" is uniform throughout the film. The peak integration method cannot be used for analysis of these data in the simple, direct manner described in Example 4.4 because the peaks overlap. However, a modified version of the peak integration method, which utilizes peak heights and widths, may be used for analysis of films of uniform composition. This method is described in 4.3.1.3 and applied to the data of Fig. 4.6.

One procedure for analyzing data of the type shown in Fig. 4.6 is to use computer fitting to separate the contributions that the different elements make to the total counts in the overlapping peaks. Peaks of specified shape are least-squares-fitted to the experimental data; the results of the fit provide the A_i with appropriate uncertainties. For a uniform film that is not too thin, a specified peak shape represented by two half-gaussians and a trapezoid works well (McIntyre *et al.*, 1987). Figure 4.7 shows the result of application of this fitting procedure to the data of Fig. 4.6. The total integrated counts $A_{FeMo} = 564{,}380$; the fitting procedure divides the counts as $A_{Fe} = 181300 \pm 2700$ and $A_{Mo} = 383100 \pm 3300$. Insertion of these values, along with appropriate values of Q′, DTR, C_{Bi}, etc., into Eq. (4.35) yields $(Nt)_{Fe} = (142 \pm 5) \times 10^{16}$ Fe atoms/cm^2, $(Nt)_{Mo} = (114 \pm 4) \times 10^{16}$ Mo atoms/cm^2, and $N_{Fe}/N_{Mo} = (1.24 \pm 0.02)$.

Of course, the simplest procedure (and often the quickest) for analysis of the film of Fig. 4.6 would be to produce a spectrum with separated peaks so the peak integration method may be applied in the simple manner described in Example 4.4. Such a spectrum (Fig. 4.8) with separated

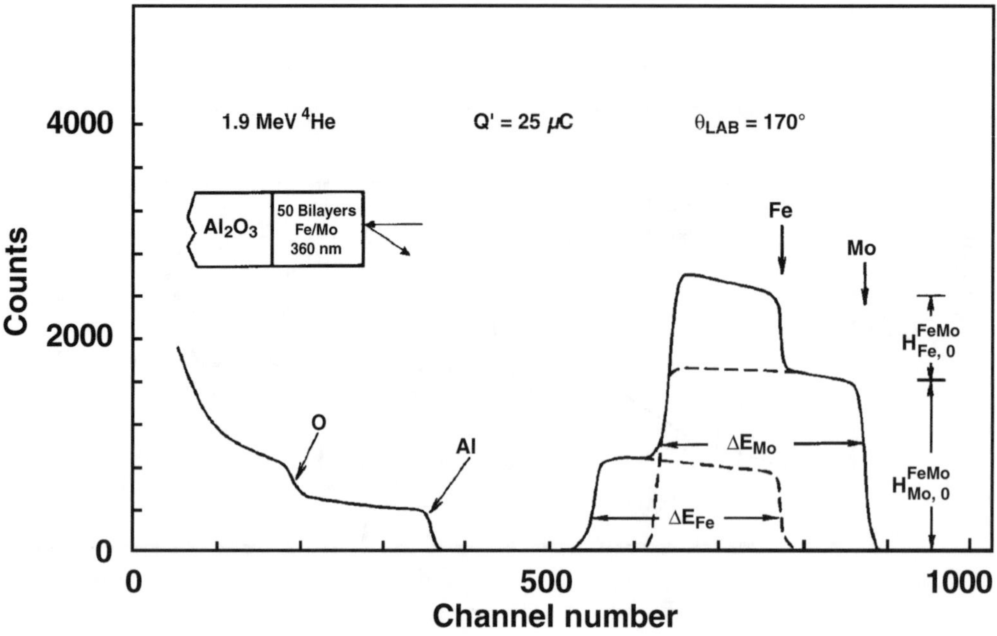

FIG. 4.6. The 1.9 MeV ^{4}He backscattering spectrum of a 360 nm Fe/Mo multilayer (50 bilayers) film on a sapphire substrate. The backscattering peaks for Fe and Mo overlap.

peaks was produced simply by increasing the ^{4}He beam energy to 3.8 Mev. The ^{4}He-Fe and ^{4}He-Mo cross sections are still Rutherford at the higher analysis energy. The ^{4}He-O cross section is not Rutherford at the higher energy (see the resonance near channel 130), but analysis of the substrate is not of interest in this case. Analysis of the data shown in Fig. 4.8 yields $(Nt)_{Fe} = (142 \pm 4) \times 10^{16}$ Fe atoms/cm^2; $(Nt)_{Mo} = (115 \pm 3) \times 10^{16}$ Mo atoms/cm^2; and $N_{Fe}/N_{Mo} = (1.236 \pm 0.006)$. Comparison of these results with those produced by computer-assisted separation of the Fig. 4.6 peak data indicates that the $(Nt)_i$ values and their uncertainties are about the same, but the accuracy of the average stoichiometric ratio, N_{Fe}/N_{Mo}, is substantially improved by use of the higher-energy data. This comparison not only points out a definite advantage to use of higher energy analysis beams, but also instills confidence in the reliability of the computer-assisted peak separation procedure.

4.3.1.2 Stoichiometry by surface spectrum heights

The stoichiometry of thin films whose composition as a function of depth is uniform may be obtained from the spectrum heights of the edges corresponding to scattering from the surface atoms. The stoichiometric ratio for the two-element film, A_mB_n, is [from Eq. (4.28)]

$$\frac{m}{n} = \frac{N_A}{N_B} = \frac{H_{A,0}^{AB}}{H_{B,0}^{AB}} \cdot \frac{\sigma_B}{\sigma_A} \cdot \frac{[\varepsilon_0]_A^{AB}}{[\varepsilon_0]_B^{AB}} \tag{4.37}$$

where the explicit expression for $[\varepsilon_0]_A^{AB}$ is [from Eq. (4.25) and Eq. (4.17)]

$$[\varepsilon_0]_A^{AB} = m\left[\frac{K_A\varepsilon^A(E_0)}{\cos\theta_1} + \frac{\varepsilon^A(K_AE_0)}{\cos\theta_2}\right] + n\left[\frac{K_A\varepsilon^B(E_0)}{\cos\theta_1} + \frac{\varepsilon^B(K_AE_0)}{\cos\theta_2}\right] . \tag{4.38}$$

The symbols are defined in Tables 4.2 and 4.3. A similar expression for $[\varepsilon_0]_B^{AB}$ may be obtained by exchanging A and B and m and n.

Note that the ratio, $[\varepsilon_0]_A^{AB}/[\varepsilon_0]_B^{AB}$, contains the ratio m/n, which is, of course, the object of the calculation.

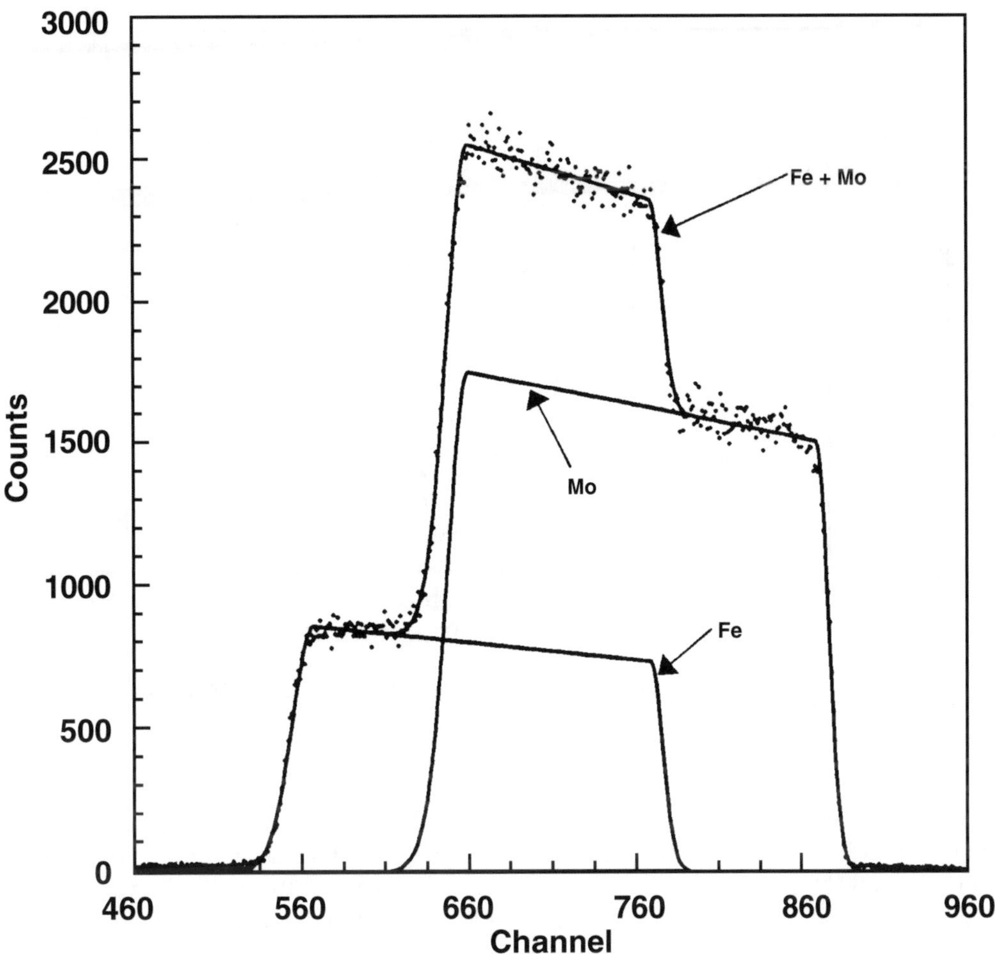

FIG. 4.7. The least-squares-fit of specified peak shapes (two half gaussians and a trapezoid) to the data (dots) of Fig. 4.6 provides integrated peak counts, A_{Fe} and A_{Mo}, for use with the peak integration analysis procedure. This fitting procedure is applicable to films that have uniform composition.

This $[\varepsilon_0]$ ratio seldom departs from 1.00 by more than a few percent. An iterative procedure may be used to obtain the m/n ratio. For the first iteration, m/n is calculated from appropriate $H^{AB}_{A,0}$, $H^{AB}_{B,0}$, σ_A, σ_B values using Eq. (4.37) with the $[\varepsilon_0]$ ratio taken as 1.00. This value of m/n is then used in Eq. (4.38) to calculate the $[\varepsilon_0]$ ratio, which, in turn, is used in Eq. (4.37) to obtain a more accurate value of m/n. The procedure is iterated until the m/n ratio ceases to change. Note that this technique may be used to determine the stoichiometry of bulk material as well as that of thin films providing the composition of the material is uniform as a function of depth. (See Example 4.5.)

4.3.1.3 Peak width methods

Data of the type shown in Fig. 4.6 may also be analyzed using the *peak width* method. Areal density calculation by this method does not require knowledge of either Q', the charge collected, or Ω, the detector solid angle. The method is usually not as accurate as the peak integration method, but can be used in cases where the efficiency of charge collection is suspect.

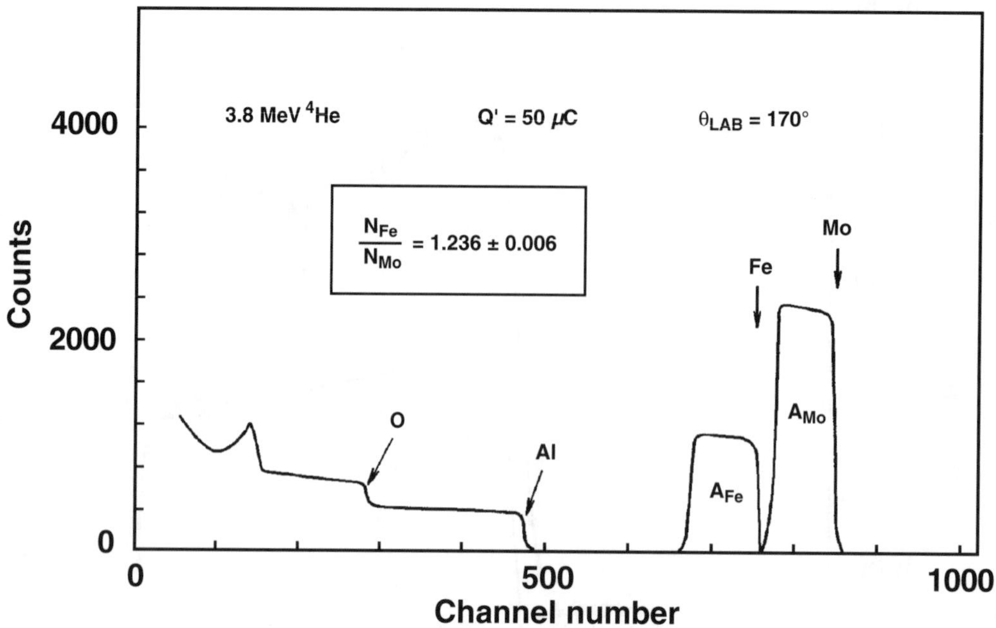

FIG. 4.8. The 3.8 MeV ^{4}He backscattering spectrum of the film whose 1.9 MeV spectrum is shown in Fig. 4.6. The backscattering peaks for Fe and Mo are separated.

For the case of the two-element compound film (A_mB_n), Eq. (4.24), with x = t, yields the following expressions for areal densities $(Nt)_A$ and $(Nt)_B$

$$(Nt)_A = \frac{m\Delta E_A}{[\varepsilon]_A^{AB}}, \quad (Nt)_B = \frac{n\Delta E_B}{[\varepsilon]_B^{AB}}, \tag{4.39}$$

where ΔE_A and ΔE_B are the energy widths (FWHM) of the peaks. The quantities $[\varepsilon]_A^{AB}$ and $[\varepsilon]_B^{AB}$ may be approximated using Eq. (4.38) with E_0 replaced by $\overline{E}$, the mean energy of the analysis ions in the film. Note that factors in Eq. (4.39) such as $m/[\varepsilon]_A^{AB}$ depend on the stoichiometric ratio m/n. This ratio may be evaluated using a combination of step-height and peak width data. From Eq. (4.39) or Eq. (4.24)

$$\frac{[\varepsilon]_A^{AB}}{[\varepsilon]_B^{AB}} = \frac{\Delta E_A}{\Delta E_B} . \tag{4.40}$$

Hence, from Eq. (4.37)

$$\frac{m}{n} = \frac{N_A}{N_B} = \frac{H_{A,0}^{AB}}{H_{B,0}^{AB}} \cdot \frac{\sigma_B}{\sigma_A} \cdot \frac{\Delta E_A}{\Delta E_B} . \tag{4.41}$$

For films of uniform composition, a useful analysis alternative which avoids much of the labor associated with the peak width method as applied in Example 4.6 is a combination of the peak width and peak integration methods. Note that combination of Eqs. (4.36), (4.37), and (4.40) yields

$$A_A = \frac{A_{AB}}{1 + \frac{H_{B,0}^{AB}}{H_{A,0}^{AB}} \cdot \frac{\Delta E_B}{\Delta E_A}} \tag{4.42}$$

which may be used to obtain the integrated peak counts necessary for use of the peak integration method. In essence, once the film stoichiometry is obtained from the data with Eq. (4.42), Eq. (4.36) may be used to find the individual element contributions to the combined peak for use in the peak integration method.

In summary, a comparison of results of the various methods for analyzing the Fe/Mo multilayer of Figs. 4.6, 4.7, and 4.8 is shown in Table 4.4. Clearly, the procedure that produces results of highest accuracy is use of the peak integration method on data with separated peaks. It is particularly important that the peaks be separated if

EXAMPLE 4.5. This example illustrates determination of film stoichiometry from measured values of spectral step heights (for the uniform film of Fig. 4.6).

• *A priori* acquisition parameters (the symbols are defined in Table 4.2) are (the analysis beam is ^{4}He)

$E_0 = 1892$ keV	$\theta = 170°$
$\theta_1 = 0°$	$\theta_2 = 10°$
$\Omega = 0.78$ msr	$C_{Bi} = 1.05 \pm 0.03$
K_{Fe} (170)° = 0.7520	K_{Mo} (170)° = 0.8471 (from Appendix 5)
$K_{Fe} \times E_0 = 1423$ keV	$K_{Mo} \times E_0 = 1603$ keV
$\mathcal{E} = (1.821 \pm 0.004)\ \dfrac{\text{keV}}{\text{ch}}$	$E' = (10 \pm 3)$ keV
$\dfrac{\sigma_R^{Fe}(E_1, 170°)}{\sigma_R^{Mo}(E_1, 170°)} = \dfrac{3.521}{9.251}$	(from Appendix 6)
$\left(\dfrac{\sigma}{\sigma_R}\right)_{Fe} = 0.996$	$\left(\dfrac{\sigma}{\sigma_R}\right)_{Mo} = 0.992$ [from Eq. (4.12)].

From Appendix 3 interpolation provides the values

ε^{Fe} (1892 keV) = 72.0×10^{-15} eVcm^2; ε^{Mo} (1892 keV) = 93.0×10^{-15} eVcm^2

ε^{Fe} (1603 keV) = 76.5×10^{-15} eVcm^2; ε^{Mo} (1603 keV) = 98.7×10^{-15} eVcm^2

ε^{Fe} (1423 keV) = 79.1×10^{-15} eVcm^2; ε^{Mo} (1423 keV) = 102×10^{-15} eVcm^2

• Relevant parameter values directly associated with the spectrum shown in Fig. 4.6 (the symbols are defined in Table 4.3)

DTR = 1.002

$H_{Fe,0}^{FeMo} = (700 \pm 40)$ cts	$H_{Mo,0}^{FeMo} = (1520 \pm 40)$ cts
$n_{Fe} = (777 \pm 1)$	$n'_{Fe} = (549 \pm 1)$
$n_{Mo} = (878 \pm 1)$	$n'_{Mo} = (643 \pm 1)$

$A_{FeMo} = 564{,}380 - (8 \pm 3)(371) = 561{,}412 \pm 1343$ cts

where (8 ± 3) cts/ch is the background on the high energy side of the FeMo signal.

• Calculation of the stoichiometric ratio, $\frac{N_{Fe}}{N_{Mo}}$.

For the first iteration, use of Eq. (4.37), with the $[\varepsilon_0]$ ratio taken as 1.00, provides

$$\frac{N_{Fe}}{N_{Mo}} = \frac{m}{n} \doteq \frac{(700 \pm 40)}{(1520 \pm 40)} \cdot \frac{9.251}{3.521} \cdot \frac{0.992}{0.996}$$

$$= 1.21 \pm .08 \ .$$

Then, use of Eq. (4.38) and the interpolated stopping cross sections given above, with m/n = 1.21 to calculate the $[\varepsilon_0]$ ratio, gives

$$\frac{[\varepsilon_0]_{Mo}^{FeMo}}{[\varepsilon_0]_{Fe}^{FeMo}} = \frac{\left((1.21)\left[(0.8471)(72.0) + \dfrac{76.5}{0.985}\right] + \left[(0.8471)(93.0) + \dfrac{98.7}{0.985}\right]\right)}{\left((1.21)\left[(0.7520)(72.0) + \dfrac{79.1}{0.985}\right] + \left[(0.7520)(93.0) + \dfrac{102}{0.985}\right]\right)}$$

$$= (1.03 \pm 0.05) \ .$$

Finally, from Eq. (4.37), the stoichiometric ratio, N_{Fe}/N_{Mo}, is

$$\frac{N_{Fe}}{N_{Mo}} = \frac{(1.21 \pm 0.08)}{(1.03 \pm 0.05)} = (1.17 \pm 0.10)$$

since further iteration of this procedure produces no change in N_{Fe}/N_{Mo}. Note that this value is in agreement with that (1.24 ± 0.02) obtained by use of the peak integration method with the computer-assisted peak separation shown in Fig. 4.7 and also that (1.236 ± 0.006) obtained using 3.8 Mev He ions shown in Fig. 4.8. The use of step heights to obtain stoichiometric ratios usually produces results of rather limited accuracy.

EXAMPLE 4.6. This example illustrates determination of the elemental areal densities by the peak width method for a uniform two-element film (for the data shown in Fig. 4.6).

- The *a priori* acquisition parameters and parameter values directly associated with the spectrum shown in Fig. 4.6 are explicitly listed in Example 4.5. Symbols are defined in Tables 4.2 and 4.3.
- Calculation of peak energy widths, ΔE_{Fe} and ΔE_{Mo}.

$$\Delta E_{Fe} = (n_{Fe} - n'_{Fe})\,\mathcal{E}$$
$$= [(777 \pm 1) - (549 \pm 1)]\,(1.821 \pm .004)$$
$$= (415 \pm 3)\ \text{keV}$$
$$\Delta E_{Mo} = (428 \pm 3)\ \text{keV}\ .$$

- Calculation of the stoichiometric ratio [Eq. (4.41)]

$$\frac{N_{Fe}}{N_{Mo}} = \frac{(700 \pm 40)}{(1520 \pm 40)} \cdot \frac{(9.251)}{(3.521)} \cdot \frac{(0.992)}{(0.996)} \cdot \frac{(415 \pm 3)}{(428 \pm 3)}$$
$$= (1.17 \pm 0.08)\ .$$

- Calculation of areal densities $(Nt)_{Fe}$ and $(Nt)_{Mo}$.

In the surface energy approximation, use of Eq. (4.38) and (4.39) yields

$$(Nt)^{SEA}_{Fe} = \frac{\Delta E_{Fe}}{[\varepsilon_0]^{FeMo}_{Fe}/m} = \frac{(415 \pm 3)}{(278 \pm 14)} \times 10^{18}$$
$$= (1.47 \pm 0.07) \times 10^{18}\ \text{atoms/cm}^2\ .$$

$$(Nt)^{SEA}_{Mo} = \frac{\Delta E_{Mo}}{[\varepsilon_0]^{FeMo}_{Mo}/n} = \frac{(428 \pm 3)}{(347 \pm 17)} \times 10^{18}$$
$$= (1.26 \pm 0.06) \times 10^{18}\ \text{atoms/cm}^2\ .$$

The energy loss of the analysis ions in the film is [Eq. (4.29)], to first order

$$\Delta E^{SEA}_{in} = (1.49)\,(72.0) + (1.23)\,(93.0) = 224\ \text{keV}$$

so the mean analysis ion energy is [Eq. (4.30)], to first order $\overline{E}^{(1)} = 1892 - (224/2) = 1780$ keV. This mean energy, $\overline{E}^{(1)}$, should be used in the evaluation of the $[\varepsilon]^{AB}_{A}$ and $[\varepsilon]^{AB}_{A}$ factors in Eq. (4.38) to obtain more accurate values of $(Nt)_{Fe}$ and $(Nt)_{Mo}$ through use of Eq. (4.39). It is, of course, necessary to interpolate additional values of ε_{Fe} and ε_{Mo} for the relevant energies from data given in Appendix 3. The results are

$$\frac{\left[\varepsilon\left(\overline{E}^{(1)}\right)\right]^{FeMo}_{Fe}}{m} = (288 \pm 14) \times 10^{-15}\ \text{eV cm}^2$$

$$\frac{\left[\varepsilon\left(\overline{E}^{(1)}\right)\right]^{FeMo}_{Mo}}{n} = (348 \pm 17) \times 10^{-15}\ \text{eV cm}^2\ .$$

The areal densities in the mean energy approximation are [Eq. (4.39)]

$$(Nt)^{MEA}_{Fe} = (1.44 \pm 0.07) \times 10^{18}\ \text{atoms/cm}^2$$

$$(Nt)^{MEA}_{Mo} = (1.23 \pm 0.06) \times 10^{18}\ \text{atoms/cm}^2$$

since further iteration of this procedure results in no further change in the (Nt) values.

the film composition varies with depth. However, it is frequently not possible to produce spectra whose peaks are separated. If the film composition is uniform, then computer-assisted separation of overlapping peaks frequently produces areal densities that have accuracies comparable to those above. However, the accuracy of the stoichiometric ratio is usually degraded by a factor of 2 or 3 compared to that above. The use of the peak width method (Example 4.6) produces results of inferior accuracy because this method depends on the accuracy of stopping cross sections (estimated to be about 5%). However, it is a method that can be carried out "by hand" for the case of overlapping peaks. Further, it may be used as a crude (5%) check on the accuracy of charge collection and solid angle measurement. If the peak width method is used in its pure form [Eq. (4.39)], it can be used to check for the presence of accidental channeling. The combination peak width and peak integration method (Example 4.7) produces results of accuracy comparable to the peak

EXAMPLE 4.7. This example illustrates use of a combination of the peak width and peak integration methods to obtain areal densities in a two-element thin film of uniform composition (Fig. 4.6).

With the data from Fig 4.6 listed in Example 4.5, Eq. (4.42) yields

$$A_{Fe} = \frac{A_{FeMo}}{1+\left(\frac{H^{FeMo}_{Mo,0}}{H^{FeMo}_{Fe,0}}\right)\left(\frac{\Delta E_{Mo}}{\Delta E_{Fe}}\right)}$$

$$= \frac{561412 \pm 1343}{1+\frac{(1520 \pm 40)}{(700 \pm 40)} \cdot \frac{(428 \pm 3)}{(415 \pm 3)}}$$

$$= (173000 \pm 7500) \text{ cts}$$

$$A_{Mo} = (388000 \pm 7500) \text{ cts} \; .$$

Substitution of these values into Eq. (4.35), with appropriate values of the other required parameters (listed in Example 4.5), the areal densities in the surface energy approximation are

$$(Nt)^{SEA}_{Fe} = (1.52 \pm 0.08) \times 10^{18} \text{ atoms/cm}^2$$

$$(Nt)^{SEA}_{Mo} = (1.31 \pm 0.05) \times 10^{18} \text{ atoms/cm}^2 \; .$$

Two further iterations, utilizing Eqs. (4.29) to (4.33), yield the areal densities in the mean energy approximation

$$(Nt)^{MEA}_{Fe} = (1.36 \pm 0.07) \times 10^{18} \text{ atoms/cm}^2$$

$$(Nt)^{MEA}_{Mo} = (1.16 \pm 0.05) \times 10^{18} \text{ atoms/cm}^2 \; .$$

width method. Its principal advantage is that it can easily be carried out "by hand," particularly if peak integration calculations have been computerized to carry out the mean energy approximation. The determination of stoichiometry from spectral step-heights (Example 4.5) produces results of only modest accuracy; the calculation may be done "by hand." This method is also applicable to bulk materials of uniform composition. It is worth noting, again, that all the above analysis methods (except the first) apply only to the case of films whose composition does not vary with depth.

Table 4.4. Comparison of results of analyses of the Fe/Mo film (of Figs. 4.6, 4.7, and 4.8) by various analysis methods.

Analysis method	$(Nt)^{*}_{Fe}$	$(Nt)^{*}_{Mo}$	$\frac{N_{Fe}}{N_{Mo}}$
Peak integration method with separated peak data, Fig. 4.8	142 ± 4	115 ± 3	1.236 ± 0.006
Peak integration method, with computer-assisted separation of overlapping Fe and Mo peaks, Fig. 4.7.	142 ± 5	114 ± 4	1.24 ± 0.02
Peak width method (with spectrum heights), Example 4.6, Fig. 4.6.	144 ± 7	123 ± 6	1.17 ± 0.08
Combination of peak width and peak integration methods, Example 4.7, Fig. 4.6.	136 ± 7	116 ± 5	1.17 ± 0.08
Stoichiometry by spectrum height, Example 4.5, Fig. 4.6	–	–	1.17 ± 0.10

*(Nt) values are in units of 10^{16} atoms/cm^2.

4.3.2 Surface layers of bulk materials

4.3.2.1 Stoichiometry by surface spectrum heights

The stoichiometry of materials whose composition does not vary with depth may be determined from measured surface spectrum heights and use of generalizations of Eqs. (4.37) and (4.38). If elements denoted by $A_mB_n \ldots D_p$ are present in the material, terms such as $H^{AB}_{A,0}$ and $[\varepsilon_0]^{AB}_A$ in Eqs. (4.37) and (4.38) should be replaced by terms $H^{AB\ldots D}_{A,0}$ and $[\varepsilon_0]^{AB\ldots D}_A$, where, for example

$$[\varepsilon_0]^{AB\ldots D}_A = m\left[\frac{K_A\varepsilon^A(E_0)}{\cos\theta_1} + \frac{\varepsilon^A(K_AE_0)}{\cos\theta_2}\right] + n\left[\frac{K_A\varepsilon^B(E_0)}{\cos\theta_1} + \frac{\varepsilon^B(K_AE_0)}{\cos\theta_2}\right] + \ldots + p\left[\frac{K_A\varepsilon^D(E_0)}{\cos\theta_1} + \frac{\varepsilon^D(K_AE_0)}{\cos\theta_2}\right] \quad (4.43)$$

etc. The symbols are defined in Tables 4.2 and 4.3. Note that the ratio $[\varepsilon_0]^{AB\ldots D}_A/[\varepsilon_0]^{AB\ldots D}_B$ in the generalized Eq. (4.37) contains the ratios m/n,...m/p, which are, of course, the objects of the calculation. As before, these $[\varepsilon_0]$ ratios seldom depart from 1.00 by more than a few percent. An iterative procedure may be used to obtain the m/n,...m/p ratios. For the first iteration, m/n is calculated from Eq. (4.37) with the $[\varepsilon_0]$ ratio taken as 1.00. These ratios may then be substituted into Eq. (4.43) to calculate the $[\varepsilon_0]$ ratios, which are used in Eq. (4.37) to obtain more accurate values of m/n, etc.

The backscattering spectrum of a ceramic glass sample is exhibited in Fig. 4.9. Cross sections for ^{4}He on all indicated elements are nearly Rutherford except for small corrections [Eq. (4.12)] due to electron shells. Substitution of $H_{i,0}^{AB...D}$ step-height values (obtained from the spectrum) into Eq. (4.37) with the $[\varepsilon_0]$ ratio taken as 1.00 yields the stoichiometry listed in Table 4.5. Atomic fractions may be obtained from the stoichiometric ratios provided by Eq. (4.37). For example, the O atomic fraction, F_O, is given by

$$\frac{1}{F_O} = 1 + \frac{N_{Na}}{N_O} + \frac{N_{Si}}{N_O} + \frac{N_K}{N_O} + \frac{N_{Zn}}{N_O} + \frac{N_{Cd}}{N_O} \quad . \tag{4.44}$$

The O atomic percent is $100F_O$. Note that the uncertainties in the atomic percents are rather large. The principal source of uncertainty is the difficulty in reading precise values of $H_{i,0}^{AB...D}$ from the graph (see Table 4.5). In view of these large uncertainties, calculation of the $[\varepsilon_0]$ ratios using Eq. (4.43) is not worthwhile, in this instance.

Table 4.5. Summary of analysis results obtained from the spectrum of the ceramic glass sample shown in Fig. 4.9.

Element	$H_{i,0}^{AB...D}$ *(counts)*	σ_R^i (1 MeV, 170°) ($\times 10^{-24}$ cm^2/sr)	Atomic percent
O	365 ± 30	0.2965	63.8 ± 1.6
Na	125 ± 15	0.5993	10.8 ± 1.0
Si	410 ± 15	0.9905	21.5 ± 1.3
K	70 ± 7	1.861	2.0 ± 0.2
Zn	175 ± 6	4.701	1.9 ± 0.1
Cd	9 ± 1	12.09	0.039 ± 0.004

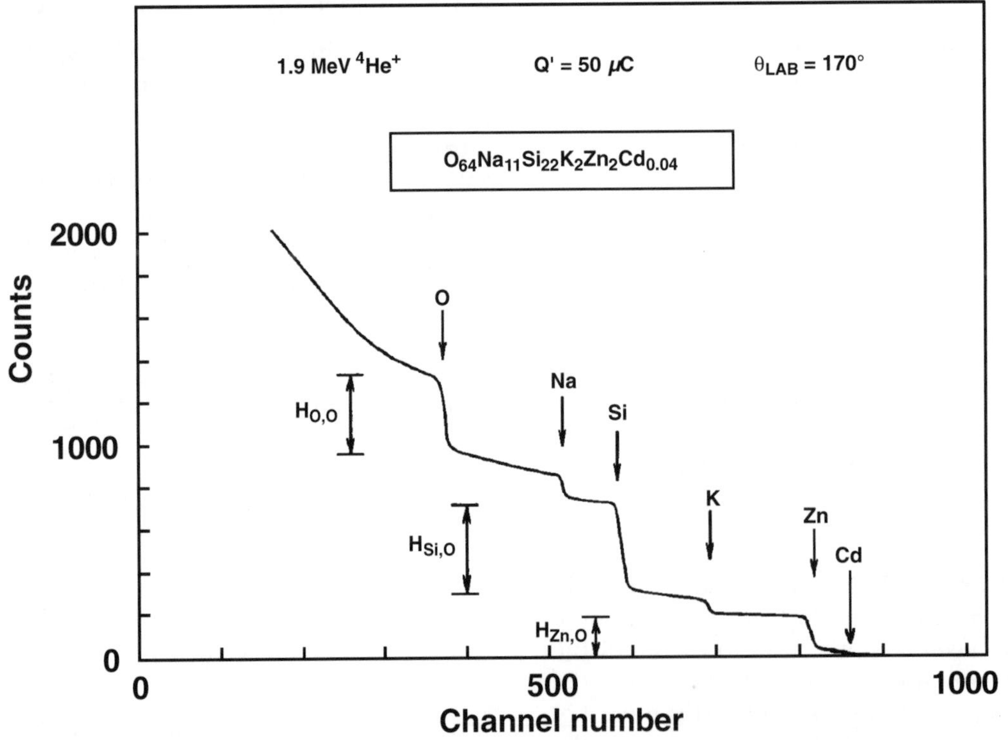

FIG. 4.9. The 1.9 MeV ^{4}He backscattering spectrum of a ceramic glass. The indicated stoichiometry was determined from the step heights. For uncertainties in the subscripts (atomic percent), see Table 4.5.

4.3.2.2 A deficiency method

A deficiency method may be used to analyze near-surface two-element materials even when the steps for both elements are not visible on the spectrum, provided the identity of the lighter element is known. The spectrum for the heavier *pure* element is compared with that for the two-element material. The deficiency in the heavy element signal caused by the presence of the light element is noted and used to obtain the material stoichiometry at the sample surface. Application of this method to a case of oxidation of tungsten is illustrated in Figure 4.10. The spectra of pure W and WO_x samples are compared. The number of analysis ions incident on both samples is the same, of course. The signal from O in the WO_x spectrum (edge near channel 370, not shown) is not visible on top of the W signal. If it is assumed that the deficiency in the W signal above channel 800 is caused by the presence of O in WO_x, then the surface stoichiometry may be calculated by inserting measured W step heights, $H^W_{W,0}$ and $H^{WO_x}_{W,0}$ from Fig. 4.10 into

$$x = \frac{N_O^{WO_x}}{N_W^{WO_x}} = \frac{[\varepsilon_0]_W^W}{[\varepsilon_0]_W^O}\left(\frac{H_{W,0}^W}{H_{W,0}^{WO_x}} - 1\right) \qquad (4.45)$$

to obtain the near-surface stoichiometric ratio. Equation (4.45) is obtained from Eqs. (4.27), (4.28), and (4.38), with

$$[\varepsilon]_A^B = \frac{K_A \varepsilon^B (E_0)}{\cos\theta_1} + \frac{\varepsilon^B (K_A E_0)}{\cos\theta_2} \qquad (4.46)$$

and $x = n/m$, $A = W$, and $B = O$. Interpolated values of the stopping cross sections ε^W and ε^O (Appendix 3) must be used in the calculation of x. The resulting near-surface stoichiometry is WO_3; the 6% uncertainty is due to uncertainties in step heights and energy loss data. From the appearance of the spectrum, it is apparent that the WO_3 layer extends to a depth corresponding to about channel 890. The WO_3 film thickness, t, may be estimated by inserting the integrated counts, A_W, between channels 890 and 960 into Eq. (4.3) to calculate $(Nt)_W$ and then use

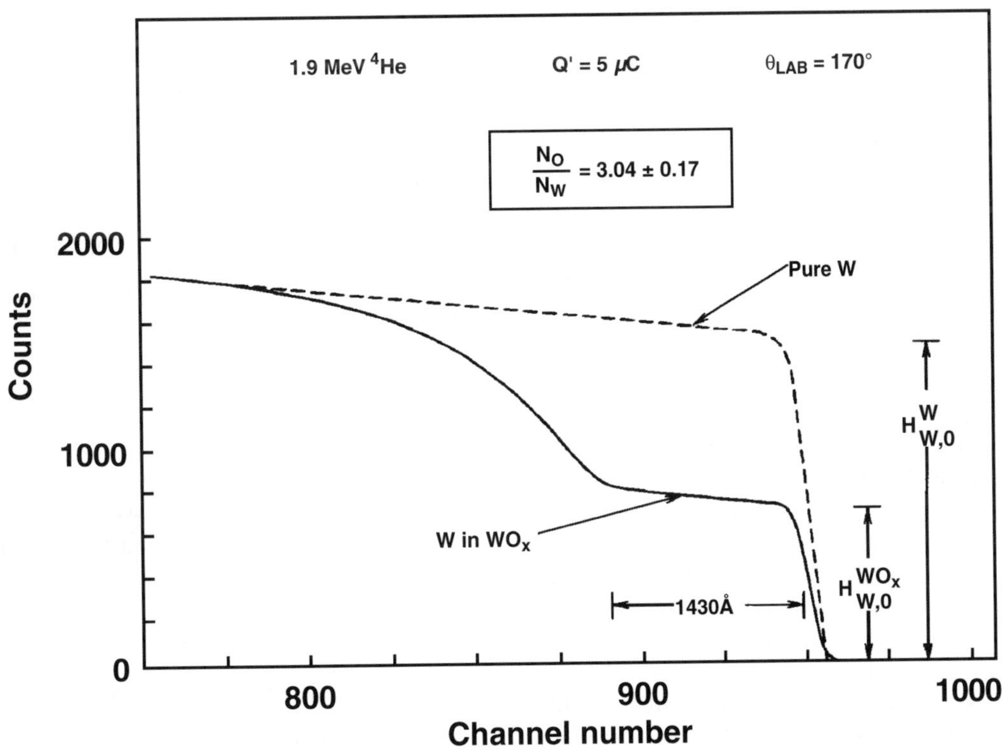

FIG. 4.10. Portions of 1.9 MeV ^{4}He backscattering spectra for a pure tungsten sample (dashed line) and for an oxidized tungsten sample (solid line). The spectra were obtained under identical experimental conditions (same integrated charge, θ_{Lab}, etc.). Analysis by the deficiency method (see 4.3.2.2) produced the WO_3 stoichiometry and film thickness shown.

Eq. (4.7) to estimate t. Of course, bulk density is *assumed* for the WO_3 film, consequently no uncertainty is quoted for the thickness estimate.

4.3.2.3 Depth profiling

All detailed examples of sample analysis presented so far involve films or materials whose compositions do not vary with depth below the surface. Since variable composition is characteristic of many samples, determination of sample stoichiometry as a function of depth is of considerable interest. In this section, two examples of depth profiling the impurity O in two Mo/Si multilayer samples are discussed. The samples, 46A and 46B, are identical multilayers consisting of 11 layers of Mo (~60 Å ea) and 10 layers of Si (~100 Å ea) on Si and C substrates, respectively. Rutherford backscattering spectra for these samples are shown in Figs. 4.11(a) and 4.12(a). The individual layers are not resolved so the films may be regarded as mixtures of Mo and Si. For sample 46B, the impurity O signal is clearly separated from the C, Si, and Mo signals in the spectrum, so the O atomic fraction may be depth profiled using a generalized version of Eq. (4.28) to convert observed O peak spectral heights to concentration ratios at appropriate depths. For sample 46A, the O signal is essentially unobservable in the Rutherford backscattering spectrum [Fig. 4.12(a)], so the O depth profile cannot be obtained from these data. In fact, sample 46B was intended to "witness" the amount of O in 46A. The purpose of the measurements (described below) utilizing the strong non-Rutherford resonance in the ^{4}He-O cross section to compare O depth profiles in 46A and 46B was to investigate the validity of this "witness" procedure.

An approximate O depth profile for sample 46B may be quickly obtained by the following procedure. The peak integration method (Section 4.3.1.1) may be used to analyze the data of Fig. 4.11. The results are shown in Table 4.6; note that the total film thickness is estimated to be ~170 nm. Of course, the Mo and Si are assumed to be distributed in pure elemental layers of bulk density, and the presence of O is ignored in making this thickness estimate. Since the average O atomic percent for the film is 15% and the concentration appears to vary linearly with depth, the spectrum height (53 cts) near the center of the quadrilateral sketched on the O peak data [Fig. 4.11(b)] corresponds to 15% atomic percent O. This establishes the vertical scale as 3.5 counts per O atomic percent

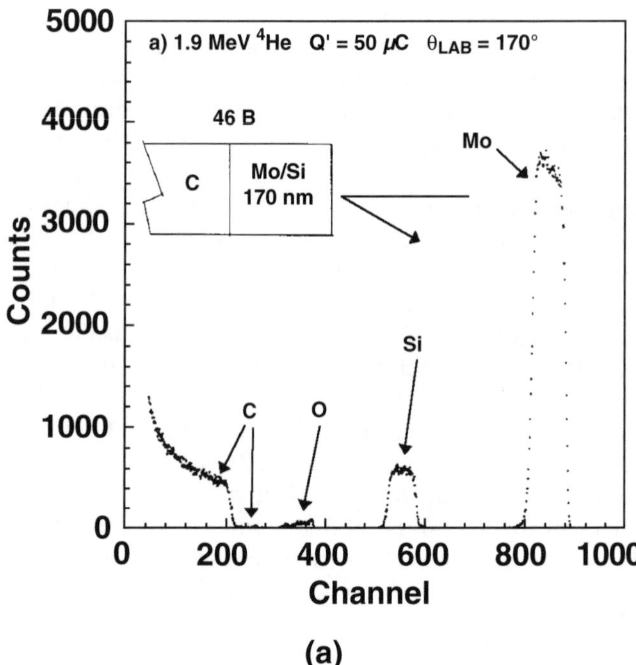

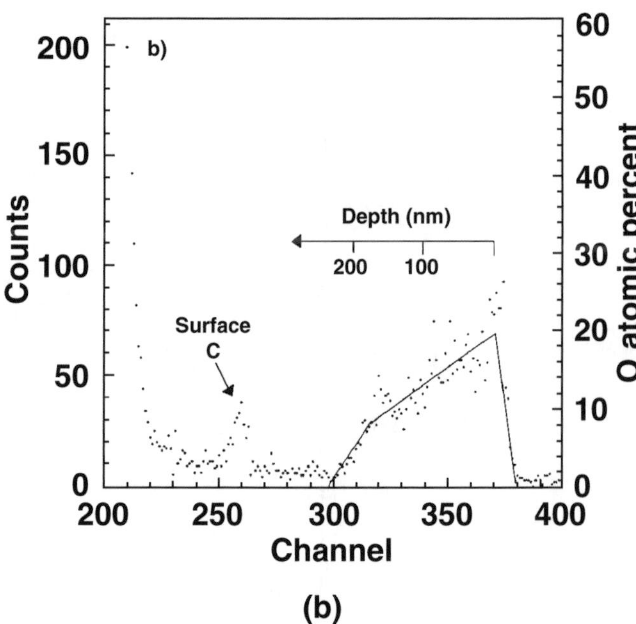

FIG. 4.11. (a) The 1.9 MeV ^{4}He Rutherford backscattering spectrum of Sample 46B, a 170 nm Mo/Si multilayer film on a C substrate. (b) An expanded view of the O peak region used for depth profiling O, with the O atomic percent as a function of depth below the surface indicated.

Table 4.6. Results of a peak integration analysis of the RBS data for Sample 46B shown in Fig. 4.11.

Element	Areal density ($\times 10^{15}$ atoms/cm^2)	Atomic percent	t (nm)
Mo	406 ± 19	37.0 ± 0.4	63
Si	531 ± 25	48.4 ± 0.4	106
O	161 ± 12	14.7 ± 0.5	

since spectrum height is proportional to concentration for low impurity concentrations. If it is assumed that the two peaks near channels 370 and 320 [Fig. 4.11(b)] are due to O on the front and back surfaces of the film, then the horizontal scale is 3.4 nm per channel. The O atomic percent varies nearly linearly from about 9% at the substrate interface to about 20% at the film surface, as indicated in Fig. 4.11(b).

The strong non-Rutherford resonance in the ^{4}He-O cross section at ^{4}He energy 3.04 MeV (Cameron, 1953) may be used to depth profile O in sample 46A. The resonance is narrow (~10 keV wide), strong ($\sigma/\sigma_R \sim 23$ at the maximum), and moderately isolated. For ^{4}He energies slightly greater than 3.04 MeV, the O signal from the film region occupied by the resonance is strongly enhanced compared to the Si signal. Figure 4.12(b) shows the strong O peak due to O at the sample 46A surface. The net integrated counts in the O peak were obtained by subtracting the large Si background. Backscattering spectra for both 46A and 46B were taken at several ^{4}He energies above and below 3.04 MeV, and the integrated counts in the O peak obtained in each case. The net O peak counts vs incident ^{4}He energy are plotted in Fig. 4.13 for both samples 46A and 46B.

The vertical scale, in terms of O atomic percent, was determined by requiring that the O atomic percent at the sample 46B surface be 20% (as was determined from the RBS spectrum of sample 46B as described above). The horizontal scale, in terms of nm, was determined from the calculated 57 keV energy loss in sample 46B (using appropriate stopping cross sections with areal densities given in Table 4.6) and the estimated 170 nm thickness. So the energy to depth conversion is 3.0 nm/keV. The

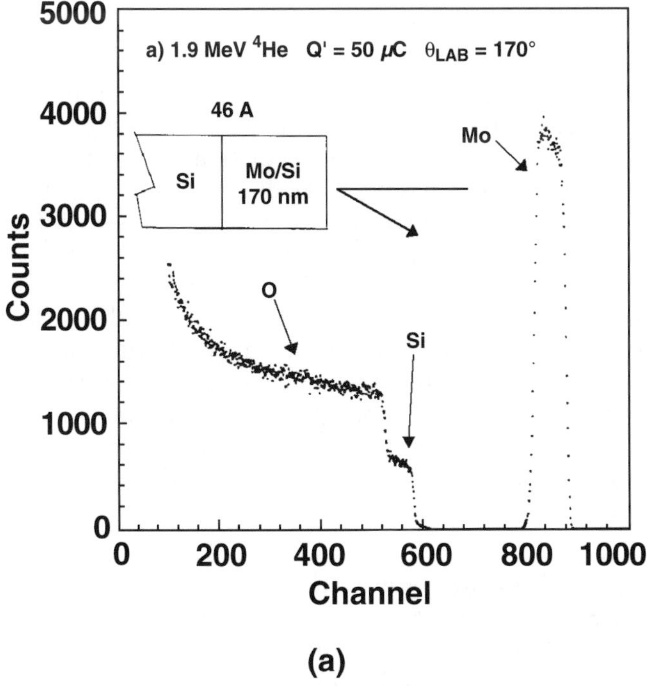

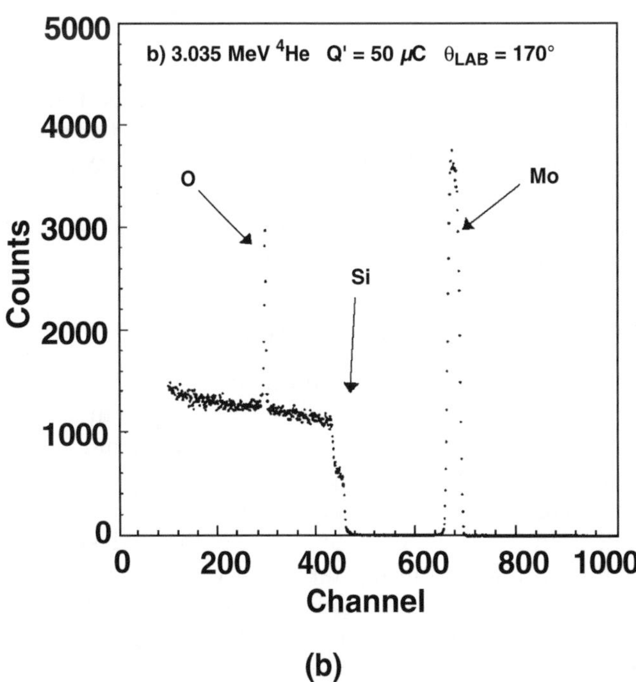

FIG. 4.12. (a) The 1.9 MeV ^{4}He Rutherford backscattering spectrum of Sample 46A, a 170 nm Mo/Si multilayer film on a Si substrate. (b) The 3.035 MeV ^{4}He backscattering spectrum of Sample 46A. The enhanced O peak in (b) is due to the strong non-Rutherford resonance in the ^{4}He-O cross section at the ^{4}He laboratory energy of 3.04 MeV.

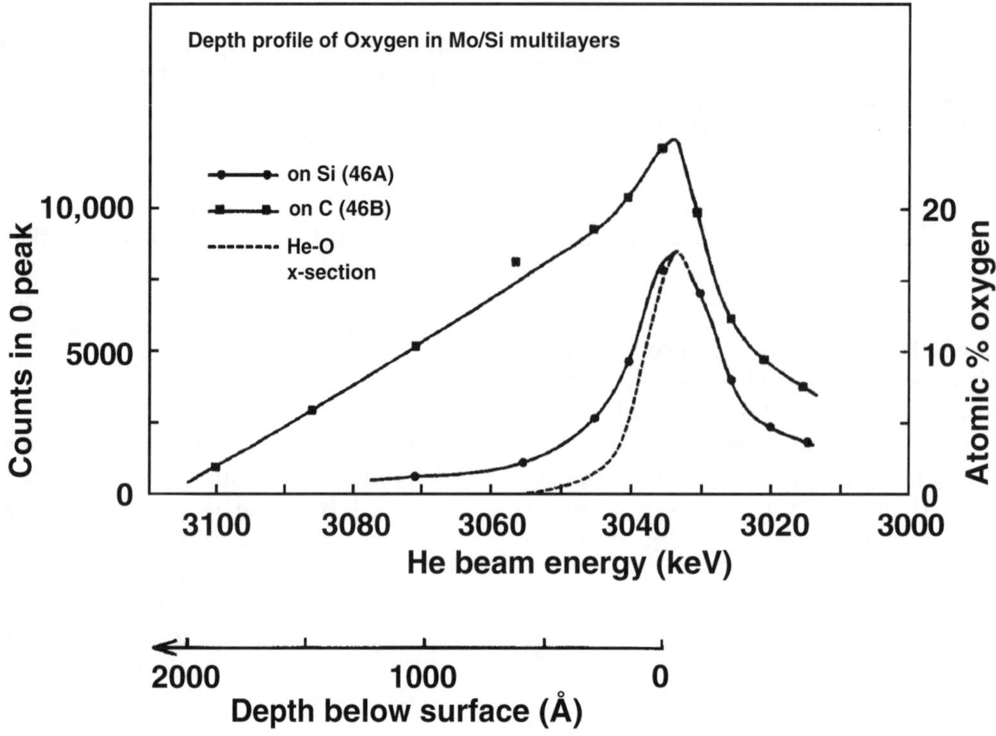

FIG. 4.13. Oxygen depth profiles for Samples 46A and 46B obtained from the data shown in Figs. 4.11 and 4.12.

resulting horizontal depth scale is shown at the bottom of Fig. 4.13. Inspection of Fig. 4.13 indicates that there is considerably more O in sample 46B than in 46A. Most of the O in 46A is very near the surface and sample 46B does *not* serve as a good "witness" sample for 46A as far as O is concerned.

4.3.3 Numerical calculations

4.3.3.1 Areal density calculations

A basic calculation in backscattering analysis involves the use of Eq. (4.35) in determining elemental areal densities (Nt values) from peak areas and experimental parameters. This is the peak integration method described in 4.3.1.1. Due to energy loss of the analysis ions in the sample, scattering takes place over a range of energies. Since this calculation involves the scattering cross section, which is a function of energy, an integration should be performed. One approach is to divide the target into thin layers for the numerical integration.

In the mean energy approximation, an iterative procedure is used as described in 4.2.4.6 and Example 4.4. This procedure is usually carried out with a microcomputer. The algorithm used is essentially that demonstrated in Example 4.4, however, the calculations often involve samples containing several layers, each with several elements. A parameterization of stopping cross sections, as described in 4.2.4.1, is an essential ingredient of programs using this procedure, since the scattering cross section in Eq. (4.35) must be evaluated at the mean energy of the beam in a particular layer. Another important function of programs calculating areal densities from peak areas is the propagation of uncertainties in the experimental quantities to obtain reasonable uncertainties in the final elemental area densities and stoichiometric ratios.

4.3.3.2 Computer simulation

Another approach to analyzing backscattering spectra with overlapping peaks is by computer simulation involving the calculation of the energy spectrum of backscattered particles using a specified target composition and experimental parameters. The procedure is to alter the target composition until the calculated and measured energy spectra are closely matched. This can be done by manually changing the target composition for each iteration or by using a least-squares fitting procedure to find the target composition that best fits the measured spectrum. An early study of this approach is given in Ziegler *et al.* (1976). Simulation programs must be used with care since the correspondence between sample composition and backscattering spectrum may not be unique (Rauhala, 1987).

A well-known backscattering analysis program, RUMP, which originated at Cornell University, is used in many laboratories (Doolittle, 1985; 1986; 1990). As an example, a RUMP simulation of the spectrum of Fig. 4.6 is shown in Fig. 4.14. Information about this program can be obtained from Computer Graphic Service, Ltd., 52 Geung Circle, Ithaca, NY 14850-8716, (607) 277-4913. A commercial software package, SCATT, and a more recent version HYPRA, are available from Charles Evans & Associates, 301 Chesapeake Drive, Redwood City, CA 94063, (415) 369-4567. A recent addition to the growing list of analysis programs is GISA, written by J. Saarilahti of the Semiconductor Laboratory, VTT, Technical Research Centre of Finland, P.O.B. 34, SF-02151 Espoo (Saarilahti and Rauhala, 1992). Many similar computer programs and variations in approaches are described in the literature (Müller *et al.*, 1978; Eskildsen, 1982; Borgesen *et al.*, 1982; Morris, 1982; Petrov *et al.*, 1983; Saunders and Ziegler, 1983; Rauhala, 1984; Kido and Youkio, 1985; Eridon and Was, 1985; Kido *et al.*, 1985; Simpson and Earwaker, 1986; Butler, 1986; Li and Al-Tamimi, 1986; Kido and Kawamoto, 1987; Rauhala, 1987; Butler, 1990).

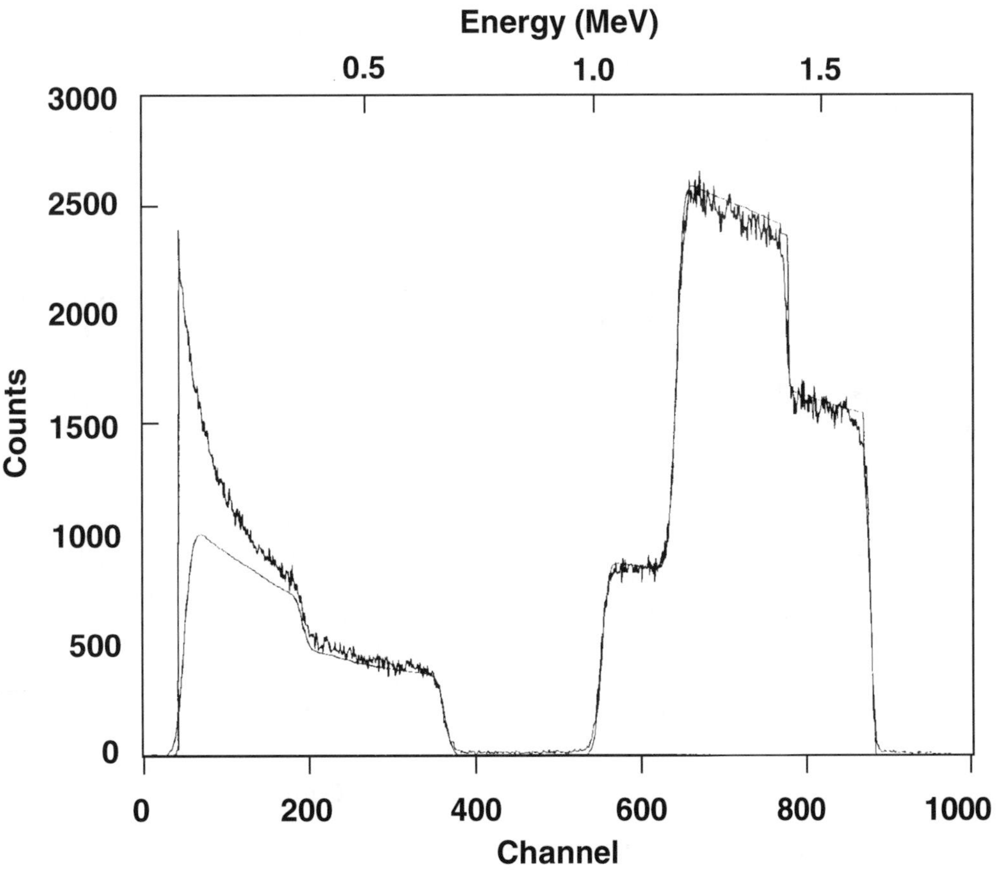

FIG. 4.14. A RUMP simulation (smooth line) of the spectrum shown in Fig. 4.6 (rough line). The experimental parameters given in Example 4.5 and the Fe and Mo areal densities given in the first row of Table 4.4 were input to the program.

4.4 EXPERIMENTAL CONSIDERATIONS

4.4.1 Design factors

Choices about ion beams, ion energies, and detectors for backscattering must be made in the context of the measurement planned. In the following sections, we discuss some common choices for ion beams and energies, considering the advantages and disadvantages of each. Detector considerations are also discussed briefly. For a more detailed discussion of detector properties, see Chapter 11.

The primary factors which must be taken into consideration in the design of a backscattering measurement are the sensitivity, mass resolution, depth of analysis, and depth resolution required to obtain the desired information. For a given measurement, the achievable limit for each of these quantities depends upon the choice for a number of experimental parameters such as the ion beam and the ion energy. The type of detector used for analysis also influences the ultimate quality of the data obtained. The final data generally result from a complex interplay of effects arising from each decision.

4.4.1.1 Sensitivity

The most obvious factor affecting the sensitivity of a given backscattering measurement is the scattering cross section which clearly varies with the mass, charge, and energy of the incoming ion. However, the ultimate sensitivity is also influenced by the level of counts from other sources (for example, pulse pile-up or heavy-atom, thick-target scattering) underlying the region of interest. An additional consideration is the practical limit on the time available to do the experiment. If a different beam provides a higher cross section but is only available at (or limited to) a much lower beam current, the achievable sensitivity may be less than that obtainable using ions with a smaller scattering cross section. Acceptable beam currents may be limited by detector considerations. Finally, sensitivity is limited by beam effects on the target such as beam heating and sputtering which can destroy fragile targets.

4.4.1.2 Mass resolution

The achievable mass resolution for a given measurement depends on factors involving the ion beam, its energy, the target composition and properties of the detector. The actual separation ΔE in energy for scattering from atoms having adjacent values of M_2 is given by Eq. (4.8) and depends on the masses of both the incoming ions and the target atoms as well as the ion energy. The ability to distinguish between adjacent values of M_2, however, depends also on the overall energy resolution δE of the experiment [Eq (4.9)]. The overall energy resolution δE depends on the detector energy resolution, energy straggling of ions within the target and the energy spread of the incoming ion beam. Other experimental parameters such as the size of the beam spot and the solid angle of the detector also influence the energy resolution.

4.4.1.3 Depth resolution

When an experiment requires depth profiling of a sample, one must consider both the accessible depth and the depth resolution when choosing an experimental technique. Both are influenced by the same factors. Clearly, if an ion loses energy rapidly, the surface thickness producing energy-resolved particles will be smaller. The energy loss factor, [S], depends on the incoming ion and its energy, the target composition, and the orientation of the target. The achievable depth resolution δx for a measurement [Eq. (4.26)] is also proportional to δE, so δx depends on the detector resolution, the energy stability of the beam, and energy straggling in the target as well as the parameters which determine [S].

4.4.2 Ion beams and ion energies

Until recently, backscattering analyses were performed almost exclusively using protons or alpha particles with energies in the Rutherford scattering regime for most target elements. While this analytical technique is convenient, it does not provide the mass resolution, sensitivity, or depth resolution required for some measurements. Light ion beams of both higher and lower energies are currently being used in specialized applications which require high sensitivity and high depth resolution. Heavy ion beams have also been shown to offer enhanced results over conventional techniques for some applications. The main features of each technique are compared in Table 4.7.

4.4.2.1 Comparison of beams and energies

Recently, several backscattering techniques were compared for the analysis of a high-temperature superconducting film, $Y_1Ba_2Cu_3O_{7-x}$ deposited on an Al_2O_3 substrate (Rauhala *et al.*, 1991). Figure 4.15 shows backscattering spectra obtained using 2.5 MeV protons and alpha particles of varying energies. Figure 4.16 shows similar spectra obtained using more massive analysis

Table 4.7. A comparison of some of the main features of some common backscattering techniques.

	Conventional backscattering	High energy, light ion backscattering	High energy, heavy ion backscattering	Medium energy backscattering
Beam/energy/ detectors	^{4}He with 0.5 MeV < E < 2.5 MeV, ^{1}H with 0.5 MeV < E < 1.0 MeV. Surface barrier detectors commonly used.	^{4}He with E > 2.5 MeV, 1H with E > 1 MeV. Surface barrier detectors are commonly used.	Any beam more massive than ^{4}He. Common beams are ^{12}C, ^{16}O, ^{19}F, and ^{35}Cl. Beam energies >5 MeV. Surface barrier and time-of-flight detectors are typical.	Any ion beam may be used. Common beams are ^{4}He, ^{7}Li, ^{12}C. Time-of-flight detectors are most useful.
Cross section	Rutherford except for isolated resonances involving low mass target atoms; negligible screening.	Rutherford for higher masses. Non-Rutherford for some lighter target species. *Experimentally determined cross sections must be used.* Negligible screening.	Beam energies are chosen to give Rutherford cross sections. Screening corrections required particularly for heavy target species.	Rutherford, but screening corrections required.
Sensitivity	Limited for low target masses due to underlying spectrum from heavy substrate or matrix species. Also may be limited by detector effects such as pulse pile-up.	Enhanced sensitivity to low mass species due to relative enhancement in cross section relative to that for higher mass species.	Increased beam energy cancels potential increase in sensitivity arising from higher Z_1. Less sensitive for lighter species than conventional backscattering due to decrease in cross section relative to heavier species. For surface barrier detectors, effects of pulse pile-up on spectrum are reduced.	Limited sensitivity to low mass species due to underlying spectra from heavier substrate or matrix species. Time-of-flight analysis minimizes effects of random coincidence background. Ultimate sensitivity limited by sputtering of target.
Mass resolution	Optimum for lower target masses, poor for heavier species. Limited by intrinsic detector resolution	Mass resolution improves linearly with increasing beam energy.	Mass separation improves with increasing beam mass. However, for surface barrier detectors, degradation of detector resolution with ion mass may result in over-all decrease in resolution.	Mass separation is smaller than for other techniques, but energy resolution of time-of-flight detector yields comparable mass resolution. Resolution may be limited by energy stability of ion beam.
Depth profiling	Relatively small [S] results in large profiling depth, but limited resolution. Resolution is limited at higher energies by [S] and at lower energies by increased energy straggling.	For broad resonances, limitations are similar to those of conventional backscattering. For narrow resonances, depth of profile is limited by proximity of other resonances while depth resolution is limited by width of resonance.	Increased [S] relative to light ions leads to shallower profiling depths and superior potential depth resolution. Depth resolution may be limited by detector resolution.	Large [S] and low beam energy results in very shallow analysis depth. Time-of-flight depth resolutions of order 1 nm exceed that available with other techniques. Resolution may be limited by energy stability of beam.
Other considerations	All target species can be analyzed simultaneously with an appropriate detector. Insignificant radiation damage for both detector and target surface. Large accelerator not required.	Depth profiles for all target species may be obtained simultaneously if resonances are broad, but narrow resonances require a series of measurements. Nuclear reaction products may complicate spectrum. Radiation effects on detectors and target surfaces unimportant. Requires a larger accelerator than conventional backscattering.	All target species heavier than the beam ion may be analyzed simultaneously. Detectors and targets subject to significant radiation damage. Pulse height defects in surface barrier detectors can complicate analysis of spectra. Production of high energy, heavy ions requires a large accelerator.	All target species may be analyzed simultaneously. Reduced beam heating of targets permits analysis of fragile targets. Sputtering may damage targets, particularly for heavier ions. Analysis may be performed with small low energy accelerators such as those used for ion implantation.

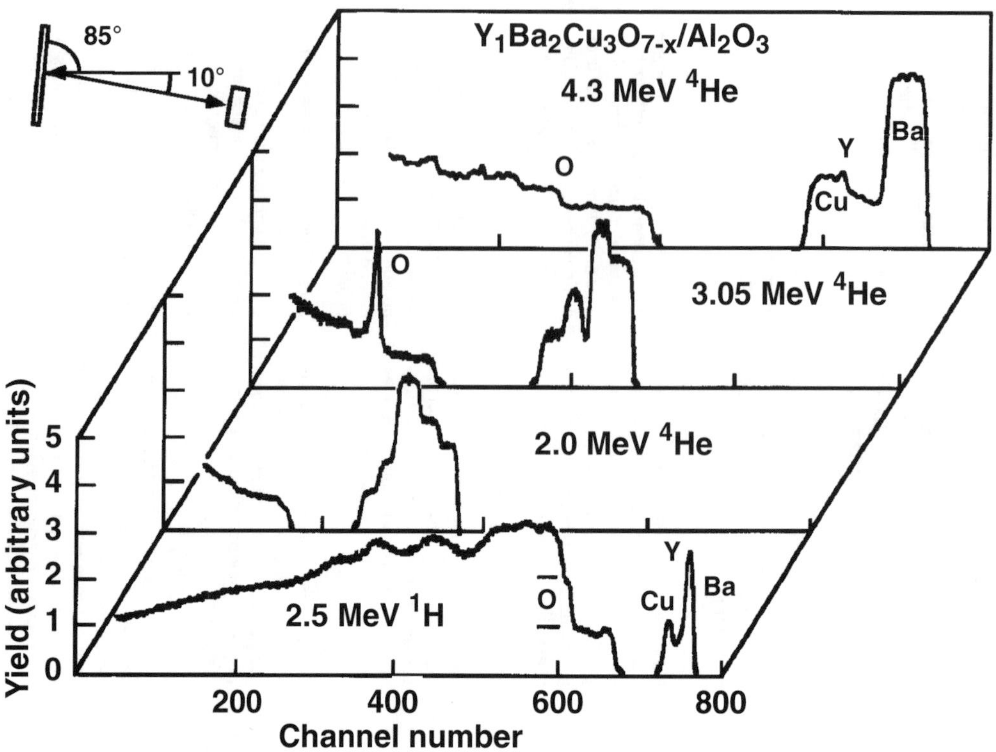

FIG. 4.15. Comparison of backscattering spectra obtained at various beam energies for ^1H and ^4He analysis of $Y_1Ba_2Cu_3O_{7-x}$. Taken from Rauhala *et al.* (1991).

ions. In each case, the analysis ions were scattered into a surface barrier detector at a scattering angle of 170°. The incident and exit angles were both 5° with respect to the surface normal. From Figure 4.15, it should be noted that, while none of the light ions provide good mass resolution for the more massive species in the target, the sensitivity for oxygen detection in the target is greatly enhanced for 3.05 MeV ^4He scattering where the elastic cross section is resonant. For scattering with heavier ions, the mass resolution improves with optimum results obtained for a 22 MeV ^{12}C analysis beam where the high energy edges resulting from scattering from the two isotopes of Cu are clearly separated. Resolution worsens for higher beam masses. Depth resolution for high mass target atoms also improves with increasing beam mass, the ^{12}C spectrum distinctly indicating an increase in Y concentration with depth.

4.4.2.2 Medium energy backscattering

Considerable work has also been done using lower energy ions for backscattering analysis. This technique, called Medium Energy Backscattering (MEBS), commonly employs analysis ions ranging from ^4He to ^{12}C with typical incident energies between 100 keV and 500 keV. Because of the relatively poor energy resolution of surface barrier detectors, other analysis systems such as time-of-flight (TOF) detectors (Mendenhall and Weller, 1989, 1990a, 1990b, 1990c) are used to detect the scattered ions.

One of the obvious advantages to MEBS is the larger elastic scattering cross section. It should be noted however that screening corrections to the cross section are generally necessary in this energy range. As for any technique, the sensitivity of MEBS is limited by background levels. Thus, MEBS is relatively insensitive to light

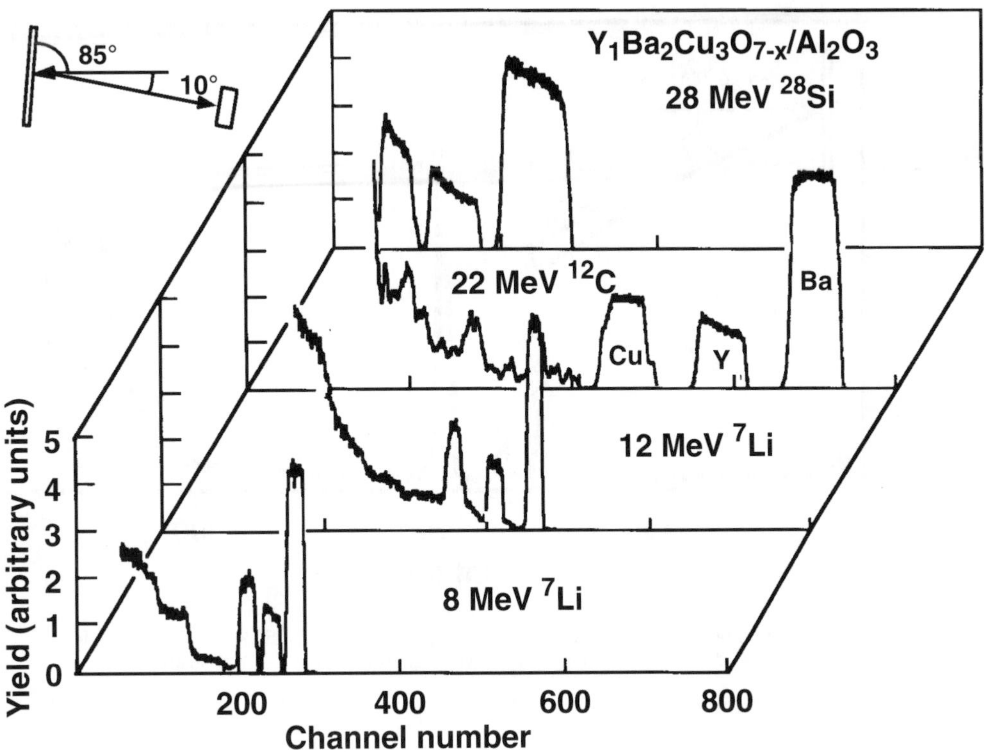

FIG. 4.16. Comparison of backscattering spectra for heavy ion analysis of $Y_1Ba_2Cu_3O_{7-x}$. Taken from Rauhala *et al.* (1991).

target constituents such as oxygen in the presence of heavier target constituents. However, MEBS, with TOF detection, can provide improved sensitivity for heavier species because there is no background arising from pulse pile-up as in a surface barrier detector. (The background arising from high count rates in a TOF detector is flat and distributed uniformly across the entire spectrum, permitting easier and less ambiguous subtraction.)

The sensitivity of this technique is also influenced by beam effects on the target. In particular, sputtering of the target is an important consideration in this energy range and often is the limiting factor in sensitivity of this technique. As a practical matter, the achievable sensitivity for any technique may be limited by the time available for the analysis and the total integrated charge which can be deposited in the target during that time. The lower energy ion beam used for MEBS results in a lower level of target heating as well as a higher cross section, thus permitting use of a higher beam current and lower total analysis doses. In some instances, this combination makes it possible to analyze fragile targets which are destroyed by higher energy beams.

The mass resolution obtained with MEBS is comparable to that of conventional backscattering, but only because the detectors used with this technique have a much better energy resolution than do surface barrier detectors. Figure 4.17 shows a MEBS spectrum obtained using a time-of-flight detector with 450 keV ^{12}C as the analysis ion. The target consisted of a sputter-deposited film containing Cu (2.7 × 10^{15} atoms/cm^2), Ag (10^{14} atoms/cm^2), and Au (3 × 10^{13} atoms/cm^2) on an Al substrate. For this measurement, the beam current was 10 nA, the total integrated charge was 10 μC and the scattering angle was 150°. The beam was incident along the surface normal. Figure 4.18 shows a simulated spectrum for 2 MeV ^4He scattered from the same target also at 150° into a surface barrier detector having a resolution

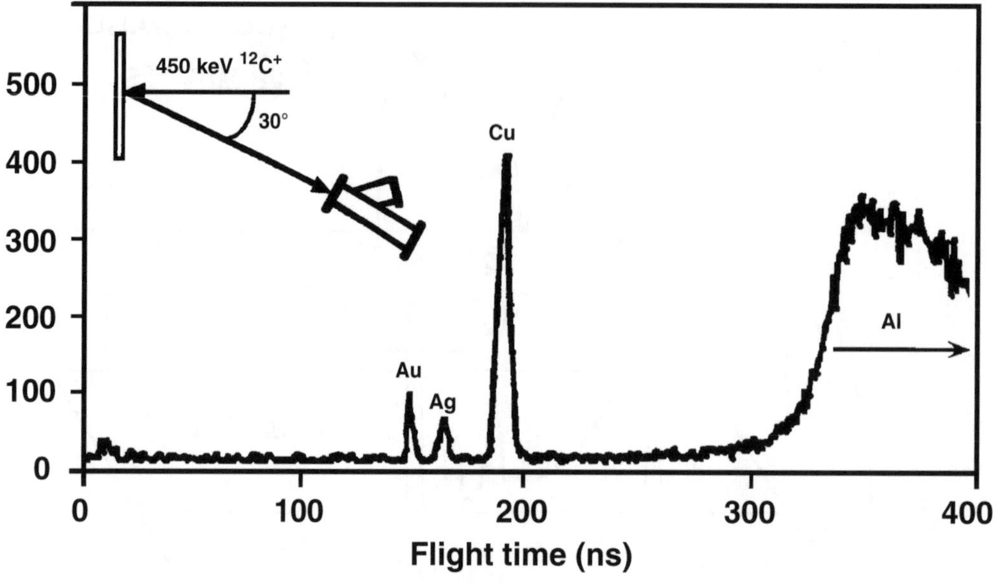

FIG. 4.17. Time-of-flight spectrum for backscattering analysis using 450 keV ^{12}C analysis ions. Target was a sputter-deposited film containing Au, Ag, and Cu on an Al substrate.

of 15 keV. The nature of the spectra is different in that the former is a plot of yield vs. flight time while the latter shows yield vs. energy, but it is clear that comparable mass resolution is obtained for both cases.

MEBS is a particularly useful technique for obtaining depth profiles at the surface of samples. Stopping powers are relatively high at the energies used and the excellent detector resolution of time-of-flight systems allows composition variations to be detected with a resolution of a few nm. Figure 4.19 (Mendenhall and Weller, 1991b) shows a comparison between two spectra obtained using a 270 keV ^7Li analysis beam. The shift of 2 keV in the Ti edge for a $TiSi_2$ film relative to that for Ti metal indicates the presence of a layer of SiO_2 only 3.5 nm thick. The accessible depth of analysis is much smaller than that available for higher energy beams however.

Finally, accelerator requirements for this technique are considerably different from those of higher energy analysis methods. In particular, the beam energy used for MEBS measurements is available from the same types of low-energy accelerators used for ion implantation and for other work in materials science. It should be noted, however, that the energy resolution of this technique is strongly influenced by the energy stability of the incident beam and special attention may need to be given to this aspect of accelerator performance.

4.4.3 Cautions

As with any experimental procedure, there are a number of possible sources of problems both in setting up the experiment and in analyzing the results. Many of these problems will be discussed in some detail in Chapter 12, but we will mention a few briefly here as well.

4.4.3.1 Sample charging

Many of the possible sources of problems in backscattering analysis arise from details of the experimental setup. For insulating targets, sample charging can alter the incident energy of the analysis beam by several keV (significant for MEBS analysis) and also can lead to inaccuracies in current integration when the sample discharges. Another source of potential error in current integration is inadequate charge collection. Both of these difficulties, together with possible preventive measures, are discussed in Chapter 12.

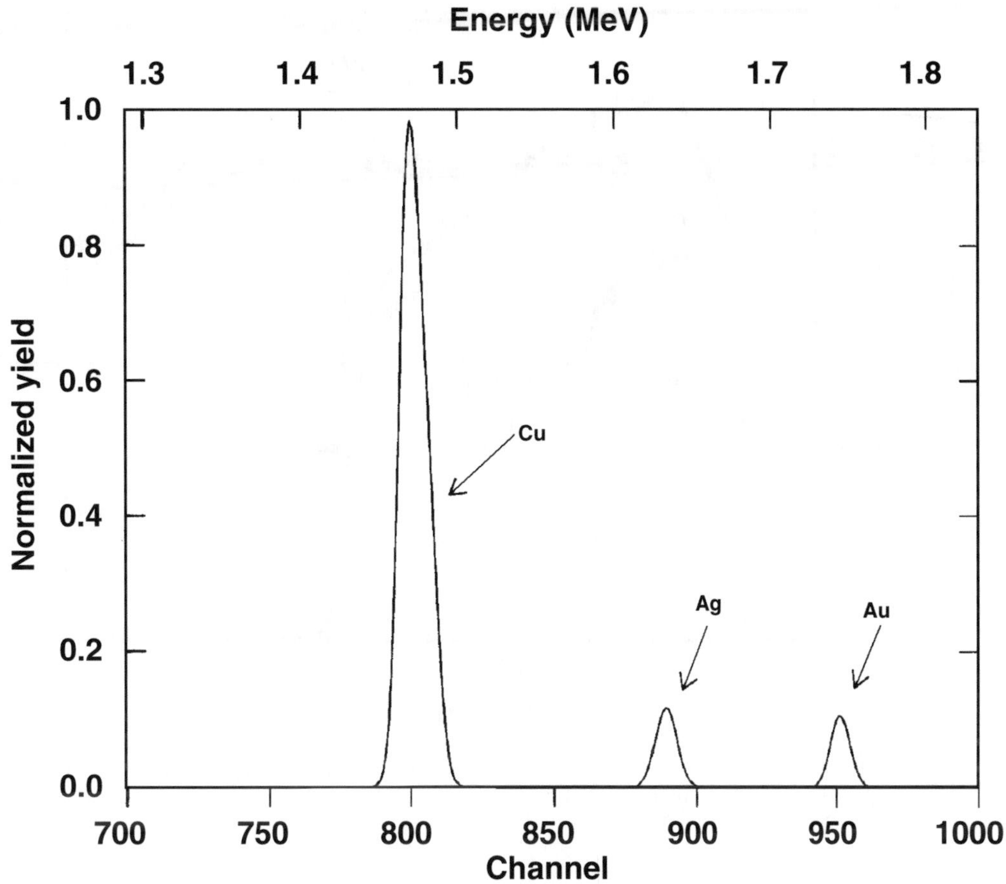

FIG. 4.18. Simulated energy spectrum for backscattering analysis using 2 MeV ^{4}He analysis ions with a surface barrier detector. Target was the same as for Figure 4.17.

4.4.3.2 Channeling

Experimental parameters such as the beam energy and the target orientation can lead to spectra which are difficult to interpret. For crystalline targets, channeling (Chapter 10) can occur if the beam is incident normal to the target surface, resulting in surface peaks and reduced sub-surface yields in the backscattering spectrum. While channeling can also occur for other angles depending on the material, it is possible to orient the target in a non-channeling position relative to the beam. Rotation of the target during bombardment is also used to avoid channeling.

4.4.3.3 Extraneous peaks

Backscattering spectra can also be complicated by processes other than elastic backscattering which occur simultaneously in the target. Inelastically scattered analysis ions and nuclear reaction products may both generate extraneous peaks in the backscattering spectrum.

4.4.3.4 Radiation exposure

Finally, we note that consideration should be given to the radiation levels associated with any backscattering technique. High-energy beams can present serious radiation hazards and even beams with keV energies generate X-rays. Appropriate precautions against radiation exposure should be taken (see Chapter 13).

4.4.4 Examples

The selection of detector, ion beam, and ion energy depends on the specific requirements of a given measurement. Some examples are given in this section.

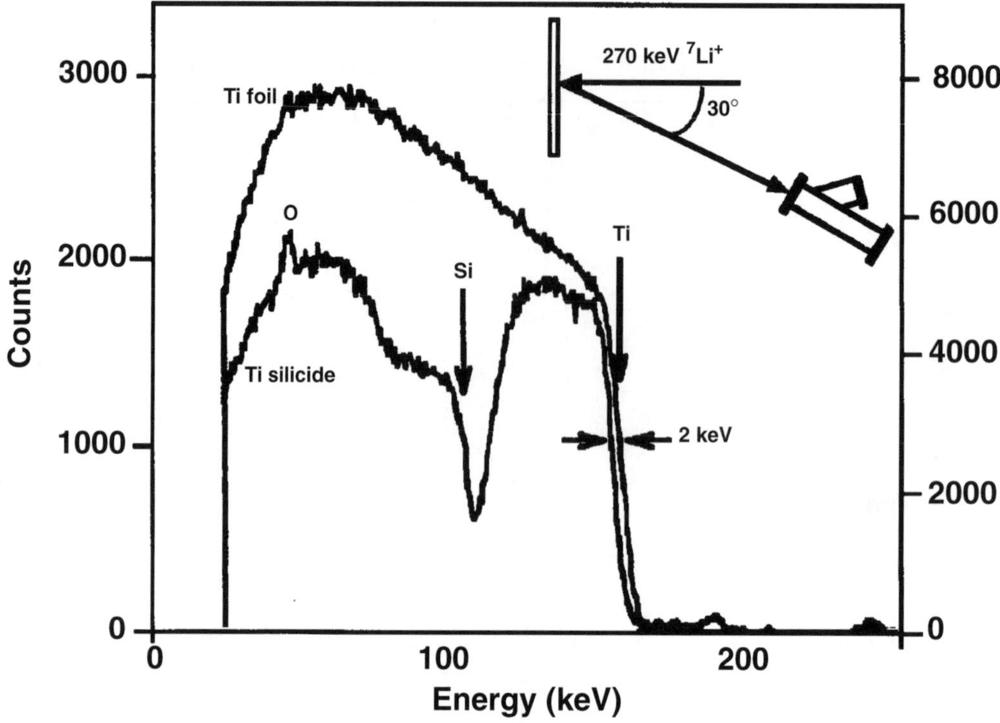

FIG. 4.19. Backscattering spectrum obtained using 270 keV $^7Li^+$ ions with a ToF analyzer. The superposed spectra correspond to a titanium foil target and a titanium silicide film. Taken from Mendenhall and Weller (1991).

4.4.4.1 Depth profile — 1.8 MeV ^{4}He

Conventional backscattering using MeV alpha particles detected by a surface barrier detector is the most convenient and appropriate analysis technique for many measurements. Figure 4.20 shows the results of a typical series of measurements performed using this technique. In this experiment, Yu *et al.* (1991) used alpha backscattering to study the formation of $IrSi_3$ layers by high-dose iridium ion implantation into silicon. For this application, good mass resolution is not necessary because the identity of the target constituents is already known. However, depth resolution was important since depth profiles of the implanted iridium and of the silicon concentration were needed. By orienting the target and detector as shown in the figure, a depth resolution of approximately 5 nm was obtained. This was sufficient to demonstrate clearly the effects of increasingly greater Ir implantation doses on both the Ir depth profile and the Si distribution.

4.4.4.2 Mass resolution — 15 MeV ^{16}O

Heavy ion backscattering has been used for a number of applications where improved mass resolution for heavy elements was needed. Weller (1983) used this technique to measure angular distributions of material sputtered from a Ag-Au alloy. Figure 4.21 shows a spectrum similar to those obtained in that work. This spectrum was obtained using 15 MeV ^{16}O ions combined with a surface barrier detector. The peaks identified as Ag and Au correspond to areal densities of $(3.63 \pm 0.10) \times 10^{13}$ atoms/cm^2 and $(3.73 \pm 0.19) \times 10^{12}$ atoms/cm^2, respectively. Although, the widths associated with each peak are large, each element is clearly resolved. Note particularly that, even with the relatively poor resolution associated with surface barrier detection of heavy ions, a Cu contaminant peak is clearly resolved from a surface peak of Fe. Also, there is no background in the region of interest arising from scattering from the Al foil substrate. (The low level background arises from bulk Fe

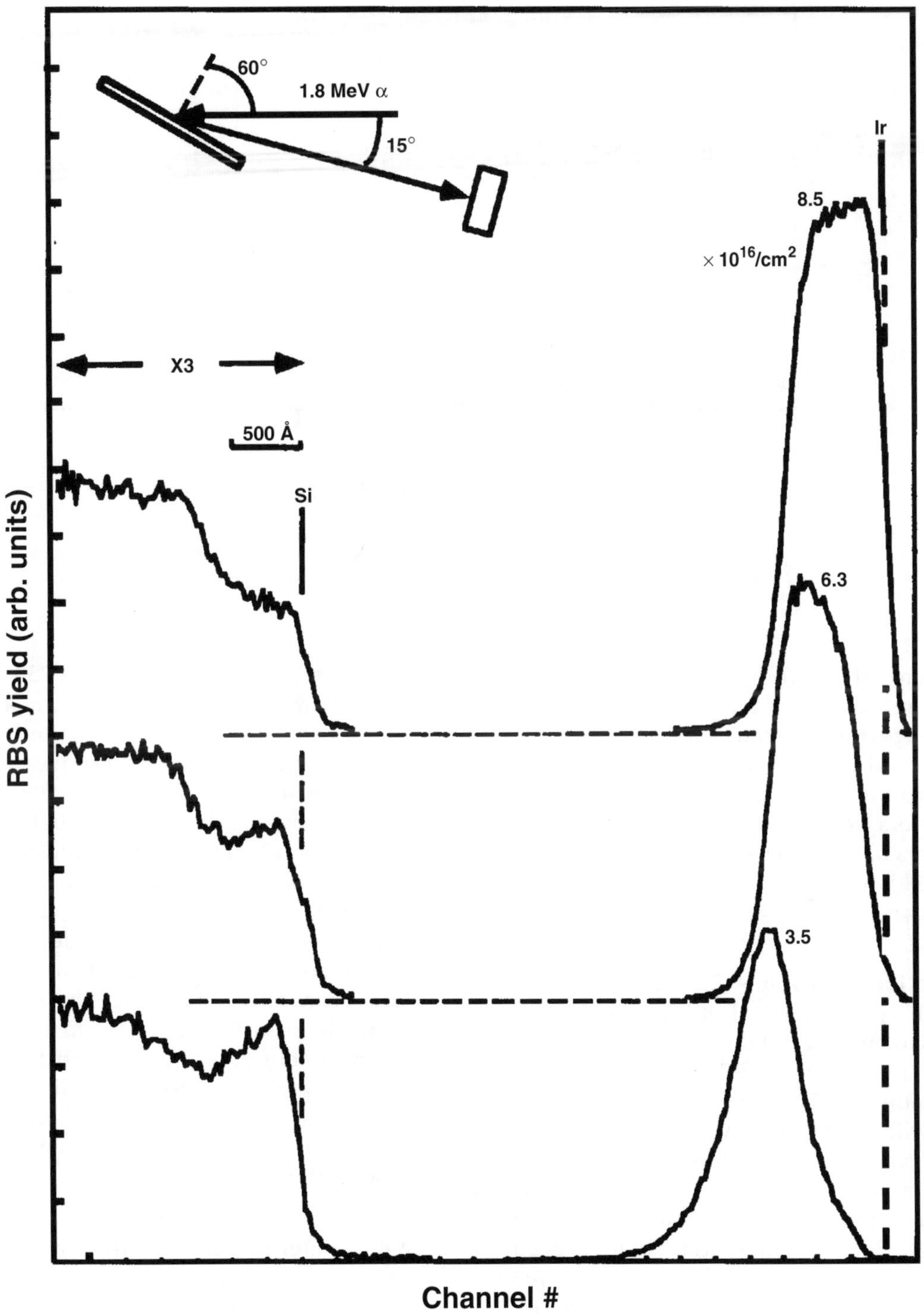

FIG. 4.20. RBS spectra of iridium-implanted silicon for three different implant doses. 1.8 MeV alpha backscattering and surface barrier detectors were used by Yu *et al.* (1991).

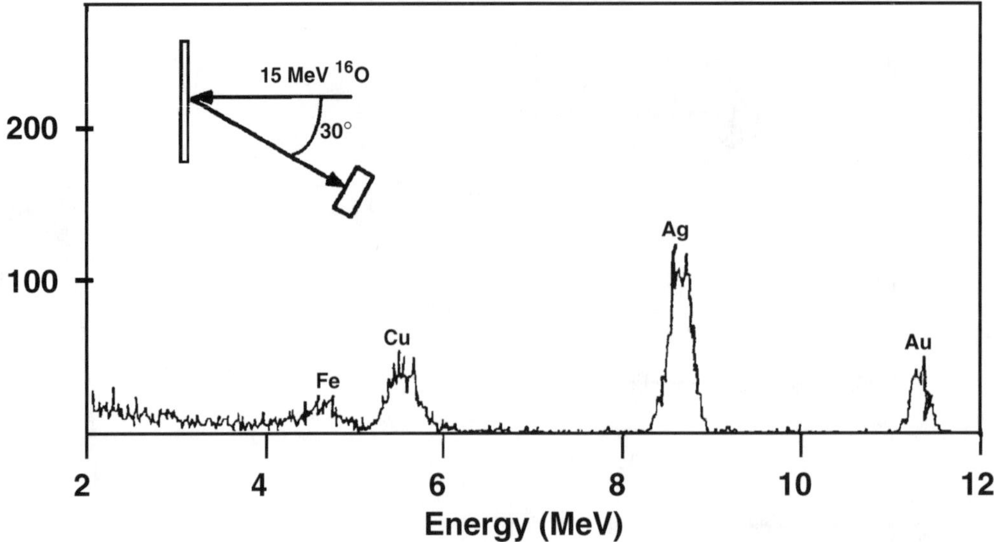

FIG. 4.21. RBS spectrum of sputter-deposited Ag and Au on an aluminum foil substrate. 15 MeV ^{16}O ions were scattered into a surface barrier detector. Taken from Weller *et al.* (1990).

contamination of the foil.) The absence of any pulse pile-up in the vicinity of the Au and Ag peaks permitted the analysis to be performed with beam currents as high as 200 nA. It should be noted that the pulse height defect associated with the detection of heavy ions by surface barrier detectors would have made it difficult to identify these elements had they been unknown species. In that case, it would have been appropriate to use a different type of detector.

4.4.4.3 Depth resolution — 25 MeV ^{35}Cl

Heavy ion backscattering has been combined with TOF analysis to obtain good depth and/or mass resolution. Döbeli *et al.* (1991) used 25 MeV ^{35}Cl to profile a sample consisting of 10 layers, 8 nm thick, of $In_{0.18}Ga_{0.82}As$ alternating with 20 nm thick layers of GaAs. Figure 4.22 shows the In portion of the spectrum obtained by scattering into (a) a surface barrier detector and (b) a TOF detector. While little information about the layered structure can be obtained from the surface barrier spectrum, the TOF spectrum clearly resolves the first eight layers of the structure. Energy dispersion, which worsens progressively with depth into the target, limits the analyzable depth in this case.

4.4.4.4 Beam modification — 250 keV ^{4}He

Medium energy backscattering analysis is most useful for the analysis of very thin films and for measurement of depth profiles very near the surface of materials. Mendenhall and Weller (1990b, 1990c) used this technique to investigate the destruction of CaF_2 and MgF_2 films by ion bombardment, irradiating targets with a relatively high current of 250 keV ^{4}He. The beam current was periodically reduced to a much smaller value for analysis so that the dose dependence of the target composition was measured. Results shown in Figure 4.23 for CaF_2, converted from a TOF to an energy spectrum, clearly indicate the loss of F from the sample for increasing irradiation doses. As seen with this application, MEBS can be particularly useful in the study of ion beam modification of materials since the same accelerator and sometimes even the same ion beam can be used for both modification and analysis.

4.4.4.5 Depth profile with elastic resonance — 3 MeV ^{4}He

Non-Rutherford elastic scattering is often used to obtain concentrations of low Z target constituents which are difficult to analyze when the cross section is

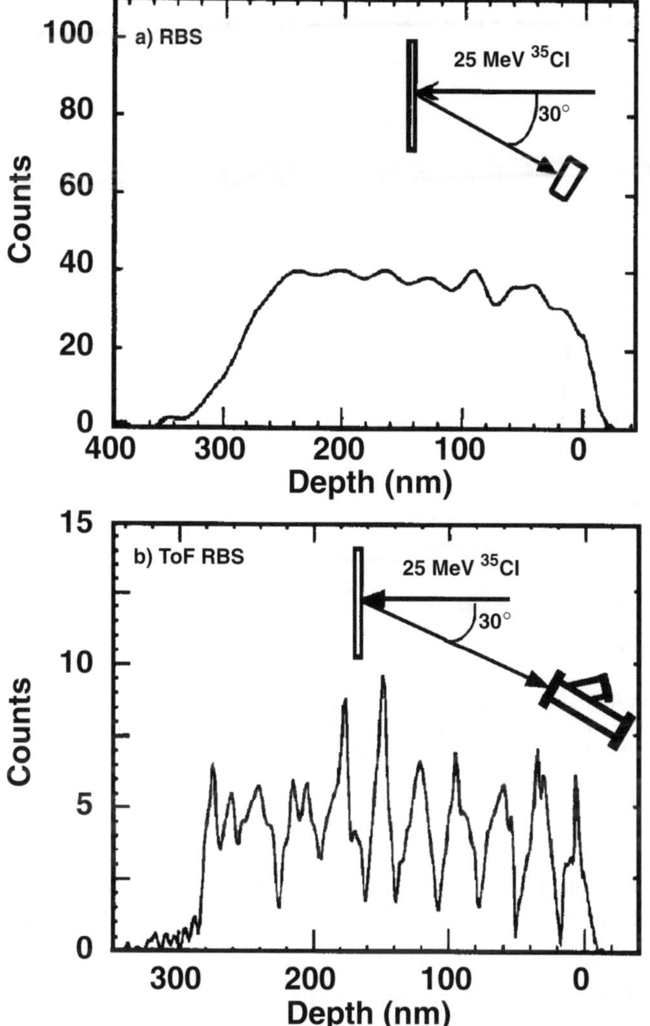

FIG. 4.22. RBS spectra obtained from target consisting of 10 layers, 8 nm thick, of $In_{0.18}Ga_{0.82}As$ alternating with 20 nm thick layers of GaAs. Figure shows In region of the spectra only for 25 MeV ^{35}Cl ions scattered into (a) a surface barrier detector and (b) a TOF detector. Taken from Döbeli *et al.* (1991)

Rutherford. Vizkelethy *et al.* (1990) demonstrated the use of this technique for profiling oxygen in buried layers. Figures 4.24 and 4.25 show spectra obtained by scattering alphas with energy 3.054 MeV and 3.254 MeV, respectively, from a target consisting of a 20 nm SiO_2 layer sandwiched between two 300 nm layers of Cu deposited on a graphite substrate. In Figure 4.24, the backscattering yield from a surface oxide layer (channel 210) is enhanced by the 3.04 MeV resonance in the $^{16}O(\alpha,\alpha)^{16}O$ elastic scattering cross section while the much smaller yield from the buried oxide layer (channel 150) results from the smaller off-resonance cross section. In Figure 4.25, however, scattering from the buried layer is resonant, yielding a more accurate estimate of the oxygen concentration at that location.

It is not possible in an article of this length to describe or give an example of every variation of backscattering analysis. Two sources for references on current backscattering techniques and applications are the proceedings of two biannual international conferences, *Applications of Accelerators in Research and Industry* (1989, 1991, 1993) and *Ion Beam Analysis* (1990, 1992, 1994).

REFERENCES

Andersen, H. H., Besenbacher, F., Loftager, P., and Möller, W. (1980), *Phys. Rev.* **A21**, 1891.

"Applications of Accelerators in Research and Industry '88" (1989), (J.L. Duggan and I.L. Morgan, ed.), *Nucl. Instr. and Meth.* **B40/41.**

"Applications of Accelerators in Research and Industry '90" (1991), *Nucl. Instr. and Meth.* **B56/57.**

"Applications of Accelerators in Research and Industry '92." (1993), *Nucl. Instr. and Meth.* **B79.**

Barbour, J. C., Doyle, B. L., and Myers, S. M. (1988), *Phys. Rev.* **B38**, 7005.

Bittner, J. W. and Moffat, R. D. (1954), *Phys. Rev.* **96**, 374.

Boerma, D.O., Labohm, F., and Reinders, J.A. (1990), *Nucl. Instr. and Meth.* **B50**, 291.

Børgesen, P., Behrisch, R., and Scherzer, B.M.U. (1982), *Appl. Phys.* **A27**, 183.

Bozoian, M., (1991a), *Nucl. Instr. and Meth.* **B56/57**, 740.

Bozoian, M., (1991b), *Nucl. Instr. and Meth.* **B58**, 127.

Bozoian, M., Hubbard, K. M., and Nastasi, M. (1990), *Nucl. Instr. and Meth.* **B51**, 311.

Bragg, W. H., and Kleeman, R. (1905), *Phil. Mag.* **10**, 5318.

Butler, J.W. (1986), *Nucl. Instr. and Meth.* **B15**, 232.

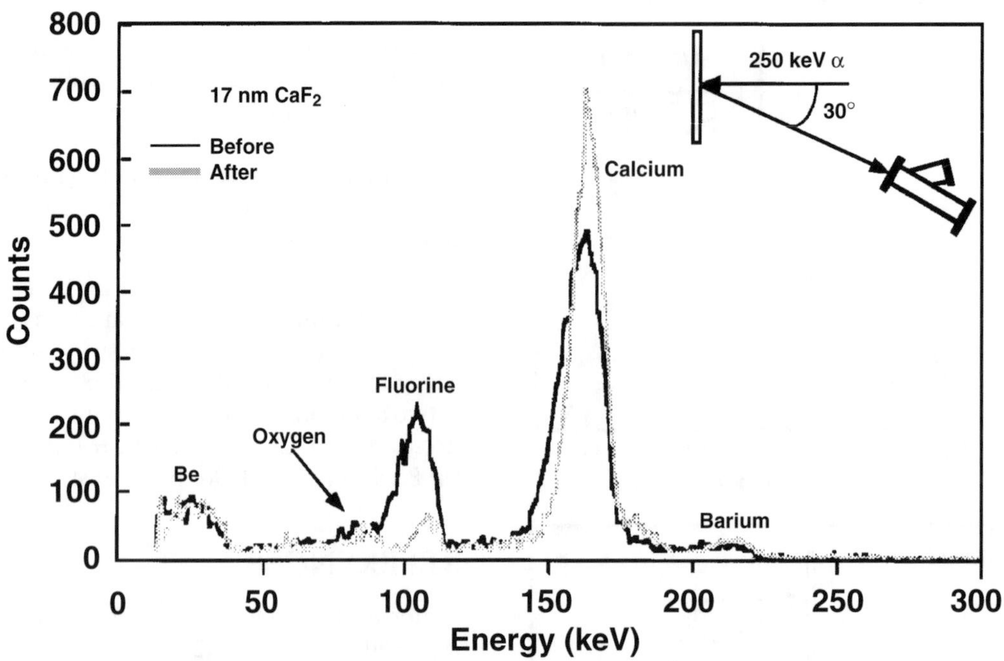

FIG. 4.23. Energy spectra showing relative concentrations of Ca and F in a CaF2 film before and after irradiation with 250 keV ^{4}He. The CaF_2 film was analyzed with 250 keV ^{4}He using a TOF detector. The TOF spectrum was then converted to the more familiar energy spectrum. Taken from Mendenhall and Weller (1990b).

Butler, J.W. (1990), *Nucl. Instr. and Meth.* **B45**, 160.

Cameron, J. R. (1953), *Phys. Rev.* **90**, 839.

Chu, W- K., Mayer, J. W., and Nicolet, M- A. (1978), "Backscattering Spectrometry," Academic Press, New York.

Döbeli, M., Haubert, P.C., Livi, R.P., Spicklemire, S.J., Weathers, D.L., and Tombrello, T.A. (1991), *Nucl. Instr. and Meth.* **B56/57**, 764.

Doolittle, L.R. (1985), *Nucl. Instr. and Meth.* **B9**, 344.

Doolittle, L.R. (1986), *Nucl. Instr. and Meth.* **B15**, 227.

Doolittle, L.R. (1990), in "High Energy and Heavy Ion Beams in Materials Analysis" (J.R. Tesmer, C.J. Maggiore, M. Nastasi, J.C. Barbour, and J.W. Mayer, eds.), p. 175. Materials Research Society, Pittsburgh, Pennsylvania.

Endt, P. M., and van der Leun, C. (1967), *Nucl. Phys.* **A105**,1.

Eridon, J.M. and Was, G.S. (1985), *Nucl. Instr. and Meth.* **B12**, 505.

Eshbach, H. L. (1983), The Bi-implanted-in-Si RBS standard was supplied by H. L. Eshbach, Central Bureau of Nuclear Measurements, Steenweg op Retie, 2440 Geel, Belgium.

Eskildsen, S.S. (1982), in "Ion Implantation into Metals" (V. Ashworth, W.A. Grant, and R.P.M. Procter, eds.), p. 315. Pergamon, Oxford.

Häusser, O., Ferguson, A. J., McDonald, A. B., Szöghy, I. M., Alexander, T. K., and Disdier, D. L. (1972), *Nucl. Phys.* **A179**, 465.

Hautala, M. and Luomajärvi, M. (1980), *Rad. Effects* **45**, 159.

Herring, D. F. (1958), *Phys. Rev.* **112**, 1217.

Hubbard, K. M., Tesmer, J. R., Nastasi, M., and Bozoian, M. (1991), *Nucl. Instr. and Meth.* **B58**, 121.

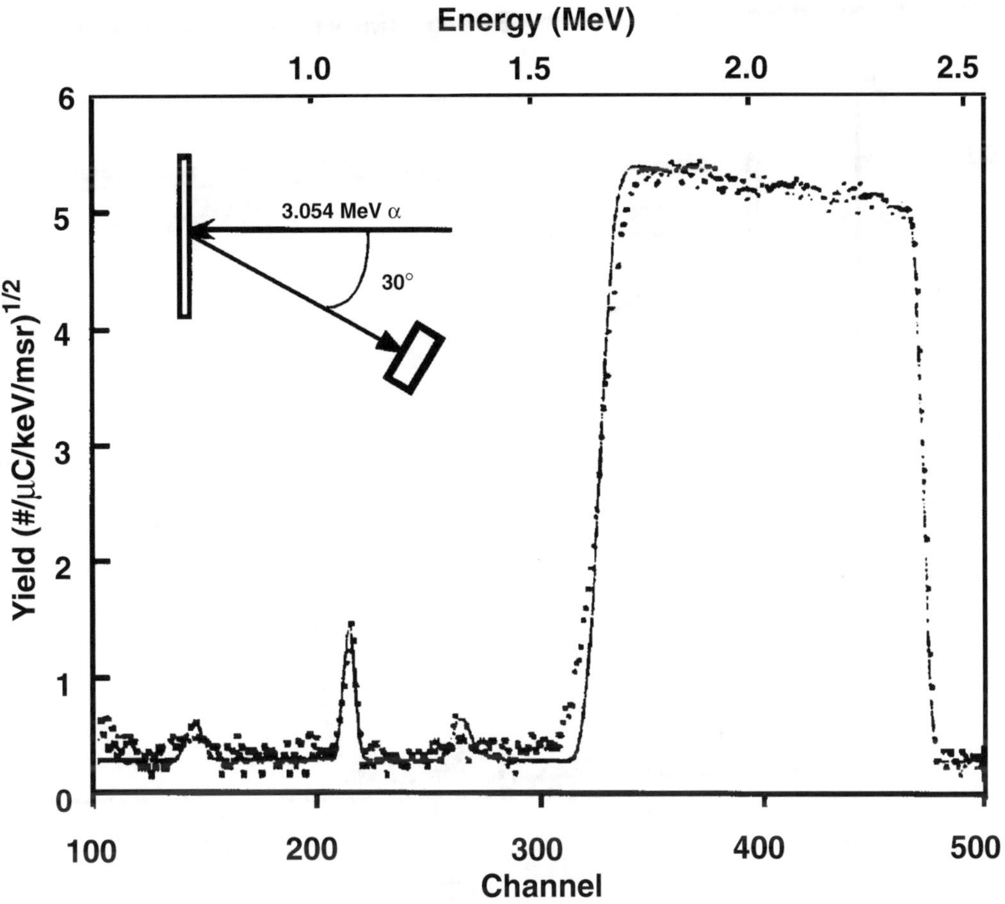

FIG. 4.24. Backscattering spectrum of a target consisting of a 20 nm SiO_2 layer sandwiched between two 300 nm layers of Cu deposited on a graphite substrate. The incident beam was 3.054 MeV alphas for which the $^{16}O(\alpha, \alpha)^{16}O$ elastic cross section is resonant. Taken from Vizkelethy *et al.* (1990).

"Ion Beam Analysis 9." (1990), *Nucl. Instr. and Meth.* **B45**.

"Ion Beam Analysis 10." (1992), *Nucl. Instr. and Meth.* **B64**.

"Ion Beam Analysis 11." (1994), *Nucl. Instr. and Meth.* **B85**.

Jarjis, R. A. (1979), "Nuclear Cross Section Data for Surface Analysis" (Department of Physics, University of Manchester) unpublished.

Kido, Y., Kakeno, M., Yamada, K., Ohsawa, H., and Kawakami, T. (1985), *J. Appl. Phys.* **58**, 3044.

Kido, Y. and Youkio, Y. (1985), *Nucl. Inst. and Meth.* **B9**. 291.

Kido, Y. and Kawamoto, J. (1987), *J. Appl. Phys.* **61**, 956.

Leavitt, J. A. (1987), *Nucl. Instr. and Meth.* **B24/25**, 717.

Leavitt, J. A., McIntyre, L. C. Jr., Stoss, P., Ashbaugh, M. D., Dezfouly-Arjomandy, B., Hinedi, M. F., and Van Zijll, G. (1988), *Nucl. Instr. and Meth.* **B35**, 333.

Leavitt, J. A., Stoss, P., Edelman, C. R., Davis, R. E., Gutierrez, S., Jubb, N. J., and Reith, T. M. (1985), *Nucl. Instr. and Meth.* **B10/11**, 719.

Leavitt, J. A., McIntyre, L. C. Jr., Champlin, R. S., Stoner, J. O. Jr., Lin, Z., Ashbaugh, M. D., Cox, R. P., and Frank, J. D. (1994), *Nucl. Instr. and Meth.* **B85**, 37.

L'Ecuyer, J., Davies, J. A., and Matsunami, N. (1979), *Nucl. Instr. and Meth.* **160**, 337.

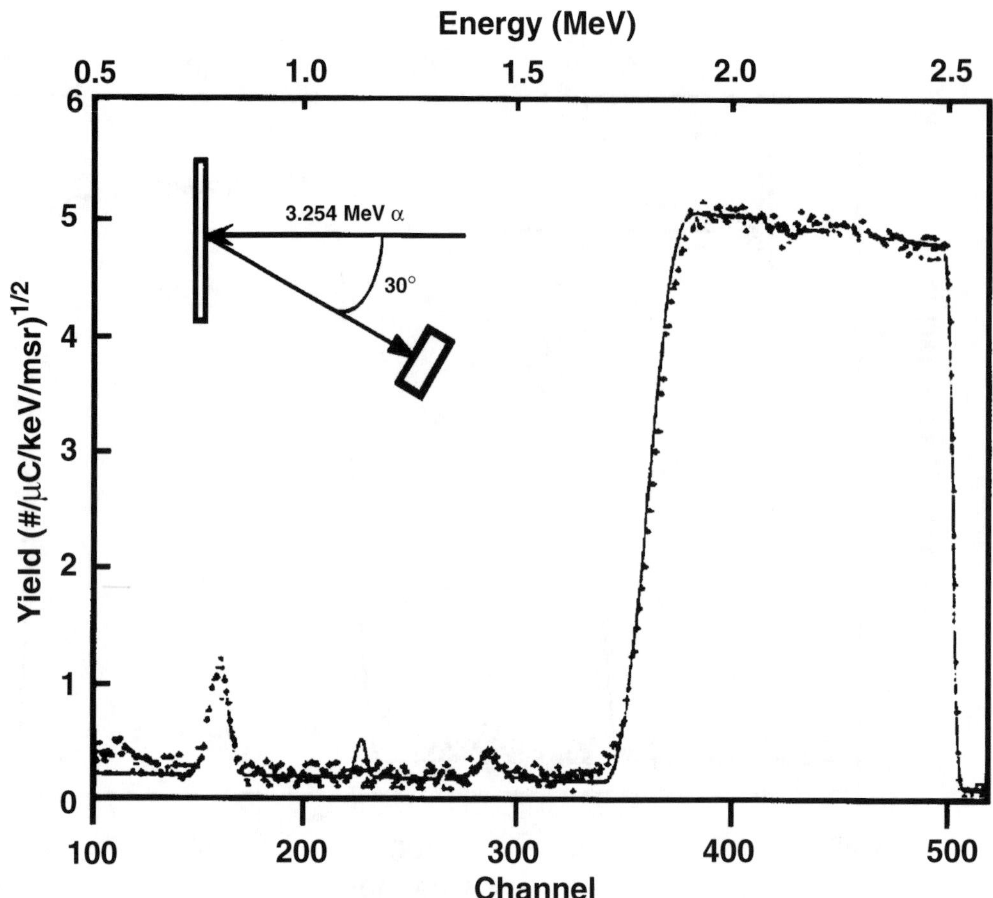

FIG. 4.25. Analysis of the same target used in Fig. 4.24 with a 3.254 MeV alpha beam. Taken from Vizkelethy *et al.* (1990).

Li, W.Z. and Al-Tamimi, Z. (1986), *Nucl Instr. and Meth.* **B15**, 241.

Macdonald, J. R., Davies, J. A., Jackman, T. E., and Feldman, L. C. (1983), *J. Appl. Phys.* **54**, 1800.

Martin, J. C., Nastasi, M., Tesmer, J. R., and Maggiore, C.J. (1988), *Appl. Phys. Lett.* **52**, 2177.

McIntyre, L.C. Jr., Ashbaugh, M.D., and Leavitt, J.A. (1987), *Mat. Res. Soc. Symp. Proc.* **93**, 401.

Mendenhall, M.H. and Weller, R.A. (1989), *Nucl. Instr. and Meth.* **B40/41**, 1239.

Mendenhall, M.H. and Weller, R.A. (1990a), *Nucl. Instr. and Meth.* **B47**, 193.

Mendenhall, M.H. and Weller, R.A. (1990b), *Nucl. Instr. and Meth.* **B51**, 400.

Mendenhall, M.H. and Weller, R.A. (1990c), *Appl. Phys. Lett.* **57**, 1712.

Mendenhall, M.H. and Weller, R. A. (1991a), *Nucl. Instr. and Meth.* **B58**, 11.

Mendenhall, M.H. and Weller, R.A. (1991b), *Nucl. Instr. and Meth.* **B59/60**, 120.

Morris, J.R. (1982), *Nucl. Inst. and Meth.* **197**, 147.

Mozer, F. W. (1956), *Phys. Rev.* **104**, 1386.

Müller, P., Szymczak, W., and Ischenko, G. (1978), *Nucl. Instr. and Meth.* **149**, 239.

Northcliffe, L. C., and Schilling, R. F. (1970), *Nuclear Data Tables* **A7**, 233.

O'Connor, D. J., and Chunyu, T. (1989), *Nucl. Instr. and Meth.* **B36**, 178.

Petrov, I., Braun, M., Fried, T., and Sätherblom (1983), *J. Appl. Phys.* **54**, 1358.

Rauhala, E. (1984), *J. Appl. Phys.* **56**, 3324.

Rauhala, E. (1987), *J. Appl. Phys.* **62**, 2140.

Rauhala, E., Saarilahti, J. and Nath, N. (1991), *Nucl. Instr. and Meth.* **B61**, 83-90.

Saarilahti, J. and Rauhala, E. (1992), *Nucl. Instr. and Meth.* **B64**, 734.

Saunders, P.A. and Ziegler, J.F. (1983), *Nucl. Instr. and Meth.* **218**, 67.

Simpson, J.C.B. and Earwaker, L.G. (1986), *Nucl. Instr. and Meth.* **B15**, 502.

Vizkelethy, G., Revesz, P., Li, J., Matienzo, L.J., Emmi, F., and Mayer, J.W. (1990), in "High Energy and Heavy Ion Beams in Materials Analysis" (J.R. Tesmer, C.J. Maggiore, M. Nastasi, J.C. Barbour, and J.W. Mayer, eds.), p. 235, Materials Research Society, Pittsburgh, Pennsylvania.

Weller, M.R. (1983), *Nucl. Instr. and Meth.* **212**, 419.

Weller, M.R., Mendenhall, M.H., Haubert, P.C., Döbeli, M., and, Tombrello, T.A. (1990), in "High Energy and Heavy Ion Beams in Materials Analysis" (J.R. Tesmer, C.J. Maggiore, M. Nastasi, J.C. Barbour, and J.W. Mayer, eds.), p. 139. Materials Research Society, Pittsburgh, Pennsylvania.

Wenzel, W. A., and Whaling, W. (1952), *Phys. Rev.* **87**, 499.

Williams, J.S. and Möller, W. (1978), *Nucl. Instr. and Meth.* **157**, 213.

Yu, K.M., Katz, B., Wu, I.C., and Brown, I.G., (1991), *Nucl. Instr. and Meth.* **B58**, 27.

Ziegler, J. F., Biersack, J. P., and Littmark, U. (1985), "The Stopping and Range of Ions in Solids," Volume 1 of The Stopping and Ranges of Ions in Matter, Pergamon Press, New York.

Ziegler, J.F., Lever, R.F., and Hirvonen, J.K. (1976), in "Ion Beam Surface Layer Analysis" (O. Meyer, G. Linker, and F. Kappler, eds.), p. 457. Plenum Press, New York.

Ziegler, J. F., and Manoyan, J. M. (1988), *Nucl. Instrum. Methods* **B35**, 215.

ELASTIC RECOIL DETECTION: ERD

(or Forward Recoil Spectrometry: FRES)

J. C. Barbour and B. L. Doyle

Sandia National Laboratories, Albuquerque, New Mexico

Contributors: G. W. Arnold, J. E. E. Baglin, P. F. Green, A. J. Kellock, J. A. Knapp, and H. J. Whitlow

CONTENTS

5.1 INTRODUCTION: ERD ANALYSIS WITH RANGE FOIL - CONCEPT AND HOW TO READ A SPECTRUM

One of the most useful modern ion-beam analysis techniques for easy depth profiling of light elements is Elastic Recoil Detection (ERD), which was first introduced in 1976 (L'Ecuyer *et al.*, 1976). Before the introduction of ERD, Nuclear Reaction Analysis (NRA) was the predominant ion-beam technique for obtaining light element depth profiles, but the ease and simplicity of ERD has led to its extensive use. Although for many cases ERD has poorer sensitivity and depth resolution than NRA, experiments which do not involve trace element analysis can be done with ERD more simply than with NRA. Compared to depth profiling of heavier atoms (Z≥11) using Rutherford Backscattering Spectrometry (RBS) with protons or alpha particles (see Chap. 4), ERD uses high-energy (≅1 MeV/amu) heavy-ion beams to kinematically recoil and depth profile low atomic number target atoms (1≤Z≤9). However, the heavy ion projectile need only have a mass greater than the target atom; and in fact alpha particles are commonly used to obtain recoil spectra for hydrogen and its isotopes (Doyle and Peercy, 1979b). This latter form of ERD with alpha particles has also been referred to as Forward REcoil Spectrometry (FRES) (Mills *et al.*, 1984; Green and Doyle, 1990). ERD has become the technique of choice for many light element profiling applications because of (1) the simplicity, (2) the possibility of simultaneous multi-element profiling, (3) fair sensitivity, and (4) good depth resolution.

Although the use of ERD is becoming a relatively common tool in materials analysis, most researchers use ERD only qualitatively (i.e., for elemental identification or profile shapes), in contrast to the quantitative analyses done with RBS. Yet, both in its concept and in its elementary execution, ERD is a simple experiment. The physical concepts governing ERD and RBS are identical; and through the use of a computer, ERD spectra are easily manipulated to obtain full quantitative analyses. In order to hasten the use of ERD in a quantitative fashion, this chapter will give the basic physical concepts of ERD and step through some simple examples of ERD analyses showing the manipulation of the spectrum at each step in order to give a quantitative elemental depth profile.

Much like RBS (Chu *et al.*, 1978), ERD depends on only four physical concepts:

- The kinematic factor describes the energy transfer from a projectile to a target nucleus in an elastic two-body collision.
- The differential scattering cross section gives the probability for the scattering event to occur.
- The stopping powers give the average energy loss of the projectile and the recoiled target atom as they traverse the sample (thereby giving a method for establishing a depth scale).
- The energy straggling gives the statistical fluctuation in the energy loss.

By applying these four physical concepts, the elastic recoil spectrum can be transformed into a quantitative concentration profile as a function of depth. The main difference in the treatment of ERD data and RBS data is in accounting for the difference in stopping between the incident projectile and the recoiled target atom. Analytical expressions analogous to those developed for RBS (but also including the effects of post-recoil range foils) will be given in this chapter.

A comparison of ERD and RBS is shown in Fig. 5.1 with ~1 MeV/amu projectiles. At the top of the figure is a schematic of the experimental set-up, which is essentially the same for ERD as RBS but with the addition of a range foil in front of the detector. In fact, RBS and ERD can be set-up in the same vacuum chamber by adding a detector (with range foil) in the forward scattering position, and by giving the sample holder the rotation capability to allow for grazing incidence of the projectile and exit of the recoiled atoms, as shown.

For the ERD analysis shown in Fig. 5.1, consider a Si^{+6} ion with $E_0(Si)$ = 28 MeV incident upon a Si_3N_4 sample containing a large amount of H. (The definition of energies can also be seen in Fig. 5.5.) The Si^{+6} loses energy ($E_0 \rightarrow E'_0$, stopping power $S_{Si} \cong 480$ eV/A) as it enters the sample, kinematically recoils an H atom with energy $E_2(H) = KE'_0 \cong 2.8$ MeV ($K \cong 0.1$), and then the H atom loses energy ($E_2 \rightarrow E_3$, with $S_H \cong 3$ eV/A) as it traverses the sample out to a Si surface barrier detector (SBD). A particle filter (range foil) is placed in front of the SBD in order to stop the high energy Si, reflected out of the incident beam, from entering the detector and overwhelming the signal or damaging the detector. (The energy loss in this range foil must also be taken into account when analyzing the data.) The differential Rutherford-scattering cross section ($d\sigma/d\Omega$) giving the

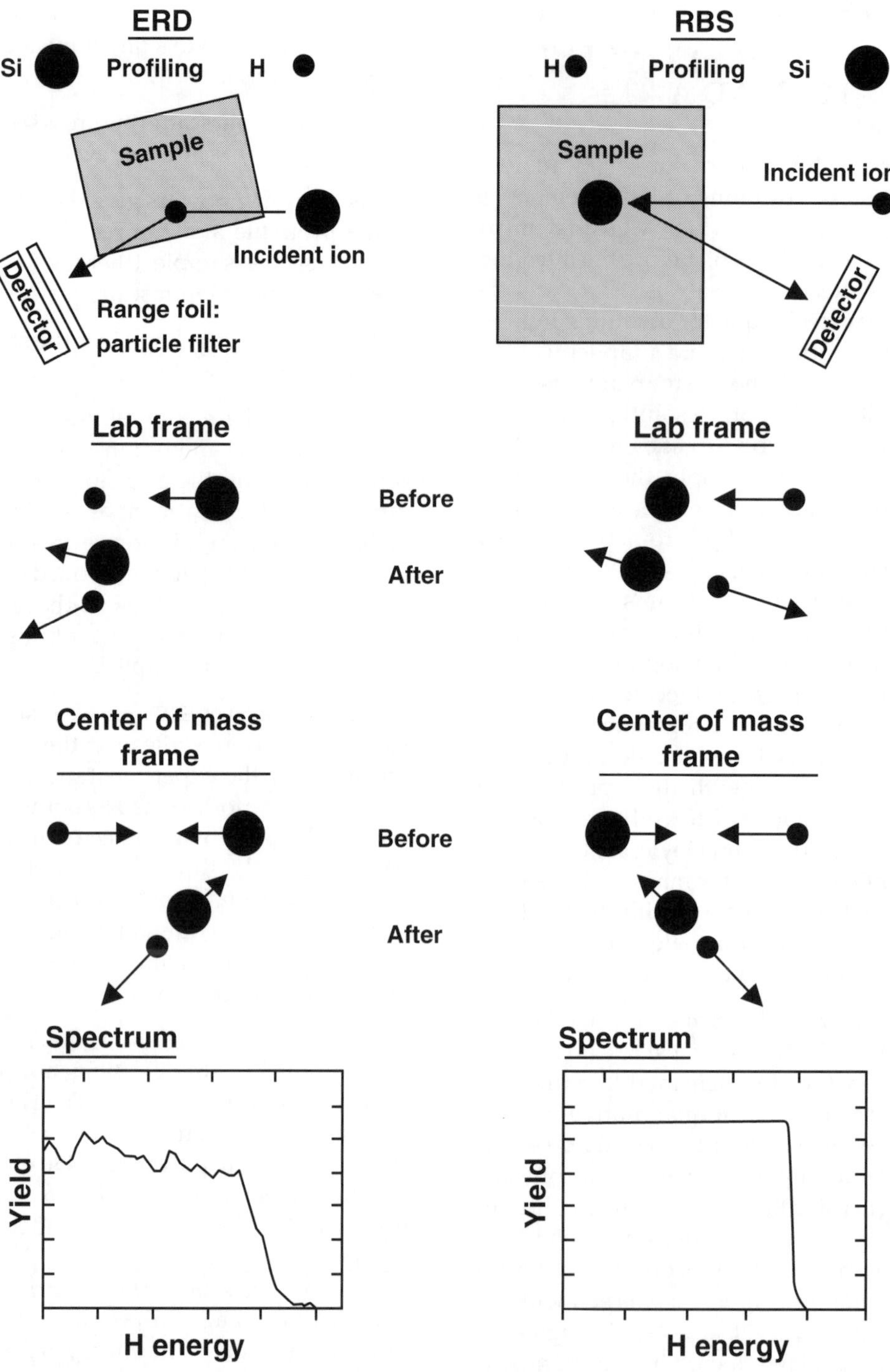

FIG. 5.1. A schematic comparison of elastic recoil detection (ERD) and Rutherford backscattering spectrometry (RBS) scattering geometries in the lab and center-of-mass (COM) frames of reference. The Si atom is represented by the larger filled circle and the H atom is represented by the smaller filled circle. At top is shown the lab-frame scattering geometry and the lab-frame scattering event (before and after the collision). In the COM frame, the scattering events for ERD and RBS appear to be equivalent. The representative spectra shown for the collection of scattered H atoms (which could be collected in a Si surface barrier detector) appear qualitatively similar for both ERD and RBS.

probability for the forward recoil scattering event is approximately 1.7 barns/sr. For RBS, a proton of energy $E_o = 1$ MeV is incident upon a Si sample. The proton loses energy ($E_o \rightarrow E'_o$, stopping power $S \cong 14$ eV/A) as it enters the sample, is kinematically backscattered with an energy $E_2 = KE'_o \cong 0.87$ MeV ($K \cong 0.87$) from a Si atom and then loses energy ($E_2 \rightarrow E_3$) as it traverses the sample out to a Si SBD. The differential Rutherford-scattering cross section giving the probability for such an event is approximately 0.26 barns/sr.

The middle portion of Fig. 5.1 schematically shows the transformation of the scattering events, described above, from the laboratory reference frame to the center-of-mass (COM) reference frame. In the COM reference frame, ERD and RBS appear as symmetrical scattering. The yield of H as a function of detected H-energy is shown in the bottom portion of Fig. 5.1, demonstrating the similarity in the spectra obtained for RBS and ERD. In fact, the ERD spectrum can be read in a manner similar to that for an RBS spectrum as shown at the bottom of Fig. 5.1. Both spectra give the yield of detected hydrogen particles as a function of the hydrogen energy, but the important point to note is that for RBS the detected particle is the same species as the incident ion, whereas for ERD the detected particle is coming from the sample. In fact, for samples containing several light element species, many of these elements can be detected simultaneously in a single spectrum. In this latter case, the detected particle energy will be that for several different recoiled particles. Figure 5.2 shows such an example where a 24 MeV Cl^{+5} beam is used for ERD analysis of an oxidized porous silicon (OPS) layer containing primarily Si and O, but with small amounts of F and some hydrocarbon contaminant on the surface. The beam is incident at an angle of 75° from the sample normal and the scattered recoil atoms (H, C, O, and F) are collected at a scattering angle of 30°.

The atomic concentration for oxygen is shown schematically in the middle portion of Fig. 5.2 with the depth scale increasing to the left. The concentration profile at the surface has a slope resulting from the rough surface, but at the bottom of the OPS layer the profile is sharp at the Si-OPS layer interface. The concentration profiles for the oxygen and other recoiled atoms reveal themselves in the ERD spectrum (open circles) as the number of detected counts as a function of channel. Also shown (solid line) is the spectrum from an nitrided oxidized porous silicon layer (NOPS). Standards can be used to calibrate the channel to energy conversion, but for these spectra only the position of the surface hydrogen peak and the known kinematic factor for scattering from the surface was used to determine the detected energy (E_d) scale shown along the top axis. Similar to RBS, the scattering from the surface of the sample is recorded at higher energies, whereas scattering from deeper in the sample is recorded at lower energies. The surface scattering positions for each detected element are marked with solid arrows in the spectra of Fig. 5.2. The energy scale of the ERD spectrum can be directly related to depth scales for each element through a knowledge of the stopping power for the incident Cl^{+5} ion and the different recoiled atoms. Note that the OPS sample was partially porous and had a rough surface which contributed to a decreased depth resolution at depth, which for the case of ERD shows a greater dependence on sample roughness than for the case of RBS because of the "grazing" angle of incidence used in ERD. Ion beam analysis techniques are insensitive to sample density and therefore porosity and surface roughness give rise to a slope in the ERD spectrum at lower energies, whereas the edge width in the spectrum at the surface energy position is a reflection of the inherent depth resolution.

The yield for a given element can be related to the atomic concentration of that element. For samples with limited depth resolution, a measure of the integrated peak area which is the areal density (number of atoms/cm^2 throughout the layer thickness) is the better analysis approach. Therefore, in the spectra of Fig. 5.2, the O concentration from the OPS sample is best determined in terms of the areal density, but the N concentration from the NOPS sample exhibits a flat-top peak in which there is sufficient depth resolution to determine a concentration profile (number of atoms/cm^3 as a function of the layer thickness). A description of the method converting an ERD spectrum to either areal density or a concentration profile will be given later in this chapter.

5.2 FUNDAMENTALS

Atoms recoiled from the surface appear at different energies for the different elements in Fig. 5.2, and therefore the masses may be separated in the spectrum as shown for F and O. However, in contrast to RBS where light masses produce a signal at low energies with a low yield and heavy masses produce a signal at high energies with high yield, the detected energy and yield of the ERD signals for the different masses depend strongly on the stopping power and thickness of the range foil. A heavier mass may or may not be detected at a higher energy than a light mass depending on the choice of projectile, projectile energy, type of absorber foil, and thickness of the absorber foil (as discussed below, see Fig. 5.3). Also, each

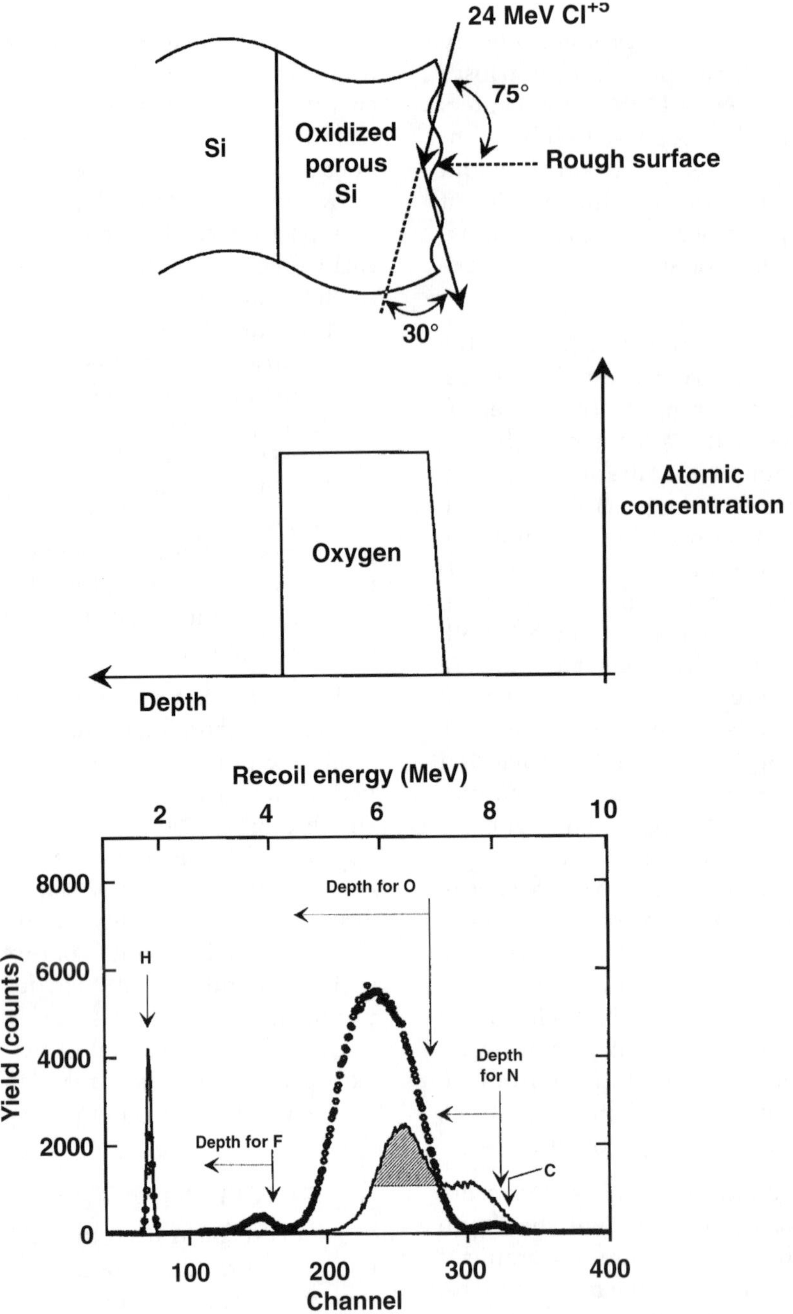

FIG. 5.2. Conversion of an oxygen concentration profile (center, with depth increasing to the left) into an ERD spectrum (bottom, open circles) for an oxidized porous silicon layer, shown schematically at top. The incident projectile is a 24 MeV Cl ion beam which impinges on the sample at an angle of 75° measured from the sample normal. The scattering angle is measured from the incident beam direction and is equal to 30°. The surface energy positions for the O, and other elements contained in this sample (which were collected in the spectrum) are shown as arrows in the figure. Each element has its own energy-to-depth conversion, but all show depth increasing to the left toward lower energy. The spectrum from a nitrided oxidized porous silicon layer is also shown in this figure (solid line) to demonstrate the mass resolution problem when signals from different recoiled atoms begin to overlap (O shown hatched sits on N background signal).

element in the spectrum has its own depth scale with a unique energy to depth transformation because of the difference in stopping powers for the different recoiled atoms. Yet similar to RBS, recoiled atoms coming from deeper in the sample have a lower energy than those atoms from the surface, and so the depth scales increase to the left (decreasing energy). Some overlap in the yield from different elements can occur such as with O signal (shaded area) on top of the N peak in the NOPS spectrum because mass and energy are not completely separable in this type of ERD using a single surface barrier detector with a range foil (in contrast to the technique of time-of-flight ERD described below). This overlap and the limited depth resolution can sometimes make element identification with ERD difficult.

Figure 5.3 shows the detected energy for the recoiled atom as a function of mass of the recoiled atom for 24 MeV Cl^{+5} ERD. The solid line (calculated) is given to show the surface energies of the recoils without the use of a range foil, and the filled circles show the surface energies determined experimentally from standards for recoils which traversed a 6 μm mylar range foil (the dashed line was drawn to guide the eye). This curve shows the strong dependence of the recoil energy for the heavier recoiled atoms (amu≥10) due to increased stopping in the range foil. In addition, this curve shows that surface energies do not simply depend on the kinematic factor as is the case for RBS. Although the curve describing the recoil energy as a function of mass could be calculated based on the known scattering parameters, stopping powers, and thickness of the range foil, the uncertainty in the thickness of the range foil and the stopping powers in mylar show that this curve is best determined experimentally and then the stopping powers (and/or range foil thickness) scaled appropriately to match the experimental curve.

The description given above for the ERD technique requires the use of a range foil in front of the particle detector. However, several examples will be given later in the chapter of other types of ERD analysis which do not require a range foil, but the basic analytical expressions developed in the following sections can be used. One rare example of the ERD technique without the use of a range foil is when a very heavy projectile is used with moderate to light mass substrates. For this case, if the geometry and beam are chosen such that recoil scattering angle, ϕ, is greater than $\Theta_{rbs}^{critical} = \sin^{-1}(M_{substrate}/M_{incident\ ion})$, then the incident ion can not be kinematically scattered into the detector and no range foil is required. This condition is particularly useful for high depth resolution ERD analysis with a Au ion.

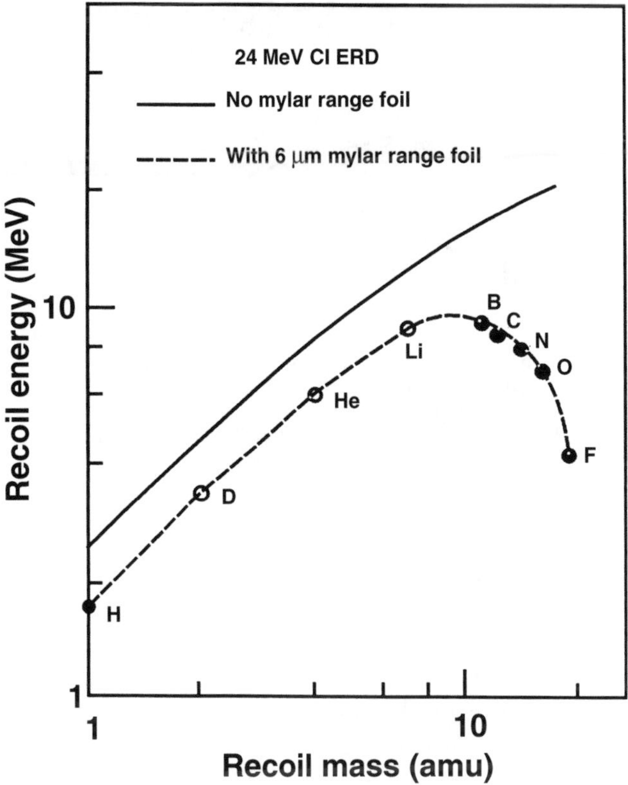

FIG. 5.3. The detected energy determined for recoiled atoms as a function of recoil mass for 24 MeV Cl^{+5} ERD. The filled circles are the experimentally determined surface energies and the open circles are the calculated surface energies for recoil atoms which have traversed a 6 μm mylar range foil (the dashed line was drawn to guide the eye). The solid line is calculated in order to show the surface energies of recoils which do not traverse a range foil.

5.2.1 Analytical expressions: ERD with a range foil and surface barrier detector

The analytical expressions which relate the observed ERD energy spectra to recoil atom type and concentration with depth are quite similar to those presented for RBS analysis in Chap. 4. Subtle differences do exist, however, and therefore the relevant expressions for ERD will be presented here.

The kinematics of the experiment in the laboratory reference frame are shown in Fig. 5.4 and a simple geometry for ERD analysis in the surface-energy approximation is shown in Fig. 5.5. The following analysis is in

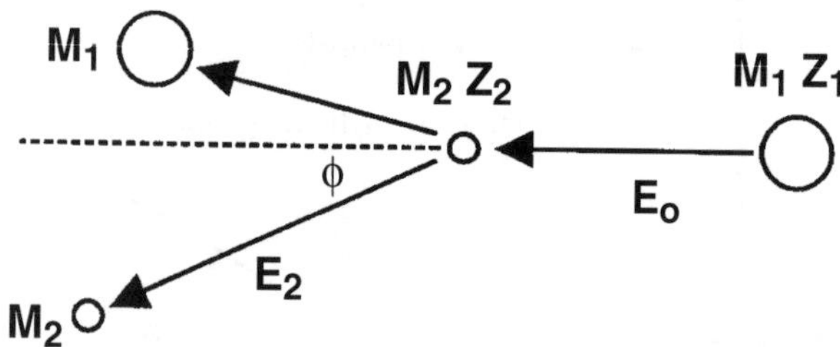

FIG. 5.4. Schematic representation of an elastic collision between a projectile of mass M_1, atomic number Z_1, and energy E_o and a target of mass M_2 (atomic number Z_2) which is at rest before the collision. After the collision, the target mass is recoiled forward at an angle ϕ with an energy E_2.

the laboratory frame of reference in which the incident beam, the direction of detection, and the sample normal are coplanar. A projectile with energy E_o, mass M_1, and atomic number Z_1 is incident upon the sample at an angle Θ_1 measured from the surface normal. Figure 5.5 shows that at a depth of x, measured along the target normal, the projectile has an energy E'_o just prior to scattering with a target atom of mass M_2 and atomic number Z_2. The recoiled atom is scattered at an angle $\phi=\pi-(\Theta_1+\Theta_2)$, where Θ_2 is the angle of detection measured from the surface normal (note that $\Theta_1+\Theta_2 > \pi/2$). As shown in Fig. 5.4, the energy of the recoiled atom, E_2, can be related to the projectile energy, E_o, by the kinematic factor, K, such that

$$E_2 = K\,E_o \tag{5.1}$$

where

$$K = \frac{4\,M_1\,M_2\cos^2\phi}{(M_1+M_2)^2}\ . \tag{5.2}$$

Therefore, the initial energy of the recoiled atom shown in Fig. 5.5 is $E_2=KE'_o$, and after traversing the sample the recoiled atom exits with energy E_3. The recoil ion is then incident normal to a 12 μm thick mylar range foil which is assumed to be parallel to the detector surface. This atom then loses more energy in the range foil and is detected at an energy E_d. The yield (Y_r) for the recoil atoms detected in energy channel E_d with a channel width of δE_d is given by

$$Y_r(E_d) = \frac{Q\,N_r(x)\,\sigma_r(E'_o,\phi)\,\Omega\,\delta E_d}{\cos\Theta_1\;dE_d/dx} \tag{5.3}$$

where Q is the incident projectile fluence, $N_r(x)$ is the atomic number density of the recoiled atom at depth x

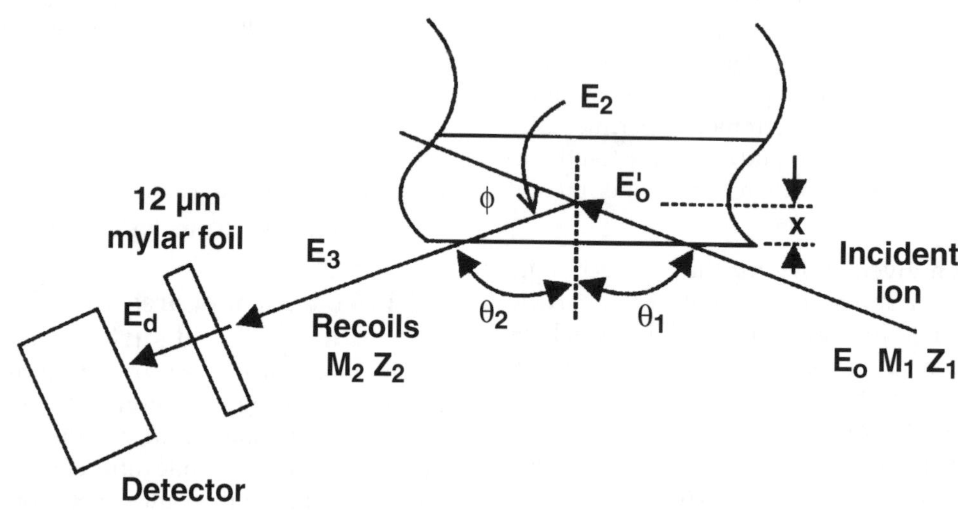

FIG. 5.5. Schematic configuration of an ERD experiment showing: the incident and exit angles (Θ_1 and Θ_2) measured relative to the sample normal, the loss of energy (E_o-E'_o) by the incident beam due to penetrating a distance x into the sample, the energy (E_2) of the recoiled atom at depth x, the energy of the recoiled atom (E_3) after it has exited the sample, and the energy of the recoiled atom (E_d) detected in a surface barrier detector after the recoil has traversed a 12 μm thick mylar range foil.

measured normal to the surface, $\sigma_r(E'_0,\phi)$ is the recoil differential scattering cross section in laboratory coordinates, Ω is the detector solid angle, and dx is the increment of depth at x corresponding to an increment of energy dE_d. The quantities Q, Ω, and Θ_1 are fixed by the experimental set-up. Rigorously, Ω is the integral of the differential surface area element in spherical coordinates but to a good approximation Ω equals the detector area divided by z^2 (z = the sample to detector distance). This approximation is good to better than 1% for a circular area and $\Omega < 125$ msr. In practice, Ω is often calibrated from the height of a hydrogen profile obtained from a standard with a known quantity of hydrogen. A defining aperture is usually placed in front of the detector, in order to maximize the depth resolution which can be degraded by the trajectory of the particles entering the detector at different scattering angles. Thus, a typical solid angle for an ERD analysis is 5 msr (similar to RBS analyses which often use 1 msr $\leq \Omega \leq$ 10 msr). In order to transform the yield, $Y_r(E_d)$, in the ERD spectrum to a concentration profile, $N_r(x)$, analytical expressions for $\sigma_r(E'_0,\phi)$, δE_d and dE_d/dx must be determined.

First, the cross section for many instances is governed by coulombic scattering as modeled by Rutherford (1911), and expressed in the laboratory reference frame is Marion and Young, (1968):

$$\sigma_r(E'_0,\phi) = \frac{[Z_1\,Z_2\,e^2\,(M_1+M_2)]^2}{[2\,M_2\,E'_0]^2\cos^3\phi} \tag{5.4}$$

where $(e^2/E'_0)^2$ = 0.020731 barns for E'_0 = 1 MeV, and 1 barn = 1×10^{-24} cm^2. For those cases in which the cross section is non-Rutherford (as discussed below) then the simplest solution for the thin film approximation is to multiply σ_r by a scaling factor, f′, in order to model the true cross section, i.e., $\sigma = f'\,\sigma_r$.

Next, consider the determination of the channel to energy conversion, δE_d. Figure 5.6 (from Arnold, 1994) shows an ERD spectrum from a $LiNbO_3$ sample with hydrogen in the surface layer of the sample. A surface hydrogen signal is found in the ERD spectrum from nearly all samples. The spectrum is plotted as yield versus detected energy with the energy channels given along the top axis. In order to determine the E_d scale, δE_d (in keV/channel) was determined from the channel corresponding to the surface of the hydrogen peak ($C_{surface}$ = channel 121, marked H) and from the calculated surface scattering energy, $E_d(H_{surface})$: $E_d(H_{surface}) = K_H E_o - \Delta E_{foil}$. The energy loss in the range foil, ΔE_{foil}, is easily determined on a computer through a slab analysis (for example, see Appendix Tables A9.4 and A9.6). In the example of Fig. 5.6, the surface recoiled hydrogen passes through a 12 µm thick mylar foil ($C_{10}H_8O_4$)which is approximated as a pure carbon layer with a density of 1.58 g/cm^3. (This effective density for carbon was derived experimentally and differs from the actual density of mylar, 1.397 g/cm^3.) Table 5.1 shows the values for the H atom energy and stopping power as it passes through the mylar foil. Initially, the surface hydrogen is recoiled (using a 24 MeV Si projectile) with an energy $K_H E_o$ = 2.3971 MeV. (Appendix Table A9.4a shows that the range of a 2.4 MeV H ion in mylar is ≅ 77 µm.) The hydrogen energy shown for each depth increment in the mylar was determined from the energy in the previous layer minus the stopping power times the slab thickness (1 µm). The final detected energy is shown in the last line of Table 5.1 as E_d = 2.1518 MeV, and therefore ΔE_{foil} is 245.3 keV. An important point demonstrated in this table is that a complete slab analysis for the mylar range foil was needed to get an accurate ΔE_{foil}, because the H stopping power changes continuously (and in a non-linear fashion) throughout the mylar. If the

Table 5.1. Stopping of hydrogen ions in mylar foil.

Depth in mylar foil *(nm)*	**Hydrogen energy, in mylar *[E (MeV)]***	**Stopping power of H in mylar at energy E *(keV/nm)***
0	2.3971	0.019729
1000	2.3774	0.019850
2000	2.3576	0.019974
3000	2.3376	0.020099
4000	2.3175	0.020228
5000	2.2973	0.020359
6000	2.2769	0.020492
7000	2.2564	0.020629
8000	2.2358	0.020769
9000	2.2150	0.020911
10000	2.1941	0.021057
11000	2.1730	0.021206
12000	2.1518	

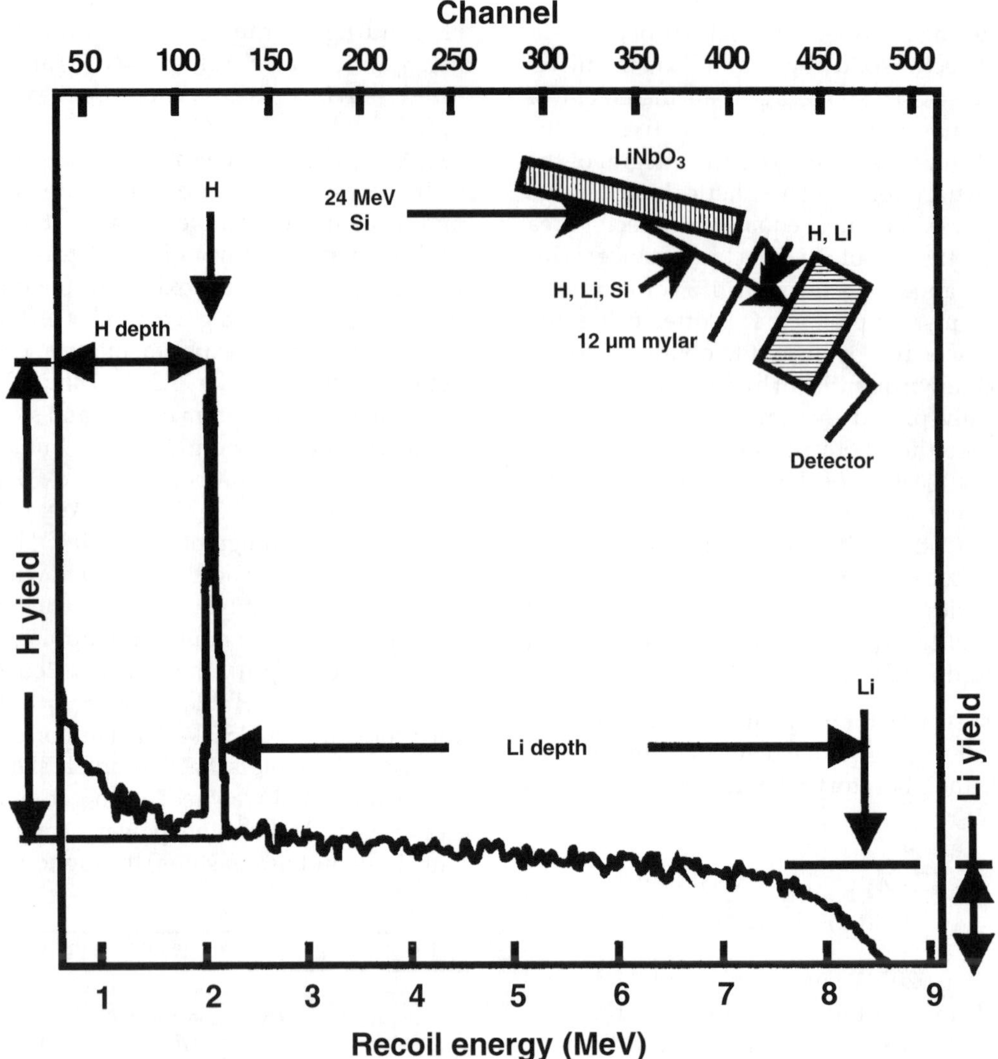

FIG. 5.6. An ERD spectrum from a thick $LiNbO_3$ crystal which has accumulated some hydrogen into the surface layer. The H and Li surface edge positions are as noted by the arrows. The projectile and detection conditions are as shown in the inset with Θ_1=75° and ϕ=30°. [From Arnold, 1994].

mylar foil was taken as a single layer 12 μm thick and the stopping power as that for the energy incident on the mylar foil (0.019729 keV/nm), then the final detected energy would be calculated as 2.1604 MeV rather than 2.1518 MeV. The magnitude of this error may be small for an energetic recoiled hydrogen (near the surface), but the error is extremely large for lower energy hydrogen or heavier recoiled atoms which have larger stopping powers. The best method to determine if the incremental depth used for the mylar film is sufficiently small is to iteratively perform an analysis of the sample while varying the incremental depth until the analysis is only slightly affected by decreasing the incremental depth.

The channel to energy conversion factor is then given by $\delta E_d = E_d(H_{surface})/C_{surface}$ = 2151.8 keV/Channel 121 = 17.78 keV/Channel, assuming non-linearities and off-sets in the energy scale have not been introduced by the electronics of the detection system. The occurrence of a non-linear energy scale or an off-set in the energy scale can be determined by measuring the surface channel position for more than one element. However, care must be taken in determining E_d(surface) for heavier elements because of the caution given above concerning the incremental mylar depth, because the calculated stopping powers for heavier elements in mylar often have errors larger than those introduced by non-linearities in the electronics, and because of heavy-ion pulse-height defects in the SDB (see Chapters 11 and 12).

The energy scale for the spectrum in Fig. 5.6 is now determined. In order to obtain a concentration depth profile for the Li, the next step in the analysis would be to calculate a depth scale for the sample and the corresponding recoil energies and effective stopping powers for each increment of depth in the sample. This type of thick film slab analysis will be given in detail in Sec. 5.2.3, and a simpler thin layer analysis will be given in the next section to quantitatively determine the areal density of hydrogen in the surface layer of this sample. Still, several qualitative features evident in Fig. 5.6 should be considered first.

Two large effects appear in Fig. 5.6 resulting from the difference in recoil atom cross sections and stopping powers as is evident in a comparison of the Li and H signals. First, the energy scale for Li is greatly extended in comparison to that for H for a given depth of analysis as a result of the difference in H and Li stopping in the mylar range foil (0.0197 keV/nm for H entering the mylar and 0.2197 keV/nm for Li entering the mylar). Second, the yield for hydrogen is significantly greater than that for Li even though the H is not a major constituent of the $LiNbO_3$. This latter effect can be understood by examining the yield equation $Y=N\ (Q\ \Omega\ \delta E/\cos\Theta_1)\ [\sigma/(dE_d/dx)]$. The first term in parentheses $(Q\Omega\,\delta E/\cos\Theta_1)$ should be the same for all elements in the spectrum, and so the enhanced yield for H in comparison to Li is primarily due to the difference in σ and dE_d/dx for the two elements. First consider σ, where the ratio of the H cross section to Li cross section is 3.7 as shown in Table A9.2. Table A9.2 also shows that in general the H Rutherford cross section for forward scattering is larger than the cross sections for other elements which may appear in the spectrum. This behavior is in marked contrast to that for RBS where heavy elements have the greater scattering cross sections. Next, the factor dE_d/dx for H (0.5337 keV/nm) is nearly a factor 6 times smaller than for Li (3.0106 keV/nm), and therefore the yield for H is enhanced by more than an order of magnitude when comparing similar concentrations of H and Li.

5.2.2 Thin layer analysis

This section will demonstrate how to compute the areal density of H in the near-surface layer of the $LiNbO_3$ sample of Fig. 5.6. (The formalism of the following approach is sufficiently general to extend to analyses of other layers which are not at the surface.) The areal density (AD_H) of hydrogen is the integrated concentration of hydrogen (N_H) over the layer thickness (τ). From the yield Eq. (5.3)

$$AD_H = \int_0^\tau N_H(x)\,dx = \frac{\cos\Theta_1}{Q\,\Omega}\int_0^\tau \frac{S^{eff}}{\sigma\,\delta E_d}\,Y(E_d)\,dx$$

$$= \frac{\cos\Theta_1}{Q\,\Omega}\int_{E_d(0)}^{E_d(x)} \frac{Y(E_d)}{\sigma}\,\frac{dE_d}{\delta E_d} \qquad (5.5)$$

where S^{eff} is the effective stopping power ($=dE_d/dx\,|_x$). Equation (5.5) shows that the effective stopping power term is eliminated by the integration when determining the areal density. In practice, the final integral of Eq. (5.5) is evaluated as a summation. For a surface peak, the cross section (σ_o) is evaluated at the specific energy E_o

$$AD_H = \frac{\cos\Theta_1}{Q\,\Omega\,\sigma_o}\sum_{\mathrm{Channel}[E_d(0)]}^{\mathrm{Channel}[E_d(x)]} Y\,(\mathrm{Channel})\ . \qquad (5.6)$$

Strictly speaking, for atoms which are present only on the surface, the summation over the peak shown in Eq. (5.6) is a result of the finite system-resolution as opposed to a summation over yield at different energies, but this equation shows the summation from Channel[$E_d(0)$] to Channel[$E_d(x)$] for generality to non-surface layers.

For the spectrum in Fig. 5.6, the incident beam is Si^{+5} at 24 MeV and is incident at an angle of 15° from the sample surface (Θ_1 = 75°). The recoiled H atoms are scattered at an angle of 30°, travel through a 12 µm mylar foil, and are detected in a 100 mm^2 surface barrier detector. An aperture in front of the SBD defines a solid angle of 5.36 msr. The total amount of charge collected from the incident beam was 10 µC, giving a total of 1.248×10^{13} Si atoms incident upon the sample [1.248×10^{13}Si = (10 µC/1.602×10^{-13} µC/Charge)/(5 Charge/Si)]. The number of integrated counts in the H signal above background from this fluence of Si was 5518. Finally, the cross section evaluated using Eq. (5.4) at E_o = 24 MeV is 2.284 b. Therefore the areal density of H in the near-surface layer of the $LiNbO_3$ is 9.3×10^{15} cm^{-2}.

5.2.3 Thick layer slab analysis: energies and effective stopping powers

The flow chart in Fig. 5.7 outlines the procedure for determining a composition depth profile in a thick layer using an iterative slab analysis. The formalism of this procedure (given below) follows that given by Doyle and Brice (1988), and example analyses will be demonstrated in Sec. 5.3. As stated above, expressions for δE_d and dE_d/dx are needed to convert the yield Y_r to a concentration profile, $N_r(x)$. These expressions can be derived in terms of the stopping powers (dE/dx) evaluated at energies corresponding to uniform increments of depth in the sample. The most simple method to determine the apparent energy loss in the detected beam as a function of depth is to divide the sample into a series of slabs of equal thickness and then create a table of the corresponding energies and stopping powers for each layer or slab of the material, e.g., as shown schematically in Fig. 5.8.

The energy and depth relationship can be expressed by the equations given below and using the following convention for superscripts and subscripts. The subscripts in the following expressions refer to the atom type, except for the notation given above for the angles. For example, the projectile energy and stopping powers are $E_o^{(n)}$ and $S_p^{(n)}$, respectively; and the recoil energy and stopping powers are either $E_2^{(n)}$ or $E_3^{(n)}$, and $S_r^{(n)}$, respectively. The superscripts refer to the layer of the target (1,2 ...) or to the range foil (superscript=0). This notation is similar to that for RBS analysis where the superscript refers to the stopping medium and the subscript refers to the scattering atom (i.e., the recoil species). For a comparison to RBS slab analysis notation, see Brice (1973), Foti *et al.* (1977), and Chap. 3 in Chu *et al.* (1978). As shown in Fig. 5.8, the projectile is incident with energy E_o at the target surface and has energy $E_o^{(1)}$, after penetrating to the interface between layers 1 and 2, and finally, has energy $E_o^{(2)\prime}$ at a variable depth x in slab 2. Thus, with $C_1 = \cos\Theta_1$ and $X^{(1)}$ the thickness of slab 1

$$X^{(1)} = C_1 \int_{E_o^{(1)}}^{E_o} dE / S_p^{(1)}(E) \quad (5.7)$$

and

$$x = C_1 \int_{E_o^{(2)\prime}}^{E_o^{(1)}} dE / S_p^{(2)}(E) \quad (5.8)$$

which relate the projectile energies to the corresponding depths within the two layers on the inward path. Similarly, the recoil atom is created at depth x within the second layer with energy $E_2 = K E_o^{(2)\prime}$, where K is given by Eq. (5.2). At the interface between layer 2 and layer 1 the recoil atom has energy $E_3^{(2)}$ and at the surface $E_3^{(1)}$. The atom incident upon the range foil has an energy $E_3 = E_3^{(1)}$ and finally after traversing the range foil the detected energy is E_d.

Similar to the equations for the projectile, the recoil atom energies on the outward path are related to depths by

$$x = C_2 \int_{E_3^{(2)}}^{E_2} dE / S_r^{(2)}(E) \quad (5.9)$$

$$X^{(1)} = C_2 \int_{E_3^{(1)}}^{E_3^{(2)}} dE / S_r^{(1)}(E) \quad (5.10)$$

and

$$X^{(0)} = \int_{Ed}^{E_3} dE / S_r^{(0)}(E) \quad (5.11)$$

with $C_2 = \cos\Theta_2$ and $X^{(0)}$ = range foil thickness.

In order to determine the effective stopping power (dE_d/dx) in terms of calculable stopping powers at different energies, the derivative of Eqs. (5.7) through (5.11) must be determined. Leibniz's rule (see Boas, 1966) for differentiation of an integral is used in this step and is given for completeness in two representative forms below

$$\frac{d}{dE'} \int_{E}^{E_o} f(E)\, dE = -f(E')$$

and

$$\frac{d}{dE'} \int_{a(E')}^{b(E')} f(E)\, dE = f(b)\frac{db}{dE'} - f(a)\frac{da}{dE'} .$$

The range foil thickness is a constant and therefore using the second form of Leibniz's rule with Eq. (5.11) gives an expression for dE_d as

$$dE_d = dE_3 \frac{S_r^{(o)}(E_d)}{S_r^{(o)}(E_3)} \quad (5.12)$$

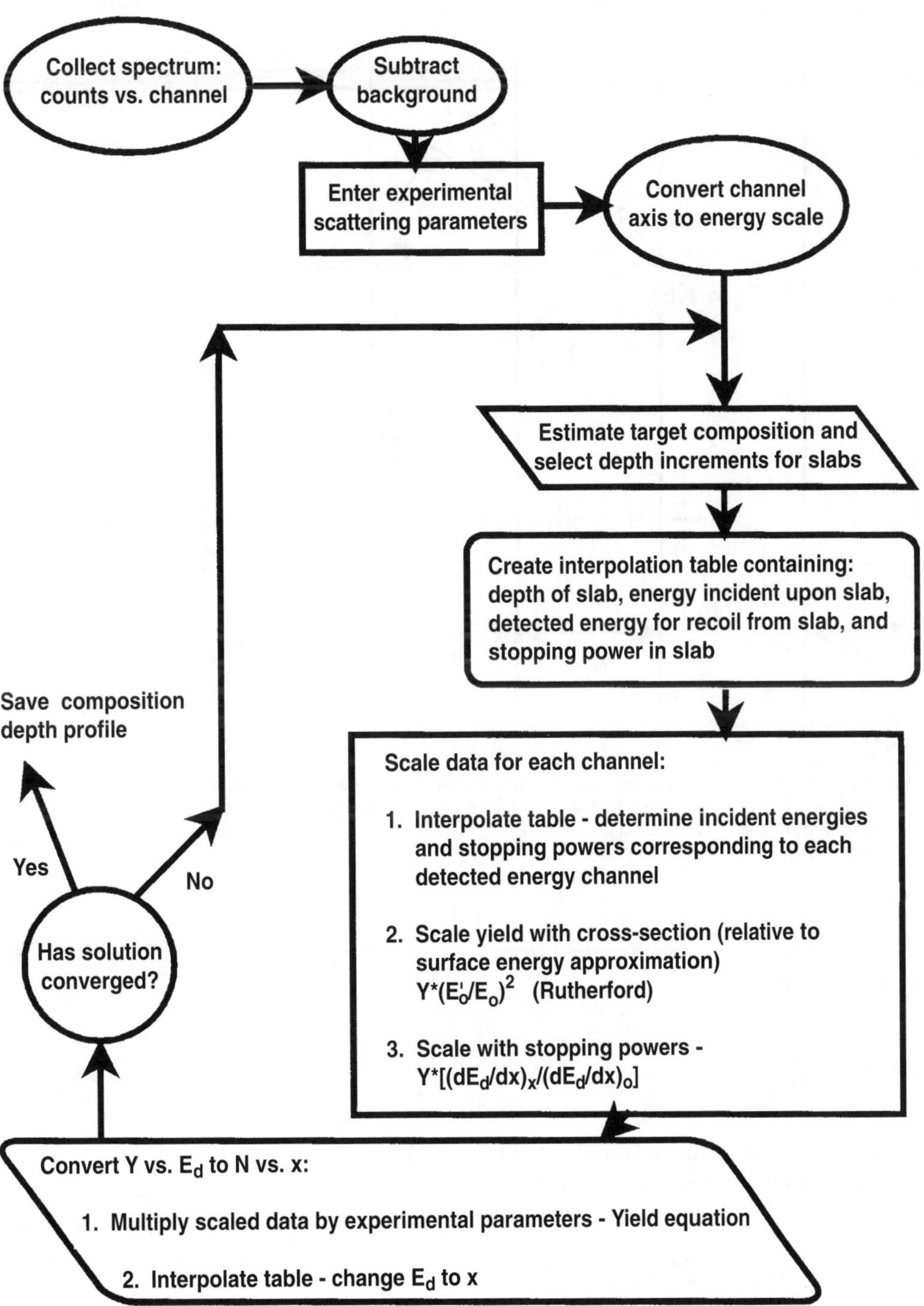

FIG. 5.7. A flow chart to demonstrate the ERD analysis of a sample. The iterative process involved with a slab analysis is shown in this figure, but most often only one analysis is needed from samples prepared for a specific materials science experiment.

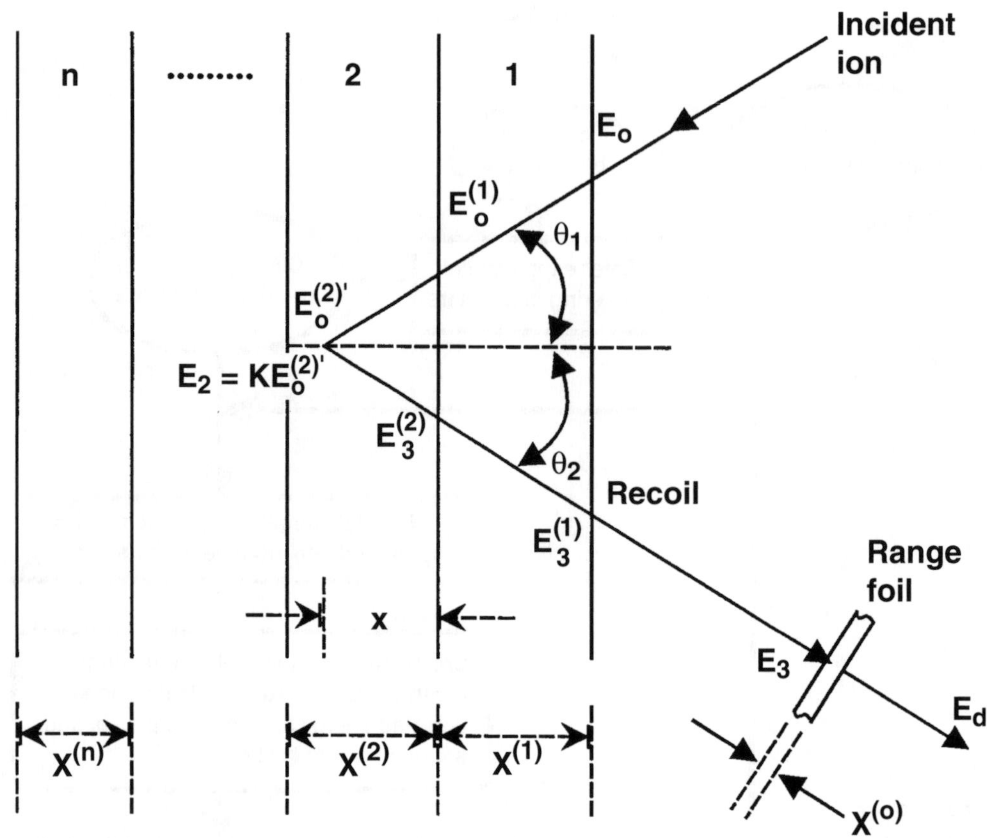

FIG. 5.8. Concept and symbols used in the derivation of the analytical equations for a slab analysis. The projectile ion penetrates to a depth x in the second layer of the sample, recoils a target atom which is detected at an energy E_d. The ′ refers to a variable depth, and the superscripts in parentheses refer to layer number.

An expression for dE_3 (= $dE_3^{(1)}$) can be determined from Eq. (5.10) in a similar manner

$$dE_3 = dE_3^{(2)} \frac{S_r^{(1)}(E_3)}{S_r^{(1)}\left(E_3^{(2)}\right)} \quad . \tag{5.13}$$

An expression for $dE_3^{(2)}$ can be determined from Eq. (5.9), but now x is a variable distance which is not constant within the second layer

$$dx = C_2 \left[dE_2/S_r^{(2)}(E_2) - dE_3^{(2)}/S_r^{(2)}\left(E_3^{(2)}\right)\right] \tag{5.14}$$

The first form of Leibniz's rule applied to Eq. (5.8) gives

$$dx = -C_1 \, dE_o^{(2)\prime} \,/\, S_p^{(2)}\left(E_o^{(2)}\right) \tag{5.15}$$

and reduction of Eqs. (5.14) and (5.15) using $dE_2 = KdE_o^{(2)\prime}$ yields

$$\frac{dE_3^{(2)}}{dx} = \frac{S_r^{(2)}\left(E_3^{(2)}\right)}{S_r^{(2)}(E_2)} [S]_{p,r}^{(2)} \tag{5.16}$$

where

$$[S]_{p,r}^{(2)} = \frac{K\, S_p^{(2)}\left(E_o^{(2)\prime}\right)}{C_1} + \frac{S_r^{(2)}(E_2)}{C_2} \tag{5.17}$$

The value $[S]_{p,r}^{(2)}$ is similar to the energy loss factor given in Chap. 2, but the square bracket requires two subscripts because the stopping powers on the inward and outward paths are for two different species: projectile and recoil.

The above equations can then be reduced to an expression for dE_d/dx in terms of calculable stopping powers

$$\frac{dE_d}{dx} = \frac{S_r^{(2)}\left(E_3^{(2)}\right)}{S_r^{(2)}\left(E_2\right)} \frac{S_r^{(1)}\left(E_3\right)}{S_r^{(1)}\left(E_3^{(2)}\right)} \frac{S_r^{(o)}\left(E_d\right)}{S_r^{(o)}\left(E_3\right)} [S]_{p,r}^{(2)} \quad . \tag{5.18}$$

The derivation for the effective stopping power given above was done for an arbitrary depth x in slab 2, however this derivation is easily extended to any slab n, as shown in Fig. 5.8. Defining $E_3^{(n)}$ as the energy of the recoil atom as it emerges from the nth slab (numbered from surface) on its outbound path then gives

$$\left.\frac{dE_d}{dx}\right|_{n\text{th}} = [S]_{p,r}^{(n)} \prod_n \tag{5.19}$$

where the product Π_n is given by

$$\prod_n = \frac{S_r^{(n)}\left(E_3^{(n)}\right)}{S_r^{(n)}\left(E_2\right)} \frac{S_r^{(n-1)}\left(E_3^{(n-1)}\right)}{S_r^{(n-1)}\left(E_3^{(n)}\right)} \cdots \frac{S_r^{(o)}\left(E_d\right)}{S_r^{(o)}\left(E_3^{(1)}\right)} \quad . \tag{5.20}$$

This derivation assumed that the stopping powers $S_r^{(n)}$ and $S_p^{(n)}$ are constant throughout the nth slab layer. But when these stopping powers are functions of depth in the layer (e.g., varying composition) as well as energy, then a more extended form of Leibniz rule must be used and $dE_d/dx^{(n)}$ must be evaluated numerically. In practice, with modern computers, the slab thickness can always be reduced to a level where this complication is eliminated. Also, an iterative process must be used if the recoil atom composition is sufficiently large to effect the stopping powers.

The relationship between thickness of the nth slab, $X^{(n)}$, measured normal to the sample surface, and the energy incident upon that slab, $E_o^{(n)}$, is given by

$$E_o^{(n)} = E_o^{(n-1)} - X^{(n)} S_p^{(n)}\left(E_o^{(n-1)}\right) \tag{5.21}$$

and the yield equation for detecting a recoiled atom from the nth slab in the material is given by

$$Y_r^{(n)}(E_d) = \frac{Q\,\Omega\,N_r^{(n)}(x)\,\sigma_r\left(E_o^{(n)},\phi\right)\delta E_d}{\cos\Theta_1\, dE_d/dx\,|_n} \quad . \tag{5.22}$$

The cross section is shown in Eq. (5.4) and the effective stopping power in Eq. (5.19). In practice, the depth in the sample is related to the detected energy measured in a spectrum. Therefore the following equation relates the detected energy for the nth slab, $E_d^{(n)}$, to the thickness $X^{(n)}$

$$E_d^{(n)} = KE_o - X^{(n)} \times \left[K \sum_{j=1}^{n} \frac{S_p^{(j)}\left(E_o^{(j-1)}\right)}{\cos\Theta_1} + \frac{S_r^{(n)}\left(KE_o^{(n)}\right)}{\cos\Theta_2} + \sum_{j=n-1}^{1} \frac{S_r^{(j)}\left(E_3^{(j+1)}\right)}{\cos\Theta_2} \right] - \Delta E_{foil} \tag{5.23}$$

The energy lost by the recoil atom in traversing the foil (ΔE_{foil}) is also calculated from a slab analysis of the foil containing NF slabs of thickness δx (i.e., $NF = X^{(0)}/\delta x$) as

$$\Delta E_{foil} = \delta x \sum_{j=1}^{NF} S_r^{(o)}\left(E_f^{(j-1)}\right)$$

where the stopping power is evaluated at the energy $E_f^{(j-1)}$ after traversing the (j–1)th slab. Although these analytical expressions appear complicated, their implementation through the use of a computer is simple and fast, and detailed examples of ERD slab analyses will be given later sections.

5.2.4 Spectral scaling

Now that the analytical equations describing the ERD spectrum are derived, several approaches can be followed to calculate composition profiles from the spectrum. In general, as shown in Fig. 5.7, an iterative process is used in which the material composition is estimated and depth increments for slabs are selected in order to make an initial guess at the stopping powers, energies, and recoil cross sections for each layer. Then an ERD spectrum could be calculated and compared to the data. A concentration depth profile would be determined by iteratively changing the slab analysis composition, recalculating the simulated spectrum, and comparing to the experimental data. This type of approach is used in RBS analysis programs such as RUMP (Doolittle, 1986, 1987).

Another approach, termed spectral scaling, is to create a table containing the sample depth, $E_o^{(n)}$, $\sigma(E_o^{(n)})$, E_d, and dE_d/dx as a function of slab layer. The data are then scaled one channel at a time to produce a concentration depth profile by interpolating between values in the table in order to determine the energies, cross sections, and stopping powers corresponding to each detected energy channel. The energy scale is transformed to depth and the yield scale converts to concentration. Simply, this scaling of the yield per channel data removes from the spectrum the effects of changes with depth into the sample of both σ and dE_d/dx. The yield Eq. (5.22) is thereby converted to appear as though the cross section and effective stopping power are constant through the sample. The conversion of this scaled thick target data to a depth profile therefore becomes as simple as the analysis of thin target data. Analytically, the scaling can be written as

$$N_r^{(n)}(x) = \frac{\cos\Theta_1}{Q\,\Omega\,\delta E_d} \frac{dE_d/dx|_n}{\sigma_r\left(E_o^{(n)},\phi\right)} Y_r^{(n)}(E_d)$$

$$= \frac{\cos\Theta_1}{Q\,\Omega\,\delta E_d} \frac{S_o^{eff}}{\sigma_o} \times \left[\frac{\sigma_r(E_o,\phi)}{\sigma_r\left(E_o^{(n)},\phi\right)} \frac{dE_d/dx|_n}{dE_d/dx|_o} Y_r^{(n)}(E_d)\right] \quad (5.24)$$

where S_o^{eff} and σ_o are evaluated at the incident energy E_o. The cross section scaling factor for Rutherford cross sections is $[\sigma_r(E_o)/\sigma_r(E_o^{(n)})] = [(E_o^{(n)})/E_o]^2$, and the stopping power scaling factor, $(dE_d/dx\,|_n)/(dE_d/dx\,|_o)$, are both determined by interpolating from tables. Examples of spectral scaling and slab analysis of ERD spectra will be given in the next sections.

5.3 EXAMPLES

We will focus on two examples in the following sections in order to demonstrate thick layer slab analyses (including tables of calculations for comparison to calculations by the reader), and to demonstrate the effect of σ and dE_d/dx scaling on the yield spectrum. The first example is a spectrum taken from a set of experiments for measuring the diffusion of hydrogen in polystyrene. This sample contained a 400-nm-thick polystyrene (PS) layer on a thin layer of deuterated PS-PMMA (polymethyl methacrylate) on a 400-nm-thick polyvinylchloride (PVC) layer. Elastic recoil detection was used to measure the interdiffusion of H from the top layer with D from the middle layer as a function of annealing experiments (Green, 1994). However, for the purpose of demonstration, an example ERD analysis will be given only for the sample before annealing treatments were performed. The second example demonstrates the analysis of both low Z and high Z elements. This analysis will measure the H and N profiles from a 300-nm-thick silicon nitride sample grown by atmospheric chemical vapor deposition at 800°C. Hydrogen is expected to be a minor constituent in this film, and therefore the matrix will be approximated as Si_3N_4.

These examples will also be used to demonstrate the following conditions. First, density exclusive units are often used for samples with varying composition (or unknown density) such that the depth scale and concentration scale are independent of the density. Therefore, the use of density exclusive units and the conversion to units which include the sample density will be shown. This approach often has accuracies of a few percent if the compositions vary little in Z such that the stopping powers change only moderately; however, large errors may occur when the composition changes from a matrix containing low Z elements to a matrix containing high Z elements.

For example, Fig. 5.9 gives the variation in effective stopping powers in density exclusive units [dE_d/dx in keV/(μg/cm^2)] corresponding to a depth of 1 μg/cm^2 for different recoil atom species from several different targets. A 12 μm mylar range foil was used for these calculations. The first row of H recoil atoms was calculated for the case of 2.8 MeV He^+ ERD analysis and the other three rows for H, C, and N recoil atoms were calculated for the cases of 28 MeV Si^{+6} ERD analyses. This figure shows that the difference in dE_d/dx for Si, SiO_2, and Si_3N_4 is small enough that only a few percent error occurs when approximating a multilayer sample as a uniform composition of just one of these materials. Yet the variation in dE_d/dx between polystyrene [$(C_8H_8)_n$] and gold is large enough to create 400% errors for all of the recoil atoms, and therefore samples containing multilayers of gold and polystyrene should be treated in a detailed slab analysis which allows for variations in sample composition as a function of depth. The more simple treatment will be used in the following analyses where the compositions are approximated as only one of the constituent layers. Other conditions which are important to consider in the following examples are the use of non-Rutherford cross sections in ERD analyses and the effect of foil slab thickness.

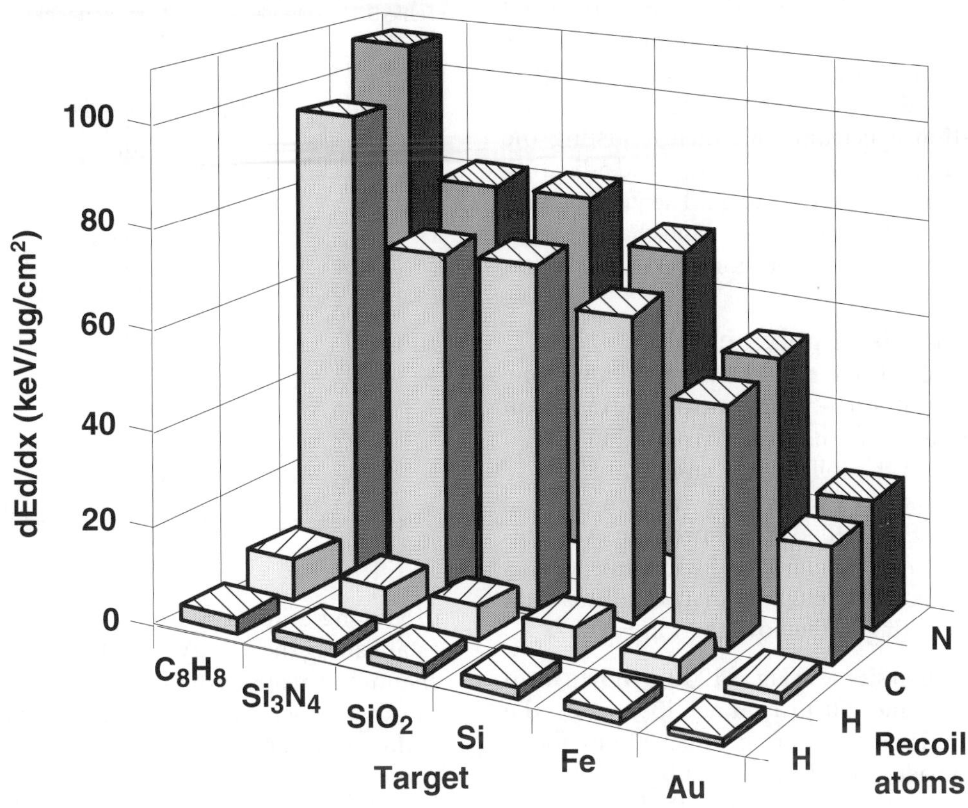

FIG. 5.9. Calculated effective stopping powers (dE_d/dx), in density exclusive units, at a depth of 1 $\mu g/cm^2$ in several different targets. The first row of H recoil atoms is for 2.8 MeV He^+ ERD, and the other rows are for 28 MeV Si^{+6} ERD. A 12 µm mylar range foil was used for these calculations.

5.3.1 Calibration of system geometry with movable detector

Before presenting detailed examples of data analysis, this section will briefly explain how the ERD data-collection geometry can be calibrated. Calibration of the system geometry is particularly important for ERD because of the high sensitivity of the technique to the angles Θ_1 and ϕ. The geometry can be calibrated once and then checked periodically (as with RBS system geometry) for a system with fixed detector angle, fixed detector aperture, and fixed sample tilt. In contrast, for a system with movable detectors and sample tilt, the geometry should be remeasured to better than 1° accuracy each time the detector or sample angles are changed.

The easiest method of calibrating the geometry is to set up the scattering angle and angle of incidence such that a spectrum collected from a standard matches that of a pre-measured spectrum from the same standard. A hydrogen spectrum is often used for this set-up procedure because H stopping is the least sensitive to variations in range foil properties and because the initial calibration of the standard can be done easily using nuclear reaction analysis [e.g., see Chap. 8 concerning the $p(^{15}N,\alpha\gamma)^{12}C$ reaction]. The standard must be one in which the H can be analyzed throughout the layer and in which H is not liberated by ion irradiation or slight heating. For example, a 0.3 µm thick Si_3N_4 film grown at a high temperature (e.g., 800°C) is a stable sample with a visible front and back edge in the H spectrum when analyzed using 2.8 MeV He^+ or 28 MeV Si^{+6} ERD. The scattering angle can then be calibrated from the front edge position of the H signal, because this edge corresponds to KE_o and K is dependent upon ϕ and not upon Θ_1. The calibration of Θ_1 is obtained from the back edge of the H signal and the height of the yield. Once the geometry has been well established, this geometry is reproducibly obtained by collecting a spectrum from the standard sample and having it match the spectrum used to calibrate the geometry (for the same amplifier settings and range foil). This is easily done by collecting a spectrum from the standard while simultaneously overlaying the calibration spectrum, and then adjusting the detector geometry to match the front edge and adjusting the sample tilt to match the back edge. The yield should also match if the detector aperture slits and charge collection are the same as in the calibration spectrum.

5.3.2 ERD analysis of polymer films

The ERD analysis of polymer films is generally done using high-energy He ion beams because of the radiation sensitivity of these films to heavy-ion irradiation (Green and Doyle, 1986). In the following example, a 2.8 MeV He^+ ion beam was used to analyze the polystyrene sample shown in Fig. 5.10(a). The spectrum was collected [as it appears in Fig. 5.10(b)] as counts versus channel number (bottom axis) and analysis of this data will progress according to the flow chart in Fig. 5.7.

First, the hydrogen signal in Fig. 5.10(b) is shown from approximately channel 70 to channel 135, while the deuterium signal is shown from channel 135 to above channel 180. No background exists to subtract from the H or D peaks in this spectrum, although the back edge of the D signal will be abruptly terminated in this analysis in order to eliminate the appearance of the H signal. The following experimental scattering parameters were used for this analysis: a 2.8 MeV $^4He^+$ ion beam was incident at 75° from the sample normal, the recoil atoms were transmitted through a 12 μm aluminized-mylar foil and collected in a SBD over a solid angle of 3.6 msr at a scattering angle of 30°, and a total incident fluence of 3 μC of charge was collected. An aluminized-mylar foil was used to keep light from entering the detector, but the thickness of the Al film had a negligible effect on the stopping of the recoil atoms and therefore will not be considered in the analysis. The slab analysis for stopping in the foil used 12 1-μm-thick layers. The composition and density of this mylar foil were $C_{10}O_4H_8$ and 1.397 g/cm^3, respectively, but for simplicity the mylar, was approximated as a pure carbon foil with a density of 1.58 g/cm^3.

The channel axis was then converted to an energy axis as shown on the top axis of Fig. 5.10(b). Calculations give K=0.480 from Eq. (5.2), and give a detected H surface edge energy equal to 949.1 keV from the stopping of 1344 keV H (=KE_0) in the 12 μm range foil as shown in Table 5.2. The measured surface edge is at channel 131 producing an energy calibration of 7.245 keV/Ch and establishing the energy axis at the top of Fig. 5.10(b). The channel corresponding to the surface edge position of D can then be calculated using this energy scale. The D surface edge energy after exiting the range foil is calculated from Table 5.3 to be 1371 keV, giving the corresponding surface channel of 189, but the D peak appears at channel 157 (1138 keV) because the deuterated-polystyrene layer is buried beneath a 400-nm-thick polystyrene layer. If the upper PS layer had been twice as thick, the D and H signals would have overlapped and a change in scattering parameters or a background subtraction would have been required.

Table 5.2. H stopping in a 12 μm range foil.

Depth in range foil *(nm)*	**Hydrogen energy in range foil *[E (MeV)]***	**Stopping power of H in range foil at energy E *(keV/nm)***
0	1.3440	0.029821
1000	1.3142	0.030283
2000	1.2839	0.030768
3000	1.2531	0.031280
4000	1.2218	0.031820
5000	1.1900	0.032391
6000	1.1576	0.032997
7000	1.1246	0.033641
8000	1.0910	0.034328
9000	1.0567	0.035062
10000	1.0216	0.035849
11000	0.9858	0.036696
12000	0.9491	

Table 5.3. Stopping of deuterons in a 12 μm range foil.

Depth in range foil *(nm)*	**Deuteron energy in range foil *[E (MeV)]***	**Stopping power of D in range foil at energy E *(keV/nm)***
0	1.8667	0.038021
1000	1.8286	0.038528
2000	1.7901	0.039059
3000	1.7511	0.039614
4000	1.7114	0.040195
5000	1.6712	0.040806
6000	1.6304	0.041448
7000	1.5890	0.042124
8000	1.5469	0.042838
9000	1.5040	0.043593
10000	1.4604	0.044394
11000	1.4160	0.045246
12000	1.3708	

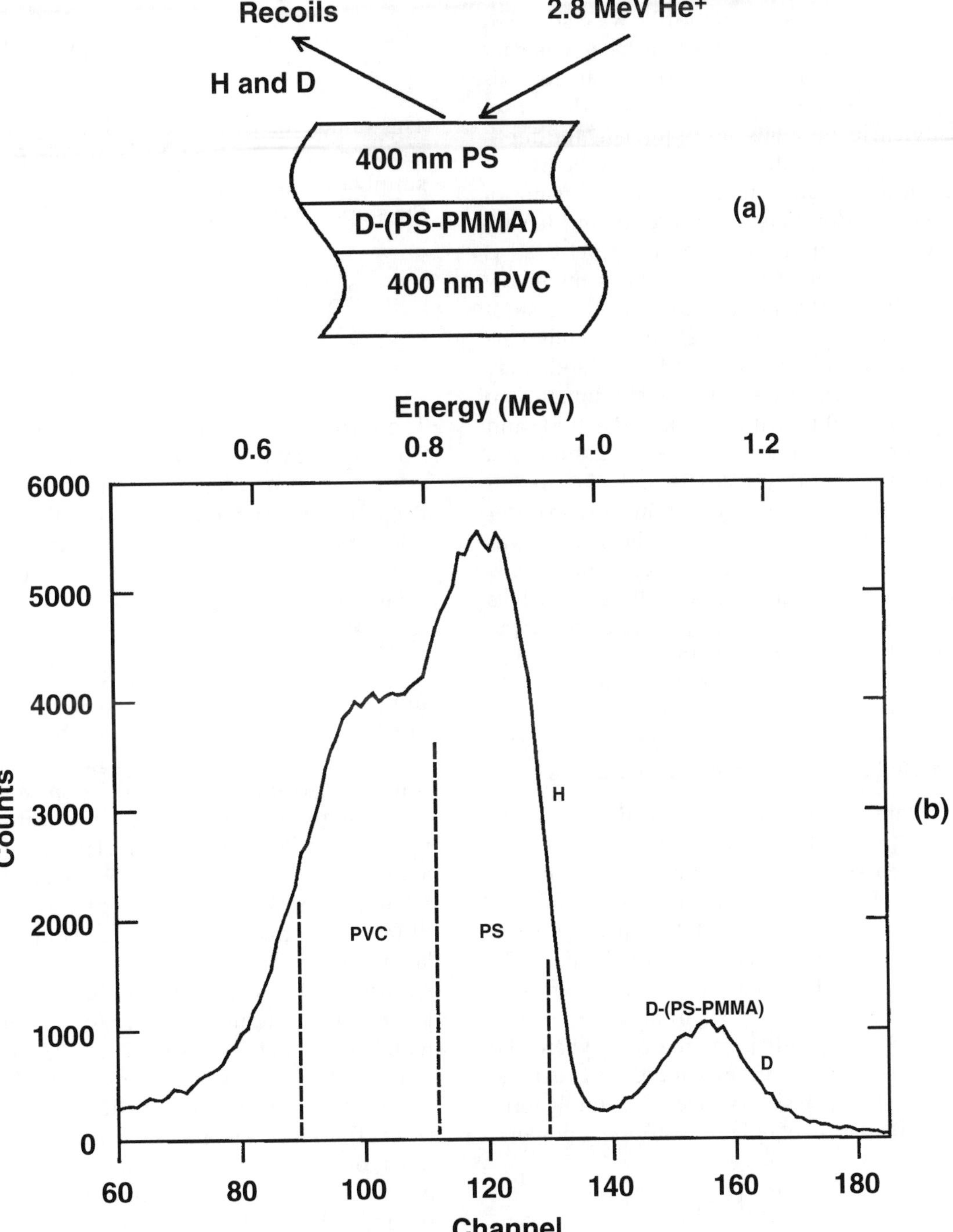

FIG. 5.10. An ERD spectrum (b) collected from the sample shown in (a). The 2.8 MeV He beam was incident at Θ_1=75°, and the recoiled atoms are collected at ϕ=30°. The H and D recoil signal are almost completely separated with their approximate surface peak positions as labeled. (From Green, 1994).

Next, the target composition of the film was estimated to be that for polystyrene [$(C_8H_8)_n$], which has a density ranging from 1.04 g/cm^3 to 1.08 g/cm^3. But for this analysis, the density was set equal to 1, and the subsequent density exclusive units are in μg/cm^2 for depth and number/g for concentration. Although the hydrogen concentration for the bottom PVC layer was not of primary importance for this study, the following density of PVC is given for completeness: ρ[PVC=$(C_2H_3Cl)_n$] is 1.68 g/cm^3. Further, the PVC layer contains the heavy element of chlorine, which can greatly affect the stopping power for that layer, and therefore the depth scale and concentration scale at the PVC layer depth may contain some error (even in density exclusive units). Two sets of interpolation tables will be presented in the H and D analyses below to compare the effect of assuming a uniform composition of PS (Tables 5.4 and 5.5) versus a uniform composition of PVC (Tables 5.6 and 5.7). In order to calculate these interpolation tables, the sample was considered to be broken into 30 equal depth increments of 4 μg/cm^2 each for a total depth of 120 μg/cm^2. This large depth is used to ensure that the tables cover the entire depth of the film. If the density of the sample was 1.04 g/cm^3, like that for PS, then this total depth in nanometers is equal to [(120 μg/cm^2)/(1.04g/cm^3)] (10^{-6}g/μg)(10^{7}nm/cm) = 1154 nm, which is 40% greater than the expected thickness of the polymer layers.

The interpolation Table 5.4 was used to determine the scaling factors for the H signal and the interpolation Table 5.5 was used to determine the scaling factors for the D signal. These tables were calculated using the following equations: $E_o^{(n)}$ from Eq. (5.21), $S_p^{(n)}$ and $S_r^{(n)}$ from Ziegler (1980), $E_2^{(n)} = KE_o^{(n)}$ where K is given by Eq. (5.2), $[S]_{p,r}^{(n)}$ from Eq. (5.17), Π_n from Eq. (5.20), the effective stopping power $dE_d/dx = [S]_{p,r}^{(n)}\Pi_n$ from Eq. (5.19), and E_d from Eq. (5.23). (Note that for the case where the sample is assumed to be a uniform composition, then the product term Π_n for each slab reduces to a simple term requiring the calculation of only four stopping powers

$$\prod_n = \frac{S_r^{target}\left(E_3^{(1)}\right)}{S_r^{target}\left(E_2^{(n)}\right)}\ \frac{S_r^{(o)}\left(E_d\right)}{S_r^{(o)}\left(E_3^{(1)}\right)}$$

where $S_r^{(o)}$ is the stopping in the foil.) Finally, an even easier method for calculating the effective stopping power is given in the last column of the tables. In this column, the effective stopping power at the *n*th layer interface, $dE_d/dx\,|_n$, is based on the average detected energy loss in that layer divided by the layer thickness, i.e., the average effective stopping power is

$$\frac{dE_d/dx\,\big|_{n-1} + dE_d/dx\,\big|_n}{2} = \frac{E_d(X^{(n-1)}) - E_d(X^{(n)})}{X^{(1)}}\ .$$

The values for E_d at each slab layer are calculated in earlier columns of the table and therefore $dE_d/dx\,|_n$ is simply calculated from the detected energies and from the stopping power of the previous layer:

$$dE_d/dx\,\big|_n = 2\,\frac{E_d(X^{(n-1)}) - E_d(X^{(n)})}{X^{(1)}} - dE_d/dx\,\big|_{n-1} \qquad (5.25)$$

The error in calculating the effective stopping power from Eq. (5.25) is at most 3% and generally less than 1%, as evidenced by a comparison of the column labeled $[S]_{p,r}^{(n)}\Pi_n$ to the last column labeled dE_d/dx. This error in calculating the dE_d/dx only affects the scaling of the yield, and since the accuracy of ion beam analysis is generally ~ 5%, this level of inaccuracy in dE_d/dx is often acceptable.

A greater error, however, can be caused by the assumption of a uniform composition throughout the sample equal to that composition for polystyrene. The difference in calculating E_d and dE_d/dx, assuming a constant PS composition versus a constant PVC composition, is slightly greater for the H recoil atom calculations than for the D recoil atom calculations, as evidenced by a comparison of Tables 5.4 and 5.6 for H recoil atoms (and 5.5 and 5.7 for D recoil atoms). For the H calculations, the difference in calculated E_d when comparing a pure PS layer to a pure PVC layer varies from a 0% difference at the surface to a 12.5% difference at the back edge of the PVC layer (corresponding to 4-5 μg/cm^2, which is approximately equal to the depth resolution). The difference in calculated dE_d/dx varies from a 20% difference at the surface to 27% difference at the back edge of the PVC layer. Therefore, an analysis based on the PS composition is only accurate for the upper PS layer of the sample, but an estimate of the PVC composition can be made by reducing the dE_d/dx scaling factor by 27% for the bottom half of the sample.

The next step in the flow chart (Fig. 5.7) for the analysis of the H and D profiles is to determine for each energy channel (i.e., channel in E_d), the corresponding values of x, E_o' at that depth, and the effective stopping power for that channel. These values are determined by interpolating between the calculated values given in Tables 5.4 (for the H profile) and 5.5 (for the D profile). For example,

Table 5.4. Interpolation table for recoil H assuming PS composition to calculate stopping powers.

Slab no.	Depth ($\mu g/cm^2$)	$E_o^{(n)}$ (keV)	$S_p^{(n)}$ ($keV/\mu g/cm^2$)	$E_2^{(n)}$ (keV)	$S_r^{(n)}$ ($keV/\mu g/cm^2$)	$[S]_{p,r}^{(n)}$ ($keV/\mu g/cm^2$)	Π_n	$[S]_{p,r}^{(n)} \Pi_n$ ($keV/\mu g/cm^2$)	E_d (keV)	dE_d/dx ($keV/\mu g/cm^2$)
surface	0	2800.000	1.21840	1344.000	0.21591	3.09382	1.26127	3.90213	949.065	3.90213
1	4	2781.170	1.22252	1334.962	0.21694	3.10545	1.26902	3.94089	933.483	3.88880
2	8	2762.276	1.22668	1325.892	0.21799	3.11722	1.2771	3.98102	917.747	3.97897
3	12	2743.318	1.23089	1316.793	0.21905	3.12913	1.28553	4.02259	901.854	3.96784
4	16	2724.295	1.23515	1307.661	0.22013	3.14119	1.29432	4.06570	885.796	4.06128
5	20	2705.206	1.23945	1298.499	0.22122	3.15339	1.3035	4.11046	869.567	4.05289
6	24	2686.050	1.24380	1289.304	0.22233	3.16574	1.31311	4.15695	853.162	4.14953
7	28	2666.828	1.24820	1280.077	0.22346	3.17824	1.32316	4.20532	836.575	4.14441
8	32	2647.537	1.25264	1270.818	0.22460	3.19091	1.3337	4.25571	819.796	4.24505
9	36	2628.177	1.25714	1261.525	0.22576	3.20373	1.34476	4.30825	802.818	4.24362
10	40	2608.749	1.26169	1252.199	0.22694	3.21671	1.35638	4.36309	785.634	4.34832
11	44	2589.249	1.26629	1242.840	0.22813	3.22987	1.36862	4.42045	768.236	4.35108
12	48	2569.679	1.27094	1233.446	0.22935	3.24319	1.38151	4.48050	750.611	4.46133
13	52	2550.037	1.27565	1224.018	0.23058	3.25669	1.39513	4.54349	732.751	4.46857
14	56	2530.322	1.28042	1214.555	0.23183	3.27036	1.40952	4.60965	714.645	4.58462
15	60	2510.533	1.28524	1205.056	0.23311	3.28422	1.42478	4.67930	696.278	4.59842
16	64	2490.670	1.29011	1195.522	0.23440	3.29827	1.44098	4.75276	677.640	4.72079
17	68	2470.732	1.29505	1185.951	0.23572	3.31251	1.45823	4.83039	658.716	4.74146
18	72	2450.717	1.30005	1176.344	0.23705	3.32694	1.47662	4.91261	639.488	4.87222
19	76	2430.625	1.30511	1166.700	0.23841	3.34158	1.49628	4.99995	619.941	4.90116
20	80	2410.455	1.31023	1157.018	0.23980	3.35642	1.51738	5.09295	600.055	5.04217
21	84	2390.206	1.31541	1147.299	0.24120	3.37147	1.54007	5.19229	579.808	5.08121
22	88	2369.876	1.32066	1137.541	0.24264	3.38674	1.56456	5.29877	559.178	5.23379
23	92	2349.466	1.32598	1127.744	0.24409	3.40223	1.59111	5.41331	538.137	5.28693
24	96	2328.973	1.33137	1117.907	0.24557	3.41795	1.61999	5.53704	516.655	5.45370
25	100	2308.397	1.33683	1108.031	0.24708	3.43390	1.65157	5.67133	494.699	5.52440
26	104	2287.737	1.34235	1098.114	0.24862	3.45009	1.68629	5.81784	472.231	5.70976
27	108	2266.991	1.34796	1088.156	0.25018	3.46652	1.72468	5.97863	449.205	5.80320
28	112	2246.158	1.35364	1078.156	0.25178	3.48321	1.76743	6.15633	425.571	6.01382
29	116	2225.238	1.35939	1068.114	0.25340	3.50016	1.81543	6.35428	401.267	6.13785
30	120	2204.229	1.36522	1058.030	0.25505	3.51737	1.86981	6.57680	376.225	6.38352

Table 5.5. Interpolation table for recoil D assuming PS composition to calculate stopping powers.

Slab no.	Depth ($\mu g/cm^2$)	$E_o^{(n)}$ (keV)	$S_p^{(n)}$ ($keV/\mu g/cm^2$)	$E_2^{(n)}$ (keV)	$S_r^{(n)}$ ($keV/\mu g/cm^2$)	$[S]_{p,r}^{(n)}$ ($keV/\mu g/cm^2$)	Π_n	$[S]_{p,r}^{(n)}\,\Pi_n$ ($keV/\mu g/cm^2$)	E_d (keV)	dE_d/dx ($keV/\mu g/cm^2$)
surface	0	2800.000	1.21840	1866.667	0.27776	4.21151	1.21392	5.11243	1370.799	5.11243
1	4	2781.170	1.22252	1854.113	0.27902	4.22702	1.21983	5.15626	1350.367	5.10370
2	8	2762.276	1.22668	1841.517	0.28030	4.24271	1.22597	5.20142	1329.761	5.19928
3	12	2743.318	1.23089	1828.878	0.28161	4.25858	1.23234	5.24800	1308.975	5.19366
4	16	2724.295	1.23515	1816.196	0.28293	4.27464	1.23895	5.29607	1288.005	5.29157
5	20	2705.206	1.23945	1803.470	0.28427	4.29090	1.24583	5.34573	1266.844	5.28875
6	24	2686.050	1.24380	1790.700	0.28562	4.30735	1.25299	5.39705	1245.486	5.39026
7	28	2666.828	1.24820	1777.885	0.28700	4.32400	1.26044	5.45015	1223.925	5.39050
8	32	2647.537	1.25264	1765.024	0.28840	4.34086	1.26821	5.50514	1202.152	5.49555
9	36	2628.177	1.25714	1752.118	0.28982	4.35792	1.27633	5.56212	1180.163	5.49920
10	40	2608.749	1.26169	1739.166	0.29126	4.37520	1.28480	5.62125	1157.948	5.60846
11	44	2589.249	1.26629	1726.166	0.29272	4.39270	1.29366	5.68267	1135.498	5.61615
12	48	2569.679	1.27094	1713.119	0.29421	4.41042	1.30294	5.74653	1112.806	5.72986
13	52	2550.037	1.27565	1700.025	0.29571	4.42838	1.31267	5.81301	1089.862	5.74224
14	56	2530.322	1.28042	1686.881	0.29724	4.44656	1.32289	5.88231	1066.655	5.86139
15	60	2510.533	1.28524	1673.689	0.29880	4.46499	1.33363	5.95464	1043.175	5.87860
16	64	2490.670	1.29011	1660.447	0.30038	4.48366	1.34494	6.03024	1019.410	6.00406
17	68	2470.732	1.29505	1647.155	0.30199	4.50258	1.35686	6.10938	995.346	6.02774
18	72	2450.717	1.30005	1633.811	0.30362	4.52175	1.36946	6.19237	970.971	6.15967
19	76	2430.625	1.30511	1620.417	0.30528	4.54119	1.38280	6.27954	946.270	6.19073
20	80	2410.455	1.31023	1606.970	0.30696	4.56090	1.39694	6.37130	921.227	6.33112
21	84	2390.206	1.31541	1593.470	0.30868	4.58088	1.41197	6.46806	895.823	6.37075
22	88	2369.876	1.32066	1579.918	0.31042	4.60114	1.42798	6.57033	870.041	6.52033
23	92	2349.466	1.32598	1566.310	0.31220	4.62170	1.44507	6.67870	843.857	6.57132
24	96	2328.973	1.33137	1552.649	0.31400	4.64255	1.46338	6.79381	817.251	6.73169
25	100	2308.397	1.33683	1538.931	0.31584	4.66370	1.48304	6.91644	790.197	6.79571
26	104	2287.737	1.34235	1525.158	0.31771	4.68517	1.50422	7.04752	762.663	6.97092
27	108	2266.991	1.34796	1511.327	0.31961	4.70696	1.52712	7.18807	734.622	7.04971
28	112	2246.158	1.35364	1497.439	0.32155	4.72907	1.55197	7.33936	706.035	7.24399
29	116	2225.238	1.35939	1483.492	0.32352	4.75152	1.57905	7.50288	676.864	7.34134
30	120	2204.229	1.36522	1469.486	0.32554	4.77432	1.60870	7.68045	647.062	7.55985

Table 5.6. Interpolation table for recoil H assuming PVC composition to calculate stopping powers.

Slab no.	Depth ($\mu g/cm^2$)	$E_o^{(n)}$ (keV)	$S_p^{(n)}$ ($keV/\mu g/cm^2$)	$E_2^{(n)}$ (keV)	$S_r^{(n)}$ ($keV/\mu g/cm^2$)	$[S]_{p,r}^{(n)}$ ($keV/\mu g/cm^2$)	Π_n	$[S]_{p,r}^{(n)}\,\Pi_n$ ($keV/\mu g/cm^2$)	E_d (keV)	dE_d/dx ($keV/\mu g/cm^2$)
surface	0	2800.000	0.98753	1344.000	0.17788	2.51872	1.26127	3.17678	949.065	3.17677
1	4	2784.738	0.99014	1336.674	0.17854	2.52610	1.26751	3.20185	936.388	3.16188
2	8	2769.435	0.99277	1329.329	0.17920	2.53356	1.27396	3.22764	923.612	3.22600
3	12	2754.092	0.99542	1321.964	0.17988	2.54109	1.28062	3.25418	910.734	3.21288
4	16	2738.708	0.99809	1314.580	0.18056	2.54869	1.28752	3.28150	897.752	3.27806
5	20	2723.283	1.00079	1307.176	0.18126	2.55636	1.29467	3.30963	884.662	3.26677
6	24	2707.816	1.00351	1299.752	0.18195	2.56411	1.30207	3.33864	871.462	3.33357
7	28	2692.307	1.00626	1292.307	0.18266	2.57194	1.30974	3.36856	858.148	3.32350
8	32	2676.755	1.00903	1284.843	0.18338	2.57985	1.31769	3.39945	844.716	3.39253
9	36	2661.161	1.01183	1277.357	0.18410	2.58783	1.32596	3.43135	831.162	3.38423
10	40	2645.523	1.01465	1269.851	0.18484	2.59590	1.33455	3.46435	817.483	3.45506
11	44	2629.842	1.01750	1262.324	0.18558	2.60405	1.34348	3.49848	803.676	3.44860
12	48	2614.117	1.02038	1254.776	0.18633	2.61228	1.35277	3.53382	789.734	3.52232
13	52	2598.347	1.02328	1247.207	0.18709	2.62060	1.36246	3.57046	775.655	3.51747
14	56	2582.532	1.02621	1239.616	0.18786	2.62901	1.37256	3.60847	761.431	3.59431
15	60	2566.673	1.02917	1232.003	0.18864	2.63750	1.38311	3.64794	747.059	3.59175
16	64	2550.767	1.03215	1224.368	0.18943	2.64609	1.39413	3.68899	732.533	3.67131
17	68	2534.815	1.03517	1216.711	0.19022	2.65476	1.40566	3.73170	717.847	3.67144
18	72	2518.817	1.03821	1209.032	0.19103	2.66354	1.41775	3.77623	702.995	3.75471
19	76	2502.772	1.04129	1201.330	0.19185	2.67240	1.43043	3.82267	687.969	3.75838
20	80	2486.679	1.04439	1193.606	0.19268	2.68137	1.44374	3.87121	672.763	3.84465
21	84	2470.538	1.04753	1185.858	0.19352	2.69043	1.45775	3.92198	657.368	3.85259
22	88	2454.349	1.05070	1178.087	0.19437	2.69960	1.47252	3.97521	641.777	3.94319
23	92	2438.110	1.05390	1170.293	0.19524	2.70887	1.4881	4.03107	625.980	3.95543
24	96	2421.823	1.05713	1162.475	0.19611	2.71824	1.50458	4.08981	609.966	4.05137
25	100	2405.485	1.06040	1154.633	0.19700	2.72772	1.52204	4.15169	593.727	4.06798
26	104	2389.097	1.06370	1146.766	0.19790	2.73732	1.54057	4.21704	577.250	4.17079
27	108	2372.657	1.06703	1138.876	0.19881	2.74702	1.5603	4.28618	560.521	4.19362
28	112	2356.167	1.07040	1130.960	0.19973	2.75684	1.58135	4.35952	543.527	4.30314
29	116	2339.624	1.07381	1123.019	0.20067	2.76678	1.60386	4.43752	526.254	4.33345
30	120	2323.028	1.07725	1115.054	0.20162	2.77684	1.62801	4.52073	508.683	4.45202

Table 5.7. Interpolation table for recoil D assuming PVC composition to calculate stopping powers.

Slab no.	Depth (μg/cm²)	$E_o^{(n)}$ (keV)	$S_p^{(n)}$ (keV/μg/cm²)	$E_2^{(n)}$ (keV)	$S_r^{(n)}$ (keV/μg/cm²)	$[S]_{p,r}^{(n)}$ (keV/μg/cm²)	Π_n	$[S]_{p,r}^{(n)}\,\Pi_n$ (keV/μg/cm²)	E_d (keV)	dE_d/dx (keV/μg/cm²)
surface	0	2800.000	0.98753	1866.667	0.22657	3.41909	1.21392	4.15050	1370.799	4.15050
1	4	2784.738	0.99014	1856.492	0.22738	3.42892	1.21865	4.17867	1354.221	4.13881
2	8	2769.435	0.99277	1846.290	0.22819	3.43883	1.22353	4.20751	1337.530	4.20641
3	12	2754.092	0.99542	1836.061	0.22901	3.44884	1.22856	4.23709	1320.726	4.19575
4	16	2738.708	0.99809	1825.805	0.22984	3.45894	1.23374	4.26742	1303.805	4.26488
5	20	2723.283	1.00079	1815.522	0.23069	3.46914	1.23908	4.29855	1286.763	4.25599
6	24	2707.816	1.00351	1805.210	0.23154	3.47945	1.24460	4.33051	1269.598	4.32634
7	28	2692.307	1.00626	1794.871	0.23240	3.48985	1.25029	4.36333	1252.308	4.31898
8	32	2676.755	1.00903	1784.503	0.23327	3.50035	1.25618	4.39706	1234.886	4.39165
9	36	2661.161	1.01183	1774.107	0.23415	3.51096	1.26226	4.43175	1217.332	4.38551
10	40	2645.523	1.01465	1763.682	0.23504	3.52168	1.26855	4.46744	1199.641	4.46007
11	44	2629.842	1.01750	1753.228	0.23595	3.53250	1.27507	4.50419	1181.809	4.45601
12	48	2614.117	1.02038	1742.744	0.23686	3.54344	1.28182	4.54205	1163.832	4.53221
13	52	2598.347	1.02328	1732.231	0.23778	3.55449	1.28882	4.58109	1145.705	4.53120
14	56	2582.532	1.02621	1721.688	0.23872	3.56565	1.29608	4.62137	1127.425	4.60881
15	60	2566.673	1.02917	1711.115	0.23967	3.57693	1.30362	4.66295	1108.986	4.61079
16	64	2550.767	1.03215	1700.511	0.24063	3.58833	1.31145	4.70593	1090.384	4.69048
17	68	2534.815	1.03517	1689.877	0.24160	3.59984	1.31960	4.75036	1071.612	4.69526
18	72	2518.817	1.03821	1679.211	0.24258	3.61149	1.32809	4.79637	1052.666	4.77806
19	76	2502.772	1.04129	1668.514	0.24358	3.62325	1.33693	4.84402	1033.538	4.78560
20	80	2486.679	1.04439	1657.786	0.24459	3.63515	1.34615	4.89346	1014.224	4.87154
21	84	2470.538	1.04753	1647.025	0.24561	3.64718	1.35578	4.94478	994.716	4.88255
22	88	2454.349	1.05070	1636.232	0.24664	3.65934	1.36585	4.99811	975.006	4.97225
23	92	2438.110	1.05390	1625.407	0.24769	3.67164	1.37639	5.05360	955.088	4.98665
24	96	2421.823	1.05713	1614.548	0.24876	3.68407	1.38743	5.11140	934.953	5.08092
25	100	2405.485	1.06040	1603.656	0.24983	3.69665	1.39903	5.17171	914.592	5.09947
26	104	2389.097	1.06370	1592.731	0.25093	3.70937	1.41121	5.23469	893.996	5.19899
27	108	2372.657	1.06703	1581.772	0.25203	3.72224	1.42403	5.30058	873.154	5.22178
28	112	2356.167	1.07040	1570.778	0.25316	3.73526	1.43755	5.36963	852.055	5.32766
29	116	2339.624	1.07381	1559.749	0.25429	3.74843	1.45183	5.44208	830.688	5.35588
30	120	2323.028	1.07725	1548.685	0.25545	3.76177	1.46694	5.51828	809.040	5.46816

channel 125 in the H profile is at detected energy 905.625 keV which corresponds to a scattering event between layers 2 and 3. The interpolated depth corresponding to the detected energy of 905.625 keV is [(917.747-905.625)/(917.747-901.854)](4 μg/cm^2) + 8 μg/cm^2 = 11.05 μg/cm^2. Table 5.8 gives the values obtained from Table 5.4 and the interpolated values needed for scaling the data point in channel 125. In addition, the cross section was calculated using the formulation of Baglin *et al.* (1992) at the interpolated value for the projectile energy $E_0^{(n)}$ = 2747.816 keV. This calculation of the cross section was used rather than the Rutherford formula in Eq. (5.4) because the measured cross section for recoiling ^{1}H by ^{4}He is non-Rutherford above a ^{4}He projectile energy of approximately 1.2 MeV. In fact, the cross section is approximately constant from an energy of 2.3 MeV to 2.9 MeV at a value of σ=255 mb/sr, which is 2.5 times greater than the Rutherford cross section calculated at 2.8 MeV. Similarly, Kellock and Baglin (1993) showed that the cross section for recoiling D by ^{4}He at a scattering angle of 30° is non-Rutherford above a ^{4}He projectile energy of 1.0 MeV, and is 12.5 times greater than the Rutherford cross section at 2.8 MeV. This cross section is also approximately constant over an energy range from 2.3 MeV to 2.9 MeV at a value of σ=454 mb/sr. A more detailed treatment of the use of non-Rutherford cross sections with ERD analysis will be given in the next section.

The final step to achieve H and D depth profiles from the data in Fig. 5.10 is to scale the yield according to Eq. (5.24) for each channel which falls within the range of E_d calculated in the interpolation tables. First take note that the cross section scaling factors, $\sigma_o(E_o)/\sigma(E_0^{(n)})$, are simply equal to 1 for both the H and D signals because the non-Rutherford cross section is constant for the entire depth of interest (2.3 MeV < $E_0^{(n)}$ < 2.8 MeV). Therefore, only the effective stopping power scaling factors (interpolated from Tables 5.4 and 5.5) need to be applied to the data. The data of Fig. 5.10 have been scaled to give the depth profiles shown in Fig. 5.11 using these scaling factors, the interpolated values of depth for each channel, and the following constant values for Eq. (5.24): $\Theta_1 = 75°$, Q = (1 He$^+$/Charge)(3 μC)/(1.602×10^{-13} μC/Charge) = 1.873×10^{13} He$^+$, Ω = 3.6 msr, δE_d = 7.245 keV/Ch, σ_o(H) = 255 mb/sr, σ_o(D) = 454 mb/sr, S_o^{eff}(H) = 3.902 keV/(μg/cm^2), and S_o^{eff}(D) = 5.112 keV/(μg/cm^2). The depth scale in this figure increases from right to left in order to preserve the connection between the profile and the ERD spectrum. Further, the left and bottom axes of this figure give the concentration and depth in density exclusive units, while the right and top axes give the concentration and depth in density inclusive units based on a density of 1.04 g/cm^3 for polystyrene.

5.3.3 Non-Rutherford cross sections in ERD analysis

Elastic recoil detection is a technique which inherently uses high energy ions for analysis, and therefore the possibility of penetrating the Coulomb barrier and the resultant use of non-Rutherford cross sections must be considered. A specific example of this type of analysis was shown in the previous section for both ^{1}H(^{4}He,^{1}H)^{4}He and ^{2}H(^{4}He,^{2}H)^{4}He scattering events. However, a more general method is needed for determining the conditions where the recoil scattering cross sections may become non-Rutherford.

Deviations from Rutherford-scattering cross sections resulting from Coulomb barrier penetration occur most often for low Z projectiles at high energies, while screening effects which can cause deviations from the Rutherford formula occur most often for high Z projectiles at low energies. Bozoian *et al.* (1990) have modeled the case of Coulomb barrier penetration and determined the energies where the cross sections may become non-Rutherford. In the lab reference frame and for a scattering angle of 30°, their formulation of this energy boundary is given by (also see Appendix 8)

$$E(\text{MeV}/\text{amu}) = \frac{-1.1934(Z_1Z_2)(M_1+M_2)}{M_1M_2A_1^{1/3}\ln(.001846\,Z_1A_2/Z_2)} \qquad (5.26)$$

Further, Andersen *et al.* (1980) have modeled the case where screening effects cause significant deviations from the Rutherford formula, and in the lab reference frame for 1% screening effects (i.e., σ/σ_r = 0.99) the energy boundary is given by

$$E(\text{MeV}/\text{amu}) = 99\,V_{LJ}\frac{(M_1+M_2)}{M_1M_2} \qquad (5.27)$$

where $V_{LJ} = 48.73Z_1Z_2\sqrt{Z_1^{2/3}+Z_2^{2/3}}$ is the Lenz-Jensen potential.

Table 5.8. Interpolated values for scaling the H data point in channel 125.

Channel no.	E_d *(keV)*	Slab no.	Depth *(μg/cm²)*	$E_o^{(n)}$ *(keV)*	$\sigma(E_o^{(n)})$ *(mb/sr)*	dE_d/dx *(keV/μg/cm²)*
	917.747	2	8	2762.276		3.97897
125	905.625		11.05	2747.816	255.2	3.97048
	901.854	3	12	2743.318		3.96784

The lowest boundary energy for possible barrier penetration can be determined for each projectile by examining the energy where barrier penetration may occur for all relevant target atoms. Deuterium atoms are thus found to be the target atoms which determine this barrier-penetration boundary energy for most projectiles. Similarly, the largest boundary energy where screening effects begin to occur for all relevant target atoms can be determined, and H target atoms are found to determine this boundary for screening effects. The determination of these projectile energy boundaries for non-Rutherford scattering are shown graphically in Fig. 5.12 as a function of the projectile atomic number. (Note that the classical Coulomb barrier value was used for the barrier penetration energy when using a Au ion beam because the model of Bozoian *et al.* (1990) predicts a value greater than the classical coulomb barrier value.) For reference, the regions corresponding to the use of 2.8 MeV He ERD, 28 MeV Si ERD, and 12 MeV Au ERD are also given in this figure. The use of a He ion beam for ERD is nearly always in the region where penetration of the Coulomb barrier can occur, and the use of a Au ion beam is nearly always in the region where significant screening effects can occur (for current ion-beam analysis accelerator

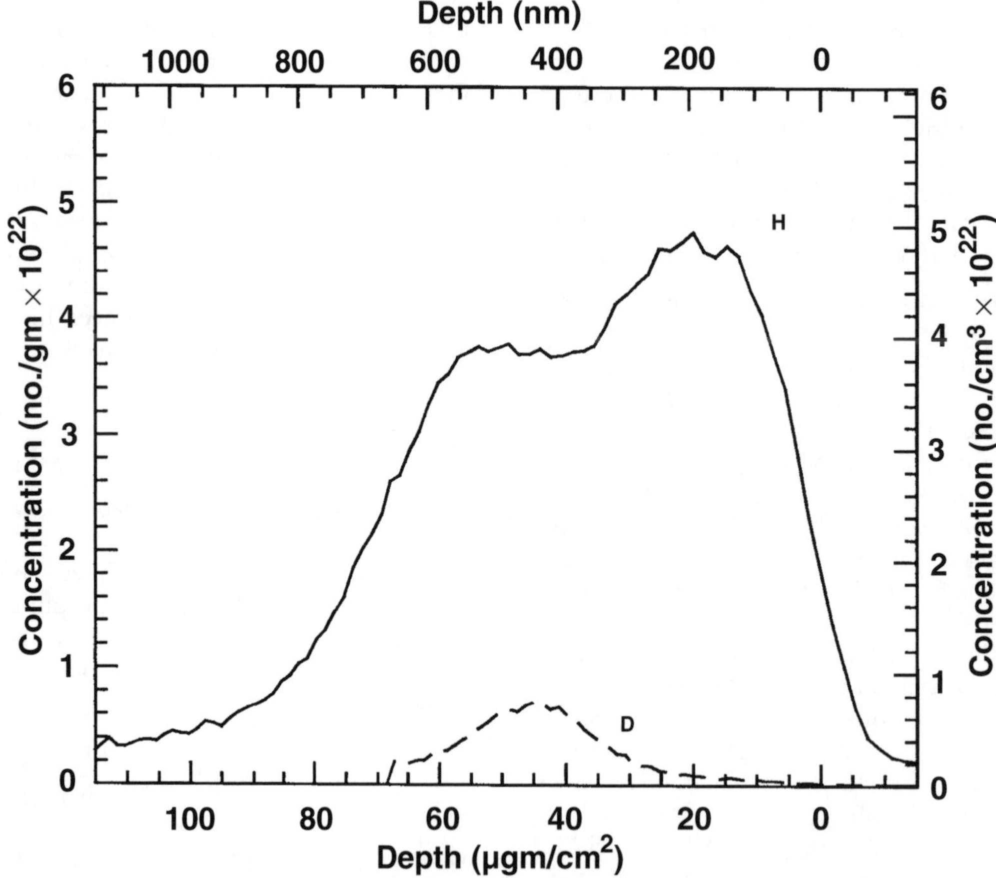

FIG. 5.11. The concentration-depth profiles for H and D determined from the spectrum shown in Fig. 5.10. The concentration and depth scales are shown in density-exclusive units along the left and bottom axes, respectively. The right and top axes were determined by assuming the density of the sample is 1.04 g/cm³.

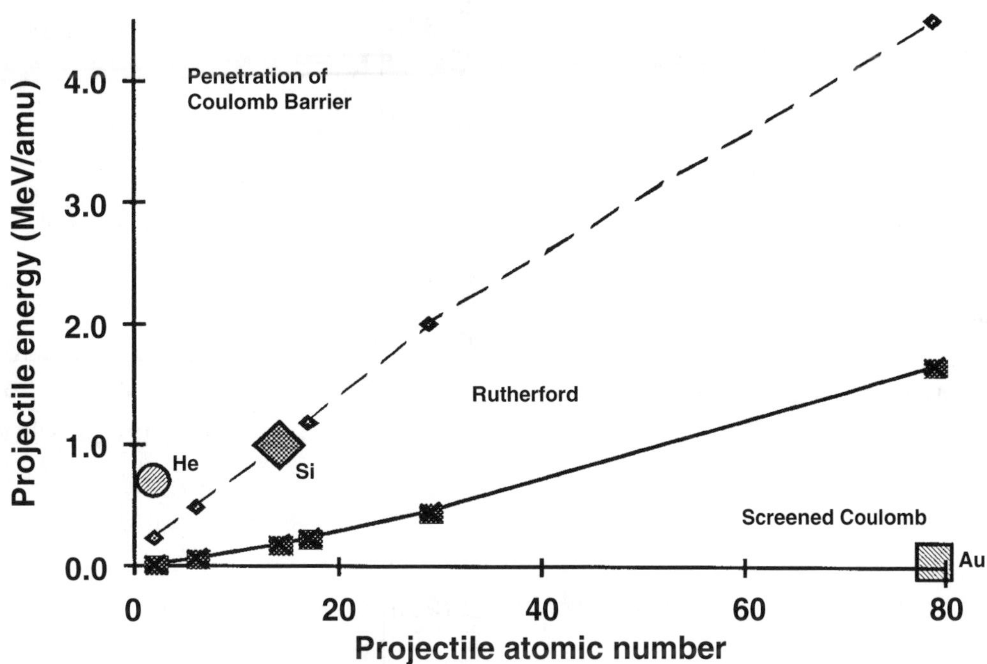

FIG. 5.12. The energy at which the ERD cross section becomes non-Rutherford can be estimated from this projectile-energy versus atomic-number plot. Above the dashed line, the projectile may have sufficient energy to penetrate the coulomb barrier and possibly cause deviations from the Rutherford formula [Eq. (5.4)]. Below the solid line, the projectiles have such low energy that significant screening of the nuclear charge occurs and again the cross sections will deviate from the Rutherford formula. The values for 2.8 MeV He ERD, 28 MeV Si ERD, and 12 MeV Au ERD are also shown for comparison.

technology). The use of 28 MeV Si for ERD analysis is close to the boundary where Coulomb barrier penetration may occur, but in practice, significant deviations from the Rutherford formula have not been measured.

If projectiles with energies which fall outside of the Rutherford region in Fig. 5.12 are to be used for ERD analyses, then an accurate determination of their scattering cross sections should be made. Baglin *et al.* (1992) and later Kellock and Baglin (1993) have made such determinations as a function of scattering angle for the important cases of He ERD used in profiling hydrogen and deuterium. A simple polynomial expression was determined for calculating these cross sections over the ^{4}He projectile energy range from 1 to 3 MeV: $\sigma(E_0^{(n)},\phi) = aE_0^{(n)}+b+(c/E_0^{(n)})+(d/[E_0^{(n)}]^2)$. (For the case of D profiling, a resonance occurs in the cross section at a ^{4}He energy of 2.135 ± 0.01 MeV, with ~ 60 keV FWHM, and therefore the polynomial fit is valid only outside of this resonance region.) Energy is expressed in MeV for this polynomial expression, and the coefficients a,b,c, and d are given in Tables 5.9 (for H recoil scattering) and 5.10 (for D recoil scattering). The measured cross section values and the polynomial fit of the data are also given in Figs. 5.13 through 5.21 for reference.

Table 5.9. Coefficients for polynomial fit of H cross sections at scattering angles ϕ.

ϕ	a	b	c	d
20	−0.143	6.043	−0.898	1.476
25	0.451	2.849	4.736	−1.503
30	0.133	4.383	2.196	−0.042
35	0.097	4.511	2.588	−0.278

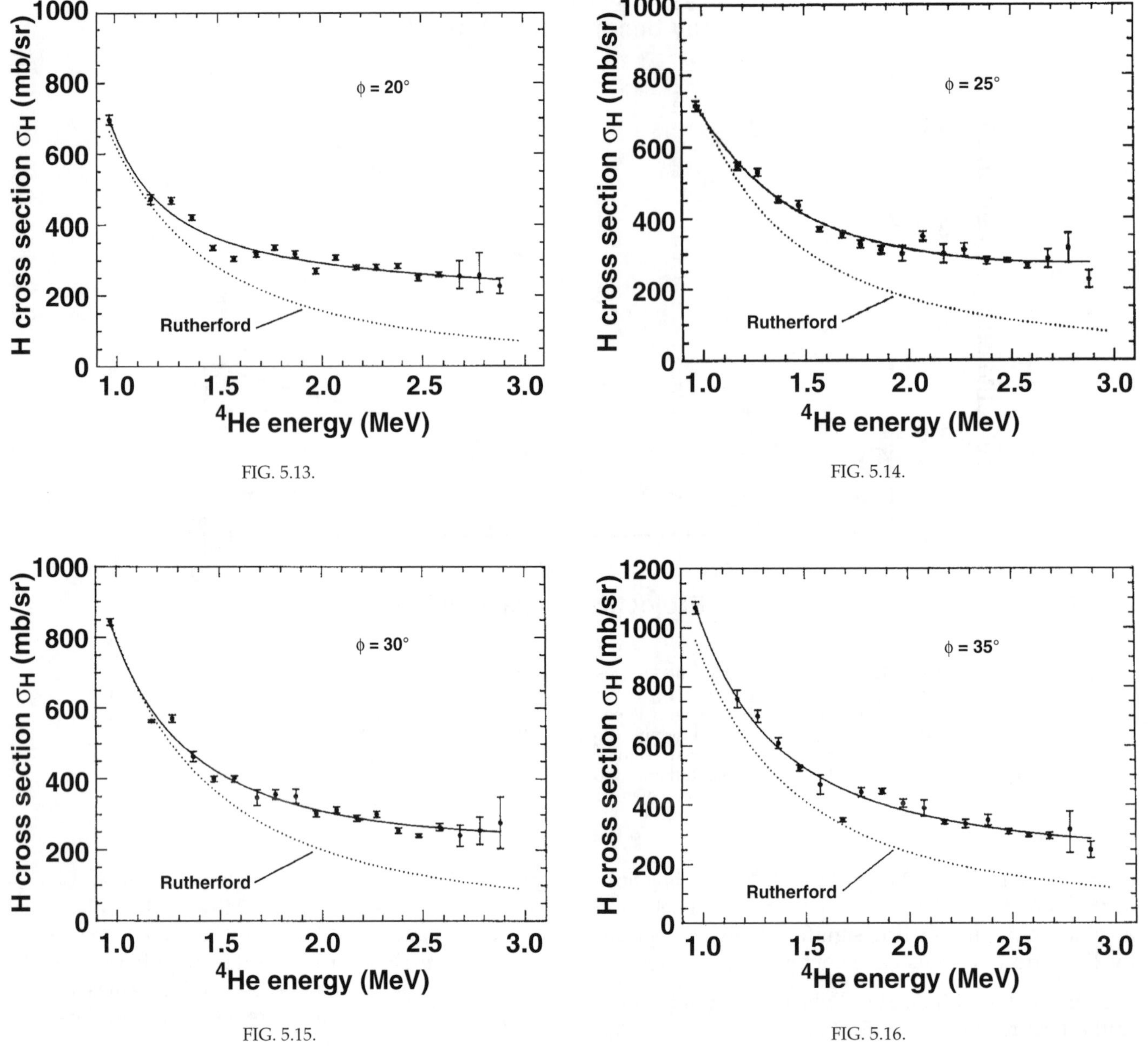

FIG. 5.13.

FIG. 5.14.

FIG. 5.15.

FIG. 5.16.

FIGS. 5.13-5.16. Hydrogen forward-recoil scattering cross sections (σ_H), plotted as a function of ^{4}He ion energy (lab frame of reference) for ϕ equals: 20° (Fig. 5.13), 25° (Fig. 5.14), 30° (Fig. 5.15), and 35° (Fig. 5.16). The solid lines through the points are polynomial fits to the data, determined by Baglin *et al.* (1992). For comparison, the cross section given by the Rutherford formula is shown as a dashed line.

5.3.4 ERD analysis of hydrogen and nitrogen in silicon nitride films

The second slab-analysis example is the determination of both the H and N depth profiles from a 300-nm-thick silicon nitride film on a Si substrate as shown in the inset to Fig. 5.22. As stated earlier, hydrogen is expected to be a minor constituent in this film, and therefore the matrix will be approximated as Si_3N_4 with a density of

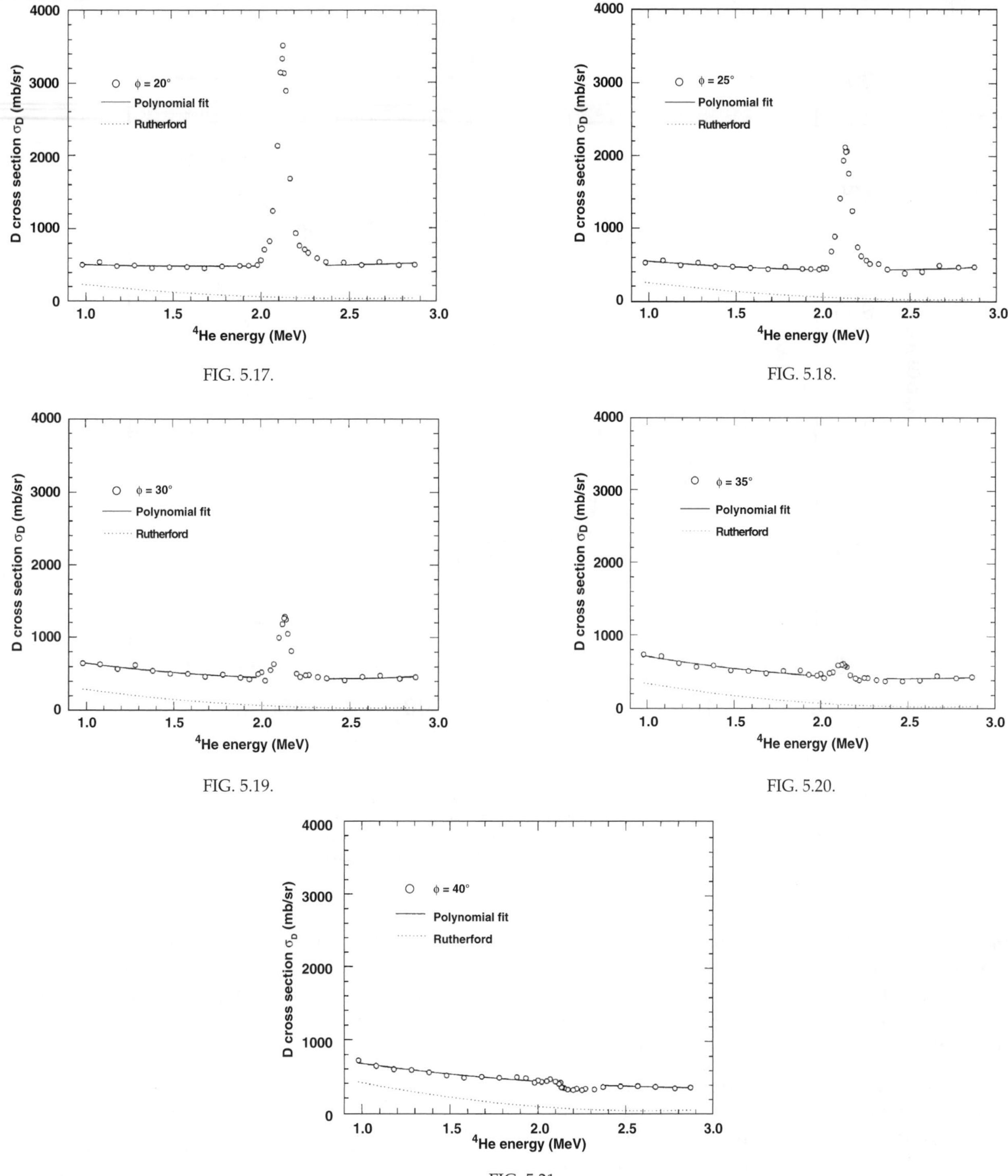

FIG. 5.17.

FIG. 5.18.

FIG. 5.19.

FIG. 5.20.

FIG. 5.21.

FIGS. 5.17-5.21. The deuteron (D) forward-recoil scattering cross sections (σ_D), plotted as a function of ^{4}He ion energy (lab frame of reference) for ϕ equals: 20° (Fig. 5.17), 25° (Fig. 5.18), 30° (Fig. 5.19), 35° (Fig. 5.20), and 40° (Fig. 5.21). The uncertainty in the cross section scale is ±5%. The solid lines through the points are polynomial fits to the data, excluding the resonance region, as determined by Kellock and Baglin (1993). The dashed lines are the Rutherford cross section given by Eq. (5.4).

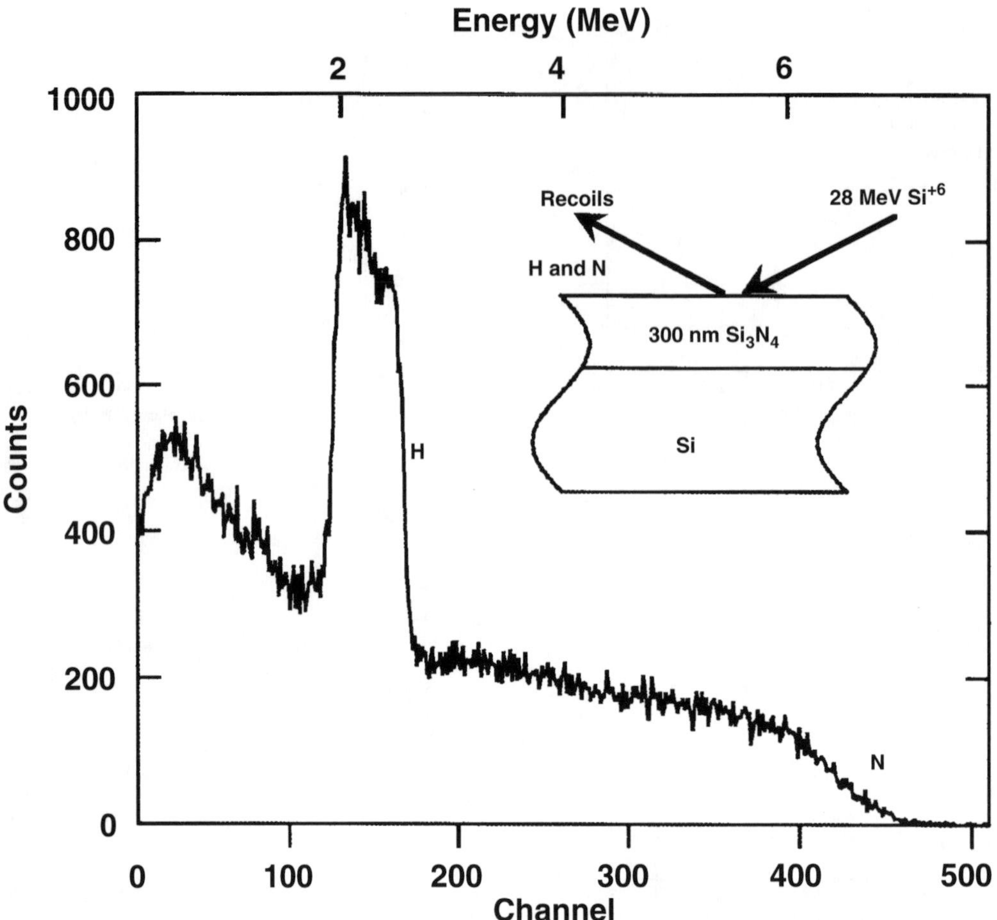

FIG. 5.22. An ERD spectrum collected from a silicon nitride layer which is approximately 300 nm thick. A 28 MeV Si beam was used for the analysis with ϕ=30° and Θ_1=75°. The hydrogen is present as a minor constituent of the film remaining after the atmospheric-pressure chemical vapor deposition at 800°C. The H signal is sitting on the N signal background.

3 g/cm^3. A normal multi-slab analysis will be presented below, and the effects on the yield spectrum of σ and dE_d/dx scaling will be demonstrated in Figs. 5.23, 5.24, and 5.25. In addition, the effect of choosing a foil slab thickness which is too thin will also be demonstrated.

The ERD spectrum from this sample is plotted in Fig. 5.22, and the following experimental scattering parameters were used for the analysis. The analysis beam was a 28 MeV $^{28}Si^{+6}$ ion beam incident at an angle Θ_1 = 75°. The recoil atoms were collected at a scattering angle ϕ = 30°, using a solid angle Ω = 5.263 msr, and a 12-μm-thick aluminized-mylar range foil. The total incident fluence was 0.5 μC of charge, which yields a value for Q of Q = (1 Si^{+6}/6 Charges)(0.5 μC)/(1.602×10^{-13} μC/Charge) = 5.202×10^{11} Si^{+6}. A slab analysis for stopping in the foil used 12 1000-nm-thick layers, and as before, the composition and density of this mylar foil were approximated as a pure carbon foil with a density of 1.58 g/cm^3. The energy scale along the top axis of Fig. 5.22 was

Table 5.10. Coefficients for polynomial fit of D cross sections at scattering angles ϕ.

ϕ	a	b	c	d
20	−0.06	6.84	−1.59	1.07
25	0.75	2.27	5.92	−2.66
30	0.58	2.99	4.94	−2.04
35	0.15	5.05	1.72	−0.34
40	−0.42	7.54	−1.77	1.19

determined from the H surface edge at channel 169 and the calculated surface edge E_d = 2579.802 keV, as shown in Table 5.11. The resulting energy scale conversion factor was δE_d = 15.265 keV/Ch.

5.3.4.1 Scaling factors for the H signal

At channels greater than ≈ 180, the ERD spectrum contains counts for nitrogen recoils alone, whereas both H and N recoils contribute to the yield for channels between 110 and 180. Therefore, the first step in analyzing the H content of this sample is to subtract the nitrogen background. A simple linear fit was used to remove the N background between channels 110 and 182, and the resultant hydrogen signal is plotted as a dashed line in Fig. 5.23. At this point, a slab analysis was used to determine the Si energy, effective stopping power, and detected recoil H energy as a function of depth into the

Table 5.11. Slab analysis for H surface-edge detected energy after traversing the stopping range-foil (divided into 12 slab layers).

Depth in range foil *(nm)*	Hydrogen energy in range foil *[E (keV)]*	Stopping power of H in range foil at energy E *(keV/nm)*
0	2796.670	0.017588
1000	2779.082	0.017672
2000	2761.411	0.017756
3000	2743.654	0.017843
4000	2725.812	0.017930
5000	2707.882	0.018019
6000	2689.863	0.018109
7000	2671.754	0.018201
8000	2653.553	0.018294
9000	2635.259	0.018389
10000	2616.870	0.018485
11000	2598.385	0.018583
12000	2579.802	

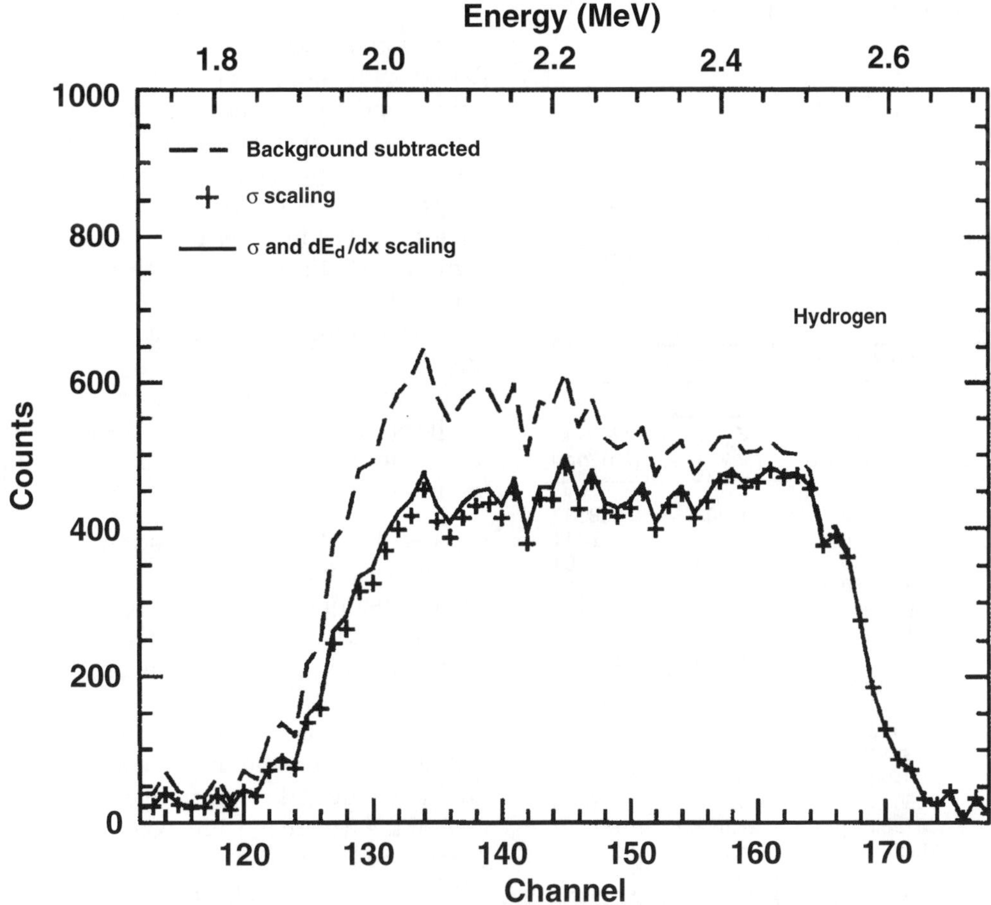

FIG. 5.23. Scaled H signal from the spectrum shown in Fig. 5.22. First the background N signal was subtracted (dashed line), then the data were scaled to make the cross section appear to be constant throughout the film and equal to σ at E_0 (+ symbols). The σ-scaled data were then scaled to remove the effect of the effective stopping power dE_d/dx (solid line). See text for a description of dE_d/dx.

sample. The resulting slab analysis is summarized in Table 5.12 for 20 slabs of equal thickness (20 nm each).

The next step in the analysis is to scale the data for the increasing recoil cross section as the Si beam penetrates the sample. Equation (5.4) shows that, for Rutherford scattering, this recoil cross section scales with the inverse square of the projectile energy. The H recoil data were normalized to remove this dependence by multiplying the yield in each channel by the ratio of $(E_o^{(n)}/E_o)^2$. The energy of the Si at depth was determined by calculating the detected energy for each channel and then interpolating Table 5.12 for the corresponding value of $E_o^{(n)}$. The results of this cross section scaling reduce the yield as shown by the +'s in Fig. 5.23. If areal density of hydrogen was needed, then it could be obtained at this point in the analysis by integrating channels. The final scaling of the data was then done to normalize the data as if the effective stopping power remained constant with depth. Each data point was multiplied by the ratio

$$\frac{dE_d/dx|_n}{dE_d/dx|_o}$$

and as with the cross section scaling, the values for $dE_d/dx\,|_n$ were interpolated from Table 5.12. However, in contrast to the σ scaling, the effective stopping power scaling increases the yield as shown by the solid line in Fig. 5.23. The application of the concentration and depth scales to the spectra will be shown below, after first calculating the scaling factors for the N signal.

Table 5.12. Interpolation table for recoiled H from Si_3N_4.

Slab no.	Depth *(nm)*	$E_o^{(n)}$ *(keV)*	E_d *(keV)*	dE_d/dx *(keV/nm)*
Surface	0	28000.000	2579.802	2.068332
1	20	27632.824	2538.395	2.072341
2	40	27265.240	2496.833	2.083884
3	60	26897.258	2455.110	2.088357
4	80	26528.889	2413.225	2.100120
5	100	26160.143	2371.171	2.105325
6	120	25791.033	2328.945	2.117234
7	140	25421.574	2286.543	2.123025
8	160	25051.777	2243.957	2.135593
9	180	24681.656	2201.183	2.141800
10	200	24311.229	2158.215	2.154978
11	220	23940.508	2115.047	2.161770
12	240	23569.512	2071.672	2.175803
13	260	23198.256	2028.082	2.183169
14	280	22826.760	1984.268	2.198190
15	300	22455.041	1940.225	2.206119
16	320	22083.121	1895.940	2.222421
17	340	21711.020	1851.404	2.231179
18	360	21338.758	1806.605	2.248740
19	380	20966.359	1761.532	2.258475
20	400	20593.848	1716.172	2.277622

5.3.4.2 Scaling factors for the N signal

This section will concentrate on the determination of the nitrogen signal scaling factors, and therefore the H signal was removed by the straight line background fit as described above for the H signal analysis. The resulting N signal is plotted in Fig. 5.24 as a dashed line. The previous hydrogen analysis gave the energy conversion for the ERD spectrum as δE_d = 15.265 keV/Ch, based on E_d for the surface channel of H. This value for the energy-scale conversion would yield a nitrogen surface channel of 289, but the measured surface channel for N is 430. The discrepancy between these two channel values is large enough to cause concern and is most likely a result of inaccurately calculating the stopping power of N in the range foil. In fact, the calculations for this analysis were based on the stopping power equations given by Ziegler (1980), and differ by about 8% from the stopping powers calculated from his later formulation (Ziegler *et al.*, 1985). In order to correct for this inaccuracy, the calculated stopping power of N in the range foil was scaled by a factor of 0.864 in order to force the calculated E_d(N) to agree with both δE_d and the measured surface channel of 430. A check on the accuracy of this scaling will be shown in determining the layer thickness from the H and N depth profiles, separately. If the scaling of the stopping power in the range foil causes errors in the determination of the N depth scale, then the H and N depth profiles for the same thickness layer will appear to have different depths.

A slab analysis was performed on this data and the results are tabulated in Table 5.13. The cross section and effective stopping power scalings were applied as shown in +'s and the solid line of Fig. 5.24, respectively. The depth and concentration scales can then be applied to the scaled spectra, as described in the next section. An interesting facet of ERD analysis with range foils is shown in Table 5.13, where the energy width of the N recoils from the back edge of the nitride layer is actually sharper than that of the front edge. For example, the difference in E_d for the top 20 nm of the sample is ≈ 480 keV, whereas the difference in E_d for 20 nm of sample at a depth of 300 nm is ≈ 100 keV. This unusual condition is caused because the effective stopping power first increases and then decreases as a function of depth, as indicated in the table.

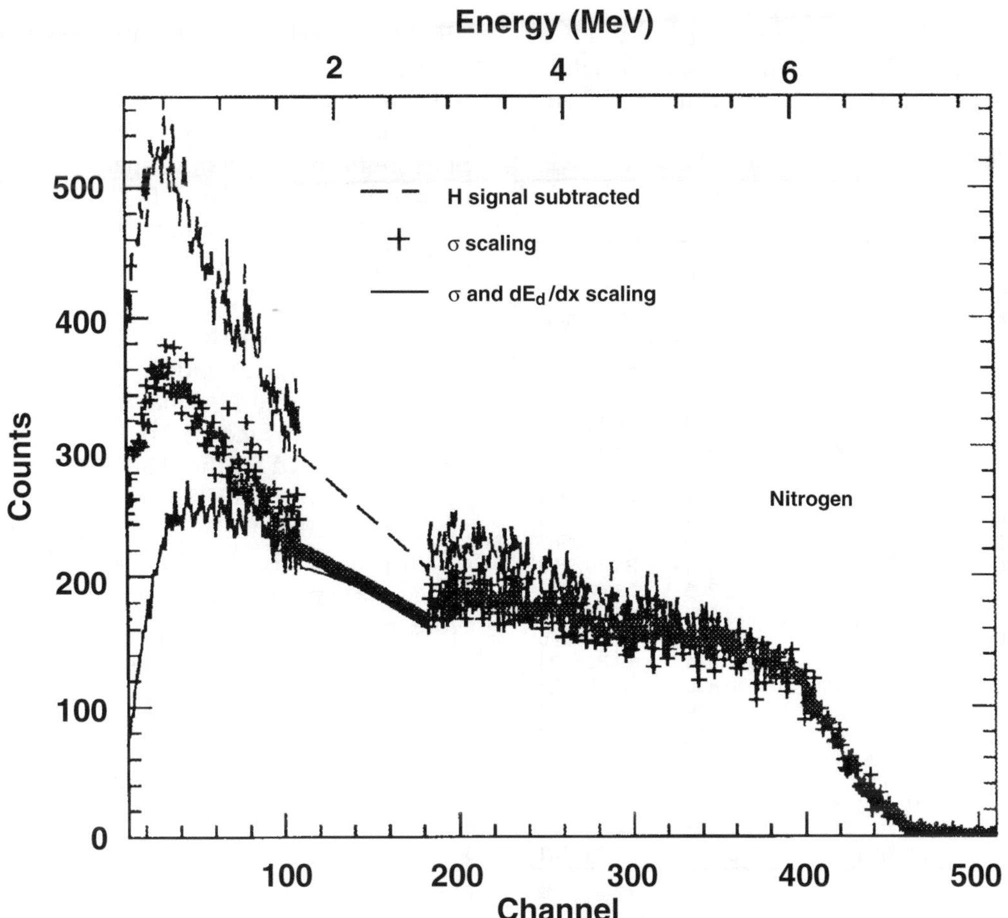

FIG. 5.24. Scaled N signal from the spectrum shown in Fig. 5.22. First the H signal was removed (dashed line), which left the artificial straight line section of the data between channels ≈100 to 200. Then the N data were scaled to make the cross section appear to be constant throughout the film and equal to σ at E_o (+ symbols). The σ-scaled data were then transformed to remove the effect of the effective stopping power dE_d/dx (solid line). The σ and dE_d/dx scalings have less effect on the N signal shown here than on the H signal shown in the previous figure.

This effect, which sharpens the back edge of the N signal, is due primarily to the range foil, and the fact that the stopping-power ratio for the range foil decreases quickly from unity

$$\left(\prod_n (\text{range foil}) = \frac{S_r^{(o)}(E_d)}{S_r^{(o)}\left(E_3^{(1)}\right)} < 1\right) \text{for } X^{(n)} \geq 220 \text{ nm} \;.$$

Before proceeding to the application of the concentration and depth scales for the H and N signals, consideration of the range-foil slab thickness should be addressed. A common mistake made in slab analyses is choosing range-foil slab layers which are too thick for an accurate determination of E_d and dE_d/dx. This mistake often results from inadvertently assuming the stopping power in the range foil increases linearly across the foil because the stopping-power ratio for the foil only depends on the stopping powers at the entrance and exit energies, $E_3^{(1)}$ and E_d. A similar assumption is used for calculating the interpolated stopping powers in the sample from the tables above, and this interpolation method is acceptable for sample layers which are very thin. However, a full multilayer slab analysis of the foil is needed to accurately determine the exit energy, E_d, and thereby the stopping power at that energy. Table 5.14 shows the error obtained when only one range-foil layer, 12-µm-thick, is used for calculating E_d and dE_d/dx (in comparison to Table 5.13 which uses 12 range-foil layers). For this sample, which is approximately 300-nm-thick, the errors in E_d and dE_d/dx increase with depth to more than 130% error at the back edge of the Si_3N_4 layer, as a result of choosing only one slab layer for the range foil. In general, a verification that the range-foil slab layer thickness does not significantly affect the analysis should be done by decreasing the slab layer thickness by half, and then recalculate the interpolation table to ensure that $E_o^{(n)}$, E_d, and dE_d/dx remain unchanged.

Table 5.13. Interpolation table for recoiled N from Si_3N_4 (using 12 layers for range-foil).

Slab no.	Depth *(nm)*	$E_o^{(n)}$ *(keV)*	E_d *(keV)*	dE_d/dx *(keV/nm)*
Surface	0	28000.000	6566.707	23.942120
1	20	27632.824	6086.800	24.048507
2	40	27265.240	5604.984	24.133084
3	60	26897.258	5121.685	24.196899
4	80	26528.889	4637.444	24.227126
5	100	26160.143	4152.970	24.220335
6	120	25791.033	3669.181	24.158523
7	140	25421.574	3187.290	24.030588
8	160	25051.777	2709.011	23.797266
9	180	24681.656	2236.866	23.417261
10	200	24311.229	1774.881	22.781236
11	220	23940.508	1329.973	21.709549
12	240	23569.512	914.614	19.826424
13	260	23198.256	550.830	16.551921
14	280	22826.760	271.684	11.362727
15	300	22455.041	106.714	5.134274
16	320	22083.121	43.870	1.150135
17	340	21711.020	24.300	0.806824
18	360	21338.758	13.716	0.251536
19	380	20966.359	7.291	0.391015
20	400	20593.848	3.951	− 0.057061

5.3.4.3 Depth profiles

The concentration and depth scales can now be applied to the H and N signals, separately, using Tables 5.12 and 5.13 and Eq. (5.24). The following calculated values were used for the H and N spectra: $\sigma_o(H) = 1678$ mb/sr, $S_o^{eff}(H) = 2.0683$ keV/nm, $\sigma_o(N) = 880$ mb/sr, and $S_o^{eff}(N) = 23.942$ keV/nm. The N and H concentration-depth profiles are thereby obtained directly from the scaled ERD spectrum and shown together in Fig. 5.25. Note that the depth scale given in Fig. 5.25 increases to the left, in units of nanometers. The N-concentration scale is shown along the left axis in density inclusive units, N/cm^3, and the H-concentration scale (which is not equal to the N-concentration scale) is shown along the right axis in units of H/cm^3. This figure also gives a check on the accuracy of the scaled nitrogen stopping power in the mylar range foil. Independently, the H and

Table 5.14. Interpolation table for recoiled N from Si_3N_4 (using 1 layer for range-foil).

Slab no.	Depth *(nm)*	$E_o^{(n)}$ *(keV)*	E_d *(keV)*	dE_d/dx *(keV/nm)*	Error (E_d) *(percent)*	Error (dE_d/dx) *(percent)*
Surface	0	28000.000	7844.692	23.469965	18	−2
1	20	27632.824	7385.229	22.476323	19	−7
2	40	27265.240	6922.469	23.799703	21	−1
3	60	26897.258	6456.356	22.811530	23	−6
4	80	26528.889	5986.857	24.138372	25	0
5	100	26160.143	5513.923	23.155086	28	−4
6	120	25791.033	5037.528	24.484417	31	1
7	140	25421.574	4557.625	23.505867	35	−2
8	160	25051.777	4074.192	24.837468	40	4
9	180	24681.656	3587.210	23.860700	46	2
10	200	24311.229	3096.668	25.193498	54	10
11	220	23940.508	2602.559	24.217389	65	11
12	240	23569.512	2104.895	25.549066	79	25
13	260	23198.256	1603.699	24.570501	98	39
14	280	22826.760	1099.003	25.899055	121	78
15	300	22455.041	590.874	24.913909	139	132
16	320	22083.121	79.394	26.234019	58	183
17	340	21711.020	− 435.325	25.237892	224	188
18	360	21338.758	−953.147	26.544292	206	196
19	380	20966.359	−1473.873	25.528357	202	194
20	400	20593.848	−1997.275	26.811864	201	201

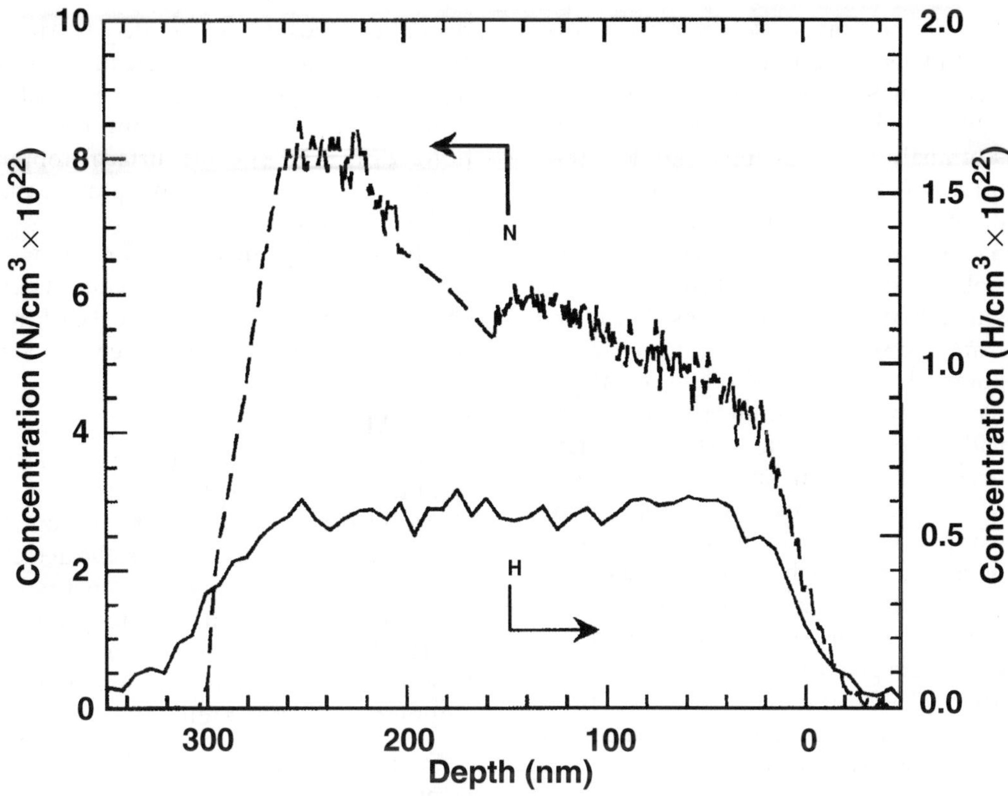

FIG. 5.25. The H (solid line) and N (dashed line) concentration-depth profiles determined from the scaled data of Figs. 5.23 and 5.24. The H concentration axis is shown at right and the N at left. Note that the back N-edge appears sharper than the front edge. See text for details.

N depth profiles yield approximately the same Si_3N_4 layer thickness, and therefore the scaling factor used for the N stopping power is acceptable for the depth scale.

The uncertainty in the N stopping power may, however, cause inaccurate scaling of the N concentration at depth. The N profile in Fig. 5.25 should remain constant with depth but appears to increase dramatically beyond 160 nm. This increase may result from inaccurately calculating the N stopping power or from multiple scattering in the sample. Further work is needed to accurately determine the N stopping power in mylar for the energy range from 1 to 7 MeV N. Several other features of the N profile are also of interest. First, as discussed above, the back edge of the N profile is sharper than the front edge because the stopping-power ratio for N in the range foil decreases quickly from unity, and the sharp drop at a depth of 300 nm results from the practical depth of analysis for N when using a 28 MeV Si projectile. If the practical minimum E_d in a surface barrier detector is 50 keV, then Table 5.13 shows the practical depth of analysis for nitrogen in Si_3N_4 is only 300 nm. In comparison, the practical depth of analysis for H in Si_3N_4 is 960 nm, when using a 28 MeV Si projectile. The practical depths of analysis for ERD performed using a surface barrier detector with a range foil will be explored in more detail in the next section on ERD capabilities.

5.4 MASS RESOLUTION, SENSITIVITY, DEPTH RESOLUTION, ANALYSIS DEPTH, AND HOW TO CHOOSE A RANGE FOIL

In previous sections, the formalism for ERD analysis was developed and examples were given for typical ERD analyses using high-energy Si and He ion beams in a reflection geometry with range foils in front of the detector. These examples showed the many complicating factors which must be taken into account when calculating a concentration profile from an ERD spectrum. The following sections will examine the system and sample parameters which affect the mass resolution, sensitivity, depth resolution, and depth of analysis. Most ERD analyses are performed using surface barrier detectors with range foils and the resolution is typically 20 to 60 nm. Typical depths of analysis in Si samples are less than 1 μm, but up to 1.4 μm of H in Si can be profiled with a depth resolution of ≅ 30-40 nm using 28 MeV ^{28}Si or 35 MeV ^{35}Cl and a scattering angle of 30°. Also, the hydrogen content in polymer films of up to 1 μm thickness can be analyzed using 2.8 MeV ^{4}He ion beams. Turos and Meyer (1984) examined several factors which influence the sensitivity, depth resolution, and depth of analysis for the specific case of hydrogen depth profiling using a 2.5 MeV ^{4}He incident beam. They showed that the maximum depth of analysis (≅ 600 nm) and greatest sensitivity were reached for $\phi = 30°$. However, the depth resolution (≅ 70 nm) was poorer at $\phi = 30°$ than at $\phi = 4°$, where the best depth resolution was achieved and equal to 23 nm. Similar to RBS analysis, a tradeoff between depth resolution and sensitivity also exists in ERD analysis. The best sensitivity for hydrogen using an ERD analysis was shown by Nagai *et al.* (1987) to be $10^{12}/cm^2$ or 1 wt. ppm, but a typical value for the sensitivity of most detectable elements in ERD analysis is 0.1 at. %.

In order to gain maximum depth of analysis, best depth resolution, and highest mass separation for ERD, the optimal combination of geometry, ion beam, and energy must be determined. An *a priori* computation to optimize these parameters is made difficult primarily because of two factors: (1) the use of a range foil and (2) the fact that the detected beam may be composed of many different atoms (with different stopping powers which also differ from that of the incident beam). Therefore, several guidelines will be given to help with this optimization and to demonstrate the tradeoffs incurred when varying the different parameters. Also, examples of transmission ERD for increased sensitivity and time-of-flight ERD (TOF-ERD) for increased mass and depth resolution will be given in later sections. Both of these techniques can avoid the complication of using range foils. In fact, Barbour *et al.* (1991) showed that TOF-ERD yields a depth resolution of 1.5 nm in the O signal at the surface for a SiO_xN_y film, and a depth resolution of 2.0 nm was reported in 1986 by Gossett using a magnetic spectrometer for ERD analysis of C in a graphite matrix.

The system parameters which influence the limits to sensitivity, depth resolution, and depth of analysis are often controllable, but sample characteristics may also affect these limits and are generally uncontrollable. System parameters which influence the sensitivity, depth resolution, and analysis depth of ERD include type of ion beam, projectile energy, angle of incidence and scattering angle, and type of range foil and its thickness. Several sample characteristics which influence the depth resolution are smoothness of the surface, stopping power of the incident and recoiled ions in the target and the range foil, and straggling of recoiled ions in the range foil. Note well that the stopping powers and straggling are also directly affected by the type of ion beam, its energy, and the choice of range foil. The influence of these various parameters will now be considered in detail.

5.4.1 Mass resolution

In conventional ERD with the use of a range foil, mass resolution is the ability to separate the signal from different recoiled atoms in the spectrum along the energy axis. For scattering from the surface, the detected energy for a recoil atom depends upon the energy imparted to that atom and its stopping in the range foil. Consequently, the selectivity for different masses is determined by the kinematic factor, K, given in Eq. (5.2), and the effect of stopping in the range foil.

First consider the mass resolution dependence on K. K is much less than $\cos^2\phi$ for $M_1 \gg M_2$, and K equals $\cos^2\phi$ for $M_1 = M_2$. Therefore, in the absence of a range foil, the detected energy would increase continuously for increasing M_2 ($\leq M_1$) and the mass resolution of the recoil atoms could be improved by increasing M_1 (i.e.,

$$\left.\frac{\partial K}{\partial M_2}\right|_{M_1}$$

approaches zero as M_1 decreases toward M_2). The effect of continuously increasing E_d was demonstrated in Fig. 5.3 by the solid line for the theoretical detected energy when no mylar foil was present. Figure 5.3 also showed the detected energy for surface recoil atoms (scattered at 30° by a 24 MeV ^{35}Cl beam) transmitted through a 6 μm mylar foil, which is just thick enough to stop the scattered ^{35}Cl beam. Recoil atoms heavier than Li have a significant amount of their energy lost in the range foil, and therefore the curve for the detected energy as a function of recoil mass exhibits a maximum. The height and position of this maximum for a given incident projectile depend on ϕ and the range foil thickness.

Another example is shown in Fig. 5.26, where the detected energy of surface recoil atoms is plotted as a function of recoil mass for 1 MeV/amu projectiles of ^{4}He, ^{12}C, ^{28}Si, and ^{35}Cl. A 10 μm Al range foil was used for

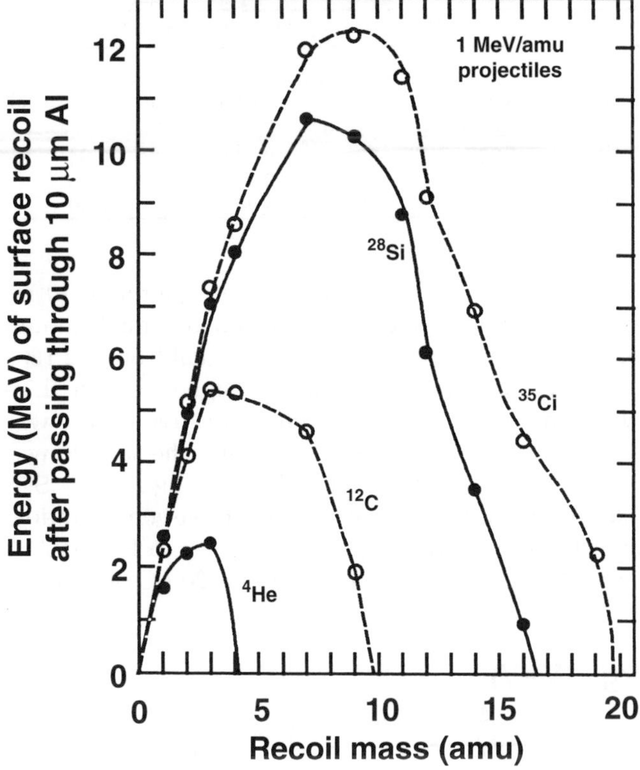

FIG. 5.26. A plot to help determine mass resolution for different beam-target combinations. Detected energies of surface-recoil atoms (ϕ=30°) as a function of recoil mass for different 1 MeV/amu projectiles: ^{4}He, ^{12}C, ^{28}Si, and ^{35}Cl. A 10 μm Al range foil was used for stopping the projectile beams. The data are shown as open and closed circles and the lines are drawn to guide the eye.

stopping the scattered incident beams, and the recoil atoms were collected at a scattering angle of 30°. The measured surface energy positions for the different recoil masses are shown as open and closed circles, and the lines are drawn to guide the eye. This figure shows that higher-mass, higher-energy projectiles yield higher-energy recoils which can then penetrate the stopping foil, often with greater mass resolution. For example, the mass resolution between H and D corresponds to an energy difference of approximately 0.8 MeV when using a 4 MeV ^{4}He projectile, but the mass resolution increases giving an energy difference of ≅ 2.5 MeV when using a 28 MeV ^{28}Si projectile. However, little mass resolution between H and D is gained by increasing the projectile mass and energy from 28 MeV ^{28}Si to 35 MeV ^{35}Cl. In some instances the mass resolution may actually be increased by decreasing M_1, because of the effect of the range foil. The energy separation between Li and Be, which are both near the peak of the curves for the Si and Cl incident beams, is greater on the 12 MeV ^{12}C curve than on either the 28 MeV ^{28}Si or 35 MeV ^{35}Cl curves. Finally, Fig. 5.26 shows that the use of higher mass projectiles gives measurability to elements like N, O, and F which are not detectable when using the lower mass projectiles, thereby directly increasing the sensitivity for these elements.

5.4.2 Sensitivity

One possible definition for sensitivity is a measure of the smallest detectable areal density (number of atoms per square centimeter). Equation (5.6) shows that the smallest detectable areal density depends on the smallest measurable ΣY = (Areal Density)(Ω Q σ)/cosΘ_1. In order to maximize the sensitivity, ΣY may be maximized by increasing Ω, Q, or σ, or by decreasing cosΘ_1 (i.e., increasing Θ_1). However, increasing Ω may also cause a reduction in depth resolution; yet if Ω is too small, then valuable beam current is wasted and sensitivity is lost without improving the depth resolution. In general, Ω should be as large as possible to increase sensitivity, but not so large that it degrades the depth resolution. Determination of the optimum solid angle is accomplished, experimentally, by varying the size of the defining aperture in front of the detector, collecting a spectrum for each size aperture, and establishing the largest aperture which causes no apparent decrease to the depth resolution in the spectrum. Increasing Q can also increase sensitivity, but possible ion-beam damage of the sample may occur and the background signal may also increase (which affects the counting statistics).

The cross section can be evaluated as a measure of the sensitivity for a fixed Ω and Q. In the approximation that $M_1 \gg M_2$ and $M \approx 2Z$, the cross section reduces to a function which depends upon the projectile atomic number, Z_1, the projectile energy, E_o, and the scattering angle, ϕ

$$\sigma_r\,(E_o,\phi) = \left(\frac{Z_1^2\,e^2}{2\,E_o}\right)^2 \frac{1}{\cos^3\phi}\ .$$

The dominant terms in the cross section which affect the sensitivity result from the dependence on Z_1 and E_o rather than the dependence on scattering geometry. The cross section increases with Z_1 as Z_1^4, and decreases with E_o as $1/E_o^2$. The cross section also increases with ϕ as $1/\cos^3\phi$, but the cosΘ_1 term in ΣY tends to cancel this increase. Table 5.15 shows that $1/\cos^3\phi$ increases by a factor of 1.85 as ϕ increases from 20° to 40°, but 1/(cosΘ_1

$\cos^3\phi$) decreases by 7% (assuming $\Theta_1 = \Theta_2$). The importance of the incidence angle and the scattering angle on the depth resolution and analysis depth will be discussed in later sections.

Table 5.15. The effect of scattering geometry on sensitivity.

ϕ	$1/\cos^3\phi$	Θ_1	$\cos\Theta_1$	$1/(\cos\Theta_1 \cos^3\phi)$
20	1.205	80	0.174	6.94
30	1.540	75	0.259	5.95
40	2.225	70	0.342	6.50

As a practical example for determining the sensitivity of ERD to hydrogen and carbon, consider the spectrum shown in Fig. 5.27. A surface hydrogen peak is centered at channel 140, a smaller H peak is sitting on the background from channel 108 to 124, and a surface carbon signal is at channel 335. The history of the sample corresponding to this spectrum is as follows. A silicon wafer was oxidized in dry O_2 to form a 57-nm-thick SiO_2 layer, a 200-nm-thick polycrystalline Si layer was deposited by CVD, and then a 600-nm-thick SiO_2 layer was deposited by pyrolysis of tetraethylorthosilicate (TEOS). The TEOS layer was densified at 850°C for 30 minutes in a flowing steam atmosphere. Interest in this sample evolved around whether H reached the vicinity of the bottom thin-oxide layer. ERD analysis (not shown) revealed that the top TEOS-oxide layer had a small amount of H (0.1 at. %), but the level of H signal from the TEOS could obscure a possible smaller H signal near the thinner 57-nm-thick oxide. Therefore, the TEOS layer was stripped in a buffered oxide etch and the sample was analyzed for the H content at depth using 24 MeV Si^{+5} ERD. A schematic drawing of this sample configuration, after the top-oxide etch, is shown inset in Fig. 5.27. A 35 nA beam was used to collect Q = 20 µC of incident

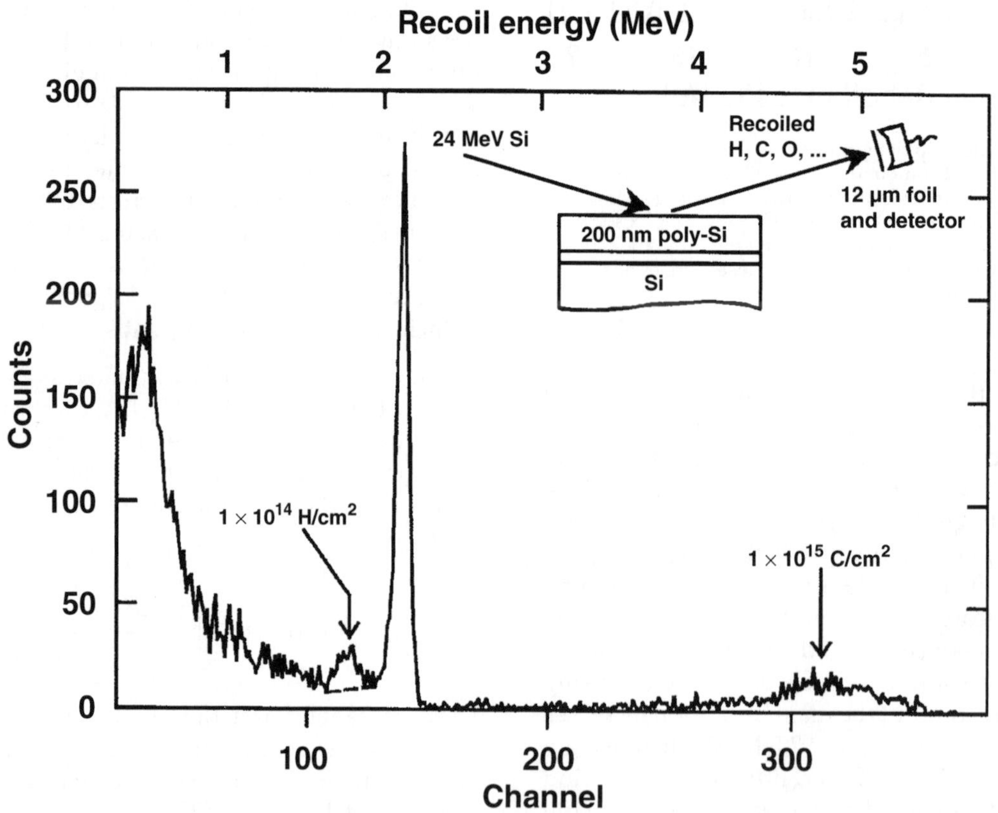

FIG. 5.27. An example ERD analysis to help estimate the sensitivity of ERD to different light elements. This ERD spectrum was taken from a sample and under the analysis conditions indicated in the inset. Two H signals are of interest: the signal at depth in the sample, indicated near channel 120; and the surface signal at channel 140. The C peak is a result of contamination (either from air or from wet chemistry). The peak near channel 20 is oxygen as a result of surface oxidation.

charge. A 12-µm mylar range foil was used in this analysis with $\phi = 30°$, $\Theta_1 = 75°$, and $\Omega = 6.1$ msr. For a given spectrum, $\cos\Theta_1/(Q\,\Omega)$ is constant. Therefore, the sensitivity for C in this spectrum can be compared to the sensitivity for H by scaling with the cross sections. In fact, the sensitivities for other elements in this type sample can be scaled from the estimated H sensitivity, determined experimentally from this spectrum.

At depth, the integrated number of counts in the H signal above background is $\Sigma Y_H = 185$, which corresponds to an areal density of 1×10^{14} H/cm^2. The total number of counts including background is $\Sigma Y_{Tot} = 346$. The standard deviation in ΣY_H is S_H, which depends on the standard deviation in total counts [$S_{Tot} = (\Sigma Y_{Tot})^{1/2} = \sqrt{346}$] and the standard deviation in the background counts [$S_{Bkg} = (\Sigma Y_{Bkg})^{1/2} = \sqrt{161}$]. The percent standard deviation ($\%S_H$) is (Bird and Williams, 1989): $100\%\ (S_H/\Sigma Y_H) = 100\%[(S_{Tot})^2 + (S_{Bkg})^2]^{1/2}/\Sigma Y_H = 100\%\ [346+161]^{1/2}/185 = 12\%$. An estimate of the H sensitivity at depth is given when the size of the standard deviation approaches the integrated number of counts above background, $\%S_H = 100\%$. (See Chapter 11 for a discussion of counting statistics and error propagation.)

If only 1×10^{13} H/cm^2 were in this sample, the signal would be only 10% of that shown in Fig. 5.27, but it would be just distinguishable from background, and of course S_H would increase tremendously: $\%S_H = 100\%\ (180+161)^{1/2}/18.5 = 100\%$. Thus the estimated sensitivity for H (ε_H) is 1×10^{13} H/cm^2. The carbon signal in Fig. 5.27 comes from a surface contaminant which was only found after the top-oxide etch. The integrated count in the C signal is 728, without any visible background (counting statistics: $\%S_C = 4\%$). This signal gives an areal density of 1×10^{15} C/cm^2. However, if only a few percent of this contaminant were present, it would still be detectable (counting statistics for 15 counts in C peak: $\%S_C = 26\%$), giving an estimated sensitivity for the carbon (ε_C) of 2×10^{13} C/cm^2. Note that this value for ε_C, estimated from the spectrum, is approximately equal to the sensitivity if it were calculated from scaling the measured H sensitivity by the relative cross sections (i.e., $\varepsilon_C \cong \varepsilon_H\sigma_H/\sigma_C$).

The measured sensitivities for H and C (solid squares) are shown in Fig. 5.28 in comparison to the calculated sensitivities (solid line) for recoil atoms up to oxygen. The solid-line calculation is based on scaling relative to the measured H sensitivity for 24 MeV Si ERD and a scattering angle of 30°. The calculated sensitivities were determined using the following assumptions: the minimum H areal density is 1×10^{13} /cm^2, the minimum ΣY is similar for each element, the cross sections obey the Rutherford formula [Eq. (5.4)], and the minimum detectable areal density scales with $1/\sigma$. In addition, calculations are shown in this figure for minimum detectable areal densities where Z_1 was increased to 17 for 24 MeV Cl ERD (small-dashed line), and where E_o was lowered to 12 MeV for Si ERD (dash-diamond). The greater increase in sensitivity arises from decreasing the energy by a factor of 2, although the depth of analysis decreases from 1.12 µm to 0.3 µm.

5.4.3 Depth resolution

Not only is the sensitivity improved by using lower energy projectiles, but the depth resolution can also be improved. Depth resolution is the ability to separate, in the spectrum along the energy axis, the signal coming from recoiled atoms at different depths in the sample. Therefore, the smallest resolvable detected energy, δE_d, determines the smallest resolvable depth interval δx

$$\delta x = \frac{\delta E_d}{dE_d / dx} \ . \tag{5.28}$$

Experimentally, the energy width δE_d is taken from the energies corresponding to the 12% and 88% of full signal height for an abrupt change in sample composition (e.g., at the sample surface). Theoretically, Eq. (5.28) is simple, but determining the theoretical δE_d can be complicated because δE_d is affected by detector resolution, straggling in the range foil and sample, nonuniformities in the range-foil thickness, sample roughness, and kinematics.

For a single layer target, and in the Gaussian approximation, the various contributions to the energy resolution can be added in quadrature

$$(\delta E_d)^2 = (\Delta E_d)^2_{det} + (\delta E_d)^2_{\substack{foil\\str}} + \left[S_r^{(o)}(E_d)\,\delta x^{(o)}\right]^2 + \left[\frac{S_r^{(o)}(E_d)}{S_r^{(o)}(E_3)}\right]^2 (\delta E_3)^2 \ . \tag{5.29}$$

The first and second terms in Eq. (5.29) result from the limited detector resolution and from straggling within the range foil. These terms are more difficult to control experimentally, and in conventional ERD analysis (with a solid state detector and range foil) the effect of detector resolution and foil straggling are often smaller than the effect from the range-foil thickness nonuniformity. The third term on the right side of Eq. (5.29) results from range-foil thickness nonuniformities, and this term can

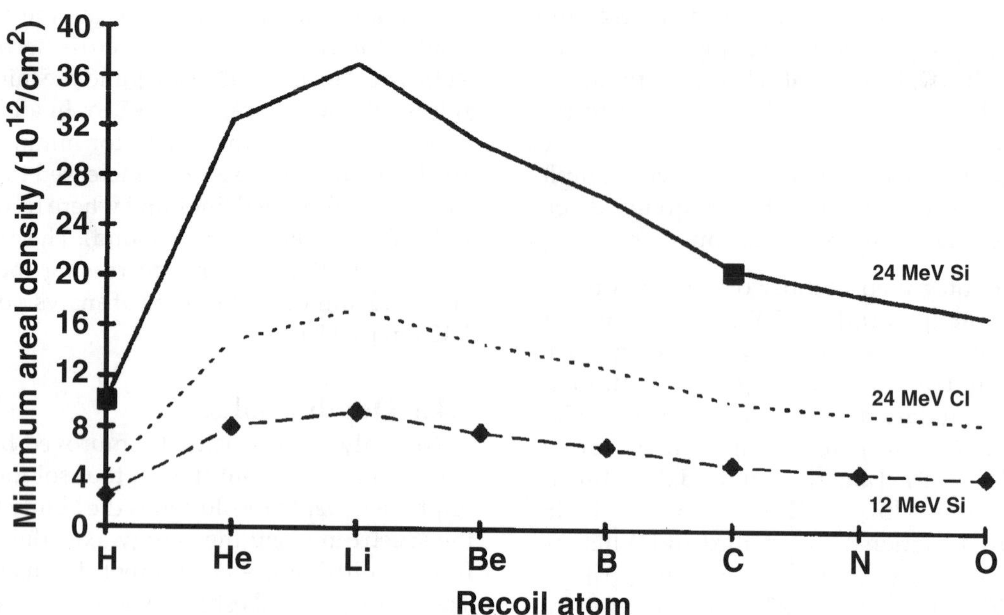

FIG. 5.28. The minimum-detectable areal density is a measure of sensitivity, shown here for recoil atoms from H to O. The ■ symbols are data determined from the ERD spectrum shown in Fig. 5.27. The solid line was calculated assuming the minimum detectable areal density scales with $1/\sigma$ and using the measured data point for H as a fixed reference. The calculated sensitivity for the 24 MeV Si beam fits to the C data point well. The other two calculated lines show a marked increase in sensitivity assuming that only the incident beam species was changed (---) or the incident energy was changed (-◆-), as indicated.

be the single most important contributor to depth resolution using conventional ERD. This effect of thickness nonuniformity on depth resolution is particularly true for the analysis of elements with $3 \leq Z \leq 9$ where over half of the energy for these elements is lost in the range foil. Mylar and other polymers are commonly used range foil materials because of their superior thickness uniformity.

The final term in Eq. (5.29) results from the uncertainty in the energy of the recoils just prior to entering the foil, and $(\delta E_3)^2$ can be evaluated as

$$(\delta E_3)^2 = (\delta E_3)^2_{\substack{sam\\str}} + \left[\frac{S_r^{(1)}(E_3)\,\delta x^{(1)}}{\cos\Theta_2}\right]^2 + \left[\frac{S_r^{(1)}(E_3)\,x^{(1)}\,\delta(\cos\Theta_2)}{\cos^2\Theta_2}\right]^2 + \left[\frac{S_r^{(1)}(E_3)}{S_r^{(1)}(E_2)}\right]^2 (\delta E_2)^2 \ . \qquad (5.30)$$

The first term on the right of Eq. (5.30) results from recoil-energy straggling, the second from sample thickness nonuniformities, the third from path length variation due to the angular acceptance width, and the last term from uncertainties in the initial outbound energy, E_2. Again, for a given projectile energy, straggling in the target is difficult to control, but the term arising from thickness nonuniformity may be controlled by preparing samples with minimum surface roughness prior to ERD analysis. The effect of sample roughness on δx tends to be more important for ERD than for RBS because of the extremely large incident and exit angles used in ERD.

The resolution in E_2 can be expressed in terms of the kinematic broadening (δK) and the uncertainty in the initial inward energy $E_o^{(1)}$

$$(\delta E_2)^2 = (E_o^{(1)}\,\delta K)^2 + (K\,\delta E_o^{(1)})^2 \qquad (5.31)$$

where

$$\left(\delta E_o^{(1)}\right)^2 = \left(\delta E_o^{(1)}\right)^2_{\substack{sam\\str}} + \left[\frac{S_p^{(1)}\left(E_o^{(1)}\right)\delta x^{(1)}}{\cos\Theta_1}\right]^2 + \left[\frac{S_p^{(1)}\left(E_o^{(1)}\right)x^{(1)}\,\delta(\cos\Theta_1)}{\cos^2\Theta_1}\right]^2 + \left[\frac{S_p^{(1)}\left(E_o^{(1)}\right)}{S_p^{(1)}(E_o)}\right]^2(\delta E_o)^2 \,. \tag{5.32}$$

The first term on the right of Eq. (5.32) results from projectile-energy straggling, the second from sample thickness nonuniformities, the third from path length variation due to a finite acceptance angle, while the last is the uncertainty in the incident energy. Highly collimated incident beams should be used to minimize the loss of resolution caused by angular spread shown in Eqs. (5.30) and (5.32). However, improvements in δx are rarely realized for collimation to angles less than the $\delta\phi$ accepted by the detector, particularly if equal angles of incidence and exit are used.

For smooth samples and a uniformly thick range foil, the kinematic broadening term can dominate the energy resolution in Eqs. (5.29) - (5.31). In this case, δE_d reduces to

$$\delta E_d \cong \frac{S_r^{(o)}(E_d)\,S_r^{(1)}(E_3)}{S_r^{(o)}(E_3)\,S_r^{(1)}(E_2)}\,E_o^{(1)}\,\delta K$$

and using Eq. (5.18) for dE_d/dx, $\delta x = E_o^{(1)}\,\delta K/[S]_{p,r}$. The depth resolution at the surface, as a result of kinematic broadening, can then be evaluated by determining the resolution in K, δK. The geometry of the scattering process is shown in the upper portion of Fig. 5.29 in which the recoiled atoms are scattered at a nominal angle ϕ relative to the incident projectile direction such that $K = 4M_1M_2\cos^2\phi/(M_1+M_2)^2$. Kinematic broadening can result because of the finite size of the detector aperture, which has an opening of size w at a distance z from the sample. The angle ϕ defines a cone of possible scattering directions and constant recoil energy, $KE_o^{(1)}$, where the incident projectile direction is the cone axis. A small variation, $d\phi \cong w/z$, in the scattering angle causes a variation in K such that $dK = 2K\tan\phi\,d\phi$, for fixed M_1 and M_2. If the detector aperture in plane AB is made to exactly match the shape of the conic section projected onto plane AB (a curved elliptically shaped slit), then $\delta\phi = d\phi$ and $\delta K = dK$. (The aperture is assumed to lie half above and half below the plane of the paper.) In this case, the depth resolution at the surface of the sample is $\delta x \cong 2KE_o w\tan\phi/(z[S]_{p,r})$. However, the most commonly used and most simple aperture to make is a rectangular slit with width w and length h, which then gives $\delta\phi > d\phi$ because the corners of the aperture permit collection of some recoiled atoms at angles greater than $\phi+d\phi/2$.

The example of the rectangular slit can be understood more easily by mapping the slit (in the plane AB, perpendicular to z) onto the basal plane of the cone (i.e., along CD). This mapping is shown in the bottom portion of Fig. 5.29, along with the cone radii corresponding to the possible angles ϕ, $\phi \pm d\phi/2$, and $\phi+(d\phi/2)+\Delta d\phi$. In the cone basal plane, the projected length of the slit remains equal to h but the projected width of the slit increases to $w/\cos\phi$ (for small $d\phi/2$). The angle $\Delta d\phi$ is the difference between $\delta\phi$ and $d\phi$, and corresponds to the shaded area (see Fig. 5.29) which allows the collection of additional recoil atoms with smaller energies. The experimental energy resolution stated above is based upon measuring the energy difference between the yield at 12% and 88% of the full spectrum height. But for simplicity in determining the theoretical energy resolution, the full difference in energy (0% yield to 100% yield) will be considered here, corresponding to a full difference in angle $\delta\phi$. (The reader is referred to Brice and Doyle [1990] for a more detailed treatment where the angular dependence of the cross section and the calculation of the standard deviation are folded into a determination of δE from the energies at 12% and 88% of the full spectrum yield.) The full difference in allowed K defined by the rectangular slit is given by

$$\delta K = \frac{4M_1M_2}{(M_1+M_2)^2}\times\left[\cos^2\left(\phi - \frac{d\phi}{2}\right) - \cos^2\left(\phi + \frac{d\phi}{2} + \Delta d\phi\right)\right]$$

$$\cong 2K\tan\phi\,(d\phi + \Delta d\phi) = 2K\tan\phi\;\delta\phi$$

$$\cong 2K\tan\phi\left[\frac{w}{z} + \frac{1}{2}\sqrt{\left(\sin 2\phi + \frac{w}{z}\right)^2 + \left(\frac{h\cos\phi}{z}\right)^2} - \frac{1}{2}\left(\sin 2\phi + \frac{w}{z}\right)\right]$$

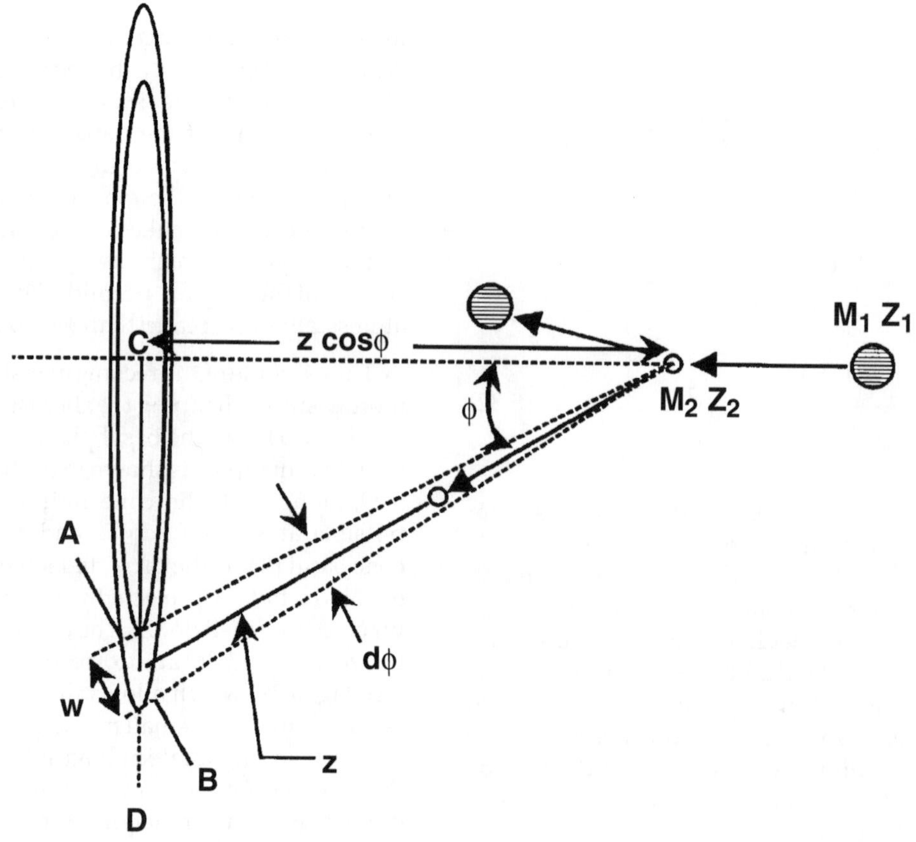

Map rectangular slit onto cone plane

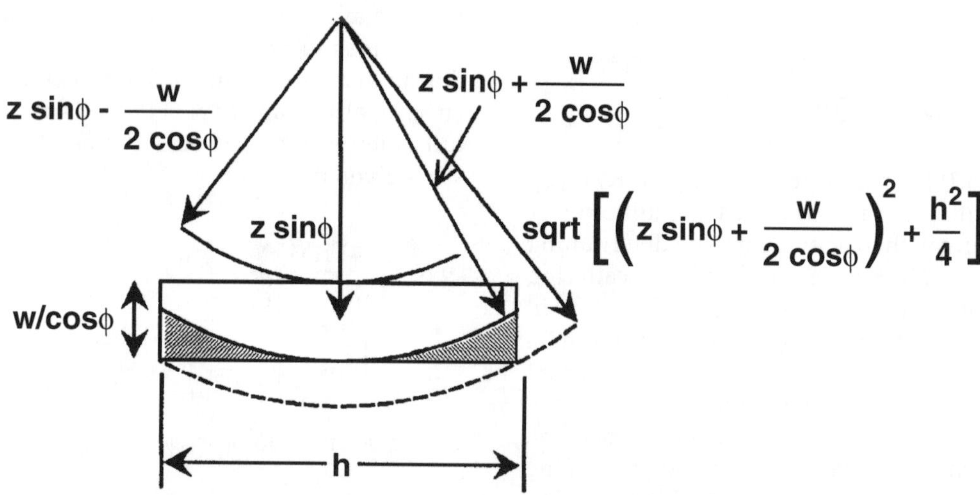

FIG. 5.29. The scattering geometry forms a cone of constant K for a given nominal scattering angle ϕ. Depth resolution in ERD can be affected by kinematic broadening which results from a finite acceptance angle (dϕ) for the detector aperture of size w at AB. If the detector aperture is a rectangular slit at a distance z from the sample, then its projection onto the cone basal plane will appear as shown at bottom. The shaded area corresponds to scattering angles greater than ϕ+(dϕ/2), and this extra area allows even lower energy recoil atoms than those defined by the cone at top to be detected.

which gives, for the depth resolution at the surface of the sample

$$\delta x \cong E_o K \tan\phi \times \frac{\sqrt{\left(\sin 2\phi + \frac{w}{z}\right)^2 + \left(\frac{h\cos\phi}{z}\right)^2} - \left(\sin 2\phi - \frac{w}{z}\right)}{[S]_{p,r}} . \quad (5.33)$$

Consider, as a practical example, determination of the optimum length and width of a rectangular detector aperture for a system with a fixed sample-detector distance. The optimum solution (Doyle and Peercy, 1979a) is to minimize δx resulting from kinematic broadening, given in Eq. (5.33), while maintaining the same sensitivity (i.e., a constant rectangular area). Energy spread of the recoiled beam is minimized for the detection condition given by

$$h = 2\sqrt{zw\tan\phi + \frac{3w^2}{4\cos^2\phi}} \cong 2\sqrt{zw\tan\phi} \quad (5.34)$$

The energy resolution is improved by ~50% when rectangular slits are replaced by curved slits with the same area, and in the case of curved slits the optimum length of the slits as a function of w and z is (Brice and Doyle, 1990)

$$h = 4\sqrt{zw\tan\phi} . \quad (5.35)$$

Figure 5.30 shows the depth resolution at the surface of a sample for H recoils through N recoils as a result of kinematic broadening for a system with a rectangular defining aperture under the constraint of a fixed rectangular area ($hw = 11\ mm^2$). The experimentally measured values of the depth resolution for H and N (■) were determined from a 300-nm-thick Si_3N_4 sample with approximately 6 at. % H throughout the film. A 24 MeV Si projectile was used for the analysis with a 12-µm-thick mylar range foil, a nominal scattering angle of 30°, z = 50 mm, w = 1 mm, h = 11 mm, $d\phi \cong 20$ mrad, and $\Delta d\phi \cong 10$ mrad. The theoretical δx (solid line) given by Eq. (5.33) is in good agreement with the measured values for these ERD conditions, but δx may be improved by using a heavier projectile such as a 24 MeV Cl projectile (dashed line). Theoretically, further improvement in the kinematic broadening of δx can be achieved, when using a Si projectile, by decreasing the scattering angle to 20° (-☒-) or by decreasing E_o to 12 MeV (-◆-). Both of these approaches increase the sensitivity, but the depth of analysis is decreased and a different thickness range foil would be required to detect the recoil atoms with Z > 2.

5.4.4 How to choose a range foil

The optimum thickness and type of stopping foil used in conventional ERD analyses depends on both the range of the projectile in the foil and the range of the recoiled atom in the foil. The stopping foil must be uniformly thick, free of pin-holes, and its thickness must be greater than or equal to the projectile range in the foil and less than the recoil range in the foil. Tables A9.3(a) and A9.3(b) are guides for determining the thickness of mylar and aluminum needed to stop different projectiles at energies commonly used in ERD analysis. The projectiles were assumed to be incident normal to the mylar foil. These tables show that a 12-µm-thick mylar foil or a 10-µm-thick Al foil may be used for stopping several different scattered projectile beams: 2.8 MeV He, 24-28 MeV Si, and 30-36 MeV Cl. Tables A9.4(a) and A9.4(b) can be used to help determine the ability of the recoiled atom to get through the range foil with sufficient energy to be detected. These latter tables give the range in mylar and Al of possible recoiled atoms. The recoil-energy ranges correspond to the possible energies for atoms coming from the surface or at depth from a sample analyzed with the incident projectiles given in Tables A9.3(a) and A9.3(b). However, a full slab analysis would more accurately determine the ability of the recoil atom to exit the range foil with sufficient energy for detection.

5.4.5 Depth of analysis

The depth of analysis in conventional ERD (using a range foil in front of a solid state detector) depends on the range of the incident projectile in the sample, the stopping of the scattered recoil atom in the sample, and the stopping of the recoil in the range foil. The range of the projectile in the sample can be estimated using a computer code, such as the RSTOP program of Ziegler (1987), but the incident angle should be taken into account when determining the projected range, R_p (measured normal to the sample surface). Typically, the projectile energy is chosen such that the sample depth is less than $1/2\ R_p$. Table A9.5 is given in the appendices to help quickly estimate the range of several different projectiles (He, Si, Cl, and Au) incident upon the following samples at Θ_1=75°: C, Si, Fe, Ge, Sn, and Pb.

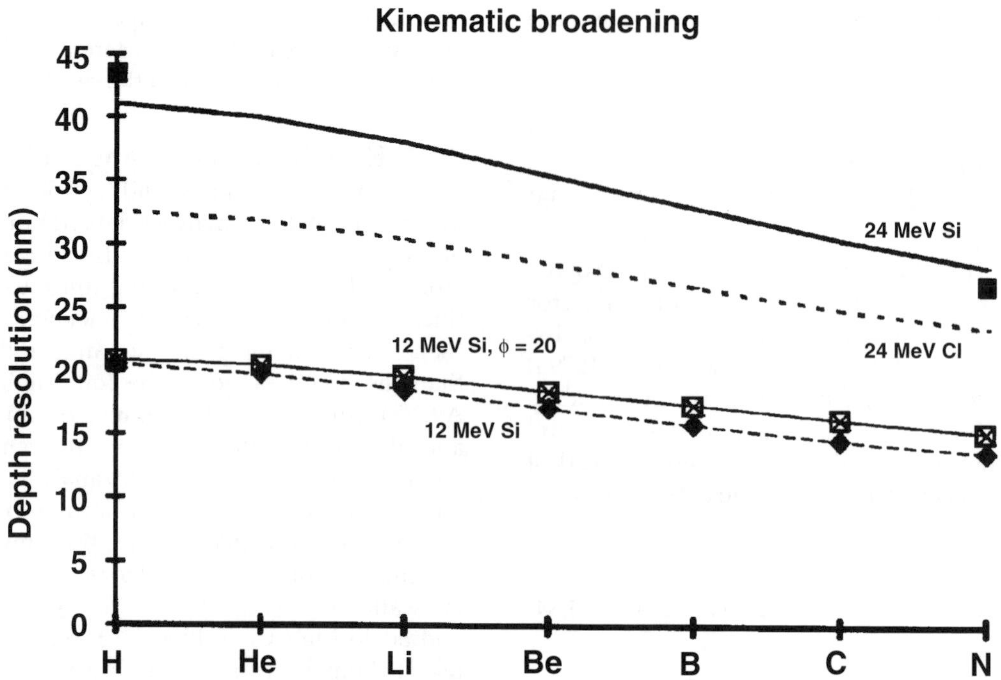

FIG. 5.30. Depth resolution resulting from kinematic broadening due to the finite rectangular-detector-aperture area: height = 11 mm and width = 1 mm. The ■ symbols are measured data for H and N, determined from the front edges of a Si_3N_4 sample, using 24 MeV Si ERD. The solid line was calculated using Eq. (5.33) for 24 MeV Si ERD with z = 50 mm and ϕ= 30°. The dashed line was calculated assuming similar conditions, but for a 24 MeV Cl projectile. The line marked -☒- was calculated for 24 MeV Si with ϕ=20°, and the line marked -◆- was calculated for 12 MeV Si with ϕ=30°. The last two calculations show that the depth resolution can be increased substantially over that obtained in the experimental measurement, but the depth of analysis would also decrease under these conditions.

Table A9.5 shows that R_p is often greater than 1 μm, but also the depth of analysis may be dominated by the stopping of the scattered recoil atom in the sample and in the range foil. In this case, the analysis depth may be much less than R_p. For example, R_p for 28 MeV Si in silicon is 2.67 μm. The practical depth of analysis for H in silicon using 28 MeV Si is 1.4 μm, but the practical depth of analysis for N in silicon using 28 MeV Si is only 0.4 μm [see Appendix A9.6(a)]. A practical limit to the analysis depth can be estimated from the minimum detectable energy, which for a surface barrier detector is ≈ 50 keV. Therefore, Tables A9.6(a) and A9.6(b), which present slab analyses using the computer code SERDAP (Barbour, 1994) to determine E_d, can be used to estimate the practical depth of analysis for several projectile-target combinations: projectiles – 2.8 MeV He, 28 MeV Si, 35 MeV Cl, and 12 MeV Au; recoils – H, C, and N; and samples – C, Si, Fe, Ge.

5.5 TRANSMISSION ERD

The previous sections of this chapter have concentrated on conventional ERD analysis geometries, but another possible geometry for ERD is the transmission mode. For transmission ERD, the sample must be thinner than the range of the recoiled atom to be profiled. The projectile beam usually impinges on the sample at or near normal incidence, and the detector is placed at a recoil scattering angle of 0°. Additional range foils may be used to stop the high intensity projectile beam, or the sample itself may be sufficient to stop the projectile. The main advantage of transmission ERD is increased sensitivity, by as much as two orders of magnitude in comparison to conventional reflection-geometry ERD. This increase in sensitivity stems from several factors: (1) since the analysis range scales with $\cos\Theta_1$, more material is probed with transmission ERD; (2) larger solid angles can be used without significant kinematic broadening in transmission ERD since $dK/d\phi=0$ at $\phi=0°$; and finally, (3) the background caused by surface hydrogen is reduced considerably in the transmission mode due to smaller multiple scattering cross sections. Wielunski *et al.* (1986) have demonstrated a 1 at. ppm sensitivity for detecting H in Ni using 4-6 MeV He projectiles in the transmission mode.

Elastic Recoil Coincidence Spectrometry (ERCS) is a variation of transmission ERD and is a fairly new technique which has had limited application to the field of materials analysis. Two detectors are used for ERCS – one to detect the scattered incident particle and the other to detect the recoiled target atom. Three-dimensional regions of counts versus projectile-energy and recoil-energy are obtained for each target element recoiled, when the spectra from these two detectors are placed in coincidence. Since thin films must be used in ERCS for transmission, this technique is especially well suited to the study of polymeric materials which can be easily made as thin self-supporting films. As indicated above, the main advantage of transmission ERD over reflection ERD is an improvement in sensitivity, but for samples where sensitivity is less of an issue (e.g., profiling major constituent elements) and where ion-beam damage is more of an issue, then ERCS can be used. In this case, the advantage of ERCS over conventional ERD is one of rapidity of analysis which helps to reduce the sample's exposure to the ion beam. The reduction in ion-beam damage is especially important when materials are being studied which are extremely sensitive to elemental loss during beam irradiation, such as polymers and biological samples. Films of polystyrene are just one example of materials which are subject to elemental loss under heavy-ion irradiation (Green and Doyle, 1986).

An example ERCS analysis of hydrogen in a 1-μm-thick mylar film is shown in Fig. 5.31 where the three-dimensional data is projected onto the recoil-energy projectile-energy plane (E_{dr} = detected energy of recoil H and E_{dp} = detected energy of projectile He). The insert depicts the scattering geometry used to collect this ERCS data. For this experiment, a 4.0 MeV He^+ beam impinged onto a self-supporting 2-μm-thick mylar film at normal incidence, and 100 nC exposures were required to obtain suitable statistics. Two standard 25 mm^2 surface barrier detectors were used to detect, in coincidence, both the scattered He and recoiled H at the angles θ and ϕ, respectively. The scattering angle θ was restricted by slits in front of the detector to define an angular acceptance of 1° (i.e., $\theta \pm 0.5°$), whereas, the recoil angle was not restricted and its angular acceptance was 16° (i.e., $\phi \pm 8°$).

Figure 5.31 displays four islands of coincidence which correspond to four different combinations of θ and ϕ for (θ,ϕ) equal: (13°,27°), (14°,35°), (12°,52°), and (10°,65°). The number of counts in each island is indicated by the size of the filled squares, with the larger squares representing greater intensity. The data in Fig. 5.31 have not been converted to a depth profile (as has been done recently by Hofsass *et al.*, 1990) but the data do display some of the benefits attributed to ERCS. For example, the coincidence islands lie along a line which satisfies the conservation of energy, as is expected for a thin film in which very little energy is lost by either the projectile or the recoil. Thicker targets would cause the line to both broaden and be shifted toward the origin. Chu and Wu (1988) suggested the use of two annular detectors for ERCS because of the cylindrical symmetry of the scattering geometry, which would then give larger detection solid angles and shorter times for data acquisition. The data collected in Fig. 5.31 produced a 2D energy spectrum similar to that which would be obtained from two annular detectors with angular ranges of $9.5°<\theta<14.5°$ and $19°<\phi<73°$. However, if annular detectors had been used in this experiment, then the counting rate would have increased by a factor of 50 and only 40 pC of integrated beam current would have been required to obtain the areal density of H with 10% statistics.

The decreased time of ion-beam exposure suggests that nuclear microscopy of both polymer and biological samples would also benefit from the use of transmission ERD or ERCS. Such instruments use beams focused to ~ μm dimensions with beam currents typically ~ 100 pA. The beams are scanned over the sample in a fashion quite analogous to that of a scanning electron microscope. The analysis given above indicates that only a 0.4 second exposure for each pixel of the scanned beam would be required to image H in a 1 μm polymer film. Typical scans are 128x128 pixels, which indicates that such an image could be accomplished in less than two hours.

5.6 TIME-OF-FLIGHT ERD (TOF-ERD)

Earlier, this chapter showed that the scattering dynamics of ERD and RBS are quite similar, especially when viewed in the center-of-mass frame of reference. Some of these similarities are good, such as the ease of acquiring

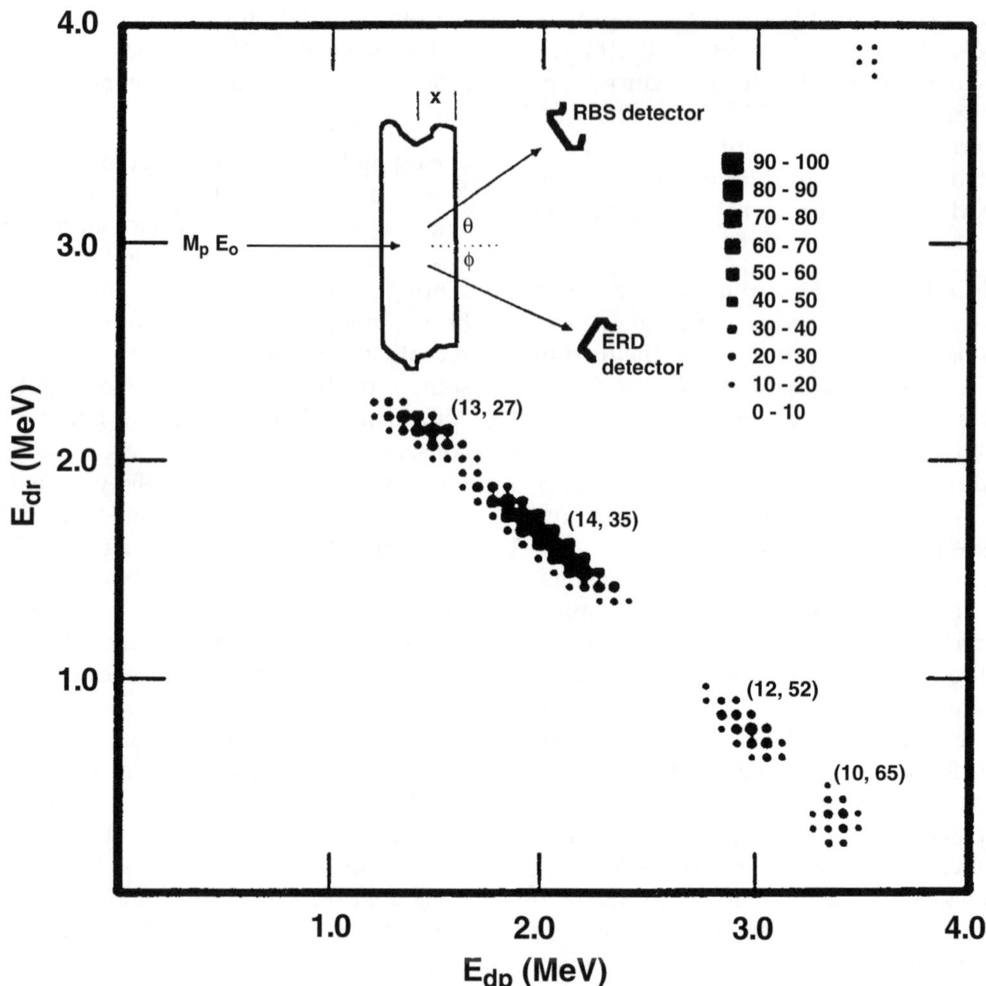

FIG. 5.31. A three-dimensional plot of projectile-recoil coincidence data used in Elastic Recoil Coincidence Spectrometry (ERCS). An insert shows the scattering geometry used in this experiment. The incident beam was a 4 MeV He^+ and the target was 2 μm Mylar foil. The four regions of data labeled as (θ,φ) correspond to different scattered projectile angles (θ) and recoil angles (φ).

both types of spectra, but other shared aspects are actually bad, such as the built-in mass-depth ambiguity which can complicate the interpretation of both ERD and RBS energy spectra (see Figs. 5.2 and 5.3 for examples). This ambiguity results because the energy of the backscattered or recoiled ions depends both on the mass of the target atom (through scattering kinematics) and on the depth in the sample where the scattering occurred (through the energy loss suffered by both the incident and exiting ion). Further, in the case of ERD, a mass ambiguity for scattering from the surface of a sample due to the energy loss suffered in the range foil also exists (note that the mass is a double-valued function of the detected energy as in Fig. 5.3). Avoiding this ambiguity is nearly impossible in RBS because the detected ion is always the same as the incoming ion. However, in ERD, the recoil ion energy and mass can be measured independently by employing either an E-ΔE particle telescope (Gebauer *et al.*, 1990) or by combining a measurement of the ions flight-time (i.e., velocity) to that of total energy (Thomas *et al.*, 1986 and Whitlow, 1990a). This latter technique is referred to as time-of-flight ERD (TOF-ERD), and both TOF-ERD and the particle telescope techniques result in measurements which are energy and mass-dispersive.

An additional benefit of TOF-ERD is that the depth resolution is improved considerably over conventional ERD, particularly for profiling atoms heavier than He or Li. This improvement in depth resolution results because the timing resolution involved with TOF-ERD is generally much better than the energy resolution involved with surface barrier detectors used in conventional ERD. Traditional ERD uses semiconductor detectors to measure the final energy of the recoil atom and such measurements are sensitive to an energy broadening effect called the nuclear or pulse height deficit, i.e., some ion energy is deposited in the detector as atomic displacements rather than through the process of producing electron-hole pairs (see 11.3.4). Timing detectors are not sensitive to this effect, and therefore possess nearly the same timing resolution for all ions. This section will describe some of the fundamentals of TOF-ERD and give typical examples of its use.

5.6.1 Methodology of TOF-ERD

Time-of-flight ERD involves the simultaneous measurement of both the velocity and total energy of the atoms recoiled from the target. Detailed descriptions of this technique have been given in papers by Thomas *et al.*, (1986) and Whitlow (1990a and 1990b), and an example of a typical TOF-ERD set-up is shown in Fig. 5.32, where ϕ=45° and Θ_1=67.5°. Several beam-target-detector geometries have been reported, but most use equal angles of incidence and exit with laboratory recoil angles between 30° and 45°. As with traditional ERD, the recoil atom energy is collected in a semiconductor detector such as a surface barrier Si detector (labeled E), but only after the atom traverses the time-of-flight detection telescope. The signal from the surface barrier detector is amplified and fed into one of two ADC inputs of a dual parameter multichannel analyzer (MCA). The other ADC input is the signal from the flight-time measurement. (See Chapter 11 for more information on TOF.)

The recoil atom velocity is determined by measuring the time required to pass between two thin foils at T1 and T2 with a corresponding flight-path length of L. Typically, L is 0.5 to 1 m and is shown equal to 0.738 m in this figure. As the atom passes through the timing foil, secondary or delta electrons are ejected and accelerated into

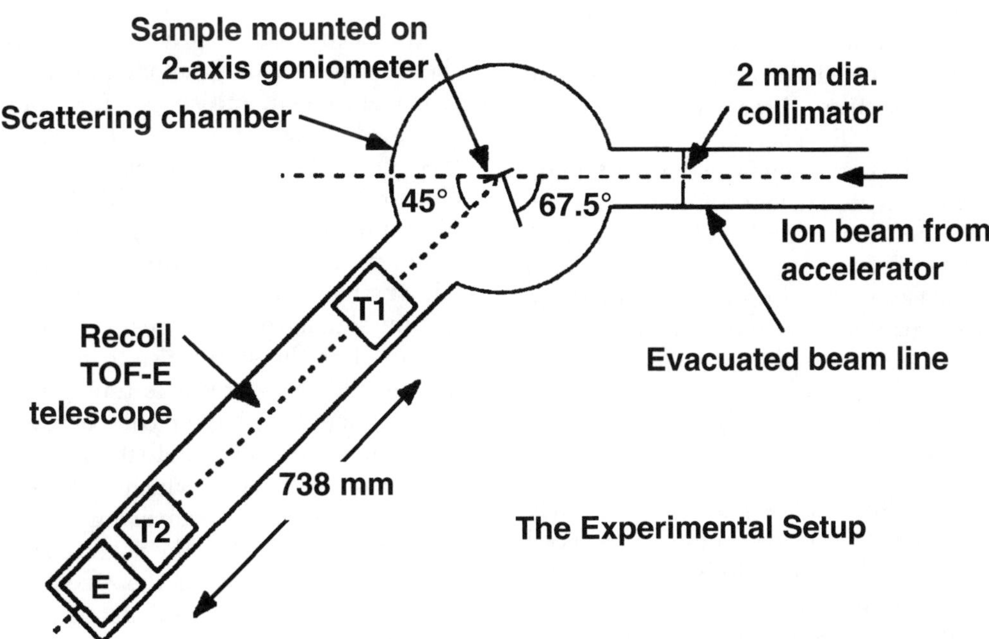

FIG. 5.32. A typical experimental configuration for mass and energy dispersive time-of-flight (TOF) ERD. The scattering angle is 45°, T1 and T2 are carbon foil time detectors which are separated by a flight path of 0.738 m, and E is a silicon surface barrier detector. [From Whitlow, 1990a].

an electron detector to mark a timing event. Further description of a typical electron detection scheme will be given below. A bilayer of formvar and carbon (2-5 μg/cm^2 thick) is commonly used for the timing foils in order to produce a large electron signal with minimal stopping of the recoil beam. Preparation and placement of these foils is a very delicate matter. The relation between E_d, L, and the flight time (t) is given by

$$E_d = \frac{M_2 L^2}{2t^2} . \quad (5.36)$$

(For $E_o < 100$ MeV and $\phi \geq 30°$, relativistic corrections to this equation can be neglected because the maximum recoil velocity is ≈ 0.07c.) Therefore, the raw data collected in the dual parameter MCA is in an energy-time space, and the recoil ions with different masses will appear in this space with loci of E vs. t points defined by Eq. (5.36). The different loci of points for each mass removes the mass-energy ambiguity from the ERD data.

Various incident ions have been used for TOF-ERD, ranging from O to Au at tens of MeV, but several advantages are obtained by using fairly low energy Au ions (10-20 MeV). First, the TOF-ERD telescope defines a small solid angle, and since Au is an extremely easy and bright beam to accelerate on a tandem accelerator, the data collection time can be reduced. Also, Eq. (5.4) shows that for the case of Au projectiles where $M_1 \gg M_2$, the recoil cross section varies as Z_1^4/E_2. Therefore, the extremely large recoil cross sections resulting from the use of high-Z/low-E ions help mitigate the low count rates associated with the extremely small solid angles required for TOF-ERD experiments. Second, if ϕ=30°, Au ions will not scatter into the TOF-ERD system unless the sample contains atomic masses greater than ~100 amu. This advantage greatly reduces accidental counts and increases the lifetime of the surface barrier detector. Last, the stopping power of Au in materials is extremely high, which improves the depth resolution.

Channel electron multiplier (CEM) plates are ultimately used as the detectors for electrons produced by recoil atoms passing through the timing foils, but there are several different approaches and designs used to guide the electrons to the CEM plates. The reader is referred to Whitlow (1990b) for a more detailed discussion of the various designs to detect and measure the arrival time of these electrons. One commonly used approach is a half-turn cyclotron time detector (Zebelman *et al.*, 1977; Thomas *et al.*, 1986) shown schematically in Fig 5.33. The secondary electrons ejected from the carbon foil are accelerated by a closely spaced (1-3 mm) grid or harp. Subsequently, these electrons enter a magnetic field and are guided through a semicircular orbit to a pair of CEM plates. A diaphragm, at right angles to the carbon foil, selects only those electrons which exit the acceleration grid within 5-10° of normal. Therefore, the electron path length is independent of the position on the foil where the electrons were created, which gives isochronous transport of the electrons to the CEM plates.

The half-turn cyclotron design has the advantage that the foil is perpendicular to the recoil trajectory, the acceleration distance is small (acceleration time ≈200 ps), and the detected electrons have traveled the same length of time to reach the CEM plates. Time resolutions of better than 100 ps FWHM have been reported (Thomas *et al.*, 1986) using half-turn cyclotron detectors. The principal disadvantages for these detectors are that they are bulky, sensitive to high voltage supply instabilities, and accurately supplying the magnetic field is complicated. Electron-cyclotron configurations where the electrons are guided by the action of crossed electric and magnetic fields have also been reported using a 1000 mm^2 detector with 109 ps FWHM (Kraus *et al.*, 1988; Bowman and Heffner, 1978).

The electron pulses produced in the channel plates are then transferred to an anode impedance matching antenna, amplified, and shaped for input into electronics designed to measure the time difference between the passage of the recoil ion through the two foils. These electronics consists of fast (200 MHz) constant fraction discriminators (CFDs) feeding the start and stop inputs of a time-amplitude converter (TAC). The signal from the "downstream" CEM should be used as the start pulse, and the signal from "upstream" CEM should be delayed and used as the stop pulse to reduce random coincidences. This also produces timing spectra which, like conventional ERD, have recoil energy increasing to the right. The TAC signal goes to the second ADC input of the dual parameter MCA, which collects a 3-dimensional spectrum with recoil beam energy channel on the x-axis, time-of-flight (offset by the delay) channel on the y-axis, and number of counts on the z-axis. The timing resolution of this measurement depends on the quality of the electronics used to make the measurement, the straggling in the start and stop foils, the geometrical variation of flight distance, and, perhaps most important, the care taken to reduce ringing in the CEM signals and the adjustments of the CFDs. Timing resolution down to 0.2 ns have been reported (Whitlow, 1990b), but values of 0.5 ns are more typical.

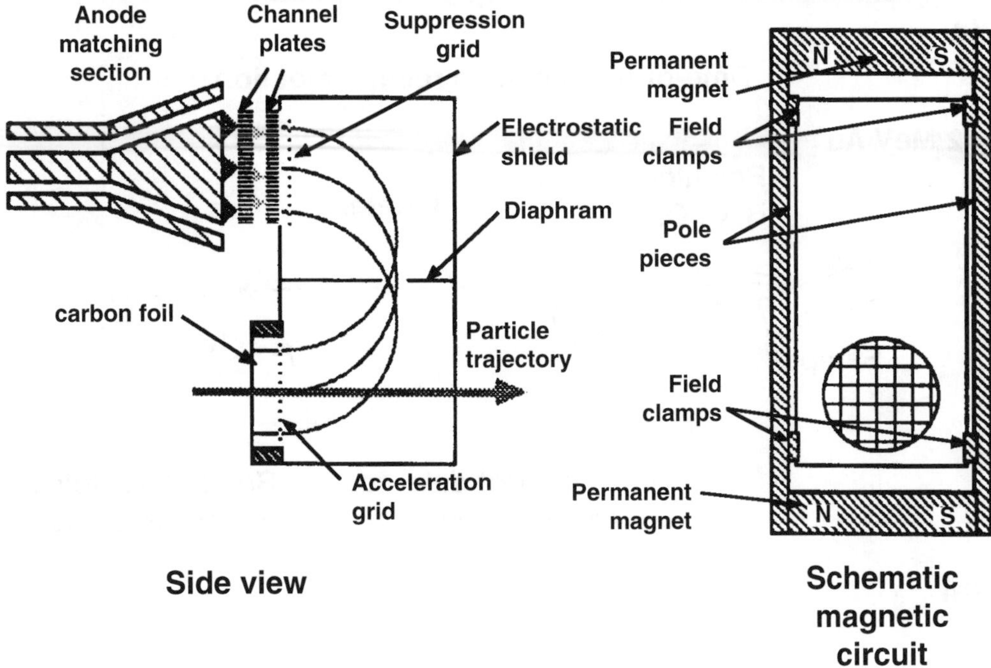

FIG. 5.33. Schematic diagram of a half-turn cyclotron electron detector used for marking the start and stop time of the recoiled atom traversing the flight path. This type of detector would be placed at T1 and another at T2, as shown in Fig. 5.32.

An example of a three-dimensional TOF-ERD spectrum is shown in Fig. 5.34(b), for 12 MeV Au projectiles incident on a sample containing a C-B-C-B-C... multilayer on a Si wafer substrate. The experimental geometry is shown schematically in Fig. 5.34(a). A recoil angle of 30° with a sample tilted at 75° was used in this experiment to be identical to the geometry commonly used for conventional ERD analysis. In the spectrum of Fig. 5.34, the time axis has been reinverted so that increasing time channels reflect lower velocities. The "front" edges of the two boron isotopes, the carbon, the silicon substrate, and even an oxygen contaminant are easily identified in the three-dimensional spectrum as islands, loci of E vs. t data points. Notice that the locus of counts in this energy-time space follows the $E \propto 1/t^2$ dependence shown in Eq. (5.36). This figure can be easily understood:

- Recoils from the surface of the sample (B+C+O) have the highest energies and lowest flight times, and are therefore positioned furthest away from the reader.
- Recoils which occur at the deepest region of the BCO layer, ≈100 nm thick, have lower energies and therefore higher flight times.
- The B, C, and O signals terminate with increasing flight time and decreasing energy corresponding to the thickness of the BCO layer.
- The deepest region of the BCO layer also represents the uppermost region of the Si substrate, and Si recoil atoms from this interface region have greater energy and lower flight times than the Si deep in the substrate.
- Since the Si wafer is much thicker than the range of analysis for the Au beam, the Si recoil counts continue to lower energies and higher flight times beyond the limits set for acceptance by the detection system.

5.6.2 TOF-ERD data analysis

The three-dimensional data collected in TOF-ERD contains considerable information and is quite useful in setting up the experiment or in identifying contaminant elements, but these data must be processed further to provide concentration depth profiles. The manipulation of these data require the definition of regions of interest around each mass island, and then the projection of this

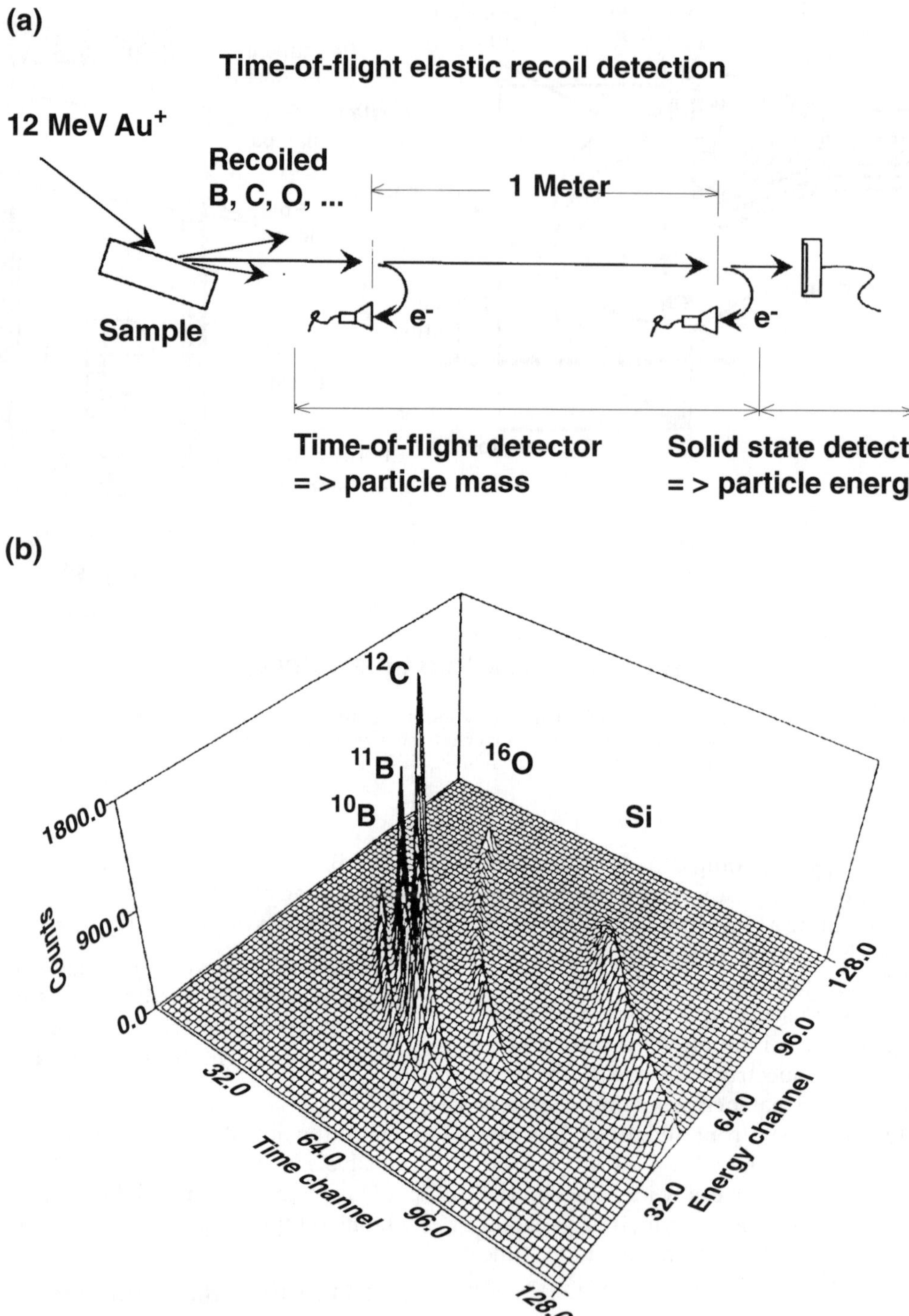

FIG. 5.34. (a) Schematic of a TOF-ERD experiment. (b) Yield vs. energy channel (from a surface barrier detector) and flight time (from a time to amplitude converter) for atoms recoiled from a BC/Si target with a 12 MeV Au beam. Each curved island is the mass-resolved yield for the mass indicated. [From Knapp *et al.*, 1992].

region of interest onto either the energy or flight-time axis. Since the recoil ions of different masses separate themselves so nicely, these projections eliminate the mass-energy ambiguity present in most traditional ERD spectra. If these data are projected onto the energy axis (Oxorn *et al.*, 1990), then they can be directly converted to concentration profiles using the methods described above. This approach is quite sufficient for many TOF-ERD experiments where the mass resolution is more important than the depth resolution. However, if better depth resolution is required, these data must be projected onto the time axis because the energy resolution obtained from the good timing resolution is greater than that achievable from the surface barrier detector.

The approach that most TOF-ERD practitioners use in the analysis of the projected data (i.e., counts vs. time) is to first transform the data back into an energy spectrum, and then analyze the data as shown in Secs. 5.2 and 5.3. The reader is cautioned that care must be taken to avoid artifacts associated with aliasing of the data by this transformation. A simple way to avoid this problem is to use a digital approach which has been developed by Knapp (1994) to transform the data. This procedure is to first calibrate the time domain data using known surface edges in the sample or standards and then convert the data into a time-event array. The time-event array is made by rebinning the channels of the time-domain spectrum so that each channel contains just one count. This is done by dividing each time channel into N subchannels where N was the number of counts in the original channel. For example, assume that channel 57 contains 21 counts in the original time-domain spectrum, and that following the calibration procedure, channel 57 represents time events with durations between 122 and 124 ns. Using Knapp's approach, the 21 events in channel 57 would be subdivided into one event at 122.0 ns, one event at 122.1 ns, one event at 122.2 ns ...... and the last event at 124.0 ns. This procedure is then repeated for all channels in the time-domain spectrum. The resulting time-event array, which is just a one-dimensional array containing individual flight-times, is then transformed into an energy-event array through Eq. (5.36). This latter array is rebinned into equal energy-width channels giving an energy-domain spectrum which can be analyzed with the procedures described in 5.2 and 5.3.

The result of transforming the data plotted in Fig. 5.34, first to a time-domain spectrum and then to an energy-domain, is shown in Fig. 5.35. While these data are still in

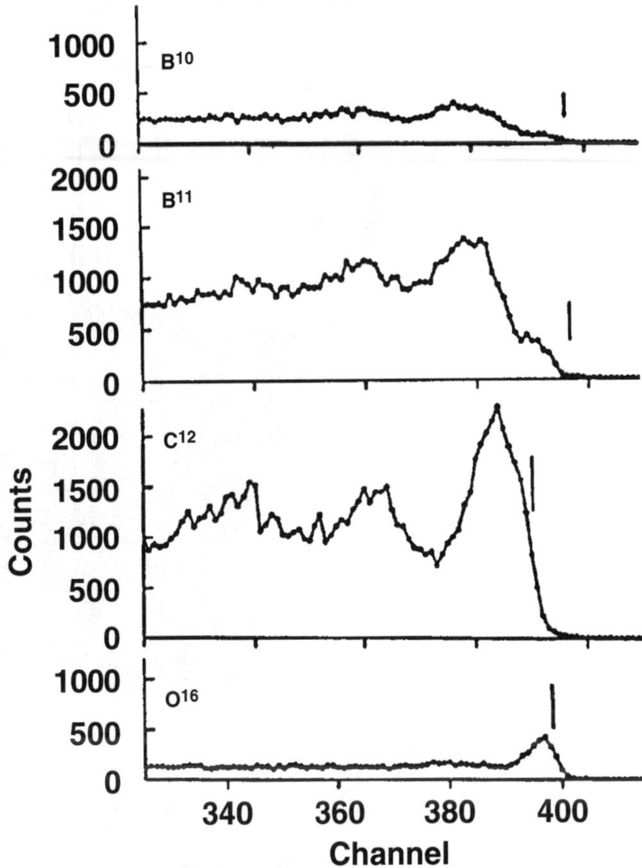

FIG. 5.35. The TOF-ERD spectra for the four masses (B^{10}, B^{11}, C^{12}, and O^{16}) in the film analyzed in Fig. 5.34 were transformed first to a time-domain spectrum and then to this energy-domain spectrum, i.e., the channels shown are energy channels. The surface edge position for each mass is indicated by the vertical lines. [From Knapp, 1994].

the form of counts vs. energy, details of the depth profiles of the B isotopes, C, and O are quite apparent. The topmost layer is a BC-oxide, followed by a C-rich layer which contains a small amount of B and O, followed by a B-rich layer containing small amounts of C and O, and so on. Note that the energy resolution deteriorates very quickly with Au-beam TOF-ERD as a result of energy straggling in the sample.

As an example of a full TOF-ERD analysis where the time-projected data is converted to a concentration-depth profile, consider the nitrogen signal in Fig. 5.36. A 10 MeV Au beam, incident at 75° off the sample normal, was used to profile a ≈12-nm-thick plasma-grown silicon nitride layer on a 4-nm-thick tunnel oxide on Si. The

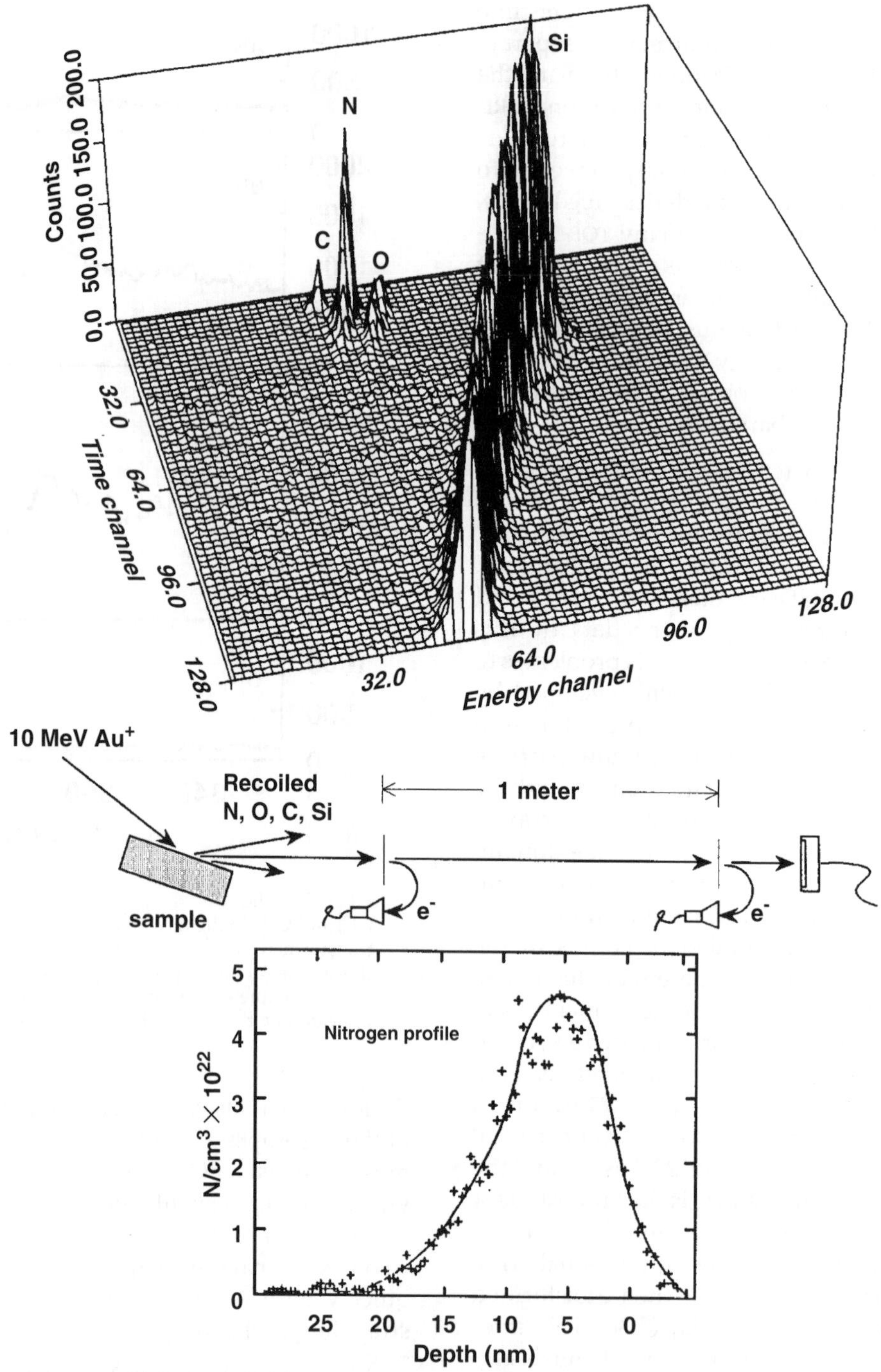

FIG. 5.36. A three-dimensional yield-time-energy TOF-ERD spectrum (top) from a plasma-deposited $SiN_x/SiO_2/Si$ sample. A 10 MeV Au beam, incident at Θ_1=75°, was used for the analysis with ϕ=30° and a flight path of 1 m. The different elements separate into islands of constant mass and form arcs in the yield-time-energy space. The data in the N island were projected onto the time axis, converted to a yield-energy plot, and then analyzed to give the N concentration-depth profile shown at bottom. The (+) are the data point and the line is drawn to guide the eye.

surface of the silicon nitride became oxidized in air after removal from the plasma growth chamber. The recoiled atoms were collected at a scattering angle of 30°, and half-turn cyclotron carbon foil detectors were used for determining the flight time across a 1-m length. The upper portion of Fig. 5.36 shows a three-dimensional plot of the yield vs. energy channel (from the SBD) and yield vs. time channel (from the TAC). Distinct islands, or regions of interest, are obtained for each element in the sample (C, N, O, and Si). The carbon signal is due to surface contamination as a result of transporting the sample through the air. (Also, carbon contamination can arise from a poor vacuum in the analysis chamber, and since a thick carbon layer can degrade the depth resolution, the TOF-ERD analysis chamber should have a vacuum $\leq 5\times10^{-8}$ Torr.)

The separate mass spectra from the three-dimensional plot were projected onto the time axis and analyzed for concentration as a function of depth. The nitrogen profile is shown in the bottom portion of Fig. 5.36, where the (+) are the scaled data and the line is drawn to guide the eye. The low level of counting statistics in this figure point out one of the difficulties encountered in TOF-ERD measurements. The data were collected for approximately 1 hour (1.2 particle-microcoulombs of charge on target), but the small solid angle (approximately 10^{-2} msr) limits the sensitivity. However, the high timing resolution of TOF-ERD yields a greater depth resolution than could be obtained from the energy scale determined from the surface barrier detector. The front edge of the nitrogen (at the sample surface) is sloped with a FWHM of ≈ 3 nm. This sloped edge is partly a result of the depth resolution of TOF-ERD (≈ 1.5 nm on O at the surface) and partially a result of the surface oxidation. While the depth resolution for N decreases rapidly at depths greater than ≈15 nm, interest is growing in the semiconductor industry in silicon oxide and nitride films which are less than 15 nm thick and which require high depth-resolution techniques for determining concentration-depth profiles of the light element constituents. Analysis of the N and O signals revealed that this sample was a Si rich nitride with an approximate composition of $SiN_{1.2}$.

5.7 SUMMARY

Elastic recoil detection, ERD, is a simple and fast method for quantitative light-element depth profiling in thin films (≤ 1 μm thick), and compliments the heavy-element depth profiling obtained from RBS analysis. Since it is complimentary to RBS analysis, ERD should be a part of every MeV ion-beam analysis system. When it is employed on small Van de Graaff accelerators, ERD can be used for determining concentration depth profiles or areal densities of H and D. When ERD is employed on larger Van de Graaffs with the ability to accelerate heavy ions, this technique can be used for concentration depth profiling even heavier elements out to fluorine.

The many attributes of RBS analysis are also shared by ERD analysis. Collection of ERD spectra can be done on the same accelerator and in the same analysis chamber as RBS. Similar to RBS, conventional ERD uses a simple detector scheme with a surface barrier detector and a stopper foil to obtain spectra. The analysis of the data is fully analytical, which may at first appear to be difficult because of the requirement for slab analyses, but with modern computers and analysis codes such as SERDAP (Barbour, 1994), the spectrum can be turned into a concentration profile within minutes. Once the system parameters are calibrated, this technique is standardless, and the analysis is independent of the matrix material and unhampered by sputtering effects which can harm other types of analyses. Finally, conventional ERD gives adequate depth resolution (tens of nanometers) and sensitivity (~ 0.1 at. %) for many types of thin film analyses performed in materials science. If greater depth resolution or greater sensitivity is required, then more exotic detection schemes can be used — such as those used with TOF ERD, which has a depth resolution an order of magnitude greater than that for conventional ERD.

USEFUL EQUATIONS FOR ERD ANALYSIS

$$K = \frac{4\, M_1\, M_2 \cos^2\phi}{(M_1 + M_2)^2} \tag{5.2}$$

$$\sigma_r(E,\phi) = \frac{\left[Z_1\, Z_2\, e^2\, (M_1 + M_2)\right]^2}{\left[2\, M_2\, E\right]^2 \cos^3\phi} \tag{5.4}$$

$$E_o^{(n)} = E_o^{(n-1)} - X^{(n)} S_p^{(n)}\left(E_o^{(n-1)}\right) \tag{5.21}$$

$$E_d^{(n)} = KE_o - X^{(n)} \left[K \sum_{j=1}^{n} \frac{S_p^{(j)}\left(E_o^{(j-1)}\right)}{\cos\Theta_1} + \frac{S_r^{(n)}\left(KE_o^{(n)}\right)}{\cos\Theta_2} + \sum_{j=n-1}^{1} \frac{S_r^{(j)}\left(E_3^{(j+1)}\right)}{\cos\Theta_2} \right] - \Delta E_{foil} \tag{5.23}$$

$$AD_H = \int_0^{\tau} N_H(x)\, dx = \frac{\cos\Theta_1}{Q\,\Omega} \int_0^{\tau} \frac{S^{eff}}{\sigma\,\delta\, E_d}\, Y(E_d)\, dx = \frac{\cos\Theta_1}{Q\,\Omega} \int_{E_d(0)}^{E_d(x)} \frac{Y(E_d)}{\sigma} \frac{dE_d}{\delta\, E_d} \tag{5.5}$$

$$AD_H = \frac{\cos\Theta_1}{Q\,\Omega\,\sigma_o} \sum_{Channel[E_d(0)]}^{Channel[E_d(x)]} Y(Channel) \tag{5.6}$$

$$Y_r^{(n)}(E_d) = \frac{Q\,\Omega\, N_r^{(n)}(x)\, \sigma_r\left(E_o^{(n)}, \phi\right) \delta\, E_d}{\cos\Theta_1\, dE_d / dx|_n} \tag{5.22}$$

$$N_r^{(n)}(x) = \frac{\cos\Theta_1}{Q\,\Omega\,\delta\, E_d} \frac{dE_d / dx|_n}{\sigma_r\left(E_o^{(n)}, \phi\right)} Y_r^{(n)}(E_d) = \frac{\cos\Theta_1}{Q\,\Omega\,\delta\, E_d} \frac{S_o^{eff}}{\sigma_o} \left[\frac{\sigma_r(E_o, \phi)}{\sigma_r\left(E_o^{(n)}, \phi\right)} \frac{dE_d / dx|_n}{dE_d / dx|_o} Y_r^{(n)}(E_d) \right] \tag{5.24}$$

$$\left.\frac{dE_d}{dx}\right|_{nth} = [S]_{p,r}^{(n)} \prod_n \tag{5.19}$$

$$[S]_{p,r}^{(2)} = \frac{K\, S_p^{(2)}\left(E_o^{(2)'}\right)}{C_1} + \frac{S_r^{(2)}(E_2)}{C_2}, \quad C_1 = \cos\theta_1, \quad C_2 = \cos\theta_2 \tag{5.17}$$

$$\prod_n = \frac{S_r^{(n)}\left(E_3^{(n)}\right)}{S_r^{(n)}(E_2)} \frac{S_r^{(n-1)}\left(E_3^{(n-1)}\right)}{S_r^{(n-1)}\left(E_3^{(n)}\right)} \cdots\cdots \frac{S_r^{(o)}(E_d)}{S_r^{(o)}\left(E_3^{(1)}\right)} \tag{5.20}$$

$$\left. {}^{dE_d}\!/_{dx} \right|_n = 2\, \frac{\left[E_d(X^{(n-1)}) - E_d(X^{(n)})\right]}{X^{(1)}} - \left. {}^{dE_d}\!/_{dx} \right|_{n-1} \tag{5.25}$$

$$\delta x = \frac{\delta E_d}{dE_d / dx} \tag{5.28}$$

$$\delta x \cong E_o K \tan\phi \; \frac{\sqrt{\left(\sin 2\phi + \frac{w}{z}\right)^2 + \left(\frac{h \cos\phi}{z}\right)^2} - \left(\sin 2\phi - \frac{w}{z}\right)}{[S]_{p,r}} \tag{5.33}$$

$$h = 2\sqrt{zw \tan\phi + \frac{3w^2}{4\cos^2\phi}} \cong 2\sqrt{zw \tan\phi} \tag{5.34}$$

$$h = 4\sqrt{zw \tan\phi} \tag{5.35}$$

$$E_d = \frac{M_2 L^2}{2t^2} \tag{5.36}$$

REFERENCES

Andersen, H. H., Besenbacher, F., Loftager, P., and Moller, W. (1980), Phys. Rev. A **21**, 1891.

Arnold, G. W. (1994), private communication.

Baglin, J. E. E., Kellock, A. J., Crockett, M. A., and Shih, A. H. (1992), Nucl. Instrum. Methods **B64**, 469.

Barbour, J. C. (1994), SERDAP computer program (Sandia National Laboratories, Dept. 1111, Albuquerque, New Mexico).

Barbour, J. C., Stein, H. J., and Outten, C. A. (1991), in *Low Energy Ion Beam and Plasma Modification of Materials,* J. M. E. Harper, K. Miyake, J. R. McNeil, and S. M. Gorbatkin, (eds.) Materials Research Society, Pittsburgh, Pennsylvania, p. 91.

Bird, J. R., and Williams, J. S. (1989), in *Ion Beams for Materials Analysis,* J. R. Bird and J. S. Williams, (eds.), Academic Press Australia, 30-52 Smidmore St., Marrickville, NSW 2204, p. 571.

Boas, M. L. (1966), *Mathematical Methods in the Physical Sciences,* John Wiley & Sons, Inc., New York, p. 159.

Bowman, J. D., and Heffner, R. H. (1978), Nucl. Instrum. Methods **148**, 503.

Bozoian, M., Hubbard, K. M., and Nastasi, M. (1990), Nucl. Instrum. Methods **B51**, 311.

Brice, D. K. (1973), Thin Solid Films **19**, 121.

Brice, D. K., and Doyle, B. L. (1990), Nucl. Instrum. Methods **B45**, 265.

Chu, W.-K., and Wu, D.T. (1988), Nucl. Instrum. Methods **B35**, 518.

Chu, W.-K., Mayer, J. W., and Nicolet, M.-A. (1978), *Backscattering Spectrometry,* Academic Press., New York.

Doolittle, L. R. (1986), Nucl. Instrum. Methods **B9**, 344.

Doolittle, L. R. (1987), Ph.D Thesis, Cornell University, Ithaca, New York.

Doyle, B. L. and Brice, D. K. (1988), Nucl. Instrum. Methods **B35**, 301.

Doyle, B. L. and Peercy, P. S. (1979a), Internal Sandia Report SAND79-02106.

Doyle, B. L. and Peercy, P. S. (1979b), Appl. Phys. Lett. **34**, 811.

Foti, G., Mayer, J. W., and Rimini, E. (1977), in *Ion Beam Handbook for Material Analysis,* J. W. Mayer and E. Rimini, (eds.), Chapter 2. Academic Press, New York.

Gebauer, B., Fink, D., Goppelt, P., Wilpert, M., and Wilpert, Th. (1990), in *High Energy and Heavy Ion Beams in Materials Analysis,* J. R. Tesmer, C. J. Maggiore, M. Nastasi, J. C. Barbour, and J. W. Mayer, (eds.), Materials Research Society, Pittsburgh, Pennsylvania, p. 257.

Gossett, C. R. (1986), Nucl. Instrum. Methods **B15**, 481.

Green, P. F. (1994), private communication.

Green, P. F., and Doyle, B. L. (1986), Nucl. Instrum. Methods **B18**, 64.

Green, P. F., and Doyle, B. L. (1990), in *New Characterization Techniques for Thin Polymer Films,* H.-M. Tong and L. T. Nguyen, (eds.), John Wiley & Sons, Inc., New York, p. 139.

Hofsass, H.C., Parich, N.R., Swanson, M.L., and Chu, W.-K. (1990), Nucl. Instrum. Methods **B45**, 1518.

Kellock A. J., and Baglin, J. E. E. (1993), Nucl. Instrum. Methods **B79**, 493.

Knapp, J. A. (1994), private communication.

Knapp, J. A., Barbour, J. C., and Doyle, B. L. (1992), J. Vac. Sci. Technol. **A10**, 2685.

Kraus, R. H., Vieira, D. J., Wollnik, H., and Wouters, J. M. (1988), Nucl. Instrum. Methods **A264**, 327.

L'Ecuyer, J., Brassard, C., Cardinal, C., Chabbal, J., Deschenes, L., Labrie, J. P., Terrault, B., Martel, J. G., and St.-Jacques, R. (1976), J. Appl. Phys. **47**, 881.

Marion, J. B. and Young, F. C. (1968), *Nuclear Reaction Analysis*, North-Holland Publishing Co., Amsterdam, p. 142.

Mills, P. J., Green, P. F., Palmstrom, C. J., Mayer, J. W., and Kramer, E. J. (1984), Appl. Phys. Lett. **45**, 958.

Nagai, H., Hayashi, S., Aratani, M., Nozaki, T., Yanokura, M., Kohno, I., Kuboi, O., and Yatsurugi, Y. (1987), Nucl. Instrum. Methods **B28**, 59.

Oxorn, K., Gujrathi, S. C., Bultena, S., Cliche, L., and Miskin, J. (1990), Nucl. Instrum. Methods **B45**, 166.

Rutherford, E. (1911), Philos. Mag. **21**, 669.

Thomas, J. P., Fallavier, M., and Ziani, A. (1986), Nucl. Instrum. Methods **B15**, 443.

Turos, A., and Meyer, O. (1984), Nucl. Instrum. Methods **B4**, 92.

Whitlow, H. J. (1990a), in *High Energy and Heavy Ion Beams in Materials Analysis,* J. R. Tesmer, C. J. Maggiore, M. Nastasi, J. C. Barbour, and J. W. Mayer, (eds.), Materials Research Society, Pittsburgh, Pennsylvania, p. 73.

Whitlow, H. J. (1990b), in *High Energy and Heavy Ion Beams in Materials Analysis,* J. R. Tesmer, C. J. Maggiore, M. Nastasi, J. C. Barbour, and J. W. Mayer, (eds.), Materials Research Society, Pittsburgh, Pennsylvania, p. 243.

Wielunski, L., Benenson, R., Horn, K., and Lanford, W.A. (1986), Nucl. Instrum. Methods **B15**, 469.

Zebelman, A. M., Meyer, W. G., Halbach, K., Poskanzer, A. M., Sextro, R. G., Gabor, G., and Landis, D. A. (1977), Nucl. Instrum. Methods **B141**, 439.

Ziegler, J. F. (1980), *Handbook of Stopping Cross-Sections for Energetic Ions in All Elements,* Pergamon Press, New York.

Ziegler, J. F. (1987), RSTOP computer code from TRIM-91 (J. F. Ziegler, IBM Research Center, Yorktown Heights, NY 10598).

Ziegler, J. F., Biersack, J. P., and Littmark, U. (1985), *The Stopping and Range of Ions in Solids,* Pergamon Press, New York.

CHAPTER

6

NUCLEAR REACTION ANALYSIS: PARTICLE-PARTICLE REACTIONS

G. Vizkelethy

Idaho State University, Pocatello, Idaho

Contributors: M. Nastasi and J. R. Tesmer

CONTENTS

6.1 GENERAL PRINCIPLES AND EQUIPMENT

Rutherford backscattering spectroscopy (RBS) is a special case of the nuclear reaction analysis (NRA) by charged particle detection. In this chapter, therefore, we will discuss only the main differences. Several review papers discuss this method in detail (Amsel and Lanford, 1984; Bird, 1980; Ziegler, 1975; Deconninck, 1978; Mayer and Rimini, 1977; Proceedings of IBA conferences 1973-1991). An excellent discussion on the fundamental physics involved in NRA can be found in Feldmann and Mayer (1986).

The technique of NRA has been studied in detail using scientometric methods (Bujdoso *et al.*, 1982). Above a certain energy of incident beam, which depends on the incident particle and the target nucleus (0.3 - 1.0 MeV for light elements), the backscattered (forward scattered or recoiled) particles and other energetic particles appear in the detected spectrum. When these particles are detected, they usually provide information that is not available by RBS. When a layer of a light element is positioned on top of a heavy substrate, then the RBS spectrum of the light element is also positioned on top of the heavy element's spectrum. Because the Rutherford cross section increases with the square of the atomic number, the light element's signal will be seen against a huge background. For example, when an oxide layer is on top of tantalum, the ratio of the tantalum-to-oxygen signal is 170; therefore, the oxygen can hardly be seen. If the $^{16}O(d,p_1)^{17}O$ reaction is used, then we can measure the oxygen without background (see Example 6.1).

There are differences between RBS and NRA that imply differences in the methods and the necessary equipment. The required equipment, however, is basically the same — accelerator up to 5 MeV energy, scattering chamber, high or ultrahigh vacuum, surface barrier detector, standard nuclear electronics, and multichannel analyzer, although the parameters for some of the equipment are different.

The nuclear reactions are isotope specific with no direct relationship between the mass of the target nucleus and the energy of the detected particles.

- There is no natural background from the high Z components of the target.
- Isotope tracing is very easy.
- The composition of the target usually cannot be measured by one measurement.

The energy of the reaction products is usually higher than the energy of the incident beam.

- The backscattered particles are well separated from the reaction products in the spectrum.
- Special detectors with large sensitive depth should be used to detect high-energy protons.

Usually the cross section of the nuclear reactions is much lower than that of RBS.

- Much higher current must be used to obtain the same data (statistics).
- Because of these high currents, the backscattered particles cause pulse pile-up, which would give high background, and could kill the detector itself. But as the energy of the reaction products is generally higher than that of the backscattered particles, these particles could be eliminated by some filtering technique, e.g., absorber foil, magnetic or electrostatic deflection, time-of-flight (TOF) technique, or special thin detector, all of which will be discussed later.

Usually there is no analytical form of the nuclear cross sections.

- Some earlier data on nuclear cross sections are available, but since the measurements were made by nuclear physicists, they are not always at the angles in which we are interested, and they are usually not precise enough for NRA (the measurements were made usually to obtain information on the nuclear levels in the nuclei).
- These cross sections should be remeasured at the optimum angle of each reaction (see Appendix 10 for some of the cross sections as the result of such measurements).

More than one reaction is possible on a certain nucleus (mainly in deuteron-induced reactions), which results in different particles (or the same particles with different energies when the target nucleus has a number of excited states close to the ground state and part of the energy is taken up by γ-ray production).

- The spectra are sometimes difficult to interpret because peaks of different particles (or same particles with slightly different energies) can overlap. The choice of appropriate filtering parameters (e.g., absorber thickness) could help to overcome this problem.

In some cases relatively narrow resonances exist [e.g., $^{18}O(p,\alpha)^{15}N$] that allow high-resolution depth profiling.

- High-resolution depth profiling requires an accelerator with high energy stability and small energy spread (usually accelerators with a slit feedback system are satisfactory).
- Automatic energy scanning is required to make this technique effective (the best method for automatic energy scanning can be found in Amsel *et al.*, 1983).

Nuclear reactions could generate neutrons (mainly deuteron-induced nuclear reactions)

- The deuteron beam is more hazardous than a standard beam, necessitating stricter safety rules. As a general rule, the energy of the deuteron beam should be as low as possible because the number of generated neutrons increases exponentially with the deuteron energy. Additional details concerning radiation safety can found in Chapter 13 and Appendix 18.

6.2 GEOMETRY AND FILTERING METHODS OF UNWANTED PARTICLES

The geometry in NRA can be either backward or forward, depending on the circumstances. The scattering geometry is the same as that in RBS except that in NRA some filtering equipment is placed in front of the detector. Figure 6.1 shows a typical NRA geometry with an absorber.

The energy of the particles leaving the target is expressed by

$$E_{out}(x) = E[E_{in}(x), Q, \theta] - \int_0^{x/\cos\alpha_{out}} S_{out}(E)dx \qquad (6.1)$$

$$E_{in}(x) = E_0 - \int_0^{x/\cos\alpha_{in}} S_{in}(E)dx \qquad (6.2)$$

where x is the depth at which the nuclear reaction occurs, E_0 is the energy of the incident beam, Q is the Q value of the reaction (several Q values for deuteron-induced reaction can be found in Appendix 10), $S(E)$ is the stopping power of the incident ion and the outcoming reaction product, α_{in} and α_{out} are the angles of the incident beam and of the detector relative to the surface normal of the sample surface, $E[E_{in}(x),Q,\Theta]$ is the energy of reaction product (see Chapter 3 and Appendix 4), and Θ is the traditional scattering angle, i.e., the angle between the incident and outcoming beam (Θ in this chapter is identical to θ in Fig. 4.1). For calculation of the stopping powers see Chapter 2 and Appendix 3.

Five filtering methods can be used.

6.2.1 Absorber foil technique

The simplest method to stop the scattered beam is to place an absorber foil in front of the detector. Usually Mylar is used for the foil because it is pinhole free, although aluminum foil is sometimes used. The energy of the particles after passing through the foil is

$$E_{abs}(x) = E_{out}(x) - \int_0^{x_{abs}} S_{abs}(E)dx \qquad (6.3)$$

where x_{abs} and $S_{abs}(E)$ are the thickness and the stopping power of the absorber foil, respectively. In Appendix 10 the energies of protons and alpha particles are calculated for deuteron-induced reaction from D to ^{19}F after passing through frequently used thicknesses of Mylar. Because of the different stopping power of alpha particles and protons, the interfering α peaks can usually be filtered out by the appropriate filter thickness. The disadvantage of this method is that it results in poor depth resolution because of large energy straggling in the absorber foil.

6.2.2 Electrostatic or magnetic deflection (also electrostatic detector and magnetic spectrometers) technique

These techniques are based on the effects of the magnetic and electrostatic fields on energetic charged particles. The deflection distance depends on the energy, mass, and charge of the particle at a constant electrostatic or magnetic field so that the different particles are separated in space. This method gives better depth resolution than use of an absorber, but because of its complicated measurement setup, it is rarely used, and we will not discuss it in detail here (Möller, 1978; Möller *et al.*, 1977; Chaturvedi *et al.*, 1990; Hirvonen and Luecke, 1978).

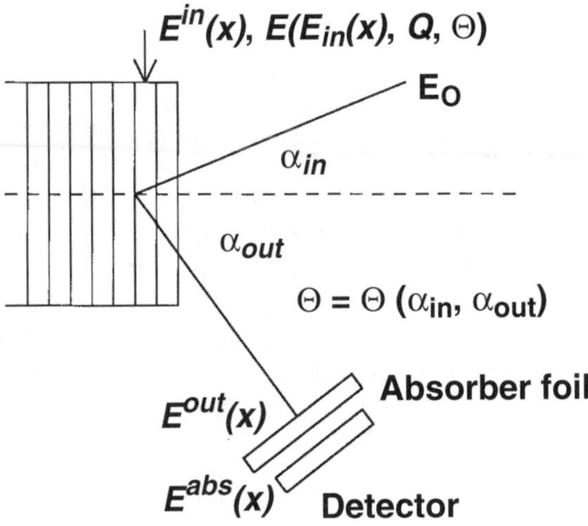

FIG. 6.1. Typical scattering geometry used in NRA experiments. E_0 is the energy of the incident beam, $E_{in}(x)$ is the energy of the incident particle at depth x, $E[E_{in}(x),Q,\Theta]$ is the energy of the reaction product in the sample at a depth x, Q is the Q value of the reaction (see Appendix 10), $E^{out}(x)$ is the energy of the reaction product, which was produced at a depth x in the sample, after traveling out of the sample but prior to passing through the absorber foil, and $E^{abs}(x)$ the energy of the reaction product after passing through the absorber foil. The angles α_{in} and α_{out} are the angles of the incident beam and of the detector relative to the surface normal of the sample surface, and Θ is the traditional scattering angle, i.e., the angle between the incident and out coming beam (Θ in this chapter is identical to θ in Fig. 4.1).

6.2.3 Time-of-flight (TOF) technique

This method is frequently used by nuclear physicists when they have to distinguish between a large number of different particles that are generated simultaneously. The technique is based on the simultaneous measurement of the energy and velocity of the detected particle. From these two values the mass of the particle can be calculated. The velocity is measured by measuring the time elapsed between the detection of the particle in two sequential detectors that are placed at a fixed distance from each other. This method gives the same depth resolution as described in 6.2.2, but requires sophisticated electronics. The full TOF measurement also requires a two-dimensional multichannel analyzer, which is not common in every ion beam laboratory. A simplified version of TOF is obtained when the measured velocity is not stored with the energy, but is used to reject the particles with velocity below a certain threshold level (Sokolov *et al.*, 1989). Another disadvantage of this method is that although the multichannel analyzer does not detect all the low-energy particles, those particles reach the detector and this huge scattered ion flux can damage or kill the detector. This technique is used in ERDA (Elastic Recoil Detection Analysis) where the difference between the flux of the forward scattered alphas and the recoiled hydrogen isotopes is not so large. Additional details concerning TOF can be found in Chapters 5 and 11.

6.2.4 Thin detector technique

This technique is used when proton and α peaks overlap and the α spectrum contains more information than the proton spectrum. The opposite of this situation is when the alpha particles are filtered out using absorber foil. Because of the difference in the stopping powers between a particles and protons, filtering can be achieved with a well-chosen detector thickness: while the α particles fully stop in the sensitive thickness of the detector, the protons lose only a fraction of their energy. This measurement can be made using either dE/dx detectors, which are especially designed to measure energy loss and are quite expensive, or with a low resistivity detector with low bias voltage (Abel *et al.*, 1990). A detailed study has been carried out by Amsel *et al.* (1992). This technique is usually combined with that of the absorber foil technique.

6.2.5 Coincidence technique

The detection of backscattered particles can also be suppressed by using a coincidence arrangement. The two reaction products are detected in different detectors in coincidence (i.e., the event is registered only if both detectors give a signal). One of the detectors should be placed at an angle that is unreachable for one of the reaction products (this angle can be calculated from the kinematics). In this case only the other detector can detect the particle, which cannot reach the first detector because of the coincidence requirement. The residual background is due only to the accidental coincidences. An application of this technique is described in Pretorius and Peisach (1978) and Wielunski and Möller (1990). This technique is limited to transmission geometries (thin foil) because of the conservation of momentum. As in the TOF technique, all of the particles (scattered and reaction products) hit the detector, which means that the large ion current can destroy it.

6.3 MEASUREMENT METHODS

6.3.1 Overall near-surface contents

If the cross section of a given nucleus changes slowly with energy in the vicinity of an energy E_0, the absolute value of nuclei per cm^2 can be determined in thin layers independently of the concentration profile and of the other components of the target. Let E_0 be the energy of the incident beam, and ΔE the energy loss in the thin layer to be measured. If $\sigma(E) \sim \sigma(E_0)$ for $E_0 > E > E - \Delta E$, then the number of counts in the peak area A from the thin layer is

$$A = \frac{Q_c \Omega \sigma(E_0) Nt}{\cos \alpha_{in}} \tag{6.4}$$

where Q_C is the collected charge, Ω is the solid angle, and Nt is the number of nuclei per cm^2. If all the quantities in Eq. (6.4) were known precisely, the absolute value of the nuclei per cm^2 would be easy to calculate. Unfortunately, usually none of the above values (except the number of counts) are known precisely enough and the measurement arrangement is changed between measurements. (For issues on absolute current measurement problems, see Chap. 12.) As the peak area A depends only on the number of nuclei per cm^2 in different matrices, the measurement can be compared to a well-known standard, i.e., $Nt = (A/A_{standard})Nt_{standard}$. (See Example 6.1.) The accuracy of the measurement is then determined only by the statistics of the measurement and the accuracy of the standard. The problem of standards and their usage is discussed later. The thickness of these thin layers is on the order of thousands of Å.

6.3.2 Nonresonant depth profiling

As the thickness of the sample becomes larger, the peak of the reaction product becomes broader. The peak shape is the convolution of the concentration profile with the cross section and the depth resolution. The depth scale can be calculated from Eqs. (6.1), (6.2), and (6.3) by calculating the detected energy of the reaction product for different depth values. An alternate technique in obtaining the depth scale is to use an expression similar to that used in ion backscattering (see Eq. 4.22)

$$\Delta E = x\, N[\varepsilon]_{nr} \tag{6.5}$$

where N is the atomic density and $[\varepsilon]_{nr}$ is the nuclear reaction stopping cross-section factor, defined as

$$[\varepsilon]_{nr} = \left(\alpha_E \frac{\varepsilon_{in}}{\cos \alpha_{in}} \right) + \frac{\varepsilon_{out}}{\cos \alpha_{out}} \tag{6.6}$$

The reaction factor α_E is taken from Eq. (6.8) (also see Table 6.1) and weights the energy loss of the incident particle in the same fashion as the kinematic factor weights the backscattering stopping cross-section factor $[\varepsilon]_{in}$ in Eq. 4.23. Values of ε_{in} and ε_{out} can be calculated assuming either a surface energy or mean energy approximation. When either the standard or unknown contain more than one element, the Bragg rule must be applied to Eq. (6.6). (See Example 6.2.)

The depth resolution in x depth is given by

$$\Delta x = \frac{\sqrt{\Delta E_d^2 + \Delta E_\gamma^2 + \Delta E_{ms}^2 + \Delta E_{abs}^2 + x\left[\frac{\sigma_{out}^2}{\cos \alpha_{out}} + \left(\frac{\partial E_1}{\partial E_2}\right)^2 \frac{\sigma_{in}^2}{\cos \alpha_{in}} \right]}}{\left(\frac{S_{in}}{\cos \alpha_{in}} \times \frac{\partial E_1}{\partial E_2} \right) + \frac{S_{out}}{\cos \alpha_{out}}} \tag{6.7}$$

where ΔE_d is the detector resolution, ΔE_g is the geometrical energy spread, ΔE_{ms} is the energy spread coming from the multiple scattering, ΔE_{abs} is the energy straggling in the absorber foil, σ_{in} and σ_{out} are the Bohr straggling for the incident ion and for the reaction product, and E_1 and E_2 are the energies of the incident ion and the reaction products in depth x. The only difference from the RBS depth resolution is that the k kinematic

Table 6.1. Values of fitting parameters α_E and β_E in the expression $E_3 = \alpha_E E_1 + \beta_E$ for $\theta = 165°$.

Reaction	Q (MeV)	α_E (MeV)	β_E (MeV)	Fitting range (MeV)
$^{14}N(d,\alpha)^{12}C$	13.579	0.122 ± 0.013	9.615 ± 0.016	0.8 - 1.4
$^{13}C(d,p)^{14}C$	5.947	0.571 ± 0.005	5.339 ± 0.006	0.6 - 2.0
$^{15}N(d,\alpha)^{13}C$	7.683	0.273 ± 0.010	5.487 ± 0.014	0.6 - 2.0
$^{9}Be(d,p)^{10}Be$	4.585	0.456 ± 0.006	3.923 ± 0.008	0.6 - 2.0
$^{12}C(d,p)^{13}C$	2.719	0.591 ± 0.003	2.384 ± 0.004	0.6 - 2.0
$^{16}O(d,p)^{17}O$	1.919	0.685 ± 0.002	1.722 ± 0.003	0.6 - 2.0

EXAMPLE 6.1. Consider a SiO_2/Si sample where the thickness of the oxide layer is unknown. We want to measure this thickness by measuring the oxygen content of the sample. The best reaction for this measurement is $^{16}O(d,p)^{17}O$. This reaction has a plateau below 900 keV down to 800 keV, shown in Fig. 6.2 (between arrows).

In this example we use $E_0 = 834$ keV deuteron beam and detect the protons at $\Theta = 135°$. To suppress the backscattered deuterons we use a 12-μm-thick Mylar foil in front of the detector (all the deuterons with energy below 900 keV are stopped by this foil). This reaction produces two groups of protons, p_0 ($Q = 1.919$ MeV) and p_1 ($Q = 1.048$ MeV). The energies of the protons can be calculated from the general kinematic formula

$$E_3^{1/2} = B \pm (B^2 + C)^{1/2} \cong (\alpha_E \, E_1 + \beta_E)^{1/2} \qquad (6.8)$$

where

$$B = \frac{(M_1 \, M_3 \, E_1)^{1/2}}{(M_3 + M_4)} \cos\theta$$

and

$$C = \frac{M_4 \, Q + E_1 \, (M_4 - M_1)}{M_3 + M_4}$$

The variables M_1, M_2, M_3, M_4, E_1, E_3, and θ in Eq. (6.8) are defined in Fig. A4.1, in Appendix 4. The second form of E_3 given in Eq. (6.8) is a linear approximation to the first form. The variables α_E and β_E are specific to the reaction under study and depend on the detection angle θ. Values for α_E, β_E, and Q are given in Table 6.1 for some specific reactions.

In this example M_1, M_2, M_3, and M_4, are the masses of the particles M_d, M_{16O}, M_p, and M_{17O}, respectively. The energy $E_1 \equiv E_0$, and E_3 is the proton energy, which in this example are 2.352 and 1.556 MeV. The energies after passing through the Mylar foil can be calculated from the data in Appendix 10. Although those energies were calculated at $\theta = 150°$, the energy of the protons in our case differs only slightly (energies in Appendix 10 are 2.36 and 1.58 MeV), so we can assume that the protons will lose the same amount of energy. Data from Appendix 10 show that p_0 loses 0.24 MeV in the absorber foil and p_1 loses 0.35 MeV. Therefore, the energies of the protons after passing through the absorber foil are 2.312 and 1.206 MeV.

Let us estimate the maximum thickness of SiO_2 that can be measured at this energy. The stopping power of deuterons in SiO_2 changes slowly between 800 and 900 keV, so we can use an average, which is 7.6 eV/Å. The energy range that we can use is approximately 30 keV, so the maximum thickness is

$$x_{max} = \frac{\Delta E}{S(E)} = \frac{30 \text{ keV}}{7.6 \frac{\text{eV}}{\text{Å}}} \approx 4000 \text{Å}$$

In Fig. 6.3 the measured spectrum (Vizkelethy, 1990) can be seen. There are two extra peaks which need some explanation. The peak at high energies comes from the $^{12}C(d,p)^{13}C$ reaction. During ion bombardment a hydrocarbon layer is always deposited on the surface. The small broad peak between the p_0 and p_1 peaks comes from the D(d,p)T reaction, because deuterium has been implanted into the sample as a result of the deuterium bombardment.

The next step is to calculate the total number of oxygen atoms from Eq. (6.4). Usually the absolute value of the cross section and the solid angle of the detector is not precisely known so standards must be used. The best oxygen standard is Ta_2O_5 made by anodic oxidation. In this example we use an 80-V-thick Ta_2O_5 standard in which the oxygen content is 6.69×10^{17} O/cm^2. We do the same measurement with the standard and calculate the peak integrals. Because we collect the same charge in both cases, the oxygen content is obtained by the comparison of the peak integrals [see Eq. (6.13)].

$$Nt_{SiO_2} = Nt_{Ta_2O_5} \frac{Y_{SiO_2}}{Y_{Ta_2O_5}}$$

$$= 6.69 \times 10^{17} \, \frac{21631}{31875} = 4.54 \times 10^{17}$$

which corresponds to 1000 Å of SiO_2.

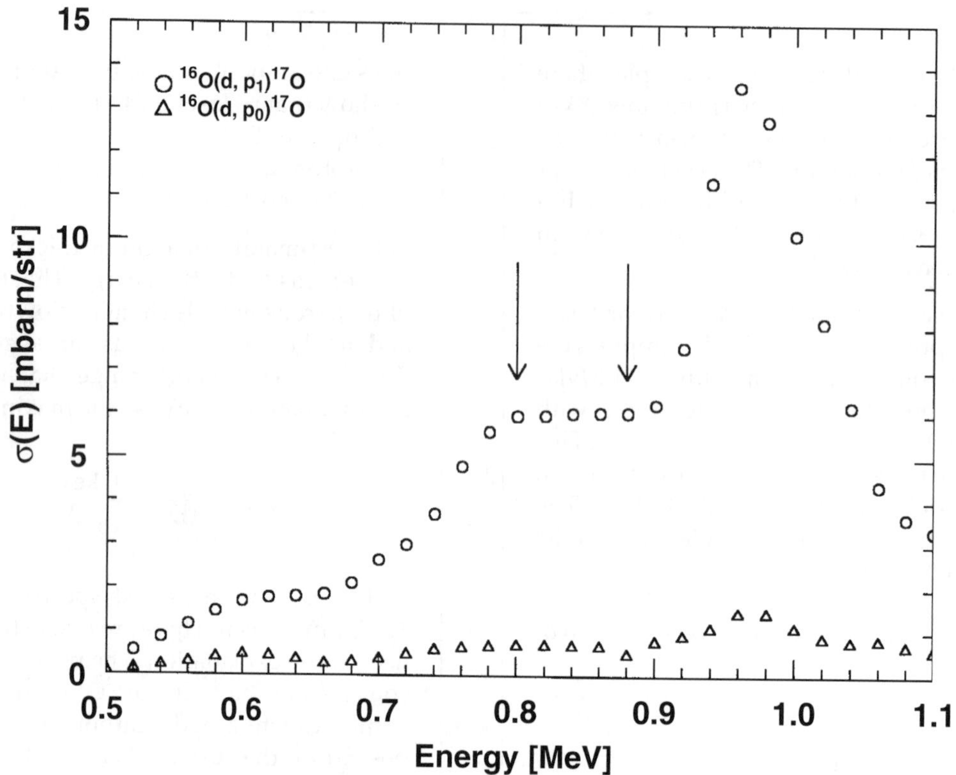

FIG. 6.2. The $^{16}O(d,p)^{17}O$ reaction for both p_0 and p_1 reaction products. This reaction has a plateau below 900 keV down to 800 keV (from Jarjis, 1979).

factor is replaced by the $\partial E_1/\partial E_2$ derivative. For some reactions this derivative is larger than one, and the stopping power of the incident ion is also large because of the usually low incident energy. The depth resolution (if no absorber is used), therefore, can be very good near the surface.

For very light nuclei at backward angles this derivative is smaller than zero. Therefore, the energy of the reaction product increases with increasing depth [D(d,p)T, $D(^3He,\alpha)^1H$, and $D(^3He,p)^4He$]. In this case (and if the stopping power of the reaction product decreases with increasing energy or changes only very little), the contribution of the deep-lying nuclei extends the spectrum toward higher energies. At a certain depth, the depth resolution becomes infinite as the denominator becomes zero.

For a slowly changing cross section, the depth profile can be calculated in the same way as in the case of RBS, and the Rutherford cross section should be replaced by the actual cross section. Where the cross section changes significantly along the measured thickness, the depth profiles may best be calculated by using simulation programs to fit the measured spectrum (Simpson and Earwaker, 1984; Vizkelethy, 1990).

6.3.3 Resonant depth profiling

As Eq. (6.7) shows, the depth resolution of the NRA is generally very poor because of the absorber. The depth resolution can be improved if resonances exist. The method is very similar to the resonance profiling by (p,γ) resonances described in Chapter 8. The detector signal goes through a windowed single channel analyzer—where the energy window is set to include only the reaction products from the resonance reaction—to a multiscaler. The energy can either be changed manually or by an automatic energy scanning system. The depth scale is given by

$$x = \frac{E_0 - E_{resonance}}{S(\bar{E}) / \cos \alpha_{in}} \tag{6.9}$$

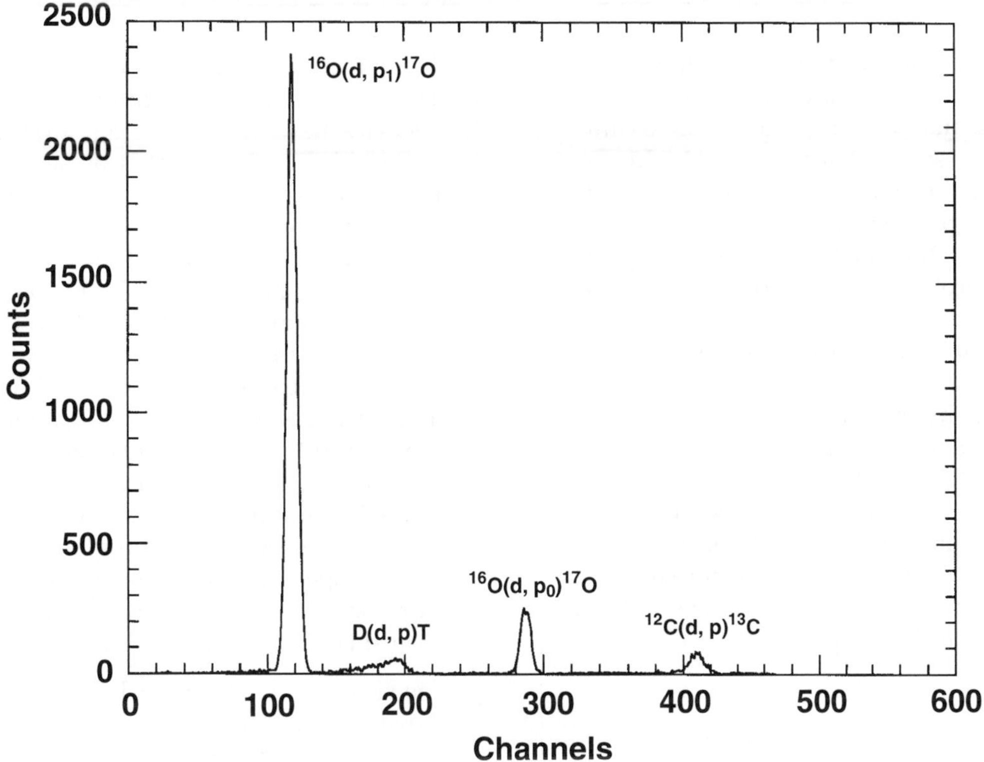

FIG. 6.3. The measured spectrum from 834 keV deuterons on a SiO_2/Si sample (from Vizkelethy, 1990). There are two extra peaks, which need some explanation. The peak at high channels (i.e., high energies) comes from the $^{12}C(d,p)^{13}C$ reaction; during ion bombardment a hydrocarbon layer is always deposited on the surface. The small broad peak between the p_0 and p_1 peaks comes from the D(d,p)T reaction, because deuterium has been implanted into the sample as a result of the deuterium bombardment.

where $E_{resonance}$ is the resonance energy. The energy loss integral can be calculated by using the average energy approximation, i.e., $\bar{E} = (E_0 - E_{resonance})/2$, which is similar to RBS average energy approximation (Chapter 4), as the energy change is usually small. The depth resolution is

$$\Delta x = \frac{\sqrt{\Gamma^2 + \frac{x}{\cos \alpha_{in}} \sigma_{st}^2 + \Delta E_{beam}^2}}{S_{in}(E_{resonance}) / \cos \alpha_{in}} \tag{6.10}$$

where Γ is the width of the resonance, σ_{st} is the Bohr straggling of the incident particle, and ΔE_{beam} is the energy spread of the incident beam. As Eq (6.10) shows, the depth resolution close to the surface can be improved by tilting the sample, but as the beam goes deeper, the energy straggling becomes the dominating effect, and the resolution becomes the same as in the nontilted case.

The measured spectrum represents the convolution of the concentration profile and the energy spread of the detection system, which is the same as the denominator in Eq. (6.10). To extract the information from the spectrum, we should deconvolve it with the depth-dependent resolution function. (See Example 6.3.) The method for the concentration profile calculation is the same as that presented for the (p,γ) reactions in Chapter 7. The theory of narrow resonances (instrumentation function, straggling close to the surface, simulation of resonance spectra, and so on.) is discussed in detail in Maurel (1980), Maurel *et al.* (1982), and Vickridge (1990). To calculate the spectrum, see Eqs. (6.11) and (6.12).

EXAMPLE 6.2. Consider the application of the $^{12}C(d,p)^{13}C$ reaction in analyzing the amount of carbon in a diamond-like-carbon (DLC) coating which contains a substantial amount of hydrogen. We will take $\alpha_{in} = 0$ and $\alpha_{out} = 15$. From Appendix 10 we see that a relatively flat plateau exists in the region between 0.9 and 1.0 MeV for the $^{12}C(d,p)^{13}C$ reaction. The proton energy following a reaction between ^{12}C and a 1 MeV deuteron at the sample surface to be 3.0 MeV is calculated using Eq. (6.8) and Table 6.1. Values of ε_{in} for the 1 MeV deuteron in carbon and hydrogen are 7.19×10^{-15} and 1.99×10^{-15} eV-cm^2/atom, respectively. Values of ε_{out} for the 3.0 MeV proton in carbon and hydrogen are 2.15×10^{-15} and 0.47×10^{-15} eV-cm^2/atom, respectively. From Table 6.1, $\alpha_E = 0.59$. From these data we can calculate the nuclear reaction stopping cross-section factor for each of the elements, carbon and hydrogen, in the sample for the $^{12}C(d,p)^{13}C$ reaction

$$[\varepsilon]^{C}_{nr} = \left(0.59 \times 7.17 + \frac{2.15}{0.97}\right) \times 10^{-15}$$

$$= 6.45 \times 10^{-15} \left(\frac{\text{eV} - \text{cm}^2}{\text{C atoms}}\right)$$

and

$$[\varepsilon]^{H}_{nr} = \left(0.59 \times 1.99 + \frac{0.47}{0.97}\right) \times 10^{-15}$$

$$= 1.22 \times 10^{-15} \left(\frac{\text{eV} - \text{cm}^2}{\text{H atoms}}\right) .$$

Assuming that the DLC coating is composed of 40 atomic % hydrogen and 60 atomic % carbon, the Bragg rule gives

$$[\varepsilon]^{C-H}_{nr} = 0.4 \times [\varepsilon]^{H}_{nr} + 0.6\,[\varepsilon]^{C}_{nr}$$

$$= 4.36 \times 10^{-15}\ (\text{eV} - \text{cm}^2/\text{atom}) .$$

EXAMPLE 6.3. Here we give an example where the $^{18}O(p,\alpha)^{15}N$ 629 keV resonance was used to determine oxygen exchange in high T_c superconductor material (Cheang Wong *et al.*, 1992). First let us estimate the depth resolution of the method and calculate the depth scale using zero tilt angle. The composition of the sample was measured by RBS, $Y_1Ba_{1.8}Cu_{2.9}O_7$. Because the high T_c layer was relatively thin, we can use an average stopping power of ~10.87 eV/Å, calculated using the Bragg rule (see Chapter 2). Equation (6.9) shows the depth scale to be 1 keV/10.87 eV/Å = 92 Å/keV. Because the resonance is 2.1 keV wide, the depth resolution on the surface is approximately 200 Å (the energy spread was assumed negligible relative to the resonance width). At first an ^{18}O reference sample was measured to calibrate the accelerator energy more precisely. The spectrum is shown in Fig. 6.4. This resonance sits on a continuous background, therefore, the spectrum does not go to zero even when the resonance is not inside the ^{18}O-containing layer. This explains the counts below the resonance energy.

The spectrum of the annealed high T_c sample is shown in Fig. 6.5. A peak can be found at the resonance energy that indicates an enriched surface layer. To calculate the depth profile from the spectrum, we must deconvolute the spectrum. The α yield (counts per channel) at E_0 beam energy is given by

$$Y(E_i) = K \sum_{j=1}^{n} c_j \Delta x \int_0^{E_i} f(E_i, x_j, E)\sigma(E)dE \qquad (6.11)$$

where K is a scaling factor, c_j is the concentration of ^{18}O in the jth layer, $\sigma(E)$ is the reaction cross section, and

$$f(E_i, x_j, E) = \frac{1}{\sqrt{2\pi\sigma_j^2}} \exp - \frac{\left[E - E(E_i, x_j)\right]^2}{2\sigma_j^2} \qquad (6.12)$$

is the energy distribution of the incoming beam with initial energy E_i in depth x_j, where $E(E_i,x_j)$ is the beam energy and σ_j is the straggling at x_j depth. There are several methods to carry out the deconvolution which are described in Chapter 7. In this case we used the Levenberg-Marquardt method (Press *et al.*, 1986) to fit the concentration profile, $\chi^2=(Y_i^{mes}-Y(E_i))^2$. The best fit was achieved by using a 150-Å-thick surface layer enriched to 30%, followed by constant volume enrichment to 3.5%.

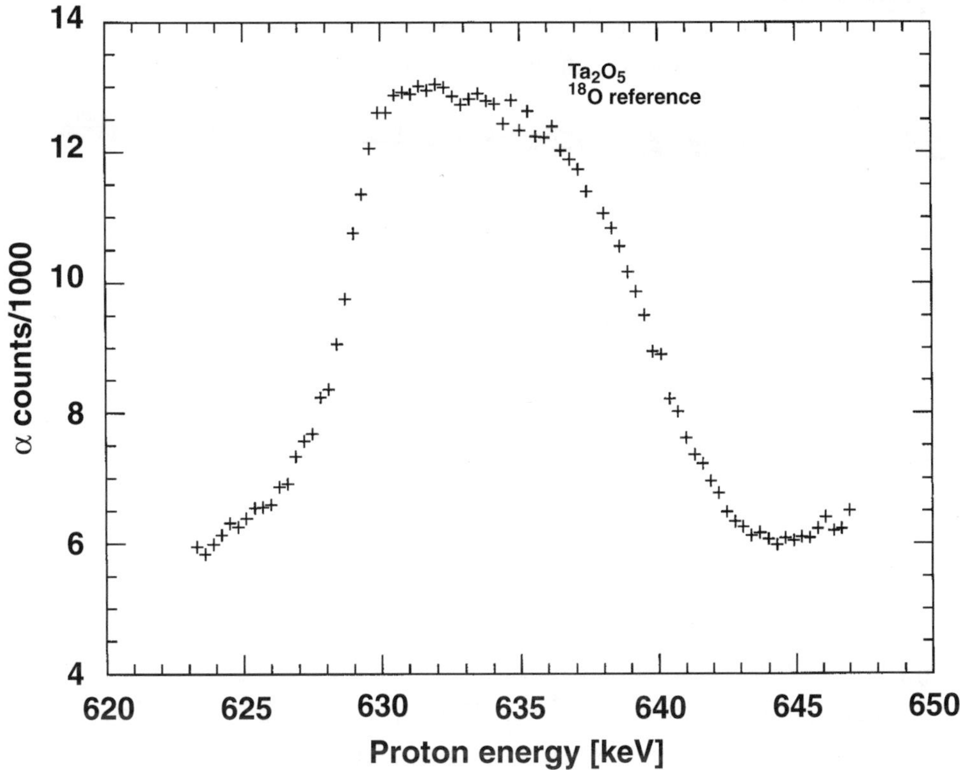

FIG. 6.4. The $^{18}O(p,\alpha)^{15}N$ 629 keV resonance from a Ta_2O_5 reference sample. This resonance sits on a continuous background and therefore the spectrum does not go to zero outside of the reaction energy.

6.4 USAGE OF STANDARDS

As mentioned previously, there are uncertainties in the calculation of the absolute overall surface content of an element. The best way to overcome these uncertainties is to use reference targets. A detailed discussion of the preparation and use of reference targets can be found in Amsel and Davies (1983).

Requirements for good standards:

- The thickness of the layer should be thin enough not to cause significant change in the cross section.
- There must be high lateral uniformity over the beam area.
- The standard should be amorphous to avoid channeling effects.
- The targets should have long-term stability in air, vacuum, and under ion bombardment.
- The preparation of the target should be highly reproducible.

Requirements for the measurement apparatus:

- Current measurement must be reproducible, which means high stability current integrator, efficient secondary electron suppression, and very good vacuum to avoid charge exchange after the beam analysis (for He or heavier beams).
- Energy calibration must be precise and reproducible.
- Detection angle (if the detector is not in a permanent position) must be reproducible.
- Dead-time correction and pile-up rejection must be accurate.

These requirements are the same as those required by high-precision RBS, so they will not be discussed here (see Chapters 4 and 12). The frequently used standards are LiF, CaF_2 (Dieumegard *et al.*, 1980), NbN (Maurel and Amsel, 1983), TaN (Amsel and David, 1969) and Ta_2O_5, (Priegle, 1972; Phillips and Pringle, 1976; Amsel *et al.*, 1978a and 1978b). The measurements of the standard

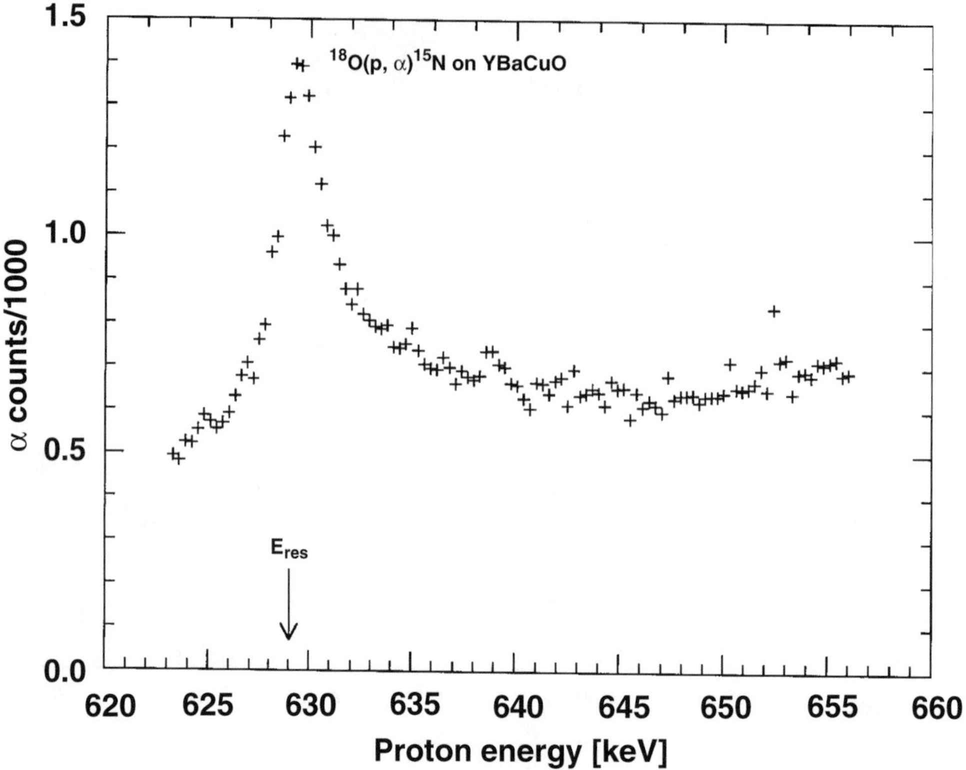

FIG. 6.5. The $^{18}O(p,\alpha)^{15}N$ 629 keV resonance spectrum from a high T_c sample following an ^{18}O oxygen exchange anneal. A peak can be found at the resonance energy that indicates an enriched surface layer.

and the unknown sample should be performed under the same condition, i.e., same solid angle, same incident energy, same absorber, and also the same uncertainty of the current measurement. In this case (if the above requirements are fulfilled), the total amount of the measured nuclei in the unknown sample is

$$Nt = Nt_{ref} \frac{A}{A_{ref}} \frac{Q_C^{ref}}{Q_C} \qquad (6.13)$$

where A and A_{ref} are the peak areas in the spectra, Q_C and Q_C^{ref} are the collected charge, and Nt and Nt_{ref} are the total amounts of the measured nuclei for the unknown sample and the reference standard, respectively. Eq. (6.13) assumes that the thickness of both the reference sample and the unknown sample is small enough to use an energy independent cross section and that their respective peaks can be fully integrated. If we collect the same amount of charge in both cases we can eliminate the Q_{ref}/Q term.

Another way to calculate the absolute overall surface content of an element is to measure the cross section ratios of different elements at a given energy. With this method only one precise reference target is needed. (See Example 6.4.) The disadvantage of the method is that we are very unlikely to find a wide energy region where the cross sections of different elements have plateaued. This limits the measurable thickness. For ^{16}O, ^{14}N, and ^{12}C at $\Theta = 150°$ and at 972 keV, the values are given in Table 6.2 (Amsel and Davies, 1983). This table permits unknown standards to be calibrated to a known standard (e.g., high precision Ta_2O_5 reference standard can be prepared easily by anodic oxidation).

In case of the $^{18}O(p,\alpha)^{15}N$ resonance, the use of a standard is also useful. It helps to determine the precise energy of the resonance and the energy spread of the beam. See Chapters 7 and 8 for more about the usage of standards with resonances.

Table 6.2. Cross section ratios.

Isotope	Nuclear reaction	Cross section ratio (σ_x/σ_0)
^{16}O	(d,p_1)	1
^{14}N	(d,p_0)	0.032
	(d,p_1+p_2)	0.078
	(d,p_4)	0.085
	(d,p_5)	0.456
	(d,p_6)	0.057
^{14}N	(d,α_0)	0.0080
	(d,α_1)	0.119
^{12}C	(d,p_0)	1.91

EXAMPLE 6.4. Let us take a sample contaminated on the surface by oxygen and carbon. Because it is quite difficult to prepare a carbon standard, a Ta_2O_5 reference sample will be used. A good oxygen standard is too thick for us to consider the oxygen cross section to be constant at 972 keV, where the $\sigma(^{12}C)/\sigma(^{16}O)$ ratio is known. Therefore, we first measure the oxygen content of the sample with the $^{16}O(d,p_1)^{17}O$ reaction at 850 keV. The cross section is quite constant down to 800 keV. Comparing the counts with a Ta_2O_5 reference standard from Eq. (6.13) we can calculate the oxygen content. Change the energy to 972 keV and measure the oxygen again and the carbon with the $^{12}C(d,p_0)^{13}C$ reaction. Because we want to measure only surface contamination, assume that both the $\sigma(^{12}C)$ and $\sigma(^{16}O)$ can be considered constant. We already know the oxygen content from the 850 keV measurement, and using data from Table 6.2, $\sigma(^{12}C)/\sigma(^{16}O) = 1.91$, we can easily calculate the carbon content

$$Nt_C = \frac{Y_C}{1.91\, Y_O} Nt_O \tag{6.14}$$

where Nt_C and Nt_O are the absolute content and Y_C and Y_O are the counts of carbon and oxygen.

In many instances the standard is a bulk specimen and the unknown sample can be either a bulk sample or a thin film. For the situation where the standard and the unknown are both bulk, a surface energy approximation to the nuclear reaction analysis can be used to obtain the composition at the surface.

Consider the binary-element unknown sample composed of elements A and B with a composition A_xB_y, where x and y are the atomic fractions of A and B, respectively (i.e., $x + y = 1$). Since we are interested in the amount of element A in our unknown, we will choose a standard containing element A, and will assume that the standard is composed of the elements A and C with the composition A_zC_w. Following the development of Eq. (4.28), the surface heights of element A in our two samples, in the unknown, $H^U_{A_0}$, and in the standard or known, $H^K_{A_0}$, will be given by

$$H^U_{A_0} = \frac{\sigma_A(E_0)\,\Omega_U\, Q_U\, \mathcal{E}_U\, x}{[\varepsilon]^{AB}_{A_{nr}} \cos\alpha_{in}}$$

and (6.15)

$$H^K_{A_0} = \frac{\sigma_A(E_0)\,\Omega_K\, Q_K\, \mathcal{E}_K\, z}{[\varepsilon]^{AC}_{A_{nr}} \cos\alpha_{in}}$$

where $\sigma_A(E_0)$ is the nuclear reaction cross section for element A, $[\varepsilon]^{AB}_{A_{nr}}$ and $[\varepsilon]^{AC}_{A_{nr}}$ are the nuclear reaction stopping cross-section factors for element A in the unknown and the known, respectively (as defined in Eq. 6.6), and the variables Ω, Q, and $\mathcal{E}$ correspond to the detector solid angle, the number of incident ions, and the energy width per channel, respectively. We can rewrite the second part of Eq. (6.15) in terms of $\sigma_A(E_0)$ and substitute into the first part of Eq. (6.15) to obtain

$$x = \frac{H^U_{A_0}}{H^K_{A_0}} \times \frac{[\varepsilon]^{AB}_{A_{nr}}}{[\varepsilon]^{AC}_{A_{nr}}} \times R \times z \tag{6.16}$$

where R is given by

$$R = \frac{\Omega_K\, Q_K\, \mathcal{E}_K}{\Omega_U\, Q_U\, \mathcal{E}_U}\ . \tag{6.17}$$

When the unknown and the known are examined under identical conditions—the same geometry and detector, integrated charge, and amplifier settings, R = 1. In the use of Eq. (6.16), one must make a first guess of the composition of the unknown so that an input value for $[\varepsilon]^{AB}_{A_{nr}}$ can be obtained. As new values for x are calculated, the value of $[\varepsilon]^{AB}_{A_{nr}}$ must be changed using the Bragg rule. Usually several iterations are required before the value of x calculated using Eq. (6.16) and the value of x used in the calculation of $[\varepsilon]^{AB}_{A_{nr}}$ converge.

For the conditions where the standard is a bulk specimen and the unknown is a thin film, we must compare the surface height of the known to the area of the unknown. Again, we will assume a known with composition A_zC_w, and an unknown with composition A_xB_y. The surface height from the known is given by the second part of Eq. (6.15). Assuming that the unknown is thin enough to have a constant cross section, the total yield, or peak integral is given by

$$A_A = \frac{\sigma_A(E_0)\,\Omega_U\,Q_U\left(N_A^{AB}\,t\right)}{\cos\alpha_{in}} \tag{6.18}$$

where N_A^{AB} is the atomic density of element A in the unknown, t is the thickness, and $\left(N_A^{AB}\,t\right)$ is the areal density. Rewriting the second part of Eq. (6.15) in terms of $\sigma_A(E_0)$, substituting into Eq. (6.18), and solving for $\left(N_A^{AB}\,t\right)$ gives

$$\left(N_A^{AB}\,t\right) = \frac{A_A}{H^K_{A_0}} \times \frac{\mathcal{E}_K}{[\varepsilon]^{AC}_{A_{nr}}} \times G \times z \tag{6.19}$$

where

$$G = \frac{\Omega_K\,\Omega_K}{\Omega_U\,\Omega_U}\ . \tag{6.20}$$

6.5 USEFUL REACTIONS

In this section we will list some useful reactions and their parameters. The detailed cross sections can be found in Appendix 11, but only their interesting parts are mentioned here. A large compilation of cross sections was prepared by Jarjis (1979), which provides more information. In addition, on-line data bases such as NNDC and SIGBASE (see references) should be consulted for cross section information.

6.5.1 (p,α) reactions [also (p,d) and (p,^{3}He)]

The (p,α) reactions can be divided into two groups, depending on their Q value. The first group has a relatively low Q value (^{6}Li, ^{9}Be, ^{10}B, ^{27}Al) and the second group has a much higher Q value (^{7}Li, ^{11}B, ^{18}O, ^{19}F, ^{23}Na, ^{31}P). In Figs. 6.6 and 6.7, the energy dependence of the reaction products on the incident energy can be seen at $\Theta = 150°$. It is useful to choose an energy with no overlap between two different reactions, for example, Fig. 6.7 demonstrates that the simultaneous detection of ^{23}Na and ^{31}P by (p,α) is impossible because the energies of the emerging a particles are the same for both reactions almost over the entire range.

As we noted previously, problems arise when the absorber foil technique is used. This is the simplest filtering method but it cannot always be applied. Above a certain proton energy, the absorber, which stops the backscattered (or forward scattered) protons, also stops the α particles. In Fig. 6.8 we calculated the minimum α (d, ^{3}He) energy, above which the α (d, ^{3}He) can penetrate the Mylar absorber but stops the protons. These data were calculated using TRIM (see Chapter 2) and chose the absorber thickness to be the range of the proton beam; then we calculated the α (d, ^{3}He) energy, which gave the same range as that of the proton beam. Usually a little thicker absorber is used to stop all the protons. In case of very light substrate, this absorber thickness can be less, depending on the maximum energy of the scattered protons.

The rough relationship between the proton and α (d,^{3}He) energy is linear, and the slopes are 1.35, 3.6, and 4.15 for d, ^{3}He, and α, respectively. These slopes can provide a good guideline to the appropriate energy in a given situation.

6.5.1.1 ^{6}Li(p,α)^{3}He and ^{6}Li(p,^{3}He)^{4}He Q = 4.02 MeV

This reaction is not useful because of its low Q value.

Cross section measurements were done at $\Theta = 20°$ and $\Theta = 60°$ between 0.5 and 3 MeV for both reactions (Marion *et al.*, 1956). Very broad resonance exists around 2 MeV, and plateaus can be found between 0.8 - 1.2 MeV and 2.2 - 3.0 MeV. A cross section measurement was also done at $\Theta = 164°$ between 0.5 - 3.5 MeV (Baskhin and Richards, 1951), and it indicated a plateau between 0.5 - 1.0 MeV. The detailed angular dependence of the cross section was measured by Elwyn *et al.* (1979). Unfortunately, absorber foil can be used only at low energies (at $\Theta = 150°$ the energy limit is 0.370 MeV for α and 0.575 MeV for ^{3}He).

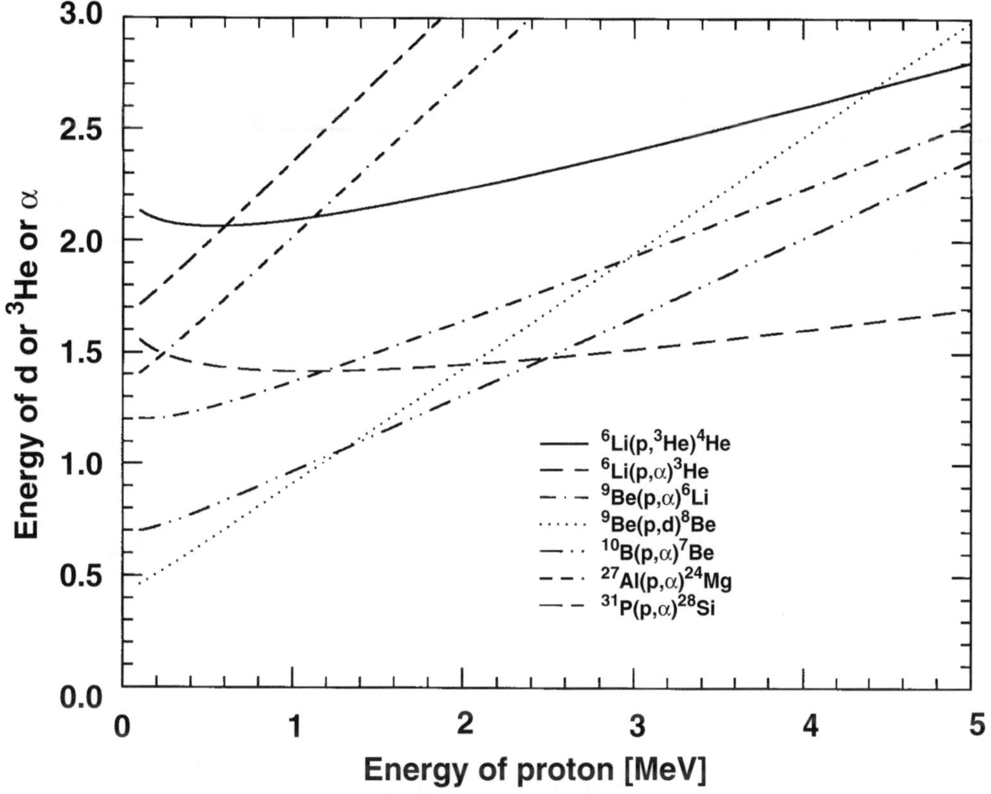

FIG. 6.6. The energy of the reaction products (with relatively low Q values) for several (p,α), (p, ^{3}He), and (p,d) reactions as a function of proton energy.

6.5.1.2 ^{7}Li(p,α)^{4}He Q = 17.347 MeV

This reaction is excellent for ^{7}Li detection. The cross section was measured at Θ = 90° between 1.4 - 2.0 MeV and Θ = 120° between 0.4 - 2.5 MeV (Sweeney and Marion, 1969). The cross section increases very smoothly up to 3.26 mb/sr. A detailed cross section measurement was carried out by Cassagnou *et al.* (1962) for Θ = 85°, 109°, 119°, 128°, 138°, 147°, 155°, and 164° between 1.5 - 3.5 MeV and by Maurel (in Mayer and Rimini, 1977) at Θ = 150° in the energy range of 0.5 - 2.0 MeV. The cross section has a broad maximum around 3 MeV. A recent application of the reaction can be found in Sagara *et al.* (1988). The (p,α) reaction that can interfere is the ^{19}F(p,α)^{16}O, but if one chooses the proton energy carefully, it can be avoided.

6.5.1.3 ^{9}Be(p,α)^{6}Li Q_{α} = 2.125 MeV
^{9}Be(p,d)^{8}Be(2α) Q_d = 0.5592 MeV

Both reactions have very low Q value, therefore, the use of an absorber is limited to very low energies. The cross sections of both reactions were measured at approximately Θ = 90° between 0.1 - 0.7 MeV (Sierk and Tombrello, 1973). Both reactions show broad resonances at approximately 330 keV, and the (p,α) reaction cross section is almost isotropic up to 0.36 MeV. Above 450 keV, both cross sections vary slowly so that they can be used for overall surface content measurement (absorber cannot be used at this energy). Accessible depth is limited by the disturbing effect of the deuterons coming from the ^{9}Be(p,d)^{8}Be(2α) reaction. For examples of depth profiling using the ^{9}Be(p,α)^{6}Li reaction, see Shaanan *et al.* (1986) and Reichle *et al.* (1990).

6.5.1.4 ^{10}B(p,α)^{7}Be Q = 1.147 MeV

This reaction has an extremely low Q value; therefore, it is not well suited for ^{10}B measurement. Cross section of both the ^{10}B(p,α_0)^{7}Be and the ^{10}B(p,α_1)^{7}Be were measured at Θ = 50° and Θ = 90° between 2 - 11 MeV (Jenkin *et al.*, 1964).

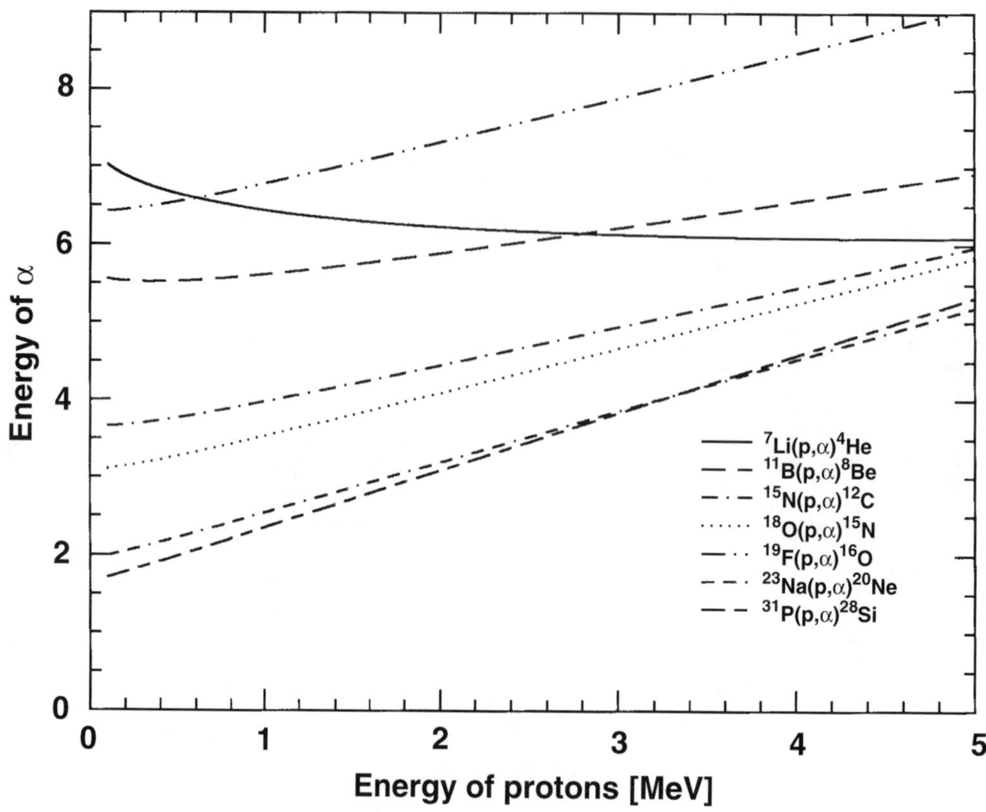

FIG. 6.7. The energy of the reaction products (with relatively high Q values) for several (p,α) reactions as a function of proton energy.

6.5.1.5 ^{11}B(p,α)^{8}Be Q = 8.582 MeV

The cross section of the (p,α_0) reaction is much smaller than that of the (p,α_1) reaction, therefore, the latter is usually used. The cross section of the (p,α_1) reaction was measured at Θ = 85.4° and Θ = 151.85° between 0.8 - 6.0 MeV (Symons and Treacy, 1963). The reaction changes very slowly between 1 - 2 MeV. For α_0, broad resonances can be found around 2.0, 2.5, and 3.8 MeV at Θ = 151.85°, but σ is very small even at the resonances as compared with the cross section of α_1.

At lower energies α_1 has a very broad resonance (Γ=300 keV) at 650 keV, σ=90 mb/sr and a narrower resonance (Γ=6.2 keV) at 163 keV, σ=5 mb/sr between Θ = 140 - 160° (Ligeon and Bontemps, 1972).

For recent applications see Lu *et al.* (1989) (high energy plateau), Scanlon *et al.* (1986), and Moncoffre *et al.* (1990) (low energy resonances).

6.5.1.6 ^{15}N(p,α)^{12}C Q = 4.966 MeV

Instead of this reaction, the ^{15}N(p,αγ)^{12}C reaction is usually used where the γ-ray is detected (see Chapter 8). The cross section was measured at Θ = 15°, 25°, 40°, 60°, 75°, 90°, 105°, and 120° between 0.9 - 1.3 MeV, and Θ = 140° between 0.9 - 3.0 MeV (Hagendorn and Marion, 1957). A broad resonance exists around 1 MeV and a narrower one is at 1.21 MeV. Cross section at Θ = 120° and Θ = 150° between 2.5 - 5.0 MeV can be found in Jausel-Hüsken and Freiesleben (1977). Interference with ^{18}O(p,α)^{15}N is possible.

6.5.1.7 ^{18}O(p,α)^{15}N Q = 3.9804 MeV

This is the most frequently used (p,α) reaction. The cross section was measured at Θ = 68°, 128°, and 170° between 0.9 - 2.0 MeV (Amsel, 1964), and at Θ = 165° between 0.5 - 1.0 MeV (Amsel and Samuel, 1967). Recently the cross section was remeasured at Θ = 155°

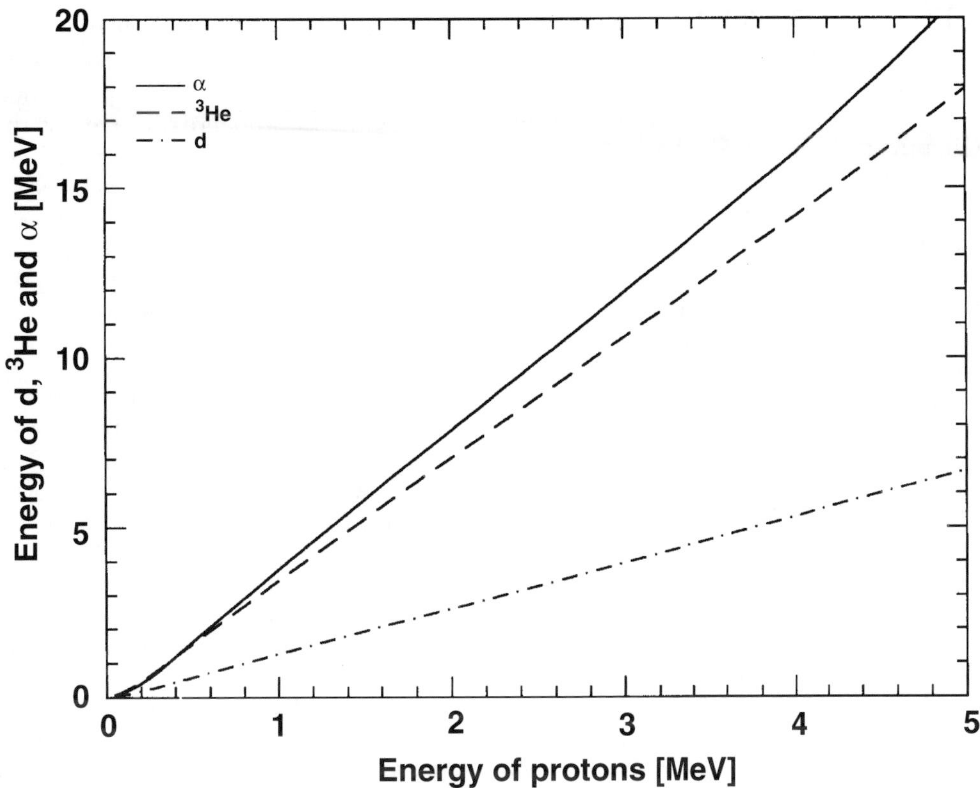

FIG. 6.8. The minimum α, d, or ^{3}He energy, above which the α, d, or ^{3}He can penetrate the Mylar absorber but stops the protons. These data were calculated using TRIM with thickness set equal to the range of the proton beam. The α, d, and ^{3}He energies are those which gave the same range as that of the proton beam in the Mylar. The relationship between the proton and α, d, or ^{3}He energy is linear, and the slopes are 1.35, 3.6, and 4.15 for d, ^{3}He, and α, respectively.

between 1.7 - 1.775 MeV (Alkemade *et al.*, 1988). Overall surface content measurements can be performed at $\Theta = 165°$ and E_p = 730 keV, where the cross section changes very slowly. The cross section has several resonances, which can be used for depth profiling. Values for the width of the resonance, Γ, and the cross section σ as a function of proton energy are given in Table 6.3. The 1165 keV resonance is very narrow, but it is weak and sits on a large continuum in the cross section. The most commonly used resonance is that of 629 keV, and the excitation curve (cross section versus proton energy) was measured very precisely (Maurel, 1980). The 152 keV resonance has not been used up to now because of its low cross section. But the fact that the cross section is isotropic in the vicinity of the resonance allows us to use an annular detector with a large solid angle. This technique has been recently investigated in details (Battistig *et al.*, 1991, 1992). For applications, see Siejka *et al.* (1971), Bird *et al.* (1981), and Chater *et al.* (1990).

Table 6.3. $^{18}O(p,\alpha)^{15}N$ parameters.

Energy *(MeV)*	Γ *(keV)*	σ *(mb/sr)*
152	0.05	
629	2.1	60
1165	0.05	
1766	4.5	135

6.5.1.8 $^{19}F(p,\alpha)^{16}O$ Q = 8.1137 MeV

In this case the γ portion of the $(p,\alpha\gamma)$ reaction is used more frequently (see Chapter 8). The α portion of the cross section was measured at $\Theta = 150°$ and 90° between 0.5 - 2.0 MeV (Dieumegard *et al.*, 1980). For overall surface content determination, E = 1260 keV can be used because of the stationary part that extends down to 1100 keV. At this energy range, this reaction is absolutely background free, and there are no interfering reactions. Several resonances exist in this energy range and values for the width of the resonance, Γ, and the cross section σ as a function of proton energy are given in Table 6.4. For resonance measurements the $^{19}F(p,\alpha\gamma)^{16}O$ reaction is highly recommended.

6.5.2 Deuteron-induced reactions

Almost all of the light elements have deuteron-induced reactions, most of them having positive Q values. We will not mention all of them here, but will cite only those that are routinely used. The reactions are mainly (d,p), but others, (d,α) and (d,^{3}He), are also used. The compound nuclei in these reactions usually have many excited states and, therefore, have many groups of emitted particles (e.g., $^{14}N(d,p_{1-7})^{15}N$, $^{19}F(d,p_{1-16})^{20}F$, etc.). More complete references may be consulted for the target mass A=3 (Ajzenberg-Selove, 1987a) , for A=5-10 (Ajzenberg-Selove, 1988), for A=11-12 (Ajzenberg-Selove, 1990), for A=13-15 (Ajzenberg-Selove, 1986a), for A=16-17 (Ajzenberg-Selove, 1986b), for A=18-20 (Ajzenberg-Selove, 1987b), and for A=21-44 (Endt and van der Leun, 1978). The large number of emitted particle groups can easily cause interference between different elements. Here we will not present the reaction product's energy versus deuteron energy curves for the deuteron-induced reaction as we did for proton reactions because researchers can calculate for their special detection angle and elements. In Appendix 10 we list the energies of the particles generated in (d,p) and (d,α) reactions at $\Theta = 150°$ and E_d=0.9 and 1.2 MeV (Amsel, 1991). Appendix 10 also contains the energies after passing through a Mylar absorber foil whose thickness was chosen to stop the deuterons. The energies of the emitted protons at $\Theta = 150°$ and E_d=1.8 MeV from elements A = 2-31 can be found in Vickridge (1988).

Table 6.4. $^{19}F(p,\alpha)^{16}O$ parameters.

Energy (Θ) *(MeV)*	Γ *(keV)*	σ *(mb/sr)*
0.777(90)	7	0.075
0.777(150)	10	0.17
0.840(150)	32	0.22
0.843(90)	17.5	0.13
1.347(150)	36	3.2
1.354(90)	36	1.05
1.652(150)	90	1.4
1.713(90)	72	2.9
1.842(150)	122	3.2
1.901(90)		1.6

The minimum detectable energy of p, ^{3}He, and α as a function of the deuteron energy using Mylar absorber is shown in Fig. 6.9. As in the case of protons, the rough relationship is linear, and the slopes are 0.76, 2.65, and 3.05 for p, ^{3}He, and α, respectively.

6.5.2.1 D(d,p)T Q = 4.033 MeV

This reaction is always present when a deuteron beam is used. The reaction occurs between the implanted deuterium and the beam. In frequently used standards the proton peak from this reaction can become significant. (for example, in ^{16}O measurements this peak appears between the p_0 and p_1 peaks). A compilation of cross section data can be found in Jarmie and Seagrove (1957).

6.5.2.2 $^3He(d,p)^4He$ and $^3He(d,\alpha)^1H$ Q = 18.352 MeV

See $D(^3He,p)^4He$ and $D(^3He,\alpha)^1H$ reactions.

6.5.2.3 $^6Li(d,\alpha)^4He$ Q = 22.374 MeV

The cross section for this reaction was measured at $\Theta = 150°$ between 0.5 - 2.0 MeV (Maurel *et al.*, 1981). It has a quasi plateau at approximately 0.65 MeV, which can be used for overall surface content measurement.

6.5.2.4 $^9Be(d,t)^8Be$ Q = 4.496 MeV

The cross section for this reaction was measured at $\Theta = 90°$ from 1.5 - 2.5 MeV (Biggerstaff *et al.*, 1962). Plateau can be found above 2 MeV.

6.5.2.5 $^9Be(d,\alpha)^7Li$ Q = 7.153 MeV

The cross sections of the α_0 and α_1 reactions were measured at Q = 90° and Q = 165° (Biggerstaff *et al.*, 1962). The cross section of α_0 shows plateaus between 1.5 -2.5 MeV at $\Theta = 90°$, and between 1.0 - 1.5 MeV at $\Theta = 165°$.

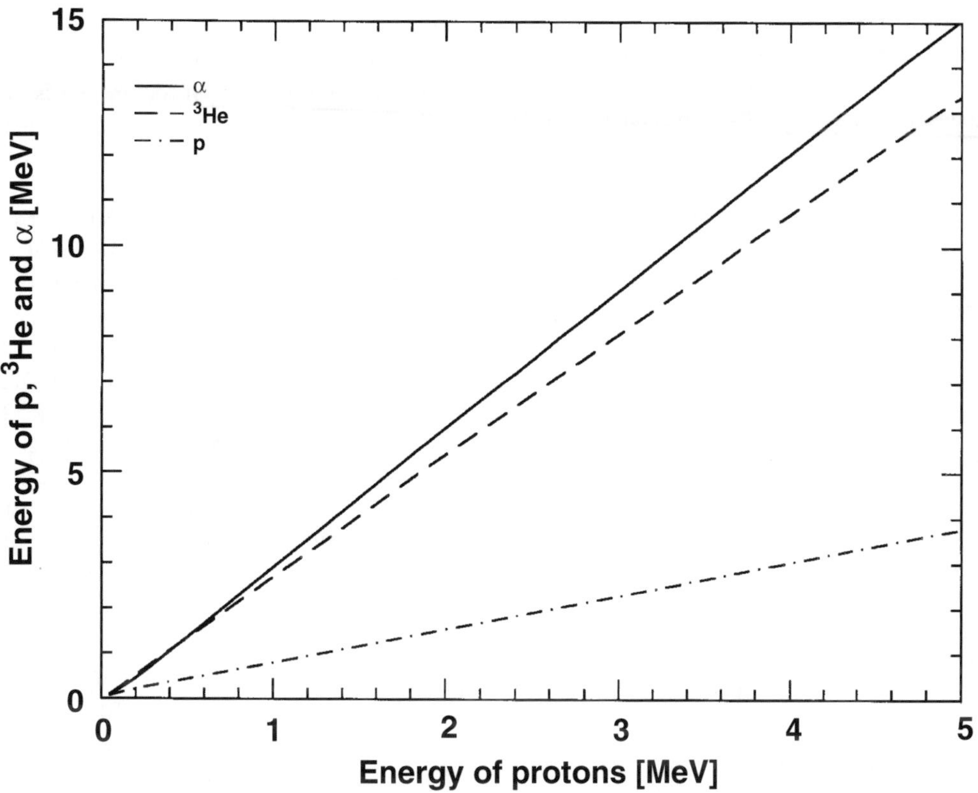

FIG. 6.9. The minimum detectable energy of p, ^{3}He, and α particles as a function of the deuteron energy using a Mylar absorber. The procedures used to calculate these energies are the same as used in Fig. 6.8. The relationship between the deuteron energy and reaction particle is linear, and the slopes are 0.76, 2.65, and 3.05 for p, ^{3}He, and α, respectively.

6.5.2.6 $^{10}B(d,\alpha)^8Be$ Q = 17.818 MeV

The cross section was measured at Θ = 156° between 1.0 - 1.8 MeV for both α_0 and α_1 (Purser and Wildenthal, 1963). Both cross sections have slowly changing regions—1.0 - 1.8 MeV for α_0 and 1.4 -1.6 MeV for α_1—but the cross section of α_1 is 10 times higher than that of the α_0 and the Q value (14.878 MeV) for α_1 is still high, therefore, the α_1 reaction is recommended.

6.5.2.7 $^{12}C(d,p)^{13}C$ Q = 2.722 MeV

A detailed cross section measurement was carried out at different angles from 1.8 MeV to 3.5 MeV by McEllistrem *et al.* (1956). The cross section of p_0 and p_1 (Q = –0.37 MeV) was measured by Kashy *et al.* (1960) at Θ = 47.6°, 80.5°, 158.4°, and 165° for p_0 and Θ = 80.5° for p_1 in the 0.7 - 2.0 MeV energy range. The p_0 cross section has a plateau at backward angles above 0.9 MeV. This plateau is usually used to measure overall carbon content in the surface layer. Thin oxygen standards can be used to determine the absolute amount of carbon, using the cross section ratio of the (d,p) reactions on ^{16}O and ^{12}C (Amsel and Davies, 1983) as shown in Table 6.2.

6.5.2.8 $^{13}C(d,p)^{14}C$ Q = 5.951 MeV

This cross section was measured at Θ = 20° and Θ = 135° in the energy range of 0.6 - 3.0 MeV (Marion and Weber, 1956). The cross section has a plateau around 1.1 MeV at Θ = 165°, which can be used for overall surface content measurement. The peaks from ^{12}C and ^{13}C are well separated because of the large difference in the Q value.

6.5.2.9 $^{14}N(d,p)^{15}N$ Q = 8.610 MeV

The cross section for this reaction was measured at Θ = 150° from 0.6 to 1.4 MeV for p_0, p_1+p_2, p_3, p_4, and p_5 (Amsel and David, 1969). At higher energies the cross section was measured at $\Theta = 60°, 90°, 120°$, and 150° in the energy range of 2.6 - 5.0 MeV for p_0-p_{10} (Koenig *et al.*, 1977). The cross section has been recently measured at $\Theta = 150°$ from 0.5 - 0.65 MeV for p_5 (Berti and Drigo, 1982) and from 0.5 - 2.0 MeV for p_0 (Simpson and Earwaker, 1984). There is a plateau in the p_1+p_2 cross section at $\Theta = 150°$ between 1.05 and 1.25 MeV. The ratio of $^{14}N(d,p_0)/[^{28}Si(d,p_1)+^{29}Si(d,p_2)]$ was measured at $\Theta = 150°$ between 1.72 and 1.82 MeV (Vickridge, 1988). This ratio is almost constant and can be used for silicon-nitride stochiometry measurement.

Because of the large Q value, very thick detectors (~ 300-500 μm) should be used to stop the protons. Some of the proton groups can interfere with the $^{12}C(d,p)^{13}C$ and $^{16}O(d,p)^{17}O$ reactions. Also the peaks from the different proton and α groups can overlap in the spectrum. This effect can be avoided by careful choice of the absorber foil.

6.5.2.10 $^{14}N(d,\alpha)^{12}C$ Q = 13.574 MeV

The cross section was measured at $\Theta = 90°$ from 1.0 to 3.5 MeV (α_0 and α_1) (Ishimatsu *et al.*, 1962), at $\Theta = 150°$ from 0.6 to 1.4 MeV (α_0 and α_1) (Amsel and David, 1969), and at $\Theta = 60°, 90°, 120°$, and 150° between 2.6 and 5.0 MeV (α_0-α_8) (Koenig *et al.*, 1977). The α_1 cross section shows a quasi plateau between 1.05 and 1.4 MeV. The use of a thick detector is recommended in this case, too.

6.5.2.11 $^{15}N(d,p)^{16}N$ Q = 0.266 MeV

Because of the low Q value this reaction is not useful.

6.5.2.12 $^{15}N(d,\alpha)^{13}C$ Q = 7.687 MeV

The cross section was measured for both α_0 and α_1 at $\Theta = 90°$ from 1.2 to 2.5 MeV (Mansour *et al.*, 1964). The paper also presents angular distribution for α_0, α_1, and $\alpha_{2,3}$ at various energies. A more precise measurement has been recently made by Sawicki *et al.* (1986) at $\Theta = 150°$ in the 0.8 -1.3 MeV energy range. These cross sections have a slowly changing part between 1.0 and 1.2 MeV.

6.5.2.13 $^{16}O(d,p)^{17}O$ Q = 1.917 MeV

This is the most frequently used reaction for ^{16}O detection. The cross section for p_0 and p_1 (Q = 1.046 MeV) was measured at $\Theta = 164.25°$ in the 0.7 - 2.2 MeV energy range (Seiler *et al.*, 1963), at $\Theta = 10°$ and 87° from 0.8 MeV to 2.0 MeV (Amsel, 1964), and at $\Theta = 135°$ from 0.2 MeV to 3.0 MeV (p_1 from 0.52 MeV) (Jarjis, 1979). Because the D(d,p)T reaction may interfere with p_0 peak (also $^{18}O(d,p)^{19}O$, but the interference is significant only in ^{18}O-enriched targets) and the cross section of p_1 is larger than that of p_0, the p_1 peak is usually used for ^{16}O content determination. The cross section of p_1 was measured with high precision at $\Theta = 150°$, between 0.4 -1.1 MeV (Amsel and Samuel, 1967). In most cases the plateau between 0.8 and 0.9 MeV is used, the usual deuteron energy is 0.83 MeV. In case of low Z backings (Al, Si) the deuteron energy should be kept low enough so that the effect of the Coulomb barrier favors the cross section for the $^{16}O(d,p)^{17}O$ reaction. The peaks from $^{11}B(d,p)^{12}B$ and $^{14}N(d,p)^{15}N$ reactions overlap the p_1 peak. Therefore, if the target contains a significant amount of ^{11}B or ^{14}N, the p_0 peak should be used (the geometry and absorber foil should be optimized to avoid the interference of the protons from the D(d,p)T reaction with the p_0 peak).

6.5.2.14 $^{16}O(d,\alpha)^{14}N$ Q = 3.11 MeV

This reaction is used for high depth resolution ^{16}O profiling. The cross section was measured at $\Theta = 164.25°$ from 0.7 MeV to 2.2 MeV (Seiler *et al.*, 1963), $\Theta = 90°, 135°$, and 165° in the 0.8 - 2.0 MeV energy range (Amsel, 1964), and $\Theta = 145°$ between 0.75 MeV and 0.95 MeV (Turos *et al.*, 1973). Usually only a thin absorber foil or no absorber is used, because an absorber that would stop the deuteron beam would also stop the alpha particles. Therefore, this reaction cannot be used on high Z targets, where the energy of the backscattered deuterons is almost equal to the beam energy. Because of the thin absorber (or no absorber), the p_0 peak from the $^{16}O(d,p)^{17}O$ reaction can interfere with the α peak. In order to avoid this interference, one should use a thin detector (~ 30 μm). This technique (depth resolution, optimization, etc.) is discussed in detail by Turos *et al.* (1973).

6.5.2.15 $^{18}O(d,p)^{19}O$ Q = 1.732 MeV

This reaction can interfere with the $^{16}O(d,p_0)^{17}O$ reaction if the target contains a significant amount of ^{18}O. To resolve the peaks from the different isotopes one should use a high-resolution detection technique. The target thickness is limited by the energy difference between the protons from ^{16}O and ^{18}O.

6.5.2.16 $^{18}O(d,\alpha)^{16}N$ Q = 4.247 MeV

The cross section of the alphas from the ground state and from the first three excited states was measured at $\Theta = 165°$ from 0.8 MeV to 2.0 MeV (Amsel, 1964). The α peaks from ^{18}O are well separated from the α peak from ^{16}O. This reaction is useful only for simultaneous ^{16}O and ^{18}O measurement; otherwise, the $^{18}O(p,\alpha)^{15}N$ should be used because of its higher cross section.

6.5.2.17 $^{19}F(d,p)^{20}F$ Q = 4.374 MeV

This reaction is important because it can interfere with the $^{16}O(d,p_1)^{17}O$ reaction. With thick targets the peak of the $^{19}F(d,p_{11}+p_{12})^{20}F$ overlaps with the p_1 peak from ^{16}O.

6.5.2.18 $^{19}F(d,\alpha)^{17}O$ Q = 10.031 MeV

The cross section was measured at $\Theta = 150°$ between 0.8 MeV and 2.0 MeV for both α_0 and α_1 (from 1 MeV) (Maurel *et al.*, 1981). The plateau in the α_1 cross section from 1.3 to 1.6 MeV is used for ^{19}F determination, using E_d=1.6 MeV. Because of the large difference in the Q values of the (d,p) and (d,α) reactions, they do not interfere with each other (at least with α_0 and α_1). The α_0 from $^{7}Li(d,\alpha)^{5}He$ and the almost continuous a spectrum from the $^{10}B(d,\alpha)^{8}Be$ reaction can overlap with the alphas from ^{19}F. The simultaneous detection of light elements and the interference is discussed in detail in Maurel *et al.* (1981).

For data (energy levels, cross section measurement, etc.) derived from the heavier elements, see Endt and van der Leun (1978).

6.5.3 ^{3}He-induced reactions

^{3}He-induced reactions have not often been used up to now, except for the $D(^{3}He,p)^{4}He$ and $D(^{3}He,\alpha)^{1}H$ reactions. Recently, ^{3}He has been used for detection of ^{12}C (Gossett, 1981) and ^{16}O (Li *et al.*, 1991). In Fig. 6.10 we show the relationship between the energy of ^{3}He and the energy of the reaction products after passing through an absorber foil that stops the primary beam. These reactions offer us an alternative reaction to detect ^{12}C and ^{16}O, mainly where deuteron beams cannot be used.

6.5.3.1 $D(^{3}He,p)^{4}He$ and $D(^{3}He,\alpha)^{1}H$ Q = 18.352 MeV

The cross section and the angular distribution of the protons were measured by several groups in the early 1950s—$^{3}He(d,p)^{4}He$ at $\Theta = 0°$ from 0.2 MeV to 1.6 MeV (Bonner *et al.*, 1952), at $\Theta = 86°$ from 0.24 MeV to 0.94 MeV (Yarnell *et al.*, 1953), and $D(^{3}He,p)^{4}He$ at various angles in the 0.1 - 0.8 MeV range (Kunz, 1955). All three measurements show almost no angular dependence; therefore, in Appendix 11 we show the total cross section instead of differential cross section. Recently the $^{3}He(d,p)^{4}He$ cross section has been remeasured by Möller and Besenbacher (1980). The total cross section shows a broad resonance around 0.64 MeV for $D(^{3}He,p)^{4}He$ (400 keV for $^{3}He(d,p)^{4}He$). This reaction is used for both the detection of deuterium and ^{3}He either detecting the protons at backward angles (Pronko and Pronko, 1974; Pronko, 1974; Dieumegard *et al.*, 1979; Qiu *et al.*, 1989; Payne *et al.*, 1989) or the alphas (Möller, 1978; Chatuverdi *et al.*, 1990; Langley *et al.*, 1974) at forward angles, or for detecting both particles in coincidence (Wielunski and Möller, 1990). The high energy protons should be detected either by a very thick (1500 μm) surface barrier detector (Dieumegard *et al.*, 1979; Payne *et al.*, 1989) or by a plastic scintillator (Qiu *et al.*, 1989). The backscattered deuterium particles should be rejected by an absorber, or electrostatic (Möller, 1978) or magnetic (Chatuverdi *et al.*, 1990) deflection, or a coincidence measurement can be used (Wielunski and Möller, 1990). It should be noted that the reaction at backward angles has inverse kinematics, i.e., the lower the energy of the incident ^{3}He, the higher the energy of the protons and alphas.

6.5.3.2 $^{6}Li(^{3}He,p)^{8}Be$ Q = 16.7863 MeV

The cross section for p_0 and p_1 was measured at $\Theta = 0°$, and 150° between 1.0 MeV and 5 MeV (Schiffer *et al.*, 1956).

6.5.3.3 $^{9}Be(^{3}He,p)^{11}B$ Q = 10.3219 MeV

The cross section was measured for p_0 and p_1 at $\Theta = 0°$, 7°, 30°, 60°, 90°, 120°, and 150° from 1.8 MeV to 5.1 MeV (Wolicki *et al.*, 1959).

6.5.3.4 $^{9}Be(^{3}He,\alpha)^{8}Be$ Q = 18.9124 MeV

Angular distribution and differential cross section were measured at Θ = 40°, 90°, 125°, and 150° in the energy range of 2.5 - 10.0 MeV (Bilwes *et al.*, 1978).

6.5.3.5 $^{10}B(^{3}He,p)^{12}C$ Q = 19.6931 MeV

Cross section measurement was done for p_0 and p_1 at Θ = 0° and 90° from 0.5 MeV to 5 MeV (Schiffer *et al.*, 1956).

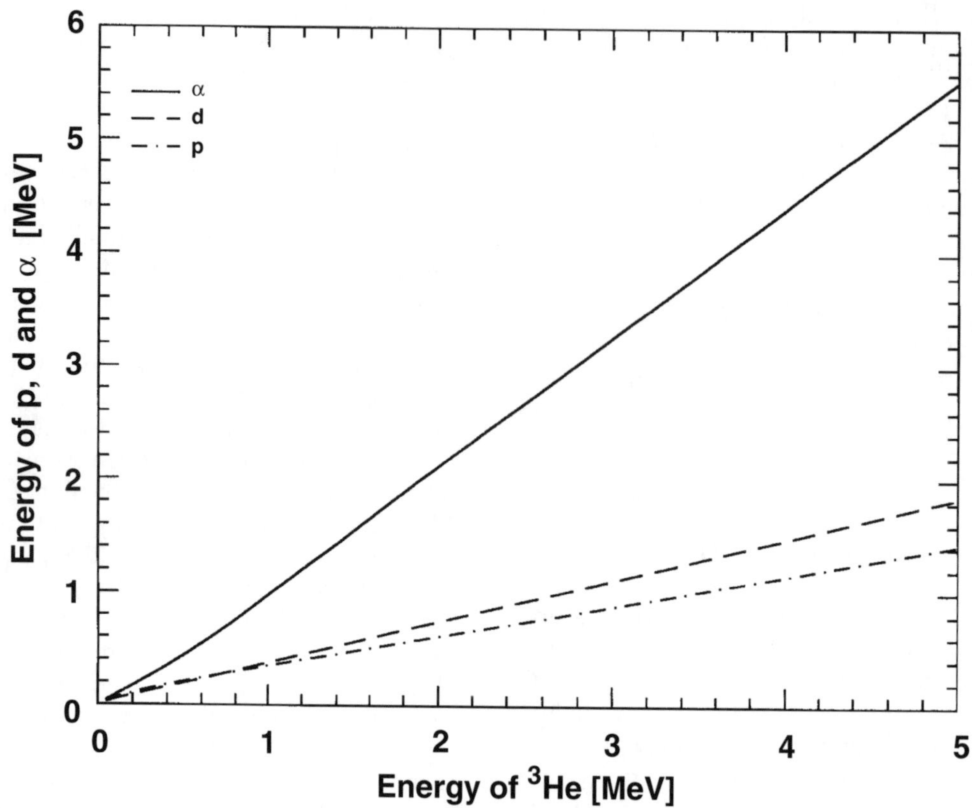

FIG 6.10. The relationship between the energy of ^{3}He and the energy of the reaction products after passing through an absorber foil that stops the primary beam.

6.5.3.6 $^{11}B(^3He,p)^{13}C$ Q = 13.1853 MeV
$^{11}B(^3He,d)^{12}C$ Q = 10.4635 MeV

The cross sections of d_0, p_0, and p_1+p_2+p_3 were measured at Θ = 60°, 90°, and 163° between 3.0 MeV and 5.5 MeV (Holmgren *et al.*, 1959).

6.5.3.7 $^{12}C(^3He,p)^{14}N$ Q = 4.7789 MeV
$^{12}C(^3He,\alpha)^{11}C$ Q = 1.8563 MeV

The cross section of α_0 and p_0 - p_{11} was measured at Θ = 76° and 159.4° in the energy range of 2.0 - 5.5 MeV (Kuan *et al.*, 1964). Recently the cross section of p_0 - p_2 was measured at Θ = 90° from 2.1 - 2.4 MeV (Tong *et al.*, 1990).

6.5.3.8 $^{16}O(^3He,p)^{18}F$ Q = 2.0321 MeV
$^{16}O(^3He,\alpha)^{15}O$ Q = 4.9139 MeV

Angular dependence and differential cross section at Θ = 90° were measured from 2.1 MeV to 3.1 MeV for α and p_0 - p_7 (Bromley *et al.*, 1959). The cross section for α was measured at Θ = 70°, 80°, 90°, 100°, 120°, 130°, 140°, and 150° in the energy range of 3.2 - 6.2 MeV (Ott and Weller, 1972). The cross section around a broad resonance at 2.37 MeV at 90° was remeasured with high precision by Abel *et al.* (1990) and Lennard *et al.* (1989). The application of this reaction is discussed in Abel *et al.*, (1990).

6.5.4 Alpha-induced reactions

The α-induced reactions are not often used for two reasons. First, only a few of the light elements have positive Q values (^{10}B, ^{11}B, ^{19}F, ^{23}Na, ^{27}Al, ^{31}P, and ^{35}Cl). This makes it possible to detect protons in an almost background-free condition. Second, the cross sections are high enough only at relatively high energies, usually above 2 MeV. Most of these cross sections have the same feature, that is they contain many narrow resonances, which might allow high resolution depth profiling. Unfortunately in most cases these resonances are not isolated enough nor are they narrow enough. Therefore, it is usually better to look for some flat places in the cross section and to measure the total areal density in one run. These cross sections have not been included in Appendix 11.

The (α,p) reactions with negative Q value can interfere with RBS measurement at high enough energies. This interference could be important in case of elastic resonance measurements. For example, the resonances of the $^{14}N(\alpha,\alpha)^{14}N$ reaction are above 3 MeV, and the peak of the protons coming from the $^{14}N(\alpha,p)^{17}O$ reaction (Q = –1.1914 MeV) can overlap with the peak of the elastically scattered α particles in a certain energy range.

6.5.4.1 $^{10}B(\alpha,p)^{13}C$ Q = 4.0626 MeV; $^{11}B(\alpha,p)^{14}C$ Q = 0.7839 MeV

These two reactions were studied mainly for nuclear physics (for details, see Dayras *et al.*, 1976; Shire *et al.*, 1953; Shire and Edge, 1955; Wilson, 1975; Lee and Schiffer, 1959; Mani *et al.*, 1963). For boron detection the $^{10}B(\alpha,p)^{13}C$ reaction was used around the 1.567 MeV resonance (Armigliato *et al.*, 1978) and around the 2.67 MeV resonance (Olivier *et al.*, 1976). The $^{11}B(\alpha,p)^{14}C$ reaction was used at 2.67 MeV (Olivier *et al.*, 1976). Recently these reactions have been investigated McIntyre *et al.* (1992).

6.5.4.2 $^{14}N(\alpha,p)^{17}O$ Q = –1.1914 MeV

Although this reaction has a negative Q value it may be worthwhile to mention here. The cross section was measured by Herring *et al.* (1958) and Kashy *et al.* (1958) and recently by Lin *et al.* (1993). The $^{28}Si(\alpha,p)^{31}P$ reaction (Q = –1.916 MeV) can interfere with this reaction. To get an estimate for the region of interference, see Fig. 6.11, which shows the energies of the protons from the $^{14}N(\alpha,p)^{17}O$ and $^{28}Si(\alpha,p)^{31}P$ reactions and also the energy of the elastically-scattered α particles from ^{14}N. An example of an application can be found in Doyle *et al.* (1985).

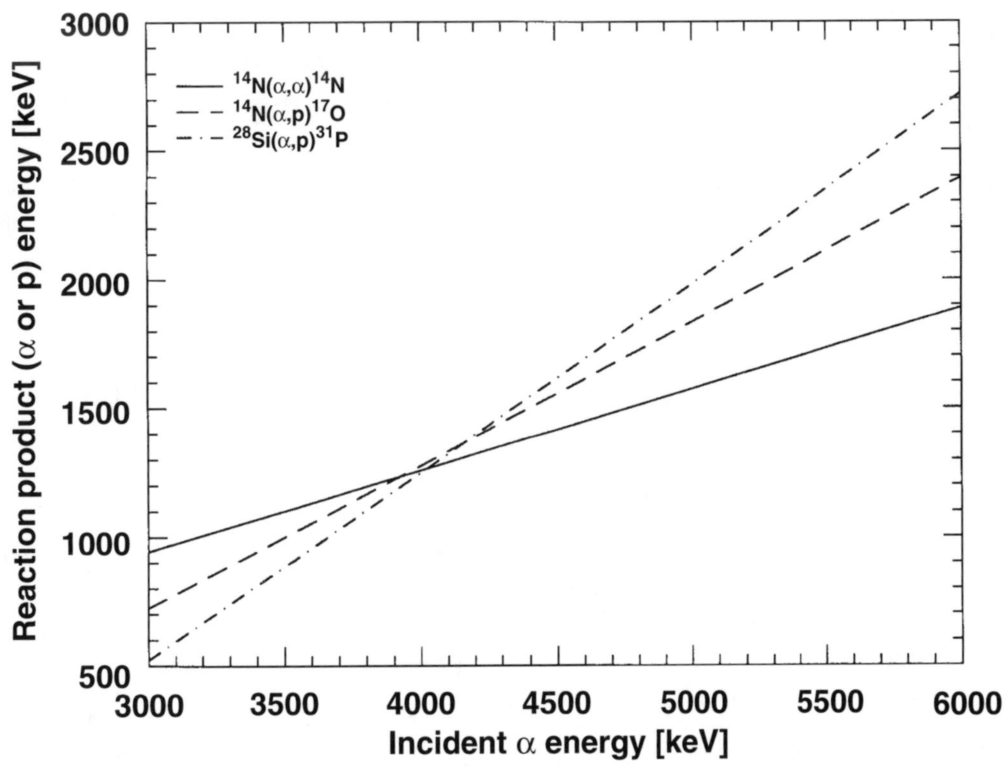

FIG. 6.11. A comparison of the reaction product energies for the $^{28}Si(\alpha,p)^{31}P$, $^{14}N(\alpha,p)^{17}O$, and $^{14}N(\alpha,\alpha)^{14}N$ reactions as a function of incident a energy.

6.5.4.3 $^{19}F(\alpha,p)^{22}Ne$ Q = 1.6746 MeV

For details on this reaction, see Endt and van der Leun (1978). The resonance around 2.443 MeV was used for depth profiling by Smulders (1986) and Kuper *et al.* (1986). Recently McIntyre *et al.* (1989) studied the depth profiling capabilities of this reaction.

6.5.4.4 $^{31}P(\alpha,p)^{34}S$ Q = 0.6316 MeV

The data of this resonance can be found in Kuperus (1964), McMurray *et al.* (1971), and Schier *et al.* (1975). The application of the resonance for depth profiling has recently been examined by Segeth *et al.* (1984), who used the 2.79 MeV and 3.048 MeV resonances, and by McIntyre *et al.* (1988) who used the 3.64 MeV and the 3.97 MeV resonances. They also investigated the 4.43 MeV resonance (which is stronger than the others), but it was not sufficiently isolated to be of easy use in depth profiling. It is worthwhile to note that the resonance energies measured in Segeth *et al.* (1984) and McIntyre *et al.* (1988) (application-based measurements) are systematically 15-20 keV less than those given in Kuperus (1964), McMurray *et al.* (1971), and Schier *et al.* (1975) (nuclear physics measurements).

REFERENCES

Abel, F., Amsel, G., d'Artemare, E., Ortega, C., Siejka, J., and Vízkelethy, G. (1990), *Nucl. Instrum. Methods* **B45**, 100.

Ajzenberg-Selove, F. (1986a), *Nucl. Phys.*, **A449**, 1.

Ajzenberg-Selove, F. (1986b), *Nucl. Phys.*, **A460**, 1.

Ajzenberg-Selove, F. (1987a), *Nucl. Phys.*, **A474**, 1.

Ajzenberg-Selove, F. (1987b), *Nucl. Phys.*, **A475**, 1.

Ajzenberg-Selove, F. (1988), *Nucl. Phys.*, **A490**, 1.

Ajzenberg-Selove, F. (1990), *Nucl. Phys.*, **A506**, 1.

Alkemade, P.F.A., Stap, C.A.M., Habraken, F.H.P.M., and van der Weg, W.F. (1988), *Nucl. Instrum. Methods*, **B35**, 135.

Amsel, G. (1964), Thesis, *Ann. Phys.*, **9**, 297.

Amsel, G. (1991), private communication.

Amsel, G., and David, D. (1969), *Rev. Phys. Appl.*, **4**, 383.

Amsel, G., and Davies, J. (1983), *Nucl. Instrum. Methods*, **218**, 177.

Amsel, G., and Lanford, W.A. (1984), *Ann. Rev. Nucl. Part. Sci.*, **34**, 435.

Amsel, G., and Samuel, D. (1967), *Anal. Chem.*, **39**, 1689.

Amsel, G., Nadai, J., Ortega, C., Rigo, S., and Siejka, J. (1978a), *Nucl. Instrum. Methods*, **149**, 705.

Amsel, G., Nadai, J.P., Ortega, C., and Siejka, J. (1978b), *Nucl. Instrum. Methods*, **149**, 713.

Amsel, G., d'Artemare, E., and Girard, E. (1983), *Nucl. Instrum. Methods*, **205**, 5.

Amsel, G., Pászti, F., Szilágyi, E., and Gyulai, J. (1992), *Nucl. Instrum. Methods* **B63**, 421.

Armigliato, A., Bentini, G.G., Ruffini, G., Battaglin, G., Della Mea, G., and Drigo, A.V. (1978), *Nucl. Instrum. Methods*, **149**, 653.

Baskhin, S., and Richards, H.T. (1951), *Phys. Rev.*, **84**, 1124.

Battistig, G., Amsel, G., and d'Artemare, E. (1991), *Nucl. Instrum. Methods* **B61**, 369.

Battistig, G., Amsel, G., and d'Artemare, E. (1992), *Nucl. Instrum. Methods* **B66**, 1.

Berti, M., and Drigo, A.V. (1982), *Nucl. Instrum. Methods*, **201**, 473.

Biggerstaff, J.A., Hood, R.F., Scott, H., and McEllistrem, M.T. (1962), *Nucl. Phys.*, **36**, 631.

Bilwes, B., Bilwes, R., Ferrero, J.L., and Garcia, A. (1978), *J. Physique*, **39**, 805.

Bird, J.R. (1980), *Nucl. Instrum. Methods*, **168**, 85.

Bird, J.R., Duerden, P., and Clapp, R.A. (1981), *Nucl. Instrum. Methods*, **191**, 345.

Bonner, T.W., Conner, J.P., and Lillie, A.B. (1952), *Phys. Rev.*, **88**, 473.

Bromley, D.A., Kuhner, J.A., and Almquist, E. (1959), *Nucl. Phys.*, **13**, 1.

Bujdoso, E., Lyon, E.S., and Noszlopi, I. (1982), *J. Radioanal. Chem.* **74**, 197.

Cassagnou, Y., Jeronymo, J.M.F., Sadeghi, A., and Forsyth, P.D. (1962), *Nucl. Phys.*, **33**, 449.

Chater, R.J., Kilner, J.A., Reson, K.J., Robinson, A.K., and Hemment, P.L.F. (1990), *Nucl. Instrum. Methods*, **B45**, 110.

Chaturvedi, U.K., Steiner, U., Zak, O., Krausch, G., Shatz, G., and Klein, J., (1990), *Appl. Phys. Lett.* **56**, 1228.

Cheang Wong, J.C., Ortega, C., Siejka, J., Trimaille, I., Sacuto, A., Balkanski, M., and Vizkelethy, G. (1992), *Nucl. Instrum. Methods*, **B64**, 179.

Dayras, R.A., Switowski, Z.E., and Tombrello, A.T. (1976), *Nucl. Phys.*, **A261**,365.

Deconninck, G. (1978), *Introduction to Radioanalytical Chemistry*, Elsevier, Amsterdam.

Dieumegard, D., Dubreuil, D., and Amsel, G. (1979), *Nucl. Instrum. Methods*, **166**, 431.

Dieumegard, D., Maurel, B., and Amsel, G. (1980), *Nucl. Instrum. Methods* **168**, 93.

Doyle, B.L., Follstaedt, D.M., Picraux, S.T., Yost, F.G., Pope, L.E., and Knapp, J.A. (1985), *Nucl. Instrum. Methods*, **B7/8**, 166.

Elwyn, A.J., Holland, R.E., Davids, C.N., Meyer-Shützmeister, L., Mooring. F.P., and Ray, Jr. W. (1979), *Phys. Rev.*, **C20**, 1984.

Endt, P.M. and van der Leun, C. (1978), *Nucl. Phys.*, **A310**, 1.

Feldman, L.C., and Mayer, J.W. (1986) *Fundamentals of Surface and Thin Film Analysis*, North-Holland, New York, Chapter 12.

Gossett, C.R. (1981), *Nucl. Instrum. Methods*, **191**, 335.

Hagendorn, F.B., and Marion, J.B. (1957), *Phys. Rev.*, **108**, 1015.

Herring, D.F., Chiba, R., Gasten, B.R., and Richards, H.T. (1958), *Phys. Rev.*, **112**, 1210.

Hirvonen, J.K., and Luecke, W.H. (1978), *Nucl. Instrum. Methods* **149**, 295.

Holmgren, H.D., Wolicki, E.A., and Johnston, R.L. (1959), *Phys. Rev.*, 114, 1281.

Ishimatsu, T., Nakashima, T., Kato, N., and Morita, S., (1962), *J. Phys. Soc. Japan*, **17**, 1189.

Jarjis, R.A. (1979), Internal Report, University of Manchaster.

Jarmie, N., and Seagrove, J. (1957), Los Alamos Scientific Laboratory Report, LA-2014.

Jausel-Hüsken, S., and Freiesleben, H. (1977), *Z. Phys.*, **A283**, 363.

Jenkin, J.G., Earwaker, L.G., and Titterton, E.W. (1964), *Nucl. Phys.*, **50**, 517.

Kashy, E., Miller, P.D. and Risser, J.R. (1958), *Phys. Rev.*, **112**, 547.

Kashy, E., Perry, R.R., and Risser, J.R. (1960), *Phys. Rev.*, **117**, 1289.

Koenig, W., Riccato, A., Stock, R., Cuzzocrea, P., Perillo, E., Sandolini, M., and Spadaccini, G. (1977), *Il Nuovo Cimento*, **A39**, 9.

Kuan, H-M., Bonner, T.W., and Risser, J.R. (1964), *Nucl. Phys.*, **51**, 481.

Kunz, W.E. (1955), *Phys. Rev.*, **97**, 456.

Kuper, F.G., De Hosson, J.Th.M., and Verwey, J.F. (1986), *J. Appl. Phys.*, **60**, 985.

Kuperus, J.(1964), *Physica* (Utrecht), **30**, 899.

Langley, R.A., Picraux, S.T., and Vook, F.L. (1974), *J. Nucl. Mat.*, **53**, 257.

Lee, Jr. L.L., and Schiffer, J.P. (1959), *Phys. Rev.*, **115**, 160.

Lennard, W.N., Tong, S.Y., Mitchell, I.V., and Massoumi, G.R. (1989), *Nucl. Instrum. Methods*, **B43**, 187.

Li, J., Vizkelethy, G., Revesz, P., Mayer, J.W., Matienzo, L.J., Emmi, F., Ortega, C., and Siejka, J. (1991), *Appl. Phys. Lett.*, **58**, 1344.

Ligeon, E., and Bontemps, A. (1972), *J. Radioanal. Chem.*, **12**, 335.

Lin, Z., McIntyre, Jr. L.C., Leavitt, J.A., Ashbaugh, M.D., and Cox, R.P. (1993), *Nucl. Instrum. Methods* **B79**, 498.

Lu, X., Yuan, Y., Zheng, Z., Jiang, W., and Liu, J. (1989), *Nucl. Instrum. Methods*, **B43**, 565.

Mani, G.S., Forsyth, P.D., and Perry, R.R. (1963), *Nucl. Phys.*, **44**, 625.

Mansour, N.A., Saad, H.R., Saleh, Z.A., Sayed, E.M., Zaloubovsky, I.I., and Gontchar, V.I., (1964), *Nucl. Phys.*, **65**, 433.

Marion, J.B., and Weber, G. (1956), *Phys. Rev.*, **103**, 167.

Marion, J.B., Weber, G., and Mozer, F.S. (1956), *Phys. Rev.* **104**, 1402.

Maurel, B. (1980), Thesis, Université Paris VII.

Maurel, B., and Amsel, G. (1983), *Nucl. Instrum. Methods*, **218**, 159.

Maurel, B., Amsel, G., and Dieumegard, D. (1981), *Nucl. Instrum. Methods*, **191**, 349),

Maurel, B., Amsel, G., and Nadai, J.P. (1982), *Nucl. Instrum. Methods*, **197**, 1.

Mayer, J.W., and Rimini, E. (eds.). (1977), *Ion Beam Handbook*, Academic Press, New York.

McEllistrem, M.T., Jones, K.W., Chiba, R., Douglas, R.A., Herring, D.F., and Silverstein, E.A. (1956), *Phys. Rev.*, **104**, 1008.

McIntyre, Jr. L.C., Leavitt, J.A., Ashbaugh, M.D., Dezfouly-Arjomady, B., Lin, Z., Oder, J., Farrow, R.F.C., and Parkin, S.S.P. (1989), *Nucl. Sci. and Techn.* (China), **1**, 56.

McIntyre, Jr. L.C., Leavitt, J.A., Ashbaugh, M.D., Lin, Z., and Stoner, Jr. J.O. (1992), *Nucl. Instrum. Methods*, **B66**, 221.

McIntyre, Jr. L.C., Leavitt, J.A., Dezfouly-Arjomandy, B., and Oder, J. (1988), *Nucl. Instrum. Methods*, **B35**, 446.

McMurray, W.R., Holtz, D.M., and van Heerden, I.J. (1971), *Z. Phys.*, 247, 453.

Moncoffre, N., Millard, N., Jaffrezic, H., and Tousset, J. (1990), *Nucl. Instrum. Methods*, **B45**, 81.

Möller, W. (1978), *Nucl. Instrum. Methods* **157**, 223.

Möller, W., and Besenbacher, F. (1980), *Nucl. Instrum. Methods*, **168**, 111.

Möller, W., Hufschmidt, M., and Kamke, D. (1977), *Nucl. Instrum. Methods* **140**, 157.

NNDC (National Nuclear Data Center), Building 197D, Brookhaven National Laboratory, Upton, NY 11973, USA. Several on-line data bases are available. Telephone (516) 282-2901, Fax (516) 282-2806, Internet: nndc@bnl.gov.

Olivier, C., McMillan, J.W., and Pierce, T.B. (1976), *J. Radioanal. Chem.*, **31**, 515.

Ott, W.R., and Weller, H.R. (1972), *Nucl. Phys.*, **A179**, 625.

Payne, R.S., Clough, A.S., Murphy, P., and Mills, P.J. (1989), *Nucl. Instrum. Methods*, **B42**, 130.

Phillips, D., and Pringle, J.P.S. (1976), *Nucl. Instrum. Methods*, **135**, 389.

Press, W.H., Falnnery, B.P., Teukolsky, S.A., and Vetterling, W.T. (1986), *Numerical Recipes*, Cambridge University Press, Cambridge.

Pretorius, R., and Peisach, M., (1978), *Nucl. Inst. Meth.* **149**, 69.

Priegle, J.P.S. (1972), *J. Electrochem. Soc.*, **119**, 482.

Proceedings of IBA Conferences, 1973 Yorktown Heights (*Thin Solid Films* **19**), 1975 Karlsruhe (Plenum Press, New York), 1978 Washington (*Nucl. Instrum. Methods* **149**), 1980 Aarhus (*Nucl. Instrum. Methods* **168**), 1981 Sydney (*Nucl. Instrum. Methods* **191**), 1983 Tempe (*Nucl. Instrum. Methods* **218**), 1985 Berlin (*Nucl. Instrum. Methods* **B15**), 1988 Johannesburg (*Nucl. Instrum. Methods* **B35**), 1989 Kingston (*Nucl. Instrum. Methods* **B45**), 1991 Eindhoven (*Nucl. Instrum. Methods* **B64**).

Pronko, P. (1974), *J. Nucl. Mat.*, **53**, 252.

Pronko, P.P., and Pronko, J.G. (1974), *Phys. Rev.*, **B9**, 2870.

Purser, K.H., and Wildenthal, B.H. (1963), *Nucl. Phys.*, **44**, 22.

Qiu, Q., Arai, E., Aratani, M., Yanokura, M., Nozaki, T., Ohji, Y., and Imura, R. (1989), *Nucl. Instrum. Methods*, **B44**, 179.

Reichle, R., Berisch, R., and Roth, J. (1990), *Nucl. Instrum. Methods* **B50** 68.

Sagara, A., Kamada, K., and Yamaguchi, S. (1988), *Nucl. Instrum. Methods*, **B34**, 465.

Sawicki, J.A., Davies, J.A., and Jackman, T.E. (1986), *Nucl. Instrum. Methods*, **B15**, 530.

Scanlon, P.J., Farell, G., Ridgway, M.C., and Valizadeh, R. (1986), *Nucl. Instrum. Methods*, **B16**, 479.

Schier, W.A., Barnes, B.K., Couchell, G.P., Egan, J.J., Harihar, P., Mathir, S.C., Mittler, A., and Sheldon, E. (1975), *Nucl. Phys.*, **A254**, 80.

Schiffer, J.P., Bonner, T.W., Davis, R.H., and Prosser, Jr. F.W. (1956), *Phys. Rev.*, **104**, 1064.

Segeth, W., Bakema, D., and Boerma, D.O. (1984), Proc. Int. Symp. on Three Day in Depth Review on the Nuclear Accelerator Impact in the Interdisciplinary Field, Lanoratori Nazionali di Legnaro, Padova, Italy, 30 May - 1 June, 1984, Mazzoldi, P., and Moschini, G. (eds.), 66.

Seiler, F., Jones, C.H., Anzick, W.J., Herring, D.F., and Jones, K.W. (1963), *Nucl. Phys.*, **45**, 647.

Shaanan, M., Kalish, R., and Richter, V. (1986), *Nucl. Instrum. Methods*, **B16**, 56.

Shire, E., and Edge, R.D. (1955), *Phil. Mag.*, **46**, 640.

Shire, E.S., Wormald, J.R., Lindsay-Jones, G., Lunden, A., and Stanley, A.G. (1953), *Phil. Mag.*, **44**, 1197.

Siejka, J., Nadai, P., and Amsel, G. (1971), *J. Electrochem. Soc.*, **118**, 727.

Sierk, A.J., and Tombrello, T.A., (1973), *Nucl. Phys.*, **A210**, 341.

SIGBASE. A service supported on a trial basis by I.C. Vickridge, Institute of Geological and Nuclear Sciences, Ltd., P.O. Box 31312, Lower Hutt, New Zealand; and G. Vizkelethy, Department of Physics, Idaho State University, Campus Box 8106, Pocatello, ID 83209-8106, USA. Internet: srlnicv@lhn.gns.cri.nz or vizkel@physics.isu.edu. Anonymous ftp to lhn.gns.cri.nz or physics.isu.edu.

Simpson, J.C.B., and Earwaker, L.G. (1984), *Vacuum*, **34**, 899.

Smulders, P.J.M. (1986), *Nucl. Instrum. Methods*, **B14**, 234.

Sokolov, J., Rafailovich, M.H., Jones, R.A.L., and Kramer, E.J. (1989), *Appl. Phys. Lett.* **54**, 590.

Sweeney, Jr. W.E., and Marion, J.B. (1969), *Phys. Rev.*, **182**, 1007.

Symons, G.D., and Treacy, P.B. (1963), *Nucl. Phys.*, **46**, 93.

Tong, S.Y., Lennard, W.N., Alkemada, P.F.A., and Mitchell, I.V. (1990), *Nucl. Instrum. Methods*, **B45**, 91.

Turos, A., Wielunski, L., and Batcz, A. (1973), *Nucl. Instrum. Methods*, **111**, 605.

Vickridge, I.C. (1988), *Nucl. Instrum. Methods*, **B34**, 470.

Vickridge, I.C. (1990), Thesis, Université Paris VII.

Vizkelethy, G., (1990), *Nucl. Instrum. Methods*, **B45**, 1.

Wielunski, M., and Möller, W. (1990), *Nucl. Instrum. Methods*, **B50**, 23.

Wilson, S.J. (1975), *Phys. Rev.*, **C11**, 1071.

Wolicki, E.A., Holmgren, H.D., Johnston, R.L., and Geer Illsley, E. (1959), *Phys. Rev.*, **116**, 1585.

Yarnell, J.L., Lovberg, R.H. and Stratton, W.R. (1953), *Phys. Rev.*, **90**, 292.

Ziegler, J.F. (1975), *New Uses of Ion Accelerators*, Plenum, New York.

CHAPTER 7

NUCLEAR REACTION ANALYSIS: PARTICLE-GAMMA REACTIONS

J.-P. Hirvonen

Technical Research Centre of Finland, Espoo, Finland

Contributor: R. Lappalainen

CONTENTS

7.1 INTRODUCTION

The unique feature of any technique for nuclear reaction analysis is that it is sensitive only to specific isotopes. This of course restricts its general applicability, but makes it the most powerful tool for those cases in which it can be utilized. To understand the limitations and advantages of nuclear reaction analysis with γ-ray emission, and to enable the reader to decide when these techniques are worth using, it is necessary to compare this technique to other methods performed with energetic ion beams.

Nuclear reaction analysis with γ-ray emission is a less frequently used ion beam analysis technique than Rutherford backscattering (RBS) or particle-induced x-ray emission (PIXE). This is understandable because elements or isotopes appear both in RBS and PIXE spectra in a simple, systematic, and predictable way. This makes the interpretation and analysis of the measurements quite straightforward. In nuclear reactions, even different isotopes of the same element behave in different ways, and no systematics can be found. For this reason, a researcher who wants to utilize nuclear reaction has to know, isotope by isotope, the energies of resonances, the energies of γ-rays, etc. In many cases, the knowledge of the properties of the isotope involved is not sufficient; possible interfering reactions in the sample matrix must be known as well. This results in huge tables full of numerical data, as can be seen in Appendix 12.

RBS and PIXE are considered as reasonably general techniques which can also be used to probe the chemical composition of completely unknown samples. Contrary to this, nuclear reaction analyses with γ-ray emission are mostly used when the chemical composition of the sample is qualitatively known and a particular γ-reaction is used to quantify the concentration or detect changes in the concentration profiles. Furthermore, there are only a few isotopes which have a potential for depth profiling. These can be found at the low atomic number section of the table of isotopes. Isotopes such as ^{13}C, ^{15}N, ^{18}O, ^{19}F, ^{22}Ne, ^{23}Na, ^{24}Mg, ^{26}Mg, ^{27}Al, ^{29}Si, and ^{30}Si are normally used. Hydrogen profiling is also widely used, and is discussed in Chapter 8. Special expertise and highly trained researchers are required to try depth profiling of isotopes that are not on the above list.

Besides depth profiling, γ-ray reactions can be used in elemental analysis in very much the same manner as PIXE. In this case, the number of possible isotopes and elements which can be detected is much higher than with depth profiling.

In summary, nuclear reaction analysis with γ-ray emission provides a quantitative and efficient technique in those cases in which it can be utilized. Its unique feature is a sensitivity for isotopes which resembles secondary ion mass spectrometry (SIMS). However, compared to SIMS, nuclear reaction analysis with γ-ray emission is non-destructive and accurate. Moreover, it is normally quite rapid and the number of samples to be measured is not a limiting factor.

7.2 PARTICLE-GAMMA REACTIONS

When a beam of charged particles, such as protons, helium ions, or deuterons hits the sample surface layer, nuclear reactions are induced and gamma radiation is emitted. Here the discussion is restricted to prompt gamma emission (Deconninck *et al.*, 1983), meaning that radiation is detected during irradiation, in contrast to activation techniques where the radiation is detected after irradiation. Activation analysis is discussed in detail in Chapter 9. It has been used widely to measure bulk concentrations of the elements with medium or heavy atomic weight. Activation methods have also been used for profiling by etching the sample surface, layer by layer.

When a sample is bombarded with particles at a fixed energy, the total gamma-ray yield from a certain reaction can be used to deduce the average concentration of the corresponding element in surface layers. With light ions of an energy of a few MeV, the maximum depth analyzed is typically well below 50 μm. When resonance reactions are used for profiling, the gamma-ray yield of a resonance reaction within a proper gamma-ray energy range is measured vs. bombarding energy starting just below the resonance. The profiled depth is typically limited to a few microns. The measurement of profiles and average concentrations using nuclear reactions leading to gamma-ray emission will be discussed separately.

7.2.1 Depth profiling by using narrow resonances

7.2.1.1 General principles

Most of the light nuclei ($Z<30$) have strong, sharp resonances in the cross section $\sigma(E)$ of nuclear reactions induced by light ions at low bombarding energies (< 3 MeV). Here the discussion is limited to resonances leading to gamma-ray emission.

Consider a flat, laterally uniform sample containing a homogeneous distribution N(x) of an isotope of the element under study. The sample is bombarded with an energy-analyzed particle beam perpendicular to the surface, and the induced γ-ray emission is detected. If the energy of the beam corresponds to the energy of a narrow resonance E_r, resonance reaction can take place only on the surface. At higher bombarding energies, the particles must slow down in the sample to the depth at which E_r is reached, when the reaction is possible. The yield detected at this bombarding energy E_b reflects the concentration at this depth. Hence the excitation curve of the resonance reaction $Y(E_b)$, i.e., the yield of the reaction vs bombarding energy gives an estimate of the concentration distribution in the surface layer.

In practice, the experimental width of the resonance profile is broadened due to the natural width of the resonance, the energy resolution of the beam, and the energy straggling of the beam particles during the energy loss E_b-E_r. In order to deduce the precise shape of the actual concentration profile $N(x)$, the shape of the resonance cross section $\sigma(E)$, the energy distribution of the beam $g(E_b, E)$, and the energy straggling of particles at depth x $f(E, E', x)$ have to be taken into account

$$Y(E_b) = C\int_0^\infty dx \int_0^\infty dE \int_0^\infty dE'\, N(x) g(E_b, E) f(E, E', x) \sigma(E') \ . \quad (7.1)$$

Here C is a constant for given detection conditions (for example, same geometry and collected charge). If the resonance is narrow and beam energy resolution is good, the functions g, f, and σ are sharp and therefore the yield curve corresponds well to the actual distribution.

As an example of the effect of experimental resolution, a measured range profile of 100 keV ^{22}Ne ions is illustrated in Fig. 7.1 (represented by the dots). The solid line is a calculated yield curve [Eq. (7.1)] for the optimized concentration profile (dashed line). This split-Gaussian concentration profile was obtained by a fitting procedure comparing the calculated and the measured yield curve. In this case, the experimental resolution is determined mainly by the energy straggling (function f), especially in the tail part of the distribution.

The actual distribution N(x) can be deconvoluted from the integral Eq. (7.1) using numerical methods. However, this problem is ill-defined and there are several mathematically correct solutions. The solution is rarely reasonable or correct, unless physically reasonable additional boundary conditions, e.g. positivity and

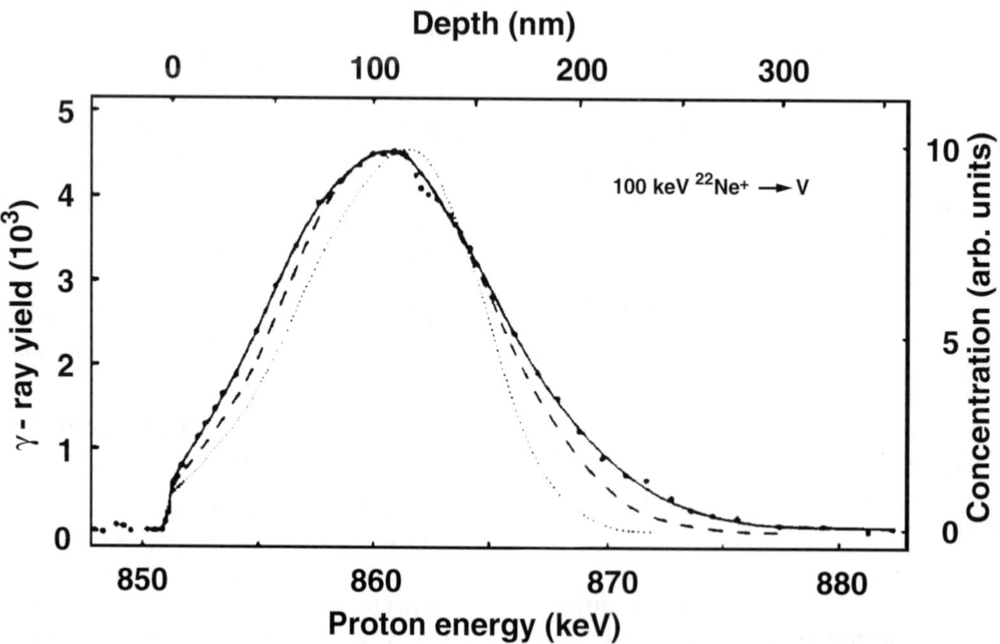

FIG. 7.1. Gamma-ray yield curve for a 100 keV ^{22}Ne distribution in vanadium. The solid line is the yield calculated for the split-Gaussian concentration profile (dashed line) obtained using a fitting procedure. The dotted line is the Monte Carlo simulation for amorphous backing (Lappalainen, 1986b).

smoothness of solution, are used in solving the equation. Several computer programs (Land *et al.*, 1976; Maurel *et al.*, 1982; Deconninck and Oystaeyen 1983; Smulders, 1986; Lappalainen, 1986a; Vickridge and Amsel 1990; Rickards, 1991) have been developed to extract an accurate depth profile from the measured yield curve. Mostly applied methods use either an iteration procedure starting from the measured curve as the first approximation or the fitting of the parameters of a certain distribution function. The first approach is suitable for all kinds of profiles. On the other hand, for example, ion implanted or diffusional profiles can often be described with split-Gaussian, Pearson, or spline functions which are easy to fit. In both cases, a complete profile can normally be obtained in less than a half an hour.

7.2.1.2 Equipment

The instrumentation is discussed in detail in Chapter 10. Here the discussion is limited to some special features relevant for profiling measurements. The necessary equipment for profiling with narrow resonances is an accelerator and a gamma-ray detector with a suitable data collection and analyzing setup, typically a multichannel analyzer or a microcomputer with an ADC board.

Accelerator

The main requirements of the accelerator for profiling are reasonable energy resolution and the possibility to change beam energy easily. A magnetic field is used to analyze the energy of the beam and to separate particles of only one type from the original beam containing several ions with different charge states. The bombarding energy is generally increased stepwise by adjusting magnetic field and terminal voltage. However, more sophisticated energy scanning systems (Amsel *et al.*, 1983; Meier and Richter, 1990) have been developed to increase the energy of the beam while keeping the analyzing magnetic field constant and selecting the energy by electrostatic deflection plates just before and after the analyzing magnet. In profiling, the advantage of the high energy resolution of the beam is restricted to very narrow resonances (<0.2 keV) and the outermost surface layer (<50 nm) since energy straggling due to the slowing down process starts to dominate the total depth resolution deeper in the sample. High energy resolution is necessary in some novel applications of narrow resonances such as studies on the Lewis effect and Doppler broadening of the resonance width due to the vibrational motion of the detected atoms (Fujimoto, 1990; Mitchell and Rolfs, 1991).

Gamma-ray detector

In the particle-induced gamma measurements, the gamma radiation is generally detected either by a NaI, BGO, or Ge(Li) detector. In all these detectors, single- and double-escape peaks are observed besides the full-energy peak corresponding to the gamma-ray energy. These escape peaks are observed 0.511 MeV and 1.022 MeV below the main peak and are prominent and especially useful with high energy gamma rays. The NaI scintillation counter is conventionally used for profiling when high efficiency is needed and energy resolution is not critical. This is normally the case when light elements are profiled in samples of high Z value using well-isolated resonances. The BGO detectors have become available quite recently. They have higher efficiency for a given size than NaI detectors and a larger fraction of pulses in the full-energy peak (Kuhn *et al.*, 1990). Furthermore, the BGO detectors offer a better signal-to-background ratio and a compact setup which are necessary, especially when profiling with weak resonances. The Ge(Li) detectors are generally used for the analysis of total average concentrations of elements in the surface layer, since these detectors have high enough energy resolution to distinguish nearby gamma rays from competing reactions. Before profiling it is helpful to measure a gamma spectrum from the sample with a Ge(Li) detector to identify possible interfering peaks from competing reactions or laboratory background and to determine the suitable energy window (region-of-interest or ROI) to sum the total yield.

Gamma ray and beam energy calibration

One standard procedure at the beginning of a set of measurements is the calibration of the gamma-ray detection system as well as the beam energy. The gamma-ray detection system is typically calibrated for energy and efficiency using a radioactive source for low energy gamma rays and well known gamma-ray resonance transitions for high energy gamma rays. Suitable radioactive sources such as ^{56}Co, ^{226}Ra, and ^{152}Eu cover the range 0.1-3.0 MeV. Figure 7.2 shows a gamma-ray calibration spectrum measured using ^{60}Co, ^{208}Tl, and ^{207}Bi. For higher gamma-ray energies, resonances such as the 992 keV ^{27}Al(p,γ)^{28}Si resonance (Anttila *et al.*, 1977a) and

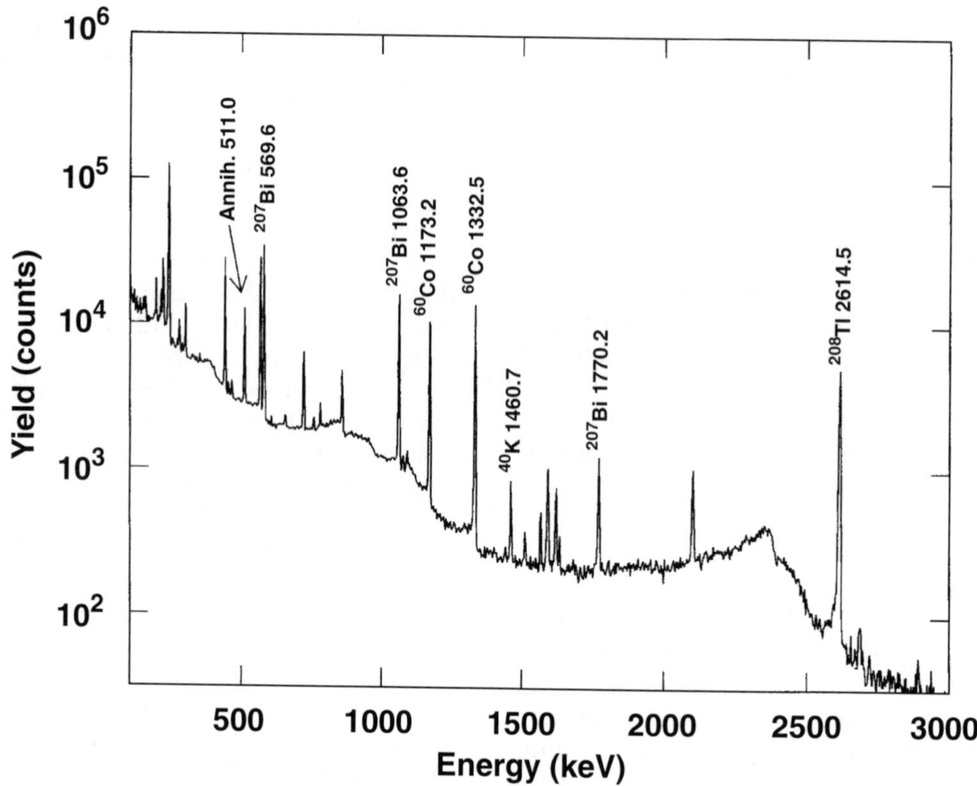

FIG. 7.2. Gamma-ray energy calibration spectrum measured using ^{60}Co, ^{208}Tl, and ^{207}Bi radioactive sources and a 110 cm^3 Ge(Li) detector.

the 441 keV $^7Li(p,\gamma)^8Be$ resonance can be used. Figure 7.3 shows a spectrum measured for the 992 keV $^{27}Al(p,\gamma)^{28}Si$ resonance. The gamma rays that correspond to the transitions from the states with a very short lifetime are emitted while the residual nuclei are slowing down in the sample. Therefore, these gamma peaks shift due to the Doppler effect unless the gamma-ray detector is at an angle of 90° with respect to the beam (Fig. 7.3). Since the Doppler shift is typically less than 1% of the gamma-ray energy, the effect can only be seen using a high resolution detector such as Ge(Li) detectors.

The energy calibration and resolution of accelerators is usually determined utilizing the same narrow, strong resonances which are used for profiling (see Appendix 17). In the case of protons, 991.90±0.04 keV $^{27}Al(p,\gamma)^{28}Si$ resonance and 429.57±0.09 keV $^{15}N(p,\alpha\gamma)^{12}N$ are generally used. The energy distribution of the beam is generally assumed to be a Gaussian

$$g(E_b, E) = \left(\frac{1}{(2\pi)^{1/2} \times \sigma_b}\right) \exp\left\{-[(E_b - E)/\sigma_b]^2 / 2\right\} \quad (7.2)$$

where E_b is the mean energy and σ_b is the standard deviation of the energy distribution. σ_b can be measured, for example, by using a resonance yield curve for a thick sample, which will be explained in 7.2.1.8.

7.2.1.3 Possible isotopes and reactions

Among light ion-induced resonance reactions, those induced by protons have been used mostly for profiling. In fact, most of the light nuclei have (p,γ) or (p,αγ) resonances at bombarding energies below 2 MeV. The cross section of an isolated resonance is given by the Breit-Wigner formula

$$\sigma_{ab}(E) = \frac{\pi\gamma\lambda^2\Gamma_a\Gamma_b}{(E - E_r)^2 + \Gamma^2 / 4} \quad (7.3)$$

where a and b refer to the incident and outgoing particle, γ is a statistical factor including spin numbers, λ is the de Broglie wavelength ($\lambda^2 \sim 1/E_b^2$), Γ is the resonance width, and Γ_a and Γ_b are partial widths. Resonance cross sections are usually tabulated either as a cross section at

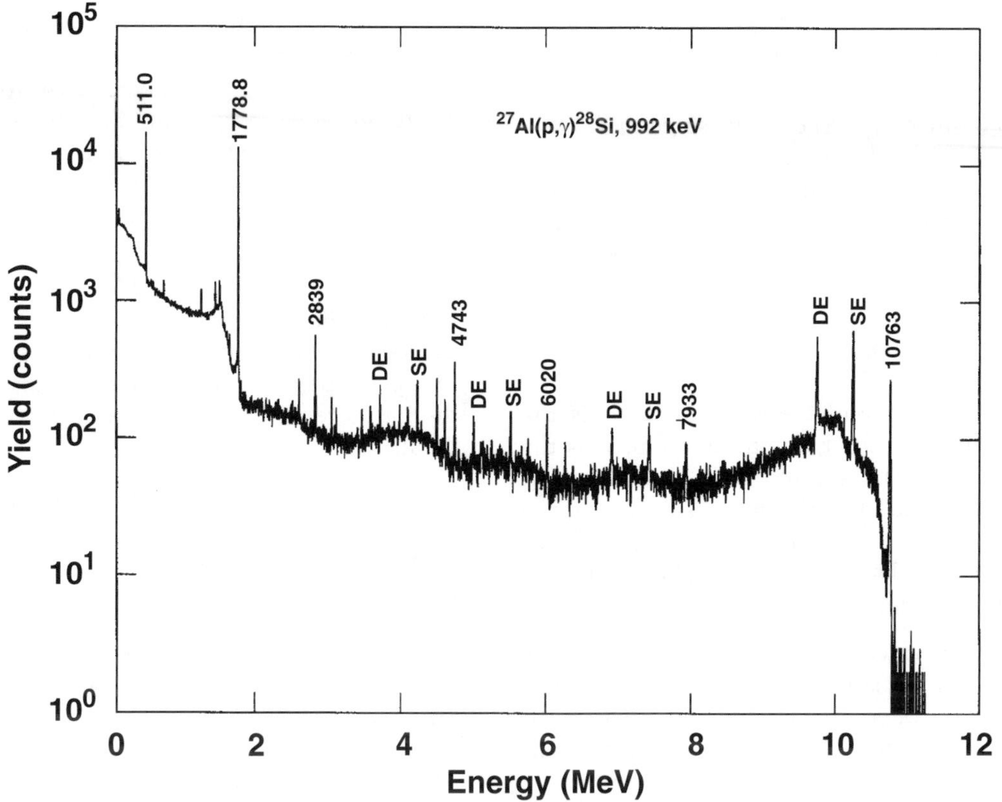

FIG. 7.3. Gamma-ray spectrum for the 992 keV $^{27}Al(p,\gamma)^{28}Si$ resonance measured using a thin film Al sample and a Ge(Li) detector. Notice that each high-energy gamma ray gives three peaks in the spectrum — full-energy photo peak (P.P.), single escape (S.E.), and double escape (D.E.) peaks. The photo peak corresponds to the case when the total energy of the gamma ray is absorbed in the detector. The escape peaks are observed when one or both of the annihilation gamma rays (511 keV) are lost from the detector.

the resonance energy [$\sigma(E_r) = \pi\gamma\lambda^2 4\Gamma_a\Gamma_b/\Gamma^2$ in barns = 10^{-24} cm^2] or as a resonance strength [$S = (2J+1)\Gamma_a\Gamma_b/\Gamma$ in eV, where J is a spin quantum number of the resonance level]. Both of these parameters are directly related to the measured gamma-ray yield and also included in Tables 1–4. If the energy resolution of the beam is smaller than the resonance width ($\sigma_b << \Gamma_r/2.355$), the total gamma-ray yield from the resonance is proportional to the maximum of the cross section. On the other hand, if $\sigma_b >> \Gamma_r/2.355$, the total yield is proportional to the integral of the cross section and resonance strength S.

Table A12.1 (in Appendix 12) lists most of the known (p,γ) resonances by proton energy below 3 MeV (Butler, 1959). Table A12.2 shows the main useful resonances of this type for most of the light elements. This table is based on the earlier compilation by Golicheff *et al.* (1972), with some useful resonances added for elements heavier than fluorine.

In principle, (α,γ), (α,nγ), and (α,pγ) resonances could offer higher depth resolution than proton resonances, due to the higher stopping power of He ions relative to H ions. However, only a few well-isolated (α,γ) resonances exist and are suitable for profiling. Some of these are collected in Table A12.3.

The same resonances that have been used for profiling with protons have been extensively utilized with inverse resonance reactions to profile hydrogen. These resonances are listed in Table A13.2. In particular, the 6.4 MeV $^{1}H(^{15}N,\alpha\gamma)^{12}C$ resonance with the detection of 4.43 MeV gamma rays is widely used for hydrogen detection, due to the easy availability of nitrogen beams and good resolution at the surface (see Chapter 8).

Some factors affecting the efficiency of measurements

The main factors which determine the efficiency of the measurement are the strength of the resonance, beam current, and gamma-ray detection efficiency. These factors can be expressed in a simplified relation for the measured gamma-yield

$$Y \sim \varepsilon w(\Theta)\Omega ItS/E_r \qquad (7.4)$$

where ε is the efficiency of the detector, $w(\Theta)$ is the angular distribution of the gamma radiation, Ω is the solid angle, I is the beam current, t is the measurement time, S is the resonance strength, and E_r is the resonance energy. For a given isotope, both the S/E_r dependence and the amount of background radiation favors the use of low bombarding energies.

The beam current can usually be increased up to a few μAs when the sample is cooled with water or liquid nitrogen and a large beam spot is used. In some cases, it might be necessary to move the beam spot during profiling due to deterioration of sample surface, for example, hydrogen beams form blisters in tungsten at high doses. The profiled element may also diffuse due to the heating of the sample and radiation-enhanced diffusion. Sometimes the counting rate of the detection system becomes a limiting factor with high beam currents. When the high counting rate is due to the very intensive low-energy gamma rays (<1 MeV) from reactions in the sample, lead absorber plates can be used to decrease the intensity of low gamma-ray energies.

Since the effective solid angle of the detector varies roughly as $1/r^2$, the sample-to-detector distance should be taken into account in the design of measurement chamber, detector shielding, and absorber attachment. The intensity of gamma radiation depends on Θ, the detection angle with respect to the beam direction as

$$w(\Theta) = 1 + a_2P_2(\cos\Theta) + a_4P_4(\cos\Theta) \qquad (7.5)$$

where $P_i(\cos\Theta)$ is the Legendre polynomial of order i, and coefficients a_i depend on the type of gamma transition. Usually, the third term is small and a good average yield can be obtained at the angles of 55° and 125°, which are zeroes for P_2.

Generally, the efficiency of a detector decreases quickly with increasing gamma-ray energy, as illustrated in Fig. 7.4 for a 110 cm^3 Ge(Li) detector (Anttila *et al.*, 1977a). Therefore, low-energy gamma rays from a resonance reaction should be included in the energy window unless interfering gamma-ray yield reduces signal-to-background ratio significantly.

Subtraction of background and interfering gamma-yield

Since reliability and accuracy of the measured depth profile depend on the correct identification of the interfering resonances and gamma rays, and on minimizing the interferences, these will be discussed next in more detail.

The background of the measured gamma-ray yield curve is mainly due to natural background radiation, competing reactions caused by contaminants or other components of the sample, reactions in slits collimating the beam, and noise in the electronics. The most common source is the natural radioactivity in the building materials of the laboratory. Figure 7.5 shows a typical laboratory background measured with a Ge(Li) detector. The energy, intensity, and origin of the strongest peaks are denoted. The strongest peaks are due to ^{208}Tl (2614.6 keV) and ^{40}K (1461 keV), which are also often used for energy calibration. The natural background occurs mainly below 3 MeV (see Table A12.8 in Appendix 12) and can be significantly reduced by a proper shielding of the detector and reducing the data collection time using higher beam current. Quite often the desired yield curve undergoes interference from resonances or continuous gamma-ray yield due to contaminants or other components of the sample. Table A12.1 lists most of the known (p,γ) and (p,αγ) resonances by proton energy below 3 MeV (Butler, 1959). For each resonance, the table includes the energy of three gamma rays which are of the greatest aid in identifying the resonance, the total cross section in millibarns at the resonance peak (or integrated cross section in evb, electron-volt barn), and the full width at half maximum of the resonance in the laboratory system of coordinates. In principle, all the resonances below and even some broad resonances above the bombarding energy can cause interfering gamma yield. The effect of contamination depends on its distribution. If the contaminant is on the surface only, the gamma yield due to a contaminant corresponds to the resonances near the bombarding energy. On the other hand,

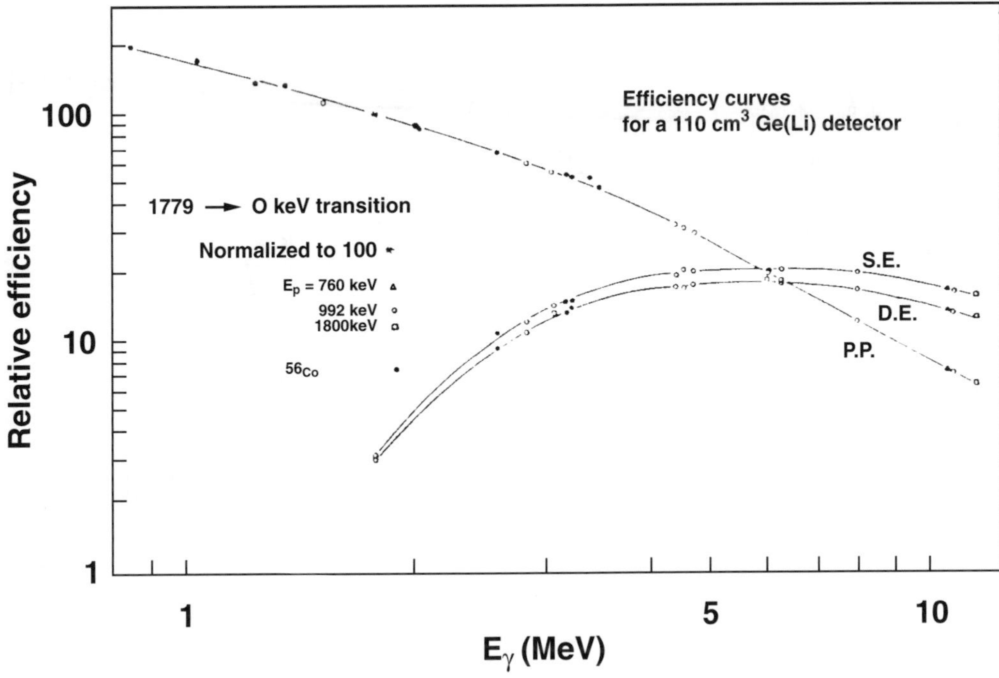

FIG. 7.4. Experimental full-energy photo peak, single escape, and double escape efficiency curves for the 110 cm^3 Ge(Li) detector measured using $^{27}Al(p,\gamma)^{28}Si$ resonances and ^{56}Co radioactive source (Anttila *et al.*, 1977a).

if the interfering yield is due to the contaminant uniformly distributed in the sample or due to a certain component of the sample, the interfering yield corresponds to the thick target gamma-ray yield, and steps in the yield will occur when the energy of a specific resonance of the interfering isotope is achieved. Table A12.6a, which will be explained later, lists the gamma rays and yields from thick targets by proton bombardment. These tables help to select a proper resonance for profiling as well as to interpret possible interferences.

The most common interfering resonances are those of fluorine, nitrogen, and carbon. These elements have several strong resonances. Some of those resonances are very broad (50-500 keV), which makes them even more likely candidates for interference even if these elements are located in the surface layer only. However, fortunately the gamma-ray yield from the resonance reactions of each of these elements corresponds to a separated energy window (region-of-interest). The most prominent gamma rays for fluorine resonances are in the energy range 6.13-7.12 MeV, 4.43 MeV for nitrogen resonances, and for carbon resonances either at the low energies 2-3 MeV or around 9 MeV. The use of the resonance tables and the selection of the proper energy windows will be explained with specific examples in 7.2.1.8.

If the profile is limited to the surface layer, the gamma-ray yield can be measured on both sides of the distribution and background subtraction can generally be done using a linear approximation. In some cases the right background level (especially above the resonance) is not that obvious, for example, when the measured profile is extending beyond the maximum analyzable depth or when the interfering gamma yield is changing significantly with increasing bombarding energy. In these cases, the right background level and the correct subtraction of interferences can be confirmed by measuring a few points of the distribution profile using a Ge(Li) detector which typically allows the resolution of the characteristic gamma rays of the separated resonances. In some cases, it is possible to determine the background and interferences by measuring the yield curve from a similar sample but without the measured isotope, or with a very thin layer containing the measured isotope. This is generally possible when ion-implanted profiles are measured.

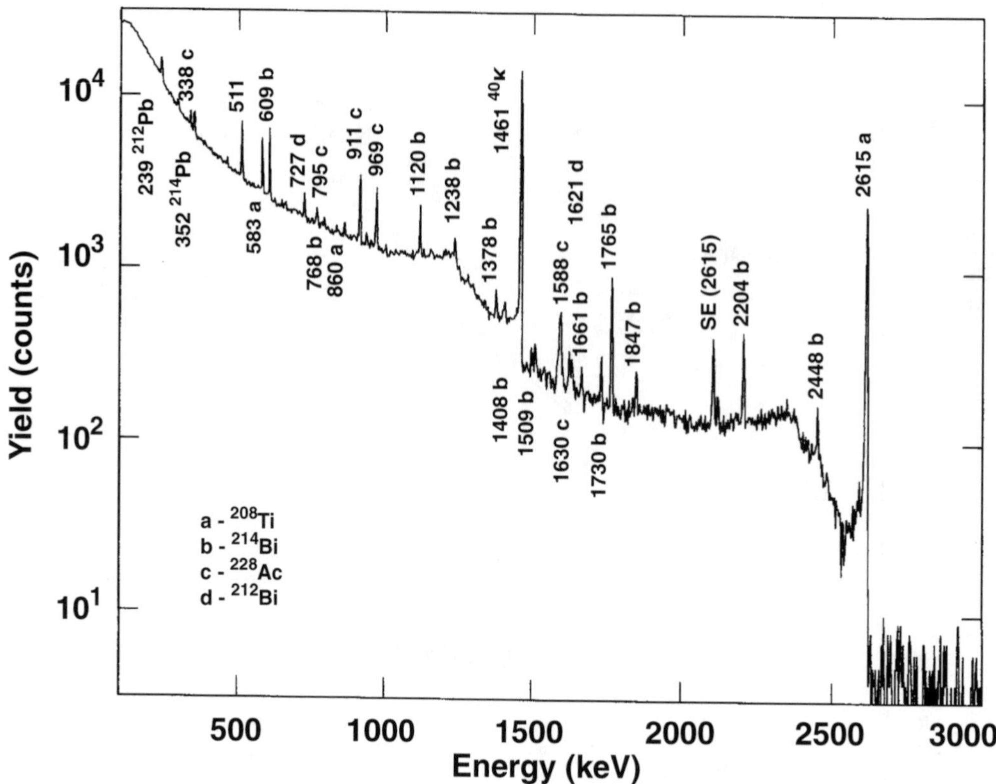

FIG. 7.5. Typical laboratory background spectrum measured with a non-shielded Ge(Li) detector. Origin of the strongest peaks are denoted (R. Lappalainen, private communication).

7.2.1.4 Concentration and depth scales

Let us assume that the background and possible interferences have been subtracted from the measured yield curve. As discussed in 7.2.1.1, the measured gamma-ray yield curve gives a fairly good representation of the actual concentration profile of atoms studied. Then the gamma-ray yield vs bombarding energy curve can be easily converted to a concentration vs depth plot. When the gamma yield for an element *A* from a sample A_mB_n (m+n=1) is compared to that of a reference standard containing a known atomic fraction *f* of the element A, the difference in the stopping cross sections in these two media must be taken into account. Then the atomic fraction of the measured element in the sample can be obtained as

$$m = \frac{fY_{A_mB_n}\varepsilon_B}{Y_{st}\varepsilon_{st} + fY_{A_mB_n}(\varepsilon_B - \varepsilon_A)} \qquad (7.6)$$

with *st* as the standard, *A* is the measured element, and *B* is the rest of the sample atoms. The Y_{st} is the gamma-ray yield step height of the resonance measured for the standard. The stopping cross sections ε of different atoms at the resonance energy can be calculated using tables, for example, Andersen and Ziegler (1977) for protons and Ziegler (1977) for helium (see Appendix 3). Stopping cross sections of the mixtures are calculated using Bragg's rule, $\varepsilon_{AmBn} = m\varepsilon_A + n\varepsilon_B$, unless experimental stopping cross section values are available.

The depth *x* which is needed to slow down bombarding particles with a mean energy E_b to an energy E′ is given by

$$x = \int_{E'}^{E_b} \frac{dE}{S(E)} \qquad (7.7)$$

where $S(E)$ is the stopping power of the sample. In practice, the depths x_i corresponding to successively increasing bombarding energies E_i can be obtained, for example, using a simple approximation

$$x(E_1) = \frac{E_1 - E_r}{S(E_1)}$$

$$x(E_i) = x_{i-1} + \frac{2(E_i - E_{i-1})}{S(E_i) + S(E_{i-1})}, \quad i \geq 2 \tag{7.8}$$

where $S(E_i)$ is the stopping power of the sample at the energy E_i. This is calculated using Bragg's rule and taking into account the fraction m at this point. Typically, the energy is in keV, S in keV/nm, and depth in nm. However, if the density of the sample is not known, stopping cross section can be used instead of stopping power and then the depth is in atoms/cm^2 or μg/cm^2.

As an example, the simple formula given above is used to calculate concentration and depth scales for the measured data given in Table 7.1. Nitrogen profile was measured for a NbN coating on steel using 429 keV $^{15}N(p,\alpha\gamma)^{12}C$ resonance. Thick TiN sample was used as a standard and the measured yield step Y_{st} was 1471. The standard was measured using the same settings as with the NbN samples. The stopping power values needed in Eq. (7.6) are calculated at the resonance energy (429 keV), and are $\varepsilon_N = 9.01$, $\varepsilon_{Ti} = 17.64$, and $\varepsilon_B = \varepsilon_{Nb} = 25.46$ in units eV/(10^{15} at/cm^2). Therefore, the stopping power of TiN is 13.32 eV/(10^{15} at/cm^2). Then, for example, the concentration corresponding to the first data point can be obtained using Eq. (7.6) as

$$m = \frac{0.50 \times 500 \times 25.46}{1471 \times 13.32 + 0.5 \times 500 \times (25.46 - 9.01)}$$

$$= 0.268 = 26.8 \text{ at. \%}$$

The depth scale was calculated using the stopping power for NbN and the density 8.4 g/cm^3. At the resonance energy

$$\varepsilon_{NbN}(429) = \frac{17.24 \text{ eV}}{10^{15} \text{ at/cm}^2}$$

$$= \frac{17.24 \times 10^{-15} \times 2 \times 6.023 \times 10^{23} \times 8.41 \text{ eV / cm}}{106.92}$$

$$= 0.163 \text{ keV / nm} \ .$$

Table 7.1. Example of the calculation of depth and concentration scales for a NbN coating on steel.

Energy *(keV)*	**Yield** *(counts)*	**Depth** *(nm)*	**Concentration** *(at. %)*
429.00	500	0	26.8
429.33	710	2.0	35.5
429.66	1190	4.0	51.5
430.97	1455	12.1	58.7
432.29	1485	20.1	59.4
434.93	1495	36.4	59.7
450.94	1510	135.8	60.0
452.29	1425	144.3	57.9
453.63	1200	152.8	51.8
456.34	945	169.9	43.9
459.05	185	187.0	11.1
461.77	25	204.3	1.61
464.50	10	221.7	0.65
467.92	0	243.6	0

Since the stopping power increases only slightly with the energy, the depth corresponding to the second point is x = 0.33/0.163 nm = 2.0 nm.

7.2.1.5 Standards

In the analysis with nuclear reactions, quantitative results are generally obtained using standard samples containing the element under determination. Different aspects of selection and preparation of standard samples for nuclear reaction analysis are discussed in the Ion Beam Analysis Conference Workshop (Amsel and Davies, 1983). The standard is measured in the same set-up as the samples and the yield is scaled with the total collected charge of bombarding particles. In the case of resonances, the resonance yield step height is determined by measuring the gamma-ray yield from the standard in the bombarding energy range E_r-5*Γ and E_r+5*Γ. This will be illustrated in 7.2.1.8.

There are several requirements for a good standard such as homogeneity, stability under beam heating, composition similar to the samples, and not too many interfering resonances or gamma-ray yield. Some typical standards are listed in Table A12.4. In many cases, high-purity elemental standards can be used as well as stabile, non-volatile compounds, e.g. oxides, carbides, nitrides, and carbonates. Standards can also be made by mixing high-purity powders in known ratios. However, special care must be taken to assure uniform mixing and stability under irradiation; this limits the choice to anhydrous materials with a reasonably high melting point. Furthermore, surface roughness and even thin impurity films can degrade depth resolution for a resonance step. In the case of insulators, it is necessary to use a thin conductive film, mask or grid, or other neutralizing arrangement to prevent charging of the sample surface. High-purity heavy metals (Ta, Mo, etc.) are suitable for this purpose.

7.2.1.6 Depth resolution and deconvolution

The depth resolution is determined by energy resolution of the beam Ω_b, Doppler broadening due to the vibration of target atoms Ω_D, the resonance width Γ_r, and energy straggling of bombarding particles $\Omega_{str}(x)$ as they penetrate the sample. The total depth resolution $d(x)$ can be obtained using Gaussian approximation

$$d(x) = \frac{\Omega_{tot}(x)}{S(x)} \tag{7.9}$$

where

$$\Omega_{tot}(x) = [\Omega_b^2 + \Omega_D^2 + \Gamma_r^2 + \Omega_{str}(x)^2]^{1/2} \; .$$

This definition of resolution corresponds to the experimental width for the infinitely narrow distribution located at depth x, where the full width, half maximum of the energy distribution is Ω_{tot}.

Energy resolution of the beam is typically in the range 0.05-1.0 keV and can be improved, for example, by collimating the beam and stabilizing the terminal voltage and magnetic field of the analyzing magnet. The Doppler broadening follows a Gaussian law and the width can be obtained from

$$\Omega_D = 2.355\left(\frac{2M_1E_bkT}{M_2}\right)^{1/2} \tag{7.10}$$

where M_1 and M_2 are the mass number of ion and target nuclei, k is the Boltzman constant, and T is the Kelvin temperature of the sample. For a 1 MeV proton beam at room temperature, Ω_D is ~ 100 eV, that is the same order of magnitude as the widths of the narrowest proton resonances. Due to the M_1/M_2 factor this term becomes more important when heavy ions are used to profile hydrogen. On the other hand, Ω_D can be minimized by cooling the sample with liquid nitrogen.

The energy straggling can be approximated by the Lindhard-Scharff model (Lindhard and Scharff, 1953)

$$\Omega_{str}(x) = 0.2942CZ_1 \times \left((Z_2/M_2)\, x[\text{nm}] r[\text{g/cm}^3]\right)^{1/2} [\text{keV}] \tag{7.11}$$

$$C = (L(\kappa)/2)^{1/2} \text{ for } \kappa \le 3, \text{ and } 1 \text{ for } \kappa \ge 3$$

$$L(\kappa) = 1.36\kappa^{1/2} - 0.016\kappa^{3/2}$$

$$\kappa = 4.0321 * 10^{-2} E_b\,[\text{keV}]/(Z_2M_1)$$

where C is the correction factor to Bohr's formulation, Z_1 and M_1 are the atomic and mass number of bombarding particles, and (Z_2/M_2) is the average atomic/mass number of the target. Although this approximation is supposed to be valid typically at depths of a few hundred nm, it agrees fairly well with experimental results even closer to the surface. See Chapter 2 for additional discussion about energy straggling.

Figure 7.6 illustrates the depth resolution of 1 MeV proton and 2 MeV $^4He^+$ beams in silicon and tantalum at room temperature calculated using the approximation given above. Ω_b and Γ_r are assumed to be 0.1 keV for both beams. This corresponds to profiling with very narrow resonances and good beam resolution. The depth resolution with protons is 4 nm and 1 nm at the surface in silicon and tantalum, respectively. However, the straggling dominates the resolution at larger depths, even at depths of a few nanometers. Therefore, it is evident that the high beam energy resolution and the minimizing of the Doppler broadening are necessary only in special surface studies with narrow resonances, since the energy straggling of the beam soon becomes the main factor of the depth resolution below the surface. The resolution with alpha resonance reactions is higher mainly due to the relatively large stopping power of $^4He^+$ ions compared to protons. Unfortunately, only a few reasonably strong and well-isolated alpha resonances exist (see Table A12.3).

The unfolding of the real profile from the experimental, broadened gamma-ray yield curve is a typical deconvolution problem. Several computer programs have

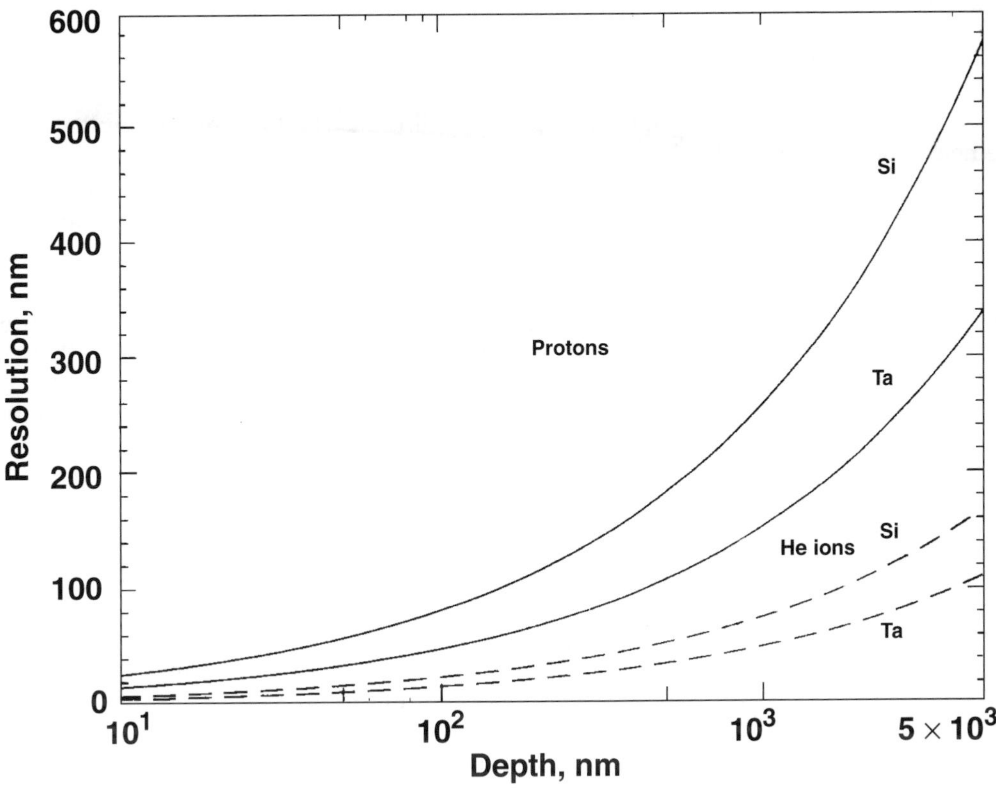

FIG. 7.6. Calculated depth resolution curves for 1 MeV protons and 2 MeV $^4He^+$ ions.

been developed especially for profiling purposes (Land *et al.*, 1976; Maurel *et al.*, 1982; Deconninck and Oystaeyen, 1983; Smulders, 1986; Lappalainen, 1986a; Vickridge and Amsel, 1990; Rickards, 1991). The main difference between these programs is the calculation of straggling. The most accurate method is the stochastic calculation of a discrete-energy-loss process (Maurel *et al.*, 1982). This type of calculation is necessary in high resolution surface studies, for example, to study the Lewis effect or the thermal vibration of surface atoms (Mitchell and Rolfs, 1991). When the profile is measured with a broad resonance ($\Gamma_r > 1$ keV), low beam energy resolution, or the accurate profile shape at the outermost surface is not of interest, straggling can be calculated accurately enough using the Landau-Vavilov theory (Landau, 1944; Vavilov, 1957). The values of these distributions can be obtained by interpolation from tabulated values or by new fast computer algorithms (Rotondi and Montagna, 1990). When the resonance is wide ($\Gamma = 4$ keV for 1 MeV protons), the straggling distribution can be obtained using a simple Gaussian approximation similar to the one for beam resolution [Eq. (7.2)].

There are basically two ways to solve the deconvolution problem: by least-square fitting of a certain analytical function describing the real profile or by iterative solution of the distribution using a suitable algorithm which takes into account proper physical boundary conditions. These programs can easily be used to simulate the yield curve for a certain user-given concentration distribution. This kind of user input is fairly easy, using modern microcomputers or graphic terminals. However, it must be kept in mind that no matter what unfolding method is used, there is never a unique solution determined by the measured yield curve only. Extra conditions such as smoothness, positivity or boundary values are always needed. The effect of experimental resolution on the profiles will be discussed with the examples.

7.2.1.7 Sensitivity

Since the detection sensitivity depends on several factors such as experimental arrangement, gamma-ray detection efficiency, interfering gamma radiation, beam current, measurement time, and substrate, the sensitivity is specific to a certain measurement. However, by taking into account the resonance strength and the gamma ray emitted, order of magnitude estimates for the sensitivity of each resonance can be calculated. Table A12.5a in Appendix 12 gives these estimates for light elements. The estimates are for the total concentration of element (unless a particular isotope is specified), and for the strongest, narrow resonance of the element.

7.2.1.8 Examples

This section gives some typical profiling examples. The best resonance choices, interferences, and some references are given for each element. Additional data can be found in Appendix 12.

Hydrogen and helium

Heavy ion accelerators have been used with ^{7}Li, ^{15}N, ^{19}F, and ^{27}Al beams for profiling hydrogen and helium (see Chapter 8). Of the reactions listed in Table A13, $H(^{7}Li,\gamma)^{8}Be$ offers the highest sensitivity (~1 ppm) and a profiling range up to several microns with 3-7 MeV beam (Padawer *et al.*, 1974; Adler and Schulte, 1978). ^{7}Li beams have also been used for helium profiling, using the 1.43 MeV resonance of the reaction $^{4}He(^{7}Li,\gamma)^{11}Be$ (Schulte, 1978). Nitrogen and fluorine beams are used conventionally for high-resolution hydrogen profiling in surface layers (<1.5 μm) (Maurel and Amsel, 1983; Damjantschitsch *et al.*, 1983; Kuhn *et al.*, 1990).

Lithium, beryllium, boron

These elements are generally profiled with nuclear reactions by detection of charged particles. All (p,γ) resonances are very broad, limiting the resolution to a few tens of nm at best. Nevertheless, low energy (p,γ) resonances of ^{7}Li (441 keV) and ^{11}B (163 keV) have been used for profiling up to a few microns with almost zero background due to the high energy of gamma rays and the low bombarding energy (Toivanen *et al.*, 1985). $^{7}Li(\alpha,\gamma)^{11}B$ resonance at 950 keV offers higher resolution and has been utilized, for example, in profiling of lithium in silicon (Zinke-Allmang *et al.*, 1986).

Carbon

$^{13}C(p,\gamma)^{14}N$ resonance at 1748 keV is the only narrow and strong resonance of carbon. This is very useful for profiling distributions of ion implanted ^{13}C (Hirvonen *et al.*, 1990). However, it must be kept in mind that at such a high bombarding energy most of the light elements have several broad resonances leading to high background at low gamma-ray energies, and even at the energy of primary gamma-rays (9.17 MeV), interferences are likely. For example, ^{13}C-implanted profiles can be measured in silicides. In this case, the main interfering gamma-ray yield is coming from Si. This is demonstrated in Fig 7.7, which shows the gamma-ray spectra taken from ^{13}C-implanted Ta plate and a Si plate at a bombarding energy of 1750 keV. All Si isotopes, ^{28}Si, ^{29}Si, and ^{30}Si, have several resonances below 1750 keV. The energy range for gamma-ray yield from each isotope is shown. In this case, the energy window for the yield curve has to be restricted to include only the 9.17 MeV gamma-ray photopeak due to the interfering gamma-ray yield from Si (Lappalainen and Hirvonen, personal communication).

Nitrogen

Both ^{14}N and ^{15}N isotopes can be profiled with several resonances. Those used most often are the 429 and 898 keV $^{15}N(p,\alpha\gamma)^{12}C$ and 1059 keV $^{14}N(p,\gamma)^{15}O$ resonances. The best depth resolution can be obtained with alpha particles, using the 1531 keV $^{14}N(\alpha,\gamma)^{18}F$ resonance (Gossett, 1985). However, generally the 429 keV (p,αγ) resonance offers a superior counting rate, a longer profiling range and good depth resolution (Γ~0.10 keV) (Maurel and Amsel, 1983). Since this resonance is one of the most used, the selection of a suitable energy window is discussed briefly.

At such a low bombarding energy, the main interferences are generally due to fluorine, carbon, and ^{14}N (Fig 7.8). The gamma-ray spectra were measured from CaF_2, graphite, and TiN with 500 keV protons and Ge(Li) or NaI detector. The spectrum measured with a Ge(Li) detector is essential for identification of a source of interference. The suitable energy range for nitrogen is about 3.1-4.8 MeV, which is just above the natural background (Fig. 7.5). Note that the energy window is selected to take into account significant broadening of gamma-ray peaks detected with a NaI detector. Since

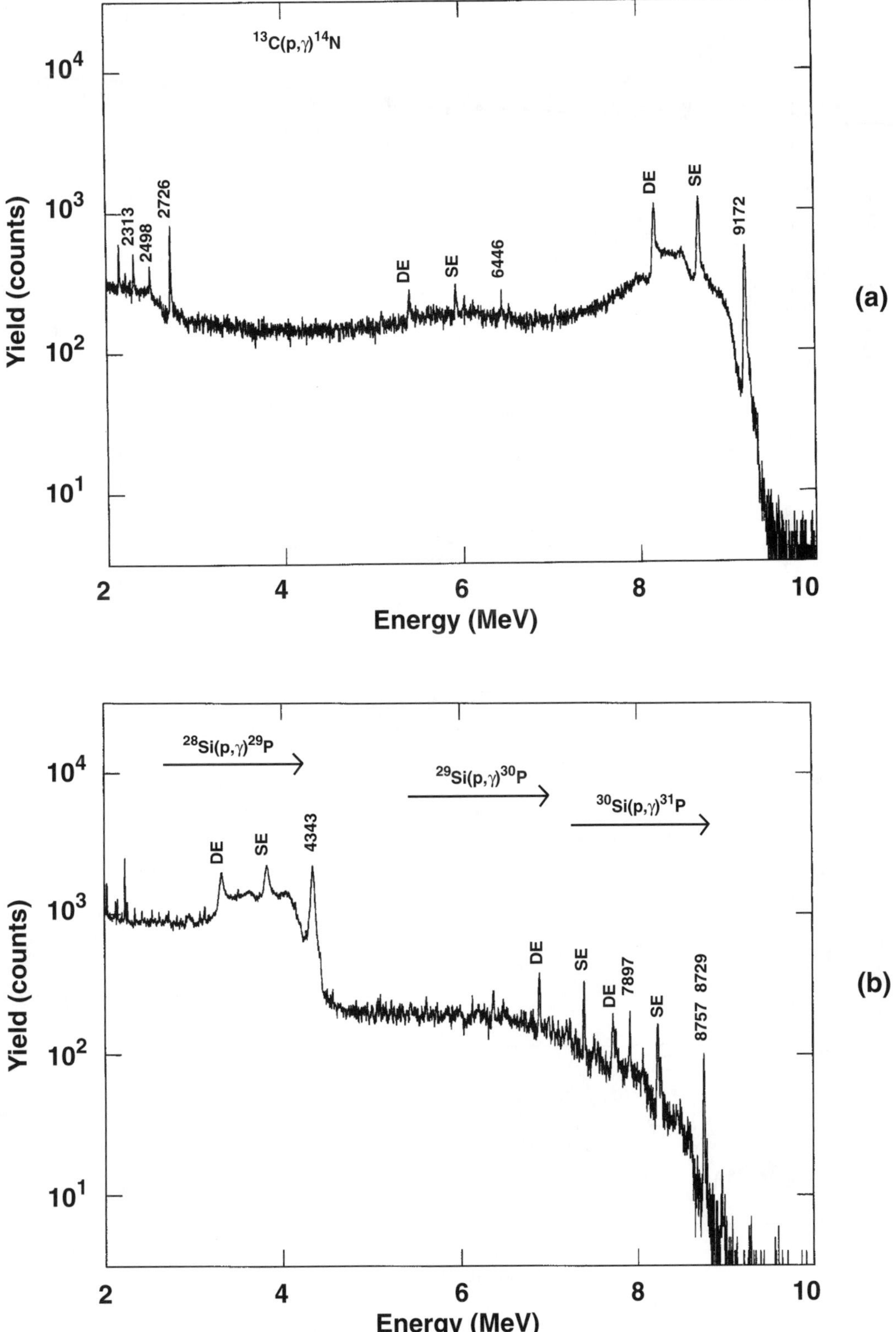

FIG. 7.7. Measured gamma-ray spectrum for a ^{13}C-implanted Ta sample (a) and a Si sample (b) at the bombarding energy of 1750 keV.

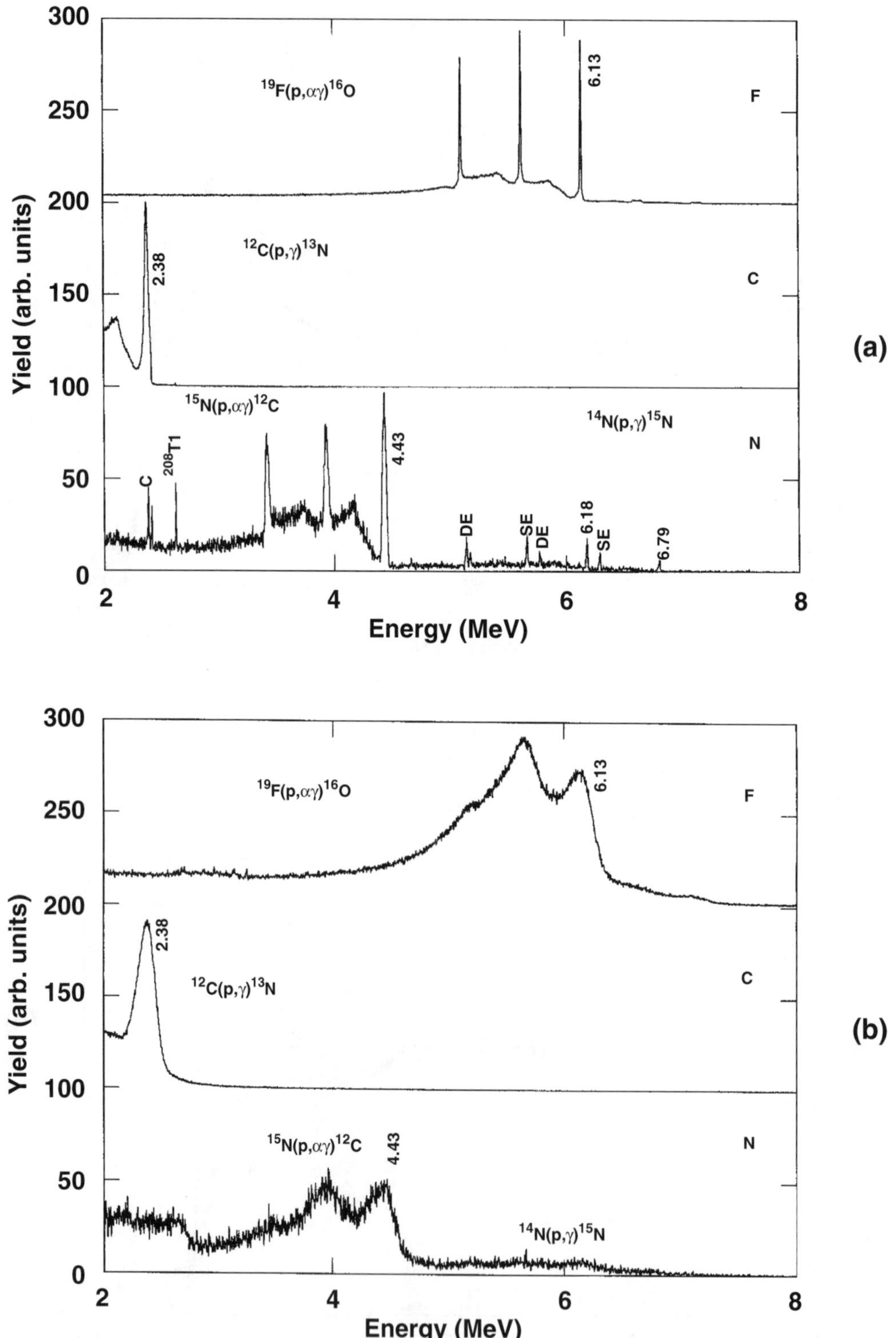

FIG. 7.8. Gamma-ray spectra for carbon, nitrogen, and fluorine measured using a 500 keV proton beam and a Ge(Li) detector (a) or a NaI detector (b).

fluorine background includes the same gamma rays over a wide bombarding energy range, the ratio of gamma-ray yield of fluorine in energy range 5.0-6.2 MeV and 3.1-4.8 MeV remains a constant. This can be used to subtract the fluorine interference changing vs bombarding energy. The same procedure can be used to subtract the background due to ^{14}N resonances since their gamma-ray energies are mainly 5-7 MeV. This interference contribution becomes even more significant when the bombarding energy is above 700 keV, which is typical, for example, in the case of thick nitride coatings.

Nitrogen resonances have been used frequently for nitrogen profiling in nitride coatings (Penttinen *et al.*, 1988; Hirvonen *et al.* 1989; Pietersen and Strydom 1988). Figure 7.9a shows an example of a nitrogen profile in a TiN coating which contains both nitrogen isotopes in natural abundances (^{14}N 99.63%, ^{15}N 0.37%). The profile was measured with the 429 keV resonance and a NaI detector with an energy window 3.1-4.8 MeV to include the 4.43 MeV gamma ray with both escape peaks. The main background comes from the natural radioactivity of building materials, especially the 2.614 MeV gamma ray of ^{208}Tl. In thick nitride coatings, the main interfering yield comes from $^{14}N(p,\gamma)$ and $^{15}N(p,\gamma)$ reactions and minor fluorine and carbon contamination. However, these can be subtracted using several energy windows as discussed above. The energy range 429-898 allows typically ~2.5 µm profiling range in nitride coatings. Resonances at higher energies can be applied as well although the background increases significantly. For example, 1.65 MeV resonance can be used up to 6 µm, but the overlapping with 1.21 and 1.98 MeV resonances must be taken into account.

The usefulness of $^{15}N(p,\alpha\gamma)^{12}C$ reaction becomes even more evident in ^{15}N-implanted samples. Detection limits up to a few ppm can be reached using a low background detection system and reasonably high currents (Kuhn *et al.*, 1990). This method has been used frequently for systematic range measurements (Land *et al.*, 1976; Anttila *et al., 1984*) and diffusion studies (Hirvonen and Anttila, 1985). Figure 7.9b shows an example of diffusion of 40 keV ^{15}N profiles in evaporated Ni (Lappalainen and Anttila, 1987). Nitrogen diffuses to the surface and to the Ni/Ta interface where the nitrogen distribution is very narrow and the broadening is mainly due to the experimental resolution at this depth. Due to the implanted profiles, it is possible to measure this kind of profile using low currents without excessive sample heating.

Oxygen

Oxygen profiling is possible using the $^{18}O(p,\gamma)$ resonances and enriched samples. All these resonances have complicated decay schemes leading to many intense gamma peaks. Table A12.2 lists only the strongest, high-energy gamma rays, although all the resonances decay through the low energy states of ^{19}F corresponding to 110 and 197 keV gamma rays which can also be due to the inelastic scattering from ^{19}F. The resonance at 1167 keV has been used for high resolution surface studies by Prof. Amsel's group. Figure 7.10a gives ^{18}O profiles in 90% ^{18}O enriched Ta_2O_5 layers (Maurel *et al.*, 1982). The prominent surface peak is due to the Lewis effect, that is, the discrete nature of the energy loss of bombarding particles (Lewis, 1981). Simulations were calculated using the stochastic theory of fast ion slowing down. The Lewis peak is very sensitive to the resolution and surface contamination. Figure 7.10b shows the effect of surface impurity layers or oxides to the simulated yield step of an Al sample. In these surface studies, special care must be taken to minimize oxidation and hydrocarbon buildup on the sample surface.

Fluorine

Fluorine is one of the most profiled elements with $(p,\alpha\gamma)$ resonances and a typical interfering contaminant due to many strong and broad resonances. The mostly used resonances are those at E_p = 340, 484, and 872 keV. They mainly decay with 6129 keV fluorine gamma ray, which is a typical characteristic peak of fluorine contamination. Figure 7.11 illustrates the profiling of fluorine in glass using the 340 keV resonance (Deconninck and Oystaeyen, 1983). The calibrated experimental resolution function is plotted vs depth for the 340 keV resonance and the closest upper resonance at 484 keV. The background yield is very low and profiles up to 1.5 µm can be measured without interferences. In this case, interferences from the other fluorine resonances cannot be resolved using a Ge(Li) detector since all the resonances decay with the same gamma rays. This is typical for all resonances which decay with the emission of a light particle and gamma rays.

Neon

In principle, the three neon stable isotopes can be profiled with (p,γ) resonances but only ^{20}Ne (natural abundance 90.51%) and ^{22}Ne (9.22%) resonances are sufficiently narrow and well isolated. The 1169 keV

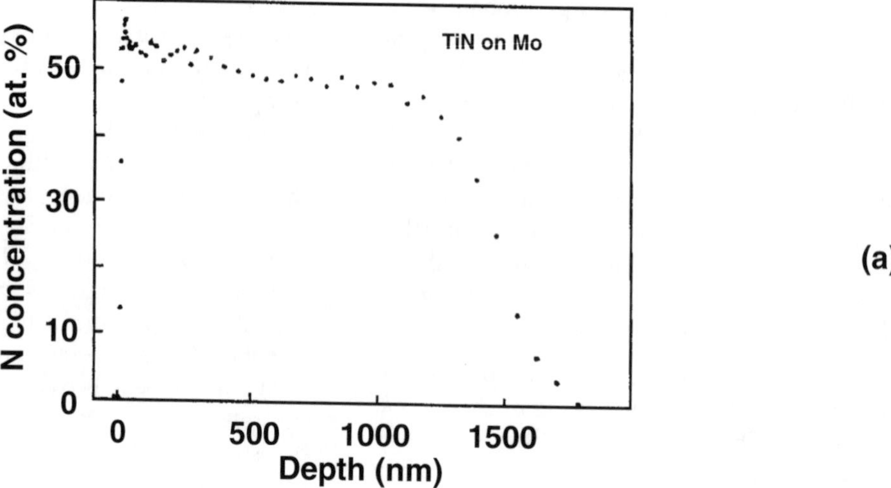

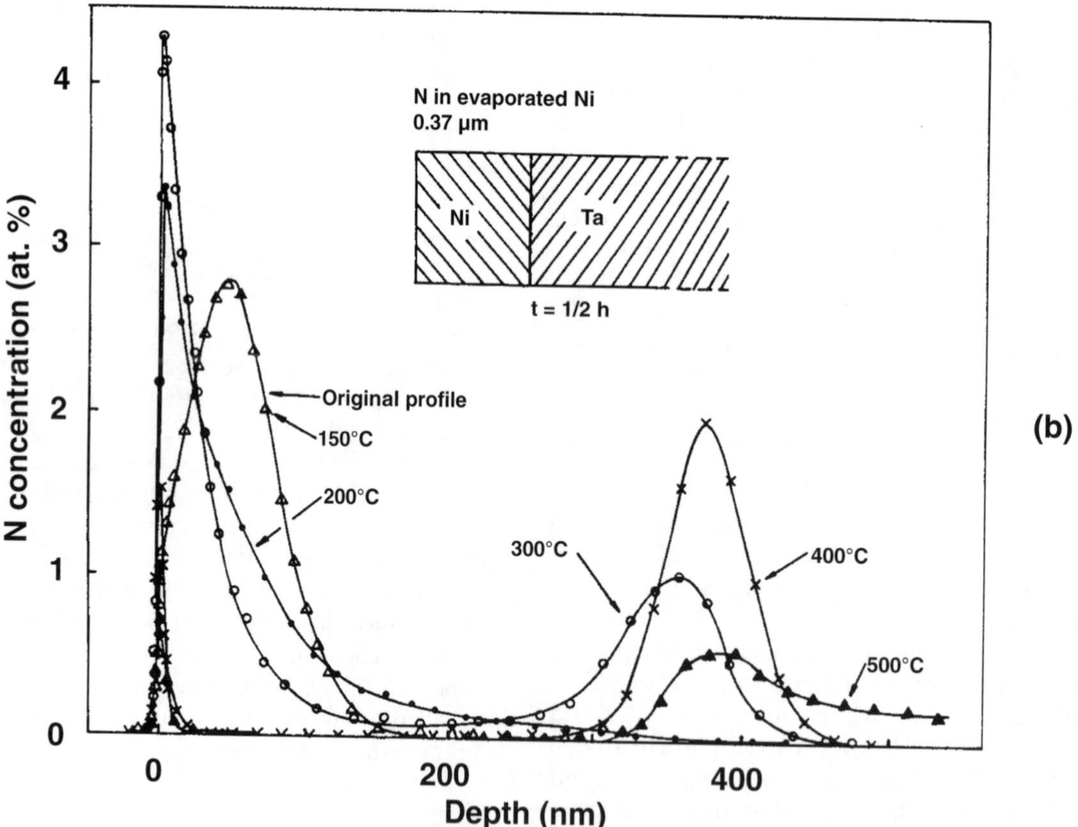

FIG. 7.9. a) Nitrogen distribution of the TiN film on Mo (Hirvonen *et al.*, 1989). b) ^{15}N profiles in evaporated Ni on Ta after isochronal annealing. The original profile is that of the 40 keV ^{15}N implants (Lappalainen and Anttila, 1987).

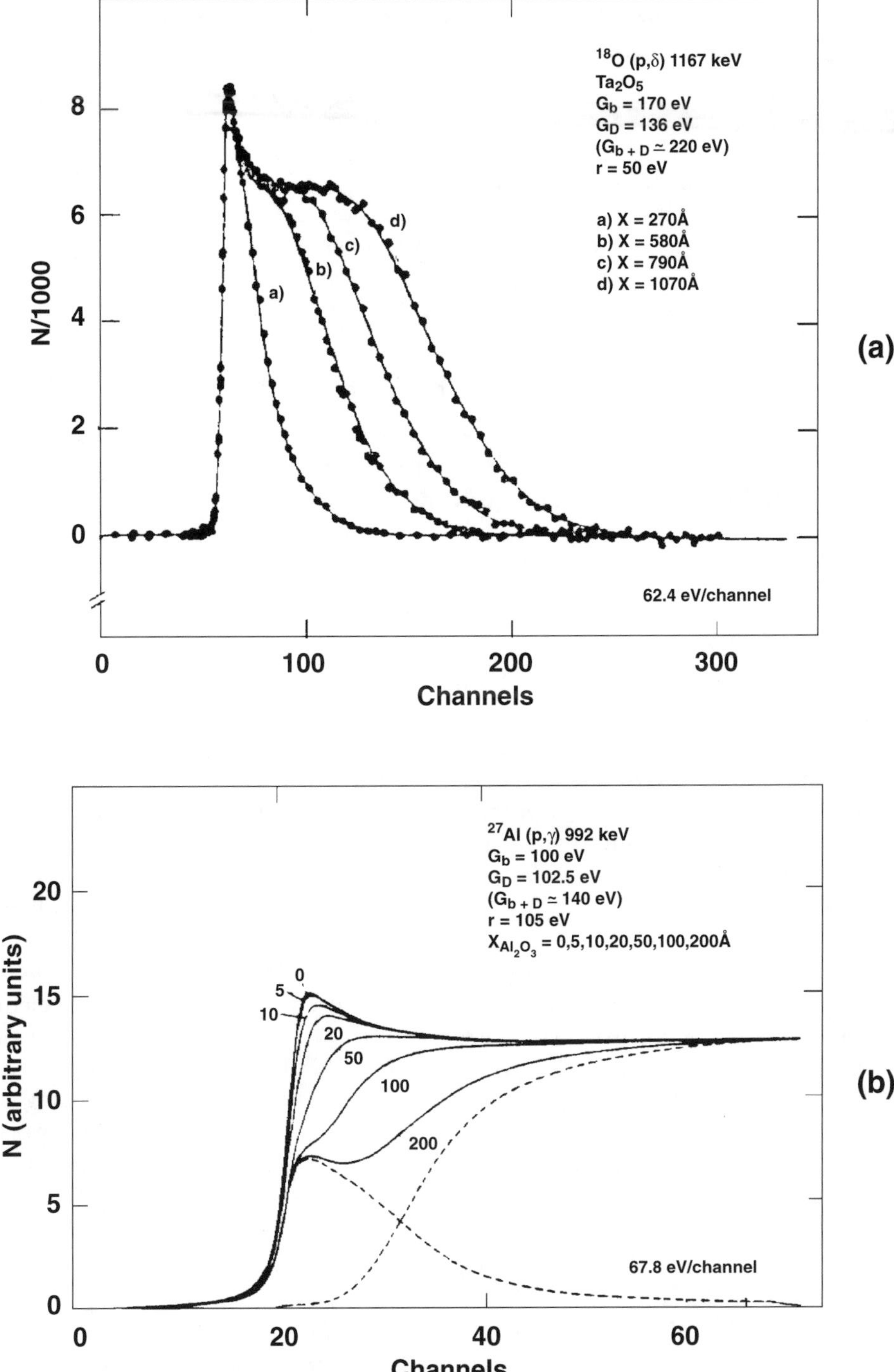

FIG. 7.10. a) Excitation curves for 90% ^{18}O-enriched Ta_2O_5 layers of increasing thickness X. The lines are best fits obtained using stochastic simulations with the parameters given in the figure. b) Calculated excitation curves for a thick aluminum sample covered with a Al_2O_3 layer of thickness X. The dotted lines represent the theoretical contributions of a 20-nm-thick oxide layer and an aluminum sample covered by an absorber equivalent to that oxide layer (Maurel *et al.*, 1982).

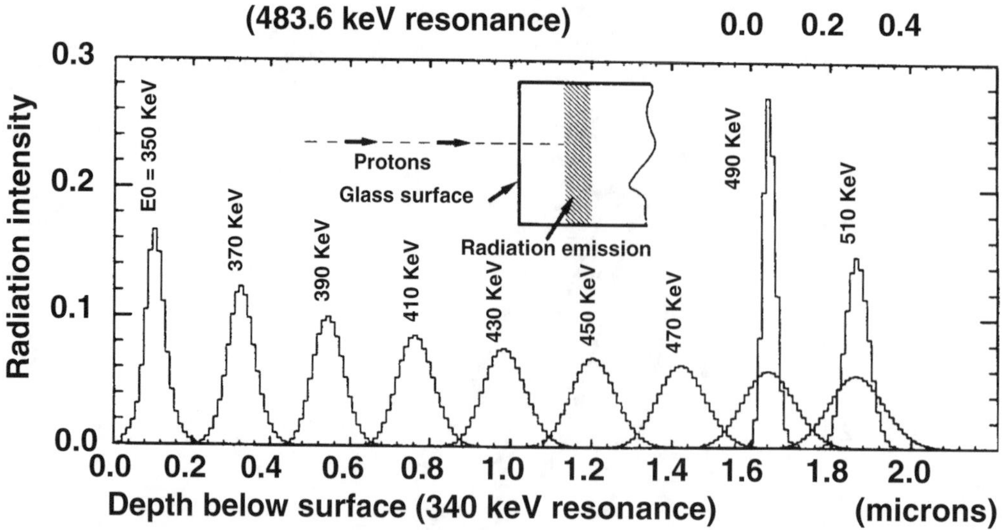

FIG. 7.11. Calculated yield initiated by the 340 keV resonance in fluoridated glass at different beam energies. The curve corresponds to the experimental resolution. Interference from the next resonance at 483 keV is also represented (Deconninck and Oystaeyen, 1983).

$^{20}Ne(p,\gamma)^{21}Na$ resonance is especially well isolated but the relatively low gamma-ray energy (3.54 MeV) is easily overlapped by interfering gamma-ray yields. However, the $^{22}Ne(p,\gamma)^{21}Na$ resonances with high energy gamma rays can be used for high accuracy profiling even in light backings (Lappalainen, 1986b).

Sodium

Sodium distributions in glasses after different treatments have been profiled using the 1011 keV $^{23}Na(p,\alpha\gamma)^{20}Ne$ resonance (Della Mea, 1986; Trocellier *et al.*, 1982). In this case, charging of the sample surface can be prevented by using a metal grid (Mo or other heavy metal). The current density must be kept low (<0.5 $\mu A/cm^2$) to avoid migration of sodium due to the probing beam.

Magnesium

Magnesium can be profiled with the 1548 keV $^{26}Mg(p,\gamma)^{27}Al$ resonance (Anttila *et al.*, 1977b) or the 2010 keV $^{24}Mg(p,p'\gamma)^{24}Mg$ resonance (Paltemaa *et al.*, 1983). In both cases, a Ge(Li) detector and effective cooling are necessary.

Aluminum

Aluminum has only one stable isotope, ^{27}Al, which has at least 22 (p,γ) resonances below 1 MeV. They are all very narrow (Γ<200 eV) but none of them offers a wide energy range free of other resonances both above and below the resonance. The two strongest resonances are those at E_p = 632 and 992 keV given in Table A12.2. Both of these resonances decay with primary high-energy gamma rays which can be well separated from the gamma-ray yield of other light elements such as fluorine, nitrogen, and carbon. In the case of deep profiles (up to several microns), interfering nearby $^{27}Al(p,\gamma)^{28}Si$ resonances can be separated using primary high-energy gamma rays and Ge(Li) detector. These resonances have been used widely, for example, for calibration (Anttila *et al.*, 1977a), oxidation (Amsel *et al.*, 1971), and diffusion studies (Hirvonen, 1982; Lappalainen, 1986a).

Measurement of a resonance yield curve for a thick homogeneous sample is typically used to determine the exact position of the resonance, the energy resolution at the surface, and the calibration factor for conversion of measured yields to concentrations. Figure 7.12a illustrates this kind of a yield curve for a 992 keV $^{27}Al(p,\gamma)^{28}Si$ resonance measured from a bulk aluminum plate. The primary gamma-ray yield (10.76 MeV) was detected with a 12.7 × 10.2 cm^2 NaI(Tl) detector using an energy window 8.0-11.5 MeV.

The yield curve for a thick sample is a smoothed step function, the resonance energy corresponding to the point where the yield is half of the step height. Figure 7.12b shows a Gaussian function and its integral. These correspond to the energy distribution of the beam on the surface and approximately the experimental yield curve for a thick sample when the resonance is narrow compared to the beam energy resolution (Fig 7.12a). The energy interval which corresponds to the yield increase from 12–88% is the experimental resolution (FWHM). The experimental beam resolution was determined for the yield step in Fig. 7.12a by fitting the yield curve using the beam resolution as a fitting parameter. In this case, the simple procedure described above would give the same result. However, when the resonance width is larger, the thick target yield curve is not an integrated Gaussian due to the longer tail of the Breit-Wigner formula than the Gaussian with the same Γ.

Silicon

Silicon has been profiled mainly by the 620 keV $^{30}Si(p,\gamma)^{31}P$ resonance, for example, to study silicon self diffusion (Hirvonen and Anttila, 1979) and diffusion in silicides. Profiles can be measured with high depth resolution and using a NaI detector with 7.0-8.3 MeV energy window.

Heavy elements

Elements heavier than silicon are seldom profiled with (p,γ) resonances. This is because resonances are typically broad, not well isolated, and at high bombarding energies. Phosphorus and sulphur are the main exceptions both having narrow resonances for profiling (Anttila *et al.*, 1977b; Kido *et al.*, 1982; Pruppers *et al.*, 1986). The (p,γ) resonances have been used frequently for nuclear lifetime measurement (Keinonen, 1985) using implanted targets in heavy backings (Ta, Mo). In these measurements, many implanted profiles of rare isotopes (M_2=30-60) have been measured.

The (p,γ) resonances of Ti, Cr, Ni, and Fe can be used for profiling, but generally it is necessary to use a Ge(Li) detector to measure primary characteristic gamma rays (Gossett, 1980; Riihonen and Keinonen, 1977).

7.2.2 Determination of concentrations in thick samples by using particle-induced gamma-ray emission (PIGE)

In principle, any ion could be used with proper bombarding energy to induce gamma-ray emission. Protons and alpha particles are mostly used, but some sparse measurements have been made using light ions such as deuterons, tritons, ^{3}He, Li, B, N, F, and Cl. Here the discussion is limited to proton- and alpha-induced gamma-ray emission.

The PIGE method is typically less sensitive than the similar methods utilizing x-ray detection (PIXE). However, the gamma-ray peaks are generally well isolated and the energy is high enough that correction for absorption will not be necessary. In addition, the high penetrability of gamma rays simplifies the experimental arrangements and calculations. Generally, PIGE is used with other complementary ion beam analysis methods such as PIXE (particle-induced x-ray emission), BS (backscattering), and NRA (nuclear reaction analysis with particle detection) (Räisänen, 1990; Lappalainen *et al.*, 1989; Bird, 1990).

The main improvements in the use of PIGE during the last decades include large, high resolution Ge(Li) detectors, external beams, and the nuclear microprobe. External beams are now a standard method in PIGE and PIXE facilities (Räisänen, 1989). The external beams offer several advantages, such as the ability to measure organic samples (even volatile) or large specimens (e.g., paintings), easy sample preparation, effective cooling, and easy, reliable beam current integration, even for insulators. Nuclear microprobes are mainly used with BS, PIXE, and other methods which have high yield with low beam currents, for example, 100 pA on a 1 μm^2 spot. However, they have occasionally been applied with the most sensitive PIGE reactions, e.g. those of fluorine. (NIM, 1991)

7.2.2.1 Thick target gamma-ray yields induced by protons and α-particles

More or less systematic measurements of gamma-ray yield vs bombarding energy under proton or alpha bombardment have been made in several laboratories. Tables A12.6a to A12.6d in Appendix 12 list the absolute thick target gamma-ray yields for elemental analysis by 1 to 9 MeV protons. Tables A12.7a and A12.7b list the thick

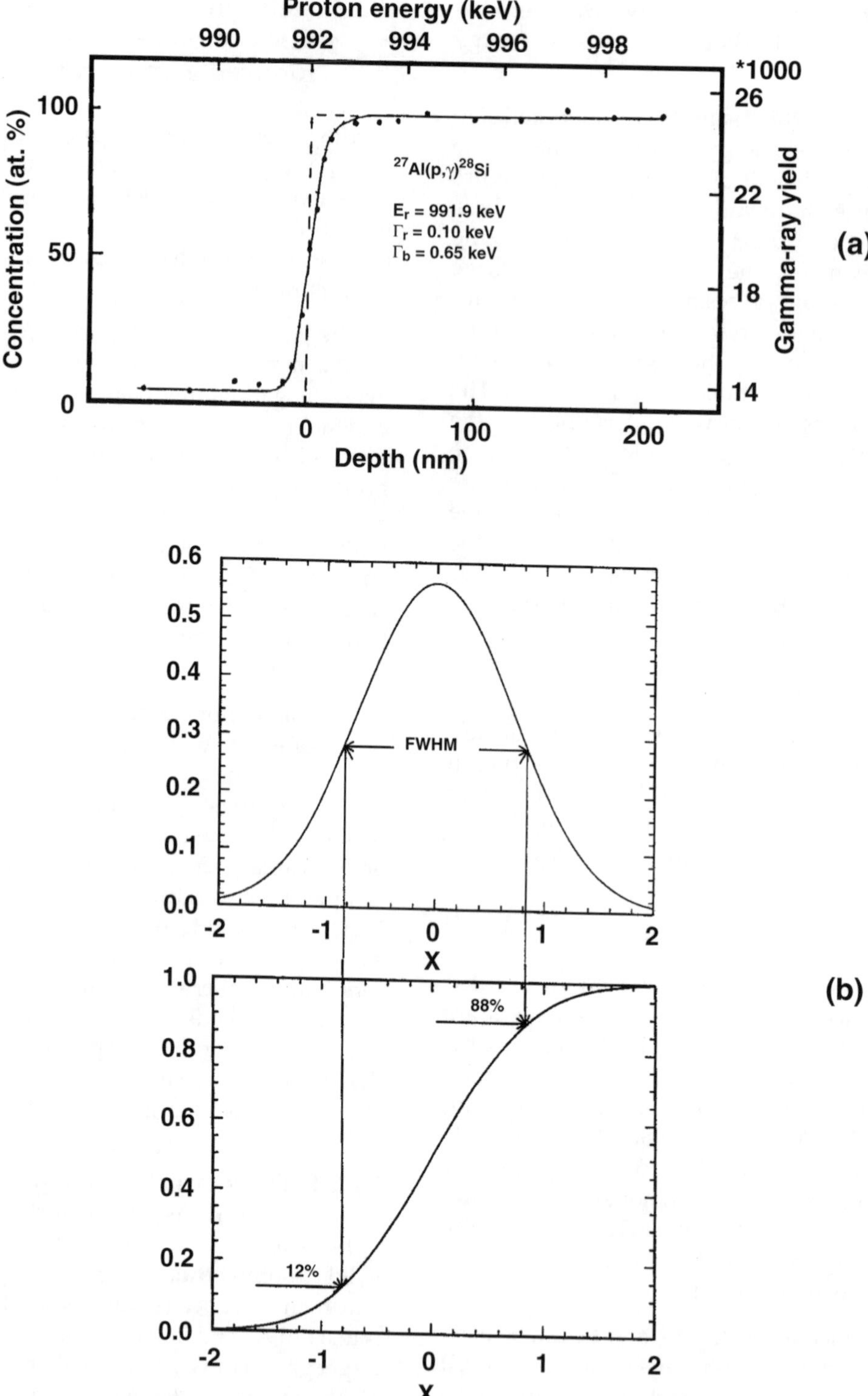

FIG. 7.12. a) Experimental resonance yield step for 992 keV $^{27}Al(p,\gamma)^{28}Si$ resonance. The solid line is a fit obtained with the program using the Vavilov theory for straggling. b) A Gaussian distribution and its integral.

target gamma-ray yields for elemental analysis by 2.4 and 5.0 MeV alpha particles. The yields given in tables A12.6 and A12.7 are absolute gamma-ray yields of the pure elemental samples at the detection angle of 55° with respect to the beam direction and are given per solid angle (sr) and total collected beam charge (μC). No error limits are given, however. The main source of error is the stopping power, which can be as much as 20% in the compound cases. On the other hand, these absolute yields are seldom used for the high accuracy analysis since the comparison with proper standards with a similar composition to that of the sample is generally possible.

The gamma-ray peaks from certain reactions are significantly broader than the resolution of a Ge(Li) detector. This broadening is mainly due to the short lifetimes of de-excitation states and the high recoil of the light nucleus (Doppler effect) or due to the large resonance width. The Doppler broadening is evident with protons for $E\gamma = 478$ keV (Li), 3562 (Be), 429 (B), 4429 (N), and 1634 (Na) and with $^4He^+$ ions for $E\gamma = 478$ (Li), 4439 (Be), 3684 (B), 1275 (F), and 1809 (Na). The broadening of peaks is useful in identifying the origin of the gamma rays, especially for elements which have only one peak.

Another source of broad peaks in the gamma-ray spectrum are the neutron-induced reactions, especially in the Ge isotopes of the Ge(Li) detector. These neutrons are from (p,n) or (α,n) reactions in the sample. In the case of 7-9 MeV proton, relative neutron yields are given in Table A12.6d. The neutron-induced peaks are mainly in the low energy region (< 1 MeV). Most typical of these are the 596 and 691 keV peaks. Table A11.8 lists most of the interfering, low-energy gamma rays and their origin. However, the neutron flux to the detector can be reduced significantly with a thick neutron absorber, for example, 10-cm-thick boron doped paraffin, plastics, or water.

7.2.2.2 Calculation of concentrations using standards

Homogeneous concentrations of elements in thick samples detected by prompt gamma-ray emission can be obtained by comparing to standards. Some typical standards are given in Table A12.4, although many multi-component elemental standards exist for the analysis of geological, biological, and medical samples. Standards are also discussed in 7.2.1.5.

The concentration of the element (isotope) c_s can be calculated from the gamma-ray yield for a certain gamma-ray peak in sample Y_s and the yield from a standard Y_{st} using this simple formula

$$c_s = \frac{c_{st} S_s Y_s}{S_{st} Y_s} \tag{7.12}$$

where the stopping powers have to be calculated at the energy $E_{1/2}$ where the thick target yield has fallen to half of its value at the bombarding energy E_b. The stopping power values can be obtained from tables in Appendix 3 or Andersen and Ziegler (1977) for protons and Ziegler (1977) for helium. If E_b is used instead of $E_{1/2}$, the error is less than a few percent. If the excitation curve is known vs bombarding energy, more accurate calculation can be made using a simple computer program, although this is not generally necessary. In the above formula, if the stopping is in atoms/cm^2 or μg/cm^2, the concentrations c_s and c_{st} are in at.% and wt.%, respectively.

7.2.2.3 Detection sensitivity

Sensitivities obtainable with protons (E < 9 MeV) and alpha particles (E = 5 MeV) are summarized in Tables A12.5c and A12.5b. These should be taken as order of magnitude values, because the actual sensitivities are better than those given in the tables. This is true especially when experimental arrangements and bombarding energy are optimized for the detection of a specific element. On the other hand, sensitivities can be much worse if the sample contains major amounts of other elements with high gamma-ray yield or if the sample can only withstand low beam currents. Therefore, possible interferences from both radioactive and prompt background, as well as from other sample components, should be investigated using Tables A12.6 to A12.8 in Appendix 12.

7.2.2.4 Examples

Figure 7.13 gives an example of the analysis of a hafnium plate with 2.4 MeV protons. The total accumulated charge was 10 mC, and a tin lining was used to reduce background. In addition to the light elements O (150 ppm), Na (~0.3 ppm), Al (~30 ppm), and P (~5 ppm), typical heavier steel components Zr (2.8%), Fe (100 ppm), and Cu (< 50 ppm) can be observed. The sensitivity of protons and helium ions for these medium mass number elements is somewhat better with higher bombarding energies. This fact and interferences in PIXE measurements make the PIGE method preferable for the analysis of samples with several of these typical steel components.

In Fig. 7.14 the spectrum taken with 2.4 MeV $^4He^+$ ions on volcanic stone (Kilimanjaro) is illustrated (Lappalainen *et al.*, 1983). This example demonstrates the high sensitivity of the method for Li, Be, B, and F which makes

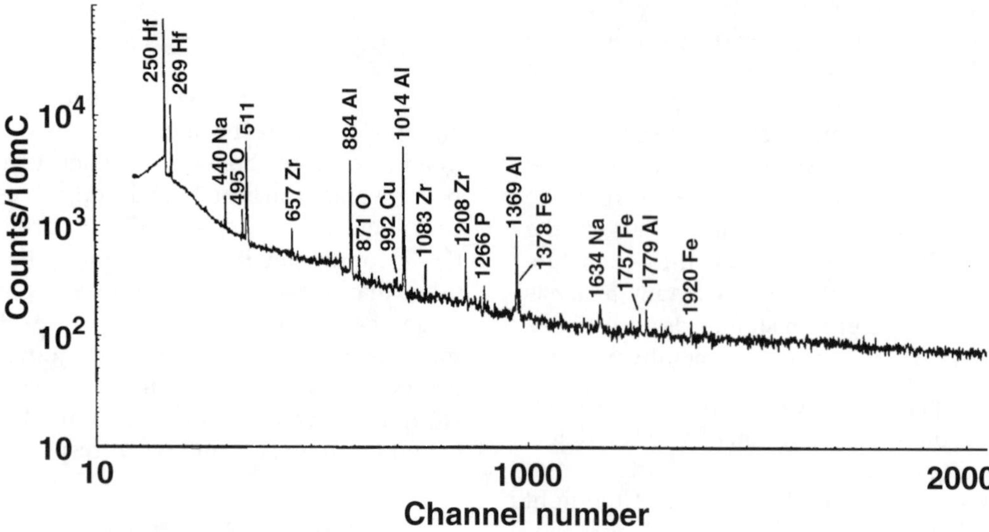

FIG. 7.13. Gamma-ray spectrum taken from a hafnium plate with 2.4 MeV protons. A lead absorber was used in front of the detector to decrease the intensity of low energy gamma rays (Räisänen and Hänninen, 1983).

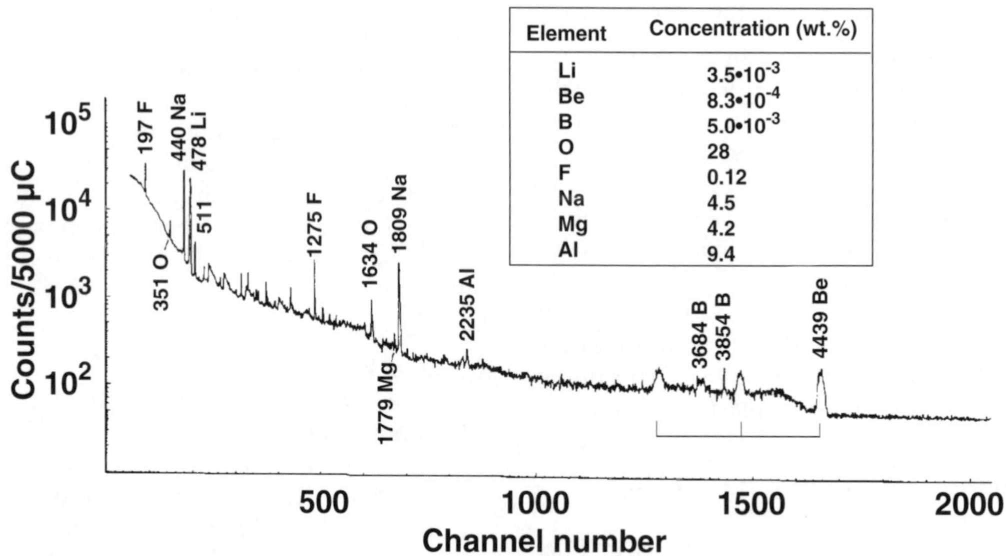

Element	Concentration (wt.%)
Li	$3.5 \cdot 10^{-3}$
Be	$8.3 \cdot 10^{-4}$
B	$5.0 \cdot 10^{-3}$
O	28
F	0.12
Na	4.5
Mg	4.2
Al	9.4

FIG. 7.14. Gamma-ray spectrum taken from a volcanic stone at with 2.4 MeV $^4He^+$ ions with 10 μA current. The insert gives the concentrations of the elements observed (Lappalainen *et al.*, 1983).

$^{4}He^{+}$ ions at low energies a preferable choice, especially when other sample components limit the sensitivity of protons. At higher bombarding energies, $^{4}He^{+}$ ions can be used also for the detection of heavier elements with detection limits down to 0.01-0.1% (Giles and Peisach, 1979).

REFERENCES

Adler, P.N. and Schulte, R.L. (1978), *Scripta Metall.* **12**, 669.

Amsel, G. and Davies, J.A. (1983), *Nucl. Instrum. Methods* **218**, 177.

Amsel, G., d'Artemare, E. and Girard, E. (1983), *Nucl. Instrum. Methods* **205**, 5.

Amsel, G., Nadai, J.P., d'Artemare, E., David, D., Girard, E. and Moulin, J. (1971), *Nucl. Instrum. Methods* **92**, 481.

Anderson, H.H. and Ziegler, J.F. (1977), *Hydrogen Stopping Powers and Ranges in All Elements*, Pergamon Press, New York.

Anttila, A., Keinonen, J., Hautala, M. and Forsblom, I. (1977a), *Nucl. Instrum. Methods* **147**, 501.

Anttila, A., Bister, M., Fontell, A. and Winterbon, K.B. (1977b), *Radiat. Eff.* **33**, 13.

Anttila, A., Hänninen, R. and Räisänen, J. (1981), *J. Radioanal. Chem.* **62**, 441.

Anttila, A., Paltemaa, R., Varjoranta, T. and Hentelä, R. (1984), *Radiat. Eff.* **86**, 179.

Bird. J.R. (1990), *Nucl. Instrum. Methods* **B45**, 516.

Butler, J.W. (1959), U.S. Naval Research Laboratory, NRL Report 5282.

Damjantschitsch, H., Weiser, M., Heusser, G., Kalbitzer, S. and Mannsperger, H. (1983), *Nucl. Instrum. Methods* **218**, 129.

Deconninck, G. and van Oystaeyen, B. (1983), *Nucl. Instrum. Methods* **218**, 165.

Deconninck, G., Demortier, D. and Bodart, F. (1983), *Atom. Energy Rev. Suppl.* 2, 151.

Della Mea, G. (1986), *Nucl. Instrum. Methods* **B15**, 495.

Fujimoto, F. (1990), *Nucl. Instrum. Methods* **B45**, 49.

Giles, I. and Peisach, M. (1979), *J. Radioanal. Chem.* **50**, 307.

Golicheff, I., Loeuillet, M., Engelmann, Ch. (1972), *J. Radioanal. Chem.* **12**, 233.

Gossett, C.R. (1980), *Nucl. Instrum. Methods* **168**, 217.

Gossett, C.R. (1985), *Nucl. Instrum. Methods* **B10/11**, 722.

Hirvonen, J.-P. (1982), Studies on Diffusion and Trapping of Atoms in Some Ion-Implanted Metals, PhD thesis, University of Helsinki, Finland.

Hirvonen, J.-P. and Anttila, A. (1979), *Appl. Phys. Lett.* **35**, 703.

Hirvonen, J-P., and Anttila, A. (1985), *Appl. Phys. Lett.* **46**, 835.

Hirvonen, J.-P., Lappalainen, R., Anttila, A. and Sirviö, E. (1989), Proc. First Int. Conf. Plasma Surface Eng., Sept. 1988, Garmisch-Partenkirchen, FRG, DGM Informationsgesellschaft, Oberursel, 1989.

Hirvonen, J.-P., Nastasi, M., Lappalainen, R. and Sickafus, K. (1990), *Mat. Res. Soc. Symp. Proc.* **157**, 203.

Keinonen, J. (1985), Capture gamma-ray spectroscopy and related topics-1984, AIP Conf. Proc. No. 125, S. Raman (ed.), AIP, New York; p 557.

Kido, Y., Kakeno, M., Yamada, K., Hioki, T. and Kawamoto, J. (1982), *J. Appl. Phys.* **53**, 4812.

Kiss, A.Z., Koltay, E., Nyako, B. Somorjai, E., Anttila, A. and Räisänen J. (1985), *J. Radioanal. Chem.* **89**, 123.

Kuhn, D., Rauch, F. and Baumann, H. (1990), *Nucl. Instrum. Methods* **B45**, 252.

Land, D.J., Simons, D.G., Brennan, J.G. and Brown, M.D. (1976), *Ion Beam Surface Layer Analysis*, ed. O. Meyer, G. Linker and F. Käppeler (eds.), Plenum, New York, vol. 2, p. 851.

Landau, L. (1944), *J. Phys. USSR* **8**, 201.

Lappalainen, R. (1986a), Application of NRB method in range, diffusion and lifetime measurements, PhD thesis, University of Helsinki, Finland.

Lappalainen, R. (1986b), *Phys. Rev.* **B34**, 3076.

Lappalainen, R. and Anttila, A. (1987), *Appl. Phys.* **A42**, 263.

Lappalainen, R., Anttila, A. and Räisänen, J. (1983), *Nucl. Instrum. Methods* **212**, 441.

Lappalainen, R., Hirvonen, J.-P., Pokela, P.J. and Alanen, J. (1989), *Thin Solid Films* **181**, 259.

Lewis, M.B. (1981), *Nucl. Instrum. Methods* **190**, 605.

Lindhard, J. and Scharff, M. (1953), *Mat. Fys. Medd. Dan. Vid. Selsk.* **27**, No. 15.

Maurel, B. and Amsel, G. (1983), *Nucl. Instrum. Methods* **218**, 159.

Maurel, B., Amsel, G. and Nadai, J.P. (1982), *Nucl. Instrum. Methods* **197**, 1.

Meier, J.H. and Richter, F.W. (1990), *Nucl. Instrum. Methods* **B47**, 303.

Mitchell, G.E. and Rolfs, C. (1991), *Nucl. Instrum. Methods* **B56/57**, 473.

NIM. (1991), Proc. of the Second Int. Conf. on Nuclear Microprobe Technology, Melbourne, Australia, Feb. 5-9, 1990. *Nucl. Instrum. Methods* **B54** (1-3).

Padawer, G.M., Kamynowski, E.A., Stauber, M.C. and D'Agostino, M.D. (1974), *Geoch. Cosmochim. Acta, Suppl.* **5-2**, 1919.

Paltemaa, R., Räisänen, J., Hautala, M. and Anttila, A. (1983), *Nucl. Instrum. Methods* **218**, 785.

Penttinen, I., Molarius, J., Korhonen, A.S. and Lappalainen, R. (1988), *J. Vac. Sci. Technol.* **A6**, 2158.

Pietersen, D. and Strydom, W.J. (1988), *Nucl. Instrum. Methods* **B35**, 467.

Pruppers, M.J.M., Zijderhand, F., Maessen, K.M.H., Bezemer, J., Habraken, F.H.P.M. and van der Weg, W.F. (1986), *Nucl. Instrum. Methods* **B15**, 512.

Räisänen, J. (1989), *Nucl. Instrum. Methods* **B40/41**, 638.

Räisänen, J. (1990), *Nucl. Instrum. Methods* **B49**, 39.

Räisänen, J. and Hänninen, R. (1983), *Nucl. Instrum. Methods* **205**, 259.

Räisänen, J., Witting, T. and Keinonen, J. (1987), *Nucl. Instrum. Methods* **B28**, 199.

Rickards, J. (1991), *Nucl. Instrum. Methods* **B56/57**, 812.

Riihonen, M. and Keinonen, J. (1977), *Nucl. Instrum. Methods* **144**, 323.

Rotondi, A. and Montagna, P. (1990), *Nucl. Instrum. Methods* **B47**, 215.

Schulte, R.L. (1978), *Nucl. Instrum. Methods* **149**, 65.

Smulders, P.J.M. (1986), *Nucl. Instrum. Methods* **B14**, 234.

Toivanen, R.O., Hirvonen, J-P. and Lindroos, V.K. (1985), *Nucl. Instrum. Methods* **B7/8**, 200.

Trocellier, P., Nens, B. and Engelmann, Ch. (1982), *Nucl. Instrum. Methods* **197**, 15.

Vavilov, P.V. (1957), *Sov. Phys. JETP* **5**, 749.

Vickridge, I. and Amsel, G. (1990), *Nucl. Instrum. Methods* **B45**, 6.

Ziegler, J.F. (1977), *Helium Stopping Powers and Ranges in All Elements*, Pergamon Press, New York.

Zinke-Allmang, M., Kössler, V. and Kalbitzer, S. (1986), *Nucl. Instrum. Methods* **B15**, 563.

NUCLEAR REACTIONS FOR HYDROGEN ANALYSIS

W. A. Lanford

State University of New York at Albany, Albany, New York

CONTENTS

8.1 INTRODUCTION

MeV ion beam analysis for hydrogen is important for the following reasons:

- Hydrogen is probably the most common elemental contaminant, especially in thin film materials.
- The presence of hydrogen can have dramatic effects on the electrical, mechanical, and chemical properties of some materials.
- Hydrogen is invisible in most modern analytical probes.

In addition, hydrogen is chemically versatile, having a valence of both +1 and –1, hence, it can react with most elements. For example, in semiconductors hydrogen can passify both acceptors and donors. Further, hydrogen is many orders of magnitude more mobile than other common contaminants. This means that if hydrogen is present in a system even at room temperature, it has the mobility to reach critical regions in short times.

Because of this situation, MeV ion beam techniques have been developed to detect hydrogen. These include nuclear reaction techniques (Leich and Tombrello, 1973 and many others) discussed in this chapter, and elastic recoil detection methods (Cohen *et al.*, 1972; L'Ecuyer *et al.*, 1976; and many others) discussed in Chapter 5.

To put this field in some perspective, over the past 10 years or so, hundreds of papers have been published where MeV ion beams have been used to measure H concentration profiles in a vast variety of materials, including electronic materials (amorphous Si, diamond, silicon nitride, boron nitride, silicon dioxide, etc.), metals (diffusion of H in metals, electromigration of H in metals, stress-induced transport of H in metals, and others), glasses (weathering of glasses, obsidian hydration, solar wind implants on lunar glasses, and many more), and superconductors. Recently, a book has been written on this subject (Khabibullaev and Skorodumov, 1989).

The purpose of this chapter is to provide practical handbook information that can be used to allow someone to quickly setup a H profiling system and to analyze data taken with such a system. This purpose is to be contrasted with that of another review recently written by the present author (Lanford, 1992), which emphasizes the general considerations in hydrogen analysis but does not present the useful tables or other practical information given here.

8.2 REACTIONS USED TO PROBE FOR ^{1}H

In principal, any proton-induced nuclear reaction can be used "in reverse" to probe for ^{1}H by bombarding the sample to be analyzed with a beam of the appropriate heavy ions and by measuring the number of nuclear reactions that occur with hydrogen in the target. However, some reactions provide better analytic characteristics than others, in terms of sensitivity, depth resolution, maximum depth of profile, interference reactions, and ease of data analysis. Another consideration is ease (expense) of producing the necessary heavy ion beam.

Perhaps the first point to make is that essentially only resonant nuclear reactions have been used to profile for ^{1}H. One reason for this is that (as outlined below), resonant nuclear reactions provide a natural way to measure depth profiles by measuring the nuclear reaction yield as a function of beam energy. Depth information could be obtained using non-resonant nuclear reactions by measuring the energy of the charged particles produced in the nuclear reactions. However, in many cases (most?), the kinematics are very unfavorable if these particles are detected at backward angles. It often turns out that the energy of the detected particle (because of the large center of mass velocity) is low and almost independent of the depth at which this particle is produced. If the detector is placed at forward angles where particle energies are higher, one may as well look directly at the elastically recoiling protons. This is a useful approach and is referred to as Elastic Recoil Detection (ERD) and is discussed in Chapter 5.

The three resonant reactions that have been used far more than others are those induced by ^{15}N (Lanford *et al.*, 1976), ^{19}F (Leich and Tombrello, 1973), and ^{7}Li (Adler *et al.*, 1974). Parameters for each of these reactions are given in Table A13.1 (Trocellier and Engelmann, 1986; Xiong *et al.* 1987; and Barnes *et al.*, 1977). Each of these reactions has its own advantages. For most applications, the ^{15}N reaction has the advantage of having the best combination of analytic characteristics (depth resolution and sensitivity). The ^{19}F reaction has the advantage that it can be conducted using natural (as opposed to isotopically enriched) F in the accelerator ion source. The ^{7}Li reaction has the advantage of allowing profiling to much greater depths in a sample than either of the other reactions mentioned above.

Below, the experimental details for using the ^{15}N resonant nuclear reaction to determine H concentration profiles will be discussed in detail. While there are some differences, in general, the same considerations apply in the use of the ^{19}F and ^{7}Li reactions.

8.3 RESONANT NUCLEAR REACTIONS

Figure 8.1 shows a schematic representation of the ^{15}N resonant nuclear reaction profiling method. This method makes use of the nuclear reaction:

$$^{15}N + {}^{1}H \rightarrow {}^{12}C + {}^{4}He + \text{gamma-ray}.$$

This reaction has a large cross section at the resonance energy (6.385 MeV ^{15}N laboratory energy) with the cross section only a few keV away from being 4 orders of magnitude smaller (Horn and Lanford, 1988). To use this reaction for analysis, the sample to be analyzed is bombarded with ^{15}N with an energy at or above the resonant energy and the number of characteristic gamma rays produced in the target is measured with a scintillation detector. When the sample is bombarded with ^{15}N at the resonance energy, the gamma-ray yield is proportional to H on the surface of the sample. When the sample is bombarded with ^{15}N above the resonance energy, there are negligible reactions with surface H because the energy is above the resonance energy. However, as the ^{15}N ions penetrate the sample, they lose energy and reach the resonance energy at some depth. Now the gamma-ray yield is proportional to H at this depth. Hence, by measuring the gamma-ray yield as a function of beam energy, the H concentration as a function of depth is determined.

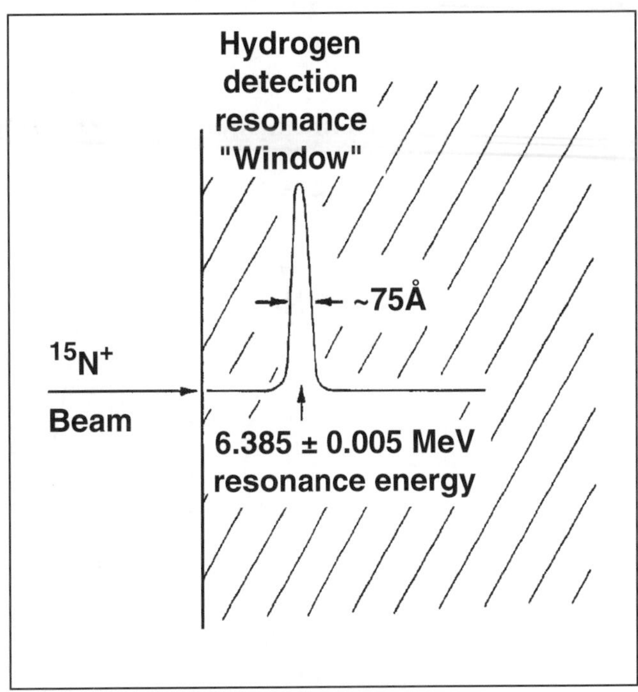

FIG. 8.1. A schematic representation of the ^{15}N resonant nuclear reaction profiling technique. The ^{15}N bombards the sample and reacts with H within a "resonance detection window" at a depth x in the target. Measuring the reaction yield vs beam energy determines the H concentration vs depth.

8.4 EXPERIMENTAL DETAILS

Figure 8.2 shows a schematic of the experimental chamber used for H profiling at the State University of NY at Albany. The ^{15}N beam enters from the right and bombards the samples which are mounted on the circumference of a rotatable sample wheel. The gamma rays from the reaction are detected by a scintillation detector (NaI crystal or a bismuth germinate crystal, BGO) located about 2 cm behind the sample (see Chapter 11). The following points should be considered.

8.4.1 Beam current integration

Because this technique relies on making an absolute measurement of gamma-ray yield per incident ion, it is important to integrate beam current carefully. Most importantly, this means suppressing any electrons that might enter or leave the Faraday cage. In the case of the chamber shown in Fig. 8.2, in addition to the –300 volt electrostatic suppression (shown), there is also (not shown) a permanent magnet mounted near the insulated coupling in the beam line. This combination of electric and magnetic suppression is very effective.

It is also important that the vacuum conditions be good enough so that no charge exchange occurs between the ion beam and the residual gas in the beam line. Otherwise, the charge carried to the Faraday cage by the ^{15}N ions may not be the same as that selected by the analyzing magnet, and the relationship between the integrated beam current to the number of incident ions is no longer known.

The chamber shown in Fig. 8.2 also has a filament to neutralize charge on insulating samples. This filament is a low voltage, low power light bulb with the glass broken off and the filament coated to lower the work function of

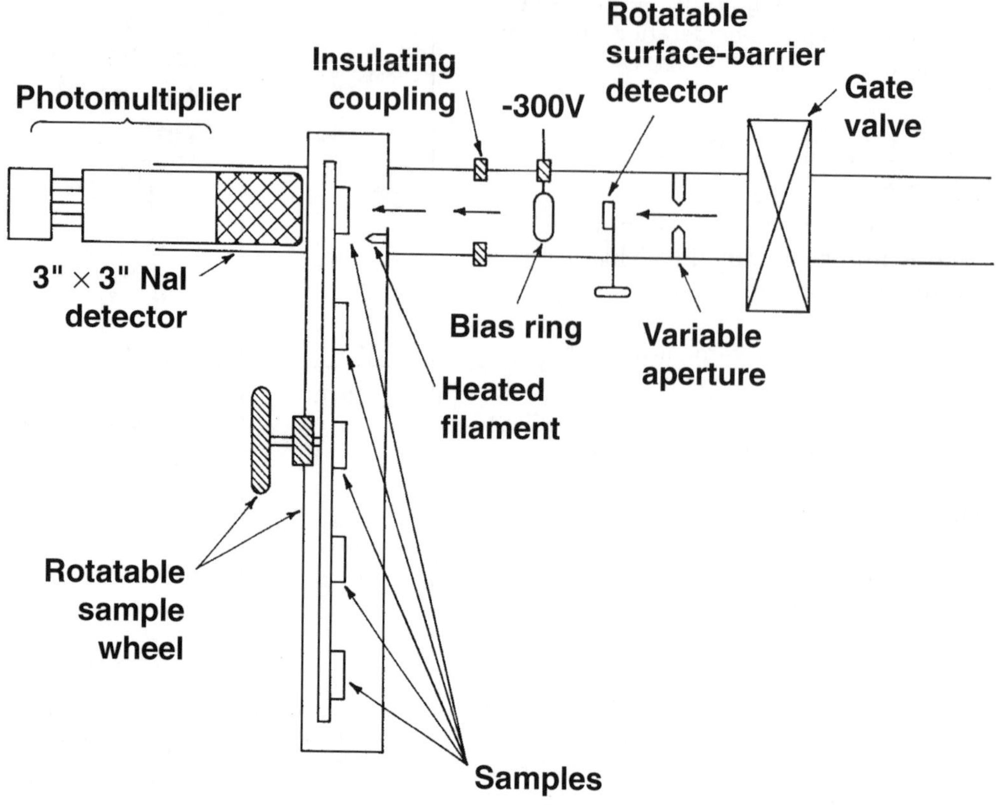

FIG. 8.2. A schematic of the chamber used for ^{15}N hydrogen profiling at the University at Albany. The ^{15}N beam bombards samples mounted on the circumference of the sample wheel. A gamma-ray detector is mounted outside the vacuum about 2 cm behind the samples. The chamber is a carefully designed Faraday cage with no slits near the targets. The filament provides electrons to stop insulating samples from charging in the beam.

the tungsten. The coating we use is a commercial product made by General Electric for this purpose. The filament is powered by a battery and is electrically enclosed in the Faraday cage. Hence, turning the filament on for insulating samples has no effect on the beam current integration system. This procedure allows analysis of insulators or conductors with equal ease.

For a more general discussion of beam current integration, see Chapter 12, Pitfalls in Ion Beam Analysis.

8.4.2 Beam geometry

It is also important to make certain that the incident beam hits the sample *and only the sample* once it enters the Faraday cage. There can be no beam-defining slits near the target, or hydrogen on these slits will cause reactions that will be detected by the gamma-ray detector, thereby causing a false signal. In the case of the setup shown in Fig. 8.2, the beam is viewed on a scintillator (a glass microscope cover slide in one of the sample positions) from the back (gamma-ray detector removed), and the ion beam is focused and steered in such a way that the beam will cleanly hit the sample to be analyzed when the sample is rotated into the beam.

8.4.3 H loss during analysis

Some materials lose H during analysis. The loss can be very rapid (e.g., CH_2) or it can be negligible (e.g., TiH_2) depending on the nature of the material. One of the great advantages of nuclear reaction analysis (NRA) for H is that it is easy to repeat measurements to see if the H content is changing during the analysis. Because of the possibility of hydrogen loss, it is useful to be able to spread the analyzing beam over a large area on the sample. The setup shown in Fig. 8.2, for example, has a beam raster system which allows the beam (after it is focused) to be rastered over a rectangular area up to a cm or more on a side. With such a system, it is possible to measure even very delicate materials, such as CH_2, with negligible loss during a run.

For some materials hydrogen can be lost simply by placing them in a vacuum. An example of this is hydrated glasses that lose water if placed in vacuum at room temperature. For such materials, freezing the samples to –30 or –40 degrees C before evacuation (and keeping them cold during analysis) can effectively stop this hydrogen loss (Schnatter *et al.*, 1988). In such a procedure, one has to be concerned that atmospheric water will condense on the samples before evacuation. In practice, this is not a serious problem. While a small amount of frost may form on the samples, it pumps off in a few minutes following evacuation.

There are also situations where it might be important to measure the hydrogen contents of samples while the samples are held in a hydrogen gas. For examples, some metals take up hydrogen from the gas phase but will give this hydrogen off when placed in vacuum. With differential pumping, it is possible to carry out this analysis with the samples held in H_2 gas at pressures up to a few Torr (Horn *et al.*, 1987).

8.4.4 Backgrounds

The large scintillation detectors used to measure the gamma-ray yield for NRA are also sensitive to other sources of radiation, most notably cosmic rays and natural radioisotopes (1.46 MeV gamma rays from decay of ^{40}K and 2.61 MeV from decay of ^{232}Th to ^{208}Pb). In some situations, there can also be background radiation produced by the accelerator. When measuring H at low levels, it is important to measure background carefully. This can be done, for example, by recording the gamma-ray counts from a "H free" sample such as a clean Si wafer which has a very low level of intrinsic hydrogen.

In a few situations, there can also be background nuclear reactions between the incident ^{15}N ions and isotopes in the target. In general, this is not expected to be a serious problem because the ^{15}N beam is far below the Coulomb barrier for every element except hydrogen. However, it is not below the Coulomb barrier for deuterium and there is a background reaction observed for deuterated samples. One reaction that occurs is the $D + {}^{15}N \rightarrow {}^{16}N + H$. The ^{16}N beta-decays (7.2 seconds half-life) to ^{16}O with accompanying high energy (6.13, 7.11, 8.87 MeV) gamma rays. The rate of this reaction can be monitored because it is not a resonance reaction so it varies smoothly with beam energy. Hence, by measuring the background as a function of energy from below the resonance energy to above it, the background yield can be estimated. Also, this background reaction produces some gamma rays of higher energy than the ^{15}N gamma rays. Hence, this background can be monitored by measuring the number of gamma rays above 4.4 MeV. Care should be taken to note that this background is not a prompt gamma-ray emitter, but the gamma ray is delayed by the beta decay of ^{16}N (7.2 seconds half-life).

The other background reaction that has been observed with a ^{15}N beam is with Li in the sample. Above about 7 MeV incident energy, high energy gamma rays are produced by reactions between ^{15}N and Li in the target.

Detailed investigations of the reactions between ^{15}N and deuterium and Li do not seem to be available in the literature. Fortunately, neither deuterium or Li are commonly present in samples, so this is usually not a problem. However, one needs to be aware of this possible problem and to take precautions when these backgrounds might be present.

8.4.5 Warning for those running N^{++} beam from a single-ended machine

When running $^{15}N^{++}$ beams from a single-ended accelerator without a system in the terminal that selects the charge to mass ratio of the particle being accelerated, there will likely also be a 1/4 energy "contaminant" beam. This beam comes from the acceleration of N_2^+ ions which break up after acceleration, but before the magnetic analysis system. This 1/4 energy beam and the full energy beam follow identical trajectories in a magnet field (they have the same ME/q^2). To separate this beam from the full energy beam requires electrostatic analysis (e.g., deflecting the beam with electric steerers and bringing it back with magnet steerers).

8.4.6 Warning for those running N beams from a tandem accelerator

Nitrogen does not form a stable negative ion. Hence, attempts to obtain an N^- beam from the ion source of a tandem accelerator will not be successful. However, there are a number of suitable negative ion complexes, such as NH^-, NH_2^-, and CN^-, which are easily available using direct extraction negative ion sources. Which complex to utilize depends on several factors that depend on the characteristic of a particular accelerator, such as is most easily produced in its ion source. However, in general, NH^- is preferable to CN^- because after stripping in the terminal, for a given terminal voltage the N ions from the former have higher energy and hence strip to a higher charge state. Any Coulomb explosion that might occur during stripping is also less for NH^- compared to CN^-. It should also be noted that when extracting such complexes from ion sources, there may be other complexes with the same atomic mass, e.g., OH^- and NH_2^-. Hence, it may be important to optimize ion source conditions monitoring the analyzed beam on target, not just the beam measured at the ion source.

8.5 ANALYSIS OF RESONANT NUCLEAR REACTION DATA

Figure 8.3 shows NRA data for measurements of H in a sample of Si implanted with 10^{16} H/cm^2 at 40 keV, and a sample of plasma-deposited amorphous Si on a crystal Si substrate. These figures show both raw data (i.e., gamma-ray counts as a function of beam energy) and final results (H concentration as a function of depth). The conversion of the raw data to final results involves multiplying the two raw data scales (axes) by constants, as outlined below.

The stopping power, dE/dx, for ^{15}N in most materials is essentially constant for beam energies between 6.4 and 10 MeV, the energies of interest for NRA profiling. Because of this, the conversion of the beam energy scale to a depth scale (x) is accomplished simply by

$$x = \frac{E - E_{res}}{dE/dx} \tag{8.1}$$

where E is the beam energy and E_{res} is the resonance energy (6.385 MeV). In cases where dE/dx is not constant, one would have to integrate dE/dx (or use range differences) to relate the energy above the resonance energy to depth within the sample.

The conversion of the yield data (Y) to hydrogen concentration, n(x), is almost as easy. Formally, the gamma-ray yield, Y, is given by

$$Y = \int Qn(x)s(x)dx \tag{8.2}$$

where Q is the number of incident ions, n is the hydrogen concentration, and σ is the cross-section. By changing variables to an integral over E, using dx = dE/(dE/dx), this becomes

$$Y = Q \int [n(E)\ \sigma(E)]/(dE/dx)dE\ . \tag{8.3}$$

Realizing that over the energies at which the cross section is non-zero, n and dE/dx are constant and can be taken through the integral sign, and $\sigma(E)$ is given by the Breit-Wigner formula

$$\sigma(E) = \frac{\sigma_0 \frac{\Gamma^2}{4}}{(E - E_{res})^2 + \frac{\Gamma^2}{4}} \tag{8.4}$$

where σ_0 is the cross section at the resonance energy and Γ is the full width at half maximum of the cross section.

Integrating this cross section over energy yields

$$Y = \frac{Q\, n\, \pi / 2\, \sigma_0\, \Gamma}{dE/dx} \tag{8.5}$$

or

$$n(x) = K\,(dE/dx)\ Y(x) \tag{8.6}$$

where K is a constant that reflects all the cross section parameters, and in practice, also the detector efficiency.

The gamma-ray yield, Y, is therefore converted to concentration using Eq. (8.6). The constant K is determined for an experimental chamber by making a measurement on a sample of known H content (see below for an example), although, in principle, it can also be determined from the known reaction parameters and an evaluation of the detector efficiency. It is important to note, however, that K is independent of the material being analyzed. All of the matrix dependent effects are in the dE/dx of Eq. (8.6).

In summary, NRA profile data, such as those in Fig. 8.3, are converted to H concentration as a function of depth using Eq. (8.1) for the depth scale and Eq. (8.6) for the concentration scale.

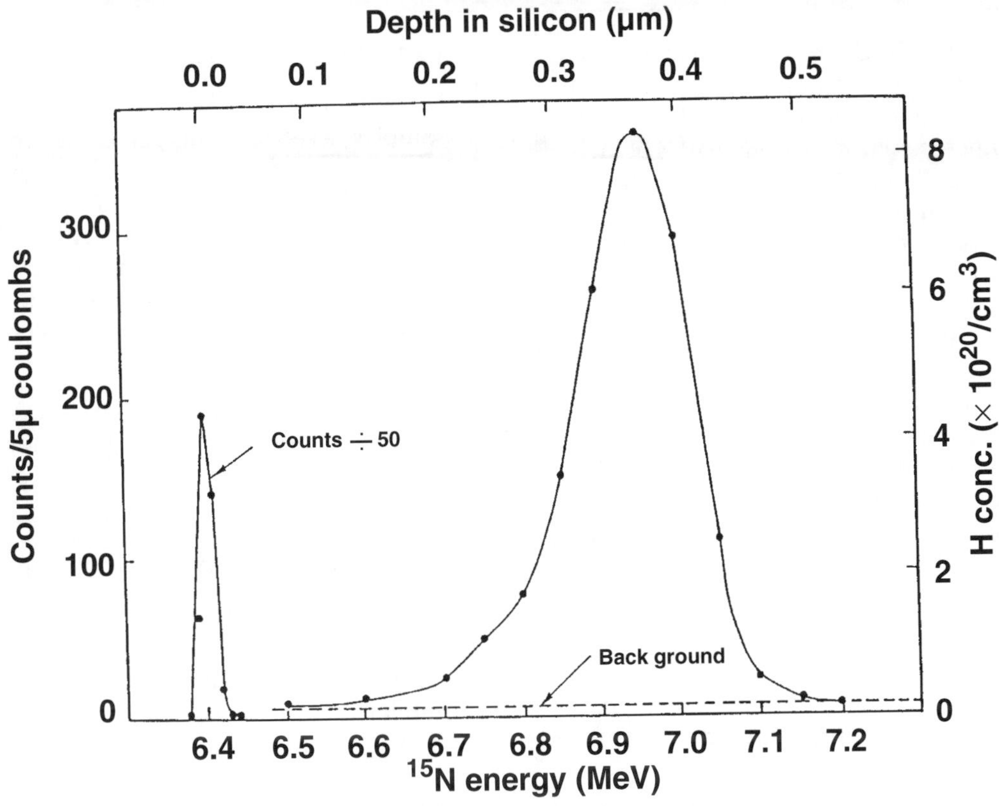

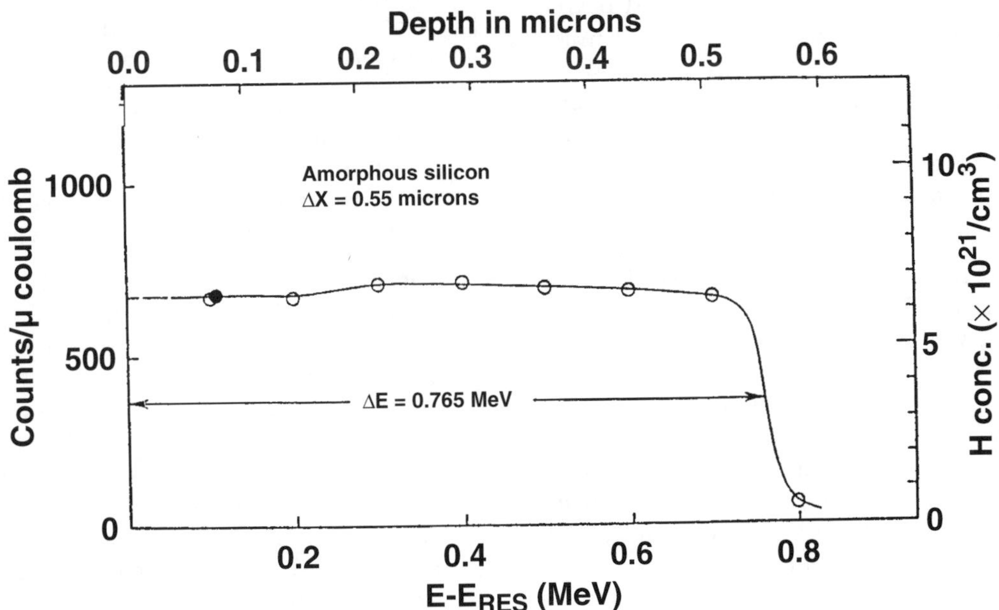

FIG. 8.3. Typical ^{15}N nuclear reaction profile data for (top) a Si sample implanted with 10^{16} H/cm^2 at 40 keV, and (bottom) a thin film of hydrogenated amorphous Si on a crystalline Si substrate. Both raw data (gamma-ray counts vs beam energy) and final results (H concentration vs depth) are given. Note that the solid datum point (bottom figure) was measured after the profile had been completed to check the H was not lost during analysis.

EXAMPLE 8.1. For the chamber shown in Fig. 8.2, using a 3-inch-diameter by 3-inch-long cylindrical BGO detector 2 cm behind the sample, with Y being counts per 1 microcoul (2 microcoul $^{15}N^{++}$) of incident ^{15}N particles, integrating all events in the 4.4 MeV gamma-ray full energy peak and the 3.9 MeV single escape peak, the constant K = 0.45 10^{19} in units of (micron/MeV cm^3), or Eq. (8.6) becomes

$$n = 0.45\ 10^{19}\ (dE/dx)\ (Y) \qquad (8.7)$$

where *dE/dx* is in units of MeV/micron, *Y* is in units of counts per 2 microcoul, and *n* is in units of H atoms/cm^3. Note that the data in Fig. 8.3 used a slightly different chamber with a different detector so the constant K for those data is different from that given here. Data using this setup will be discussed further below under calibration procedures.

8.5.1 Chamber calibration: Determination of K

While in principle the chamber constant K can be determined from the reaction cross section, resonance width, and detector efficiency, it is dangerous to trust the literature for absolute cross sections to the accuracy needed for analysis. The usual procedure is to record H profile data from a sample of known H content and determine K by comparing these data with the known H content. There are two types of standard samples that can be used, one has a known amount of H/cm^2, and one has a known H/cm^3.

The usual way to make samples of a known H/cm^2 is by ion implantation. The calibration of the chamber shown in Fig. 8.2 was established this way. This procedure has the advantage of allowing the determination of K for a chamber with an accuracy limited only by the precision with which the experiment can be carried out, i.e., the precision of the implantation and the precision of the H profile measurement. Determination of K by using samples of known H/cm^3 relies on knowing the dE/dx for the sample. Since dE/dx is generally known only to 5 - 10%, this limits the accuracy with which K can be determined by this procedure to 5 - 10%, at best.

EXAMPLE 8.2. Measurements on CH_2.

Measurement of Y for a variety of polyethylene samples gave 522 counts per 0.06 microcoulombs of $^{15}N^{++}$ beam. Note that this material can lose H very rapidly in the analysis beam. To minimize this effect, the beam current was very low (few namps) and it was rastered over several mm by several mm. Note also that different sources of "polyethylene" gave different count rates (by as much as 5% from the average).

The dE/dx for H is 28.4 MeV/mg/cm^2 and for carbon is 8.4 MeV/mg/cm^2. See Table A13.2. The Bragg rule for combining stopping powers gives the stopping power of CH_2

$$dE/dX_{CH2} = [(2)(28.4)+(12)(8.4)]/[2+12]$$
$$= 11.26\ MeV/mg/cm^2$$

Assuming a density of 1 gm/cm^3 for polyethylene implies that the dE/dx [in the units used for Eq. (8.6)] is 1.13 MeV/micron, and that the number of C atoms/cm^3 in polyethylene is 4.3 10^{22} atoms/ cm^3.

(Note, to convert dE/dx from units of MeV/mg/ cm^2 to MeV/micron, just multiply the former by the specific gravity in gm/cm^3 and divide by 10.)

Using this dE/dx, along with the measured Y (522 counts/0.06 uC = 17,400 count/2 uC) in Eq. (8.7), gives the hydrogen content of n = 8.84 10^{22} H atom/ cm^3, or the ratio of H/C atoms = 8.84/4.3 = 2.05, in good agreement with the expected value of 2.

Note that in calculating the ratio of H/C atoms, the assumed density of the polyethylene cancels out. This cancellation can be confirmed by the reader by carrying out the calculation for the H/C ratio assuming a different density for the polyethylene. (For example, assuming the density is 2 gram/cm^3 doubles the dE/dx in units of MeV/micron, which doubles the deduced absolute H concentration. Doubling the assumed density also doubles the deduced absolute C content. Hence, taking the ratio, the assumed density cancels.)

Profiles of implant data, such as those shown in Fig. 8.3, can be used to determine K. In this case, Eq. (8.6) is integrated over the yield curve giving nt (atoms/cm^2) = k $\int$ Y(E)dE.

Probably the more common way to determine K is to record data for samples of known H/cm^3. There are a large variety of possible materials, including plastics and metal hydrides that could be used (see Table A12.3). To illustrate this procedure for the chamber shown in Fig. 8.2, the use of polyethylene as a "standard" will be presented. This is not the procedure used to establish K for this chamber, but is used as a check of the ion-implantation-based calibration and as an easy way to check that nothing in the analysis chamber has changed over time.

It is perhaps worth emphasizing that unlike some procedures which require standards for each class of materials to be analyzed, for nuclear reaction analysis one has only to establish the chamber constant K once. Given K, using the procedure outlined above, any material for which dE/dx is known can be analyzed.

8.5.2 Energy loss values

Table A13.2 in the appendices gives a list of elements with their densities (specific gravities) and ^{15}N, ^{19}F and ^{7}Li stopping powers at their resonance energy. These stopping powers are from the program STOP written by James Ziegler of IBM Watson Research Laboratory.

8.5.3 Depth resolution

There exists some confusion about the depth resolution of these methods in the literature. Commonly, the near surface depth resolution was considered to be given simply by the width of the resonance divided by dE/dx. Deeper into the sample, the depth resolution would be dominated by straggle of the incident beam. For the ^{15}N case, the resonance is so narrow (1.8 keV) that the motion of H bonded in a solid will shift the reaction on and off resonance (Zinke-Allmang *et al.*, 1985). As a result, the reaction behaves as if the cross section is not Breit-Wigner (Lorentzian), but Gaussian with the width determined by how the H is bonded (in the range of 5 -15 keV). For the ^{15}N reaction, this Doppler effect dominates the near surface depth resolution.

In general, the evaluation of the depth resolution of resonant nuclear reaction profiling requires the convolution of a Gaussian width (from beam energy straggle and Doppler effects) with a Breit-Wigner (Lorentzian) width (from the resonance cross section). This has to be done numerically. Amsel and Maurel (1983) have done this for us and presented the general result in graphical form (Fig. 8.4).

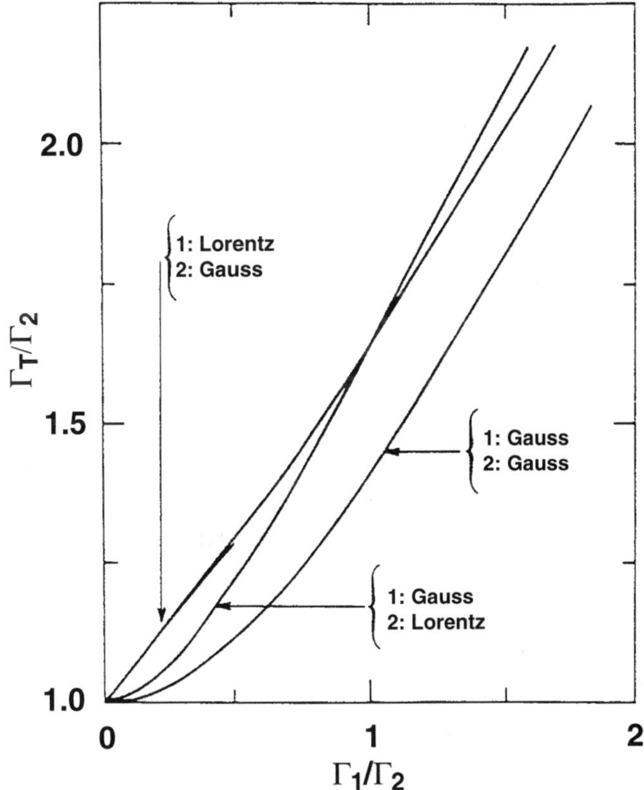

FIG. 8.4. This figure (Amsel and Maurel, 1983) gives the result in graphical form of the width that results from the convolution of a Gaussian with a Lorentzian (Breit-Wigner), as well as the result for a Gaussian convoluted with a Gaussian. The results are expressed as the ratio of the Total (T) width over the width of one of the functions. See the text for an example.

8.5.4 Off-resonance contributions

When profiling through a surface layer with a high H content into a material with a low H content, the off-resonance cross section interacting with H in the surface layer can contribute to the count rate. This is shown schematically in Fig. 8.5. It is clearly important to know this contribution. For the case of the ^{15}N reaction, this off-resonance contribution has been carefully measured (at least for beam energies below 8 MeV) (Horn and Lanford, 1988). The counts from the off-resonance cross section can be determined by integrating the area under the surface layer peak (high hydrogen). The off-resonance contribution is then

$$Y_{\text{off res.}} = (1.28 \times 10^{-5})\ I \qquad (8.8)$$

where *I* is the integrated area (in count·keV) of the surface layer peak.

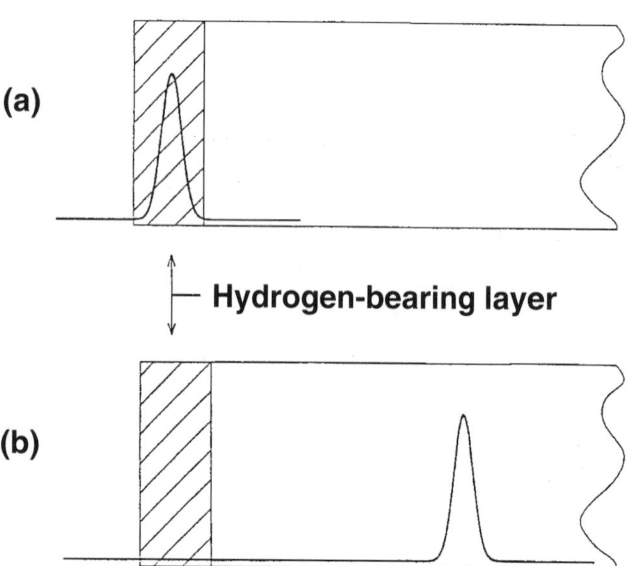

FIG. 8.5. A schematic representation of profiling through a hydrogen-bearing surface layer into a low H region. In situation (b), the off resonance cross section can interact with the H in the surface layer contributing to the count rate. The counts from the process are (to a good approximation) just proportional to the integrated amount of H in the surface layer times the off resonance cross section. See the text for an example.

EXAMPLE 8.3. Convolution of a Gaussian and a Lorentzian.

As an example, let us evaluate the depth resolution (FWHM) of a resonant nuclear reaction (Breit-Wigner width of 25 keV) at a depth in a target where the straggle is 20 keV (a Gaussian distribution with a FWHM of 20 keV). In the notation of Fig. 8.4, $\Gamma_1/\Gamma_2 = 20/25 = 0.8$. From the Fig. 8.4, reading the Gaussian/Lorentz at $\Gamma_1/\Gamma_2 = 0.8$ gives $\Gamma_T/\Gamma_2 = 1.45$ or the total width is (1.45)(25) = 36 keV. The depth resolution is then given by this energy width divided by the dE/dx at the resonance energy.

EXAMPLE 8.4. Evaluation of the Off-Resonance Yield.

Figure 8.6 shows the raw data for a hydrogen profile of a sample which consisted of a plasma-deposited silicon nitride (very high hydrogen content), on aluminum (low hydrogen content), on a silicon substrate. The surface nitride gave a very high count rate (about 5500 counts/4.8 microcoul). The question was whether the low count rate observed in the aluminum (energies 6.7 and 7.3 MeV) was due to real hydrogen in the aluminum or due to the off-resonance yield with hydrogen in the surface nitride. Integration of counts in the surface nitride layer needs to be done carefully, which is not possible for the data plotted in the log plot of Fig. 8.6. Roughly, one can see in this figure that there are about 5500 counts from the resonance energy (6385 keV) to about 6550 keV. Careful integration of these data gives $I = 0.96\ 10^6$ (keV)(counts). Using Eq. (8.8), the off-resonance yield should be

$$Y_{\text{off res}} = (1.28\ 10^{-5})(0.96\ 10^{6}) = 12.3 \text{ counts.}$$

This is indeed about the count rate observed in the aluminum layer, and the conclusion is that there is no measurable hydrogen in this layer. The large peak at the back edge of the aluminum (around 7.5 MeV) is real and clearly not due to off-resonance reactions. Rather, this peak results from the accumulation of hydrogen at the Al/Si interface during sample annealing.

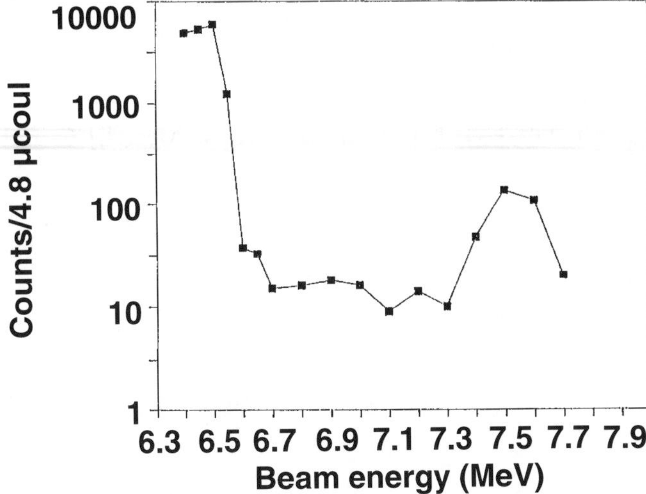

FIG. 8.6. The raw data for a hydrogen profile of a sample consisting of a thin layer of plasma silicon nitride on aluminum on a silicon substrate. The very high count rate surface peak (6.385-6.55 MeV) corresponds to the H in the silicon nitride. The low count rate region (6.7-7.3 MeV) corresponds to the aluminum layer and the peak at 7.5 MeV corresponds to the Al/Si interface. The approximately 10 counts in the Al region are attributed to the off-resonance reaction with the surface H layer. See text.

8.5.5 High sensitivity measurements

The ^{15}N profiling technique outlined above will give a sensitivity on the order of hundreds of ppm (atomic), with beam currents on the order of 50 namp and run times on the order of a few minutes. To measure samples with lower H content, it is necessary to (1) increase the beam current, or (2) decrease the background, or (3) increase the detector efficiency, or (4) use a reaction with a larger cross section. All of these are done. The increase in beam current depends on the accelerator available and how much beam a sample will take (with cooling and beam rastering). The decrease in gamma-ray detector background is accomplished with material and electronic shielding of the gamma-ray detector to reduce the number of cosmic ray events recorded (Damjantschitsch *et al.*, 1983; Kuhn *et al.*, 1990; and Horn and Lanford, 1990). New well-type BGO detectors provide a high detector efficiency compared to the commonly used 3-inch by 3-inch cylindrical detector (Kuhn *et al.*, 1991). Going to the 13.35 MeV ^{15}N resonance increases the gamma ray yield by an order of magnitude (at a sacrifice in depth resolution and more complicated off-resonance contributions). Hence there are many ways to increase the sensitivity of this method and all of these approaches have been explored. However, experience indicates that many research problems in this field can be effectively studied without such changes.

8.6 SUMMARY AND CHECK LIST OF NUCLEAR REACTION PROFILING PROCEDURES AND CONCERNS

Given the availability of suitable energy beams of ^{15}N, ^{19}F, or ^{7}Li and a gamma-ray detector (e.g. 3-inch by 3-inch NaI or BGO), NRA for H profiles requires only the ability to bombard the sample cleanly with the analyzing beam, to integrate the incident beam effectively, and to measure the resultant gamma rays with a fixed efficiency. Probably the two most common difficulties encountered revolve around the questions: "Is the beam hitting the sample and only the sample?" and "Is the H stable in the sample during analysis?" The first of these can be managed by taking some care in designing the beam geometry (no slits near the target but ability to monitor beam position, beam size, and any possible beam halos). The second pitfall can be managed by taking care to minimize beam intensity on the target by rastering the beam, by running minimum total number of ions required to get good statistical accuracy, and by cooling the target.

Analysis of the data is done by using Eqs. (8.1) and (8.6), with the chamber calibration determined by running one or more samples of known H content.

REFERENCES

Adler, P. N., Kamykowski, E. A. , and Padwer, G. M. (1974), in *Hydrogen in Metal*, American Society of Metals, Materials Park, Ohio. p. 623.

Amsel, G. and Maurel, B. (1983), *Nucl. Instrum. Methods* **218**, 183.

Barnes, C. A., Overley, J. C., Switzkowski, Z. E. and Tombrello, T. A. (1977), *Appl. Phys. Lett.* **31**, 239.

Cohen, B. L., Fink, C. L., and Degnan, J. H. (1972), *J. Appl. Phys.* **43**, 19.

Damjantschitsch, H., Weiser, M., Heusser, G. Kalbitzer, S., Mannsperger, H. (1983), *Nucl. Instrum. Methods* **218**, 129.

Horn, K. M. and Lanford, W. A. (1988), *Nucl. Instrum. Methods* **B29**, 609.

Horn, K. M. and Lanford, W. A. (1990), *Nucl. Instrum. Methods* **B45**, 256.

Horn, K. M., Lanford, W. A., Rodbell, K. and Ficalora, P. (1987), *Nucl. Instrum. Methods* **B26,** 559.

Khabibullaev, P. K. and Skorodumov, G. G. (1989), *Determination of Hydrogen in Materials, Nuclear Physics Methods*, Springer-Verlag, Berlin, Germany.

Kuhn, D., Rauch, F., and Baumann, H. (1990), *Nucl. Instrum. Methods* **B45,** 253.

Kuhn, D., Sturm, H. and Rauch, F., (1991), *10th International Conference on Ion Beam Analysis (1-5 July, 1991, Eindhoven, The Netherlands) Book of Abstracts*, p. 73.

Lanford, W. A. (1992), *Nucl. Instrum. Methods in Phys.* Res. **B66,** 65.

Lanford, W. A., Trautvetter, H. P., Zieger, J. F. , and Keller, J. (1976), *Appl. Phys. Lett.* **28,** 566.

L'Ecuyer, J., Brassand, C., Cardinal, C., Chabbal, J., Deschenes, L., Labrie, J. P., Terreault, B. , Martel, J. G. and St. Jacques, R., *J. Appl. Phys.* (1976), **47**, 381 and many others.

Leich, D. A. and Tombrello, T. A. (1973), *Nucl. Instrum. Methods* **108,** 67, and many others.

Schnatter, K. H., Doremus, R. H. and Lanford, W. A. (1988), *J. Non-Cryst. Solids* **102**, 11.

Trocellier, P., and Engelmann, Ch. (1986), *J. Radioanalytical and Nuclear Chemistry* **100,** 117.

Xiong, F., Rauch, F. , Shi, C. , Zhou, Z., Livi, R. P., and Tombrello, T. A. (1987), *Nucl. Instrum. Methods in Phys.* Res. **B27**, 432.

Zinke-Allmang, M., Kalbitzer, S. and Weiser, M. (1985), *Z. Phys.* **A230**, 697.

CHAPTER

9

CHARGED PARTICLE ACTIVATION ANALYSIS

G. Blondiaux and J.-L. Debrun

CNRS-CERI, Orleans, France

C.J. Maggiore

Los Alamos National Laboratory, Los Alamos, New Mexico

CONTENTS

9.1 INTRODUCTION

Charged particle activation analysis (CPAA) can in principle be used to determine almost any element, in solids or liquids, over a wide range of concentrations (from less than the part per billion level to the percent level). However, taking into account practical considerations and the availability of numerous other analytical techniques, the important application of CPAA is to analyze light elements such as C, N, and O at trace levels in solids. For this application, CPAA is the best reference method because:

- Chemical contaminations do not introduce significant errors in the measurement, which is especially important in the case of ubiquitous elements like C, N, and O.
- Detection limits are good because, unlike "in-beam" analysis, it uses radioisotopes which can be separated from disturbing backgrounds and then measured.
- As in other nuclear methods, absolute calibration is direct, easy, and free from matrix effects.

This handbook chapter will present only essential information. More details on activation analysis, radiochemical procedures, counting, etc. can be found in De Soete (1972). For more details on charged particle activation itself, readers should consult Vandecasteele (1988). Appendix 14 of this book contains a compilation of cross-sections and yield curves used in CPAA.

9.2 THEORY

9.2.1 General principle of radioactivation

The sample to be analyzed is "activated" by irradiation with an ion beam of suitable type and energy. The artificial radioisotopes such as in Example 9.1 are created by nuclear reactions on the matrix itself and on the impurities of interest. Qualitative analysis is based on the identification of the created radioisotopes, mainly using high resolution gamma-ray spectroscopy. Quantitation is based on the correspondence between the stable isotope present in the sample and the radioisotope produced by the beam. It depends on the direct relationship that can be established between the concentration of the element (isotope) of interest, and the counting rate (activity) of the corresponding radioisotope.

EXAMPLE 9.1. Activation for analysis.

Chromium can be analyzed via the (p, n) reaction on the stable isotope, ^{52}Cr, which produces ^{52}Mn. This nuclide is radioactive and emits characteristic gamma rays (1434, 935, and 744 keV) with a half-life of 5.6 days. Detection of ^{52}Mn after irradiation of the sample with protons (< 15 MeV) indicates the presence of chromium because below 15 MeV no other element can produce ^{52}Mn by a proton-induced reaction.

9.2.2 Creation of radioisotopes by irradiation

During irradiation, radioisotopes are created via nuclear reactions and immediately begin to decay so that the instantaneous rate of production for a *thin target* is given by

$$d\mathcal{N}/dt = n\sigma i dx - \lambda\mathcal{N} \qquad (9.1)$$

where

$\mathcal{N}$ = number of radioactive atoms at time t,

n = number of atoms to be analyzed/mass,

σ = nuclear cross section for the reaction of interest,

i = number of incident ions on target/unit time,

dx = target thickness (g/cm^2), and

λ = decay constant of the radioisotope of interest.

The decay constant is $\lambda = \ln(2)/\tau$, where τ is the half-life of the radioisotope.

$\mathcal{N}_0$, the number of radioactive atoms at the end of an irradiation time, t_i, is obtained by integration

$$\mathcal{N}_0 = (n\sigma i/\lambda)\, dx\, [1 - e^{-\lambda t_i}] \qquad (9.2)$$

The activity at the end of the irradiation, the total number of disintegrations per second, is

$$A_0 = \lambda\mathcal{N}_0 \text{ (becquerels)} \ . \qquad (9.3)$$

The term $[1 - e^{-\lambda t_i}]$ is called the "saturation factor." After a rapid growth, it approaches 1 asymptotically as shown in Fig. 9.1. The activity at time, t_d, after the end of irradiation (the origin of time is now the end of irradiation) is given by

$$A_{td} = A_0 e^{-\lambda t_d} \tag{9.4}$$

where *d* stands for decay.

Cross sections are functions of energy, and because the incident ions lose energy as they traverse the sample, σ can also be thought of as a function of depth, *x*. The activity for a *thick target* at the end of irradiation is then given by

$$A_0 = ni\left(1 - e^{-\lambda t_i}\right) \int_0^{E_i} \frac{\sigma(E)}{S(E)} dE \tag{9.5}$$

where E_i is the energy of the incident ions in the sample and *S(E)* is the stopping power, *dE/dx*. See Chapter 2 for a discussion of stopping power.

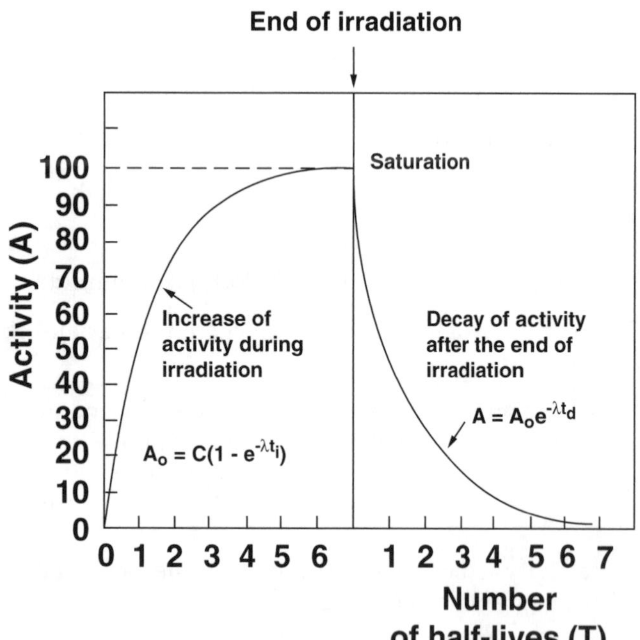

FIG. 9.1. Creation and increase of the activity during irradiation, and decay after the end of the irradiation.

9.2.3 Standardization for quantitation

In principle, the concentration, *n* (atoms/g) of the element to be analyzed can be obtained using Eq. (9.5), measuring both the absolute activity of the radionuclide of interest and the integrated beam intensity. The absolute cross section and the stopping power of the incident ion for numerical integration of the integral must also be known. It is not easy to measure absolute activities and intensities, and the absolute cross sections are not very well known either. Thus, for more accurate analysis, it is common to measure relative to the activity of a standard. The sample and the standard are irradiated at the same energy with appropriate total charges, Q_i, and the ratio of the activities after irradiation is compared. The ratio of the concentrations is then

$$\frac{n_x}{n_s} = \frac{A_x}{\left(1 - e^{-\lambda t_i}\right)_x} \frac{\left(1 - e^{-\lambda t_i}\right)_s}{A_s} \frac{Q_s}{Q_x} \frac{\int_0^{E_i} \frac{\sigma(E)dE}{S_s(E)}}{\int_0^{E_i} \frac{\sigma(E)dE}{S_x(E)}}$$

$$\frac{n_x}{n_s} = \frac{A_x}{\left(1 - e^{-\lambda t_i}\right)_x} \frac{\left(1 - e^{-\lambda t_i}\right)_s}{A_s} \frac{Q_s}{Q_x} \mathcal{R} \tag{9.6}$$

The subscripts *x* and *s* refer to the unknown sample and the standard, respectively. The end of irradiation is usually taken as the reference time. Instead of the absolute activity, *A*, the measured counting rate corrected for the decay during counting time, t_c, and the delay time before measurement (Example 9.2), t_d, are used. These corrections are given by

$$C_0 = \frac{\lambda}{\left(1 - e^{-\lambda t_c}\right)} e^{\lambda t_d} C \tag{9.7}$$

where C is the integrated count acquired during the counting time. The corrected C_0's from the standard and the unknown are used in place of the activities in Eq. (9.6). By knowing the concentration, n_s, in the standard and measuring the ratio of the counting rates in the standard and the unknown, the concentration in the unknown, n_x, can be calculated if the ratio of the integrals, $\mathcal{R}$, in Eq. (9.6) is calculated. Several methods have been proposed to calculate $\mathcal{R}$. The choice of an appropriate method depends on the precision required and the information available on the activation cross section versus energy.

EXAMPLE 9.2. Corrections for counting times and delays.

Consider the case with no interferences and an activation product with a half-life of 10 minutes [$\lambda = \ln(2)/\tau = \ln(2)/10 = .693/10 = .0693$]. The sample activation begins at 1:17 and ends at 1:39 for a sample irradiation time, t_i = 22 minutes. If the beam current is 0.1 μA, then the integrated charge is 132 μC. If the counting of the sample begins at 1:44 and ends at 1:55 then the counting time, t_c = 11 minutes, and the delay time, t_d = 5 minutes. Then according to Eq. (9.7) the count rate for the unknown at the end of the irradiation is

$$C_0 = \frac{\lambda}{\left(1 - e^{-\lambda t_c}\right)} e^{\lambda t_d} C_x$$

$$= \frac{0.0693}{\left(1 - e^{-0.0693 \times 11}\right)} e^{0.0693 \times 5} C_x = 0.184\ C_x$$

where C_x is the total count in the peak of interest during the 11-minute counting time. If the activation of the standard begins at 1:57 and ends at 1:58 with a beam current of 0.01 μA, the integrated charge is 0.6 μC. If the counting of the standard begins at 2:15 and ends at 2:18, the counting time, t_c = 3 minutes, and the delay time, t_d = 17 minutes. Equation (9.7) gives the corrected count rate for the standard

$$C_0 = \frac{\lambda}{\left(1 - e^{-\lambda t_c}\right)} e^{\lambda t_d} C_s$$

$$= \frac{0.0693}{\left(1 - e^{-0.0693 \times 3}\right)} e^{0.0693 \times 17} C_s = 1.199\ C_s$$

where C_s is the total count in the peak during the 3-minute counting time. Then substituting the corrected counting rates for the absolute activities in Eq. (9.6) we have

$$\frac{n_x}{n_s} = \frac{A_x}{\left(1 - e^{-\lambda t_i}\right)_x} \frac{\left(1 - e^{-\lambda t_i}\right)_s}{A_s} \frac{Q_s}{Q_x} \frac{\int_0^{E_i} \frac{\sigma(E)dE}{S_s(E)}}{\int_0^{E_i} \frac{\sigma(E)dE}{S_x(E)}}$$

$$\frac{n_x}{n_s} = \frac{0.184 C_x \left(1 - e^{-0.0693 \times 1}\right)_s \times 0.6}{\left(1 - e^{-0.0693 \times 22}\right)_x \times 1.199\ C_s \times 132} \mathcal{R}$$

$$\frac{n_x}{n_s} = 5.97 \times 10^{-6} \frac{C_x}{C_s} \mathcal{R}$$

The substitution of the corrected measured count rates for the absolute activities in Eq. (9.6) assumes that the sample and the standard are counted under exactly the same conditions.

9.2.3.1 Approximations based on range

The following approximations require no detailed knowledge of the cross sections, but only relative stopping powers or ranges. Detailed justifications can be found in Ricci and Hahn (1965, 1967), Ishii (1978a), and Chaudri *et al.* (1976). The error introduced is less than 10% if the irradiation energy is reasonably higher than the threshold energy. The higher the incident energy and the higher the reaction threshold, the smaller the error introduced by these approximations. The error also becomes smaller as the average mean atomic numbers of the matrix and the sample are closer.

Activated range approximation $\mathcal{R} = \frac{R_s(E_i) - R_s(E_{th})}{R_x(E_i) - R_x(E_{th})}$ Ricci (1965)

Total range approximation $\mathcal{R} = \frac{R_s(E_i)}{R_x(E_i)}$ Ricci (1967)

In these approximations E_i is the incident beam energy and E_{th} is the threshold energy for the reaction.

9.2.3.2 Numerical integration

If the integrals in Eq. (9.6) are evaluated numerically, only relative cross-sections and relative stopping powers are needed. Vandecasteele (1988) has shown through calculations that the result is quite insensitive to large variations in the shape of the cross-section curve.

9.2.3.3 Approximations based on stopping power

It has been shown by Ishii *et al.* (1978a) that $\mathcal{R}$ can be approximated with a negligible error by the ratio of the stopping powers at a calculable average energy. If the average stopping power, $\langle S \rangle$, is defined such that

$$\int_0^{E_i} \frac{\sigma(E)dE}{S(E)} = \frac{1}{\langle S \rangle} \int_0^{E_i} \sigma(E)dE \qquad (9.8)$$

then $\langle S \rangle = S(E_m)$ where E_m is the average energy

$$E_m = \frac{\int_0^{E_i} E\sigma(E)dE}{\int_0^{E_i} \sigma(E)dE} \ . \qquad (9.9)$$

E_m is matrix independent to a good approximation (< 1%) and the ratio of the integrals, $\mathcal{R}$, is then given by

$$\mathcal{R} = \frac{S_x(E_m)}{S_s(E_m)} \ . \qquad (9.10)$$

The systematic error from the approximation is small and can be calculated (Ishii *et al.*, 1978a), and just as for the numerical integration, large variations in the shape of the cross section have little effect.

A simple method of approximating $\mathcal{R}$ suggested by Chaudri *et al.* (1976) is to use Eq. (9.10) at the mean energy, $E_m = (E_i + E_{th})/2$, where E_i is the incident energy and E_{th} is the threshold energy for the reaction. This approximation is usually adequate for most applications. Example 9.3 shows the results for different approximation methods.

EXAMPLE 9.3. Calculating the ratio of integrals.

The reaction $^{14}N(p, \alpha)^{11}C$ can be used to measure nitrogen in materials, and nylon can be used as a standard. To measure the nitrogen in a sample of vanadium with 6 MeV protons, the ratio of integrals in Eq. (9.6) can be calculated. The total range approximation of Ricci gives

$$\mathcal{R} \approx \frac{R_s(E_i)}{R_x(E_i)} = \frac{R_{Nylon}(6)}{R_V(6)}$$

$$= \frac{420.3\ \mu m \times 1.13\ gm/cm^3}{135.5\ \mu m \times 6.10\ gm/cm^3} = 0.575 \ .$$

The reaction has a threshold energy of 5 MeV, so the mean energy of Chaudri is

$$E_m = (5 + 6)/2 = 5.5\ MeV \ .$$

Then the mean stopping power approximation gives

$$\mathcal{R} \approx \frac{S_x(E_m)}{S_s(E_m)} = \frac{S_V(5.5)}{S_{Nylon}(5.5)}$$

$$= \frac{4.55 \times 10^{-2}\ keV/\mu g/cm^3}{7.66 \times 10^{-2}\ keV/\mu g/cm^3} = 0.594 \ .$$

According to Vandecasteele (1988), the value of $\mathcal{R}$ for this case by numerical integration is 0.603. Figure 9.2 plots the deviations for the range and stopping power approximations versus energy and Z of the matrix where

$$\Delta = \frac{(\mathcal{R} - \mathcal{R}_{approx})}{\mathcal{R}} \times 100\% \ .$$

Notice that the higher the incident energy and the higher the reaction threshold, the smaller the error introduced by these approximations. The error also becomes smaller as the average mean atomic numbers of the matrix and the standard are closer.

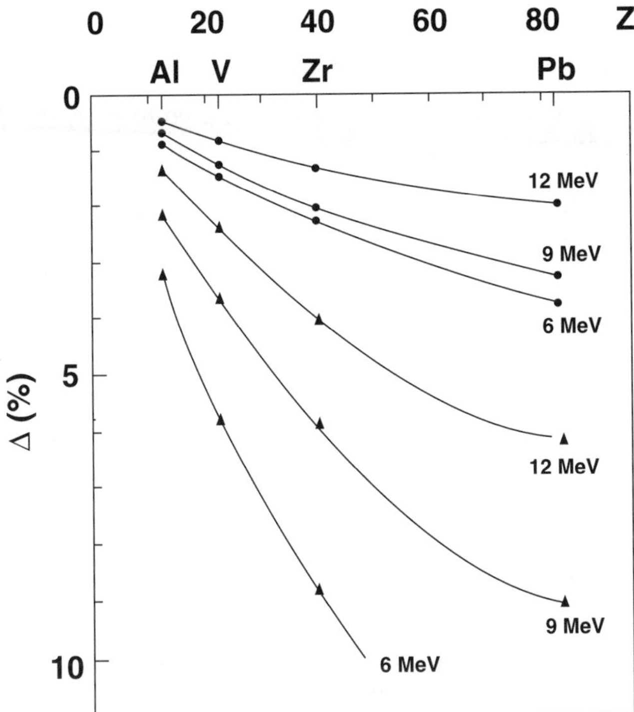

FIG. 9.2. Difference (%) between the ratio, $\mathcal{R}$, of the "activation integrals," obtained by numerical integration and by different approximations as a function of the atomic number of the matrix for different irradiation energies. The standard is nylon and this is for the $^{14}N(p, \alpha)^{11}C$ reaction (Vandecasteele, 1988).

- • stopping power approximation at the mean energy of Chaudri *et al.* (1976).
- ▲ total range approximation of Ricci and Hahn (1967).

Note that E_m can also be calculated from a thick target yield curve (Ishii, 1978b). Using the average stopping power method, Blondiaux *et al.* (1981) showed that *no measurable error* is introduced when chemical compounds are used for standards, or when a compound matrix is analyzed. Bragg-Kleeman's rule on the additivity of stopping powers is valid in the energy domain encountered in charged particle activation.

9.3 EXPERIMENTAL TECHNIQUES

Activation analysis usually consists of these independent steps: (1) irradiation with the appropriate beam to activate the sample; (2) post-irradiation etching to remove activated surface contamination; (3) radiochemical separation to isolate the radioisotope of interest when it cannot be detected directly in the sample; and (4) measurement of the activity.

9.3.1 Reaction choice

Cyclotrons and electrostatic accelerators are used primarily with protons, deuterons, and ^{3}He for activation analysis. Alpha particles are of limited interest and tritons are interesting but rarely available. Heavy ions ($Z > 2$) are of little analytical use, and difficult to use in practice because of sample deterioration due to high stopping power. Table 9.1 lists the ions, energies, reactions, and sampling depths commonly used in activation analysis.

Because there is a large choice of ions and energies, a given analytical problem may sometimes be solved in various ways. The strategy is to obtain an adequate sensitivity for the elements to be analyzed while keeping the matrix activity to a minimum. The nuclear interferences, production of the same reaction product by a different reaction on a different chemical element, must

Table 9.1. Ions, reactions, and sampling depths.

Incident ion	Reaction	Energy *(MeV)*	Sampling depth
p	(p, n) (p, 2n) (p, pn) (p, α)	10 - 30	100 μm - mm's
d	(d, n) (d, 2n) (d, p) (d, α)	3 - 20	10 μm - 2 mm
t	(t, n) (t, p) (t, d)	3 - 15	10 μm - 100 μm
^{3}He	(^{3}He, n) (^{3}He, 2n) (^{3}He, p) (^{3}He, α)	3 - 20	few μm - 100 μm
α	(α, n) (α, 2n) (α, 3n) (α, p) (α, pn) (α, αn)	15 - 45	10 μm - 100's μm

also be avoided. In practice, this determines the reaction that is used. Example 9.4 is typical of the possibilities. When analyzing an element of atomic number Z, one can use radioisotopes of elements Z-1, Z, Z+1 and perhaps Z+2 and Z-2. Figure 9.3 shows the Z versus nucleon number, N, plot of the chart of the nuclides, and the nuclear reactions that can be used to reach adjacent nuclei which may be of analytical interest. Notice that there are several different ways to produce the same final radionuclide from the same starting isotope. Similarly there are possible interferences for a particular radionuclide due to other reactions from isotopes of other elements. Most of the time, thick targets are irradiated, where thick means thicker than the range of the incident particles. In that case, the higher the energy, the higher the sensitivity because you activate a larger volume of material.

EXAMPLE 9.4. Activation of nickel.

Several different reactions can be used to analyze for nickel, among them: $^{60}Ni(p, n)^{60}Cu$, $^{58}Ni(p, pn)^{57}Ni$, $^{58}Ni(p, \alpha)^{55}Co$, $^{60}Ni(d, n)^{61}Cu$, and $^{60}Ni(^{3}He, pn)^{61}Cu$. If you wanted to use the (p, α) reaction that produces ^{55}Co, you would need to be concerned about the interference from the $^{56}Fe(p, 2n)^{55}Co$ reaction. Figure 9.4 shows the yield of ^{55}Co from Ni and Fe as a function of incident proton energy. Clearly the activation should be carried out below 15 MeV, the threshold for the (p, 2n) reaction on ^{56}Fe.

The nuclear interferences increase in number and importance with the energy because more reaction channels become available. Figure 9.5 is an example of the possibilities for a single isotope of copper, ^{63}Cu, and an incident ^{3}He beam. Few of the radioisotopes have suitable half-lives, not too long and not too short, and suitable decay schemes. Furthermore, only a fraction of the "suitable" radioisotopes is free of nuclear interference.

Van de Graaff accelerators are preferred to cyclotrons because of higher beam quality and because they are easier to operate, particularly when an energy change is needed. For the higher energies, however, one needs a large tandem, which may not be easily accessible; a small cyclotron may then be an easier solution.

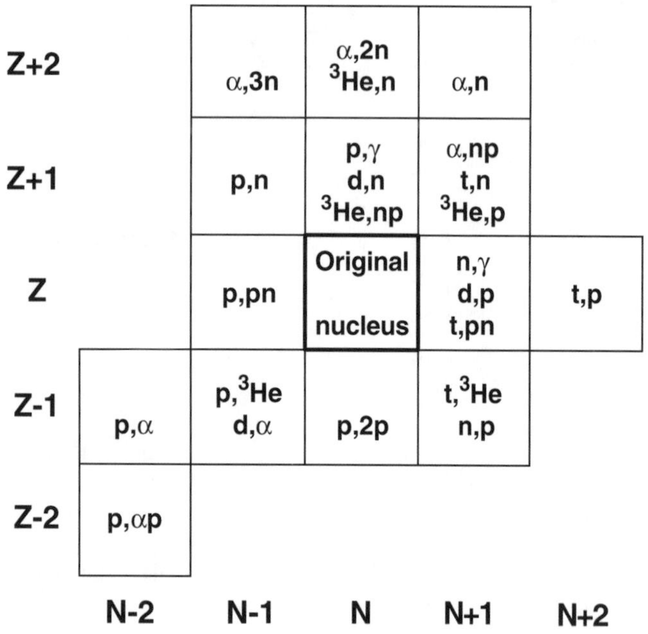

	N-2	N-1	N	N+1	N+2
Z+2		α,3n	α,2n ^{3}He,n	α,n	
Z+1		p,n	p,γ d,n ^{3}He,np	α,np t,n ^{3}He,p	
Z		p,pn	**Original nucleus**	n,γ d,p t,pn	t,p
Z-1	p,α	p,^{3}He d,α	p,2p	t,^{3}He n,p	
Z-2	p,αp				

FIG. 9.3. Reactions to obtain radioisotopes with atomic numbers Z - 1, Z - 2, Z, Z + 1, and Z + 2, starting from a target of atomic number Z.

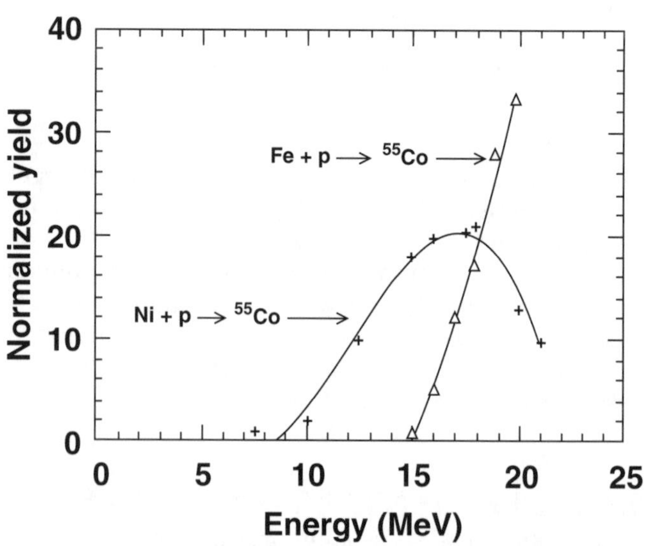

FIG. 9.4. Interference of Fe by the $^{56}Fe(p, 2n)^{55}Co$ reaction above 15 MeV, when analyzing Ni by the $^{58}Ni(p, \alpha)^{55}Co$ reaction.

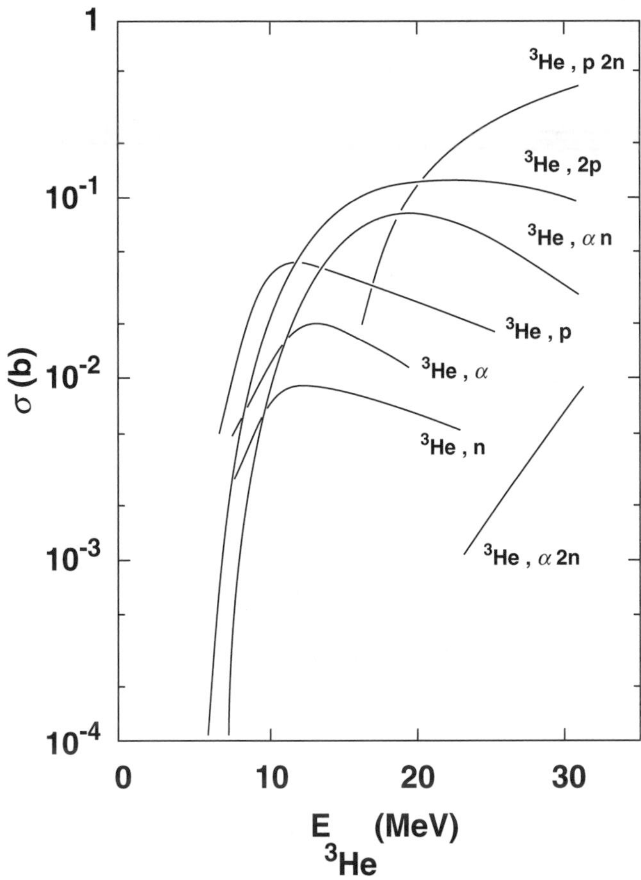

FIG. 9.5. Excitation functions for ^{3}He-induced nuclear reactions on ^{63}Cu (Lamb, 1969).

9.3.2 Irradiation procedures

There are two basic problems that must be solved for accurate activation analysis: current integration and specimen cooling. Figures 9.6 and 9.7 show typical irradiation setups. Direct current integration and air cooling are usually preferred, but may not always be feasible. Irradiation in air or in a controlled atmosphere such as helium simplifies the cooling of the samples. At high energies, particularly for irradiations in air, the activation in a thin foil may be used for monitoring the beam current. Note particularly the safety considerations given in Chapter 13. The induced activity in the monitor foil serves as a measure of the integrated charge, Q = it, where *i* is the beam current and *t* is the irradiation time. Whenever possible, direct current integration should be used because it is faster, simpler, and more accurate than a thin foil monitor.

There is no rule for the irradiation times and intensities because all cases are different. It is clear however that it is not profitable to irradiate more than one or two half-lives of the radioisotope of interest because of the existence of the saturation factor. Beam intensities for the standards are generally of the order of nanoamperes. For the samples, they may reach several microamperes and in some cases tens of microamperes. Note that the samples may easily be fused or vaporized, even below 1 microampere at several MeV, if they are insulating or not well cooled.

9.3.3 Post-irradiation chemistry

9.3.3.1 Etching

To demonstrate the importance of surface contamination, Table 9.2 gives the range of surface concentrations for C, N, and O for various samples after chemical etching under the best conditions, i.e., the cleanest surface that can be obtained (Quaglia, 1976). This minimum surface contamination is activated at the same time as the trace element in the bulk and will correspond to a minimum detectable bulk concentration. The table also gives the equivalent of ppm/weight in the bulk that it would represent if not removed. This equivalent is a function of the nature of the sample, the incident energy, and the reaction used. It is clear that etching is an absolute necessity at low levels, at least for C and O.

Cleaning by etching *before* irradiation is not consistently as good as indicated in Table 9.2, and etching *after* irradiation is delicate when using low energies. Even though surface contamination may be very thin (tens of Angstroms), microns of material must be removed by etching because of the phenomenon of *recoil implantation*. For instance, ^{18}F nuclei from the ^{16}O(^{3}He, p)^{18}F reaction

Table 9.2. Surface contamination.

Contaminant	Minimum surface concentration *(μg/cm²)*	Equivalent bulk concentration *(ppm)*
C	0.05 - 0.5	0.1 - 10
N	<0.01	<0.01
O	0.05 - 0.5	0.1 - 10

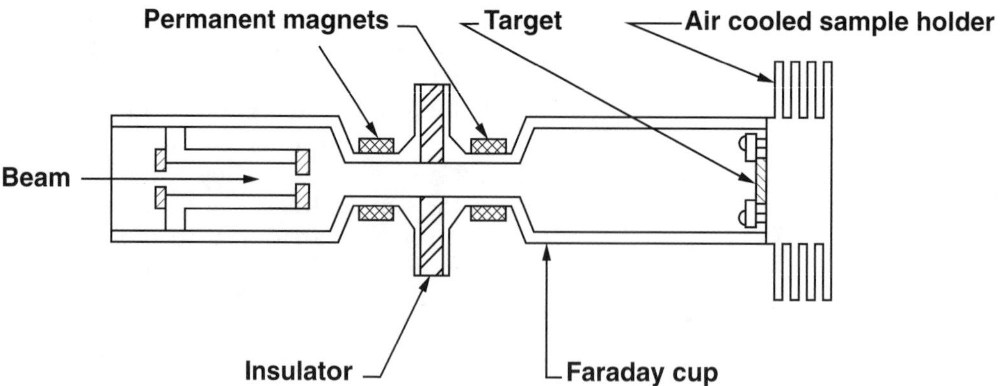

FIG. 9.6. Setup for irradiation of samples in the vacuum at low power using direct beam current integration.

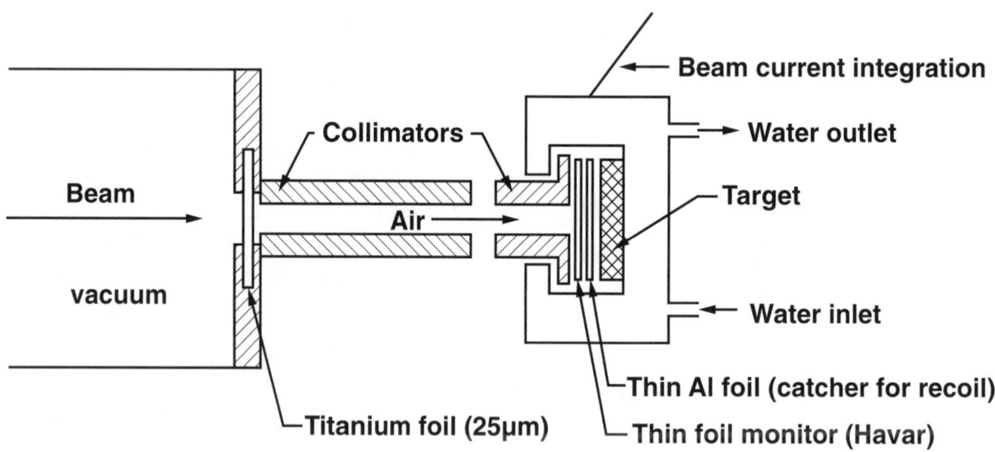

FIG. 9.7. Setup for irradiation of samples at atmospheric pressure, either air or controlled atmosphere, at high power. Irradiation dose measured by activation of a thin foil such as Havar; beam current value is only indicative.

travel ~ 0.7 μm in GaAs for an incident energy of 3 MeV, and 4 μm in Al for an incident energy of 18 MeV. A limited and known number of microns must be removed after irradiation to minimize the effects of surface contamination. The etching must be studied for each particular type of sample. To measure the etching, one can use mechanical means, or sometimes use the activation of the matrix (Valladon *et al.*, 1980a). The etching itself can be purely chemical if the vacuum is very good in the accelerator. Usually it must be a combination of mechanical etching with diamond paste followed by chemical etching because of the polymerization of hydrocarbons at the beam spot (Blondiaux *et al.*, 1980). This polymerized layer inhibits the chemical etching and contains huge quantities of carbon. For elements other than C, N, and O, surface contamination is not a severe problem. However, for trace analysis care should be taken to clean the sample surface before irradiation because of possible diffusion under irradiation (Blondiaux *et al.*, 1984). It is always wise to control the sample temperature during irradiation (Wauters *et al.*, 1986).

Table 9.3 lists recipes used for etching semiconductor matrices. The composition of the etching baths may be varied depending on the needs of any particular analysis. Etching is not always a straightforward process. The polymerized carbon layer must always be removed before chemical etching and the material should be removed in several different baths to avoid recontamination problems. Furthermore, etching is a non-linear process (the amount of material removed is not proportional to the time).

Table 9.3. Semiconductor etches.

Matrix	Composition of the bath (acids are commercial concentrated acids)	Approximate etching speed (μm/min)
Si	CH_3COOH (2 vol.) + HNO_3 (3 vol.) + HF (1 vol.)	7
GaAs	HF (1 vol.) + HNO_3 (1 vol.) + H_2SO_4 (1 vol.)	13
InP	Br_2 (0.5%) in methanol	3

9.3.3.2 Radiochemical separations

In favorable cases when there is low matrix activation or the production of only short-lived radioisotopes from the matrix compared to the half-life of the nuclide of interest, the radioactivity can be measured directly in the sample. This is often the case for the analysis of a light element in a high Z matrix where the Coulomb barrier effect can be used. Generally speaking, the radioisotope of interest cannot be measured directly, and it has to be isolated from unwanted activities using radiochemical procedures. Numerous procedures based on solvent extraction, precipitation, ion exchange, distillation, or volatilization/fusion have been described (De Soete *et al.*, 1972). The methods (Example 9.5) are not particularly difficult, but they do require experience with the chemical techniques. This chapter will not discuss the details of the methods. If you need to use radiochemical separation, find a chemist to collaborate. With good chemical procedures, the separation of the radionuclide of interest from interfering radionuclides makes activation analysis a very sensitive method compared to the other "prompt methods" based on the use of ion beams.

One should always be aware of the possible danger arising from the handling of radioactive samples. This is of course the case for most matrices at high energy (> 10 MeV), but it is also the case even at a few MeV for low Z matrices (B, C, N, O, F...). Be sure to check the samples with a monitor after irradiation. See Chapter 13 on Radiation Safety.

EXAMPLE 9.5. Radiochemical separation for oxygen in InP.

The reaction used was $^{16}O(t, n)\,^{18}F$ at 6 MeV (Bordes *et al.*, 1987). Although indium itself is little activated at 6 MeV by tritons, it is strongly activated via (n, n') and (n, γ) reactions from secondary neutrons produced by triton-induced reactions on phosphorous. It is impossible to detect ^{18}F directly because of the overwhelming ^{113}In, ^{114}In, ^{115}In, and ^{116}In activities. After irradiation and etching ~5 μm, the InP sample is dissolved in HCl after adding a few milligrams of fluoride (NaF) as an inactive fluorine carrier. ^{18}F is then extracted using diphenyldichlorosilane in di-isopropylether. The organic phase is rinsed and directly counted with a Ge detector. The oxygen concentration varied from 6 to 16×10^{15} at/cm^3, with a detection limit of 10^{14} at/cm^3. The complete post-irradiation treatment takes less than 30 minutes.

9.3.4 Measurement of the radioactivity

Standard procedures are used for the analysis of the induced activity of the sample (De Soete *et al.*, 1972). Gamma ray spectroscopy with a germanium detector and multichannel analyzer is all that is required for detection of specific gamma rays. For the case of pure positron emitters where the 511-keV photons must be detected at very low levels, γ-γ coincidence counting with sodium iodide is used. Figure 9.8 shows a typical coincidence detector geometry used to detect the simultaneous back-to-back 511-keV gamma-rays emitted when a positron annihilates. Computer programs to analyze the sometimes quite complex spectra exist in most laboratories. The exponential decay in the counting rate as shown in Example 9.6 should always be followed to confirm that the proper activity is being measured. This is particularly important in the case of positron emitters where the 511 keV transition is not specific.

9.4 SPECIFIC APPLICATIONS

As mentioned previously, CPAA is the method of choice for the trace analysis of C, N, and O because it does not depend on the availability of blanks or standards with known concentrations of these ubiquitous elements. Other methods such as neutron activation or photon activation are not effective for the analysis of C, N, or O at trace levels because there is no suitable reaction or the cross section is too small for adequate sensitivity. CPAA is one of the reference methods accepted by the European Community Bureau of Reference for the trace analysis of C, N, and O.

EXAMPLE 9.6. Following the decay curve.

The usual method of measuring the decay constant is to plot the count per time interval versus time on a semi-log plot. If the data can be fit with a single straight line whose slope is the decay constant of desired activity, then you are following the correct activity. Suppose that you intend to activate a sample by producing a radioisotope with a 3-minute half-life and you measure the following activity in consecutive 1-minute intervals: 10,100, 7,847, 6,378, 4,929, 4,030, 3,092, 2,550, 1,939, 1,614, 1,214, 1,023, 758, 650, 437, 412, 294. Figure 9.9 is a plot of the activity versus time and its exponential fit. The decay constant, $\lambda = 0.2321 = \ln(2)/\tau = 0.6931/\tau$, implies $\tau = 2.98$ minutes. More complex situations with multiple activation products will produce data with breaks in the curve and can be fit with non-linear least squares procedures.

The following sections discuss the reactions that have been used for the analysis of these light elements and their interferences. Cross-section curves and thick-target yields for reactions of interest and for interfering reactions are needed when looking for the best accuracy when calibrating by the numerical integration and average stopping power methods. They also are valuable for locating resonances and maxima to control the analysis and evaluate sensitivities. Relative thick target yield curves are easy to measure; they are useful to calculate the average energy and directly evaluate sensitivities versus energy.

9.4.1 Carbon

The two main reactions are given in Table 9.4. They allow the determination of carbon down to the ppb level (10^{-9}) if a post-irradiation radiochemical separation is used. Non-destructive analysis is possible well below the ppm level in matrices with $Z > 30$ where it is possible to take advantage of the coulomb barrier effect. In the case of radiochemical separations, the methods mostly used are based on reducing fusion for ^{13}N and trapping N_2 on hot titanium, or on oxidizing fusion for ^{11}C with absorption of $^{11}CO_2$ on soda asbestos. The usual standard is graphite.

Table 9.4. Carbon reactions.

Reaction	Threshold *(MeV)*	Interfering reactions	Threshold *(MeV)*
$^{12}C(d, n)^{13}N$	0.3	$^{14}N(d, t)^{13}N$	4.9
		$^{14}N(d, dn)^{13}N$	12
		$^{16}O(d, \alpha n)^{13}N$	8.4
$^{12}C(^3He, a)^{11}C$	0	$^{9}Be(^3He, n)^{11}C$	0
		$^{10}B(^3He, d)^{11}C$	0
		$^{11}B(^3He, t)^{11}C$	2.5
		$^{14}N(^3He, \alpha d)^{11}C$	10
		$^{16}O(^3He, 2\alpha)^{11}C$	6.3

9.4.1.1 $^{12}C(d, n)^{13}N$

The $^{12}C(d, n)^{13}N$ reaction produces a pure positron emitter with a half-life of 9.97 minutes. It is free of interferences below 4.9 MeV and in practice can usually be used up to the 8.4 MeV threshold for the oxygen reaction. The reasons are that the cross sections for the deuteron-induced reactions on nitrogen are small and the concentration of nitrogen in the sample is usually less than the carbon concentration. The corresponding cross sections are shown in Appendix 14. Figure A14.1 shows the cross section versus energy for the (d, n) reaction and Fig. A14.2 shows the ratio of ^{13}N produced from carbon to ^{13}N produced from nitrogen as a function of deuteron energy. The thick target yield curves for the reaction on carbon and the interfering reactions on N and O are shown in Figs. A14.3, A14.4, and A14.5. They are taken from Krasnov *et al.* (1970). Example 9.7 shows the use of this reaction in metals and Example 9.8 shows its use with a radiochemical separation.

9.4.1.2 $^{12}C(^3He, \alpha)^{11}C$

This reaction also produces a pure positron emitter, but with a 20.3 minute half-life. Several measured experimental cross-section curves are shown in Fig. 9.10 (Lamb, 1969). The differences between the curves are typical of total cross-section measurements, but they are not important for activation analysis using standards. Beryllium and boron interfere at all energies. The cross section curves for these interferences are given in Figs. A14.7 and A14.8. The thick target yield curves for the reaction on carbon and the interfering reactions on Be and B are given in Figs. A14.9, A14.10, and A14.11.

It is clear that the $^{12}C(^3He, \alpha)^{11}C$ reaction can be used only if the concentration levels for Be and B are much

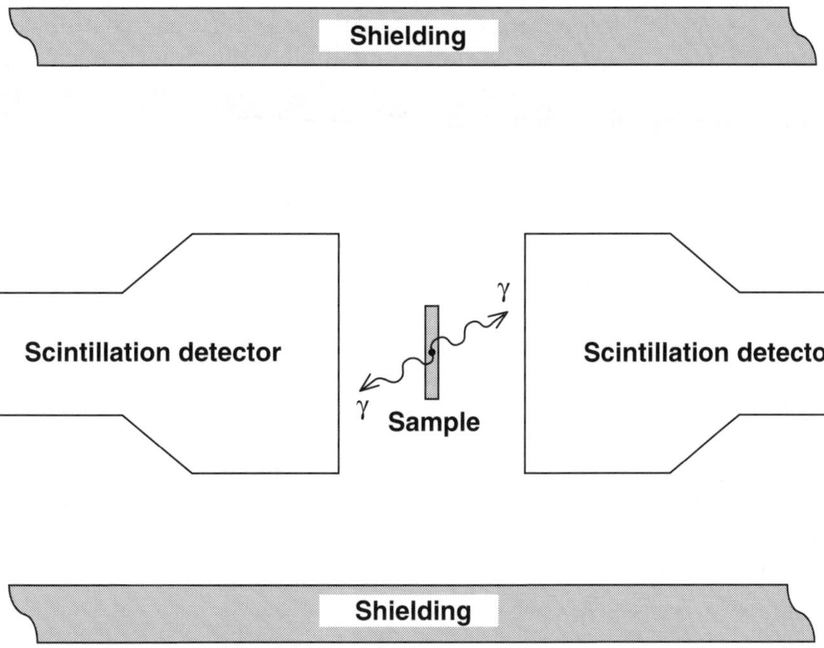

FIG. 9.8. Typical detector geometry for coincidence detection of the 511-keV annihilation radiation from a positron emitter.

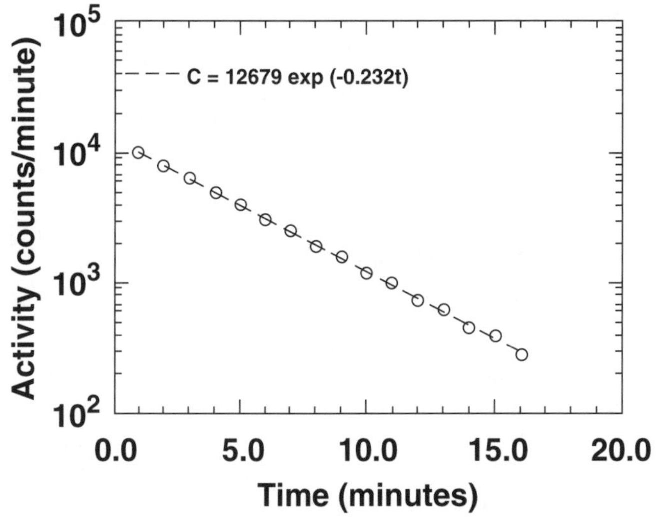

FIG. 9.9. Semi-log plot of activity versus time for a radionuclide with a 3-minute half-life.

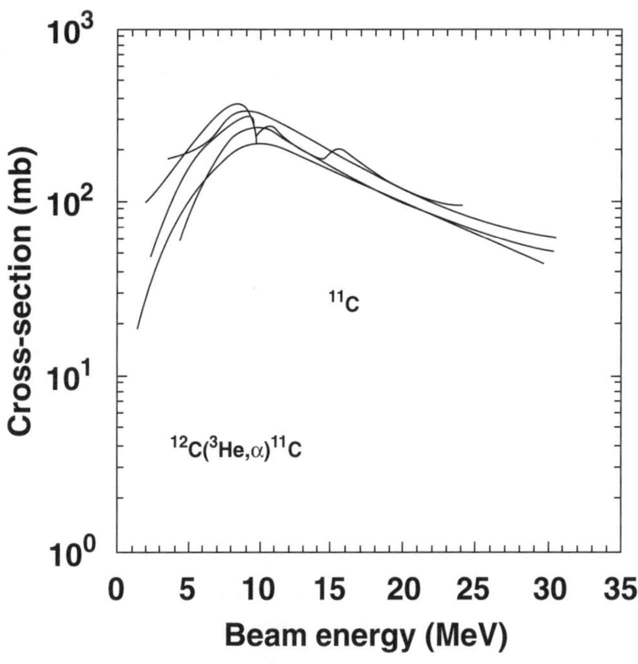

FIG. 9.10. Cross-sections for the $^{12}C(^{3}He, \alpha)^{11}C$ reaction compiled by Lamb (1969).

lower than those for carbon. Note that this reaction should not be used for the analysis of carbon in semiconductors such as GaAs or InP grown by the liquid encapsulated Czochralski (LEC) method because these crystals are grown in the presence of molten B_2O_3 and present a relatively high boron concentration. Nitrogen and oxygen may also interfere above ~10 and ~12 MeV respectively. The thick target yield curves for these higher energy interferences are also given in Krasnov *et al.* (1970). For oxygen see Fig. A14.22.

9.4.1.3 Other reactions for carbon

There are other reactions for carbon, but they are of little analytical interest. The cross-section curves for $^{12}C(p, pn)^{11}C$, $^{12}C(^{3}He, pn)^{13}N$, $^{12}C(^{3}He, n)^{14}O$ and $^{12}C(\alpha, \alpha n)^{11}C$ are compiled in an International Atomic Energy Agency handbook (Albert, 1987).

References for actual analysis of carbon in various matrices are given in Table 9.5.

9.4.2 Nitrogen

For nitrogen, there is no obvious choice for the nuclear reaction that should be used. The four reactions that we can recommend are listed in Table 9.6. Other possible reactions are: $^{14}N(d, \alpha n)^{11}C$, $^{14}N(^{3}He, \alpha)^{13}N$, $^{14}N(^{3}He, d)^{15}O$, and $^{14}N(\alpha, \alpha n)^{13}N$. Data for most of these reactions can be found in Albert (1987). The preferred reaction is $^{14}N(p, \alpha)^{11}C$, because the ppb level can be reached and ^{11}C has a half-life of 20.3 minutes, easily compatible with radiochemical separations. The other radioisotopes have much shorter half-lives (^{14}O: 70.6 s, ^{15}O: 122 s, ^{16}N: 7.1 s).

9.4.2.1 $^{14}N(p, n)^{14}O$

As shown in Fig. A14.12, the cross section is not very high (Nozaki and Iwamoto, 1981; Kuan and Risser, 1964). The sensitivity is therefore limited to about 1 ppm/weight. The interesting point is that the reaction is interference-free up to ~22 MeV and that ^{14}O possesses a characteristic high energy gamma ray (2,313 keV). It is thus possible to analyze nitrogen without radiochemical separation and without complex decay curve analysis of the 511-keV peak if the matrix does not activate too much. The drawback of this reaction is its high threshold energy of 6.3 MeV.

9.4.2.2 $^{14}N(p, \alpha)^{11}C$

If the sample does not contain too much boron, this is the most suitable reaction for nitrogen analysis at levels of a ppm or greater. Figure A14.13 shows the cross-section curve for the $^{14}N(p, \alpha)^{11}C$ reaction (Jacobs *et al.*, 1974). A detailed curve up to 7 MeV has recently been redetermined by Köhl *et al.* (1990). Figure A14.14 shows the cross-section for the interfering $^{11}B(p, n)^{11}C$ reaction (Anders, 1981). The thick target yield curves for the

EXAMPLE 9.7. Carbon in Zr, Nb, Ta, and W.

Taking advantage of the coulomb barrier effect, Vandecasteele *et al.* (1979a) could analyze carbon in Zr, Nb, Ta, and W without chemical separation. The $^{12}C(d, n)^{13}N$ reaction was used. The samples were irradiated 2 to 5 minutes at 5 MeV for Zr or 7 MeV for Nb, Ta, and W, with an intensity of 1 to 2 microamperes. After irradiation and etching, the 511-keV annihilation peak was measured in coincidence between two (3 × 3 inches) NaI crystals. Concentrations ranged from a few tens of ppm/weight in Zr and Nb, down to 1 ppm in Ta, and less than 15 ppb/weight in W.

EXAMPLE 9.8. Carbon in GaAs with radiochemical separation.

Wei *et al.* (1987) analyzed carbon in LEC grown GaAs by the $^{12}C(d, n)^{13}N$ reaction at 8 MeV. After irradiation for 20 minutes with a 3-μa beam of deuterons, the samples were mechanically polished and etched to remove 4 μm from the surface. Then the sample was vaporized in a graphite crucible (reducing fusion) with the ^{13}N trapped on titanium grains (sponge) at 900°C. The detection limit was 8×10^{13} at/cm^3 (0.3 ppb/weight). By measuring the carbon content from various parts of the crystal in the 10 to 30 ppb range, it was possible to determine the segregation coefficient of carbon in a "Gallium rich" crystal of GaAs grown by the LEC method.

Table 9.5. References for carbon analyses.

Matrix	Reaction used	Reference
Al	$^{12}C(d, n)^{13}N$	Goethals *et al.*, 1979
Si	$^{12}C(^{3}He, \alpha)^{11}C$	Nozaki *et al.*, 1970, 1974a, 1976; Engelmann, 1971b; Marschal *et al.*, 1971; Endo *et al.*, 1972; Vandecasteele *et al.*, 1974a; Böttger *et al.*, 1980
	$^{12}C(d, n)^{13}N$	Engelmann *et al.*, 1971b; Böttger *et al.*, 1980; Martin and Haas, 1972; Sanni *et al.*, 1984; Böttger *et al.*, 1975
Fe	$^{12}C(\alpha, n)^{15}O$	Mayolet *et al.*, 1972
Ni	$^{12}C(d, n)^{13}N$	Strijckmans *et al.*, 1980
Zr	$^{12}C(d, n)^{13}N$	Vandecasteele *et al.*, 1979a
Zircalloy	$^{12}C(d, n)^{13}N$	Mortier *et al.*, 1981
Nb	$^{12}C(d, n)^{13}N$	Shikano *et al.*, 1987a; Vandecasteele *et al.*, 1979a
Mo	$^{12}C(d, n)^{13}N$	Vandecasteele *et al.*, 1981
W	$^{12}C(d, n)^{13}N$	Vandecasteele *et al.*, 1979a, 1981
Ta	$^{12}C(\alpha, \alpha n)^{11}C$	Engelmann, 1971b; Vandecasteele *et al.*, 1979a
Au	$^{12}C(d, n)^{13}N$	Vandecasteele and Hoste, 1985
GaAs	$^{12}C(d, n)^{13}N$	Wei *et al.*, 1987; Blondiaux and Debrun, 1988; Valladon *et al.*, 1980b; Meyer and Bethge, 1990; Shikano *et al.*, 1987b
Ga Al As	$^{12}C(d, n)^{13}N$	Misdaq *et al.*, 1986

Table 9.6. Nitrogen reactions.

Reaction	Threshold *(MeV)*	Interfering reaction(s)	Threshold *(MeV)*
$^{14}N(p, n)^{14}O$	6.3	$^{16}O(p, t)^{14}O$	21.7
$^{14}N(p, \alpha)^{11}C$	3.1	$^{11}B(p, n)^{11}C$ $^{12}C(p, d)^{11}C$	3 17.9
$^{14}N(d, n)^{15}O$	0	$^{16}O(d, t)^{15}O$	10.6
$^{14}N(t, p)^{16}N$	0	$^{13}C(t, \gamma)^{16}N$ $^{17}O(t, \alpha)^{16}N$	0 0

$^{14}N(p, \alpha)^{11}C$ and the $^{11}B(p, n)^{11}C$ reaction are shown in Figs. A14.15 and A14.16 (Krasnov *et al.*, 1970). The use of this reaction is shown in Example 9.9.

9.4.2.3 $^{14}N(d, n)^{15}O$

This reaction is interference-free up to ~11 MeV. Although it is intrinsically sensitive enough to reach the ppb level, it cannot reach it in practice because ^{15}O must be radiochemically separated, which takes time, and a large amount of activity is lost because of the short half-life of ^{15}O (122 s). The best use of this reaction is probably for concentrations > 0.1 ppm/weight. The cross-section curve is shown in Fig. A14.17 (Köhl *et al.*, 1990), and the thick target yield curve is shown in Fig. A14.18 (Engelmann, 1971a).

9.4.2.4 $^{14}N(t, p)^{16}N$

The interferences from ^{13}C and ^{17}O should be negligible in most cases because of the low isotopic abundances and low cross-sections. The drawback of this reaction is the requirement for access to a beam of tritons, but it has very good potential for the analysis of nitrogen. It is the equivalent of the analysis of oxygen with 14 MeV neutrons [$^{16}O(n, p)^{16}N$], except that one can get a much higher flux on the sample and using 3 MeV tritons results in low matrix activity compared nitrogen. Preliminary tests (unpublished results from the group in Orleans) have shown that sensitivities of less than 0.1 ppm can be obtained in practice in less than one minute (half-life of ^{16}N is 7.1 s).

References for actual analysis of nitrogen in various matrices are given in Table 9.7.

9.4.3 Oxygen

The main reactions for the analysis of oxygen are given in Table 9.8.

9.4.3.1 $^{16}O(p, \alpha)^{13}N$

This reaction is a pure positron emitter with a 10 minute half-life. The cross-section is shown in Fig. A14.19 (Furukawa and Tanaka, 1960). The thick target yield curve is shown in Fig. A14.20 (Krasnov *et al.*, 1970). This reaction is not selective at any energy, and should be used only in selected cases (high fluorine concentration in the sample, and low carbon concentration). The importance of interfering reactions is shown in Fig. A14.21 for carbon and in Fig. A14.15 for nitrogen.

EXAMPLE 9.9. Nitrogen in silicon, with radiochemical separation.

Nozaki *et al.* (1970) analyzed nitrogen in silicon by the $^{14}N(p, \alpha)^{11}C$ reaction. This was, of course, undoped silicon (boron-free). The samples were irradiated for 20 minutes at 13 MeV with a current of few microamperes (no precise value is given). Nylon was used as the standard but Kapton is also often used. After etching, the samples were dissolved in HF + KIO_4; ^{11}C was converted to $^{11}CO_2$ and absorbed in NaOH. $BaCO_3$ was precipitated and counted (total time ~30 minutes).

The lowest concentration actually seen was ~1 ppb/weight, the most frequent concentration observed was a few ppb, and the maximum 20 ppb. This technique was used to study the segregation of nitrogen during zone melting (clearly towards the tail end of the ingot), and the purification during this process (good purification). Other publications in which carbon and oxygen were also analyzed (Nozaki *et al.*, 1974a; 1976) complete this study.

Table 9.7. References for nitrogen analyses.

Matrix	Reaction used	Reference
C	$^{14}N(p, \alpha)^{11}C$	Nozaki *et al.*, 1966
Si	$^{14}N(p, \alpha)^{11}C$	Nozaki *et al.*, 1970, 1974a, 1976; Köhl *et al.*, 1990; Krogner, 1975
Ti	$^{14}N(p, n)^{14}O$	Strijckmans *et al.*, 1977
Al-Ti	$^{14}N(p, n)^{14}O$	Vandecasteele *et al.*, 1979b
Fe	$^{14}N(d, n)^{15}O$	Mayolet *et al.*, 1972
Ni	$^{14}N(p, \alpha)^{11}C$	Strijckmans *et al.*, 1980
Zr	$^{14}N(p, \alpha)^{11}C$	Petit, 1978; Strijckmans *et al.*, 1982
	$^{14}N(d, n)^{15}O$	Giovagnoli *et al.*, 1979
	$^{14}N(\alpha, \alpha n)^{13}N$	Strijckmans *et al.*, 1982
Zircalloy	$^{14}N(\alpha, \alpha n)^{13}N$	Mortier *et al.*, 1981
Nb	$^{14}N(p, n)^{14}O$	Strijckmans *et al.*, 1977
	$^{14}N(p, 2\alpha)^{7}Be$	Sastri and Krivan, 1981
Mo	$^{14}N(d, n)^{15}O$	Mayolet *et al.*, 1972
	$^{14}N(p,\alpha)^{11}C$	Vandecasteele *et al.*, 1981
Ta	$^{14}N(p, n)^{14}O$	Strijckmans *et al.*, 1977
	$^{14}N(p, \alpha)^{11}C$	Strijckmans *et al.*, 1978
W	$^{14}N(p, \alpha)^{11}C$	Vandecasteele *et al.*, 1981

9.4.3.2 $^{16}O(^{3}He, p)^{18}F$

Helium 3 activation is perhaps the most widely used activation method for the analysis of oxygen. ^{18}F is a pure positron emitter with a 110-minute half-life. The interferences of fluorine and sodium exist at all energies and are only < 5% for concentrations identical to that of oxygen (Engelmann and Marschal, 1971). Also, the thresholds of the reactions are low, which limits the activation of the matrix, and the cross section is quite good (see Fig. A14.22; Lamb, 1969). The thick target yield is shown in Fig. A14.23 and the thick target yield curve for the interfering reaction on fluorine, (Krasnov *et al.*, 1970) is shown in Fig. A14.24.

9.4.3.3 $^{16}O(\alpha, d)^{18}F$

Activation with alpha particles has been used in the past when ^{3}He was rare and expensive. The use of alpha particles should be avoided because of the high value of the fluorine interference and because the high threshold values cause high matrix activation. The cross-section curve (Nozaki *et al.*, 1974a) and the thick target yield curve (Krasnov *et al.*, 1970) are shown in Fig. A14.25 and

Table 9.8. Oxygen reactions.

Reaction	Threshold *(MeV)*	Interfering reactions	Threshold *(MeV)*
$^{16}O(p, \alpha)^{13}N$	5.5	$^{12}C(p, \gamma)^{13}N$	0
		$^{13}C(p, n)^{13}N$	3.2
		$^{14}N(p, d)^{13}N$	8.9
$^{16}O(^{3}He, p)^{18}F$	0	$^{19}F(^{3}He, \alpha)^{18}F$	0
$^{16}O(^{3}He, n)^{18}Ne$ +e.c $\longrightarrow ^{18}F$	3.8	$^{23}Na(^{3}He, 2\alpha)^{18}F$	0.4
$^{16}O(\alpha, d)^{18}F$	20.4	$^{15}N(\alpha, n)^{18}F$	8.1
$^{16}O(\alpha, pn)^{18}F$	23.2	$^{19}F(\alpha, \alpha n)^{18}F$	12.6
$^{16}O(\alpha, 2n)^{18}Ne$ β^{+} ^{18}F	29.7	$^{23}Na(\alpha, 2\alpha n)^{18}F$	25
$^{16}O(t, n)^{18}F$	0	$^{20}Ne(t, \alpha n)^{18}F$	4
		$^{19}F(t, tn)^{18}F$	12.1

A14.26 respectively. The thick target yield curve for the interference of fluorine is shown in Fig. A14.27.

9.4.3.4 $^{16}O(t, n)^{18}F$

This is certainly the best reaction for the analysis of oxygen at trace level. It has a good cross section at low energy, as shown in Fig. A14.28 (Revel *et al.*, 1977), and is interference-free up to 4 MeV. In practice it can be considered as interference-free up to 12 MeV because neon is a rare and inert gaseous element whose concentration in most samples is negligible. The thick target yield curve (Bordes *et al.*, 1987), is shown in Fig. A14.29 along with yields for other triton-induced reactions. Example 9.10 is typical of the use of this reaction.

References for the analysis of oxygen mostly in metals and semiconductors are given in Table 9.9.

9.4.4 Channeled activation for C and O

Channeling of the incident ions, in combination with RBS, PIXE, or prompt nuclear reactions, has been widely used to study the location and role of dopants or impurities in single crystals (Feldman *et al.*, 1982). These methods are usually applied at high concentrations, typically above 0.1%. RBS and PIXE cannot be used for light elements like C and O, and prompt nuclear reactions are intrinsically limited in sensitivity and subject to surface contamination problems. Chapter 10 is a general discussion of channeling.

The combination of channeling with charged particle activation, although not free of problems, allows studies on the lattice location of C and O at levels <1 ppm/weight. The reasons for this are the high intrinsic sensitivity at low energy and the possibility of removing surface contamination. Also note that *accidental channeling*, as in other methods, can be a source of large errors in charged particle activation analysis. This has been demonstrated up to factors of ±2 in the work of Hanna Bakraji *et al.* (1991a).

EXAMPLE 9.10. Oxygen analysis in GaAs using the $^{16}O(t, n)^{18}F$ reaction.

The samples were irradiated for 2 hours at 3 MeV and 0.5 microampere. After irradiation, 5 μm were removed by etching and the samples could then be directly measured on a Ge(Li) detector. Oxygen was not detected in these samples and the experimental detection limit was 6 ppb/weight. This limit could have been increased by a factor of 5 to 10 by separating ^{18}F because there is some background from the activation of GaAs.

Examples 9.11 and 9.12 demonstrate that charged particle activation combined with channeling can be used as a tool to study the lattice position of some light elements at the sub-ppm level, probably down to 0.1 ppm/weight. The interest of the method is that it is applicable to a number of as-grown pure materials, not especially doped for the purpose of characterization. Also, the elements that can be investigated — B, C, and O — are most of the time the major residual impurities of the semiconductor materials (B for Czochralski crystals prepared under liquid B_2O_3). A considerable amount of work remains to be done to fully understand the

EXAMPLE 9.11. Carbon in $Ga_{1-x}Al_xAs$.

The detailed work is published in Misdaq *et al.* (1986). $Ga_{1-x}Al_xAs$ samples were prepared by vapor phase epitaxy with organo-metallic compounds. The aluminum concentration x was varied between 25 and 50% by varying the flow of trimethylaluminum relative to that of trimethylgallium. Thick (20 to 30 μm) layers were prepared especially for the needs of the analysis. Carbon was analyzed using the $^{12}C(d, n)^{13}N$ reaction at 2.7 MeV (1 μA, 15 minutes). ^{13}N could be detected directly without chemical separation. Irradiations were performed in random, <100>, and <110> directions. The carbon concentration increased with the Al content at fixed arsine flow, and decreased with the arsine flow at fixed Al concentration. Figure 9.11 represents the C/R ratio of the channeled yield (C) to the random yield (R), versus the carbon concentration for the <100> direction, and for three different Al concentrations (52%, 32%, and 25%). C/R ratios were also measured in the <110> direction for samples before and after annealing. Detailed analysis of these data were interpreted as evidence of interstitial carbon and of carbon in substitution on arsenic sites. Arsenic vacancies seemed to play a major role concerning the proportion of carbon present on the different sites. The thermal treatment revealed a change of site, from interstitial to substitutional.

Table 9.9. References for oxygen analyses.

Matrix	Reaction used	Reference
Na	$^{18}O(p, n)^{18}F$	Engelmann , 1970a
Al	$^{16}O(^3He, p)^{18}F$	Vialatte, 1973; Vandecasteele and Hoste, 1975a; Petri and Sastri, 1975; Fedoroff *et al.*, 1981
	$^{16}O(t, n)^{18}F$	Vialatte, 1973; Valladon *et al.*, 1978
	$^{18}O(p, n)^{18}F$, $^{16}O(p, \alpha)^{13}N$, $^{16}O(\alpha, pn)^{18}F$	Vialatte, 1973
Al Mg	$^{16}O(^3He, p)^{18}F$	Fedoroff *et al.*, 1974
Al Si	$^{16}O(^3He, p)^{18}F$	Vandecasteele and Hoste, 1975a
Si	$^{16}O(^3He, p)^{18}F$	Nozaki *et al.*, 1970; Nozaki *et al.*, 1976; Sanni *et al.*, 1984; Nozaki *et al.*, 1966; Schweikert and Rook, 1970; Engelmann *et al.*, 1970; Kim, 1971; Vandecasteele *et al.*, 1974b; Rook and Schweikert, 1969; Kim, 1969
	$^{16}O(\alpha, pn)^{18}F$	Saito *et al.*, 1963; Vandecasteele *et al.*, 1974b
	$^{18}O(p, n)^{18}F$	Schweikert *et al.*, 1974
	$^{16}O(t, n)^{18}F$	Valladon and Debrun, 1977
Si N	$^{16}O(^3He, p)^{18}F$	Nozaki *et al.*, 1979
Ti	$^{16}O(t, n)^{18}F$	Valladon and Debrun, 1977
Cr	$^{16}O(^3He, p)^{18}F$	Debrun *et al.*, 1969a
Fe	$^{16}O(\alpha, pn)^{18}F$	Fedoroff *et al.*, 1974; Debrun *et al.*, 1969a
Co	$^{16}O(\alpha, pn)^{18}F$	Kohn *et al.*, 1974
Ni	$^{16}O(^3He, p)^{18}F$	Strijckmans *et al.*, 1980; Debrun *et al.*, 1969a
Cu	$^{16}O(^3He, p)^{18}F$	Kohn *et al.*, 1974; Lee *et al.*, 1970; Debrun *et al.*, 1975; Vandecasteele and Hoste, 1975b
	$^{16}O(\alpha, pn)^{18}F$	Vandecasteele *et al.*, 1975; Defebve *et al.*, 1976
Cu Zn	$^{16}O(^3He, p)^{18}F$	Debrun and Barrandon, 1975
Zn Te	$^{16}O(^3He, p)^{18}F$	Caneau *et al.*, 1981
Ga P	$^{16}O(^3He, p)^{18}F$	Kim, 1971
Ga As	$^{16}O(^3He, p)^{18}F$	Misaelides *et al.*, 1987
	$^{16}O(t, n)^{18}F$	Valladon *et al.*, 1980b; Valladon and Debrun, 1977; Huber *et al.*, 1978; Huber et al., 1979
Ga Al As	$^{16}O(^3He, p)^{18}F$	Hanna Bakraji *et al.*, 1991a
Ge	$^{16}O(^3He, p)^{18}F$	Revel *et al.*, 1977; Vandecasteele and Hoste, 1975c
	$^{16}O(t, n)^{18}F$	Valladon and Debrun, 1977
Zr	$^{16}O(\alpha, pn)^{18}F$	Revel and Albert, 1968
Mo	$^{16}O(t, n)^{18}F$	Valladon and Debrun, 1977
	$^{16}O(^3He, p)^{18}F$	Fedoroff *et al.*, 1974; Revel *et al.*, 1968
	$^{16}O(\alpha, pn)^{18}F$	Faure *et al.*, 1972a
Ag	$^{16}O(\alpha, pn)^{18}F$	Kohn *et al.*, 1974
InP	$^{16}O(t, n)^{18}F$	Bordes *et al.*, 1987
Hf	$^{16}O(\alpha, pn)^{18}F$	Revel and Albert, 1968
Ta	$^{16}O(^3He, p)^{18}F$	Debrun and Barrandon, 1975
W	$^{16}O(^3He, p)^{18}F$	Vandecasteele *et al.*, 1981; Debrun and Barrandon, 1975
	$^{16}O(\alpha, pn)^{18}F$	Revel and Albert, 1968
Pb	$^{16}O(\alpha, pn)^{18}F$	Faure *et al.*, 1972b; Vandecasteele and Hoste, 1975d
Pt	$^{16}O(^3He, p)^{18}F$	Petri and Erdtmann, 1981
Au	$^{16}O(\alpha, pn)^{18}F$	Defebve *et al.*, 1976
Fluoride glasses	$^{18}O(p, n)^{18}F$	Mitachi *et al.*, 1985
	$^{16}O(d, n)^{17}F$	Barthe *et al.*, 1990
Diamond	$^{16}O(^3He, p)^{18}F$	Maggiore *et al.*, 1989

EXAMPLE 9.12. Oxygen in $Ga_{1-x}Al_xAs$

Work similar to that on carbon was performed on oxygen using the $^{16}O(^3He, p)^{18}F$ reaction at 3 MeV (Hanna Bakraji *et al.*, 1991a). The samples were irradiated 45 minutes at 0.6 μA. Oxygen (^{18}F) could be measured directly without chemical separation. The ^{18}F yield for the $^{16}O(^3He, p)^{18}F$ reaction was measured after orientation along the <100> axis. Figure 9.12 gives the ratio (C/R) of this "channeled yield," C, to that of the random yield, R. The results, as for carbon, could be explained in terms of arsenic vacancies.

The detection limit of interstitial oxygen under these conditions is estimated to be ~100 ppb/weight. It would be ~10 ppb/weight using 2.5. MeV tritons and the $^{16}O(t, n)^{18}F$ reaction. In the present case, abnormally high (C/R) ratios have in a few cases been measured. These high values, as indicated earlier in this chapter, are due to different stopping powers for the channeled and random ions. The effect is particularly large here because 1.5 μm are removed by etching, while 90% of the induced activity is confined within 4 mm of the surface.

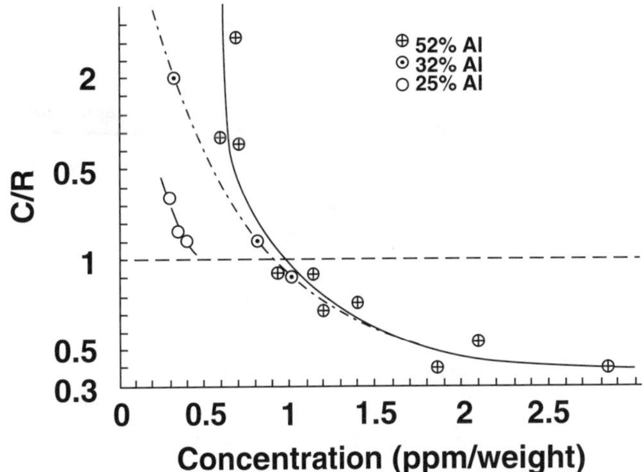

FIG. 9.11. Variation of the ratio C/R of the channeled yield (C) of ^{13}N in the <100> direction to the yield of ^{13}N in a random direction (R) versus the carbon concentration in Ga Al As, for three different Al concentrations (Misdaq *et al.*, 1986).

relationship between the light elements cited above and the properties of semiconductors. Charged particle activation is an interesting tool for these types of studies.

Activation has several drawbacks however. The first one is that activation, in the case of carbon, takes place over 20 to 30 μm (although the activation of the last microns is negligible). As the ions penetrate the sample, dechanneling becomes more and more important, so that the final observed ^{13}N yield is a mixed response which is the sum of a "channeled activation yield," a "partly channeled activation yield," and a random yield. Furthermore, the best channeled part of the sample (first 2 μm) is eliminated by etching. The second drawback is that etching the same number of microns for the channeled sample and the random sample leads to distorted C/R ratios because the energy loss is not the same for channeled and random ions. The method can be improved by trying to reduce the etching after irradiation and by using thinner layers. Surface cleaning under ultra-high vacuum, either using ion sputtering or laser ablation/desorption so that etching after irradiation would become unnecessary, would be a major improvement.

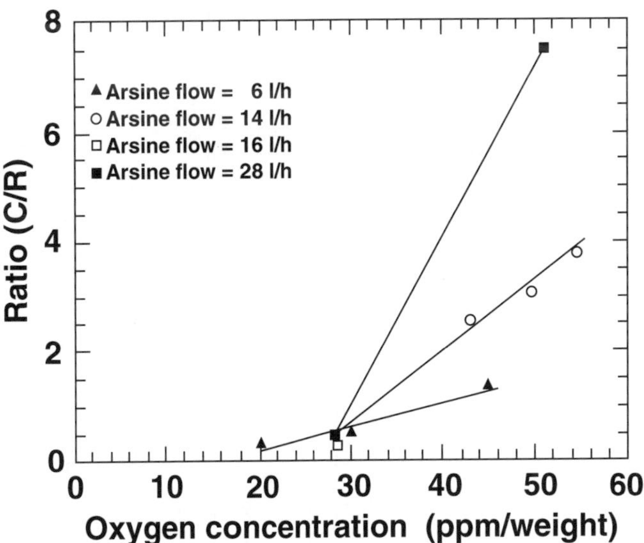

FIG. 9.12. Ratio C/R of the ^{18}F yield when channeling along the <100> axis (C) to that of the random ^{18}F yield (R) versus the oxygen concentration in Ga Al As, for different arsine flows (Hanna Bakraji *et al.*, 1991a).

9.4.5 Analyses of other elements

Table 9.10 is a list of the use of proton activation for the analysis of various matrices (Vandecasteele, 1988).

Figures A14.30 and A14.31 show the calculated "best detection limits" for a number of elements in an aluminum matrix, under standard conditions (Debrun *et al.*,

Table 9.10. References for proton activation analyses.

Matrix	Reference	Chemical separation
Na	Nordmann *et al.*, 1975	yes
Al	Debrun *et al.*, 1969b; Dabney *et al.*, 1973; Vandecasteele *et al.*, 1980	yes
	Debrun *et al.*, 1973; Debrun *et al.*, 1976; Barrandon *et al.*, 1976; Shibata *et al.*, 1979	no
Si	Debrun *et al.*, 1976; Barrandon *et al.*, 1976	no
Fe	Dabney *et al.*, 1973; Dams *et al.*, 1986	yes
	Goethals *et al.*, 1986	no
Co	Debrun and Barrandon, 1973; Debrun *et al.*, 1976; Barrandon *et al.*, 1976; Benaben *et al.* 1975; Krivan, 1975a, 1975b, 1976	no
Ni	Vandecasteele *et al.*, 1980	yes
Cu	Vandecasteele *et al.*, 1980; Benaben *et al.*, 1979	yes
Zn	De Brucker *et al.*, 1987	yes
Nb	Debrun *et al.*, 1976; Barrandon *et al.*, 1976; Barrandon *et al.*, 1974; Krivan, 1975c	no
	Faix *et al.*, 1979; Faix and Krivan, 1982	yes
Rh	Debrun *et al.*, 1976; Barrandon *et al.*, 1976; Debrun *et al.*, 1975	no
Ag	Debrun and Barrandon, 1973; Debrun *et al.*, 1976; Barrandon *et al.*, 1976	no
Cd Te	Delmas and Debrun, 1976	yes
InP	Lacroix *et al.*, 1984	no
Pr	Debrun *et al.*, 1976; Barrandon *et al.*, 1976	no
Biological	Bonardi *et al.*, 1982; Zikovsky and Schweikert, 1977	
Tb	Debrun *et al.*, 1976; Barrandon *et al.*, 1976	no
Dy	Debrun *et al.*, 1976; Barrandon *et al.*, 1976	no
Ho	Debrun *et al.*, 1976; Barrandon *et al.*, 1976	no
Ta	Krivan, 1975a; Barrandon *et al.*, 1976; Barrandon *et al.*, 1974; Krivan *et al.*, 1974	no
	Sastri and Krivan, 1982; Schmid *et al.*, 1987	yes
W	Sastri and Krivan, 1982; 1983	yes
Ir	Debrun *et al.*, 1976; Barrandon *et al.*, 1976	no
Au	Riddle and Schweikert, 1973	yes
	Debrun *et al.*, 1976; Barrandon *et al.*, 1976	no
Glass	Riddle and Schweikert, 1973	yes
Zr F4	Barthe *et al.*, 1989	yes
Coal	Landsberger, 1984	no
Minerals	Delmas *et al.*, 1976	no
Environ-mental	Wauters *et al.*, 1987a	no
	Wauters *et al.*, 1987b; Wauters *et al.*, 1987c	yes
Airborne particulate matter	Krivan *et al.*, 1974; Priest *et al.*, 1980; Desaedeleer and Ronneau, 1976a; Parsa *et al.*, 1974	no
Fly ash	Strijckmans *et al.*, 1985	

1976; Barrandon *et al.*, 1976), using 10 MeV protons. The evolution of these sensitivities which follows that of the thick target yields with the proton energy (assuming there are no nuclear interferences), is shown in Figs. A14.32, A14.33, and A14.34 for a number of elements (Albert, 1987).

Table 9.11 shows actual results of analysis for impurities in a silicon matrix, which is a favorable matrix.

9.4.6 Thin layer activation (TLA) for studies on wear and corrosion

The sample to be studied is irradiated with an ion beam in order to induce activation over a well-defined area and to a limited depth. The loss of material due to corrosion or wear is related to the loss of radioactivity. The advantages of the method are its speed, sensitivity, and the possibility of achieving continuous measurements. For instance, the method is 100 times faster than

Table 9.11. Proton activation analyses in Si.

	"Clean" sample *(at/cm³)*	Contaminated sample* *(at/cm³)*
B	1.2×10^{16}	10^{16}
C	$\leq 5 \times 10^{12}$	3×10^{15}
Ti	$\leq 2 \times 10^{12}$	7×10^{13}
Cr	$\leq 6 \times 10^{12}$	$\leq 6 \times 10^{12}$
Fe	$\leq 10^{13}$	2×10^{13}
Ni	2×10^{14}	2×10^{14}
Cu	$\leq 4 \times 10^{13}$	$\leq 4 \times 10^{13}$
Ge	2×10^{13}	10^{13}
Zr	$\leq 10^{12}$	$\leq 10^{12}$

* Contamination by the "wetting agent" used during the metallurgy.

methods based on mechanical measurements and 1,000 times faster than those based on ultrasonic measurements. Important financial savings can be made in industry using TLA.

The potential of the thin layer activation method and its different applications have been described in detail in review articles, for instance in Conlon (1982). We will limit ourselves here to only one example (9.13): corrosion-erosion in heat exchangers. Note that most applications deal with studies on wear in engines, turbines, etc.

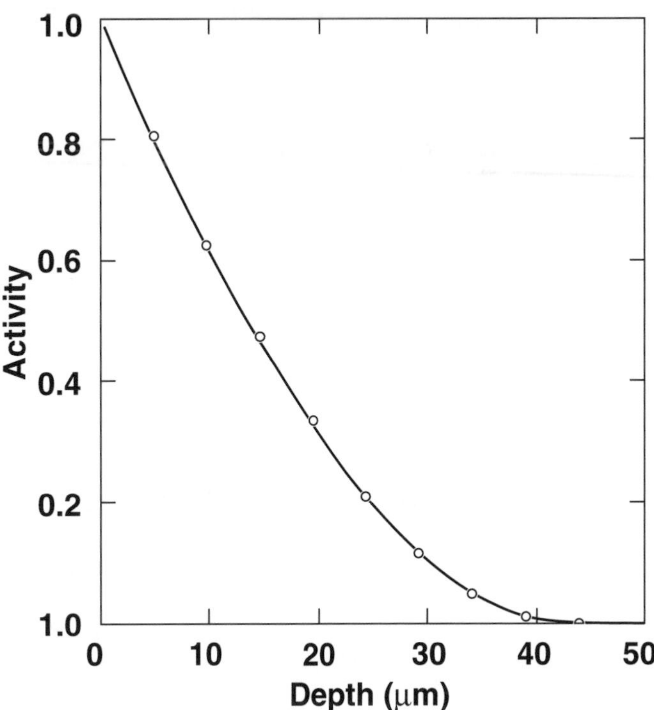

FIG. 9.13. Relative total activity versus depth (total activity at depth zero equal to 1) for a steel sample irradiated with 9 MeV protons at an angle of 22°, and for the ^{56}Co activity from the ^{56}Fe(p, n)^{56}Co reaction.

EXAMPLE 9.13. Corrosion-erosion phenomenon in nuclear power plants.

The materiel studied was unalloyed steel and the reaction used was ^{56}Fe(p, n)^{56}Co (Bouchacourt *et al.*, 1989). The half-life of ^{56}Co is 77.3 days so that the loss of activity due to the natural decay of the radioisotope is small compared to the loss of activity due to corrosion. The irradiation energy was 9 MeV; the corresponding activated depth was 40 μm as shown in Fig. 9.13. The sensitivity is 2×10^{-3} mm/year, for a 100-hour test. Examples of results are given in Figs. 9.14 and 9.15. The variation of the corrosion-erosion with the velocity and the temperature of the cooling water and the effects of alloying with a small percentage of chromium were readily demonstrated

9.5 CONCLUSION

Charged particle activation can in principle be used to analyze almost any element at the ppb (10^{-9}) level. H, He, and Si are notable exceptions. Except for C, N, and O at very low levels, there are many excellent methods to achieve trace analysis: Inductively Coupled Plasma (ICP), ICP with mass spectrometry (ICP-MS), secondary ion mass spectrometry (SIMS), glow discharge mass spectrometry (GDMS), etc. Therefore, charged particle activation should be used only in difficult cases, or as an independent method for the calibration of standards.

Proton activation at E ~ 11 MeV provides a multi-elemental and nondestructive analysis in the case of matrices which are the following elements or compounds and alloys of these elements: H, Be, C, N, O, F, Al, Si, P, S, Cl, Mn, Co, Nb, Rh, Ag, In, Pr, Tb, Ho, Ta, Au (see Table 9.10). Proton activation is a good complement of thermal neutron activation because elements which are very sensitive in one method are less sensitive in the other. Activation with other particles such as ^{3}He and ^{4}He has been very seldom used, and is of little interest in practice.

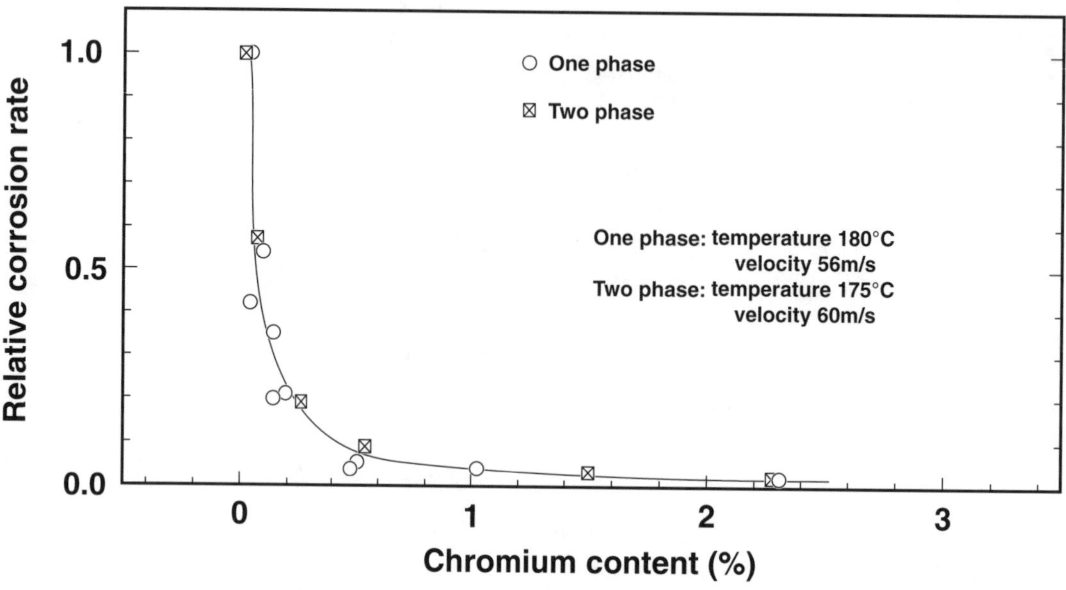

FIG. 9.14. Corrosion rate for unalloyed steel for different speeds of the circulating water, versus the water temperature (Bouchacourt *et al.*, 1989).

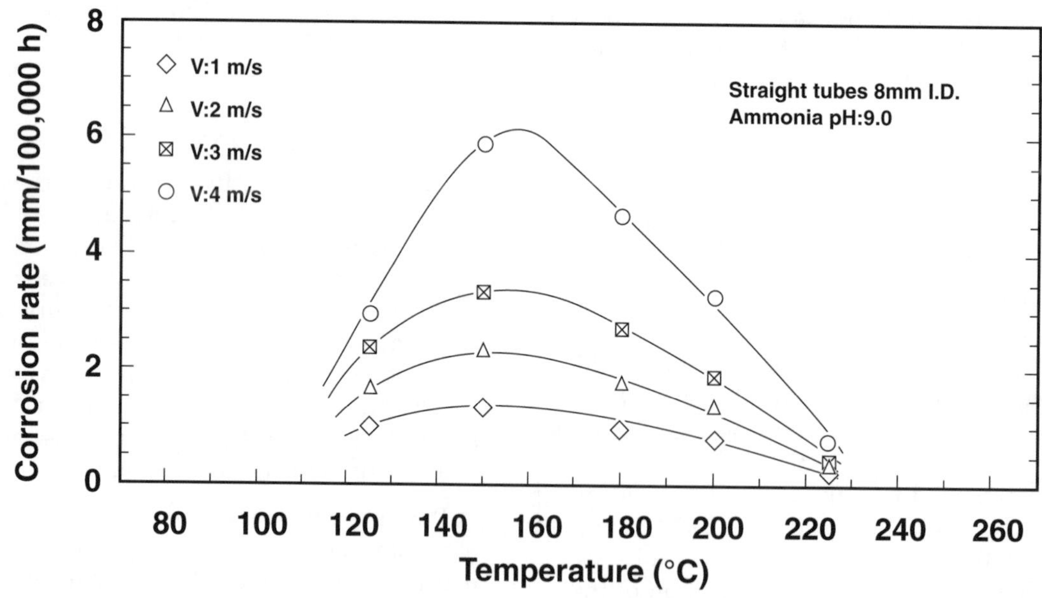

FIG. 9.15. Relative corrosion rate versus the chromium content of steel, for two experimental conditions (Bouchacourt *et al.*, 1989).

REFERENCES

Albert, P. (1987), Handbook on nuclear activation data, IAEA Technical reports series No. 273, 479-628.

Anders, B. (1981), *Z. Phys.* **A301**, 353.

Barrandon, J.N., Benaben, P., Debrun, J.L., and Valladon, M. (1974), *Anal. Chim. Acta* **73**, 39.

Barrandon, J.N., Benaben, P., and Debrun, J.L. (1976), *Anal. Chim. Acta* **83**, 157.

Barthe, M.F., Bajard, M.T., Giovagnoli, A., Blondiaux, G., Debrun, J.L., Lesergent, M., Tregoat, Y., and Barraud, J.Y. (1989), *Nucl. Instr. and Methods* **B40/41**, 1202.

Barthe, M.F., Giovagnoli, A., Blondiaux, G., Debrun, J.L., Tregoat, Y., and Barraud, J.Y. (1990), *Nucl. Instr. and Methods* **B45**, 105.

Benaben, P., Barrandon, J.N., and Debrun, J.L. (1975), *Anal. Chim. Acta* **78**, 129.

Benaben, P., Barrandon, J.N., and Debrun, J.L. (1979), *Radiochem. Radioanal. Lett* **37**, 241.

Blondiaux, G., and Debrun, J.L. (1988), Proceedings of the 12th International Symposium on the Application of Ion Beams in Materials Science, Tokyo. (1987), Hosei University Press, p. 85.

Blondiaux, G., Sastri, C.S., Valladon, M., and Debrun, J.L. (1980), *J. Radioanal. Nucl. Chem.* **56**, 163.

Blondiaux, G., Valladon, M., Koemmerer, C., Giovagnoli, A., and Debrun, J.L. (1981), *Analysis of Non-metals in Metals*, G. Kraft (ed.), Walter de Gruyter and Co., Berlin-New York, p. 233.

Blondiaux, G., Albert, P., Giovagnoli, A., and Debrun, J.L. (1984), *Anal. Chim. Acta* **160**, 289.

Bonardi, M., Birattari, C., Gilardi, M.C., Pietra, R., and Sabbioni, E. (1982), *J. Radioanal. Chem.* **70**, 337.

Bordes, N., Blondiaux, G., Maggiore, C.J., Valladon, M., Debrun, J.L., Coquille, R., and Gauneau, M. (1987), *Nucl. Instr. and Methods* **B24/25**, 722.

Bottger, M.L., and Birnstein, D. (1975), *Isotopenpraxis* **11**, 127.

Bottger, M.L., Birnstein, D., Helbig, W., and Niese, S. (1980), *J. Radioanal. Chem.* **58**, 173.

Bouchacourt, M., Marsigne, C., Dubail, A., Blondiaux, G., and Debrun, J.L. (1989), *Nucl. Instr. and Methods* **B40/41**, 1199.

Caneau, C., Schneider, M., and Moreau, R. (1981), *J. Crypt. Growth* **53**, 605.

Chaudri, M.A., Burns, G., Reen, E., Rouse, J.L., and Spicer, B.M. (1976), Proceedings of the International Conference on Modern Trends in Activation Analysis, Munich.

Conlon, T.W. (1982), *Contemp. Phys.* **23**, 353.

Dabney, S.A., Swindle, D.L., Beck, J.N., Francis, G., and Schweikert, E.A. (1973), *J. Radioanal. Chem.* **16**, 375.

Dams, R., Alluyn, F., Vanloo, B., Wauters, G., and Vandecasteele, C. (1986), Z. *Anal. Chem.* **325**, 163.

De Brucker, N., Strijckmans, K., and Vandecasteele, C. (1987), *Anal. Chim. Acta* **195**, 323.

Debrun, J.L., and Barrandon, J.N. (1973), *J. Radioanal. Chem.* **17**, 291.

Debrun, J.L., and Barrandon, J.N. (1975), Proceedings of the 7th Int. Conf. on Cyclotrons and their Applications, 507, Birkhauser.

Debrun, J.L., Barrandon, J.N., and Albert, P. (1969a), *Bull. Soc. Chim. France* **3**, 1011.

Debrun, J.L., Barrandon, J.N., and Albert, P. (1969b), *Bull. Soc. Chim. France* **3**, 1017.

Debrun, J.L., Barrandon, J.N., Benaben, P., and Kouxel, C. (1975), *Anal. Chem.* **47**, 637.

Debrun, J.L., Barrandon, J.N., and Benaben, P. (1976), *Anal. Chem.* **48**, 167.

Defebve, P., Lerch, P., and Vandecasteele, C. (1976), *Radiochem. Radioanal. Lett* **24**, 51.

Delmas, R., and Debrun, J.L. (1976), *Radiochem. Radioanal. Lett.* **27**, 91.

Delmas, R., Barrandon, J.N., and Debrun, J.L. (1976), *Analysis* **4**, 339.

De Soete, D., Gijbels, R., and Hoste, J. (1972), *Neutron Activation Analysis*, Wiley Intertscience, London.

Desaedeleer, G., and Ronneau, C. (1976a), *J. Radioanal. Chem.* **32**, 117.

Desaedeleer, G., and Ronneau, C. (1976b), *Anal. Chem.* **48**, 572.

Endo, Y., Yatsurugi, Y., Akiyama, N., and Nozaki, T. (1972), *Anal. Chem.* **44**, 2258.

Engelmann, C. (1970), *J. Radioanal. Chem.* **6**, 227.

Engelmann, C. (1971a), *J. Radioanal. Chem.* **7**, 89.

Engelmann, C. (1971b), *J. Radioanal. Chem.* **7**, 281.

Engelmann, C., and Marschal, A. (1971), *Radiochem. Radioanal. Lett* **6**, 189

Engelmann, C., Gosset, J., and Rigaud, J.M. (1970), *Radiochem. Radioanal. Lett.* **5**, 319.

Faix, W.G., and Krivan, V. (1982) *Talanta* **29**, 285.

Faix, W.G., Mitchell, J.W., and Krivan, V. (1979), *J. Radioanal. Chem.* **53**, 97.

Faure, L., Boissier, M., and Tousset, J. (1972a), *J. Radioanal. Chem.* **10**, 213.

Faure, L., Giroux, J., and Tousset, J. (1972b), *J. Radioanal. Chem.* **10**, 223.

Fedoroff, M., Debove, L., and Loos-Neskovic, C. (1974), *J. Radioanal. Chem.* **21**, 331.

Fedoroff, M., Loos-Neskovic, C., Rouchaud, J.C., and Revel, G. (1981), *Analysis of non-metals in metals*, G. Kraft (ed.), De Gruyter, Berlin, p.243.

Feldman, L.C., Mayer, J.W., and Picraux, S.T. (1982), *Materials Analysis by Ion Channeling*, Academic Press, New York.

Furukawa, M., and Tanaka, S. (1960), *J. Phys. Soc. Japan* **15**, 2167.

Giovagnoli, A., Valladon, M., Koemmerer, C., Blondiaux, G., and Debrun, J.L. (1979), *Anal. Chim. Acta* **109**, 411.

Goethals, P., Vandecasteele, C., and Hoste, J. (1979), *Anal. Chim. Acta* **108**, 367.

Goethals, P., Bosman, B., and Vandecasteele, C. (1986), *Bull. Soc. Chim. Belg.* **95**, 331.

Hanna Bakraji, E., Ducouret, E., Blondiaux, G., and Debrun, J.L. (1991a), *Nucl. Instr. and Methods* **B56/57**, 819.

Hanna Bakraji, E., Blondiaux, G., Ducouret, E., and Debrun, J.L. (1991b), *Nucl. Instr. and Methods* **B56/57**, 896.

Huber, A., Morillot, G., Linh, N.T., Debrun, J.L., and Valladon, M. (1978), *Nucl. Instr. and Methods* **149**, 543.

Huber, A., Linh, N.T., Valladon, M., Debrun, J.L., Martin, G.M., Mitouneau, A., and Mircea, A. (1979), *J. Appl. Phys.* **50**, 4022.

Ishii, K., Valladon, M., and Debrun, J.L. (1978a), *Nucl. Instr. and Methods* **150**, 213.

Ishii, K., Valladon, M., Sastri, C.S., and Debrun, J.L. (1978b), *Nucl. Instr. and Methods* **153**, 503.

Jacobs, W.W., Bodansky, D., Chamberlin, D., and Oberg, D.L. (1974), *Phys. Rev.* **C9**, 2134.

Jaszczak, R.J., Macklin, R.L., and Gibbons, J.H. (1969), *Phys. Rev.* **181**, 1428.

Kim, C.K. (1969), *Radioanal. Lett.* **2**, 53.

Kim, C.K. (1971), *Anal. Chim. Acta* **54**, 407.

Kohl, F., Krauskopf, J., Misaelides, P., Michelmann, R., Wolf, G., and Bethge, K. (1990), *Nucl. Instr. and Methods* **B50**, 19.

Kohn, A., Barrandon, J.N., Debrun, J.L., Valladon, M., and Vialette, B. (1974), *Anal. Chem.* **46**, 1737.

Krasnov, N., Dmitriyev, P.P., Dmitriyeva, S.P., Konstantinov, I.O., and Molin, G.A. (1970), *Uses of Cyclotrons in Chemistry, Metallurgy and Biology*, C.B. Amphlett (ed.), Butterworths, London.

Krivan, V. (1975a), *J. Radioanal. Chem.* **26**, 151.

Krivan, V. (1975b), *Anal. Chim. Acta* **79**, 161.

Krivan, V. (1975c), *Anal. Chem.* **47**, 469.

Krivan, V. (1976), *Talanta* **23**, 621.

Krivan, V., Swindle, D.L., and Schweikert, E.A. (1974), *Anal. Chem.* **46**, 1626.

Krogner, K. (1975), *Isotopenpraxis* **11**, 278.

Kuan, H.M., and Risser, J.R. (1964), *Nucl. Phys.* **51**, 518.

Lacroix, R., Blondiaux, G., Giovagnoli, A., Valladon, M., Debrun, J.L., Coquille, R., and Gauneau, M. (1984), *J. Radioanal. Nucl. Chem.* **83**, 91.

Lamb, J.F. (1969), Lawrence Berkeley Laboratory, University of California, Report UCRL-18981.

Landsberger, S. (1984), *Int. J. Environ. Anal. Chem.* **16**, 337.

Lee, D.M., Stauffacher, C.V., and Markowitz, S.S. (1970), *Anal. Chem.* **42**, 994.

Liebler, V., Bethge, K., Krauskopf, J., Meyer, J.D., Misaelides, P., and Wolf, G. (1989), *Nucl. Instr. and Methods* **B36**, 7.

Maggiore, C.J., Blacic, J.D., Blondiaux, G., Debrun, J.L., Hage Ali, M., Mathez, E., and Misdaq, M.A. (1989), *Nucl. Instr. and Methods* **B40/41**, 1193.

Marschal, A., Gosset, J., and Englemann, C. (1971), *J. Radioanal. Chem.* **8**, 243

Martin, J., and Haas, E. (1972), *Z. Anal. Chim.* **259**, 97

Mayolet, F., Reimers, P., and Englemann, C. (1972), *J. Radioanal. Chem.* **12**, 115.

Meyer, J.D., and Bethge, K. (1990), Proceedings of the 6th conference on semi-insulating III-V Materials, Toronto, Canada, ch. 4.

Michelmann, R.W., Krauskopf, J., Meyer, J.D., and Bethge, K. (1990), *Nucl. Instr. and Methods* **B51**, 1

Misaelides, P., Krauskopf, J., Wolf, G., and Bethge, K. (1987), *Nucl. Instr. and Methods* **B18**, 281.

Misdaq, M.A., Blondiaux, G., Andre, J.P., Hage Ali, M., Valladon, M., Maggiore, C.J., and Debrun, J.L. (1986), *Nucl. Instr. and Methods* **B15**, 329.

Mitachi, S., Sakaguchi, S., Yonezawa, H., Shikano, K., Shigematsu, T., and Takahashi, S. (1985), *Jpn. J. Appl. Phys.* **24**, L827.

Mortier, R., Strijckmans, K., and Vandecasteele, C. (1981), *Bull. Soc. Chim. Belg.* **90**, 297.

Nordmann, F., Fluhr, A., Tinelli, G., and Engelmann, C. (1975), *Analysis* **3**, 171.

Nozaki, T.M., and Iwamoto, M. (1981), *Radiochim. Acta* **29**, 57.

Nozaki, T., Okuo, T., Akutsu, H., and Furukawa, M. (1966), *Bull. Chem. Soc. Japan* **39**, 2685.

Nozaki, T., Yatsurugi, Y., and Akiyama, N. (1970), *J. Radioanal. Chem.* **4**, 87.

Nozaki, T., Yatsurugi, Y., Akiyama, N., Endo, Y., and Makide, Y. (1974a), *J. Radioanal. Chem.* **19**, 109.

Nozaki, T., Iwamoto, M., and Ido, T. (1974b), *Int. J. Appl. Rad. Isot.* **25**, 393.

Nozaki, T., Yatsurugi, Y., and Endo, Y. (1976), *J. Radioanal. Chem.* **32**, 43.

Nozaki, T., Iwamoto, M., Usami, K., Mukai, K., and Hiraiwa, A. (1979), **52**, 449.

Parsa, B., and Markowitz, S.S. (1974), *Anal. Chem.* **46**, 186.

Petit, J. (1978), *Mem. Sci. Rev. Met.* **75**, 395.

Petri, H., and Sastri, C.S. (1975), *Z. Anal. Chim.* **277**, 25.

Petri, H., and Erdtmann, G. (1981), *Analysis of non-metals in metals*, G. Kraft (ed.), De Gruyter, Berlin, p. 253.

Priest, P., Devillers, M., and Desaedeleer, G. (1980), *Radiochemical analysis*, W.S. Lyon (ed.), Ann Arbor Science, Ann Arbor, p. 191.

Quaglia, L. (1976), Eurisotop report ITE No. 90.

Revel, G., and Albert, P. (1968), *J. Nucl. Mat.* **25**, 87.

Revel, G., Da Cunha Belo, M., Linck, I., and Kraus, L. (1977), *Rev. Phys. Appl.* **12**, 81.

Ricci, E., and Hahn, R.L. (1965), *Anal. Chem.* **37**, 742.

Ricci, E., and Hahn, R.L. (1967), *Anal. Chem.* **39**, 794.

Riddle, D.C., and Schweikert, E.A. (1973), *J. Radioanal. Chem.* **16**, 413.

Rook, H.L., and Schweikert, E.A. (1969), *Anal. Chem.* **41**, 958.

Saito, K., Nozaki, T., Tanaka, S., Furukawa, M., and Cheng, H. (1963), *Int. J. Appl. Rad. Isot* **14**, 357.

Sanni, A.O., Roche, N.G., Dowell, N., Schweikert, E.A., and Ramsey, T.H. (1984), *J. Radioanal. Nucl. Chem.* **81**, 125.

Sastri, C.S., and Krivan, V. (1981), *Anal. Chem.* **53**, 2242.

Sastri, C.S., and Krivan, V. (1982), *Anal. Chim. Acta* **141**, 399.

Sastri, C.S., and Krivan, V. (1983), *J. Trace. Microprobe Techn.* **1**, 293.

Schmid, W., Egger, K.P., and Krivan, V. (1987), *J. Radioanal. Nucl. Chem.* **119**

Schweikert, E.A., and Rook, H.L. (1970), *Anal. Chem.* **42**, 1525.

Schweikert, E.A., McGinley, J.R., Francis, G., and Swindle, D.L. (1974), *J. Radioanal. Chem.* **19**, 89.

Shibata, S., Tanaka, S., Suzuki, T., Umezawa, H., Lo, J.G., and Yeh, S.J. (1979), *Int. J. Appl. Rad. Isot.* **30**, 563.

Shikano, K., Katoh, M., Shigematsu, T., and Yonezawa, H. (1987a), *J. Radioanal. Nucl. Chem.* **119**, 237.

Shikano, K., Katoh, M., Shigematsu, T., and Yonezawa, H. (1987b), *J. Radioanal. Nucl. Chem.* **119**, 433.

Strijckmans, K., Vandecasteele, C., and Hoste, J. (1977), *Anal. Chim. Acta* **89**, 255.

Strijckmans, K., Vandecasteele, C., and Hoste, J. (1978), *Anal. Chim. Acta* **96**, 195.

Strijckmans, K., Vandecasteele, C., and Hoste, J. (1980), *Z. Anal. Chim.* **303**, 106.

Strijckmans, K., Motier, R., Vandecasteele, C., and Hoste, J. (1982), *Mikrochim. Acta* **II**, 321.

Strijckmans, K., DeBrucker, N., and Vandecasteele, C. (1985), *J. Radioanal. Nucl. Chem. Lett.* **96**, 389.

Valladon, M., and Debrun, J.L. (1977), *J. Radioanal. Chem.* **39**, 385

Valladon, M., Giovagnoli, A., and Debrun, J.L. (1978), *Analysis* **6**, 452.

Valladon, M., Blondiaux, G., Giovagnoli, A., Koemmerer, C., and Debrun, J.L. (1980a), *Anal. Chim. Acta* **116**, 25.

Valladon, M., Blondiaux, G., Koemmerer, C., Hallais, J., Poiblaud, G., Huber, A., and Debrun, J.L. (1980b), *J. Radioanal. Chem.* **58**, 169.

Vandecasteele, C. (1988), *Activation Analysis with Charged Particles*, Ellis Horwood Ltd., Chichester.

Vandecasteele, C., Adams, F., and Hoste, J. (1974a), *Anal. Chim. Acta* **72**, 269.

Vandecasteele, C., Adams, F., and Hoste, J. (1974b), *Anal. Chim. Acta* **71**, 67.

Vandecasteele, C., and Hoste, J. (1975a), *Bull. Soc. Chim. Belg.* **84**, 673.

Vandecasteele, C., and Hoste, J. (1975b), *Anal. Chim. Acta* **79**, 302.

Vandecasteele, C., and Hoste, J. (1975c), *Anal. Chim. Acta* **78**, 121.

Vandecasteele, C., and Hoste, J. (1975d), *J. Radioanal. Chem.* **27**, 465.

Vandecasteele, C., and Hoste, J. (1985), *J. Radioanal. Nucl. Chem. Lett.* **89**, 167.

Vandecasteele, C., Adams, F., and Hoste, J. (1975), *Anal. Chim. Acta* **76**, 27.

Vandecasteele, C., Strijckmans, K., and Hoste, J. (1979a), *Anal. Chim. Acta* **108**, 127.

Vandecasteele, C., Strijckmans, K., Kuffer, R., and Hoste, J. (1979b), *Radiochem. Radioanal. Lett.* **38**, 261.

Vandecasteele, C., Dewaele, J., Esprit, M., and Goethals, P. (1980), *Anal. Chim. Acta* **119**, 121.

Vandecasteele, C., Strijckmans, K., Engelmann, C., and Ortner, H.M. (1981), *Talanta* **28** 19.

Vialatte, B. (1973), *J. Radioanal. Chem.* **17**, 301.

Wauters, G., Vandecasteele, C., and Hoste, J. (1986), *J. Radioanal. Nucl. Chem.* **98**, 345.

Wauters, G., Vandecasteele, C., and Hoste, J. (1987a), *J. Trace Microprobe Techn.* **5**, 169.

Wauters, G., Vandecasteele, C., Strijckmans, K., and Hoste, J. (1987b), *J. Radioanal. Nucl. Chem.* **112**, 23.

Wauters, G., Vandecasteele, C., and Hoste, J. (1987c), *J. Radioanal. Nucl. Chem.* **110**, 477.

Wei, L.C., Blondiaux, G., Giovagnoli, A., Valladon, M., and Debrun, J.L. (1987), *Nucl. Instr. and Methods* **B24/25**, 999.

Wilkinson, D.H. (1955), *Phys. Rev.* **100**, 32.

Wohllebran, K., and Schuster, E. (1967), *Radiochimica Acta* **8**, 78.

Zikovsky, L., and Schweikert, E.A. (1977), *J. Radioanal. Chem.* **37**, 571.

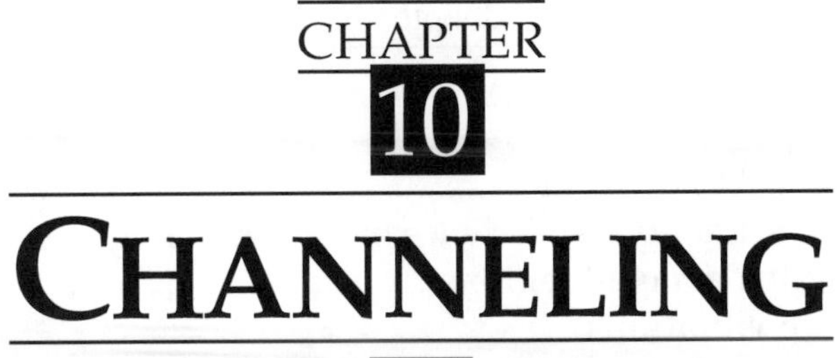

CHANNELING

M. L. Swanson

University of North Carolina, Chapel Hill, North Carolina

CONTENTS

10.1 INTRODUCTION

In this chapter we shall describe the phenomenon of channeling, its measurement and its applications. General references are Morgan (1973), Gemmell (1974), Feldman *et al.* (1982), Swanson (1982), Howe *et al.* (1983), and Mayer and Rimini (1977) (the "Catania Handbook"). We shall not discuss high energy channeling (e.g., above a few MeV per nucleon), low energy channeling (below 50 keV), positron or electron channeling, or emission channeling (i.e., channeling of emitted particles from nuclear reactions). Good references for these topics are: high energy channeling, Uggerhøj (1980) and Andersen (1980); low energy channeling, Buck *et al.* (1983) and Aono *et al.* (1989); positron channeling, Andersen *et al.* (1971) and Schultz *et al.* (1988); and electron and emission channeling, Gemmell (1974) and Hofsäss and Lindner (1991).

Channeling is the steering of a beam of energetic ions into open spaces (channels) between close-packed rows or planes of atoms in a crystal, as shown schematically in Fig. 10.1 (Morgan, 1973; Gemmell, 1974). A ball and stick model of a diamond cubic structure (e.g., Si, Ge, or diamond), showing a random direction, a {110} planar channel, and a <110> axial channel, is illustrated in Fig. 10.2 (Chu *et al.*, 1978). The steering is the result of a correlated series of small-angle screened Coulomb collisions between an ion and the atoms bordering the channel. Thus, channeled ions do not penetrate closer than the screening distance of the vibrating atomic nuclei, and the probability of large-angle Rutherford collisions (backscattering), nuclear reactions, or inner-shell x-ray excitation is greatly reduced compared with the probability of such interactions from a non-channeled (random) beam of ions. Consequently, for ions incident at small angles ψ to a close-packed direction, a large reduction in yield from such interactions with host atoms is observed (Fig. 10.1). The normalized yield χ_h from host atoms for such interactions is defined by the ratio of the yield for ions incident at an angle ψ to the yield for a "randomly" directed beam (see Section 10.42, step 4). In the best axial channels for a highly perfect crystal, χ_h is of the order of 0.01 at low temperatures.

When both the incident beam and the detector are aligned with channels, "double alignment" occurs, and χ_h is even smaller. This is because the backscattered ions are "blocked" or shadowed by the rows of atoms bordering the channel, and thus are prevented from reaching the detector. (See also Figs. 10.4 and 10.44.)

Thermal vibrations basically cause χ_h, which is increased further by any perturbation of the crystal lattice, including point defects, dislocations, stacking faults, mosaic structure, precipitates, and elastic strain (Sections 10.5 and 10.6). Thus channeling has often been used to measure implantation damage, dislocation networks, quality of epitaxial layers, and strain in strained layer superlattices.

Figure 10.3 illustrates the effect of crystalline imperfections on channeling (Bøgh, 1968; Gemmell, 1974; Swanson, 1982; Feldman *et al.*, 1982). For small concentrations of displaced host atoms caused, for example, by point defects or dislocations, channeled ions are gradually deflected out of channels (dechanneled) by multiple scattering with the displaced atoms [Fig. 10.3(b)]. The normalized yield χ_h is a measure of the fraction of dechanneled ions, and the rate of change of χ_h with depth is proportional to the concentration of displaced host atoms. For large concentrations ($\gtrsim$10%) of displaced host atoms, the probability of single large-angle collisions with displaced host atoms becomes important. If a near-surface layer of the crystal is completely disordered (amorphized), as shown in Fig. 10.3(c), χ_h becomes equal to unity for that layer. Since the ions lose energy as they penetrate into the crystal, the energy increment over which $\chi_h = 1$ corresponds to the thickness of the amorphous region.

The position of solute atoms in a crystal lattice can be determined quite directly and precisely from channeling experiments by measuring the normalized yields χ_h from host atoms and χ_s from solute atoms for the same depth increment in the crystal (Swanson, 1982; Feldman *et al.*, 1982; see Section 10.7). If solute atoms are in the same lattice sites as host atoms, then $\chi_s \cong \chi_h$ for any angle ψ. However, if solute atoms project into a given channel <*uvw*>, the channeled ions interact with them, causing an increased yield χ_s. If, for example, 50% of solute atoms project into a given channel, the value of χ_s for that channel would be 0.5 if the ion flux distribution were uniform in a channel (Fig. 10.1). Actually, because of the steering action involved in channeling, the ion flux is peaked near the center of a channel. Thus, if the solute atoms are displaced to positions near the center of a channel, a peaking in χ_s occurs near perfect alignment ($\psi = 0$). Such a peaking effect is unambiguous evidence that the solute atoms lie near the center of the channel. By comparing yields for different channels, the position of solute atoms can be determined rather accurately (to ~0.005 nm in favorable cases) by a triangulation procedure.

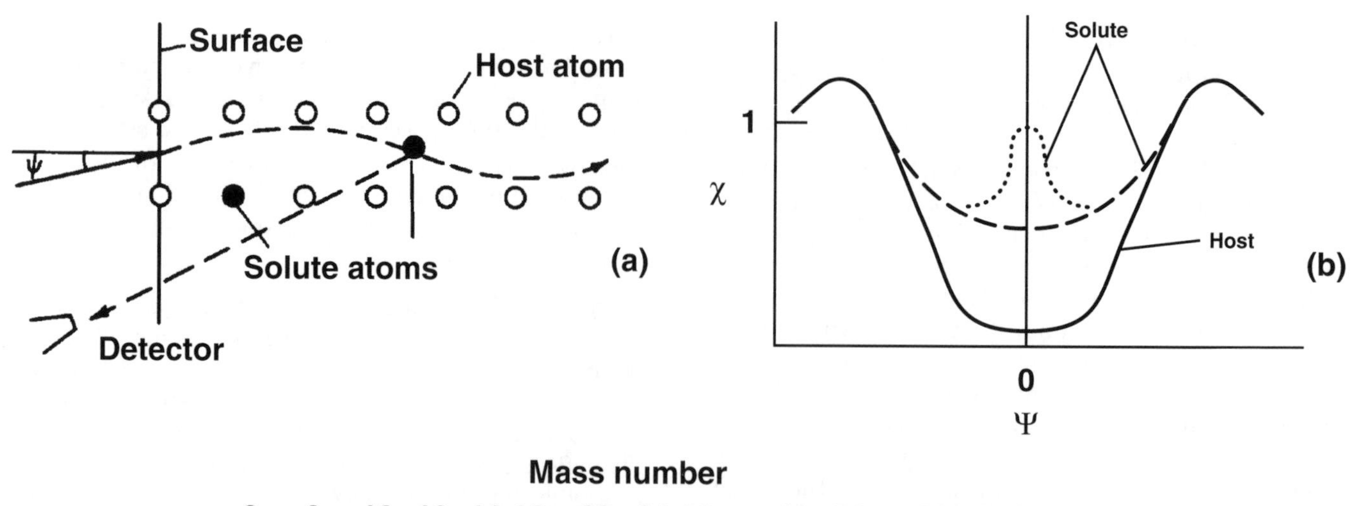

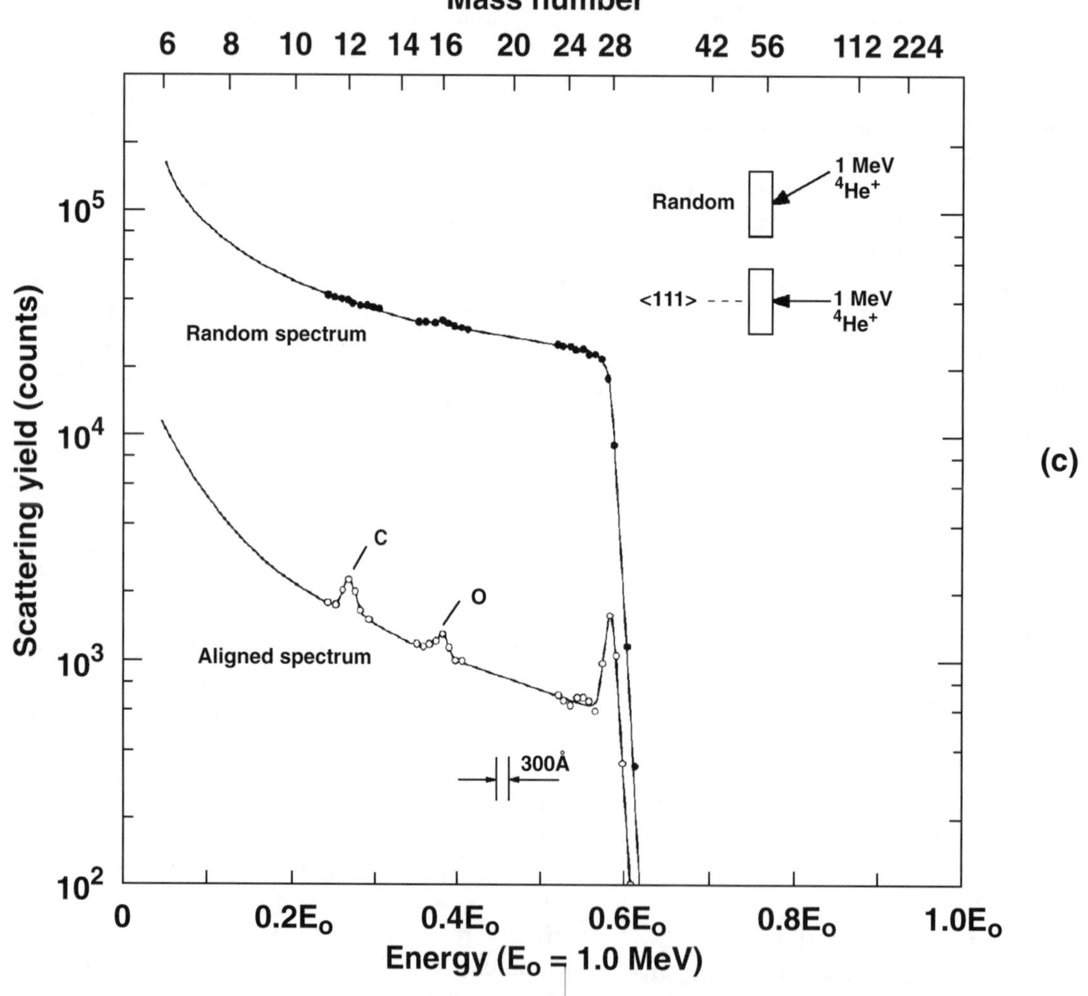

FIG. 10.1. (a) Schematic view of the channeling of ions directed at an angle ψ to a close-packed row of atoms in a crystal. (b) The normalized yield χ_h of ions that are backscattered from host atoms (the RBS yield) shows a strong dip at $\psi = 0$. If 50% of solute atoms are displaced into the channel, the normalized yield χ_s of ions backscattered from the solute atoms is approximately half the random yield; i.e., $\chi_s = 0.5$ at $\psi = 0$ (broken curve). If the displaced solute atoms are located near the center of the channel, a peak in yield may occur (dotted line). (c) The energy spectra of backscattered He ions from Si are shown for good <111> alignment ($\psi = 0$) and for random alignment. Small peaks caused by surface O and C atoms show up in the aligned spectrum.

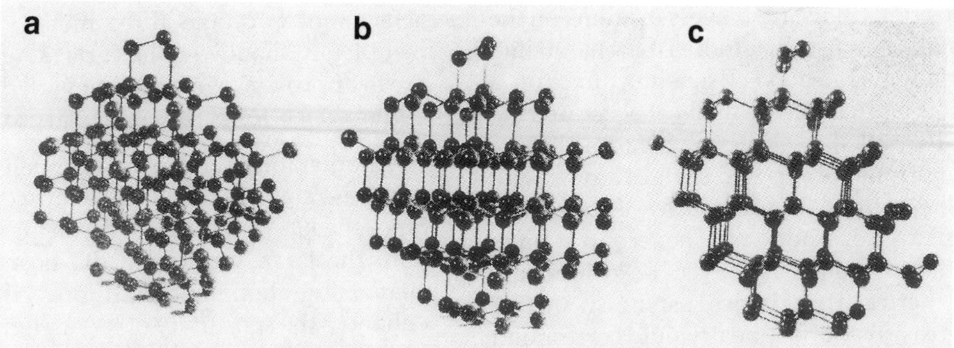

FIG. 10.2. Model of lattice atoms, showing the atomic configuration in the diamond cubic structure, viewed along (a) random, (b) {110} planar, or (c) <110> axial directions (from Chu *et al.*, 1978).

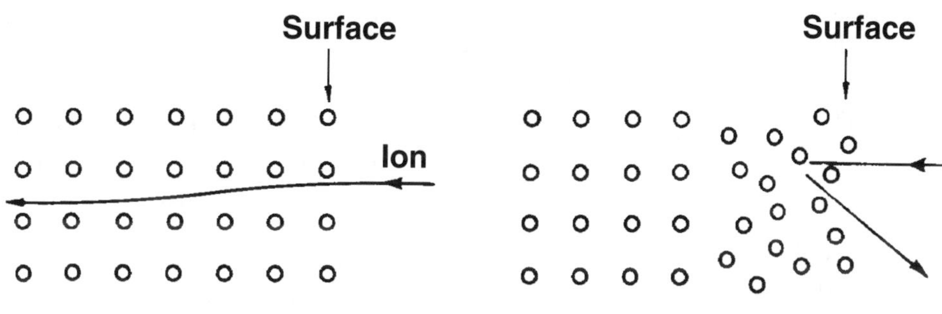

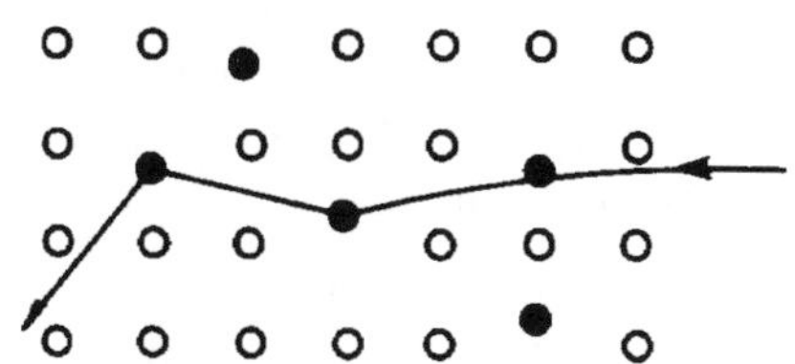

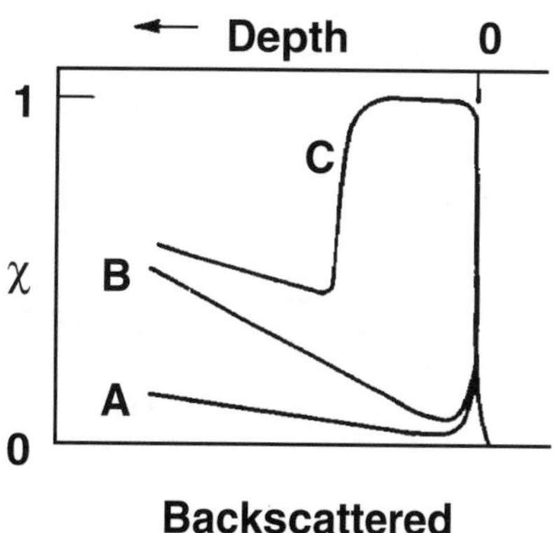

FIG. 10.3. Schematic diagram illustrating ion scattering from defects: (a) little scattering for shallow depths in a nearly perfect crystal, (b) dechanneling by multiple scattering from point defects, (c) direct backscattering from an amorphous surface layer.

The lattice positions of the first layers of atoms on the surface of crystals can also be studied by channeling methods (Section 10.8). In a good channeling direction, a clearly resolvable surface peak in the backscattering yield [Figs. 10.1(c), 10.3] is observed because the ion beam is scattered from the exposed surface atoms with the same intensity as from a random array of atoms, i.e., the surface atoms are not shadowed. The second atom in each string of atoms could, however, be completely shadowed in a perfect rigid lattice, as shown in Fig. 10.4(a), in which case the surface peak would correspond to the density of the top atoms of each row, or approximately 10^{15} atoms cm^{-2} (usually denoted as one atom per string). The second atom in a string is no longer shadowed if it is displaced out of the shadow cone of the incident ions; e.g., by thermal vibrations or by a shift in lattice positions caused by surface reconstruction or relaxation (see Section 10.8). Thus channeling can be used to measure lateral reconstruction, surface relaxation, vibrational amplitudes, and premelting of atoms in the surface and near-surface planes. The lattice positions of adsorbate atoms can also be determined.

An example of the utilization of the shadowing concept for the analysis of epitaxial growth is given by the Rutherford backscattering (RBS) spectra of Fig. 10.4(b), where the surface peaks for thin layers of Au deposited on (111)Ag are shown. If the coverage of the initial monolayers of Au is uniform, then along <110> directions the underlying Ag atoms will be shadowed by Au atoms even for one monolayer coverage [Fig. 10.4(b), inset]. As shown in this figure, for <110> alignment, one monolayer of Au reduced the Ag surface peak by almost a factor of two, and three monolayers of Au almost completely shadowed the surface Ag atoms. These data are in approximate agreement with the calculated shadowing effects for perfect epitaxial growth of Au onto Ag.

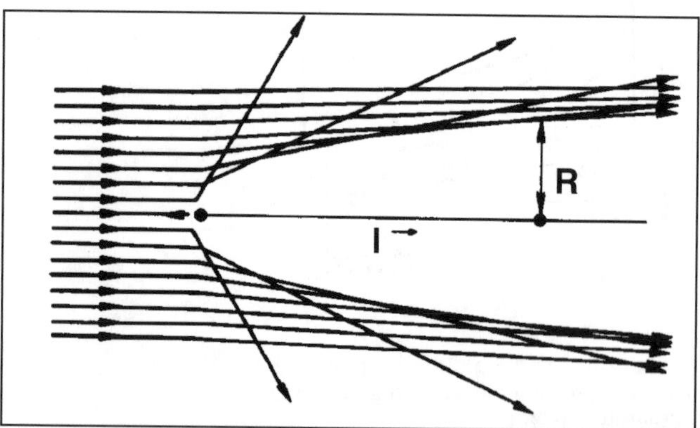

FIG. 10.4(a). An example of a shadow cone behind a surface atom bombarded by energetic ions. R is the radius of the cone at the next atom (Feldman *et al.*, 1982; Swanson, 1982).

Another important use of channeling is for analysis of low mass elements in a heavier host lattice (e.g., O in Si). Since the RBS yield of the light element is superimposed onto the large yield from the host atoms, the use of channeling alignment to suppress the host atom yield enhances the sensitivity of RBS for light element analysis (see Fig. 10.46).

The experimental equipment necessary for channeling is basically the same as described in Chapters 4–8, 11, and 12 for backscattering or nuclear reaction analysis, with the additional requirement of a goniometer for sample alignment. (See Fig. 10.10 and Sections 10.3 and 10.4.)

10.2 THEORY OF CHANNELING

10.2.1 Basic concepts of channeling

Ion channeling was predicted many years ago (Stark, 1912), but was first detected only in 1963, both from the experimental range studies of Davies and co-workers (Piercy *et al.*, 1963) and from computer simulations of ion trajectories by Robinson and Oen (1963a, 1963b). Channeling can occur for neutral, positive, or negative particles, but we shall discuss only the more common case of positive ions. Channeling has been observed for a wide range of ion masses and for energies from the keV to GeV range. We shall be concerned mostly with $^1H^+$, $^2H^+$, and $^4He^+$ ions of energy 0.3-2.5 MeV, which can be obtained from a small Van de Graaff accelerator, and are used for most channeling analyses.

An energetic ion which is directed at a small angle ψ to close-packed rows or planes of atoms in a crystal is steered by a series of gentle collisions with the atoms so that it is channeled into the regions between these rows or planes (Figs. 10.1 and 10.5). The channeling effect is easily understood in terms of a correlated series of elastic two-body collisions (Lehman and Leibfried, 1963; Nelson and Thompson, 1963; Erginsoy, 1965; Lindhard, 1965). The validity of the classical collision model has been shown by Lindhard (1965). The assumptions necessary for this treatment of channeling are:

- for a channeled ion, the scattering angles are small;
- successive collisions are strongly correlated (because ions passing close to an atom in a string must also come close to the next atom);

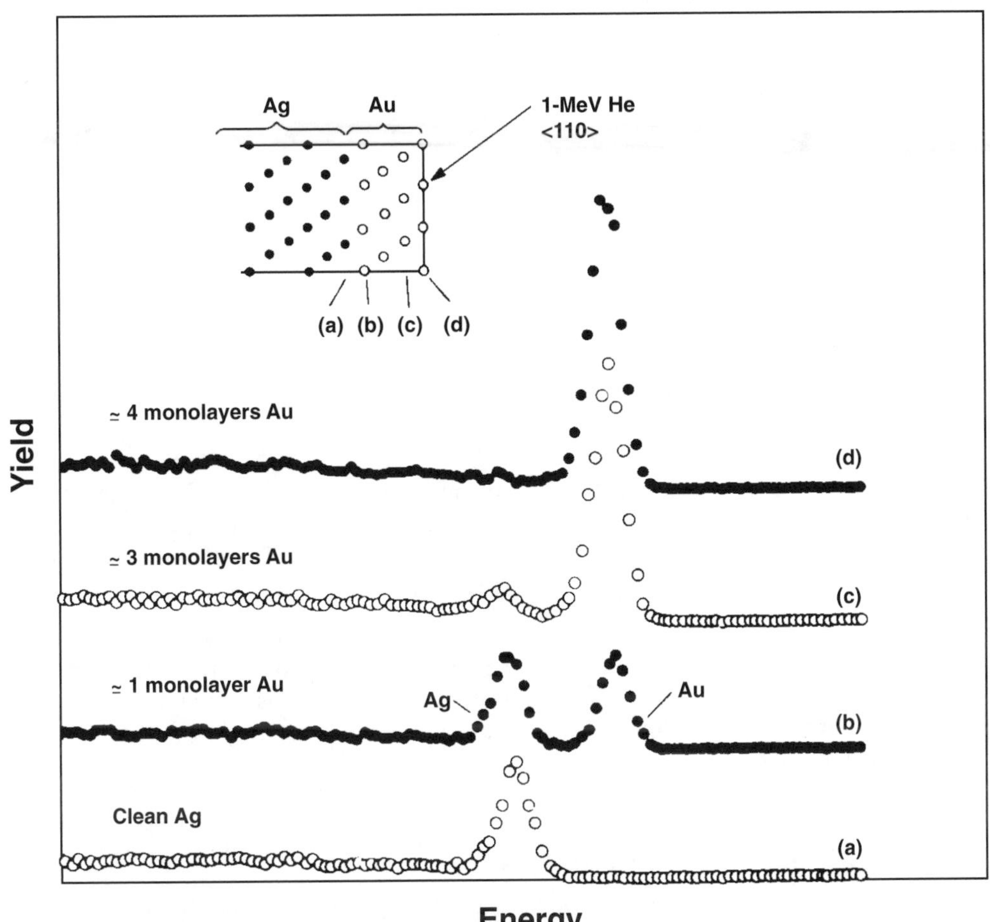

FIG. 10.4(b). Backscattering spectra for 1.0 MeV He ions incident along a <110> axis of (a) a clean Ag (111) surface and (b), (c), (d) Au-covered surfaces. The Ag surface peak decreases because of the Au adsorption (from Culbertson *et al.*, 1981).

- the collisions are elastic two-body encounters; and
- for the description of channeling, a real crystal can be approximated by perfect strings of atoms in which the atomic spacing d is uniform (Fig. 10.5).

For a series of correlated collisions, it has been shown by Lindhard (1965) that the ion can be considered to move in a transverse potential, V_T, which results from an averaging of the potentials of each atom in a string. This average (continuum) potential is thus

$$V_T(\rho) = \frac{1}{d}\int_{-\infty}^{\infty} V[(\rho^2 + x^2)^{1/2}]\,dx \qquad (10.1)$$

where ρ is the distance of the ion from the string of atoms, x is the distance travelled along the string, and d is the spacing between atoms in the string (Fig. 10.5). If the string consists of different atomic species, the potentials for each atom are different, and an average potential is used.

The two-body potential $V(r)$ is generally taken to have the Thomas-Fermi form

$$V(r) = (Z_1 Z_2 e^2/r)\,\varphi(r/a) \qquad (10.2)$$

where Z_1 and Z_2 are the atomic numbers of the ion and an atom in the string, respectively, e is the electronic charge, r is the nuclear separation distance, $\varphi(r/a)$ is the Thomas-Fermi screening function, and 'a' is the screening distance. The use of such a potential is appropriate for the intermediate impact parameters involved in ion channeling, as discussed by Lindhard *et al.* (1968). It will be noted that such a two-body potential is independent of the ion velocity, i.e., inelastic scattering is neglected.

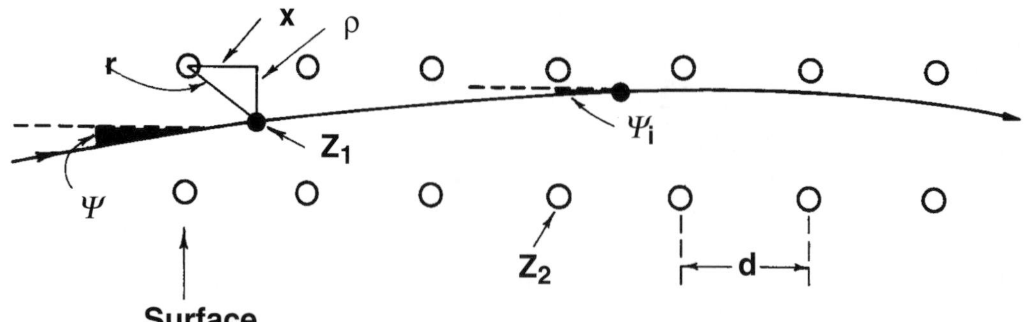

FIG 10.5. String model of channeling.

An analytical approximation for φ(r/a), developed by Molière (1947), is often used in calculating continuum potentials

$$\varphi(r/a) = \sum_{i=1}^{3} \alpha_i \exp(-\beta_i r/a) \tag{10.3}$$

where α_i and β_i are constants. The screening length a is given by an equation of the form (Firsov, 1958; Gemmell, 1974)

$$a = 0.8853 a_0 \left(Z_1^{1/2} + Z_2^{1/2}\right)^{-2/3} \tag{10.4}$$

where a_0 is the Bohr radius, 5.292×10^{-11} m.

Integration of Eq. (10.1), using a potential of type (10.2), gives the string potential (Example 10.1)

$$V_T(\rho) = \frac{2Z_1Z_2e^2}{d} f(\rho/a) . \tag{10.5}$$

An approximation for *f(ρ/a)*, called the standard potential, has been given by Lindhard (1965) (Example 10.1)

$$f_L(\rho/a) = 0.5 \ln[(Ca/\rho)^2 + 1] \tag{10.6}$$

where C is a constant, $C \cong 3^{1/2}$.

When using Molière's screening function, Eq. (10.3), f(ρ/a) becomes

$$f_M(\rho/a) = \sum_{i=1}^{3} \alpha_i K_0(\beta_i \rho/a) \tag{10.7}$$

where α_i and β_i are constants as in Eq. (10.3), and K_0 is a modified Bessel function. It is seen that the ρ dependence of such a potential [Eq. (10.5)] is not so strong as that of the Thomas-Fermi potential [Eq. (10.2)]. Note that $(f_M)^{1/2}$ is given by F_{RS} of Fig. A15.2, in Appendix 15 (Example 10.1).

Under the high-energy conditions which we shall consider, the approximation of the continuum string potential [Eq. (10.1)] is valid if the incident angle ψ is less than a characteristic angle ψ_1, given by

$$\psi_1 = \left(\frac{2Z_1Z_2e^2}{Ed}\right)^{1/2} \tag{10.8}$$

where E is the incident energy of the ion (Example 10.2). This condition results from the requirement that many collisions are involved in the scattering of an ion as it approaches a string of atoms.

EXAMPLE 10.1. Calculation of f_L and f_M for $^4He^+$ incident along a <110> channel in Si.

$Z_1 = 2$, $Z_2 = 14$, $a = 0.0157$ nm from Eq. (10.4).

(Note that the approximation of Table A15.3 gives a = 0.0194 nm.) Then, from Eq. (10.6), $f_L(\rho/a) = 0.69$, 0.28 and 0.086 for ρ/a = 1, 2 and 3, respectively. From Fig. A15.2, using $f_M = (F_{RS})^2$, $f_M = 0.66$, 0.31 and 0.12 for the same values of ρ/a.

These values of f_M or f_L can be inserted into Eq. (10.5) to find the string potential. Using d = (0.5431/1.414) nm = 0.384 nm from Tables A15.1 and A15.3, and $e^2 = 1.44 \times 10^{-13}$ cm MeV, we get $2Z_1Z_2e^2/d = 210$ eV.

EXAMPLE 10.2. Calculation of ψ_1 for 2 MeV $^4He^+$ along a <110> channel in Si.

$Z_1 = 2, Z_2 = 14, e^2 = 1.44 \times 10^{-13}$ cm MeV, E = 2 MeV, d = 0.384 nm (Example 10.1). Thus $\psi_1 = 0.0102$ rad $= 0.584°$.

For mixed strings of different atoms, Z_2 is an arithmetic average of the atoms in the string. If the atoms are not uniformly spaced along the string, d is also taken as an average spacing (Example 10.3). Table A15.2 gives the values by which the lattice constant d_o must be multiplied to give the different interatomic spacings d and interplanar spacings d_p in common diatomic compounds.

The implication of this picture of channeling is that the channeled ion oscillates between strings (or planes) of atoms, never penetrating closer to the strings than the screening distance a (≅0.01 nm). Thus, any close encounter processes, such as back-scattering at large angles, production of characteristic x-rays or nuclear reactions, are prohibited for channeled ions. Consequently, as outlined in the introduction, a strong reduction in yield from such processes occurs for small incident angles ψ (see Figs. 10.1 and 10.3).

The normalized yield $\chi_h^{<uvw>}$ for alignment along an axial channel <uvw>, according to the continuum model (Lindhard, 1965), is given simply by the fraction of the channel area which is blocked by the vibrating strings of atoms (Example 10.4), or

$$\chi_h^{<uvw>} = N\, d\, \pi \left(u_2^2 + a^2\right) . \qquad (10.9)$$

Here N is the atomic density and u_2 is the transverse vibrational amplitude, $u_2 = <z^2 + y^2>^{1/2}$. This result agrees remarkably well with experimental data for channeling of protons in a material like tungsten where u_2 is small, but is generally less than experimental values in other cases (Barrett, 1971). The experimental $\chi_h^{<uvw>}$ is increased appreciably if the divergence of the ion beam exceeds ≅0.1°, or if oxides or other impurities are present on the surface of the crystal.

EXAMPLE 10.3. Interatomic spacings d and interplanar spacings d_p for GaP (zinc blende structure).

$Z = (Z_{Ga} + Z_P)/2 = (15 + 31)/2 = 23$

$d_o = 0.545$ nm (Table A15.4)

<111> axis: $d = d_o(3^{1/2}/4 + 3 \times 3^{1/2}/4)/2 = 0.472$ nm (Table A15.2)

{110} plane: $d_p = d_o/(2 \times 2^{1/2}) = 0.193$ nm

The half-width $\psi_{1/2}$ of the channeling dip, defined as the angle ψ at which the normalized yield χ_h from host atoms reaches a value halfway between its minimum value and the random value (unity), is related to the Lindhard characteristic angle ψ_1 by some constant $C_\psi (\cong 1)$ which depends somewhat on the vibrational amplitude of the atoms

$$\psi_{1/2} = C_\psi\, \psi_1 \ . \qquad (10.10)$$

See Eq. (10.14) and Appendix 15, Eq. (A15.4) for an evaluation of C_ψ. A comparison of experimental $\psi_{1/2}$ values with the relations [Eqs. (10.8) and (10.10)] shows that $\psi_{1/2}$ varies in the predicted way with energy E, string spacing d and atomic numbers Z_1 and Z_2. Similar considerations apply to planar channeling (see Appendix 15).

The continuum, single-string model of channeling, with the inclusion of atomic vibrations, has been used to calculate the shape of channeling dips. In Fig. 10.6, the calculations are compared with experimental results for $^1H^+$ ions in W (Andersen, 1967). In this case, the aligned value of χ_h, the angular width of the dip, and the increase in χ_h above unity in the shoulder regions agree quite well with the experimental data. The reason for the high yields in the shoulders is that ions which are incident at angles slightly larger than $\psi_{1/2}$ have a higher than average probability of colliding with atoms in the strings. These calculations are valid only near the surface of the crystal, since multiple scattering of ions by electrons and other defects, as well as the energy loss of ions as they penetrate the crystal, are neglected.

EXAMPLE 10.4. Normalized <110> yield for 2 MeV ^4He in Si.

$N = 8/(0.543\ \text{nm})^3 = 49.9\ \text{nm}^{-3}$ (Tables A15.1 and A15.3), d = 0.384 nm, as in the Example 10.1, $u_2 = 2^{1/2} u_1$ (or $u_2 = 1.414\, u_1$) = 0.0106 nm (Table A15.3), and a = 0.0157 nm from Eq.(10.4). Thus $\chi = 0.022$.

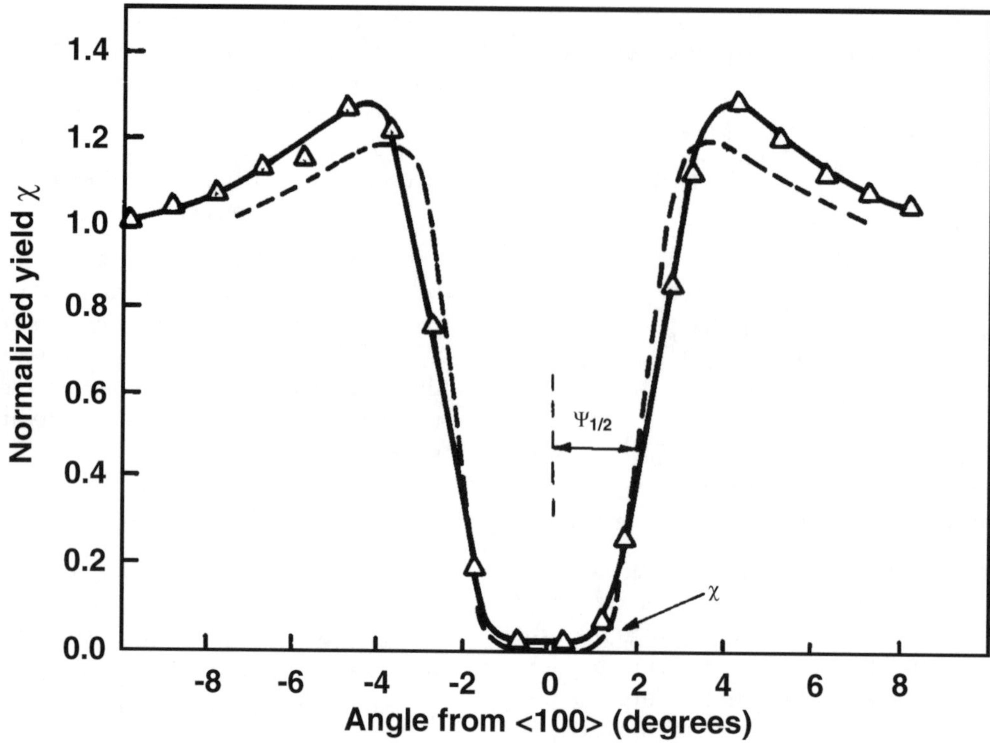

FIG. 10.6. Experimental and calculated (dashed line) backscattering yields for <100> axial channeling of 480 MeV protons in W at 309 K (from Andersen, 1967).

10.2.2 Calculations of channeling parameters

More accurate calculations of channeling parameters have been performed using continuum potentials which are the sum of single-string potentials for the nearest and next-nearest rows of atoms bordering a channel. An approximation such as the Molière screening function, Eq. (10.3), is used in each continuum string potential. Typical potentials calculated in this way are shown in Fig. 10.7 for 1 MeV He^+ ions in a <110> channel of Al (Matsunami *et al.*, 1978).

Because of the steering action of channeling, ions are directed towards the center of a channel, resulting in an enhanced ion flux there. In order to calculate ion trajectories, and thus the spatial distributions of channeled ions within a given channel, it is usually assumed that the transverse component $E_\perp$ of the ion energy is conserved (energy losses are neglected). Thus, the ion trajectories are described only by the transverse energies

$$E_\perp = E\psi_i^2 + V_T(\rho) \qquad (10.11)$$

where ψ_i is the angle between the ion's direction of motion and the atomic strings (see Fig. 10.5).

In analytical calculations, it is further assumed that statistical equilibrium of ion positions has been reached. This means that ions striking the crystal at an initial incident angle ψ, and at a distance ρ_{in} from an atomic string, are uniformly distributed within an accessible area specified by the equation

$$E\psi^2 + V_T(\rho_{in}) = V_T(\rho_A) \ . \qquad (10.12)$$

The ions are excluded from an area $\pi\rho_A^2$ close to the string. It follows that the ion flux is enhanced near the center of a channel. Under these conditions, a simple estimate of the flux distribution within a channel can be made for a given ψ, using potential contours of the type shown in Fig. 10.7. For $\psi = 0$, the normalized flux F_j at the jth potential contour is given approximately by (Example 10.5)

$$F_j = \int_{A_j}^{A_0} dA / A = \ln(A_0 / A_j) \qquad (10.13)$$

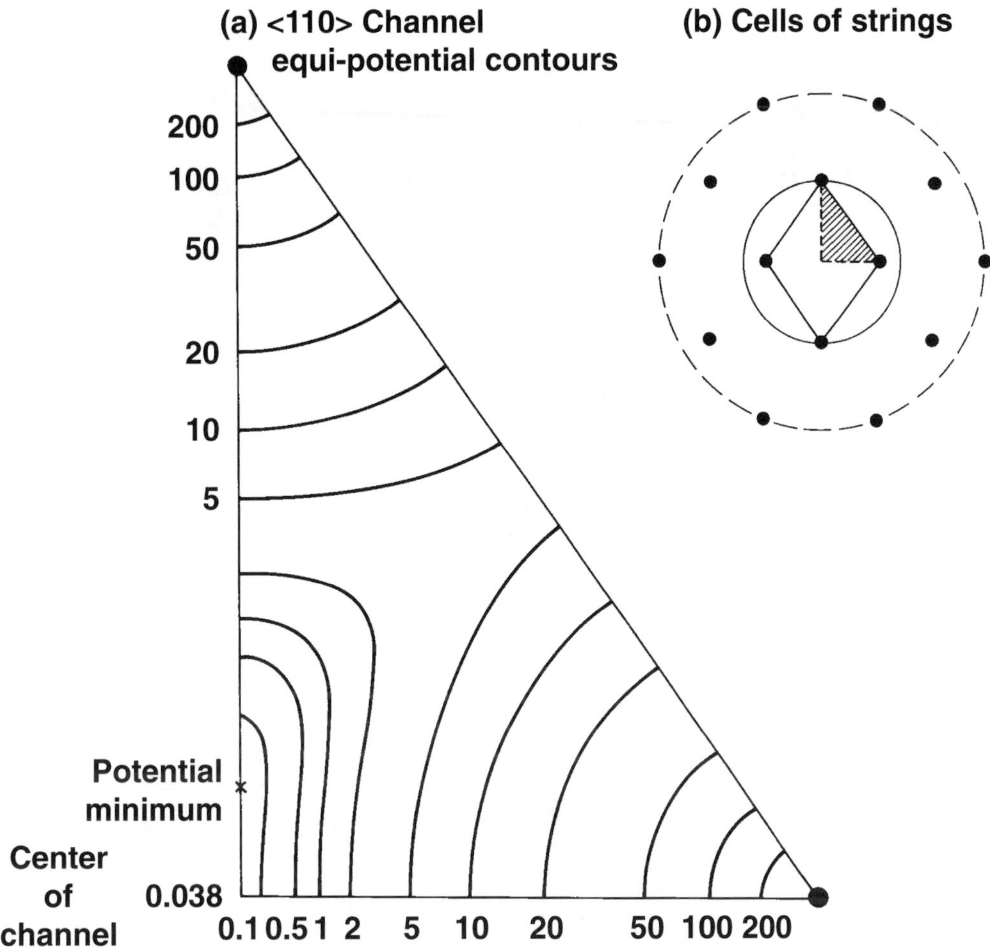

FIG. 10.7. (a) Equipotential contours (in units of electron volts) for He^+ ions in Al for one-quarter of a <110> channel, as indicated by the shaded area in (b). The results were obtained for the 14 strings of atoms indicated by the filled circles in (b). The first and second cells of strings are designated by full and broken circles, respectively (from Matsunami *et al.*, 1978).

where A_0 is the cross-sectional area of the channel and A_j is the accessible area (at a lower potential than the jth potential contour) (Andersen *et al.*, 1971).

Detailed calculations of ion flux distributions, using continuum potentials calculated by summing several string potentials, and including thermal vibrations and multiple scattering, have given flux profiles of the type shown in Fig. 10.8 (Van Vliet, 1971; Matsunami *et al.*, 1978; Nielsen, 1988).

> **EXAMPLE 10.5.** Calculation of F_j. In Fig. 10.7, the areas outside the strings enclosed by the 20 eV potential contours are ~0.6 of the total, giving $F_j = \ln(1/0.6) = 0.7$, in good agreement with Fig. 10.8.

Before statistical equilibrium is reached, the ion flux not only varies across a channel but also exhibits strong depth oscillations. This occurs because of the rather regular movement of channeled ions from one side of a channel to the other. Thus, peaks in the mid-channel flux (and minima in the flux at the channel walls) occur at $\lambda/4$ where λ is the wave length of the channeled ion's path. An analytical expression for λ, based on the harmonic bowl approximation for the continuum potential is $\lambda = 4\pi\rho_{ch}2/(3n)^{1/2}Ca\psi_1$, where n is the number of atomic strings bordering the channel, ρ_{ch} is the distance from a string to the center of the channel, C is a constant

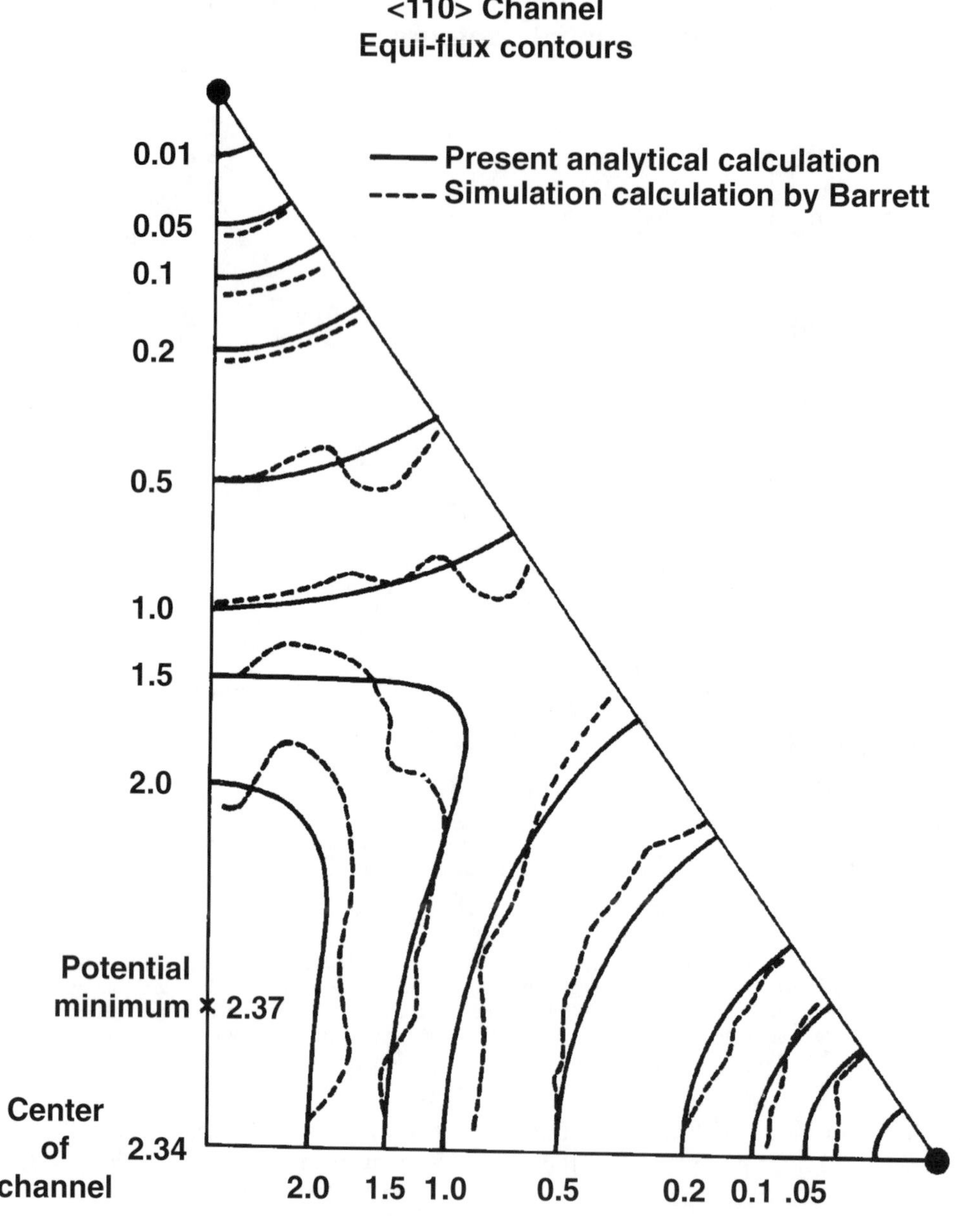

FIG. 10.8. Equi-flux contours for 1 MeV He^+ ions in Al at 30 K, shown for one quarter of a <110> channel. Results of an analytical calculation are shown as solid lines and a Monte Carlo simulation as broken lines. The flux values have been normalized to the value for random incidence (from Matsunami *et al.*, 1978).

taken as $3^{1/2}$, a is the screening distance, and ψ_1 is the Lindhard characteristic angle, Eq. (10.8) (Van Vliet, 1971).

These depth oscillations are less regular for axial channels than for planar channels, since axially-channeled ions can move to neighboring channels more easily than can planar channeled ions. With increasing depth, the oscillations are damped out rapidly because of multiple scattering and the anharmonicity of potential contours. In Fig. 10.9, the calculated axial depth oscillations of the mid-channel flux are shown for 1 MeV He^+ in a <100> channel of Cu (Van Vliet, 1971).

Barrett (1971) has used Monte Carlo computer simulations of ion trajectories to determine channeling parameters, using the impulse approximation for binary

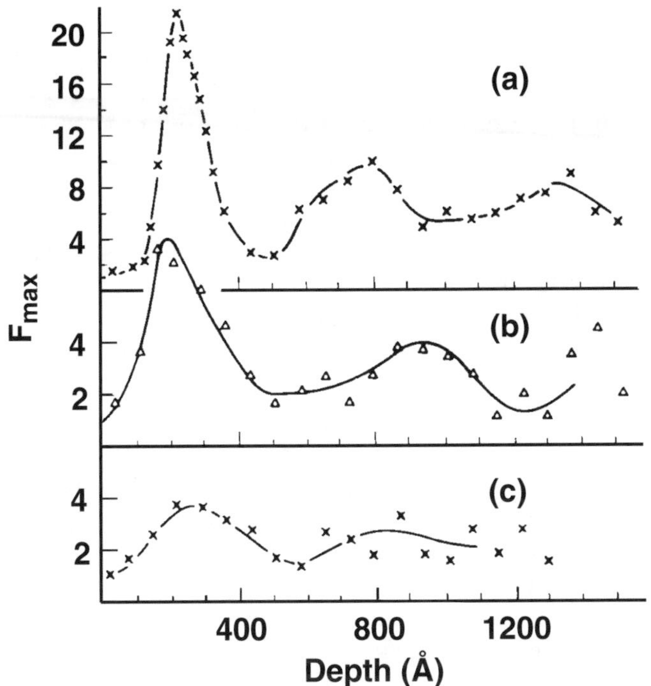

FIG. 10.9. Depth variation of the calculated mid-channel flux of 1 MeV He^+ ions in a <100> axial channel of Cu. (a) no multiple scattering included; (b) multiple scattering, energy losses, thermal vibrations (T = 273 K), and beam divergence (0.06°) included; and (c) as for (b) but with a beam divergence of 0.23°. The full curves are best fits to the computed results (from Van Vliet, 1971).

collisions, Molière's approximation for the Thomas-Fermi potential and thermal vibrations according to the Debye theory. These calculations give almost the same ion flux distributions in a channel, when averaged over the depth interval 50-150 nm, as the analytical calculations (see Fig. 10.8). Similar computer calculations were performed by Morgan and Van Vliet (1972), and Smulders and Boerma (1987).

A review of computer simulation methods for channeling has been given by Barrett (1990). References to a variety of simulation applications to channeling experiments are given, as well as references to a number of simulation programs. Copies of such programs can be obtained from J.H. Barrett (Oak Ridge National Labs) and D.O. Boerma (Univ. of Groningen, the Netherlands). They are complex, so that learning to use an existing program requires a certain amount of expertise. In addition, each experiment often has particular features that require extra effort to produce a modified version of the program. Because of these complexities in using computer simulations, they are not used as often as might be desirable.

Study of surface structures by ion scattering is one application of channeling in which computer simulation is most needed and has been almost universally used. Double alignment (simultaneous channeling and blocking) has proven much more powerful than single alignment, and is now accepted as an important technique for surface studies (see Section 10.8 and Fig. 10.44). A review with many examples has been given by van der Veen (1985). The simulation program described by Tromp and van der Veen (1983) was written for such channeling studies and is highly recommended.

The result of Barrett's Monte Carlo calculations for tungsten, averaged over a depth of 55 nm, gave angular half-widths for axial channels (Example 10.6)

$$\psi_{1/2} = k[f_M(mu_1/a)]^{1/2}\psi_1 \qquad (10.14)$$

where k and m are constants, $u_1 = <z^2>^{1/2}$ is the root mean square vibrational amplitude, and f_M and ψ_1 are defined by Eqs. (10.7) and (10.8) (Example 10.2). This equation is thus very similar to Eq. (10.10). Values of k = 0.80 and m = 1.2 gave very good agreement with experimental results for several different targets, ion masses, and ion energies. An equation similar to (10.14) was given for planar channels. Calculated values of $\psi_{1/2}$ are compared with experimental values in Appendix 15 (Table A15.5).

The minimum yield $\chi_h^{<uvw>}$ for <uvw> axial channels was also calculated by Barrett (1971). The results fit the equation (Example 10.7)

$$\chi_h^{<uvw>} = N\, d\, \pi\left(3u_2^2 + 0.2\, a^2\right) \qquad (10.15)$$

EXAMPLE 10.6. Calculation of $\psi^{1/2}$ for 1 MeV ^4He incident on <110> Ge at 293 K.

$Z_1 = 2$, $Z_2 = 32$, $e^2 = 1.44 \times 10^{-13}$ cm MeV, E = 1 MeV, $d = d_o/2^{1/2} = 0.400$ nm (Tables A15.1, A15.3); thus $\psi_1 = 1.23°$ [Eq. (10.8)]. $f_M^{1/2} = F_{RS} = 0.95$ (see Eq. A15.4 and Fig. A15.2, and using $u_1 = 0.0085$ nm and a = 0.0148 nm from Table A15.3). Thus $\psi_{1/2} = 0.8\, F_{RS}\psi_1 = 0.93°$.

EXAMPLE 10.7. Calculation of the minimum yield $\chi_h^{<uvw>}$ using Eq. 10.15 for ^4He ions along <110> axial channels in Si.

$\chi = 0.023$, using the values of N, d, u_2, and a in Example 10.4. Without the a^2 term, $\chi = 0.020$. These values are close to the value 0.022 obtained from Eq. (10.9).

where the symbols are as defined for Eq. (10.9). The term in *a* is sufficiently small that it can usually be neglected.

In Eq. (10.15), χ has a somewhat larger dependence on the vibrational amplitude than in Eq. (10.9), presumably because account is taken of ions having large transverse energies, which have an enhanced probability of a nuclear collision with lattice atoms. In addition, the Monte Carlo simulations were done (in tungsten) for a depth range of 34-137 nm, so that some dechanneling effects could rise from multiple collisions. In general, Barrett's Eq. (10.15) agrees well with experimental results (see Appendix 15), although in some cases the experimental values are lower than the calculated ones (Howe *et al.*, 1976). $\chi_h^{<uvw>}$ is in fact somewhat dependent on energy; this dependence is expressed by Eq. A15.8 in Appendix 15, and some comparisons with experimental results are shown in Fig. A15.4.

The normalized yield χ^1 under double alignment conditions is expected to be considerably less than the yield χ for single alignment. Feldman and Appleton (1969) have derived an expression for χ^1 from Lindhard's theory

$$\chi^1 = \left(2 - \frac{1}{2}\sin^2\alpha\right)\chi^2 \tag{10.16}$$

where α is the angle between the detector and beam directions. Barrett's (1971) calculations yielded

$$\chi^1 = \nu(\alpha)\,\chi^2 \tag{10.17}$$

where $\nu(90°) = 1.1 \pm 0.1$ and $\nu(180°) = 1.2 \pm 0.1$. It is apparent that the use of double alignment can considerably increase the sensitivity of the channeling method for defect studies.

10.2.3 Dechanneling

In the foregoing discussion of channeling, two important effects were ignored, the energy loss of ions as they penetrate a crystal and the gradual increase of the transverse energy $E_\perp$ of the ions because of multiple collisions with electrons and displaced atoms. The latter effect eventually causes ions to be deflected out of channels (dechanneled), producing an increase in minimum yield $\chi_h^{<uvw>}$ and a decrease in the channel half-width $\psi_{1/2}$ as the ions penetrate deeper into the material (Fig. 10.3). For shallow depths, where $\chi<0.2$, dechanneling is caused mainly by nuclear multiple collisions with thermally vibrating atoms and with lattice defects, whereas multiple electronic collisions become increasingly important at greater depths (Gemmell, 1974).

In backscattering experiments, the (rate of) dechanneling is measured by $d\chi/dx$, where $\chi(x)$ is the normalized yield at a distance *x* along the beam path. The dechanneled fraction of ions is given by $\chi(x)$-$\chi(0)$, where the normalized yield at the surface, $\chi(0)$, includes the probability of direct backscattering events from displaced atoms. One of the difficulties in comparing such experimental results with theory is the determination of the actual depth *x*. It is conveniently found by assuming that the stopping power dE/dx in an aligned direction is equal to that in a random direction (the Aarhus convention; see Section 10.3.5).

In a commonly used phenomenological dechanneling model, the rate at which the channeled fraction $(1\text{-}\chi)$ of ions becomes dechanneled by a concentration N_d of irradiation-produced defects is given by (Bøgh, 1968; Merkle *et al.*, 1973)

$$d(1-\chi)/dx = -(1-\chi)(N_d\sigma_d + \sigma_{th}) \tag{10.18}$$

where σ_d is an effective defect cross section for dechanneling and σ_{th} is the effect of thermal vibrations. Thus for small χ, the slope $d\chi/dx$ is proportional to the defect concentration N_d. This is a very simple way of measuring the defect concentration versus depth, although absolute values of defect concentrations are dependent on the accuracy of theoretical predictions for cross sections of different defects. An assumption inherent in Eq. (10.18) that the effects of lattice defects and thermal vibrations are independent is in general not true (Howe *et al.*, 1976; Matsunami and Itoh, 1973).

Integration of Eq. (10.18) gives

$$\chi(x) = 1 - [1 - \chi_v(x)]\left[\exp\left(-\int_0^x N_d\sigma_d dx\right)\right] \tag{10.19}$$

where χ_v is the value for the undamaged crystal.

In a more fundamental treatment of dechanneling, the diffusion equation for the multiple electronic and nuclear scattering of ions is solved to find the distribution $g(E_\perp, x)$ of the transverse ion energy $E_\perp$ as a function of depth (Lindhard, 1965; Bonderup *et al.*, 1972; Gemmell, 1974). An ion is assumed to be dechanneled when its transverse energy exceeds a critical value $E_\perp^c$. Rechanneling is neglected. Then the dechanneled fraction of ions is

$$\chi(x) - \chi(0) = \int_{E_\perp^c}^{\infty} g(E_\perp, x) dE_\perp \quad . \qquad (10.20)$$

Using this approach, it has been shown by Matsunami and Itoh (1973) that there is an 'interference' between lattice defects and thermal vibrations. This means that thermal vibrations increase the transverse ion energy $E_\perp$ so that the effect of lattice defects on dechanneling becomes larger (or the channel becomes effectively narrower). It is thus useful to perform dechanneling studies of defects at low temperatures where thermal vibrations are minimized.

One problem with the random walk (diffusion) approach, other than the computational difficulties, is the difficulty of specifying an initial distribution $g(E_\perp, 0)$ at the surface. In fact, for small values of $\chi(<0.2)$, it can be assumed that most of the ions which are dechanneled had large values of $E_\perp$ at the surface. Thus, $dE_\perp/dx$ for those ions is much larger than for well-channeled ions and is determined largely by nuclear scattering. This leads to the 'steady-increase' approximation for dechanneling, in which diffusion is neglected, and the transverse energy of a given ion at a depth x is determined solely by its initial transverse energy at the surface (Foti *et al.*, 1971). This approach leads to dechanneling results that are similar to the diffusion equation values when $\chi<0.2$ (Bonderup *et al.*, 1972).

10.3 EXPERIMENTAL METHODS

10.3.1 Channeling equipment

The experimental equipment and methods are basically as described in Chapters 4, 6, 7, and 11, but some details that are specific to channeling will be discussed here.

To obtain collimated beams of light ions in the 0.3-2.5 MeV energy range, an electrostatic accelerator is normally used. A beam divergence of 0.1° or less (achieved by placing apertures at least 1-2 m apart) is suitable, and the beam at target is approximately 1 mm in diameter. Micrometer-sized beams can be used, but then ion damage can become a severe problem (see Section 10.3.5). A single-crystal sample is mounted on a two- or three-axis goniometer in a target chamber containing suitable solid-state detectors (Fig. 10.10). Usually, backscattering geometry is used, with the scattering angle approximately 150° (Chu *et al.*, 1978). Increased depth resolution can be obtained by using beams which are incident at glancing angles to the sample surface, or by placing the detector at a glancing angle. Double-alignment geometry is most easily achieved by placing an annular detector in the beam path. For single-alignment measurements, the solid angle of the detector is large enough to avoid accidental double alignment.

Ideally, the sample is surrounded by a cold shield at liquid nitrogen temperatures or lower to avoid condensation of hydrocarbons and other contaminants on the sample. This shield is biased negatively with respect to the sample in order to repel secondary electrons, and the ion fluence is best monitored by measuring the total charge accumulated on the sample holder and shield. For defect studies, it is useful to cool the sample also. This is conveniently done by connecting it to a nitrogen bath or a cryogenic cooler by means of a copper braid (Bøttiger *et al.*, 1973). A small electrical heater attached to the sample holder is then used to vary the sample temperature between about 20 and 400 K. For higher temperatures, the braid is disconnected. An ultra-high vacuum target chamber is necessary for surface studies. Associated equipment is desirable for complementary studies using low-energy electron diffraction (LEED), Auger spectroscopy, and other techniques (Davies and Norton, 1980).

The goniometer is obviously a crucial part of the channeling equipment. Early channeling experiments were performed with manual two axes goniometers (Davies *et al.*, 1967). (For non-computerized goniometers, tilt and azimuthal rotations are the most convenient). Although angular scans are tedious using manual angular manipulations, the superior flexibility for control of the angular spacings and ion dose during the scans is a decided advantage. In addition, alignment can be done more rapidly manually, thus introducing less ion damage to the sample.

Commercial computerized goniometers are now readily available, but because the sample is usually mounted on a massive plate, in general they offer less flexibility in the sample geometry (e.g., compensation

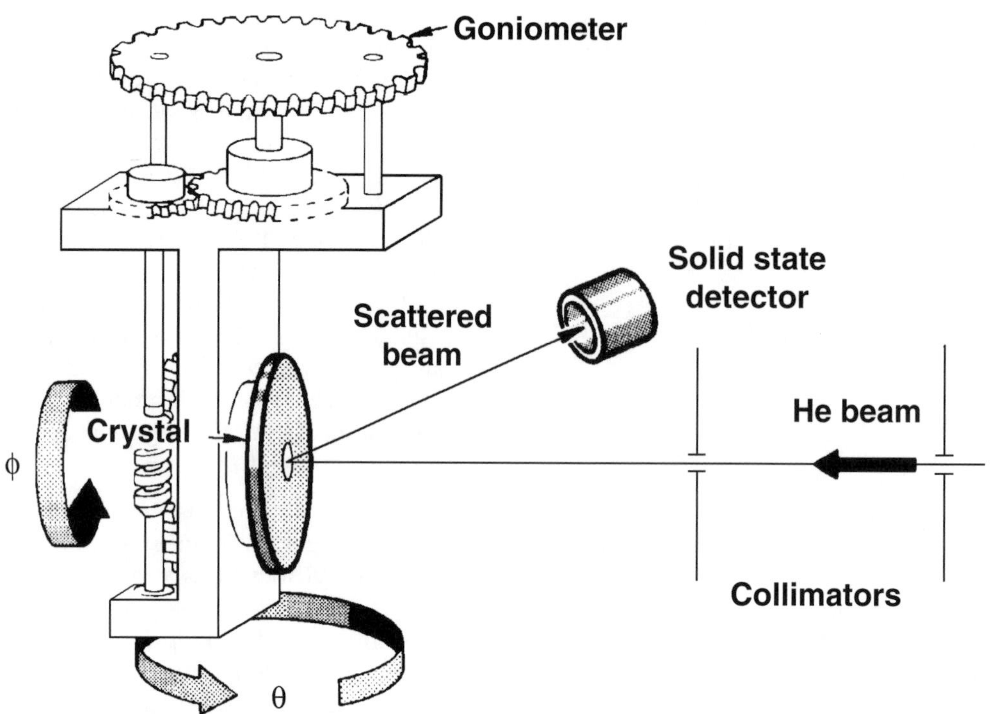

FIG 10.10. Schematic view of the setup for channeling experiments. The ion beam impinges on a crystal mounted on a two axis goniometer, with tilt angle θ (rotation about the vertical axis), and azimuthal angle ϕ (rotation about the crystal normal). The nuclear particle detector measures the energy of the backscattered ions (from Feldman *et al.*, 1982).

for sample thickness to retain the axes of rotation at the same spot on the sample surface); in sample heating and cooling; and in use of cold shields for reduction of contamination and as partial Faraday cups for better current integration. Thus, a self-designed manual or computerized instrument may be better (and cheaper) for many applications (Gemmell, 1971). When designing the goniometer, provision should be made for thermal and spatial separation of a low mass target holder, so that it may conveniently be heated and cooled and surrounded by a cooling shield. Such a design was used in the manual goniometer of Bøttiger *et al.* (1973). An angular precision of 0.01° is desirable for the tilt angle.

10.3.2 Choice of ion species and energy

For most channeling applications, the use of $^4He^+$ beams (at 1-2 MeV) is suitable because of the moderate mass resolution, beam damage, channel width, and depth resolution. If one desires to probe more deeply into the sample, $^1H^+$ beams are suitable, but the mass resolution is poor. Thus $^1H^+$ beams are suitable for probing the distribution of deep damage; e.g., caused by high energy implantations to ~1 μm (see Fig. 10.26). As discussed in Chapter 4, the use of heavier ions like C increases the mass resolution and the sensitivity for detection of small quantities of heavy elements in a light element host material. However, the depth resolution is not improved when surface barrier-Si detectors are used because of the decrease in detector resolution. In addition, the damage introduced by heavier ions is much greater, varying approximately as $Z_1^2Z_2^2M_1/(M_2E)$, or as the square of the ion's atomic number and as the mass M_1 of the ion (Swanson and Maury, 1975; Schiøtt and Thomsen, 1972).

The choice of ion energy is determined by considerations of cross section and mass resolution. The cross section for backscattering generally decreases with increasing energy, according to the Rutherford relationship (Chapters 3 and 4), however, at high energies (e.g., above 2 MeV $^4He^+$ ions onto light elements like O, N, C), the cross section for backscattering may increase by a factor of up to 10 because of non-Rutherford behavior. In addition, several resonances in (α, α) reactions occur at higher He energies. Thus higher energies can be used to good advantage in order to increase sensitivity, but care must be taken to avoid unwanted resonances, for example when studying Si. Since channeling studies are dependent on the channel width, $\psi_{1/2}$, it must also be remembered that the width varies as $E^{-1/2}$ so that detailed angular scans can be more difficult to perform at higher energies.

For the use of nuclear reactions (see Chapters 6–8), the choice of both ion species and energy are clearly crucial. Consideration of the nuclear cross sections, depth of analysis, resonance versus non-resonance reactions, and uniqueness of the nuclear reaction Q values are dictated by the specific goal. For example, the $^{11}B(p, \alpha)\,^{8}Be$ reaction has a high cross section but gives no depth resolution because the emitted alphas have a variety of energies. Thus, when using that reaction for lattice location of B, the depth of B must already be known, for example, by using implantation, epitaxial deposition, etc. (Fig. 10.37). Alternatively, resonant reactions can be used for ^{11}B analysis (Chapters 6 and 7).

10.3.3 Choice of RBS, x-rays, or nuclear reactions (NRA)

For channeling studies of crystal perfection, RBS in the Rutherford scattering regime is the best choice, although in some specialized cases such as study of the lattice vibrations of the constituents of high T_c superconductors, use of non-Rutherford RBS, x-rays, or nuclear reactions might be appropriate.

For surface studies, a combination of RBS and nuclear reactions is often chosen. Both of these methods can be used simultaneously with the same ion species and energy, or a combination of RBS with one particle and NRA with another can be employed. For example, the surface coverage of O has been studied using the $^{16}O(d,p)^{17}O$ reaction, while the reconstruction or relaxation of the surface caused by the O layer has been followed using ^{4}He backscattering (Section 10.8).

For lattice location of impurities, RBS is the best choice when the impurity mass is sufficiently larger than the host atom mass; i.e., the energy separation of the ions backscattered from impurity atoms and host atoms will give enough depth of analysis. For example, to measure the lattice position of As in Si using 1 MeV $^{4}He^{+}$ ions at a scattering angle of 150°, the energies of backscattered He from surface Si and As atoms are $E_{Si} = 0.58$ MeV and $E_{As} = 0.82$ MeV so that the separation in energy is 240 keV. This corresponds to a depth of 400 nm, since the energy loss is about 600 eV/nm (Chapter 2). Thus He^{+} RBS is suitable for As lattice location up to a depth of 400 nm. However, in less favorable cases, e.g., looking at Ag impurities in Zr, the difference in backscattering energies for 1 MeV He^{+} is only 20 keV, giving a depth region of only 20 nm. This can be increased two-fold by using 2 MeV He^{+}.

For light impurity elements in a heavy element host lattice, accurate measurement of backscattering yields from the impurity element is difficult because the signal is buried in the signal from the heavier host element. In such cases, analysis via nuclear reactions or characteristic x-rays is used. There are often several ion energies and species that give suitable nuclear reaction or x-ray yields for a given impurity species. Important considerations for lattice location studies are the amounts of damage produced and the depth of analysis (10.3.5). The damage depends on the reaction cross section, the ion species, and the ion energy. Nuclear reactions emitting alphas (of a unique energy) give much better depth resolution than those producing protons. Because of the large penetration of characteristic x-rays, the depth from which they emanate is difficult to determine. Thus, since lattice location studies require a comparison of yields from impurity atoms and host atoms at the same depth, there have been few such studies done using x-rays (Ecker, 1983; Lennard *et al.*, 1988). However, there are ways of overcoming this problem, such as introducing the impurity into a shallow layer near the surface.

In general, for lattice location studies, heavy impurities in a light matrix are done by RBS, very light impurities (H to S) in a heavier lattice are done by NRA, and intermediate mass impurities in a heavy host lattice are done by characteristic x-ray yields.

10.3.4 Choice of channels and angular scans

The choice of specific angular scans depends of course on the particular application. Although detailed discussion of this topic for specific cases is given in Sections 10.5–10.8, some general comments are appropriate here. For studies of surface reconstruction, surface relaxation, and strained layer superlattices, it is clearly of importance to measure channeling both at normal incidence and at some oblique angle. Thus the precision of orientation of the surface plane is of some concern. In some cases of surface studies, a combination of channeling and blocking is very useful (Section 10.8).

For studies of lattice damage, it is often necessary to measure channeling yields for more than one axial or planar channel because defects may present different distortions (strain) along different channels; e.g., dislocation loops lying along specific crystal planes. In the case of a network of misfit dislocations caused by P doping of Si, Picraux *et al.* (1980) found that the sensitivity is enhanced by using planar channeling (Fig. 10.27).

For lattice location of impurities, angular scans along only two axial channels are often sufficient if the impurities lie on high symmetry interstitial sites (Section 10.7). In more difficult cases, however, specific planar scans are necessary, for example, in the location of ^2H in body-centered cubic metals (Picraux, 1981). Axial angular scans done along specific planar channels not only give more information than scans done randomly, but also are necessary for precise lattice location studies using Monte Carlo methods (Barrett, 1978; Smulders and Boerma, 1987).

10.3.5 Problems

10.3.5.1 Ion damage

The damage caused by energetic ions is caused by elastic collisions with target atoms. Since damage from the analyzing beam can interfere with the results of channeling analysis, especially for studies of low levels of implantation damage or lattice location studies of low concentrations of impurities, it is advisable to minimize that damage by:

- increasing the detector solid angle;
- analyzing only near channeling directions;
- moving the analyzing spot frequently; and
- using high energy and low mass ions.

The simplest ways of increasing detector solid angle are to use larger detectors and to move the detector closer to the target. These methods cause some loss of depth and mass resolution. There are two methods of reducing that loss. One method is to use multiple detectors, or specially constructed detectors. A second method is to use coincidence techniques, where the energies of the scattered ion and the recoiled target atom are measured in coincidence (Hofsäss *et al.*, 1991; Chu and Wu, 1988). The coincidence method is only suitable for low mass elements and requires a uniform thin film (approximately 1 μm thick) because the detectors are mounted on opposite sides of the sample.

The amount of ion damage is obtained directly from the fraction of the incident ion energy that goes into nuclear collisions. This is obtained best by a Monte Carlo simulation method, such as the widely used TRIM program (Ziegler, 1991). Approximations such as the Kinchin-Pease method or the Sigmund approach (Swanson and Maury, 1975) can also be used. In these cases, the concentration of Frenkel pairs (vacancy-interstitial pairs) is proportional to the squares of the ion and target atomic numbers (Z_1 and Z_2), and to the maximum knock-on energy, and is inversely proportional to the ion energy.

Since the damage is caused by nuclear collisions, it is caused only by the dechanneled part of the beam and is proportional to $\chi_h^{<uvw>}$. This has been shown to be true experimentally (Swanson *et al.*, 1979). Thus if channeling analyses are done only near good channeling directions, damage is reduced. In practice, this may reduce the extent of angular scans for lattice location of impurities. Alternatively, the beam spot can be moved often, but this may cause problems if the sample is not uniform in crystal quality, impurity concentration, or defect distribution.

Reducing damage by using high energy or low mass ions in order to lower the collision cross section also reduces the yield of backscattered ions so that the benefit is questionable.

10.3.5.2 Depth of analysis

There are two basic problems with determining the depth of analysis for channeling studies. The first is to determine the stopping powers of the analyzing beam while entering the sample and leaving the sample. The stopping powers of He and H in most materials are well known, although the Bragg law (additivity of stopping powers of different elements in a compound or alloy) is less certain (Chapter 2).

In the case of a channeled beam, the stopping power is as much as 50% less than that of a randomly directed beam because of the lowered stopping from close collisions (Gemmell, 1974). However, according to the so-called Aarhus convention used for channeling analysis, the aligned and random stopping powers are assumed to be equal because of the uncertainty in establishing the different stopping powers in each experiment. In practice, this is not a bad approximation in many cases because that part of the channeled beam that interacts with atoms to be backscattered is the part with a large transverse momentum, and thus is close to the channel walls.

The second problem with depth of analysis is that for lattice location of impurities it is necessary to compare channeling yields from the impurity atoms and from the host atoms at the same depth in the crystal. For backscattering, this is relatively easy to do when using He ions which have a depth of resolution of typically 10 nm. For nuclear reactions, if protons are used or are the reaction product, it is more difficult to compare yields at the same depth. Similarly, if the $^{15}N(p,\alpha)\ ^{12}C$ reaction is used to determine the lattice position of N atoms in Fe, we want to compare α yields from this reaction with proton backscattering yields from the Fe host atoms. Since the depth resolution for backscattered protons is only about 100 nm, it may be difficult to compare yields at the same depth, especially if the N is in a shallow implanted layer. In this case, it may be necessary to use He ions to determine χ_{Fe} in the desired depth interval.

10.3.5.3 Scattering cross sections

Although the Rutherford collision cross section is accurate for most ion/target combinations when using $^4He^+$ ions, care must be taken when using protons because nuclear reactions become significant for proton interactions with light elements at energies greater than 1 MeV (see Chapters 4 and 12 and Appendices 7 and 8). For impurity lattice location using nuclear reactions or characteristic x-rays, cross sections as a function of particle energy are often not accurately known so that standards may be required.

For many channeling studies, the accurate knowledge of scattering cross sections is not necessary, since only normalized yields rather than absolute yields are used.

10.3.5.4 Sensitivity

Channeling is a relatively insensitive method to measure lattice damage. Although it is very effective for a quick, non-destructive measure of the quality of epitaxy and gross ion damage or amorphization, it is not sensitive to small concentrations of defects or dislocations, nor can it easily distinguish types of defects (Section 10.6). Thus it is very useful to compare results from other measurements such as electron microscopy.

For lattice location of impurities, the impurity concentration must be quite high, typically 10^{-3} to 10^{-4} atomic fraction. This is determined by folding the typical areal concentration sensitivity, 10^{14} atoms/cm^2, with the depth over which analysis is desired, typically 10-100 nm. For deeper impurities, the inevitable dechanneling will complicate the analysis.

10.4 CRYSTAL ALIGNMENT

10.4.1 Stereographic projection

In order to align a crystal for channeling measurements, it is necessary to understand the essentials of the stereographic projection. Good introductory treatments of the method are given by Barrett (1952), Cullity (1967), and Amoros *et al.* (1975). Standard stereographic projections of cubic crystals (<100>, <110>, and <111> normals) and hexagonal crystals are shown in Appendix 15, Figs A15.7–A15.11. Stereographic projections of other crystal structures such as tetragonal, orthorhombic, and monoclinic are given in the book by Smaill (1972). Table 10.1 shows the generally accepted crystallographic notation for axes and planes. The general notation refers to a set of axes or planes, and the specific notation refers to a single, specific axis or plane.

The stereographic projection provides an accurate representation of the angles between planes and axes of a crystal. For simplicity, consider first a cubic crystal, for which each axis [uvw] in the Miller index notation is perpendicular to the plane (hkl) of the same index. Consequently, that axis is also perpendicular to every direction or axis contained in that plane. (If we denote one of the axes contained in the plane as $[h_ik_il_i]$, then the sum of products $h_iu+k_iv+l_iw = 0$. This is just the criterion that the sum of the products of direction cosines of perpendicular lines equals zero). As the simplest example of this notation, the [100] axis is defined as the direction from the origin in cartesian coordinates to a point one positive unit along the x-axis, and zero units along the y and z axes. The [100] axis is perpendicular to the (100) plane, which is the plane intersecting the x-axis at unit distance, and intersecting the y- and z-axes at infinity; i.e., the (hkl) planar notation corresponds to intersections of the plane at 1/h, 1/k and 1/l points on the x-, y-, and

Table 10.1. Bracket notation for crystallographic axes and planes.

	Axis	Plane
General notation	<uvw>	{hkl}
Examples	<100>, <210>	{100}, {210}
Specific notation	[uvw]	(hkl)
Examples	[010], [102]	(001), (120)

z-axes. In Fig. 10.11, the normals to {100} planes of a cubic crystal are shown as they intersect a surrounding sphere at points called poles. The positions of the poles determine the orientations of the {100} planes. Note that the normal to a (100) plane intersects the sphere in two diametrically opposed poles, the [$\bar{1}$00] and [100] poles. The planes are also represented by great circles on the sphere, as shown. Note that the (100) planar great circle cuts the [0$\bar{1}$0], [010], [001], and [00$\bar{1}$] poles. The angle between two planes equals the angle between their normals or between their great circles. These angles can be measured precisely by the distance between the poles along a great circle connecting the poles.

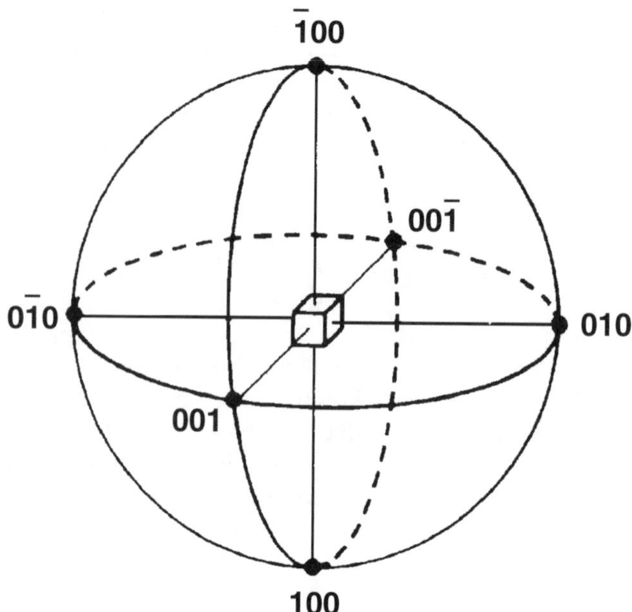

FIG. 10.11. {100} poles of a cubic crystal (from Cullity, 1967).[2]

In order to transfer this spherical crystallographic information to a planar representation, the stereographic projection is used (Fig. 10.12). A projection plane is placed at the end of any diameter of the sphere, and the point of projection is chosen as the other end of that diameter (the line AB in Fig. 10.12). If a light source is placed at B, the shadow of the pole P is projected onto the projection plane at P′. This gives the equiangular stereographic projection, preserving angular relationships but distorting areas. The pole A represents the plane NESW, which is normal to the diameter AB. The great circle NESW projects to form the basic circle N′E′S′W′ on the projection. The plane sketched at C has a pole at P which is projected to the point P′. Thus the angle between the planes C and NESW is given by the distance between P′ and A.

The Wulff net, shown in Fig. 10.13, is marked with equal angular spacings and is used to measure angular distances of a stereographic projection. To measure angles on a stereographic projection, the projection is placed on a Wulff net and rotated until the poles of interest lie on a great circle (the longitudes of the Wulff net), since angles can be measured only along great circles.

To determine the orientation of a crystal, the Laue back reflection x-ray method is often used. In this method, a beam of white x-rays is diffracted back from the crystal onto a film placed between the crystal and the x-ray source. We wish to translate the Laue x-ray pattern to a stereographic projection of the crystal.[1] This is done using the Greninger chart, Fig. 10.14. The chart is placed over the x-ray film with its center coinciding with the film center and with the edges of the chart and film parallel. The angular coordinates γ and δ of the normal to the plane causing a diffraction spot are then read directly on the Greninger chart, as indicated in Fig. 10.14. In this case, the spot indicated has $\gamma = 18°$ and $\delta = -14°$, and its position is marked on a transparent sheet placed over the Wulff net, as shown in Fig. 10.13. Note that the Wulff net must be placed with its great circles (or meridians) running from side to side. The x-ray film must be read from the back side to view the crystal as it appears to the diffracting x-ray beam or to an ion beam. Normally the upper right hand corner of the film as viewed from the crystal is cut away (Fig. 10.15), so that when the film is read, this cut corner must be at the upper left, as shown in Fig. 10.16.

As an example, consider the back-reflection Laue pattern for a randomly oriented Al crystal, shown in Fig. 10.16, and plotted on the stereographic projection of Fig. 10.17. Note that only a small part of the complete projection is seen by this typical set of diffraction spots. The great circles (A-E) joining these spots are drawn by suitable rotation of the Wulff net. These great circles

1. In many cases, such as commercially available Si slices, the orientation of the surface plane is known, and a flat on the disk's circumference indicates a low index crystallographic direction. In such cases, a comparison of channeling planar dips with the stereographic projections of Appendix 15 suffices to establish the sample orientation, and x-ray data are not required.

2. B.D. Cullity, *Elements of X-Ray Diffraction*, © 1978, 1956 by Addison-Wesley Publishing Company, Inc. Reprinted with permission of the publisher.

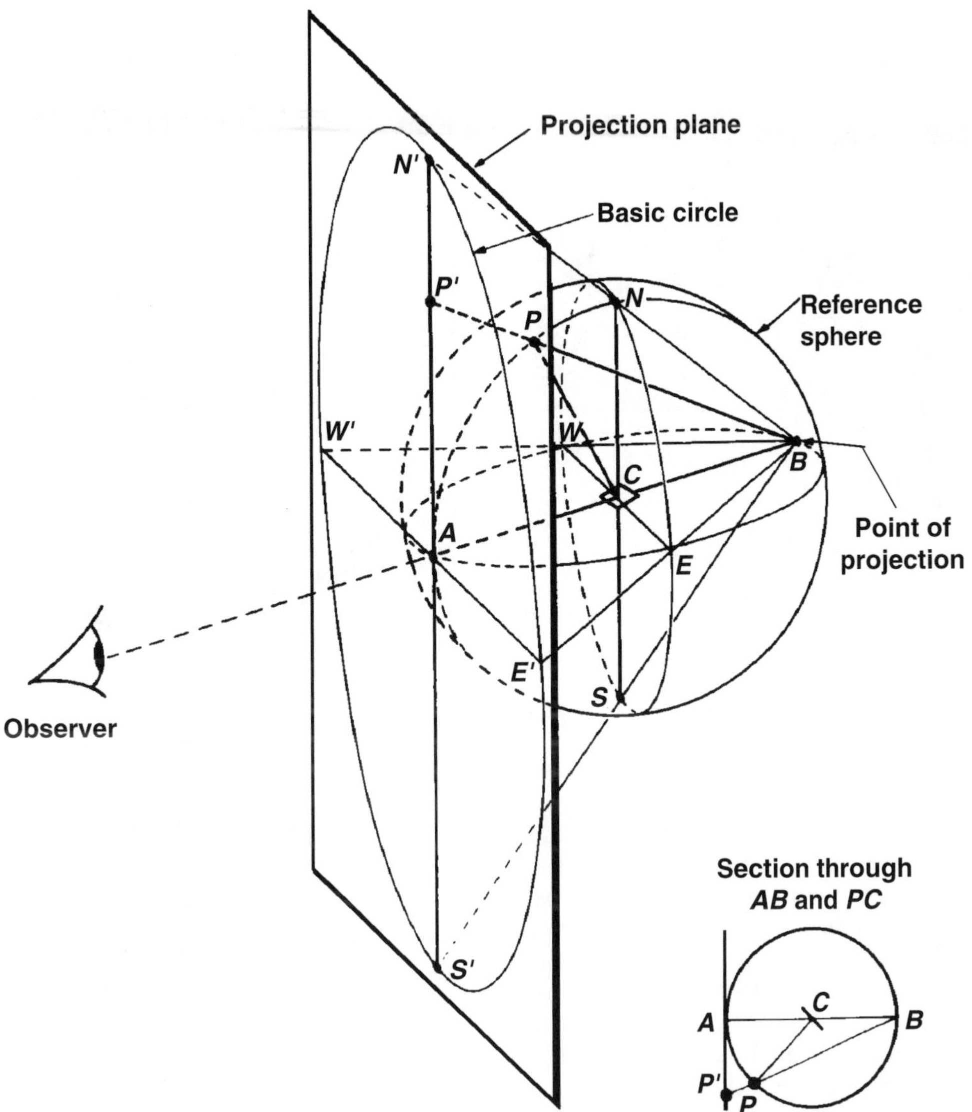

FIG. 10.12. Illustration of the concept of stereographic projections (from Cullity, 1967).

represent major crystallographic planes which intersect in poles representing low index crystallographic directions. Insertion of poles P_A-P_E lying at 90° to the great circles, representing the normals to those planes, is helpful. Then, the stereographic projection is laid over a standard projection (Appendix 15) to assist in the final determination of the crystal orientation. It is also helpful to check the angles between low index directions; i.e., 35° between <110> and <111> axes, 45° between <110> and <100>, and 55° between <100> and <111> (see Appendix 15, Table A15.8, and Figs. A15.7–15.9), and to look for the spherical triangle joining those axes.

In this example, we see that the crystal normal lies roughly equidistant from these major axes, which is often convenient when one wants to measure channeling along all of these axes with a minimum of crystal tilt.

10.4.2 Obtaining an RBS channeling spectrum

Step 1. Use a single crystal sample

The most convenient way to check whether the sample is a single crystal is by etching to look for grain boundaries.

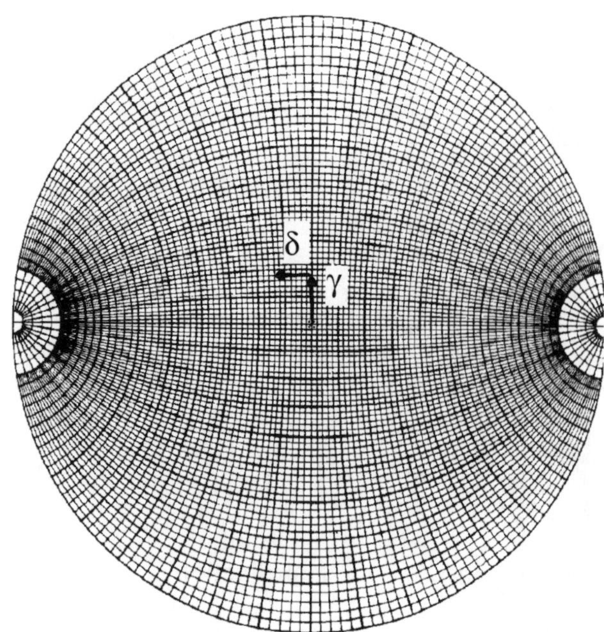

FIG. 10.13. Wulff net drawn to 2° intervals. The angular coordinates γ and δ for a single diffraction spot on a Laue x-ray pattern are obtained by using the Greninger chart, as shown in Fig. 10.14 (from Cullity, 1967).

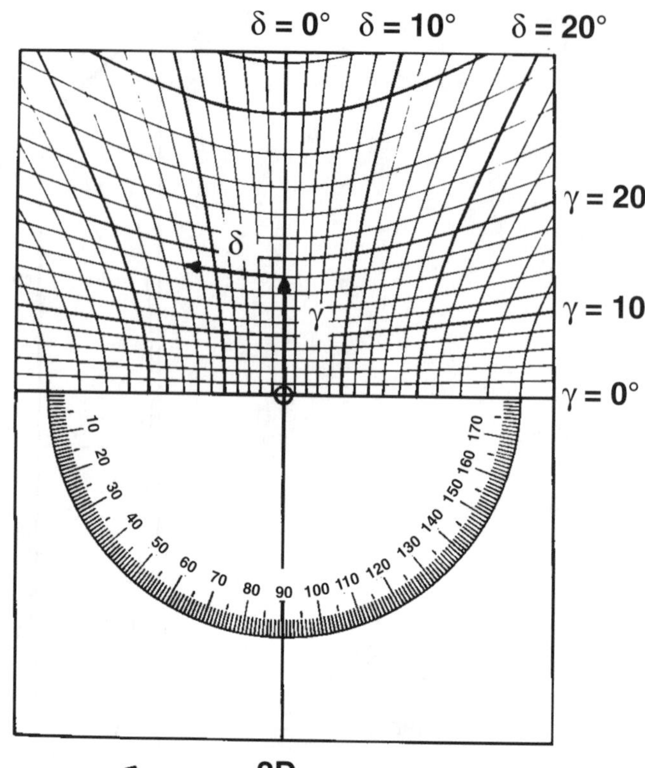

FIG. 10.14. Greninger chart for the solution of back-reflection Laue patterns (from Cullity, 1967).

A better way is to record Laue back-reflection x-ray patterns at different parts of the sample (Section 10.4.1). Electron diffraction patterns are also suitable. If the spots on the x-ray patterns are sharp, if the patterns are identical for different parts of the crystal, and if the pattern indexes as a single crystal (Section 10.4.1), then the crystal will be suitable for channeling. If double or multiple spots appear close to each other on the x-ray pattern [Fig. 10.18(a)], considerable mosaic spread (Section 10.5.3) is indicated, which means that parts of the crystal (crystallites) have small misorientations of one to several degrees with respect to each other. Such crystals are not suitable for channeling if the ion beam hits more than one such crystallite since the half width for channeling is typically 1° or less.

Step 2. Use well-polished, etched or cleaved surfaces

It is very important for channeling that the surface be relatively clean, oxide-free, and (above all), absent of plastic deformation. If the crystal is prepared by cutting with a diamond saw, spark cut, or even polished on diamond paste wheels, there will be a severely damaged layer extending at least 1 μm deep (see Whitton, 1965; 1969). This layer will show up in x-ray patterns as smeared spots [Fig. 10.18(c)]. If the x-ray pattern looks sharp, the surface is normally suitable for channeling.

The usual way of preparing metal surfaces, which are often much softer and thus more susceptible to deformation than semiconductor or insulator surfaces, is by mechanical polishing followed by etching, chemical polishing, or electropolishing (see Whitton, 1965; 1969). The mechanical polishing first uses emery paper, then diamond paste on a polishing wheel, followed by vibratory polishing with 0.05 μm alumina slurry. If thick layers (about 1 μm) are removed by the chemical polishing or electropolishing procedures, then the last mechanical polishing steps can be deleted, but the surface may be rather wavy.

Semiconductor surfaces like Si are often obtained in a prepolished state with a fine mirror finish, and no further treatment is necessary for channeling. In some cases, like GaAs, a combined mechanical-chemical polishing procedure is used.

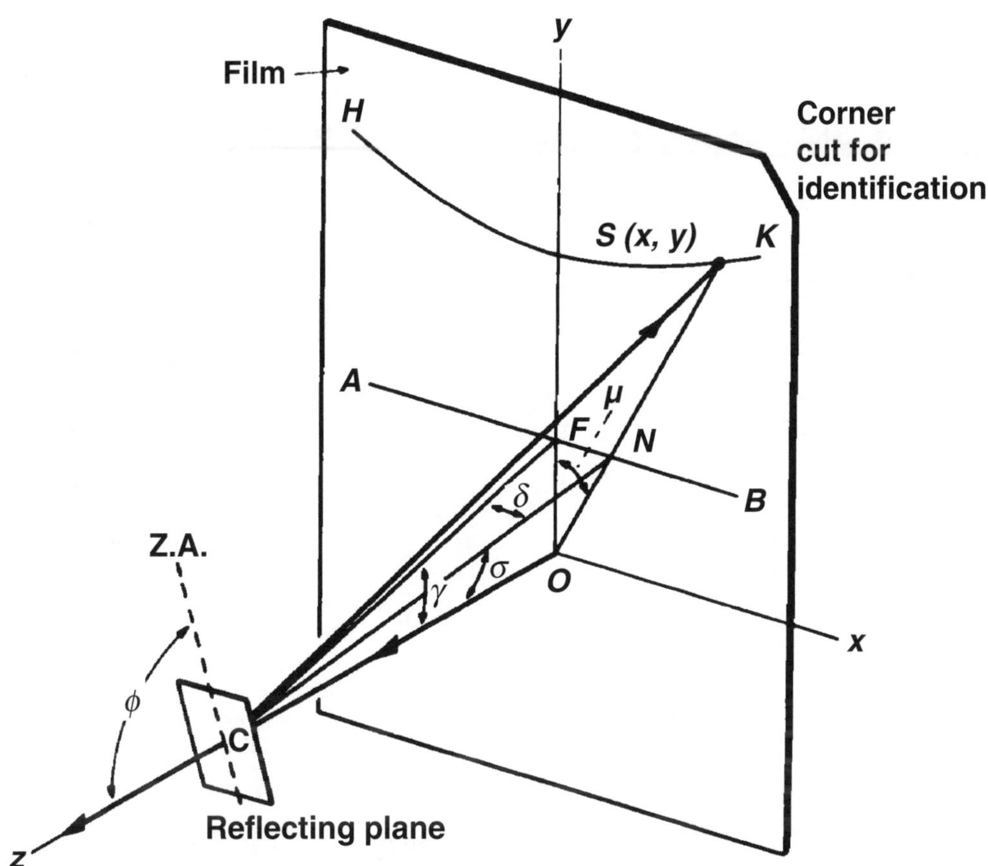

FIG. 10.15. Location of back-reflection Laue spot (from Cullity, 1967).

Some brittle materials like NaCl can be cleaved, leaving a deformation-free, clean surface.

The thin oxide that forms on many materials like Si and Al on exposure to air affects channeling very little [Eq. (A15.11)].

Step 3. Mount the sample and find a suitable channel

The crystal is mounted on a goniometer in an accelerator target chamber. The experimental equipment required to obtain a channeled RBS spectrum is the same as described in Chapter 4, Fig. 4.1, with the necessary addition of a goniometer to align crystallographic channels with the ion beam. A two-axis goniometer with two tilts, or preferably one tilt θ and one azimuthal rotation ϕ is suitable, as indicated in Fig. 10.10.

As an example of aligning a crystal for channeling, consider a (110) Si slice. (Note that this is the common notation for a Si crystal with a {110} planar surface, or a <110> axial normal; see Section 10.4.1 for the bracket notations.) The sample is mounted on the goniometer which is tilted to a small angle, typically 6°, so that the beam is incident on the target at approximately 6° from a <110> axial channel (which is typically 1–2° from the normal to the surface; see Appendix 15 for the stereographic projection of a <110> aligned crystal, and Section 10.4.1 for an explanation of the stereographic projection). A random RBS spectrum is then obtained, as shown in Fig. 10.19(a), top curve, and using a single channel analyzer, an energy window is set at just below the maximum energy of the He ions backscattered from the Si crystal. The RBS counts in this window are then collected for a given ion dose as the azimuthal angle is incremented in steps of 1-2°. The RBS counts are recorded as a function of the azimuthal angle φ, as shown in Fig. 10.19(b). The angles φ at which major dips in yield occur correspond to alignment along planar channels. For good crystals like Si, these channels give dips having minima of about 20-30% of the random yield. The speed of this alignment

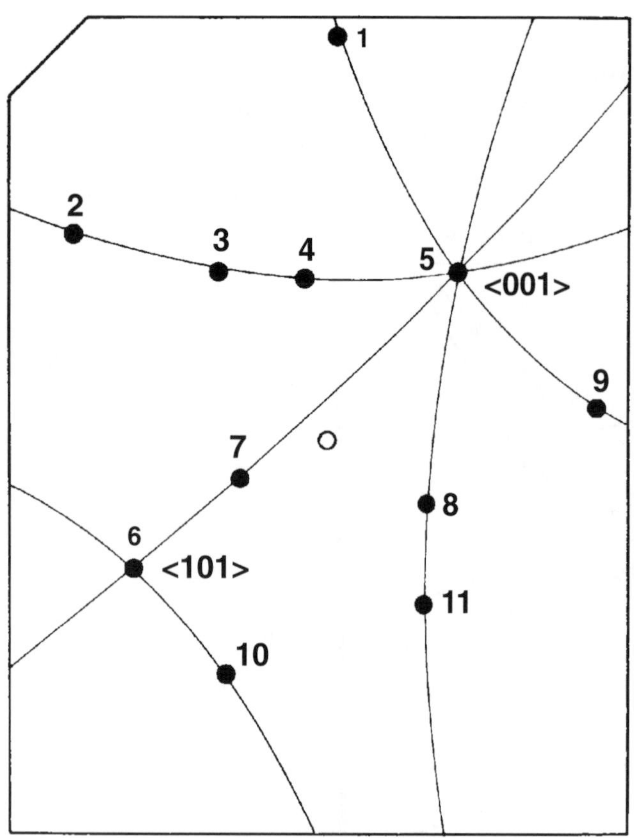

FIG. 10.16. Selected diffraction spots of a back-reflection Laue pattern of an Al crystal (from Cullity, 1967).

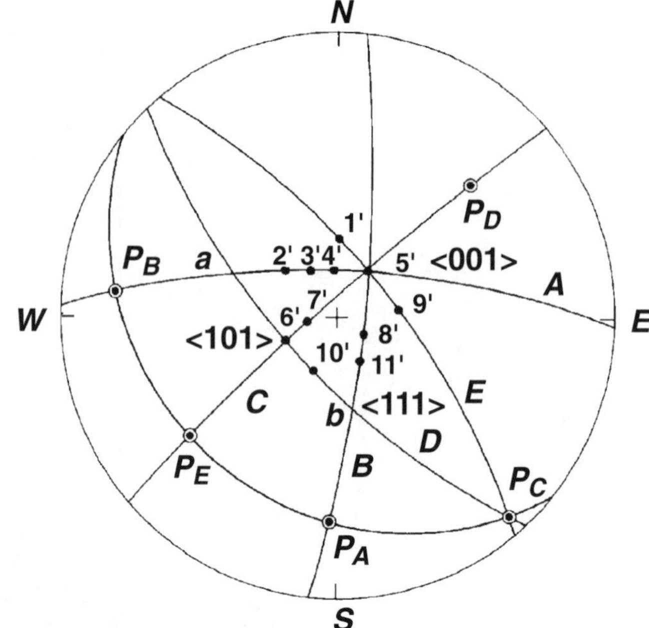

FIG. 10.17. Stereographic projection corresponding to the back-reflection pattern of Fig. 10.16 (from Cullity, 1967). The crystal normal is indicated by the cross (+).

process can be enhanced by using a broad energy window in the single-channel analyzer, but then a deeper layer of the crystal is probed where the dips are reduced because of dechanneling (see Fig. 10.3). Note that the energy window is normally set just below the surface to avoid the inevitable surface peak (see Section 10.8 and Fig. 10.3).

The angles at which these planar dips occur are then plotted on a polar net for the given tilt angle θ, as shown in Fig. 10.19(c). The procedure is repeated for one or more different tilts, e.g., θ = 5°, the angles for the planar dips are plotted on the polar diagram, and the points are connected by straight lines. These lines represent {110}, {100}, and two {111} crystal planes, and they intersect at a <110> crystallographic axis located near the origin, θ = 0. The angles φ and θ at this intersection point are the goniometer settings for <110> alignment. It is sometimes difficult to distinguish between <110> and <100> crystal alignment, since in both cases major planes intersect at 90°. However, a close inspection of Fig. A15.6, which shows the major planes for cubic crystals having different alignment, is helpful. In the case of a <110> surface normal, the {110} and {100} planes are perpendicular, but the {111} planes are about 35° from the {110} planes, as is also the case in Fig. 10.19(c).

To find the precise angles θ and φ for <110> alignment, an angular scan similar to Fig. 10.19(b) is performed, but over a smaller angular range of about 2-4°, by varying the tilt angle θ in small increments (about 0.05°) with φ set at the approximate value 60° from Fig 10.19(c). The result is a curve like that shown in Fig. 10.19(d). Fine tuning of the aligned value of φ can then be done by setting θ at the minimum value 0.90° of Fig. 10.19(d) and doing a similar scan with varying φ.

In practice, for crystals already cut along a major crystallographic plane like our (110) Si slice, it is convenient to insert a small wedge below the sample in order to tilt the crystal by about 2° from perfect alignment; this simplifies the alignment fine tuning because the alignment is almost independent of φ when the tilt angle is very small.

In order to find a non-normal channel, it is convenient to tilt the crystal along a good planar channel the required amount to reach the desired channel. The amount

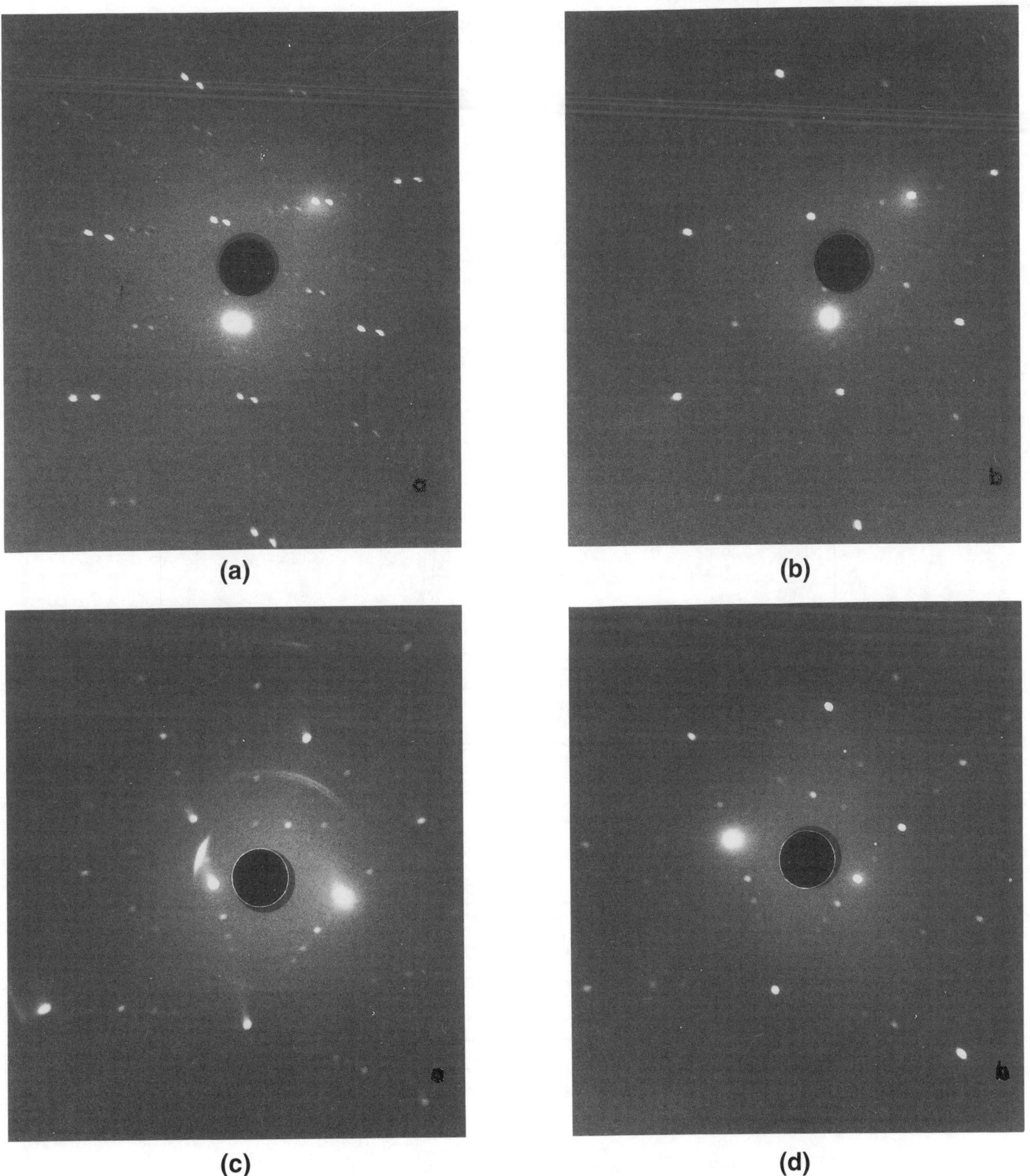

FIG. 10.18. Back-reflection x-ray patterns for (a) a bicrystal of Al, showing double diffraction spots; (b) a single crystal of Al; (c) a deformed surface of Al after 0.25 µm diamond paste polish; (d) the same Al surface, after vibratory polishing a further 15 hours with 0.05 µm Al_2O_3 slurry, and then electropolishing, showing a strain-free surface (from Quenneville, 1979).

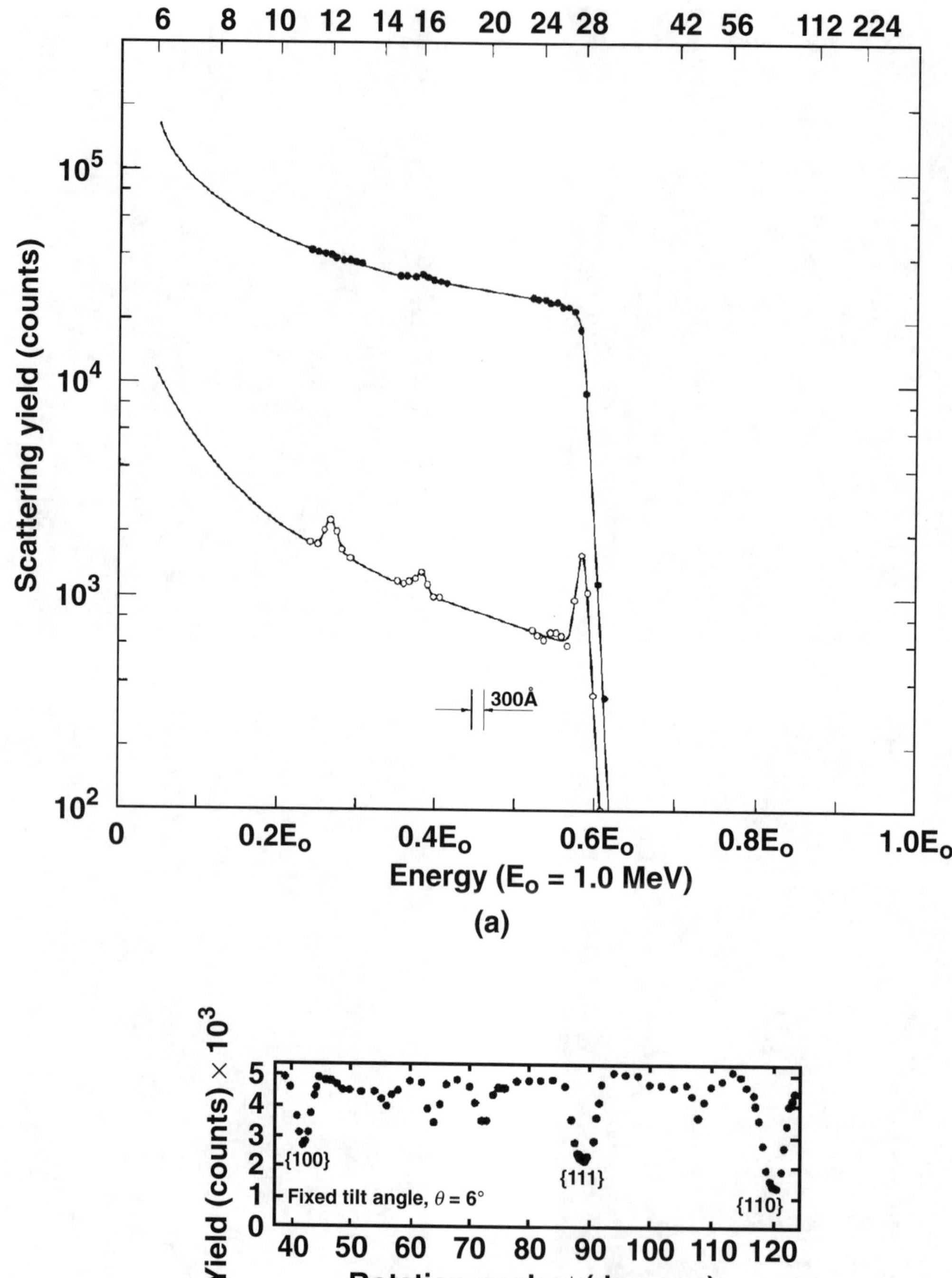

FIG. 10.19. How to align a Si (110) crystal for channeling. (a) RBS spectra for a randomly-oriented crystal (top) and an aligned crystal (bottom). A narrow energy window is shown below the surface peak. (b) RBS yield as a function of the azimuth angle ϕ for a tilt angle of 6°. (c) Polar plot of the planar minima from panel (b). (d) Precise determination of the value of θ for <110> alignment, with ϕ set at 60°.

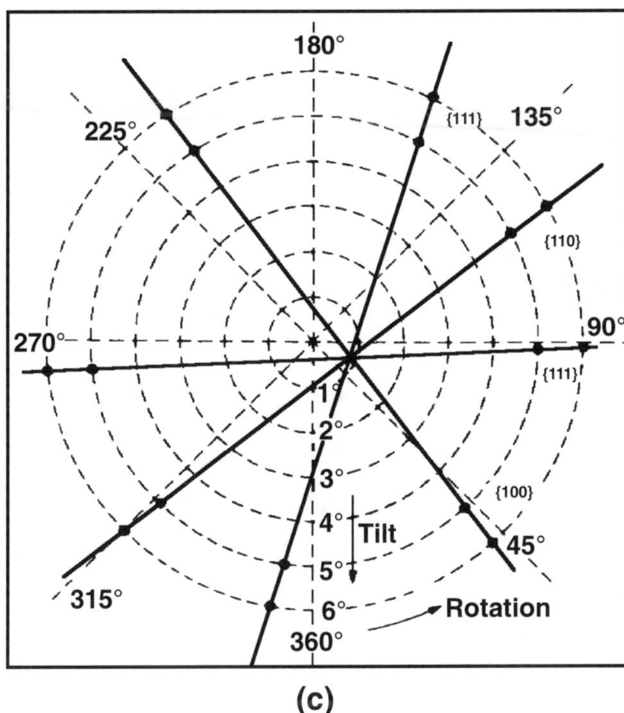

(c)

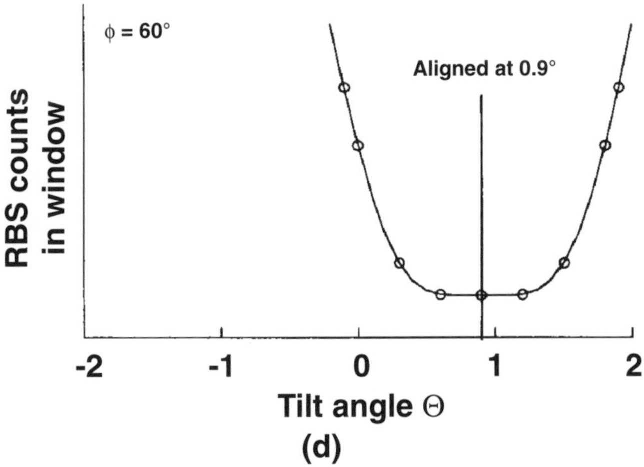

(d)

FIG. 10.19. (continued).

of tilt is given by the known angles between axes in a crystal. For cubic crystals these angles are listed in Table A15.8 in Appendix 15. For example, to find the <100> axis in our (110) Si crystal, the tilt from <110> is 45°. It is of course necessary to tilt along the correct plane joining these axes, which is a {100} plane. This plane can be chosen by the strength of its dip compared with the other planar dips. Note that in diamond cubic crystals, the {100} planes are weaker (give smaller dips) than the {110} planes, whereas the opposite is true for face-centered cubic crystals (see Appendix 15, Fig. A15.6).

Step 4. Find the normalized yield

The normalized yield $\chi^{<uvw>}$ (x) is the ratio of the RBS yield for good <uvw> alignment to the yield for a randomly-oriented crystal as a function of the depth x along the beam path. After an aligned RBS spectrum is recorded for a given integrated current, the sample is tilted about 5° from the aligned tilt, and then rotated continuously to obtain a good random spectrum for the same ion dose. A random spectrum is one for which the ion beam is directed in a random crystallographic direction, which does not exist except for an amorphous material. Since the use of a fixed off-axis alignment normally gives a yield that deviates by 5-10% from the random or average yield [Fig. 10.19(b)], it is usual to rotate the crystal to get an accurate random spectrum.

Finally, the normalized yield can be obtained by dividing the aligned yield by the random yield and converting the energy scale to a depth scale. Problems with the possible difference of depth scale for aligned and random spectra were discussed in Section 10.3.5. Problems with current integration are discussed in Chapters 11 and 12. Normally, the beam intensity (ion flux) should be reduced for the random spectrum in order to avoid pulse pile-up or dead time, since the backscattering yield is much greater for random alignment.

The alignment procedure outlined in steps 1–3 can be done manually or using alignment programs with computer-controlled goniometers (Section 10.3.1). The alignment can also be done by a different quick manual method by rotating the sample continuously and observing the planar dips on a ratiometer, which displays the ratio of the number of counts in the energy window to the beam current. If the beam current is very stable, it is sufficient to observe the RBS counts in the energy window. This method is often the best in complicated cases where the crystal orientation is uncertain because both the tilt and azimuth angles can be adjusted to follow a given planar dip. This avoids the problem of the false assignment of different individual dips in yield with a given plane.

Step 5. How to do angular scans

An angular scan through a planar channel is best done by traversing the plane at right angles, and it should be done at an angle of about 5° from a major axis in order to avoid high index crystallographic directions. For example, to scan through the $(0\bar{1}1)$ plane in a <100>-oriented Si crystal, the crystal would first be aligned with the [011] axis, then rotated about 5° from that axis along the $(0\bar{1}1)$ plane, and then tilted to scan across the plane, as shown in Fig. 10.20(a). Before doing the scan, it is best to measure the $(0\bar{1}1)$ planar yield for several positions near 5° from the [011] axis in order to find a flat region in the planar yield.

Axial scans can be done in several ways: (1) along planar channels, (2) along arbitrary (non-planar) paths, and (3) along a random path, by rotating about the axis for each angular increment. Computer programs must be used for this approach.

For comparison with calculations, methods (1) and (3) are preferable. To scan through an axis along a plane it is sufficient to note the θ and ϕ settings for the axial alignment and for two points in the plane on either side of the axis, then interpolate the intermediate θ and ϕ settings for the scan. If the alignment is slightly off the axial setting, the plane will be encountered on one side of the axis, giving an asymmetric scan [Fig. 10.20(b)].

10.5 EPITAXIAL LAYERS AND STRAIN

The quality of epitaxial layers and the strain in such layers or in multilayers can be determined by channeling measurements at normal incidence or at oblique angles. In general, good epitaxy can only be achieved when the lattice parameters of the substrate and overlayer are almost the same. For a small lattice parameter mismatch, perfect epitaxy can be accommodated by vertical elastic strain in the overlayer. When the stress in the strained epilayer exceeds the elastic modulus, the strain will be accommodated by the creation of "misfit" dislocations.

10.5.1 Elemental layers and strained layer superlattices

10.5.1.1 Elemental layers

A good example of the use of channeling to study the initial registry of epitaxial layers on the substrate atoms was given by the evaporation of Au onto (111)Ag [Fig. 10.4(b), Culbertson *et al.* (1981)].

An example of epitaxial layers with a considerable lattice mismatch is the Ge/Si system, where the lattice parameter of Ge is 4.4% greater than that of Si. The

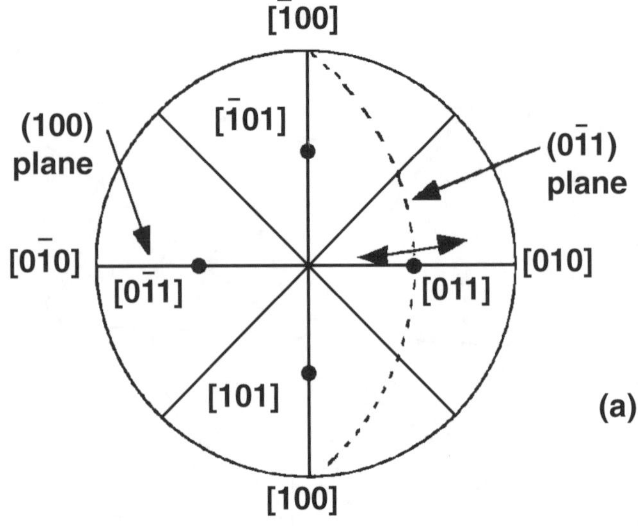

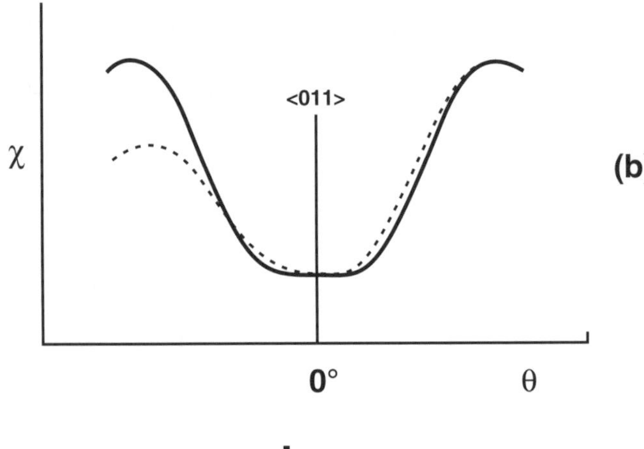

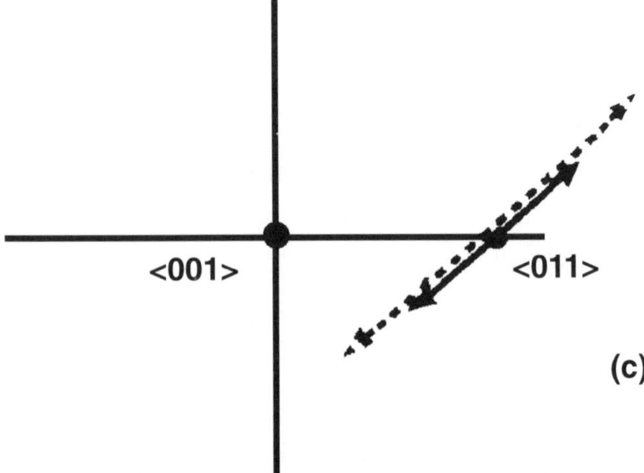

FIG. 10.20. (a) Illustration of an angular scan path for a $(0\bar{1}1)$ planar scan; (b) symmetric axial angular scan and asymmetric scan caused by a slight misalignment; and (c) scan paths for symmetric (—) and asymmetric (- - -) axial scans.

epitaxy of Ge on Si is easily measured by RBS. It is predicted theoretically that only about six monolayers of (100) Ge could be grown on Si without creation of mismatch dislocations or twins. However, since growth, for example, via molecular beam epitaxy (MBE) is not an equilibrium process, considerable latitude exists in the film quality and the critical thickness for creation of mismatch dislocations.

A thin epitaxial overlayer of Ge on (100) Si with a capping layer of Si is shown schematically in Fig. 10.21 (Feldman *et al.*, 1987). For a lattice match in planes parallel to the surface, a considerable strain is introduced in the normal direction. In the case of a (100) surface, this strain should not create an increase in $\chi_{Ge}^{<100>}$, but does cause Ge atoms to project into <110> channels at 45° to the surface. This will produce some ion beam bending in these channels as the ions are steered into the Ge <110> channels. However, for very thin layers (Fig. 10.21), the projection of Ge atoms into the paths of ions that have been well channeled by the Si overlayer causes unusual asymmetric channeling effects (Fig. 10.22). An ion beam directed at an angle slightly greater than that required for <110> alignment (>45°) is steered more towards the displaced Ge atoms giving a large Ge yield, while a beam at <45° is steered away from the displaced Ge atoms, giving a lower than normal Ge yield in the shoulder region. As control experiments, Fig. 10.22(c) shows a <100> angular scan, which gives a symmetric scan, as expected, and Fig. 10.22(d) shows the <110> angular scan for 100 monolayers of a $Si_{0.99}Ge_{0.01}$ epitaxial film, which shows very little strain, again as expected.

Computer simulations of the asymmetric angular scans of Figs. 10.22(a) and (b) give good agreement when the perpendicular displacement of Ge atoms in the *n*th layer of Ge at an off-normal angle θ is given by

$$\Delta x_n = \frac{(n - 1/2)(d_{o\perp} - d_{oSi})\sin\theta}{4} \qquad (10.21)$$

where d_{oSi} is the Si lattice constant and

$$d_{o\perp} = d_{oSi} + \frac{(1+\upsilon)(d_{oGe} - d_{oSi})}{1-\upsilon} \qquad (10.22)$$

is the perpendicular lattice constant of the Ge. This equation follows from application of Poisson's equation. The Poisson ratio is $\upsilon = 0.273$ for Ge.

Transmission channeling is also a very useful way to study thin epitaxial layers; for example a few monolayers of Ge on Si (Lyman *et al.*, 1991).

10.5.1.2 Strained-layer superlattices

Epitaxial multilayer structures, consisting of alternating layers of materials with a small lattice mismatch, are called strained-layer superlattices if the layers are thin enough that the lattice mismatch is accommodated by uniform elastic strains. In practice, for mismatches of ~1%, the layer thicknesses can be up to a few tens of nanometers. For thicker layer or greater mismatches, the strain creates dislocations. Typical strained layer superlattice materials are $GaAs_xP_{1-x}$/GaP, $Al_xGa_{1-x}As$/GaAs, $In_xGa_{1-x}As$/GaAs, Si_xGe_{1-x}/Si, and AlSb/GaSb. These materials exhibit interesting quantum well effects.

Channeling measurements can be used to measure the onset of dislocation creation at a critical layer thickness by measurement of the normalized yield at normal incidence. In addition, the tilt angle $\Delta\psi$ at each interface between layers, or the lattice strain, can be determined using channeling measurements for axes inclined to the surface, as illustrated in Fig. 10.23(a) (Picraux *et al.*, 1983, 1986). Since alignment for each layer is different, measurement of the angular difference for alignment in each layer gives a measure of the tilt angle. However, because the ion beam is steered by the first layer, such measurements of the shift in channeling angle using only the first and second layers gives a value smaller than $\Delta\psi$, especially for planar channels. A better measurement is obtained by comparing the alignment of the first layer with the average alignment from several deeper layers or with the substrate alignment.

An example of the influence of strain in such superlattices is shown in Figs. 10.23(b) and (c) for an $In_xGa_{1-x}As$/GaAs superlattice. Fig 10.23(b) shows the RBS data for random alignment and for [100] and [110] alignment. Because of the steering of the ion beam as it passes from one layer to the next for non-normal incidence, the dechanneling along the [110] axis is much greater than along the [100] axis (normal to the surface). However, it is difficult to calculate the strain from such measurements because it is dependent on many variables; e.g., the thickness of the layers relative to the wave length of the channeled ion's depth oscillations (Fig. 10.9).

Fig 10.23(c) shows the [110] angular scan data for the first layer (using the As yield), and for the fifth plus sixth layers (using the Ga and As yields, respectively). The angle between the first layer and the average of layers 5 + 6 equals $\Delta\psi/2$. This measurement gives $\Delta\psi = 0.64°$, which is close to that calculated for this superlattice.

A further, more sophisticated measure of $\Delta\psi$ is obtained by measurements of "catastrophic dechanneling,"

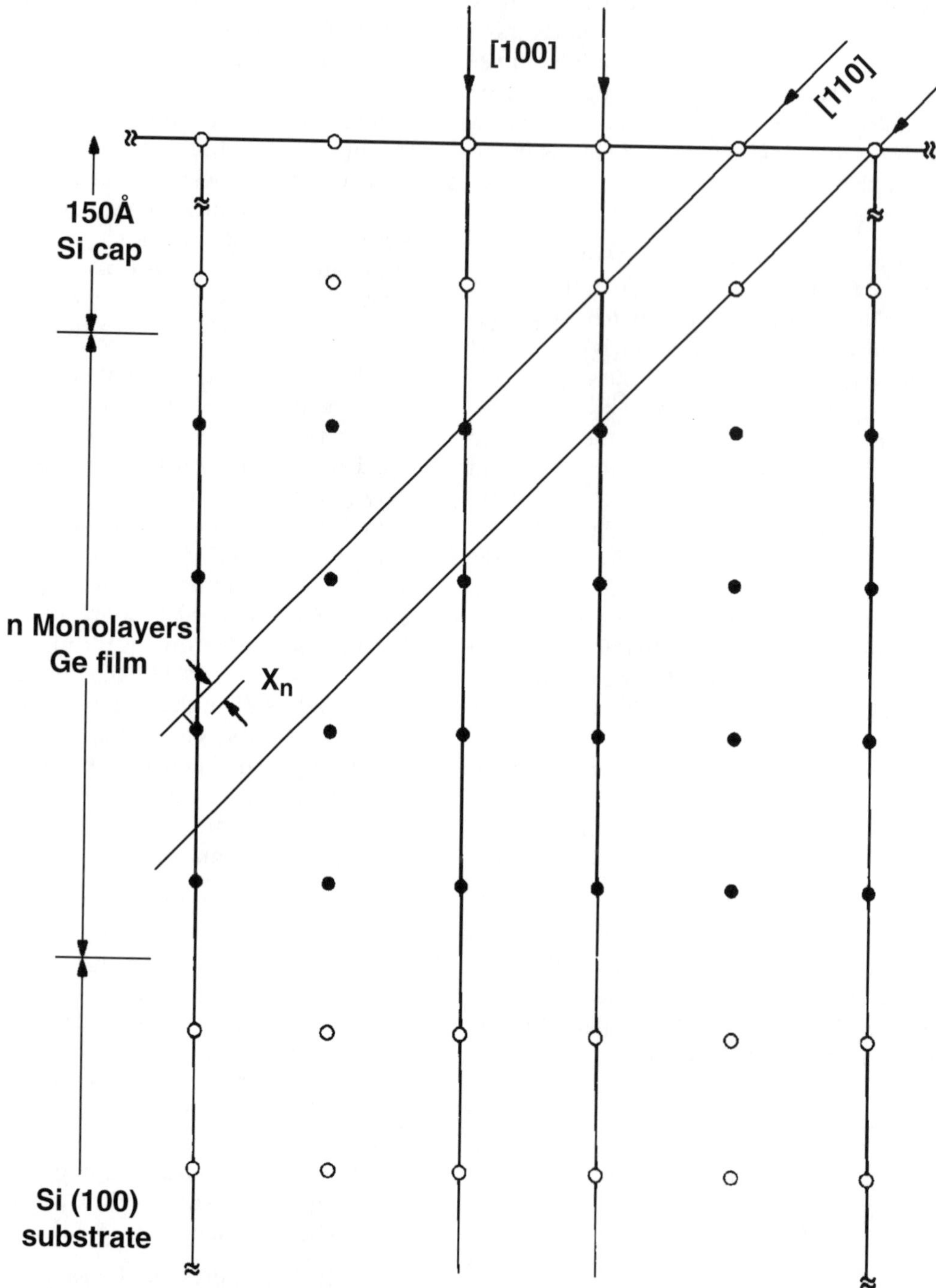

FIG. 10.21. Schematic diagram of the structure of a strained epitaxial film of Ge capped with the substrate material, Si. The channeling directions [100] and [110] are shown. The displacements x_n from substitutional sites of the Ge atoms are indicated (Feldman *et al.*, 1987).

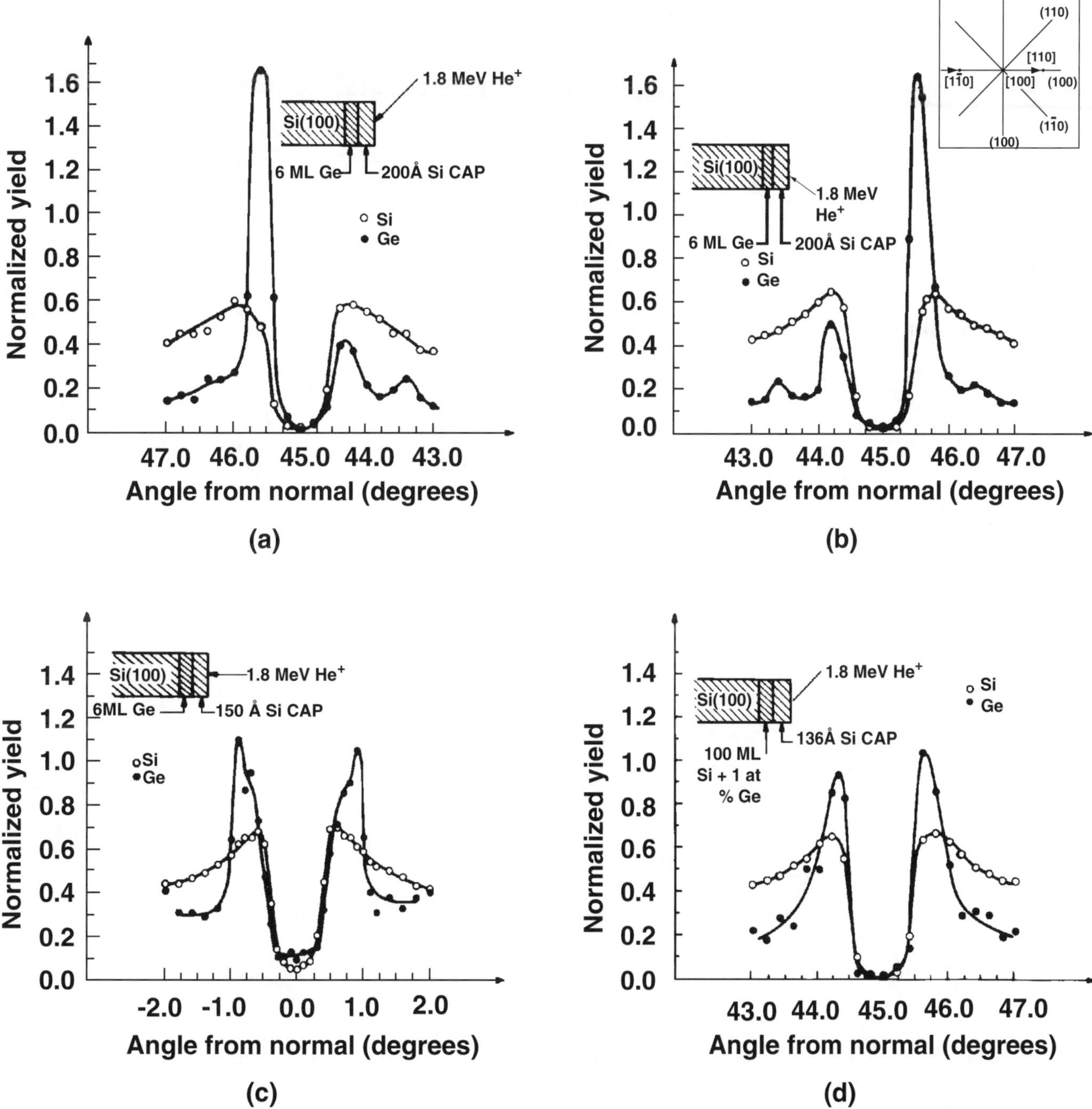

FIG. 10.22. Channeling angular scan of Ge/Si structures, using 1.8 MeV He ions (Feldman *et al.*, 1987). (a) Scan through the [1$\bar{1}$0] axial direction in the (100) plane for an epitaxial structure consisting of Si(100)/Ge (6ML)/Si (20nm); (b) scan through the [110] axial direction in the (100) plane for the same sample as in (a); (c) scan through the [100] (crystal normal) channeling direction for the same sample as in (a); (d) scan through the [1$\bar{1}$0] channeling direction in a sample of Si(100)/$Si_{0.99}Ge_{0.01}$ (100ML)/Si (13.6 nm). The stereogram in the inset indicates the direction of the angular scan for (a) and (b).

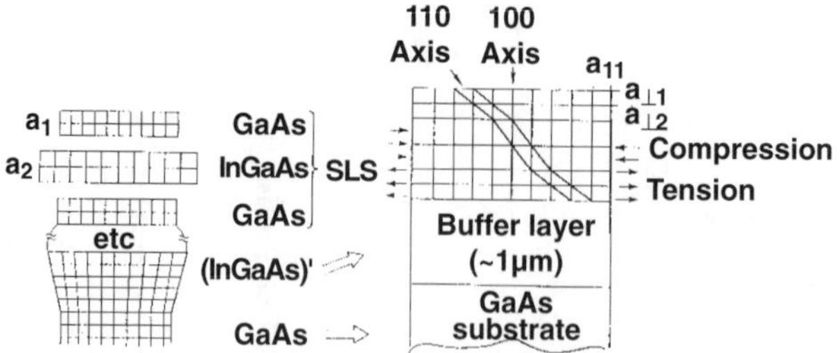

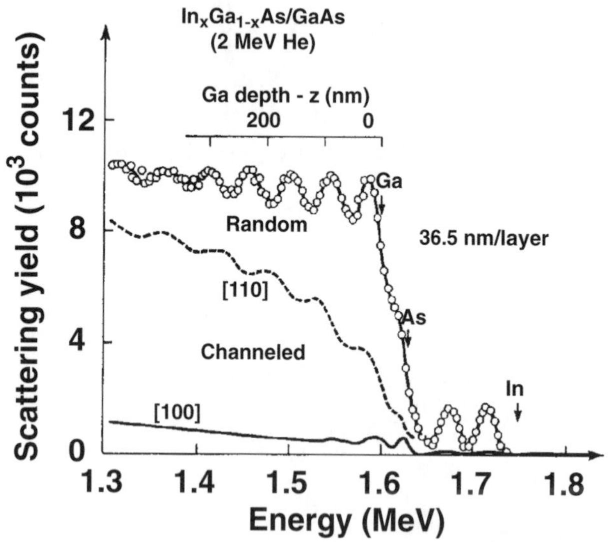

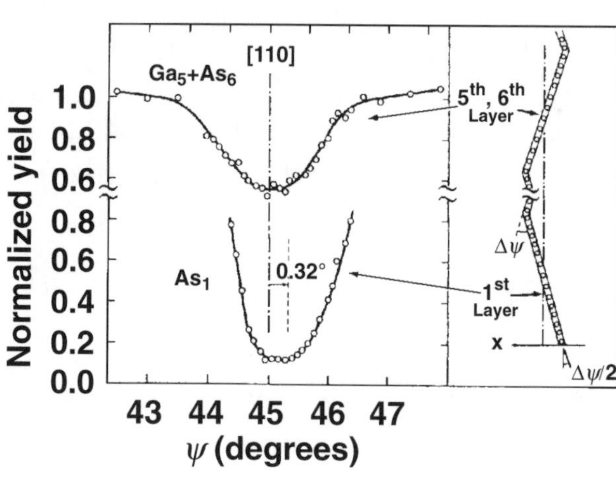

FIG. 10.23. Channeling in superlattices. (a) Illustration of a superlattice with channeling beam paths. (b) 2 MeV He axial channeling results for [110] and [100] dechanneling in an $In_{0.22}Ga_{0.78}As$ (36.5nm)/GaAs (36.5nm) strained layer superlattice (SLS). (c) Angular [110] scans in an $In_{0.15}Ga_{0.85}As$ (38nm)/GaAs (38nm) SLS with the smaller lattice constant (GaAs) material corresponding to the top layer. The RBS yields for the first layer (As scattering signal), and the fifth plus sixth layers (Ga and As signals, respectively) are shown. The angle ψ is measured with respect to the [100] axis which is normal to the surface. A schematic representation of [110] rows for each layer is also shown (from Picraux *et al.*, 1983, 1986).

which is caused by a resonance between the wavelength of the planar channeled trajectory and the superlattice period (Ellison *et al.*, 1988). It results from the channeled ions in the first layer being steered into the walls of the second layer. Thus an ion beam directed slightly to one side of the channeling alignment is strongly dechanneled, while a beam directed to the other side experiences resonant channeling.

10.5.2 Compound layers

A convenient way to measure the quality of compound epitaxial layers (pseudomorphic films), whether on the surface or buried, is via channeling. In the case of silicide layers on Si, which are often used in semiconductor device processing, the silicide layers are usually produced by depositing a thin layer of the desired metallic component, for example Ni, on the Si surface and then heating to form the desired alloy. The rate of evolution of the compound formation and the uniformity of the stoichiometry can be determined simultaneously with the quality of the epitaxy.

An example of silicide epitaxy is shown in Fig. 10.24 for a 72 nm layer of $NiSi_2$ on (111) Si (Chiu *et al.*, 1980). The ratio of <111> to random yields indicates $\chi_{Ni}^{<111>} = 0.04$, a value close to that expected for a perfect $NiSi_2$ crystal. The random spectrum also gives the stoichiometry of the $NiSi_2$ from the ratio of the yield from Si in the plateau region to the yield from Ni atoms in the large peak. Note that $(Z_{Ni}/Z_{Si})^2 = 4$, so that for the $NiSi_2$ stoichiometry the ratio of Ni yield to Si yield in the random spectrum should be approximately 2, as observed (see Chapter 4).

The absence of disorder at the interface between the silicide and the Si substrate is also indicated by the absence of a peak in the Si <111> yield at the interface (near 1.1 MeV).

Buried layers can also be studied in the same way.

10.5.3 Mosaic spread

Crystals are normally imperfect. They contain not only point defects, impurities, and dislocations, but also regions having slightly differing orientations. These regions can be considered as a mosaic structure of crystallites separated by low-angle grain boundaries. Mosaic spread refers to the angular distribution of crystallite alignment. The mosaic spread in a crystal can be determined by measurements of the energy dependence of the normalized yield χ or the half width $\Psi_{1/2}$, assuming a Gaussian distribution of the crystallite orientations. The yield at a given incident energy E is given by (Ishiwara and Furukawa, 1976)

$$\chi(\psi, E) = 1 - \left(\frac{1 - \chi_0}{1 + \sigma^2 / \Delta_0^2} \right) \exp\left(\frac{-\psi^2}{\sigma^2 + \Delta_0^2} \right) \qquad (10.23)$$

where ψ is the tilt angle of the beam from the most preferred axis in the polycrystals, σ is the standard deviation of the spread in crystallite orientation, and χ_o is the minimum yield as $\sigma \to 0$, a measure of the density of scattering centers. Δ_0 is the standard deviation of the angular yield in single crystals, which is known to satisfy the relation

$$\Delta_0(E) = C_1(Z_1/E)^{1/2} \qquad (10.24)$$

where C_1 is a constant. Fig. 10.25 shows some results for the angular dependence of the yield of 1.0- and 2.0-MeV $^4He^+$ ions backscattered from a PtSi film. From Eqs. (10.23) and (10.24), the energy dependence of $\psi_{1/2}$ may be expressed as

$$\psi_{1/2}^2 = \ln 2\left(\Delta_0^2 + \sigma^2\right) = \sigma^2 \ln 2(1 + E_c / E) \qquad (10.25)$$

where *ln* 2 is the conversion factor of the standard deviation to the half-width at half-maximum value and $E_c = C_1 Z_1/\sigma^2$. Using Eq. (10.25), the standard deviation σ can be obtained from the plot of $\psi_{1/2}$ versus Z_1/E, as shown in Fig. 10.25 for PtSi films on {111} Si. The estimated values of σ for the samples A and B (annealed for 2 h at 400°C and 750°C, respectively) were 0.87° and 0.59°, respectively. Information about χ_o and σ can also be obtained from the energy dependence of χ

$$(1 - \chi)^{-1} = (1 - \chi_0)^{-1} (1 + E/E_c) \ . \qquad (10.26)$$

10.6 DAMAGE MEASUREMENTS

10.6.1 Basics

Damage to a crystal involves displaced lattice atoms, and broadly includes strain, point defects, dislocations, stacking faults, twins, defect clusters, small precipitates, and amorphous regions. Damage produced by an ion beam consists of simple Frenkel pairs (a vacant lattice site plus the self-interstitial atom ejected from that site), and clusters of vacancies and self-interstitials, some of which may condense to form voids or dislocation loops. High concentrations of defects caused by ion irradiation can produce amorphous regions.

As outlined in Section 10.1 and Fig. 10.3, displaced atoms can affect channeling in two qualitative ways: by dechanneling (gradual deflection of the channeled ions out of channels) or by direct backscattering of channeled ions. Of course in practice a combination of these two processes usually occurs. Limiting cases are (1) "distortion"-type defects (e.g., a high density of dislocations), which cause mostly dechanneling because they involve small displacements of host atoms into the channels; and (2) "obstruction"-type defects (e.g., amorphous regions), which cause mostly direct backscattering because they occupy central positions in the channels.

Lattice vibrations can be considered to be another example of distortion-type defects and it should be noted

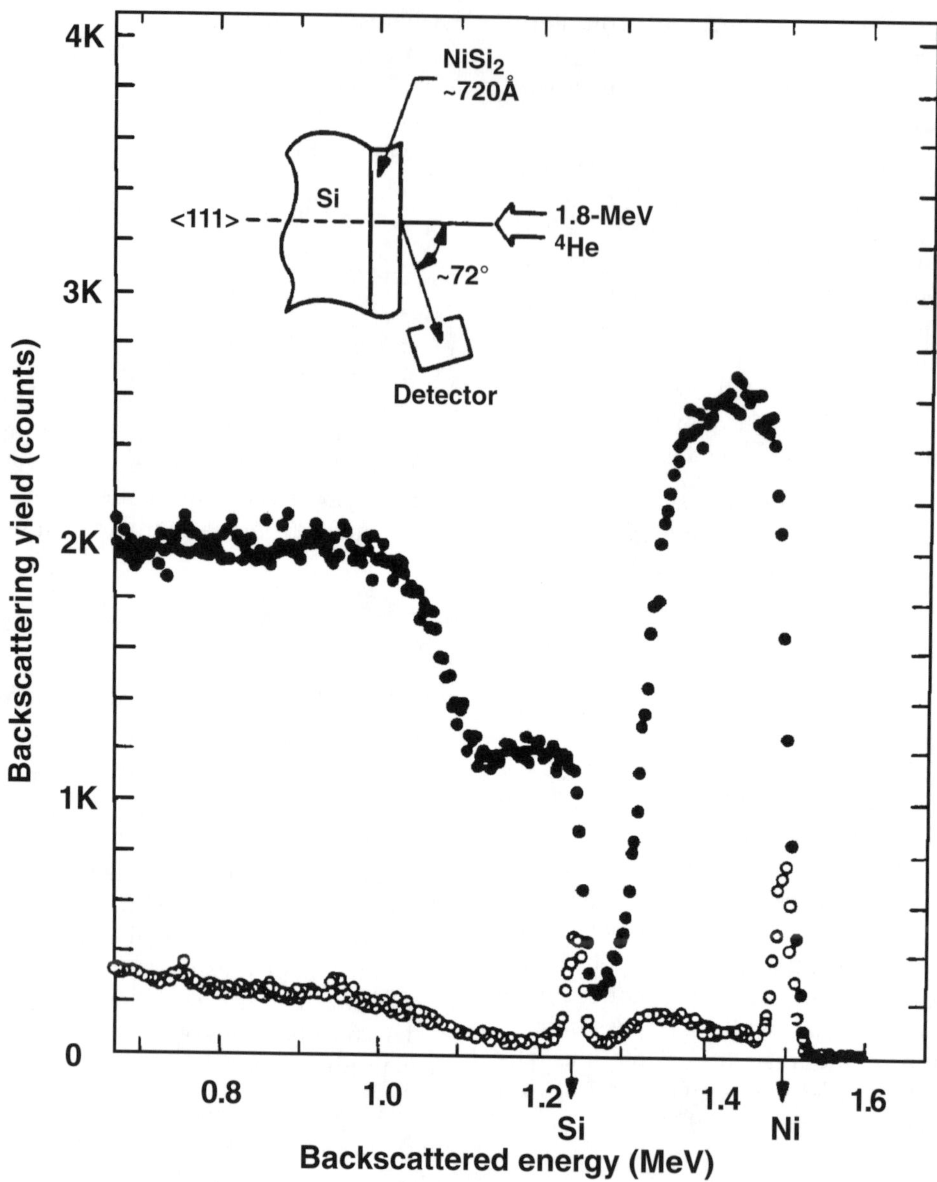

FIG. 10.24. Random and <111> channeling spectra for a 72 nm $NiSi_2$ epitaxial film on a (111) Si substrate (from Chiu *et al.*, 1980).

that the amount of dechanneling caused by dislocations or point defects is dependent on the temperature of the sample; that is, the amounts of dechanneling caused by defects and by thermal vibrations are not in general additive (Matsunami and Itoh, 1973; Howe *et al.*, 1976).

The effect of lattice defects on channeling can be described most simply by considering that the ion beam consists of a channeled fraction which is backscattered only from displaced atoms, and a dechanneled fraction which is backscattered from all atoms in the same way as a randomly directed beam (Bøgh, 1968; Merkle *et al.*, 1973; Feldman *et al.*, 1982; Swanson, 1982). Channeled ions gradually become dechanneled (deflected out of the channel) by multiple collisions with electrons and with displaced atoms, including thermally vibrating atoms and lattice defects (Gemmell, 1974). In backscattering geometry, the normalized yield χ_h is equal to the dechanneled fraction of the beam plus the yield arising

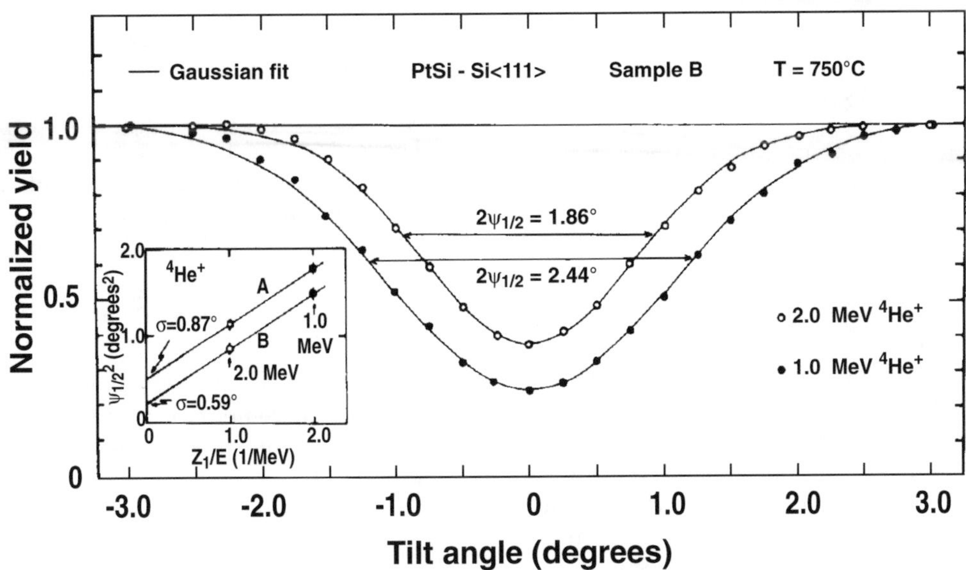

FIG. 10.25. Angular dependence of the backscattering yield of He ions around a <111> axis in PtSi films 28 nm thick (annealed at 750°C). The solid lines represent a Gaussian fit to the experimental data points. Also shown, in the inset, is a plot of $(\Psi_{1/2})^2$ versus Z_1/E (from Ishiwara *et al.*, 1979).

from the backscattering of the channeled ions from displaced atoms. For alignment along a channel *<uvw>*, the normalized yield from host atoms, $\chi_h^{<uvw>}$ can thus be written as

$$\chi_h^{<uvw>} = \chi_{hD}^{<uvw>} + \left(1 - \chi_{hD}^{<uvw>}\right) \times \sum C_{hi} g_i^{<uvw>} F_i^{<uvw>} \quad (10.27)$$

where $\chi_{hD}^{<uvw>}$ is the dechanneled fraction of ions and $\left(1 - \chi_{hD}^{<uvw>}\right)$ the channeled fraction of ions. The total fraction C_h of displaced atoms can be divided into components C_{hi} which are distributed among various sites "*i*" in the channel where the normalized ion flux has values $F_i^{<uvw>}$. (This normalization is with respect only to the channeled part of the beam.) In Eq. (10.27), $g_i^{<uvw>}$ is a geometric factor giving the fraction of each site which projects into the given channel.

Two limiting cases of Eq. (10.27) are shown schematically in Fig. 10.3:

- For small fractions C_h of displaced atoms, only dechanneling is important [Fig. 10.3(b)] and thus

$$\chi_h^{<uvw>} \cong \chi_{hD}^{<uvw>} \quad . \quad (10.28)$$

The methods of calculating the dechanneling component $\chi_{hD}^{<uvw>}$ and relating it to the concentration and cross section of the defects were described briefly in Section 10.2.3.

- If dechanneling is small compared with direct backscattering of the channeled beam from displaced atoms [Fig. 10.3(c)], then

$$\chi_h^{<uvw>} \cong \sum C_{hi} g_i^{<uvw>} F_i^{<uvw>} \quad . \quad (10.29)$$

This situation occurs, for example, when the near-surface region of semiconductors is made amorphous by ion bombardment, in which case $g_i^{<uvw>} F_i^{<uvw>} = 1$. It is usually assumed that for single, randomly-displaced atoms this defect scattering factor $g_i F_i$ is unity. In general, however, detailed knowledge about the spatial distribution of the displaced atoms is required in order to carry out a rigorous analysis. This distribution is usually difficult to obtain, especially for relatively low fractions of displaced host atoms.

In practice, for the case where only dechanneling is important [Eq. (10.28)], it is easiest to estimate defect concentrations at a given depth from Eq. (10.18). An equivalent formulation is (Feldman *et al.*, 1982)

$$N_d(x) = \frac{1}{\sigma_d} \frac{d}{dx} \ln\left(\frac{1 - \chi_v(x)}{1 - \chi_h(x)}\right) \quad (10.30)$$

where it is assumed that the effects of thermal vibrations and defects are independent. From Eq. (10.18), for small defect concentrations, $1 - \chi \cong 1$ and

$$N_d(x) = \frac{1}{\sigma_d}\left(\frac{d\chi}{dx} - \sigma_{th}\right) = \frac{1}{\sigma_d}\left(\frac{d\chi}{dx} - \frac{d\chi_v}{dx}\right) . \quad (10.31)$$

10.6.2 Dechanneling by defects

10.6.2.1 Point defects

An example of the effect of irradiation-induced point defects on dechanneling is given in Fig. 10.26 for an Al-0.08 at. % Ag crystal, where χ is plotted as a function of depth x before and after irradiation with 0.5 MeV He ions at 70 K, and after annealing at 293 K. After irradiation, the slope $d\chi/dx$ was almost constant up to a depth of about 700 nm, indicating an almost uniform concentration of defects, according to Eq. (10.18), for small values of χ_h. Near the end of range of the 0.5 MeV He ions (1300 nm), the damage increased strongly because most of the ion energy was deposited into nuclear collisions there. After the 293 K anneal, the near-surface damage (up to about 1200 nm) recovered completely, indicating that this damage consisted mostly of simple point defects, but the end-of-range damage recovered very little because of the more complex defects produced in that densely damaged layer. The cross section for defect scattering was calculated from the data to be 4×10^{-18} cm^2, which is an order of magnitude larger than the Rutherford cross section.

10.6.2.2 Dislocations

A beautiful example of the dechanneling caused by dislocations was given by Picraux *et al.* (1980). In this experiment, a network of misfit dislocations was created in Si at a well-defined depth by implanting the Si with P to a fluence of 10^{16} cm^{-2} and then annealing at 1323 K to diffuse the P into the crystal. At the tail of the P concentration profile, a regular network of misfit edge dislocations was created because of the lattice contraction in the P-diffused region. The dislocation array is seen clearly in the transmission electron micrograph of Fig. 10.27 (Picraux *et al.*, 1980). The depth of these dislocations (~450 nm) is easily determined from the position of the step in the backscattering spectra for both <111> axial and {110} planar channels (Fig. 10.27). Note that the step is much more pronounced for the planar channel because of the smaller channel half width for planar channels. The rates of dechanneling are similar above and below the dechanneling steps, indicating that the crystalline perfection was similar on either side of the dislocation network.

The total dislocation concentration can be estimated from the magnitude of the step height. In the case of Fig. 10.27, where the dechanneling occurs over only a small depth interval, from Eq. (10.18) the step height is

$$\Delta\chi_h = (1 - \chi_h)N_d\sigma_d\Delta_x \quad (10.32)$$

and the effective total dislocation length $N_d\Delta x$ (where Δx is the depth increment of the dislocation-rich layer) can be estimated from theoretical values of σ_d (Mory and Quéré, 1972).

TEM analysis showed that the dislocations were mainly loops of edge character lying on {111} and {110} planes. The mean dislocation-loop diameter was 27.5 nm and the total projected length of dislocation lines per unit area was 2.4×10^5 cm^{-2}. For dislocation lines the planar dechanneling cross section per unit dislocation length is λ_p (given in units of distance) and is given approximately by (Mory and Quéré, 1972)

$$\lambda_p = \left(\frac{Eb}{\alpha Z_1 Z_2 e^2 N_p}\right)^{1/2} \quad (10.33)$$

where b is the magnitude of the Burgers vector of the dislocation loops, α is a constant dependent on dislocation orientation and type, and N_p is the atomic density in the plane. Assuming negligible direct scattering and the additivity of dechanneling, the density of disorder as a function of x is given by Eq. (10.30) or (10.32) with $\sigma_d = \lambda_p$ and N_d equal to the total projected length of dislocation lines per unit volume at a depth x. Actually, λ_p should be a function of the diameter D of the dislocation loops, and Eq. (10.33) can be considered as an asymptotic value λ_∞ when D tends to infinity (Quéré, 1978).

10.6.2.3 Energy dependence of dechanneling

The energy dependence of dechanneling from different classes of defects should differ markedly. Thus, in principle dechanneling can be used to identify the type of defects, although in most practical cases where a mixture of defect types occurs, the utilization of other techniques such as electron microscopy is required.

The energy dependence is expected to be E° for stacking faults, $E^{1/2}$ for dislocations (see Eq. 10.33), and E^{-1} to

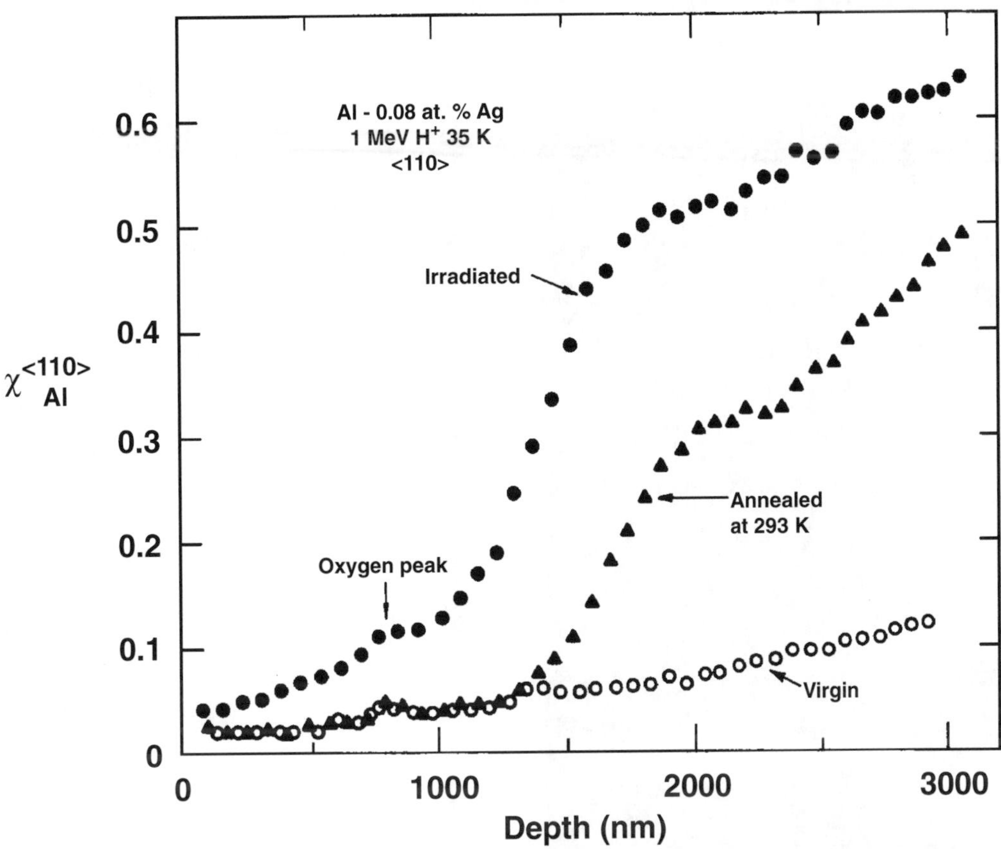

FIG. 10.26. Normalized backscattering yields as a function of depth for 1 MeV H ions aligned along a <110> axis of an Al-0.08 at. % Ag crystal . The effects of an irradiation at 70 K with 0.5 MeV He ions to a fluence of 4×10^{15} cm^{-2}, as well as subsequent annealing for 16 h at 293 K are shown. The depth scale was calculated assuming that aligned and random stopping powers were equal (from Swanson *et al.*, 1982).

$E^{-1/2}$ for point defects and defect clusters (Quéré, 1974). In the simple case of Fig. 10.27, the dechanneling step exhibited the expected $E^{1/2}$ dependence, as shown in Fig. 10.27(c).

10.6.3 Heavily damaged regions

Channeling is commonly used to measure the epitaxial regrowth of semiconductors that are amorphized by ion implantation. A classic example of this is shown in Fig. 10.28, where Si was implanted with Si in order to amorphize the surface layer without introducing foreign atoms. The implantation energy was varied from 50 to 250 keV, which amorphized the crystal to a depth of 460 nm. This depth was determined directly from the energy interval over which the backscattering yield of 2 MeV He ions reached the random yield (the plateau region). A series of furnace anneals caused the amorphous layer to recrystallize epitaxially onto the substrate. The epitaxial growth was linear with time for a (100) surface orientation of the crystal, and is easily measured from the data of Fig. 10.28. The regrowth rate was slower and nonlinear for a crystal having a (111) surface, and considerable residual damage (largely in the form of twinned regions) remained after complete recrystallization.

Ion-induced buried damaged layers have also been measured in metals and semiconductors by channeling in cases where the crystal has not been amorphized, using an analysis based on Eq. (10.27). The variation of damage with depth has been measured and compared with theoretical range distributions of the implanted ions.

An example of such an analysis is shown in Fig. 10.29 for implanted Si (Eisen, 1973). In order to obtain the depth profile of the displaced atoms [i.e., $N_d(x)$ versus x],

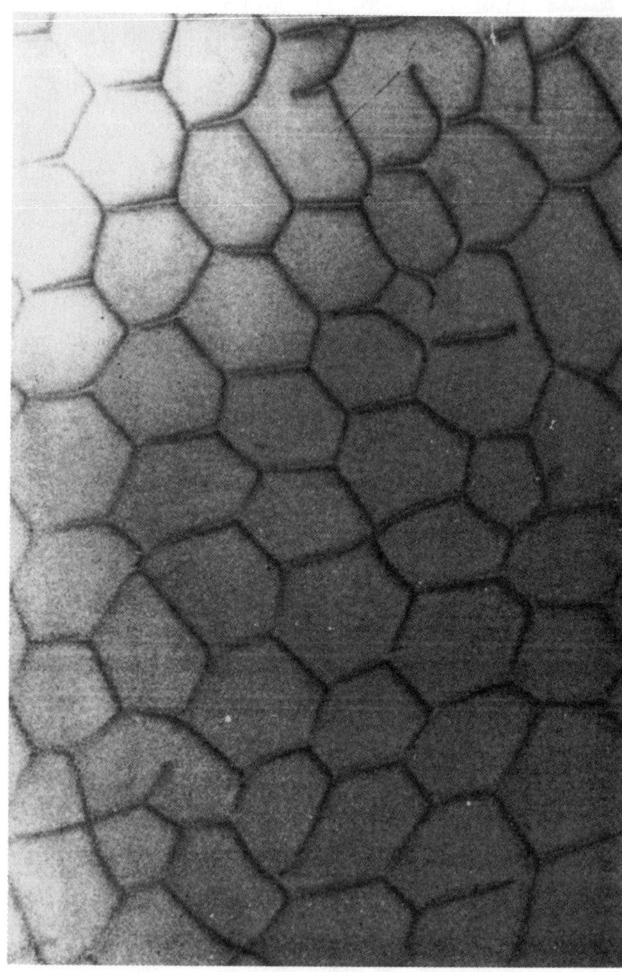

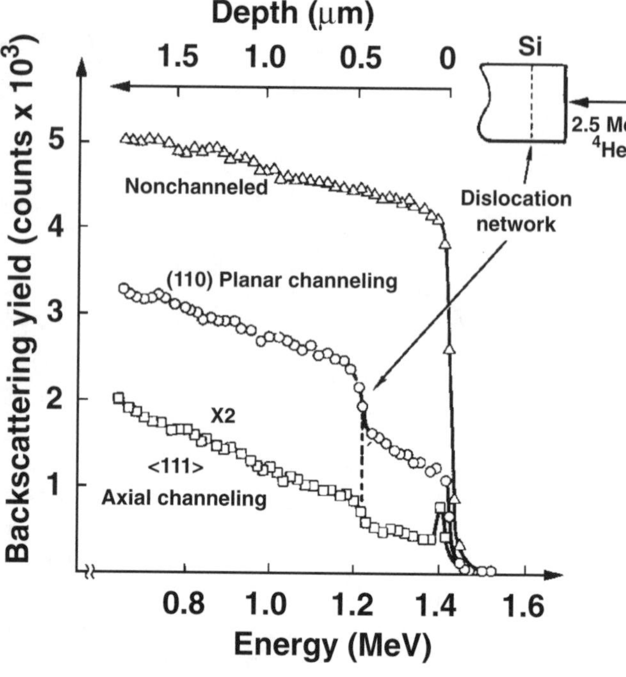

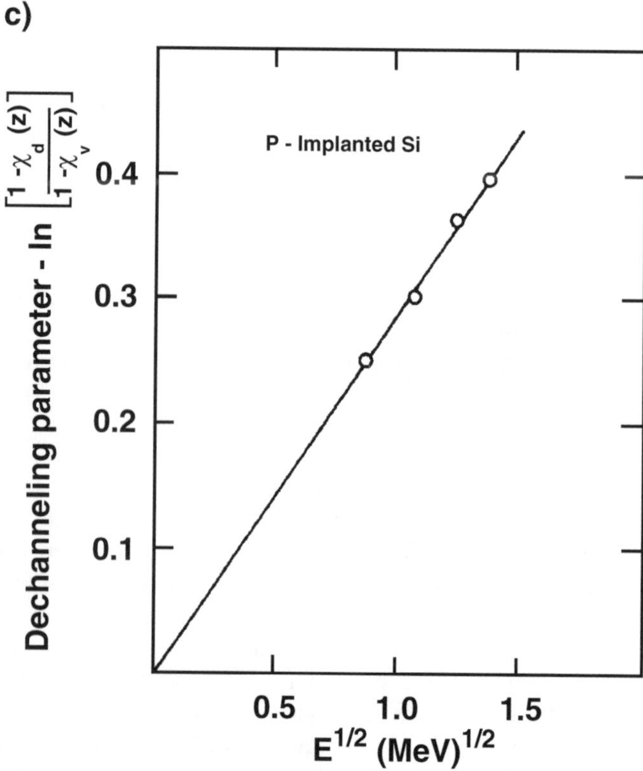

FIG. 10.27. Analysis of phosphorous-diffused Si by (a) transmission electron microscopy; (b) backscattering spectra for 2.5 MeV He ions incident along {110} planar and <111> axial channels; and (c) energy dependence of the planar dechanneling. A misfit dislocation network is observed at a depth of 450 nm (from Picraux *et al.*, 1980).

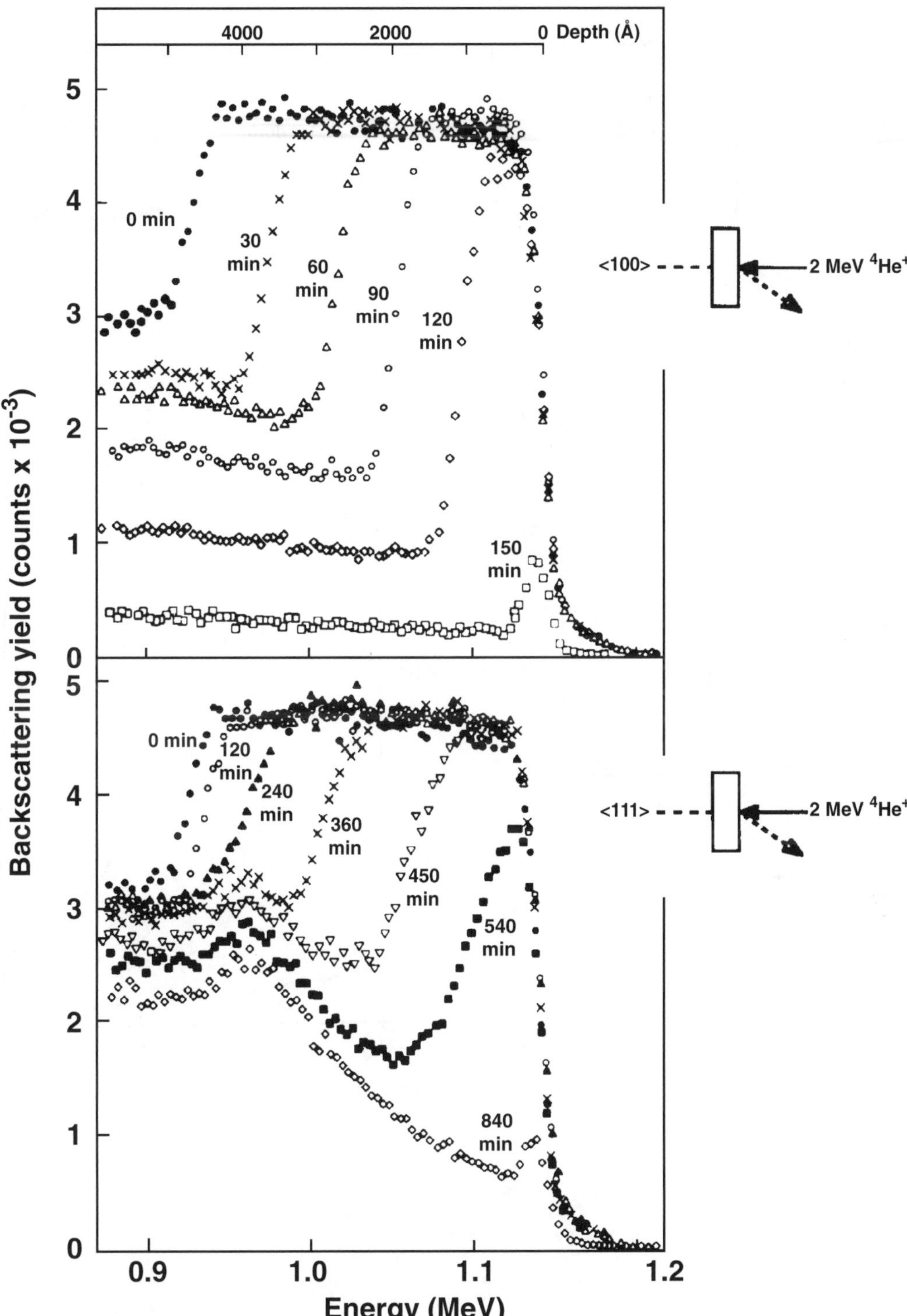

FIG. 10.28. Backscattering spectra for 2 MeV He ions channeled in silicon samples that were implanted at 80 K, preannealed at 673 K for 60 min and annealed at 823 K for the indicated times. The upper portion shows <100> spectra for a sample with the surface normal 0.3° off a <100> axis. The lower portion shows <111> spectra for a sample with the surface normal 0.3° off a <111> axis. The depth scale was calculated assuming the bulk density of Si (from Csepregi *et al.*, 1976).

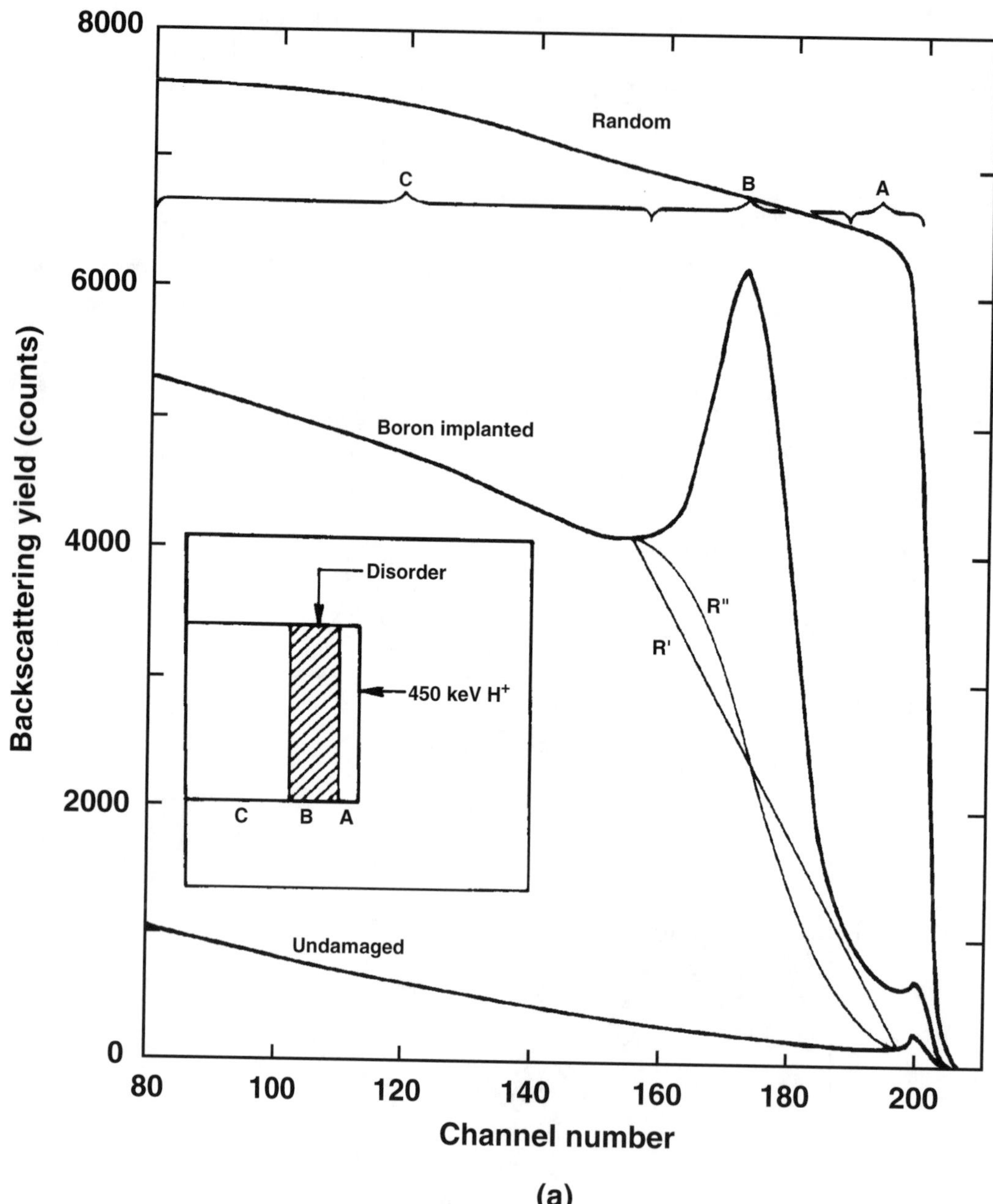

FIG. 10.29. Analysis of ion-induced damage in Si. (a) Aligned <110> and random backscattering spectra for a 450 keV proton beam incident on boron-implanted Si. The spectrum showing a damage peak was obtained from a sample implanted with 8×10^{15} B ions per cm^2. The line labeled R′ is a linear approximation to the dechanneled yield χ_{hD} [Eq. (10.27)], and the curve labeled R″ was calculated using the Keil plural scattering theory. The labeled regions of the spectrum for the B-implanted sample correspond to the regions with the same label in the schematic disorder distribution shown in the inset.
(b) Comparison of disorder distributions calculated using the straight-line approximation (•) and the Keil plural scattering theory (o) (from Eisen, 1973).

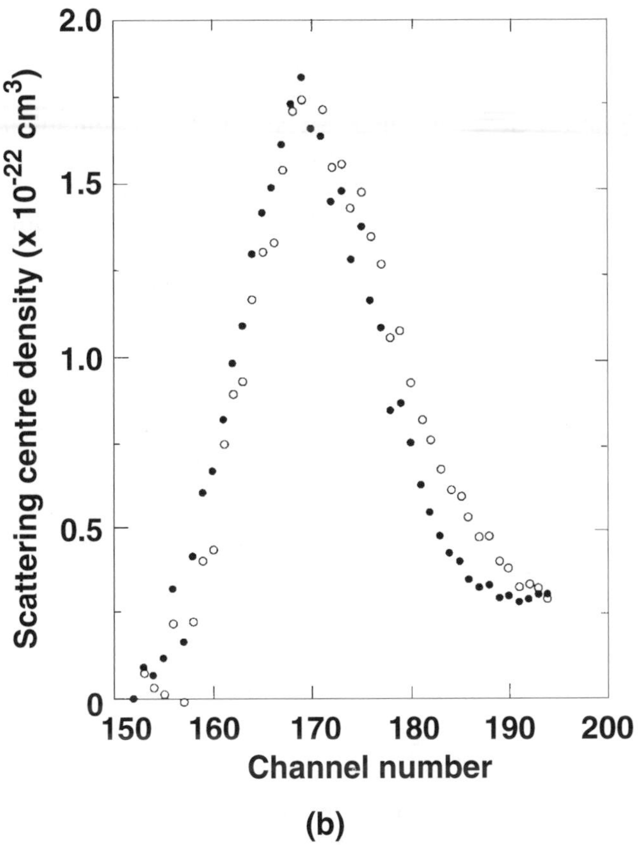

FIG. 10.29. (continued).

it is necessary to determine the dechanneled component χ_{hD} as a function of depth. The analysis for a disordered region situated inside the crystal usually involves using an iterative procedure starting at the surface and moving in succeeding increments of depth to determine the dechanneled fraction χ_{hD} in the next layer, which then enables the direct backscattering contribution from displaced atoms and thus the defect density $N_d(x)$ to be determined at that layer [see Chu *et al.* (1978) and Rimini (1978) for details]. If the damage is located near the surface and there is a well-defined peak in the aligned spectrum, as in Fig. 10.29, then $\chi_{hD}(x)$ can be approximated by drawing a straight line from a point on the aligned curve for the undamaged crystal near the beginning of the damaged region, to a point on the spectrum for the aligned damaged crystal just below the damaged region. The number $N_d\Delta x$ of displaced atoms per cm^2, where Δx is the thickness of the damaged layer, can then be obtained from the area A_d of the disordered peak (in integrated counts) from the relationship

$$\Delta x N_d = A_d \delta E_l H^{-1}[K_{M2}\,\varepsilon(E) + \varepsilon(K_{M2}E)/\cos\theta]^{-1} \qquad (10.34)$$

where δE is the energy width of a channel in the backscattered spectrum, H is the height of the random spectrum (counts per channel) near the crystal surface, K_{M2} is the kinematic factor giving the ratio of the projectile energy after the elastic collision (with a crystal atom of mass M_2) to that before the collision, θ is the angle through which the incident projectile is scattered, and ε is the stopping cross section given by $\varepsilon = (1/N)(dE/dx)$ where N is the atomic density of the crystal and dE/dx is the energy loss per unit path length experienced by the projectile as it penetrates the crystal.

Although single scattering accounts for dechanneling in a reasonable manner for low amounts of disorder, plural or multiple scattering of the channeled ions becomes important as the amount of disorder increases. Fig. 10.29 compares a disorder distribution obtained using the straight-line approximation, as referred to above, to one obtained using a plural scattering treatment developed by Keil *et al.* (1960).

The defect concentration extracted in the above treatments represents an average of the defect distribution weighted by the channeled-ion flux. As discussed in Section 10.2, the channeled beam is not uniformly distributed across the channel between the channel rows. The flux distributions can be changed by tilting the crystal axis to an angle ψ with respect to the beam direction (where $\psi<\psi_{1/2}$) and the variations in the measured yields as a function of ψ may then be related to the distribution across the channel of the displaced atoms [Eq. (10.27), Baeri *et al.*, (1976)].

10.6.4 Lattice vibrations

10.6.4.1 Gas atoms

Carstanjen (1980, 1989) has studied the effect of the measuring temperature on the position of deuterium in Pd. In this case, D was diffused into Pd to a concentration of 0.7 at. % and analyzed by the D(^{3}He,p)^{4}He reaction. The presence of a large peak in yield from D for a <110> axial channel and {111} planar channel, together with complete shadowing for <100> axial and {100} planar channels showed that the D was in octahedral interstitial sites (Fig. A15.12). The vibrational amplitude of D was estimated by measurement of the rate of increase in aligned yield for the <100> channeling dip as the temperature was raised in the range 295-446 K.

10.6.4.2 High T_c superconductors

Measurements of axial <001> channeling in $YBa_2Cu_3O_{7-\delta}$ and $ErBa_2Cu_3O_{7-\delta}$ showed a discontinuous increase in the half width $\psi_{1/2}$ as the temperature was lowered through the transition temperature T_c (= 90 K) (Sharma *et al.*, 1988, 1989), indicating a "stiffening" of the lattice or a reduction in lattice vibrations. In the earlier experiments, which were done with $YBa_2Cu_3O_{7-\delta}$, it was difficult to separate the Cu signal from the Y signal so that it was not known whether this anomalous change in $\psi_{1/2}$ was due to the Y-Ba sublattice or to the Cu-O sublattice. When replacing Y with the heavier Er atoms (Fig. 10.30), it became clear that the change in $\psi_{1/2}$ was due almost entirely to the Cu-O rows.

These results illustrate the power of the RBS-channeling method for direct studies of vibrational properties of crystals. In this case, neutron diffraction studies of the phonon spectra did not show a clear discontinuity at T_c. It can be concluded that the anomaly was due to long wave length phonons (correlated vibrations) which are detected by channeling but not by neutron diffraction.

10.7 LATTICE SITES OF SOLUTE ATOMS

10.7.1 Principles

One of the most often used applications of channeling has been to determine lattice sites of impurity atoms in crystals (Davies, 1973; Picraux, 1975; Swanson, 1982; Feldman *et al.*, 1982). As shown in Fig. 10.1, impurity atoms that project into axial or planar channels are "seen" by the ion beam, causing an increased backscattering yield χ_s from the solute atoms as compared to χ_h from the host atoms. By comparing yields for different channels, the positions of solute atoms can be determined quite accurately using a triangulation procedure. The channeling method is one of the most direct ways of determining actual lattice positions of solute atoms, in contrast to other methods like the Mössbauer effect and electron paramagnetic resonance, which determine the electronic environment of solute atoms rather than giving direct spatial information.

The easiest lattice site to study by channeling is a high symmetry interstitial site such as the tetrahedral interstitial site in a face-centered cubic lattice (Fig. 10.31). This position is the $(1/4, 1/4, 1/4)d_0$ site, where d_0 is the lattice parameter. It lies exactly at the center of <100> channels, is one-half the distance to the center of <110> channels, and is completely shadowed in <111> channels (that is, it lies along <111> strings of atoms). The projection of tetrahedral sites into each of these channels is shown in Fig. 10.31. The normalized yields of backscattered ions (or other close encounter reaction products) from solute atoms which are exclusively in tetrahedral sites will be $\chi_s^{<111>} \cong \chi_h^{<111>}$, $\chi_s^{<110>} \cong 1$, and $\chi_s^{<100>} > 1$. In fact, from <100> angular scans, a clear peak in the yield from solute atoms should be seen at perfect <100> alignment because the ion flux is high at the center of the channel.

The angular scans of backscattering yields show characteristic profiles when solute atoms project various distances into the channels (Davies, 1973; Gemmell, 1974; Feldman *et al.*, 1982; Swanson, 1982), as shown schematically for a low-index axial channel in Fig. 10.32.

- (a) For solute atoms in substitutional lattice sites, the yields from solute and host atoms are identical for all channels (assuming the same vibrational amplitude).
- (b) For solute atoms projecting a few hundredths of nanometers into a channel, the dip in yield from solute atoms is narrower than that from host atoms.
- (c) For large projections (~0.1 nm), small double peaks or a small central peak may be observed.
- (d) For displacements into the center of a channel, a large, narrow peak in yield from solute atoms is seen.
- (e) For randomly located solute atoms, a flat solute profile is observed.

Combinations of substitutional plus small displacements or substitutional plus interstitial sites are also shown in Fig. 10.32. For high-symmetry interstitial positions, the effect of the crystal symmetry is to give profiles of types (a), (c), and (d) for different channels — such an interstitial position often lies in the center of one set of channels, off-center in another set, and completely shadowed in another set (see Fig. 10.31). Lower-symmetry positions project into all channels. The projections of tetrahedral and octahedral interstitial sites into major channels for bcc, fcc, diamond-cubic, and hcp structures are shown in Appendix 15. For example, the octahedral interstitial site in an hcp structure would show profiles of type (a) for $\{11\bar{2}0\}$ planar channels, type (c) for $<10\bar{1}0>$ axial channels, and type (d) for <0001>, $<11\bar{2}0>$, {0001}, and $\{10\bar{1}0\}$ channels.

The general procedure for locating solute atoms is to determine the type of displacement that predominates, using selected angular-scan data of the type shown in Fig. 10.32, and then to compare the channeling data with analytical or computer calculations to determine precise

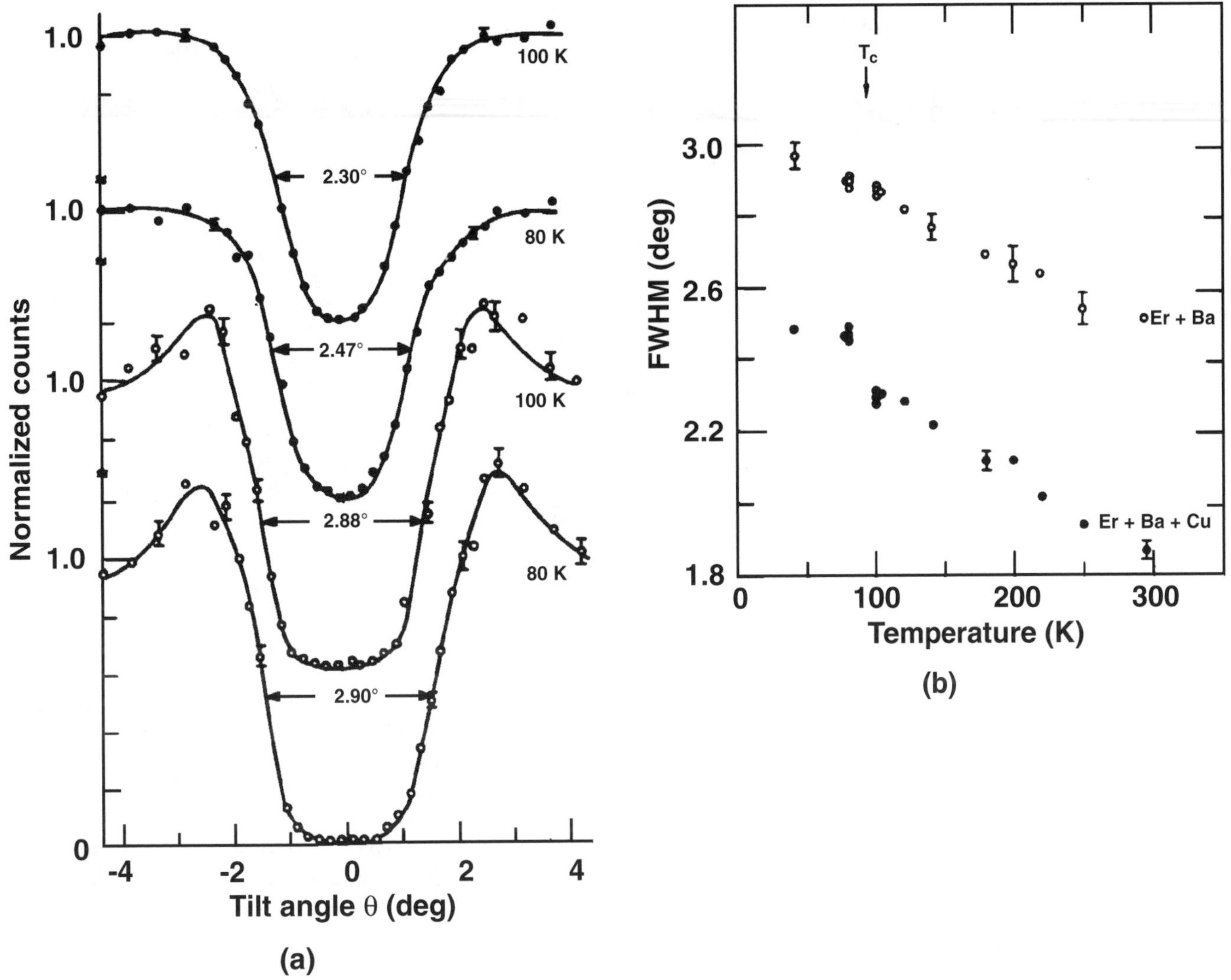

FIG. 10.30. Change in channel width at the superconducting transition temperature of $ErBa_2Cu_3O_{7-d}$. (a) [001] axial channeling scans obtained from the combined Er-Ba-Cu RBS yields (closed symbols), and from the Er-Ba yields (open symbols) at 80 K and 100 K. (b) Full width at half maximum (FWHM) for the axial scans of (a) (from Sharma *et al.*, 1989).

lattice positions. Symmetry considerations dictate how much information of this type is required. A considerable number of scans may be necessary when a low-symmetry position or several different solute atom positions occur. However, if the solute atoms are in a high-symmetry position, channeling data along only two or three axes usually are sufficient to determine the specific position (Fig. 10.31).

To determine exact lattice sites of solute atoms, it is necessary to compare angular scans of the type shown in Fig. 10.32 with those calculated from flux contours (Fig. 10.8). Analytical calculations of <110> angular scans obtained from the results of Fig. 10.8 are shown in Fig. 10.33 for various displacements of solute atoms into the <110> channel of Al (Matsunami *et al.*, 1978). Displacement of 0.2 nm (2Å) along <100> corresponding to the tetrahedral site of Fig. 10.31 gives a peak to valley ratio of 2.2 in the <100> angular scan. When unique, high-symmetry positions are involved, it is possible in this way to specify the position of the solute atom to less than 0.01 nm. If a variety of sites occurs, this procedure inevitably leads to a variety of possible combinations of lattice sites, and considerable model fitting may be

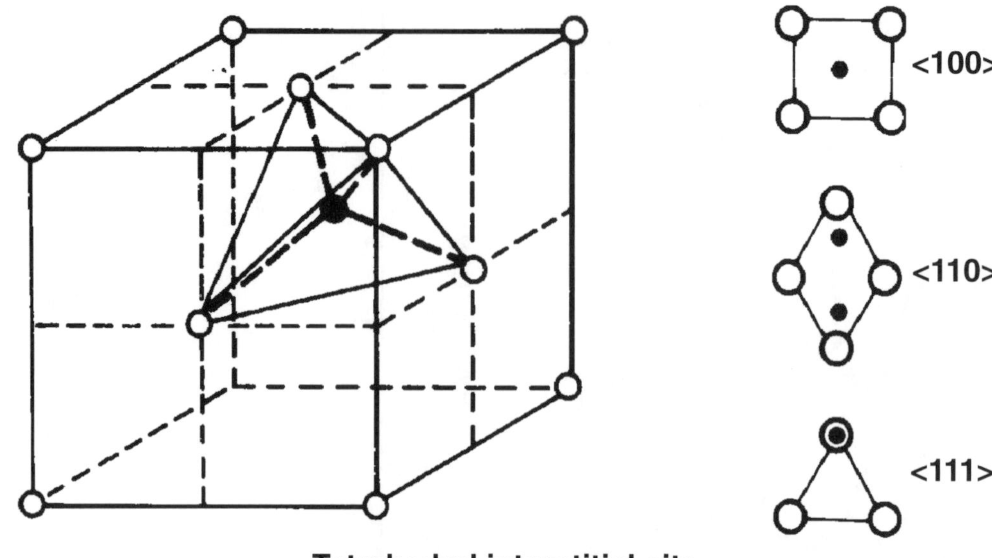

FIG. 10.31. Tetrahedral interstitial (•) in a face-centered-cubic structure. Projections of this site into <100>, <110>, and <111> channels are shown (from Swanson, 1982).

required. In such cases, the physical insight gained by performing a systematic series of experiments to alter the lattice sites is helpful.

Both qualitative and quantitative information on lattice location of solute atoms is obtained from measurements of the normalized yields $\chi_h^{<uvw>}$ from host atoms and $\chi_s^{<uvw>}$ from solute atoms for perfect alignment along various <uvw> channels. A useful empirical quantity from such measurements is the apparent fraction of solute atoms $f_{ds}^{<uvw>}$ that are displaced from lattice sites into the given channel (Swanson, 1982)

$$f_{ds}^{<uvw>} = \frac{\chi_s^{<uvw>} - \chi_h^{<uvw>}}{1 - \chi_h^{<uvw>}} . \tag{10.35}$$

This would be the actual fraction of solute atoms that project into the channel if the (normalized) ion flux at the solute atom's position were unity. However, because the ion flux varies greatly across a channel (Fig. 10.8), $f_{ds}^{<uvw>}$ is larger than the true displaced fraction when the solute atoms are near the center of the channel, and $f_{ds}^{<uvw>}$ is smaller than the true displaced fraction when the solute atoms are near the edge of the channel. If a fraction C_{si} of solute atoms is present in each of several different lattice sites labeled "i", then

$$f_{ds}^{<uvw>} = \sum C_{Si} g_i^{<uvw>} F_i^{<uvw>} \tag{10.36}$$

where $g_i^{<uvw>} F_i^{<uvw>}$ is the effective ion flux at each of those sites [compare Eq. (10.27)]. Here $g_i^{<uvw>}$ is a geometric factor equal to the fraction of solute atom sites which project into the given <uvw> channel, and $F_i^{<uvw>}$ is the normalized ion flux for each site calculated with respect to only the average channeled flux. In other words, $F_i^{<uvw>}$ is the ratio of the channeled beam intensity at the solute atom site to the total channeled beam intensity. This is the quantity normally calculated, as in Fig. 10.8, where dechanneling is neglected.

As an example, consider that one-half of the solute atoms are in substitutional lattice sites ($C_{s1} = 0.5$ and $F_{s1} = 0$) and one-half are in interstitial sites (s2) where the flux $F_{s2} = 2$; then $f_{ds}^{<uvw>} = 1$. From flux calculations of the type shown in Fig. 10.8, empirical values of $f_{ds}^{<uvw>}$ for various channels can be compared with calculated values of $F_i^{<uvw>}$ for different sites. If the solute atoms are in only substitutional lattice sites and one unique interstitial site, then

$$\frac{f_{ds}^{<uvw>}}{f_{ds}^{<uvw>*}} = \frac{g_i^{<uvw>} F_i^{<uvw>}}{g_i^{<uvw>*} F_i^{<uvw>*}} \tag{10.37}$$

where *<uvw>* and *<uvw>** are two different channeling directions or planes. A comparison of the observed ratio on the left-hand side of Eq. (10.37) with calculated ratios on the right-hand side of the equation for various possible sites "i" enables the position of the solute atoms to

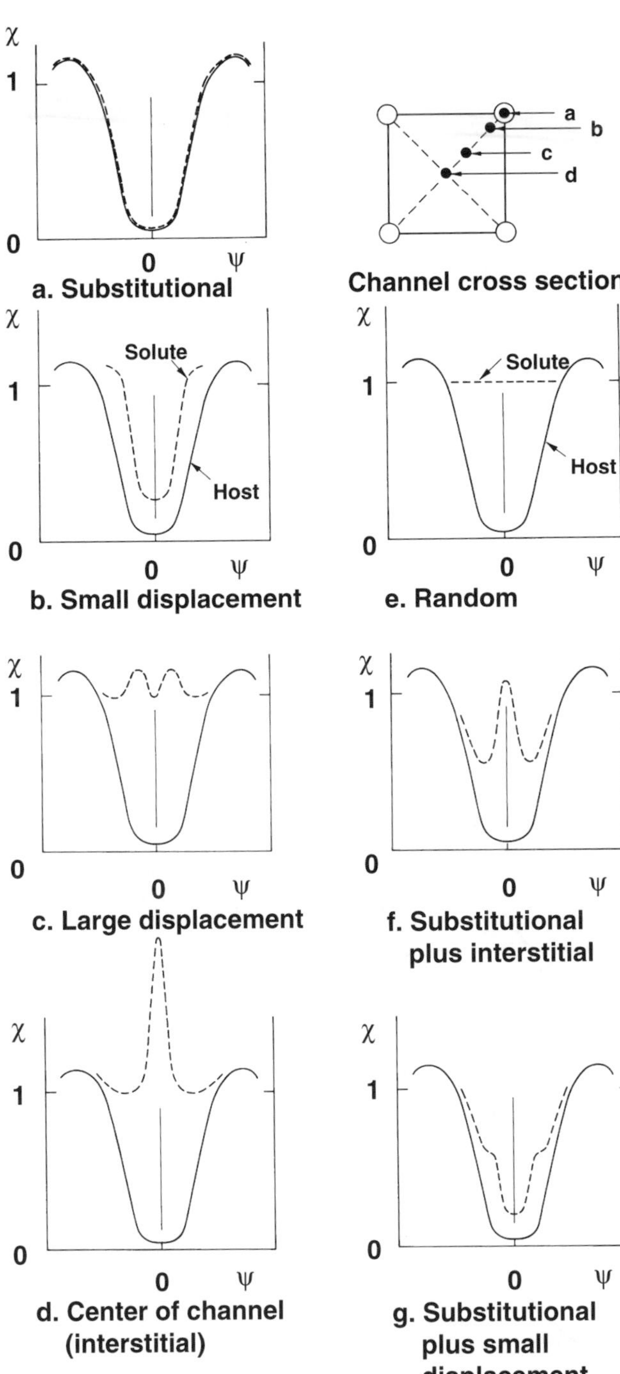

FIG. 10.32. Angular scans for different projections of a solute atom into an axial channel (from Howe *et al.*, 1983).

be determined (Matsunami *et al.*, 1978). It is important to realize, however, that $F_i^{<uvw>}$ is very sensitive to factors such as surface oxides, beam divergence, and depth of analysis.

10.7.2 Analysis methods

10.7.2.1 Comparison of RBS, x-ray, and nuclear reaction methods

As outlined in Section 10.3.3, the choice of RBS, x-ray analysis, or nuclear reaction analysis (NRA) (or a combinations of these) for lattice-site measurements is dictated by the particular element to be analyzed.

Careful attention must be given to the cross section for a particular reaction and to the depth information available. Some examples of the use of these methods will be given later.

10.7.2.2 Analytical versus Monte Carlo analysis

There are four degrees of sophistication for analysis of lattice site data, in order of increasing accuracy:

- Examine the angular scans (or just the apparent displaced fraction $f_{ds}^{<uvw>}$) to estimate lattice sites from Fig. 10.32.
- Use the statistical equilibrium approximation Eq. (10.13) for the ion flux at different positions in each channel, and compare with angular scan or $f_{ds}^{<uvw>}$ data (Andersen *et al.*, 1971; Swanson *et al.*, 1973).
- Use analytical calculations with multi-string Molière-type potentials to calculate equipotential contours and ion fluxes, and then calculate angular scan yields [Eq. (10.13), Figs. 10.7, 10.8, 10.33] (Matsunami *et al.*, 1978; Nielsen, 1988).
- Use Monte Carlo calculations of flux profiles and angular scans (Barrett, 1990; Smulders and Boerma, 1987; Smulders *et al.*, 1990).

The choice of analysis depends on the amount of information required and the complexity of the lattice sites and crystal structures. There are many programs available for both analytical and Monte Carlo calculations. J.H. Barrett (Oak Ridge National Labs) and D.O. Boerma (University of Groningen, the Netherlands) can be contacted about the Monte Carlo programs.

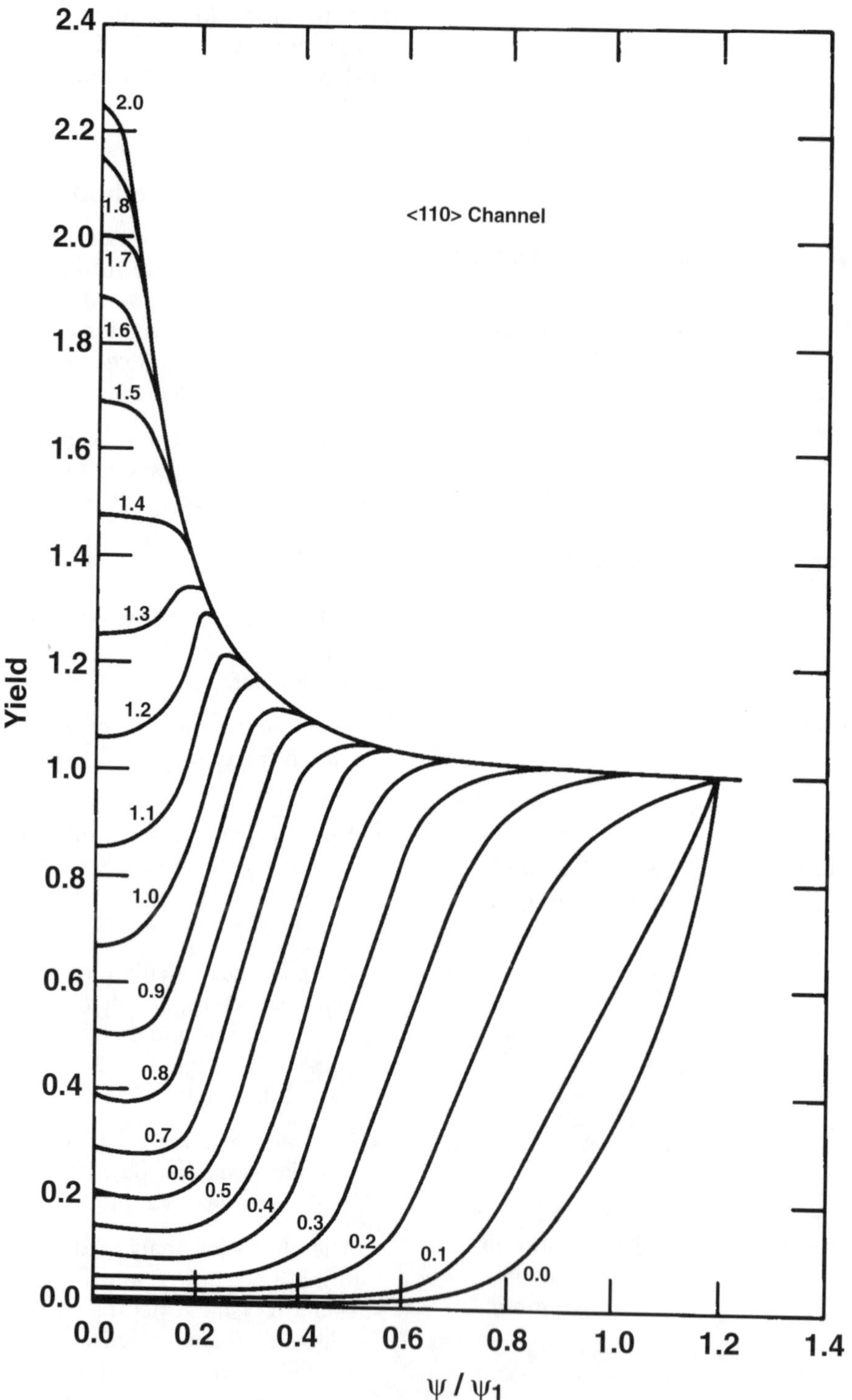

FIG. 10.33. Calculated yields in a <110> axial channel for atoms displaced along <100> directions as a function of the incident angle ψ, normalized to the Lindhard characteristic angle ψ_1. The numbers on the curves are the displacement values in Angstroms. The calculation was for 1 MeV He ions in a <110> channel of Al at 30 K at a depth of 100 nm and for a beam divergence of 0.048° (from Matsunami *et al.*, 1978).

10.7.3 Substitutional sites in elements and compounds

10.7.3.1 Elements

As an example of substitutional lattice sites (position "a" of Fig. 10.32), we consider the Cu-Au system, which exhibits continuous solid solubility, with short- and long-range ordering. For small concentrations (~1%) of Au in Cu, it is expected that the Au atoms are on random substitutional lattice sites. Evidence for this is obtained from x-ray data as well as from many other physical properties. Channeling data (Alexander and Poate, 1972) also indicate for a Cu - 2 at. % Au alloy that the Au atoms occupy substitutional lattice sites, as shown in Fig 10.34. Here the <110> angular dependences of χ_{Cu} and χ_{Au} are almost identical, as expected for this alloy. However, Barrett (1988) has analyzed these data by computer simulations to show that the slightly lower values of χ_{Au} compared with χ_{Cu} up to an angle of almost 2° can be explained by a 14% lower vibration amplitude for Au atoms as compared with Cu atoms in this alloy.

The segregation of solute atoms to grain boundaries or to precipitates may or may not produce a decrease in the substitutional component of the solute atoms. An interesting study of segregation was performed in a low-alloy steel (Knapp and Follstaedt, 1981). An Fe crystal was implanted with both Ti and Sb and then annealed 1 h at 873 K. This treatment produced fcc TiC precipitates by the gettering of C impurity atoms. One <100> axis of each precipitate was aligned with a <100> axis of the host bcc Fe crystal, but the other two <100> TiC axes were aligned with <110> Fe axes. Thus in the channeling experiment, attenuation of Ti yields occurred only along <100> Fe axes. However, attenuation of Sb yields was observed for <100>, <110>, and <111> Fe channels, thus showing that an appreciable fraction of the Sb atoms occupied substitutional Fe lattice sites adjacent to TiC precipitates.

10.7.3.2 Ordered structures and compounds

In substitutional solid solutions, ordered structures often occur (e.g., Cu_3Au, CuAu). In an ordered lattice or compound, a given channel may be bordered by strings of only one element or by mixed strings of both elements. In the former case, a difference in angular half-width $\psi_{1/2}$ occurs for the different elements because of the relation [Eq. (10.10)], $\psi_{1/2} \propto (2Z_1Z_2e^2/Ed)^{1/2}$ where Z_1 and Z_2 are the projectile and target atomic numbers, E is the projectile energy, and d is the atomic spacing along the strings of atoms bordering the channel. This difference has also been observed (Eriksson and Davies, 1969; Picraux *et al.*, 1969) in many compounds containing elements of very different Z_2 values such as UO_2 and GaP. Such an effect has been observed in ordered Cu_3Au (Fig. 10.35). Note that in ordered Cu_3Au, the <100> axial channels are bordered by three strings of pure Cu atoms and 1 string of pure Au atoms. The effect of short-range ordering, which may also occur for more dilute solid solutions, has yet to be investigated.

The lattice positions of impurity atoms in compounds having complex structures have recently been studied by channeling. A good example is the study of the lattice sites of Ta and Hf solute atoms in $LiNbO_3$ (Rebouta *et al.*, 1990, 1992). Since the angular scans for Ta impurity atoms in $LiNbO_3$ were identical to those of Nb [Fig. 10.36(a)], it was concluded that Ta atoms replaced Nb atoms in the lattice. However, as shown in Fig. 10.36(b), the angular scans for Hf impurity atoms were completely different and indicated that the Hf atoms occupied Li sites (Appendix 15, Fig. A15.17). This result was unexpected because the atomic radii of Ta and Hf are very similar.

A question of some interest in semiconductor research is whether amphoteric elements like Si or Sn occupy group III or group V lattice sites in compounds like GaAs or GaP. Channeling methods can be applied to study this problem. Both axial and planar channeling have been used to measure the sites of In and Sn in GaP and GaAs. The principle of the planar method (Swanson *et al.*, 1989) is that a (111) plane in GaP is bounded on one side by Ga atoms and on the other side by P atoms. Thus a beam of ions directed towards the Ga wall of the channel is steered so that it encounters only Ga atoms at a specific depth (equal to approximately $\lambda/4$, where λ is the wavelength for planar depth oscillations) corresponding to the critical angle ψ_c with respect to the perfectly aligned direction. Similarly, a beam directed towards the P wall of the channel encounters only P atoms at a similar depth. Thus oscillations in yield from Ga atoms and P atoms vary with depth in a complementary way. The ion beam will interact with implanted dopant atoms in the same way as with the host atoms which the dopant atoms replace, so that the sites occupied by the dopant atoms can be identified at any depth where the oscillations are measurable. The tilt angle and beam energy can be "tuned" to a specific depth. The axial channeling method (Andersen *et al.*, 1982) uses the same principle but since axial depth oscillations damp out much faster than planar oscillations, the axial method is applicable only for solute atoms distributed very close to the surface.

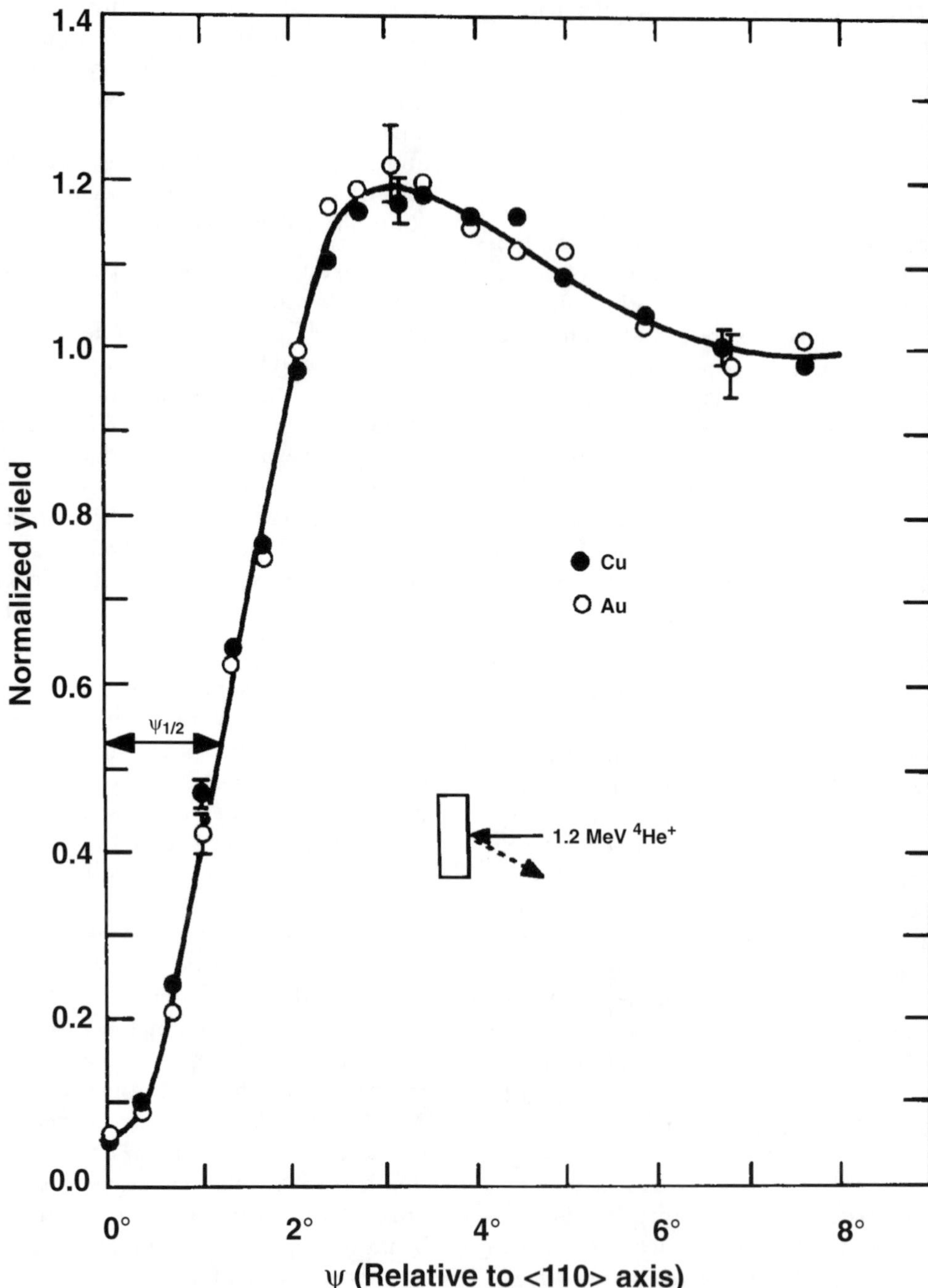

FIG. 10.34. Angular dependence of the normalized backscattering yields of 1.2 MeV He ions from Cu and Au atoms in a single crystal of Cu containing 2 at. % Au. The yields were measured for the depth interval 40 - 90 nm below the surface (from Alexander and Poate, 1972).

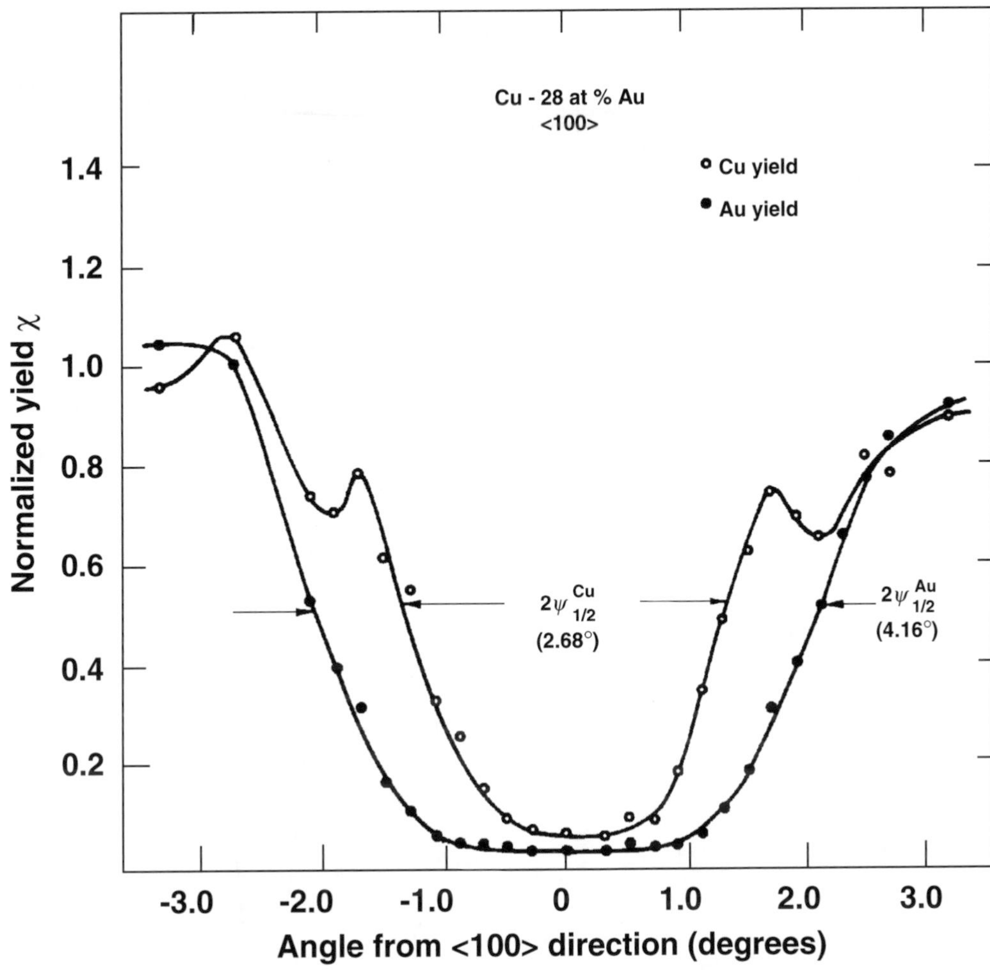

FIG. 10.35. Angular dependence of the normalized yield χ of backscattered 1 MeV He ions from Cu and Au atoms for a scan through a <100> axial channel in a $Cu_{0.72}Au_{0.28}$ crystal. The measurements were taken at 35 K (from Howe *et al.*, 1983).

10.7.4 Low symmetry interstitial sites

Low symmetry interstitial sites can involve small projections [Fig. 10.32(b)] and/or large projections [Fig. 10.32(c)] of solute atoms into different channels. When narrowing of the channeling dip occurs for several channels, it can be concluded that the solute atoms are displaced only small distances (≤0.05 nm) from normal lattice sites. On the other hand, if the projections are small into some low index channels and large into others, one can conclude that the solute atoms are displaced larger distances into well-defined interstitial sites.

10.7.4.1 Small displacements

A good example of small displacements was given by the case of Bi implanted into Si (Picraux *et al.*, 1972), where narrowing occurred for both <110> and <111> channels. The data were in good agreement with 50% of the Bi atoms lying on perfect lattice sites and 50% of the Bi atoms displaced ~0.045 nm in <110> directions. Double-alignment data in this case provided additional information concerning the lattice sites of Bi. Because the yield from Bi atoms in <110> double alignment was reduced by a factor of 3 as compared with the yield in single alignment, it appears that at least 95% of the Bi atoms were located within 0.06 nm of <110> lattice rows.

In principle, enhanced thermal vibrations of the solute atoms could also cause narrowing, but such effects have not been observed as yet for substitutional solutes.

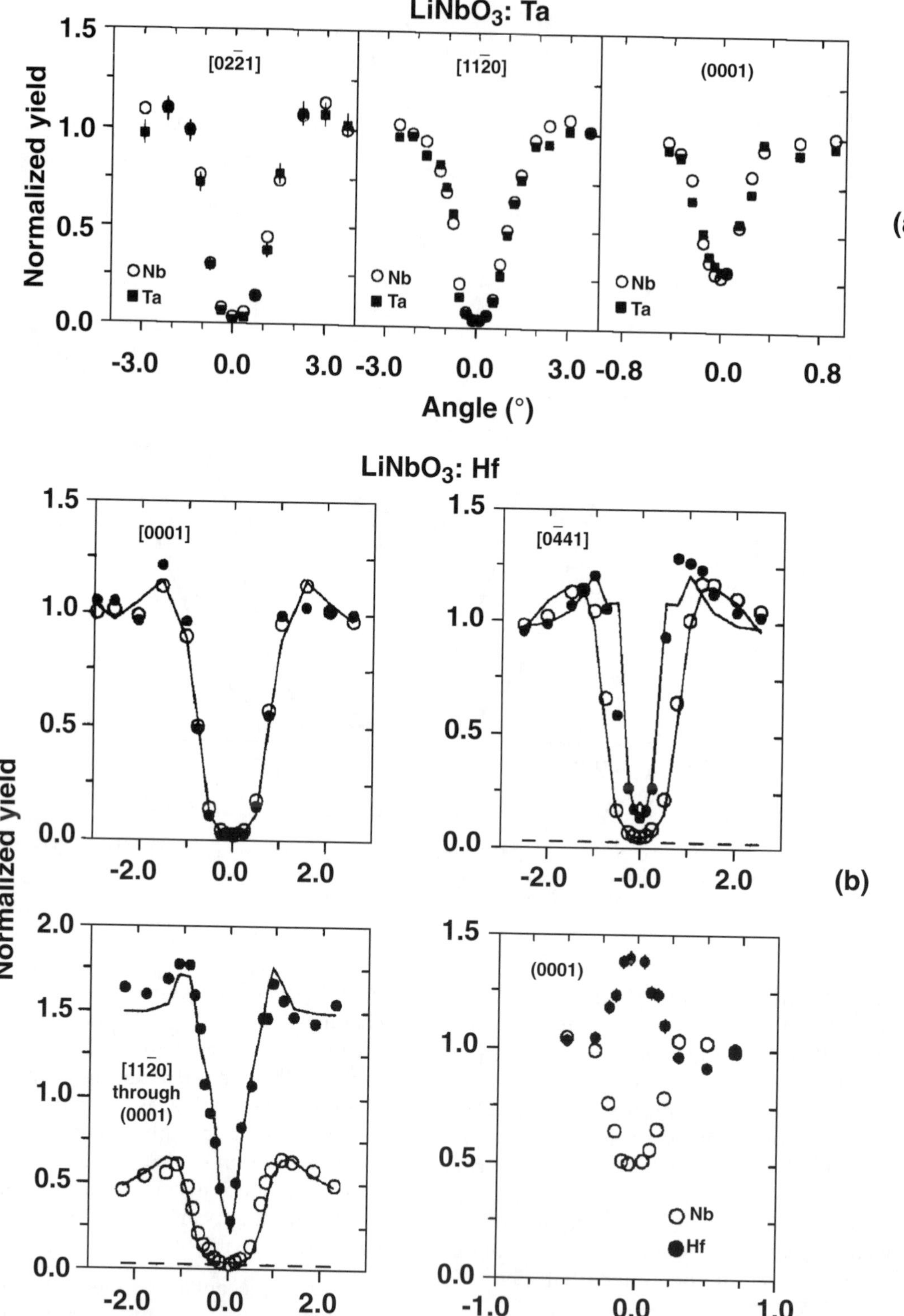

FIG. 10.36. Angular scans for $LiNbO_3$ containing Ta or Hf. (a) Normalized backscattering yields from Ta and Nb for the < 11$\bar{2}$0 > and < 02$\bar{2}$1 > axes, and {0001} plane. (b) Normalized backscattering yields from Hf and Nb for the <1000>, < 11$\bar{2}$0 >, and < 04$\bar{4}$1 > axes, and {0001} plane. The lines were drawn to guide the eye (from Rebouta *et al.*, 1992).

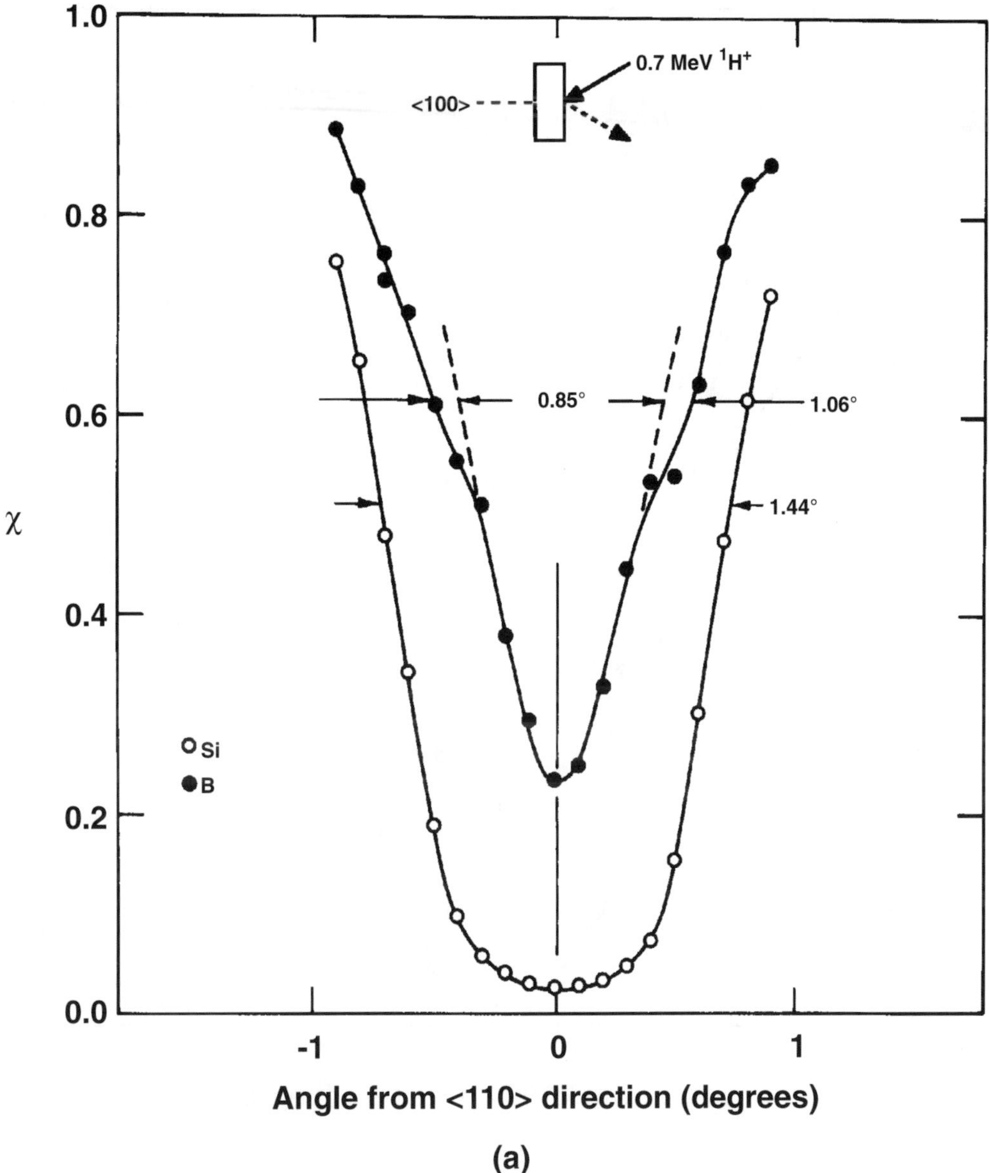

(a)

FIG. 10.37. Lattice location of B atoms in Si. Angular scans for a laser annealed Si - 0.1 at. % B crystal after irradiation at 35 K with 0.7 MeV He ions to a fluence of 5×10^{16} cm^{-2}, followed by a 20 hours anneal at 293 K. The normalized yields χ_{Si} of H ions (incident energy 0.7 MeV) backscattered from Si atoms and the normalized yields χ_B of α particles from the $^{11}B(p,\alpha)^8Be$ nuclear reaction are plotted as a function of the angle from the indicated axes. The measurements were taken at 35 K (from Swanson *et al.*, 1981). (a) <110> scan; (b) <100> scan; and (c) projections of the bond-centered and split <100> interstitial positions into <110> and <100> channels in Si. The average normalized ion fluxes F_i calculated for the indicated displacements are shown.

10.7.4.2 Large displacements

An example of interstitial sites displaced more than 0.05 nm from lattice positions is given by B in Si as shown in Fig. 10.37 (Swanson *et al.*, 1981), where narrowing occurred for <110> axial channels but not for <100> channels. Also, the value of $\chi_B^{<100>}$ was greater than $\chi_B^{<110>}$. Thus the displaced B atoms projected a smaller distance into <110> channels than into <100> channels. Possible positions for the B atoms, as seen along these channels, are given in Fig. 10.37(c), showing the calculated

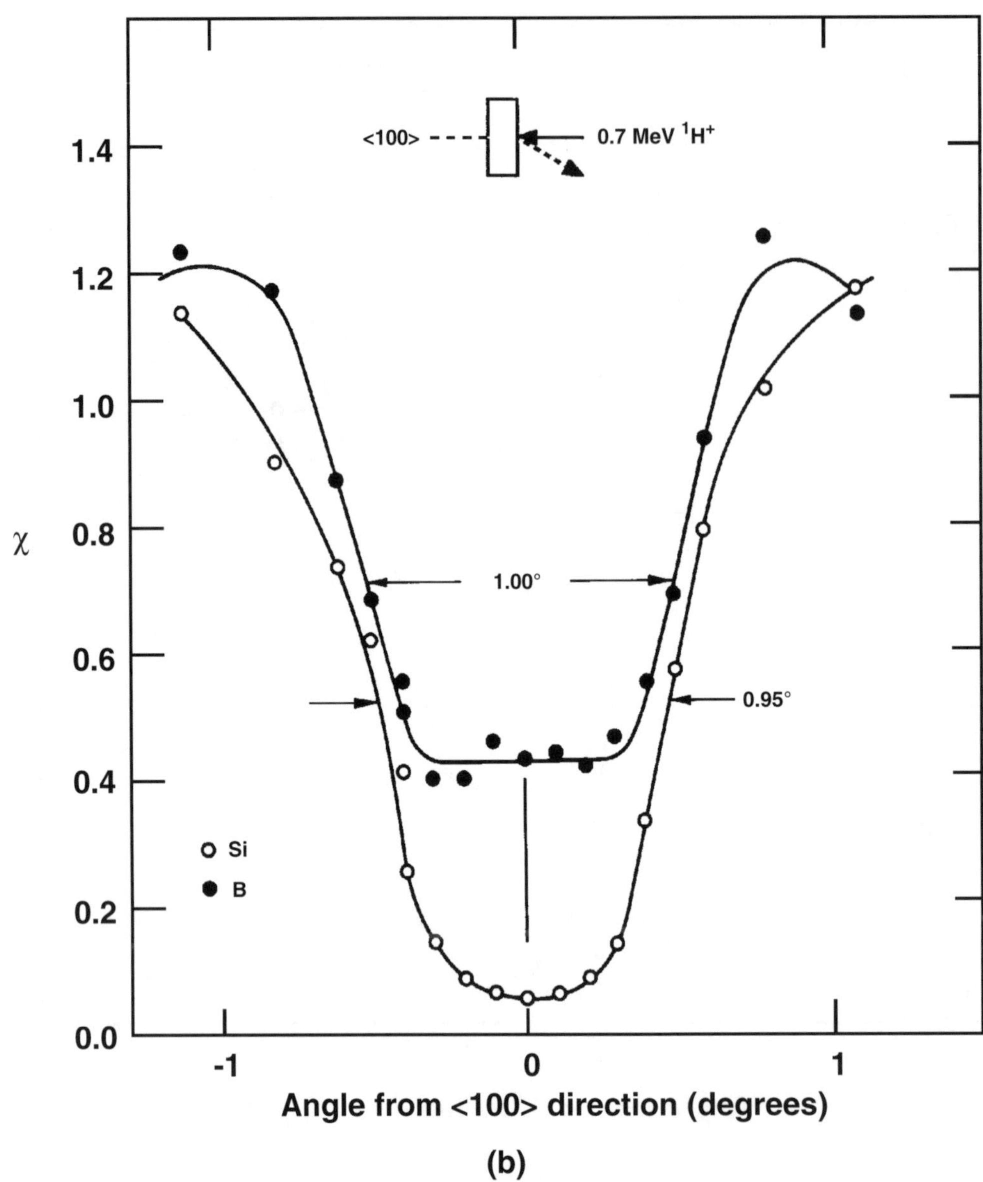

FIG. 10.37. (continued).

ion fluxes for displacements of 0.079 nm along <100> directions (split <100> position), or 0.12 nm along <111> directions (bond-centered position). Although either displacement would fit the <110> and <100> channeling data of Fig. 10.37, the bond-centered position is excluded by the <111> data (not shown), which would have a peak in yield for that position of the B atoms.

In this case, the B was implanted at an energy of 25 keV to a fluence of 10^{15} cm^{-2}, and the sample was ruby laser annealed (1.6 J cm^{-2}). The crystal was then irradiated with 0.7-MeV $^{1}H^{+}$ at 35 K and isochronally annealed up to 293 K. The major displacement of B atoms occurred during annealing near 240 K. It had been determined previously by EPR measurements that this annealing stage corresponded to migration of a defect caused by the interaction of interstitial Si atoms with B atoms (Watkins, 1975). This defect was either a Si-B split interstitial or a B interstitial. Thus in this channeling example, the displacement of B atoms from lattice sites was due to clustering of B interstitials.

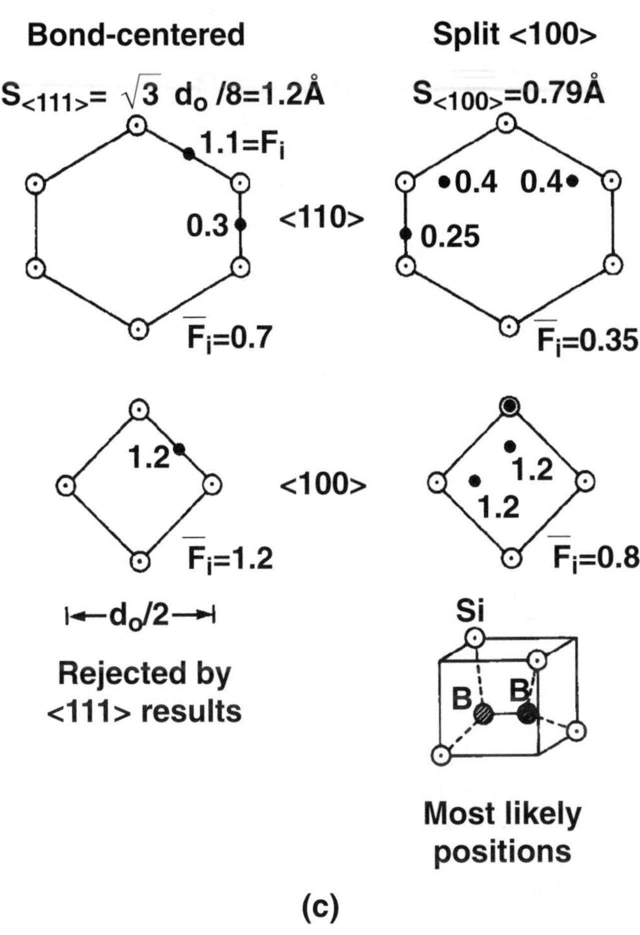

FIG. 10.37. (continued).

This analysis of B in Si is an example of the use of nuclear reactions to determine lattice positions of light solute atoms. The yield of α particles from the $^{11}B(p, \alpha)^{8}Be$ nuclear reaction was compared with the yield of backscattered 0.7-MeV $^{1}H^{+}$ ions from the Si atoms. Because of the many alpha energies emitted by this particular nuclear reaction, it is difficult to extract depth information from an energy analysis of the emitted α particles. However, the laser-annealing treatment distributes the B atoms almost uniformly up to a depth of only 200-300 nm. When the yields from B atoms and Si atoms are compared for this shallow depth interval, the problem of increased yields at greater depths because of dechanneling is eliminated.

The data of Fig. 10.37, as well as <111> angular scans, were analyzed by both analytical and Monte Carlo methods (Smulders *et al.*, 1990). Although some differences in the two methods were apparent, both methods indicated that the B atoms were displaced 0.9 ± 0.1 Å along either <100> or <110> directions. A likely model for the B configuration is the B-B <100> split interstitial, as shown in the inset of Fig. 10.37(c).

It is interesting to note that even with several angular scans and sophisticated analysis methods, the B sites in this case of low symmetry interstitial sites could not be positively established. This difficulty is of course not unique to channeling studies in complicated cases like this.

10.7.5 Gas atom sites

Small gas atoms often occupy octahedral or tetrahedral interstitial sites in metals, as shown by channeling experiments (Bugeat and Ligeon, 1979; Carstanjen, 1980; Yagi *et al.*, 1985). One example that illustrates the use of the channeling method for determining these high-symmetry interstitial positions is the lattice location of deuterium in Pd (Fig. 10.38). The nuclear reaction $^{2}H(^{3}He,p)^{4}He$ was used to detect the deuterium (Carstanjen *et al.* 1978). The angular scans of Fig. 10.38 clearly demonstrate that the deuterium atoms occupied octahedral interstitial sites since those sites are completely shadowed in <100> axial and {100} planar channels, but are exactly in the center of <110> axial and {111} planar channels. (See Appendix 15 for projections of different interstitial sites into various channels.)

Small gas atoms like H or D often are retained in metals only in a trapped state. Picraux (1981) showed that the position of D that was implanted into Fe was ~0.04 nm from octahedral sites. This is an example of sophisticated analysis involving computer fitting of both axial and planar channeling data. The results indicated that the D was trapped by single vacancies.

Many experiments with ion-implanted Si have shown that the implanted solutes can occupy a variety of lattice sites, but it is often not clear whether these are equilibrium or defect-associated sites. One example is Yb implanted into Si — the solute atoms lie near the tetrahedral interstitial sites, producing strong peaking in Yb yields for <110> and small dips in Yb yields for <100> and <111> channels (Andersen *et al.*, 1971).

10.7.6 Solute atoms associated with point defects

In general, the presence of nearby point defects will pull solute atoms out of their normal lattice sites, resulting

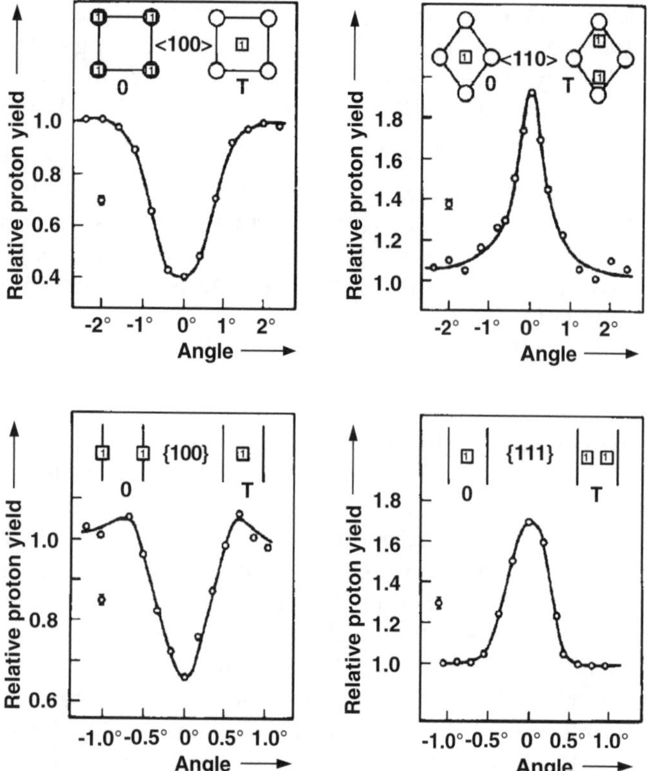

FIG. 10.38. Analysis of D in Pd. Angular scans for D, using proton yields from the nuclear reaction $D(^3He,p)^4He$ for <100> and <110> axes, and {100} and {111} planes in $Pd_{0.993}D_{0.007}$ at 295 K, using 0.75 MeV ^3He ions (from Carstanjen *et al.*, 1978).

in either small or large displacements. An example of large displacements into low symmetry interstitial sites is that of the small solute atom Cu in Al, which can trap self-interstitial Al atoms that are produced by particle bombardment. The predominant trapping configuration is found by channeling to be the <100> mixed dumbbell, in which a solute atom and a host atom straddle a lattice site. Because the displacement of solute atoms in such mixed dumbbells is large (0.1-0.14 nm), the solute atoms project considerably into <110> channels, about 2/3 of the distance to the octahedral sites, giving a peaking in backscattering yield from the solute atoms in these channels (Matsunami *et al.*, 1978). The results of Monte Carlo studies agreed with these analytical calculations (Barrett, 1978).

In some cases, normally substitutional solute atoms can be injected into high symmetry interstitial sites by the trapping of point defects. We consider an alloy of Al containing 0.2% In which was irradiated to produce vacancies. Subsequent annealing caused the vacancies to cluster about the In atoms, thus causing a large fraction of the In atoms to occupy tetrahedral sites surrounded by four vacant lattice sites (Fig. 10.39 inset).

The growth of the peak in yield from In atoms for a <100> angular scan as a function of irradiation fluence is shown in Fig. 10.39 (Swanson and Howe, 1983). This peak is strong evidence that the In atoms occupy the tetrahedral interstitial site, which lies in the center of <100> channels (Fig. 10.31). For the highest fluence, the peak-to-valley ratio was 2.3, which is close to that calculated for tetrahedral sites. Thus a large fraction (~0.60) of the In atoms were in tetrahedral interstitial sites. Most of the remaining In atoms occupied substitutional or random sites. Only the central part of the angular scan was measured (Fig. 10.39) in order to reduce the effect of damage from the analyzing beam.

10.7.7 Random sites

When solute atoms are in random sites, the apparent displaced fractions $f_{ds}^{<uvw>}$ for all channeling alignments are equal, and no structure is seen in angular scans (Fig. 10.32). In this case, the solutes may be present in precipitates of random alignment or in association with large defect clusters. The solutes may also have diffused to the surface. If the solute atoms are incorporated in precipitates with some preferred orientation with respect to the host crystal, attenuation of yields occurs in some channels (Section 10.7.3.1).

10.8 SURFACE STUDIES

10.8.1 Basic considerations

Channeling is used to study many surface properties, such as surface reconstruction, relaxation of the outermost layer of atoms, surface melting, and lattice positions of adsorbate atoms (van der Veen, 1985; Swanson, 1982; Howe *et al.*, 1983). As outlined in Section 10.1, a surface peak is seen in channeling spectra even for a perfect, uncontaminated surface because the outermost layer of atoms is not shadowed (Figs. 10.3 and 10.4). An example of a surface peak for clean Pt is shown in Fig. 10.40 (Davies *et al.*, 1978). The presence of a surface oxide of course will enhance the surface peak, so that measurements are usually done in UHV. Depending on the radius *R* of the shadow cone [Fig. 10.4(a)], more than one atom per string may be seen. For Coulomb scattering, the radius R_c of the cone at the second atom is

$$R_c = 2(Z_1Z_2e^2d/E)^{1/2} \qquad (10.38)$$

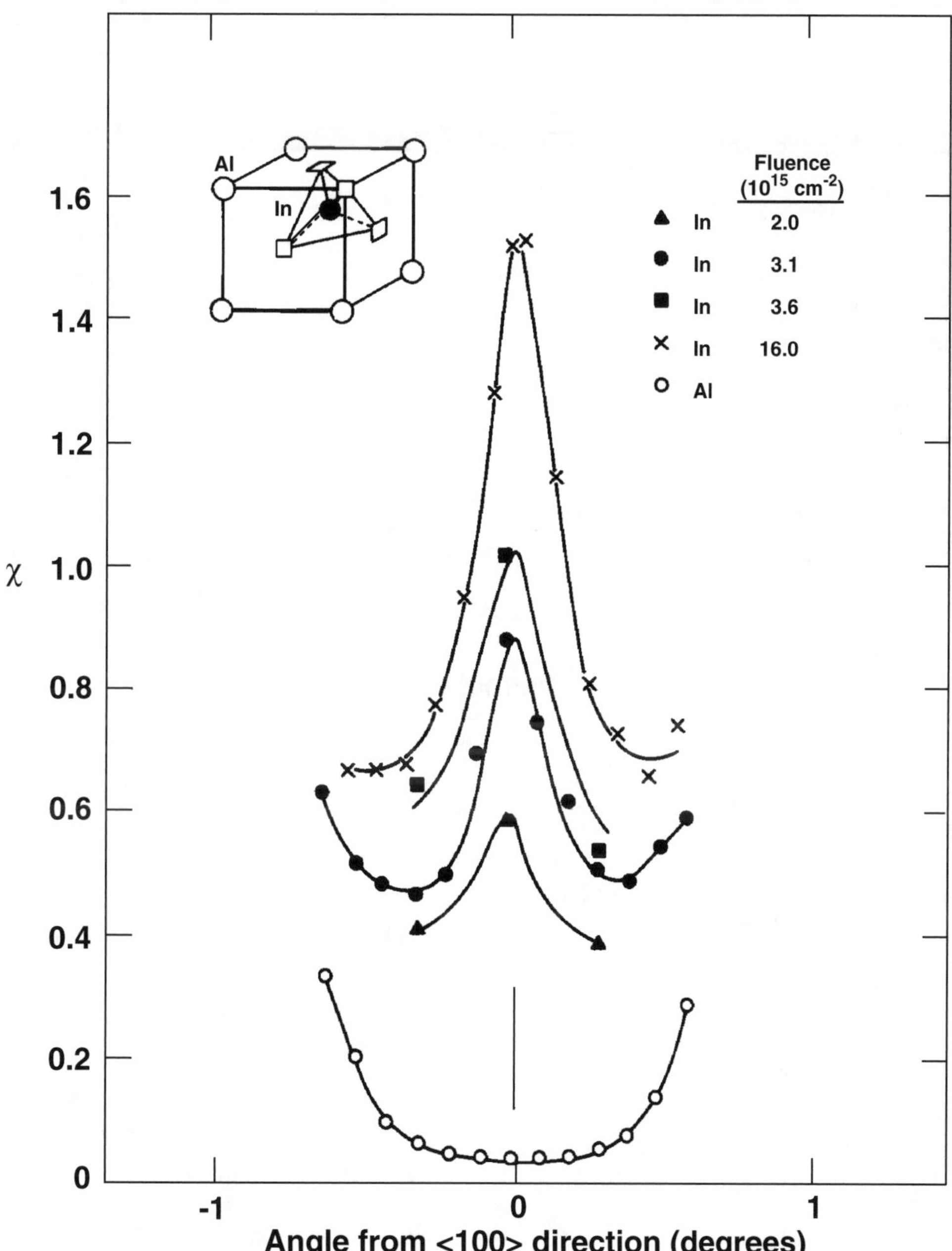

FIG. 10.39. <100> angular scans for irradiated Al-0.02 at. % In. Backscattering yields χ of 1 MeV He ions from Al and In atoms are shown as a function of the angle from a <100> axis. The sample was irradiated with 1 MeV He ions to the indicated fluences at 35 K, and then annealed for 600 s at 220 K before each of the angular scans. The yields were measured at 35 K for a depth interval of 50-280 nm from the surface. Since the Al yields were almost the same after each irradiation, only one set of Al data is shown (from Swanson and Howe, 1983). The inset shows the deduced tetrahedral interstitial position of an In atom (●), surrounded by four vacancies (□).

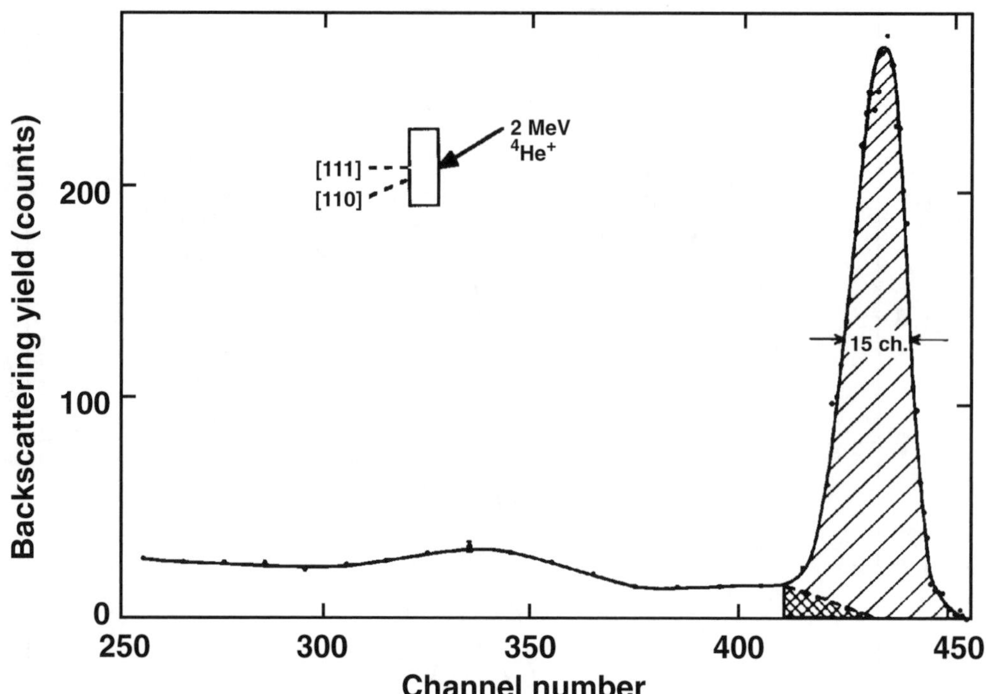

FIG. 10.40. Backscattered energy spectra for 2 MeV He ions incident on a Pt(111) surface along a <110> direction at 60 K. The energy scale has been expanded with a large zero offset. The width (FWHM) of the surface peak is equivalent to a depth resolution of 3.5 nm. Its area is equivalent to 1.39 Pt atoms for a <110> row (from Davies *et al.*, 1978).

where *d* is the spacing of atoms in the string and *E* is the energy of the incident ion (Example 10.8) (Bøgh, 1973; Feldman *et al.*, 1982). If the thermal vibrational amplitude exceeds R_c, a second atom is seen by the ion beam.

Fig. 10.41 shows schematically how the yield of the surface peak can be increased beyond one atom per string:

- by displacement of the surface laterally, i.e., in a reconstructed or melted surface [Fig. 10.41(b)];
- by relaxation of the surface layer inwards or outwards (observable only for non-perpendicular incidence [Fig. 10.41(c)]; and
- by an increase in the thermal vibrational amplitude u_2 beyond R_c.

Each of these effects is more pronounced with increasing ion energy, which reduces R_c according to Eq. (10.38).

EXAMPLE 10.8. The shadow cone radius for a Au layer on Ag [Fig. 10.4(b)].

For a 1 MeV He beam on Au, $Z_1 = 2$, $Z_2 = 79$, $d^{<110>} = 0.288$ nm (Tables A15.1 and A15.3), $e^2 = 1.44 \times 10^{-13}$ cm MeV, E = 1 MeV. Thus $R_c = 0.0162$ nm, which can be compared with $u_2 = 0.012$ nm.

In order to calculate the actual number of atoms per string which contribute to the surface peak, Monte Carlo computer simulations of the type described in Section 10.2 (Barrett, 1971; Tromp and van der Veen, 1983) are required. In such calculations, a Thomas-Fermi screened Coulomb potential and the Debye approximation to vibrational amplitudes are often used. One refinement is the inclusion of correlations between thermal vibrations of adjacent atoms, which increases the shadowing effect (Jackson and Barrett, 1979).

The effect of thermal vibrations on the surface peak can be reduced by using materials with a high melting point. For example, if 1 MeV He^+ ions are incident along

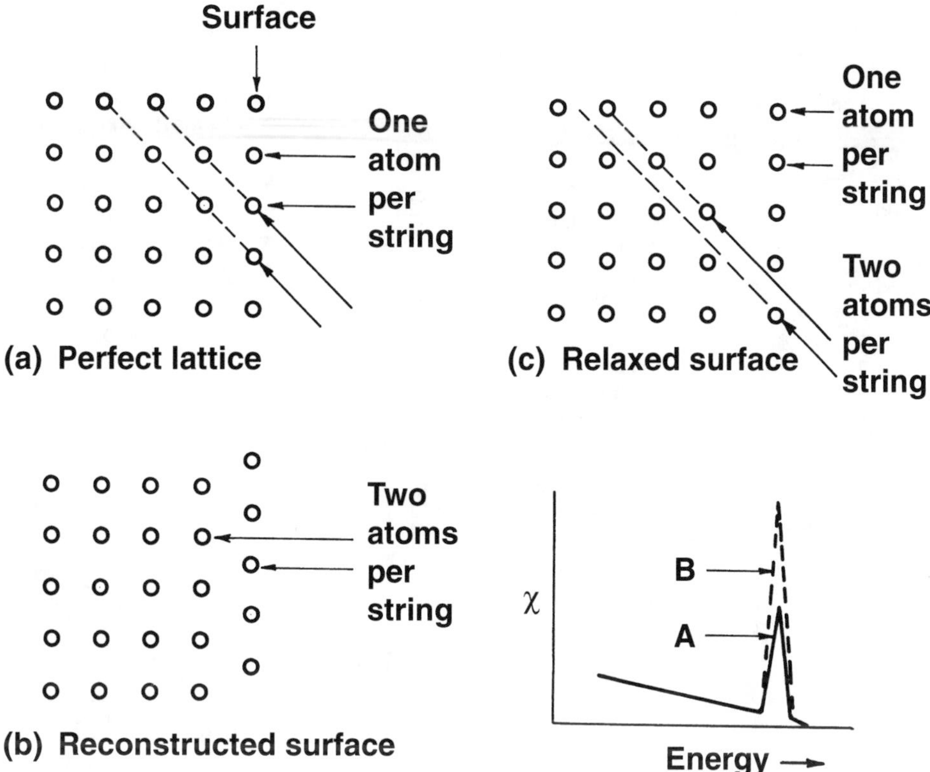

FIG. 10.41. Schematic illustration of measurements of channeling surface peaks to study displacements of surface atoms. (a) perfect lattice; (b) reconstructed surface; and (c) relaxed surface (from Swanson, 1982).

a <100> direction of Pt, the radius of the shadow cone at the second atom in a <100> string, as calculated using the Molière approximation to the Thomas-Fermi potential, is R_M = 0.0154 nm, which is considerably larger than the vibrational amplitude at room temperature, u_2 = 0.0089 nm.

Since RBS cross sections are accurately known, this surface-peak area provides a quantitative measure of the number of unshadowed lattice atoms per square centimeter. This number in turn contains useful information about structural changes in the surface region, such as the outward relaxation shown in Fig. 10.42. For perpendicular incidence (<111> in Fig. 10.42), the second atom in each row is shadowed, and thus the observed surface-peak area will correspond to 1 atom per row (4.5×10^{15} Pt atoms cm^{-2}) at all energies, provided the shadow cone radius R at the second atom is much larger than u_2. Note that a surface relaxation Δd, or an enhanced vibrational amplitude perpendicular to the surface, has no effect on this <111> shadowing, but any lateral displacement within the surface plane would be clearly detected.

For nonperpendicular axes, such as the <110> in Fig. 10.42, an outward (or inward) displacement Δd shifts the shadow cone relative to the underlying row of atoms, thus causing the surface peak to increase towards a value of 2 atoms per row. Since R varies approximately as $E^{-1/2}$ [Eq. (10.38)], this <110> surface peak (in a static lattice) would increase from a value of 1 atom per row at low E to approximately 2 atoms per row at high E. Thus the magnitude of the surface relaxation Δd (and also of any lateral displacements) can be obtained by measuring the surface-peak area for various low-index directions as a function of energy. In practice, a more sensitive experimental procedure is to perform detailed angular scans through each of the chosen close-packed axes.

In most cases of practical interest, thermal vibration effects are not negligible, and the observed surface peak

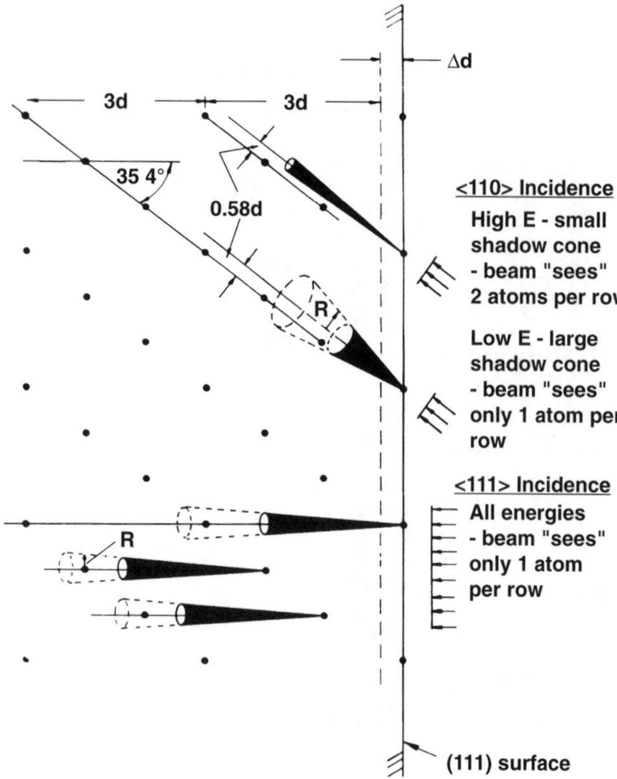

FIG. 10.42. Atomic configurations near the surface of a Pt(111) crystal, illustrating the use of channeling to investigate the surface relaxation Δd (from Howe *et al.*, 1983).

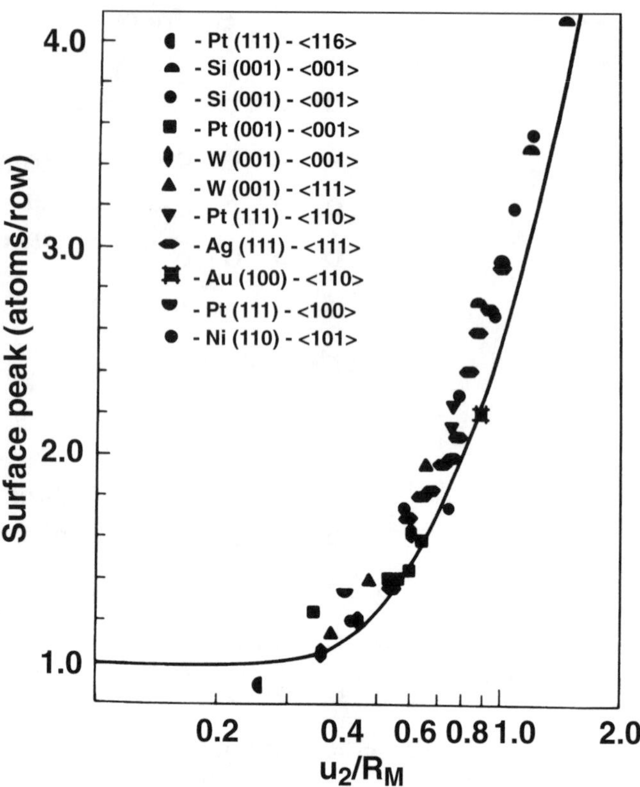

FIG. 10.43. Comparison of the "universal" curve [Eq. (10.39)] with experimental values for a number of different "bulk-like" surfaces. The experimental values were determined from backscattering measurements. The notation Pt(111)-<116> indicates a Pt crystal with a (111) surface plane using backscattering measurements in the <116> axial direction. u_2 is the vibrational amplitude and R_M is the shadow cone radius at the second atom, using the Molière potential [compare Eq. (10.38)] (from Feldman *et al.*, 1982).

may include significant contributions from several underlying atoms per row. Quantitative analytical estimates are then no longer possible, except in the two-atom limit of Fig. 10.4, where Feldman *et al.* (1977) have shown that the surface-peak area N in atoms per row is given by (Example 10.9)

$$N = 1 + \left(1 + R^2 / 2u_2^2\right) \exp\left(-R^2 / 2u_2^2\right) . \qquad (10.39)$$

The first term in Eq. (10.39) is the contribution from the first atom in the row, and the second term is the temperature-dependent contribution from the second atom. Although this two-atom approximation is often not adequate for detailed comparison with an experiment, it has provided a very useful scaling parameter (u_2/R) for correlating various surface peak intensity measurements (Fig. 10.43).

In addition to the single alignment method which was illustrated in Fig. 10.41, a double-alignment technique is extremely useful for study of surface structure (Turkenburg *et al.*, 1976). In this approach [Fig. 10.44(a)], the incident beam is directed along one crystallographic axis and the detector is aligned with another axis. The backscattering yields from both surface atoms and atoms well below the surface [the bulk yield, Fig. 10.44(b)] are

EXAMPLE 10.9. 1 monolayer of Au on Ag, 1 MeV He^+, <110> [Fig. 10.4(b) and Example 10.8].

$R_c = 0.0162$ nm for Au, and $u_2 = 0.0132$ for Ag. Thus $u_2/R_c = 0.81$ and from Fig. 10.43, the surface peak (for Au plus Ag) corresponds to 2 atoms per row. This means that one atom per row of Ag is still visible, in approximate agreement with Fig. 10.4(b).

measured. Along the incident beam direction in Fig. 10.44(a), only 50% of the atomic rows originate in the surface plane; the other 50% originate in the second plane. Consequently, at least 50% of the backscattered ions contributing to the surface peak can be blocked along certain emission directions by the surface plane of atoms, thus producing a blocking dip in the surface peak yield. Any relaxation Δd of the surface plane will rotate the position of this blocking dip through an angle Δθ relative to the underlying bulk blocking direction [Fig. 10.44(c)]. Thus the magnitude and sign of Δd may be obtained from purely geometrical considerations. Another big advantage of this blocking technique is that it can be readily adapted to provide information on the location of adsorbate atoms since these too may produce observable blocking dips in the backscattered yield from the underlying crystal.

10.8.2 Surface relaxation

The measurement of the relaxation of surface Ni atoms by the blocking method is shown in Fig. 10.44(d) (van der Veen *et al.*, 1979a). In this case, the (110) surface atoms for a clean surface relaxed inwards by 4 ± 1% of the interplanar spacing. When the surface was contaminated by oxygen, an outward relaxation of −1% was observed. Similarly, a relaxation of ≤0.003 nm (1.5%) was observed for (111) Pt surface atoms both by this method (van der Veen *et al.*, 1979b) and by single alignments of the Pt surface peak as a function of the incident angle of the probing ion beam (Davies *et al.*, 1980). These results are in general agreement with LEED data, but the accuracy of the channeling method is considerably greater. The technique of Davies *et al.* (1980) has been extended to provide information about surface Debye temperatures.

10.8.3 Surface impurities

One special feature of the combined shadowing and blocking technique is the ability to locate the position of impurity atoms adsorbed on a crystal surface. An excellent example of this is the study of sulfur on Ni(110) by van der Veen *et al.* (1979c). Comparison of the clean and sulfur-covered spectra revealed that the sulfur atoms lie 0.087 ± 0.003 nm above the Ni surface, in good agreement with the LEED analysis of 0.093 ± 0.01 nm.

10.8.4 Lateral reconstruction

As indicated in Fig. 10.41, reconstruction of the surface layer can be studied by channeling measurements of the surface peak area.

An example of single-alignment measurements of the backscattering surface peak to determine parameters for the reconstruction of a Pt(100) surface is shown in Fig. 10.45. The clean surface was in the reconstructed state, characterized by a (5 × 20) LEED pattern and by a large surface peak. As CO molecules were adsorbed on the surface, the magnitude of the surface peak decreased at the same rate as the disappearance of the (5 × 20) LEED pattern. At a CO coverage of 6.4×10^{14} molecules cm^{-2}, or 0.50 ± 0.03 monolayers, the transformation to a (1 × 1) surface corresponding to the bulk structure was complete (Davies and Norton, 1980). Since the reduction in magnitude of the surface peak corresponded to 1.65 ± 0.05 $\times 10^{14}$ Pt atoms cm^{-2}, which is more than the density of the Pt(100) plane, the reconstruction must involve displacements of Pt atoms in more than one plane. The CO coverage was measured using nuclear reaction yields, which gave an absolute accuracy of ±0.03 monolayers. Similar studies of the effect of surface contaminants on reconstruction have been reported by Feldman (1979).

10.8.5 Interface structures

The nature of the interface between a crystal and an overlying oxide can also be studied by channeling. In the case of Si which was covered by an oxide grown either by the $CO_2 + SiH_4$ process or by thermal oxidation at 1270 K, it was shown that the oxide was stoichiometric SiO_2 and one or two monolayers of Si at the Si-SiO_2 interface were displaced from normal lattice sites (Feldman *et al.*, 1978; Jackman *et al.*, 1980). For these experiments, the total areas of the surface silicon peak and oxygen peak were measured as the oxide was gradually removed by etching in an HF solution. Typical backscattering spectra are shown in Fig. 10.46 for a detector placed at a grazing angle and for a detector placed at a scattering angle of 180°. The former detector geometry yielded greater accuracy because of the reduced background and was used for the thinner oxides. The surface Si peak in this case corresponded to Si atoms contained in the oxide, Si atoms in a disordered Si-SiO interface region, and non-shadowed Si atoms in the outermost layers of the perfect Si crystal. The variation of the total areal density of Si atoms in the surface peak as a function of the oxygen coverage [Fig. 10.46(b)] demonstrated that the oxide was stoichiometric SiO_2. In addition, the density of Si atoms extrapolated to zero oxygen atom density was 8.6×10^{15} atoms cm^{-2}, which was greater than the density of non-shadowed Si atoms, 6.4×10^{15} cm^{-2}, in the surface region of a clean Si{110} crystal. The excess 2.2×10^{15} atoms cm^{-2} were attributed to displaced Si atoms at the Si-SiO_2

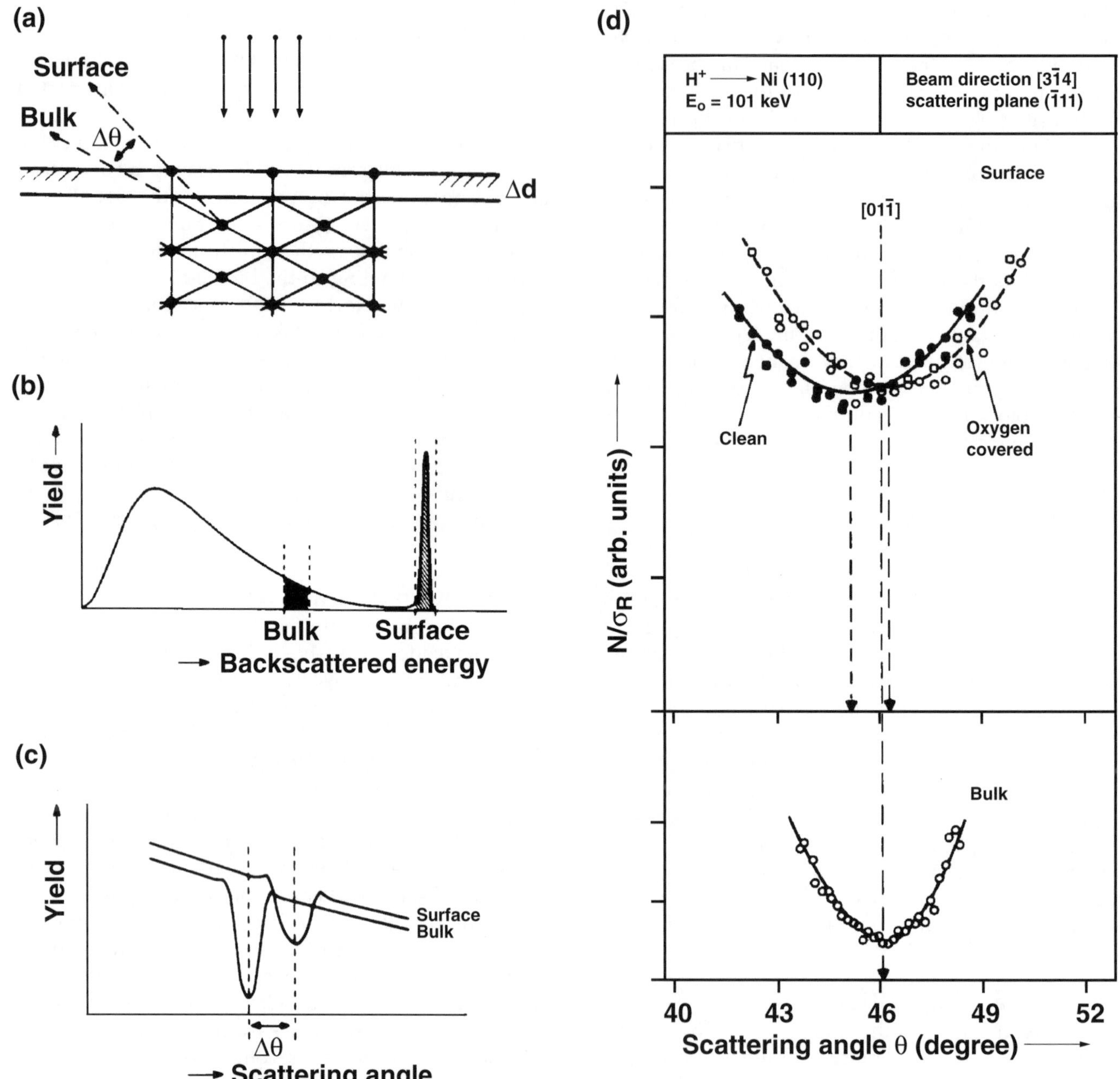

FIG. 10.44. An illustration of the method to study surface relaxation experimentally, based on channeling and blocking.
(a) The scattering plane coincides with the crystallographic plane. A displacement of the first layer results in different angular positions of the blocking cone for surface and bulk atoms.
(b) An energy scan of backscattered ions detected at an angle θ by means of an electrostatic energy analyzer.
(c) Angular scans of backscattered intensities from surface and bulk atoms. The displacement caused by the relaxation effect causes a shift Δθ of the surface blocking minimum (from Turkenburg *et al.*, 1976).
(d) Blocking minima for 0.1 MeV protons scattered from bulk and surface Ni atoms for a clean (110) surface and for an oxygen coverage of 1/3 monolayer. The yield N has been normalized by the Rutherford cross section σ_R. The filled symbols represent the clean surface and the open symbols the oxygen-covered surface. Measurements were taken in the order ●,○,■,□ (from van der Veen *et al.*, 1979a).

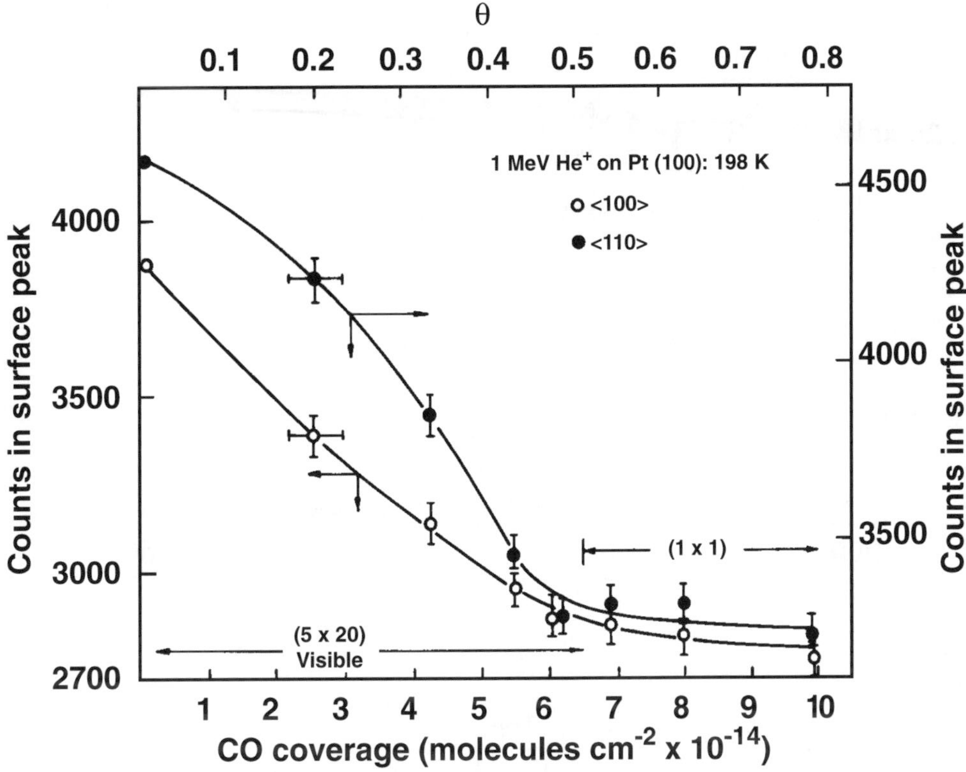

FIG. 10.45. Correlation between LEED observations and backscattering measurements of the surface peak on Pt(100), as a function of CO coverage. Note the gradual transition from the reconstructed (5 × 20) to the (1 × 1) structure. The backscattering measurements were made with 1 MeV He ions at 198 K for <100> (o) and <110> (•) directions (from Davies and Norton, 1980).

interface. The data could not distinguish whether these displaced Si atoms consisted of two monolayers of unregistered Si or one monolayer of non-registered Si atoms plus ~0.5 nm of SiO.

Measurements using thin Si crystals, in which the yields of 1.2 MeV He^+ ions scattered from both front and back surfaces were compared, indicated that the displaced Si atoms at the Si-SiO_2 interface were 0.01-0.1 nm from lattice sites rather than being in an amorphous layer (Feldman *et al.*, 1978).

Fischer *et al.* (1988) studied the bonding at a $CoSi_2$/Si interface by using the thin $CoSi_2$ overlayer as an "ion lens" to focus the ion beam onto the interface atoms of the Si(111) substrate. They found that the Si atoms at the interface bonded to the Co atoms of the silicide.

10.8.6 Surface melting

Channeling has been used to study surface pre-melted layers, using double alignment to reduce the backscattered yield from subsurface atoms. As shown in Fig. 10.47, a huge pre-melting surface peak was observed in Pb (Frenken and van der Veen, 1985; Frenken *et al.*, 1986). This peak was already anomalously high at 500 K, which is 100° below the melting point (600.7 K). The measurements were taken using 97 keV protons aligned along the [$\bar{1}$01] axis, and the backscattered protons emerging along the [011] axis were detected using an electrostatic energy analyzer. The depth resolution was 4.1 monolayers.

A Monte Carlo computer simulation of the experiment was performed. Lattice vibrations were modeled by Gaussian probability densities of the atoms around their equilibrium positions (quasiharmonic approximation). The bulk thermal-vibration amplitude varies smoothly from 0.018 nm at room temperature to 0.028 nm just below T_m. This causes the surface peak area calculated for a bulk-like solid surface to follow curve I in Fig. 10.47(b). Curve II is obtained by also accounting for enhanced surface-vibration amplitudes and relaxations of the first two interlayer distances. At 500 K the

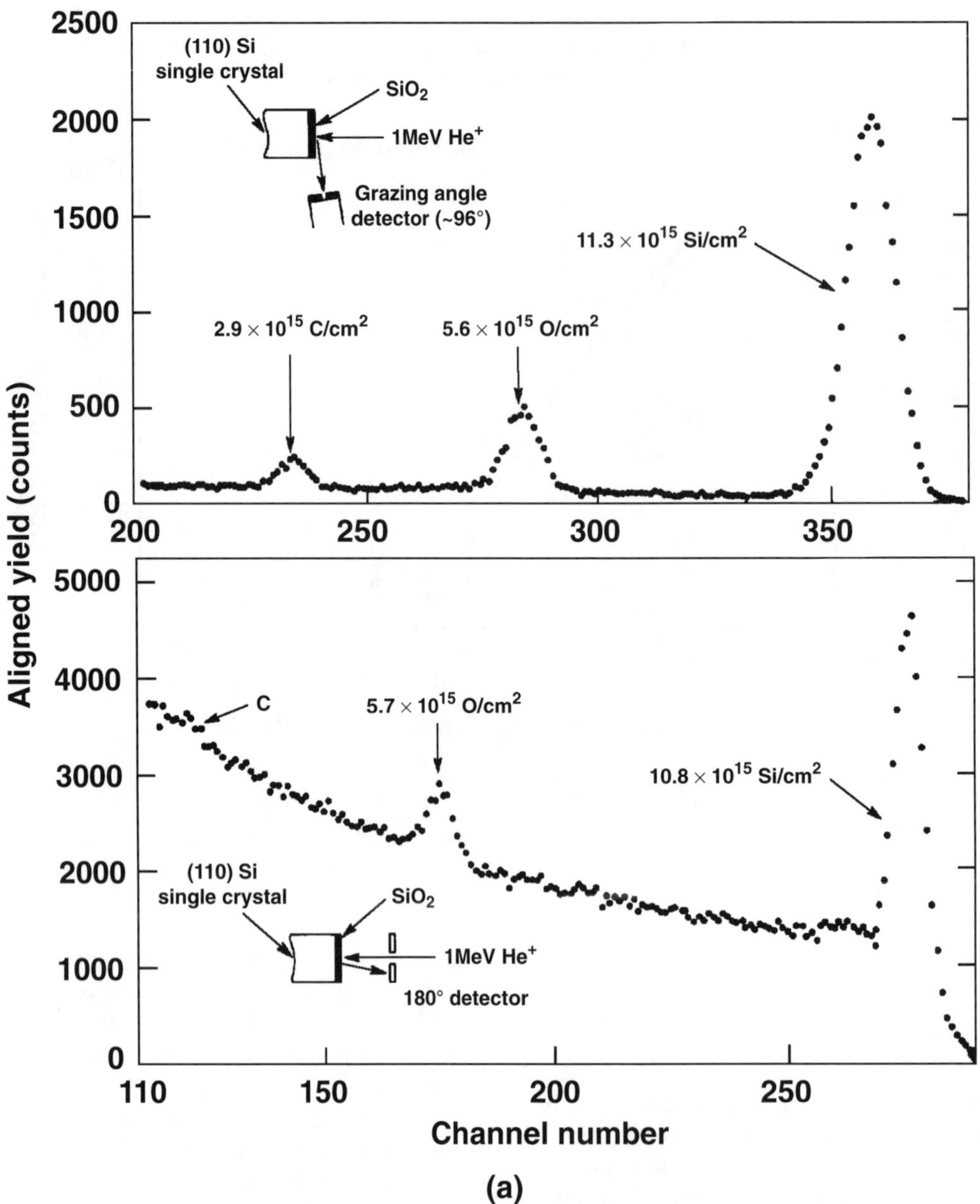

FIG. 10.46. (a) Backscattering yields of 1 MeV He ions for <110> alignment of a Si crystal (having a {110} planar surface) that was covered by 1.5 nm of oxide. Results from both a grazing angle detector and a 180° backscattering detector are shown.
(b) Measured densities of Si and O atoms for a Si(110) surface that was covered by oxides of varying thicknesses. The densities were determined from the Si and O peak areas as shown in (a). The straight line corresponds to stoichiometric SiO_2 plus 8.6×10^{15} atoms cm^{-2} of excess Si. A contribution of 6.4×10^{15} non-shadowed Si atoms per cm^2 from the Si substrate is indicated (from Jackman *et al.*, 1980).

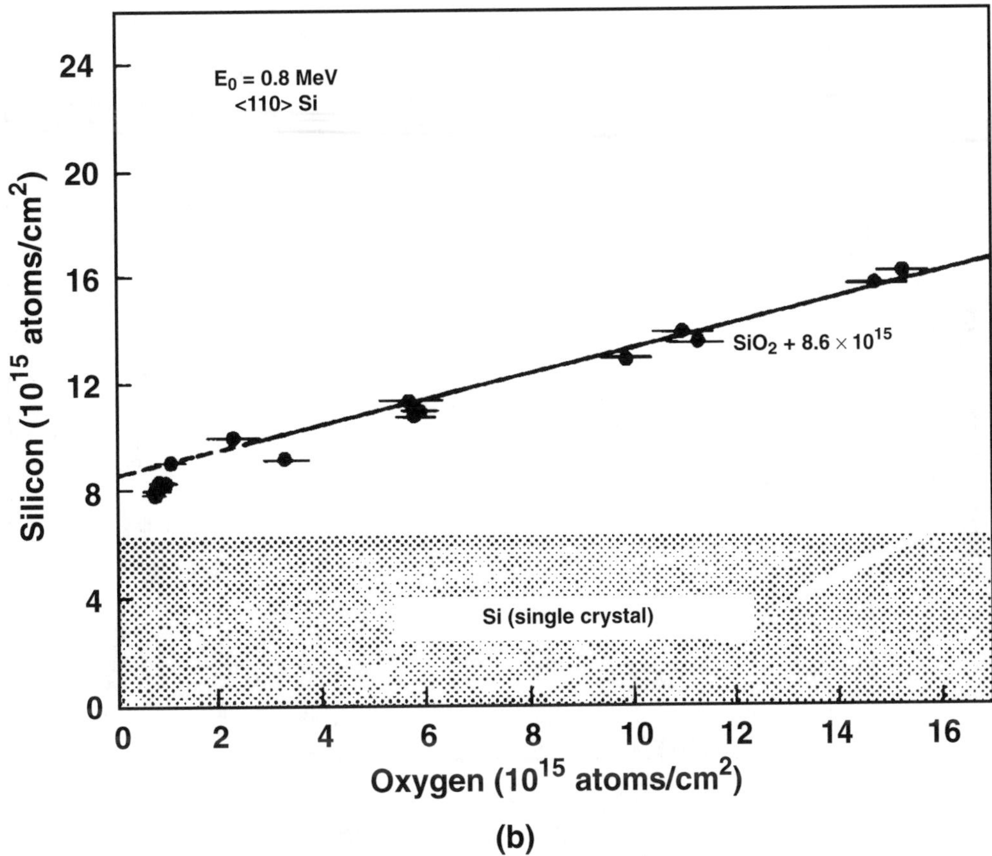

FIG. 10.46. (continued).

measured surface peak areas already exceed those in curve II. At 600.5 K the discrepancy amounts to as much as 20 extra visible lead layers. All data above 560 K, e.g., curve d in Fig. 10.47(a), can be fitted by a liquid surface layer (curve M) plus a partially ordered three monolayer-thick transition layer (curve I) lying between the liquid layer and the crystalline substrate. These were the first direct measurements of surface melting. Later measurements showed that the surface melting was strongly dependent on the orientation of the surface.

ACKNOWLEDGMENTS

The author is grateful to J.A. Davies, L.C. Feldman, L.M. Howe, S.T. Picraux, R.P Sharma, J.C. Soares, and D.O. Boerma for supplying figures. D.O. Boerma and P.J.M. Smulders provided useful comments and offered to supply computer programs. Special thanks to J.H. Barrett for helpful comments and suggestions for improving the manuscript. The help of N.R. Parikh, J.C. Austin, and W.C. Hughes is greatly appreciated. Comments from the book and chapter editors were also extremely useful.

NOTATION FOR CHANNELING

a = Thomas-Fermi screening length

a_o = Bohr radius = 5.292×10^{-11} m

d = atomic spacing along axial direction

d_o = lattice constant

d_p = interplanar spacing

e^2 = electronic charge squared = 1.44×10^{-13} cm MeV

E = ion energy

dE/dx = energy loss along the ion beam path

F_{RS} = square root of the Molière string potential [Eq. (10.7)].

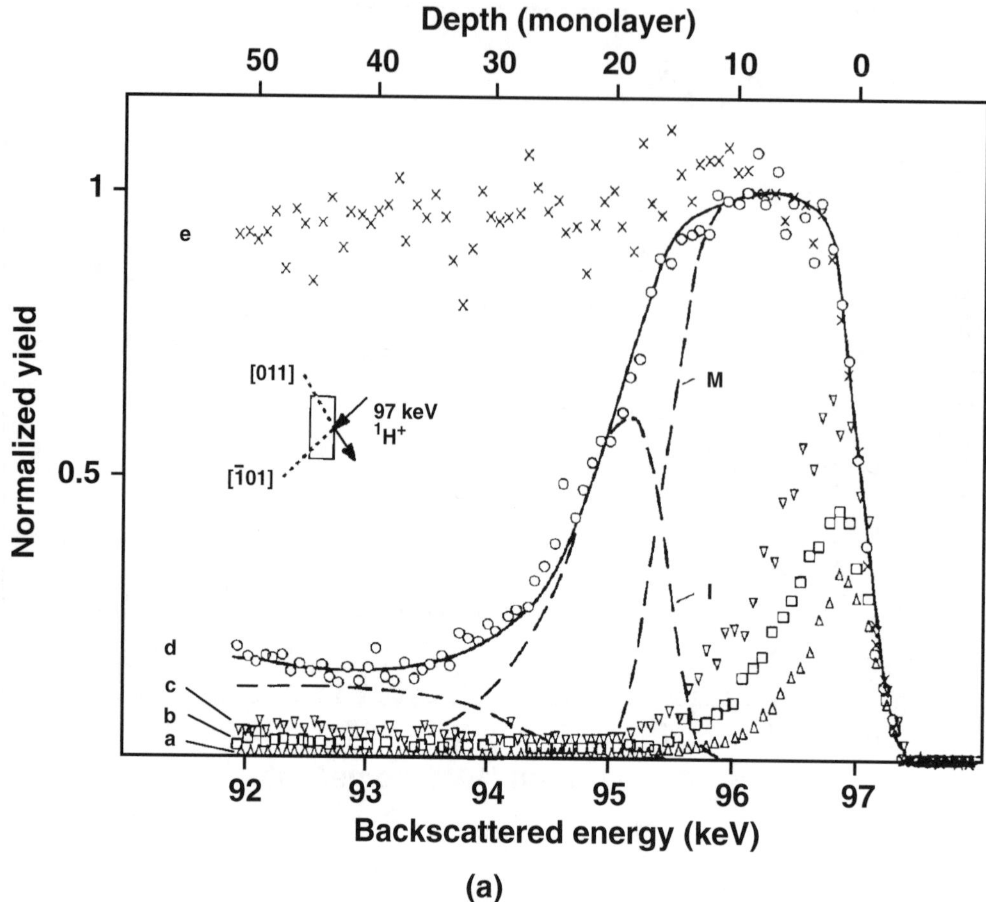

FIG. 10.47. (a) Normalized backscattering yields of 97 keV protons from Pb near the melting point, with the incident proton beam directed along the $[\bar{1}01]$ axis and the backscattered beam along the [011] axis: curve a, 295 K; curve b, 506 K; curve c, 561 K; curve d, 600.5 K; and curve e, 600.8 K. Spectrum d is fitted by a sum of contributions M from a liquid surface layer, and I from a partially ordered transition layer.
(b) Surface peak area from (a) as a function of temperature. The vertical line indicates the bulk melting point T_m. The inset is an expanded view of the highest 10 K interval. The shaded band in it corresponds to the calibration uncertainty in T_m. The arrow indicates the surface melting point. Curve I is calculated for thermal vibrations, and curve II includes enhanced surface vibrations and relaxations of the first two layers. The right hand vertical axis gives the number of molten layers from the fits using curves like M in (a). (From Frenken and van der Veen, 1985.)

F_{PS} = square root of the planar potential (Fig. A15.3).

H = number of counts per unit energy loss in an RBS spectrum [Eq. (10.34)]

M_1, M_2 = atomic masses of ions and target atoms

N = atomic density per unit volume

N_d = defect concentration per unit volume

T = crystal temperature

u_1 = one dimensional vibrational amplitude

u_2 = two dimensional vibrational amplitude (= 1.414 u_1)

x = distance along ion beam path

Z_1, Z_2 = atomic numbers of ions and target atoms

χ = normalized yield = (aligned yield)/(random yield)

χ_h = normalized yield from host atoms

χ_s = normalized yield for solute atoms

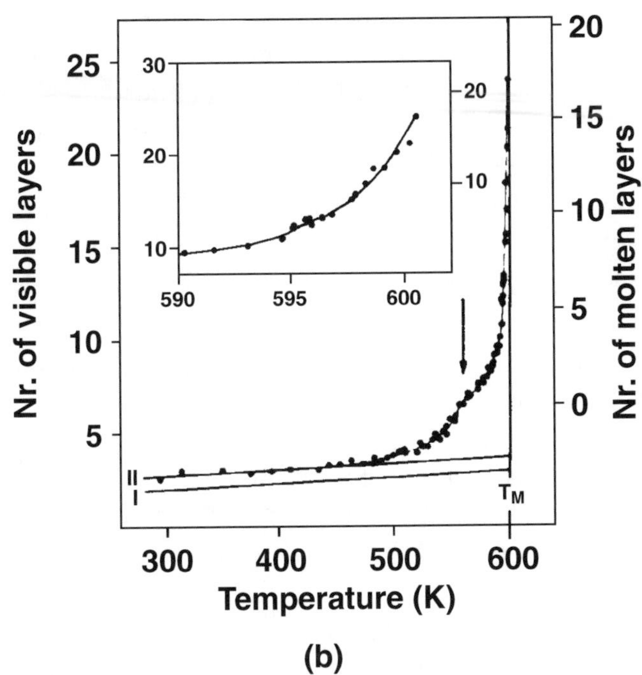

FIG. 10.47. (continued).

χ_v = normalized yield for undamaged crystal

$\chi_h^{<uvw>}$, $\chi_s^{<uvw>}$ = normalized yields for host and solute atoms for alignment along <uvw> axial channels.

$\chi_h^{\{hkl\}}$, $\chi_s^{\{hkl\}}$ = normalized yields for host and solute atoms for alignment along {hkl} planar channels.

χ_{hD} = dechanneled fraction of ions

ϕ = crystal azimuthal angle

ϕ_D = the Debye function

λ_p = dechanneling length caused by dislocations [Eq. (10.33)]

θ = crystal tilt angle

θ_D = Debye temperature (K)

σ_d = cross section for dechanneling by defects

σ_{th} = cross section for dechanneling by thermal vibrations

ψ_1 = characteristic angle for channeling

$\psi_{1/2}$ = half-width of the channeling dip

REFERENCES

GENERAL

Aono, M., Katayama, M., Nomura, E., Chasse, T., Choi, D., and Kato, M. (1989), *Nucl. Instr. Meth.* **37/38**, 264. Coaxial ion scattering spectroscopy (CAICISS).

Andersen, J.U., Augustyniak, W.M., and Uggerhøj, E. (1971), *Phys. Rev.* **B3**, 705. MeV positron channeling.

Andersen, J.U. (1980), *Nucl. Instr. Meth.* **170**, 1 and following papers. Channeling radiation.

Bøgh, E. (1968), *Canad. J. Phys.* **46**, 653. Defect studies.

Buck, T.M., Wheatley, G.H., and Jackson, D.P. (1983), *Nucl. Instr. Meth.* **218**, 257. Low energy ion scattering from Cu_3Au (see also adjacent articles in that journal).

Chu, W.K., Mayer, J.W., and Nicolet, M.-A. (1978), *Backscattering Spectrometry*, Academic Press, New York.

Culbertson, R.J., Feldman, L.C., Silverman, P.J., and Boehm, H. (1981), *Phys. Rev. Lett.* **47**, 657. Epitaxial growth (see Fig. 10.4b).

Feldman, L.C., Mayer, J.W., and Picraux, S.T. (1982), *Materials Analysis by Ion Channeling*, Academic Press, New York.

Gemmell, D.S. (1974), *Rev. Mod. Phys.* **46**, 129. Good survey of channeling.

Hofsäss, H., and Lindner, G. (1991), *Phys. Reports* **201**, 121. Emission channeling.

Howe, L.M., Swanson, M.L., and Davies, J.A. (1983), *Methods of Experimental Physics*, Vol. 21, Academic Press, New York, p. 275. Review of defect and surface studies.

Mayer, J.W., and Rimini, E. (eds.). (1977), *Ion Beam Handbook for Material Analysis*, Academic Press, New York. This is the old Ion Beam Handbook, which also includes a section on x-rays ("Catania Handbook").

Morgan, D.V. (ed.). (1973), *Channeling: Theory, Observations and Applications*, Wiley (Interscience), New York.

Schultz, P.J., Logan, L.R., Jackman, T.E., and Davies, J.A. (1988), *Phys. Rev.* **B38**, 6369. Positron channeling at <50 keV.

Swanson, M.L. (1982), *Rep. Prog. Phys.* **45**, 47. Channeling studies of lattice defects.

Uggerhøj, E. (1980), *Nucl. Instr. Meth.* **170**, 105. Channeling at GeV energies.

THEORY AND COMPUTER SIMULATIONS

Andersen, J.U. (1967), *Mat.-Fys. Medd.: K. Dan. Vidensk. Selsk.* **36**, No. 7. Comparison of experimental channeling yields with simple theory. (H in W; Fig. 10.6).

Andersen, J.U., Andreasen, O., Davies, J.A., and Uggerhøj, E. (1971), *Rad. Effects* **7**, 25. First observation of flux peaking — for lattice location of Yb in Si.

Barrett, J.H. (1971), *Phys. Rev.* **B3**, 1527. Monte Carlo computer simulations of channeling [see Eqs. (10.14), (10.15), (A15.4–A15.6), (A15.8)].

Barrett, J.H. (1990), *Nucl. Instr. Meth.* **B44,** 367. Short review of computer simulation of channeling.

Bøgh, E. (1968), *Canad. J. Phys.* **46**, 653. Study of lattice damage.

Bonderup, E., Esbensen, H., Andersen, J.U., and Schiøtt, H.E. (1972), *Rad. Effects* **12**, 261. Theory of dechanneling.

Erginsoy, C. (1965), *Phys. Rev. Lett.* **15**, 360. Early theory.

Feldman, L.C., and Appleton, B.R. (1969), *Appl. Phys. Lett.* **15**, 305. Double alignment.

Firsov, O.B. (1958), *Soviet Phys.* **JETP 6**, 534. Screening functions.

Foti, G., Grasso, F., Quattracchi, R., and Rimini, E. (1971), *Phys. Rev.* **B3**, 2169. Dechanneling.

Gemmell, D.S. (1974), *Rev. Mod. Phys.* **46**, 129. A good general review of channeling.

Howe, L.M., Swanson, M.L., and Quenneville, A.F. (1976), *Nucl. Instr. Meth.* **132**, 241. Channeling yields and dechanneling experiments in Al, Si, Cu, Ge.

Lehmann, C., and Leibfried, G. (1963), *J. Appl. Phys.* **34**, 2821. Early theory.

Lindhard, J. (1965), *Mat.-Fys. Medd.: K. Dan. Vidensk. Selsk.* **34**, No. 14. Comprehensive theory of channeling — the classic in the field.

Lindhard, J., Nielsen, V., and Scharff, M. (1968), *Mat.-Fys. Medd.: K. Dan. Vidensk. Selsk.* **36**, No. 10.

Matsunami, N., and Itoh, N. (1973), *Phys. Lett.* **43A**, 435. Theory of dechanneling.

Matsunami, N., Swanson, M.L., and Howe, L.M. (1978), *Canad. J. Phys.* **56**, 1057. Analytical calculations of equipotential contours and ion flux distributions (Figs. 10.7, 10.8).

Merkle, K.L., Pronko, P.P., Gemmell, D.S., Mikkelson, R.C., and Wrobel, J.R. (1973), *Phys. Rev.* **B8**, 1002. Simple dechanneling theory.

Molière, G. (1947), *Z. Naturf.* **a 2**, 133. Formulation of interatomic potential.

Morgan, D.V., and Van Vliet, D. (1972), *Rad. Effects* **12**, 203. Computer simulations.

Nelson, R.S., and Thompson, M.W. (1963), *Phil. Mag.* **8**, 1677. Early theory.

Nielsen, B. Bech. (1988), *Phys. Rev.* **B37**, 6353. Analytical calculations of ion flux distributions.

Piercy, G.R., Brown, F., Davies, J.A., and McCargo, M. (1963), *Phys. Rev. Lett.* **10**, 399. First experimental evidence for channeling — from range studies in Al.

Robinson, M.T., and Oen, U.S. (1963a), *Appl. Phys, Lett.* **2**, 30.

Robinson, M.T., and Oen, U.S. (1963b), *Phys. Rev.* **132**, 2385. First predictions of channeling by computer simulations.

Smulders, P.J.M., and Boerma, D.O. (1987), *Nucl. Instr. Meth.* **B29**, 471. Computer simulations of channeling.

Stark, J. (1912), *Phys. Z.* **13**, 973. First prediction of ion channeling.

Tromp, R.M., and van der Veen, J.F. (1983), *Surf. Sci.* **133**, 137. Computer simulations for surface studies.

van der Veen, J.F. (1985), *Surf. Sci. Rept* **5**, 199. Computer simulations and review of surface studies.

Van Vliet, D. (1971), *Rad. Effects* **10**, 137. Analytical calculations of ion flux distributions, and depth oscillations of ion flux.

EXPERIMENTAL METHODS

Barrett, J.H. (1978), *Nucl. Instr. Meth.* **149**, 341. Monte Carlo computer simulations for precise lattice location of impurity atoms.

Bøttiger, J., Davies, J.A., Lori, J., and Whitton, J.L. (1973), *Nucl. Instr. Meth.* **109**, 579. Low temperature setup for channeling measurements.

Chu, W.K., Mayer, J.W., and Nicolet, M.-A. (1978), *Backscattering Spectrometry*, Academic Press, New York. Experimental setup and alignment description.

Chu, W.K., and Wu, D.T. (1988), *Nucl. Instr. Meth.* **B35**, 518. Coincidence methods of analysis.

Davies, J.A., Denhartog, J., Eriksson, L., and Mayer, J.W. (1967), *Canad. J. Phys.* **45**, 4053. Early channeling studies in implanted Si using a manual two-axis goniometer.

Davies, J.A., and Norton, P.R. (1980), *Nucl. Instr. Meth.* **168**, 611. Combination of channeling and nuclear reaction analysis with other surface science methods.

Ecker, K.H. (1983), *J. Nucl. Mat.* **119**, 301. Lattice location in metals using characteristic x-rays produced by channeled protons.

Feldman, L.C., Mayer, J.W., and Picraux, S.T. (1982), *Materials Analysis by Ion Channeling*, Academic Press, New York. General review and Fig. 10.10.

Gemmell, D.S. (1974), *Rev. Mod. Phys.* **46**, 129. Channeled versus random stopping powers.

Gemmell, D.S. (1971), *Nucl. Instr. Meth.* **91**, 15. Early computerized goniometer.

Hofsäss, H.C., Parikh, N.R., Swanson, M.L., and Chu, W.K. (1991), *Nucl. Instr. Meth.* **B58**, 49. Coincidence method for enhancing the sensitivity for elastic recoil analysis.

Lennard, W.N., Swanson, M.L., Eger, D., Springthorpe, A.J., and Shepherd, F.R. (1988), *J. Electr. Mater.* **17**, 1. Lattice positions of Zn in InP using characteristic x-ray yields.

Picraux, S.T. (1981), *Nucl. Instr. Meth.* **182/183**, 413. Lattice location of gas atoms in metals using axial and planar angular scans.

Picraux, S.T., Follstaedt, D.M., Baeri, P., Campisano, S.U., Foti, G., and Rimini, E. (1980), *Rad. Effects* **49**, 75. Dechanneling effect of dislocations using planar channeling. (See Fig. 10.27.)

Schiøtt, H.E., and Thomsen, P.V. (1972), *Rad. Effects* **14**, 39. Calculation of ion damage.

Smulders, P.J.M., and Boerma, D.O. (1987), *Nucl. Instr. Meth.* **B29**, 471. Computer simulations of channeling for precise lattice location of impurities.

Swanson, M.L., Offermann, P., and Ecker, K.H. (1979), *Canad. J. Phys.* **57**, 457. Reduced damage from channeled ions.

Swanson, M.L., and Maury, F. (1975), *Canad. J. Phys.* **53**, 1117. Calculation of ion damage in Al, and channeling studies of defect-induced solute atom displacements.

Ziegler, J. (1991), *Transport of Ions in Matter* (TRIM), IBM Corp. Software for Monte Carlo calculations of ion ranges and damage.

CRYSTAL ALIGNMENT

Amoros, J.L., Buerger, M.J., and de Amoros, M.C. (1975), *The Laue Method*, Academic Press, New York. Description of Laue x-ray diffraction patterns.

Barrett, C.S. (1952), *Structure of Metals*, McGraw-Hill, New York. Elementary crystallography.

Cullity, B.D. (1967), *Elements of X-Ray Diffraction*, Addison-Wesley, London. Description of stereographic projection and crystal orientation using Laue back reflection x-ray patterns.

Quenneville, A.F. (1979), Atomic Energy of Canada Report CRNL #1926. X-ray patterns illustrating mosaic spread and surface deformation.

Smaill, J.S. (1972), *Metallurgical Stereographic Projections*, Wiley (Halsted Press), New York. Many stereographic projections of non-cubic structures.

Whitton, J.L. (1965), *J. Appl. Phys.* **36**, 3917. Vibratory polishing methods.

Whitton, J.L. (1969), *Proc. Roy. Soc.* **A311**, 63. Polishing methods.

EPITAXIAL LAYERS AND STRAIN

Chiu, K.C.R., Poate, J.M., Feldman, L.C., and Doherty, C.J. (1980), *Appl. Phys. Lett.* **36**, 544. Epitaxy of $NiSi_2$ on Si (see Fig. 10.24).

Culbertson, R.J., Feldman, L.C., Silverman, P.J., and Boehm, H. (1981), *Phys. Rev. Lett.* **47**, 657. Measurement of the initial registry of epitaxial layers.

Ellison, J.A., Picraux, S.T., Allen, W.R., and Chu, W.K. (1988), *Phys. Rev.* **B37**, 7290. Also Picraux, S.T., Biefeld, R.M., Allen, W.R., Chu, W.K., and Ellison, J.A. (1988), *Phys. Rev.* **B38**, 11086; Allen, W.R., Chu, W.K., Picraux, S.T., Biefeld, R.M., and Ellison, J.A. (1989), *Phys. Rev.* **B39**, 3954; Chu, W.K., Allen, W.R., Picraux, S.T., and Ellison, J.A. (1990), *Phys. Rev.* **B42**, 5923. Planar channeling in superlattices; the second paper gives experimental details of catastrophic dechanneling.

Feldman, L.C., Bevk, J., Davidson, B.A., Gossmann, H.-J., and Mannaerts, J.P. (1987), *Phys. Rev. Lett.* **59**, 664. Epitaxy of Ge on Si.

Ishiwara, H., and Furukawa, S. (1976), *J. Appl. Phys.* **47**, 1686. Mosaic spread.

Ishiwara, H., Hikosaka, K., Nagatomo, M., and Furukawa, S. (1979), *Surf. Sci.* **86**, 711. Channeling study of the mosaic spread of silicides (see Fig. 10.25).

Lyman, P.F., Thevuthasan, S., and Seiberling, L.E. (1991), *J. Crystal Growth* **113**, 45. Transmission channeling studies of Ge layers grown on Si.

Picraux, S.T., Chu, W.K., Allen, W.R., and Ellison, J.A. (1986), *Nucl. Instr. Meth.* **B15**, 306. Strain in superlattices.

Picraux, S.T., Dawson, L.R., Osbourn, G.C., and Bielfeld, R.M. (1983), *Appl. Phys. Lett.* **43**, 1020. Strain in superlattices.

DAMAGE MEASUREMENTS

Baeri, P., Campisano, S.U., Ciavola, G., and Rimini, E. (1976), *Nucl. Instr. Meth.* **132**, 237. Defect displacements from channeling data.

Bøgh, E. (1968), *Canad. J. Phys.* **46**, 653. Study of lattice damage.

Bonderup, E., Esbensen, H., Andersen, J.U., and Schiøtt, H.E. (1972), *Rad. Effects* **12**, 261. Theory of dechanneling.

Carstanjen, H.-D. (1980), *Phys. Stat. Sol.* (a)**59**, 11.

Carstanjen, H.-D. (1989), *Z. Physikal. Chemie* **165**, 141. Effect of lattice vibrations on channeling yields from deuterium.

Chu, W.K., Mayer, J.W., and Nicolet, M.-A. (1978), *Backscattering Spectrometry*, Academic Press, New York.

Csepregi, L., Mayer, J.W., and Sigmon, T.W. (1976), *Appl. Phys. Lett.* **29**, 92. RBS analysis of epitaxial growth of amorphized Si (see Fig. 10.28).

Eisen, F.H. (1973), *Channeling: Theory, Observations and Applications,* D.V. Morgan, (ed.), Wiley (Interscience), New York, p 415. Dechanneling by buried damaged layers (see Fig. 10.29).

Feldman, L.C., Mayer, J.W., and Picraux, S.T. (1982), *Materials Analysis by Ion Channeling*, Academic Press, New York.

Gemmell, D.S. (1974), *Rev. Mod. Phys.* **46**, 129. Review.

Howe, L.M., Swanson, M.L., and Davies, J.A. (1983), *Methods of Experimental Physics*, Vol. 21, Academic Press, New York, p. 275. Review.

Howe, L.M., Swanson, M.L., and Quenneville, A.F. (1976), *Nucl. Instr. Meth.* **132**, 241. Channeling yields and dechanneling experiments in Al, Si, Cu, Ge.

Keil, E., Zeitler, E., and Zinn, W. (1960), *Z Naturforsch.* **A15**, 1031. Plural scattering theory.

Matsunami, N., and Itoh, N. (1973), *Phys. Lett.* **43A**, 435. Theory of dechanneling.

Merkle, K.L., Pronko, P.P., Gemmell, D.S., Mikkelson, R.C., and Wrobel, J.R. (1973), *Phys. Rev.* **B8**, 1002. Dechanneling experiments.

Mory, J., and Quéré, Y. (1972), *Rad. Effects* **13**, 57. Theory of dechanneling by dislocations.

Picraux, S.T., Follstaedt, D.M., Baeri, P., Campisano, S.U., Foti, G., and Rimini, E. (1980), *Rad. Effects* **49**, 75. Dechanneling effect of dislocations, using planar channeling (see Fig. 10.27).

Quéré, Y. (1974), *J. Nucl. Mat.* **53**, 262. Energy dependence of dechanneling.

Quéré, Y. (1978), *Rad. Effects* **38**, 131. Dechanneling by dislocations.

Rimini, E. (1978), *Material Characterization Using Ion Beams* J.P. Thomas and A. Cachard (eds.), Plenum, New York, p. 455. Damage distributions from dechanneling data.

Sharma, R.P., Rehn, L.E., Baldo, P.M., and Liu, J.Z. (1988), *Phys. Rev.* **B38**, 9287. (1989), *Phys. Rev. Lett.* **62**, 2869. Discontinuity of channeling widths in high Tc superconductors at the transition temperature. (See Fig. 10.30.)

Swanson, M.L. (1982), *Rep. Prog. Phys.* **45**, 47. Review of channeling studies of damage.

Swanson, M.L., Howe, L.M., Jackman, T.E., and Moore, J.A. (1982), *Nucl. Instr. Meth.* **194**, 165. Dechanneling by irradiation-induced defects (Fig. 10.26).

LATTICE SITES OF SOLUTE ATOMS

Alexander, R.B., and Poate, J.M. (1972), *Rad. Effects* **12**, 211. Angular scans for substitutional impurities: 2% Au in Cu (Fig. 10.34).

Andersen, J.U., Andreasen, O., Davies, J.A., and Uggerhøj, E. (1971), *Rad.Effects* **7**, 25. First observation of flux peaking–for lattice location of Yb in Si.

Andersen, J.U., Chechenin, N.G., and Zhang, Z.H. (1982), *Nucl. Instr. Meth.* **194**, 129. Lattice sites of solute atoms in GaP.

Barrett, J.H. (1978), *Nucl. Instr. Meth.* **149,** 341. Monte Carlo computer simulations for precise lattice location of impurity atoms.

Barrett, J.H. (1988). *Phys. Rev.* **B38,** 5069. Analysis of lattice vibrations in Cu - 2% Au (Fig. 10.34).

Barrett, J.H. (1990), *Nucl. Instr. Meth.* **B44**, 367. Short review of computer simulation of channeling.

Bugeat J.P., and Ligeon, E. (1979), *Phys. Letters* **71A**, 93. Lattice sites of H in f.c.c. metals.

Carstanjen, H.D. (1980), *Phys. Stat. Sol.* (a)**59**, 11. Review of lattice location of gas atoms.

Carstanjen, H.D., Dunstl, J., Lobl, G., and Sizmann, R. (1978), *Phys. Stat. Sol.* (a)**45**, 529. Lattice location of D in Pd. (See Fig. 10.38.)

Davies, J.C. (1973), *Channeling: Theory, Observations and Applications*, D.V. Morgan (ed.), Wiley (Interscience), New York, p. 391. General lattice location methods.

Eriksson, L., and Davies, J.A. (1969), *Ark. Fys.* **39**, 439. Variation of channel width with the atomic mass of constituent elements in compounds.

Feldman, L.C., Mayer, J.W., and Picraux, S.T. (1982), *Materials Analysis by Ion Channeling*, Academic Press, New York. General review.

Knapp, J.A., and Follstaedt, D.M. (1981), *Nucl. Instr. Meth.* **182/183**, 1017. Channeling study of solute atom location in precipitates in steel.

Gemmell, D.S. (1974), *Rev. Mod. Phys.* **46**, 129. General procedures.

Howe, L.M., Swanson, M.L., and Davies, J.A. (1983), *Methods of Experimental Physics*, Vol. 21, Academic Press, New York, p. 275. General review and Figs. 10.32 and 10.35.

Matsunami, N., Swanson, M.L., and Howe, L.M. (1978), *Canad. J. Phys.* **56**, 1057. Analytical calculations of equipotential contours and ion flux distributions (Figs. 10.7 10.8 and 10.33).

Nielsen, B. Bech (1988), *Phys. Rev.* **B37**, 6353. Analytical calculations of ion flux distributions.

Picraux, S.T. (1975), *New Uses of Ion Accelerators*, J.F. Ziegler (ed.), Plenum Press, New York, p. 229. General procedures.

Picraux, S.T. (1981), *Nucl. Instr. Meth.* **182/183**, 413. Lattice position of trapped D in Fe.

Picraux, S.T., Brown, W.L., and Gibson, W.M. (1972), *Phys. Rev.* **B6**, 1382. Lattice position of Bi implanted into Si.

Picraux, S.T., Davies, J.A., Eriksson, L., Johansson, N.G.E., and Mayer, J.W. (1969), *Phys. Rev.* **180**, 873. Variation of channel width with constituent elements in compounds.

Rebouta, L., Soares, J.C., daSilva, M.F., Sanz-Garcia, J.A., Dieguez, E., and Agullo-Lopez, F. (1990), *Nucl. Instr. Meth.* **B50**, 428.

Rebouta, L., Soares, J.C., daSilva, M.F., Sanz-Garcia, J.A., Dieguez, E., and Agullo-Lopez, F. (1992), *J. Mater. Res.* **7**, 130. Lattice positions of Ta and Hf in $LiNbO_3$ (Fig. 10.36).

Smulders, P.J.M., and Boerma, D.O. (1987), *Nucl. Instr. Meth.* **B29**, 471. Computer simulations of channeling.

Smulders, P.J.M., Boerma, D.O., Nielsen, B. Bech, and Swanson, M.L. (1990), *Nucl. Instr. Meth.* **B45**, 438. Lattice location of B in Si using analytical and computer simulations of channeling.

Swanson, M.L. (1982), *Rep. Prog. Phys.* **45**, 47. Review, including effects of lattice defects.

Swanson, M.L., and Howe, L.M. (1983), *Nucl. Instr. Meth.* **218**, 613. Tetrahedral interstitial sites of In atoms in Al caused by trapped vacancies (Fig. 10.39).

Swanson, M.L., Howe, L.M., Saris, F.W., and Quenneville, A.F. (1981), *Defects in Semiconductors*, J. Narayan and T.Y. Tan (eds.), North-Holland, Amsterdam, p. 71. Lattice sites of B in Si.

Swanson, M.L., Maury, F., and Quenneville, A.F. (1973), *Phys. Rev Lett.* **31**, 1057. Lattice sites of "mixed dumbbell" interstitials in Al.

Swanson, M.L., Parikh, N.R., Sandhu, G.S., Frey, E.C., Zhang, Z.H., and Chu, W.K. (1989), Mat. Res. Soc. Symp. Proc. **144**, Materials Research Society, Pittsburgh, Pennsylvania, p. 409. Using depth oscillations of planar channeling yields to locate solute atoms in compound semiconductors.

Yagi, E., Kobayashi, T., Nakamura, S., Fukai, Y., and Watanabe, K. (1985), *Phys. Rev.* **B31**, 1640. Lattice position of H in V — effect of stress.

Watkins, G.D. (1975), *Phys. Rev.* **B12**, 5824. EPR studies of interstitial B atoms in Si.

SURFACE AND INTERFACE STUDIES

Barrett, J.H. (1971), *Phys. Rev.* **B3**, 1527. Monte Carlo computer simulations of channeling [see Eqs. (10.14), (10.15), (A15.4–A15.6), (A15.8)].

Bøgh, E. (1973), *Channeling: Theory, Observations and Applications,* D.V. Morgan (ed.), Wiley (Interscience), New York, p. 435. Review.

Culbertson, R.J, Feldman, L.C., Silverman, P.J., and Boehm, H. (1981), *Phys. Rev. Lett.* **47**, 657. Epitaxial growth of Au on Ag (see Fig. 10.4).

Davies, J.A., Jackson, D.P., Norton, P.R., Posner, D.E., and Unertl, W.N. (1980), *Solid State Commun.* **34**, 41. Surface relaxation of Pt.

Davies, J.A., Jackson, D.P., Matsunami, N., Norton, P.R., and Andersen, J.U. (1978), *Surf. Sci.* **78**, 274. Surface relaxation of Pt.

Davies, J.A., and Norton, P.R. (1980), *Nucl. Instr. Meth.* **168**, 611. Reconstruction of Pt surfaces (Fig. 10.45).

Feldman, L.C. (1979), *ISISS 1979 Surface Science*, CRC Press, Cleveland. Effect of surface impurities on reconstruction.

Feldman, L.C., Kauffman, R.L., Silverman, P.J., Zuhr, R.A., and Barrett, J.H. (1977), *Phys. Rev. Lett.* **39**, 38. Effect of thermal vibrations on channeling surface peak.

Feldman, L.C., Mayer, J.W., and Picraux, S.T. (1982), *Materials Analysis by Ion Channeling*, Academic Press, New York. A good review.

Feldman, L.C., Silverman, P.J., Williams, J.S., Jackman, T.E., and Stensgaard, I. (1978), *Phys. Rev. Lett.* **41**, 1396. Si-SiO_2 interfaces in thin Si crystals.

Fischer, A.E.M.J., Gustafsson, T., and van der Veen, J.F. (1988), *Phys. Rev.* **B37**, 6305. Silicide/Si interface bonding.

Frenken, J.W.M., and van der Veen, J.F. (1985), *Phys. Rev. Lett.* **54**, 134. Surface melting of Pb, Fig. 10.47.

Frenken, J.W.M., Maree, P.M.J., and van der Veen, J.F. (1986), *Phys. Rev.* **B34**, 7506. Surface melting of Pb (Fig. 10.47).

Howe, L.M., Swanson, M.L., and Davies, J.A. (1983), *Methods of Experimental Physics*, Vol. 21, Academic Press, New York, p. 275.

Jackman, T.E., MacDonald, J.R., Feldman, L.C., Silverman, P.J., and Stensgaard, I. (1980), *Surf. Sci.* **100**, 35. Study of thin layers of SiO_2, and the Si-SiO_2 interface (Fig. 10.46).

Jackson, D.P., and Barrett, J.H. (1979), *Phys. Lett.* **71A**, 359. Effect of correlated thermal vibrations on the surface channeling peak.

Swanson, M.L. (1982), *Rep. Prog. Phys.* **45**, 47. Review.

Tromp, R.M., and van der Veen, J.F. (1983), *Surf. Sci.* **133**, 137. Computer simulations for surface studies.

Turkenburg, W.C., Soszka, W., Saris, F.W., Kersten, H.H., and Colenbrander, B.G. (1976), *Nucl. Instr. Meth.* **132**, 587. Double alignment for surface studies (Fig. 10.44).

van der Veen, J.F. (1985), *Surf. Sci. Rept.* **5**. 199. Review of surface studies.

van der Veen, J.F., Smeenk, R.G., Tromp, R.M., and Saris, F.W. (1979a), *Surf. Sci.* **79**, 212. Relaxation of Ni surfaces [Fig. 10.44(d)].

van der Veen, J.F., Smeenk, R.G., Tromp, R.M., and Saris, F.W. (1979b), *Surf. Sci.* **79**, 219. Relaxation of Pt surfaces.

van der Veen, J.F., Tromp, R.M., Smeenk, R.G., and Saris, F.W. (1979c), *Surf. Sci.* **82**, 468. Position of S on Ni surfaces.

INSTRUMENTATION AND LABORATORY PRACTICE

R. A. Weller

Vanderbilt University, Nashville, Tennessee

CONTENTS

11.1 INTRODUCTION

Although there are many aspects of instrumentation which are important for specific tasks which arise in connection with ion beam analyses, five general areas are of special significance. These are ion beam production, transport, and diagnostics; particle detectors; pulse electronics; vacuum technology and thin film fabrication; and statistics and data handling. Anyone who is active in the field of ion beam analysis can expect to have to deal with some or all of these areas routinely. This chapter consists of brief overviews of each of these areas of technology emphasizing the relevant topics for ion beam analytical work and suggesting more detailed references for further study.

Most of the technology which is in current use by materials scientists for ion beam analysis was originally developed for the study of nuclei and may be accessed bibliographically through such key words as "nuclear electronics" and "radiation detection" and in journals such as *Nuclear Instruments and Methods in Physics Research*, *Nuclear Physics*, and the *IEEE Transactions on Nuclear Science*. Because much of the instrumentation needed for ion beam analysis is also widely used for fundamental nuclear studies, environmental monitoring, and nuclear medicine, it is possible to build a laboratory almost exclusively with commercial equipment. The manufacturers of this equipment are often excellent sources of technical information on both the theory and applications of their products. Students beginning careers in materials analysis should be especially encouraged to seek out manufacturer's catalogs and other literature for the educational value which they can derive from them.

11.2 ION SOURCES, ACCELERATORS, AND BEAMS

The steps in the generation of an ion beam which is suitable for analytical work are:

- production of ions from neutral atoms or molecules,
- acceleration of these ions by electric fields,
- selection of ions with specific mass and energy, and,
- the transport of the ions to the specimen to be analyzed.

11.2.1 Ion sources

Many ion sources are in common use in ion beam work; they depend upon a wide range of physical principles for their operation (Brown, 1989). The polarity of ions which is needed depends upon the type of accelerator used. In general, a positive ion can be produced from any ion in the periodic table and some ion sources such as the Penning source are capable of producing multiply-charged ions of high order. This can be quite useful since, for a given accelerating voltage, the energy of a projectile increases with its charge state. Negative ions are more difficult to obtain and in some cases, such as nitrogen, are so unstable that for practical purposes they cannot be produced. Single-ended accelerators, those with only one stage of acceleration, always use positive ions. If they did not, they would produce a parasitic beam of electrons that would generate a very large background of x-ray radiation. Tandem accelerators of the kind often used for Rutherford backscattering analysis require ions to be initially negative.

Most ion sources used for ion beam analytical work depend upon collisions between particles to generate ions. In sources in which the original atoms or molecules are gaseous, electron bombardment is the primary process that leads to ionization. These sources can be used, however, only for elements which are volatile or have volatile compounds. Sources which produce ions by sputtering can be used to produce a wider assortment of beams but often at reduced current. Sputter sources can be designed to produce either positive or negative ions. Plasma sources are generally much more efficient at producing positive ions. One gas-fed source which can produce large quantities of both polarities is the duoplasmatron. This general purpose source is in widespread use. Another commonly encountered gas-fed source is the radio frequency or rf source. In these devices a plasma is produced in a bottle of low pressure gas by the application of a radio frequency signal of around 100 MHz. These "rf" sources are often found in the terminals of single-ended electrostatic accelerators.

An interesting special purpose ion source that is distinguished by its simplicity and ease of use is the surface ionization source[1]. It consists of a porous tungsten slug which has been loaded with an alkali metal salt. When heated, the device emits alkali metal ions. The ionization

1. Spectra-Mat, Inc., 1240 Highway 1, Watsonville, CA 95076.

occurs because of the large difference in work function between tungsten and the alkali metals. The only input required to operate the source is power to the heater and, of course, a bias to accelerate the ions away from the surface. Surface ionization sources produce beams with very narrow energy dispersion and are useful for a wide range of tests and calibrations.

11.2.2 Accelerators

The next stage in beam production is the accelerator. Exhaustive discussions of the types of accelerators now used widely for materials analysis are found in England (1974). Voltages below approximately 500 kV can be produced and maintained with reasonable success using only air insulation. Above this value, it becomes necessary to surround the terminal of the main high voltage power supply with a pressurized gas. Sulfur hexafluoride, SF_6, is perhaps the best insulating gas, although various mixtures of CO_2 and N_2 are also used. The vessels that contain the insulating gas give accelerators their characteristic tank-like look. Voltages up to about 400 kV can be generated with relatively conventional power supplies which use voltage multiplication circuitry similar to the original design of Cockcroft and Walton (England, 1974). Above about 500 kV other techniques must be used to generate high voltage. By far the most widely used device for the production of MeV beams is the Van de Graaff generator, named for its inventor, Robert J. Van de Graaff. An engaging history of this ingenious device has been compiled by Bromley (1974). We will limit our further discussion of accelerators to electrostatic accelerators of this type. Discussions of the principles by which other accelerators operate can be found in England's book or the more recent monograph (Humphries, 1986).

Van de Graaff generators operate by mechanically moving electrical charge from ground potential to the high voltage terminal. They depend crucially for their operation upon the fact often encountered in a first course on electricity and magnetism that charge always resides on the exterior surface of a conductor. In older Van de Graaffs, charge is sprayed from sharp needles raised to a potential of a few tens of keV onto a wide rubber-impregnated endless belt that extends from ground to the high voltage terminal. This belt moves continuously on rollers in such a way that charge is carried up to and physically inside the high voltage terminal. The energy to do this is supplied by an electric motor which propels the belt. Once the charge on the belt has entered the terminal, it no longer feels a force. It is removed from the belt by another set of needles which are electrically connected to the terminal and flows to the terminal's surface. The "pelletron" accelerator[2] is a variant of the Van de Graaff design that uses chains of alternating metallic and dielectric links in place of continuous belts to transport the charge. This scheme avoids the problem of patchy charging of the belt and has reduced the associated voltage ripple on the terminal substantially.

The capacitance, C, of a Van de Graaff terminal is typically of order 100 pF. The charge, q, required to establish a given voltage, V, follows from the usual relation $q = CV$. The voltage is distributed uniformly between the terminal and ground by a chain of high voltage resistors and specially engineered equipotential surfaces. These are designed to minimize sparks which can be quite destructive in accelerators with large stored energy ($CV^2/2$). A movable electrode covered with needles called corona points is provided to stabilize the voltage. The voltage of the terminal can be measured by a device called a generating voltmeter. This device is a variation on the concept of a vibrating reed electrometer which is mounted inside the accelerator tank. It is useful for measurements at the few percent level but not accurate enough to serve as the principal reference for the terminal voltage, although it is sometimes used in this way.

All accelerators which are routinely used for materials analysis have terminals which are positively charged. In single-ended accelerators (Fig. 11.1a), the ion source is placed inside the high voltage terminal. A modest voltage (typically from a few kV to perhaps 150 kV) is used to inject ions into the accelerating column. The energy with which ions arrive at ground potential is just nV in electron volts where n is the ion's charge. The design of single-ended accelerators is complicated by the need to provide power to the ion source while it is thousands to millions of volts above ground potential. In lower voltage accelerators, the ion source is powered directly from the AC power grid through an isolation transformer or

2. "Pelletron" is a trade name of National Electrostatics Corporation, Middleton, WI 53562. National Electrostatics was founded by R.G. Herb of the University of Wisconsin, one of the pioneers in the development of electrostatic accelerators. Van de Graaff's company, High Voltage Engineering Corporation (known universally as HVEC), no longer designs new accelerators in the United States but does have a European offspring, HVEC Europa, that is an active participant in the small accelerator market.

by a generator in the terminal driven by a motor at ground potential through an insulating shaft. In higher voltage machines the terminal generator is driven directly by the belt or charging chains. Access to the ion source is also difficult because it is inside the pressure vessel. As a result, single-ended accelerators are usually limited in the range of ions that can be conveniently produced. Additional flexibility can be realized by installing an insulated gas feed line to the ion source so that gas changes can be made at a manifold external to the pressure vessel.

Tandem accelerators (Fig. 11.1b) use a clever trick to multiply the energy of the beam. Negative ions are produced at ground potential and accelerated to the positive terminal located midway along the beam tube structure (Rose and Galejs, 1967). Inside the terminal, they encounter either a thin carbon foil (typically 2-5 µg/cm^2) or a tenuous gas (usually N_2 or H_2 to minimize scattering). Because of their high velocity, they are stripped of electrons through collisions in this region and emerge from the terminal with positive charge. Since the terminal polarity is also positive, they are again accelerated. The total energy which they acquire is (n+1)V where n is the final charge state of the ion and V the terminal voltage. Ordinarily, there will be a range of final charge states produced at a given terminal voltage. To obtain a desired final beam energy one must choose the terminal voltage and select a specific charge state for the final ion. Tandems can produce ion beams that are far more energetic than those available from single-ended accelerators with conventional ion sources, because n can be much larger. As a result, tandems are more appropriate for techniques such as nuclear reaction analysis where high energies and heavy ion beams such as ^{15}N or ^{19}F may be needed.

11.2.3 Beams

The ion beam that emerges from the accelerator typically contains a range of energies and even a multiplicity of species. These result from the imperfect vacuum of the beam tube, poor quality of the beam injected into the accelerator, and processes such as scattering and charge exchange that occur during acceleration. To extract a beam suitable for materials analysis, it is necessary to reject all ions that do not have the desired mass and energy. This is almost universally accomplished by passing the beam into the field of a magnet, called an analyzing magnet, whose field has been tuned to permit only the desired species to pass through its narrow entrance and exit slits. Under most circumstances, this means that it is possible to select a unique beam with the desired mass and energy. However, this is not always the case. Magnets actually filter the ions by momentum per unit charge or, equivalently, by the magnetic rigidity

$$B\rho = \left(\frac{2mE}{q^2}\right)^{1/2} \tag{11.1}$$

where B is the magnetic induction, m is the particle's mass, E its energy, q its charge, and ρ its radius of curvature, which is established by the magnet's entrance and exit slits. As a result, interferences can occur. Perhaps the most common of these is encountered when trying to accelerate He^{2+} ions in a single-ended accelerator. In this case, the hydrogen molecular ion H_2^+, with half the charge, half the energy and half the mass of the He^{2+} ion, cannot be distinguished by a magnet. It is sometimes hard to detect this condition by looking at backscattering spectra alone, since backscattered protons can be interpreted as ordinary structure in an otherwise unremarkable He spectrum. A particularly insidious example occurs when analyzing a specimen that is expected to have isolated peaks whose masses are not known *a priori*. See Chapter 12 for more details.

The careful calibration of analyzing magnets is usually done using resonant nuclear reactions if the accelerator's energy is adequate (Marion and Young, 1968; Overly *et al.*, 1969). See Appendix 17 for a listing of many of these reactions and techniques for energy calibration. For lower energy accelerators, it is necessary to rely on a precision voltage divider or a combination of partial knowledge of the expected beam and Eq. (11.1). One proceeds by sweeping the magnetic field B and noting the values at which beams are found. The most intense beams usually are the ones expected on the basis of, for example, the source gas composition. By making some initial assumptions and iteratively assigning masses to other beams, it is possible to arrive at a consistent relationship between B and m at a given energy. All this is necessary because ion sources almost universally produce many more beams than the uninitiated operator would reasonably expect. These arise from impurities in the feed material and sometimes from the materials comprising the source itself. In addition, molecules can be manufactured in the source. For example, chlorine feed gas is especially effective at generating volatile chlorides from solid constituents within the source. The process is so efficient, in fact, that it is used to produce beams of some species that are otherwise very difficult to vaporize.

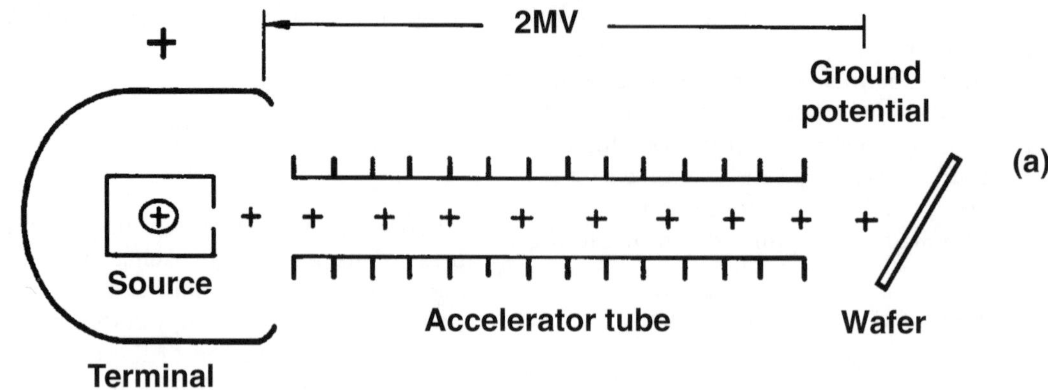

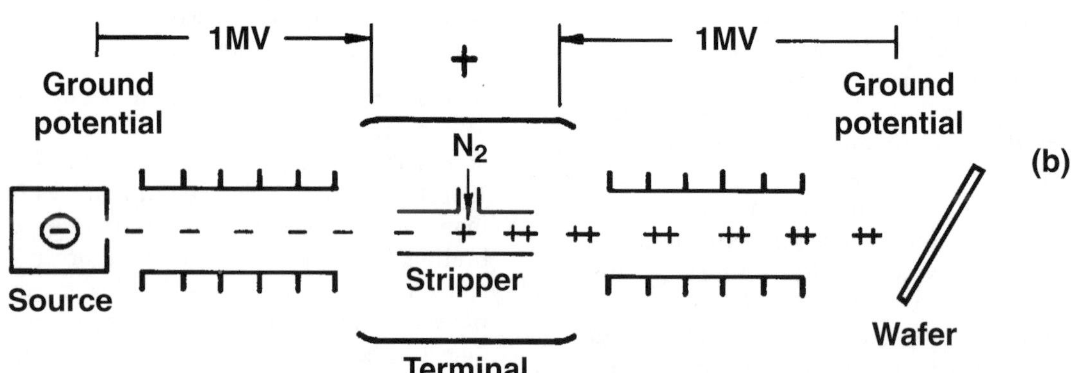

FIG. 11.1. Schematic diagram of single-ended and tandem Van de Graaff electrostatic accelerators. Terminal voltages up to approximately 20 MV have been achieved with tandem accelerators. Voltages of 1-3 MV are typical for machines designed for ion beam analytical applications. Figure courtesy of National Electrostatics Corporation, Middleton, WI 53562.

The strength of the field of the analyzing magnet is measured using either a Hall effect or nuclear magnetic resonance probe. Both are accurate enough to enable the operator to generate an accurate calibration. In some smaller accelerators, the magnetic field probe is omitted on account of its cost. In this case, one obtains an approximate measure of B by inference from the current in the magnet to which B is nominally proportional. However, using current in place of a direct measure of B can lead to larger errors because of the hysteresis that is characteristic of large electromagnets. We note finally one additional characteristic of analyzing magnets that has surprised more than one unsuspecting operator. Always slowly reduce the current in a large electromagnet to zero before switching off the power supply. To do otherwise is to court disaster, because of the magnet's huge inductance. A high capacity, large-area diode placed (reverse-biased) across the power supply's terminals is good insurance against the potentially catastrophic damage that can be done to a supply that has not been specifically engineered to drive large inductive loads.

The beam is transported to the target chamber through an evacuated pipe. It is steered to its destination by a series of electric and magnetic fields produced by devices called lenses. This nomenclature is appropriate because a powerful analogy can be drawn between the effect of ordinary glass lenses on light and the effect of these carefully designed electromagnetic fields on charged particles whose motions in phase space are constrained by Liouville's theorem (Lawson, 1977). The technology of designing and placing electromagnetic lenses is called ion optics and has been elevated to a high art by the designers of electron microscopes. An introduction to the subject may be found in Moore *et al.* (1989) and more detailed treatments in Lawson (1977), England (1974), or Humphries (1986).

Most of the charged particle lenses encountered by materials scientists will be either magnetic or electrostatic quadrupole lenses or electrostatic gap lenses (Harting and Read, 1976). A detailed discussion of the operation of these devices is beyond the scope of this chapter. In general, however, both behave in a manner analogous to (thick) convex glass lenses and so are capable of focusing a parallel beam or producing the "image" of an "object". Quadrupole and gap lenses differ for practical purposes mostly in the ranges of energies that they can focus. Quadrupole lenses are generally stronger, in the usual sense, than gap lenses and magnetic lenses are required for high energy particles. For low-energy ion work, a commercial vendor has created a do-it-yourself kit for electrostatic optics that retains much of the romance of the "Erector Set" that launched many scientific careers[3].

11.2.4 Beam diagnostics

The manner in which one uses lenses to transport and shape an ion beam follows quite closely the analogous procedure for visible-light optics. However, to do this operation you must be able to see the ion beam's position and its cross sectional shape. This is done using a series of beam viewers that fluoresce when the beam strikes them. Fused quartz is the most common viewer material. It should be coated with a thin layer of graphite to provide a path for the electrical charge to reach ground. Otherwise electric potentials of truly astonishing proportions may be generated along with significant quantities of x-rays when sparks discharge the surface. (Colloidal graphite, available commercially in spray cans[4], is also useful in eliminating patch potentials on electrostatic lenses and as an ultra-high vacuum compatible solid lubricant.) The quartz viewer emits a blue glow when struck by most ion beams of more than a few nanoamps and energies exceeding about 100 keV. For lower currents, conventional phosphors can be used. In many cases, mechanical sensors, such as wires sweeping through the beam, are used to generate an "electrical" view of the beam.

By placing a series of viewers along the beam line, it is possible to propagate an image of the aperture from which the ion source emits the beam, all the way to the plane of the target. For proper behavior, the trajectory of the beam should be along the optical axes of the lenses. Accelerator beam lines have electrostatic or electromagnetic deflectors to accomplish this. When a beam is properly steered into a lens, adjusting the lens will alter the focus but not the average position of the beam. For best results, an ion beam should be controlled primarily with steerers and lenses. It is common practice to use slits to define the cross section of an ion beam but the practice should be used with considerable discretion. The role of slits in ion beam transport should be restricted to eliminating that portion of the beam that is manifestly aberrant. (Slits are, of course, an integral part of analyzing magnets and other dispersive elements.) If used to define the beam area, scattering from the slits can reduce the quality of the delivered beam. In any event, slits should be very thin at their edges, but not thinner than the range of the ion beam.

The final size of the beam on the target can be increased or decreased by the lens system subject to the restrictions imposed by Liouville's theorem. Conventional accelerator beam spots, when well focused, are typically from 0.1-3.0 mm in diameter. To obtain beam spots smaller than about 0.1 mm, it is necessary to use special optical elements and ion sources. As an example, see Jamieson *et al.* (1989).

3. Kimball Physics, Inc., Kimball Hill Rd., Wilton, NH 03086.
4. Acheson Colloids Co., Port Huron, MI 48060.

The quantity of beam which is obtained on the target is measured by carefully measuring the electrical current which flows between the target and ground. From the point of view of circuit theory, the particle accelerator is a constant current source with one side grounded. The ions in the beam are the charge carriers and the circuit is completed when the ions are intercepted by a material object connected to ground. A simple ammeter can be used to measure the current. However, there are a number of physical phenomena which are initiated by the ion bombardment which complicate the measurement. The most important is secondary electron emission. When the ions in the beam strike a surface, electrons are emitted with energies from a few eV to over 1 keV. This flow of electrons away from the target constitutes a current of the same polarity as the beam current and is indistinguishable from it by a current measurement alone. One might then suggest that by biasing the target at a comfortable level of say +2 kV these electrons could be restrained and accuracy of the measurement restored. However, it would then be noticed that every stray electron in the vacuum chamber streamed to the target. Thus arises the subject of Faraday cups. These are electrodes specially designed to intercept ion beams with minimal interference from secondary processes. The general theory of Faraday cups is given in England (1974). Some of the problems with their use in materials work are discussed in Chapter 12.

Since most ion beam analytical work depends upon the total number of ions which strike a target rather than their rate of arrival, charge rather than current is usually the quantity of most importance. Since the ion beams produced by accelerators vary in intensity (often as a result of the action of the system used to stabilize the energy) on a time scale that is very short compared with the typical time for an experiment, it is necessary to measure the charge directly rather than to measure current and time separately and multiply them after the fact. Instruments to make this measurement are called current integrators and are manufactured by several suppliers of general nuclear instrumentation. They produce a train of output pulses each one of which corresponds to a fixed amount of input charge. Current integration is, thus, reduced to pulse counting. Alternatively, one can purchase a linear voltage-to-frequency converter (sometimes called a voltage controlled oscillator) from any of a number of electronic component manufacturers, and with the addition of an operational amplifier wired as a current-to-voltage transducer, produce a simple current integrator quite inexpensively.

11.3 DETECTORS

11.3.1 Overview

When an ion beam strikes the surface of a specimen, it initiates a series of processes which usually result in the emission of radiation. The detection and measurement of this radiation is the crucial step in ion beam analysis. An exhaustive treatment of the subject is found in Knoll (1989). Additional information on devices such as magnetic spectrographs and other specialized systems may be found in England (1974) and in the reviews contained in the special issue of *Nuclear Instruments and Methods* edited by Bromley (1979).

In this introduction, we will restrict our attention to three kinds of devices, semiconductor detectors, scintillators, and electron multipliers. Gas-filled detectors and neutron detectors are discussed in Knoll (1989). Historically, the development of detectors with a potential for application in materials analysis has been driven mostly by the demands of nuclear physics. Reports of recent advances in detector technology are found, for the most part, in the literature of that field. Current information on commercial detector properties may be found in the literature of manufacturers.

11.3.2 Electron multipliers

Electron multipliers are among the simplest yet most versatile of particle detectors. A schematic is shown in Fig. 11.2. An electron multiplier consists of a cathode which is sensitive to radiation and a set of electrodes called dynodes which are biased at progressively more positive voltages. The last electrode is called the anode, as in all vacuum tubes. When radiation in the form of photons, electrons, or other particles strikes the cathode, electrons are emitted. These are accelerated toward the first dynode where they strike and liberate additional electrons. In commercial electron multipliers, the process is repeated as many as a dozen times, giving gains up to 10^9 for single events at the cathode. Pulses from electron multipliers are typically very short, with 5-10 ns being common. Rise times in the range 100-200 ps are obtained with the best devices designed for timing applications.

Electron multipliers are used in two distinct applications in materials analysis. In one, they are called photomultipliers and are used in combination with a scintillator to measure γ- and x-ray radiation. In the other, they are used directly to detect photons or particles in, for example, time-of-flight measurements of particle velocity.

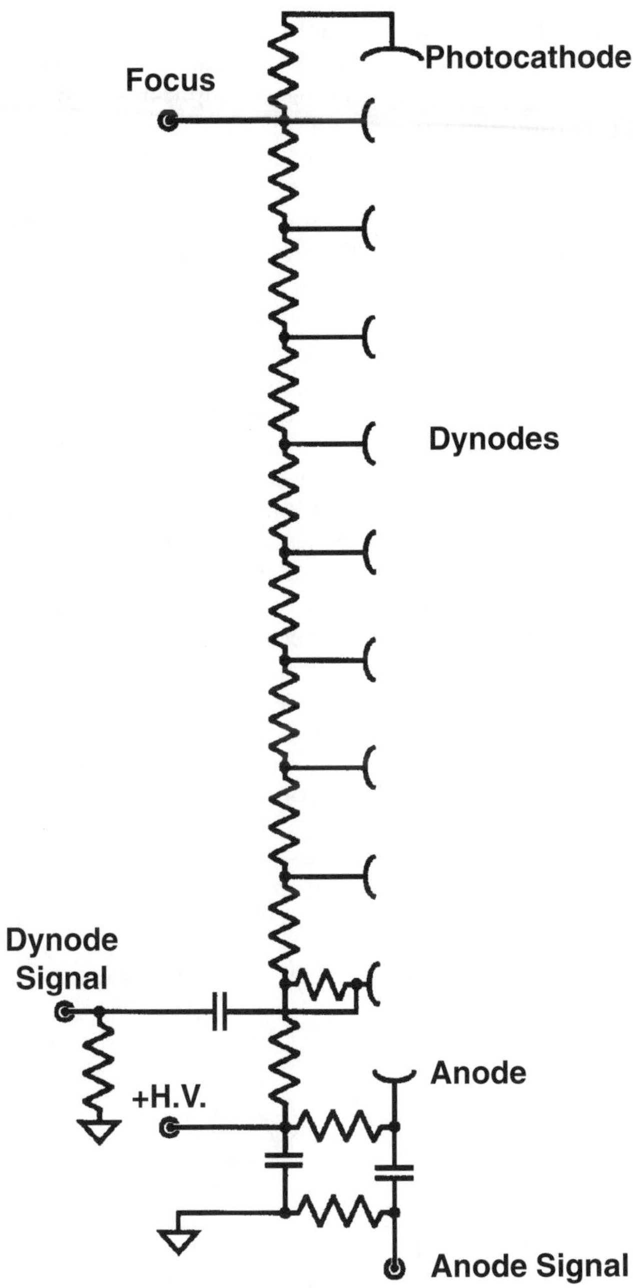

FIG. 11.2. A typical bias arrangement for a photomultiplier tube showing the resistive divider chain for the dynodes and signal outputs at both the last dynode and the anode.

The primary characteristic that distinguishes the device that is chosen for each application is the composition of the cathode. For measurements of light, the cathode is covered with a material that emits electrons when irradiated and is called a photocathode. For particle detection, high sensitivity to light would yield an unnecessarily high level of background counts and so cathode materials with relatively high work functions are chosen. The best source of information on photomultipliers is the literature of manufacturers[5]. Knoll (1989) gives a summary table of data on a number of commonly encountered photomultiplier tubes, but in some cases, higher performance replacements are now to be preferred.

For work inside a vacuum system a specialized form of electron multiplier is now in widespread use. It is termed a continuous dynode or channel electron multiplier and is often generically called a "channeltron" after the trade name used by the original manufacturer, Galileo Electro-Optics Corp.[6] These devices do not have discrete dynodes, but rather have curved internal surfaces of a resistive material (a reduced Pb glass) which both grades the potential and acts as a source of stored charge. Pulses from channel electron multipliers are extremely fast, giving these devices better characteristics than any other detector for timing applications such as time-of-flight spectrometry. They are also manufactured in the form of flat plates with a high density of microchannels each of which is an active multiplier (Wiza, 1979).

Originally developed as light amplifiers for use in night vision devices, microchannel plates provide large-sensitive-area detection and extremely fast response times. They are widely used as the active detecting elements for particle timing applications. Recently they have also been combined with photocathodes in conventional vacuum tubes giving photomultiplier tubes with remarkable characteristics. The Hamamatsu microchannel-plate, photomultiplier tube, R2566U, has a rise time of 100 ps, a fall time of 120 ps, and a total pulse width of 200 ps when operated under standard conditions. Its frequency response is 3 GHz.

5. Burle Industries, Inc. (formerly RCA), 1000 New Holland Ave., Lancaster, PA 17601. Other large suppliers of photomultiplier tubes include Hamamatsu Corporation, 360 Foothill Road, P.O. Box 6910, Bridgewater, NJ 08807; Thorn EMI Electron Tubes, Inc. 100 Forge Way Unit F, Rockaway, NJ 07866; Philips Components (formerly Amperex but now part of the Philips conglomerate) 100 Providence Pike, Slatersville, RI 02876; and, finally, if you need to lower your photomultiplier tube into an oil well, EMR Photoelectric, Princeton, NJ 08542.
6. Galileo Electro-Optics Corp., Galileo Park, P.O. Box 550, Sturbridge, MA 01566.

11.3.3 Scintillators

Some materials emit brief pulses of light when they are struck by radiation. An example which has already been mentioned is fused quartz. However, there are much better materials and these form the basis for the detection of γ-rays and x-rays, and less frequently, electrons and particles. The qualities that distinguish a good scintillator are a large absorption cross section for the incident radiation, efficient generation of scintillation photons, and a long mean free path for these photons to be reabsorbed. The most common scintillators used in materials work are inorganic crystals such as sodium iodide activated with thallium, denoted NaI(Tl), and bismuth germanate, $Bi_4Ge_3O_{12}$, which is usually referred to as BGO[7]. There is also a large variety of plastic and liquid scintillators, but these are used less frequently in materials work. One exception that should be noted, however, is in timing or coincidence studies where certain plastics such as NE111[8] are distinguished by their extremely fast response times. The properties of some common scintillators are included in Tables 11.1 and 11.2.

The physical principles by which scintillators operate are relatively well known. Energetic photons interact with all matter through the photoelectric effect, Compton scattering, and pair production (Fig. 11.3). Materials containing high Z elements such as iodine and bismuth are, thus, favored in scintillators because of their large photoelectric cross sections. The electrons produced by these processes lose energy by further collisions until their energy is fully dissipated. In scintillators such as sodium iodide, some of this energy is absorbed by intentionally created defects (thallium impurities in the case of NaI) called activators whose energy states lie within the band gap of the host crystal. When these excited states decay, they do so by the emission of photons whose energies are lower than the band gap of the crystal. These photons, which are usually in the visible region of the spectrum, are thus able to travel significant distances in the crystal with minimum likelihood of reabsorption. Photomultiplier tubes outside the crystal collect the scintillation photons and generate electrical pulses from them.

Ideally the number of scintillation photons is proportional to the energy of the initial radiation. However, a number of processes intervene to complicate this simple proportionality. Consider the case of an energetic gamma ray produced, for example, in the nuclear reaction $^{15}N(p,\alpha\gamma)^{12}C$ which is sometimes used for the measurement of surface hydrogen. Suppose that the γ-ray interacts with the detector through the production of an electron-positron pair and that each of these particles comes to rest within the crystal depositing all of its kinetic energy. At this stage, all of the energy except that required to create the particles' rest masses has been accounted for. Immediately upon being stopped, the positron will annihilate with an electron from the crystal generating two γ-rays each with an energy of 511 keV. There are now three possibilities. Either both of these γ-rays are absorbed by the crystal or one or both escape. It turns out that for practical detectors each of these possibilities is reasonably likely. Therefore, one may get three different deposited energies and three different photomultiplier pulse amplitude groups from the absorption of a flux of monoenergetic γ-rays. Other physical processes lead to additional idiosyncrasies in gamma ray spectra (Fig. 11.4). For a complete discussion of these, see Knoll (1989) and Heath (1964).

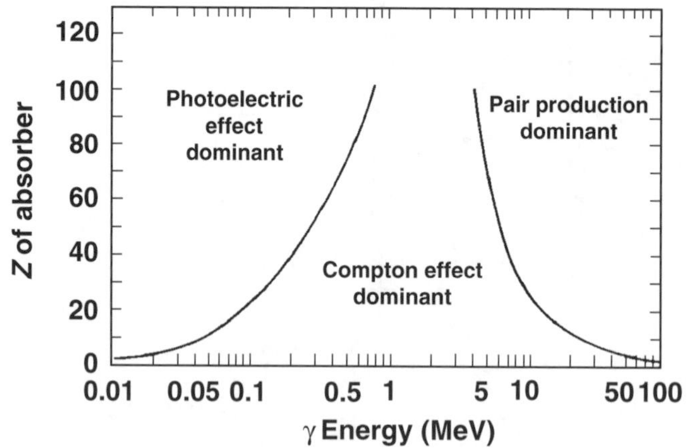

FIG. 11.3. Relative importance of the photoelectric effect, Compton effect, and pair production as a function of the γ-ray energy and atomic number of the absorber. From Robley D. Evans, The Atomic Nucleus (McGraw Hill, New York, 1955), p. 712. Copyright McGraw Hill, used by permission.

7. Solon Technologies, Inc., 6801 Cochran Road, Solon, OH 44139. This is the new name of the Harshaw Chemical Company.
8. NE Technology, Inc., 9 Deer Park Drive, Monmoth, NJ 08852. Also, Bicron Corporation, 12345 Kinsman Road, Newbury, OH 44065. Many of the scintillators in use for nuclear spectrometry carry the "NE" designation. Bicron sells equivalent materials under less familiar part numbers.

Table 11.1. Properties of inorganic scintillators. Sources: Radiation Detectors Catalog, Solon Technologies, Inc.; Sakai (1988), Holl *et al.* (1988), Knoll (1989).

Scintillator	Density *(g/cm³)*	Light output relative to NaI(Tl) *(%)*	Wave-length of max. emission λ_{max} *(nm)*	Index of refrac-tion at λ_{max}	Principal decay time constant *(μs)*	Pulse rise time 10–90% *(μs)*	Light output *(photons/ MeV)*	Absolute scintilla-tion efficiency *(%)*	Hygro-scopic?
Na(Tl)	3.67	100	415	1.85	0.23	0.5	38000	11.3	Yes
CsI(Tl)	4.51	45	550	1.80	1.0	4	52000	11.9	Slightly
CsI(Na)	4.51	85	420	1.84	0.63	4	39000	11.4	Yes
$Bi_4Ge_3O_{12}$	7.13	12-14	480	2.15	0.3	0.8	8200	2.1	No
$CdWO_4$	7.9	40	540	2.2-2.3	5.0	–	15000	3.8	No
CaF_2(Eu)	3.18	50	435	1.44	0.94	4	24000	6.7	No
BaF_2 Fast Slow	4.88	4 20	225 310	1.49@ 325 nm	0.0008 0.62	 3	10000	4.5	No
CsF	4.64	5	390	1.48 @ 589 nm	0.005	–	–	–	Yes
LiI(Eu)	4.08	35	470	1.96	1.4 n radiation 0.6 γ radiation	–	11000	2.8	Yes

11.3.4 Semiconductor detectors

Ordinarily, a reverse biased semiconductor diode does not conduct electricity because it contains a region from which charge carriers have been depleted. However, when electron-hole pairs are formed randomly in the depletion region of such a diode, following the usual laws of thermodynamics, the holes and electrons are swept in opposite directions by the established electric field and produce the small, temperature-dependent leakage current that is a characteristic of the device. It is well known that light can also cause the generation of electron-hole pairs by directly promoting electrons from the valence to the conduction band of the material. This is the principle of operation of the photodiode. Semiconductor detectors for particle and high-energy photon detection, most of which are fabricated from Si or Ge (Table 11.3), work in exactly the same way. See McKenzie (1979).

When an energetic ion enters the depletion region of a semiconductor detector, it begins to lose energy through collisions with the electrons and nuclei of the constituent atoms. Energy deposited in collisions with nuclei generates phonons (heat) and is effectively lost, but energy used to excite electrons, the so-called electronic component of stopping power, is available to produce charge carriers at a rate of a few eV per pair. The ion comes to rest in the detector (provided, of course, that it is sufficiently thick) in a matter of a few picoseconds and leaves in its wake a large number of electron-hole pairs. As with thermally-generated charge carriers, these are swept from the depletion region and appear as a pulse of current at the detector's terminals. The length of this pulse is typically a few ns. It is collected by a charge-sensitive preamplifier for further processing.

The most common particle detector used for ion beam materials analysis is the Si surface barrier detector. For light ions such as protons or α particles, the quantity of collected charge is quite linear (about 3.6-3.7 eV per electron-hole pair) in the total energy deposited in electronic processes. The overall resolution is typically in the range 15-20 keV. For all ions, the energy lost to nuclear collisions gives rise to a non-linearity known as pulse-height defect which can become quite large. These phenomena also worsen the energy resolution of the detector. Some energy is also lost as the ion passes through the Au entrance window and dead layer beneath it. For

Table 11.2. Physical constants of plastic and liquid scintillators. Also included are glass scintillators, and for comparison, anthracene and NaI(Tl). Source: Table of Physical Constants of Scintillators, NE Technology, Inc., 9 Deer Park Drive, Monmouth, NJ 08852. Used by permission.

	Scintillator	Type	Density	Refractive index	Melting softening or boiling point (°C)	Light output (% anthracene)	Decay constant, main component (ns)	Wavelength of maximum emission (nm)	Content of loading element (% by wt.)	H/C (no. H atoms/ no. of C atoms)	Principal applications
Plastic	NE 102A	Plastic	1.032	1.581	75°	65	2.4	423	- - - -	1.104	γ, α, β, fast n
	NE 104	Plastic	1.032	1.581	75°	68	1.9	406	- - - -	1.100	ultra-fast counting
	NE 108	Plastic	1.032	1.58	75°	65	15	545	- - - -	1.103	with photodiodes
	NE 110	Plastic	1.032	1.58	75°	60	3.3	434	- - - -	1.104	γ, α, β, fast n, etc.
	NE 111A	Plastic	1.032	1.58	75°	55	1.6	370	- - - -	1.103	ultra-fast timing
	NE 114	Plastic	1.032	1.58	75°	50	4.0	434	- - - -	1.109	as for NE 110
	NE 115	Plastic	1.032	1.582	75°	35	320	395	- - - -	1.108	phoswiches
	NE 118	Plastic	1.020	1.580	99°	60	3.3	434	- - - -	1.104	high temperatures
	NE 125	Plastic	1.14	1.594	80°	55	2.7	425	D 13.8%	0.957 D/C	n
	Pilot F	Plastic	1.032	1.581	75°	65	2.1	425	- - - -	1.103	γ, α, β, fast n
	Pilot U	Plastic	1.032	1.58	75°	67	1.36	391	- - - -	1.100	ultra fast timing
	Pilot 425	Plastic	1.19	1.49	100°	- - -		425	- - - -	1.6	Cherenkov detector
Liquid	NE 213	Liquid	0.874	1.508	141°	78	3.7	425	- - - -	1.213	fast n (P.S.D.)
	NE 224	Liquid	0.877	1.505	169°	80	2.6	425	- - - -	1.330	γ, fast n
	NE 226	Liquid	1.61	1.38	80°	20	3.3	430	- - - -	0	γ, insensitive to n
	NE 230	Deuterated liquid	0.945	1.50	81°	60	3.0	425	D 14.2%	0.984 D/C	Special applications
	NE 232	Deuterated liquid	0.89	1.43	81°	60	4	430	D 24.5%	1.96 D/C	Special applications
	NE 235	Liquid	0.858	1.47	350°	40	4	420	- - - -	2.0	large tanks
	NE 236	Liquid	0.796	1.444	192°	48	3.5	425	- - - -	1.92	low temperatures
	NE 237	Liquid	0.813	1.459	192°	61	3.15	425	- - - -	1.82	high flash point, general
Loaded liquid	NE 311 & 311A	B loaded liquid	0.91	1.411	85°	65	3.8	425	B 5%	1.701	n, β
	NE 314A	Pb loaded liquid	0.96	1.53	141°	25	2.0	425	Pb 7.5%	1.261	γ, x-rays
	NE 316	Sn loaded liquid	0.93	1.496	148.5°	35	4.0	425	Sn 10%	1.411	γ, x-rays
	NE 320	^{6}Li loaded liquid	0.906	1.497	160°	34	2.19	425	^{6}Li 0.15%	1.428	n
	NE 343	Gd loaded liquid	0.884	1.502	168°	65	3.0	425	Gd 0.5%	1.360	n
Neutron (ZnS-type) and glass	NE 422 & 426	^{6}Li-ZnS(Ag)	2.36	- - - -	110°	300	200	450	Li 5%	- - - -	slow n
	NE 451	ZnS(Ag) plastic	1.443	- - - -	110°	300	200	450	- - - -	- - - -	fast n
	NE 901, 902, 903	Glass	2.64	1.58	c.1200°	28	20 & 60	395	Li 2.3%	- - - -	n, β
	NE 904, 905, 906	Glass	2.5	1.55	c.1200°	25	20 & 58	395	Li 6.6%	- - - -	n
	NE 907, 908	Glass	2.42	1.566	c.1200°	20	18 & 62	399	Li 7.5%	- - - -	n
	NE 912, 913	Glass	2.42	1.55	c.1200°	25	18 & 55	397	Li 7.7%	- - - -	n, β (low background)
Crystal	Anthracene	Crystal	1.25	1.62	217°	100	30	447	- - - -	0.715	γ, α, β, fast n
	NaI(Tl)	Crystal	3.67	1.775	650°	230	230	413	- - - -	- - - -	γ, x-rays

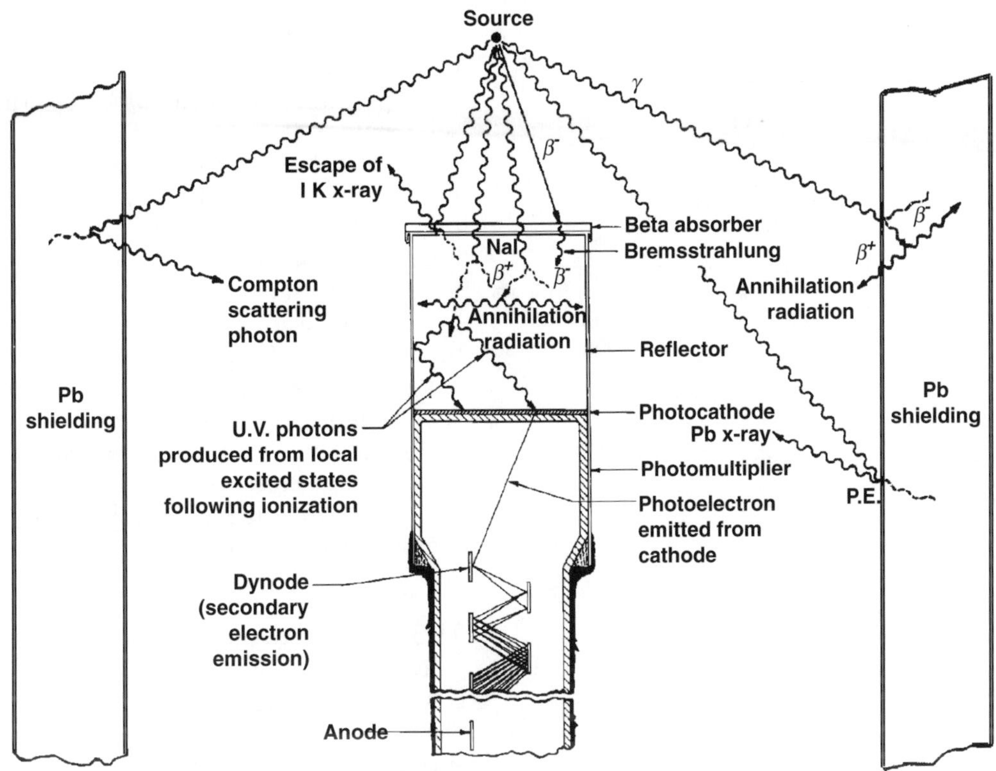

FIG. 11.4. Physics processes affecting a γ-ray spectrum. From Heath (1964).

dense ionization, recombination can also occur before the electrons and holes are fully separated, but this effect is usually quite small. For very precise work with surface barrier detectors, it is necessary to consider carefully the physics of their operation in order to understand the circumstances under which their linearity breaks down (Lennard *et al.*, 1986). Caution should also be exercised when detecting heavy ions ($A > 4$) because of their larger pulse-height defect and because of the irreversible radiation damage that such ions cause to the detector. For example, when the $\sqrt{E_r}$, the reduced energy, [see Eq. (2.3a) and elsewhere in Chapter 2] is ≈ 3 the nuclear stopping is ≈36% of the electronic stopping and the pulse height defect will be a significant. On the other hand, when $\sqrt{E_r} = 10$, the nuclear stopping is $\approx 10^{-3}$ of the electronic stopping.

The largest supplier of Si surface barrier detectors is EG&G Ortec[9]. The reference section of the Ortec detector catalog gives a concise but complete description of surface barrier detectors. It also provides advice on the style and operating characteristics to chose for specific applications such as high resolution spectroscopy, fast timing or rugged service. Be sure to specify a detector with a depletion region thick enough to stop the highest energy particle that you expect to encounter in your experiments. Otherwise the energy of these particles cannot be measured. The nomogram shown in Fig. 11.5 can be used to determine the depletion layer depth in an Si surface barrier detector which is required to stop α particles, protons, and electrons of various energies and to estimate (as a function of the resistivity of the Si) the detector bias which is required to obtain this depletion depth.

Surface barrier detectors are used primarily to detect electrons and ions. For detecting x and γ radiation, lithium-drifted detectors and high purity or intrinsic germanium are the solid-state detectors of choice. The principal advantage that these solid-state photon detectors have over scintillation detectors is high resolution. Some are conspicuously less easy to use, however, because they must be cooled to liquid nitrogen temperatures. They are

9. EG&G Ortec, 100 Midland Road, Oak Ridge, TN 37831.

Table 11.3. Properties of pure silicon and germanium. Sources: England (1974), Knoll (1989), Elmsley (1989).

	Si	Ge
Atomic number	14	32
Atomic mass (u)	28.0855	72.61
Stable Isotopes (u) *(%)*	28 – 92.23 29 – 4.67 30 – 3.10	70 – 20.5 72 – 27.4 73 – 7.8 74 – 36.5 76 – 7.8
Density *(g/cm^3)*	2.329	5.323
Atomic density ($10^{22}/cm^3$)	4.96	4.41
Linear thermal expansion coefficient *(K^{-1})*	4.2×10^{-6}	5.57×10^{-6}
Crystal structure	diamond	diamond
Lattice constant @ 300 K *(nm)*	0.5430	0.5658
Melting point *(K)*	1683	1210.6
Dielectric constant	12	16
Forbidden energy gap *(eV)*		
300 K	1.115	0.665
0 K	1.165	0.746
Intrinsic carrier density @ 300 K *(cm^{-3})*	1.5×10^{10}	2.4×10^{13}
Intrinsic resistivity @ 300 K *($\Omega\cdot cm$)*	2.3×10^{5}	47
Electron mobility *($cm^2/V\cdot s$)*		
300 K	1350	3900
77 K	2.1×10^{4}	3.6×10^{4}
Hole mobility *($cm^2/V\cdot s$)*		
300 K	480	1900
77 K	1.1×10^{4}	4.2×10^{4}
Carrier saturation velocity *(cm/s)*		
300 K	8.2×10^{6}	5.9×10^{6}
77 K	10^{7}	9.6×10^{6}
Energy per electron-hole pair *(eV)*		
300 K	3.62	Not applicable
77 K	3.76	2.96

also more costly. The detector in this class that is most commonly used for materials work is the lithium-drifted silicon x-ray detector, Si(Li), which is pronounced "silly". They are frequently found on scanning electron microscopes where they are used for dispersive x-ray analysis of the elemental constituents of the sample. Si(Li) detectors operate on the same physical principles as other solid state detectors and interact with electromagnetic radiation in the same way as scintillators described above. For γ-ray work, an intrinsic germanium detector is the preferred solution. Lithium drifted germanium detectors, Ge(Li), pronounced "jelly", require continuous refrigeration by liquid nitrogen (77 K) to avoid destruction by lithium precipitation and have been made obsolete as the technology for producing high purity germanium has been perfected. [Si(Li) detectors are also subject to some deterioration when warmed to room temperature and should be continuously maintained at 77 K for best results and longest life.] The catalogs of both EG&G Ortec and Canberra Industries[10] are useful references for γ- and x-ray detectors.

11.3.5 Detector systems

There are two complex detector systems that are becoming increasingly important for ion beam analytical work. These are particle telescopes and time-of-flight spectrometers. A particle telescope is a composite detector that simultaneously measures the energy lost by a particle as it first passes through a thin Si detector (or a gas detector) and then stops in a conventional solid state detector (England, 1974). This configuration returns two parameters for each event. The energy lost in the thin detector is approximately proportional to the stopping power of the ion, while the sum of the energies recorded in the two detectors is the ion's total energy. By extending the conventional concept of an energy spectrum to two dimensions and plotting the number of counts against both the energy and stopping power, it is possible to separate events generated by ions of different species.

The time-of-flight spectrometer, one variation of which is shown in Fig. 11.6, is also a composite device composed of two discrete particle detectors. Its function is to measure the time that is required for a single particle to pass between two points. Such systems had their genesis in detectors developed during the 1970s for heavy ion nuclear spectroscopy (Gabor *et al.*, 1975; Betts, 1979). Their application to ion beam analysis has been reviewed recently (Whitlow, 1990), and continues to be refined. In all of the designs, the energetic particle first passes through a carbon foil about 1-5 $\mu g/cm^2$ thick which is either

10. Canberra Industries, Inc., One State Street, Meriden, CT 06450.

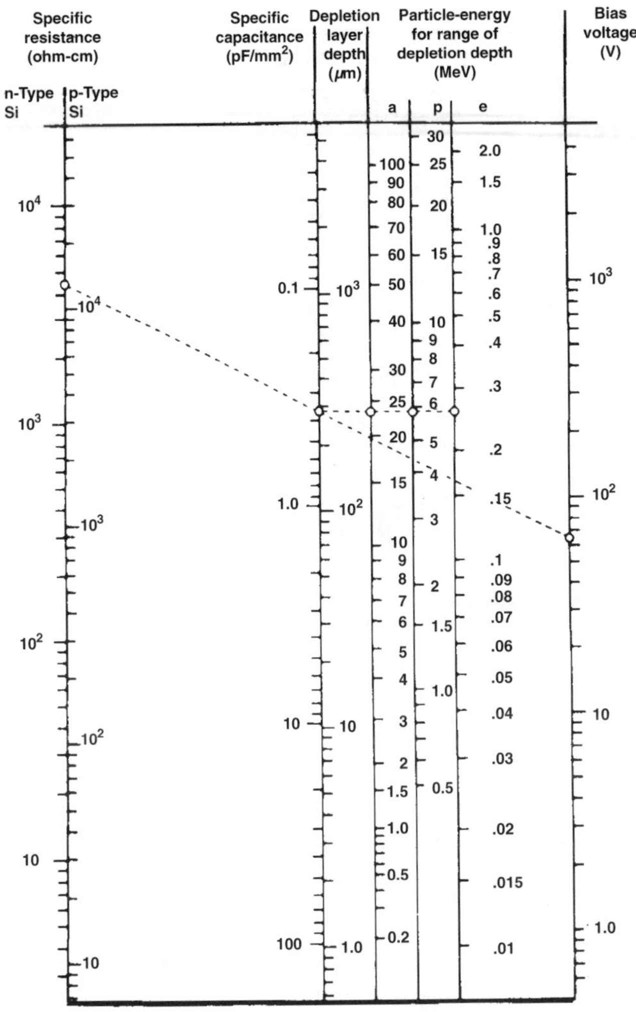

1 µm of Si = 0.2325 mg/cm²
1 mg/cm² of Si = 4.3 µm

A straight edge intersecting the center vertical line at the required depletion depth will give combinations of resistivity and detector bias that may be used to achieve that depth. (Shown, for example, is the voltage that must be applied to a 13,000 ohm-cm p-type or 4500 ohm-cm n-type silicon detector to stop a 23-MeV alpha, a 6-MeV proton, or a 250-keV electron within the depletion depth.)

FIG. 11.5. Silicon surface barrier nomogram, showing the relationship between bias, depletion depth, resistivity, and specific capacitance. Similar to the nomogram reported by J. L. Blankenship, IEEE Trans. Nucl. Sci. **NS7** (2-3) (1960) 190. From the Ortec Catalog, copyright EG&G Ortec, used by permission.

self-supporting or mounted on a high-transmission mesh[11]. Electrons emitted from the carbon foil are collected by a channel electron multiplier or a microchannel plate, and a "start" pulse is generated. After drifting a fixed distance, the particle is again examined by another carbon foil, by a semiconductor detector, or by direct impact on the cathode of a channel electron multiplier. This generates a "stop" pulse and the interval between the start and stop pulses is measured electronically. To understand the details of this process, consider an example.

Let us suppose that a time-of-flight system is being designed to give an energy resolution for a 2 MeV α particle which is comparable with the 20 keV being obtained using a silicon surface barrier detector. Assume that the total length between the carbon foil and stop detector is 1 m. The flight time required for the α particle to cover this distance at an energy of 2 MeV is about 100 ns. An energy resolution of 20 keV is equivalent to 1% so that a timing resolution of 0.5% is required ($E \propto t^{-2}$). This means that timing resolution of 500 ps is needed. Consider first the implications of this resolution for the start and stop pulses. These pulses are produced by very fast electron multipliers, and although there is no directly useful information in their amplitudes, these amplitudes do vary. The pulses may also need to be amplified by a voltage-sensitive preamplifier. As a result, they can have durations of up to several ns and rise times approaching 1 ns. These relatively large values complicate the electronic task of consistently deciding just when the antecedent pulse occurred.

The problem of generating consistent time markers is solved with a device known as a constant-fraction discriminator. It analyzes the analog pulse from the detector or preamplifier and emits a standard fast negative logic pulse when the amplitude has reached a specific fraction of its maximum value. It does this by splitting the input signal into two components, attenuating one component, inverting and delaying the other (typically with a few centimeters of cable), and then recombining the two into one fast bipolar pulse. The zero-crossing of this bipolar pulse occurs at a time which is independent of the input pulse amplitude and is used as the trigger event for the generation of the fast-negative-logic time marker. Since logic pulses have a consistent amplitude and shape, the time of the original analog pulse can be established with the required precision even though the amplitude of the original pulse may vary by an order of magnitude.

Consider now the method by which the interval is timed. If 500 ps timing precision is required, then to time the interval directly, a clock ticking with this period is

11. Buckbee-Mears, St. Paul, 245 East 6th Street, St. Paul, MN 55101. Purveyors of electroformed meshes.

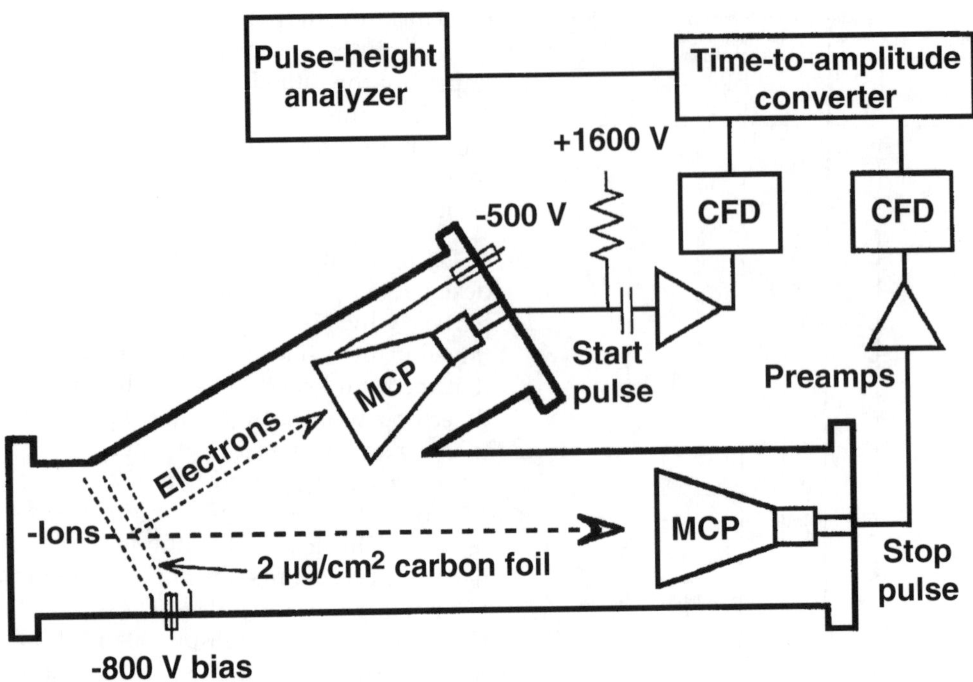

FIG. 11.6. A time-of-flight detector system.

needed. This is 2 GHz and to count this frequency directly is rather a formidable task. More commonly, another method is used, which has the added virtue that a multichannel analyzer can then be used for time-of-flight spectrometry in the same way that it is used with the detectors discussed above for energy spectrometry. The start and stop pulses are connected to a device called a time-to-amplitude converter or TAC. A TAC is a specialized, single-shot pulse generator whose output is proportional to the time interval between its "start" and "stop" inputs. The TAC converts the problem of accurately measuring a very short time interval to the problem of accurately measuring a pulse amplitude. Since pulse height analysis is a well-developed art, the time-of-flight problem is solved. Timing resolutions of about 150 ps are obtainable with present technology. This is more than adequate for most ion beam analytical applications.

Although they are less convenient and much more expensive than surface barrier detectors, time-of-flight systems have been shown to be of particular value when used to measure backscattering of low energy ions (Mendenhall and Weller, 1991) or of heavy ions at high energies (Döbeli *et al.*, 1991, and references therein). Time-of-flight spectrometers are superior to surface barrier detectors in low energy work because they offer superior energy resolution. For high energy heavy ions they are substantially more robust than solid state detectors whose lifetimes may be extremely short (see Chapter 4). Time-of-flight spectrometry is also useful for (forward) elastic recoil detection, since a time-dispersive detector is fundamentally more appropriate than a single, energy-dispersive detector in this application (see Chapter 5). A mathematical description of the effect of random coincidences in a time-of-flight spectrometer on observed spectra is given in 11.6.4.

11.4 ELECTRONICS

11.4.1 Introduction

The vast majority of all modern scientific measurements follow the same general form. The physical quantity of interest is related to an electrical quantity through a transducer, the electrical quantity is processed using the standard techniques of analog electronics, the analog signal is digitized with an appropriate analog-to-digital converter, and the resulting values are stored digitally by a computer, possibly after additional digital processing. The procedures used for ion beam materials analysis follow this scenario quite closely.

The transducers used in ion beam work are the various detectors which were discussed in the previous section. The purpose of this section is to introduce the remainder of the chain. All of the significant signal processing and analysis tools necessary for ion beam studies of materials are available from commercial vendors. However, this does not imply that it is possible to treat these tools as "black boxes". Quite the contrary is true. If one is to obtain best results in routine measurements and have the flexibility to push into new areas, at least the functioning of standard electronic modules must be understood.

In addition, it is important to have enough understanding of elementary analog electronics to be able to recognize the need for simple devices and build them. Perhaps the best reference to guide the reader to achieving this skill is the excellent and aptly named text *The Art of Electronics* (Horowitz and Hill, 1989). The approximate minimum level of proficiency in electronics that is desirable for materials scientists active in ion beam work is the ability to build simple active circuits with operational amplifiers and passive components. Some knowledge of discrete transistor circuits is also useful, but has become less important as the power and variety of integrated, analog electronic components has increased. Horowitz and Hill provide a number of tables that are very useful for selecting proper components for simple electronic projects around the laboratory.

11.4.2 Nuclear electronics

The best places to learn about the fundamental elements of nuclear electronics are the tutorial sections of the manufacturers catalogs and their applications notes (Ortec, 1984; Canberra, 1978; Duggan, 1988). The text by Nicholson is comprehensive but more advanced (1974). In this introduction, we will discuss only the simplest energy spectroscopy system for use with surface barrier or photon detectors. Our principle aim will be to introduce the nomenclature of the field and to place it in context.

Commercial nuclear instrumentation was standardized by the United States Atomic Energy Commission, the forerunner of today's Department of Energy (USDOE, 1990). The standard is basically for analog electronics, although there are also definitions for logic signals. The basic concept is of the nuclear instrument module (NIM) from which the standard gets its name[12]. Table 11.4, adapted from manufacturers' catalogs, lists a number of standard modules, their inputs, outputs, and functions. Signal and power inputs to these modules and their outputs have been clearly defined. NIM modules may occupy one or more slots in any position of a NIM bin. Bins may stand alone or be mounted in 19 inch "relay" racks. NIM bins contain DC power supplies (±12 V, ±24 V, and now often ±6 V) for modules which can operate within the standard current limitations established for the number of slots which they occupy, as well as AC power for such items as detector bias supplies. The modules connect to the bin via a standard connector on the rear of the module. This connection is intended primarily for power and does not provide digital data paths as do the more recent, computer based, CAMAC (IEEE-583; IEEE, 1976; Clout, 1982) and FASTBUS (IEEE-960; USDOE, 1977) modular instrumentation systems. The pin assignments for the NIM bin connector are shown in Fig. 11.7 and for the preamplifier power connector in Table 11.5.

NIM analog pulses may be of either polarity. The preferred range for positive signals is 0-10 V. Fast analog signals are preferred to be negative and in the range 0 to –1 V or 0 to –5 V. There are two logic families, medium-speed positive logic and fast negative logic. These logic pulses have rigorous specifications which are given in Table 11.6. NIM positive logic has states which are defined by voltage and is typically useful for rates from DC up to about 20 MHz. Fast negative logic has states which are defined by current. With rise times of ≈ 2 ns, it is capable of handling rates up to and beyond 100 MHz. Fast timing applications use fast negative logic exclusively.

Because of the age of the NIM standard, its logic and voltage levels do not correspond with any standard logic family used today or with the ±15 V power supplies that are now commonly used to power operational amplifiers for analog circuit functions. For the occasional analog construction project, it is usually easiest to simply operate op-amp circuitry at the reduced ±12 V that is available from the NIM bin. Interfacing logic signals presents a bit more of a challenge. By far the most common integrated logic in use today is transistor-transistor logic

12. The original NIM standard was drafted by the AEC's Committee on Nuclear Instrument Modules. By the time of the CAMAC standardization, the name National Instrumentation Methods Committee appeared on documents. Now, since NIM is international, it should simply be considered to be a name. "CAMAC" was originally conceived by the Europeans as a name, not an acronym, but, by popular demand, in the U.S. it was embellished to "Computer Automated Measurement and Control". Strike a blow for English; use words, not acronyms in your writing.

Table 11.4. Summary of some commonly available NIM modules, their typical inputs, outputs, and functions.

NIM module	Input	Output	Comments
Preamplifier	Linear pulse from a detector	Linear tail pulse or voltage amplified pulse	Initial amplification and shaping for spectroscopy or timing. Preamps mount near the detector, not in a NIM bin.
Spectroscopy amplifier	Linear tail pulse	Shaped linear pulse	Final amplification and shaping for analog-to-digital conversion.
Biased amplifier	Shaped linear pulse	Shaped linear pulse	Output pulse amplitude proportional to input amplitude less bias level. Used to expand a region of a spectrum.
Timing amplifier	Fast linear pulse	Amplified fast linear pulse	Broadband voltage amplification for timing applications
Delay-line amplifier	Linear tail pulse	Delay-line shaped linear pulse	A spectroscopy amplifier with delay-line pulse shaping. Used for high-count-rate spectroscopy.
Summing amplifier	Shaped linear pulses and often positive logic pulses	Sum of pulses in two or more channels	Sums linear or logic signals. Useful for mixing signals.
Delay or delay amplifier	Linear or logic pulses	Delayed linear or logic pulses of approximately the same amplitude and shape	Introduces an adjustable delay up to a few μs. Short delays are achieved by measured lengths of cable. Longer delays use delay lines with buffer amplifiers.
Single channel analyzer	Shaped linear pulse	NIM positive logic pulse	Generates an output pulse only if the input pulse amplitude lies between adjustable upper and lower thresholds.
Constant fraction discriminator	Fast linear pulse	NIM fast negative logic pulse	Generates the output pulse a fixed period after the input pulse reaches a specific fraction of its maximum amplitude. Used for highest resolution timing applications.
Linear gate	Linear pulse at the "input". NIM positive logic pulse at the "gate"	Linear pulse of the same amplitude and shape as the input	Linear pulse is passed only if the gate is enabled.
Gate and delay generator	NIM positive or fast negative logic pulse	Logic pulses with adjustable amplitude, width and delay, often with fast negative logic marker pulses	A handy tool for working with logic signals. Can be used for conversion between NIM positive and fast negative logic or between NIM and other logic families.
Coincidence / anticoincidence unit	Two or more NIM positive logic pulses	NIM positive logic pulse	Generates an output pulse if input pulses occur coincidentally in user-designated logic states.
Scaler, counter (timer)	NIM positive or fast negative logic pulses and shaped linear pulses with amplitudes typically ≥ 100 mV	Visual or computer output of accumulated number of pulses	Counts pulses. Commercial units often contain internal time bases for slow timing applications and deliver logic output pulses in proportion (e.g., 0.1) to the number of input pulses.
Ratemeter	NIM positive or fast negative logic pulses or linear pulses with amplitudes typically ≥ 100 mV	Visual indication of pulse rate. Analog voltage or current proportional to input pulse rate	Gives an analog readout of pulse rate. Units typically generate a fixed charge pulse internally for each input pulse edge then filter and deliver the resulting current (or a proportional voltage) as the output.
Pulser	None required but analog pulse amplitude input and logic trigger are optional on some units	Shaped linear pulse, tail pulse, NIM positive or fast negative logic pulses	Pulsers generate test pulses for tests and calibration. Some even generate randomly occurring pulses (by triggering on an internal noise source).
Current integrator	Ion beam or other current	NIM positive logic pulses at a rate proportional to charge accumulation	Typically used to integrate accelerator beam. An analog output proportional to input current is usually also available.

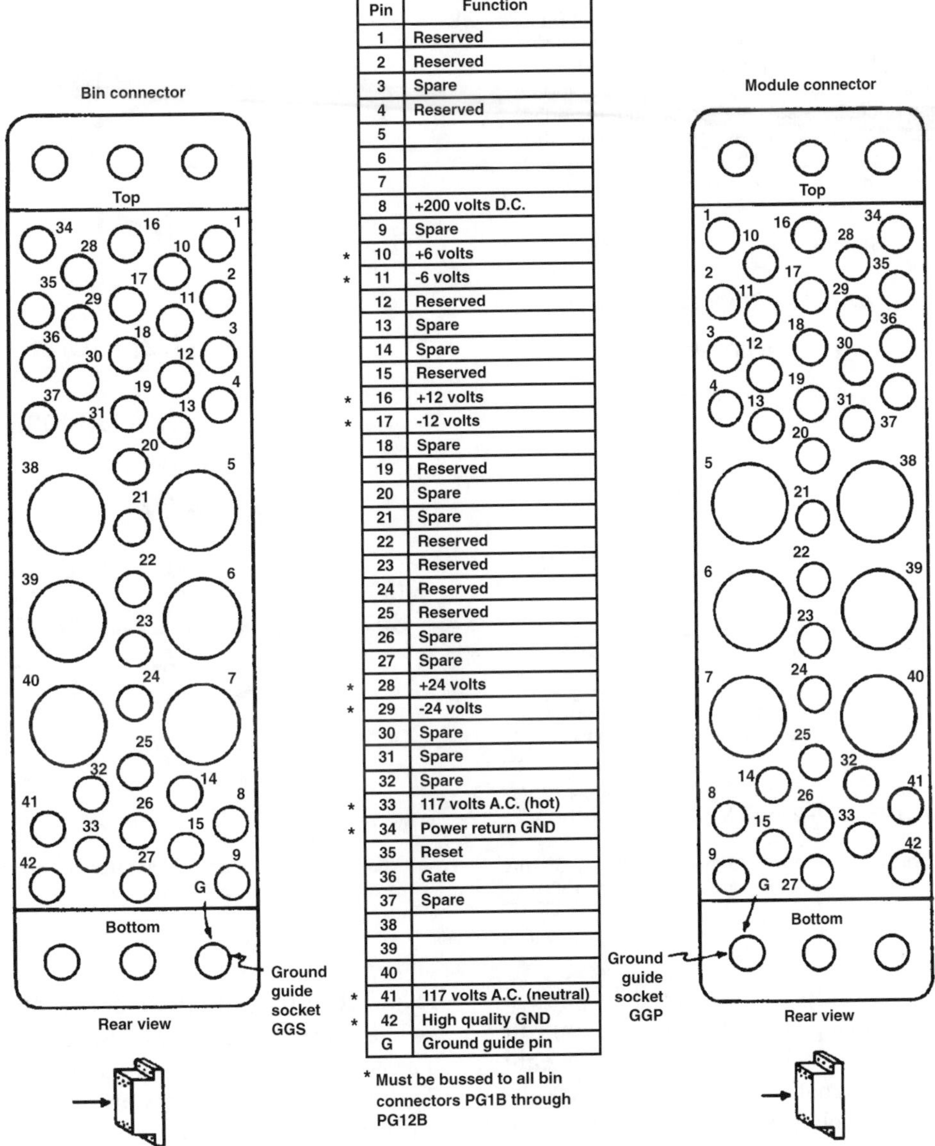

	Pin	Function
	1	Reserved
	2	Reserved
	3	Spare
	4	Reserved
	5	
	6	
	7	
	8	+200 volts D.C.
	9	Spare
*	10	+6 volts
*	11	-6 volts
	12	Reserved
	13	Spare
	14	Spare
	15	Reserved
*	16	+12 volts
*	17	-12 volts
	18	Spare
	19	Reserved
	20	Spare
	21	Spare
	22	Reserved
	23	Reserved
	24	Reserved
	25	Reserved
	26	Spare
	27	Spare
*	28	+24 volts
*	29	-24 volts
	30	Spare
	31	Spare
	32	Spare
*	33	117 volts A.C. (hot)
*	34	Power return GND
	35	Reset
	36	Gate
	37	Spare
	38	
	39	
	40	
*	41	117 volts A.C. (neutral)
*	42	High quality GND
	G	Ground guide pin

* Must be bussed to all bin connectors PG1B through PG12B

Notes:

1. Reserved pins are for future assignment by the committee and shall not be used until such assignments are made.

2. GP-1 = Guide Pin GGP = Ground Guide Pin
   GS-1 = Guide Socket GGS = Ground Guide Socket

3. The POWER RETURN GROUND (bin pin 34) is the return bus for all dc supplies. The HIGH QUALITY GROUND (bin pin 42) is intended as zero potential reference. High quality ground current load by any module should be limited to 1 mA. Also, pulses or varying loads should be limited to no more than 100 microamperes. Care should be taken not to capacitively couple high quality ground to the local ground. Any capacitive coupling to local ground should be isolated by a resistance of at least 1000 ohms. The CHASSIS GROUND is normally connected to the building ground. Connections between the HIGH QUALITY GROUND bus (bin pin 42), the power RETURN GROUND BUS (bin pin 34), and the chassis are made at the ground guide pin of PG1B.

4. Signals on pins 35 and 36 shall utilize the "logic levels for transmission of digital data down to DC" of the nim specifications. A logic "1" gates ON and RESETS.

GP-1 GS-1 GP-1 GS-1 Note C GP-1 GS-1 GGS GGP
Bin conn. Mod. conn.
Front faces
Showing guide pins and sockets

FIG. 11.7. The pin assignments for the NIM standard connector. From USDOE (1990) p. 25.

Table 11.5. Power to NIM preamps is typically supplied from 9-pin "D" connectors which are installed on spectroscopy amplifiers for this purpose. The pin assignments shown are for Ortec and Canberra amplifiers. This connector is not part of the NIM standard but is used by several manufacturers. However, be sure to verify that your equipment uses compatible pin assignments. When this connector was first introduced, incompatible pin assignments used by different manufacturers led to much inconvenience.

Pin	Signal	Description
1	GND	Power supply ground
2	GND	High quality ground
4	+12	+12 volt dc supply
6	−24	−24 volt dc supply
7	+24	+24 volt dc supply
9	−12	−12 volt dc supply

Table 11.6. NIM standard logic levels.

NIM logic family	Logic level	Output (must deliver)	Input (must accept)
Positive logic	Logic 1 Logic 0	+4 to +12 V −2 to +1 V	+3 to +12 V −2 to 1.5 V
Fast negative logic (50Ω)	Logic 1 Logic 0	−14 to −18 mA +1 to −1 mA	−12 to −36 mA +20 to −4 mA

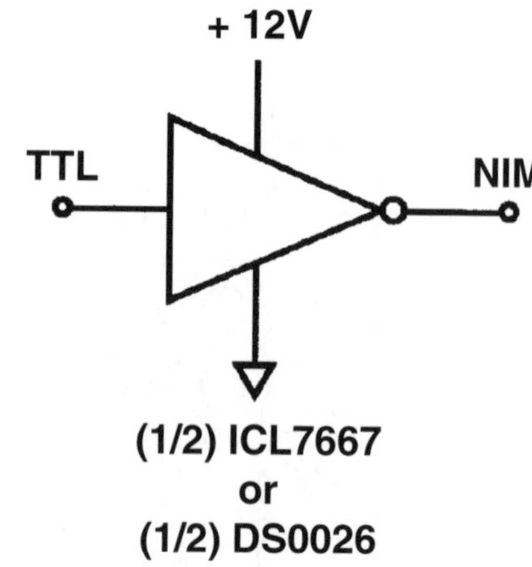

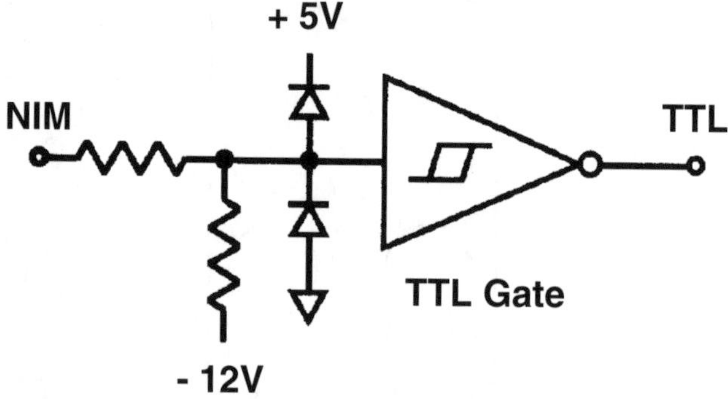

FIG. 11.8. Simple circuits to interface medium speed positive NIM logic and TTL.

(TTL) and its compatible CMOS successors. To drive a standard TTL circuit from a positive NIM output you can use a commercial module such as the Ortec 416A gate and delay generator. Alternatively, you can use a comparator or simply resistively couple the signal to a TTL gate with Schmitt trigger input which has been diode-clamped to ground and the +5 V supply as shown in Fig. 11.8. To boost a TTL output to drive a positive NIM input, use a driver with TTL compatible input such as the ICL7667 (an updated DS0026) connected to the +12 V supply (Fig. 11.8). This circuit is also useful as a simple 50Ω cable driver for TTL logic. NIM fast negative logic most closely resembles emitter-coupled logic (ECL) and modules to interface fast negative NIM to ECL are commercially available (e.g., the Ortec EC1600 16 channel ECL-NIM-ECL converter). If you need to build such an interface, an (aging) example is given in Althaus and Nagel (1972). Before beginning, become familiar with the latest high speed interface circuitry from the leading manufacturers of integrated circuits.

11.4.3 Basic energy spectroscopy

The simplest form of an energy spectroscopy system that is useful for ion beam analytical work is shown in Fig. 11.9, a block diagram of the modules (all commercially available) that are needed. The detector is indicated by a stylized diode symbol.

The voltage required to establish the reverse bias of the detector is supplied by a bias power supply that is conventionally a separate module, often with several independent outputs. The high voltage is usually physically applied to the preamplifier module which contains circuitry for limiting the current that can flow in the detector and for capacitively shunting the pulses into the charge-sensitive preamplifier circuitry which operates with ground reference. The pulse from the preamplifier is further amplified and shaped by the main spectroscopy amplifier and digitized and displayed by a multichannel analyzer. Schematics of the shape of the pulse are shown in Fig. 11.9 at various stages of processing.

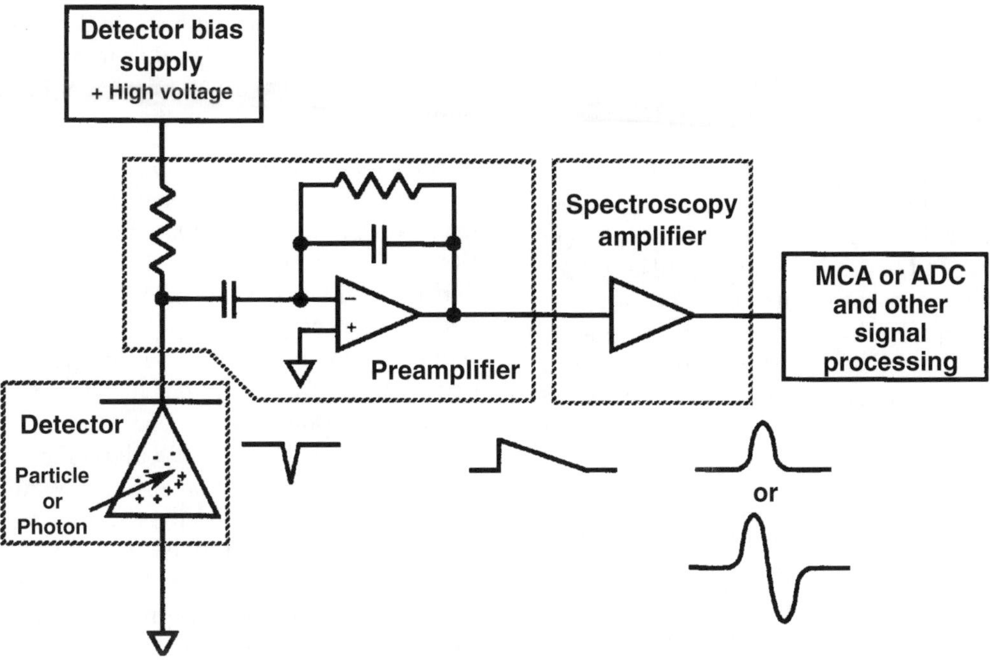

FIG. 11.9. A simple energy spectrometry system.

It is very important to emphasize that the preamplifiers and amplifiers that are used for nuclear spectroscopy are very different from those that are encountered, for example, in high-fidelity audio applications. Although both are, in the strictest sense, linear, the audio amplifier is designed to have essentially flat frequency response from near DC to several times the maximum frequency that humans can hear. Spectroscopy amplifiers, on the other hand, have frequency responses that are carefully tailored to preserve the information content which resides in the amplitude of the detector's pulse.

11.4.4 Preamplifiers

Preamplifiers come in two types, charge sensitive and voltage sensitive. Energy spectroscopy systems, such as the one in Fig. 11.9, use charge sensitive preamplifiers. As the diagram implies, these devices integrate the pulse from the detector. The reason for this is that the information about the radiation's energy is encoded in the number of electron-hole pairs that are created by the detector. By integrating the pulse from the detector, the energy is measured directly. The RC time constant in the preamplifier's feedback circuit is chosen to be of order 100 µs, which is vastly longer than the pulse generated by the detector. For this reason, the preamplifier's output resembles the graph of e^{-t} for $t > 0$ and is called a tail pulse. The information is contained in the height of the step at $t = 0$.

During the setup and calibration of the system, it is desirable to have a way to produce pulses without having to provide a source of radiation. This is done with a special pulse generator. Commercial spectroscopy preamplifiers provide an input for this signal. If the pulser output is very linear with its amplitude control, then the overall linearity of the spectroscopy system can be measured. This is particularly important for high resolution γ-ray systems because of the large number of very narrow peaks that appear in spectra. The resolution of the electronic system can also be observed by disconnecting the detector and recording the width of the pulser line.

Voltage sensitive preamplifiers are used for timing applications such as the time-of-flight spectrometer discussed previously. These preamplifiers simply increase the amplitude and power of the pulse without integrating it or materially altering its shape (except for a possible polarity reversal). It is most important to match the frequency response of the preamplifier and its rise time with the detector signal being processed. For example, microchannel plates need faster preamplifiers than surface barrier detectors. Manufacturers' literature provides guidance on the proper selections for different applications. The pin assignment for the preamp power connector is shown in Table 11.5.

11.4.5 Spectroscopy amplifiers

Modern spectroscopy amplifiers are extremely versatile and powerful tools. Their basic function is to accept tail pulses as input and to produce output pulses which are approximately Gaussian in shape, proportional in amplitude to the input pulse step, and approximately 1-2 µs in duration. They accomplish this by first differentiating and then integrating the input signal. A diagram of a modern semi-Gaussian shaping amplifier is shown in Fig. 11.10.

The first stage of a spectroscopy amplifier is shown schematically in Fig. 11.10 as a CR differentiator. This configuration of capacitor and resistor (ignoring R_2 for the moment) is referred to as a differentiating circuit because for signals which vary more slowly than the time constant, the voltage at point 2 is approximately the derivative of the input. For the tail pulse whose rise time is much shorter than the time constant, the resulting output is a much more quickly decreasing exponential as shown. This signal is amplified by A_1 and fed to an integrator circuit, which in its simplest form, is just another capacitor and resistor wired as shown. This circuit slows the very rapid rise of the pulse and yields the shape shown at point 3. A second amplifier A_2 provides isolation for this stage and adds additional gain. In a modern amplifier, the simple RC integrator is replaced by a more sophisticated active integrator in order to improve the signal to noise ratio for low level signals from semiconductor detectors. However, the principle is the same.

Most modern spectroscopy amplifiers have additional circuitry that aids in improving performance at high count rates. Three such features are pole-zero cancellation, baseline restoration, and pile-up rejection. The pole-zero circuit is also shown in Fig. 11.10. It derives its name from the cancellation of a pole and a zero in the complex plane in the Laplace transform analysis of the circuit. Its operation is quite simple. If R_2 were not present in the circuit, then point 2 would be AC coupled to the input and the voltage there would always average to zero. At moderate to high count rates, pulses would be shifted by the undershoots of previous pulses and, most disastrously, this shift would be rate dependent. R_2 allows a small fraction of the input pulse to circumvent the blocking capacitor C. When R_2 is properly adjusted, the undershoot is exactly canceled and the rate-dependent error is eliminated. This is what you are doing when adjusting the pole-zero cancellation potentiometer of a spectroscopy amplifier.

Another feature of good amplifiers is a baseline restorer. Because modern amplifiers are multi-staged and DC coupled, the later stages amplify DC offsets of previous stages. The baseline restorer deals with this source of offset. At low count rates, a simple CR differentiator (which is equivalent to AC coupling) would suffice. However, this circuit has problems at high pulse rates as

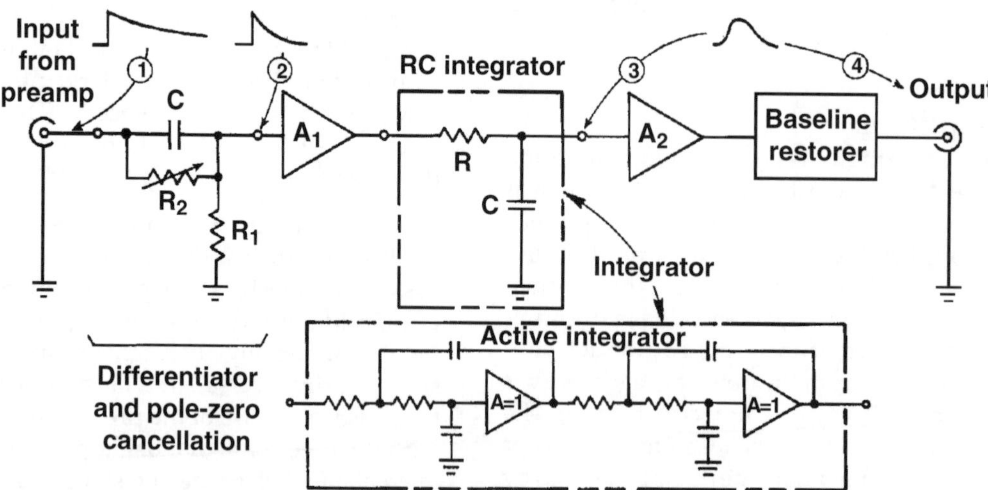

FIG. 11.10. A simplified schematic of a spectroscopy amplifier. Adopted from the EG&G Ortec catalog. Used with permission.

just described. The problem is solved at the expense of more complex circuitry. You should use your amplifier's baseline restoration circuit when trying to do precise spectroscopy at high count rates.

Pile-up rejection is also very useful when high count rates are encountered. If two pulses arrive within a time interval which is less than the approximate half width of the main amplifier's output pulse, they are summed and are subsequently indistinguishable from a single pulse with larger amplitude. This leads to significant distortion of a spectrum when the count rate is high and these overlaps are frequent. In order to reduce pile-up, many spectroscopy amplifiers have pile-up rejection circuitry. The principle is very simple. Within the amplifier, a parallel signal path is established from the input. By differentiating the input signal, it is possible to create very narrow pulses that correspond to the leading edges of the input tail pulses. When a pulse arrives at the amplifier's input, a flag is set in digital logic indicating that the amplifier is busy. If a second input pulse arrives during the time (equal to the width of the shaped output pulse) that this flag remains set, a second digital logic pulse is generated coincident with the output pulse. This digital pulse is provided as an output of the amplifier and can be used to gate a multichannel analyzer or other modules and prevent the piled-up analog pulse from being processed. Pile-up is thus reduced at the expense of additional "dead time" in the spectroscopy system. Figure 11.11 shows a gamma ray spectrum taken with and without pile-up rejection. The procedure for correcting for dead time is discussed below. For additional discussion of pile-up see Knoll (1989) and references therein.

Many spectroscopy amplifiers permit you to select either unipolar (semi-Gaussian) or bipolar output pulses (Fig. 11.12). The latter are typically obtained by an additional differentiation of the former. Consult the manual for your multichannel analyzer to determine which it prefers to receive as input. If you cannot find a preference stated, use unipolar pulses. Adjust the amplifier using an oscilloscope and keep the oscilloscope connected and turned on throughout the experiment. Seeing badly shaped pulses is often the first clue that something has gone wrong. For most additional processing, the amplifier's output pulses should be positive and have amplitudes in the range of 1-10 V. Adjust the gain to make it so. Preamplifier pulses are sometimes inverted relative to the case shown in Fig. 11.9. If your final pulse is negative, use the amplifier's input polarity switch to invert it. Adjust the shaping constants according to the instructions given in your multichannel analyzer manual or, if the manual is not available, to give an aesthetically pleasing, bell-shaped pulse with a width of about 1-2 μs as illustrated in Fig. 11.12. If you need to use bipolar pulses, be sure that they swing positively first, then negatively. Read your amplifier's manual and follow its directions to adjust the pole-zero potentiometer. Finally, if any of the connectors or controls on your amplifier become loose, fix them. There is little else to go wrong with a good spectroscopy amplifier.

11.4.6 Multichannel analyzers

A properly adjusted spectroscopy amplifier's output pulse is ready for analog-to-digital conversion. This is most often carried out in a multichannel analyzer (MCA) (Fig. 11.13). The MCA has been evolving rapidly in the last few years. It was originally invented by D. H. Wilkinson for nuclear studies and his analog-to-digital conversion technique is still widely used in commercial equipment (Wilkinson, 1950). Both stand-alone MCAs and personal computer plug-in cards are now widely used. In addition, for high count rate applications, or multiple data channels, computer based systems such as CAMAC, which was originally developed for high-energy physics, are used. Here, we shall describe a stand-alone MCA with Wilkinson analog-to-digital converter for the sake of simplicity. Data acquisition and handling are the most individualized aspects of nuclear spectroscopy. The best advice that can be given is to understand your own system, whatever its details, and not to leave those details to someone else.

The heart of the MCA is the analog-to-digital converter (ADC). The ADC analyzes the maximum amplitude of each pulse with a precision (called conversion gain) ranging typically from one part in 256 up to one in 8192. Fig. 11.14 illustrates the principle of the Wilkinson ADC. The input pulse is used to charge a capacitor until the peak of the pulse is detected (by a zero crossing of its derivative). At this time, or with a small delay, a constant current source is turned on and begins to linearly discharge the capacitor. At the time that the discharging begins, a high frequency clock is also started and its ticks are counted until the voltage on the capacitor crosses zero. This stops the discharge, stops the clock and makes the total counts available as the digitized pulse amplitude.

Good Wilkinson ADCs use clock frequencies of 100 MHz or higher and, therefore, can do an 8k channel

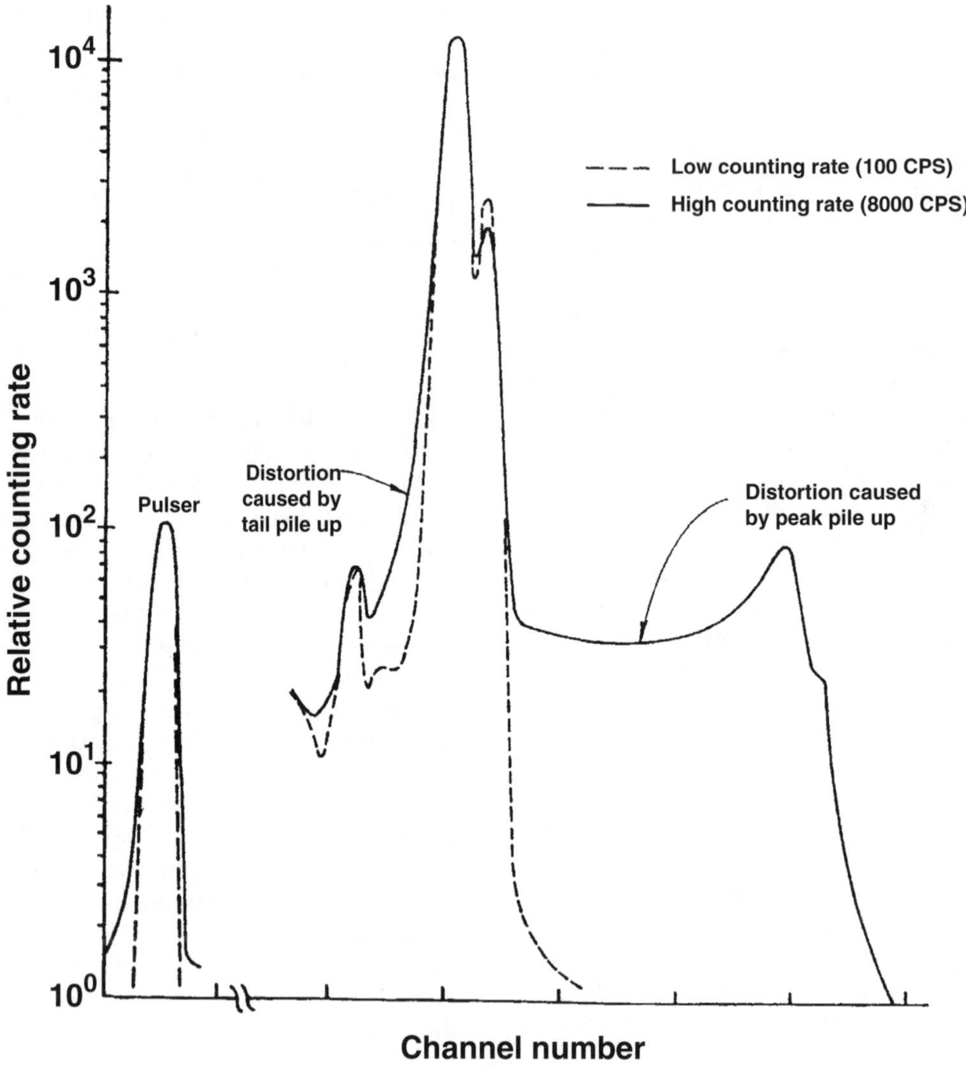

FIG. 11.11. Spectra of ^{55}Fe showing distortion caused by pulse pile-up. From L. Wielopolski and R.P. Gardner, "Prediction of the Pulse-height spectral distortion caused by the peak pile-up effect", *Nucl. Instr. and Methods* **133** (1976) 303. Used with permission.

conversion in a time of order 100 μs and a lower precision conversion proportionally more quickly. During the conversion, the analyzer is unavailable for processing additional pulses. It is necessary to take this dead time along with pile-up dead time into account when doing quantitative work. One way to measure the ADC dead time is to gate an oscillator using the circuitry in the ADC which inhibits additional pulses while one is being processed. By inhibiting counts while the ADC is "dead" a direct measure of "live time" is produced. A better way is to leave a stable pulser attached to the preamplifier and adjust the amplitude to insert test pulses into a portion of the spectrum that does not contain real features of interest. By comparing the number of these pulses observed in the MCA spectrum with the total number generated, a direct measure of counting losses due to both ADC dead time and pile-up rejection is obtained. (See Chapter 12.) For most applications a rate of test pulses as low as 60 Hz is satisfactory. Some authors argue (see Nicholson, 1974) that random rather than constant frequency test pulses should be used. However, for practical ion beam analytical work, the difference is irrelevant if the pulser rate is below about 10% of the real count rate. In modern MCA designs, the total real time and live time are usually made

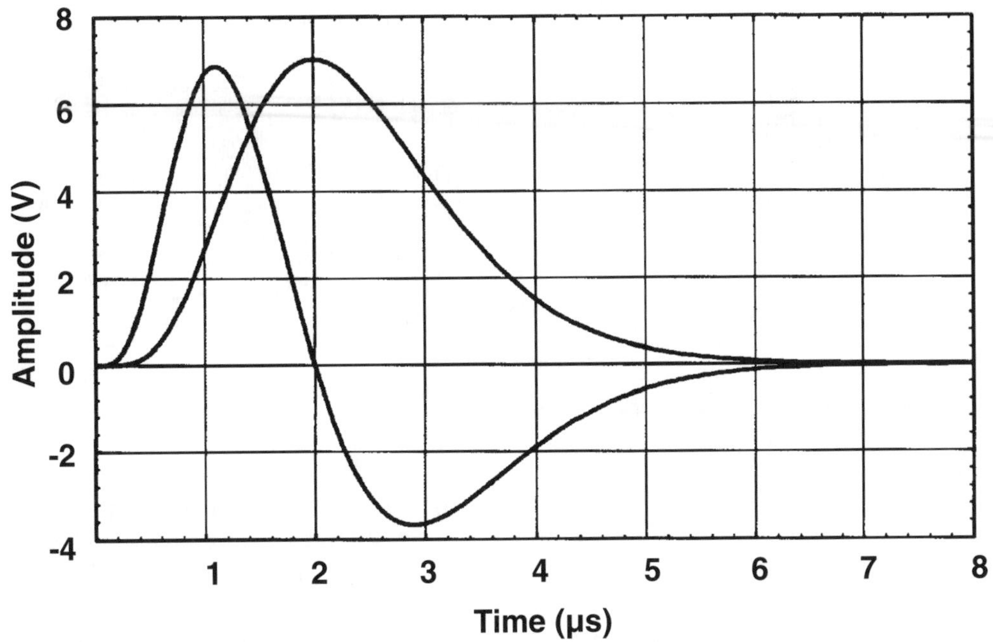

FIG. 11.12. Characteristic shape of unipolar and bipolar NIM linear pulses from a spectroscopy amplifier. The bipolar pulse is obtained as the derivative of the unipolar pulse.

available to the user. However, these are often less accurate measures than can be obtained using a pulser. As a result, for the most precise work, you should do your own dead time estimation. The use of the measured dead time to correct data is given in 11.6.2.

The output of the ADC is a number which is proportional to the height of the input pulse. An output event is produced, of course, only when a pulse appears at the ADC input. The MCA takes the number which is computed by the ADC and uses it as an address to access a memory bank (Fig. 11.13). When a pulse arrives, the MCA addresses the indicated memory register and increments it by one to record that one pulse of that amplitude has been measured. In this way, a histogram of measured pulse amplitudes is accumulated. Commercial standalone MCAs display this histogram in real time on a video monitor and offer several display options, and often an assortment of analytical capabilities. They usually also provide some mechanism for down-loading the information to a computer for storage and further processing. Plug-in cards for personal computers use the computer's display under control of specialized software.

It is assumed that the quantity being measured, e.g., energy E or time T, is related to the MCA channel n via a linear relationship such as

$$E = a(n - z) \tag{11.2}$$

where a and z are constants. The constant z is called the zero offset. Both z and a must be obtained by calibrating the analyzer. This is done by using standards in which spectral features with known energies can be identified. Tables 11.7, 11.8, and 11.9 contain useful x-ray, γ-ray, and α calibration sources[13]. Clearly two spectral features with known energies are sufficient to calibrate a linear analyzer. However, it is better to have more so that a and z can be obtained with probable errors by linear regression. Subsequently one obtains E corresponding to a channel number n from Eq. (11.2). For very precise γ-ray work with germanium detectors it is necessary to verify the linearity of the analyzer. For routine ion beam analysis, all commercial equipment is sufficiently linear that Eq. (11.2) can be assumed to hold.

Although multichannel analyzers will suffice for most ion beam analytical work, there are some instances where

13. For an electron source with discrete lines you can use a radioactive source that decays by internal conversion (the nuclear analog of Auger electron emission). See Jardine and Lederer (1974).

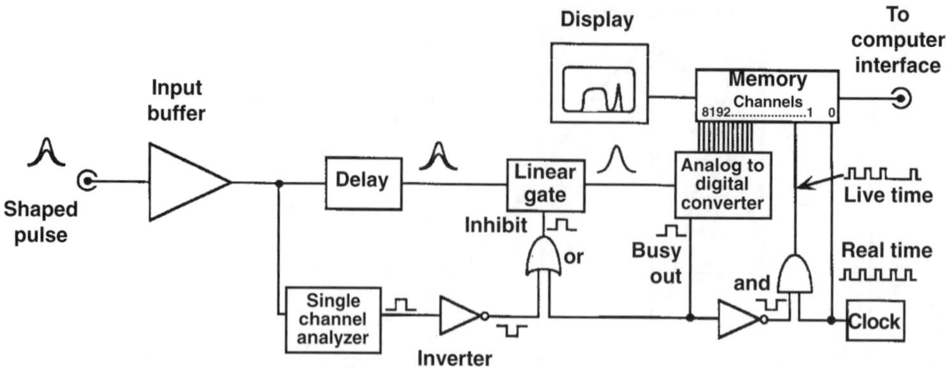

FIG. 11.13. Schematic diagram of a multichannel analyzer.

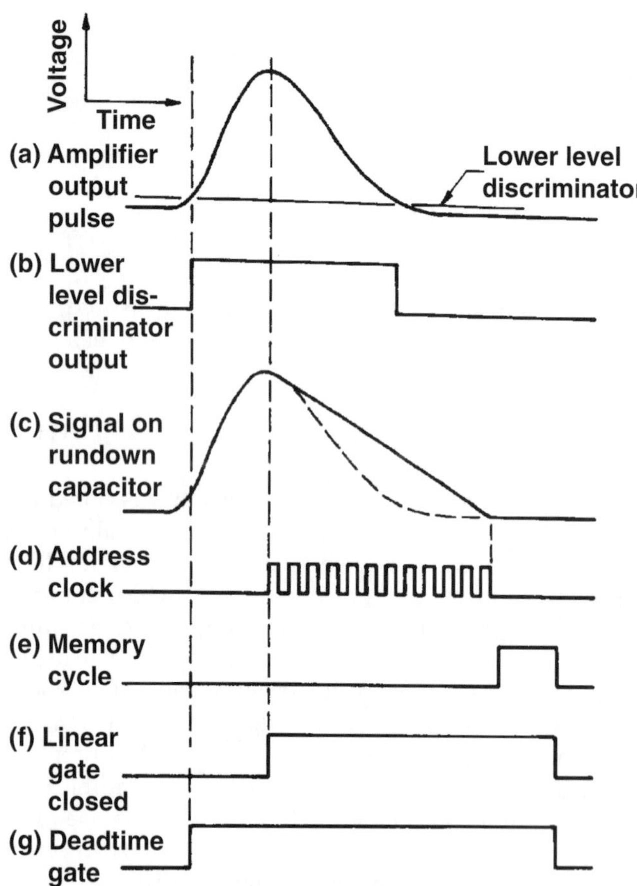

FIG. 11.14. Timing diagram of a Wilkinson ADC. From the Ortec Catalog. Used by permission.

the histogramming of a single parameter to produce a spectrum is inadequate. Experiments using position-sensitive detectors (Knoll, 1989) or multi-parameter particle detection systems (England, 1974), such as particle telescopes (which simultaneously measure total energy and stopping power of an ion by using a thin surface barrier detector in combination with a thick one) are two such cases. When more than one parameter must be stored in connection with each event, the best general-purpose approach is to use a CAMAC data acquisition system, since these were developed for this purpose. Commercial hardware and software are available to implement multi-dimensional histogramming or to store data on an event-by-event basis for later computer analysis.

11.4.7 Cables and connections

The analog and logic signals, collected beam current, detector biases, and most other electrical signals are routed from module to module and throughout the laboratory using coaxial cables (Bryant, 1984). These cables are used so widely for electronic equipment that their specifications have been standardized by the United States Government. Table 11.10 contains a detailed listing of cables and their specifications by their "RG" numbers (Radio Guide #/Universal). The three most widely used cables for nuclear spectroscopy are RG-62/U, RG-58A/U, and RG-174/U. In addition RG-8A/U and RG-59A/U are often used as high voltage cables. Analog and positive logic NIM signals are routed using RG-62/U cable which has a characteristic impedance of 93 Ω. Module inputs for these signals typically have an impedance of around 1 kΩ so that several inputs may be daisy-chained together driven in parallel from the same output signal. Because of the mismatch between the impedance of the cable and the instrument input, it is necessary to terminate the cable in its characteristic impedance. This is very important. Improperly

Table 11.7. Calibration sources for solid state x-ray detectors, including the intensity ratio of specified x-rays to the indicated γ-ray. Uncertainties in the final digits are shown in parentheses. Quoted x-ray energies should not be regarded as standards because of uncertainty caused by line shape (many of the lines have complex structure), chemical shifts, and other shifts. Sources: Campbell and McNelles (1975), Helmer *et al.* (1979), Tuli (1985), Lederer and Shirley (1978).

Isotope	Half-life	E_X *(keV)*	E_γ *(keV)*	Intensity *(X/γ)*
^{54}Mn	312.5 d	5.47 ($K_{\alpha\beta}$)	834.84	0.251 (1)
^{57}Co	270.9 d	6.4 (K_α) 7.1 (K_β) 14.41 (γ)	122.06	0.57 (1) 0.79 (2) 0.112 (2)
^{65}Zn	243.9 d	8.04 (K_α) 8.9 (K_β)	1,115.5	0.660 (5) 0.091 (2)
^{85}Sr	64.84 d	13.38 (K_α) 15.0 (K_β)	514.00	0.502 (3) 0.088 (1)
^{88}Y	106.6 d	14.12 (K_α) 15.85 (K_β)	898.04	0.549 (7) 0.989 (2)
^{109}Cd	462.9 d	22.10 (K_α) 25.0 (K_β)	88.037	22 (1) 4.7 (2)
^{113}Sn	115.1 d	24.14 (K_α) 27.4 (K_β)	391.69	1.22 (4) 0.27 (1)
^{137}Cs	30.17 y	32.1 (K_α) 36.6 (K_β)	661.66	0.067 (2) 0.0159 (5)
^{139}Ce	137.7 d	33.29 (K_α) 38.0 (K_β)	165.85	0.81 (9) 0.20 (2)
^{198}Au	2.696 d	70.15 (K_α) 80.7 (K_β)	411.80	0.0229 (5) 0.0064 (2)
^{203}Hg	46.60 d	72.11 (K_α) 83.0 (K_β)	279.20	0.125 (3) 0.0348 (8)

terminated cables, especially long ones, are one of the most common sources of poor or erroneous signals.

For routing fast negative logic and high speed analog signals, one typically uses either RG-58A/U or RG-174/U cable. The former is more appropriate for long cable runs. The latter is a much smaller diameter cable, but has greater attenuation per unit length. Even so, RG-174/U is more convenient for short, dense interconnections. It is not usually necessary to provide external termination for a standard NIM fast negative input since it is normally terminated internally in 50 Ω. However, if fast signals are being routed through equipment that may not be internally terminated, it is vital that proper termination be provided externally. Signals propagate in coaxial cable at about 60% of the speed of light (propagation velocity = $(Z_oC)^{-1}$ where Z_o is the characteristic impedance and *C* is the capacitance/meter). Fast negative logic pulses can be as short as 5 ns. As a result, it is possible even in relatively short cable runs to get distinct, multiple pulses from improper termination. Do not step on coaxial cables or bend them in arcs shorter than about 10 of their own diameters. If you need to make a 90° bend at a junction, do it with a 90° "elbow" connector, not by kinking the cable.

Coaxial cables are attached to NIM modules (and most other laboratory equipment) by standard interlocking connectors (Table 11.11). These too have standard designations ("UG" in this case, for Union Guide/Universal) but these are not used as frequently as the "RG" numbers of cables. The three most important connectors are the BNC (Baby Neill-Concelman), the SHV (safe high voltage), and the LEMO (the company's name)[14]. BNC connectors are rated for voltages up to 500 VDC and are by far the most common connector. SHVs are used for handling voltages above the limit for BNCs (up to 5 kVDC) and for ultra-low level currents where the leakage at a BNC connection could be significant. Another high-voltage connector, the MHV, is also still encountered on some equipment. However, it is possible to force an MHV plug onto a BNC socket with potentially disastrous consequences for equipment and people. For this reason, do not use MHV connectors. If you have a piece of equipment that has MHV connectors on it, treat it with special care and replace the MHVs with SHVs the next time that maintenance on the unit is required. LEMOs are elegant little connectors that make possible very high connection densities, especially when used with RG-174/U cables. They are found most frequently on designs such as timing discriminators where large numbers of 50 Ω cable connections are needed on narrow modules.

The most common problems that can occur in interconnecting systems are incorrect routing of signals, poor quality or improperly maintained cables and connectors, improper termination of cables, and ground loops. The last is the most insidious. All "grounds" are not equal. In an accelerator laboratory, some of the equipment is

14. LEMO USA, Inc., P.O. Box 11006, Santa Rosa, CA 95406. Only the 00 shell size of LEMO is used for nuclear instrumentation. Compatible connectors are also manufactured in the United States by Kings Connectors, Tuckahoe, NY 10707 under the trade name K-Loc. The official ANSI/IEEE designation of this connector is 50CM.

Table 11.8. Suggested γ-ray calibration sources including half-lives and principal γ-ray energies. Uncertainties in the final digits are shown in parentheses. Sources: Helmer *et al.* (1979), Tuli (1985).

Source	Half-life	E_γ (keV)
^{7}Be	53.29 d	477.605 (3)
$^{13}C + \alpha$	α source	6,129.27 (5)
^{22}Na	2.602 y	1,274.54 (1)
^{24}Na	15.02 h	1,368.63 (1)
		2,754.03 (1)
^{46}Sc	83.83 d	889.277 (3)
		1,120.545 (4)
^{51}Cr	27.704 d	320.084 (1)
^{54}Mn	312.5 d	834.843 (6)
^{59}Fe	44.496 d	1,099.251 (4)
		1,291.596 (7)
^{56}Co	78.76 d	846.764 (6)
		1,037.844 (4)
		1,175.10 (1)
		1,238.29 (1)
		1,360.21 (1)
		1,771.35 (2)
		1,810.72 (2)
		1,963.71 (1)
		2,015.18 (1)
		2,034.76 (1)
		2,113.11 (1)
		2,212.92 (1)
		2,598.46 (1)
		3,009.60 (2)
		3,201.95 (1)
		3,253.42 (1)
		3,273.00 (1)
		3,451.15 (1)
^{57}Co	270.9 d	122.0614 (3)
		136.4743 (5)
^{60}Co	5.271 y	1,173.238 (4)
		1,332.502 (5)
^{65}Zn	243.9 d	1,115.546 (4)
^{88}Y	106.64 d	898.042 (4)
		1,836.06 (1)
^{95}Zr	64.02 d	724.199 (5)
^{94}Nb	2.03×10^{4} y	702.645 (6)
		871.119 (4)
^{108m}Ag	127 y	433.936 (4)
		614.281 (4)
		722.929 (4)
^{110m}Ag	249.76 d	446.811 (3)
		620.360 (3)
		657.762 (2)
		677.623 (2)
		687.015 (3)
		706.682 (3)
		744.277 (3)
		763.944 (3)
		818.031 (4)
		884.685 (3)
		937.493 (4)
		1,384.300 (4)
		1,475.788 (6)
		1,505.040 (5)
		1,562.302 (5)
^{124}Sb	60.20 d	602.730 (3)
		645.855 (2)
		713.781 (5)
		722.786 (4)
		790.712 (7)
		968.201 (4)
		1,045.131 (4)
		1,325.512 (6)
		1,368.164 (7)
		1,436.563 (7)
		1,690.980 (6)
		2,090.942 (8)
^{137}Cs	30.17 y	661.660 (3)
^{144}Ce	284.4 d	696.510 (3)
		1,489.160 (5)
		2,185.662 (7)
^{152}Eu	13.33 y	121.7824 (4)
		244.699 (1)
		344.281 (2)
^{153}Gd	241.6 d	69.6734 (2)
		97.4316 (3)
		103.1807 (3)
^{170}Tm	128.6 d	84.2551 (3)
^{182}Ta	114.5 d	67.7500 (2)
		84.6808 (3)
		100.1065 (3)
		113.6723 (4)

Table 11.8. Suggested γ-ray calibration sources including half-lives and principal γ-ray energies. Uncertainties in the final digits are shown in parentheses. Sources: Helmer *et al.* (1979), Tuli (1985). (continued).

Source	Half-life	E_γ *(keV)*
^{182}Ta (Cont)		116.4186 (7)
		152.4308 (5)
		156.3874 (5)
		179.3948 (5)
		198.353 (1)
		222.110 (1)
		229.322 (1)
		264.076 (1)
		1,121.301 (5)
		1,189.050 (5)
		1,221.408 (5)
		1,231.016 (5)
		1,257.418 (5)
		1,273.730 (5)
		1,289.156 (5)
		1,373.836 (5)
		1,387.402 (4)
^{192}Ir	73.831 d	136.3434 (5)
		205.7955 (5)
		295.958 (1)
		308.457 (1)
		316.508 (1)
^{192}Ir (Cont)		416.472 (1)
		468.072 (1)
		484.578 (1)
		588.585 (2)
		604.415 (2)
		612.466 (2)
		884.542 (2)
^{198}Au	2.696 d	411.804 (1)
		675.888 (2)
		1,087.691 (3)
^{203}Hg	46.6 d	279.197 (1)
^{207}Bi	32.2 y	569.702 (2)
		1,063.662 (4)
		1,770.24 (1)
^{228}Th	1.9131 y	238.632 (2)
		583.191 (2)
		860.564 (5)
		893.408 (5)
		1,620.74 (1)
		2,614.53 (1)

usually located in the target room and some is in the control room. With such distributed systems, it is possible, even likely, to encounter situations where there are large loops in the ground wiring. These can be the sources of offsets up to several volts. Get to know ground loops and design your systems to avoid them. For help, consult Morrison (1977). Amazingly, he has made the subject interesting. If you are building a laboratory, consider specifying a high quality signal ground as part of the construction. Such a system consists of a large conducting mesh that is planted in the earth near the laboratory and is connected to various strategic points via heavy conductors. Do not underestimate the time and effort that can go into eliminating ground loops or the damage to your data that they can do.

11.5 VACUUM TECHNOLOGY

11.5.1 Vacuum practice

All ion beam analyses are conducted in vacuum because vacuum is required for the production and transport of ion beams and because surface properties are highly dependent upon surface cleanliness. In addition, vapor-deposited thin films must often be manufactured in support of the routine practice of ion beam analysis. A full discussion of vacuum systems is beyond the scope of this chapter. However, some graphs have been included because of their general applicability to problems arising in the course of ion beam analysis. For a complete discussion of all aspects of vacuum technology see Roth (1982). For ideas on producing atomically clean surfaces on elemental targets consult Musket *et al.* (1982).

Probably the most useful general advice that could be given about vacuum practice would be to always design systems conservatively and to aim for the lowest pressure and cleanest system that time and resources permit. Here "clean" refers to the quantity of hydrocarbon vapors that enter the system from oil diffusion pumps and mechanical roughing pumps. If present in sufficient quantities in the sample chamber, molecules of these hydrocarbons can be cracked by the action of the beam while they temporarily reside on the sample surface. This leaves behind a film of non-volatile fragments which can interfere with surface analyses or surface modifications.

Table 11.9. Alpha particle energy standards, including half-lives, energies, and relative intensities. Uncertainties in the last digit of the energies are shown in parentheses. Very thin ^{212}Bi and ^{212}Po sources can be prepared by collecting ^{220}Rn on a sample surface in the vicinity of a thorium source. The controlling half-live is 10.64 h, that of ^{212}Pb from which ^{212}Bi forms by β^- decay. Sources: Rytz (1979), Lederer and Shirley (1978), Tuli (1985).

Isotope	Half-life	E_α (MeV)	I_α (%)
^{146}Sm	1.03×10^8 y	2.470 (6) (calculated)	100
^{150}Gd	1.8×10^6 y	2.719 (8) (calculated)	100
^{148}Gd	75 ± 30 y	3.18271 (2)	100
^{232}Th	1.41×10^{10} y	4.013 (3) 3.954 (8)	77 23
^{238}U	4.468×10^9 y	4.197 (5) 4.150 (5)	77 23
^{235}U	7.038×10^8 y	4.599 (2) 4.400 (2) 4.374 (4) 4.368 (2) 4.218 (2)	1.2 57 6.1 12.3 6.2
^{236}U	2.342×10^7 y	4.494 (3) 4.445 (5)	74 26
^{230}Th	7.54×10^4 y	4.688 (2) 4.621 (2)	75.5 24.5
^{234}U	2.45×10^5 y	4.775 (1) 4.723 (1)	72.5 27.5
^{209}Po	102 y	4.881 (2)	99.52
^{231}Pa	3.28×10^4 y	5.059 (1) 5.028 (1) 5.014 (1) 4.952 (1) 4.736 (1)	11.0 20.0 25.4 22.8 8.4
^{208}Po	2.898 y	5.115 (2)	100
^{239}Pu	2.412×10^4 y	5.1566 (4) 5.144 (1) 5.105 (1)	73.3 15.1 11.5
^{240}Pu	6570 y	5.1682 (2) 5.1237 (2)	73.5 26.4
^{243}Am	7380 y	5.350 (1) 5.2753 (5) 5.2330 (5)	0.22 87.4 11.0
^{210}Po	138.376 d	5.3044 (1)	99+
^{228}Th	1.9131 y	5.4232 (2) 5.3403 (2)	72.7 26.7
^{241}Am	432.2 y	5.544 (1) 5.4856 (1) 5.4429 (1)	0.34 85.2 13.1
^{238}Pu	87.74 y	5.4991 (2) 5.4563 (2)	71.5 28.5
^{244}Cm	18.10 y	5.80482 (5) 5.76270 (5)	76.4 23.6
^{243}Cm	28.5 y	6.066 (1) 6.058 (1) 5.992 (1) 5.7850 (5) 5.7411 (5)	1.0 5.0 5.8 73.54 10.65
^{242}Cm	162.8 d	6.1128 (1) 6.0694 (1)	73.8 26.2
^{254}Es	275.5 d	6.429 (1)	93.2
^{253}Es	20.47 d	6.63257 (5) 6.5914 (5)	90.0 6.6
^{212}Bi	60.55 m	6.08994 (4) 6.05083 (4)	27.2 69.9
^{212}Po	0.298 μs	8.7842 (1)	100

To avoid this problem, the sample area should be pumped by an oil-free pump. Ion pumps are a good choice but not necessarily the best choice for all situations. It is essential to always remember that the monolayer formation time at 10^{-6} torr is about 2 s (see Appendix 16, Fig. A16.1) so that significant experiments on surface properties require ultra-high vacuum ($<10^{-9}$ torr). It is also important to remember when designing vacuum systems and internal components that there are some materials which work well in vacuum and some which cause problems. An example of the latter is cadmium, which has a high vapor pressure (see Appendix 16, Fig. A16.2). Although not used for structural purposes, cadmium is commonly used to plate steel fasteners to suppress corrosion. If these fasteners are heated, cadmium can be spread throughout the vacuum chamber contaminating samples and hardware alike. Develop the habit of using stainless steel fasteners exclusively inside your chamber and you will avoid this. If you have a diffusion pumped system, you can still create a localized hydrocarbon-free volume with a liquid-nitrogen-cooled shield around the sample. For best results, use

Table 11.10. Properties of some common coaxial cables. Principal source: *Master Catalog 885*, Belden Wire and Cable, Richmond, Indiana.

Cable Designation (Radio Guide–# /Universal)	Center conductor style (AWG stranding)	Insulation	Cable O.D. (mm)	Maximum voltage (V_{RMS})	Nominal impedence Z_o (Ω)	Nominal capacitance C (pF/m)	Signal propagation velocity (% of speed of light)	(MHz)	Nominal signal attenuation (dB/100·m)	Governing specification	Comments
RG–8/U RG–8A/U	13 (7 × 21)	Polyethylene	10.29	5,000	52	96.8	66	100 400 1000	7.2 15.4 29.2	JAN-C-17A MIL-C-17D	Black PVC jacket. RG-8A/U has non-contaminating jacket.
RG–9/U	13 (7 × 21)	Polyethylene	10.67	5,000	51	98.4	66	100 400 1000	7.2 15.4 29.2	JAN-C-17A	Gray, non-contaminating PVC jacket. Double shielded.
RG–11/U RG–11A/U	18 (7 × 26)	Polyethylene	10.29	5,000	75	67.3	66	100 400 1000	6.6 13.8 23.3	JAN-C-17A MIL-C-17D	Black PVC jacket. RG-11A/U has non-contaminating jacket.
RG–58/U	20 (solid)	Polyethylene	4.95	1,900	53.5	93.5	66	100 400 1000	14.8 32.8 55.8	JAN-C-17A	Black PVC jacket.
RG–58A/U	20 (19 × 32)	Polyethylene	4.95	1,900	50	101.0	66	100 400 1000	16.1 37.7 70.5	JAN-C-17A	Black PVC jacket. Used for fast NIM signals.
RG–58C/U	20 (19 × .0071)	Polyethylene	4.95	1,900	50	101.0	66	100 400 1000	16.1 37.7 70.5	MIL-C-17F	Black, non-contaminating PVC jacket.
RG–59/U	22 (solid)	Polyethylene	6.15	2,300	73	68.9	66	100 400 1000	11.2 23.3 39.4	JAN-C-17A	Black PCV jacket.
RG–59B/U	23 (solid)	Polyethylene	6.15	2,300	75	67.3	66	100 400 1000	11.2 23.0 39.4	MIL-C-17D	Black, non-contaminating PVC jacket.
RG–62/U	22 (solid)	Semi-solid polyethylene	6.04	700	93	44.3	84	100 400 1000	8.9 17.7 28.5	JAN-C-17A	Black PVC jacket. Used for NIM linear and positive logic signals.
RG–62B/U	24 (7 × 32)	Semi-solid polyethylene	6.15	700	93	44.3	84	100 400 1000	9.5 20.0 36.1	MIL-C-17D	Black, non-contaminating PVC jacket. Also used for NIM applications.

Table 11.10. Properties of some common coaxial cables. Principal source: *Master Catalog 885*, Belden Wire and Cable, Richmond, Indiana (continued).

Cable Designation (Radio Guide–# /Universal)	Center conductor style (*AWG stranding*)	Insulation	Cable O.D. (*mm*)	Maximum voltage (V_{RMS})	Nominal impedence Z_o (Ω)	Nominal capacitance C (*pF/m*)	Signal propagation velocity (*% of speed of light*)	(*MHz*)	Nominal signal attenuation (*dB/100·m*)	Governing specification	Comments
RG–174/U	26 (7 × 34)	Polyethylene	2.56	1,500	50	101.0	66	100 400 1000	29.2 57.4 98.4	MIL-C-17D	Black PVC jacket. Used for high speed NIM signals, often with LEMO connectors.
RG-178B/U	30 (7 × 38)	Teflon	1.83	1,000	50	95.10	69.5	100 400 1000	45.9 91.9 150.9	MIL-C-17D	Brown, fluorinated ethylene-propylene jacket. High temperature.
RG–213/U	13 (7 × 21)	Polyethylene	10.29	5,000	50	101.0	66	100 400 1000	7.2 15.4 29.2	MIL-C-17D	Black, non-contaminating PVC jacket.
RG–223/U	19 (solid)	Polyethylene	5.38	1,900	50	101.0	66	100 400 1000	14.8 30.2 53.5	MIL-C-17F	Black, non-contaminating PVC jacket. Double shielded.
RG–316/U	26 (7 ×. 0067)	Teflon	2.49	1,200	50	95.2	69.5	100 400 1000	34.1 54.1 101.7	MIL-C-17D	Brown, fluorinated ethylene-propylene jacket. High temperature.
2077†	4 (solid)	High molecular weight polyethylene	38.1	300,000 VDC	67	83.0	60	–	–	–	Very high voltage coaxial cable.

†Manufactured by Dielectric Sciences, Inc., Chelmsford, MA.

Table 11.11. Properties of common coaxial connectors. Sources: *Catalog 500-89*, Kings Electronics Co., Inc., Tuckahoe, NY; Malco Division, Microdot, Inc., South Pasadena, CA; *S Series Connectors Catalog*, LEMO USA, Santa Rosa, CA.

Series	Nominal impedance (Ω)	Nominal cable size range *(inches)*	Coupling style	Maximum frequency *(GHz)*	Voltage rating (V_{RMS})	Dielectric withstanding voltage (V_{RMS})	Governing specification	Comments
BNC	50	0.150 – 0.250	Bayonet	4	500	1500	MIL-C-39012	Universal, inexpensive. "Baby Neill-Concelman."
SHV	Not constant	0.054 – 0.440	Bayonet	–	3500	10,000 VDC	MIL-C-39012 NBS-ND-545	Preferred HV connector.
MHV	Not constant	0.054 – 0.440	Bayonet	–	3500	6500 VDC	MIL-C-39012	Potentially dangerous. Avoid rigorously.
UHF	Not constant	0.199 – 0.432	Threaded	0.2	500	–	None	The original pre-WWII rf connector. Inexpensive, but of low quality for signals.
Microdot	93, 50, 70	< 0.154	Threaded	2	750	1000		93Ω version used frequently on surface barrier detectors.
50CM LEMO (series 00) K-Loc	50	0.150 – 0.250	Locking	2 (typical)	500	1500	NBS-ND-549	Very elegant when used with RG-174/U cable.
TNC	50	0.150 – 0.250	Threaded	11	500	1500	MIL-C-39012	Essentially a threaded high frequency BNC.
N	50	0.350 – 0.450	Threaded	11	1000	2500	MIL-C-39012	Patriarch of the family of rf connectors. (Born, 1942. Named for Paul Neill.)
HN	50	0.195 – 0.945	Threaded	4	1500	5000	MIL-C-39012	High voltage version of type N rf connector.
C	50	0.350 – 0.450	Bayonet	11	1000 3000 (HV type)	3000	MIL-C-39012	Common. High voltage type useable to 2 GHz. (Designer: Carl Concelman.)
SC	50	0.350 – 0.450	Threaded	11	1000	3000	MIL-C-39012	A threaded C for hostile environments.
SMA	50	0.047 – 0.250	Threaded	12 – 18 (some versions to 27 GHz)	350 – 500	500 – 1500	MIL-C-39012	Most widely used microwave connector. Sub-Miniature (A), very high frequency. Maximum ratings depend upon cable.

metal-to-metal external seals such as the now standard "conflat"[15] design or compatible alternatives from other suppliers. For exhaustive information on the properties of elastomer seals, consult the literature of manufacturers[16].

A final observation. When your vacuum system leaks, as it will eventually, find the leak and fix it properly. Do not use wax, glue or other temporary measures because you are too busy or lazy to do the job correctly. Now having said this, we know that you will do something like this anyway. Therefore, obtain a small bottle and spray can of Vacseal[17], and a tube of a solventless epoxy such as Torr-Seal from Varian while your system is still leak-free so that if you must cast good vacuum practice to the wind, at least your crude patches will work. You should also know that, if you *must* glue something that will be used in an ultra-high vacuum system, the thermal-curing epoxies H21D (electrically conducting) and H72 (electrically insulating) from Epo-tek[18] are useful alternatives to Torr-Seal.

11.5.2 Thin films

Thin films for use in ion beam work can be produced electrochemically, by electron-beam evaporation, resistive-heater evaporation, sputtering, ion implantation, molecular-beam epitaxy, or by laser-assisted deposition. They are used in applications as diverse as beam stripper foils for tandem accelerators, entrance windows for gas detectors and occasionally even as exit windows for bringing ion beams into the air to perform analyses on samples that are incompatible with vacuum. The classic reference on thin films is Maissel and Glang (1970) but there are more recent comprehensive reviews (Schuegraf, 1988). Much recent work has been undertaken in conjunction with the development of semiconductor technology and is covered in conference proceedings such as Poate *et al.* (1978) and Lucovsky (1988). The most common technique for making thin films is by evaporation. An extensive listing of the physical properties of compounds including their vapor pressure has been compiled by Graper (1990) to assist in this task. This table is presented in Appendix 16, Table A16.1

There are several factors which must be considered in selecting a foil for a specific purpose. These include the required thickness, constraints on the foil's composition such as the likelihood of nuclear reactions, the area, if the foil is self supporting, the differential pressure which the foil must support, and the level of radiation that the foil will experience. Beam stripper foils mounted in the terminal of a tandem Van de Graaff accelerator are universally composed of carbon and are usually about 3-5 $\mu g/cm^2$ thick. At the other extreme are vacuum-system exit windows which must support atmospheric pressure on one side and yet permit an ion beam to penetrate. In the early days of nuclear physics, these windows were almost always made of Ni. Now, however, Al, Ti, and the radiation-resistant organic material, kapton, are also used. The required thickness depends, of course, on the area of the window, but it is typically a few mg/cm^2. Ti is also commonly used for exit windows in high-current, electron accelerators which manufacture plastic films by radiation-induced polymerization. An analysis of the mechanical behavior of thin films subjected to a differential pressure has been given by Spivack (1972). If the required thickness of a film is incompatible with the area that must be covered, it is often possible to mount the film on a high-transmission mesh for support[11].

One of the most serious complications which arises in the production of thin films by deposition is the phenomenon of islanding (Feldman and Mayer, 1986). If deposited atoms are mobile on the surface, they may coalesce into non-contiguous clumps called islands which locally can be much thicker than would be assumed on the basis of the average coverage of the surface as a whole. The presence of islanding can be detected by high-resolution backscattering by, for example, insufficient shifting of the thick target edge attributable to the atoms of the substrate. See also Fig. 12.10. Islanding can diminish the value of submonolayer films intended for use as backscattering energy-resolution standards. For better results, use the thick target edges of mono-isotopic elements in this application.

15. Varian Vacuum Products, 121 Hartwell Avenue, Lexington, MA 02173.
16. For example, Parker Seal Group, O-Ring Division, 2360 Palumbo Drive, P.O. Box 11751, Lexington, KY 40512. See, in particular, the catalog entitled *Parker O-Ring Handbook, ORD-5700* or its most current successor.
17. Space Environment Laboratories, P.O. Box 1061, Boulder, CO 80306.
18. Epoxy Technology, Inc., 14 Fortune Drive, Billerica, MA 01821.

11.6 STATISTICS AND DATA HANDLING

11.6.1 Counting statistics

The data which are obtained in the course of ion beam analysis of materials have intrinsic variability, and this is reflected in uncertainty in the final result of the analysis. Errors come in two classes, systematic and random. Systematic errors are errors which affect each data point in the same way. Examples are improper calibration of a multichannel analyzer or other instrument, a poorly designed Faraday cup that suffers from significant secondary electron emission, an uncertainty in the thickness of a carbon foil used in a time-of-flight spectrometer system, a poorly constructed aperture defining the solid angle subtended by a surface barrier detector, long-term drifts in electronic instruments that are not accounted for, and human errors such as reading the wrong scale on a meter with two range scales. Some systematic errors can be eliminated by proper design of an experiment and some cannot. Careful attention to detail usually minimizes these errors. If a measurement has small systematic errors it is said to be accurate.

Most of this section will be devoted to a discussion of random errors. Although these can include such things as errors in judgment in reading analog meters, line voltage fluctuations, mechanical vibrations, and other small disturbances, we will confine our attention to the most important random error in ion beam analysis, counting statistics. Experiments with small random errors are said to be precise.

The counting of events generated by radiation quanta is central to all ion beam analyses, and the fundamental statistical property of these events is that they occur with constant probability per unit time. As a result, they obey Poisson statistics, which is to say, that if the events occur at a mean rate of λ per second, then the probability $P(n)$ of observing exactly n events in a time t is given by

$$P(n) = \frac{(\lambda t)^n e^{-\lambda t}}{n!} = \frac{m^n e^{-m}}{n!} \tag{11.3}$$

where $m = \lambda t$ is the mean number of counts expected in the interval.

The mean and variance of this distribution are defined as

$$m = \sum_{n=0}^{n=\infty} nP(n) \text{ and } \sigma^2 = \sum_{n=0}^{n=\infty} (n-m)^2 P(n) \; . \tag{11.4}$$

According to a mathematical result known as the central limit theorem (Mathews and Walker, 1970), if m is large, then P(n) assumes the approximate form

$$P(n) = \frac{e^{(n-m)^2/2m}}{\sqrt{2\pi m}} = \frac{e^{(n-m)^2/2\sigma^2}}{\sqrt{2\pi\sigma^2}} \; . \tag{11.5}$$

Thus, for large m, the probability of obtaining exactly n counts is normally distributed with variance $\sigma^2 = m$. This is an extremely important result, because it provides the rule for assigning the statistical error to a measurement. For example, if the number of counts measured in a backscattering experiment is 100, then the standard deviation $\sigma = \sqrt{100} = 10$, or 10%. If a 1% result is desired, then the required number of counts is 10^4. Since one way to get more counts is to run an experiment longer, and since getting one more decade of precision requires 100 times more effort, there is a certain tyranny implied by this result. Usually, it is satisfactory to get about 10^3 counts in the spectral feature of interest. The resulting 3% precision is as good or better than most other sources of error in a typical ion beam experiment.

11.6.2 Error propagation

The final result of an experiment is usually obtained by mathematical manipulation of individually measured quantities. For example, the true number of counts in a peak is the difference between the total measured counts in the region of the peak and the background upon which these are superimposed. To get the correct result, you must subtract the background. Alternatively, one may wish to compute a cross section obtained by dividing a measured number of counts by the product of target thickness, integrated beam current, and detector solid angle, each of which is itself uncertain. How good is the final answer? To find out, you need to consider the subject of error propagation.

Rules for error propagation are found in standard texts that discuss data handling such as Bevington (1969). They are obtained by observing that if z=f(x,y) where x and y are random variables and f is some function

$$dz = \frac{\partial f}{\partial x} dx + \frac{\partial f}{\partial y} dy \; . \tag{11.6}$$

The quantities dx, dy, and dz are then interpreted as differences of individual measured values from the mean, and the expression is squared and averaged over the complete data set. For example, $dz \equiv \delta z_i = z_i - \langle z \rangle$, and similarly for dx and dy. Here, the angle brackets $\langle\ \rangle$ imply the average value of the quantity enclosed and i denotes the *i*th data point. Thus, since $\langle z_i - \langle z \rangle \rangle^2 \equiv \sigma_z^2$, etc., one obtains

$$\sigma_z^2 = \sigma_x^2 \left(\frac{\partial f}{\partial x} \right)^2 + \sigma_y^2 \left(\frac{\partial f}{\partial y} \right)^2 + 2\sigma_{xy}^2 \left(\frac{\partial f}{\partial x} \right)\left(\frac{\partial f}{\partial y} \right) \quad (11.7)$$

where

$$\sigma_{xy}^2 \equiv \langle (x - \langle x \rangle)(y - \langle y \rangle) \rangle \ . \quad (11.8)$$

If $\sigma_{xy}^2 = 0$, then the errors in x and y are said to be uncorrelated. In charged particle spectrometry, errors are usually either known to be uncorrelated or are assumed to be. Some counter examples are given in Nicholson (1974).

Table 11.12 gives the rules for propagating errors for several specific functional forms of f. Notice, in particular, the form z=x−y which is used for background subtraction of y counts from a total of x. In this case $\sigma_z^2 = \sigma_x^2 + \sigma_y^2 = x + y$. Thus, if a peak contains 400 total counts of which 100 are background, one obtains a net count of 300 with an error of $\sqrt{500}$ or about 7.5% error.

Table 11.12. Error propagation formulae for some simple functions of the random variables x and y. The parameters a and b are constants.

Function	Uncorrelated errors	Correlated errors
$z = ax \pm by$	$\sigma_z^2 = a^2\sigma_x^2 + b^2\sigma_y^2$	$\sigma_z^2 = a^2\sigma_x^2 + b^2\sigma_y^2 + 2ab\sigma_{xy}^2$
$z = \pm\, axy$	$\frac{\sigma_z^2}{z^2} = \frac{\sigma_x^2}{x^2} + \frac{\sigma_y^2}{y^2}$	$\frac{\sigma_z^2}{z^2} = \frac{\sigma_x^2}{x^2} + \frac{\sigma_y^2}{y^2} + 2\frac{\sigma_{xy}^2}{xy}$
$z = \pm \frac{a\,x}{y}$	$\frac{\sigma_z^2}{z^2} = \frac{\sigma_x^2}{x^2} + \frac{\sigma_y^2}{y^2}$	$\frac{\sigma_z^2}{z^2} = \frac{\sigma_x^2}{x^2} + \frac{\sigma_y^2}{y^2} - 2\frac{\sigma_{xy}^2}{xy}$
$z = a\,x^{\pm b}$	$\frac{\sigma_z^2}{z^2} = b^2\frac{\sigma_x^2}{x^2}$	
$z = a\,e^{\pm b\,x}$	$\frac{\sigma_z^2}{z^2} = b^2\sigma_x^2$	
$z = a\ln(\pm bx)$	$\sigma_z^2 = a^2\frac{\sigma_x^2}{x^2}$	

It is clear from this example that whenever possible, steps should be taken to physically reduce background rather than relying on manipulation of data after the experiment to subtract it.

If the complete probability distributions $P_1(x)$ of x, and $P_2(y)$ of y, are known, then the distribution of values of $z = f(x,y)$ can be obtained from

$$P(z) = \int\int P_1(x)\cdot P_2(y)\cdot \delta(z - f(x,y))\, dx\, dy \quad (11.9)$$

where $\delta(x)$ is the Dirac delta function and the integral covers all of (x,y) space.

11.6.3 Dead time correction

The problem of dead time has been discussed in the section on multichannel analyzers and in conjunction with pile-up. Here we quantify the discussion. Suppose that λ is the actual rate of events and that λ' is the measured rate in the presence of dead time. Let the dead time per pulse be τ. Then in one second, the total dead time is $\lambda'\tau$. The number of counts that were missed during this time is just the true rate times this interval. Since the number of missed counts in one second can also be expressed as $\lambda - \lambda'$ we have

$$\lambda - \lambda' = \lambda\lambda'\tau \ . \quad (11.10)$$

From this expression it follows that

$$\lambda = \frac{\lambda'}{1 - \lambda'\tau} \quad \text{or} \quad \lambda' = \frac{\lambda}{1 + \lambda\tau} \ . \quad (11.11)$$

Notice that for a measurement that takes time t to accumulate n' counts, the rate $\lambda' = n'/t$. Therefore, the quantity $\lambda'\tau = n'\tau/t$. Since τ is the (mean) dead time per count, we see that $\lambda'\tau$ is just the fractional dead time of the analyzer, which is, of course, equal to the fractional number of missed pulses that one obtains using the pulser technique discussed in 11.4.6.

The problem of dead time in pulse conversion is made worse by the randomness of pulses (England, 1974). Accurate counting can occur at a much higher rate if pulses are regularly spaced. Some advanced systems have been designed to regularize the rate by buffering the analog pulses until the ADC can accept them, but this is not common practice. If you have measurable dead time, you should simply plan to make the above correction. If the dead time exceeds about 5%, it is excessive and should be reduced by, for example, reducing the ion beam current.

11.6.4 Random coincidences

Another place that the randomness of pulse arrival has important consequences is in time-of-flight measurements. For low count rates, the start and stop pulses are almost all correlated. However, as the count rate increases, events overlap. For example, a fast particle can overtake a slower one which has initiated the start pulse and cause a premature stop pulse. Similarly, a start pulse can be initiated just before an earlier particle generates a stop, thus producing an event of impossibly short duration. Before considering the specific problem of time-of-flight spectrometers, let us consider more generally the rate of random coincidences which accompany the measurement of true events in a coincidence experiment.

Suppose that the rate of events in one detector is λ_1 and that the rate in another is λ_2 and that each of these values is somewhat less than the true rate, λ, of correlated events because of detector efficiency or other experimental considerations. Let τ be the resolving time of the coincidence apparatus. From Eq. (11.3) the probability of at least one event in detector 1 during the time τ is $P_1 = 1 - \exp(-\lambda_1\tau)$ and the probability of at least one event in detector 2 is $P_2 = 1 - \exp(-\lambda_2\tau)$. The probability of a coincidence event during τ is thus $P = P_1P_2 \approx \lambda_1\lambda_2\tau^2$. In time T there are T/τ such intervals so that the total number of random events is $B = \lambda_1\lambda_2T\tau$. Now in many situations we can assume that $\lambda_1 \approx \lambda_2 \approx \lambda$, so that $B \approx \lambda^2T\tau$. However, λT is just the total number of truly coincident events. Therefore, one finds that $B/(\lambda T) \approx \lambda\tau$, which says that the ratio of background to true counts goes up with the rate λ of events. In other words, the signal-to-noise ratio in coincidence experiments (such as time-of-flight measurements) improves as the beam current, and therefore the event rate, is reduced. Obviously a measurement entails a compromise between data quality and running time.

Time-of-flight spectra can be significantly altered by random coincidences between start and stop detectors at high counting rates. It is possible to compute this effect if one makes some assumptions about the apparatus (Weller, 1993). Here we present a formula applicable to a spectrometer such as the one shown in Fig. 11.6 which has distinct start and stop detectors. We assume that the true rate of particles entering the detector is λ, that the normalized intrinsic distribution of flight times t is $f(t)$, and that the probabilities of triggering the start and stop detectors (the intrinsic efficiencies) as a function of flight time are $\eta_1(t)$ and $\eta_2(t)$. We further define the mean intrinsic efficiencies of the detectors to be

$$\eta_1 = \int \eta_1(t)\, f(t)\, dt \tag{11.12}$$

and

$$\eta_2 = \int \eta_2(t)\, f(t)\, dt \ . \tag{11.13}$$

With these definitions, it is possible to express the mathematical form of the probability, $P(t)$, that an ion entering the time-of-flight spectrometer generates an event with flight time t, in terms of the intrinsic distribution $f(t)$ and the true event rate λ. This formulation includes the effect of random pulses in the stop detector. The expression is

$$P(t) = pe^{-\eta_2\lambda t}\begin{bmatrix} \eta_1\eta_2\lambda + \eta_1(t)\eta_2(t)f(t) \\ -\,\eta_2\lambda\displaystyle\int_0^t \eta_1(t')\eta_2(t')f(t')\cdot dt' \end{bmatrix} . \tag{11.14}$$

Here p is the probability that the system is not busy with a previous pulse at the time that a start pulse arrives. Note that this equation predicts a component of additive random background proportional to λ and a multiplicative exponential damping factor $e^{-\eta_2\lambda t}$ which tends to attenuate the spectrum at long arrival times. In time-of-flight backscattering spectra, this attenuation is ordinarily negligible in the portion of the spectrum that has features of interest and so can be ignored. Moreover, the principal constituent of the additive background, at least for the common case of backscattering analysis of heavy constituents on a lighter substrate, is constant and can be measured easily by examining regions of the spectrum which do not contain kinematically allowable backscattering events. Thus, background subtraction can be made quite easily. Notice too that as the rate of events is increased, the background increases linearly. This is in accordance with the expectation gleaned above by considering a simple coincidence apparatus. Note finally, that even at low counting rates, it is necessary to measure $\eta_1(t)\cdot\eta_2(t)$ in order to recover f(t) from the measured spectrum. This is a disadvantage of time-of-flight spectrometry when compared with the near unit quantum efficiency of energy spectrometry using surface barrier detectors.

Equation 11.14 can be converted to a differential equation and formally solved to obtain the best achievable approximation, $\eta_1(t)\eta_2(t)f(t)$, to the true distribution, $f(t)$, from the measured distribution $P(t)$. The result is interesting but is not useful for practical work, since in a real experiment, one does not usually sample the distribution for a large enough range of t values. The formal solution of Eq. (11.14) is

$$\eta_1(t)\eta_2(t)f(t) = p^{-1}e^{\eta_2\lambda t}\int_0^t [\eta_2\lambda P(t') + P'(t')]dt' \quad (11.15)$$

where P' denotes the derivative. Using the simple assumption that the time-of-flight spectrometer's mean dead time τ is the mean duration of observed events, one obtains for τ

$$\tau = \frac{\int_0^\infty t \cdot P(t)dt}{\int_0^\infty P(t)dt} \quad (11.16)$$

and, as in 11.6.3, for the live time factor, p

$$p = \frac{1}{1+\eta_1\lambda\tau} . \quad (11.17)$$

The random coincidence background sets a lower limit on the sensitivity of time-of-flight measurements but it should be noted from Eq. (11.14) that this background depends linearly upon the rate of events $\eta_2\lambda$ in the stop detector. If a significant portion of these counts is a result of uninteresting, low-energy particles, then the signal-to-noise performance of a spectrometer can be improved by placing a range foil (see Chapter 5) just before the stop detector. In this location, the foil causes negligible loss of timing resolution for the particles which pass through it and pin holes in the foil only produce additional background, not distortions in the spectrum.

11.6.5 Advanced data manipulation

The nearest thing to a cult classic in the literature of statistics for scientists is the set of notes of Orear (1958) which he based upon unpublished work of Enrico Fermi. Most of the essential contents of these notes, as well as many other topics, are now contained in the chapters on statistical handling of data in Press *et al.* (1992). This reference is particularly valuable because it provides executable C language functions. A FORTRAN and Pascal edition of the book is also available. For most analytical tasks, curve fitting is the most appropriate approach to reduce ion beam data. The routines in Press *et al.* have been designed to propagate errors properly, even through complicated non-linear fitting procedures, and are the preferred approach even for such mundane tasks as background subtraction. Functions are also presented for performing sophisticated statistical tests on data sets as well as such useful procedures as Fourier transforms (by the fast Fourier transform algorithm).

The reader is cautioned, however, to use these tools with discretion. For example, it is sometimes desirable to "smooth" data for presentation. There are many ways to do this but our recommendation is to use the procedure of Marchand and Marmet (1983), or perhaps even better, to perform a Gaussian convolution. The Gaussian function is well known to have the same functional form in both real and reciprocal space. As a result, it is easy to do a convolution by simply taking the Fourier transform of the data, multiplying by the transform of a Gaussian with a width of a few channels, and (Fourier) inverting the product. We find this result superior to the smoothing procedure suggested in Press *et al.* (1992).

While on the subject of convolutions, and in closing, we issue a warning. Be very careful about trying to "deconvolve" data by manipulating the Fourier transform. The process has practical difficulties and should be avoided. Instead, create a fitting function based upon an analysis of your experiment and fit your data using the procedures presented in Press *et al.* (1992). This is stable and will lead to satisfactory results if you have developed a good model.

REFERENCES

Althaus, R.F., and Nagel, L.W. (1972), "NIM Fast Logic Modules Utilizing MECL III Integrated Circuits," *IEEE Trans. Nucl. Sci.* **NS-19**, 1, p. 520.

Betts, R.R. (1979), "Time of Flight Detectors for Heavy Ions," *Nucl. Instr. and Methods* **162**, 531.

Bevington, Phillip R. (1969), *Data Reduction and Error Analysis for the Physical Sciences*, McGraw Hill, New York. A standard introductory text to the subject of data analysis.

Blankenship, J.L. (1960), *Trans. Nucl. Sci.* **NS7** (2-3), 190.

Bromley, D. Allan. (1974), "The Development of Electrostatic Accelerators," *Nucl. Instr. and Methods* **122,** 1. See also D.A. Bromley, "The Development of Heavy-Ion Nuclear Physics," in D.A. Bromley (ed.), (1984), *Treatise on Heavy-Ion Science*, Vol. 1 Plenum, New York, p. 3. A highly readable historical article by a person with broad knowledge of both the field and the personalities who made it.

Bromley, D.A. (ed.). (1979), "Detectors in Nuclear Science," *Nucl. Instr. and Methods* **162**. This entire issue is devoted to invited reviews of detector technology. The articles contain much valuable technical and historical information.

Brown, Ian G. (1989), *The Physics and Technology of Ion Sources*, John Wiley, New York.

Bryant, John H. (1984), "Coaxial Transmission Lines, Related Two-Conductor Transmission Lines, Connectors, and Components: A U.S. Historical Perspective," *IEEE Transactions on Microwave Theory and Techniques*, **MTT-32**, 9, p. 970. (This volume also contains a number of fascinating review articles on the history of microwave electronics.) See also M.A. Maury, "Microwave Coaxial Connector Technology: A Continuing Revolution," (1990), in *1990 State of the Art Reference*, a special issue of *Microwave Journal*, Horizon House Publishing, Norwood, MA, p. 39.

Campbell, J.L., and McNelles, L.A. (1975), An intercomparison of efficiency-calibration techniques for semiconductor x-ray detectors, *Nucl. Instr. and Methods* **125,** 205.

Canberra. (1978), *Canberra Laboratory Manual*, Canberra Industries, Meriden, Connecticut.

Clout, Peter. (1982), *A CAMAC Primer*, Los Alamos National Laboratory, LA-UR-82-2718. This document may also be obtained from a CAMAC module manufacturer, KineticSystems Corporation, 11 Maryknoll Drive, Lockport, Illinois 60441.

Döbeli, M., Haubert, P.C., Livi, R.P., Spicklemire, S.J., Weathers, D.L., and Tombrello, T.A. (1991), "A time-of-flight detector for heavy ion RBS," *Nucl. Instr. and Methods* **B56/57,** 764.

Duggan, Jerome L. (1988), *Laboratory Investigations in Nuclear Science* (Applications Note TN-2000), The Nucleus, Tennelec Inc., 761 Emory Valley Road, Oak Ridge, Tennessee. Tennelec, now a division of Oxford Instruments, is also a significant supplier of NIM instrumentation.

Elmsley, John. (1989), *The Elements*, Oxford University Press, Oxford.

England, J.B.A. (1974), *Techniques in Nuclear Structure Physics*, Halstead Press, New York. A two-volume treatise on all aspects of experimental nuclear physics. Very useful reference at an advance level.

Evans, Robley D. (1955), *The Atomic Nucleus*, McGraw Hill, New York.

Feldman, L.C., and Mayer, J.W. (1986), *Introduction to Surface and Thin Film Analysis*, North-Holland, New York, p. 136.

Gabor, G., Schimmerling, W., Greiner, D., Bieser, F., and Lindstrom, P. (1975), "High Resolution Spectrometry for Relativistic Heavy Ions," *Nucl. Instr. and Methods* **130**, 65.

Graper, E. (1990), *Thin Film Handbook*, Lebow Corporation, 5960 Mandarin Avenue, Goleta, CA 93117.

Harting, E., and Read, F.H. (1976), *Electrostatic Lenses*, Elsevier, New York. Many figures and tables for persons who need to design electrostatic optics.

Heath, R.L. (1964), "Scintillation Spectrometry—Gamma Ray Spectrum Catalog," U.S. Atomic Energy Commission Idaho Operations Office Research and Development Report IDO-16880-1. This reference is dated but it contains an instructive introduction which is interesting from a historical perspective. It is the origin of the often-copied figure shown in this chapter as Fig. 11.4.

Helmer, R.B., Van Assche, P.H.M., and VanDer Leun, C. (1979), Recommended standards for gamma-ray energy calibration, *Atomic Data and Nuclear Data Tables* **24,** 39.

Holl, I., Lorenz, E., and Mageras, G. (1988), A measurement of the light yield of common inorganic scintillators, *Trans. Nucl. Sci.* **35** (1), 105.

Horowitz, Paul and Hill, Winfield. (1989), *The Art of Electronics*, 2nd Ed., Cambridge University Press, Cambridge. An essential text and reference for physical scientists. Readable, wide ranging, full of practical experience and insight.

Humphries, Jr., Stanley. (1986), *Principles of Charged Particle Acceleration*, John Wiley, New York.

IEEE. (1976), Institute of Electrical and Electronics Engineers, *CAMAC Instrumentation and Interface Standards*, Institute of Electrical and Electronics Engineers, distributed by John Wiley, New York. See also, U.S. Energy Research and Development Administration NIM Committee (1976), "CAMAC Tutorial Articles/US NIM Committee, TID-26618," Supt. of Documents, U.S. Government Printing Office, Washington.

Jamieson, D.N., Grime, G.W., and Watt, F. (1989), "The New Oxford Scanning Proton Microprobe Analytical Facility," *Nucl. Instr. and Methods* **B40/41,** 669. Other articles on this subject are found frequently in this and other proceedings of the Conference on the Application of Accelerators in Research and Industry, which has been held biennially at the University of North Texas for many years. For a number of years, these proceedings have been published as special volumes of *Nuclear Instruments and Methods in Physics Research.*

Jardine, L.J., and Lederer, C.M. (1974), "Relative Efficiency Calibration of a Si(Li) Electron Spectrometer," *Nucl. Instr. and Methods* **120**, 515.

Knoll, Glenn F. (1989), *Radiation Detection and Measurement,* 2nd Ed., John Wiley, New York. An essential reference for persons who work with radiation.

Lawson, J.D. (1977), *The Physics of Charged-Particle Beams*, Oxford University Press, Oxford, p. 172.

Lederer, C.M., and Shirley, V.S. (eds.). (1978), *Table of Isotopes,* 7th Ed., John Wiley & Sons, New York.

Lennard, W.N., Geissel, H., Winterbon, K.B., Phillips, D., Alexander, T.K., and Forster, J.S. (1986), *Nucl. Instr. and Methods* **A248,** 454.

Lucovsky, Gerald (ed.). (1988), *Deposition and Growth: Limits for Microelectronics,* American Institute of Physics, Conference Proceedings No. 167, New York.

Maissel, Leon I., and Glang, Reinhard. (1970), *Handbook of Thin Film Technology,* McGraw Hill, New York.

Marchand, P., and Marmet, L. (1983), "Binomial smoothing filter: A way to avoid some pitfalls of least-squares polynomial smoothing," *Reviews of Scientific Instruments,* **54,** 1034.

Marion, J.B., and Young, F.C. (1968), *Nuclear Reaction Analysis, Graphs and Tables,* North Holland, Amsterdam, p. 145. A classic handbook for experimental nuclear physicists. Has many tables (some out of date, of course) including a listing of accelerator calibration reactions.

Mathews, Jon and Walker, R.L. (1970), *Mathematical Methods of Physics,* 2nd Ed., W.A. Benjamin, New York, p. 383. Graduate text in mathematical physics with an introduction to many useful topics in statistics and other areas.

McKenzie, J.M. (1979), "Development of the Semiconductor Radiation Detector," *Nucl. Instr. and Methods* **162,** 49.

Mendenhall, Marcus H., and Weller, Robert A. (1991), "High-Resolution Medium-Energy Backscattering Spectrometry," *Nucl. Inst. and Methods* **B59/60,** 120.

Moore, John H., Davis, Christopher C., and Coplan, Michael A. (1989), *Building Scientific Apparatus,* 2nd Ed., Addison Wesley, New York, Chapter 5, p. 305. An essential text and reference for students of experimental physical sciences.

Morrison, Ralph. (1977), *Grounding and Shielding Techniques in Instrumentation,* 2nd Ed., John Wiley, New York. A highly readable little book on a subject that might at first glance seem to be very dull. Recommended for anyone who is serious about making precise electrical measurements.

Musket, R.G., McLean, W., Colmenares, C.a., Makowiecki, D.M., and Siekhaus, W.J. (1982), "Preparation of Atomically Clean Surfaces of Selected Elements: A Review," *Applications of Surface Science* **10**, 143.

Nicholson, P.W. (1974), *Nuclear Electronics,* John Wiley, New York. Specialized volume on nuclear electronics. A good review.

Orear, Jay. (1958), "Notes on Statistics for Physicists," University of California Radiation Laboratory, Berkeley, publication UCRL-3417, August 13.

Ortec. (1984), *Experiments in Nuclear Science, AN34, Laboratory Manual,* 3rd Ed., Revised, EG&G Ortec, Oak Ridge, Tennessee.

Overley, J.C., Parker, P.D., and Bromley, D.A. (1969), "The Calibration of Tandem Accelerators," *Nucl. Instr. and Methods* **68,** 61. Contains reactions that are particularly useful for calibrating large accelerators.

Poate, J.M., Tu, K.N., and Mayer, J.W. (eds.). (1978), *Thin-film interdiffusion and reactions,* John Wiley & Sons, New York.

Press, William H., Teukolsky, Saul A., Vetterling, William T., and Flannery, Brian P. (1992), *Numerical Recipes in C,* 2nd Ed., Cambridge University Press,

Cambridge. See especially Chapter 14 p. 609 and Chapter 15, p. 656. An instant classic in the field of numerical analysis, written by and for practicing scientists. An essential reference.

Rose, P.H., and Galejs, A. (1967), "The Production and Acceleration of Ion Beams in the Tandem Accelerator," in F.J.M. Farley (ed), *Progress in Nuclear Techniques and Instrumentation,* North Holland, Amsterdam, pp. 1-116. A classic article to introduce serious students to accelerators.

Roth, A. (1982), *Vacuum Technology,* 2nd Ed., North Holland, Amstedam. Excellent reference on all aspects of vacuum technology.

Rytz, A. (1979), New catalogue of recommended alpha energy and intensity value, *Atomic Data and Nuclear Data Tables* **23,** 507.

Sakai, Eiji. (1988), Recent measurements on scintillator-photodetector systems, *Trans. Nucl. Sci.* **43** (1), 418.

Schuegraf, Klaus K. (ed.). (1988), *Handbook of thin-film deposition processes and techniques,* Noyes Publications, Park Ridge, New Jersey.

Spivack, M.A. (1972), "Mechanical Properties of Very Thin Polymer Films," *Rev. Sci. Instr.* **43**, p. 985.

Tuli, J.K. (1985), *Nuclear Wallet Cards,* National Nuclear Data Center, Brookhaven National Laboratory, Upton, NY.

USDOE. (1977), U.S. NIM Committee. Advanced Study Group, *Future data bus requirements for laboratory high speed data acquisition systems. TID-27621,* Energy Research and Development Administration (now U.S. Department of Energy), Washington.

USDOE. (1990), *Standard NIM Instrumentation System,* DOE/ER-0457T, U.S. Department of Energy, Office of Energy Research, Washington, D.C. This document supersedes *Standard Nuclear Instrumentation Modules,* Report No. TID20893, Rev. 4 (1974), U. S. Atomic Energy Commission.

Weller, Robert A. (1993), "Instrumental effects on time-of-flight spectra," *Nucl. Instr. and Methods* **B79**, 817.

Wielopolski, L., and Gardner, R.P. (1976), Prediction of the pulse-height spectral distortion caused by the peak pile-up effect, *Nucl. Instr. and Methods* **133,** 303.

Whitlow, Harry J. (1990), "Time-of-flight Spectroscopy Methods for Analysis of Materials with Heavy Ions: A Tutorial," in J.H. Tesmer, C.J. Maggiore, M. Nastasi, J.C. Barbour, and J.W. Mayer (eds.). *High Energy and Heavy Ion Beams in Materials Analysis,* Materials Research Society, Pittsburgh, Pennsylvania, p. 243.

Wilkinson, D.H. (1950), "A Stable Ninety-Nine Channel Pulse Amplitude Analyzer for Slow Counting," *Proc. Cambridge Phil. Soc.* **46 Pt. 3**, 508.

Wiza, J.L. (1979), "Microchannel Plate Detectors," *Nucl. Instr. and Methods* **162,** 587.

PITFALLS IN ION BEAM ANALYSIS

J. A. Davies

McMaster University, Hamilton, Ontario, Canada

W. N. Lennard and I. V. Mitchell

University of Western Ontario, London, Ontario, Canada

CONTENTS

12.1 INTRODUCTION

The purpose of this chapter is to publicize the more common pitfalls that often prevent IBA techniques from achieving their full quantitative accuracy. Various problem areas are identified and some guidelines are provided which enable the reader either to minimize the problem or at least estimate its approximate magnitude. Where helpful, a brief discussion of the underlying physics is included, but the chapter does not attempt to present a comprehensive or quantitatively accurate treatment of the individual phenomena involved.

Ion beam analysis (IBA) at MeV energies generally involves the yield, Y_i, of a specific beam/target-atom interaction for an accurately known dose, Q, of bombarding ions having atomic number Z_1 and energy E. From such a measurement of Y_i/Q, the number N_i (atoms/cm^2) of 'i atoms' in the target can then be obtained from the appropriate **differential** cross section, $\sigma(Z_1,E,\theta) = d\sigma/d\Omega$, and the detector solid angle, $\Delta\Omega$

$$\frac{Y_i}{Q} = N_i\ \sigma(Z_i, E, \theta)\ \Delta\Omega\ . \tag{12.1}$$

In the first three sections of the chapter, we will discuss pitfalls associated with a measurement of the quantity Y_i/Q in Section 12.2, cross section evaluation in Section 12.3, and solid angle calibration in Section 12.4. Section 12.5 briefly describes various undesirable target modifications that may result during the ion bombardment, for example, target heating, sputtering, radiation damage, and surface charging (insulators). Miscellaneous other effects (surface roughness, non-linear detector response) are treated in Section 12.6.

12.2 LOST BEAM AND EVENTS

12.2.1 Faraday cup

Accurate measurement of the parameter Y_i/Q requires, above all, the ability to measure the integrated beam dose Q — and hence some knowledge of Faraday cup principles. This has been one of the biggest pitfalls among IBA practitioners. It is rather ironic that our nuclear physics forefathers recognized 40 years ago the importance of good Faraday cup design better than we do, even though at that time they did not have available nearly such a detailed understanding of the relevant particle/surface interaction mechanisms (Auger, secondary electron, photon emission, sputtering, etc.).

A typical example of a 'good' Faraday cup design is shown in Fig. 12.1. It consists of an electrostatically shielded cup surrounding the target, so that only a small

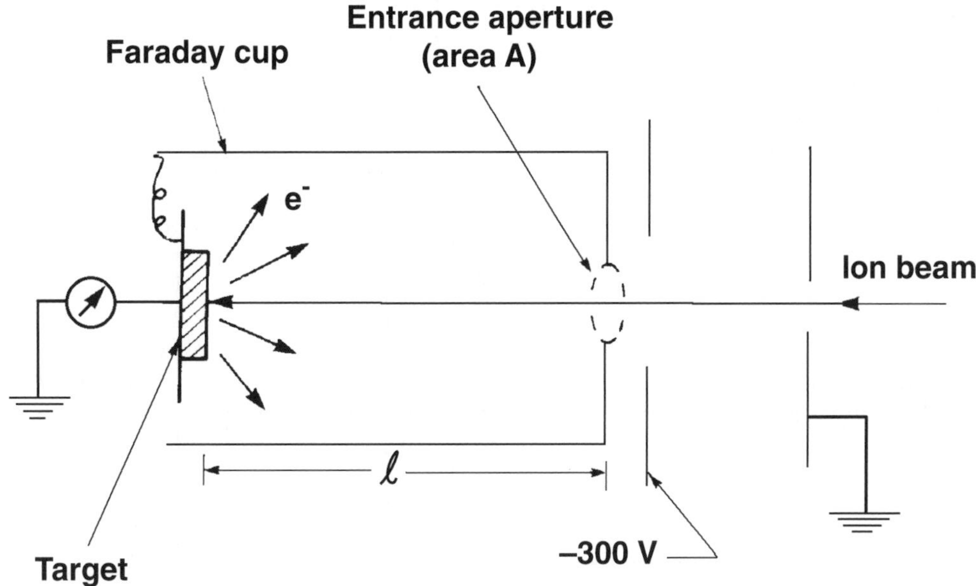

FIG. 12.1. Good Faraday cup design. The Faraday cup/target assembly must be mounted within a grounded electrostatic shield (not shown).

fraction of any emitted secondary electrons can escape. For isotropic emission, this fraction is given by $A/2\pi l^2$, where A is the aperture area and l is the aperture-to-target distance. In addition, at the entrance aperture, a –300 V electric field (or alternatively a weak magnetic field) is placed to prevent any low-E electrons from leaving or entering the cup (weak suppression fields have negligible effects on the trajectories of beam particles).

Such a large Faraday cup is not readily compatible with many IBA applications, such as channeling measurements, large area implants, high-temperature or low-temperature studies. Hence, we often need to modify the design without seriously compromising basic Faraday cup concepts.

One widely used (but very unreliable) modification is shown in Fig. 12.2, where a solid metal cylinder (or alternatively a cylindrical metal mesh) at –300 V potential surrounds the target to suppress low-E secondary electron emission. Often this metal cylinder also serves as a liquid nitrogen trap to minimize target surface contamination by hydrocarbons, etc. Such a modification effectively accomplishes the major objective of a Faraday cup: it prevents the loss of the numerous low energy (<300 eV) secondary electrons produced by ion bombardment.

However, there are at least three minor processes which are often overlooked—and which this simple suppression system does not adequately control. In many ion/target combinations, one or more of these minor processes can contribute a significant error unless a better Faraday cup system is employed (Table 12.1).

12.2.1.1 Higher energy electrons

A tail of higher energy electrons, extending up to several keV, arises from two sources: direct Coulomb ejection of inner-shell target electrons by the incident beam, and Auger electrons resulting from the subsequent filling of these K- or L-vacancies. This effect reaches its maximum value when the ion velocity roughly matches the inner-shell electron velocity in the target atom (as shown in Fig. 12.3). Experimental tests show that, with MeV He ions, the effect is usually <2%. However, since the ionization cross section, σ_I, scales with Z_1^2 (Z_1 is the projectile atomic number), this correction becomes quite serious when using heavier ion beams in the same velocity regime. For example, the σ_I value for 20 MeV ^{16}O in Si is ~5×10^{-17} cm^2, compared to only 1-2 $\times 10^{-18}$ cm^2 for 2 MeV H^+ and $^4He^+$ ions, respectively. In the simple suppression system of Fig. 12.2, such high energy electrons strike the surrounding cylinder, producing low energy secondary electrons which are then accelerated towards the target, causing an underestimate in the true beam flux.

Table 12.1. Problems requiring **good** Faraday cup design.

Cause of error	Maximum effect
High-E electrons, including Auger electrons	1-2% with MeV He. Much larger with heavier ions
Photons, x-rays (→ secondary e^- on impact)	~0.1% in metal targets. >100% in polymers, etc.
Excited sputtered atoms, M* (→ secondary e^- on impact)	Negligible at MeV energies in metals. ≥ 50% in some insulators and at keV energies

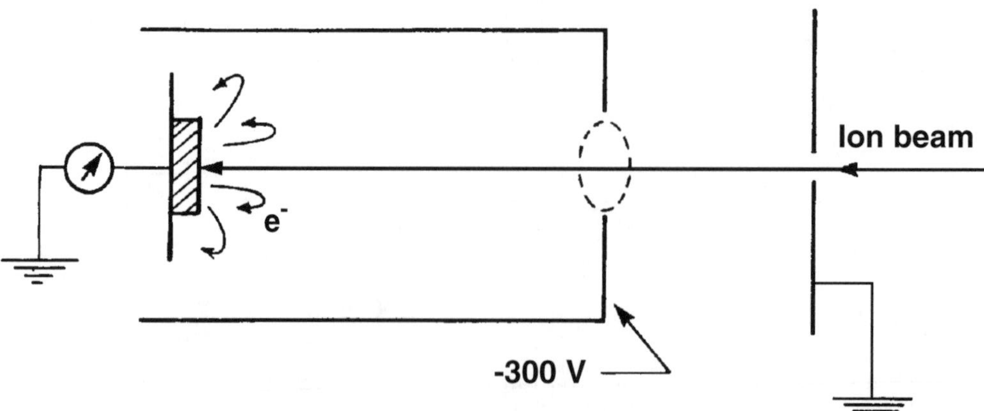

FIG. 12.2. Poor Faraday cup design.

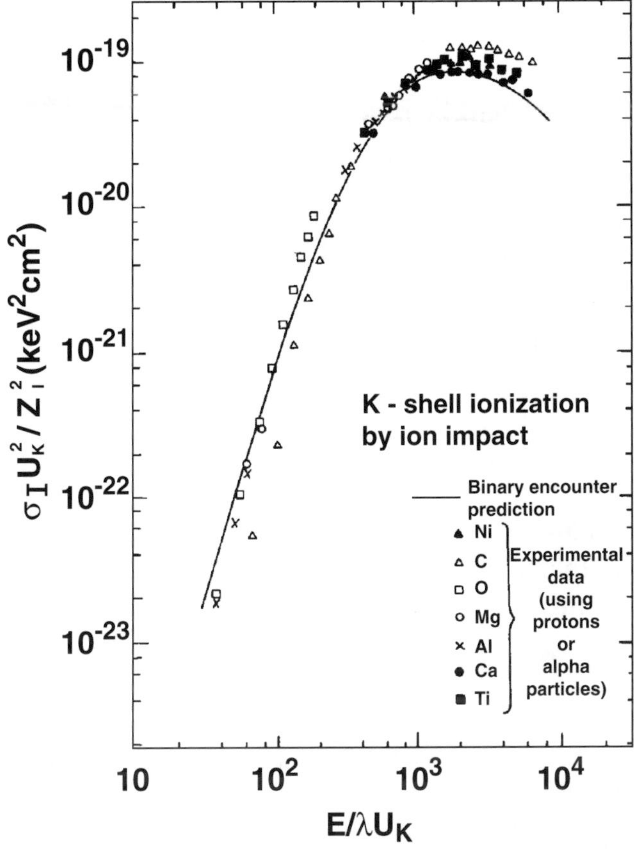

FIG. 12.3. Energy dependence of inner-shell ionization cross sections (σ_I) for projectiles (Z_1) of similar velocity, E/λ. U_K is the K-shell binding energy and $\lambda = M_1/M_p$, where M_p is the proton mass (taken from Garcia, 1971). Note the strong Z_1^2-dependence of σ_I.

12.2.1.2 Inner-shell vacancies

Inner-shell vacancies produced in (12.2.1.1) can alternatively decay by photon emission. These photons (x-rays, UV, etc.) also produce a flow of secondary electrons toward the target on striking the metal cylinder. Since the x-ray emission fluorescence yield is not large, <<1 (except for high-Z targets), this effect is usually very small in metallic and semiconductor targets. However, many insulators such as polymers and alkali halides are highly transparent to visible and UV emission. In such cases, tens or even hundreds of such photons per incident ion may be emitted and the resulting flow of negative electrons from the surrounding cylinder can readily exceed the incident positive ion current, as was reported by Venkatesan *et al.*, (1984) in an MeV He bombardment of polymers. Again, the beam fluence is underestimated.

12.2.1.3 Sputtered atoms ejected

A significant fraction of the sputtered atoms ejected from the target consist of excited neutral particles which can then de-excite on collision with the surrounding cylinder, again producing low-energy secondary electrons. See Section 12.5.1.3 for a discussion of sputtering at MeV energies. In metals and semiconductors, such sputtering effects are negligible, but in insulators there can again be a significant contribution.

Good Faraday cup design (Fig. 12.1) reduces all three of the above effects by the geometrical solid-angle factor ($A/2\pi l^2$) of the cup — a technique that is usually sufficient to reduce their contributions to negligible levels. Furthermore the –300V suppression electrode shown in Fig. 12.1 is completely shadowed by the slightly smaller apertures of the Faraday cup and its surrounding electrostatic shield. Consequently, even the few high energy electrons, photons, etc., which do escape from the cup are unable to generate any flow of low-energy electrons back to the target. Note that this –300V suppression electrode is important, even in the presence of a small ($A/2\pi l^2$) geometrical factor, in order to prevent electrons and negative ions generated by the beam at upstream apertures, etc., from drifting along the beam axis into the Faraday cup.

For channeling studies, where tilting and angular rotation of the target are necessary, one suitable modification is to disconnect mechanically the target/goniometer system from the surrounding Faraday cup, as in Fig. 12.1.

A particularly common and useful solution to the Faraday cup problem is to use the so-called 'transmission' design, shown in Fig. 12.4 (Sitter *et al.*, 1982), in which a rotating or oscillating paddle intercepts the beam several times per second. The duty cycle of such a cup can be measured by placing a normal Faraday cup in the target position. However, this is usually not necessary, since the use of a calibrated RBS standard (Section 12.4.3) simultaneously evaluates the combined effect of the detector solid angle and the duty cycle. Note that the paddle should be located roughly in the middle of such a transmission cup and not near the downstream end. As the edge of the paddle traverses the beam, secondary emission can readily occur in the forward as well as in the backward direction. The Chalk River group discovered this particular 'pitfall' quite by accident and as a result, their first UHV transmission Faraday cup had to be rebuilt (Sitter *et al.*, 1982).

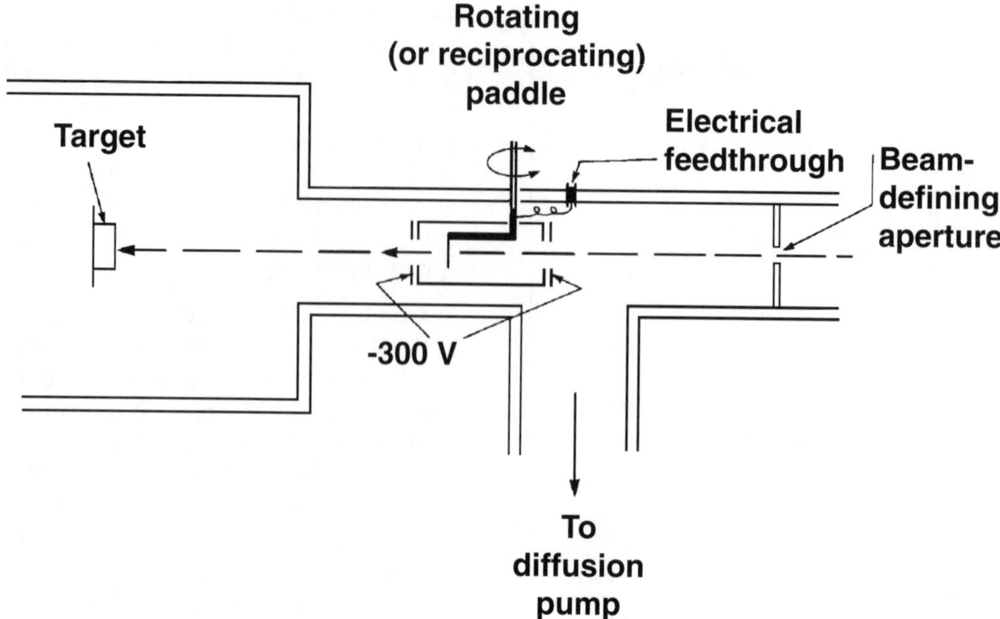

FIG. 12.4. Transmission-type of Faraday cup (Sitter *et al.*, 1982).

One major advantage of the transmission concept is that the integrated current measurement Q is completely independent of target material, channeling, etc. Also, it greatly simplifies experimental studies in UHV and at both high and low temperatures. Furthermore, since its only function is to give a reproducible measurement of **relative** beam current, rather than an accurate absolute value, the Faraday cup design is not quite as critical.

12.2.2 Beamline charge exchange

The use of current integration to record the number of beam particles incident on target presumes that the charge state (q_i) of the incident beam is known. Any process that allows q_i to alter downstream of the analyzing magnet can introduce errors. The most obvious source of such a problem is a charge-changing collision between the ion and residual gas atoms or molecules that comprise the vacuum. These are 'soft' collisions which need not deflect the ion significantly from its original direction so that they are carried with the beam. While there exist methods for estimating the magnitude of the effect, it is best to suppress it via improvement to the vacuum.

The probability, *P(L)*, for a charge-changing collision through electron capture or loss is given by

$$P(L) = 1 - e^{-nL\sigma_{CE}} \tag{12.2}$$

where L is the path length (cm), σ_{CE} is the sum of all cross sections for charge-changing collisions that convert the initial charge state, i, to a different charge state, j (cm^2/atom), and n is the gas target density (atoms/cm^3). For diatomic gases at room temperature, the relationship between n and the pressure p (in torr) is $n/p = 7 \times 10^{16}$ atoms/cm^3/torr. Experimental values for σ_{CE} can be found in Allison's paper (1958) for He ions in air. For example, for 1.7 MeV $^4He^+$ (q=1+) ions, a pressure of 10^{-6} torr in a 2 meter pathlength, $\sigma_{12} = 3.7 \times 10^{-17}$ (cm^2) yielding P = 0.051%. The probability of a second charge-changing collision converting a He^{++} ion back to a He^+ ion ($\sigma_{21} = 0.50 \times 10^{-17}$ cm^2) need not be considered since the number of He^{++} formed is so small.

As a rule-of-thumb, if we wish P to be <1% for a charge-changing process with a cross section $\sigma_{CE} = 5 \times 10^{-15}$ cm^2 (a conservatively large value!), then the approximate condition becomes $pL < 3 \times 10^{-7}$, where the beamline pressure, p, is in torr and the pathlength, L, is in meters.

12.2.3 Particle identification

A precise knowledge of the incident particle parameters is essential for all ion beam experiments. Scientists in the field of accelerator mass spectrometry (AMS) have worked very hard to achieve discrimination levels of $\sim 10^{15}$ between, for example, ^{14}N and ^{14}C species in carbon dating research using particle telescopes, time-of-flight detection, etc. (Purser *et al.*, 1978). These sophisticated setups are generally only available in ion beam laboratories dedicated to AMS research. We assume here that the accelerator has only an analyzing magnet (with relatively poor mass resolution) to perform momentum selection of the beam, with a magnet constant proportional to ME/q^2, where M is the ion mass, E is the kinetic energy, and q is the ion charge. Most accelerators suffer from periodic problems due to unwanted contaminant beams that arrive at the target together with the desired species. Such problems are caused by:

- Atomic or molecular particles whose kinematic properties mimic the acceleration and deflection kinematics of the desired particle;
- Unwanted particles that reach the target because of a low probability charge exchange sequence occurring upstream of the target; or
- Particles that have completely incorrect kinematics but arrive at the target because of (wall) scattering.

One of the simplest ways to test for a mixture of 'equal-mass' beams in single-ended machines is to detect elastically scattered projectiles from a thin self-supporting Au target. An energy spectrum analysis with a standard surface barrier detector system will readily distinguish the fragments of molecular impurities, such as mixtures of H_3^+, HD^+ and $^3He^+$, or $^4He^+$ and D_2^+, which may arise through ion source memory effects.

The consequence of scattering the three ion species, H_3^+, HD^+, and $^3He^+$, from a thin Au scattering foil, when the incident species have equal energy (1 MeV), is shown in Table 12.2. We assume that the detector is located at 90° (lab. angle) and that all kinematic scattering factors are unity. The three beams would give rise to three distinct peaks:

Table 12.2. Relative particle yields and energies for several "mass 3" ions incident at 1.0 MeV.

	$E_{product}$ *(MeV)*		
Beam	**0.33**	**0.67**	**1.00**
H_3^+	27	–	–
HD^+	9	2.25	–
$^3He^+$	–	–	4

- For H_3^+, a single peak at 0.33 MeV energy;
- For HD^+, two peaks corresponding to 0.33 MeV protons and 0.67 MeV deuterons; and
- For $^3He^+$, a single peak at 1 MeV energy.

The relative intensities are distinctive. The entries in Table 12.2 are based on a simple $(Z_1/E)^2$ dependence of the elastic (Rutherford) cross section for the dissociated fragments.

Another example of the use of this Au foil scattering technique is the identification of ^{16}O ions in the presence of ^4He ions — a curiosity observed by the authors when poor beamline vacuum developed, thereby allowing a $^{16}O^+$ impurity beam co-accelerated to 2 MeV with the $^4He^+$ beam to charge-exchange to $^{16}O^{++}$ in the drift section between the accelerator and the analyzing magnet. This process allowed two beams of 'identical' rigidity, $ME/q^2 = 8$ (MeV-amu), to be passed by the magnet.

One can also study the charge state purity distribution for the beam after passage through a thin foil, or test for beam impurities by examining the x-ray spectrum produced by the beam when impacting a solid target. These techniques are not easy to quantify, and require additional experimental apparatus (Lennard, 1989).

The molecular interferences are not problematic for tandem-type accelerators, where there is an injection magnet at the source, an analyzing magnet following acceleration, and a (collisional) stripping event at the high voltage terminal.

Wall scattering can be minimized using a series of apertures; however, care must be exercised so as not to restrict beamline pumping speed excessively.

12.2.4 Deadtime

We restrict the comments to 'singles' experiments (as opposed to coincidence experiments), which are the most common type for ion beam analysis applications. Deadtime and/or pileup can arise in any of the electronic components: preamplifier, amplifier, ADC, etc., due to high counting rates. Solid state detectors should be operated at pulse rates separated by a time interval that is more than the deadtime inherent in the rest of the electronic circuitry. The temporal response of the system can be improved through the use of baseline restoration, pulsed optical feedback amplifiers, pulse pileup rejection circuitry, to name a few.

It is a good practice to *always* use a test pulse at the appropriate preamplifier input, that is, introducing a signal which is periodic with known frequency, ν (*e.g.*, ν=60 Hz), and which has an amplitude larger than any other feature of the pulse height spectrum. Thus, the total number of 'missing' pulses due to count rate losses during an acquisition time, Δt, can be calculated to first order from the *measured* area, A_m, under the pulser peak. The correction then amounts to $\nu\Delta t/A_m$ for all channels of the spectrum. Caution should be exercised when using a Faraday cup that rotates into the beam periodically, because there are no pulser losses when the beam is hitting the cup. Here the precise duty cycle of the cup must be measured and its effect factored into the correction. Another way to measure counting losses using a pulser is to trigger the pulser with an output (usually scaled) from the current integrator. This way, if the beam goes off, so do the pulses, and since one usually counts for a preset charge, the number of pulses is also known.

In general, counting rate losses will be non-negligible for total pulse counting rates exceeding 10^3 s^{-1}, which applies to almost all RBS measurements. Raising the lower level discriminator of the amplifier and/or ADC can decrease the deadtime problem, provided that the low energy region of the spectrum is extraneous to the experiment. Care should also be taken to consider the effect of large, rapid fluctuations in the incident beam intensity. The reader is referred to Knoll's excellent text (1989) for more detail. Chapter 11 discuses this topic further.

12.3 CROSS SECTION EVALUATION

12.3.1 Rutherford window

How accurate is the Rutherford scattering law? This is often the key question in achieving quantitative RBS analysis, since its main advantage is the assumed existence of a universal and predictable scattering cross section, σ_R(cm^2/sr), whose dependence on beam energy E(MeV), scattering angle θ, and atomic numbers Z_1 and Z_2 of the beam and target atoms, respectively, may be accurately described by

$$\sigma_R = 1.296 \times 10^{-27} \left(\frac{Z_1 Z_2}{E}\right)^2 \sin^{-4}\left(\frac{\theta}{2}\right) \qquad (12.3)$$

where E and θ are expressed in center-of-mass coordinates. To transform Eq. (12.3) into the normal laboratory coordinates, three separate corrections are involved — to convert E, θ, and the solid angle $\Delta\Omega$ [Eq. (12.1)], respectively — and the resulting expression is rather complex. However, since $M_1 << M_2$ is the usual RBS condition, all three terms can be combined quite accurately into a single correction term, given by Marion and Young (1968): replacing the $\sin^{-4}(\theta/2)$ term in Eq. (12.3) by $[\sin^{-4}(\theta/2) - 2(M_1/M_2)^2]$.

The validity of this Rutherford scattering law [Eq. (12.3)] requires the distance of closest approach (or 'collision diameter'), b, between the projectile and target nuclei to fall within the so-called Rutherford 'window' in order that this point-charge scattering formula be quantitative. This means that b must be considerably larger than the nuclear radius, r_n, of the target atom. At the same time, b must also be smaller than the atomic K-shell radius, r_K, in order to minimize electron screening effects.

Let us consider briefly the factors controlling this Rutherford window. The three appropriate distances involved (in Å) may be estimated from the following simple formulae

$$b = Z_1 Z_2 e^2/E, \text{ where } e^2 = 14.4 \text{ [eV Å]} .$$

$$r_n = 1.4\, Z_2^{1/3} \times 10^{-5}$$

$$r_K = 0.5/Z_2 .$$

As suggested below, a quick and simple estimate of the upper and lower energy limits within which the scattering cross section should be within ±4% of the σ_R value may be derived by requiring

$$0.5\, r_K > b > 3\, r_n .$$

The resulting lower and upper energy limits are then given by $0.03\, Z_1 Z_2^2$ keV and $0.3\, Z_1 Z_2^{2/3}$ MeV, respectively. Representative values are listed in Table 12.3, together with some E_{max} values derived by Bozoian *et al.* (1990) from a detailed analysis of the nuclear force perturbation on σ_R (see Section 3.3.3 in Chapter 3 and Appendix 8).

Upper E limit for RBS: Unless $b > 3r_n$, the point-charge approximation inherent in the Rutherford law is not satisfied. The resulting cross section, σ, falls below σ_R in a predictable way. Even more serious, short-range nuclear forces also become significant at small values of b, producing large and not so readily predicted fluctuations, both above and below σ_R. In special cases, the resulting nuclear reactions, elastic scattering resonances, etc. may be used to advantage in the analysis of low-Z ions, as discussed in Chapters 3, 6, 7, 8, and 9.

In summary, provided the condition $b > 3r_n$ is fulfilled, the nuclear force perturbation is usually much less than ±4%.

Lower E limit for RBS: Provided $b < 0.5r_K$, then the major part of the collision takes place after the projectile has penetrated all the electron shells of the target atom, that is, in the unscreened Coulomb field assumed in Rutherford scattering. At the MeV energies involved in IBA, this condition is almost always fulfilled, even for the heaviest target nuclei (Table 12.3).

Note, however, that a small screening correction to σ_R is always necessary, as discussed in 4.2.2.3, even though most of the collision is completely unscreened. This arises because the initial part of each scattering trajectory is fully screened (i.e., from ∞ to ~0.5 Å from the target nucleus). Consequently, the incident projectile penetrates into the unscreened region with somewhat higher energy than would occur if the target atom was a bare nucleus. The resulting decrease in cross section below σ_R is usually much less than 4%. Furthermore, experimental studies by L'Ecuyer *et al.* (1979), Besenbacher *et al.* (1980), and Hautala and Luomajarvi (1980) show that its magnitude is predicted with reasonable accuracy by the relationship

$$\sigma = \sigma_R \left[1 - \frac{0.049\, Z_1\, Z_2^{4/3}}{E} \right] \qquad (12.4)$$

where E is the projectile energy in keV. Values of this correction term (denoted by F) for a ^{4}He beam are given as a function of Z_2 and E in Fig. 12.5. Correction factors are given in Appendix 6.

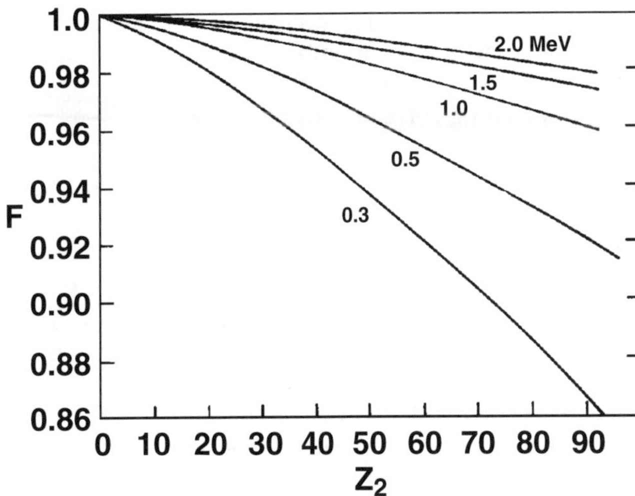

FIG. 12.5. Screening-correction factor, *F*, to the Rutherford cross section for ^{4}He backscattering as a function of Z_2 and E (L'Ecuyer *et al.*, 1979).

Strictly speaking, the center-of-mass energy should be inserted in Eq. (12.4), but the screening correction is usually small enough that the use of laboratory coordinates introduces a negligible error.

Backscattering studies in the so-called Medium Energy Ion Scattering (MEIS) regime often involve energies well below the E_{min} value for Rutherford scattering (Table 12.3). In such cases, the entire scattering process occurs in a screened field and hence requires an appropriate screened-Coulomb potential to obtain the classical cross section. For quantitative work, suitable algorithms are given by Mendenhall and Weller (1991); however,

Table 12.3. RBS energy window (MeV) for quantitative (±4%) Rutherford scattering.

Ion	Target	E_{min}	E_{max}
H^+	Si	0.006	1.7
	Au	0.2	5.5
$^4He^+$	Si	0.01	3.5 (3.5)*
	Au	0.4	11 (19.0)*
$^{16}O^+$	Si	0.05	14
	Au	1.5	45

*Values derived by Bozoian *et al.* (1990).

they point out that a fairly accurate (2-3%) approximation to the screened Coulomb cross section can still be obtained directly from the σ_R value by applying the simple screening correction of Eq. (12.4).

12.3.2 Scattering angle problems

12.3.2.1 Double/plural scattering

One major assumption in RBS, and indeed in all IBA studies, is that the incoming and outgoing trajectories are completely linear, that is, only one significant angular deflection, namely the Rutherford backscattering event, is occurring. However, the mean free path, λ, for unscreened[1] scattering events $[(N\pi a_{TF}^2)^{-1}]$ is only ~1000 Å, where N (atoms Å^{-3}) is the atom density of the target and a_{TF} (~0.1 Å) is the appropriate Thomas-Fermi screening length. Hence, even at quite shallow depths, a fraction of the beam undergoes significant secondary deflections along the incoming/outgoing trajectories.

Such secondary deflections obviously change θ, the scattering angle involved in the main RBS collision (as illustrated in the schematic of Table 12.4), thus affecting the magnitudes of both σ_R and the kinematic energy loss factor. Furthermore, it also affects the depth-to-energy conversion scale.

Quantitative treatment of such secondary scattering corrections would require a full Monte Carlo simulation

Table 12.4. Probability, P_{crit}, of a secondary collision with scattering angle, $\Delta\theta$, greater than θ_{crit} (for a total path length of 10^{18} atoms cm^{-2}).

Energy *(MeV)*	Ion	Target	P_{crit} *(% of trajectories)* Normal incidence (θ_{crit} = 20°)	Grazing (ERD) geometry (θ_{crit} = 2°)
2.0	$^4He^+$	Si	0.01	1.0
2.0	$^4He^+$	Au	0.3	33

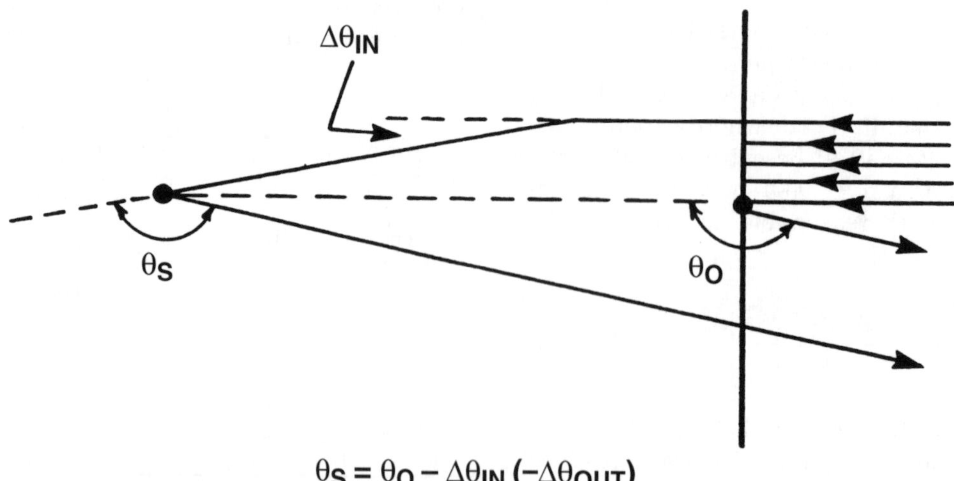

1. Only unscreened scattering can contribute significantly to the mean deflection angle at MeV energies.

of each particle trajectory and is not normally feasible. Nevertheless, a simple figure-of-merit exists for quickly checking in any given case the approximate depth beyond which such a correction is expected to become serious.

First, one must estimate the maximum acceptable deflection, θ_{crit}, along the ion trajectory (see below) and then use the integral form [Eq. (12.5)] of the Rutherford scattering law to obtain σ_{crit}, the total cross section for scattering angles larger than θ_{crit}

$$\sigma_{crit} = 1.6 \times 10^{-26} \left(\frac{Z_1 Z_2}{E}\right)^2 \cot^2\left(\frac{\theta_{crit}}{2}\right) . \qquad (12.5)$$

Multiplying this value of σ_{crit} (cm^2) by the path length (atoms cm^{-2}) gives the overall probability, *P*, of at least one significant deflection occurring along either the incoming or outgoing path.

The crucial question, obviously, is to estimate the appropriate value of θ_{crit}, because this depends strongly on the experimental geometry involved. For near-perpendicular incidence, symmetry considerations would suggest that secondary deflections would have to be at least 20-30° in order to change significantly either the mean value of σ_R or the depth-to-energy conversion scale. On the other hand, in the grazing-incidence/exit geometry used in ERD (see Chapter 5) or for enhancing the depth resolution of RBS, a θ_{crit} value of 2°, or even less, may influence the depth-to-energy conversions.

Representative values of P for the above two cases are given in Table 12.4 for MeV ^{4}He ions in various targets after traversing a path length of 10^{18} atoms cm^{-2} (~0.1 μm). Clearly, for grazing-angle geometry, secondary scattering effects are always a serious problem, even at quite shallow depths. [Note that at 10° incidence to the surface, the chosen path length of 10^{18} atoms cm^{-2} (Table 12.4) corresponds to a depth of only ~160 Å normal to the surface.]

Even in the case of normal incidence, a significant correction may occur in high-Z targets, such as Au, at depths greater than ~10^{19} atoms cm^{-2} (~1 μm), as shown in Fig. 12.6. A detailed discussion of this effect is given by Moore (1980).

Included in Fig. 12.6 are the corresponding computer simulations and (in Fig. 12.6c) the ratio between the measured and the simulated RBS yields as a function of depth. In both cases, significant deviations (~5-20%) appear as the backscattered energy falls below channel 400, that is, at a depth of ~1.2×10^{19} Au atoms cm^{-2}. In a low-Z target (for example, Si), Moore (1980) finds the observed deviations to be roughly a factor of two smaller than in Au for the same channel (E) in the RBS energy spectrum.

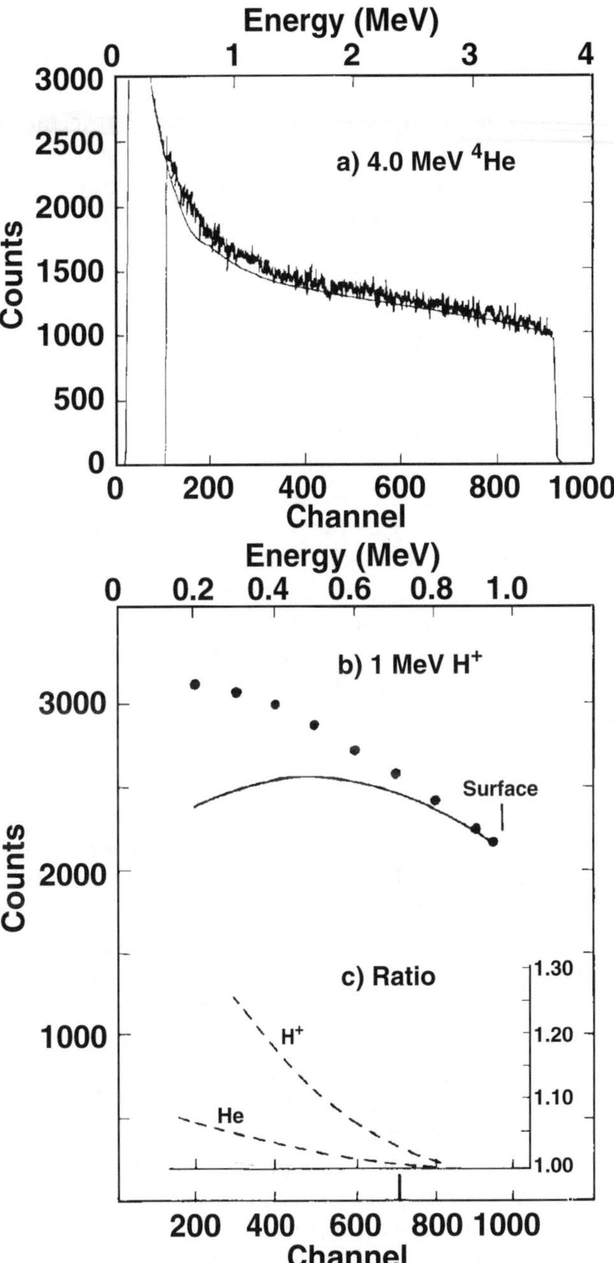

FIG. 12.6. Comparison of predicted (smooth curve) and experimental backscattered energy spectra from a gold target at a scattering angle of 150°: a) 4 MeV ^{4}He (Almeida and Macaulay-Newcombe, 1991); b) 1 MeV H$^+$ (Moore, 1980); c) Ratio of experimental to theoretical values.

12.3.2.2 Limitations of analysis codes

Note that all existing computer simulations of RBS profiles (RUMP, etc. - see Section 4.3.3.2) assume that the incoming/outgoing trajectories are linear. Furthermore, at present, no simple mechanism exists for modifying those simulations to include secondary scattering events because this would require some sort of Monte Carlo treatment for each particle trajectory.

12.4 SOLID ANGLE CALIBRATION

12.4.1 General remarks

These comments apply to particle detectors, where the intrinsic detector efficiency is known to be 100% for energies exceeding ~30 keV. For γ-ray and neutron measurements, it is more difficult to achieve an accuracy of ~1% for the overall detection efficiency. In practice, the solid angle for charged-particle detectors is almost never measured. It is difficult to determine the detector-to-target distance, l, with a precision of 1% (generating a 2% uncertainty in the solid angle). The use of a calibrated aperture on the detector, with an area, A, less than the active area, also presents some problems at the 1% level for relatively small area detectors. If the solid angle ($\sim A/l^2$ sr) is to be measured accurately, then the detector must be positioned at a rather large distance from the target to minimize the uncertainty in the determination of l, as was done in the Chalk River calibration (Davies *et al.*, 1986) of the Bi RBS standard (Section 12.4.3). In this situation, one generally compensates for the loss in counts by the use of higher fluences, thereby creating additional problems due to beam damage, etc. (see Section 12.5.1). Thus, most groups prefer to use suitable standards to calibrate the combined geometry and current integration constant for their experimental arrangements. It should be noted that if the solid angle and current integration are calibrated on one beam, then care should be taken that there is no relative error introduced via current integration when performing an experiment in the same geometry with a **different** beam.

Whenever the cross section varies significantly across the finite solid angle of the detector, as is the case for many nuclear reactions (Chapters 5–8), care must be used to properly integrate over the solid angle with respect to the angular dependence of the cross section. Care should also be taken in choosing an aperture which is not only opaque to MeV α particles, but also for more penetrating particles, for example, high energy protons. If there is an accurately calibrated radioactive α source of small area (for example, ^{241}Am), then the solid angle may be determined by placing the source at the position of the target. However, it is usually difficult to reduce the uncertainty ($\Delta l/l$) in positioning the α-source to the desired 1-2% level.

12.4.1 Standards for nuclear reaction analysis (NRA)

In discussing reference standards, we are also talking about those nuclear reactions for which the corresponding cross sections are well known at the level of a few percent. To date, probably the most widely used standard for NRA applications has been that for ^{16}O, specifically anodically-grown thin Ta_2O_5 targets on a Ta substrate. The Paris group (Amsel and Samuel, 1967) pioneered their use through the $^{16}O(d,p_1)^{17}O$ reaction at a bombarding energy of 972 keV and a detector angle of 150°. Here, there is a broad (~85 keV) resonance in the reaction cross section. The protons are emitted with an energy of ~1.6 MeV, and the cross section of 13.3 mb sr^{-1} has been measured with a precision of ±3% (Davies and Norton, 1980). A successful 'round-robin' intercomparison of absolute measurements in different laboratories using this reaction was recently reported (Seah *et al.*, 1988).

For those not having access to deuterium beams, substitution of ^{18}O in the electrolyte allows $Ta_2{}^{18}O_5$ targets to be fabricated. The $^{18}O(p,\alpha)^{15}N$ reaction at E = 750 keV could then be used, although the cross section value is not known as well as for the $^{16}O(d,p_1)^{17}O$ reaction (Christensen *et al.*, 1990). It would be helpful to have precise cross section values for ^{3}He + ^{16}O reactions. Some work has been reported (Lennard *et al.*, 1989; Abel *et al.*, 1990), but more studies are needed to firm up the values, particularly for higher ^{3}He energies. It should be noted that oxygen from a Ta_2O_5 oxide layer will begin dissolving back into the Ta substrate at a temperature of ~700K (Smyth, 1966). Indifference to beam heating of standards is therefore not encouraged.

The cross section for the d + ^{3}He → p + ^{4}He reaction (Q-value ~18 MeV) is very well known (±2%) for a center-of-mass energy of ~250 keV (σ_{lab}=58 mb sr^{-1}) (Davies and Norton, 1980; Möller and Besenbacher, 1980). Unfortunately, ^{3}He targets fabricated via ion implantation are sometimes not stable under prolonged ion beam bombardment (Alexander *et al.*, 1984; Geissel *et al.*, 1984). As well, stable hydride or polymer targets suitable for hydrogen (or deuterium) thin-film standards are still lacking. There are problems yet to be solved in the preparation of reference targets via ion implantation (current integration, dose uniformity, depth profiles, and stability). The reader is referred to Amsel and Davies (1983) for more detailed discussions on this topic.

12.4.3 Bi calibration standard

In 1975 an RBS round-robin experiment was conducted to test whether or not the often claimed ±2% accuracy was actually being achieved. Several Si wafers were implanted at UKAEA (Harwell) with ~5 × 10^{15} Bi cm^{-2} at an energy of 40 keV, which locates the Bi distribution at a depth of ~200 Å. They were then partitioned into 1 cm^2 pieces and distributed to ~50 RBS laboratories around the world. The results reported by Baglin (1975) at the Karlsruhe IBA conference were disappointing. Even among experienced laboratories, discrepancies of ±20% were common, and some results varied by 50% or more. The individual causes were never fully resolved, but from the shape of the 1.9 MeV H^+ RBS spectra (which includes the strong ~1.6 MeV resonance in Si), Baglin concluded that several laboratories did not even have a correctly calibrated accelerator energy scale.

Subsequently, a collaboration between ion beam groups at Chalk River, l'Université de Paris, Harwell, and Geel (Belgium) developed several independent methods of calibrating the Bi content of the Harwell standards, eventually obtaining very satisfactory (±2% or better) agreement among the different laboratories (Cohen *et al.*, 1983). In the course of these studies, several of the pitfalls noted in this chapter became recognized for the first time. For example, the energy-dependence of the Bi yield was observed to deviate slightly from the expected E^{-2} Rutherford law, and this observation led to the recognition of the electron screening correction (12.4.3). Many pitfalls associated with poor Faraday cup techniques (12.2.1) were identified and corrected. Also, the problem of charge exchange along the beam line (12.2.2) was found to be responsible for significant fluctuations (±4%) in some of the Harwell Series II Bi implantations (Davies *et al.*, 1986).

At Chalk River, an absolute measurement of the Bi RBS scattering yield was carried out using an accurately calibrated beam energy, a proper Faraday cup system, and a carefully measured scattering angle and detector solid angle. Thus, all the necessary quantities in Eq. (12.1) were measured directly without the use of an independently calibrated standard.

In Paris, the Bi RBS yield was compared to that obtained from a thin Ta film whose thickness had been measured independently by weighing. Again, as in the Chalk River study, proper Faraday cup techniques were needed, since two separate RBS measurements were involved.

In Geel, the Bi-implanted wafer was placed in an evaporation chamber, and coated with a known weight of V or Cu (measured *in situ* with a quartz microbalance). In the subsequent RBS analysis, the Bi/Cu or Bi/V yield ratio was measured in a simultaneous measurement, so that accurate current integration was not required in this case.

The resulting comparison, with the same Bi-implanted wafer being used in all three laboratories, is shown in Table 12.5. The ±2% agreement shows that with appropriate care, RBS can achieve the claimed absolute accuracy without the use of any calibrated standard. In practice, since the detector solid angle and the scattering angle are difficult quantities to measure accurately, it is usually simpler to use a previously calibrated standard instead. For this purpose, Harwell's Series I and Series II Bi-implanted wafers are an excellent and widely available choice.

Series I wafers were distributed worldwide in 1974/75; their Bi content is 4.87 (± 0.08) × 10^{15} atoms cm^{-2} (Table 12.5). Series II wafers are commercially available from Geel laboratories (contact U. Watjen, Central Bureau for Nuclear Measurements, Geel, Belgium); they

Table 12.5. Intercomparison of RBS standards (units of 10^{15} atoms cm^{-2}), adopted from Cohen *et al.*, 1983.

Sample	CRNL value	Paris value (Amsel)	Geel value (Irving Mitchell)
Ta film	30.7 ± 0.4	31.0 ± 0.4	
Bi-implant (Harwell Series I)	4.77 ± 0.07	4.88 ± 0.05	4.92 ± 0.10
Bi-implant (Harwell Series III)	3.60 ± 0.05	3.65 ± 0.05	

have the same nominal Bi content as Series I, but fluctuations as large as ±4% have been found (as noted above). Hence, whenever greater accuracy is required, one should obtain a wafer that has been cross-calibrated against a Series I standard.

12.5 UNWANTED TARGET-BEAM INTERACTIONS

12.5.1 Beam-induced damage

12.5.1.1 Sample heating

The power deposited in a target under beam impact is expressible in watts through the product of the particle energy (MeV) and the number of particles per second (μA), where the latter represents particle current, not electrical current. If the cross sectional area of the beam is known, then the power density (watts per unit area) is also readily calculated. Basic heat flow calculations can then be made.

For a target of semi-infinite thickness, the validity of this calculation rests on the fact that the fraction of incident particles reflected or scattered out of the target is negligible (< 1%), which is satisfied for all beams at MeV energies. It further requires that secondary particle emission (sputtered particle flux, nuclear reaction product flux, electron emission, etc.) produces no secondary heating effects. Again, for MeV particles, this assumption is usually satisfied. For thin targets, the heat input is determined by the energy deposited by the beam within the target; this may be much less than the incident energy. For each particle, incident with energy E_0 and exiting with energy E_1, the energy loss is

$$\Delta E = E_0 - E_1 = \int_0^t S(E)\,dz \qquad (12.6)$$

where *t* is the target thickness and *S(E)* is the (energy dependent) projectile stopping power along the particle *(z)* direction. While it is usually of no interest in ion beam analysis applications, heat generation in the target will proceed via the same fundamental interactions that have been identified in discussions of 'spike' phenomena. Beam energy expended through electronic stopping is coupled to electrons, that is, the electrons are 'heated' and this excitation must then be coupled to the lattice through electron-phonon interactions, thereby producing macroscopic sample heating with a characteristic time constant of $\sim 10^{-12}$ seconds. The route to lattice heating is more direct in the nuclear stopping regime, which dominates near the end of the particle range. As noted above, however, the quantity of interest is usually the mean power deposited. The variation in rate of thermalization of the lattice with depth is rarely of concern.

The sample response to beam heating depends on experimental conditions. If the beam spot is small in area, radial symmetry may be assumed. At temperatures above a few hundred degrees centigrade, cooling will be dominated by the radiative mechanism (T^4) rather than by conduction. (When conductivity within the sample is good, temperature profiles will be rather unremarkable.)

However, mechanical contact does not ensure good **thermal** contact. The existence of a hot spot can lead to differences in

- Electron emission;
- Sticking probabilities for adsorbates (e.g. O_2, CO, H_2O, etc.) and thus lower coverages relative to cooler surrounding material;
- Varying decomposition rates for adsorbed hydrocarbons; and
- Annealing of beam-induced disorder.

Thermal gradients can result in redistribution of mobile impurities, including embedded gas atoms from the beam, and can decompose materials. The question of how best to determine the target temperature under the beam lies outside the scope of these comments. Recent work has shown how important temperature control can be: large changes (up to an order of magnitude) in the level of residual disorder in MeV energy self-irradiation of Si over the temperature range 270K-320K for otherwise identical irradiation conditions have been observed (Schultz *et al.*, 1992). Hence, significant changes in beam-induced disorder may result from beam heating of even a few degrees.

12.5.1.2 Radiation damage

In many IBA applications, radiation damage produced by the ion beam is a key limitation. In general, such effects depend not only on the primary energy loss process involved, but also on subsequent solid state diffusion effects. Hence, the nature and temperature of the target, the dose rate, and the total fluence are all relevant parameters. The review literature on radiation damage and sputtering is extensive and we give here only some simple guidelines to assist newcomers in the IBA field.

Radiation damage processes arise from two widely different mechanisms of energy transfer:

- Nuclear (or atomic) stopping, $(dE/dx)_n$, in which scattering of the incident ion by the (partially-) screened target nucleus results in significant momentum being transferred to the whole atom. This contribution is the dominant energy-loss process at low (keV) energies, but at IBA energies it is only a very small fraction ($\sim 10^{-3}$) of the total stopping process.
- Electronic stopping, $(dE/dx)_e$, in which energy is lost to various electronic excitation and ionization processes. At IBA energies, this is the dominant energy loss process (> 99%).

In metals and most semiconductors, electronic excitation and ionization decay almost instantaneously without producing permanent damage effects. In such materials, the major source of radiation damage is the small nuclear stopping component (first mechanism, above). An upper limit to the resulting defect density may be obtained by multiplying the nuclear stopping power, $(dE/dx)_n$ (eV/10^{15} atoms cm^{-2}), by the total ion fluence and dividing by twice the displacement energy, E_d. At IBA energies, the nuclear stopping power is usually between 0.03 and 0.3 eV/10^{15} atoms cm^{-2} and E_d is typically ~30 eV. Hence, a beam fluence of 10^{15} ions cm^{-2} produces at most a defect density in the 0.05-0.5% range.

On the other hand, in insulators and other molecular compounds, damage production rates are often considerably greater because here the electronic stopping power (second mechanism, above) may also cause bond breakage and hence permanent chemical and structural changes in the target. Radiation chemistry studies in a wide variety of inorganic and organic molecular solids, using e^-, γ-ray, and MeV ion bombardment, have shown that the number of bonds broken per 100 eV of deposited energy (the so-called g-factor) is roughly 10, indicating that in such materials electronic stopping processes may be even more effective in breaking bonds than the nuclear collision cascades. Since the electronic stopping power at IBA energies is two to three orders of magnitude greater than the nuclear stopping power, it is evident that the total rate of damage creation in insulators can be as much as 10^3 greater than the nuclear collision estimates in the previous paragraph.

In practice, the observed levels of damage vary widely from one type of insulator to another. We have presented here only a rough estimate of the maximum rate at which damage is created; in many materials, self-annealing reduces the damage. For example, most ceramic materials (WC, BN) and certain inorganic oxides (MgO, Al_2O_3, SiO_2, UO_2) are fairly resistant, whereas alkali halides, polymers, and most organic and biological materials are rapidly and permanently decomposed by electronic stopping processes.

Damage creation due to electronic stopping is not restricted to insulators, but may occur in any polyatomic molecular solid, including high-T_c superconductors.

Finally, it should be emphasized that even when a large amount of damage is created by the analyzing beam, this does not necessarily introduce significant error into the quantities being analyzed. However, if some of the resulting defect species are mobile (radiation-enhanced diffusion) or volatile (for example, H_2), then significant changes in depth distribution or stoichiometry may result. In some cases, low-temperature analysis may reduce the problem.

Single crystal targets, which involve RBS/channeling analyses, are particularly sensitive to all types of radiation damage, even when the resulting defect species are non-mobile. Interaction between solute atoms and point defects (vacancies or interstitials), for example, can cause solute atoms to move into a completely different lattice site configuration (see Chapter 10, Section 10.7.6).

12.5.1.3 Sputtering

As part of the radiation damage process, near-surface atoms occasionally receive sufficient kinetic energy or electronic excitation to be ejected — sputtered — from the target surface. Obviously, if the amount sputtered during IBA becomes significant (10^{15} cm^{-2} ≈ one monolayer), then the surface structure, stoichiometry, film thickness, etc. could be irreversibly changed. Again, as in the radiation damage section (12.5.1.2), we divide the discussion into two widely different types of behavior.

In metals and most semiconductors, only the nuclear stopping component contributes to sputtering and the observed yield is directly proportional to $(dE/dx)_n$. The relationship between the sputtering yield, Y (atoms/incident ion), and $(dE/dx)_n$ (eV/10^{15} atoms cm^{-2}) can be expressed as

$$Y = 0.1 \frac{\left(\frac{dE}{dx}\right)_n}{U_s \cos\phi} \; . \qquad (12.7)$$

The numerator $[0.1\ (dE/dx)_n]$ is a rough estimate of the total energy (in ev) contributing to the sputtering process, U_s (the surface binding energy of the target) is usually between 2 and 10 eV/atom, and ϕ is the angle between the incident beam direction and the surface normal. Fig. 12.7, taken from Andersen (1987), depicts the sputtering behavior of a typical metal (Ni) as a function of the energy and atomic number of the incident beam. The exceptionally good fit of the predicted $(dE/dx)_n$ dependence (solid curves) to the experimental Ni data is somewhat fortuitous. In general, the agreement between predicted and measured sputtering yields is no better than a factor of two.

In the low-E regime (10-100 keV) where $(dE/dx)_n$ reaches its maximum value, sputtering yields as large as 10 atoms/ion are often encountered. But at IBA energies, $(dE/dx)_n$ is in the range of 0.03-0.3 eV/10^{15} atoms/cm^{-2}; hence, the sputtering yield is less than 10^{-2}.

In insulators, the collision cascade mechanism of sputtering (described in the preceding paragraph) still occurs and hence sputtering yields of at least 10^{-3} atoms/ion occur in all IBA studies. In insulators, however, the much larger $(dE/dx)_e$ process can also produce significant sputtering through a variety of mechanisms: bond breaking, 'Coulomb explosion' effects, formation of volatile

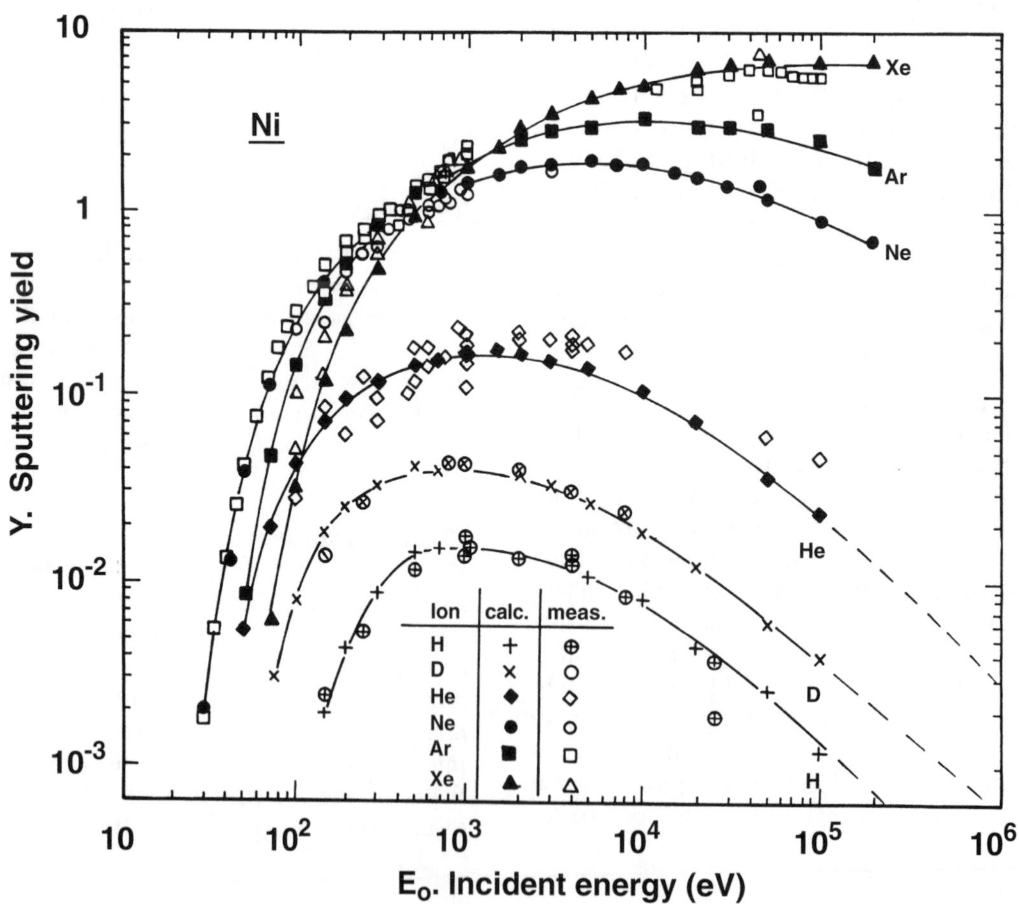

FIG. 12.7. Experimental data and predicted (using TRIM, see Ziegler *et al.*, 1985) sputtering yields for various ions in Ni (Andersen, 1987).

products (H_2, O_2, etc.). The exact mechanisms are, in general, not well characterized and they depend markedly on the type of insulating material involved. Nevertheless, experimental evidence (Tombrello, 1984) shows that the sputtering yield in insulators can be as high as 10 (in rare cases, even 100) atoms/incident ion. In such materials, the electronic stopping process has enhanced the sputtering rate by as much as 10^4.

Note that in polyatomic targets, preferential sputtering of certain atomic species can often occur. Hence, whenever the total amount sputtered is greater than one monolayer (~10^{15} cm^{-2}), significant changes in near-surface stoichiometry may also result. An extreme example is the rapid dehydrogenation (or graphitization) of polymers that occurs during IBA with MeV He^+ beams [see, for example, the review of polymer damage by Brown (1986)].

12.5.2 Insulators

Compared to metals and semiconductors, most insulating materials present a series of special problems in regard to IBA. In the preceding section (12.5.1) on beam-induced damage, the greatly enhanced sensitivity of insulators to beam-induced effects such as radiation damage and sputtering has already been emphasized. Here, we briefly note two additional complications that arise in certain types of insulating materials: target charging and photon emission.

12.5.2.1 Charging effects

Depending on the geometry, thickness and resistivity of the target material, surface charging under MeV ion bombardment can in severe cases reach several tens of kV. (This value is, of course, still small compared to the incident beam energy, unlike the insulator situation encountered in Auger and secondary ion mass spectroscopy.) However, since the charge states of the incident and backscattered beam are not necessarily the same, surface charging can seriously distort the energy and hence the shape of the observed RBS spectrum, as shown for the case of quartz in Fig. 12.8.

Furthermore, surface charging sometimes produces sufficiently high electric fields to interfere with the performance of the Faraday cup system. Also, in certain insulators (for example, $BaTiO_3$), excessive surface charging may even cause the target to disintegrate.

One effective way to neutralize surface charging effects is to provide a supply of low-E electrons from a small, hot filament located nearby. If such a filament is powered via an isolation transformer and is electrically connected to the target holder through a suitable (+50V) bias to prevent electrons escaping to the chamber walls, then quantitative current integration is still achievable.

Generally, the use of a transmission Faraday cup is a better way to solve these current integration problems.

Other possible methods that have been used to reduce surface charging are:

- Coating the surface at least partially with a very thin layer of conducting material, such as graphite (for example, rubbing a pencil lightly across the surface) or an evaporated metal.
- Sweeping the beam so that it also bombards the adjacent (metal) target holder or a suitably placed metal grid in front of the target. The latter process generates a supply of secondary electrons to neutralize the positively charged surface of the insulator.

In many insulators, charging effects are small enough to be ignored.

12.5.2.2 Photon emission

In insulators, much of the electronic excitation resulting from the $(dE/dx)_e$ stopping mechanism is eventually converted into optical emission in the visible and ultra violet region. Polymers, alkali halides, etc. are transparent to such radiation. Consequently, photons emitted even from a depth of several microns can readily escape from the target. In such cases, hundreds or even thousands of photons may be emitted per incident ion. We have already discussed (Faraday cup design, 12.2.1) how these photons can generate a flux of secondary electrons at the chamber walls. Furthermore, the energy resolution of charged-particle detectors can be seriously degraded by photogeneration of carriers in the depleted region. (The standard surface barrier detector has a Au electrode of thickness ~200 Å which is transparent to visible light.)

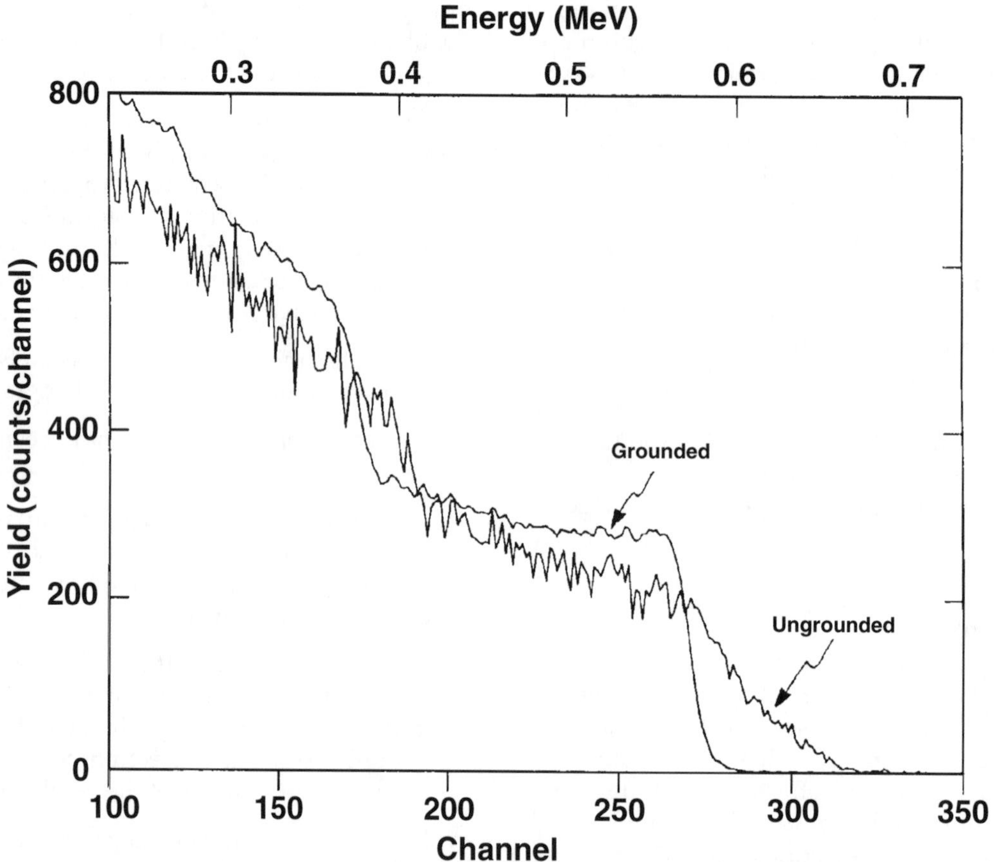

FIG. 12.8. Surface charging effect. Comparison of RBS spectra from a quartz target using 1 MeV ^{4}He: a) ungrounded; b) grounded via a thin conductive surface layer of graphite by rubbing a pencil lightly across the surface (Almeida and Macauley-Newcombe, 1991).

12.6 OTHER EFFECTS

12.6.1 Surface roughness

When performing experiments where the incident or emergent charged particle makes a small angle (for example, < 15°) with a target surface which is not smooth, problems in the shape of the energy distribution may arise that are caused by the surface topography. This is very often the case for ERDA measurements (see Chapter 5). Even at near-perpendicular incidence, porous targets, sintered powders, etc. can introduce very serious surface roughness complications.

12.6.2 Target non-uniformity

Care must be taken in resting the interpretation of RBS spectra on limited data. Apparently simple features may be misleading if it is assumed that the target has lateral uniformity under the analyzing beam. As shown by Campisano *et al.* (1975), the shape of a Pb spectrum recorded from a non-uniform surface film of Pb on Si may closely resemble that of a laterally uniform diffusion profile of Pb in Si.

This confusion is removed by supplementary measurements and analysis (Campisano *et al.*, 1978). As shown in Fig. 12.9a, the consequences of varying the target angle, ϕ, to the incident beam for a Pb film of thickness 50 nm overlaying an Si substrate are to give a familiar broadening $[(\cos \phi)^{-1}]$ of the Pb profile in the RBS spectrum, and a correlated shift in the leading edge of the Si continuum. Totally different profiles are obtained for samples which have been annealed to 280°C for 20 minutes, after deposition. In this case (Fig. 12.9b), the decrease in the scattered particle yield, H_{Si}, from Si atoms at the surface as the tilting angle (ϕ) increases rules out interdiffusion. The results have been modeled in terms of raised Pb features (islands) and good agreement found

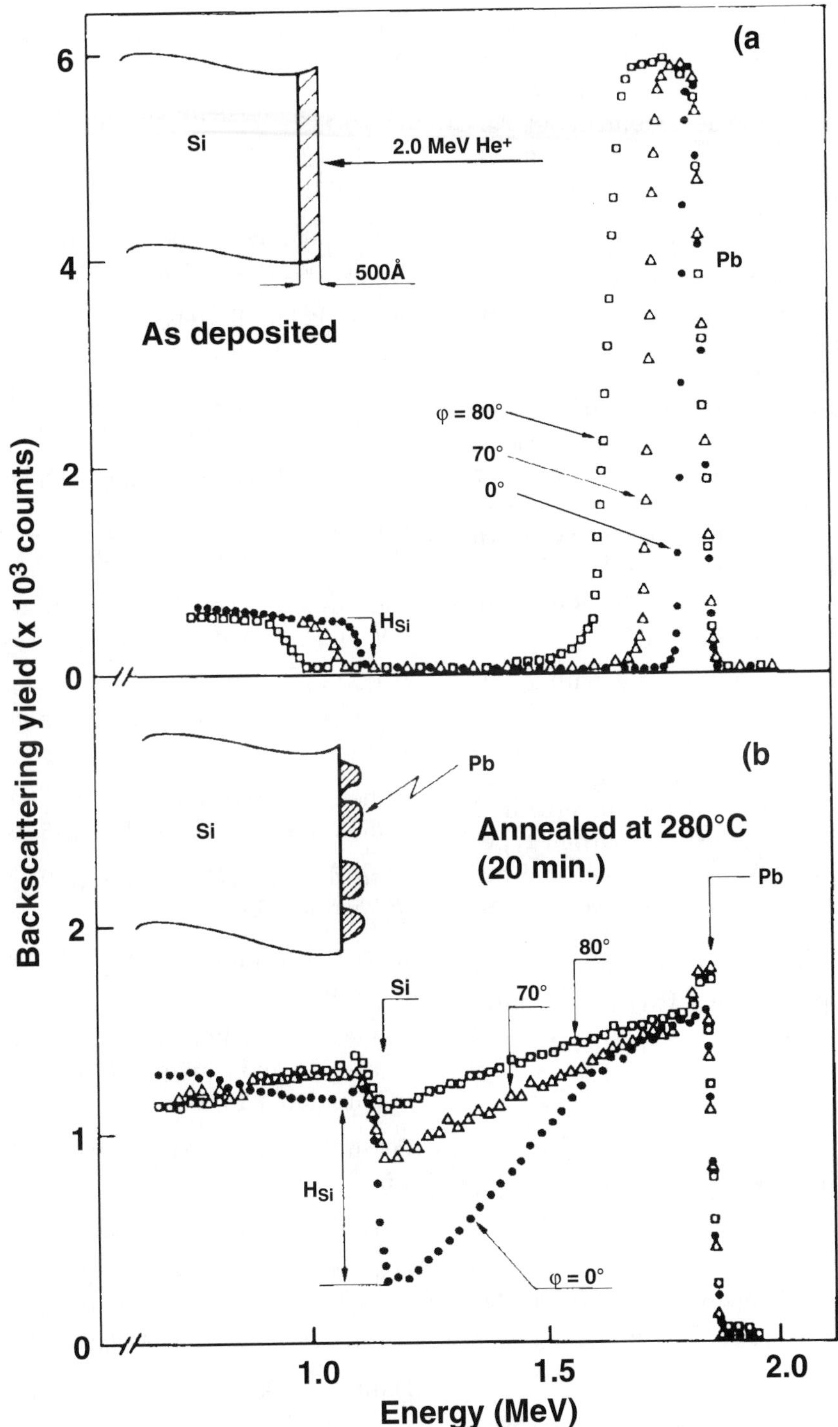

FIG. 12.9. Backscattered energy spectra for several tilting angles. a) From a uniform Pb layer (500 Å) on a Si substrate; and b) from the same layer after annealing for 20 minutes at 280°C (Campisano *et al.*, 1978).

with island dimensions and spacings measured by SEM analysis which displays the altered surface morphology. RBS methods have been used very effectively (Zinke-Allmang *et al.*, 1989) to study the development of layer-by-layer and island growth (Stranski-Krastanov films).

12.6.3 Non-linear detector response

Semiconductor radiation detectors are known to respond in a non-linear fashion to particle energy. Normally, the deviations are lumped into the so-called 'pulse height defect', which includes the effects of detector entrance-window/dead-layer and also the nuclear (non-ionizing) energy loss of the projectile arising from elastic collisions with the detector lattice atoms. There is a further effect that has been studied in some detail (Lennard *et al.*, 1986; Bauer and Bortels, 1990; Comedi and Davies, 1991) in Si for light ions (^{1,2}H, ^{3,4}He, ^{7}Li,...) which arises as a consequence of the differing ionization densities produced by the incident particle.

So long as experiments are only concerned with counting the number of events, these effects are unimportant. However, in those few spectrometric measurements where one is concerned with the **precise energy** (to better than 1-2%), care must be taken to interpolate measured peak energies using particles (of the **same particle type**) whose energies are well known, as is done in γ-ray spectrometry. This procedure poses a problem for all particles except ^{4}He due to the lack of suitable sources. Most precise particle energy determinations are performed using magnetic spectrometers (Rytz, 1979); however, the event rate is dismal.

REFERENCES

Abel, F., Amsel, G., d'Artemare, E., Ortega, C., Siejka, J. and Vizkelethy, G. (1990), *Nucl. Instr. and Meth. in Phys. Res.* **B45**, 100.

Alexander, T.K., Ball, G.C., Lennard, W.N., Geissel, H. and Mak, H.-B (1984), *Nucl. Phys.* **A427**, 526.

Allison, S.K. (1958), *Rev. Mod. Phys.* **30**, 1137.

Almeida, F. and Macaulay-Newcombe, R.G. (1991), private communication, McMaster University, Hamilton, Canada.

Amsel, G. and Davies, J.A. (1983), *Nucl. Instr. and Meth.* **218**, 177.

Amsel, G. and Samuel, D. (1967), *Anal. Chem.* **39**, 1689.

Andersen, H.H. (1987), *Nucl. Instr. and Meth. in Phys. Res.* **B18**, 321.

Andersen, H.H., Besenbacher, F., Loftager, P. and Möller, W. (1980), *Phys. Rev.* **A21**, 1891.

Bauer, P. and Bortels, G. (1990), *Nucl. Instr. and Meth. in Phys. Res.* **A299**, 205.

Bozoian, M., Hubbard, K.M. and Nastasi, M. (1990), *Nucl. Instr. and Meth. in Phys. Res.* **B51**, 311.

Brown, W.L. (1986), *Rad. Eff.* **98**, 115.

Campisano, S.U., Ciavola, G., Costanzo, E., Foti, G. and Rimini, E. (1978), *Nucl. Instr. and Meth.* **149**, 229.

Campisano, S.U., Foti, G., Grasso, F. and Rimini, E. (1975), *Thin Solid Films* **25**, 431.

Christensen, N.S., Jensen, F., Besenbacher, F. and Stensgaard, I. (1990), *Nucl. Instr. and Meth. in Phys. Res.* **B51**, 97

Cohen, C., Davies, J.A., Drigo, A.V. and Jackman, T.E. (1983), *Nucl. Instr. and Meth.* **218**, 147.

Comedi, D. and Davies, J.A. (1992), *Nucl. Instr. and Meth. in Phys. Res.* **B67**, 93.

Davies, J.A. and Norton, P.R. (1980), *Nucl. Instr. and Meth.* **168**, 611.

Davies, J.A., Jackman, T.E., Eschbach, H.L., Domba, W., Watjen, U. and Chivers, D. (1986), *Nucl. Instr. and Meth. in Phys. Res.* **B15**, 238.

Feldman, L.C. and Appleton, B.R. (1969), *Appl. Phys. Lett.* **15**, 305.

Garcia, J.D. (1971), *Phys. Rev.* **A4**, 955.

Geissel, H., Lennard, W.N., Ball, G.C., Forster, J.S., Lone, M.A., Milani, L., Phillips, D. and Plattner, H.H. (1984), *Nucl. Instr. and Meth. in Phys. Res.* **B2**, 770.

Hautala, M. and Luomajarvi, M. (1980), *Rad. Eff.* **45**, 159.

Knoll, G.F. (1989), *Radiation Detection and Measurement*, 2nd edition, John Wiley and Sons, New York.

L'Ecuyer, J., Davies, J.A. and Matsunami, N. (1979), *Nucl. Instr. and Meth. in Phys. Res.* **160**, 337.

Lennard, W.N. (1989), *Vacuum* 39, 413.

Lennard, W.N., Geissel, H., Winterbon, K.B., Phillips, D., Alexander, T.K. and Forster, J.S. (1986), *Nucl. Instr. and Meth. in Phys. Res.* **A248**, 454.

Lennard, W.N., Tong, S.Y., Mitchell, I.V. and Massoumi, G.R. (1989), *Nucl. Instr. and Meth. in Phys. Res.* **B43**, 187.

Marion, J.B. and Young, F.C. (1968), *Nuclear Reaction Analysis: Graphs and Tables*, North Holland, New York, p.154.

Mendenhall, M.H. and Weller, R.A. (1991), *Nucl. Instr. and Meth. in Phys. Res.* **B58**, 11.

Möller, W. and Besenbacher, F. (1980), *Nucl. Instr. and Meth.* **168**, 111.

Moore, J. (1980), *Nucl. Instr. and Meth.* **174**, 577.

Pronko, P.P., Appleton, B.R., Holland, O.W. and Wilson, S.R. (1979), *Phys. Rev. Lett.* **43**, 779.

Purser, K.H., Litherland, A.E. and Gove, H.E. (1979), *Nucl. Instr. and Meth.* **162**, 637.

Rytz, A. (1979), *At. Data Nucl. and Data Tables* **23**, 507.

Schultz, Peter J., Jagadish, C., Ridgway, M.C., Elliman, R.G. and Williams, J.S. (1991), *Phys. Rev.* **B44**, 9118.

Seah, M.P., David, D., Davies, J.A., Jeynes, C., Ortega, C., Read, P.M., Sofield, C.J. and Weber, G. (1988), *Nucl. Instr. and Meth. in Phys. Res.* **B30**, 140.

Sitter, C., Davies, J.A., Jackman, T.E. and Norton, P.R. (1982), *Rev. Sci. Instr.* **53**, 797.

Smyth, D.M. (1966), *J. Electrochem. Soc.* **113**, 1271.

Tombrello, T.A. (1984), *Nucl. Instr. and Meth. in Phys. Res.* **B2**, 555.

Venkatesan, T., Brown, W.L. and Wilkens, B.J. (1984), *Nucl. Instr. and Meth. in Phys. Res.* **B1**, 605.

Ziegler, J.F., Biersack, J.P. and Littmark, U. (1985), *The Stopping and Range of Ions in Solids*, Pergamon Press, New York.

Zinke-Allmang, M., Feldman, L.C., Nakahara, S. and Davidson, B.A. (1989), *Phys. Rev.* **B39**, 7848.

CHAPTER 13

RADIOLOGICAL SAFETY

P. M. DeLuca, Jr.

University of Wisconsin, Madison, Wisconsin

J. R. Tesmer

Los Alamos National Laboratory, Los Alamos, New Mexico

CONTENTS

13.1 INTRODUCTION

Although a complete discussion of the many aspects of the radiation environment encountered near low-energy particle accelerators is beyond the scope of this chapter, general information is provided here to guide the user. More detailed information is available in the references listed at the end of this chapter.

The following sections discuss the various agencies involved in regulation and development of standards, as well as briefly review the more common radiological parameters, the typical radiation environment encountered and its potential hazards, and general guidance in the development of suitable radiation protection for an ion-beam analysis facility.

13.2 REGULATORY AGENCIES

Unfortunately, current research is conducted in a complex and diverse regulatory environment. Frequently, the same aspect of radiological safety is controlled by two or more agencies, and some effort is required to determine which is actually the controlling agency. The following discussion outlines some of the agencies involved in the control of radiation near accelerators.

Although special nuclear materials (SNM), i.e., plutonium, ^{233}U, ^{235}U, uranium enriched in ^{233}U or ^{235}U, and tritium, are not usually found in most ion-beam analysis laboratories, for the purpose of a more complete overview they are included in this discussion. In the United States, the Nuclear Regulatory Commission (USNRC) has regulatory control of the use of all SNM and by-product materials—material produced as a result of using special nuclear materials. The rules and regulations by which the USNRC controls by-product and SNM are contained in Chapter 10 (Energy) of the Code of Federal Regulations, Parts 0 through 50. Most of the essential aspects are discussed in Part 20 (10CFR20, 1993). USNRC does not usually control the use of radiation-producing apparatus other than that involved with the aforementioned materials, nor do they control naturally-occurring or accelerator-produced radionuclides.

Control of radiation-producing apparatus and naturally-occurring or accelerator-produced radionuclides is usually provided by state agencies. The most common of these devices are radiographic apparatus used in medical procedures. Some states, called *Agreement States*, have assumed the regulatory activities of USNRC, and those states control radiation consistent with the regulations of USNRC. Beyond these regulatory bodies, the United States Environmental Protection Agency (USEPA) regulates radiation in the environment generally occupied by the public. Obviously, USEPA control overlaps USNRC and Agreement State activities.

Further, the United States Department of Energy (USDOE) provides the regulatory environment for those institutions which it operates. USDOE generally adopts the position of USNRC on matters of radiation control. However, USDOE has recently taken a position of greater conservatism with respect to radiation than USNRC. Two regulations are of particular interest to operators and users of ion-beam analysis facilities: USDOE order 5480.25 (DOE 5480.25, 1992), Safety of Accelerator Facilities, and Part 835 of Chapter 10 (Energy) of the Code of Federal Regulations (10CFR835, 1993), Occupational Radiation Protection (formerly USDOE order 5480.11) and their associated guidance documents.

The United States Department of Transportation provides the regulatory environment for transportation of radioactive materials, both those controlled by USNRC and State Agencies. USEPA is also involved in regulating transportation insofar as it affects the environment occupied by the general public.

While this description is specific for the United States, it suggests a rather bleak and imposing situation. Readers are cautioned to carefully determine who are the controlling parties for their facilities and resolve any potential areas of conflicting or overlapping interest.

Although the regulatory environment is fraught with confusion and some uncertainty, the scientific community that provides guidance to the several regulatory agencies does so in a coherent and comprehensive manner. In the U.S., the National Council of Radiation Protection (NCRP), Baltimore, MD, was organized in the early 1930s to provide crucial guidance on the proper and safe use of radiation. As such, the NCRP has promulgated these findings in a series of reports organized by topic and use. Considerable information is contained in these brief reports. On an international scale, the International Commission on Radiological Protection and the International Commission of Radiation Units and Measurements (ICRU) provide similar guidance. All these agencies coordinate their activities to provide a comprehensive and cohesive presentation to the scientific community.

13.3 RADIOLOGICAL QUANTITIES

Various radiological units and quantities are used to describe the effects of radiation. The radiation found near low-energy accelerators can be classified as either directly- or indirectly-ionizing radiation. Protons and alphas are directly ionizing, while x-rays, γ-rays, and neutrons are not. Hence, except for beams brought into the air for use (external beams), most of the radiation hazards near low-energy accelerators are from indirectly-ionizing radiation.

Despite several alternate parameterizations, the most accepted and well understood parameter is *absorbed dose*. As stated by the ICRU (ICRU 51, 1992), absorbed dose D in a material is the quotient of dε by dm, where *dε* is the mean energy imparted by ionizing radiation to matter of mass *dm*

$$D = \frac{d\varepsilon}{dm} \ . \qquad (13.1)$$

The unit of absorbed dose is one *joule* per kilogram and is denoted as the gray (Gy). Note that absorbed dose is defined for a specific material, leading to the usage one tissue gray. Absorbed dose is a non-stochastic quantity. Absorbed dose is the physical result of ionizing radiation. The unit Gy replaces the older unit rad, where 100 rad equals 1 Gy.

For a better description of the effects of indirectly ionizing radiation, an additional quantity is defined, the pre-cursor to absorbed dose, called *kerma*, an acronym for Kinetic Energy Released in Matter. The ICRU (ICRU 33, 1980) defines *kerma* as the quotient of dE_{tr} by dm, where dE_{tr} is the sum of the initial kinetic energies of all charged particles ionizing particles liberated by uncharged particles in a material of mass *dm*

$$K = \frac{dE_{tr}}{dm} \ . \qquad (13.2)$$

The unit of kerma is the same as absorbed dose, the gray. The kerma per unit particle fluence is the kerma factor. Note that kerma is not directly measurable, but can be calculated from basic particle interaction probabilities. Hence for indirectly ionizing radiation, the absorption of energy consists of a two-step process: (1) indirectly ionizing radiation converts some energy to directly ionizing charged particles (kerma) and (2) the resulting ionizing charged particles impart energy to matter by excitation and ionization (absorbed dose).

While the biological effect of 1 gray is largely similar for all types of ionizing radiation, differing by perhaps less than a factor of 10, the differences are sufficient to warrant the scaling of the absorbed dose by a biological effectiveness parameter, *Q*. This leads to the quantity *dose equivalent*. The ICRU (ICRU 51, 1992) defines dose equivalent (H) as the product of Q and D at a point in tissue, where *D* is the absorbed dose and *Q* is the quality factor at that point

$$H = QD \ . \qquad (13.3)$$

The unit of dose equivalent is the *sievert (Sv)*, where 1 Sv = J kg^{-1}. Note that dose equivalent and absorbed dose differ only by the scaling of biological effect, and hence have the same base units, i.e., J kg^{-1}. For x- or γ-radiation, absorbed dose and dose equivalent are essentially equal. For neutrons, the biological effect varies with neutron energy. Table 13.1, taken from 10CFR20.1004(b).2 (10CFR20, 1993), indicates the variation of biological effect for neutrons with neutron energy. The introduction of a biological effectiveness factor considerably complicates the determination of dose equivalent for fast neutrons because physical measurements are usually associated with energy absorbed per unit mass. Figure 13.1 plots dose equivalent per unit fluence values as a function of neutron energy. This information is useful for estimating the dose equivalent that would result from a known reaction.

As a point of information, the difference in biological effect among various radiations is in large part due to the rate of energy transfer by charge particles at a microscopic level. Specifically, an energy transfer rate of about 100 keV/µm produces the largest observed biological effect (ICRU 51, 1992). Thus, fast neutrons which produce recoil protons and other heavier charged particles are observed to have a larger biological effect than, for example, γ-rays which produce electrons with an energy transfer rate less than about 5 keV/µm .

13.4 DIRECT EXPOSURE

Exposure to the charged particle beam poses an extreme hazard because these particles are directly ionizing. This is one of the major hazards for researchers using external beams. The specific hazards associated with Particle-Induced-X-ray Emission (PIXE) in air are discussed by Doyle *et al.* (1991).

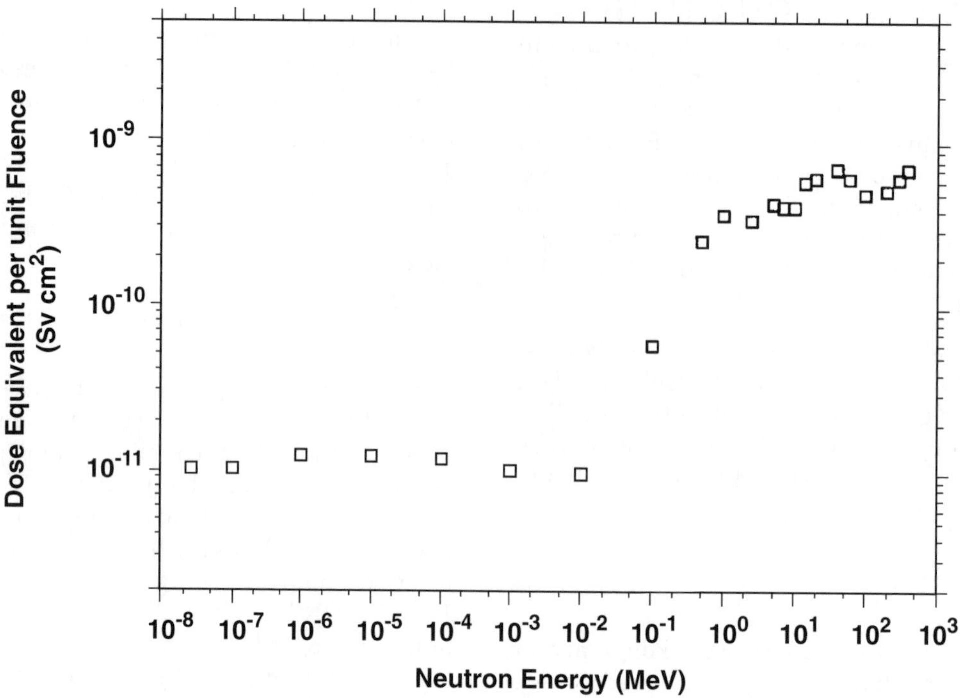

FIG. 13.1. Neutron dose equivalent per unit neutron fluence values plotted as a function of neutron bombarding energy.

For a particle flux density incident upon skin, the absorbed dose rate is given by

$$\dot{D} = {}_mS_{tiss} * \phi * 1.602\,10^{-10}\left(\frac{Gy}{sec}\right) \tag{13.4}$$

where ${}_mS_{tiss}$ is the mass collision stopping power for the particles of energy E in tissue in units of MeVcm^{2}g^{-1} and ϕ is the charged particle flux density in units of cm^{-2}s^{-1}. Figure 13.2 plots values of ${}_mS_{tiss}$ for tissue bombarded by protons and α-particles of energy less than 10 MeV. In a similar manner, Fig. 13.3 plots $\dot{D}$ values vs. charged particle energy normalized to a fluence rate of 6.24×10^{11}s^{-1}, corresponding to a beam current of 1 nA with 1 mm^2 area. ${}_mS_{tiss}$ values are taken from the tabulations of Anderson and Ziegler (1977). Such energy absorption rates are sufficient to cause massive skin damage, albeit local to the surface of the skin.

While the energy absorption rate is very large for these directly ionizing particles, the range in tissue for particles below 10-MeV energy is modest. Figure 13.4 plots proton and alpha-particle range values in tissue for a similar particle energy range. Particle range values were determined from integration of the same data used for the stopping powers (Anderson and Ziegler, 1977). While not affording complete protection, the dead layer of the skin (the epidermis) is about 50 μm thick (ICRP, 1974) corresponding to the range of approximately 1.5-MeV protons and 6.4-MeV α particles.

Clearly the hazard from direct exposure to the charged particle beam is eliminated by introduction of a modest thickness of material to stop the beam. Moreover, the direct exposure damage is confined to superficial layers of the skin, with the greatest hazard being exposure of the eyes. For example, the NCRP recommendation for superficial exposure is 0.5 Sv yr^{-1} (NCRP, 1987), which can be produced by a fluence of 1.2×10^5 mm^{-2} of 1-MeV protons. This fluence can be easily achieved from protons backscattered at 120° during bombardment of a 10-μm-thick gold layer with 200 nC of protons and at 10 cm target-to-skin distance. Exposure time would be 2 s at 100 nA beam current!

While the most important danger of direct exposure to charged particles is macroscopic organ damage, namely the skin, stochastic effects are still possible with skin cancer being the most dangerous. Such a possibility would become pronounced at exposures in the range of

Table 13.1. Mean quality factor, Q, and fluence per unit dose equivalent for monoenergetic neutrons.

Neutron energy (*MeV*)	Quality factor[1] (*Q*)	Fluence per unit dose equivalent[2] ($cm^{-2}Sv^{-1}$)
2.5×10^{-8}	2	980×10^{8}
1×10^{-7}	2	980×10^{8}
1×10^{-6}	2	810×10^{8}
1×10^{-5}	2	810×10^{8}
1×10^{-4}	2	840×10^{8}
1×10^{-3}	2	980×10^{8}
1×10^{-2}	2.5	1010×10^{8}
1×10^{-1}	7.5	170×10^{8}
5×10^{-1}	11	39×10^{8}
1	11	27×10^{8}
2.5	9	29×10^{8}
5	8	23×10^{8}
7	7	24×10^{8}
10	6.5	24×10^{8}
14	7.5	7×10^{8}
20	8	16×10^{8}
40	7	14×10^{8}
60	5.5	16×10^{8}
1×10^{2}	4	20×10^{8}
2×10^{2}	3.5	19×10^{8}
3×10^{2}	3.5	16×10^{8}
4×10^{2}	3.5	14×10^{8}

1. Value of Q at the point where the dose equivalent is maximum in a 30–cm diameter tissue-equivalent cylinder (phantom).
2. Monoenergetic neutrons incident normally on a 30–cm diameter tissue-equivalent cylinder (phantom).

0.1 to 3.0 Gy, corresponding to fluences of 2.4×10^4 mm^{-2} to 71×10^4 mm^{-2} of 1-MeV protons. However, exposure to external beams of vastly higher particle fluences and energies are more likely and are hence the principle concern.

13.5 INDIRECT EXPOSURE

For many analyses the incident particle energies are typically a few MeV and generally below the nucleon binding energy. Therefore, the potential hazard from neutron exposure is often assumed to be small. This is not necessarily the case. Significant quantities of neutrons are generated at low particle bombardment energies when low-Z projectiles strike low-Z targets. Some of the important neutron-producing reactions below 5-MeV particle bombarding energy are given in Table 13.2. A program by Drosg was used to estimate these values from thick isotopically pure targets (Drosg, 1992).

Typically, these reactions are not deliberately used, but may occur inadvertently. For example, a sample

Table 13.2. Various reactions that produce significant quantities of neutrons for bombardment by protons and deuterons below 5 MeV. Values of thick target neutron yield Y, mean neutron energy $\bar{E}_n$, and neutron dose equivalent rate H at 0° and 90° with respect to the bombardment direction are listed.

Reaction	Q (*MeV*)	Y ($sr^{-1}pC^{-1}$) (0°)	Y (90°)	$\bar{E}_n$ (*MeV*) (0°)	$\bar{E}_n$ (90°)	H ($\mu Sv\ h^{-1}nA^{-1}$) (0°)	H (90°)
$^3H(p,n)^3He$		5.93	3.35	1.2	0.42	0.64	0.30
$^2H(d,n)^3He$	2.2	2.26	0.27	5.24	2.95	0.24	0.03
$^3H(d,n)^4He$	17.8	0.59	0.35	18.3	14.8	0.10	0.06
$^2H(t,n)^4He$		0.43	0.40	18.7	14.4	0.08	0.07

bombarded with deuterons will embed deuterons at the depth of their range, and since hydrogen readily diffuses in many materials, continued bombardment will produce neutrons from the ^{2}H(d, n) reaction. Increasing the bombardment energy will produce an even more prolific source of neutrons as the bombarding deuterons can reach the embedded deuterons with sufficient energy to have a significantly increased interaction probability. Also, mixing beams of deuterons and tritons on the same sample will produce substantial amounts of 15-MeV neutrons. Note that the *sample* can be beam collimation apparatus or any other components exposed to the beam! In fact, the ^{2}H(d, n) reaction can be a significant source of neutrons in older ion-beam analysis facilities and can eventually curtail (if unshielded) the use of deuteron beams for analysis. Baking out the components with the embedded deuterium can reduce this problem.

Another significant source of neutrons is produced by deuteron breakup on beam transport equipment and samples when the deuteron bombarding energy exceeds the deuteron binding energy in the center-of-mass, i.e., 2.2-MeV lab energy for heavy targets. Fortunately, most analysis reactions use lower bombarding energies.

Elastic resonance reactions for the analysis of light elements with alphas or protons may also produce neutrons. As the beam energy is increased above the coulomb barrier (see Appendix 8) the production of neutrons and gammas increases dramatically for light targets. Data for (α,n) reactions up to 10 MeV are given in Appendix 18. Both neutrons and gammas from higher energy proton beams may require the use of shielding. Fortunately, most useful proton resonances are at lower energies. Nevertheless in these cases, a knowledge of what you are about to analyze is important for preventing

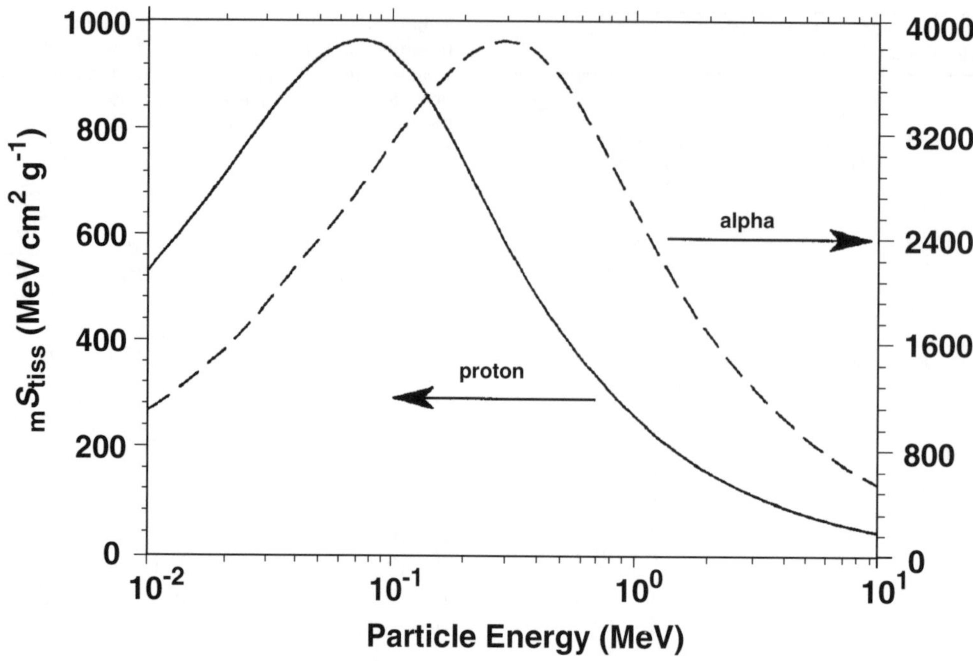

FIG. 13.2. Values for the mass collision stopping power for protons and alpha particles as a function of particle energy for interactions with muscle tissue.

radiation exposures. In addition, the sample materials may become activated with short half-life beta emitters (or internal conversion gammas). These are produced on purpose in activation analysis (see Chapter 9), but are not welcomed by the unknowing experimenter. These radionuclides may be produced by the (p,n), (α,p), and (α,n) reactions. For example, ^{48}Ti(α,n) yields ^{51}Cr with a 28-day half-life. Hence, the sample with Ti in it has now become a radioactive source with a relatively long half-life and must be treated as such! Another common example is ^{29,30}Si(α,p)^{32,33}P with half-lives of approximately 14 and 25 days respectively. It is good practice to monitor all samples for radioactivity which have been irradiated with deuterons, or any particles with energies above the coulomb barrier of the elements in the sample.

External (in air) beams may produce radionuclides such as ^{11}C, ^{15}O, ^{13}N, ^{18}F, etc., which may then be inhaled. Doyle *et al.* (1991) discusses the hazards for the case of PIXE. The allowable concentrations are listed in 10CFR835 (1993).

13.6 RADIATION DETECTORS

Most accelerators used for materials analyses are capable of producing unwanted ionizing radiation. This situation, coupled with the desirability of having the researcher near the sample analysis chamber and the accelerator controls, requires that great attention be given to reducing or limiting the possibility of exposure to radiation. In such situations radiation detectors play a leading role in keeping the working environment safe. As the beam energy available for analyses goes up, the hazard increases from both prompt radiation and activated samples, making radiation detection systems mandatory.

There are several types of detectors usually encountered in accelerator laboratories used for ion beam materials analyses. These divide into two classes of devices: (1) those to assess the absorbed dose to individuals and (2) those to survey the environment occupied by workers. The most common are personnel dosimeters. All accelerator laboratories should require personnel to wear dosimeters. These detectors come in several types that serve different purposes and have different applications. The principle difference is whether they respond actively or require subsequent readout to determine the absorbed dose. An example of a small active device is the pocket ionization chamber. Modern instruments use a small silicon diode operated in photo-voltaic mode, the response of which is interpreted by a microprocessor for active indication of absorbed dose and dose rate.

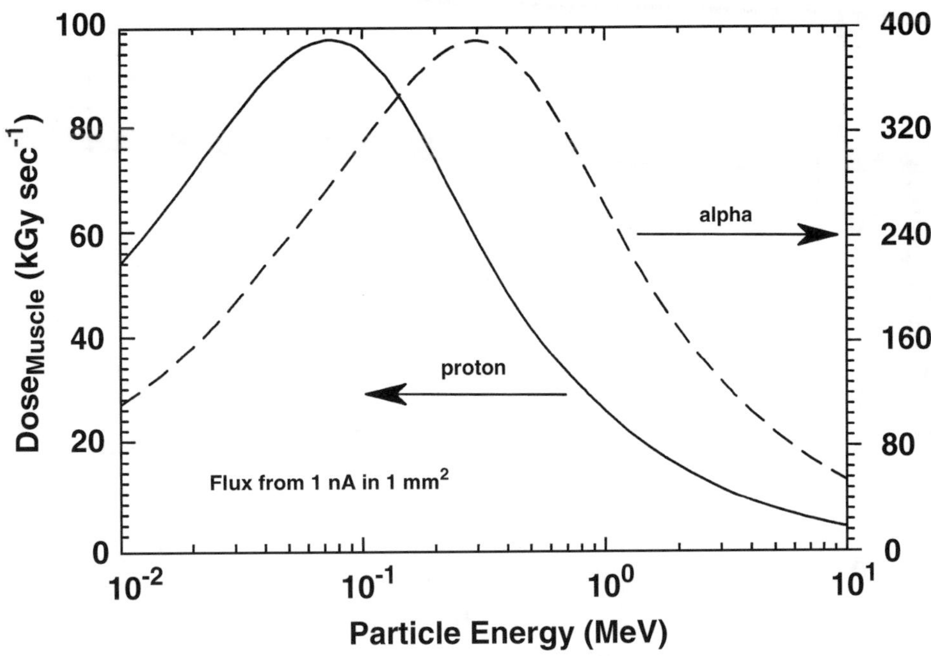

FIG. 13.3. Values of the absorbed dose rate for protons and alpha particles as a function of particle energy for interactions with muscle tissue. Values correspond to a particle flux rate of 0.6×10^{12} s^{-1} cm^{-2}.

Passive instruments include the so-called film badge and the thermoluminescent dosimeters (TLD) of the 'ring' or 'body' type. TLD dosimeters will be discussed in detail here. Such devices are usually supplied as a means of monitoring long-term (monthly) exposures to gamma, beta, and neutron radiation. Their advantages include ease of readout (by photomultiplier tubes, see Chapter 11), reasonable sensitivity, long-term stability, and environmental ruggedness. They do, however, have some disadvantages such as insensitivity to low levels of radiation, perhaps 10 mrem per month, poor sensitivity to beta radiation, and a neutron response that is not proportional to either absorbed dose or dose equivalent. Neutron response is accomplished by using a lithium-fluoride TLD element. Neutrons are moderated by the body and diffuse back into the TLD element where the ^{6}Li(n,α) reaction produces energy depositions in the TLD element. Clearly the TLD response is sensitive to the neutron energy impinging on the individual, hence requiring a knowledge of the bombarding neutron spectrum. Generally the neutron albedo from the body decreases with neutron bombarding energy and becomes very small above a few MeV neutron energy. Hence the LiF TLD neutron response falls in a similar manner.

Recently, thermoplastic materials, usually CR-39, have served as neutron-absorbed dosimeters. These materials record the radiation-absorbed dose by using an etching technique to display damage to the plastic produced by neutron-induced, heavy-charged particle recoils. Their neutron sensitivity exhibits an effective threshold of about 0.5 MeV with good response to 10s of MeV. Some suppliers of personnel dosimeters now combine track-etch materials and TLDs in the same package to more completely record the neutron-, photon-, and beta-absorbed dose. The LiF TLD neutron albedo response provides coverage below 1 MeV, while the track-etch detector extends the neutron response to 10-20 MeV.

The other common types of detectors are survey instruments that consist of a radiation-sensing element, sensitive to a particular type of radiation, connected to electronics which convert the signals from the detector to readings on the instrument. These detectors are usually much more sensitive to radiation than the dosimeters discussed above.

Portable survey detectors are obviously the ones you can carry around. Instruments are commercially available for detecting all of the types of radiation produced

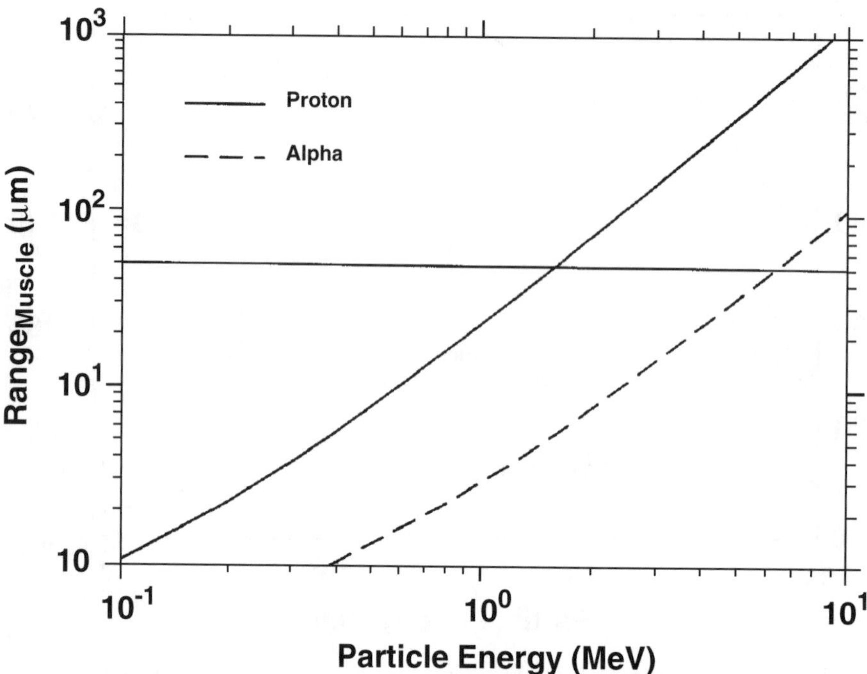

FIG. 13.4. Values of the range of protons and α particles as a function of particle energy for muscle tissue.

by an accelerated beam. Many detectors are calibrated to give readings in dose equivalent units such as rem or sievert, (Sv), while others indicate in absorbed dose units such as rad or gray (Gy). Other detectors are calibrated in counts per min (cpm), which is useful for measuring low levels of radiation such as induced activities near background. These detectors cannot, however, tell you directly the type of radiation being measured. With a little experimentation and a knowledge of the ranges of emitted particles an educated guess can be made.

Portable detectors are usually only calibrated for a certain type and energy range of radiation. Various publications give standards for the calibration and use of detectors, and the collection of pamphlets published by the National Council on Radiation Protection is an excellent source. If a detector is not routinely checked and calibrated it is of limited use and may be illegal in the United States. Training in the selection and use of a detector is extremely helpful if not a requirement, and many institutions have personnel whose responsibility is to assure the appropriate use and calibration of detectors as well as radiation generating equipment. These are usually the same people who maintain the dosimeters.

The two most common types of detectors used in accelerator laboratories are neutron and combination x- and gamma-ray detectors. Most neutron detectors give a response per unit neutron fluence that is proportional to dose equivalent. This is accomplished by surrounding a thermal-neutron-sensitive element with moderating material. For an appropriate choice of thickness of moderating material the response of the instrument per unit neutron fluence is dose-equivalent for a considerable range of neutron energy. These detectors are easily identified by their characteristic large round or cylindrical polyethylene moderator which houses the sensing detector. Because of the neutron moderation process and the practical limit to the size of the moderator, these detectors are not very efficient and have a slow response time. Unless a passive thermal neutron detector is used, they are only useful when neutrons are continuously generated, i.e., they do not work in the presence of pulsed neutron sources. Some thermal neutron-sensing-detectors are:

- LiI crystal, which uses the ^{6}Li(n,α) reaction to produce light pulses observed by a photomultiplier tube and counting system;
- a boron-trifluoride-filled proportional counter, which uses the ^{10}B(n,α) reaction to generate charge pulses; and
- Au foil, which uses the ^{197}Au(n,γ)^{198}Au reaction to produce activity that is subsequently measured by conventional γ-ray analysis.

For the active detectors, neutron dose equivalent rates near 0.2 μSv per hour can be determined.

Most x- or γ-ray dosimeters use gas ionization for detection of the radiation. The gas-filled cavity is assumed to operate as a Bragg-Grey cavity so that the energy absorbed by the gas is in radiation equilibrium with the walls and the response is proportional to absorbed dose in the wall material. If the wall material is tissue-like, then the response is directly in units of tissue absorbed dose. Greater sensitivity is possible by using a pressurized gas cavity. Some common type photon detectors are equipped with a moving metal shield covering part of the wall material. When the shield is removed, a much thinner wall is exposed allowing energetic beta particles and low energy photons to penetrate into the cavity. The source of radiation is measured with the shield in place and removed. The difference in the measurements is an indication of the energy or type of radiation being detected if the range of the radiation is known. For example, the shield would absorb low-energy γ- or x-rays, but would pass higher energy gammas.

Area detectors, sometimes called area monitors, are fixed-position detectors similar to the portable detectors, and continuously monitor the radiation at specified locations. These are commonly used as part of the accelerator safety interlock system which is designed to stop beam production if a preset level of radiation is exceeded. If your accelerator is capable of beam energies sufficient for elastic resonance backscattering, then such a system is a very worthwhile investment. Moreover, a survey of the area will indicate if such detectors are needed. The definitions of radiological areas, required posting, and the types of protective systems required in the U.S. are described in 10CFR835, 1993.

13.7 CONCLUSION

In today's environment, a careful evaluation of the hazards to personnel and the environment associated with accelerator operation is a requirement (and there certainly are other hazards in addition to radiation). The consideration of sample activation as well as prompt radiation must be addressed. Particular attention should be given to abiding by regulations, orders, and laws, which seem to be changing daily. The concept of ALARA, As Low As Reasonably Achievable, must also be taken into consideration. If the evaluation (including credible accidents) shows the hazards from radiation to be of sufficient magnitude, then radiation interlock systems by themselves may not be adequate, and some type of access control or shielding may also be necessary to keep personnel safe. It is good practice to keep a log book recording beam species, energies, and target combinations in addition to any prompt radiation or sample activation that is detected. This may be used as a reference for future analyses and for indicating when a new beam-sample combination may need to be monitored for possible radiation production.

REFERENCES

Andersen, H.H. and Ziegler, J.F. (1977), *Hydrogen: Stopping Powers and Ranges in All Elements.* Volume 3 of *The Stopping and Ranges of Ions in Matter,* Pergammon, New York.

10CFR20 (Code of Federal Regulations, Title 10, Part 20). (1993), United States Nuclear Regulatory Commission Rules and Regulations of Standards for Protection Against Radiation, Office of the Federal Register, National Archives and Records Administration, Washington, D.C.

10CFR835 (Code of Federal Regulations, Title 10, Part 835). (1993), United States Department of Energy Rules and Regulations of Occupational Radiation Protection, Office of the Federal Register, National Archives and Records Administration, Washington, D.C.

DOE 5480.25. (1992), Safety of Accelerator Facilities, U.S. Department of Energy, Office of Energy Research, Washington, D.C.

Doyle, B.L., Walsh, D.S., and Lee, S.R. (1991), *Nucl. Instrum. Methods* **B54**, 244.

Drosg, M. (1992), IAEA-NDS-87 Computer Code DROSG-87. Technical Report, Institute for Experimental Physics, University of Wein, Vienna, Austria.

ICRU. (1992), *Quantities and Units in Radiation Protection Dosimetry*, Report 51, International Commission on Radiation Units and Measurements, Bethesda, MD.

ICRU. (1980), *Radiation Quantities and Units*, Report 33, International Commission on Radiation Units and Measurements, Bethesda, MD.

ICRP. (1974), *Report of the Task Group on Reference Man*. ICRP Publication 23 of International Commission on Radiological Protection, Pergamon Press, Oxford, UK.

NCRP. (1987), *Recommendations on Limits for Exposure to Ionizing Radiation*, Report No. 91, National Council on Radiation Protection and Measurements, Bethesda, MD.

APPENDIX 1
ELEMENTS

Compiled by
M. Nastasi

Los Alamos National Laboratory, Los Alamos, New Mexico

J.C. Barbour

Sandia National Laboratories, Albuquerque, New Mexico

J.R. Tesmer

Los Alamos National Laboratory, Los Alamos, New Mexico

Table A1. The elements.

Element	Atomic number *(Z)*	Isotopic mass *(amu)*	Relative abundance	Atomic weight *(amu)*	Atomic density *(atoms/cm³)*	Mass density *(grams/cm³)*
H	1	1.0078 2.0141 3.0160*	99.985 0.015	1.00794		
He	2	3.0160 4.0026	0.0001 99.9999	4.002602		
Li	3	6.0151 7.0160	7.5 92.5	6.941	4.60E+22	0.53
Be	4	7.0169* 9.0122 10.0135*	100	9.01282	1.24E+23	1.85
B	5	10.0129 11.0093	19.9 80.1	10.811	1.30E+23	2.34
C	6	12.0000 13.0034 14.0032*	98.9 1.1	12.011	1.31E+23	2.62
N	7	14.0031 15.0001	99.63 0.37	14.00674	5.38E+22	1.251

Table A1. The elements (continued).

Element	Atomic number (Z)	Isotopic mass (amu)	Relative abundance	Atomic weight (amu)	Atomic density ($atoms/cm^3$)	Mass density ($grams/cm^3$)
O	8	15.9949	99.762	15.9994	5.38E+22	1.429
		16.9991	0.038			
		17.9992	0.2			
F	9	18.9984	100	18.9984032	5.38E+22	1.696
Ne	10	19.9924	90.51	20.1797	2.69E+22	0.901
		20.9938	0.27			
		21.9914	9.22			
Na	11	21.9944*				
		22.9898	100	22.989768	2.54E+22	0.97
Mg	12	23.9850	78.99	24.305	4.31E+22	1.74
		24.9858	10.			
		25.9826	11.01			
Al	13	25.9869*				
		26.9815	100	26.981539	6.03E+22	2.70
Si	14	27.9769	92.23	28.0855	5.00E+22	2.33
		28.9765	4.67			
		29.9738	3.1			
		31.9741*				
P	15	30.9738	100	30.973762	3.54E+22	1.82
		31.9739*				
		32.9717*				
S	16	31.9721	95.02	32.066	3.89E+22	2.07
		32.9715	0.75			
		33.9679	4.21			
		34.9690*				
		35.9671	0.02			
Cl	17	34.9689	75.77	35.4527	5.38E+22	3.17
		35.9683*				
		36.9659	24.23			
Ar	18	35.9675	0.337	39.948	2.69E+22	1.784
		36.9668*				
		37.9627	0.063			
		38.9643*				
		39.9624	99.6			
K	19	38.9637	93.2581	39.0983	1.32E+22	0.86
		39.9624	0.0117			
		40.9618	6.7302			
Ca	20	39.9626	96.941	40.078	2.33E+22	1.55
		40.9623*				
		41.9586	0.647			
		42.9588	0.135			
		43.9555	2.086			
		44.9562*				
		45.9537	0.004			
		46.9546*				
		47.9525	0.187			
Sc	21	44.9559	100	44.95591	4.02E+22	3.0

Table A1. The elements (continued).

Element	Atomic number (*Z*)	Isotopic mass (*amu*)	Relative abundance	Atomic weight (*amu*)	Atomic density (*atoms/cm³*)	Mass density (*grams/cm³*)
Ti	22	45.9526	8.0	47.88	5.66E+22	4.50
		46.9518	7.3			
		47.9479	73.8			
		48.9479	5.5			
		49.9448	5.4			
V	23	49.9472	0.25	50.9415	6.86E+22	5.8
		50.9440	99.75			
Cr	24	49.9461	4.35	51.9961	8.33E+22	7.19
		51.9405	83.79			
		52.9407	9.5			
		53.9389	2.36			
Mn	25	54.9380	100	54.93805	8.14E+22	7.43
Fe	26	53.9396	4.35	55.847	8.44E+22	7.83
		55.9349	83.79			
		56.9354	9.5			
		57.9333	2.36			
Co	27	58.9332	100	58.9332	9.09E+22	8.90
Ni	28	57.9353	68.27	58.69	9.13E+22	8.90
		59.9308	26.1			
		60.9311	1.13			
		61.9283	3.59			
		63.9280	0.91			
Cu	29	62.9296	69.17	63.546	8.49E+22	8.96
		64.9278	30.83			
Zn	30	63.9291	48.6	65.39	6.58E+22	7.14
		65.9260	27.9			
		66.9271	4.1			
		67.9248	18.8			
		69.9253	0.6			
Ga	31	68.9256	60.1	69.723	5.10E+22	5.91
		70.9247	39.9			
Ge	32	69.9243	20.5	72.61	4.41E+22	5.32
		71.9221	27.4			
		72.9235	7.8			
		73.9212	36.5			
		75.9214	7.8			
As	33	74.9216	100	74.92159	4.60E+22	5.72
Se	34	73.9225	0.9	78.96	3.66E+22	4.80
		75.9192	9.			
		76.9199	7.6			
		77.9173	23.5			
		79.9165	49.6			
		81.9167	9.4			
Br	35	78.9183	50.69	79.904	2.35E+22	3.12
		80.9163	49.31			

Table A1. The elements (continued).

Element	Atomic number *(Z)*	Isotopic mass *(amu)*	Relative abundance	Atomic weight *(amu)*	Atomic density *(atoms/cm^3)*	Mass density *(grams/cm^3)*
Kr	36	77.9204	0.35	83.8	2.69E+22	3.74
		79.9164	2.25			
		81.9135	11.6			
		82.9141	11.5			
		83.9115	57.			
		85.9106	17.3			
Rb	37	84.9118	72.17	85.4678	1.08E+22	1.53
		86.9092	27.83			
Sr	38	83.9134	0.56	87.62	1.79E+22	2.6
		85.9093	9.86			
		86.9089	7.00			
		87.9056	82.58			
Y	39	88.9059	100	88.90585	3.05E+22	4.5
Zr	40	89.9047	51.45	91.224	4.28E+22	6.49
		90.9056	11.27			
		91.9050	17.17			
		93.9063	17.33			
		95.9083	2.78			
Nb	41	92.9064	100	92.90635	5.54E+22	8.55
Mo	42	91.9068	14.84	95.94	6.40E+22	10.2
		93.9051	9.25			
		94.9058	15.92			
		95.9047	16.68			
		96.9060	9.55			
		97.9054	24.13			
		99.9075	9.63			
Tc	43	98	100			11.5
Ru	44	95.9076	5.52	101.07	7.27E+22	12.2
		97.9053	1.88			
		98.5059	12.7			
		99.9042	12.6			
		100.9056	17.0			
		101.9043	31.6			
		103.9054	18.7			
Rh	45	102.9055	100	102.9055	7.26E+22	12.4
Pd	46	101.9056	1.02	106.42	6.79E+22	12.0
		103.9040	11.14			
		104.9051	22.33			
		105.9035	27.33			
		107.9039	26.46			
		109.9052	11.72			
Ag	47	106.9051	51.84	107.8682	5.86E+22	10.5
		108.9048	48.16			

Table A1. The elements (continued).

Element	Atomic number (Z)	Isotopic mass (amu)	Relative abundance	Atomic weight (amu)	Atomic density (atoms/cm^3)	Mass density (grams/cm^3)
Cd	48	105.9065	1.25	112.411	4.63E+22	8.65
		107.9042	0.89			
		109.9030	12.49			
		110.9042	12.80			
		111.9028	24.13			
		112.9044	12.22			
		113.9034	28.73			
		115.9048	7.49			
In	49	112.9041	4.3	114.82	3.83E+22	7.31
		114.9039	95.7			
Sn	50	111.9048	1.			
		113.9028	0.7	118.71	3.70E+22	7.30
		114.9033	0.4			
		115.9017	14.7			
		116.9030	7.7			
		117.6016	24.3			
		118.9033	8.6			
		119.9022	32.4			
		121.9034	4.6			
		123.9053	5.6			
Sb	51	120.9038	57.3	121.75	3.30E+22	6.68
		122.9042	42.7			
Te	52	119.9041	0.096	127.6	2.94E+22	6.24
		121.9031	2.6			
		122.9043	0.908			
		123.9028	4.816			
		124.9044	7.14			
		125.9033	18.95			
		127.9045	31.69			
		129.9062	33.8			
I	53	126.9045	100	126.90447	2.33E+22	4.92
Xe	54	123.9061	0.10	131.29	2.70E+22	5.89
		125.9043	0.09			
		127.9035	1.91			
		128.9048	26.40			
		129.9035	4.10			
		130.9051	21.20			
		131.9041	26.90			
		133.9054	10.40			
		135.9072	8.90			
Cs	55	132.9054	100	132.90543	8.47E+21	1.87
Ba	56	129.9063	0.106	137.327	1.53E+22	3.5
		131.9050	0.101			
		133.9045	2.417			
		134.9057	6.592			
		135.9046	7.854			
		136.9058	11.230			
		137.9052	71.700			
La	57	137.9071	0.09	138.9055	2.90E+22	6.7
		138.9063	99.91			

Table A1. The elements (continued).

Element	Atomic number (*Z*)	Isotopic mass (*amu*)	Relative abundance	Atomic weight (*amu*)	Atomic density (*atoms/cm*^{3})	Mass density (*grams/cm*^{3})
Ce	58	135.9071	0.19	140.115	2.91E+22	6.78
		137.9060	0.25			
		139.9054	88.48			
		141.9092	11.08			
Pr	59	140.9077	100	140.90765	2.89E+22	6.77
Nd	60	141.9077	27.13	144.24	2.92E+22	7.00
		142.9098	12.18			
		143.9101	23.80			
		144.9126	8.30			
		145.9131	17.19			
		147.9169	5.76			
		149.9209	5.64			
Pm	61	145	100	(145)	2.69E+22	6.475
Sm	62	143.9120	3.1	150.36	3.02E+22	7.54
		146.9149	15.0			
		147.9148	11.3			
		148.9172	13.8			
		149.9173	7.4			
		151.9197	26.7			
		153.9222	22.7			
Eu	63	150.9198	47.8	151.965	2.08E+22	5.26
		152.9212	52.2			
Gd	64	151.9198	0.2	157.25	3.02E+22	7.89
		153.9209	2.18			
		154.9226	14.8			
		155.9221	20.47			
		156.9240	15.65			
		157.9241	24.84			
		159.9271	21.86			
Tb	65	158.9253	100	158.92534	3.13E+22	8.27
Dy	66	155.9243	0.06	162.5	3.16E+22	8.53
		157.9244	0.1			
		159.9252	2.34			
		160.9269	18.9			
		161.9268	25.5			
		162.9287	24.9			
		163.9292	28.2			
Ho	67	164.9303	100	164.93032	3.21E+22	8.80
Er	68	161.9288	0.14	167.26	3.26E+22	9.05
		163.9292	1.61			
		165.9303	33.6			
		166.9320	22.95			
		167.9324	26.8			
		169.9355	14.9			
Tm	69	168.9342	100	168.93421	3.33E+22	9.33

Table A1. The elements (continued).

Element	Atomic number (Z)	Isotopic mass (amu)	Relative abundance	Atomic weight (amu)	Atomic density (atoms/cm^3)	Mass density (grams/cm^3)
Yb	70	167.9339	0.13	173.04	2.43E+22	6.98
		169.9348	3.05			
		170.9363	14.30			
		171.9364	21.90			
		172.9382	16.12			
		173.9389	31.80			
		175.9426	12.70			
Lu	71	174.9408	97.4	174.967	3.39E+22	9.84
		175.9427	2.6			
Hf	72	173.9400	0.16	178.49	4.42E+22	13.1
		175.9414	5.20			
		176.9432	18.60			
		177.9437	27.10			
		178.9458	13.74			
		179.9465	35.20			
Ta	73	179.9475	0.012	180.9479	5.52E+22	16.6
		180.9480	99.988			
W	74	179.9467	0.13	183.85	6.32E+22	19.3
		181.9482	26.30			
		182.9502	14.30			
		183.9509	30.67			
		185.9544	28.60			
Re	75	184.9530	37.4	186.207	6.79E+22	21.0
		186.9557	62.6			
Os	76	183.9525	0.02	190.2	7.09E+22	22.4
		185.9538	1.58			
		186.9557	1.6			
		187.9558	13.3			
		188.9581	16.1			
		189.9584	26.4			
		191.9615	41.0			
Ir	77	190.9606	37.3	192.22	7.05E+22	22.5
		192.9629	62.7			
Pt	78	189.9599	0.01	195.08	6.61E+22	21.4
		191.9610	0.79			
		193.9627	32.9			
		194.9650	33.8			
		195.9650	25.3			
		197.9679	7.2			
Au	79	196.9666	100	196.96654	5.90E+22	19.3
Hg	80	195.9658	0.15	200.59	4.06E+22	13.53
		197.9668	10.1			
		198.9683	17.			
		199.9683	23.1			
		200.9703	13.2			
		201.9706	29.65			
		203.9735	6.8			
Tl	81	202.9723	29.524	204.3833	3.49E+22	11.85
		204.9744	70.476			

Table A1. The elements (continued).

Element	Atomic number (*Z*)	Isotopic mass (*amu*)	Relative abundance	Atomic weight (*amu*)	Atomic density (*atoms/cm*^{3})	Mass density (*grams/cm*^{3})
Pb	82	203.9730 205.9745 206.9759 207.9767	1.4 24.1 22.1 52.4	207.2	3.31E+22	11.4
Bi	83	208.9804	100	208.98037	2.82E+22	9.8
Po	84	208.9824	100	(209)	2.71E+22	9.4
At	85	210	100	(210)	0.00E+00	
Rn	86	222.0176	100	(222)	2.69E+22	9.91
Fr	87	223	100	(223)	0.00E+00	
Ra	88	226.0254	100	226.0254	1.33E+22	5.
Ac	89	227	100	227.0278	2.67E+22	10.07
Th	90	232.0381	100	232.0381	3.04E+22	11.7
Pa	91	231.0359	100	231.0359	4.01E+22	15.4
U	92	234.0410 235.0439 238.0400	0.0055 0.7200 99.2745	238.0289	4.78E+22	18.9
Np	93	237.0482	100	237.0482	5.18E+22	20.4
Pu	94	244.0642	100	(244)	4.89E+22	19.8
Am	95	243.0614	100	(243)	3.37E+22	13.6

* Radioactive, unstable isotopes of atomic number 20 or less which may be produced by nuclear reactions with analysis beams.
Data from *Nuclides and Isotopes*, revised 1989 by F. W. Walker, J.R. Parrington, and F. Feiner, published by the General Electric Company, San Jose, California.

APPENDIX 2

Physical Constants, Conversions, and Useful Combinations

Compiled by
J. R. Tesmer

Los Alamos National Laboratory, Los Alamos, New Mexico

Constants

Avogadro constant	$N_A = 6.022 \times 10^{23}$	atoms (molecules)/mole
Gas constant	$R = 8.314$	J/mole/°K
Boltzmann constant	$k = 8.617 \times 10^{-5}$	eV/°K
Planck constant	$h = 6.626 \times 10^{-34}$	J s
	$= 4.136 \times 10^{-21}$	MeV s
	$\hbar = 6.582 \times 10^{-16}$	eV s
Speed of light	$c = 2.998 \times 10^{8}$	meter/s
Electronic charge	$e = 1.602 \times 10^{-19}$	C
Barn	$b = 1 \times 10^{-24}$	cm^2
Electron rest mass	$m_e = 5.110 \times 10^{-1}$	MeV
	$= 9.109 \times 10^{-31}$	kg
Proton rest mass	$m_p = 9.383 \times 10^{2}$	MeV
	$= 1.673 \times 10^{-27}$	kg
Deuteron rest mass	$m_d = 1.876 \times 10^{3}$	MeV
	$= 3.344 \times 10^{-27}$	kg
Alpha particle rest mass	$m_\alpha = 3.727 \times 10^{3}$	MeV
	$= 6.645 \times 10^{-27}$	kg

Combinations

Bohr radius	$a_0 = \hbar^2/2me^2 = 5.292 \times 10^{-2}$ nm
Bohr velocity	$v_0 = e^2/\hbar = 2.188 \times 10^{6}$ m/s
Fine structure constant	$\alpha = e^2/\hbar c = 7.297 \times 10^{-3}$
Hydrogen binding energy	$e^2/2a_0 = 13.606$ eV
Classical electron radius	$r_e = e^2/mc^2 = 2.818 \times 10^{-6}$ nm
Electronic charge	$e^2 = 1.440$ eV nm
	$= 1.440 \times 10^{-13}$ MeV cm

Conversions

1 Å	=	10^{-8} cm
	=	10^{-1} nm
1 μm	=	1×10^4 Å
1 Ci	=	3.700×10^{10} Bq
	=	3.700×10^{10} decays/s

electrons/Coulomb	=	6.242×10^{18}
Speed of 1 MeV alpha		7×10^6 m/s
1eV	=	1.602×10^{-19} J
1 ev/particle	=	23.06 kcal/mol
	=	96.53 kJ/mol

APPENDIX

3

STOPPING AND RANGE

Compiled by

E. Rauhala

University of Helsinki, Helsinki, Finland

CONTENTS

A3.1 TABLES OF ELECTRONIC AND NUCLEAR STOPPING FOR PROTONS AND ^{4}HE

Table A3.1.1. Proton electronic stopping cross sections [eV cm^2/10^{15} atoms].

		E_p (keV)																		
Z_2	At	100	200	300	400	500	750	1000	1250	1500	2000	2500	3000	4000	5000	6000	7000	8000	9000	10000
1	H	4.690	3.924	3.028	2.402	1.987	1.415	1.125	0.945	0.821	0.656	0.549	0.474	0.374	0.309	0.264	0.231	0.205	0.185	0.168
2	He	6.123	5.559	4.570	3.789	3.225	2.374	1.906	1.607	1.396	1.116	0.935	0.808	0.638	0.530	0.455	0.399	0.356	0.322	0.294
3	Li	8.120	7.066	5.784	4.855	4.206	3.224	2.662	2.287	2.014	1.638	1.388	1.208	0.965	0.807	0.696	0.613	0.548	0.497	0.454
4	Be	10.044	7.794	6.430	5.559	4.940	3.929	3.296	2.854	2.525	2.064	1.754	1.530	1.226	1.028	0.888	0.783	0.702	0.636	0.583
5	B	13.653	11.049	8.777	7.322	6.348	4.890	4.045	3.476	3.060	2.488	2.107	1.834	1.465	1.226	1.058	0.932	0.835	0.757	0.693
6	C	14.428	12.225	9.879	8.278	7.185	5.546	4.602	3.969	3.506	2.866	2.439	2.131	1.714	1.441	1.248	1.103	0.991	0.900	0.826
7	N	16.222	14.448	11.858	9.843	8.406	6.291	5.154	4.427	3.911	3.209	2.742	2.404	1.943	1.640	1.423	1.260	1.133	1.030	0.946
8	O	16.145	14.592	12.240	10.377	9.016	6.925	5.737	4.953	4.385	3.602	3.079	2.700	2.183	1.843	1.600	1.418	1.276	1.161	1.067
9	F	13.434	13.138	11.840	10.510	9.377	7.392	6.170	5.343	4.740	3.904	3.343	2.934	2.374	2.004	1.739	1.540	1.384	1.259	1.155
10	Ne	14.373	13.784	12.249	10.866	9.738	7.753	6.483	5.601	4.951	4.049	3.450	3.019	2.436	2.056	1.787	1.586	1.429	1.302	1.198
11	Na	20.975	17.400	14.724	12.861	11.477	9.151	7.680	6.652	5.890	4.826	4.114	3.600	2.904	2.450	2.129	1.888	1.700	1.549	1.425
12	Mg	20.700	17.582	14.840	12.842	11.383	9.035	7.605	6.618	5.885	4.854	4.155	3.646	2.949	2.490	2.162	1.916	1.724	1.569	1.441
13	Al	20.460	16.771	14.302	12.611	11.347	9.185	7.786	6.794	6.048	4.994	4.280	3.760	3.049	2.581	2.248	1.997	1.801	1.643	1.513
14	Si	26.428	20.790	16.690	14.173	12.473	9.845	8.271	7.192	6.397	5.288	4.542	4.002	3.264	2.778	2.430	2.168	1.963	1.797	1.660
15	P	27.108	22.980	18.245	15.256	13.308	10.437	8.773	7.640	6.804	5.634	4.845	4.270	3.482	2.962	2.589	2.308	2.087	1.909	1.761
16	S	27.706	23.767	18.895	15.424	13.095	9.856	8.182	7.120	6.361	5.313	4.603	4.081	3.355	2.867	2.515	2.246	2.034	1.861	1.718
17	Cl	32.424	27.755	22.077	18.046	15.339	11.560	9.596	8.346	7.453	6.219	5.385	4.773	3.922	3.351	2.939	2.625	2.377	2.176	2.009
18	Ar	29.803	26.252	21.181	17.660	15.307	11.913	10.024	8.762	7.835	6.533	5.647	4.996	4.095	3.494	3.061	2.733	2.474	2.264	2.091
19	K	32.604	27.364	22.806	19.560	17.187	13.360	11.067	9.523	8.405	6.879	5.876	5.159	4.193	3.564	3.118	2.784	2.522	2.311	2.137
20	Ca	27.434	25.497	22.298	19.429	17.137	13.353	11.140	9.685	8.638	7.197	6.228	5.520	4.537	3.879	3.404	3.041	2.755	2.523	2.330
21	Sc	32.520	28.996	24.330	20.629	17.902	13.702	11.354	9.829	8.738	7.245	6.248	5.525	4.529	3.868	3.392	3.031	2.747	2.516	2.325
22	Ti	30.923	27.425	23.139	19.970	17.671	13.996	11.771	10.241	9.110	7.528	6.462	5.687	4.627	3.930	3.432	3.056	2.762	2.524	2.328
23	V	32.362	30.449	25.685	21.772	18.900	14.507	12.043	10.434	9.277	7.691	6.632	5.865	4.812	4.115	3.614	3.234	2.935	2.693	2.492
24	Cr	29.221	26.042	22.612	19.922	17.842	14.299	12.065	10.515	9.367	7.767	6.691	5.911	4.845	4.142	3.638	3.257	2.957	2.715	2.514
25	Mn	27.737	25.590	22.441	19.866	17.845	14.366	12.153	10.611	9.466	7.864	6.784	6.000	4.926	4.215	3.706	3.320	3.017	2.770	2.566
26	Fe	29.619	26.554	22.962	20.220	18.128	14.588	12.358	10.806	9.653	8.039	6.949	6.156	5.066	4.344	3.825	3.431	3.121	2.869	2.660
27	Co	27.314	25.488	22.394	19.882	17.917	14.526	12.358	10.837	9.702	8.105	7.022	6.231	5.141	4.416	3.894	3.498	3.184	2.930	2.718
28	Ni	25.010	23.114	20.982	19.112	17.532	14.579	12.557	11.087	9.968	8.366	7.266	6.456	5.335	4.587	4.047	3.637	3.312	3.049	2.830
29	Cu	23.737	22.858	20.672	18.811	17.299	14.528	12.616	11.204	10.112	8.524	7.416	6.593	5.445	4.674	4.116	3.691	3.356	3.083	2.856
30	Zn	24.168	23.234	21.369	19.515	17.912	14.910	12.869	11.392	10.268	8.657	7.548	6.729	5.589	4.823	4.268	3.844	3.508	3.234	3.005
31	Ga	25.262	24.769	22.350	19.999	18.093	14.863	12.848	11.431	10.355	8.798	7.704	6.886	5.728	4.940	4.365	3.924	3.574	3.288	3.049
32	Ge	27.976	25.582	22.637	20.329	18.570	15.570	13.602	12.165	11.051	9.413	8.252	7.377	6.137	5.291	4.672	4.197	3.820	3.512	3.255

Table A3.1.1 (continued).

Z_2	At	E_p (keV) 100	200	300	400	500	750	1000	1250	1500	2000	2500	3000	4000	5000	6000	7000	8000	9000	10000
33	As	32.061	27.841	23.789	20.865	18.759	15.397	13.326	11.865	10.754	9.145	8.017	7.172	5.977	5.163	4.568	4.111	3.747	3.450	3.202
34	Se	31.135	27.569	23.547	20.782	18.832	15.675	13.657	12.195	11.068	9.420	8.258	7.387	6.157	5.319	4.708	4.239	3.866	3.561	3.307
35	Br	33.566	30.712	25.425	21.985	19.770	16.489	14.466	12.986	11.827	10.099	8.858	7.918	6.577	5.661	4.990	4.475	4.066	3.732	3.454
36	Kr	35.989	32.580	27.175	23.330	20.725	16.866	14.608	13.030	11.829	10.079	8.843	7.912	6.590	5.688	5.027	4.519	4.115	3.785	3.509
37	Rb	40.881	34.760	29.727	26.070	23.329	18.744	15.874	13.883	12.407	10.340	8.945	7.930	6.535	5.610	4.945	4.439	4.041	3.717	3.449
38	Sr	41.169	37.473	31.631	26.871	23.439	18.390	15.654	13.857	12.537	10.656	9.342	8.355	6.954	5.995	5.292	4.752	4.322	3.971	3.678
39	Y	41.148	36.698	30.916	26.468	23.278	18.465	15.751	13.936	12.597	10.694	9.373	8.386	6.992	6.040	5.343	4.808	4.381	4.032	3.741
40	Zr	43.611	39.052	32.316	27.378	23.970	18.959	16.148	14.260	12.863	10.881	9.511	8.492	7.061	6.091	5.383	4.841	4.410	4.059	3.765
41	Nb	42.281	37.275	32.202	28.357	25.428	20.481	17.371	15.209	13.606	11.361	9.845	8.741	7.223	6.215	5.489	4.938	4.502	4.148	3.853
42	Mo	38.632	34.286	29.645	26.144	23.484	18.997	16.170	14.199	12.733	10.673	9.276	8.256	6.847	5.907	5.229	4.712	4.302	3.969	3.691
43	Tc	42.293	36.487	30.998	27.192	24.427	19.862	16.982	14.954	13.431	11.272	9.796	8.713	7.213	6.212	5.489	4.938	4.502	4.148	3.854
44	Ru	38.848	35.301	30.956	27.490	24.787	20.136	17.168	15.087	13.535	11.350	9.866	8.782	7.283	6.284	5.562	5.012	4.576	4.222	3.926
45	Rh	38.753	34.554	30.424	27.143	24.563	20.071	17.171	15.125	13.592	11.425	9.948	8.865	7.365	6.363	5.637	5.083	4.645	4.287	3.989
46	Pd	36.105	33.392	29.993	27.073	24.683	20.373	17.511	15.463	13.916	11.714	10.205	9.096	7.558	6.528	5.782	5.213	4.762	4.395	4.089
47	Ag	34.727	33.811	30.421	27.166	24.484	19.855	16.982	15.006	13.540	11.471	10.050	8.998	7.524	6.525	5.796	5.236	4.791	4.426	4.121
48	Cd	39.929	35.794	31.058	27.555	24.913	20.419	17.530	15.484	13.943	11.754	10.255	9.153	7.623	6.598	5.855	5.288	4.839	4.472	4.166
49	In	38.637	36.026	31.612	28.082	25.374	20.775	17.837	15.760	14.196	11.972	10.446	9.323	7.762	6.716	5.957	5.378	4.919	4.545	4.233
50	Sn	40.480	37.148	32.232	28.347	25.421	20.606	17.627	15.556	14.008	11.817	10.317	9.212	7.676	6.645	5.896	5.324	4.870	4.500	4.191
51	Sb	45.199	37.917	32.395	28.625	25.843	21.153	18.151	16.026	14.428	12.158	10.604	9.462	7.877	6.815	6.046	5.458	4.992	4.612	4.296
52	Te	43.369	39.074	34.022	29.898	26.733	21.564	18.467	16.356	14.790	12.569	11.031	9.885	8.267	7.164	6.354	5.731	5.235	4.828	4.488
53	I	49.486	42.827	35.234	30.235	26.896	21.892	18.923	16.849	15.279	13.014	11.430	10.246	8.574	7.435	6.601	5.961	5.450	5.032	4.683
54	Xe	49.208	42.905	35.610	30.706	27.390	22.364	19.354	17.241	15.636	13.314	11.686	10.467	8.746	7.573	6.715	6.056	5.531	5.102	4.743
55	Cs	51.609	44.552	38.322	33.723	30.247	24.391	20.703	18.135	16.225	13.546	11.734	10.412	8.593	7.384	6.514	5.852	5.330	4.906	4.553
56	Ba	56.156	47.269	39.113	33.731	30.012	24.198	20.675	18.231	16.404	13.817	12.044	10.740	8.926	7.709	6.826	6.153	5.619	5.183	4.820
57	La	54.859	46.788	39.659	34.522	30.723	24.496	20.686	18.079	16.166	13.512	11.735	10.446	8.680	7.508	6.665	6.023	5.515	5.102	4.757
58	Ce	52.058	43.997	35.740	30.722	27.529	22.835	19.988	17.937	16.346	13.992	12.311	11.039	9.225	7.981	7.067	6.364	5.803	5.344	4.961
59	Pr	48.431	41.662	35.991	32.055	29.115	24.064	20.760	18.388	16.586	14.004	12.223	10.907	9.074	7.842	6.948	6.266	5.724	5.283	4.915
60	Nd	47.221	41.423	35.650	31.643	28.703	23.744	20.531	18.226	16.472	13.950	12.204	10.910	9.101	7.881	6.993	6.314	5.775	5.334	4.966
61	Pm	46.506	42.441	36.754	32.395	29.188	23.991	20.764	18.483	16.751	14.251	12.504	11.200	9.361	8.112	7.198	6.496	5.937	5.481	5.099
62	Sm	45.281	41.887	37.534	33.901	30.941	25.578	21.989	19.409	17.455	14.671	12.764	11.364	9.426	8.132	7.198	6.486	5.924	5.466	5.085
63	Eu	39.845	37.121	33.428	30.324	27.870	23.589	20.754	18.672	17.049	14.643	12.918	11.608	9.731	8.437	7.483	6.746	6.158	5.676	5.273
64	Gd	43.820	40.931	35.835	31.836	28.858	23.953	20.849	18.629	16.932	14.465	12.732	11.433	9.595	8.340	7.421	6.712	6.148	5.685	5.298

Table A3.1.1 (continued).

Z_2	At	E_p (keV) 100	200	300	400	500	750	1000	1250	1500	2000	2500	3000	4000	5000	6000	7000	8000	9000	10000
65	Tb	41.490	38.528	34.093	30.706	28.141	23.715	20.769	18.608	16.934	14.481	12.749	11.448	9.607	8.350	7.429	6.720	6.155	5.692	5.304
66	Dy	41.907	38.331	34.049	30.848	28.380	23.994	21.012	18.813	17.108	14.613	12.857	11.542	9.684	8.420	7.494	6.783	6.215	5.751	5.362
67	Ho	37.767	34.464	30.894	28.250	26.198	22.486	19.896	17.947	16.415	14.140	12.514	11.283	9.528	8.320	7.431	6.743	6.192	5.740	5.361
68	Er	40.367	36.940	33.419	30.610	28.316	24.046	21.072	18.869	17.161	14.666	12.914	11.604	9.756	8.499	7.579	6.871	6.307	5.844	5.457
69	Tm	39.012	35.853	31.790	28.815	26.587	22.718	20.095	18.139	16.606	14.328	12.698	11.462	9.694	8.476	7.576	6.879	6.321	5.862	5.477
70	Yb	39.363	37.187	33.221	29.783	27.109	22.688	19.955	18.022	16.541	14.358	12.789	11.589	9.850	8.633	7.726	7.018	6.448	5.978	5.582
71	Lu	39.261	36.606	32.276	28.985	26.570	22.574	19.979	18.072	16.581	14.360	12.760	11.541	9.786	8.568	7.664	6.963	6.399	5.935	5.545
72	Hf	40.031	36.717	32.171	28.971	26.650	22.729	20.113	18.171	16.650	14.390	12.772	11.543	9.784	8.568	7.669	6.972	6.413	5.953	5.566
73	Ta	38.745	36.256	32.399	29.528	27.329	23.402	20.689	18.659	17.068	14.713	13.035	11.769	9.963	8.721	7.806	7.097	6.530	6.063	5.671
74	W	39.512	36.629	33.025	29.957	27.451	22.931	19.918	17.750	16.103	13.740	12.105	10.893	9.192	8.039	7.195	6.545	6.026	5.599	5.242
75	Re	38.186	35.761	32.598	29.957	27.756	23.622	20.743	18.614	16.967	14.564	12.877	11.614	9.829	8.610	7.714	7.022	6.468	6.013	5.631
76	Os	37.547	35.535	32.574	30.009	27.839	23.721	20.838	18.702	17.049	14.636	12.941	11.673	9.879	8.654	7.754	7.058	6.502	6.044	5.660
77	Ir	37.446	35.754	32.320	29.648	27.528	23.623	20.883	18.826	17.214	14.831	13.136	11.858	10.037	8.788	7.866	7.154	6.583	6.114	5.720
78	Pt	36.452	35.891	32.204	29.160	26.877	23.048	20.522	18.640	17.153	14.915	13.288	12.040	10.234	8.974	8.036	7.307	6.720	6.236	5.828
79	Au	35.401	36.506	34.037	31.101	28.511	23.889	20.968	18.921	17.367	15.094	13.465	12.217	10.404	9.131	8.178	7.434	6.833	6.336	5.917
80	Hg	37.220	38.229	34.633	31.034	28.227	23.715	20.975	19.025	17.513	15.254	13.610	12.345	10.501	9.209	8.244	7.492	6.885	6.385	5.964
81	Tl	41.050	39.960	35.338	31.599	28.848	24.386	21.555	19.496	17.891	15.504	13.785	12.474	10.584	9.271	8.297	7.540	6.931	6.430	6.008
82	Pb	44.030	42.516	37.926	33.715	30.407	25.039	21.859	19.682	18.045	15.664	13.965	12.667	10.784	9.465	8.478	7.708	7.086	6.572	6.139
83	Bi	49.414	46.138	40.496	35.669	31.972	26.074	22.628	20.291	18.547	16.035	14.258	12.908	10.961	9.603	8.591	7.803	7.167	6.643	6.202
84	Po	46.636	46.679	40.551	35.316	31.616	26.120	22.920	20.669	18.937	16.375	14.534	13.130	11.108	9.705	8.665	7.858	7.211	6.677	6.230
85	At	49.061	47.884	41.151	35.883	32.173	26.530	23.184	20.839	19.049	16.433	14.574	13.166	11.149	9.755	8.724	7.923	7.281	6.752	6.308
86	Rn	52.704	49.004	41.732	36.309	32.535	26.845	23.498	21.156	19.369	16.748	14.879	13.458	11.414	9.996	8.944	8.126	7.469	6.927	6.472
87	Fr	53.590	50.345	43.427	37.696	33.492	27.117	23.530	21.124	19.331	16.741	14.902	13.503	11.480	10.068	9.015	8.193	7.532	6.985	6.525
88	Ra	55.137	52.805	45.454	39.103	34.494	27.742	24.070	21.626	19.802	17.149	15.252	13.802	11.704	10.239	9.147	8.298	7.614	7.051	6.577
89	Ac	58.943	55.547	47.306	40.728	36.021	28.958	24.966	22.277	20.277	17.408	15.397	13.883	11.723	10.235	9.135	8.282	7.599	7.038	6.566
90	Th	54.502	52.441	45.462	39.538	35.160	28.443	24.609	22.019	20.088	17.311	15.353	13.874	11.752	10.283	9.192	8.346	7.666	7.105	6.634
91	Pa	51.092	50.540	44.945	39.570	35.397	28.809	24.983	22.375	20.417	17.582	15.576	14.057	11.878	10.372	9.257	8.392	7.699	7.128	6.650
92	U	49.883	47.584	41.398	36.707	33.311	27.815	24.349	21.862	19.952	17.165	15.196	13.714	11.605	10.157	9.088	8.262	7.599	7.055	6.597

Table A3.1.2. Helium electronic stopping cross sections [eV $cm^2/10^{15}$ atoms].

Z_2	At	E_{He} (keV) 100	200	300	400	500	750	1000	1250	1500	2000	2500	3000	4000	5000	6000	7000	8000	9000	10000
1	H	7.66	10.22	11.79	12.73	13.19	12.98	11.87	10.61	9.46	7.68	6.47	5.62	4.50	3.80	3.31	2.94	2.65	2.41	2.21
2	He	9.14	12.67	15.03	16.63	17.62	18.23	17.42	16.13	14.79	12.48	10.74	9.43	7.64	6.46	5.63	5.00	4.50	4.10	3.77
3	Li	13.51	17.72	20.37	22.04	22.97	23.22	22.02	20.44	18.89	16.29	14.32	12.82	10.67	9.20	8.12	7.28	6.61	6.06	5.60
4	Be	22.92	25.98	27.02	27.21	26.98	25.71	24.24	22.82	21.49	19.15	17.21	15.63	13.22	11.49	10.18	9.16	8.33	7.65	7.07
5	B	23.14	30.67	34.88	37.03	37.80	36.53	33.75	30.96	28.50	24.58	21.68	19.44	16.22	13.99	12.34	11.06	10.04	9.20	8.50
6	C	23.62	31.51	36.32	39.15	40.55	40.28	37.78	34.89	32.20	27.83	24.56	22.05	18.46	15.98	14.14	12.72	11.57	10.63	9.84
7	N	26.91	35.14	40.47	44.04	46.22	47.41	45.21	41.86	38.41	32.54	28.20	25.01	20.67	17.82	15.78	14.21	12.96	11.93	11.06
8	O	26.70	34.90	40.22	43.83	46.12	47.78	46.17	43.35	40.32	34.92	30.74	27.54	23.01	19.94	17.69	15.95	14.55	13.40	12.43
9	F	22.80	28.85	33.14	36.48	39.04	42.61	43.28	42.28	40.50	36.34	32.55	29.41	24.75	21.52	19.13	17.27	15.77	14.54	13.49
10	Ne	17.14	27.62	34.76	39.04	41.75	44.83	44.98	43.71	41.85	37.75	34.01	30.84	26.01	22.56	19.98	17.97	16.36	15.03	13.92
11	Na	40.46	50.00	54.65	56.87	57.74	57.14	54.99	52.36	49.63	44.49	40.08	36.41	30.81	26.79	23.77	21.40	19.50	17.93	16.61
12	Mg	44.33	50.33	53.88	56.14	57.42	57.67	55.61	52.70	49.66	44.11	39.59	35.95	30.52	26.66	23.75	21.47	19.62	18.08	16.78
13	Al	40.64	49.94	53.88	55.46	55.91	55.06	53.18	50.93	48.57	43.99	39.97	36.56	31.25	27.37	24.42	22.08	20.19	18.61	17.28
14	Si	41.84	59.22	68.03	71.66	72.32	68.78	63.71	59.03	54.95	48.33	43.22	39.17	33.20	28.98	25.83	23.36	21.37	19.73	18.35
15	P	45.21	57.35	67.07	73.59	76.88	75.92	70.21	64.36	59.32	51.56	45.89	41.53	35.21	30.79	27.47	24.88	22.78	21.04	19.57
16	S	45.49	60.17	69.42	75.19	78.30	78.33	73.02	66.48	60.35	50.69	43.98	39.20	32.84	28.70	25.69	23.37	21.49	19.92	18.60
17	Cl	53.23	70.50	81.30	88.00	91.56	91.48	85.27	77.69	70.59	59.38	51.57	45.98	38.52	33.64	30.11	27.37	25.15	23.32	21.76
18	Ar	41.68	60.44	72.95	80.93	85.34	86.41	81.18	74.73	68.76	59.29	52.49	47.41	40.25	35.32	31.65	28.77	26.43	24.47	22.82
19	K	47.14	72.33	83.39	88.44	90.54	89.93	85.90	80.88	75.78	66.59	59.12	53.15	44.41	38.38	33.94	30.53	27.82	25.59	23.74
20	Ca	50.77	61.58	68.97	74.47	78.44	83.07	82.55	79.37	75.12	66.40	59.01	53.13	44.72	39.04	34.90	31.70	29.12	26.98	25.18
21	Sc	57.46	72.38	81.89	88.26	92.26	94.97	91.73	86.18	80.16	69.34	60.92	54.51	45.58	39.62	35.30	31.98	29.31	27.11	25.26
22	Ti	49.40	66.70	77.30	83.93	87.78	89.87	86.81	82.14	77.26	68.49	61.40	55.71	47.27	41.29	36.81	33.30	30.45	28.10	26.12
23	V	45.77	64.91	78.40	87.91	94.15	99.53	96.76	90.98	84.61	73.21	64.41	57.72	48.35	42.07	37.49	33.95	31.12	28.78	26.81
24	Cr	43.02	66.19	74.38	79.29	82.40	85.08	83.70	80.57	76.79	69.18	62.50	56.93	48.46	42.41	37.86	34.30	31.43	29.05	27.05
25	Mn	40.06	60.07	69.09	75.30	79.39	83.45	82.76	80.03	76.51	69.20	62.68	57.20	48.82	42.80	38.26	34.70	31.83	29.44	27.43
26	Fe	43.68	61.90	73.48	80.40	84.19	86.82	85.12	81.79	77.93	70.29	63.63	58.09	49.65	43.59	39.03	35.44	32.54	30.14	28.11
27	Co	41.33	56.39	66.79	74.18	78.77	83.10	82.50	79.89	76.52	69.48	63.18	57.85	49.65	43.72	39.23	35.69	32.81	30.42	28.40
28	Ni	39.97	55.56	63.35	67.87	70.92	75.12	76.11	75.18	73.19	68.03	62.79	58.08	50.47	44.74	40.31	36.78	33.88	31.45	29.39
29	Cu	34.95	48.68	57.97	64.47	68.92	74.26	75.15	74.04	72.03	67.14	62.27	57.89	50.72	45.23	40.91	37.41	34.52	32.09	30.01
30	Zn	33.71	56.24	61.84	65.59	69.19	75.27	77.22	76.62	74.72	69.50	64.18	59.41	51.73	45.98	41.54	37.98	35.07	32.63	30.54
31	Ga	44.53	54.50	62.30	68.61	73.51	80.34	81.60	79.88	76.89	70.18	64.19	59.22	51.66	46.15	41.90	38.48	35.65	33.25	31.19
32	Ge	51.38	63.16	70.68	75.93	79.51	83.38	83.05	80.88	78.02	72.06	66.68	62.06	54.70	49.13	44.73	41.14	38.14	35.60	33.40

Table A3.1.2 (continued).

Z_2	At	E_{He} (keV) 100	200	300	400	500	750	1000	1250	1500	2000	2500	3000	4000	5000	6000	7000	8000	9000	10000
33	As	62.62	75.03	82.34	86.98	89.72	91.19	88.62	84.65	80.44	72.76	66.47	61.35	53.59	47.92	43.53	39.99	37.06	34.58	32.46
34	Se	49.33	66.74	77.69	84.51	88.35	90.32	87.64	83.84	79.98	73.06	67.29	62.48	54.93	49.26	44.81	41.19	38.18	35.63	33.44
35	Br	41.24	63.78	80.11	91.20	97.73	100.86	96.08	90.16	84.89	76.69	70.60	65.73	58.20	52.47	47.89	44.11	40.94	38.22	35.87
36	Kr	52.47	73.80	88.15	97.73	103.50	106.83	102.54	96.33	90.29	80.36	72.95	67.21	58.76	52.64	47.90	44.06	40.86	38.15	35.81
37	Rb	55.67	96.87	106.10	110.87	113.33	113.91	110.62	105.81	100.55	90.47	81.82	74.66	63.82	56.05	50.21	45.62	41.90	38.82	36.22
38	Sr	78.89	92.96	103.61	111.75	117.45	122.53	119.08	112.08	104.37	90.83	80.73	73.26	62.95	55.98	50.76	46.62	43.21	40.32	37.84
39	Y	73.67	91.90	103.69	111.68	116.72	120.17	116.25	109.63	102.60	90.23	80.77	73.56	63.34	56.30	51.01	46.81	43.36	40.46	37.97
40	Zr	66.95	91.49	107.71	118.41	124.79	128.13	122.50	114.36	106.30	92.91	83.00	75.53	64.94	57.61	52.09	47.71	44.12	41.11	38.53
41	Nb	76.22	94.78	107.18	114.73	119.01	121.91	119.39	114.71	109.30	98.62	89.33	81.60	69.85	61.43	55.08	50.10	46.06	42.71	39.88
42	Mo	55.02	82.87	96.85	104.85	109.22	112.13	109.86	105.62	100.73	91.09	82.69	75.70	65.03	57.36	51.56	46.99	43.28	40.19	37.58
43	Tc	65.75	91.79	106.60	114.76	118.77	119.68	115.62	110.29	104.83	94.75	86.23	79.16	68.31	60.42	54.40	49.62	45.71	42.46	39.69
44	Ru	46.62	82.85	97.38	105.44	110.42	115.18	114.09	110.44	105.80	96.15	87.53	80.25	69.05	60.96	54.82	49.97	46.03	42.76	39.98
45	Rh	45.89	84.58	98.81	105.14	108.90	112.74	111.86	108.62	104.36	95.30	87.04	80.00	69.08	61.12	55.06	50.26	46.35	43.08	40.31
46	Pd	68.64	84.55	91.99	97.98	102.57	108.61	109.38	107.31	103.87	95.79	88.01	81.22	70.46	62.49	56.38	51.51	47.52	44.19	41.36
47	Ag	56.37	73.46	85.37	94.32	100.95	109.75	111.17	108.70	104.51	94.98	86.33	79.14	68.33	60.66	54.87	50.29	46.55	43.42	40.74
48	Cd	57.80	83.68	99.21	108.39	113.51	117.03	114.85	110.74	106.02	96.66	88.39	81.40	70.54	62.59	56.50	51.66	47.70	44.39	41.57
49	In	56.85	80.10	95.02	104.92	111.27	117.47	116.53	112.77	108.05	98.45	89.97	82.82	71.78	63.71	57.53	52.61	48.59	45.22	42.35
50	Sn	64.87	86.65	100.60	109.90	115.86	121.28	119.46	114.80	109.30	98.61	89.56	82.14	70.94	62.89	56.77	51.93	47.96	44.65	41.83
51	Sb	73.40	101.80	116.03	122.59	125.12	124.41	120.38	115.40	110.20	100.27	91.62	84.34	73.05	64.80	58.48	53.46	49.35	45.92	43.00
52	Te	85.15	100.62	110.59	117.68	122.56	127.53	125.93	121.20	115.33	103.69	93.87	85.97	74.33	66.14	59.97	55.06	51.03	47.65	44.74
53	I	71.06	102.55	122.45	134.33	140.36	140.91	133.44	124.85	117.00	104.32	94.77	87.30	76.19	68.15	61.96	56.98	52.85	49.36	46.37
54	Xe	74.13	103.67	122.25	133.57	139.59	140.99	134.38	126.30	118.72	106.25	96.71	89.19	77.93	69.74	63.41	58.31	54.07	50.49	47.41
55	Cs	69.81	119.16	132.46	139.99	143.98	145.87	142.27	136.49	130.00	117.33	106.36	97.23	83.32	73.33	65.77	59.82	55.00	50.98	47.59
56	Ba	87.15	122.06	141.83	152.36	156.99	155.56	147.60	138.74	130.45	116.41	105.34	96.48	83.23	73.74	66.52	60.79	56.11	52.19	48.86
57	La	79.39	121.37	139.99	148.81	152.88	153.47	148.21	140.99	133.36	119.16	107.29	97.64	83.25	73.11	65.54	59.64	54.87	50.93	47.61
58	Ce	75.28	108.34	129.20	141.30	146.89	145.13	135.87	126.59	118.76	106.82	98.04	91.11	80.52	72.60	66.32	61.18	56.85	53.16	49.97
59	Pr	73.31	105.32	122.61	131.40	135.41	136.48	133.11	128.38	123.22	113.00	103.84	95.99	83.60	74.40	67.28	61.58	56.89	52.96	49.60
60	Nd	68.30	99.45	117.82	128.16	133.42	135.68	132.15	127.09	121.70	111.40	102.38	94.72	82.69	73.75	66.83	61.27	56.68	52.83	49.53
61	Pm	77.42	100.79	115.98	126.25	132.84	138.63	136.28	130.93	124.81	113.26	103.61	95.70	83.64	74.81	67.98	62.47	57.92	54.07	50.76
62	Sm	69.95	97.33	113.13	122.91	129.17	136.33	136.95	134.30	130.05	120.10	110.49	102.03	88.56	78.54	70.81	64.66	59.61	55.39	51.81
63	Eu	76.57	90.99	100.79	108.14	113.58	120.70	121.75	119.67	116.19	108.21	100.70	94.15	83.64	75.61	69.22	63.96	59.53	55.74	52.46
64	Gd	67.81	91.84	107.91	118.99	126.29	133.48	132.25	127.82	122.47	112.01	103.04	95.58	84.01	75.43	68.73	63.31	58.81	54.99	51.70

Table A3.1.2 (continued).

Z_2	At	E_{He} *(keV)*																		
		100	200	300	400	500	750	1000	1250	1500	2000	2500	3000	4000	5000	6000	7000	8000	9000	10000
65	Tb	63.61	87.17	102.46	112.65	119.17	125.59	125.03	121.86	117.78	109.26	101.45	94.65	83.70	75.35	68.75	63.37	58.88	55.07	51.78
66	Dy	63.27	88.50	104.07	113.77	119.60	125.00	124.52	121.81	118.19	110.20	102.56	95.77	84.69	76.19	69.46	63.99	59.42	55.55	52.22
67	Ho	61.44	82.04	94.61	102.51	107.38	112.30	112.46	110.69	108.02	101.76	95.50	89.77	80.21	72.70	66.67	61.70	57.51	53.94	50.84
68	Er	61.62	87.24	101.38	109.56	114.51	120.21	121.23	119.81	117.06	109.97	102.67	95.98	84.94	76.42	69.69	64.21	59.65	55.79	52.46
69	Tm	60.42	82.60	96.74	105.91	111.60	116.92	116.36	113.72	110.36	103.26	96.61	90.71	81.02	73.50	67.46	62.48	58.29	54.71	51.60
70	Yb	70.78	87.10	98.40	106.87	113.07	120.93	121.60	118.72	114.42	105.24	97.20	90.57	80.46	73.03	67.21	62.45	58.43	54.99	51.99
71	Lu	61.84	82.92	96.95	106.61	112.96	119.34	118.62	115.30	111.23	103.18	96.13	90.14	80.57	73.24	67.38	62.53	58.44	54.93	51.87
72	Hf	55.78	81.40	98.08	108.70	114.99	119.92	118.35	114.94	111.08	103.50	96.69	90.76	81.11	73.64	67.66	62.72	58.57	55.01	51.92
73	Ta	55.46	78.66	94.65	105.22	111.77	118.12	118.17	116.01	112.98	106.16	99.49	93.45	83.44	75.62	69.36	64.21	59.88	56.19	52.99
74	W	59.21	87.56	99.83	107.23	112.40	119.11	120.19	118.25	114.83	106.59	98.58	91.53	80.30	71.92	65.43	60.21	55.92	52.30	49.21
75	Re	45.47	76.02	93.80	103.68	109.45	116.22	117.91	116.95	114.51	107.81	100.78	94.32	83.65	75.44	68.95	63.68	59.28	55.56	52.36
76	Os	44.17	74.43	92.04	101.95	107.92	115.37	117.58	116.92	114.66	108.14	101.16	94.72	84.04	75.81	69.29	63.99	59.58	55.84	52.62
77	Ir	44.47	71.65	89.98	101.73	108.91	116.30	117.30	115.87	113.32	106.94	100.39	94.35	84.24	76.32	69.98	64.76	60.39	56.65	53.42
78	Pt	49.79	71.74	87.51	99.05	107.13	116.62	117.60	115.22	111.73	104.40	97.79	92.07	82.81	75.59	69.75	64.89	60.75	57.18	54.06
79	Au	53.96	72.09	85.48	96.20	104.70	117.89	122.53	122.12	119.14	110.71	102.44	95.39	84.59	76.73	70.63	65.66	61.48	57.90	54.78
80	Hg	55.79	74.12	88.99	101.16	110.64	123.79	126.40	123.80	119.27	109.60	101.37	94.72	84.64	77.17	71.23	66.32	62.15	58.54	55.38
81	Tl	55.68	80.75	98.63	111.54	120.39	130.04	129.94	126.18	121.39	112.03	104.05	97.41	86.98	79.07	72.77	67.56	63.16	59.38	56.09
82	Pb	72.95	93.87	108.53	119.57	127.72	138.15	139.14	135.41	129.80	118.03	108.00	99.98	88.19	79.83	73.40	68.19	63.82	60.08	56.83
83	Bi	87.18	108.90	123.50	134.15	141.70	150.30	149.54	144.34	137.56	124.08	112.92	104.10	91.29	82.30	75.44	69.93	65.34	61.42	58.02
84	Po	74.80	94.19	111.76	126.74	138.27	151.86	150.88	144.21	136.41	122.71	112.32	104.31	92.49	83.85	77.04	71.45	66.73	62.67	59.15
85	At	57.54	90.83	115.55	133.35	145.19	156.23	153.47	146.32	138.54	124.88	114.28	105.95	93.56	84.53	77.50	71.77	66.97	62.87	59.32
86	Rn	73.67	105.99	128.08	143.15	152.69	160.25	156.07	148.31	140.22	126.28	115.58	107.21	94.83	85.83	78.81	73.07	68.26	64.14	60.57
87	Fr	85.63	113.07	131.95	145.53	154.84	164.24	161.77	154.40	145.78	129.96	117.66	108.27	94.96	85.71	78.66	72.98	68.24	64.19	60.67
88	Ra	96.24	117.93	135.57	149.75	160.29	172.09	169.81	161.41	151.52	133.82	120.58	110.77	97.14	87.75	80.59	74.78	69.91	65.73	62.10
89	Ac	83.92	118.75	143.02	160.11	171.38	181.46	177.25	167.94	157.73	139.76	126.05	115.62	100.74	90.38	82.51	76.20	70.96	66.52	62.69
90	Th	78.77	110.00	132.04	148.06	159.15	170.89	169.07	161.70	152.86	136.44	123.53	113.58	99.32	89.35	81.76	75.65	70.58	66.26	62.52
91	Pa	88.49	108.38	124.97	138.79	149.60	164.06	165.45	160.25	152.68	137.38	124.85	115.05	100.84	90.80	83.10	76.88	71.69	67.26	63.43
92	U	61.59	95.36	119.20	135.53	145.93	155.26	153.20	147.59	141.24	129.35	119.38	111.12	98.29	88.73	81.22	75.10	69.99	65.64	61.89

Table A3.1.3. Proton nuclear stopping cross sections [10^{-2} eV cm^2/10^{15} atoms].

Z_2	At	E_p (keV) 10	20	30	40	50	75	100	150	200	300	500	1000
1	H	5.720	3.311	2.383	1.881	1.563	1.112	0.871	0.616	0.481	0.338	0.216	0.117
2	He	5.359	3.133	2.266	1.794	1.493	1.066	0.838	0.594	0.464	0.327	0.210	0.114
3	Li	6.330	3.754	2.732	2.172	1.813	1.301	1.024	0.729	0.571	0.404	0.260	0.142
4	Be	7.976	4.794	3.511	2.800	2.344	1.689	1.333	0.952	0.747	0.530	0.342	0.187
5	B	9.774	5.887	4.334	3.469	2.910	2.104	1.665	1.192	0.938	0.666	0.431	0.236
6	C	11.897	7.239	5.357	4.300	3.616	2.623	2.080	1.493	1.176	0.837	0.542	0.298
7	N	13.147	8.072	6.001	4.831	4.069	2.961	2.352	1.692	1.335	0.952	0.618	0.341
8	O	14.301	8.938	6.611	5.335	4.502	3.285	2.615	1.885	1.489	1.064	0.692	0.382
9	F	14.570	9.178	6.813	5.511	4.659	3.408	2.717	1.962	1.552	1.110	0.723	0.400
10	Ne	16.204	10.287	7.662	6.214	5.261	3.858	3.081	2.229	1.765	1.265	0.825	0.458
11	Na	16.541	10.576	7.902	6.423	5.446	4.004	3.201	2.321	1.840	1.320	0.862	0.479
12	Mg	17.906	11.530	8.725	7.039	5.977	4.405	3.527	2.561	2.032	1.460	0.955	0.531
13	Al	18.267	11.841	8.987	7.263	6.177	4.562	3.657	2.660	2.113	1.520	0.996	0.555
14	Si	19.649	12.820	9.759	7.900	6.729	4.980	3.998	2.913	2.316	1.668	1.094	0.610
15	P	19.796	12.994	9.920	8.120	6.859	5.088	4.090	2.984	2.375	1.712	1.124	0.628
16	S	21.067	13.913	10.652	8.734	7.386	5.490	4.418	3.228	2.572	1.856	1.221	0.683
17	Cl	20.873	13.862	10.641	8.740	7.398	5.510	4.440	3.249	2.590	1.871	1.232	0.690
18	Ar	20.178	13.474	10.370	8.530	7.296	5.393	4.351	3.188	2.544	1.840	1.212	0.680
19	K	22.303	14.977	11.557	9.522	8.154	6.037	4.876	3.578	2.858	2.069	1.365	0.766
20	Ca	23.446	15.829	12.245	10.104	8.662	6.424	5.195	3.817	3.051	2.211	1.460	0.821
21	Sc	22.454	15.235	11.813	9.762	8.377	6.223	5.037	3.706	2.964	2.150	1.421	0.800
22	Ti	22.539	15.369	11.945	9.885	8.491	6.317	5.119	3.771	3.019	2.191	1.450	0.817
23	V	22.589	15.478	12.057	9.992	8.591	6.401	5.193	3.830	3.068	2.229	1.477	0.833
24	Cr	23.506	16.184	12.636	10.487	9.025	6.735	5.469	4.038	3.238	2.355	1.561	0.881
25	Mn	23.571	16.303	12.757	10.602	9.133	6.824	5.547	4.101	3.290	2.395	1.589	0.898
26	Fe	24.494	17.020	13.347	11.107	9.577	7.234	5.831	4.316	3.465	2.525	1.677	0.948
27	Co	24.469	17.078	13.421	11.182	9.651	7.301	5.889	4.365	3.507	2.557	1.700	0.962
28	Ni	25.822	18.102	14.255	11.894	10.274	7.784	6.284	4.663	3.749	2.736	1.820	1.031
29	Cu	25.050	17.634	13.915	11.624	10.050	7.624	6.160	4.576	3.681	2.689	1.791	1.016
30	Zn	25.504	18.029	14.255	11.923	10.317	7.838	6.338	4.713	3.794	2.774	1.848	1.049
31	Ga	25.011	17.752	14.063	11.777	10.199	7.759	6.278	4.674	3.765	2.754	1.837	1.044
32	Ge	25.078	17.871	14.185	11.892	10.308	7.852	6.358	4.739	3.820	2.797	1.867	1.062
33	As	25.324	18.117	14.408	12.094	10.491	8.002	6.484	4.838	3.903	2.859	1.910	1.087
34	Se	25.012	17.963	14.311	12.027	10.441	7.974	6.529	4.830	3.898	2.858	1.911	1.089
35	Br	25.684	18.515	14.779	12.434	10.803	8.261	6.769	5.012	4.048	2.970	1.988	1.133
36	Kr	25.424	18.396	14.710	12.390	10.774	8.249	6.764	5.013	4.051	2.975	1.992	1.137
37	Rb	25.841	18.767	15.033	12.676	11.031	8.457	6.940	5.148	4.163	3.059	2.050	1.170
38	Sr	26.102	19.024	15.267	12.887	11.223	8.614	7.074	5.253	4.250	3.125	2.096	1.198
39	Y	26.608	19.462	15.645	13.220	11.522	8.855	7.277	5.408	4.378	3.222	2.162	1.237
40	Zr	26.798	19.668	15.838	13.398	11.686	8.991	7.395	5.500	4.456	3.281	2.204	1.261
41	Nb	27.161	20.003	16.135	13.663	11.926	9.187	7.561	5.628	4.562	3.362	2.260	1.294
42	Mo	27.129	20.047	16.196	13.729	11.993	9.249	7.617	5.675	4.603	3.394	2.283	1.308
43	Tc	27.093	20.086	16.255	13.793	12.057	9.309	7.672	5.721	4.643	3.426	2.306	1.323
44	Ru	27.324	20.323	16.473	13.992	12.240	9.461	7.803	5.823	4.728	3.491	2.351	1.350
45	Rh	27.608	20.600	16.724	14.220	12.447	9.632	7.950	5.937	4.824	3.564	2.402	1.380
46	Pd	27.453	20.549	16.709	14.221	12.456	9.650	7.970	5.956	4.842	3.580	2.415	1.388

Table A3.1.3 (continued).

Z_2	At	E_p (keV) 10	20	30	40	50	75	100	150	200	300	500	1000
47	Ag	27.817	20.886	17.009	14.490	12.701	9.850	8.141	6.089	4.953	3.665	2.473	1.423
48	Cd	27.409	20.642	16.836	14.356	12.593	9.776	8.085	6.110	4.925	3.647	2.462	1.417
49	In	27.526	20.793	16.985	14.497	12.724	9.889	8.184	6.189	4.992	3.698	2.499	1.439
50	Sn	27.303	20.685	16.922	14.457	12.698	9.879	8.181	6.192	4.996	3.704	2.504	1.444
51	Sb	27.274	20.723	16.978	14.518	12.760	9.938	8.235	6.238	5.035	3.735	2.527	1.458
52	Te	26.653	20.310	16.663	14.262	12.543	9.779	8.108	6.146	4.964	3.685	2.495	1.440
53	I	27.426	20.958	17.220	14.753	12.982	10.132	8.407	6.378	5.153	3.827	2.593	1.498
54	Xe	27.117	20.780	17.098	14.661	12.910	10.086	8.373	6.357	5.138	3.819	2.589	1.496
55	Cs	27.389	21.046	17.341	14.883	13.113	10.255	8.519	6.473	5.234	3.893	2.640	1.527
56	Ba	27.088	20.872	17.221	14.793	13.042	10.209	8.487	6.453	5.220	3.885	2.637	1.526
57	La	27.355	21.134	17.462	15.012	13.244	10.378	8.632	6.568	5.316	3.958	2.688	1.557
58	Ce	27.686	21.447	17.744	15.268	13.478	10.572	8.798	6.700	5.425	4.042	2.747	1.591
59	Pr	28.095	21.820	18.078	15.569	13.751	10.797	8.991	6.852	5.550	4.137	2.813	1.631
60	Nd	27.998	21.801	18.086	15.589	13.778	10.828	9.022	6.880	5.575	4.159	2.830	1.642
61	Pm	28.014	21.869	18.166	15.671	13.858	10.901	9.089	6.936	5.676	4.197	2.857	1.658
62	Sm	27.918	21.849	18.173	15.690	13.883	10.931	9.119	6.964	5.702	4.218	2.873	1.669
63	Eu	28.144	22.080	18.389	15.889	14.068	11.087	9.255	7.073	5.793	4.288	2.922	1.698
64	Gd	27.703	21.787	18.168	15.711	13.918	10.979	9.170	7.013	5.747	4.255	2.902	1.688
65	Tb	27.909	22.002	18.371	15.899	14.093	11.127	9.299	7.117	5.834	4.322	2.949	1.716
66	Dy	27.781	21.954	18.353	15.896	14.098	11.142	9.316	7.135	5.851	4.337	2.961	1.724
67	Ho	27.851	22.060	18.465	16.006	14.203	11.235	9.400	7.204	5.910	4.383	2.994	1.745
68	Er	27.934	22.178	18.585	16.123	14.316	11.334	9.488	7.276	5.972	4.431	3.029	1.766
69	Tm	28.122	22.379	18.777	16.302	14.482	11.476	9.612	7.376	6.057	4.497	3.075	1.794
70	Yb	27.910	22.260	18.699	16.247	14.441	11.454	9.599	7.371	6.055	4.497	3.078	1.796
71	Lu	28.050	22.422	18.858	16.397	14.583	11.577	9.706	7.459	6.130	4.555	3.119	1.821
72	Hf	27.936	22.380	18.845	16.399	14.592	11.594	9.726	7.479	6.149	4.571	3.132	1.830
73	Ta	27.988	22.471	18.943	16.497	14.687	11.679	9.803	7.543	6.204	4.615	3.163	1.849
74	W	27.971	22.506	18.995	16.554	14.745	11.736	9.856	7.589	6.244	4.647	3.187	1.864
75	Re	28.036	22.607	19.102	16.659	14.847	11.827	9.938	7.657	6.303	4.692	3.220	1.884
76	Os	27.856	22.509	19.041	16.618	14.818	11.814	9.932	7.657	6.305	4.697	3.225	1.888
77	Ir	27.970	22.648	19.180	16.751	14.945	11.925	10.031	7.738	6.375	4.750	3.263	1.912
78	Pt	27.952	22.681	19.229	16.806	15.002	11.980	10.082	7.783	6.414	4.782	3.287	1.927
79	Au	28.078	22.830	19.376	16.947	15.135	12.097	10.186	7.868	6.486	4.838	3.327	1.951
80	Hg	27.956	22.777	19.352	16.938	15.135	12.107	10.199	7.883	6.502	4.851	3.338	1.959
81	Tl	27.816	22.708	19.314	16.917	15.123	12.107	10.205	7.892	6.512	4.861	3.347	1.965
82	Pb	27.809	22.747	19.368	16.976	15.183	12.165	10.259	7.939	6.553	4.894	3.371	1.980
83	Bi	27.937	22.896	19.516	17.117	15.318	12.283	10.363	8.025	6.626	4.951	3.412	2.005
84	Po	28.165	23.128	19.734	17.321	15.507	12.445	10.505	8.140	6.724	5.025	3.465	2.038
85	At	28.527	23.470	20.047	17.607	15.772	12.667	10.698	8.294	6.854	5.125	3.536	2.080
86	Rn	27.333	22.529	19.263	16.930	15.172	12.195	10.305	7.994	6.608	4.990	3.412	2.009
87	Fr	27.550	22.752	19.473	17.126	15.355	12.352	10.442	8.106	6.703	5.064	3.464	2.040
88	Ra	27.520	22.769	19.508	17.168	15.400	12.398	10.486	8.144	6.737	5.093	3.485	2.054
89	Ac	27.732	22.986	19.714	17.360	15.580	12.553	10.622	8.255	6.831	5.166	3.537	2.086
90	Th	27.461	22.802	19.576	17.250	15.489	12.488	10.573	8.222	6.806	5.149	3.527	2.081
91	Pa	27.904	23.213	19.948	17.589	15.801	12.750	10.799	8.403	6.959	5.267	3.610	2.130
92	U	27.396	22.831	19.638	17.328	15.573	12.575	10.657	8.296	6.873	5.204	3.568	2.107

Table A3.1.4. Helium nuclear stopping cross sections [10^{-2} eV $cm^2/10^{15}$ atoms].

Z_2	At	E_{He} *(keV)* 10	20	30	40	50	75	100	150	200	300	500	1000
1	H	56.91	35.17	26.24	21.17	17.85	13.02	10.36	7.46	5.90	4.21	2.74	1.51
2	He	60.18	36.90	27.41	22.05	18.57	13.51	10.73	7.71	6.09	4.34	2.82	1.55
3	Li	72.46	45.42	33.63	27.17	22.94	16.75	13.34	9.62	7.61	5.44	3.54	1.96
4	Be	90.89	57.86	43.14	35.02	29.66	21.77	17.39	12.60	9.98	7.15	4.67	2.59
5	B	109.43	70.66	53.53	43.22	36.72	27.09	21.70	15.77	12.52	9.00	5.89	3.28
6	C	131.62	86.17	65.70	53.23	45.37	33.62	27.01	19.70	15.67	11.29	7.41	4.14
7	N	144.57	95.73	73.38	60.22	50.95	37.90	30.52	22.32	17.79	12.84	8.45	4.73
8	O	156.24	104.57	80.56	66.31	56.75	41.97	33.88	24.84	19.83	14.34	9.46	5.31
9	F	158.70	107.20	82.95	68.46	58.69	43.54	35.21	25.87	20.68	14.99	9.90	5.56
10	Ne	174.79	119.29	92.75	76.78	65.96	49.09	39.78	29.31	23.47	17.04	11.28	6.35
11	Na	177.61	122.26	95.45	79.21	68.17	50.86	41.30	30.50	24.45	17.78	11.79	6.65
12	Mg	190.63	132.43	103.84	86.41	74.50	56.28	45.35	33.57	26.95	19.63	13.04	7.38
13	Al	193.46	135.48	106.64	88.94	76.81	58.18	46.96	34.83	28.00	20.43	13.59	7.70
14	Si	206.38	145.75	115.19	96.32	83.33	63.29	51.16	38.04	30.62	22.38	14.91	8.46
15	P	206.91	147.21	116.75	97.83	84.77	64.54	52.25	38.92	31.37	22.95	15.32	8.71
16	S	218.51	156.67	124.73	104.76	90.92	69.40	56.80	42.00	33.89	24.84	16.60	9.45
17	Cl	215.60	155.62	124.29	104.60	90.91	69.55	57.00	42.22	34.11	25.03	16.75	9.55
18	Ar	207.76	150.90	120.88	101.93	88.71	68.01	55.81	41.40	33.48	24.60	16.49	9.41
19	K	227.54	166.51	133.88	113.15	98.63	75.81	62.31	46.31	37.50	27.60	18.52	10.59
20	Ca	237.56	175.02	141.20	119.58	104.38	80.42	66.19	49.28	39.95	29.44	19.79	11.33
21	Sc	226.89	168.09	135.98	115.36	100.82	77.82	64.13	47.81	38.80	28.62	19.26	11.05
22	Ti	226.65	168.92	137.05	116.48	101.94	78.85	65.06	48.57	39.45	29.15	19.64	11.28
23	V	226.09	169.47	137.89	117.41	102.88	79.74	65.87	49.25	40.05	29.62	19.98	11.49
24	Cr	233.85	176.33	143.90	122.76	107.70	83.65	69.20	52.31	42.17	31.23	21.10	12.15
25	Mn	233.42	176.96	144.82	123.75	108.71	84.60	70.07	53.04	42.80	31.74	21.46	12.37
26	Fe	241.15	183.86	150.89	129.17	113.61	88.60	73.47	55.70	44.99	33.40	22.61	13.05
27	Co	239.87	183.83	151.26	129.70	114.21	89.23	74.08	56.24	45.46	33.79	22.91	13.24
28	Ni	251.59	193.86	159.96	137.40	121.15	94.84	78.83	59.94	48.49	36.09	24.49	14.17
29	Cu	243.30	188.35	155.78	134.02	118.29	92.76	77.19	58.77	47.58	35.44	24.08	13.95
30	Zn	246.54	191.80	159.04	137.04	121.09	95.13	79.25	60.42	48.95	36.51	24.84	14.40
31	Ga	240.93	188.29	156.49	135.05	119.46	94.01	78.40	59.85	48.99	36.23	24.67	14.32
32	Ge	240.59	188.88	157.36	136.00	120.43	94.94	79.26	60.59	49.63	36.74	25.04	14.56
33	As	241.92	190.79	159.33	137.91	122.25	96.54	80.69	61.76	50.63	37.51	25.60	14.90
34	Se	238.09	188.58	157.83	136.81	121.40	96.03	80.34	61.57	50.51	37.46	25.59	14.91
35	Br	243.36	193.61	162.43	141.01	125.26	99.25	83.12	63.78	52.37	38.88	26.59	15.51
36	Kr	240.07	191.78	161.24	140.17	124.64	98.92	82.92	63.71	52.35	38.90	26.63	15.55
37	Rb	243.00	194.93	164.25	142.99	127.28	101.18	84.91	65.32	53.71	39.94	27.37	16.00
38	Sr	244.49	196.93	166.29	144.97	129.17	102.85	86.39	66.54	54.76	40.76	27.96	16.36
39	Y	248.22	200.73	169.87	148.30	132.26	105.48	88.70	68.40	56.33	41.96	28.82	16.88
40	Zr	249.06	202.20	171.47	149.89	133.81	106.88	89.96	69.46	57.25	42.68	29.34	17.20
41	Nb	251.49	204.95	174.16	152.45	136.22	108.98	91.82	70.97	58.54	43.68	30.06	17.64
42	Mo	250.35	204.77	174.35	152.81	136.68	109.51	92.35	71.46	58.99	44.05	30.34	17.82
43	Tc	249.20	204.56	174.51	153.15	137.10	110.01	92.86	71.94	59.42	44.41	30.62	18.01
44	Ru	250.45	206.32	176.36	154.97	138.86	111.59	94.27	73.12	60.44	45.64	31.20	18.36
45	Rh	252.18	208.48	178.54	157.08	140.88	113.38	95.88	74.45	61.58	46.54	31.84	18.76
46	Pd	250.00	207.38	177.93	156.73	140.69	113.38	95.97	74.60	61.74	46.70	31.98	18.86

Table A3.1.4 (continued).

Z_2	At	E_{He} (keV) 10	20	30	40	50	75	100	150	200	300	500	1000
47	Ag	252.45	210.12	180.62	159.30	143.12	115.51	97.85	76.15	63.07	47.74	32.72	19.32
48	Cd	248.07	207.13	178.37	157.49	141.61	114.45	97.04	75.60	62.65	47.47	32.56	19.24
49	In	248.36	208.05	179.47	158.66	142.78	115.56	98.06	76.48	63.42	48.09	33.02	19.53
50	Sn	245.67	206.43	178.39	157.87	142.20	115.24	97.88	76.41	63.41	48.11	33.06	19.57
51	Sb	244.69	206.25	178.53	158.19	142.59	115.72	98.37	76.87	63.83	48.47	33.34	19.75
52	Te	238.55	201.66	174.85	155.09	139.91	113.69	96.72	75.66	62.87	47.78	32.88	19.50
53	I	244.64	207.45	180.18	160.00	144.47	117.55	100.09	78.38	65.17	49.57	34.14	20.27
54	Xe	241.27	205.18	178.50	158.68	143.38	116.82	99.55	78.04	64.92	49.42	34.06	20.24
55	Cs	242.97	207.23	180.58	160.71	145.33	118.56	101.12	79.35	66.06	50.32	34.71	20.64
56	Ba	239.70	205.01	178.93	159.41	144.27	117.85	100.59	79.01	65.81	50.17	34.64	20.61
57	La	241.38	207.03	180.98	161.41	146.19	119.57	102.14	80.31	66.94	51.07	35.28	21.02
58	Ce	243.61	209.53	183.46	163.79	148.46	121.58	103.94	81.81	68.23	52.09	36.02	21.47
59	Pr	246.51	212.60	186.44	166.62	151.14	123.94	106.04	83.54	69.72	53.27	36.86	21.99
60	Nd	245.06	211.90	186.11	166.50	151.14	124.09	106.25	83.79	69.97	53.49	37.04	22.12
61	Pm	244.58	212.04	186.51	167.02	151.73	124.72	106.88	84.36	70.49	53.93	37.37	22.34
62	Sm	243.17	211.35	186.18	166.89	151.72	124.87	107.08	84.60	70.73	54.15	37.55	22.47
63	Eu	244.51	213.06	187.95	168.65	153.43	126.43	108.50	85.81	71.78	55.00	38.16	22.85
64	Gd	240.17	209.79	185.33	166.45	151.54	125.01	107.37	84.99	71.13	54.54	37.87	22.70
65	Tb	241.36	211.35	186.97	168.09	153.14	126.48	108.71	86.13	72.13	55.34	38.46	23.07
66	Dy	239.72	210.41	186.40	167.74	152.92	126.45	108.76	86.25	72.27	55.49	38.59	23.16
67	Ho	239.77	210.94	187.13	168.55	153.77	127.30	109.58	86.97	72.92	56.03	38.98	23.42
68	Er	239.93	211.58	187.96	169.45	154.70	128.21	110.44	87.74	73.60	56.59	39.40	23.69
69	Tm	240.99	213.01	189.48	170.99	156.21	129.61	111.73	88.84	74.57	57.37	39.97	24.05
70	Yb	238.68	211.44	188.33	170.10	155.50	129.16	111.42	88.68	74.47	57.33	40.36	24.07
71	Lu	239.34	212.50	189.53	171.33	156.73	130.33	112.51	89.62	75.30	58.02	40.87	24.40
72	Hf	237.88	211.66	189.02	171.03	156.55	130.33	112.59	89.76	75.46	58.18	41.01	24.50
73	Ta	237.83	212.07	189.63	171.73	157.29	131.09	113.32	90.42	76.06	58.68	41.39	24.74
74	W	237.20	211.95	189.77	172.00	157.65	131.52	113.78	90.86	76.47	59.03	41.67	24.93
75	Re	237.26	212.46	190.45	172.77	158.45	132.34	114.56	91.57	77.11	59.56	42.07	25.19
76	Os	235.29	211.12	189.49	172.04	157.88	132.00	114.34	91.47	77.07	59.57	42.10	25.23
77	Ir	235.77	211.98	190.50	173.10	158.96	133.04	115.32	92.33	77.83	60.20	42.58	25.53
78	Pt	235.17	211.87	190.62	173.36	159.29	133.46	115.76	92.76	78.23	60.55	42.85	25.71
79	Au	235.75	212.82	191.70	174.49	160.43	134.55	116.79	93.66	79.03	61.21	43.35	26.03
80	Hg	234.30	211.92	191.12	174.09	160.16	134.46	116.79	93.74	79.14	61.33	43.46	26.12
81	Tl	232.71	210.88	190.40	173.58	159.79	134.28	116.71	93.75	79.19	61.41	43.54	26.18
82	Pb	232.22	210.84	190.58	173.88	160.16	134.73	117.17	94.20	79.61	61.77	43.83	26.37
83	Bi	232.86	211.81	191.68	175.02	161.30	135.83	118.21	95.11	80.42	62.44	44.33	26.70
84	Po	234.31	213.53	193.45	176.78	163.02	137.42	119.66	96.36	81.52	63.33	45.00	27.11
85	At	236.86	216.25	196.14	179.38	165.51	139.66	121.69	98.08	83.01	64.53	45.88	27.66
86	Rn	226.68	207.30	188.22	172.26	159.03	134.31	117.10	94.45	79.98	62.20	44.25	26.70
87	Fr	228.06	208.95	189.92	173.95	160.68	135.84	118.51	95.66	81.04	63.07	44.90	27.11
88	Ra	227.43	208.73	189.93	174.08	160.90	136.15	118.86	96.01	81.38	63.37	45.14	27.27
89	Ac	228.77	210.33	191.58	175.74	162.51	137.65	120.24	97.21	82.43	64.23	45.78	27.68
90	Th	226.18	208.30	189.93	174.35	161.32	136.77	119.54	96.72	82.06	63.98	45.63	27.60
91	Pa	229.41	211.64	193.18	177.46	164.29	139.42	121.94	98.73	83.81	65.38	46.66	28.25
92	U	224.93	207.83	189.90	174.57	161.70	137.35	120.19	97.39	82.71	64.56	46.10	27.93

A3.2 CALCULATION OF PROTON STOPPING CROSS SECTIONS

Proton stopping cross sections ε_H [eVcm^2/10^{15}atoms] above 25 keV may be obtained (Ziegler *et al.*, 1985; Ziegler and Biersack, 1991; references in Chapter 2) from:

$$\varepsilon_H = \frac{\varepsilon_{Low} \cdot \varepsilon_{High}}{\varepsilon_{Low} + \varepsilon_{High}}$$

$$\varepsilon_{Low} = A1\, E^{A2} + A3\, E^{A4}$$

$$\varepsilon_{High} = A5 / E^{A6} \ln [(A7 / E) + A8\, E] \ ,$$

where *E* is in keV/amu. The coefficients A1 – A8 are given in Table A3.2.1.

Table A3.2.1. Proton stopping cross section coefficients.

Z_2	At	A1	A2	A3	A4	A5	A6	A7	A8
1	H	0.0121702	0.00533578	1.12874	0.364197	1120.7	1.12128	2477.31	0.009770990
2	He	0.4890013	0.0050512491	0.8613451	0.4674054	745.3815	1.0422672	7988.3889	0.033328667
3	Li	0.8583748	0.0050147482	1.6044494	0.3884424	1337.3032	1.047033	2659.2306	0.018979873
4	Be	0.8781010	0.0051049349	5.4231571	0.2031973	1200.6151	1.0211124	1401.8432	0.038529280
5	B	1.4607952	0.0048835929	2.3380238	0.4424895	1801.2741	1.0352217	1784.1234	0.020239625
6	C	2.10544	0.00490795	2.08723	0.46258	1779.22	1.01472	2324.45	0.020269400
7	N	0.645636	0.00508289	4.09503	0.33879	2938.49	1.04017	2911.08	0.010721900
8	O	0.751093	0.00503003	3.93983	0.346199	2287.85	1.01171	3997.24	0.018426800
9	F	1.30187	0.00514136	3.82737	0.28151	2829.94	1.02762	7831.3	0.020940300
10	Ne	4.7339096	0.0044505735	0.0298622	1.4940358	1825.3641	0.9789632	130.76313	0.021576591
11	Na	6.097248	0.0044291901	3.1929400	0.4576301	1363.3487	0.9518161	2380.6086	0.081834623
12	Mg	14.013106	0.0043645904	2.2641223	0.3632649	2187.3659	0.9909772	6264.8005	0.046200118
13	Al	0.0390926	0.0045416623	6.9692434	0.3297639	1688.3008	0.9594386	1151.9784	0.048981572
14	Si	2.178134	0.0044454523	2.6045162	0.6088463	1550.2068	0.9330245	1703.8459	0.031619771
15	P	17.575478	0.0038345645	0.0786935	1.2388076	2805.9699	0.9728416	1037.5875	0.012878599
16	S	3.1473	0.0044716	4.9747	0.41024	4005.8	1.0011	1898.8	0.007659200
17	Cl	3.3544	0.004474	5.9206	0.41003	4403.9	0.99623	2006.9	0.008231000
18	Ar	2.0378865	0.0044775111	3.0742856	0.5477292	3505.0123	0.9757545	1714.0455	0.011700915
19	K	0.7417072	0.0043051307	1.1514679	0.9508284	917.2098	0.8781994	389.93209	0.189257680
20	Ca	9.1315679	0.0043809163	5.4610696	0.3132704	3891.8065	0.9793344	6267.9299	0.015195719
21	Sc	7.2247467	0.0043718065	6.1016923	0.3751071	2829.2167	0.9521775	6376.1223	0.020398311
22	Ti	0.1469983	0.0048456345	6.3484619	0.4105728	2164.1297	0.9402756	5292.6185	0.050263311
23	V	5.0611377	0.0039867276	2.6173973	0.5795689	2218.8786	0.9236074	6323.0262	0.025669274
24	Cr	0.5326669	0.0042967841	0.3900533	1.2725384	1872.6784	0.9077604	64.16607	0.030106759
25	Mn	0.4769667	0.0043038251	0.3145150	1.3289335	1920.5366	0.9064879	45.57647	0.027469102
26	Fe	0.0274264	0.0035443371	0.0315632	2.1754694	1919.5468	0.9009877	23.90246	0.025362735
27	Co	0.1638265	0.0043042231	0.0734540	1.8591669	1918.4213	0.8967809	27.60974	0.023184259
28	Ni	4.2562307	0.0043736651	1.5605733	0.7206703	1546.8412	0.8795769	302.01726	0.040944280
29	Cu	2.3508344	0.0043236552	2.8820471	0.5011277	1837.724	0.8999210	2376.9522	0.049650048
30	Zn	3.1095234	0.0038454560	0.1147724	1.5037112	2184.6911	0.8930896	67.30570	0.016587522
31	Ga	15.321773	0.0040306222	0.6539090	0.6766839	3001.7127	0.9248419	3344.1849	0.016366484
32	Ge	3.6931934	0.0044813010	8.6080131	0.2763752	2982.6649	0.9276011	3166.5676	0.030873833
33	As	7.1372758	0.0043134079	9.4247048	0.2793700	2725.8278	0.9159744	3166.0773	0.025007912
34	Se	4.8979355	0.0042936681	3.7793113	0.5000385	2824.4987	0.9102752	1282.4213	0.017061405
35	Br	1.3682731	0.0043024195	2.5678732	0.6082169	6907.8291	0.9817031	628.00764	0.006805497
36	Kr	2.2352779	0.0043096475	4.8856225	0.4788315	4972.2755	0.9514553	1185.2258	0.009235595
37	Rb	0.4205605	0.0041168716	0.0169498	2.3615903	2252.7284	0.8919173	39.75195	0.027756682
38	Sr	30.779774	0.0037736081	0.5581330	0.7681590	7113.1556	0.9769719	1604.3936	0.006526788

Table A3.2.1 (continued).

Z_2	At	A1	A2	A3	A4	A5	A6	A7	A8
39	Y	11.575976	0.0042119068	7.0244432	0.3776444	4713.5219	0.9426371	2493.2196	0.011269740
40	Zr	6.2405791	0.0041916428	5.2701225	0.4945333	4234.5506	0.9323158	2063.9198	0.011844234
41	Nb	0.3307332	0.0041243377	1.7246018	1.1062067	1930.1909	0.8690703	27.41631	0.038208323
42	Mo	0.0177470	0.0041715317	0.1458581	1.7305221	1803.6188	0.8631518	29.66948	0.032122562
43	Tc	3.7228678	0.0041768103	4.6286038	0.5676889	1678.0247	0.8620204	3093.9512	0.062440177
44	Ru	0.1399838	0.0041328551	0.2557267	1.4241133	1919.2613	0.8632628	72.79694	0.032235102
45	Rh	0.2858978	0.0041385894	0.3130097	1.3423521	1954.8161	0.8617511	115.18178	0.029341706
46	Pd	0.7600193	0.0042179200	3.3859683	0.7628467	1867.3886	0.8580521	69.99413	0.036447779
47	Ag	6.395683	0.0041934620	5.4689075	0.4137814	1712.6134	0.8539727	18493.003	0.056470873
48	Cd	3.471693	0.0041343969	3.2337247	0.6378845	1116.3584	0.8195894	4766.0254	0.117895110
49	In	2.5265128	0.0042282025	4.5319769	0.5356240	1030.8484	0.8165170	16252.232	0.197218350
50	Sn	7.3682953	0.0041006764	4.6791094	0.5142830	1160.0010	0.8245361	17964.821	0.133160090
51	Sb	7.7197216	0.0043879762	3.2419754	0.6843444	1428.1143	0.8339777	1786.6706	0.066512413
52	Te	16.779901	0.0041917673	9.3197716	0.2956838	3370.9153	0.9028867	7431.7168	0.026159672
53	I	4.2132343	0.0042097824	4.6753325	0.5794508	3503.9280	0.8926145	1468.8716	0.014359044
54	Xe	4.6188934	0.0042203349	5.8164363	0.5218418	3961.2382	0.9040966	1473.2618	0.014194525
55	Cs	0.1851741	0.0036214664	0.0005878	3.5315422	2931.2998	0.8893605	26.17981	0.026392988
56	Ba	4.8248318	0.0041457784	6.0934255	0.5702562	2300.1078	0.8635875	2980.7187	0.038678811
57	La	0.4985675	0.0041054097	1.9775408	0.9587658	0786.5483	0.7850915	806.59969	0.408823790
58	Ce	3.275439	0.0042177424	5.7680306	0.5405402	6631.2873	0.9428168	744.06608	0.008302589
59	Pr	2.9978278	0.0040901358	4.5298608	0.6202474	2161.1538	0.8566884	1268.5942	0.043030519
60	Nd	2.8701111	0.0040959577	4.2567723	0.6137956	2130.4334	0.8523469	1704.1091	0.039384664
61	Pm	10.852925	0.0041148811	5.8907486	0.4683363	2857.1706	0.8754973	3654.168	0.029955419
62	Sm	3.64072	0.0041782043	4.8742398	0.5786142	1267.6986	0.8221108	3508.1718	0.241737030
63	Eu	17.645466	0.0040991510	6.5855038	0.3273433	3931.3048	0.9075401	5156.6611	0.036278412
64	Gd	7.5308869	0.0040813717	4.9389060	0.5067915	2519.6680	0.8581850	3314.6247	0.030514306
65	Tb	5.4741845	0.0040828895	4.8969573	0.5111306	2340.0736	0.8529648	2342.6752	0.035661774
66	Dy	4.266075	0.0040667385	4.5031787	0.5525674	2076.3920	0.8415133	1666.5639	0.040801264
67	Ho	6.8312811	0.0040485776	4.3986952	0.5167493	2003.0028	0.8343741	1410.4455	0.034779520
68	Er	1.2707048	0.0040553442	4.6294611	0.5742751	1626.2816	0.8185828	995.68122	0.055319240
69	Tm	5.7561274	0.0040490505	4.356999	0.5249555	2207.3232	0.8379555	1579.5099	0.027165033
70	Yb	14.127459	0.0040595861	5.8303935	0.3775487	3645.8910	0.8782317	3411.7714	0.016392116
71	Lu	6.6947551	0.0040602520	4.9361227	0.4796132	2719.0292	0.8524863	1885.8388	0.019713210
72	Hf	3.0618944	0.0040511084	3.5802967	0.5908198	2346.0989	0.8371319	1221.9881	0.020071670
73	Ta	10.810811	0.0033007875	1.3767142	0.7651179	2003.7300	0.8226856	1110.5686	0.024957749
74	W	2.7100691	0.0040960825	1.2289456	0.9859815	1232.3872	0.7906638	155.42021	0.047294287
75	Re	0.5234523	0.0040243824	1.4037997	0.8551002	1461.3907	0.7967727	503.34277	0.036789456
76	Os	0.4616005	0.0040202735	1.3014089	0.8704331	1473.5357	0.7968670	443.08542	0.036301488
77	Ir	0.9781437	0.0040374101	2.0126698	0.7225008	1890.8068	0.8174691	930.70144	0.027690448
78	Pt	3.2085673	0.0040510075	3.6657671	0.5361780	3091.1569	0.8560235	1508.1176	0.015401358
79	Au	2.0035097	0.0040430629	7.4882362	0.3560990	4464.3312	0.8883581	3966.5437	0.012838852
80	Hg	15.429952	0.0039432123	1.1237408	0.7070324	4595.7209	0.8843687	1576.4704	0.008853375
81	Tl	3.1512351	0.0040523543	4.0995555	0.5424994	3246.3125	0.8577231	1691.7661	0.015058053
82	Pb	7.1896291	0.0040587571	8.6927070	0.3584227	4760.5609	0.8883332	2888.2709	0.011029181
83	Bi	9.320869	0.0040539730	11.542820	0.3202666	4866.1620	0.8912398	3213.3794	0.011934944
84	Po	29.242217	0.0036194863	0.1686396	1.1226448	5687.9614	0.8981205	1033.2571	0.007130312
85	At	1.8522161	0.0039972862	3.1556025	0.6509577	3754.9715	0.8638291	1602.0163	0.012041676
86	Rn	3.221995	0.0040040926	5.9023588	0.5267790	4040.1546	0.8680370	1658.3527	0.011746940
87	Fr	9.3412359	0.0039660558	7.920988	0.4297687	5180.8957	0.8877259	2173.1554	0.009200702
88	Ra	36.182673	0.0036003237	0.5834109	0.8674703	6990.2108	0.9108200	1417.0974	0.006218743
89	Ac	5.9283892	0.0039694789	6.4082404	0.5212246	4619.5148	0.8808273	2323.5230	0.011627375
90	Th	5.2453649	0.0039744049	6.7968897	0.4854236	4586.3094	0.8779443	2481.5001	0.011282428
91	Pa	33.701736	0.003690143	0.4725717	0.8923500	5295.6866	0.8892973	2053.3026	0.009190849
92	U	2.7588977	0.0039805707	3.2091513	0.6612173	2505.3660	0.8286302	2065.1403	0.022815839

A3.3 PLOTS OF STOPPING CROSS SECTIONS FOR SELECTED IONS IN SOME ELEMENTAL AND COMPOUND TARGETS (according to Ziegler *et al.*, 1985).

Note: all curves maintain their relative positions. They do not cross over.

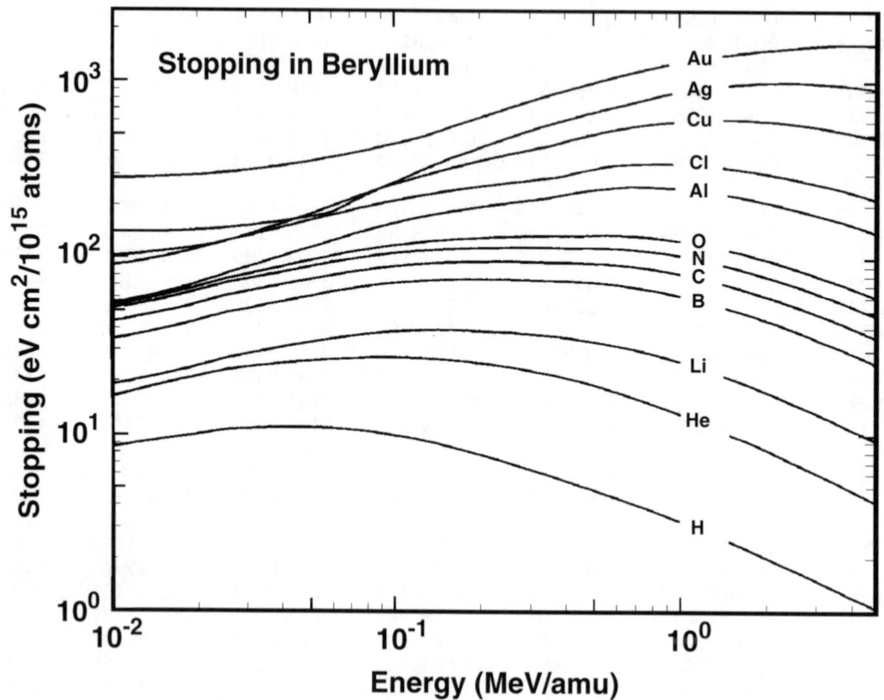

FIG. A3.3.1. Semiempirical stopping cross sections for ^1H, ^4He, ^7Li, ^{11}B, ^{12}C, ^{14}N, ^{16}O, ^{27}Al, ^{35}Cl, ^{63}Cu, ^{107}Ag and ^{197}Au ions in beryllium. Atomic weight = 9.012 amu, mass density = 1.82 g/cm^3, atomic density = $1.21 \cdot 10^{23}$ atoms/cm^3.

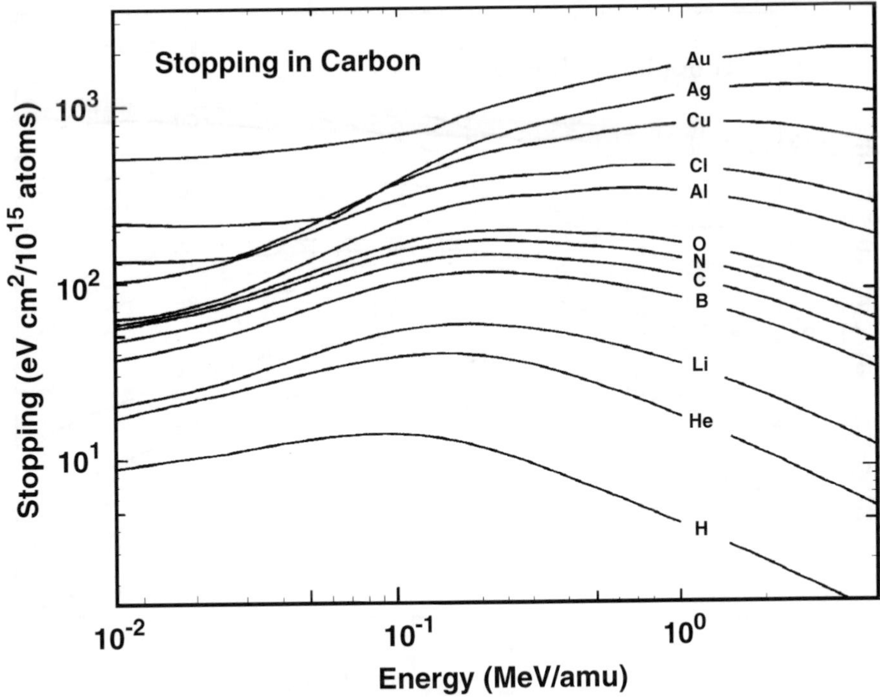

FIG. A3.3.2. Semiempirical stopping cross sections for ^{1}H, ^{4}He, ^{7}Li, ^{11}B, ^{12}C, ^{14}N, ^{16}O, ^{27}Al, ^{35}Cl, ^{63}Cu, ^{107}Ag and ^{197}Au ions in carbon. Atomic weight = 12.011 amu, mass density = 3.516 g/cm^3, atomic density = $1.76 \cdot 10^{23}$ atoms/cm^3.

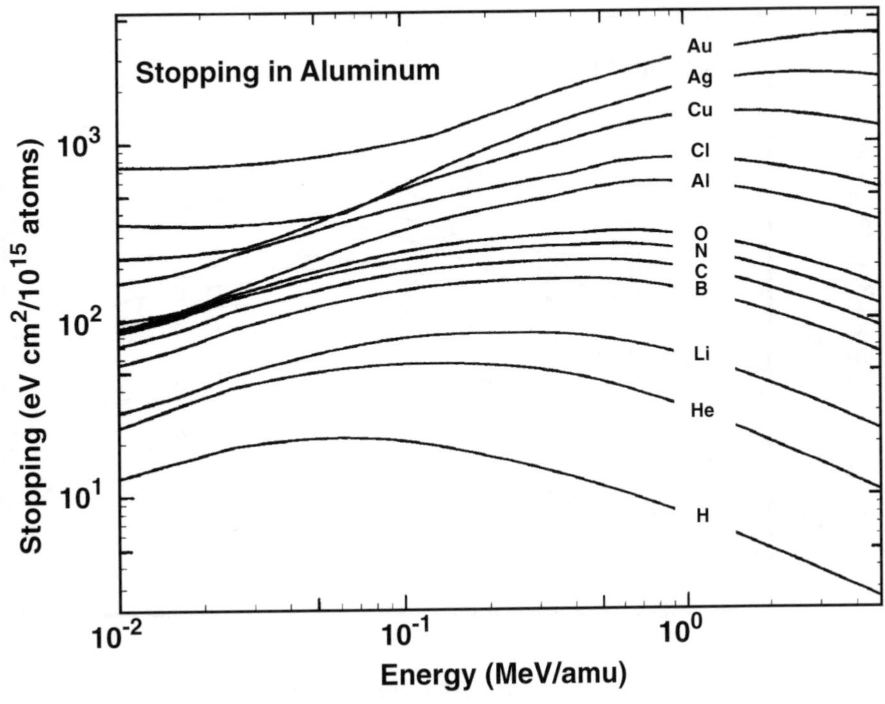

FIG. A3.3.3. Semiempirical stopping cross sections for ^{1}H, ^{4}He, ^{7}Li, ^{11}B, ^{12}C, ^{14}N, ^{16}O, ^{27}Al, ^{35}Cl, ^{63}Cu, ^{107}Ag and ^{197}Au ions in aluminum. Atomic weight = 26.982 amu, mass density = 2.7 g/cm^3, atomic density = $6.02 \cdot 10^{22}$ atoms/cm^3.

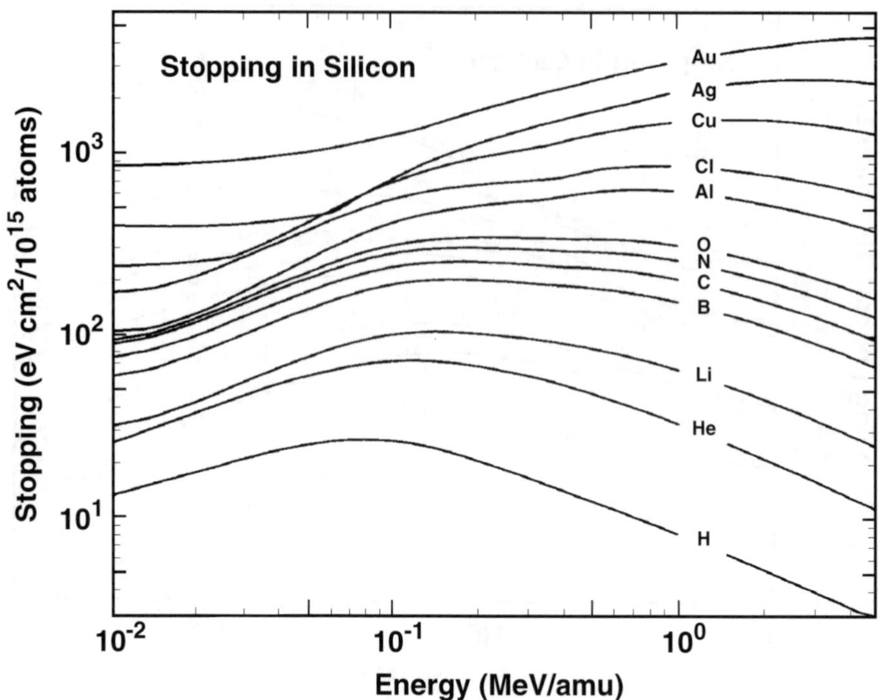

FIG. A3.3.4. Semiempirical stopping cross sections for ^{1}H, ^{4}He, ^{7}Li, ^{11}B, ^{12}C, ^{14}N, ^{16}O, ^{27}Al, ^{35}Cl, ^{63}Cu, ^{107}Ag and ^{197}Au ions in silicon. Atomic weight = 28.086 amu, mass density = 2.33 g/cm^3, atomic density = $5.00 \cdot 10^{22}$ atoms/cm^3.

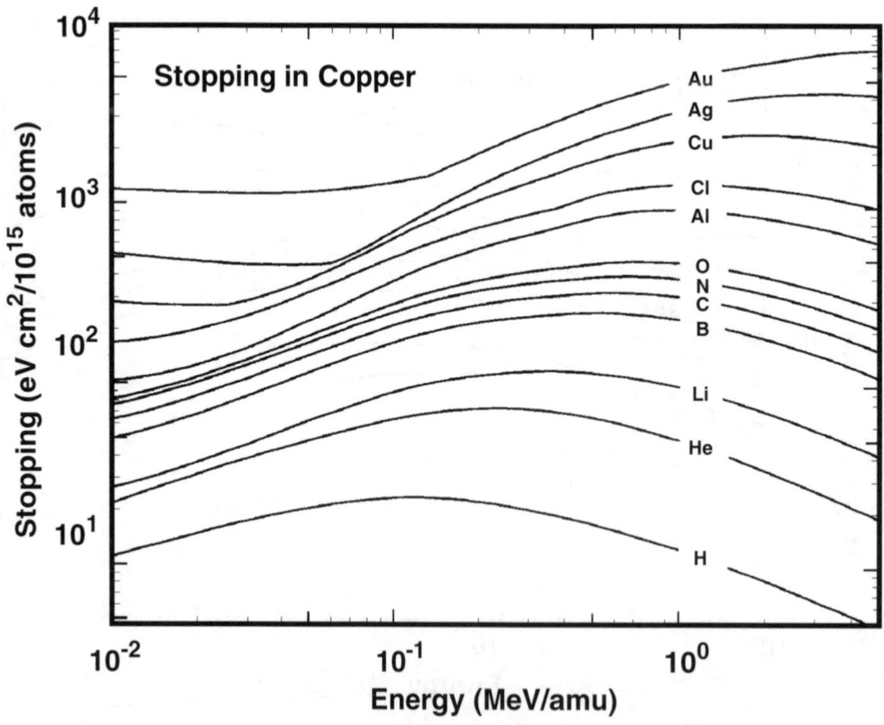

FIG. A3.3.5. Semiempirical stopping cross sections for ^{1}H, ^{4}He, ^{7}Li, ^{11}B, ^{12}C, ^{14}N, ^{16}O, ^{27}Al, ^{35}Cl, ^{63}Cu, ^{107}Ag and ^{197}Au ions in copper. Atomic weight = 63.546 amu, mass density = 8.93 g/cm^3, atomic density = $8.45 \cdot 10^{22}$ atoms/cm^3.

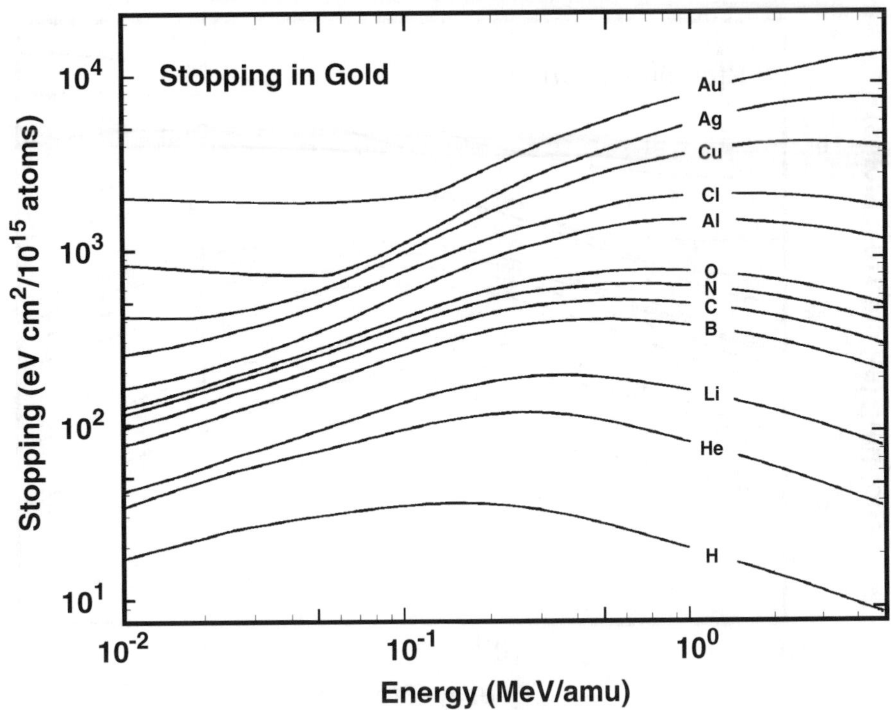

FIG. A3.3.6. Semiempirical stopping cross sections for ^{1}H, ^{4}He, ^{7}Li, ^{11}B, ^{12}C, ^{14}N, ^{16}O, ^{27}Al, ^{35}Cl, ^{63}Cu, ^{107}Ag and ^{197}Au ions in gold. Atomic weight = 196.967 amu, mass density = 19.28 g/cm^3, atomic density = $5.90 \cdot 10^{22}$ atoms/cm^3.

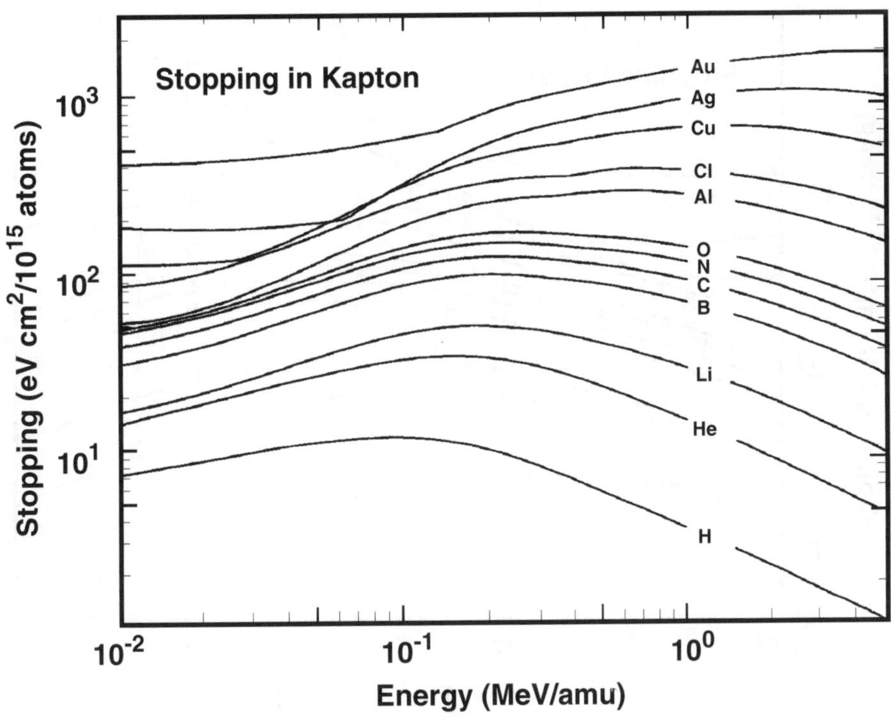

FIG. A3.3.7. Semiempirical stopping cross sections for ^{1}H, ^{4}He, ^{7}Li, ^{11}B, ^{12}C, ^{14}N, ^{16}O, ^{27}Al, ^{35}Cl, ^{63}Cu, ^{107}Ag and ^{197}Au ions in Kapton ($H_{10}C_{22}N_2O_5$). Mean atomic weight = 9.80 amu, mass density = 1.43 g/cm^3.

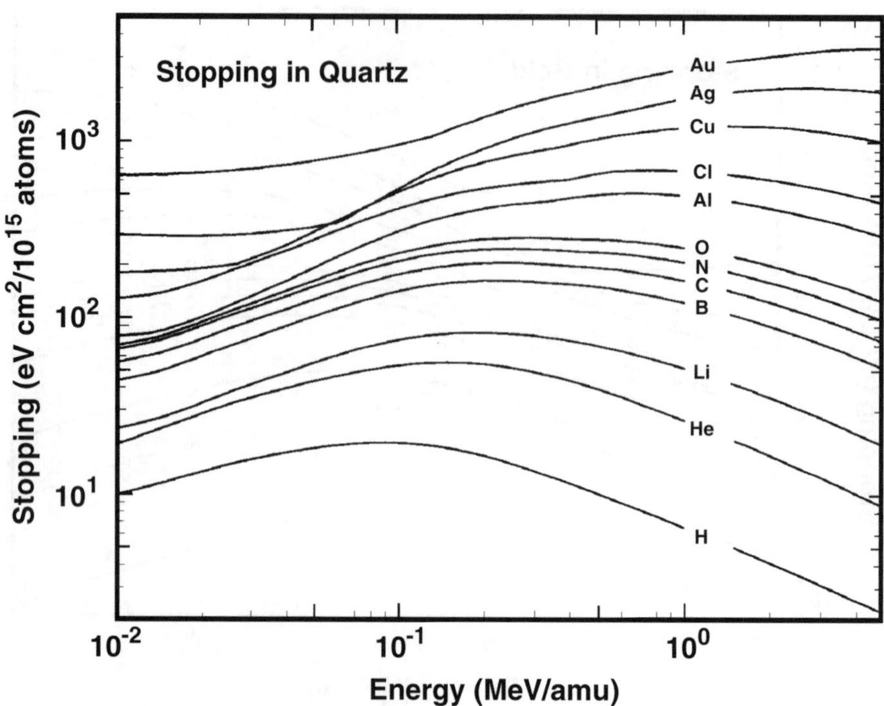

FIG. A3.3.8. Semiempirical stopping cross sections for ^{1}H, ^{4}He, ^{7}Li, ^{11}B, ^{12}C, ^{14}N, ^{16}O, ^{27}Al, ^{35}Cl, ^{63}Cu, ^{107}Ag and ^{197}Au ions in quartz (SiO_2). Mean atomic weight = 20.03 amu, mass density = 2.32 g/cm^3.

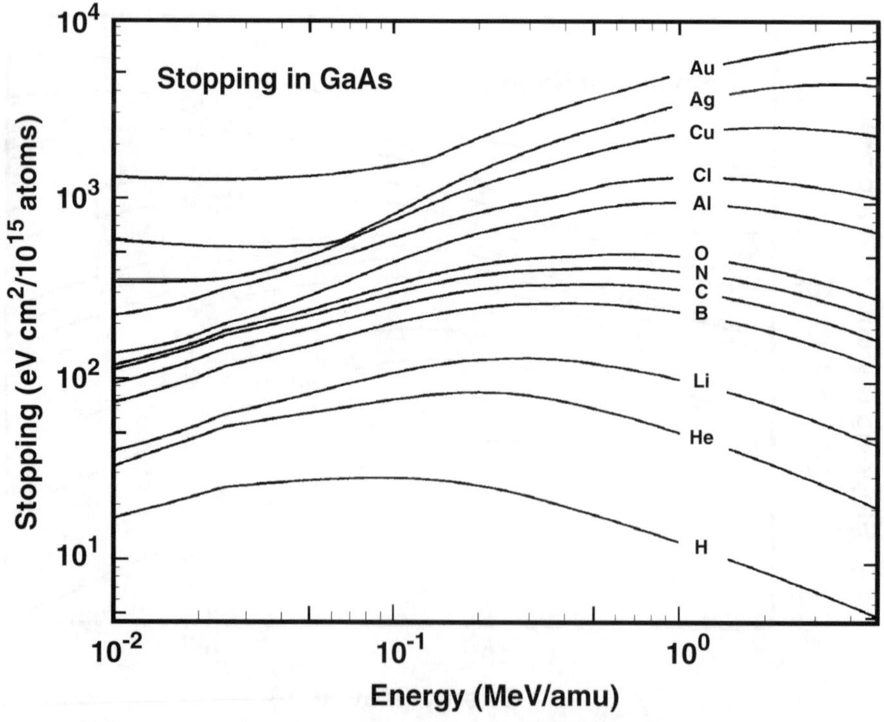

FIG. A3.3.9. Semiempirical stopping cross sections for ^{1}H, ^{4}He, ^{7}Li, ^{11}B, ^{12}C, ^{14}N, ^{16}O, ^{27}Al, ^{35}Cl, ^{63}Cu, ^{107}Ag and ^{197}Au ions in GaAs. Mean atomic weight = 72.32 amu, mass density = 5.32 g/cm^3.

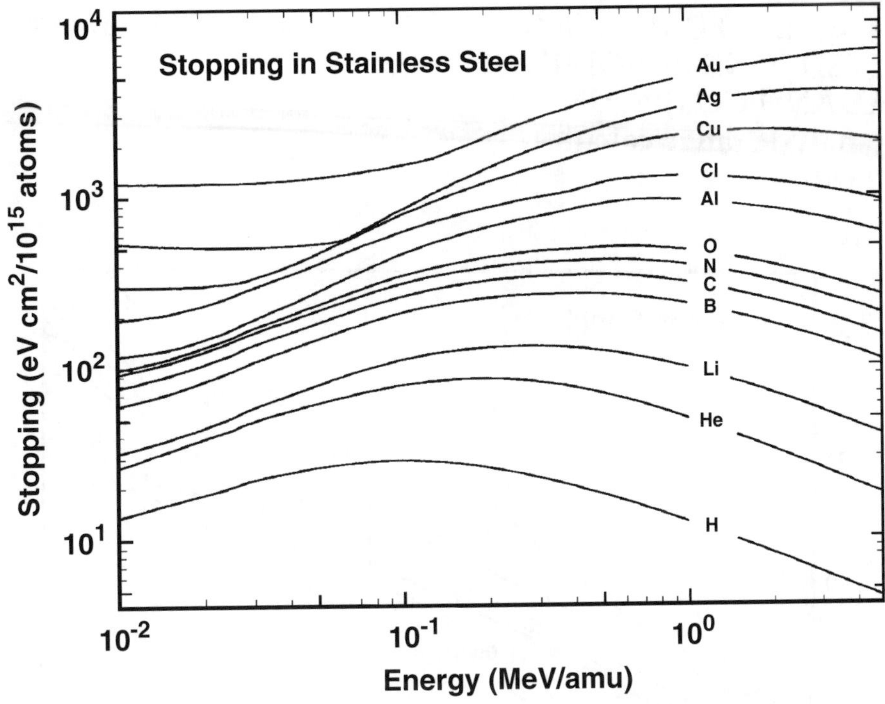

FIG. A3.3.10. Semiempirical stopping cross sections for ^{1}H, ^{4}He, ^{7}Li, ^{11}B, ^{12}C, ^{14}N, ^{16}O, ^{27}Al, ^{35}Cl, ^{63}Cu, ^{107}Ag and ^{197}Au ions in stainless steel (8% Cr, 74% Fe, 18% Ni). Mean atomic weight = 56.1 amu, mass density = 8.0 g/cm^3.

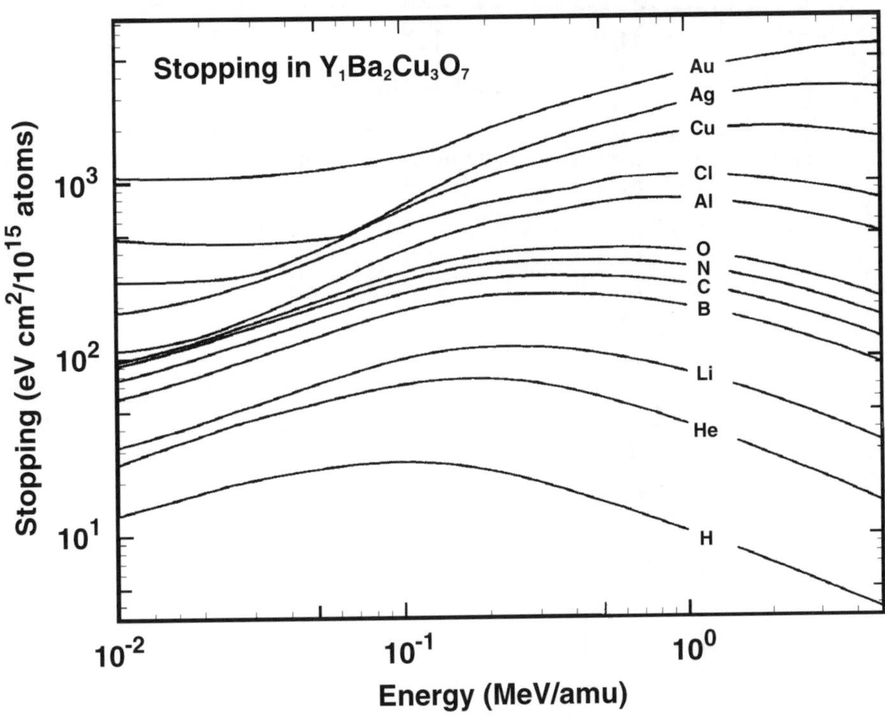

FIG. A3.3.11. Semiempirical stopping cross sections for ^{1}H, ^{4}He, ^{7}Li, ^{11}B, ^{12}C, ^{14}N, ^{16}O, ^{27}Al, ^{35}Cl, ^{63}Cu, ^{107}Ag and ^{197}Au ions in $Y_1Ba_2Cu_3O_7$. Mean atomic weight = 51.2 amu, mass density = 6.54 g/cm^3.

A3.4 PLOTS OF PROJECTED RANGES FOR SELECTED IONS IN SOME ELEMENTAL AND COMPOUND TARGETS (analytic range calculations by TRIM-91).

Note: all curves maintain their relative positions. They do not cross over.

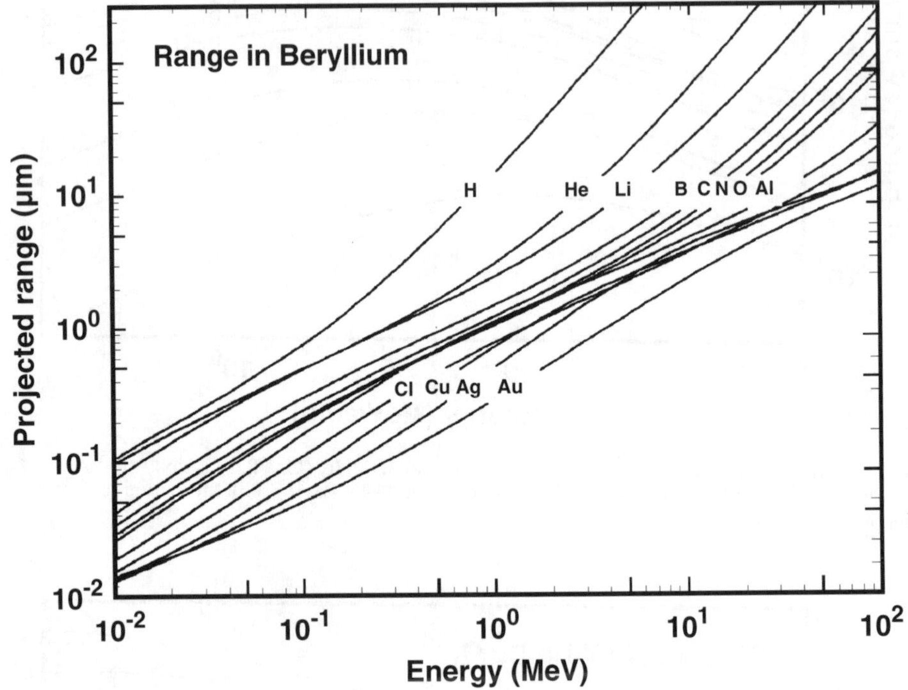

FIG. A3.4.1. Projected ranges for ^{1}H, ^{4}He, ^{7}Li, ^{11}B, ^{12}C, ^{14}N, ^{16}O, ^{27}Al, ^{35}Cl, ^{63}Cu, ^{107}Ag and ^{197}Au ions in beryllium. Atomic weight = 9.012 amu, mass density = 1.82 g/cm^3, atomic density = $1.21 \cdot 10^{23}$ atoms/cm^3.

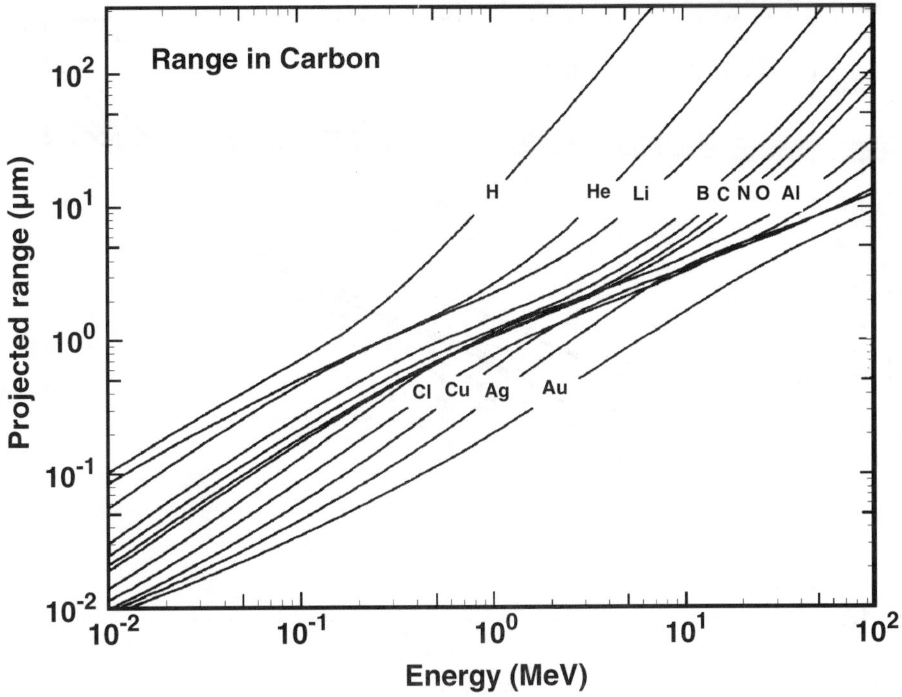

FIG. A3.4.2. Projected ranges for ^{1}H, ^{4}He, ^{7}Li, ^{11}B, ^{12}C, ^{14}N, ^{16}O, ^{27}Al, ^{35}Cl, ^{63}Cu, ^{107}Ag and ^{197}Au ions in carbon. Atomic weight = 12.011 amu, mass density = 3.516 g/cm^3, atomic density = $1.76 \cdot 10^{23}$ atoms/cm^3.

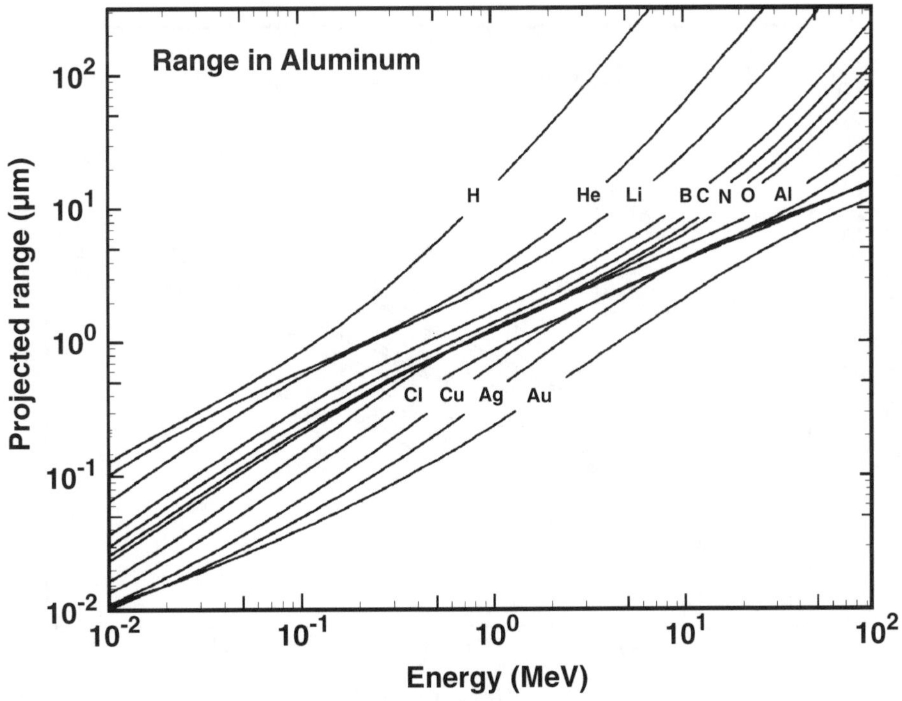

FIG. A3.4.3. Projected ranges for ^{1}H, ^{4}He, ^{7}Li, ^{11}B, ^{12}C, ^{14}N, ^{16}O, ^{27}Al, ^{35}Cl, ^{63}Cu, ^{107}Ag and ^{197}Au ions in aluminum. Atomic weight = 26.982 amu, mass density = 2.7 g/cm^3, atomic density = $6.02 \cdot 10^{22}$ atoms/cm^3.

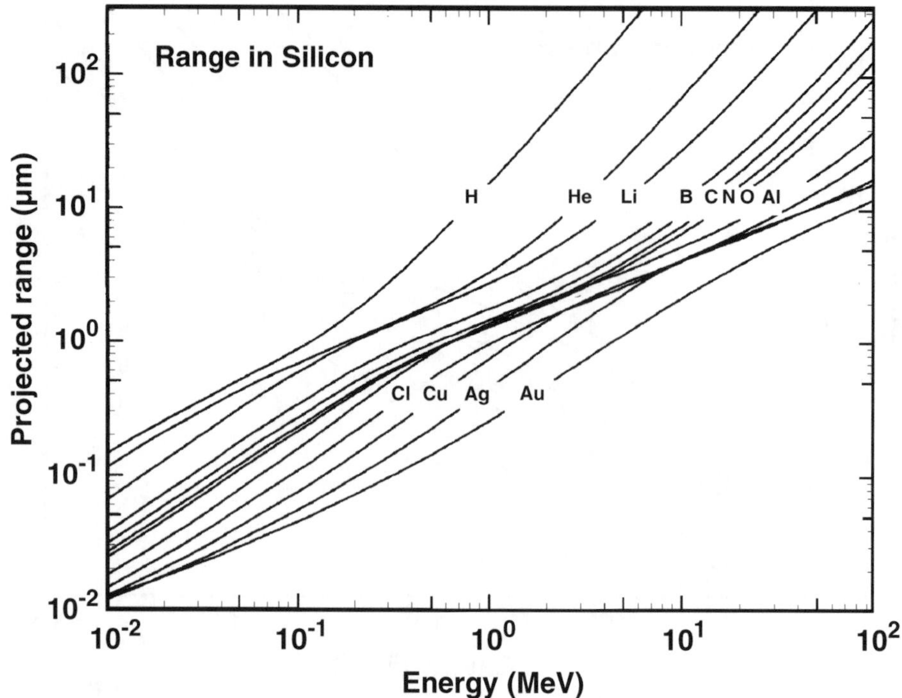

FIG. A3.4.4. Projected ranges for ^{1}H, ^{4}He, ^{7}Li, ^{11}B, ^{12}C, ^{14}N, ^{16}O, ^{27}Al, ^{35}Cl, ^{63}Cu, ^{107}Ag and ^{197}Au ions in silicon. Atomic weight = 28.086 amu, mass density = 2.33 g/cm^3, atomic density = $5.00 \cdot 10^{22}$ atoms/cm^3.

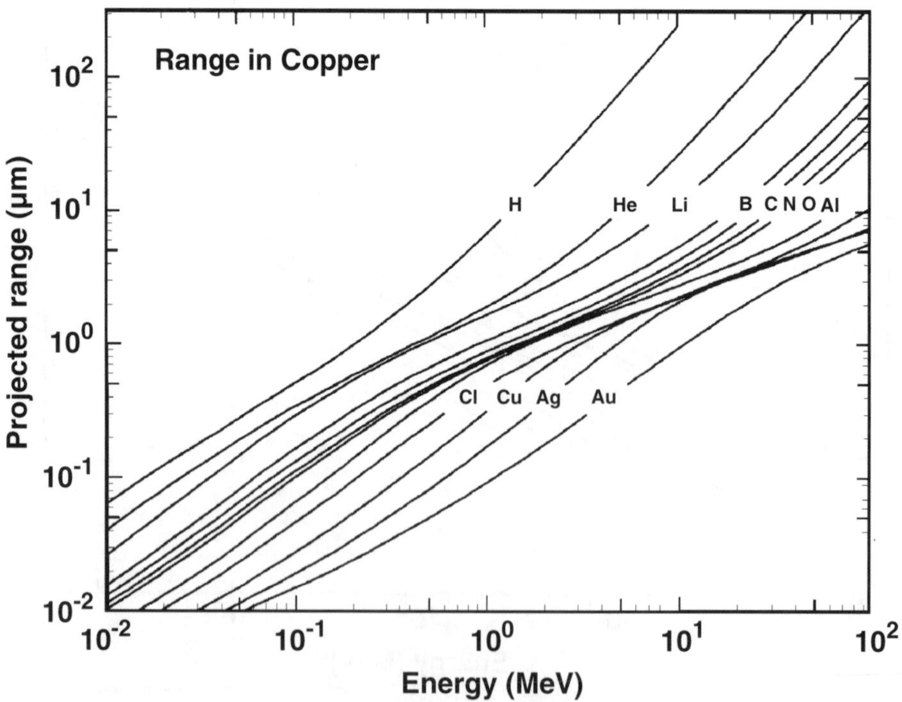

FIG. A3.4.5. Projected ranges for ^{1}H, ^{4}He, ^{7}Li, ^{11}B, ^{12}C, ^{14}N, ^{16}O, ^{27}Al, ^{35}Cl, ^{63}Cu, ^{107}Ag and ^{197}Au ions in copper. Atomic weight = 63.546 amu, mass density = 8.93 g/cm^3, atomic density = $8.45 \cdot 10^{22}$ atoms/cm^3.

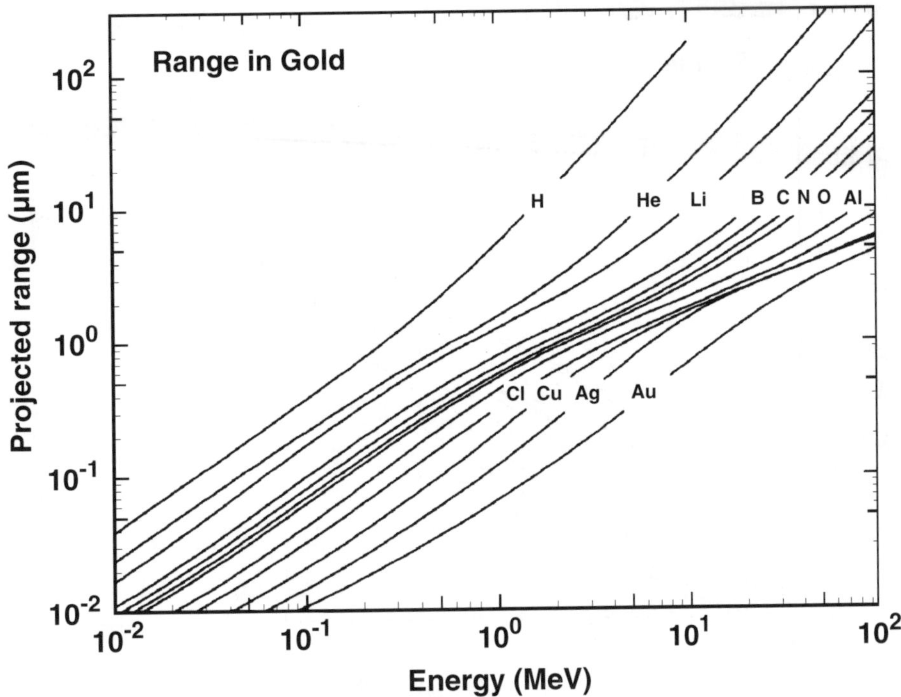

FIG. A3.4.6. Projected ranges for ^{1}H, ^{4}He, ^{7}Li, ^{11}B, ^{12}C, ^{14}N, ^{16}O, ^{27}Al, ^{35}Cl, ^{63}Cu, ^{107}Ag and ^{197}Au ions in gold. Atomic weight = 196.967 amu, mass density = 19.28 g/cm^{3}, atomic density = $5.90 \cdot 10^{22}$ atoms/cm^{3}.

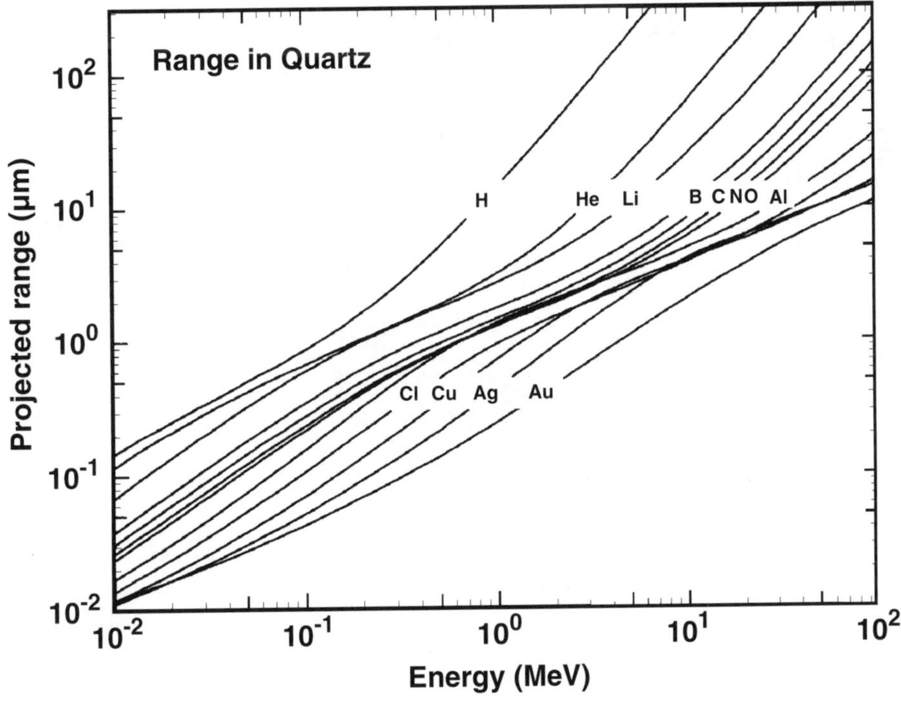

FIG. A3.4.7. Projected ranges for ^{1}H, ^{4}He, ^{7}Li, ^{11}B, ^{12}C, ^{14}N, ^{16}O, ^{27}Al, ^{35}Cl, ^{63}Cu, ^{107}Ag and ^{197}Au ions in quartz (SiO_2). Mean atomic weight = 20.03 amu, mass density = 2.32 g/cm^{3}.

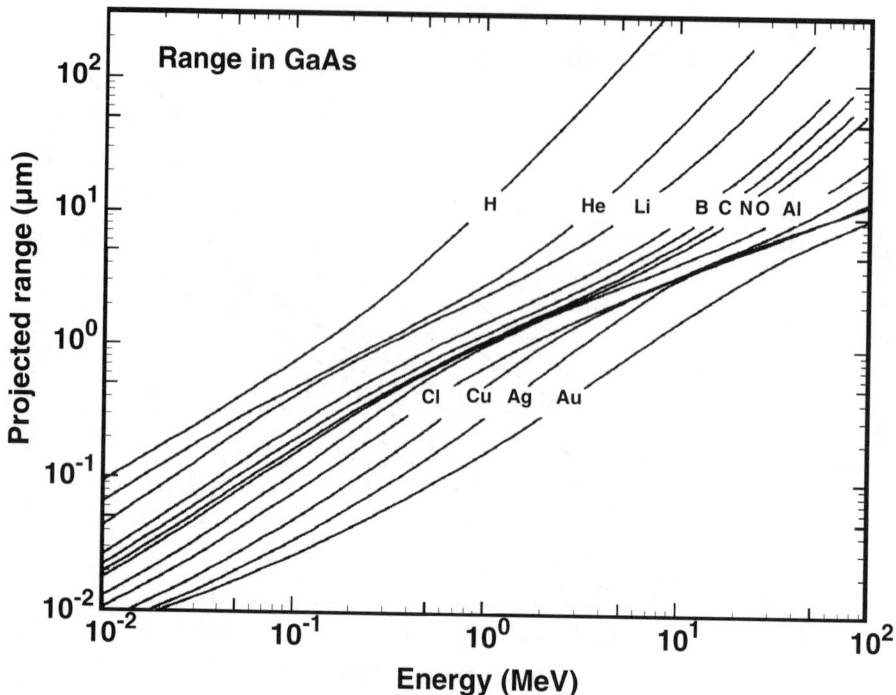

FIG. A3.4.8. Projected ranges for ^{1}H, ^{4}He, ^{7}Li, ^{11}B, ^{12}C, ^{14}N, ^{16}O, ^{27}Al, ^{35}Cl, ^{63}Cu, ^{107}Ag and ^{197}Au ions in GaAs. Mean atomic weight = 72.32 amu, mass density = 5.32 g/cm^3.

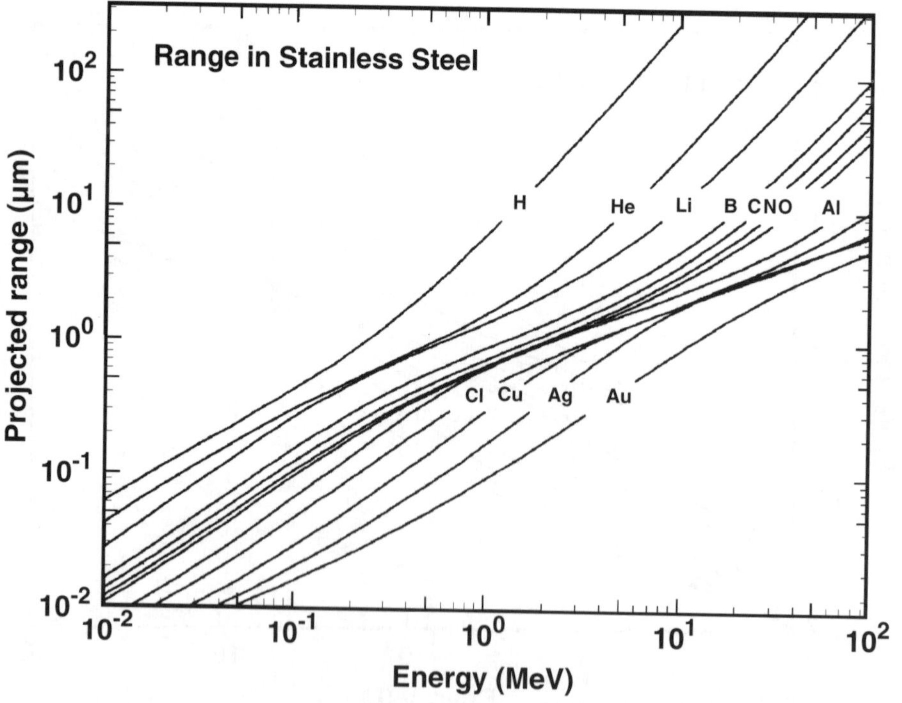

FIG. A3.4.9. Projected ranges for ^{1}H, ^{4}He, ^{7}Li, ^{11}B, ^{12}C, ^{14}N, ^{16}O, ^{27}Al, ^{35}Cl, ^{63}Cu, ^{107}Ag and ^{197}Au ions in stainless steel (8% Cr, 74% Fe, 18% Ni). Mean atomic weight = 56.1 amu, mass density = 8.0 g/cm^3.

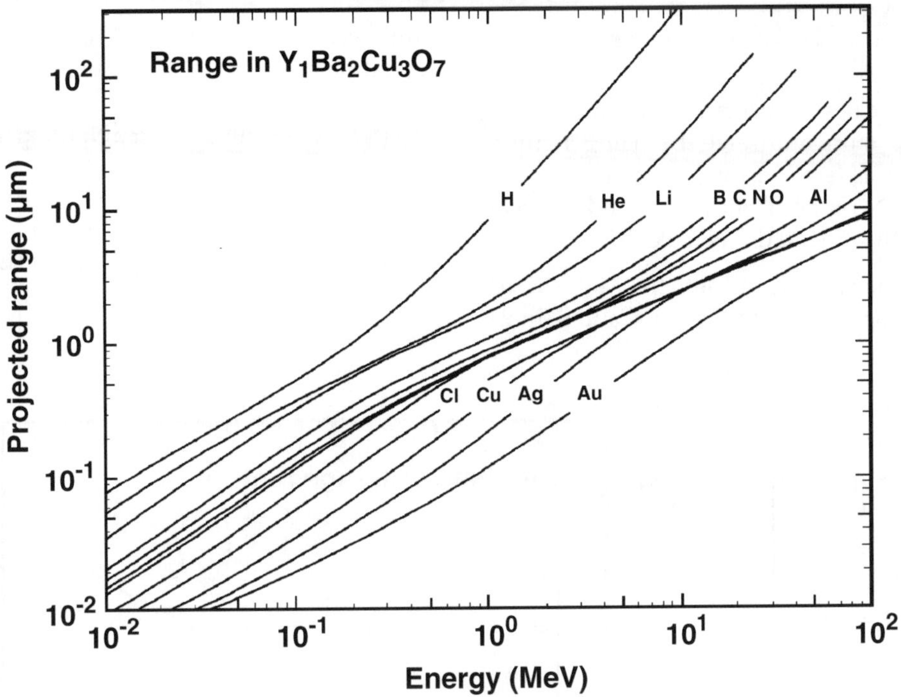

FIG. A3.4.10. Projected ranges for ^{1}H, ^{4}He, ^{7}Li, ^{11}B, ^{12}C, ^{14}N, ^{16}O, ^{27}Al, ^{35}Cl, ^{63}Cu, ^{107}Ag and ^{197}Au ions in $Y_1Ba_2Cu_3O_7$. Mean atomic weight = 51.2 amu, mass density = 6.54 g/cm^3.

A3.5 READING THE LOGARITHMIC SCALES

The logarithmic scales in the stopping and range figures may be read more accurately by using a ruler: Measure the lengths L (e.g., in mm) spanned by any number of decades N on both horizontal and vertical scales according to the schematic (FIG. A3.5.1).

The conversions between E in MeV (or MeV/amu) and E in mm, between S [eV cm^2/10^{15}atoms] and S [mm] and between R [μm] and R [mm] are then given by the equations:

Stopping: $E[mm] = (L_E/N_E)(\log(E[MeV/amu]) - a)$

$S[eVcm^2/10^{15}atoms] = 10^{((NS[mm]/L) + k)}$

Range: $E[mm] = (L_E/N_E)(\log(E[MeV]) - a)$

$R[\mu m] = 10^{((NR[mm]/L) + k)}$

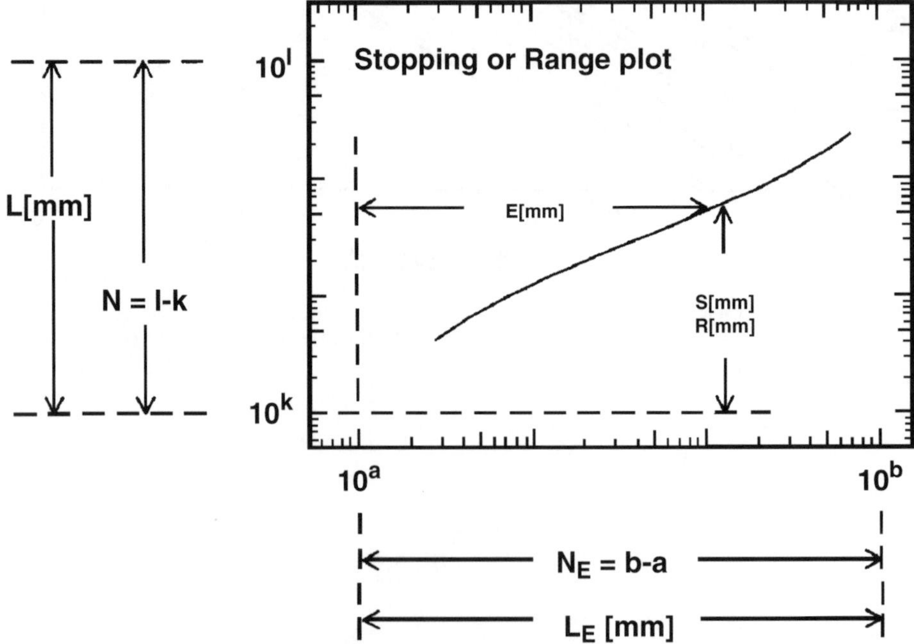

FIG. A3.5.1. Notation for reading the logarithmic scales.

APPENDIX

4

SCATTERING AND REACTION KINEMATICS

R. A. Weller

Vanderbilt University, Nashville, Tennessee

CONTENTS

A4.1 INTRODUCTION

This appendix is a compendium of formulae for the analysis of particle scattering and reactions. The forms that have been chosen for the equations are those that are most useful for the analysis of backscattering, elastic recoil, and nuclear reaction experiments performed for materials analysis. The tables are organized into two sections. Table A4.1 contains nonrelativistic expressions for the analysis of elastic scattering data and should suffice for reducing data from most Rutherford backscattering and elastic recoil experiments. Table A4.2 again presents nonrelativistic equations, but these are for the analysis of nuclear reactions where the final particles are not necessarily the same as the initial ones or where processes such as coulomb excitation of a nucleus significantly change the final kinetic energy of the system.

For high precision work, relativistic kinematics should be used. A discussion of relativistic kinematics can be found in Hagedorn (1963).

A4.2 ELASTIC COLLISIONS

You should use this section if you are analyzing nonrelativistic elastic collisions between a projectile and a target that is at rest in the laboratory. A good working definition of such collisions is that no particle's speed exceeds 10% of the speed of light, c, ($c = 3.0 \times 10^8$ m/s) and no nuclear reactions occur.

Figure A4.1 shows the scattering geometry and illustrates the various angles and energies. Table A4.1 contains the expressions for elastic scattering.

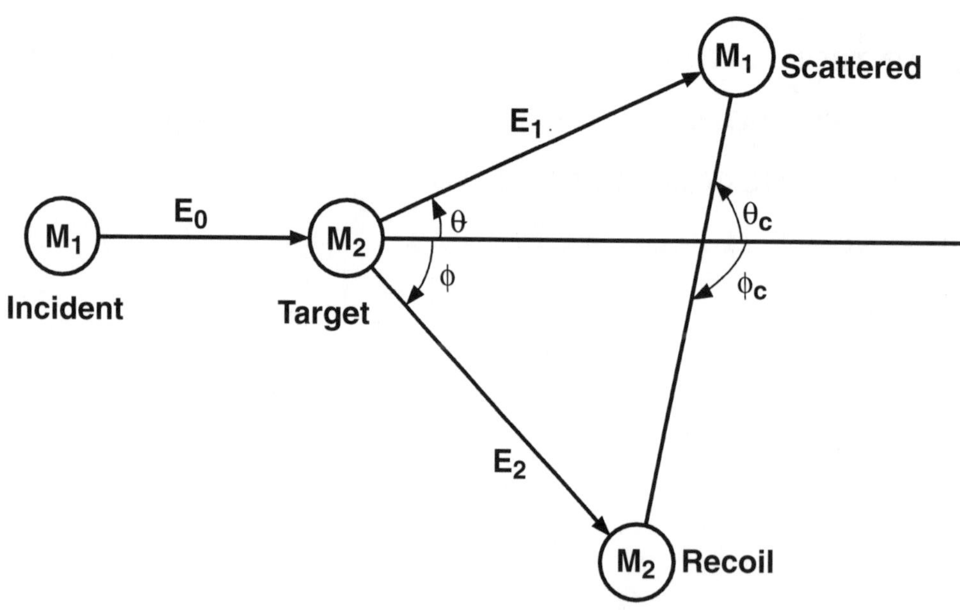

FIG. A4.1. Schematic scattering event as seen in the laboratory and center-of-mass coordinate systems illustrating the angles, energies, and velocities for nonrelativistic elastic collisions.

E_0 = Energy of the incident projectile.

E = Total kinetic energy in the center-of-mass system.

M_1 = Mass of the incident projectile.

M_2 = Mass of the target particle.

θ = Laboratory angle of the scattered projectile.

θ_c = Center-of-mass angle of the scattered projectile.

ϕ = Laboratory angle of the recoiling target.

ϕ_c = Center-of-mass angle of the recoiling target.

E_1 = Laboratory energy of the scattered projectile.

K = Backscattering kinematic factor E_1/E_0.

E_2 = Laboratory energy of the recoiling target.

x = Mass ratio M_1/M_2.

$\sigma(\theta)$ = Laboratory differential scattering cross section. $\sigma(\theta)d\Omega$ is the cross section to deflect the incident projectile by an angle θ into the solid angle $d\Omega$. $\sigma(\theta)$ is often symbolized by $\frac{d\sigma}{d\Omega}$.

$\sigma_c(\theta_c)$ = Differential scattering cross section in the center-of-mass system.

$\sigma_R(\phi)$ = Laboratory differential recoil cross section. Cross section per unit solid angle for the target nucleus to recoil at an angle ϕ with respect to the direction of the incident particle.

Scattering and Reaction Kinematics

Table A4.1. Kinematic expressions for elastic scattering.

Quantity calculated	**Expression**	**Equation number**
Center-of-mass energy	$E = \frac{M_2 E_0}{M_1 + M_2} = \frac{E_0}{1+x}$	A4.1
Laboratory energy of the scattered projectile for $M_1 \le M_2$	$K = \frac{E_1}{E_0} = \frac{\{x \cdot \cos(\theta) + [1 - x^2 \sin^2(\theta)]^{1/2}\}^2}{(1+x)^2}$	A4.2
	When $M_1 = M_2$, $\theta \le \frac{\pi}{2}$.	
Laboratory energies of the scattered projectile for $M_1 > M_2$	$\frac{E_1}{E_0} = \frac{\{x \cdot \cos(\theta) \pm [1 - x^2 \sin^2(\theta)]^{1/2}\}^2}{(1+x)^2}$ $\theta \le \sin^{-1}(x^{-1})$	A4.3
Laboratory energy of the recoil nucleus	$\frac{E_2}{E_0} = 1 - \frac{E_1}{E_0} = \frac{4M_1M_2}{(M_1+M_2)^2} \cos^2(\phi)$ $= \frac{4x}{(1+x)^2} \cos^2(\phi) = \frac{4x}{(1+x)^2} \sin^2(\theta_c/2)$	A4.4
	where $\phi \le \frac{\pi}{2}$.	
Laboratory angle of the recoil nucleus	$\phi = \frac{\pi - \theta_c}{2} = \frac{\phi_c}{2}$ $\sin(\phi) = \left(\frac{M_1E_1}{M_2E_2}\right)^{1/2} \sin(\theta)$	A4.5
Center-of-mass angle of the scattered projectile	$\theta_c = \pi - 2\phi = \pi - \phi_c$	A4.6
	When $M_1 \le M_2 \Rightarrow x \le 1$, θ_c is defined for all $\theta \le \pi$ and $\theta_c = \theta + \sin^{-1}[x \sin(\theta)]$ When $M_1 > M_2 \Rightarrow x > 1$, θ_c is double valued and the laboratory scattering angle is limited to the range $\theta \le \sin^{-1}(x^{-1})$. In this case $\theta_c = \theta + \sin^{-1}[x \sin(\theta)]$ or $\theta_c = \pi + \theta - \sin^{-1}[x \sin(\theta)]$	
Laboratory scattering cross section in terms of the center-of-mass cross section	$\sigma(\theta) = \frac{\sigma_c(\theta_c)\sin^2(\theta_c)}{\sin^2(\theta)\cos(\theta_c - \theta)}$ $= \sigma_c[\theta_c(\theta)] \times \frac{\{x \cdot \cos(\theta) + [1 - x^2 \sin^2(\theta)]^{1/2}\}^2}{[1 - x^2 \sin^2(\theta)]^{1/2}}$	A4.7
Laboratory recoil cross section	$\sigma_R(\phi) = 4\sigma_c[\theta_c(\phi)] \cos(\phi) = 4\sigma_c(\pi - 2\phi) \cos\phi$	A4.8

EXAMPLE A4.1. Laboratory scattering cross sections.

Cross sections such as the Rutherford scattering cross section are ordinarily computed in the center-of-mass coordinate system where they are expressed as functions of E and θ_c, but are needed for the evaluation of experiments in the laboratory coordinate system where the relevant parameters are E_0, θ, and the particle masses. To make the transformation proceed as follows.

Given E_0, M_1, M_2, and θ:

1) Compute $x = M_1/M_2$.

2) Compute $E = E_0/(1 + x)$.

3) Compute the center-of-mass scattering angle

$$\theta_c = \theta + \sin^{-1}[x \sin(\theta)] \ .$$

4) Compute the cross section in the center-of-mass frame for the scattering event of interest. For the Rutherford scattering cross section

$$\sigma_c(\theta_c) = \frac{Z_1^2 Z_2^2 e^4}{16 E^2 \sin^4(\theta_c/2)} \ . \tag{A4.9}$$

Here Z_1 and Z_2 are the atomic numbers of the colliding atoms and e is the electron's charge (note, $e^2 = 1.44$ eV-nm).

5) Obtain the desired result

$$\sigma(\theta) = \frac{\sigma_c(\theta_c)\sin^2(\theta_c)}{\sin^2(\theta)\cos(\theta_c - \theta)} \ . \tag{A4.10}$$

There is a relatively compact closed-form expression for the Rutherford cross section in the laboratory coordinate system (Chu *et al.*, 1978). It is

$$\sigma(\theta) = \left(\frac{Z_1 Z_2 e^2}{2E_0}\right)^2 \times \frac{\left\{\cos(\theta) + [1 - x^2 \sin^2(\theta)]^{1/2}\right\}^2}{\sin^4(\theta)[1 - x^2 \sin^2(\theta)]^{1/2}} \tag{A4.11}$$

EXAMPLE A4.2. Laboratory recoil cross sections.

In elastic recoil experiments, one ordinarily measures the number of recoiling target atoms observed in a detector with fixed angle ϕ and a known solid angle. The cross section for these events is related to the center-of-mass, scattering cross section by Eq. (A4.8). The procedure for obtaining this cross section is analogous to that outlined in Example A4.1.

Given E_0, M_1, M_2, and ϕ:

1) Compute x.

2) Compute E.

3) Compute $\theta_c = \pi - 2\phi$.

4) Compute the center-of-mass scattering cross section [Note that $\sin(\theta_c/2) = \cos(\phi)$].

5) Obtain the desired result

$$\sigma_R(\phi) = 4\,\sigma_c(\theta_c)\cos(\phi) \ . \tag{A4.12}$$

For Rutherford scattering, this result is especially simple

$$\sigma_R(\phi) = \left(\frac{Z_1 Z_2 e^2}{2E_0}\right)^2 \frac{(1+x)^2}{\cos^3(\phi)} \ . \tag{A4.13}$$

A4.3 INELASTIC COLLISIONS

The nuclear reactions that are most commonly used for materials analysis involve light ions. These reactions can usually be analyzed without invoking relativistic kinematics. This section, which is adapted from the compilation by Marion and Young (1968), contains non-relativistic equations for the analysis of two kinds of scattering events:

- when there is a net energy change (a non-zero "Q" value), or
- when nucleons are transferred, resulting in reaction products with atomic masses that differ from those of the reactants (and where also $Q \neq 0$ in general).

Table A4.2 contains expressions for the case where a particle of mass M_1 is incident upon another with mass M_2 which is at rest.

Figure A4.2 shows the scattering geometry and illustrates the various angles and energies.

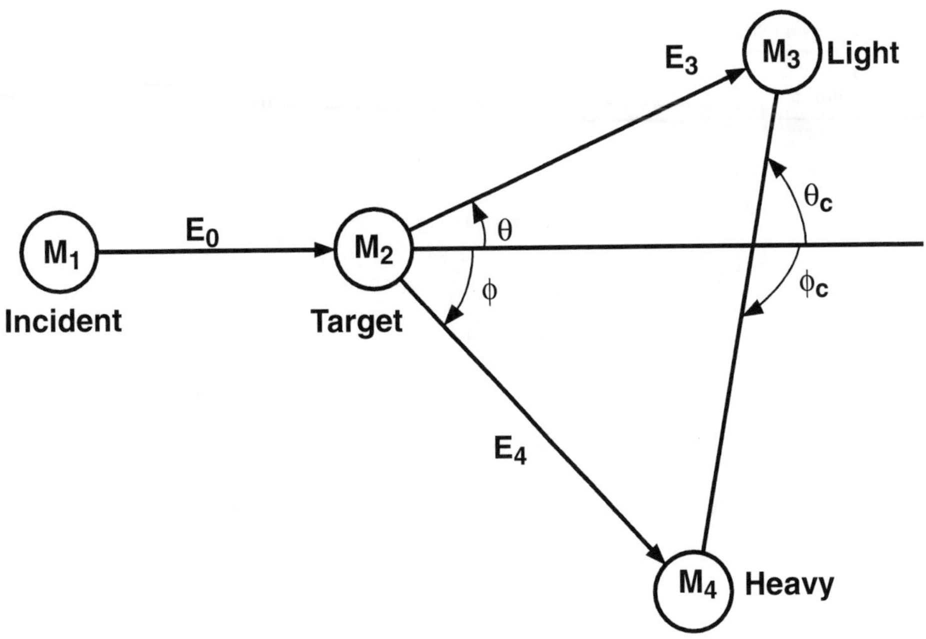

FIG. A4.2. Schematic scattering event as seen in the laboratory and center-of-mass coordinate systems illustrating the angles, energies, and velocities for nonrelativistic inelastic collisions.

E_0 = Energy of the incident projectile.

M_1 = Mass of the incident projectile.

M_2 = Mass of the target particle.

M_3 = Mass of the lighter reaction product.

M_4 = Mass of the heavier reaction product.

Q = Energy released by the reaction = $(M_1 + M_2 - M_3 - M_4)c^2$. $Q < 0 \Rightarrow$ endothermic reaction.

θ = Laboratory angle of M_3.

θ_c = Center-of-mass angle of M_3.

ϕ = Laboratory angle of M_4.

ϕ_c = Center-of-mass angle of M_4.

E_3 = Laboratory energy of M_3.

E_{3c} = Center-of-mass energy of M_3.

E_4 = Laboratory energy of M_4.

E_{4c} = Center-of-mass energy of M_4.

$\sigma_3(\theta)$ = Laboratory differential scattering cross section for production of the light product.

$\sigma_{3c}(\theta_c)$ = Center-of-mass differential cross section for the production of the light product.

$\sigma_4(\phi)$ = Laboratory differential cross section for the production of the heavy product.

$\sigma_{4c}(\phi_c)$ = Center-of-mass differential cross section for the production of the heavy product.

$$E_T \equiv E_0 + Q = E_3 + E_4$$

$$A \equiv \frac{M_1 M_4 (E_0 / E_T)}{(M_1 + M_2)(M_3 + M_4)}$$

$$B \equiv \frac{M_1 M_3 (E_0 / E_T)}{(M_1 + M_2)(M_3 + M_4)}$$

$$C \equiv \frac{M_2 M_3}{(M_1 + M_2)(M_3 + M_4)}\left(1 + \frac{M_1 Q}{M_2 E_T}\right) = \frac{E_{4c}}{E_T}$$

$$D \equiv \frac{M_2 M_4}{(M_1 + M_2)(M_3 + M_4)}\left(1 + \frac{M_1 Q}{M_2 E_T}\right) = \frac{E_{3c}}{E_T}$$

The quantities A, B, C, and D obey the constraints, A+B+C+D = 1 and AC = BD.

Table A4.2. Kinematic expressions for inelastic scattering.

Quantity calculated	Expression	Equation number
Laboratory energy of the light reaction product	$\frac{E_3}{E_T} = B + D + 2(AC)^{1/2}\cos(\theta_c)$ When $B \le D$, $\frac{E_3}{E_T} = B\{\cos(\theta) + [D/B - \sin^2(\theta)]^{1/2}\}^2$ When $B > D$, E_3 is double valued, $\frac{E_3}{E_T} = B\{\cos(\theta) \pm [D/B - \sin^2(\theta)]^{1/2}\}^2$ and $\theta \le \sin^{-1}[(D/B)^{1/2}]$	A4.14
Laboratory energy of the heavy reaction product	$\frac{E_4}{E_T} = A + C + 2(AC)^{1/2}\cos(\phi_c)$ When $A \le C$, $\frac{E_4}{E_T} = A\{\cos(\phi) + [C/A - \sin^2(\phi)]^{1/2}\}^2$ When $A > C$, E_4 is double valued, $\frac{E_4}{E_T} = A\{\cos(\phi) \pm [C/A - \sin^2(\phi)]^{1/2}\}^2$ and $\phi \le \sin^{-1}[(C/A)^{1/2}]$	A4.15
Laboratory angle of the heavy product	$\sin(\phi) = \left(\frac{M_3E_3}{M_4E_4}\right)^{1/2}\sin(\theta)$	A4.16
Center-of-mass angle of the light product	$\sin(\theta_c) = \left(\frac{E_3/E_T}{D}\right)\sin(\theta)$	A4.17
Cross-section transformation between the laboratory and center-of-mass systems for the light product	$\frac{\sigma_3(\theta)}{\sigma_{3c}(\theta_c)} = \frac{\sin(\theta_c)\,d\theta_c}{\sin(\theta)\,d\theta} = \frac{\sin^2(\theta_c)}{\sin^2(\theta)\cos(\theta_c - \theta)}$ $= \frac{E_3/E_T}{(AC)^{1/2}[D/B - \sin^2(\theta)]^{1/2}}$	A4.18
Cross-section transformation between the laboratory and center-of-mass systems for the heavy product	$\frac{\sigma_4(\phi)}{\sigma_{4c}(\phi_c)} = \frac{\sin(\phi_c)\,d\phi_c}{\sin(\phi)\,d\phi} = \frac{\sin^2(\phi_c)}{\sin^2(\phi)\cos(\phi_c - \phi)}$ $= \frac{E_4/E_T}{(AC)^{1/2}[C/A - \sin^2(\phi)]^{1/2}}$	A4.19
Cross-section ratio for associated particles in the laboratory system	$\frac{\sigma_3(\theta)}{\sigma_4(\phi)} = \frac{\sin(\phi)\,d\phi}{\sin(\theta)\,d\theta} = \frac{\sin^2(\phi)\cos(\phi_c - \phi)}{\sin^2(\theta)\cos(\theta_c - \theta)}$	A4.20

REFERENCES

Chu, W.-K., Mayer, J.W., and Nicolet, M.-A. (1979), *Backscattering Spectrometry*, Academic Press, New York, pp. 320-322.

Hagedorn, R. (1963), *Relativistic Kinematics*, W.A. Benjamin, New York, pp. 1-44.

Marion, J.B., and Young, F.C. (1968), *Nuclear Reaction Analysis*, North-Holland, Amsterdam, pp. 140-142.

K FACTORS FOR RBS

Compiled by

J. C. Barbour

Sandia National Laboratories, Albuquerque, New Mexico

CONTENTS

Table A5.1. Rutherford backscattering spectrometry kinematic factors for H as a projectile.

This table gives the RBS kinematic factors K_{M_2}, defined by Eq. (A4.2), for H as a projectile (M_1=1.0078 amu) and for the isotopic masses (M_2) of the elements. The kinematic factors are given as a function of scattering angle from 180° to 90° measured in the laboratory frame of reference. The first row for each element gives the average atomic weight of that element and the average kinematic factor (Average K_{M_2}) for that element. The subsequent rows give the isotopic masses for that element and the K_{M_2} for those isotopic masses. The average K_{M_2} is calculated as the weighted average of the K_{M_2}, in which the relative abundances (ϖ_{M_2}) for the M_2 are given in Appendix 1:

$$\text{Average } K_{M_2} = \Sigma\, \varpi_{M_2} K_{M_2}\text{ , summed over the masses } M_2.$$

Atomic no. (Z)	El.	Isotopic mass (M_2) (amu)	180°	170°	160°	150°	140°	130°	120°	110°	100°	90°
3	Li	6.941	0.5568	0.5592	0.5666	0.5789	0.5960	0.6179	0.6443	0.6748	0.7090	0.7461
		6.015	0.5084	0.5110	0.5187	0.5317	0.5500	0.5733	0.6017	0.6348	0.6721	0.7130
		7.016	0.5607	0.5632	0.5705	0.5827	0.5998	0.6215	0.6477	0.6781	0.7120	0.7488
4	Be	9.012	0.6381	0.6403	0.6468	0.6576	0.6725	0.6913	0.7139	0.7397	0.7682	0.7988
5	B	10.811	0.6877	0.6897	0.6955	0.7051	0.7184	0.7352	0.7551	0.7778	0.8027	0.8293
		10.013	0.6677	0.6697	0.6758	0.6859	0.6999	0.7175	0.7385	0.7625	0.7889	0.8171
		11.009	0.6927	0.6946	0.7004	0.7099	0.7230	0.7396	0.7592	0.7816	0.8061	0.8323
6	C	12.011	0.7143	0.7161	0.7216	0.7306	0.7429	0.7585	0.7769	0.7979	0.8208	0.8452
		12.000	0.7141	0.7159	0.7214	0.7304	0.7427	0.7583	0.7768	0.7977	0.8207	0.8450
		13.003	0.7330	0.7347	0.7399	0.7484	0.7601	0.7748	0.7921	0.8118	0.8333	0.8561
7	N	14.007	0.7495	0.7512	0.7561	0.7641	0.7752	0.7891	0.8055	0.8241	0.8443	0.8658
		14.003	0.7495	0.7511	0.7560	0.7641	0.7752	0.7891	0.8055	0.8240	0.8443	0.8657
		15.000	0.7640	0.7656	0.7702	0.7779	0.7884	0.8016	0.8172	0.8347	0.8539	0.8741
8	O	15.999	0.7770	0.7785	0.7829	0.7902	0.8003	0.8128	0.8276	0.8442	0.8624	0.8815
		15.995	0.7770	0.7785	0.7829	0.7902	0.8002	0.8128	0.8275	0.8442	0.8623	0.8815
		16.999	0.7887	0.7901	0.7943	0.8013	0.8109	0.8228	0.8369	0.8527	0.8699	0.8881
		17.999	0.7992	0.8005	0.8046	0.8112	0.8204	0.8318	0.8452	0.8603	0.8767	0.8940
9	F	18.998	0.8087	0.8100	0.8138	0.8202	0.8290	0.8399	0.8527	0.8672	0.8828	0.8993
10	Ne	20.179	0.8187	0.8199	0.8236	0.8297	0.8380	0.8484	0.8606	0.8744	0.8892	0.9048
		19.992	0.8173	0.8185	0.8222	0.8284	0.8368	0.8472	0.8595	0.8733	0.8883	0.9040
		20.994	0.8252	0.8264	0.8300	0.8359	0.8439	0.8540	0.8658	0.8790	0.8934	0.9084
		21.991	0.8324	0.8336	0.8370	0.8427	0.8505	0.8601	0.8715	0.8842	0.8979	0.9124
11	Na	22.990	0.8391	0.8402	0.8435	0.8490	0.8565	0.8658	0.8767	0.8889	0.9022	0.9160
12	Mg	24.305	0.8470	0.8481	0.8512	0.8565	0.8636	0.8725	0.8829	0.8945	0.9071	0.9203
		23.985	0.8452	0.8463	0.8495	0.8548	0.8620	0.8710	0.8815	0.8933	0.9060	0.9194
		24.986	0.8509	0.8520	0.8551	0.8602	0.8671	0.8758	0.8860	0.8973	0.9096	0.9225
		25.983	0.8562	0.8572	0.8602	0.8652	0.8719	0.8803	0.8901	0.9011	0.9129	0.9253
13	Al	26.982	0.8612	0.8621	0.8650	0.8698	0.8763	0.8845	0.8939	0.9046	0.9160	0.9280
14	Si	28.086	0.8662	0.8672	0.8700	0.8746	0.8809	0.8887	0.8979	0.9081	0.9192	0.9307
		27.977	0.8658	0.8667	0.8695	0.8742	0.8805	0.8883	0.8975	0.9078	0.9189	0.9305
		28.976	0.8701	0.8710	0.8737	0.8782	0.8844	0.8920	0.9009	0.9108	0.9216	0.9328
		29.974	0.8741	0.8750	0.8777	0.8820	0.8880	0.8954	0.9040	0.9137	0.9241	0.9349
15	P	30.974	0.8779	0.8788	0.8814	0.8856	0.8914	0.8986	0.9070	0.9163	0.9264	0.9370
16	S	32.064	0.8818	0.8826	0.8852	0.8893	0.8949	0.9018	0.9100	0.9191	0.9288	0.9390
		31.972	0.8815	0.8823	0.8849	0.8890	0.8946	0.9016	0.9097	0.9188	0.9287	0.9389
		32.972	0.8849	0.8857	0.8882	0.8922	0.8976	0.9044	0.9124	0.9212	0.9307	0.9407
		33.968	0.8881	0.8889	0.8912	0.8952	0.9005	0.9071	0.9148	0.9234	0.9327	0.9424
		35.967	0.8939	0.8947	0.8970	0.9007	0.9057	0.9120	0.9194	0.9275	0.9363	0.9455

Table A5.1. Rutherford backscattering spectrometry kinematic factors for H as a projectile (continued).

Atomic no. *(Z)*	El.	Isotopic mass (M_2) *(amu)*	180°	170°	160°	150°	140°	130°	120°	110°	100°	90°
17	Cl	35.453	0.8924	0.8932	0.8955	0.8993	0.9044	0.9108	0.9182	0.9265	0.9354	0.9447
		34.969	0.8911	0.8919	0.8942	0.8980	0.9032	0.9096	0.9171	0.9255	0.9346	0.9440
		36.966	0.8967	0.8974	0.8996	0.9032	0.9082	0.9143	0.9214	0.9294	0.9380	0.9469
18	Ar	39.948	0.9040	0.9047	0.9067	0.9101	0.9147	0.9204	0.9271	0.9345	0.9425	0.9508
		35.967	0.8939	0.8947	0.8970	0.9007	0.9057	0.9120	0.9194	0.9275	0.9363	0.9455
		37.963	0.8992	0.9000	0.9021	0.9057	0.9105	0.9165	0.9234	0.9312	0.9396	0.9483
		39.962	0.9040	0.9047	0.9068	0.9102	0.9148	0.9205	0.9271	0.9345	0.9425	0.9508
19	K	39.098	0.9020	0.9027	0.9048	0.9083	0.9129	0.9188	0.9256	0.9331	0.9413	0.9497
		38.964	0.9017	0.9024	0.9045	0.9080	0.9127	0.9185	0.9253	0.9329	0.9411	0.9496
		40.000	0.9041	0.9048	0.9069	0.9102	0.9148	0.9205	0.9272	0.9346	0.9426	0.9508
		40.962	0.9063	0.9069	0.9089	0.9123	0.9167	0.9223	0.9288	0.9361	0.9439	0.9520
20	Ca	40.078	0.9043	0.9050	0.9070	0.9104	0.9150	0.9207	0.9273	0.9347	0.9427	0.9509
		39.963	0.9040	0.9047	0.9068	0.9102	0.9148	0.9205	0.9271	0.9345	0.9425	0.9508
		41.959	0.9084	0.9090	0.9110	0.9142	0.9186	0.9241	0.9305	0.9376	0.9452	0.9531
		42.959	0.9104	0.9111	0.9130	0.9162	0.9205	0.9258	0.9320	0.9390	0.9464	0.9542
		43.956	0.9124	0.9130	0.9149	0.9180	0.9222	0.9274	0.9335	0.9403	0.9476	0.9552
		45.954	0.9160	0.9166	0.9184	0.9214	0.9255	0.9305	0.9363	0.9428	0.9498	0.9571
		47.952	0.9194	0.9199	0.9217	0.9246	0.9284	0.9333	0.9389	0.9451	0.9519	0.9588
21	Sc	44.956	0.9142	0.9148	0.9167	0.9197	0.9239	0.9290	0.9349	0.9416	0.9487	0.9561
22	Ti	47.878	0.9192	0.9198	0.9216	0.9244	0.9283	0.9332	0.9388	0.9450	0.9518	0.9588
		45.953	0.9160	0.9166	0.9184	0.9214	0.9254	0.9305	0.9363	0.9428	0.9498	0.9571
		46.952	0.9177	0.9183	0.9201	0.9230	0.9270	0.9319	0.9376	0.9440	0.9509	0.9580
		47.948	0.9194	0.9199	0.9217	0.9245	0.9284	0.9333	0.9389	0.9451	0.9519	0.9588
		48.948	0.9209	0.9215	0.9232	0.9260	0.9298	0.9346	0.9401	0.9462	0.9528	0.9597
		49.945	0.9224	0.9230	0.9247	0.9274	0.9312	0.9358	0.9413	0.9473	0.9537	0.9604
23	V	50.942	0.9239	0.9245	0.9261	0.9288	0.9325	0.9371	0.9424	0.9483	0.9546	0.9612
		49.947	0.9225	0.9230	0.9247	0.9275	0.9312	0.9358	0.9413	0.9473	0.9537	0.9604
		50.944	0.9239	0.9245	0.9261	0.9288	0.9325	0.9371	0.9424	0.9483	0.9546	0.9612
24	Cr	51.996	0.9254	0.9259	0.9275	0.9302	0.9338	0.9383	0.9435	0.9493	0.9555	0.9620
		49.946	0.9225	0.9230	0.9247	0.9275	0.9312	0.9358	0.9413	0.9473	0.9537	0.9604
		51.940	0.9253	0.9259	0.9275	0.9301	0.9338	0.9382	0.9434	0.9492	0.9555	0.9619
		52.941	0.9267	0.9272	0.9288	0.9314	0.9350	0.9394	0.9445	0.9502	0.9563	0.9626
		53.939	0.9280	0.9285	0.9301	0.9326	0.9361	0.9405	0.9455	0.9511	0.9571	0.9633
25	Mn	54.938	0.9292	0.9298	0.9313	0.9338	0.9373	0.9415	0.9464	0.9519	0.9578	0.9640
26	Fe	55.847	0.9303	0.9309	0.9324	0.9349	0.9382	0.9424	0.9473	0.9527	0.9585	0.9645
		53.940	0.9280	0.9285	0.9301	0.9326	0.9361	0.9405	0.9455	0.9511	0.9571	0.9633
		55.935	0.9305	0.9310	0.9325	0.9350	0.9383	0.9425	0.9474	0.9528	0.9586	0.9646
		56.935	0.9316	0.9321	0.9336	0.9361	0.9394	0.9435	0.9483	0.9536	0.9593	0.9652
		57.933	0.9328	0.9333	0.9347	0.9371	0.9404	0.9444	0.9491	0.9544	0.9600	0.9658
27	Co	58.933	0.9339	0.9344	0.9358	0.9382	0.9414	0.9454	0.9500	0.9551	0.9607	0.9664
28	Ni	58.688	0.9336	0.9341	0.9355	0.9379	0.9411	0.9451	0.9498	0.9549	0.9605	0.9662
		57.935	0.9328	0.9333	0.9347	0.9371	0.9404	0.9444	0.9491	0.9544	0.9600	0.9658
		59.931	0.9349	0.9354	0.9368	0.9392	0.9423	0.9462	0.9508	0.9559	0.9613	0.9669
		60.931	0.9360	0.9364	0.9378	0.9401	0.9432	0.9471	0.9516	0.9566	0.9619	0.9675
		61.928	0.9370	0.9374	0.9388	0.9411	0.9441	0.9479	0.9523	0.9573	0.9625	0.9680
		63.928	0.9389	0.9393	0.9407	0.9429	0.9458	0.9495	0.9538	0.9586	0.9637	0.9690
29	Cu	63.546	0.9385	0.9390	0.9403	0.9425	0.9455	0.9492	0.9535	0.9583	0.9634	0.9688
		62.930	0.9379	0.9384	0.9398	0.9420	0.9450	0.9487	0.9531	0.9579	0.9631	0.9685
		64.928	0.9398	0.9402	0.9416	0.9437	0.9466	0.9503	0.9545	0.9592	0.9642	0.9694

Table A5.1. Rutherford backscattering spectrometry kinematic factors for H as a projectile (continued).

Atomic no. *(Z)*	El.	Isotopic mass (M_2) *(amu)*	180°	170°	160°	150°	140°	130°	120°	110°	100°	90°
30	Zn	65.396	0.9402	0.9406	0.9419	0.9441	0.9470	0.9506	0.9548	0.9595	0.9644	0.9696
		63.929	0.9389	0.9393	0.9407	0.9429	0.9458	0.9495	0.9538	0.9586	0.9637	0.9690
		65.926	0.9407	0.9411	0.9424	0.9445	0.9474	0.9510	0.9552	0.9598	0.9648	0.9699
		66.927	0.9415	0.9420	0.9433	0.9453	0.9482	0.9517	0.9558	0.9604	0.9653	0.9703
		67.925	0.9424	0.9428	0.9441	0.9461	0.9489	0.9524	0.9565	0.9610	0.9658	0.9708
		69.925	0.9440	0.9444	0.9456	0.9476	0.9504	0.9537	0.9577	0.9621	0.9667	0.9716
31	Ga	69.723	0.9438	0.9442	0.9455	0.9475	0.9502	0.9536	0.9576	0.9619	0.9666	0.9715
		68.926	0.9432	0.9436	0.9449	0.9469	0.9497	0.9531	0.9571	0.9615	0.9663	0.9712
		70.925	0.9447	0.9452	0.9464	0.9483	0.9510	0.9544	0.9583	0.9626	0.9672	0.9720
32	Ge	72.632	0.9460	0.9464	0.9476	0.9495	0.9521	0.9554	0.9592	0.9634	0.9679	0.9726
		69.924	0.9440	0.9444	0.9456	0.9476	0.9504	0.9537	0.9577	0.9621	0.9667	0.9716
		71.922	0.9455	0.9459	0.9471	0.9490	0.9517	0.9550	0.9588	0.9631	0.9676	0.9724
		72.924	0.9462	0.9466	0.9478	0.9497	0.9524	0.9556	0.9594	0.9636	0.9681	0.9727
		73.921	0.9469	0.9473	0.9485	0.9504	0.9530	0.9562	0.9599	0.9641	0.9685	0.9731
		75.921	0.9483	0.9487	0.9498	0.9517	0.9542	0.9573	0.9610	0.9650	0.9693	0.9738
33	As	74.922	0.9476	0.9480	0.9492	0.9510	0.9536	0.9568	0.9604	0.9645	0.9689	0.9735
34	Se	78.993	0.9502	0.9506	0.9517	0.9535	0.9559	0.9589	0.9624	0.9663	0.9705	0.9748
		73.923	0.9469	0.9473	0.9485	0.9504	0.9530	0.9562	0.9599	0.9641	0.9685	0.9731
		75.919	0.9483	0.9487	0.9498	0.9517	0.9542	0.9573	0.9610	0.9650	0.9693	0.9738
		76.920	0.9489	0.9493	0.9504	0.9523	0.9548	0.9579	0.9615	0.9654	0.9697	0.9741
		77.917	0.9496	0.9499	0.9511	0.9529	0.9553	0.9584	0.9619	0.9659	0.9701	0.9745
		79.916	0.9508	0.9512	0.9523	0.9540	0.9564	0.9594	0.9629	0.9667	0.9708	0.9751
		81.917	0.9520	0.9523	0.9534	0.9551	0.9575	0.9604	0.9638	0.9675	0.9715	0.9757
35	Br	79.904	0.9508	0.9512	0.9522	0.9540	0.9564	0.9594	0.9629	0.9667	0.9708	0.9751
		78.918	0.9502	0.9506	0.9517	0.9535	0.9559	0.9589	0.9624	0.9663	0.9705	0.9748
		80.916	0.9514	0.9518	0.9528	0.9546	0.9570	0.9599	0.9633	0.9671	0.9712	0.9754
36	Kr	83.8	0.9530	0.9534	0.9544	0.9561	0.9584	0.9612	0.9646	0.9682	0.9722	0.9762
		77.920	0.9496	0.9500	0.9511	0.9529	0.9553	0.9584	0.9619	0.9659	0.9701	0.9745
		79.916	0.9508	0.9512	0.9523	0.9540	0.9564	0.9594	0.9629	0.9667	0.9708	0.9751
		81.913	0.9520	0.9523	0.9534	0.9551	0.9575	0.9604	0.9638	0.9675	0.9715	0.9757
		82.914	0.9525	0.9529	0.9539	0.9556	0.9580	0.9608	0.9642	0.9679	0.9719	0.9760
		83.911	0.9531	0.9534	0.9545	0.9562	0.9585	0.9613	0.9646	0.9683	0.9722	0.9763
		85.911	0.9542	0.9545	0.9555	0.9572	0.9594	0.9622	0.9654	0.9690	0.9728	0.9768
37	Rb	85.468	0.9539	0.9543	0.9553	0.9569	0.9592	0.9620	0.9652	0.9688	0.9727	0.9767
		84.912	0.9536	0.9540	0.9550	0.9567	0.9589	0.9618	0.9650	0.9686	0.9725	0.9765
		86.909	0.9547	0.9550	0.9560	0.9576	0.9599	0.9626	0.9658	0.9694	0.9731	0.9771
38	Sr	87.617	0.9550	0.9554	0.9564	0.9580	0.9602	0.9629	0.9661	0.9696	0.9734	0.9773
		83.913	0.9531	0.9534	0.9545	0.9562	0.9585	0.9613	0.9646	0.9683	0.9722	0.9763
		85.909	0.9542	0.9545	0.9555	0.9572	0.9594	0.9622	0.9654	0.9690	0.9728	0.9768
		86.909	0.9547	0.9550	0.9560	0.9576	0.9599	0.9626	0.9658	0.9694	0.9731	0.9771
		87.906	0.9552	0.9555	0.9565	0.9581	0.9603	0.9630	0.9662	0.9697	0.9734	0.9773
39	Y	88.906	0.9557	0.9560	0.9570	0.9586	0.9608	0.9634	0.9666	0.9700	0.9737	0.9776
40	Zr	91.221	0.9568	0.9571	0.9580	0.9596	0.9617	0.9643	0.9674	0.9708	0.9744	0.9781
		89.905	0.9561	0.9565	0.9574	0.9590	0.9612	0.9638	0.9669	0.9704	0.9740	0.9778
		90.906	0.9566	0.9569	0.9579	0.9595	0.9616	0.9642	0.9673	0.9707	0.9743	0.9781
		91.905	0.9571	0.9574	0.9584	0.9599	0.9620	0.9646	0.9676	0.9710	0.9746	0.9783
		93.906	0.9580	0.9583	0.9592	0.9607	0.9628	0.9654	0.9683	0.9716	0.9751	0.9788
		95.908	0.9588	0.9591	0.9601	0.9615	0.9636	0.9661	0.9690	0.9722	0.9756	0.9792

Table A5.1. Rutherford backscattering spectrometry kinematic factors for H as a projectile (continued).

Atomic no. (Z)	El.	Isotopic mass (M_2) (amu)	180°	170°	160°	150°	140°	130°	120°	110°	100°	90°
41	Nb	92.906	0.9575	0.9579	0.9588	0.9603	0.9624	0.9650	0.9680	0.9713	0.9749	0.9785
42	Mo	95.931	0.9588	0.9591	0.9600	0.9615	0.9636	0.9661	0.9690	0.9722	0.9756	0.9792
		91.907	0.9571	0.9574	0.9584	0.9599	0.9620	0.9646	0.9676	0.9710	0.9746	0.9783
		93.905	0.9580	0.9583	0.9592	0.9607	0.9628	0.9654	0.9683	0.9716	0.9751	0.9788
		94.906	0.9584	0.9587	0.9596	0.9611	0.9632	0.9657	0.9686	0.9719	0.9754	0.9790
		95.905	0.9588	0.9591	0.9601	0.9615	0.9636	0.9661	0.9690	0.9722	0.9756	0.9792
		96.906	0.9593	0.9596	0.9605	0.9619	0.9639	0.9664	0.9693	0.9725	0.9759	0.9794
		97.905	0.9597	0.9600	0.9609	0.9623	0.9643	0.9667	0.9696	0.9727	0.9761	0.9796
		99.908	0.9605	0.9607	0.9616	0.9631	0.9650	0.9674	0.9702	0.9733	0.9766	0.9800
43	Tc	98.000	0.9597	0.9600	0.9609	0.9623	0.9643	0.9668	0.9696	0.9728	0.9761	0.9796
44	Ru	101.019	0.9609	0.9612	0.9620	0.9634	0.9654	0.9677	0.9705	0.9736	0.9768	0.9802
		95.908	0.9588	0.9591	0.9601	0.9615	0.9636	0.9661	0.9690	0.9722	0.9756	0.9792
		97.905	0.9597	0.9600	0.9609	0.9623	0.9643	0.9667	0.9696	0.9727	0.9761	0.9796
		98.506	0.9599	0.9602	0.9611	0.9625	0.9645	0.9669	0.9698	0.9729	0.9763	0.9797
		99.904	0.9605	0.9607	0.9616	0.9631	0.9650	0.9674	0.9702	0.9733	0.9766	0.9800
		100.906	0.9608	0.9611	0.9620	0.9634	0.9653	0.9677	0.9705	0.9735	0.9768	0.9802
		101.904	0.9612	0.9615	0.9624	0.9638	0.9657	0.9680	0.9708	0.9738	0.9771	0.9804
		103.905	0.9619	0.9622	0.9631	0.9644	0.9663	0.9686	0.9713	0.9743	0.9775	0.9808
45	Rh	102.906	0.9616	0.9619	0.9627	0.9641	0.9660	0.9683	0.9710	0.9741	0.9773	0.9806
46	Pd	106.415	0.9628	0.9631	0.9639	0.9653	0.9671	0.9694	0.9720	0.9749	0.9780	0.9812
		101.906	0.9612	0.9615	0.9624	0.9638	0.9657	0.9680	0.9708	0.9738	0.9771	0.9804
		103.904	0.9619	0.9622	0.9631	0.9644	0.9663	0.9686	0.9713	0.9743	0.9775	0.9808
		104.905	0.9623	0.9626	0.9634	0.9648	0.9666	0.9689	0.9716	0.9745	0.9777	0.9810
		105.904	0.9626	0.9629	0.9638	0.9651	0.9669	0.9692	0.9719	0.9748	0.9779	0.9811
		107.904	0.9633	0.9636	0.9644	0.9657	0.9675	0.9698	0.9724	0.9752	0.9783	0.9815
		109.905	0.9640	0.9643	0.9651	0.9664	0.9681	0.9703	0.9729	0.9757	0.9787	0.9818
47	Ag	107.868	0.9633	0.9636	0.9644	0.9657	0.9675	0.9698	0.9724	0.9752	0.9783	0.9815
		106.905	0.9630	0.9633	0.9641	0.9654	0.9673	0.9695	0.9721	0.9750	0.9781	0.9813
		108.905	0.9637	0.9639	0.9647	0.9661	0.9678	0.9701	0.9726	0.9755	0.9785	0.9817
48	Cd	112.412	0.9648	0.9650	0.9658	0.9671	0.9688	0.9710	0.9735	0.9762	0.9792	0.9822
		105.907	0.9627	0.9629	0.9638	0.9651	0.9669	0.9692	0.9719	0.9748	0.9779	0.9811
		107.904	0.9633	0.9636	0.9644	0.9657	0.9675	0.9698	0.9724	0.9752	0.9783	0.9815
		109.903	0.9640	0.9643	0.9651	0.9664	0.9681	0.9703	0.9729	0.9757	0.9787	0.9818
		110.904	0.9643	0.9646	0.9654	0.9667	0.9684	0.9706	0.9731	0.9759	0.9789	0.9820
		111.903	0.9646	0.9649	0.9657	0.9669	0.9687	0.9708	0.9733	0.9761	0.9791	0.9821
		112.904	0.9649	0.9652	0.9660	0.9672	0.9690	0.9711	0.9736	0.9763	0.9793	0.9823
		113.903	0.9652	0.9655	0.9663	0.9675	0.9692	0.9713	0.9738	0.9765	0.9794	0.9825
		115.905	0.9658	0.9661	0.9668	0.9681	0.9698	0.9718	0.9743	0.9769	0.9798	0.9828
49	In	114.818	0.9655	0.9658	0.9665	0.9678	0.9695	0.9716	0.9740	0.9767	0.9796	0.9826
		112.904	0.9649	0.9652	0.9660	0.9672	0.9690	0.9711	0.9736	0.9763	0.9793	0.9823
		114.904	0.9655	0.9658	0.9665	0.9678	0.9695	0.9716	0.9740	0.9767	0.9796	0.9826
50	Sn	118.613	0.9666	0.9668	0.9676	0.9688	0.9704	0.9725	0.9748	0.9774	0.9802	0.9831
		111.905	0.9646	0.9649	0.9657	0.9669	0.9687	0.9708	0.9733	0.9761	0.9791	0.9821
		113.903	0.9652	0.9655	0.9663	0.9675	0.9692	0.9713	0.9738	0.9765	0.9794	0.9825
		114.903	0.9655	0.9658	0.9665	0.9678	0.9695	0.9716	0.9740	0.9767	0.9796	0.9826
		115.902	0.9658	0.9661	0.9668	0.9681	0.9698	0.9718	0.9743	0.9769	0.9798	0.9828
		116.903	0.9661	0.9664	0.9671	0.9683	0.9700	0.9721	0.9745	0.9771	0.9800	0.9829

Table A5.1. Rutherford backscattering spectrometry kinematic factors for H as a projectile (continued).

Atomic no. (Z)	El.	Isotopic mass (M_2) (amu)	180°	170°	160°	150°	140°	130°	120°	110°	100°	90°
		117.602	0.9663	0.9666	0.9673	0.9685	0.9702	0.9722	0.9746	0.9773	0.9801	0.9830
		118.903	0.9667	0.9669	0.9677	0.9689	0.9705	0.9725	0.9749	0.9775	0.9803	0.9832
		119.902	0.9669	0.9672	0.9679	0.9691	0.9707	0.9728	0.9751	0.9777	0.9805	0.9833
		121.903	0.9675	0.9677	0.9684	0.9696	0.9712	0.9732	0.9755	0.9781	0.9808	0.9836
		123.905	0.9680	0.9682	0.9689	0.9701	0.9717	0.9736	0.9759	0.9784	0.9811	0.9839
51	Sb	121.758	0.9674	0.9677	0.9684	0.9696	0.9712	0.9732	0.9755	0.9780	0.9808	0.9836
		120.904	0.9672	0.9675	0.9682	0.9694	0.9710	0.9730	0.9753	0.9779	0.9806	0.9835
		122.904	0.9677	0.9680	0.9687	0.9699	0.9715	0.9734	0.9757	0.9782	0.9809	0.9837
52	Te	127.586	0.9689	0.9691	0.9698	0.9709	0.9725	0.9744	0.9766	0.9790	0.9816	0.9843
		119.904	0.9669	0.9672	0.9679	0.9691	0.9707	0.9728	0.9751	0.9777	0.9805	0.9833
		121.903	0.9675	0.9677	0.9684	0.9696	0.9712	0.9732	0.9755	0.9781	0.9808	0.9836
		122.904	0.9677	0.9680	0.9687	0.9699	0.9715	0.9734	0.9757	0.9782	0.9809	0.9837
		123.903	0.9680	0.9682	0.9689	0.9701	0.9717	0.9736	0.9759	0.9784	0.9811	0.9839
		124.904	0.9682	0.9685	0.9692	0.9703	0.9719	0.9738	0.9761	0.9786	0.9812	0.9840
		125.903	0.9685	0.9687	0.9694	0.9706	0.9721	0.9740	0.9763	0.9787	0.9814	0.9841
		127.905	0.9690	0.9692	0.9699	0.9710	0.9726	0.9744	0.9766	0.9791	0.9817	0.9844
		129.906	0.9694	0.9697	0.9704	0.9715	0.9730	0.9748	0.9770	0.9794	0.9820	0.9846
53	I	126.905	0.9687	0.9690	0.9697	0.9708	0.9723	0.9742	0.9765	0.9789	0.9815	0.9842
54	Xe	131.293	0.9698	0.9700	0.9707	0.9718	0.9732	0.9751	0.9772	0.9796	0.9821	0.9848
		123.906	0.9680	0.9682	0.9689	0.9701	0.9717	0.9736	0.9759	0.9784	0.9811	0.9839
		125.904	0.9685	0.9687	0.9694	0.9706	0.9721	0.9740	0.9763	0.9787	0.9814	0.9841
		127.904	0.9690	0.9692	0.9699	0.9710	0.9726	0.9744	0.9766	0.9791	0.9817	0.9844
		128.905	0.9692	0.9694	0.9701	0.9712	0.9728	0.9746	0.9768	0.9792	0.9818	0.9845
		129.904	0.9694	0.9697	0.9704	0.9715	0.9730	0.9748	0.9770	0.9794	0.9820	0.9846
		130.905	0.9697	0.9699	0.9706	0.9717	0.9732	0.9750	0.9772	0.9795	0.9821	0.9847
		131.904	0.9699	0.9701	0.9708	0.9719	0.9734	0.9752	0.9773	0.9797	0.9822	0.9848
		133.905	0.9703	0.9706	0.9712	0.9723	0.9738	0.9756	0.9777	0.9800	0.9825	0.9851
		135.907	0.9708	0.9710	0.9716	0.9727	0.9741	0.9759	0.9780	0.9803	0.9827	0.9853
55	Cs	132.905	0.9701	0.9703	0.9710	0.9721	0.9736	0.9754	0.9775	0.9799	0.9824	0.9849
56	Ba	137.327	0.9711	0.9713	0.9719	0.9730	0.9744	0.9762	0.9782	0.9805	0.9829	0.9854
		129.906	0.9694	0.9697	0.9704	0.9715	0.9730	0.9748	0.9770	0.9794	0.9820	0.9846
		131.905	0.9699	0.9701	0.9708	0.9719	0.9734	0.9752	0.9773	0.9797	0.9822	0.9848
		133.904	0.9703	0.9706	0.9712	0.9723	0.9738	0.9756	0.9777	0.9800	0.9825	0.9851
		134.906	0.9706	0.9708	0.9714	0.9725	0.9740	0.9758	0.9778	0.9801	0.9826	0.9852
		135.905	0.9708	0.9710	0.9716	0.9727	0.9741	0.9759	0.9780	0.9803	0.9827	0.9853
		136.906	0.9710	0.9712	0.9718	0.9729	0.9743	0.9761	0.9782	0.9804	0.9829	0.9854
		137.905	0.9712	0.9714	0.9720	0.9731	0.9745	0.9763	0.9783	0.9806	0.9830	0.9855
57	La	138.905	0.9714	0.9716	0.9722	0.9733	0.9747	0.9764	0.9785	0.9807	0.9831	0.9856
		137.907	0.9712	0.9714	0.9720	0.9731	0.9745	0.9763	0.9783	0.9806	0.9830	0.9855
		138.906	0.9714	0.9716	0.9722	0.9733	0.9747	0.9764	0.9785	0.9807	0.9831	0.9856
58	Ce	140.115	0.9716	0.9719	0.9725	0.9735	0.9749	0.9766	0.9787	0.9809	0.9833	0.9857
		135.907	0.9708	0.9710	0.9716	0.9727	0.9741	0.9759	0.9780	0.9803	0.9827	0.9853
		137.906	0.9712	0.9714	0.9720	0.9731	0.9745	0.9763	0.9783	0.9806	0.9830	0.9855
		139.905	0.9716	0.9718	0.9724	0.9735	0.9749	0.9766	0.9786	0.9809	0.9832	0.9857
		141.909	0.9720	0.9722	0.9728	0.9738	0.9752	0.9769	0.9789	0.9811	0.9835	0.9859
59	Pr	140.908	0.9718	0.9720	0.9726	0.9737	0.9751	0.9768	0.9788	0.9810	0.9834	0.9858

Table A5.1. Rutherford backscattering spectrometry kinematic factors for H as a projectile (continued).

Atomic no. *(Z)*	El.	Isotopic mass (M_2) *(amu)*	180°	170°	160°	150°	140°	130°	120°	110°	100°	90°
60	Nd	144.242	0.9724	0.9726	0.9733	0.9743	0.9756	0.9773	0.9793	0.9814	0.9837	0.9861
		141.908	0.9720	0.9722	0.9728	0.9738	0.9752	0.9769	0.9789	0.9811	0.9835	0.9859
		142.910	0.9722	0.9724	0.9730	0.9740	0.9754	0.9771	0.9791	0.9812	0.9836	0.9860
		143.910	0.9724	0.9726	0.9732	0.9742	0.9756	0.9773	0.9792	0.9814	0.9837	0.9861
		144.913	0.9726	0.9728	0.9734	0.9744	0.9757	0.9774	0.9794	0.9815	0.9838	0.9862
		145.913	0.9728	0.9730	0.9736	0.9746	0.9759	0.9776	0.9795	0.9816	0.9839	0.9863
		147.917	0.9731	0.9733	0.9739	0.9749	0.9762	0.9779	0.9798	0.9819	0.9841	0.9865
		149.921	0.9735	0.9737	0.9743	0.9752	0.9765	0.9782	0.9800	0.9821	0.9843	0.9866
61	Pm	145.000	0.9726	0.9728	0.9734	0.9744	0.9757	0.9774	0.9794	0.9815	0.9838	0.9862
62	Sm	150.36	0.9735	0.9737	0.9743	0.9753	0.9766	0.9782	0.9801	0.9822	0.9844	0.9867
		143.912	0.9724	0.9726	0.9732	0.9742	0.9756	0.9773	0.9792	0.9814	0.9837	0.9861
		146.915	0.9729	0.9731	0.9737	0.9747	0.9761	0.9777	0.9796	0.9818	0.9840	0.9864
		147.915	0.9731	0.9733	0.9739	0.9749	0.9762	0.9779	0.9798	0.9819	0.9841	0.9865
		148.917	0.9733	0.9735	0.9741	0.9751	0.9764	0.9780	0.9799	0.9820	0.9842	0.9866
		149.917	0.9735	0.9737	0.9743	0.9752	0.9765	0.9782	0.9800	0.9821	0.9843	0.9866
		151.920	0.9738	0.9740	0.9746	0.9755	0.9768	0.9784	0.9803	0.9824	0.9845	0.9868
		153.922	0.9741	0.9743	0.9749	0.9759	0.9771	0.9787	0.9805	0.9826	0.9847	0.9870
63	Eu	151.965	0.9738	0.9740	0.9746	0.9756	0.9768	0.9784	0.9803	0.9824	0.9846	0.9868
		150.920	0.9736	0.9738	0.9744	0.9754	0.9767	0.9783	0.9802	0.9822	0.9844	0.9867
		152.921	0.9740	0.9742	0.9748	0.9757	0.9770	0.9786	0.9804	0.9825	0.9846	0.9869
64	Gd	157.252	0.9747	0.9749	0.9754	0.9764	0.9776	0.9792	0.9810	0.9829	0.9851	0.9873
		151.920	0.9738	0.9740	0.9746	0.9755	0.9768	0.9784	0.9803	0.9824	0.9845	0.9868
		153.921	0.9741	0.9743	0.9749	0.9759	0.9771	0.9787	0.9805	0.9826	0.9847	0.9870
		154.923	0.9743	0.9745	0.9751	0.9760	0.9773	0.9789	0.9807	0.9827	0.9848	0.9871
		155.922	0.9745	0.9747	0.9752	0.9762	0.9774	0.9790	0.9808	0.9828	0.9849	0.9872
		156.924	0.9746	0.9748	0.9754	0.9763	0.9776	0.9791	0.9809	0.9829	0.9850	0.9872
		157.924	0.9748	0.9750	0.9755	0.9765	0.9777	0.9793	0.9810	0.9830	0.9851	0.9873
		159.927	0.9751	0.9753	0.9758	0.9768	0.9780	0.9795	0.9813	0.9832	0.9853	0.9875
65	Tb	158.925	0.9750	0.9751	0.9757	0.9766	0.9779	0.9794	0.9812	0.9831	0.9852	0.9874
66	Dy	162.498	0.9755	0.9757	0.9762	0.9771	0.9783	0.9798	0.9816	0.9835	0.9855	0.9877
		155.924	0.9745	0.9747	0.9752	0.9762	0.9774	0.9790	0.9808	0.9828	0.9849	0.9872
		157.924	0.9748	0.9750	0.9755	0.9765	0.9777	0.9793	0.9810	0.9830	0.9851	0.9873
		159.925	0.9751	0.9753	0.9758	0.9768	0.9780	0.9795	0.9813	0.9832	0.9853	0.9875
		160.927	0.9753	0.9754	0.9760	0.9769	0.9781	0.9796	0.9814	0.9833	0.9854	0.9876
		161.927	0.9754	0.9756	0.9761	0.9770	0.9783	0.9798	0.9815	0.9834	0.9855	0.9876
		162.929	0.9756	0.9757	0.9763	0.9772	0.9784	0.9799	0.9816	0.9835	0.9856	0.9877
		163.929	0.9757	0.9759	0.9764	0.9773	0.9785	0.9800	0.9817	0.9836	0.9857	0.9878
67	Ho	164.930	0.9759	0.9760	0.9766	0.9775	0.9786	0.9801	0.9818	0.9837	0.9858	0.9879
68	Er	167.256	0.9762	0.9764	0.9769	0.9778	0.9789	0.9804	0.9821	0.9840	0.9860	0.9880
		161.929	0.9754	0.9756	0.9761	0.9770	0.9783	0.9798	0.9815	0.9834	0.9855	0.9876
		163.929	0.9757	0.9759	0.9764	0.9773	0.9785	0.9800	0.9817	0.9836	0.9857	0.9878
		165.930	0.9760	0.9762	0.9767	0.9776	0.9788	0.9802	0.9819	0.9838	0.9858	0.9879
		166.932	0.9761	0.9763	0.9769	0.9777	0.9789	0.9804	0.9821	0.9839	0.9859	0.9880
		167.932	0.9763	0.9765	0.9770	0.9779	0.9790	0.9805	0.9822	0.9840	0.9860	0.9881
		169.936	0.9766	0.9767	0.9773	0.9781	0.9793	0.9807	0.9824	0.9842	0.9862	0.9882
69	Tm	168.934	0.9764	0.9766	0.9771	0.9780	0.9791	0.9806	0.9823	0.9841	0.9861	0.9881

Table A5.1. Rutherford backscattering spectrometry kinematic factors for H as a projectile (continued).

Atomic no. *(Z)*	El.	Isotopic mass (M_2) *(amu)*	180°	170°	160°	150°	140°	130°	120°	110°	100°	90°
70	Yb	173.034	0.9770	0.9771	0.9777	0.9785	0.9796	0.9810	0.9827	0.9845	0.9864	0.9884
		167.934	0.9763	0.9765	0.9770	0.9779	0.9790	0.9805	0.9822	0.9840	0.9860	0.9881
		169.935	0.9766	0.9767	0.9773	0.9781	0.9793	0.9807	0.9824	0.9842	0.9862	0.9882
		170.936	0.9767	0.9769	0.9774	0.9782	0.9794	0.9808	0.9825	0.9843	0.9863	0.9883
		171.936	0.9768	0.9770	0.9775	0.9784	0.9795	0.9809	0.9826	0.9844	0.9863	0.9883
		172.938	0.9770	0.9771	0.9776	0.9785	0.9796	0.9810	0.9827	0.9845	0.9864	0.9884
		173.939	0.9771	0.9773	0.9778	0.9786	0.9797	0.9811	0.9828	0.9846	0.9865	0.9885
		175.943	0.9773	0.9775	0.9780	0.9788	0.9800	0.9814	0.9830	0.9847	0.9866	0.9886
71	Lu	174.967	0.9772	0.9774	0.9779	0.9787	0.9799	0.9813	0.9829	0.9847	0.9866	0.9885
		174.941	0.9772	0.9774	0.9779	0.9787	0.9799	0.9813	0.9829	0.9847	0.9866	0.9885
		175.943	0.9773	0.9775	0.9780	0.9788	0.9800	0.9814	0.9830	0.9847	0.9866	0.9886
72	Hf	178.49	0.9777	0.9778	0.9783	0.9791	0.9803	0.9816	0.9832	0.9850	0.9868	0.9888
		173.940	0.9771	0.9773	0.9778	0.9786	0.9797	0.9811	0.9828	0.9846	0.9865	0.9885
		175.941	0.9773	0.9775	0.9780	0.9788	0.9800	0.9814	0.9830	0.9847	0.9866	0.9886
		176.943	0.9775	0.9776	0.9781	0.9790	0.9801	0.9815	0.9831	0.9848	0.9867	0.9887
		177.944	0.9776	0.9778	0.9783	0.9791	0.9802	0.9816	0.9832	0.9849	0.9868	0.9887
		178.946	0.9777	0.9779	0.9784	0.9792	0.9803	0.9817	0.9832	0.9850	0.9869	0.9888
		179.947	0.9778	0.9780	0.9785	0.9793	0.9804	0.9818	0.9833	0.9851	0.9869	0.9889
73	Ta	180.948	0.9780	0.9781	0.9786	0.9794	0.9805	0.9819	0.9834	0.9852	0.9870	0.9889
		179.947	0.9778	0.9780	0.9785	0.9793	0.9804	0.9818	0.9833	0.9851	0.9869	0.9889
		180.948	0.9780	0.9781	0.9786	0.9794	0.9805	0.9819	0.9834	0.9852	0.9870	0.9889
74	W	183.849	0.9783	0.9785	0.9790	0.9797	0.9808	0.9821	0.9837	0.9854	0.9872	0.9891
		179.947	0.9778	0.9780	0.9785	0.9793	0.9804	0.9818	0.9833	0.9851	0.9869	0.9889
		181.948	0.9781	0.9783	0.9787	0.9795	0.9806	0.9820	0.9835	0.9852	0.9871	0.9890
		182.950	0.9782	0.9784	0.9789	0.9797	0.9807	0.9821	0.9836	0.9853	0.9872	0.9890
		183.951	0.9783	0.9785	0.9790	0.9798	0.9808	0.9822	0.9837	0.9854	0.9872	0.9891
		185.954	0.9786	0.9787	0.9792	0.9800	0.9810	0.9824	0.9839	0.9856	0.9874	0.9892
75	Re	186.207	0.9786	0.9787	0.9792	0.9800	0.9811	0.9824	0.9839	0.9856	0.9874	0.9892
		184.953	0.9784	0.9786	0.9791	0.9799	0.9809	0.9823	0.9838	0.9855	0.9873	0.9892
		186.956	0.9787	0.9788	0.9793	0.9801	0.9811	0.9824	0.9840	0.9856	0.9874	0.9893
76	Os	190.24	0.9790	0.9792	0.9797	0.9804	0.9815	0.9827	0.9842	0.9859	0.9876	0.9895
		183.952	0.9783	0.9785	0.9790	0.9798	0.9808	0.9822	0.9837	0.9854	0.9872	0.9891
		185.954	0.9786	0.9787	0.9792	0.9800	0.9810	0.9824	0.9839	0.9856	0.9874	0.9892
		186.956	0.9787	0.9788	0.9793	0.9801	0.9811	0.9824	0.9840	0.9856	0.9874	0.9893
		187.956	0.9788	0.9789	0.9794	0.9802	0.9812	0.9825	0.9840	0.9857	0.9875	0.9893
		188.958	0.9789	0.9791	0.9795	0.9803	0.9813	0.9826	0.9841	0.9858	0.9876	0.9894
		189.958	0.9790	0.9792	0.9796	0.9804	0.9814	0.9827	0.9842	0.9859	0.9876	0.9894
		191.962	0.9792	0.9794	0.9798	0.9806	0.9816	0.9829	0.9844	0.9860	0.9878	0.9896
77	Ir	192.216	0.9792	0.9794	0.9799	0.9806	0.9817	0.9829	0.9844	0.9860	0.9878	0.9896
		190.961	0.9791	0.9793	0.9797	0.9805	0.9815	0.9828	0.9843	0.9859	0.9877	0.9895
		192.963	0.9793	0.9795	0.9799	0.9807	0.9817	0.9830	0.9845	0.9861	0.9878	0.9896
78	Pt	195.08	0.9795	0.9797	0.9802	0.9809	0.9819	0.9832	0.9846	0.9862	0.9879	0.9897
		189.960	0.9790	0.9792	0.9796	0.9804	0.9814	0.9827	0.9842	0.9859	0.9876	0.9894
		191.961	0.9792	0.9794	0.9798	0.9806	0.9816	0.9829	0.9844	0.9860	0.9878	0.9896
		193.963	0.9794	0.9796	0.9800	0.9808	0.9818	0.9831	0.9845	0.9862	0.9879	0.9897
		194.965	0.9795	0.9797	0.9801	0.9809	0.9819	0.9832	0.9846	0.9862	0.9879	0.9897
		195.965	0.9796	0.9798	0.9802	0.9810	0.9820	0.9832	0.9847	0.9863	0.9880	0.9898
		197.968	0.9798	0.9800	0.9804	0.9812	0.9822	0.9834	0.9848	0.9864	0.9881	0.9899

Table A5.1. Rutherford backscattering spectrometry kinematic factors for H as a projectile (continued).

Atomic no. *(Z)*	El.	Isotopic mass (M_2) *(amu)*	180°	170°	160°	150°	140°	130°	120°	110°	100°	90°
79	Au	196.967	0.9797	0.9799	0.9803	0.9811	0.9821	0.9833	0.9848	0.9864	0.9881	0.9898
80	Hg	200.588	0.9801	0.9803	0.9807	0.9814	0.9824	0.9836	0.9850	0.9866	0.9883	0.9900
		195.966	0.9796	0.9798	0.9802	0.9810	0.9820	0.9832	0.9847	0.9863	0.9880	0.9898
		197.967	0.9798	0.9800	0.9804	0.9812	0.9822	0.9834	0.9848	0.9864	0.9881	0.9899
		198.968	0.9799	0.9801	0.9805	0.9813	0.9823	0.9835	0.9849	0.9865	0.9882	0.9899
		199.968	0.9800	0.9802	0.9806	0.9814	0.9824	0.9836	0.9850	0.9866	0.9882	0.9900
		200.970	0.9801	0.9803	0.9807	0.9815	0.9824	0.9837	0.9851	0.9866	0.9883	0.9900
		201.971	0.9802	0.9804	0.9808	0.9815	0.9825	0.9837	0.9851	0.9867	0.9884	0.9901
		203.973	0.9804	0.9806	0.9810	0.9817	0.9827	0.9839	0.9853	0.9868	0.9885	0.9902
81	Tl	204.383	0.9805	0.9806	0.9811	0.9818	0.9827	0.9839	0.9853	0.9869	0.9885	0.9902
		202.972	0.9803	0.9805	0.9809	0.9816	0.9826	0.9838	0.9852	0.9868	0.9884	0.9901
		204.974	0.9805	0.9807	0.9811	0.9818	0.9828	0.9840	0.9854	0.9869	0.9885	0.9902
82	Pb	207.217	0.9807	0.9809	0.9813	0.9820	0.9830	0.9841	0.9855	0.9870	0.9886	0.9903
		203.973	0.9804	0.9806	0.9810	0.9817	0.9827	0.9839	0.9853	0.9868	0.9885	0.9902
		205.975	0.9806	0.9808	0.9812	0.9819	0.9829	0.9841	0.9854	0.9870	0.9886	0.9903
		206.976	0.9807	0.9809	0.9813	0.9820	0.9829	0.9841	0.9855	0.9870	0.9886	0.9903
		207.977	0.9808	0.9809	0.9814	0.9821	0.9830	0.9842	0.9856	0.9871	0.9887	0.9904
83	Bi	208.980	0.9809	0.9810	0.9815	0.9822	0.9831	0.9843	0.9856	0.9871	0.9887	0.9904
84	Po	208.982	0.9809	0.9810	0.9815	0.9822	0.9831	0.9843	0.9856	0.9871	0.9887	0.9904
85	At	210.000	0.9810	0.9811	0.9816	0.9822	0.9832	0.9844	0.9857	0.9872	0.9888	0.9904
86	Rn	222.018	0.9820	0.9821	0.9825	0.9832	0.9841	0.9852	0.9865	0.9879	0.9894	0.9910
87	Fr	223.000	0.9821	0.9822	0.9826	0.9833	0.9842	0.9853	0.9865	0.9879	0.9894	0.9910
88	Ra	226.025	0.9823	0.9825	0.9829	0.9835	0.9844	0.9855	0.9867	0.9881	0.9896	0.9911
89	Ac	227.000	0.9824	0.9825	0.9829	0.9836	0.9844	0.9855	0.9868	0.9882	0.9896	0.9912
90	Th	232.038	0.9828	0.9829	0.9833	0.9839	0.9848	0.9858	0.9871	0.9884	0.9899	0.9914
91	Pa	231.036	0.9827	0.9828	0.9832	0.9839	0.9847	0.9858	0.9870	0.9884	0.9898	0.9913
92	U	238.018	0.9832	0.9833	0.9837	0.9843	0.9852	0.9862	0.9874	0.9887	0.9901	0.9916
		234.041	0.9829	0.9831	0.9834	0.9841	0.9849	0.9860	0.9872	0.9885	0.9899	0.9914
		235.044	0.9830	0.9831	0.9835	0.9841	0.9850	0.9860	0.9872	0.9886	0.9900	0.9915
		238.040	0.9832	0.9833	0.9837	0.9843	0.9852	0.9862	0.9874	0.9887	0.9901	0.9916
93	Np	237.048	0.9831	0.9833	0.9836	0.9843	0.9851	0.9861	0.9873	0.9887	0.9901	0.9915
94	Pu	244.064	0.9836	0.9837	0.9841	0.9847	0.9855	0.9865	0.9877	0.9890	0.9904	0.9918
95	Am	243.061	0.9836	0.9837	0.9840	0.9846	0.9855	0.9865	0.9876	0.9889	0.9903	0.9917

Table A5.2. Rutherford backscattering spectrometry kinematic factors for He as a projectile.

This table gives the RBS kinematic factors K_{M_2}, defined by Eq. (A4.2), for He as a projectile (M_1=4.0026 amu) and for the isotopic masses (M_2) of the elements. The kinematic factors are given as a function of scattering angle from 180° to 90° measured in the laboratory frame of reference. The first row for each element gives the average atomic weight of that element and the average kinematic factor (Average K_{M_2}) for that element. The subsequent rows give the isotopic masses for that element and the K_{M_2} for those isotopic masses. The average K_{M_2} is calculated as the weighted average of the K_{M_2}, in which the relative abundances (ϖ_{M_2}) for the M_2 are given in Appendix 1:

$$\text{Average } K_{M_2} = \Sigma\, \varpi_{M_2} K_{M_2}\ , \text{ summed over the masses } M_2.$$

Atomic no. *(Z)*	El.	Isotopic mass (M_2) *(amu)*	180°	170°	160°	150°	140°	130°	120°	110°	100°	90°
3	Li	6.941	0.0722	0.0735	0.0774	0.0845	0.0955	0.1115	0.1343	0.1662	0.2099	0.2680
		6.015	0.0404	0.0412	0.0438	0.0485	0.0560	0.0675	0.0846	0.1102	0.1476	0.2009
		7.016	0.0748	0.0761	0.0802	0.0874	0.0987	0.1151	0.1384	0.1708	0.2150	0.2735
4	Be	9.012	0.1482	0.1502	0.1564	0.1671	0.1832	0.2056	0.2356	0.2747	0.3241	0.3849
5	B	10.811	0.2111	0.2135	0.2208	0.2333	0.2516	0.2766	0.3090	0.3499	0.3999	0.4592
		10.013	0.1839	0.1861	0.1930	0.2049	0.2225	0.2465	0.2782	0.3186	0.3686	0.4288
		11.009	0.2178	0.2203	0.2277	0.2403	0.2589	0.2841	0.3167	0.3577	0.4076	0.4667
6	C	12.011	0.2501	0.2526	0.2604	0.2736	0.2928	0.3187	0.3518	0.3929	0.4423	0.5001
		12.000	0.2498	0.2523	0.2600	0.2733	0.2925	0.3183	0.3515	0.3926	0.4420	0.4998
		13.003	0.2801	0.2828	0.2908	0.3044	0.3240	0.3502	0.3835	0.4244	0.4731	0.5293
7	N	14.007	0.3086	0.3113	0.3194	0.3333	0.3531	0.3795	0.4127	0.4532	0.5009	0.5555
		14.003	0.3085	0.3112	0.3193	0.3332	0.3530	0.3794	0.4126	0.4531	0.5008	0.5554
		15.000	0.3349	0.3377	0.3459	0.3599	0.3798	0.4062	0.4392	0.4790	0.5257	0.5787
8	O	15.999	0.3597	0.3625	0.3708	0.3848	0.4047	0.4309	0.4635	0.5027	0.5483	0.5998
		15.995	0.3596	0.3624	0.3707	0.3847	0.4046	0.4308	0.4634	0.5026	0.5482	0.5997
		16.999	0.3830	0.3857	0.3940	0.4080	0.4278	0.4538	0.4860	0.5244	0.5689	0.6188
		17.999	0.4047	0.4074	0.4157	0.4296	0.4493	0.4750	0.5067	0.5443	0.5877	0.6362
9	F	18.998	0.4251	0.4278	0.4360	0.4498	0.4693	0.4946	0.5258	0.5627	0.6050	0.6520
10	NE	20.179	0.4473	0.4500	0.4582	0.4718	0.4910	0.5159	0.5464	0.5824	0.6234	0.6688
		19.992	0.4441	0.4468	0.4549	0.4686	0.4879	0.5128	0.5434	0.5795	0.6207	0.6664
		20.994	0.4621	0.4647	0.4728	0.4863	0.5053	0.5299	0.5599	0.5952	0.6354	0.6797
		21.991	0.4789	0.4816	0.4896	0.5029	0.5216	0.5458	0.5752	0.6098	0.6490	0.6920
11	Na	22.990	0.4948	0.4974	0.5053	0.5185	0.5370	0.5607	0.5896	0.6233	0.6615	0.7034
12	Mg	24.305	0.5142	0.5168	0.5245	0.5375	0.5555	0.5787	0.6069	0.6397	0.6766	0.7171
		23.985	0.5098	0.5123	0.5201	0.5331	0.5513	0.5746	0.6029	0.6360	0.6732	0.7140
		24.986	0.5240	0.5265	0.5342	0.5470	0.5649	0.5878	0.6155	0.6478	0.6842	0.7238
		25.983	0.5373	0.5399	0.5474	0.5600	0.5776	0.6001	0.6273	0.6589	0.6944	0.7330
13	Al	26.982	0.5500	0.5525	0.5600	0.5724	0.5897	0.6117	0.6384	0.6693	0.7040	0.7416
14	Si	28.086	0.5632	0.5657	0.5730	0.5852	0.6022	0.6238	0.6499	0.6801	0.7139	0.7505
		27.977	0.5620	0.5645	0.5718	0.5840	0.6010	0.6227	0.6489	0.6791	0.7130	0.7497
		28.976	0.5734	0.5759	0.5831	0.5951	0.6118	0.6331	0.6588	0.6884	0.7215	0.7573
		29.974	0.5843	0.5867	0.5938	0.6056	0.6220	0.6430	0.6681	0.6971	0.7294	0.7644
15	P	30.974	0.5946	0.5970	0.6040	0.6156	0.6318	0.6523	0.6770	0.7054	0.7370	0.7711
16	S	32.064	0.6053	0.6076	0.6145	0.6259	0.6418	0.6619	0.6861	0.7139	0.7448	0.7780
		31.972	0.6045	0.6068	0.6137	0.6251	0.6410	0.6612	0.6854	0.7132	0.7441	0.7775
		32.972	0.6139	0.6161	0.6229	0.6342	0.6498	0.6696	0.6934	0.7206	0.7509	0.7835
		33.968	0.6228	0.6250	0.6317	0.6428	0.6582	0.6776	0.7009	0.7277	0.7573	0.7892
		35.967	0.6395	0.6417	0.6482	0.6589	0.6738	0.6926	0.7150	0.7408	0.7692	0.7997

Table A5.2. Rutherford backscattering spectrometry kinematic factors for He as a projectile (continued).

Atomic no. *(Z)*	El.	Isotopic mass (M_2) *(amu)*	180°	170°	160°	150°	140°	130°	120°	110°	100°	90°
17	Cl	35.453	0.6353	0.6374	0.6440	0.6548	0.6698	0.6887	0.7114	0.7374	0.7662	0.7970
		34.969	0.6314	0.6336	0.6402	0.6511	0.6662	0.6853	0.7082	0.7344	0.7634	0.7946
		36.966	0.6474	0.6495	0.6559	0.6665	0.6811	0.6996	0.7216	0.7468	0.7747	0.8046
18	Ar	39.948	0.6689	0.6709	0.6770	0.6871	0.7010	0.7186	0.7395	0.7634	0.7897	0.8179
		35.967	0.6396	0.6417	0.6482	0.6589	0.6738	0.6926	0.7150	0.7408	0.7692	0.7997
		37.963	0.6549	0.6570	0.6633	0.6737	0.6880	0.7062	0.7279	0.7526	0.7800	0.8092
		39.962	0.6690	0.6710	0.6771	0.6872	0.7011	0.7187	0.7396	0.7635	0.7898	0.8179
19	K	39.098	0.6630	0.6651	0.6712	0.6815	0.6956	0.7134	0.7346	0.7589	0.7857	0.8142
		38.964	0.6621	0.6642	0.6703	0.6806	0.6947	0.7126	0.7339	0.7582	0.7850	0.8137
		40.000	0.6692	0.6713	0.6774	0.6874	0.7014	0.7189	0.7398	0.7637	0.7900	0.8181
		40.962	0.6756	0.6776	0.6836	0.6936	0.7073	0.7246	0.7451	0.7686	0.7944	0.8220
20	Ca	40.078	0.6697	0.6717	0.6778	0.6879	0.7018	0.7193	0.7402	0.7641	0.7903	0.8184
		39.963	0.6690	0.6710	0.6771	0.6872	0.7011	0.7187	0.7396	0.7635	0.7898	0.8179
		41.959	0.6820	0.6840	0.6899	0.6997	0.7132	0.7302	0.7504	0.7734	0.7988	0.8258
		42.959	0.6881	0.6901	0.6959	0.7055	0.7188	0.7356	0.7555	0.7781	0.8030	0.8295
		43.956	0.6940	0.6959	0.7017	0.7112	0.7243	0.7407	0.7603	0.7826	0.8070	0.8331
		45.954	0.7052	0.7071	0.7126	0.7218	0.7345	0.7505	0.7695	0.7910	0.8146	0.8398
		47.952	0.7156	0.7174	0.7228	0.7318	0.7441	0.7596	0.7780	0.7988	0.8217	0.8459
21	Sc	44.956	0.6997	0.7016	0.7073	0.7166	0.7295	0.7457	0.7650	0.7869	0.8109	0.8365
22	Ti	47.878	0.7152	0.7170	0.7224	0.7314	0.7437	0.7592	0.7776	0.7985	0.8214	0.8457
		45.953	0.7052	0.7071	0.7126	0.7218	0.7345	0.7505	0.7695	0.7910	0.8146	0.8398
		46.952	0.7105	0.7123	0.7178	0.7269	0.7394	0.7551	0.7738	0.7950	0.8182	0.8429
		47.948	0.7156	0.7174	0.7228	0.7318	0.7441	0.7596	0.7780	0.7988	0.8217	0.8459
		48.948	0.7205	0.7223	0.7276	0.7365	0.7486	0.7639	0.7820	0.8025	0.8250	0.8488
		49.945	0.7252	0.7270	0.7323	0.7410	0.7530	0.7680	0.7858	0.8060	0.8282	0.8516
23	V	50.942	0.7298	0.7316	0.7368	0.7454	0.7572	0.7720	0.7896	0.8095	0.8312	0.8543
		49.947	0.7253	0.7270	0.7323	0.7410	0.7530	0.7680	0.7858	0.8061	0.8282	0.8516
		50.944	0.7298	0.7316	0.7368	0.7454	0.7572	0.7720	0.7896	0.8095	0.8312	0.8543
24	Cr	51.996	0.7345	0.7362	0.7414	0.7498	0.7615	0.7761	0.7934	0.8129	0.8344	0.8570
		49.946	0.7252	0.7270	0.7323	0.7410	0.7530	0.7680	0.7858	0.8061	0.8282	0.8516
		51.940	0.7343	0.7360	0.7411	0.7496	0.7613	0.7759	0.7932	0.8128	0.8342	0.8569
		52.941	0.7386	0.7403	0.7454	0.7537	0.7652	0.7796	0.7967	0.8160	0.8371	0.8594
		53.939	0.7428	0.7444	0.7494	0.7577	0.7690	0.7832	0.8001	0.8191	0.8399	0.8618
25	Mn	54.938	0.7468	0.7485	0.7534	0.7615	0.7727	0.7867	0.8033	0.8221	0.8425	0.8642
26	Fe	55.847	0.7504	0.7520	0.7569	0.7649	0.7760	0.7898	0.8062	0.8247	0.8449	0.8662
		53.940	0.7428	0.7444	0.7494	0.7577	0.7690	0.7832	0.8001	0.8191	0.8399	0.8618
		55.935	0.7507	0.7524	0.7572	0.7653	0.7763	0.7901	0.8065	0.8250	0.8451	0.8664
		56.935	0.7545	0.7561	0.7610	0.7689	0.7798	0.7934	0.8095	0.8278	0.8476	0.8686
		57.933	0.7582	0.7598	0.7646	0.7724	0.7831	0.7966	0.8125	0.8305	0.8501	0.8708
27	Co	58.933	0.7618	0.7634	0.7681	0.7758	0.7864	0.7997	0.8154	0.8331	0.8524	0.8728
28	Ni	58.688	0.7608	0.7624	0.7671	0.7749	0.7855	0.7989	0.8146	0.8324	0.8518	0.8723
		57.935	0.7582	0.7598	0.7646	0.7724	0.7831	0.7966	0.8125	0.8305	0.8501	0.8708
		59.931	0.7653	0.7668	0.7714	0.7791	0.7896	0.8027	0.8182	0.8356	0.8547	0.8748
		60.931	0.7686	0.7702	0.7747	0.7823	0.7926	0.8056	0.8209	0.8381	0.8569	0.8767
		61.928	0.7719	0.7734	0.7779	0.7854	0.7956	0.8084	0.8235	0.8405	0.8590	0.8786
		63.928	0.7782	0.7797	0.7841	0.7914	0.8013	0.8138	0.8285	0.8451	0.8631	0.8822

Table A5.2. Rutherford backscattering spectrometry kinematic factors for He as a projectile (continued).

Atomic no. *(Z)*	El.	Isotopic mass (M_2) *(amu)*	180°	170°	160°	150°	140°	130°	120°	110°	100°	90°
29	Cu	63.546	0.7770	0.7785	0.7829	0.7902	0.8002	0.8128	0.8276	0.8442	0.8624	0.8815
		62.930	0.7751	0.7766	0.7811	0.7884	0.7985	0.8112	0.8260	0.8428	0.8611	0.8804
		64.928	0.7812	0.7827	0.7870	0.7942	0.8041	0.8164	0.8309	0.8473	0.8651	0.8839
30	Zn	65.396	0.7825	0.7840	0.7883	0.7955	0.8053	0.8175	0.8320	0.8482	0.8659	0.8846
		63.929	0.7782	0.7797	0.7841	0.7914	0.8013	0.8138	0.8285	0.8451	0.8631	0.8822
		65.926	0.7842	0.7856	0.7899	0.7970	0.8068	0.8189	0.8333	0.8494	0.8670	0.8855
		66.927	0.7870	0.7884	0.7927	0.7997	0.8094	0.8214	0.8356	0.8515	0.8689	0.8871
		67.925	0.7898	0.7912	0.7954	0.8024	0.8119	0.8238	0.8378	0.8535	0.8707	0.8887
		69.925	0.7952	0.7965	0.8007	0.8075	0.8168	0.8284	0.8420	0.8574	0.8741	0.8917
31	Ga	69.723	0.7946	0.7960	0.8001	0.8069	0.8162	0.8279	0.8416	0.8570	0.8738	0.8914
		68.926	0.7925	0.7939	0.7981	0.8049	0.8144	0.8261	0.8399	0.8555	0.8724	0.8902
		70.925	0.7977	0.7991	0.8032	0.8099	0.8191	0.8306	0.8441	0.8593	0.8758	0.8932
32	Ge	72.632	0.8019	0.8032	0.8073	0.8138	0.8229	0.8341	0.8474	0.8623	0.8785	0.8955
		69.924	0.7952	0.7965	0.8007	0.8074	0.8167	0.8284	0.8420	0.8574	0.8741	0.8917
		71.922	0.8002	0.8016	0.8056	0.8123	0.8214	0.8327	0.8461	0.8611	0.8774	0.8946
		72.924	0.8027	0.8040	0.8080	0.8146	0.8236	0.8348	0.8480	0.8629	0.8790	0.8959
		73.921	0.8051	0.8064	0.8104	0.8169	0.8258	0.8369	0.8499	0.8646	0.8805	0.8973
		75.921	0.8097	0.8110	0.8149	0.8212	0.8299	0.8408	0.8536	0.8679	0.8835	0.8998
33	As	74.922	0.8074	0.8087	0.8126	0.8191	0.8279	0.8389	0.8518	0.8663	0.8820	0.8986
34	Se	78.993	0.8163	0.8176	0.8213	0.8275	0.8359	0.8464	0.8588	0.8727	0.8877	0.9035
		73.923	0.8051	0.8064	0.8104	0.8169	0.8258	0.8369	0.8499	0.8646	0.8805	0.8973
		75.919	0.8097	0.8110	0.8149	0.8212	0.8299	0.8408	0.8536	0.8679	0.8835	0.8998
		76.920	0.8119	0.8132	0.8170	0.8233	0.8320	0.8427	0.8553	0.8695	0.8849	0.9011
		77.917	0.8141	0.8154	0.8192	0.8254	0.8339	0.8446	0.8570	0.8711	0.8863	0.9023
		79.916	0.8183	0.8196	0.8233	0.8294	0.8377	0.8481	0.8604	0.8741	0.8890	0.9046
		81.917	0.8223	0.8236	0.8272	0.8332	0.8414	0.8516	0.8635	0.8770	0.8916	0.9068
35	Br	79.904	0.8183	0.8195	0.8232	0.8293	0.8377	0.8481	0.8603	0.8741	0.8890	0.9046
		78.918	0.8162	0.8175	0.8212	0.8274	0.8358	0.8464	0.8587	0.8726	0.8877	0.9035
		80.916	0.8203	0.8216	0.8253	0.8313	0.8396	0.8499	0.8620	0.8756	0.8903	0.9057
36	Kr	83.8	0.8259	0.8271	0.8307	0.8366	0.8446	0.8546	0.8664	0.8796	0.8938	0.9088
		77.920	0.8141	0.8154	0.8192	0.8254	0.8339	0.8446	0.8571	0.8711	0.8863	0.9023
		79.916	0.8183	0.8196	0.8233	0.8294	0.8377	0.8481	0.8604	0.8741	0.8890	0.9046
		81.913	0.8223	0.8236	0.8272	0.8332	0.8414	0.8516	0.8635	0.8770	0.8915	0.9068
		82.914	0.8243	0.8255	0.8291	0.8350	0.8431	0.8532	0.8651	0.8784	0.8928	0.9079
		83.911	0.8262	0.8274	0.8309	0.8368	0.8448	0.8548	0.8666	0.8797	0.8940	0.9089
		85.911	0.8299	0.8310	0.8345	0.8403	0.8482	0.8580	0.8695	0.8824	0.8963	0.9110
37	Rb	85.468	0.8290	0.8302	0.8337	0.8395	0.8474	0.8573	0.8688	0.8818	0.8958	0.9105
		84.912	0.8280	0.8292	0.8328	0.8386	0.8465	0.8564	0.8680	0.8811	0.8952	0.9100
		86.909	0.8316	0.8328	0.8363	0.8420	0.8498	0.8595	0.8709	0.8836	0.8975	0.9119
38	Sr	87.617	0.8329	0.8340	0.8375	0.8431	0.8509	0.8605	0.8718	0.8845	0.8982	0.9126
		83.913	0.8262	0.8274	0.8309	0.8368	0.8448	0.8548	0.8666	0.8797	0.8940	0.9089
		85.909	0.8299	0.8310	0.8345	0.8403	0.8482	0.8580	0.8695	0.8824	0.8963	0.9110
		86.909	0.8316	0.8328	0.8363	0.8420	0.8498	0.8595	0.8709	0.8836	0.8975	0.9119
		87.906	0.8334	0.8345	0.8380	0.8436	0.8513	0.8610	0.8722	0.8849	0.8986	0.9129
39	Y	88.906	0.8351	0.8362	0.8396	0.8452	0.8529	0.8624	0.8736	0.8861	0.8996	0.9138

Table A5.2. Rutherford backscattering spectrometry kinematic factors for He as a projectile (continued).

Atomic no. *(Z)*	El.	Isotopic mass (M_2) *(amu)*	180°	170°	160°	150°	140°	130°	120°	110°	100°	90°
40	Zr	91.221	0.8389	0.8400	0.8433	0.8488	0.8563	0.8656	0.8765	0.8888	0.9020	0.9159
		89.905	0.8368	0.8379	0.8413	0.8468	0.8544	0.8638	0.8749	0.8873	0.9007	0.9148
		90.906	0.8384	0.8395	0.8429	0.8484	0.8559	0.8652	0.8762	0.8885	0.9017	0.9157
		91.905	0.8400	0.8411	0.8445	0.8499	0.8573	0.8666	0.8774	0.8896	0.9028	0.9165
		93.906	0.8432	0.8443	0.8475	0.8528	0.8601	0.8692	0.8799	0.8918	0.9047	0.9182
		95.908	0.8462	0.8472	0.8504	0.8557	0.8629	0.8718	0.8822	0.8940	0.9066	0.9199
41	Nb	92.906	0.8416	0.8427	0.8460	0.8514	0.8588	0.8679	0.8787	0.8907	0.9038	0.9174
42	Mo	95.931	0.8461	0.8472	0.8504	0.8556	0.8628	0.8718	0.8822	0.8939	0.9066	0.9199
		91.907	0.8400	0.8411	0.8445	0.8499	0.8573	0.8666	0.8774	0.8896	0.9028	0.9165
		93.905	0.8432	0.8443	0.8475	0.8528	0.8601	0.8692	0.8799	0.8918	0.9047	0.9182
		94.906	0.8447	0.8458	0.8490	0.8543	0.8615	0.8705	0.8811	0.8929	0.9057	0.9191
		95.905	0.8462	0.8472	0.8504	0.8557	0.8629	0.8718	0.8822	0.8940	0.9066	0.9199
		96.906	0.8476	0.8487	0.8519	0.8571	0.8642	0.8730	0.8834	0.8950	0.9075	0.9207
		97.905	0.8491	0.8501	0.8533	0.8584	0.8655	0.8742	0.8845	0.8960	0.9084	0.9214
		99.908	0.8519	0.8529	0.8560	0.8611	0.8680	0.8766	0.8867	0.8980	0.9102	0.9230
43	Tc	98.000	0.8492	0.8503	0.8534	0.8585	0.8656	0.8743	0.8846	0.8961	0.9085	0.9215
44	Ru	101.019	0.8533	0.8543	0.8574	0.8624	0.8693	0.8778	0.8878	0.8990	0.9111	0.9237
		95.908	0.8462	0.8472	0.8504	0.8557	0.8629	0.8718	0.8822	0.8940	0.9066	0.9199
		97.905	0.8491	0.8501	0.8533	0.8584	0.8655	0.8742	0.8845	0.8960	0.9084	0.9214
		98.506	0.8499	0.8510	0.8541	0.8592	0.8662	0.8750	0.8852	0.8966	0.9090	0.9219
		99.904	0.8519	0.8529	0.8560	0.8610	0.8680	0.8766	0.8867	0.8980	0.9102	0.9230
		100.906	0.8532	0.8542	0.8573	0.8623	0.8692	0.8777	0.8877	0.8989	0.9110	0.9237
		101.904	0.8545	0.8556	0.8586	0.8636	0.8704	0.8789	0.8888	0.8999	0.9119	0.9244
		103.905	0.8571	0.8581	0.8611	0.8660	0.8727	0.8811	0.8908	0.9017	0.9135	0.9258
45	Rh	102.906	0.8558	0.8569	0.8599	0.8648	0.8716	0.8800	0.8898	0.9008	0.9127	0.9251
46	Pd	106.415	0.8602	0.8612	0.8641	0.8689	0.8755	0.8837	0.8932	0.9039	0.9154	0.9275
		101.906	0.8545	0.8556	0.8586	0.8636	0.8704	0.8789	0.8888	0.8999	0.9119	0.9244
		103.904	0.8571	0.8581	0.8611	0.8660	0.8727	0.8811	0.8908	0.9017	0.9135	0.9258
		104.905	0.8584	0.8594	0.8624	0.8672	0.8739	0.8821	0.8918	0.9026	0.9143	0.9265
		105.904	0.8596	0.8606	0.8636	0.8684	0.8750	0.8832	0.8928	0.9035	0.9151	0.9272
		107.904	0.8620	0.8630	0.8659	0.8707	0.8771	0.8852	0.8946	0.9052	0.9166	0.9285
		109.905	0.8644	0.8653	0.8682	0.8729	0.8792	0.8872	0.8965	0.9068	0.9180	0.9297
47	Ag	107.868	0.8620	0.8630	0.8659	0.8706	0.8771	0.8852	0.8946	0.9052	0.9165	0.9284
		106.905	0.8609	0.8618	0.8647	0.8695	0.8761	0.8842	0.8937	0.9043	0.9158	0.9278
		108.905	0.8632	0.8642	0.8671	0.8718	0.8782	0.8862	0.8956	0.9060	0.9173	0.9291
48	Cd	112.412	0.8672	0.8681	0.8709	0.8755	0.8817	0.8895	0.8986	0.9088	0.9198	0.9312
		105.907	0.8596	0.8606	0.8636	0.8684	0.8750	0.8832	0.8928	0.9035	0.9151	0.9272
		107.904	0.8620	0.8630	0.8659	0.8707	0.8771	0.8852	0.8946	0.9052	0.9166	0.9285
		109.903	0.8644	0.8653	0.8682	0.8729	0.8792	0.8872	0.8965	0.9068	0.9180	0.9297
		110.904	0.8655	0.8665	0.8693	0.8739	0.8803	0.8881	0.8973	0.9076	0.9187	0.9303
		111.903	0.8666	0.8676	0.8704	0.8750	0.8813	0.8891	0.8982	0.9084	0.9194	0.9309
		112.904	0.8677	0.8687	0.8715	0.8760	0.8823	0.8900	0.8991	0.9092	0.9201	0.9315
		113.903	0.8688	0.8697	0.8725	0.8770	0.8832	0.8909	0.8999	0.9100	0.9208	0.9321
		115.905	0.8709	0.8718	0.8746	0.8790	0.8851	0.8927	0.9015	0.9114	0.9221	0.9332
49	In	114.818	0.8698	0.8707	0.8735	0.8780	0.8841	0.8917	0.9007	0.9106	0.9214	0.9326
		112.904	0.8677	0.8687	0.8715	0.8760	0.8823	0.8900	0.8991	0.9092	0.9201	0.9315
		114.904	0.8699	0.8708	0.8735	0.8780	0.8842	0.8918	0.9007	0.9107	0.9215	0.9327

Table A5.2. Rutherford backscattering spectrometry kinematic factors for He as a projectile (continued).

Atomic no. *(Z)*	El.	Isotopic mass (M_2) *(amu)*	180°	170°	160°	150°	140°	130°	120°	110°	100°	90°
50	Sn	118.613	0.8737	0.8745	0.8772	0.8816	0.8876	0.8950	0.9037	0.9133	0.9238	0.9347
		111.905	0.8666	0.8676	0.8704	0.8750	0.8813	0.8891	0.8982	0.9084	0.9194	0.9309
		113.903	0.8688	0.8697	0.8725	0.8770	0.8832	0.8909	0.8999	0.9100	0.9208	0.9321
		114.903	0.8699	0.8708	0.8735	0.8780	0.8842	0.8918	0.9007	0.9107	0.9215	0.9327
		115.902	0.8709	0.8718	0.8746	0.8790	0.8851	0.8927	0.9015	0.9114	0.9221	0.9332
		116.903	0.8720	0.8729	0.8756	0.8800	0.8860	0.8936	0.9023	0.9122	0.9227	0.9338
		117.602	0.8727	0.8736	0.8763	0.8807	0.8867	0.8942	0.9029	0.9127	0.9232	0.9342
		118.903	0.8740	0.8749	0.8775	0.8819	0.8879	0.8953	0.9039	0.9136	0.9240	0.9349
		119.902	0.8750	0.8758	0.8785	0.8828	0.8887	0.8961	0.9047	0.9143	0.9246	0.9354
		121.903	0.8769	0.8778	0.8804	0.8846	0.8905	0.8977	0.9062	0.9156	0.9258	0.9364
		123.905	0.8787	0.8796	0.8822	0.8864	0.8921	0.8993	0.9076	0.9169	0.9270	0.9374
51	Sb	121.758	0.8767	0.8776	0.8802	0.8845	0.8903	0.8976	0.9060	0.9155	0.9257	0.9363
		120.904	0.8759	0.8768	0.8794	0.8837	0.8896	0.8969	0.9054	0.9149	0.9252	0.9359
		122.904	0.8778	0.8787	0.8813	0.8855	0.8913	0.8985	0.9069	0.9163	0.9264	0.9369
52	Te	127.586	0.8820	0.8828	0.8853	0.8894	0.8950	0.9020	0.9101	0.9192	0.9290	0.9391
		119.904	0.8750	0.8758	0.8785	0.8828	0.8887	0.8961	0.9047	0.9143	0.9246	0.9354
		121.903	0.8769	0.8778	0.8804	0.8846	0.8905	0.8977	0.9062	0.9156	0.9258	0.9364
		122.904	0.8778	0.8787	0.8813	0.8855	0.8913	0.8985	0.9069	0.9163	0.9264	0.9369
		123.903	0.8787	0.8796	0.8822	0.8864	0.8921	0.8993	0.9076	0.9169	0.9269	0.9374
		124.904	0.8797	0.8805	0.8831	0.8872	0.8929	0.9000	0.9083	0.9176	0.9275	0.9379
		125.903	0.8806	0.8814	0.8839	0.8881	0.8937	0.9008	0.9090	0.9182	0.9281	0.9384
		127.905	0.8823	0.8831	0.8856	0.8897	0.8953	0.9023	0.9104	0.9194	0.9292	0.9393
		129.906	0.8840	0.8848	0.8873	0.8913	0.8969	0.9037	0.9117	0.9206	0.9302	0.9402
53	I	126.905	0.8814	0.8823	0.8848	0.8889	0.8945	0.9015	0.9097	0.9188	0.9286	0.9388
54	Xe	131.293	0.8851	0.8860	0.8884	0.8924	0.8979	0.9046	0.9125	0.9214	0.9309	0.9408
		123.906	0.8787	0.8796	0.8822	0.8864	0.8921	0.8993	0.9076	0.9169	0.9270	0.9374
		125.904	0.8806	0.8814	0.8839	0.8881	0.8937	0.9008	0.9090	0.9182	0.9281	0.9384
		127.904	0.8823	0.8831	0.8856	0.8897	0.8953	0.9023	0.9104	0.9194	0.9292	0.9393
		128.905	0.8832	0.8840	0.8865	0.8905	0.8961	0.9030	0.9110	0.9200	0.9297	0.9398
		129.904	0.8840	0.8848	0.8873	0.8913	0.8968	0.9037	0.9117	0.9206	0.9302	0.9402
		130.905	0.8848	0.8857	0.8881	0.8921	0.8976	0.9044	0.9123	0.9212	0.9307	0.9407
		131.904	0.8857	0.8865	0.8889	0.8929	0.8983	0.9051	0.9130	0.9218	0.9312	0.9411
		133.905	0.8873	0.8881	0.8905	0.8944	0.8998	0.9064	0.9142	0.9229	0.9322	0.9420
		135.907	0.8888	0.8896	0.8920	0.8959	0.9012	0.9077	0.9154	0.9240	0.9332	0.9428
55	Cs	132.905	0.8865	0.8873	0.8897	0.8937	0.8991	0.9058	0.9136	0.9223	0.9317	0.9415
56	Ba	137.327	0.8899	0.8907	0.8931	0.8969	0.9021	0.9086	0.9162	0.9247	0.9338	0.9434
		129.906	0.8840	0.8848	0.8873	0.8913	0.8969	0.9037	0.9117	0.9206	0.9302	0.9402
		131.905	0.8857	0.8865	0.8889	0.8929	0.8983	0.9051	0.9130	0.9218	0.9312	0.9411
		133.904	0.8873	0.8881	0.8905	0.8944	0.8998	0.9064	0.9142	0.9229	0.9322	0.9420
		134.906	0.8881	0.8889	0.8912	0.8952	0.9005	0.9071	0.9148	0.9234	0.9327	0.9424
		135.905	0.8888	0.8896	0.8920	0.8959	0.9012	0.9077	0.9154	0.9240	0.9332	0.9428
		136.906	0.8896	0.8904	0.8927	0.8966	0.9019	0.9084	0.9160	0.9245	0.9337	0.9432
		137.905	0.8904	0.8911	0.8935	0.8973	0.9025	0.9090	0.9166	0.9250	0.9341	0.9436
57	La	138.905	0.8911	0.8919	0.8942	0.8980	0.9032	0.9096	0.9172	0.9256	0.9346	0.9440
		137.907	0.8904	0.8911	0.8935	0.8973	0.9025	0.9090	0.9166	0.9250	0.9341	0.9436
		138.906	0.8911	0.8919	0.8942	0.8980	0.9032	0.9096	0.9172	0.9256	0.9346	0.9440
58	Ce	140.115	0.8920	0.8928	0.8951	0.8988	0.9040	0.9104	0.9178	0.9262	0.9351	0.9445
		135.907	0.8888	0.8896	0.8920	0.8959	0.9012	0.9077	0.9154	0.9240	0.9332	0.9428

Table A5.2. Rutherford backscattering spectrometry kinematic factors for He as a projectile (continued).

Atomic no. *(Z)*	El.	Isotopic mass (M_2) *(amu)*	180°	170°	160°	150°	140°	130°	120°	110°	100°	90°
		137.906	0.8904	0.8911	0.8935	0.8973	0.9025	0.9090	0.9166	0.9250	0.9341	0.9436
		139.905	0.8918	0.8926	0.8949	0.8987	0.9039	0.9103	0.9177	0.9261	0.9350	0.9444
		141.909	0.8933	0.8940	0.8963	0.9001	0.9052	0.9115	0.9188	0.9271	0.9359	0.9451
59	Pr	140.908	0.8926	0.8933	0.8956	0.8994	0.9045	0.9109	0.9183	0.9266	0.9355	0.9448
60	Nd	144.242	0.8949	0.8956	0.8979	0.9016	0.9066	0.9128	0.9201	0.9282	0.9369	0.9460
		141.908	0.8933	0.8940	0.8963	0.9001	0.9052	0.9115	0.9188	0.9271	0.9359	0.9451
		142.910	0.8940	0.8948	0.8970	0.9007	0.9058	0.9121	0.9194	0.9276	0.9364	0.9455
		143.910	0.8947	0.8954	0.8977	0.9014	0.9064	0.9126	0.9199	0.9280	0.9368	0.9459
		144.913	0.8954	0.8961	0.8984	0.9020	0.9070	0.9132	0.9205	0.9285	0.9372	0.9462
		145.913	0.8961	0.8968	0.8990	0.9027	0.9076	0.9138	0.9210	0.9290	0.9376	0.9466
		147.917	0.8974	0.8981	0.9003	0.9039	0.9088	0.9149	0.9220	0.9299	0.9384	0.9473
		149.921	0.8987	0.8994	0.9016	0.9051	0.9100	0.9160	0.9230	0.9308	0.9392	0.9480
61	Pm	145.000	0.8954	0.8962	0.8984	0.9021	0.9071	0.9133	0.9205	0.9286	0.9372	0.9463
62	Sm	150.36	0.8989	0.8997	0.9018	0.9054	0.9102	0.9162	0.9232	0.9310	0.9394	0.9481
		143.912	0.8947	0.8954	0.8977	0.9014	0.9064	0.9126	0.9199	0.9280	0.9368	0.9459
		146.915	0.8967	0.8975	0.8997	0.9033	0.9082	0.9144	0.9215	0.9295	0.9380	0.9470
		147.915	0.8974	0.8981	0.9003	0.9039	0.9088	0.9149	0.9220	0.9299	0.9384	0.9473
		148.917	0.8980	0.8988	0.9010	0.9045	0.9094	0.9155	0.9225	0.9304	0.9388	0.9477
		149.917	0.8987	0.8994	0.9016	0.9051	0.9100	0.9160	0.9230	0.9308	0.9392	0.9480
		151.920	0.9000	0.9007	0.9028	0.9063	0.9111	0.9171	0.9240	0.9317	0.9400	0.9487
		153.922	0.9012	0.9019	0.9040	0.9075	0.9122	0.9181	0.9249	0.9326	0.9408	0.9493
63	Eu	151.965	0.9000	0.9007	0.9028	0.9064	0.9111	0.9171	0.9240	0.9317	0.9400	0.9487
		150.920	0.8993	0.9001	0.9022	0.9057	0.9106	0.9165	0.9235	0.9313	0.9396	0.9483
		152.921	0.9006	0.9013	0.9034	0.9069	0.9117	0.9176	0.9245	0.9321	0.9404	0.9490
64	Gd	157.252	0.9032	0.9039	0.9059	0.9093	0.9140	0.9197	0.9265	0.9339	0.9420	0.9504
		151.920	0.9000	0.9007	0.9028	0.9063	0.9111	0.9171	0.9240	0.9317	0.9400	0.9487
		153.921	0.9012	0.9019	0.9040	0.9075	0.9122	0.9181	0.9249	0.9326	0.9408	0.9493
		154.923	0.9018	0.9025	0.9046	0.9081	0.9128	0.9186	0.9254	0.9330	0.9411	0.9496
		155.922	0.9024	0.9031	0.9052	0.9086	0.9133	0.9191	0.9259	0.9334	0.9415	0.9499
		156.924	0.9030	0.9037	0.9058	0.9092	0.9138	0.9196	0.9263	0.9338	0.9419	0.9503
		157.924	0.9036	0.9043	0.9063	0.9097	0.9143	0.9201	0.9268	0.9342	0.9422	0.9506
		159.927	0.9047	0.9054	0.9075	0.9108	0.9154	0.9210	0.9277	0.9350	0.9429	0.9512
65	Tb	158.925	0.9041	0.9048	0.9069	0.9103	0.9149	0.9206	0.9272	0.9346	0.9426	0.9509
66	Dy	162.498	0.9061	0.9068	0.9088	0.9121	0.9167	0.9222	0.9287	0.9360	0.9438	0.9519
		155.924	0.9024	0.9031	0.9052	0.9086	0.9133	0.9191	0.9259	0.9334	0.9415	0.9499
		157.924	0.9036	0.9043	0.9063	0.9097	0.9144	0.9201	0.9268	0.9342	0.9422	0.9506
		159.925	0.9047	0.9054	0.9075	0.9108	0.9154	0.9210	0.9276	0.9350	0.9429	0.9512
		160.927	0.9053	0.9060	0.9080	0.9113	0.9159	0.9215	0.9281	0.9354	0.9433	0.9515
		161.927	0.9058	0.9065	0.9085	0.9119	0.9164	0.9220	0.9285	0.9358	0.9436	0.9518
		162.929	0.9064	0.9071	0.9091	0.9124	0.9169	0.9224	0.9289	0.9362	0.9440	0.9520
		163.929	0.9069	0.9076	0.9096	0.9129	0.9174	0.9229	0.9294	0.9366	0.9443	0.9523
67	Ho	164.930	0.9075	0.9081	0.9101	0.9134	0.9178	0.9233	0.9298	0.9369	0.9446	0.9526
68	Er	167.256	0.9087	0.9094	0.9113	0.9145	0.9189	0.9244	0.9307	0.9378	0.9454	0.9533
		161.929	0.9058	0.9065	0.9085	0.9119	0.9164	0.9220	0.9285	0.9358	0.9436	0.9518
		163.929	0.9069	0.9076	0.9096	0.9129	0.9174	0.9229	0.9294	0.9366	0.9443	0.9523
		165.930	0.9080	0.9087	0.9106	0.9139	0.9183	0.9238	0.9302	0.9373	0.9449	0.9529
		166.932	0.9085	0.9092	0.9112	0.9144	0.9188	0.9242	0.9306	0.9377	0.9453	0.9532
		167.932	0.9090	0.9097	0.9117	0.9149	0.9192	0.9247	0.9310	0.9380	0.9456	0.9534
		169.936	0.9101	0.9107	0.9127	0.9158	0.9202	0.9255	0.9318	0.9387	0.9462	0.9540

Table A5.2. Rutherford backscattering spectrometry kinematic factors for He as a projectile (continued).

Atomic no. *(Z)*	El.	Isotopic mass (M_2) *(amu)*	180°	170°	160°	150°	140°	130°	120°	110°	100°	90°
69	Tm	168.934	0.9096	0.9102	0.9122	0.9154	0.9197	0.9251	0.9314	0.9384	0.9459	0.9537
70	Yb	173.034	0.9116	0.9122	0.9141	0.9173	0.9215	0.9268	0.9329	0.9398	0.9471	0.9548
		167.934	0.9090	0.9097	0.9117	0.9149	0.9192	0.9247	0.9310	0.9380	0.9456	0.9534
		169.935	0.9101	0.9107	0.9127	0.9158	0.9202	0.9255	0.9318	0.9387	0.9462	0.9540
		170.936	0.9106	0.9112	0.9131	0.9163	0.9206	0.9259	0.9321	0.9391	0.9465	0.9542
		171.936	0.9111	0.9117	0.9136	0.9168	0.9210	0.9264	0.9325	0.9394	0.9468	0.9545
		172.938	0.9116	0.9122	0.9141	0.9172	0.9215	0.9268	0.9329	0.9398	0.9471	0.9548
		173.939	0.9120	0.9127	0.9146	0.9177	0.9219	0.9272	0.9333	0.9401	0.9474	0.9550
		175.943	0.9130	0.9136	0.9155	0.9186	0.9228	0.9280	0.9340	0.9408	0.9480	0.9555
71	Lu	174.967	0.9125	0.9132	0.9151	0.9182	0.9224	0.9276	0.9337	0.9404	0.9477	0.9553
		174.941	0.9125	0.9132	0.9151	0.9181	0.9224	0.9276	0.9337	0.9404	0.9477	0.9553
		175.943	0.9130	0.9136	0.9155	0.9186	0.9228	0.9280	0.9340	0.9408	0.9480	0.9555
72	Hf	178.49	0.9142	0.9148	0.9167	0.9197	0.9238	0.9290	0.9349	0.9416	0.9487	0.9561
		173.940	0.9120	0.9127	0.9146	0.9177	0.9219	0.9272	0.9333	0.9401	0.9474	0.9550
		175.941	0.9130	0.9136	0.9155	0.9186	0.9228	0.9280	0.9340	0.9408	0.9480	0.9555
		176.943	0.9135	0.9141	0.9160	0.9190	0.9232	0.9284	0.9344	0.9411	0.9483	0.9558
		177.944	0.9139	0.9146	0.9164	0.9195	0.9236	0.9287	0.9347	0.9414	0.9486	0.9560
		178.946	0.9144	0.9150	0.9169	0.9199	0.9240	0.9291	0.9351	0.9417	0.9488	0.9562
		179.947	0.9149	0.9155	0.9173	0.9203	0.9244	0.9295	0.9354	0.9420	0.9491	0.9565
73	Ta	180.948	0.9153	0.9159	0.9178	0.9207	0.9248	0.9299	0.9358	0.9423	0.9494	0.9567
		179.947	0.9149	0.9155	0.9173	0.9203	0.9244	0.9295	0.9354	0.9420	0.9491	0.9565
		180.948	0.9153	0.9159	0.9178	0.9207	0.9248	0.9299	0.9358	0.9423	0.9494	0.9567
74	W	183.849	0.9166	0.9172	0.9190	0.9219	0.9260	0.9310	0.9368	0.9432	0.9502	0.9574
		179.947	0.9149	0.9155	0.9173	0.9203	0.9244	0.9295	0.9354	0.9420	0.9491	0.9565
		181.948	0.9158	0.9164	0.9182	0.9212	0.9252	0.9303	0.9361	0.9427	0.9497	0.9569
		182.950	0.9162	0.9168	0.9186	0.9216	0.9256	0.9306	0.9365	0.9430	0.9499	0.9572
		183.951	0.9166	0.9172	0.9190	0.9220	0.9260	0.9310	0.9368	0.9433	0.9502	0.9574
		185.954	0.9175	0.9181	0.9199	0.9228	0.9268	0.9317	0.9375	0.9439	0.9507	0.9579
75	Re	186.207	0.9176	0.9182	0.9200	0.9229	0.9269	0.9318	0.9375	0.9439	0.9508	0.9579
		184.953	0.9171	0.9177	0.9195	0.9224	0.9264	0.9314	0.9371	0.9436	0.9505	0.9576
		186.956	0.9179	0.9185	0.9203	0.9232	0.9272	0.9321	0.9378	0.9441	0.9510	0.9581
76	Os	190.24	0.9193	0.9199	0.9216	0.9245	0.9284	0.9332	0.9388	0.9451	0.9518	0.9588
		183.952	0.9166	0.9172	0.9190	0.9220	0.9260	0.9310	0.9368	0.9433	0.9502	0.9574
		185.954	0.9175	0.9181	0.9199	0.9228	0.9268	0.9317	0.9375	0.9439	0.9507	0.9579
		186.956	0.9179	0.9185	0.9203	0.9232	0.9272	0.9321	0.9378	0.9441	0.9510	0.9581
		187.956	0.9183	0.9189	0.9207	0.9236	0.9275	0.9324	0.9381	0.9444	0.9512	0.9583
		188.958	0.9187	0.9193	0.9211	0.9240	0.9279	0.9328	0.9384	0.9447	0.9515	0.9585
		189.958	0.9192	0.9197	0.9215	0.9244	0.9283	0.9331	0.9387	0.9450	0.9517	0.9587
		191.962	0.9200	0.9206	0.9223	0.9251	0.9290	0.9338	0.9394	0.9456	0.9522	0.9591
77	Ir	192.216	0.9201	0.9207	0.9224	0.9252	0.9291	0.9339	0.9394	0.9456	0.9523	0.9592
		190.961	0.9196	0.9202	0.9219	0.9247	0.9286	0.9334	0.9390	0.9453	0.9520	0.9589
		192.963	0.9204	0.9209	0.9227	0.9255	0.9293	0.9341	0.9397	0.9458	0.9525	0.9594
78	Pt	195.08	0.9212	0.9218	0.9235	0.9263	0.9301	0.9348	0.9403	0.9464	0.9530	0.9598
		189.960	0.9192	0.9197	0.9215	0.9244	0.9283	0.9331	0.9387	0.9450	0.9517	0.9587
		191.961	0.9200	0.9206	0.9223	0.9251	0.9290	0.9338	0.9394	0.9456	0.9522	0.9591
		193.963	0.9208	0.9213	0.9231	0.9259	0.9297	0.9344	0.9400	0.9461	0.9527	0.9596
		194.965	0.9212	0.9217	0.9234	0.9262	0.9300	0.9348	0.9403	0.9464	0.9529	0.9598
		195.965	0.9215	0.9221	0.9238	0.9266	0.9304	0.9351	0.9406	0.9466	0.9532	0.9600
		197.968	0.9223	0.9229	0.9246	0.9273	0.9311	0.9357	0.9411	0.9472	0.9536	0.9604

Table A5.2. Rutherford backscattering spectrometry kinematic factors for He as a projectile (continued).

Atomic no. *(Z)*	El.	Isotopic mass (M_2) *(amu)*	180°	170°	160°	150°	140°	130°	120°	110°	100°	90°
79	Au	196.967	0.9219	0.9225	0.9242	0.9270	0.9307	0.9354	0.9408	0.9469	0.9534	0.9602
80	Hg	200.588	0.9233	0.9238	0.9255	0.9282	0.9319	0.9365	0.9419	0.9478	0.9542	0.9609
		195.966	0.9215	0.9221	0.9238	0.9266	0.9304	0.9351	0.9406	0.9466	0.9532	0.9600
		197.967	0.9223	0.9229	0.9246	0.9273	0.9311	0.9357	0.9411	0.9472	0.9536	0.9604
		198.968	0.9227	0.9232	0.9249	0.9277	0.9314	0.9360	0.9414	0.9474	0.9539	0.9606
		199.968	0.9230	0.9236	0.9253	0.9280	0.9317	0.9363	0.9417	0.9477	0.9541	0.9608
		200.970	0.9234	0.9240	0.9256	0.9284	0.9321	0.9366	0.9420	0.9479	0.9543	0.9609
		201.971	0.9238	0.9243	0.9260	0.9287	0.9324	0.9370	0.9423	0.9482	0.9545	0.9611
		203.973	0.9245	0.9251	0.9267	0.9294	0.9330	0.9376	0.9428	0.9487	0.9550	0.9615
81	Tl	204.383	0.9246	0.9252	0.9268	0.9295	0.9332	0.9377	0.9429	0.9488	0.9551	0.9616
		202.972	0.9241	0.9247	0.9263	0.9290	0.9327	0.9373	0.9425	0.9484	0.9548	0.9613
		204.974	0.9249	0.9254	0.9270	0.9297	0.9333	0.9378	0.9431	0.9489	0.9552	0.9617
82	Pb	207.217	0.9256	0.9262	0.9278	0.9304	0.9340	0.9385	0.9437	0.9495	0.9557	0.9621
		203.973	0.9245	0.9251	0.9267	0.9294	0.9330	0.9376	0.9428	0.9487	0.9550	0.9615
		205.975	0.9252	0.9258	0.9274	0.9300	0.9337	0.9381	0.9434	0.9492	0.9554	0.9619
		206.976	0.9256	0.9261	0.9277	0.9304	0.9340	0.9384	0.9436	0.9494	0.9556	0.9621
		207.977	0.9259	0.9264	0.9280	0.9307	0.9343	0.9387	0.9439	0.9496	0.9558	0.9622
83	Bi	208.980	0.9262	0.9268	0.9284	0.9310	0.9346	0.9390	0.9442	0.9499	0.9560	0.9624
84	Po	208.982	0.9262	0.9268	0.9284	0.9310	0.9346	0.9390	0.9442	0.9499	0.9560	0.9624
85	At	210.000	0.9266	0.9271	0.9287	0.9313	0.9349	0.9393	0.9444	0.9501	0.9562	0.9626
86	Rn	222.018	0.9304	0.9309	0.9324	0.9349	0.9383	0.9425	0.9473	0.9528	0.9586	0.9646
87	Fr	223.000	0.9307	0.9312	0.9327	0.9352	0.9386	0.9427	0.9476	0.9530	0.9587	0.9647
88	Ra	226.025	0.9316	0.9321	0.9336	0.9360	0.9394	0.9435	0.9483	0.9536	0.9593	0.9652
89	Ac	227.000	0.9319	0.9324	0.9339	0.9363	0.9396	0.9437	0.9485	0.9538	0.9595	0.9653
90	Th	232.038	0.9333	0.9338	0.9353	0.9376	0.9409	0.9449	0.9496	0.9548	0.9603	0.9661
91	Pa	231.036	0.9330	0.9335	0.9350	0.9374	0.9406	0.9447	0.9493	0.9546	0.9601	0.9659
92	U	238.018	0.9349	0.9354	0.9368	0.9392	0.9423	0.9462	0.9508	0.9559	0.9613	0.9669
		234.041	0.9339	0.9344	0.9358	0.9382	0.9414	0.9454	0.9500	0.9551	0.9606	0.9664
		235.044	0.9341	0.9346	0.9361	0.9384	0.9416	0.9456	0.9502	0.9553	0.9608	0.9665
		238.040	0.9349	0.9354	0.9368	0.9392	0.9423	0.9462	0.9508	0.9559	0.9613	0.9669
93	Np	237.048	0.9347	0.9352	0.9366	0.9389	0.9421	0.9460	0.9506	0.9557	0.9611	0.9668
94	Pu	244.064	0.9365	0.9370	0.9384	0.9406	0.9437	0.9475	0.9520	0.9569	0.9622	0.9677
95	Am	243.061	0.9362	0.9367	0.9381	0.9404	0.9435	0.9473	0.9518	0.9568	0.9621	0.9676

Table A5.3. Rutherford backscattering spectrometry kinematic factors for Li as a projectile.

This table gives the RBS kinematic factors K_{M_2}, defined by Eq. (A4.2), for Li as a projectile (M_1=7.016 amu) and for the isotopic masses (M_2) of the elements. The kinematic factors are given as a function of scattering angle from 180° to 90° measured in the laboratory frame of reference. The first row for each element gives the average atomic weight of that element and the average kinematic factor (Average K_{M_2}) for that element. The subsequent rows give the isotopic masses for that element and the K_{M_2} for those isotopic masses. The average K_{M_2} is calculated as the weighted average of the K_{M_2}, in which the relative abundances (ϖ_{M_2}) for the M_2 are given in Appendix 1:

$$\text{Average } K_{M_2} = \Sigma\, \varpi_{M_2} K_{M_2}\text{, summed over the masses } M_2.$$

Atomic no. (*Z*)	El.	Isotopic mass (M_2) (*amu*)	180°	170°	160°	150°	140°	130°	120°	110°	100°	90°
4	Be	9.012	0.0155	0.0159	0.0171	0.0193	0.0229	0.0289	0.0386	0.0546	0.0812	0.1245
5	B	10.811	0.0455	0.0464	0.0492	0.0543	0.0625	0.0747	0.0929	0.1196	0.1582	0.2125
		10.013	0.0310	0.0316	0.0337	0.0376	0.0438	0.0535	0.0683	0.0910	0.1253	0.1760
		11.009	0.0491	0.0500	0.0530	0.0585	0.0671	0.0800	0.0990	0.1267	0.1664	0.2215
6	C	12.011	0.0689	0.0702	0.0740	0.0809	0.0916	0.1073	0.1297	0.1612	0.2046	0.2625
		12.000	0.0687	0.0699	0.0738	0.0806	0.0913	0.1070	0.1294	0.1609	0.2042	0.2621
		13.003	0.0894	0.0909	0.0955	0.1036	0.1161	0.1341	0.1592	0.1936	0.2395	0.2991
7	N	14.007	0.1106	0.1123	0.1175	0.1267	0.1407	0.1606	0.1879	0.2244	0.2721	0.3325
		14.003	0.1105	0.1122	0.1174	0.1267	0.1406	0.1605	0.1878	0.2243	0.2720	0.3324
		15.000	0.1315	0.1334	0.1392	0.1493	0.1646	0.1860	0.2149	0.2530	0.3018	0.3626
8	O	15.999	0.1523	0.1544	0.1607	0.1716	0.1879	0.2105	0.2407	0.2800	0.3295	0.3903
		15.995	0.1523	0.1543	0.1606	0.1715	0.1878	0.2104	0.2406	0.2799	0.3294	0.3902
		16.999	0.1728	0.1750	0.1817	0.1933	0.2104	0.2340	0.2652	0.3053	0.3552	0.4157
		17.999	0.1928	0.1951	0.2021	0.2142	0.2321	0.2565	0.2884	0.3290	0.3791	0.4391
9	F	18.998	0.2122	0.2145	0.2219	0.2344	0.2528	0.2778	0.3104	0.3513	0.4013	0.4606
10	Ne	20.179	0.2342	0.2367	0.2442	0.2572	0.2761	0.3017	0.3346	0.3756	0.4254	0.4838
		19.992	0.2308	0.2333	0.2409	0.2538	0.2726	0.2981	0.3310	0.3721	0.4219	0.4805
		20.994	0.2490	0.2516	0.2593	0.2725	0.2918	0.3176	0.3507	0.3918	0.4413	0.4990
		21.991	0.2665	0.2691	0.2770	0.2905	0.3099	0.3360	0.3693	0.4103	0.4593	0.5163
11	Na	22.990	0.2834	0.2860	0.2941	0.3077	0.3274	0.3536	0.3869	0.4278	0.4763	0.5324
12	Mg	24.305	0.3046	0.3073	0.3154	0.3292	0.3490	0.3754	0.4086	0.4491	0.4970	0.5518
		23.985	0.2996	0.3023	0.3104	0.3242	0.3440	0.3703	0.4036	0.4442	0.4922	0.5474
		24.986	0.3153	0.3180	0.3262	0.3401	0.3600	0.3863	0.4195	0.4598	0.5073	0.5615
		25.983	0.3304	0.3331	0.3413	0.3553	0.3752	0.4016	0.4346	0.4746	0.5215	0.5748
13	Al	26.982	0.3449	0.3476	0.3559	0.3699	0.3899	0.4161	0.4490	0.4886	0.5349	0.5873
14	Si	28.086	0.3603	0.3630	0.3713	0.3853	0.4052	0.4314	0.4640	0.5032	0.5488	0.6002
		27.977	0.3588	0.3616	0.3698	0.3839	0.4038	0.4300	0.4626	0.5018	0.5475	0.5990
		28.976	0.3723	0.3750	0.3833	0.3973	0.4172	0.4433	0.4757	0.5145	0.5595	0.6101
		29.974	0.3852	0.3880	0.3963	0.4102	0.4301	0.4560	0.4881	0.5265	0.5709	0.6207
15	P	30.974	0.3977	0.4005	0.4087	0.4227	0.4424	0.4682	0.5000	0.5380	0.5817	0.6306
16	S	32.064	0.4108	0.4135	0.4218	0.4357	0.4553	0.4809	0.5124	0.5498	0.5929	0.6409
		31.972	0.4097	0.4125	0.4207	0.4346	0.4543	0.4798	0.5114	0.5489	0.5920	0.6401
		32.972	0.4213	0.4241	0.4323	0.4461	0.4657	0.4910	0.5223	0.5593	0.6018	0.6491
		33.968	0.4325	0.4352	0.4434	0.4571	0.4766	0.5017	0.5327	0.5693	0.6111	0.6576
		35.967	0.4537	0.4564	0.4645	0.4781	0.4972	0.5219	0.5522	0.5879	0.6286	0.6735
17	Cl	35.453	0.4482	0.4509	0.4591	0.4727	0.4919	0.5168	0.5472	0.5831	0.6241	0.6695
		34.969	0.4433	0.4460	0.4541	0.4678	0.4871	0.5121	0.5427	0.5788	0.6201	0.6658
		36.966	0.4637	0.4664	0.4745	0.4880	0.5069	0.5315	0.5614	0.5967	0.6368	0.6810

Table A5.3. Rutherford backscattering spectrometry kinematic factors for Li as a projectile (continued).

Atomic no. *(Z)*	El.	Isotopic mass (M_2) *(amu)*	180°	170°	160°	150°	140°	130°	120°	110°	100°	90°
18	Ar	39.948	0.4917	0.4943	0.5022	0.5154	0.5340	0.5578	0.5868	0.6207	0.6591	0.7012
		35.967	0.4537	0.4564	0.4645	0.4781	0.4972	0.5219	0.5522	0.5879	0.6286	0.6735
		37.963	0.4734	0.4761	0.4841	0.4975	0.5163	0.5406	0.5702	0.6050	0.6445	0.6880
		39.962	0.4918	0.4945	0.5024	0.5156	0.5341	0.5579	0.5869	0.6208	0.6592	0.7013
19	K	39.098	0.4840	0.4866	0.4946	0.5079	0.5265	0.5505	0.5798	0.6141	0.6530	0.6957
		38.964	0.4828	0.4854	0.4934	0.5067	0.5254	0.5494	0.5787	0.6131	0.6520	0.6948
		40.000	0.4922	0.4948	0.5027	0.5159	0.5344	0.5582	0.5872	0.6211	0.6595	0.7015
		40.962	0.5006	0.5032	0.5111	0.5242	0.5425	0.5661	0.5948	0.6282	0.6661	0.7075
20	Ca	40.078	0.4928	0.4954	0.5033	0.5165	0.5350	0.5588	0.5878	0.6216	0.6599	0.7020
		39.963	0.4918	0.4945	0.5024	0.5156	0.5341	0.5579	0.5869	0.6208	0.6592	0.7013
		41.959	0.5091	0.5117	0.5194	0.5324	0.5506	0.5740	0.6023	0.6354	0.6727	0.7135
		42.959	0.5173	0.5199	0.5276	0.5405	0.5585	0.5816	0.6096	0.6423	0.6790	0.7192
		43.956	0.5252	0.5278	0.5354	0.5482	0.5661	0.5889	0.6166	0.6489	0.6851	0.7247
		45.954	0.5404	0.5429	0.5504	0.5630	0.5805	0.6029	0.6300	0.6614	0.6967	0.7351
		47.952	0.5546	0.5571	0.5645	0.5768	0.5940	0.6159	0.6424	0.6731	0.7074	0.7447
21	Sc	44.956	0.5329	0.5354	0.5430	0.5557	0.5734	0.5960	0.6234	0.6553	0.6910	0.7300
22	Ti	47.878	0.5540	0.5565	0.5639	0.5763	0.5935	0.6154	0.6419	0.6726	0.7070	0.7443
		45.953	0.5404	0.5429	0.5504	0.5630	0.5805	0.6029	0.6300	0.6614	0.6967	0.7351
		46.952	0.5476	0.5501	0.5576	0.5700	0.5873	0.6095	0.6363	0.6673	0.7022	0.7400
		47.948	0.5546	0.5571	0.5645	0.5768	0.5940	0.6159	0.6424	0.6731	0.7074	0.7447
		48.948	0.5614	0.5639	0.5712	0.5834	0.6004	0.6221	0.6483	0.6786	0.7125	0.7493
		49.945	0.5680	0.5704	0.5777	0.5898	0.6067	0.6282	0.6540	0.6840	0.7174	0.7537
23	V	50.942	0.5744	0.5768	0.5840	0.5960	0.6127	0.6340	0.6596	0.6891	0.7222	0.7579
		49.947	0.5680	0.5704	0.5777	0.5898	0.6067	0.6282	0.6541	0.6840	0.7174	0.7537
		50.944	0.5744	0.5768	0.5840	0.5960	0.6127	0.6340	0.6596	0.6892	0.7222	0.7579
24	Cr	51.996	0.5809	0.5833	0.5905	0.6024	0.6189	0.6399	0.6652	0.6944	0.7270	0.7622
		49.946	0.5680	0.5704	0.5777	0.5898	0.6067	0.6282	0.6540	0.6840	0.7174	0.7537
		51.940	0.5806	0.5830	0.5902	0.6021	0.6186	0.6396	0.6650	0.6942	0.7268	0.7620
		52.941	0.5867	0.5891	0.5962	0.6079	0.6243	0.6451	0.6702	0.6990	0.7312	0.7660
		53.939	0.5926	0.5949	0.6020	0.6136	0.6298	0.6505	0.6752	0.7037	0.7355	0.7698
25	Mn	54.938	0.5983	0.6006	0.6076	0.6192	0.6352	0.6556	0.6801	0.7083	0.7397	0.7735
26	Fe	55.847	0.6034	0.6057	0.6126	0.6241	0.6400	0.6602	0.6844	0.7123	0.7433	0.7768
		53.940	0.5926	0.5949	0.6020	0.6136	0.6299	0.6505	0.6752	0.7038	0.7355	0.7698
		55.935	0.6039	0.6062	0.6131	0.6245	0.6404	0.6606	0.6849	0.7127	0.7437	0.7771
		56.935	0.6093	0.6116	0.6184	0.6298	0.6455	0.6655	0.6895	0.7170	0.7476	0.7806
		57.933	0.6146	0.6168	0.6236	0.6349	0.6505	0.6703	0.6940	0.7212	0.7514	0.7840
27	Co	58.933	0.6197	0.6220	0.6287	0.6398	0.6553	0.6749	0.6983	0.7253	0.7551	0.7872
28	Ni	58.688	0.6184	0.6206	0.6274	0.6385	0.6540	0.6737	0.6972	0.7242	0.7541	0.7864
		57.935	0.6146	0.6169	0.6236	0.6349	0.6505	0.6703	0.6940	0.7212	0.7514	0.7840
		59.931	0.6247	0.6270	0.6336	0.6447	0.6600	0.6794	0.7026	0.7292	0.7587	0.7904
		60.931	0.6296	0.6318	0.6384	0.6494	0.6645	0.6837	0.7067	0.7330	0.7622	0.7935
	Ni	61.928	0.6344	0.6366	0.6431	0.6539	0.6690	0.6880	0.7107	0.7367	0.7655	0.7965
		63.928	0.6435	0.6457	0.6521	0.6628	0.6775	0.6961	0.7184	0.7439	0.7720	0.8022
29	Cu	63.546	0.6418	0.6439	0.6504	0.6611	0.6759	0.6946	0.7169	0.7425	0.7708	0.8011
		62.930	0.6390	0.6412	0.6477	0.6584	0.6733	0.6921	0.7146	0.7404	0.7688	0.7994
		64.928	0.6480	0.6501	0.6565	0.6670	0.6816	0.7001	0.7221	0.7473	0.7751	0.8050

Table A5.3. Rutherford backscattering spectrometry kinematic factors for Li as a projectile (continued).

Atomic no. (Z)	El.	Isotopic mass (M_2) (amu)	180°	170°	160°	150°	140°	130°	120°	110°	100°	90°
30	Zn	65.396	0.6499	0.6520	0.6583	0.6688	0.6834	0.7017	0.7237	0.7488	0.7765	0.8061
		63.929	0.6435	0.6457	0.6521	0.6628	0.6775	0.6961	0.7184	0.7439	0.7720	0.8022
		65.926	0.6523	0.6544	0.6607	0.6711	0.6856	0.7039	0.7257	0.7506	0.7782	0.8076
		66.927	0.6565	0.6586	0.6648	0.6752	0.6895	0.7076	0.7292	0.7539	0.7811	0.8102
		67.925	0.6606	0.6627	0.6689	0.6791	0.6933	0.7113	0.7326	0.7570	0.7840	0.8128
		69.925	0.6685	0.6706	0.6767	0.6867	0.7007	0.7183	0.7392	0.7631	0.7895	0.8176
31	Ga	69.723	0.6677	0.6697	0.6758	0.6859	0.6999	0.7175	0.7385	0.7625	0.7889	0.8171
		68.926	0.6646	0.6667	0.6728	0.6830	0.6971	0.7148	0.7360	0.7601	0.7868	0.8152
		70.925	0.6723	0.6744	0.6804	0.6904	0.7042	0.7217	0.7424	0.7661	0.7921	0.8200
32	Ge	72.632	0.6786	0.6806	0.6865	0.6964	0.7100	0.7271	0.7475	0.7708	0.7964	0.8237
		69.924	0.6685	0.6706	0.6767	0.6867	0.7007	0.7183	0.7392	0.7631	0.7895	0.8176
		71.922	0.6761	0.6781	0.6841	0.6940	0.7077	0.7250	0.7455	0.7689	0.7947	0.8222
		72.924	0.6797	0.6817	0.6877	0.6975	0.7111	0.7282	0.7485	0.7717	0.7973	0.8245
		73.921	0.6833	0.6853	0.6912	0.7009	0.7144	0.7313	0.7515	0.7744	0.7997	0.8266
		75.921	0.6902	0.6922	0.6980	0.7076	0.7208	0.7374	0.7572	0.7797	0.8045	0.8308
33	As	74.922	0.6868	0.6888	0.6946	0.7043	0.7176	0.7344	0.7544	0.7771	0.8021	0.8287
34	Se	78.993	0.7002	0.7021	0.7078	0.7171	0.7300	0.7462	0.7654	0.7873	0.8113	0.8368
		73.923	0.6833	0.6853	0.6912	0.7009	0.7144	0.7313	0.7515	0.7745	0.7997	0.8266
		75.919	0.6902	0.6922	0.6980	0.7076	0.7208	0.7374	0.7572	0.7797	0.8045	0.8308
		76.920	0.6936	0.6955	0.7013	0.7108	0.7239	0.7404	0.7600	0.7823	0.8068	0.8328
		77.917	0.6969	0.6988	0.7045	0.7139	0.7269	0.7432	0.7626	0.7847	0.8090	0.8348
		79.916	0.7032	0.7051	0.7107	0.7200	0.7327	0.7488	0.7679	0.7895	0.8133	0.8386
		81.917	0.7093	0.7112	0.7167	0.7258	0.7384	0.7541	0.7729	0.7941	0.8174	0.8422
35	Br	79.904	0.7032	0.7050	0.7106	0.7199	0.7327	0.7487	0.7678	0.7895	0.8133	0.8385
		78.918	0.7001	0.7020	0.7076	0.7170	0.7298	0.7461	0.7653	0.7872	0.8112	0.8367
		80.916	0.7063	0.7082	0.7137	0.7229	0.7356	0.7515	0.7704	0.7919	0.8154	0.8404
36	Kr	83.8	0.7148	0.7166	0.7221	0.7310	0.7434	0.7589	0.7773	0.7982	0.8211	0.8455
		77.920	0.6969	0.6988	0.7045	0.7139	0.7269	0.7432	0.7627	0.7847	0.8090	0.8348
		79.916	0.7032	0.7051	0.7107	0.7200	0.7327	0.7488	0.7679	0.7895	0.8133	0.8386
		81.913	0.7093	0.7112	0.7167	0.7258	0.7383	0.7541	0.7729	0.7941	0.8174	0.8422
		82.914	0.7123	0.7141	0.7196	0.7286	0.7411	0.7567	0.7753	0.7963	0.8194	0.8440
		83.911	0.7152	0.7170	0.7224	0.7314	0.7437	0.7592	0.7776	0.7985	0.8214	0.8457
		85.911	0.7208	0.7226	0.7279	0.7368	0.7489	0.7642	0.7822	0.8027	0.8252	0.8490
37	Rb	85.468	0.7196	0.7213	0.7267	0.7356	0.7477	0.7631	0.7812	0.8018	0.8243	0.8483
		84.912	0.7180	0.7198	0.7252	0.7341	0.7463	0.7617	0.7800	0.8006	0.8233	0.8474
		86.909	0.7235	0.7253	0.7306	0.7394	0.7514	0.7665	0.7844	0.8048	0.8270	0.8506
38	Sr	87.617	0.7254	0.7272	0.7325	0.7412	0.7531	0.7682	0.7860	0.8062	0.8283	0.8517
		83.913	0.7152	0.7170	0.7224	0.7314	0.7437	0.7592	0.7776	0.7985	0.8214	0.8457
		85.909	0.7208	0.7226	0.7279	0.7367	0.7489	0.7641	0.7822	0.8027	0.8252	0.8490
		86.909	0.7235	0.7253	0.7306	0.7394	0.7514	0.7665	0.7844	0.8048	0.8270	0.8506
		87.906	0.7262	0.7280	0.7332	0.7419	0.7539	0.7689	0.7866	0.8068	0.8288	0.8522
39	Y	88.906	0.7288	0.7306	0.7358	0.7444	0.7563	0.7711	0.7888	0.8087	0.8306	0.8537
40	Zr	91.221	0.7347	0.7364	0.7415	0.7500	0.7616	0.7762	0.7935	0.8131	0.8345	0.8571
		89.905	0.7314	0.7331	0.7383	0.7469	0.7586	0.7734	0.7908	0.8106	0.8323	0.8552
		90.906	0.7339	0.7357	0.7408	0.7493	0.7609	0.7756	0.7929	0.8125	0.8340	0.8567
		91.905	0.7364	0.7381	0.7432	0.7516	0.7632	0.7777	0.7949	0.8144	0.8356	0.8581
		93.906	0.7413	0.7429	0.7480	0.7562	0.7676	0.7819	0.7988	0.8180	0.8388	0.8610
		95.908	0.7459	0.7476	0.7525	0.7607	0.7719	0.7860	0.8026	0.8214	0.8419	0.8637

Table A5.3. Rutherford backscattering spectrometry kinematic factors for Li as a projectile (continued).

Atomic no. *(Z)*	El.	Isotopic mass (M_2) *(amu)*	180°	170°	160°	150°	140°	130°	120°	110°	100°	90°
41	Nb	92.906	0.7389	0.7406	0.7456	0.7540	0.7655	0.7799	0.7969	0.8162	0.8373	0.8596
42	Mo	95.931	0.7459	0.7475	0.7525	0.7606	0.7719	0.7859	0.8025	0.8214	0.8419	0.8636
		91.907	0.7364	0.7381	0.7432	0.7516	0.7632	0.7777	0.7949	0.8144	0.8356	0.8582
		93.905	0.7413	0.7429	0.7480	0.7562	0.7676	0.7819	0.7988	0.8180	0.8388	0.8610
		94.906	0.7436	0.7453	0.7503	0.7585	0.7698	0.7840	0.8007	0.8197	0.8404	0.8623
		95.905	0.7459	0.7476	0.7525	0.7607	0.7719	0.7860	0.8026	0.8214	0.8419	0.8637
		96.906	0.7482	0.7498	0.7547	0.7628	0.7740	0.7879	0.8044	0.8231	0.8434	0.8650
		97.905	0.7504	0.7520	0.7569	0.7650	0.7760	0.7899	0.8062	0.8247	0.8449	0.8663
		99.908	0.7548	0.7564	0.7612	0.7691	0.7800	0.7936	0.8097	0.8279	0.8478	0.8688
43	Tc	98.000	0.7506	0.7523	0.7571	0.7652	0.7762	0.7900	0.8064	0.8249	0.8451	0.8664
44	Ru	101.019	0.7570	0.7586	0.7634	0.7713	0.7820	0.7956	0.8115	0.8296	0.8493	0.8701
		95.908	0.7459	0.7476	0.7525	0.7607	0.7719	0.7860	0.8026	0.8214	0.8419	0.8637
		97.905	0.7504	0.7520	0.7569	0.7650	0.7760	0.7899	0.8062	0.8247	0.8449	0.8663
		98.506	0.7517	0.7534	0.7582	0.7662	0.7772	0.7910	0.8073	0.8257	0.8458	0.8670
		99.904	0.7547	0.7564	0.7612	0.7691	0.7800	0.7936	0.8097	0.8279	0.8478	0.8688
		100.906	0.7569	0.7585	0.7632	0.7711	0.7819	0.7954	0.8114	0.8295	0.8492	0.8700
		101.904	0.7589	0.7605	0.7653	0.7731	0.7838	0.7972	0.8131	0.8310	0.8505	0.8712
		103.905	0.7630	0.7646	0.7692	0.7769	0.7875	0.8007	0.8163	0.8340	0.8532	0.8735
45	Rh	102.906	0.7610	0.7626	0.7673	0.7750	0.7857	0.7990	0.8147	0.8325	0.8519	0.8723
46	Pd	106.415	0.7678	0.7694	0.7740	0.7815	0.7919	0.8049	0.8202	0.8375	0.8564	0.8763
		101.906	0.7589	0.7605	0.7653	0.7731	0.7838	0.7972	0.8131	0.8310	0.8505	0.8712
		103.904	0.7630	0.7646	0.7692	0.7769	0.7875	0.8007	0.8163	0.8340	0.8532	0.8735
		104.905	0.7650	0.7665	0.7712	0.7788	0.7893	0.8024	0.8179	0.8354	0.8545	0.8746
		105.904	0.7669	0.7685	0.7731	0.7807	0.7911	0.8041	0.8195	0.8369	0.8558	0.8757
		107.904	0.7707	0.7722	0.7768	0.7843	0.7945	0.8074	0.8225	0.8396	0.8583	0.8779
		109.905	0.7744	0.7759	0.7804	0.7877	0.7979	0.8105	0.8255	0.8423	0.8607	0.8800
47	Ag	107.868	0.7706	0.7721	0.7767	0.7842	0.7944	0.8073	0.8225	0.8396	0.8582	0.8779
		106.905	0.7688	0.7704	0.7749	0.7825	0.7928	0.8058	0.8210	0.8383	0.8570	0.8768
		108.905	0.7726	0.7741	0.7786	0.7860	0.7962	0.8090	0.8240	0.8410	0.8595	0.8790
48	Cd	112.412	0.7788	0.7802	0.7847	0.7919	0.8019	0.8143	0.8290	0.8455	0.8635	0.8825
		105.907	0.7669	0.7685	0.7731	0.7807	0.7911	0.8041	0.8195	0.8369	0.8558	0.8757
		107.904	0.7707	0.7722	0.7768	0.7843	0.7945	0.8074	0.8225	0.8396	0.8583	0.8779
		109.903	0.7744	0.7759	0.7804	0.7877	0.7979	0.8105	0.8255	0.8423	0.8607	0.8800
		110.904	0.7762	0.7777	0.7821	0.7894	0.7995	0.8121	0.8269	0.8436	0.8618	0.8810
		111.903	0.7779	0.7794	0.7838	0.7911	0.8011	0.8136	0.8283	0.8449	0.8630	0.8820
		112.904	0.7797	0.7811	0.7855	0.7928	0.8027	0.8151	0.8297	0.8462	0.8641	0.8830
		113.903	0.7814	0.7828	0.7872	0.7944	0.8042	0.8166	0.8311	0.8474	0.8652	0.8840
		115.905	0.7847	0.7862	0.7905	0.7976	0.8073	0.8194	0.8337	0.8499	0.8674	0.8858
49	In	114.818	0.7829	0.7844	0.7887	0.7958	0.8056	0.8179	0.8323	0.8485	0.8662	0.8848
		112.904	0.7797	0.7811	0.7855	0.7928	0.8027	0.8151	0.8297	0.8462	0.8641	0.8830
		114.904	0.7831	0.7845	0.7889	0.7960	0.8058	0.8180	0.8324	0.8486	0.8663	0.8849
50	Sn	118.613	0.7888	0.7902	0.7945	0.8014	0.8110	0.8229	0.8370	0.8528	0.8700	0.8882
		111.905	0.7779	0.7794	0.7838	0.7911	0.8011	0.8136	0.8283	0.8449	0.8630	0.8820
		113.903	0.7814	0.7828	0.7872	0.7944	0.8042	0.8166	0.8311	0.8474	0.8652	0.8840
		114.903	0.7831	0.7845	0.7888	0.7960	0.8058	0.8180	0.8324	0.8486	0.8663	0.8849
		115.902	0.7847	0.7862	0.7905	0.7976	0.8073	0.8194	0.8337	0.8498	0.8674	0.8858
		116.903	0.7864	0.7878	0.7921	0.7991	0.8088	0.8208	0.8350	0.8510	0.8684	0.8868

Table A5.3. Rutherford backscattering spectrometry kinematic factors for Li as a projectile (continued).

Atomic no. (Z)	El.	Isotopic mass (M_2) (amu)	180°	170°	160°	150°	140°	130°	120°	110°	100°	90°
		117.602	0.7875	0.7889	0.7932	0.8002	0.8098	0.8218	0.8359	0.8519	0.8692	0.8874
		118.903	0.7895	0.7910	0.7952	0.8021	0.8117	0.8236	0.8376	0.8534	0.8705	0.8886
		119.902	0.7911	0.7925	0.7967	0.8036	0.8131	0.8249	0.8388	0.8545	0.8715	0.8894
		121.903	0.7942	0.7955	0.7997	0.8065	0.8158	0.8275	0.8412	0.8567	0.8735	0.8912
		123.905	0.7971	0.7985	0.8026	0.8093	0.8185	0.8301	0.8436	0.8588	0.8754	0.8928
51	Sb	121.758	0.7939	0.7953	0.7995	0.8063	0.8156	0.8273	0.8411	0.8565	0.8733	0.8910
		120.904	0.7926	0.7940	0.7982	0.8051	0.8145	0.8262	0.8400	0.8556	0.8725	0.8903
		122.904	0.7957	0.7970	0.8012	0.8079	0.8172	0.8288	0.8424	0.8578	0.8745	0.8920
52	Te	127.586	0.8023	0.8037	0.8077	0.8142	0.8232	0.8345	0.8477	0.8626	0.8788	0.8957
		119.904	0.7911	0.7925	0.7967	0.8036	0.8131	0.8249	0.8388	0.8545	0.8715	0.8894
		121.903	0.7942	0.7955	0.7997	0.8065	0.8158	0.8275	0.8412	0.8567	0.8735	0.8912
		122.904	0.7957	0.7970	0.8012	0.8079	0.8172	0.8288	0.8424	0.8578	0.8745	0.8920
		123.903	0.7971	0.7985	0.8026	0.8093	0.8185	0.8301	0.8436	0.8588	0.8754	0.8928
		124.904	0.7986	0.7999	0.8040	0.8107	0.8199	0.8313	0.8448	0.8599	0.8763	0.8936
		125.903	0.8000	0.8014	0.8054	0.8120	0.8211	0.8325	0.8459	0.8609	0.8773	0.8944
		127.905	0.8028	0.8042	0.8081	0.8147	0.8237	0.8349	0.8481	0.8630	0.8791	0.8960
		129.906	0.8055	0.8069	0.8108	0.8173	0.8262	0.8372	0.8503	0.8649	0.8808	0.8975
53	I	126.905	0.8014	0.8028	0.8068	0.8134	0.8224	0.8337	0.8470	0.8620	0.8782	0.8952
54	Xe	131.293	0.8073	0.8087	0.8126	0.8190	0.8278	0.8388	0.8517	0.8662	0.8820	0.8985
		123.906	0.7971	0.7985	0.8026	0.8093	0.8185	0.8301	0.8436	0.8588	0.8754	0.8928
		125.904	0.8000	0.8014	0.8054	0.8120	0.8212	0.8325	0.8459	0.8609	0.8773	0.8944
		127.904	0.8028	0.8042	0.8081	0.8147	0.8237	0.8349	0.8481	0.8630	0.8791	0.8960
		128.905	0.8042	0.8055	0.8095	0.8160	0.8249	0.8361	0.8492	0.8639	0.8799	0.8968
		129.904	0.8055	0.8069	0.8108	0.8173	0.8262	0.8372	0.8503	0.8649	0.8808	0.8975
		130.905	0.8069	0.8082	0.8121	0.8185	0.8274	0.8384	0.8513	0.8659	0.8817	0.8983
		131.904	0.8082	0.8095	0.8134	0.8198	0.8286	0.8395	0.8524	0.8668	0.8825	0.8990
		133.905	0.8108	0.8121	0.8159	0.8222	0.8309	0.8417	0.8544	0.8687	0.8842	0.9004
		135.907	0.8133	0.8146	0.8184	0.8246	0.8332	0.8438	0.8564	0.8705	0.8858	0.9018
55	Cs	132.905	0.8095	0.8108	0.8147	0.8210	0.8297	0.8406	0.8534	0.8678	0.8833	0.8997
56	Ba	137.327	0.8150	0.8163	0.8201	0.8263	0.8347	0.8453	0.8578	0.8717	0.8869	0.9028
		129.906	0.8055	0.8069	0.8108	0.8173	0.8262	0.8372	0.8503	0.8649	0.8808	0.8975
		131.905	0.8082	0.8095	0.8134	0.8198	0.8286	0.8395	0.8524	0.8668	0.8825	0.8990
		133.904	0.8108	0.8121	0.8159	0.8222	0.8309	0.8417	0.8544	0.8687	0.8842	0.9004
		134.906	0.8120	0.8133	0.8171	0.8234	0.8320	0.8428	0.8554	0.8696	0.8850	0.9011
		135.905	0.8133	0.8146	0.8184	0.8246	0.8332	0.8438	0.8564	0.8705	0.8858	0.9018
		136.906	0.8145	0.8158	0.8196	0.8258	0.8343	0.8449	0.8574	0.8714	0.8866	0.9025
		137.905	0.8157	0.8170	0.8207	0.8269	0.8354	0.8459	0.8583	0.8722	0.8873	0.9032
57	La	138.905	0.8169	0.8182	0.8219	0.8281	0.8365	0.8470	0.8593	0.8731	0.8881	0.9038
		137.907	0.8157	0.8170	0.8207	0.8269	0.8354	0.8459	0.8583	0.8722	0.8873	0.9032
		138.906	0.8169	0.8182	0.8219	0.8281	0.8365	0.8470	0.8593	0.8731	0.8881	0.9038
58	Ce	140.115	0.8184	0.8196	0.8233	0.8294	0.8378	0.8482	0.8604	0.8741	0.8890	0.9046
		135.907	0.8133	0.8146	0.8184	0.8246	0.8332	0.8438	0.8564	0.8705	0.8858	0.9018
		137.906	0.8157	0.8170	0.8207	0.8269	0.8354	0.8459	0.8583	0.8722	0.8873	0.9032
		139.905	0.8181	0.8194	0.8231	0.8292	0.8375	0.8480	0.8602	0.8740	0.8889	0.9045
		141.909	0.8204	0.8217	0.8253	0.8314	0.8396	0.8499	0.8620	0.8756	0.8903	0.9058
59	Pr	140.908	0.8193	0.8205	0.8242	0.8303	0.8386	0.8490	0.8611	0.8748	0.8896	0.9051

Table A5.3. Rutherford backscattering spectrometry kinematic factors for Li as a projectile (continued).

Atomic no. *(Z)*	El.	Isotopic mass (M_2) *(amu)*	180°	170°	160°	150°	140°	130°	120°	110°	100°	90°
60	Nd	144.242	0.8230	0.8243	0.8279	0.8338	0.8420	0.8522	0.8641	0.8775	0.8920	0.9072
		141.908	0.8204	0.8217	0.8253	0.8314	0.8396	0.8499	0.8620	0.8756	0.8903	0.9058
		142.910	0.8216	0.8228	0.8265	0.8325	0.8407	0.8509	0.8629	0.8764	0.8911	0.9064
		143.910	0.8227	0.8239	0.8276	0.8335	0.8417	0.8519	0.8638	0.8772	0.8918	0.9070
		144.913	0.8238	0.8250	0.8286	0.8346	0.8427	0.8528	0.8647	0.8780	0.8925	0.9076
		145.913	0.8249	0.8261	0.8297	0.8356	0.8437	0.8537	0.8656	0.8788	0.8932	0.9082
		147.917	0.8271	0.8283	0.8318	0.8376	0.8456	0.8556	0.8673	0.8804	0.8946	0.9094
		149.921	0.8292	0.8304	0.8339	0.8396	0.8475	0.8574	0.8689	0.8819	0.8959	0.9106
61	Pm	145.000	0.8239	0.8251	0.8287	0.8347	0.8428	0.8529	0.8648	0.8781	0.8926	0.9077
62	Sm	150.36	0.8296	0.8308	0.8343	0.8400	0.8479	0.8577	0.8692	0.8822	0.8961	0.9108
		143.912	0.8227	0.8239	0.8276	0.8335	0.8417	0.8519	0.8638	0.8772	0.8918	0.9070
		146.915	0.8260	0.8272	0.8308	0.8366	0.8447	0.8547	0.8664	0.8796	0.8939	0.9088
		147.915	0.8271	0.8283	0.8318	0.8376	0.8456	0.8556	0.8673	0.8804	0.8946	0.9094
		148.917	0.8281	0.8293	0.8328	0.8386	0.8466	0.8565	0.8681	0.8811	0.8952	0.9100
		149.917	0.8292	0.8303	0.8339	0.8396	0.8475	0.8574	0.8689	0.8819	0.8959	0.9106
		151.920	0.8312	0.8324	0.8359	0.8416	0.8494	0.8591	0.8705	0.8833	0.8972	0.9117
		153.922	0.8332	0.8344	0.8378	0.8435	0.8512	0.8608	0.8721	0.8848	0.8985	0.9128
63	Eu	151.965	0.8313	0.8324	0.8359	0.8416	0.8494	0.8591	0.8706	0.8834	0.8972	0.9117
		150.920	0.8302	0.8314	0.8349	0.8406	0.8485	0.8582	0.8697	0.8826	0.8965	0.9112
		152.921	0.8322	0.8334	0.8368	0.8425	0.8503	0.8600	0.8713	0.8841	0.8978	0.9123
64	Gd	157.252	0.8364	0.8376	0.8410	0.8465	0.8541	0.8635	0.8746	0.8870	0.9005	0.9146
		151.920	0.8312	0.8324	0.8359	0.8416	0.8494	0.8591	0.8705	0.8833	0.8972	0.9117
		153.921	0.8332	0.8344	0.8378	0.8435	0.8512	0.8608	0.8721	0.8848	0.8985	0.9128
		154.923	0.8342	0.8354	0.8388	0.8444	0.8521	0.8616	0.8729	0.8855	0.8991	0.9134
		155.922	0.8352	0.8363	0.8397	0.8453	0.8530	0.8625	0.8736	0.8862	0.8997	0.9139
		156.924	0.8361	0.8373	0.8407	0.8462	0.8538	0.8633	0.8744	0.8868	0.9003	0.9144
		157.924	0.8371	0.8382	0.8416	0.8471	0.8547	0.8641	0.8751	0.8875	0.9009	0.9149
		159.927	0.8390	0.8401	0.8434	0.8489	0.8564	0.8657	0.8766	0.8888	0.9021	0.9159
65	Tb	158.925	0.8380	0.8392	0.8425	0.8480	0.8555	0.8649	0.8759	0.8882	0.9015	0.9154
66	Dy	162.498	0.8413	0.8424	0.8457	0.8511	0.8585	0.8677	0.8784	0.8905	0.9036	0.9172
		155.924	0.8352	0.8363	0.8397	0.8453	0.8530	0.8625	0.8736	0.8862	0.8997	0.9139
		157.924	0.8371	0.8382	0.8416	0.8471	0.8547	0.8641	0.8751	0.8875	0.9009	0.9149
		159.925	0.8390	0.8401	0.8434	0.8489	0.8564	0.8657	0.8766	0.8888	0.9021	0.9159
		160.927	0.8399	0.8410	0.8443	0.8497	0.8572	0.8665	0.8773	0.8895	0.9027	0.9164
		161.927	0.8408	0.8419	0.8452	0.8506	0.8580	0.8672	0.8780	0.8901	0.9032	0.9169
		162.929	0.8417	0.8428	0.8461	0.8515	0.8588	0.8680	0.8787	0.8908	0.9038	0.9174
		163.929	0.8426	0.8437	0.8469	0.8523	0.8596	0.8687	0.8794	0.8914	0.9044	0.9179
67	Ho	164.930	0.8434	0.8445	0.8478	0.8531	0.8604	0.8695	0.8801	0.8920	0.9049	0.9184
68	Er	167.256	0.8454	0.8465	0.8497	0.8550	0.8622	0.8712	0.8817	0.8934	0.9062	0.9195
		161.929	0.8408	0.8419	0.8452	0.8506	0.8580	0.8672	0.8780	0.8901	0.9032	0.9169
		163.929	0.8426	0.8437	0.8469	0.8523	0.8596	0.8687	0.8794	0.8914	0.9044	0.9179
		165.930	0.8443	0.8454	0.8486	0.8539	0.8612	0.8702	0.8808	0.8926	0.9055	0.9189
		166.932	0.8452	0.8463	0.8495	0.8547	0.8620	0.8709	0.8815	0.8933	0.9060	0.9193
		167.932	0.8460	0.8471	0.8503	0.8555	0.8627	0.8717	0.8821	0.8939	0.9065	0.9198
		169.936	0.8477	0.8488	0.8519	0.8571	0.8642	0.8731	0.8834	0.8950	0.9076	0.9207
69	Tm	168.934	0.8469	0.8479	0.8511	0.8563	0.8635	0.8724	0.8828	0.8945	0.9071	0.9203

Table A5.3. Rutherford backscattering spectrometry kinematic factors for Li as a projectile (continued).

Atomic no. *(Z)*	El.	Isotopic mass (M_2) *(amu)*	180°	170°	160°	150°	140°	130°	120°	110°	100°	90°
70	Yb	173.034	0.8502	0.8512	0.8544	0.8595	0.8665	0.8752	0.8854	0.8968	0.9092	0.9221
		167.934	0.8460	0.8471	0.8503	0.8555	0.8627	0.8717	0.8821	0.8939	0.9065	0.9198
		169.935	0.8477	0.8488	0.8519	0.8571	0.8642	0.8731	0.8834	0.8950	0.9076	0.9207
		170.936	0.8485	0.8496	0.8527	0.8579	0.8650	0.8738	0.8841	0.8956	0.9081	0.9211
		171.936	0.8493	0.8504	0.8535	0.8587	0.8657	0.8745	0.8847	0.8962	0.9086	0.9216
		172.938	0.8501	0.8512	0.8543	0.8594	0.8664	0.8751	0.8853	0.8968	0.9091	0.9220
		173.939	0.8509	0.8520	0.8551	0.8602	0.8671	0.8758	0.8860	0.8973	0.9096	0.9225
		175.943	0.8525	0.8535	0.8566	0.8617	0.8686	0.8771	0.8872	0.8984	0.9106	0.9233
71	Lu	174.967	0.8517	0.8528	0.8559	0.8609	0.8679	0.8765	0.8866	0.8979	0.9101	0.9229
		174.941	0.8517	0.8528	0.8558	0.8609	0.8678	0.8765	0.8866	0.8979	0.9101	0.9229
		175.943	0.8525	0.8535	0.8566	0.8617	0.8686	0.8771	0.8872	0.8984	0.9106	0.9233
72	Hf	178.49	0.8544	0.8555	0.8585	0.8635	0.8703	0.8788	0.8887	0.8998	0.9118	0.9244
		173.940	0.8509	0.8520	0.8551	0.8602	0.8671	0.8758	0.8860	0.8973	0.9096	0.9225
		175.941	0.8525	0.8535	0.8566	0.8616	0.8685	0.8771	0.8872	0.8984	0.9106	0.9233
		176.943	0.8533	0.8543	0.8574	0.8624	0.8692	0.8778	0.8878	0.8990	0.9111	0.9237
		177.944	0.8540	0.8550	0.8581	0.8631	0.8699	0.8784	0.8884	0.8995	0.9116	0.9241
		178.946	0.8548	0.8558	0.8588	0.8638	0.8706	0.8791	0.8890	0.9001	0.9120	0.9245
		179.947	0.8555	0.8565	0.8596	0.8645	0.8713	0.8797	0.8896	0.9006	0.9125	0.9249
73	Ta	180.948	0.8563	0.8573	0.8603	0.8652	0.8719	0.8803	0.8901	0.9011	0.9130	0.9253
		179.947	0.8555	0.8565	0.8596	0.8645	0.8713	0.8797	0.8896	0.9006	0.9125	0.9249
		180.948	0.8563	0.8573	0.8603	0.8652	0.8719	0.8803	0.8901	0.9011	0.9130	0.9253
74	W	183.849	0.8584	0.8594	0.8623	0.8672	0.8738	0.8821	0.8918	0.9026	0.9143	0.9265
		179.947	0.8555	0.8565	0.8596	0.8645	0.8713	0.8797	0.8896	0.9006	0.9125	0.9249
		181.948	0.8570	0.8580	0.8610	0.8659	0.8726	0.8809	0.8907	0.9016	0.9134	0.9257
		182.950	0.8577	0.8587	0.8617	0.8666	0.8733	0.8816	0.8913	0.9021	0.9139	0.9261
		183.951	0.8584	0.8594	0.8624	0.8673	0.8739	0.8822	0.8918	0.9026	0.9143	0.9265
		185.954	0.8599	0.8608	0.8638	0.8686	0.8752	0.8834	0.8929	0.9036	0.9152	0.9273
75	Re	186.207	0.8600	0.8610	0.8639	0.8688	0.8753	0.8835	0.8931	0.9038	0.9153	0.9274
		184.953	0.8592	0.8601	0.8631	0.8679	0.8745	0.8828	0.8924	0.9031	0.9148	0.9269
		186.956	0.8606	0.8615	0.8645	0.8693	0.8758	0.8839	0.8935	0.9041	0.9156	0.9277
76	Os	190.24	0.8628	0.8637	0.8666	0.8713	0.8778	0.8858	0.8952	0.9057	0.9170	0.9289
		183.952	0.8584	0.8594	0.8624	0.8673	0.8739	0.8822	0.8918	0.9026	0.9143	0.9265
		185.954	0.8599	0.8608	0.8638	0.8686	0.8752	0.8834	0.8929	0.9036	0.9152	0.9273
		186.956	0.8606	0.8615	0.8645	0.8693	0.8758	0.8839	0.8935	0.9041	0.9156	0.9277
		187.956	0.8612	0.8622	0.8651	0.8699	0.8764	0.8845	0.8940	0.9046	0.9161	0.9280
		188.958	0.8619	0.8629	0.8658	0.8705	0.8770	0.8851	0.8945	0.9051	0.9165	0.9284
		189.958	0.8626	0.8636	0.8665	0.8712	0.8776	0.8857	0.8951	0.9056	0.9169	0.9288
		191.962	0.8639	0.8649	0.8677	0.8724	0.8788	0.8868	0.8961	0.9065	0.9177	0.9295
77	Ir	192.216	0.8641	0.8651	0.8679	0.8726	0.8790	0.8869	0.8962	0.9066	0.9179	0.9296
		190.961	0.8633	0.8642	0.8671	0.8718	0.8782	0.8862	0.8956	0.9060	0.9173	0.9291
		192.963	0.8646	0.8655	0.8684	0.8731	0.8794	0.8873	0.8966	0.9070	0.9182	0.9298
78	Pt	195.08	0.8660	0.8669	0.8697	0.8743	0.8807	0.8885	0.8977	0.9079	0.9190	0.9306
		189.960	0.8626	0.8636	0.8665	0.8712	0.8776	0.8857	0.8951	0.9056	0.9169	0.9288
		191.961	0.8639	0.8649	0.8677	0.8724	0.8788	0.8868	0.8961	0.9065	0.9177	0.9295
		193.963	0.8652	0.8662	0.8690	0.8737	0.8800	0.8879	0.8971	0.9074	0.9186	0.9302
		194.965	0.8659	0.8668	0.8696	0.8743	0.8806	0.8884	0.8976	0.9079	0.9190	0.9305
		195.965	0.8665	0.8675	0.8703	0.8749	0.8812	0.8890	0.8981	0.9083	0.9194	0.9309
		197.968	0.8678	0.8687	0.8715	0.8761	0.8823	0.8900	0.8991	0.9092	0.9201	0.9315

Table A5.3. Rutherford backscattering spectrometry kinematic factors for Li as a projectile (continued).

Atomic no. *(Z)*	El.	Isotopic mass (M_2) *(amu)*	180°	170°	160°	150°	140°	130°	120°	110°	100°	90°
79	Au	196.967	0.8672	0.8681	0.8709	0.8755	0.8817	0.8895	0.8986	0.9088	0.9198	0.9312
80	Hg	200.588	0.8694	0.8703	0.8731	0.8776	0.8837	0.8914	0.9003	0.9103	0.9211	0.9324
		195.966	0.8665	0.8675	0.8703	0.8749	0.8812	0.8890	0.8981	0.9083	0.9194	0.9309
		197.967	0.8678	0.8687	0.8715	0.8761	0.8823	0.8900	0.8991	0.9092	0.9201	0.9315
		198.968	0.8684	0.8693	0.8721	0.8766	0.8828	0.8906	0.8996	0.9097	0.9205	0.9319
	Hg	199.968	0.8690	0.8699	0.8727	0.8772	0.8834	0.8911	0.9000	0.9101	0.9209	0.9322
		200.970	0.8696	0.8705	0.8733	0.8778	0.8839	0.8916	0.9005	0.9105	0.9213	0.9325
		201.971	0.8702	0.8711	0.8739	0.8784	0.8845	0.8921	0.9010	0.9109	0.9217	0.9329
		203.973	0.8714	0.8723	0.8750	0.8795	0.8856	0.8931	0.9019	0.9118	0.9224	0.9335
81	Tl	204.383	0.8716	0.8726	0.8753	0.8797	0.8858	0.8933	0.9021	0.9119	0.9226	0.9336
		202.972	0.8708	0.8717	0.8745	0.8789	0.8850	0.8926	0.9015	0.9114	0.9220	0.9332
		204.974	0.8720	0.8729	0.8756	0.8800	0.8861	0.8936	0.9024	0.9122	0.9228	0.9338
82	Pb	207.217	0.8733	0.8742	0.8769	0.8813	0.8872	0.8947	0.9034	0.9131	0.9236	0.9345
		203.973	0.8714	0.8723	0.8750	0.8795	0.8856	0.8931	0.9019	0.9118	0.9224	0.9335
		205.975	0.8726	0.8735	0.8762	0.8806	0.8866	0.8941	0.9028	0.9126	0.9231	0.9341
		206.976	0.8732	0.8741	0.8767	0.8811	0.8871	0.8946	0.9033	0.9130	0.9235	0.9344
		207.977	0.8737	0.8746	0.8773	0.8817	0.8876	0.8950	0.9037	0.9134	0.9238	0.9347
83	Bi	208.980	0.8743	0.8752	0.8778	0.8822	0.8881	0.8955	0.9041	0.9138	0.9242	0.9350
84	Po	208.982	0.8743	0.8752	0.8778	0.8822	0.8881	0.8955	0.9042	0.9138	0.9242	0.9350
85	At	210.000	0.8749	0.8758	0.8784	0.8827	0.8886	0.8960	0.9046	0.9142	0.9245	0.9353
86	Rn	222.018	0.8812	0.8821	0.8846	0.8887	0.8943	0.9013	0.9095	0.9187	0.9285	0.9387
87	Fr	223.000	0.8817	0.8826	0.8851	0.8892	0.8948	0.9018	0.9099	0.9190	0.9288	0.9390
88	Ra	226.025	0.8832	0.8840	0.8865	0.8906	0.8961	0.9030	0.9111	0.9200	0.9297	0.9398
89	Ac	227.000	0.8837	0.8845	0.8870	0.8910	0.8965	0.9034	0.9114	0.9204	0.9300	0.9400
90	Th	232.038	0.8860	0.8869	0.8893	0.8933	0.8987	0.9054	0.9133	0.9220	0.9315	0.9413
91	Pa	231.036	0.8856	0.8864	0.8888	0.8928	0.8983	0.9050	0.9129	0.9217	0.9312	0.9411
92	U	238.018	0.8888	0.8895	0.8919	0.8958	0.9011	0.9077	0.9153	0.9239	0.9331	0.9427
		234.041	0.8870	0.8878	0.8902	0.8941	0.8995	0.9062	0.9140	0.9227	0.9320	0.9418
		235.044	0.8874	0.8882	0.8906	0.8945	0.8999	0.9066	0.9143	0.9230	0.9323	0.9420
		238.040	0.8888	0.8896	0.8919	0.8958	0.9011	0.9077	0.9153	0.9239	0.9331	0.9427
93	Np	237.048	0.8883	0.8891	0.8915	0.8954	0.9007	0.9073	0.9150	0.9236	0.9329	0.9425
94	Pu	244.064	0.8914	0.8921	0.8944	0.8982	0.9034	0.9098	0.9173	0.9257	0.9347	0.9441
95	Am	243.061	0.8909	0.8917	0.8940	0.8978	0.9030	0.9095	0.9170	0.9254	0.9345	0.9439

Table A5.4. Rutherford backscattering spectrometry kinematic factors for C as a projectile.

This table gives the RBS kinematic factors K_{M_2}, defined by Eq. (A4.2), for C as a projectile (M_1=12 amu) and for the isotopic masses (M_2) of the elements. The kinematic factors are given as a function of scattering angle from 180° to 90° measured in the laboratory frame of reference. The first row for each element gives the average atomic weight of that element and the average kinematic factor (Average K_{M_2}) for that element. The subsequent rows give the isotopic masses for that element and the K_{M_2} for those isotopic masses. The average K_{M_2} is calculated as the weighted average of the K_{M_2}, in which the relative abundances (ϖ_{M_2}) for the M_2 are given in Appendix 1:

$$\text{Average } K_{M_2} = \Sigma\, \varpi_{M_2} K_{M_2}\text{ , summed over the masses } M_2.$$

Atomic no. *(Z)*	El.	Isotopic mass (M_2) *(amu)*	180°	170°	160°	150°	140°	130°	120°	110°	100°	90°
7	N	14.007	0.0060	0.0061	0.0066	0.0076	0.0092	0.0121	0.0170	0.0261	0.0437	0.0772
		14.003	0.0059	0.0061	0.0066	0.0076	0.0092	0.0120	0.0169	0.0261	0.0436	0.0770
		15.000	0.0123	0.0127	0.0136	0.0154	0.0185	0.0235	0.0318	0.0459	0.0702	0.1111
8	O	15.999	0.0204	0.0209	0.0224	0.0251	0.0297	0.0369	0.0485	0.0670	0.0966	0.1428
		15.995	0.0204	0.0208	0.0223	0.0251	0.0296	0.0369	0.0484	0.0669	0.0965	0.1427
		16.999	0.0297	0.0304	0.0324	0.0361	0.0422	0.0516	0.0660	0.0883	0.1222	0.1724
		17.999	0.0400	0.0408	0.0434	0.0481	0.0555	0.0669	0.0840	0.1094	0.1468	0.2000
9	F	18.998	0.0510	0.0520	0.0551	0.0606	0.0695	0.0827	0.1021	0.1302	0.1703	0.2258
10	Ne	20.179	0.0647	0.0658	0.0695	0.0761	0.0863	0.1014	0.1232	0.1539	0.1966	0.2539
		19.992	0.0624	0.0636	0.0671	0.0736	0.0836	0.0985	0.1199	0.1503	0.1926	0.2498
		20.994	0.0743	0.0756	0.0797	0.0869	0.0981	0.1144	0.1377	0.1700	0.2141	0.2726
		21.991	0.0864	0.0878	0.0923	0.1003	0.1125	0.1302	0.1550	0.1890	0.2346	0.2939
11	Na	22.990	0.0987	0.1002	0.1051	0.1137	0.1269	0.1458	0.1719	0.2073	0.2541	0.3141
12	Mg	24.305	0.1149	0.1166	0.1220	0.1314	0.1456	0.1658	0.1934	0.2303	0.2782	0.3387
		23.985	0.1109	0.1126	0.1179	0.1271	0.1411	0.1610	0.1883	0.2249	0.2726	0.3331
		24.986	0.1233	0.1251	0.1307	0.1405	0.1552	0.1761	0.2044	0.2419	0.2904	0.3511
		25.983	0.1355	0.1374	0.1433	0.1536	0.1691	0.1908	0.2200	0.2583	0.3073	0.3681
13	Al	26.982	0.1477	0.1497	0.1559	0.1667	0.1827	0.2051	0.2351	0.2741	0.3235	0.3843
14	Si	28.086	0.1610	0.1631	0.1696	0.1808	0.1974	0.2205	0.2512	0.2908	0.3406	0.4012
		27.977	0.1597	0.1618	0.1683	0.1794	0.1960	0.2190	0.2497	0.2892	0.3390	0.3997
		28.976	0.1716	0.1738	0.1805	0.1920	0.2091	0.2327	0.2639	0.3039	0.3538	0.4143
		29.974	0.1834	0.1856	0.1925	0.2044	0.2219	0.2459	0.2776	0.3179	0.3680	0.4282
15	P	30.974	0.1949	0.1973	0.2043	0.2165	0.2344	0.2589	0.2909	0.3315	0.3816	0.4415
16	S	32.064	0.2073	0.2097	0.2169	0.2294	0.2477	0.2725	0.3049	0.3458	0.3958	0.4553
		31.972	0.2063	0.2087	0.2159	0.2283	0.2466	0.2714	0.3038	0.3446	0.3947	0.4542
		32.972	0.2175	0.2199	0.2273	0.2399	0.2585	0.2836	0.3163	0.3572	0.4072	0.4663
		33.968	0.2284	0.2309	0.2384	0.2513	0.2701	0.2955	0.3283	0.3694	0.4192	0.4779
		35.967	0.2497	0.2522	0.2599	0.2732	0.2924	0.3182	0.3514	0.3925	0.4419	0.4997
17	Cl	35.453	0.2442	0.2467	0.2544	0.2675	0.2867	0.3124	0.3454	0.3865	0.4361	0.4941
		34.969	0.2391	0.2417	0.2493	0.2624	0.2814	0.3070	0.3400	0.3811	0.4308	0.4890
		36.966	0.2600	0.2625	0.2704	0.2838	0.3031	0.3291	0.3623	0.4034	0.4526	0.5099
18	Ar	39.948	0.2894	0.2921	0.3001	0.3138	0.3336	0.3598	0.3931	0.4339	0.4823	0.5380
		35.967	0.2497	0.2522	0.2599	0.2732	0.2924	0.3182	0.3514	0.3925	0.4419	0.4997
		37.963	0.2700	0.2726	0.2806	0.2941	0.3136	0.3397	0.3730	0.4139	0.4629	0.5196
		39.962	0.2896	0.2922	0.3003	0.3140	0.3337	0.3600	0.3933	0.4340	0.4824	0.5381
19	K	39.098	0.2812	0.2838	0.2918	0.3055	0.3251	0.3513	0.3846	0.4255	0.4741	0.5303
		38.964	0.2799	0.2826	0.2905	0.3042	0.3238	0.3500	0.3833	0.4242	0.4728	0.5291
		40.000	0.2899	0.2926	0.3007	0.3144	0.3341	0.3604	0.3937	0.4344	0.4828	0.5385
		40.962	0.2990	0.3017	0.3098	0.3236	0.3434	0.3697	0.4030	0.4436	0.4917	0.5468

Table A5.4. Rutherford backscattering spectrometry kinematic factors for C as a projectile (continued).

Atomic no. *(Z)*	El.	Isotopic mass (M_2) *(amu)*	180°	170°	160°	150°	140°	130°	120°	110°	100°	90°
20	Ca	40.078	0.2906	0.2933	0.3014	0.3151	0.3348	0.3611	0.3944	0.4351	0.4834	0.5391
		39.963	0.2896	0.2922	0.3003	0.3140	0.3337	0.3600	0.3933	0.4341	0.4824	0.5381
		41.959	0.3083	0.3110	0.3191	0.3330	0.3528	0.3792	0.4124	0.4528	0.5006	0.5552
		42.959	0.3173	0.3200	0.3282	0.3421	0.3620	0.3884	0.4215	0.4618	0.5092	0.5633
		43.956	0.3261	0.3289	0.3371	0.3510	0.3710	0.3973	0.4304	0.4705	0.5175	0.5711
		45.954	0.3433	0.3460	0.3543	0.3682	0.3882	0.4145	0.4474	0.4870	0.5334	0.5859
		47.952	0.3596	0.3624	0.3707	0.3847	0.4046	0.4308	0.4634	0.5026	0.5482	0.5997
21	Sc	44.956	0.3348	0.3375	0.3458	0.3597	0.3797	0.4060	0.4390	0.4789	0.5256	0.5786
22	Ti	47.878	0.3590	0.3617	0.3700	0.3840	0.4040	0.4301	0.4628	0.5020	0.5476	0.5991
		45.953	0.3432	0.3460	0.3543	0.3682	0.3882	0.4145	0.4474	0.4870	0.5334	0.5859
		46.952	0.3515	0.3543	0.3625	0.3765	0.3965	0.4227	0.4555	0.4949	0.5409	0.5929
		47.948	0.3596	0.3623	0.3706	0.3846	0.4046	0.4308	0.4634	0.5026	0.5482	0.5997
		48.948	0.3675	0.3703	0.3786	0.3926	0.4125	0.4386	0.4711	0.5100	0.5553	0.6062
		49.945	0.3752	0.3780	0.3863	0.4003	0.4202	0.4462	0.4785	0.5173	0.5621	0.6126
23	V	50.942	0.3828	0.3855	0.3938	0.4078	0.4277	0.4536	0.4858	0.5243	0.5688	0.6187
		49.947	0.3752	0.3780	0.3863	0.4003	0.4202	0.4462	0.4786	0.5173	0.5621	0.6126
		50.944	0.3828	0.3856	0.3939	0.4078	0.4277	0.4536	0.4858	0.5243	0.5688	0.6187
24	Cr	51.996	0.3906	0.3933	0.4016	0.4156	0.4354	0.4612	0.4932	0.5314	0.5755	0.6249
		49.946	0.3752	0.3780	0.3863	0.4003	0.4202	0.4462	0.4786	0.5173	0.5621	0.6126
		51.940	0.3902	0.3929	0.4012	0.4152	0.4350	0.4609	0.4929	0.5311	0.5752	0.6246
		52.941	0.3974	0.4002	0.4085	0.4224	0.4422	0.4679	0.4998	0.5377	0.5815	0.6304
		53.939	0.4045	0.4073	0.4155	0.4295	0.4492	0.4748	0.5065	0.5442	0.5876	0.6360
25	Mn	54.938	0.4115	0.4142	0.4225	0.4363	0.4560	0.4815	0.5131	0.5505	0.5935	0.6415
26	Fe	55.847	0.4176	0.4204	0.4286	0.4425	0.4620	0.4875	0.5188	0.5560	0.5987	0.6462
		53.940	0.4045	0.4073	0.4156	0.4295	0.4492	0.4748	0.5065	0.5442	0.5876	0.6360
		55.935	0.4182	0.4210	0.4292	0.4431	0.4626	0.4881	0.5194	0.5566	0.5992	0.6467
		56.935	0.4249	0.4276	0.4359	0.4497	0.4692	0.4945	0.5256	0.5625	0.6048	0.6518
		57.933	0.4314	0.4341	0.4423	0.4561	0.4755	0.5007	0.5317	0.5683	0.6103	0.6568
27	Co	58.933	0.4378	0.4405	0.4487	0.4624	0.4818	0.5068	0.5376	0.5740	0.6156	0.6617
28	Ni	58.688	0.4361	0.4389	0.4470	0.4608	0.4801	0.5052	0.5361	0.5725	0.6142	0.6604
		57.935	0.4314	0.4341	0.4423	0.4561	0.4755	0.5007	0.5317	0.5683	0.6103	0.6568
		59.931	0.4440	0.4467	0.4549	0.4686	0.4878	0.5128	0.5434	0.5795	0.6207	0.6663
		60.931	0.4501	0.4528	0.4610	0.4746	0.4938	0.5186	0.5490	0.5848	0.6257	0.6709
		61.928	0.4561	0.4588	0.4669	0.4805	0.4996	0.5243	0.5545	0.5901	0.6306	0.6754
		63.928	0.4677	0.4704	0.4785	0.4919	0.5108	0.5353	0.5651	0.6002	0.6400	0.6839
29	Cu	63.546	0.4655	0.4682	0.4762	0.4897	0.5087	0.5331	0.5631	0.5982	0.6382	0.6823
		62.930	0.4620	0.4647	0.4728	0.4863	0.5053	0.5298	0.5599	0.5952	0.6354	0.6797
		64.928	0.4734	0.4760	0.4841	0.4975	0.5163	0.5406	0.5702	0.6050	0.6445	0.6880
30	Zn	65.396	0.4758	0.4785	0.4865	0.4999	0.5187	0.5429	0.5724	0.6071	0.6465	0.6898
		63.929	0.4677	0.4704	0.4785	0.4919	0.5109	0.5353	0.5651	0.6002	0.6400	0.6839
		65.926	0.4789	0.4815	0.4895	0.5029	0.5216	0.5458	0.5752	0.6098	0.6489	0.6920
		66.927	0.4843	0.4870	0.4949	0.5082	0.5268	0.5509	0.5801	0.6144	0.6532	0.6959
		67.925	0.4896	0.4922	0.5002	0.5134	0.5319	0.5558	0.5849	0.6189	0.6574	0.6997
		69.925	0.4999	0.5025	0.5104	0.5235	0.5419	0.5655	0.5941	0.6277	0.6655	0.7070
31	Ga	69.723	0.4989	0.5015	0.5093	0.5224	0.5408	0.5645	0.5932	0.6268	0.6647	0.7063
		68.926	0.4948	0.4974	0.5053	0.5185	0.5370	0.5607	0.5896	0.6233	0.6615	0.7034
		70.925	0.5049	0.5075	0.5153	0.5284	0.5467	0.5701	0.5986	0.6319	0.6695	0.7106

Table A5.4. Rutherford backscattering spectrometry kinematic factors for C as a projectile (continued).

Atomic no. (*Z*)	El.	Isotopic mass (M_2) (*amu*)	180°	170°	160°	150°	140°	130°	120°	110°	100°	90°
32	Ge	72.632	0.5131	0.5157	0.5235	0.5364	0.5545	0.5777	0.6059	0.6388	0.6758	0.7163
		69.924	0.4999	0.5025	0.5104	0.5235	0.5419	0.5654	0.5941	0.6277	0.6655	0.7070
		71.922	0.5098	0.5124	0.5202	0.5332	0.5514	0.5747	0.6030	0.6360	0.6733	0.7140
		72.924	0.5147	0.5172	0.5250	0.5379	0.5560	0.5792	0.6073	0.6401	0.6770	0.7174
		73.921	0.5194	0.5219	0.5296	0.5425	0.5605	0.5835	0.6115	0.6440	0.6807	0.7207
		75.921	0.5286	0.5311	0.5388	0.5515	0.5693	0.5920	0.6196	0.6517	0.6877	0.7270
33	As	74.922	0.5240	0.5266	0.5342	0.5470	0.5649	0.5878	0.6156	0.6479	0.6842	0.7239
34	Se	78.993	0.5419	0.5444	0.5520	0.5645	0.5820	0.6043	0.6313	0.6627	0.6979	0.7361
		73.923	0.5194	0.5219	0.5297	0.5425	0.5605	0.5835	0.6115	0.6440	0.6807	0.7207
		75.919	0.5286	0.5311	0.5387	0.5515	0.5692	0.5920	0.6196	0.6517	0.6877	0.7270
		76.920	0.5330	0.5356	0.5432	0.5558	0.5735	0.5961	0.6235	0.6554	0.6911	0.7301
		77.917	0.5374	0.5399	0.5475	0.5601	0.5777	0.6002	0.6274	0.6590	0.6945	0.7331
		79.916	0.5460	0.5485	0.5559	0.5684	0.5858	0.6080	0.6349	0.6660	0.7009	0.7389
		81.917	0.5542	0.5567	0.5641	0.5764	0.5936	0.6156	0.6421	0.6728	0.7071	0.7445
35	Br	79.904	0.5459	0.5484	0.5559	0.5683	0.5857	0.6079	0.6348	0.6659	0.7009	0.7388
		78.918	0.5417	0.5442	0.5518	0.5643	0.5818	0.6041	0.6312	0.6625	0.6977	0.7360
		80.916	0.5501	0.5526	0.5601	0.5725	0.5897	0.6118	0.6385	0.6694	0.7041	0.7417
36	Kr	83.8	0.5617	0.5641	0.5714	0.5837	0.6007	0.6224	0.6485	0.6788	0.7127	0.7494
		77.920	0.5374	0.5400	0.5475	0.5601	0.5777	0.6002	0.6274	0.6590	0.6945	0.7331
		79.916	0.5460	0.5485	0.5559	0.5684	0.5858	0.6080	0.6349	0.6660	0.7009	0.7389
		81.913	0.5542	0.5567	0.5641	0.5764	0.5936	0.6156	0.6421	0.6728	0.7071	0.7444
		82.914	0.5582	0.5607	0.5681	0.5803	0.5974	0.6192	0.6456	0.6760	0.7101	0.7471
		83.911	0.5622	0.5646	0.5719	0.5841	0.6011	0.6228	0.6490	0.6792	0.7131	0.7498
		85.911	0.5698	0.5723	0.5795	0.5916	0.6084	0.6298	0.6556	0.6855	0.7188	0.7549
37	Rb	85.468	0.5681	0.5706	0.5778	0.5899	0.6068	0.6283	0.6542	0.6841	0.7175	0.7537
		84.912	0.5660	0.5685	0.5758	0.5879	0.6048	0.6264	0.6523	0.6824	0.7160	0.7524
		86.909	0.5736	0.5760	0.5832	0.5952	0.6119	0.6332	0.6589	0.6885	0.7216	0.7574
38	Sr	87.617	0.5762	0.5786	0.5858	0.5977	0.6144	0.6356	0.6611	0.6906	0.7235	0.7591
		83.913	0.5622	0.5646	0.5719	0.5841	0.6012	0.6228	0.6490	0.6792	0.7131	0.7498
		85.909	0.5698	0.5723	0.5795	0.5916	0.6084	0.6298	0.6556	0.6855	0.7188	0.7549
		86.909	0.5736	0.5760	0.5832	0.5952	0.6119	0.6332	0.6589	0.6885	0.7216	0.7574
		87.906	0.5773	0.5797	0.5868	0.5988	0.6154	0.6366	0.6620	0.6915	0.7243	0.7598
39	Y	88.906	0.5809	0.5833	0.5904	0.6023	0.6188	0.6399	0.6652	0.6944	0.7269	0.7622
40	Zr	91.221	0.5890	0.5913	0.5984	0.6101	0.6264	0.6472	0.6721	0.7008	0.7329	0.7674
		89.905	0.5844	0.5868	0.5939	0.6057	0.6222	0.6431	0.6682	0.6972	0.7296	0.7645
		90.906	0.5879	0.5903	0.5974	0.6091	0.6255	0.6463	0.6712	0.7000	0.7321	0.7668
		91.905	0.5914	0.5937	0.6008	0.6125	0.6287	0.6494	0.6742	0.7028	0.7346	0.7690
		93.906	0.5981	0.6004	0.6074	0.6190	0.6350	0.6555	0.6800	0.7082	0.7395	0.7734
		95.908	0.6046	0.6069	0.6138	0.6253	0.6412	0.6613	0.6855	0.7133	0.7443	0.7776
41	Nb	92.906	0.5948	0.5971	0.6041	0.6158	0.6319	0.6524	0.6771	0.7055	0.7371	0.7712
42	Mo	95.931	0.6046	0.6069	0.6138	0.6252	0.6411	0.6613	0.6854	0.7133	0.7442	0.7775
		91.907	0.5914	0.5937	0.6008	0.6125	0.6287	0.6494	0.6742	0.7028	0.7346	0.7690
		93.905	0.5981	0.6004	0.6074	0.6190	0.6350	0.6554	0.6799	0.7082	0.7395	0.7734
		94.906	0.6014	0.6037	0.6107	0.6222	0.6381	0.6584	0.6828	0.7108	0.7419	0.7755
		95.905	0.6046	0.6069	0.6138	0.6253	0.6412	0.6613	0.6855	0.7133	0.7443	0.7776
		96.906	0.6078	0.6101	0.6170	0.6283	0.6441	0.6642	0.6882	0.7159	0.7466	0.7796
		97.905	0.6109	0.6132	0.6200	0.6314	0.6471	0.6670	0.6909	0.7183	0.7488	0.7816
		99.908	0.6171	0.6193	0.6261	0.6373	0.6528	0.6725	0.6961	0.7232	0.7532	0.7855

Table A5.4. Rutherford backscattering spectrometry kinematic factors for C as a projectile (continued).

Atomic no. *(Z)*	El.	Isotopic mass (M_2) *(amu)*	180°	170°	160°	150°	140°	130°	120°	110°	100°	90°
43	Tc	98.000	0.6112	0.6135	0.6203	0.6316	0.6473	0.6673	0.6911	0.7186	0.7490	0.7818
44	Ru	101.019	0.6203	0.6225	0.6292	0.6404	0.6558	0.6754	0.6988	0.7257	0.7555	0.7876
		95.908	0.6046	0.6069	0.6138	0.6253	0.6412	0.6613	0.6855	0.7133	0.7443	0.7776
		97.905	0.6109	0.6132	0.6200	0.6314	0.6471	0.6670	0.6909	0.7183	0.7488	0.7816
		98.506	0.6128	0.6151	0.6219	0.6332	0.6488	0.6687	0.6925	0.7198	0.7502	0.7828
		99.904	0.6171	0.6193	0.6261	0.6373	0.6528	0.6725	0.6961	0.7232	0.7532	0.7855
		100.906	0.6201	0.6223	0.6290	0.6401	0.6556	0.6752	0.6986	0.7255	0.7554	0.7874
		101.904	0.6230	0.6252	0.6319	0.6430	0.6583	0.6778	0.7011	0.7278	0.7575	0.7893
		103.905	0.6287	0.6310	0.6376	0.6485	0.6637	0.6829	0.7059	0.7323	0.7616	0.7929
45	Rh	102.906	0.6259	0.6281	0.6348	0.6458	0.6610	0.6804	0.7035	0.7301	0.7595	0.7911
46	Pd	106.415	0.6357	0.6378	0.6444	0.6552	0.6702	0.6891	0.7118	0.7377	0.7665	0.7973
		101.906	0.6230	0.6252	0.6319	0.6430	0.6583	0.6778	0.7011	0.7278	0.7575	0.7893
		103.904	0.6287	0.6310	0.6376	0.6485	0.6637	0.6829	0.7059	0.7323	0.7616	0.7929
		104.905	0.6316	0.6338	0.6403	0.6512	0.6663	0.6855	0.7083	0.7345	0.7636	0.7947
		105.904	0.6343	0.6365	0.6431	0.6539	0.6689	0.6879	0.7106	0.7367	0.7655	0.7964
		107.904	0.6397	0.6419	0.6484	0.6591	0.6740	0.6928	0.7152	0.7409	0.7694	0.7998
		109.905	0.6450	0.6472	0.6536	0.6642	0.6789	0.6974	0.7196	0.7450	0.7731	0.8031
47	Ag	107.868	0.6396	0.6418	0.6483	0.6590	0.6739	0.6927	0.7151	0.7408	0.7693	0.7998
		106.905	0.6371	0.6392	0.6457	0.6565	0.6715	0.6904	0.7129	0.7388	0.7675	0.7982
		108.905	0.6424	0.6446	0.6510	0.6617	0.6764	0.6951	0.7174	0.7430	0.7712	0.8015
48	Cd	112.412	0.6513	0.6535	0.6598	0.6703	0.6847	0.7031	0.7249	0.7499	0.7775	0.8070
		105.907	0.6343	0.6365	0.6431	0.6539	0.6689	0.6879	0.7107	0.7367	0.7655	0.7964
		107.904	0.6397	0.6419	0.6484	0.6591	0.6740	0.6928	0.7152	0.7409	0.7694	0.7998
		109.903	0.6450	0.6471	0.6536	0.6642	0.6789	0.6974	0.7196	0.7450	0.7731	0.8031
		110.904	0.6476	0.6497	0.6561	0.6666	0.6813	0.6997	0.7218	0.7470	0.7749	0.8047
		111.903	0.6501	0.6522	0.6586	0.6691	0.6836	0.7020	0.7239	0.7490	0.7767	0.8063
		112.904	0.6526	0.6547	0.6610	0.6715	0.6859	0.7042	0.7260	0.7509	0.7784	0.8079
		113.903	0.6551	0.6572	0.6635	0.6739	0.6882	0.7064	0.7281	0.7528	0.7801	0.8094
		115.905	0.6599	0.6620	0.6682	0.6785	0.6927	0.7107	0.7321	0.7565	0.7835	0.8124
49	In	114.818	0.6573	0.6594	0.6657	0.6760	0.6903	0.7084	0.7299	0.7545	0.7817	0.8108
		112.904	0.6526	0.6547	0.6610	0.6715	0.6859	0.7042	0.7260	0.7509	0.7784	0.8079
		114.904	0.6575	0.6596	0.6659	0.6762	0.6905	0.7086	0.7301	0.7547	0.7818	0.8109
50	Sn	118.613	0.6659	0.6679	0.6741	0.6842	0.6983	0.7160	0.7370	0.7611	0.7877	0.8160
		111.905	0.6501	0.6522	0.6586	0.6691	0.6836	0.7020	0.7239	0.7490	0.7767	0.8063
		113.903	0.6551	0.6572	0.6635	0.6739	0.6882	0.7064	0.7281	0.7528	0.7801	0.8094
		114.903	0.6575	0.6596	0.6659	0.6762	0.6905	0.7086	0.7301	0.7547	0.7818	0.8109
		115.902	0.6599	0.6620	0.6682	0.6785	0.6927	0.7107	0.7321	0.7565	0.7835	0.8124
		116.903	0.6623	0.6644	0.6705	0.6808	0.6949	0.7128	0.7341	0.7584	0.7852	0.8138
		117.602	0.6639	0.6660	0.6722	0.6823	0.6964	0.7142	0.7354	0.7596	0.7863	0.8148
		118.903	0.6669	0.6690	0.6751	0.6852	0.6992	0.7169	0.7379	0.7619	0.7884	0.8167
		119.902	0.6692	0.6712	0.6773	0.6874	0.7013	0.7189	0.7398	0.7637	0.7900	0.8180
		121.903	0.6737	0.6757	0.6817	0.6917	0.7054	0.7228	0.7435	0.7671	0.7930	0.8208
		123.905	0.6780	0.6800	0.6860	0.6958	0.7095	0.7266	0.7471	0.7704	0.7960	0.8234
51	Sb	121.758	0.6733	0.6753	0.6814	0.6914	0.7051	0.7225	0.7432	0.7668	0.7928	0.8206
		120.904	0.6714	0.6735	0.6795	0.6896	0.7034	0.7209	0.7417	0.7654	0.7915	0.8194
		122.904	0.6758	0.6778	0.6838	0.6938	0.7075	0.7247	0.7453	0.7687	0.7946	0.8221

Table A5.4. Rutherford backscattering spectrometry kinematic factors for C as a projectile (continued).

Atomic no. *(Z)*	El.	Isotopic mass (M_2) *(amu)*	180°	170°	160°	150°	140°	130°	120°	110°	100°	90°
52	Te	127.586	0.6856	0.6876	0.6935	0.7031	0.7165	0.7334	0.7534	0.7762	0.8013	0.8280
		119.904	0.6692	0.6712	0.6773	0.6874	0.7013	0.7189	0.7398	0.7637	0.7900	0.8180
		121.903	0.6737	0.6757	0.6817	0.6917	0.7054	0.7228	0.7435	0.7671	0.7930	0.8208
		122.904	0.6758	0.6778	0.6838	0.6938	0.7075	0.7247	0.7453	0.7687	0.7946	0.8221
		123.903	0.6780	0.6800	0.6860	0.6958	0.7095	0.7266	0.7471	0.7704	0.7960	0.8234
		124.904	0.6801	0.6821	0.6881	0.6979	0.7114	0.7285	0.7488	0.7720	0.7975	0.8247
		125.903	0.6822	0.6842	0.6901	0.6999	0.7134	0.7304	0.7506	0.7736	0.7990	0.8260
		127.905	0.6863	0.6883	0.6942	0.7038	0.7172	0.7340	0.7540	0.7767	0.8018	0.8285
		129.906	0.6904	0.6923	0.6981	0.7077	0.7209	0.7375	0.7573	0.7798	0.8045	0.8309
53	I	126.905	0.6843	0.6863	0.6921	0.7019	0.7153	0.7322	0.7523	0.7752	0.8004	0.8272
54	Xe	131.293	0.6930	0.6949	0.7007	0.7102	0.7233	0.7399	0.7595	0.7818	0.8064	0.8325
		123.906	0.6780	0.6800	0.6860	0.6958	0.7095	0.7266	0.7471	0.7704	0.7960	0.8234
		125.904	0.6822	0.6842	0.6901	0.6999	0.7134	0.7304	0.7506	0.7736	0.7990	0.8260
		127.904	0.6863	0.6883	0.6941	0.7038	0.7172	0.7340	0.7540	0.7767	0.8018	0.8285
		128.905	0.6884	0.6903	0.6961	0.7058	0.7190	0.7358	0.7556	0.7783	0.8032	0.8297
		129.904	0.6903	0.6923	0.6981	0.7077	0.7209	0.7375	0.7573	0.7798	0.8045	0.8309
		130.905	0.6923	0.6942	0.7000	0.7095	0.7227	0.7392	0.7589	0.7813	0.8059	0.8321
		131.904	0.6943	0.6962	0.7019	0.7114	0.7245	0.7409	0.7605	0.7828	0.8072	0.8332
		133.905	0.6981	0.7000	0.7057	0.7151	0.7280	0.7443	0.7636	0.7856	0.8098	0.8355
		135.907	0.7018	0.7037	0.7093	0.7186	0.7314	0.7476	0.7667	0.7885	0.8123	0.8377
55	Cs	132.905	0.6962	0.6981	0.7038	0.7132	0.7262	0.7426	0.7621	0.7842	0.8085	0.8344
56	Ba	137.327	0.7044	0.7062	0.7118	0.7211	0.7338	0.7498	0.7688	0.7904	0.8141	0.8393
		129.906	0.6904	0.6923	0.6981	0.7077	0.7209	0.7375	0.7573	0.7798	0.8045	0.8309
		131.905	0.6943	0.6962	0.7019	0.7114	0.7245	0.7409	0.7605	0.7828	0.8072	0.8332
		133.904	0.6981	0.7000	0.7057	0.7150	0.7280	0.7443	0.7636	0.7856	0.8098	0.8355
		134.906	0.7000	0.7018	0.7075	0.7168	0.7297	0.7459	0.7652	0.7871	0.8111	0.8366
		135.905	0.7018	0.7037	0.7093	0.7186	0.7314	0.7476	0.7667	0.7885	0.8123	0.8377
		136.906	0.7036	0.7055	0.7111	0.7204	0.7331	0.7492	0.7682	0.7898	0.8136	0.8388
		137.905	0.7054	0.7073	0.7129	0.7221	0.7348	0.7507	0.7697	0.7912	0.8148	0.8399
57	La	138.905	0.7072	0.7091	0.7146	0.7238	0.7364	0.7523	0.7711	0.7925	0.8160	0.8410
		137.907	0.7054	0.7073	0.7129	0.7221	0.7348	0.7507	0.7697	0.7912	0.8148	0.8399
		138.906	0.7072	0.7091	0.7146	0.7238	0.7364	0.7523	0.7711	0.7925	0.8160	0.8410
58	Ce	140.115	0.7093	0.7112	0.7167	0.7258	0.7384	0.7541	0.7729	0.7941	0.8174	0.8422
		135.907	0.7018	0.7037	0.7093	0.7186	0.7314	0.7476	0.7667	0.7885	0.8123	0.8377
		137.906	0.7054	0.7073	0.7129	0.7221	0.7348	0.7507	0.7697	0.7912	0.8148	0.8399
		139.905	0.7090	0.7108	0.7163	0.7255	0.7380	0.7538	0.7726	0.7939	0.8172	0.8420
		141.909	0.7124	0.7143	0.7197	0.7288	0.7412	0.7569	0.7754	0.7965	0.8196	0.8441
59	Pr	140.908	0.7107	0.7126	0.7181	0.7271	0.7396	0.7554	0.7740	0.7952	0.8184	0.8430
60	Nd	144.242	0.7163	0.7181	0.7236	0.7325	0.7448	0.7603	0.7786	0.7994	0.8222	0.8464
		141.908	0.7124	0.7143	0.7197	0.7288	0.7412	0.7569	0.7754	0.7965	0.8195	0.8441
		142.910	0.7141	0.7160	0.7214	0.7304	0.7428	0.7583	0.7768	0.7977	0.8207	0.8451
		143.910	0.7158	0.7176	0.7231	0.7320	0.7443	0.7598	0.7782	0.7990	0.8218	0.8461
		144.913	0.7175	0.7193	0.7247	0.7336	0.7459	0.7613	0.7795	0.8003	0.8230	0.8470
		145.913	0.7191	0.7209	0.7263	0.7352	0.7474	0.7627	0.7809	0.8015	0.8241	0.8480
		147.917	0.7224	0.7242	0.7295	0.7382	0.7503	0.7655	0.7835	0.8039	0.8262	0.8499
		149.921	0.7255	0.7273	0.7326	0.7413	0.7532	0.7683	0.7861	0.8063	0.8284	0.8518

Table A5.4. Rutherford backscattering spectrometry kinematic factors for C as a projectile (continued).

Atomic no. *(Z)*	El.	Isotopic mass (M_2) *(amu)*	180°	170°	160°	150°	140°	130°	120°	110°	100°	90°
61	Pm	145.000	0.7176	0.7194	0.7248	0.7337	0.7460	0.7614	0.7796	0.8004	0.8231	0.8471
62	Sm	150.36	0.7261	0.7279	0.7332	0.7419	0.7538	0.7688	0.7866	0.8067	0.8288	0.8521
		143.912	0.7158	0.7176	0.7231	0.7320	0.7443	0.7598	0.7782	0.7990	0.8218	0.8461
		146.915	0.7208	0.7226	0.7279	0.7367	0.7489	0.7641	0.7822	0.8027	0.8252	0.8490
		147.915	0.7224	0.7241	0.7295	0.7382	0.7503	0.7655	0.7835	0.8039	0.8262	0.8499
		148.917	0.7240	0.7257	0.7310	0.7398	0.7518	0.7669	0.7848	0.8051	0.8273	0.8509
		149.917	0.7255	0.7273	0.7326	0.7413	0.7532	0.7683	0.7861	0.8063	0.8284	0.8518
		151.920	0.7286	0.7304	0.7356	0.7442	0.7561	0.7710	0.7886	0.8086	0.8304	0.8536
		153.922	0.7316	0.7334	0.7385	0.7471	0.7588	0.7736	0.7910	0.8108	0.8324	0.8554
63	Eu	151.965	0.7287	0.7304	0.7356	0.7443	0.7561	0.7710	0.7886	0.8086	0.8305	0.8536
		150.920	0.7271	0.7288	0.7341	0.7427	0.7547	0.7696	0.7873	0.8074	0.8294	0.8527
		152.921	0.7301	0.7319	0.7371	0.7456	0.7575	0.7723	0.7898	0.8097	0.8314	0.8545
64	Gd	157.252	0.7365	0.7382	0.7433	0.7517	0.7633	0.7778	0.7950	0.8144	0.8357	0.8582
		151.920	0.7286	0.7304	0.7356	0.7442	0.7561	0.7710	0.7886	0.8086	0.8304	0.8536
		153.921	0.7316	0.7334	0.7385	0.7471	0.7588	0.7736	0.7910	0.8108	0.8324	0.8554
		154.923	0.7331	0.7348	0.7400	0.7485	0.7602	0.7749	0.7922	0.8119	0.8334	0.8562
		155.922	0.7346	0.7363	0.7414	0.7499	0.7615	0.7761	0.7934	0.8130	0.8344	0.8571
		156.924	0.7360	0.7377	0.7429	0.7513	0.7629	0.7774	0.7946	0.8141	0.8354	0.8579
		157.924	0.7375	0.7392	0.7443	0.7526	0.7642	0.7786	0.7958	0.8151	0.8363	0.8588
		159.927	0.7403	0.7420	0.7470	0.7553	0.7668	0.7811	0.7981	0.8172	0.8382	0.8604
65	Tb	158.925	0.7389	0.7406	0.7457	0.7540	0.7655	0.7799	0.7969	0.8162	0.8373	0.8596
66	Dy	162.498	0.7438	0.7455	0.7505	0.7587	0.7700	0.7842	0.8009	0.8199	0.8406	0.8625
		155.924	0.7346	0.7363	0.7414	0.7499	0.7615	0.7761	0.7934	0.8130	0.8344	0.8571
		157.924	0.7375	0.7392	0.7443	0.7526	0.7642	0.7786	0.7958	0.8151	0.8363	0.8588
		159.925	0.7403	0.7420	0.7470	0.7553	0.7668	0.7811	0.7981	0.8172	0.8382	0.8604
		160.927	0.7417	0.7434	0.7484	0.7567	0.7680	0.7823	0.7992	0.8183	0.8391	0.8612
		161.927	0.7431	0.7447	0.7497	0.7580	0.7693	0.7835	0.8003	0.8193	0.8400	0.8620
		162.929	0.7444	0.7461	0.7511	0.7593	0.7705	0.7847	0.8014	0.8203	0.8410	0.8628
		163.929	0.7458	0.7474	0.7524	0.7606	0.7718	0.7858	0.8025	0.8213	0.8418	0.8636
67	Ho	164.930	0.7471	0.7488	0.7537	0.7618	0.7730	0.7870	0.8036	0.8223	0.8427	0.8644
68	Er	167.256	0.7501	0.7518	0.7567	0.7647	0.7758	0.7896	0.8060	0.8245	0.8447	0.8661
		161.929	0.7431	0.7447	0.7497	0.7580	0.7693	0.7835	0.8003	0.8193	0.8401	0.8620
		163.929	0.7458	0.7474	0.7524	0.7606	0.7718	0.7858	0.8025	0.8213	0.8418	0.8636
		165.930	0.7484	0.7501	0.7550	0.7631	0.7742	0.7881	0.8046	0.8233	0.8436	0.8651
		166.932	0.7497	0.7514	0.7563	0.7643	0.7754	0.7893	0.8057	0.8242	0.8445	0.8659
		167.932	0.7510	0.7527	0.7575	0.7655	0.7766	0.7904	0.8067	0.8252	0.8453	0.8666
		169.936	0.7536	0.7552	0.7600	0.7680	0.7789	0.7926	0.8088	0.8271	0.8470	0.8681
69	Tm	168.934	0.7523	0.7539	0.7588	0.7668	0.7777	0.7915	0.8077	0.8261	0.8462	0.8674
70	Yb	173.034	0.7574	0.7590	0.7638	0.7716	0.7824	0.7959	0.8118	0.8299	0.8495	0.8703
		167.934	0.7510	0.7527	0.7575	0.7655	0.7766	0.7904	0.8067	0.8252	0.8453	0.8666
		169.935	0.7536	0.7552	0.7600	0.7680	0.7789	0.7926	0.8088	0.8271	0.8470	0.8681
		170.936	0.7548	0.7564	0.7612	0.7692	0.7800	0.7937	0.8098	0.8280	0.8478	0.8688
		171.936	0.7561	0.7577	0.7625	0.7703	0.7812	0.7947	0.8108	0.8289	0.8486	0.8695
		172.938	0.7573	0.7589	0.7637	0.7715	0.7823	0.7958	0.8118	0.8298	0.8495	0.8702
		173.939	0.7585	0.7601	0.7649	0.7727	0.7834	0.7969	0.8127	0.8307	0.8503	0.8709
		175.943	0.7609	0.7625	0.7672	0.7749	0.7856	0.7989	0.8147	0.8325	0.8518	0.8723

Table A5.4. Rutherford backscattering spectrometry kinematic factors for C as a projectile (continued).

Atomic no. *(Z)*	El.	Isotopic mass (M_2) *(amu)*	180°	170°	160°	150°	140°	130°	120°	110°	100°	90°
71	Lu	174.967	0.7597	0.7613	0.7661	0.7738	0.7845	0.7979	0.8137	0.8316	0.8511	0.8716
		174.941	0.7597	0.7613	0.7660	0.7738	0.7845	0.7979	0.8137	0.8316	0.8511	0.8716
		175.943	0.7609	0.7625	0.7672	0.7749	0.7856	0.7989	0.8147	0.8325	0.8518	0.8723
72	Hf	178.49	0.7639	0.7654	0.7701	0.7778	0.7883	0.8015	0.8171	0.8346	0.8538	0.8740
		173.940	0.7585	0.7601	0.7649	0.7727	0.7834	0.7969	0.8127	0.8307	0.8503	0.8709
		175.941	0.7609	0.7625	0.7672	0.7749	0.7856	0.7989	0.8147	0.8325	0.8518	0.8723
		176.943	0.7621	0.7637	0.7684	0.7761	0.7867	0.8000	0.8156	0.8333	0.8526	0.8730
		177.944	0.7633	0.7648	0.7695	0.7772	0.7877	0.8010	0.8166	0.8342	0.8534	0.8736
		178.946	0.7644	0.7660	0.7706	0.7783	0.7888	0.8020	0.8175	0.8350	0.8541	0.8743
		179.947	0.7656	0.7671	0.7717	0.7794	0.7898	0.8029	0.8184	0.8359	0.8549	0.8750
73	Ta	180.948	0.7667	0.7682	0.7729	0.7804	0.7909	0.8039	0.8193	0.8367	0.8556	0.8756
		179.947	0.7656	0.7671	0.7717	0.7794	0.7898	0.8029	0.8184	0.8359	0.8549	0.8750
		180.948	0.7667	0.7682	0.7729	0.7804	0.7909	0.8039	0.8193	0.8367	0.8556	0.8756
74	W	183.849	0.7699	0.7714	0.7760	0.7835	0.7938	0.8067	0.8219	0.8391	0.8577	0.8774
		179.947	0.7656	0.7671	0.7717	0.7794	0.7898	0.8029	0.8184	0.8359	0.8549	0.8750
		181.948	0.7678	0.7694	0.7740	0.7815	0.7919	0.8049	0.8202	0.8375	0.8564	0.8763
		182.950	0.7689	0.7705	0.7750	0.7826	0.7929	0.8059	0.8211	0.8383	0.8571	0.8769
		183.951	0.7700	0.7716	0.7761	0.7836	0.7939	0.8068	0.8220	0.8391	0.8578	0.8775
		185.954	0.7722	0.7737	0.7783	0.7857	0.7959	0.8087	0.8237	0.8407	0.8592	0.8788
75	Re	186.207	0.7725	0.7740	0.7785	0.7859	0.7961	0.8089	0.8240	0.8409	0.8594	0.8789
		184.953	0.7711	0.7727	0.7772	0.7847	0.7949	0.8077	0.8229	0.8399	0.8585	0.8781
		186.956	0.7733	0.7748	0.7793	0.7867	0.7969	0.8096	0.8246	0.8415	0.8599	0.8794
76	Os	190.24	0.7767	0.7782	0.7827	0.7900	0.8000	0.8126	0.8273	0.8440	0.8622	0.8813
		183.952	0.7700	0.7716	0.7761	0.7836	0.7939	0.8068	0.8220	0.8391	0.8578	0.8775
		185.954	0.7722	0.7737	0.7783	0.7857	0.7959	0.8087	0.8237	0.8407	0.8592	0.8788
		186.956	0.7733	0.7748	0.7793	0.7867	0.7969	0.8096	0.8246	0.8415	0.8599	0.8794
		187.956	0.7744	0.7759	0.7803	0.7877	0.7978	0.8105	0.8254	0.8423	0.8606	0.8800
		188.958	0.7754	0.7769	0.7814	0.7887	0.7988	0.8114	0.8263	0.8431	0.8613	0.8806
		189.958	0.7764	0.7779	0.7824	0.7897	0.7998	0.8123	0.8271	0.8438	0.8620	0.8812
		191.962	0.7785	0.7800	0.7844	0.7917	0.8016	0.8141	0.8288	0.8453	0.8633	0.8823
77	Ir	192.216	0.7788	0.7802	0.7846	0.7919	0.8019	0.8143	0.8290	0.8455	0.8635	0.8825
		190.961	0.7775	0.7790	0.7834	0.7907	0.8007	0.8132	0.8280	0.8446	0.8627	0.8818
		192.963	0.7795	0.7810	0.7854	0.7926	0.8025	0.8150	0.8296	0.8461	0.8640	0.8829
78	Pt	195.08	0.7816	0.7831	0.7875	0.7946	0.8045	0.8168	0.8313	0.8476	0.8654	0.8841
		189.960	0.7765	0.7779	0.7824	0.7897	0.7998	0.8123	0.8271	0.8438	0.8620	0.8812
		191.961	0.7785	0.7800	0.7844	0.7917	0.8016	0.8141	0.8288	0.8453	0.8633	0.8823
		193.963	0.7805	0.7820	0.7864	0.7936	0.8035	0.8158	0.8304	0.8468	0.8647	0.8835
		194.965	0.7815	0.7830	0.7873	0.7945	0.8044	0.8167	0.8312	0.8475	0.8653	0.8840
		195.965	0.7825	0.7840	0.7883	0.7955	0.8053	0.8175	0.8320	0.8482	0.8659	0.8846
		197.968	0.7845	0.7859	0.7902	0.7973	0.8070	0.8192	0.8335	0.8497	0.8672	0.8857
79	Au	196.967	0.7835	0.7849	0.7893	0.7964	0.8062	0.8184	0.8327	0.8490	0.8666	0.8851
80	Hg	200.588	0.7869	0.7884	0.7926	0.7997	0.8093	0.8213	0.8355	0.8515	0.8688	0.8871
		195.966	0.7825	0.7840	0.7883	0.7955	0.8053	0.8175	0.8320	0.8482	0.8659	0.8846
		197.967	0.7845	0.7859	0.7902	0.7973	0.8070	0.8192	0.8335	0.8497	0.8672	0.8857
		198.968	0.7854	0.7869	0.7912	0.7982	0.8079	0.8200	0.8343	0.8504	0.8678	0.8862
		199.968	0.7864	0.7878	0.7921	0.7991	0.8088	0.8208	0.8350	0.8511	0.8685	0.8868

Table A5.4. Rutherford backscattering spectrometry kinematic factors for C as a projectile (continued).

Atomic no. *(Z)*	El.	Isotopic mass (M_2) *(amu)*	180°	170°	160°	150°	140°	130°	120°	110°	100°	90°
		200.970	0.7873	0.7887	0.7930	0.8000	0.8096	0.8216	0.8358	0.8517	0.8691	0.8873
		201.971	0.7883	0.7897	0.7939	0.8009	0.8105	0.8225	0.8365	0.8524	0.8697	0.8878
		203.973	0.7901	0.7915	0.7957	0.8027	0.8122	0.8240	0.8380	0.8538	0.8709	0.8889
81	Tl	204.383	0.7905	0.7919	0.7961	0.8030	0.8125	0.8244	0.8383	0.8540	0.8711	0.8891
		202.972	0.7892	0.7906	0.7948	0.8018	0.8113	0.8232	0.8373	0.8531	0.8703	0.8884
		204.974	0.7910	0.7924	0.7966	0.8035	0.8130	0.8248	0.8387	0.8544	0.8715	0.8894
82	Pb	207.217	0.7930	0.7944	0.7986	0.8054	0.8148	0.8265	0.8403	0.8559	0.8728	0.8905
		203.973	0.7901	0.7915	0.7957	0.8027	0.8122	0.8240	0.8380	0.8538	0.8709	0.8889
		205.975	0.7919	0.7933	0.7975	0.8044	0.8138	0.8256	0.8395	0.8551	0.8720	0.8899
		206.976	0.7928	0.7942	0.7984	0.8052	0.8146	0.8264	0.8402	0.8557	0.8726	0.8904
		207.977	0.7937	0.7951	0.7992	0.8061	0.8154	0.8271	0.8409	0.8564	0.8732	0.8909
83	Bi	208.980	0.7946	0.7960	0.8001	0.8069	0.8162	0.8279	0.8416	0.8570	0.8738	0.8914
84	Po	208.982	0.7946	0.7960	0.8001	0.8069	0.8162	0.8279	0.8416	0.8570	0.8738	0.8914
85	At	210.000	0.7955	0.7969	0.8010	0.8077	0.8170	0.8286	0.8423	0.8576	0.8743	0.8919
86	Rn	222.018	0.8054	0.8067	0.8107	0.8172	0.8260	0.8371	0.8502	0.8648	0.8807	0.8974
87	Fr	223.000	0.8062	0.8075	0.8114	0.8179	0.8267	0.8378	0.8508	0.8654	0.8812	0.8979
88	Ra	226.025	0.8085	0.8098	0.8137	0.8201	0.8288	0.8398	0.8526	0.8671	0.8827	0.8992
89	Ac	227.000	0.8092	0.8105	0.8144	0.8208	0.8295	0.8404	0.8532	0.8676	0.8832	0.8996
90	Th	232.038	0.8130	0.8143	0.8181	0.8243	0.8329	0.8436	0.8562	0.8703	0.8856	0.9017
91	Pa	231.036	0.8123	0.8135	0.8174	0.8236	0.8322	0.8430	0.8556	0.8698	0.8851	0.9012
92	U	238.018	0.8172	0.8185	0.8222	0.8283	0.8367	0.8472	0.8595	0.8733	0.8883	0.9040
		234.041	0.8144	0.8157	0.8195	0.8257	0.8342	0.8448	0.8573	0.8713	0.8865	0.9025
		235.044	0.8151	0.8164	0.8202	0.8264	0.8349	0.8454	0.8579	0.8718	0.8870	0.9029
		238.040	0.8172	0.8185	0.8222	0.8284	0.8368	0.8472	0.8595	0.8733	0.8883	0.9040
93	Np	237.048	0.8166	0.8178	0.8216	0.8277	0.8361	0.8466	0.8590	0.8728	0.8879	0.9036
94	Pu	244.064	0.8213	0.8226	0.8262	0.8322	0.8405	0.8507	0.8627	0.8763	0.8909	0.9063
95	Am	243.061	0.8207	0.8219	0.8256	0.8316	0.8398	0.8501	0.8622	0.8758	0.8905	0.9059

Table A5.5. Rutherford backscattering spectrometry kinematic factors for N as a projectile.

This table gives the RBS kinematic factors K_{M_2}, defined by Eq. (A4.2), for N as a projectile (M_1=14.003 amu) and for the isotopic masses (M_2) of the elements. The kinematic factors are given as a function of scattering angle from 180° to 90° measured in the laboratory frame of reference. The first row for each element gives the average atomic weight of that element and the average kinematic factor (Average K_{M_2}) for that element. The subsequent rows give the isotopic masses for that element and the K_{M_2} for those isotopic masses. The average K_{M_2} is calculated as the weighted average of the K_{M_2}, in which the relative abundances (ϖ_{M_2}) for the M_2 are given in Appendix 1:

$$\text{Average } K_{M_2} = \Sigma\, \varpi_{M_2} K_{M_2}\ \text{, summed over the masses } M_2.$$

Atomic no. (Z)	El.	Isotopic mass (M_2) (amu)	180°	170°	160°	150°	140°	130°	120°	110°	100°	90°
8	O	15.999	0.0044	0.0046	0.0049	0.0057	0.0070	0.0092	0.0131	0.0207	0.0359	0.0665
		15.995	0.0044	0.0045	0.0049	0.0056	0.0069	0.0091	0.0131	0.0206	0.0358	0.0664
		16.999	0.0093	0.0096	0.0103	0.0118	0.0142	0.0182	0.0251	0.0371	0.0586	0.0966
		17.999	0.0156	0.0160	0.0172	0.0194	0.0231	0.0290	0.0387	0.0548	0.0815	0.1249
9	F	18.998	0.0229	0.0234	0.0251	0.0281	0.0331	0.0410	0.0534	0.0730	0.1039	0.1514
10	Ne	20.179	0.0327	0.0334	0.0356	0.0396	0.0461	0.0561	0.0713	0.0945	0.1294	0.1805
		19.992	0.0310	0.0317	0.0338	0.0377	0.0439	0.0536	0.0684	0.0911	0.1255	0.1762
		20.994	0.0399	0.0407	0.0433	0.0480	0.0554	0.0668	0.0839	0.1093	0.1466	0.1998
		21.991	0.0493	0.0502	0.0532	0.0587	0.0673	0.0802	0.0993	0.1271	0.1668	0.2219
11	Na	22.990	0.0590	0.0601	0.0636	0.0698	0.0795	0.0939	0.1147	0.1445	0.1862	0.2429
12	Mg	24.305	0.0724	0.0736	0.0776	0.0847	0.0957	0.1118	0.1347	0.1667	0.2105	0.2687
		23.985	0.0690	0.0703	0.0741	0.0810	0.0917	0.1074	0.1299	0.1615	0.2048	0.2628
		24.986	0.0794	0.0807	0.0850	0.0925	0.1041	0.1211	0.1450	0.1780	0.2228	0.2817
		25.983	0.0898	0.0912	0.0958	0.1040	0.1165	0.1345	0.1597	0.1941	0.2400	0.2996
13	Al	26.982	0.1003	0.1019	0.1068	0.1155	0.1288	0.1478	0.1741	0.2097	0.2566	0.3167
14	Si	28.086	0.1120	0.1137	0.1189	0.1282	0.1423	0.1623	0.1897	0.2263	0.2741	0.3345
		27.977	0.1108	0.1125	0.1178	0.1270	0.1410	0.1609	0.1882	0.2247	0.2724	0.3329
		28.976	0.1214	0.1232	0.1287	0.1384	0.1531	0.1738	0.2020	0.2393	0.2877	0.3484
		29.974	0.1319	0.1338	0.1396	0.1497	0.1650	0.1864	0.2154	0.2535	0.3024	0.3632
15	P	30.974	0.1424	0.1443	0.1504	0.1610	0.1768	0.1988	0.2285	0.2672	0.3165	0.3773
16	S	32.064	0.1537	0.1558	0.1621	0.1730	0.1894	0.2121	0.2424	0.2817	0.3313	0.3920
		31.972	0.1528	0.1548	0.1611	0.1720	0.1883	0.2110	0.2412	0.2805	0.3301	0.3908
		32.972	0.1631	0.1652	0.1717	0.1830	0.1997	0.2229	0.2537	0.2934	0.3432	0.4038
		33.968	0.1732	0.1754	0.1821	0.1937	0.2108	0.2345	0.2657	0.3058	0.3557	0.4162
		35.967	0.1932	0.1955	0.2025	0.2147	0.2325	0.2569	0.2889	0.3295	0.3796	0.4395
17	Cl	35.453	0.1881	0.1903	0.1973	0.2093	0.2269	0.2512	0.2830	0.3234	0.3735	0.4336
		34.969	0.1833	0.1855	0.1924	0.2043	0.2218	0.2459	0.2775	0.3178	0.3679	0.4281
		36.966	0.2030	0.2053	0.2125	0.2249	0.2430	0.2678	0.3000	0.3408	0.3909	0.4505
18	Ar	39.948	0.2313	0.2337	0.2413	0.2542	0.2731	0.2986	0.3315	0.3725	0.4223	0.4809
		35.967	0.1932	0.1955	0.2025	0.2147	0.2325	0.2569	0.2889	0.3295	0.3796	0.4395
		37.963	0.2126	0.2150	0.2223	0.2349	0.2533	0.2783	0.3108	0.3518	0.4018	0.4611
		39.962	0.2314	0.2339	0.2414	0.2544	0.2732	0.2987	0.3316	0.3727	0.4225	0.4810
19	K	39.098	0.2233	0.2258	0.2332	0.2460	0.2647	0.2900	0.3227	0.3638	0.4136	0.4725
		38.964	0.2221	0.2245	0.2320	0.2447	0.2634	0.2887	0.3214	0.3624	0.4123	0.4713
		40.000	0.2317	0.2342	0.2418	0.2547	0.2736	0.2991	0.3320	0.3731	0.4228	0.4814
		40.962	0.2406	0.2431	0.2507	0.2638	0.2829	0.3085	0.3416	0.3827	0.4323	0.4905

Table A5.5. Rutherford backscattering spectrometry kinematic factors for N as a projectile (continued).

Atomic no. *(Z)*	El.	Isotopic mass (M_2) *(amu)*	180°	170°	160°	150°	140°	130°	120°	110°	100°	90°
20	Ca	40.078	0.2324	0.2349	0.2425	0.2554	0.2743	0.2998	0.3327	0.3738	0.4235	0.4821
		39.963	0.2314	0.2339	0.2414	0.2544	0.2732	0.2987	0.3316	0.3727	0.4225	0.4810
		41.959	0.2496	0.2521	0.2598	0.2731	0.2923	0.3181	0.3513	0.3923	0.4418	0.4996
		42.959	0.2584	0.2610	0.2688	0.2822	0.3015	0.3275	0.3607	0.4018	0.4510	0.5083
		43.956	0.2671	0.2697	0.2776	0.2910	0.3105	0.3366	0.3699	0.4109	0.4599	0.5168
		45.954	0.2840	0.2866	0.2946	0.3083	0.3280	0.3542	0.3875	0.4283	0.4769	0.5329
		47.952	0.3003	0.3029	0.3111	0.3248	0.3447	0.3710	0.4043	0.4448	0.4929	0.5480
21	Sc	44.956	0.2756	0.2782	0.2862	0.2998	0.3194	0.3455	0.3788	0.4197	0.4685	0.5250
22	Ti	47.878	0.2996	0.3023	0.3104	0.3242	0.3440	0.3703	0.4036	0.4442	0.4922	0.5473
		45.953	0.2840	0.2866	0.2946	0.3083	0.3280	0.3542	0.3875	0.4283	0.4769	0.5329
		46.952	0.2922	0.2949	0.3029	0.3166	0.3364	0.3627	0.3960	0.4367	0.4850	0.5405
		47.948	0.3002	0.3029	0.3110	0.3248	0.3446	0.3710	0.4042	0.4448	0.4928	0.5479
		48.948	0.3082	0.3108	0.3190	0.3328	0.3527	0.3791	0.4123	0.4527	0.5005	0.5551
		49.945	0.3159	0.3186	0.3268	0.3407	0.3606	0.3869	0.4201	0.4604	0.5079	0.5621
23	V	50.942	0.3235	0.3262	0.3344	0.3484	0.3683	0.3946	0.4278	0.4679	0.5150	0.5688
		49.947	0.3159	0.3186	0.3268	0.3407	0.3606	0.3870	0.4201	0.4604	0.5079	0.5621
		50.944	0.3235	0.3262	0.3344	0.3484	0.3683	0.3947	0.4278	0.4679	0.5150	0.5688
24	Cr	51.996	0.3314	0.3341	0.3423	0.3563	0.3762	0.4026	0.4356	0.4755	0.5224	0.5756
		49.946	0.3159	0.3186	0.3268	0.3407	0.3606	0.3869	0.4201	0.4604	0.5079	0.5621
		51.940	0.3310	0.3337	0.3419	0.3559	0.3758	0.4022	0.4352	0.4752	0.5220	0.5753
		52.941	0.3383	0.3410	0.3493	0.3633	0.3833	0.4096	0.4425	0.4823	0.5288	0.5817
		53.939	0.3455	0.3482	0.3565	0.3705	0.3905	0.4167	0.4496	0.4892	0.5354	0.5878
25	Mn	54.938	0.3526	0.3553	0.3636	0.3776	0.3976	0.4238	0.4565	0.4959	0.5418	0.5938
26	Fe	55.847	0.3588	0.3616	0.3699	0.3839	0.4038	0.4300	0.4627	0.5019	0.5475	0.5990
		53.940	0.3455	0.3482	0.3565	0.3705	0.3905	0.4168	0.4496	0.4892	0.5354	0.5878
		55.935	0.3595	0.3622	0.3705	0.3845	0.4045	0.4306	0.4633	0.5025	0.5481	0.5996
		56.935	0.3663	0.3690	0.3773	0.3913	0.4113	0.4374	0.4699	0.5089	0.5542	0.6052
		57.933	0.3729	0.3757	0.3840	0.3980	0.4179	0.4439	0.4763	0.5151	0.5601	0.6107
27	Co	58.933	0.3795	0.3822	0.3905	0.4045	0.4244	0.4504	0.4826	0.5212	0.5659	0.6160
28	Ni	58.688	0.3778	0.3805	0.3888	0.4028	0.4227	0.4487	0.4810	0.5196	0.5644	0.6146
		57.935	0.3729	0.3757	0.3840	0.3980	0.4179	0.4440	0.4763	0.5151	0.5601	0.6107
		59.931	0.3859	0.3886	0.3969	0.4109	0.4308	0.4567	0.4888	0.5271	0.5715	0.6212
		60.931	0.3922	0.3949	0.4032	0.4172	0.4370	0.4628	0.4948	0.5329	0.5770	0.6263
		61.928	0.3984	0.4011	0.4094	0.4233	0.4431	0.4688	0.5007	0.5386	0.5823	0.6312
		63.928	0.4104	0.4131	0.4214	0.4353	0.4549	0.4805	0.5121	0.5495	0.5926	0.6406
29	Cu	63.546	0.4081	0.4108	0.4191	0.4330	0.4527	0.4783	0.5099	0.5474	0.5906	0.6388
		62.930	0.4045	0.4072	0.4155	0.4294	0.4491	0.4747	0.5064	0.5441	0.5875	0.6360
		64.928	0.4163	0.4190	0.4272	0.4411	0.4607	0.4862	0.5176	0.5548	0.5976	0.6452
30	Zn	65.396	0.4188	0.4216	0.4298	0.4436	0.4632	0.4886	0.5199	0.5571	0.5997	0.6471
		63.929	0.4104	0.4132	0.4214	0.4353	0.4550	0.4805	0.5121	0.5495	0.5926	0.6406
		65.926	0.4220	0.4247	0.4330	0.4468	0.4663	0.4917	0.5229	0.5599	0.6024	0.6496
		66.927	0.4276	0.4304	0.4386	0.4524	0.4719	0.4971	0.5282	0.5650	0.6071	0.6539
		67.925	0.4332	0.4359	0.4441	0.4578	0.4773	0.5024	0.5333	0.5699	0.6117	0.6582
		69.925	0.4440	0.4467	0.4548	0.4685	0.4878	0.5127	0.5433	0.5794	0.6207	0.6663
31	Ga	69.723	0.4429	0.4456	0.4537	0.4674	0.4867	0.5117	0.5423	0.5784	0.6197	0.6655
		68.926	0.4386	0.4413	0.4495	0.4632	0.4826	0.5076	0.5384	0.5747	0.6163	0.6623
		70.925	0.4492	0.4519	0.4601	0.4737	0.4929	0.5177	0.5482	0.5840	0.6250	0.6702

Table A5.5. Rutherford backscattering spectrometry kinematic factors for N as a projectile (continued).

Atomic no. *(Z)*	El.	Isotopic mass (M_2) *(amu)*	180°	170°	160°	150°	140°	130°	120°	110°	100°	90°
32	Ge	72.632	0.4578	0.4605	0.4686	0.4822	0.5012	0.5259	0.5560	0.5915	0.6320	0.6766
		69.924	0.4440	0.4467	0.4548	0.4685	0.4878	0.5127	0.5433	0.5794	0.6207	0.6663
		71.922	0.4544	0.4571	0.4652	0.4788	0.4979	0.5226	0.5529	0.5885	0.6292	0.6741
		72.924	0.4594	0.4621	0.4702	0.4838	0.5028	0.5274	0.5575	0.5930	0.6333	0.6778
		73.921	0.4644	0.4671	0.4752	0.4886	0.5076	0.5321	0.5621	0.5973	0.6373	0.6815
		75.921	0.4741	0.4768	0.4848	0.4982	0.5170	0.5413	0.5709	0.6057	0.6451	0.6886
33	As	74.922	0.4693	0.4720	0.4800	0.4935	0.5124	0.5367	0.5665	0.6015	0.6413	0.6851
34	Se	78.993	0.4883	0.4909	0.4988	0.5121	0.5307	0.5546	0.5837	0.6178	0.6564	0.6987
		73.923	0.4644	0.4671	0.4752	0.4887	0.5076	0.5321	0.5621	0.5973	0.6373	0.6815
		75.919	0.4741	0.4768	0.4848	0.4982	0.5170	0.5413	0.5709	0.6057	0.6451	0.6886
		76.920	0.4788	0.4815	0.4895	0.5028	0.5216	0.5457	0.5752	0.6097	0.6489	0.6920
		77.917	0.4835	0.4861	0.4941	0.5074	0.5260	0.5501	0.5794	0.6137	0.6526	0.6953
		79.916	0.4925	0.4952	0.5031	0.5163	0.5348	0.5586	0.5875	0.6214	0.6597	0.7018
		81.917	0.5013	0.5039	0.5118	0.5248	0.5432	0.5667	0.5954	0.6288	0.6666	0.7080
35	Br	79.904	0.4924	0.4951	0.5030	0.5162	0.5347	0.5585	0.5874	0.6213	0.6597	0.7017
		78.918	0.4880	0.4907	0.4986	0.5119	0.5304	0.5544	0.5835	0.6176	0.6562	0.6986
		80.916	0.4970	0.4996	0.5074	0.5206	0.5390	0.5627	0.5915	0.6252	0.6632	0.7049
36	Kr	83.8	0.5092	0.5118	0.5196	0.5326	0.5508	0.5741	0.6025	0.6355	0.6728	0.7136
		77.920	0.4835	0.4861	0.4941	0.5074	0.5261	0.5501	0.5794	0.6137	0.6526	0.6953
		79.916	0.4925	0.4952	0.5031	0.5163	0.5348	0.5586	0.5875	0.6214	0.6597	0.7018
		81.913	0.5013	0.5039	0.5117	0.5248	0.5432	0.5667	0.5954	0.6288	0.6666	0.7080
		82.914	0.5056	0.5082	0.5160	0.5290	0.5473	0.5707	0.5992	0.6324	0.6700	0.7110
		83.911	0.5098	0.5124	0.5201	0.5331	0.5513	0.5746	0.6029	0.6360	0.6732	0.7140
		85.911	0.5180	0.5205	0.5283	0.5411	0.5591	0.5822	0.6102	0.6428	0.6796	0.7197
37	Rb	85.468	0.5161	0.5187	0.5265	0.5393	0.5574	0.5805	0.6086	0.6413	0.6782	0.7184
		84.912	0.5139	0.5165	0.5242	0.5372	0.5553	0.5785	0.6066	0.6394	0.6764	0.7169
		86.909	0.5220	0.5245	0.5322	0.5450	0.5630	0.5859	0.6138	0.6462	0.6827	0.7225
38	Sr	87.617	0.5247	0.5273	0.5350	0.5477	0.5656	0.5885	0.6162	0.6485	0.6848	0.7244
		83.913	0.5098	0.5124	0.5201	0.5331	0.5513	0.5746	0.6029	0.6360	0.6732	0.7140
		85.909	0.5180	0.5205	0.5283	0.5411	0.5591	0.5822	0.6102	0.6428	0.6796	0.7197
		86.909	0.5220	0.5245	0.5322	0.5450	0.5630	0.5859	0.6138	0.6462	0.6827	0.7225
		87.906	0.5259	0.5284	0.5361	0.5489	0.5667	0.5896	0.6172	0.6494	0.6857	0.7252
39	Y	88.906	0.5298	0.5323	0.5399	0.5526	0.5704	0.5931	0.6207	0.6527	0.6886	0.7279
40	Zr	91.221	0.5384	0.5410	0.5485	0.5611	0.5787	0.6011	0.6283	0.6598	0.6952	0.7338
		89.905	0.5336	0.5361	0.5437	0.5564	0.5740	0.5967	0.6240	0.6558	0.6916	0.7305
		90.906	0.5374	0.5399	0.5474	0.5600	0.5776	0.6001	0.6273	0.6589	0.6944	0.7330
		91.905	0.5411	0.5436	0.5511	0.5636	0.5811	0.6035	0.6306	0.6620	0.6972	0.7356
		93.906	0.5483	0.5508	0.5582	0.5707	0.5880	0.6102	0.6369	0.6679	0.7027	0.7405
		95.908	0.5553	0.5578	0.5652	0.5775	0.5947	0.6166	0.6430	0.6737	0.7080	0.7452
41	Nb	92.906	0.5447	0.5472	0.5547	0.5672	0.5846	0.6069	0.6338	0.6650	0.7000	0.7380
42	Mo	95.931	0.5553	0.5577	0.5651	0.5774	0.5946	0.6165	0.6430	0.6736	0.7079	0.7451
		91.907	0.5411	0.5436	0.5511	0.5636	0.5811	0.6035	0.6306	0.6620	0.6972	0.7356
		93.905	0.5483	0.5508	0.5582	0.5707	0.5880	0.6102	0.6369	0.6679	0.7027	0.7405
		94.906	0.5518	0.5543	0.5617	0.5741	0.5914	0.6134	0.6400	0.6708	0.7053	0.7428
		95.905	0.5553	0.5578	0.5652	0.5775	0.5947	0.6166	0.6430	0.6737	0.7080	0.7452
		96.906	0.5587	0.5612	0.5686	0.5808	0.5979	0.6197	0.6460	0.6765	0.7105	0.7475
		97.905	0.5621	0.5646	0.5719	0.5841	0.6011	0.6228	0.6489	0.6792	0.7130	0.7497
		99.908	0.5687	0.5712	0.5784	0.5905	0.6074	0.6288	0.6547	0.6846	0.7180	0.7541

Table A5.5. Rutherford backscattering spectrometry kinematic factors for N as a projectile (continued).

Atomic no. *(Z)*	El.	Isotopic mass (M_2) *(amu)*	180°	170°	160°	150°	140°	130°	120°	110°	100°	90°
43	Tc	98.000	0.5624	0.5649	0.5722	0.5844	0.6014	0.6231	0.6492	0.6795	0.7133	0.7500
44	Ru	101.019	0.5722	0.5746	0.5819	0.5939	0.6106	0.6320	0.6577	0.6874	0.7205	0.7564
		95.908	0.5553	0.5578	0.5652	0.5775	0.5947	0.6166	0.6430	0.6737	0.7080	0.7452
		97.905	0.5621	0.5646	0.5719	0.5841	0.6011	0.6228	0.6489	0.6792	0.7130	0.7497
		98.506	0.5641	0.5666	0.5739	0.5860	0.6030	0.6246	0.6507	0.6808	0.7145	0.7511
		99.904	0.5687	0.5711	0.5784	0.5905	0.6074	0.6288	0.6547	0.6846	0.7180	0.7541
		100.906	0.5720	0.5744	0.5816	0.5936	0.6104	0.6318	0.6575	0.6872	0.7204	0.7563
		101.904	0.5751	0.5775	0.5848	0.5967	0.6134	0.6346	0.6602	0.6897	0.7227	0.7584
		103.905	0.5814	0.5838	0.5909	0.6028	0.6193	0.6403	0.6656	0.6948	0.7273	0.7625
45	Rh	102.906	0.5783	0.5807	0.5879	0.5998	0.6164	0.6375	0.6629	0.6923	0.7250	0.7604
46	Pd	106.415	0.5889	0.5912	0.5983	0.6100	0.6263	0.6471	0.6720	0.7008	0.7328	0.7674
		101.906	0.5751	0.5775	0.5848	0.5967	0.6134	0.6347	0.6602	0.6897	0.7227	0.7584
		103.904	0.5814	0.5838	0.5909	0.6028	0.6193	0.6403	0.6656	0.6948	0.7273	0.7625
		104.905	0.5844	0.5868	0.5939	0.6057	0.6222	0.6431	0.6682	0.6972	0.7295	0.7645
		105.904	0.5874	0.5898	0.5969	0.6086	0.6250	0.6458	0.6708	0.6996	0.7317	0.7664
		107.904	0.5933	0.5957	0.6027	0.6143	0.6305	0.6511	0.6758	0.7043	0.7360	0.7703
		109.905	0.5990	0.6014	0.6083	0.6199	0.6359	0.6563	0.6807	0.7089	0.7402	0.7740
47	Ag	107.868	0.5932	0.5955	0.6026	0.6142	0.6304	0.6510	0.6757	0.7042	0.7359	0.7702
		106.905	0.5904	0.5927	0.5998	0.6115	0.6278	0.6485	0.6733	0.7020	0.7339	0.7684
		108.905	0.5962	0.5985	0.6055	0.6171	0.6332	0.6537	0.6783	0.7066	0.7381	0.7721
48	Cd	112.412	0.6059	0.6082	0.6151	0.6265	0.6424	0.6625	0.6866	0.7144	0.7452	0.7784
		105.907	0.5874	0.5898	0.5969	0.6086	0.6250	0.6458	0.6708	0.6996	0.7317	0.7664
		107.904	0.5933	0.5957	0.6027	0.6143	0.6305	0.6511	0.6758	0.7043	0.7360	0.7703
		109.903	0.5990	0.6014	0.6083	0.6199	0.6359	0.6563	0.6807	0.7089	0.7402	0.7740
		110.904	0.6018	0.6042	0.6111	0.6226	0.6385	0.6588	0.6831	0.7111	0.7422	0.7758
		111.903	0.6046	0.6069	0.6138	0.6252	0.6411	0.6613	0.6855	0.7133	0.7442	0.7776
		112.904	0.6073	0.6096	0.6165	0.6279	0.6437	0.6638	0.6878	0.7155	0.7462	0.7793
		113.903	0.6100	0.6123	0.6191	0.6305	0.6462	0.6662	0.6901	0.7176	0.7482	0.7810
		115.905	0.6153	0.6176	0.6243	0.6356	0.6512	0.6709	0.6946	0.7218	0.7520	0.7844
49	In	114.818	0.6125	0.6147	0.6215	0.6328	0.6485	0.6684	0.6922	0.7195	0.7499	0.7826
		112.904	0.6073	0.6096	0.6165	0.6279	0.6437	0.6638	0.6878	0.7155	0.7462	0.7793
		114.904	0.6127	0.6150	0.6218	0.6330	0.6487	0.6686	0.6924	0.7197	0.7501	0.7827
50	Sn	118.613	0.6218	0.6241	0.6308	0.6419	0.6572	0.6768	0.7001	0.7269	0.7566	0.7886
		111.905	0.6046	0.6069	0.6138	0.6252	0.6411	0.6613	0.6855	0.7133	0.7442	0.7776
		113.903	0.6100	0.6123	0.6191	0.6305	0.6462	0.6662	0.6901	0.7176	0.7482	0.7810
		114.903	0.6127	0.6150	0.6218	0.6330	0.6487	0.6686	0.6924	0.7197	0.7501	0.7827
		115.902	0.6153	0.6176	0.6243	0.6356	0.6511	0.6709	0.6946	0.7218	0.7519	0.7844
		116.903	0.6179	0.6201	0.6269	0.6381	0.6536	0.6732	0.6968	0.7238	0.7538	0.7861
		117.602	0.6197	0.6219	0.6286	0.6398	0.6552	0.6748	0.6983	0.7252	0.7551	0.7872
		118.903	0.6230	0.6252	0.6319	0.6430	0.6583	0.6778	0.7011	0.7278	0.7574	0.7893
		119.902	0.6254	0.6277	0.6343	0.6453	0.6606	0.6800	0.7032	0.7297	0.7592	0.7909
		121.903	0.6303	0.6325	0.6391	0.6500	0.6652	0.6844	0.7073	0.7336	0.7627	0.7939
		123.905	0.6351	0.6373	0.6438	0.6546	0.6696	0.6886	0.7113	0.7373	0.7661	0.7969
51	Sb	121.758	0.6300	0.6322	0.6388	0.6497	0.6648	0.6840	0.7070	0.7333	0.7624	0.7937
		120.904	0.6279	0.6301	0.6367	0.6477	0.6629	0.6822	0.7052	0.7317	0.7610	0.7924
		122.904	0.6327	0.6349	0.6415	0.6524	0.6674	0.6865	0.7093	0.7354	0.7644	0.7954

Table A5.5. Rutherford backscattering spectrometry kinematic factors for N as a projectile (continued).

Atomic no. (*Z*)	El.	Isotopic mass (M_2) (*amu*)	180°	170°	160°	150°	140°	130°	120°	110°	100°	90°
52	Te	127.586	0.6435	0.6456	0.6520	0.6627	0.6774	0.6961	0.7183	0.7438	0.7720	0.8022
		119.904	0.6255	0.6277	0.6343	0.6454	0.6606	0.6800	0.7032	0.7297	0.7592	0.7909
		121.903	0.6303	0.6325	0.6391	0.6500	0.6652	0.6844	0.7073	0.7336	0.7627	0.7939
		122.904	0.6327	0.6349	0.6415	0.6524	0.6674	0.6865	0.7093	0.7354	0.7644	0.7954
		123.903	0.6351	0.6373	0.6438	0.6546	0.6696	0.6886	0.7113	0.7373	0.7661	0.7969
		124.904	0.6374	0.6396	0.6461	0.6569	0.6718	0.6907	0.7132	0.7391	0.7677	0.7984
		125.903	0.6397	0.6419	0.6484	0.6591	0.6739	0.6927	0.7152	0.7409	0.7693	0.7998
		127.905	0.6442	0.6464	0.6528	0.6634	0.6781	0.6968	0.7190	0.7444	0.7725	0.8026
		129.906	0.6487	0.6508	0.6571	0.6677	0.6823	0.7007	0.7227	0.7478	0.7756	0.8054
53	I	126.905	0.6420	0.6442	0.6506	0.6613	0.6761	0.6948	0.7171	0.7427	0.7709	0.8012
54	Xe	131.293	0.6516	0.6537	0.6600	0.6705	0.6850	0.7033	0.7251	0.7501	0.7777	0.8072
		123.906	0.6351	0.6373	0.6438	0.6546	0.6696	0.6886	0.7113	0.7373	0.7661	0.7969
		125.904	0.6397	0.6419	0.6484	0.6591	0.6739	0.6927	0.7152	0.7409	0.7693	0.7998
		127.904	0.6442	0.6464	0.6528	0.6634	0.6781	0.6968	0.7190	0.7444	0.7725	0.8026
		128.905	0.6465	0.6486	0.6550	0.6656	0.6802	0.6987	0.7208	0.7461	0.7741	0.8040
		129.904	0.6487	0.6508	0.6571	0.6677	0.6822	0.7007	0.7227	0.7478	0.7756	0.8054
		130.905	0.6508	0.6529	0.6593	0.6698	0.6843	0.7026	0.7245	0.7495	0.7771	0.8067
		131.904	0.6530	0.6551	0.6614	0.6718	0.6862	0.7045	0.7263	0.7512	0.7786	0.8081
		133.905	0.6572	0.6592	0.6655	0.6758	0.6902	0.7082	0.7298	0.7544	0.7816	0.8107
		135.907	0.6613	0.6633	0.6695	0.6798	0.6940	0.7119	0.7332	0.7576	0.7844	0.8132
55	Cs	132.905	0.6551	0.6572	0.6635	0.6738	0.6882	0.7064	0.7280	0.7528	0.7801	0.8094
56	Ba	137.327	0.6641	0.6662	0.6723	0.6825	0.6966	0.7144	0.7356	0.7598	0.7864	0.8149
		129.906	0.6487	0.6508	0.6571	0.6677	0.6823	0.7007	0.7227	0.7478	0.7756	0.8054
		131.905	0.6530	0.6551	0.6614	0.6718	0.6863	0.7045	0.7263	0.7512	0.7786	0.8081
		133.904	0.6572	0.6592	0.6655	0.6758	0.6902	0.7082	0.7298	0.7544	0.7816	0.8107
		134.906	0.6592	0.6613	0.6675	0.6778	0.6921	0.7101	0.7315	0.7560	0.7830	0.8119
		135.905	0.6613	0.6633	0.6695	0.6798	0.6940	0.7119	0.7332	0.7576	0.7844	0.8132
		136.906	0.6633	0.6653	0.6715	0.6817	0.6958	0.7136	0.7349	0.7591	0.7859	0.8144
		137.905	0.6653	0.6673	0.6735	0.6836	0.6977	0.7154	0.7365	0.7606	0.7872	0.8156
57	La	138.905	0.6672	0.6693	0.6754	0.6855	0.6995	0.7171	0.7382	0.7622	0.7886	0.8168
		137.907	0.6653	0.6673	0.6735	0.6836	0.6977	0.7154	0.7365	0.7606	0.7872	0.8156
		138.906	0.6672	0.6693	0.6754	0.6855	0.6995	0.7171	0.7382	0.7622	0.7886	0.8168
58	Ce	140.115	0.6696	0.6716	0.6777	0.6878	0.7017	0.7192	0.7401	0.7640	0.7902	0.8183
		135.907	0.6613	0.6633	0.6695	0.6798	0.6940	0.7119	0.7332	0.7576	0.7844	0.8132
		137.906	0.6653	0.6673	0.6735	0.6836	0.6977	0.7154	0.7365	0.7606	0.7872	0.8156
		139.905	0.6692	0.6712	0.6773	0.6874	0.7013	0.7189	0.7398	0.7636	0.7900	0.8180
		141.909	0.6730	0.6750	0.6811	0.6911	0.7048	0.7222	0.7430	0.7666	0.7926	0.8204
59	Pr	140.908	0.6711	0.6731	0.6792	0.6892	0.7031	0.7206	0.7414	0.7651	0.7913	0.8192
60	Nd	144.242	0.6773	0.6793	0.6853	0.6952	0.7088	0.7260	0.7465	0.7699	0.7956	0.8230
		141.908	0.6730	0.6750	0.6811	0.6911	0.7048	0.7222	0.7430	0.7666	0.7926	0.8204
		142.910	0.6749	0.6769	0.6829	0.6929	0.7066	0.7239	0.7445	0.7680	0.7939	0.8215
		143.910	0.6768	0.6788	0.6847	0.6946	0.7083	0.7255	0.7461	0.7694	0.7952	0.8226
		144.913	0.6786	0.6806	0.6866	0.6964	0.7100	0.7272	0.7476	0.7708	0.7965	0.8238
		145.913	0.6804	0.6824	0.6883	0.6981	0.7117	0.7288	0.7491	0.7722	0.7977	0.8249
		147.917	0.6840	0.6860	0.6918	0.7016	0.7150	0.7319	0.7520	0.7750	0.8002	0.8270
		149.921	0.6875	0.6894	0.6953	0.7049	0.7182	0.7350	0.7549	0.7776	0.8026	0.8292
61	Pm	145.000	0.6788	0.6807	0.6867	0.6966	0.7102	0.7273	0.7477	0.7710	0.7966	0.8239

Table A5.5. Rutherford backscattering spectrometry kinematic factors for N as a projectile (continued).

Atomic no. *(Z)*	El.	Isotopic mass (M_2) *(amu)*	180°	170°	160°	150°	140°	130°	120°	110°	100°	90°
62	Sm	150.36	0.6882	0.6901	0.6960	0.7056	0.7189	0.7356	0.7555	0.7781	0.8030	0.8296
		143.912	0.6768	0.6788	0.6847	0.6946	0.7083	0.7255	0.7461	0.7694	0.7952	0.8227
		146.915	0.6822	0.6842	0.6901	0.6999	0.7134	0.7304	0.7506	0.7736	0.7989	0.8260
		147.915	0.6840	0.6860	0.6918	0.7016	0.7150	0.7319	0.7520	0.7750	0.8002	0.8270
		148.917	0.6857	0.6877	0.6936	0.7033	0.7166	0.7335	0.7535	0.7763	0.8014	0.8281
		149.917	0.6875	0.6894	0.6953	0.7049	0.7182	0.7350	0.7549	0.7776	0.8026	0.8291
		151.920	0.6909	0.6928	0.6986	0.7082	0.7214	0.7380	0.7577	0.7802	0.8049	0.8312
		153.922	0.6943	0.6962	0.7019	0.7114	0.7245	0.7409	0.7605	0.7828	0.8072	0.8332
63	Eu	151.965	0.6910	0.6929	0.6987	0.7083	0.7215	0.7381	0.7578	0.7803	0.8050	0.8313
		150.920	0.6892	0.6912	0.6970	0.7066	0.7198	0.7365	0.7563	0.7789	0.8038	0.8302
		152.921	0.6926	0.6945	0.7003	0.7098	0.7229	0.7395	0.7591	0.7815	0.8061	0.8322
64	Gd	157.252	0.6996	0.7015	0.7072	0.7166	0.7294	0.7457	0.7649	0.7868	0.8109	0.8364
		151.920	0.6909	0.6928	0.6986	0.7082	0.7214	0.7380	0.7577	0.7802	0.8049	0.8312
		153.921	0.6943	0.6962	0.7019	0.7114	0.7245	0.7409	0.7605	0.7828	0.8072	0.8332
		154.923	0.6959	0.6978	0.7035	0.7130	0.7260	0.7424	0.7619	0.7840	0.8083	0.8342
		155.922	0.6975	0.6994	0.7051	0.7145	0.7275	0.7438	0.7632	0.7852	0.8094	0.8352
		156.924	0.6992	0.7010	0.7067	0.7161	0.7290	0.7452	0.7645	0.7865	0.8105	0.8362
		157.924	0.7007	0.7026	0.7083	0.7176	0.7305	0.7466	0.7658	0.7877	0.8116	0.8371
		159.927	0.7039	0.7058	0.7114	0.7206	0.7334	0.7494	0.7684	0.7900	0.8138	0.8390
65	Tb	158.925	0.7023	0.7042	0.7098	0.7191	0.7319	0.7480	0.7671	0.7889	0.8127	0.8380
66	Dy	162.498	0.7078	0.7097	0.7152	0.7244	0.7370	0.7528	0.7716	0.7930	0.8164	0.8413
		155.924	0.6975	0.6994	0.7051	0.7145	0.7275	0.7438	0.7632	0.7852	0.8094	0.8352
		157.924	0.7007	0.7026	0.7083	0.7176	0.7305	0.7466	0.7658	0.7877	0.8116	0.8371
		159.925	0.7039	0.7058	0.7114	0.7206	0.7333	0.7494	0.7684	0.7900	0.8138	0.8390
		160.927	0.7054	0.7073	0.7129	0.7221	0.7348	0.7507	0.7697	0.7912	0.8148	0.8399
		161.927	0.7070	0.7088	0.7144	0.7235	0.7362	0.7521	0.7709	0.7923	0.8158	0.8408
		162.929	0.7085	0.7103	0.7159	0.7250	0.7376	0.7534	0.7722	0.7935	0.8169	0.8417
		163.929	0.7100	0.7118	0.7173	0.7264	0.7390	0.7547	0.7734	0.7946	0.8179	0.8426
67	Ho	164.930	0.7115	0.7133	0.7188	0.7278	0.7403	0.7560	0.7746	0.7957	0.8189	0.8435
68	Er	167.256	0.7148	0.7167	0.7221	0.7311	0.7434	0.7590	0.7774	0.7983	0.8212	0.8455
		161.929	0.7070	0.7088	0.7144	0.7235	0.7362	0.7521	0.7709	0.7923	0.8158	0.8408
		163.929	0.7100	0.7118	0.7173	0.7264	0.7390	0.7547	0.7734	0.7946	0.8179	0.8426
		165.930	0.7129	0.7148	0.7202	0.7292	0.7417	0.7573	0.7758	0.7968	0.8199	0.8444
		166.932	0.7144	0.7162	0.7217	0.7306	0.7430	0.7586	0.7770	0.7979	0.8209	0.8452
		167.932	0.7158	0.7176	0.7231	0.7320	0.7443	0.7598	0.7782	0.7990	0.8218	0.8461
		169.936	0.7187	0.7205	0.7258	0.7347	0.7469	0.7623	0.7805	0.8011	0.8237	0.8477
69	Tm	168.934	0.7173	0.7191	0.7245	0.7334	0.7456	0.7611	0.7793	0.8001	0.8228	0.8469
70	Yb	173.034	0.7229	0.7247	0.7300	0.7388	0.7509	0.7660	0.7840	0.8043	0.8266	0.8503
		167.934	0.7158	0.7176	0.7231	0.7320	0.7443	0.7598	0.7782	0.7990	0.8218	0.8461
		169.935	0.7187	0.7205	0.7258	0.7347	0.7469	0.7623	0.7805	0.8011	0.8237	0.8477
		170.936	0.7201	0.7219	0.7272	0.7361	0.7482	0.7635	0.7816	0.8022	0.8247	0.8486
		171.936	0.7214	0.7232	0.7286	0.7374	0.7495	0.7647	0.7828	0.8032	0.8256	0.8494
		172.938	0.7228	0.7246	0.7299	0.7387	0.7508	0.7659	0.7839	0.8042	0.8265	0.8502
		173.939	0.7242	0.7260	0.7312	0.7400	0.7520	0.7671	0.7850	0.8053	0.8274	0.8510
		175.943	0.7269	0.7286	0.7339	0.7425	0.7545	0.7694	0.7872	0.8072	0.8292	0.8526
71	Lu	174.967	0.7256	0.7273	0.7326	0.7413	0.7533	0.7683	0.7861	0.8063	0.8284	0.8518
		174.941	0.7255	0.7273	0.7326	0.7413	0.7532	0.7683	0.7861	0.8063	0.8284	0.8518
		175.943	0.7269	0.7286	0.7339	0.7425	0.7545	0.7694	0.7872	0.8072	0.8292	0.8526

Table A5.5. Rutherford backscattering spectrometry kinematic factors for N as a projectile (continued).

Atomic no. *(Z)*	El.	Isotopic mass (M_2) *(amu)*	180°	170°	160°	150°	140°	130°	120°	110°	100°	90°
72	Hf	178.49	0.7302	0.7319	0.7371	0.7457	0.7575	0.7723	0.7898	0.8097	0.8315	0.8545
		173.940	0.7242	0.7260	0.7312	0.7400	0.7520	0.7671	0.7850	0.8053	0.8274	0.8510
		175.941	0.7269	0.7286	0.7339	0.7425	0.7545	0.7694	0.7871	0.8072	0.8292	0.8526
		176.943	0.7282	0.7299	0.7352	0.7438	0.7557	0.7706	0.7882	0.8082	0.8301	0.8533
		177.944	0.7295	0.7312	0.7364	0.7450	0.7569	0.7717	0.7893	0.8092	0.8310	0.8541
		178.946	0.7308	0.7325	0.7377	0.7463	0.7580	0.7728	0.7903	0.8102	0.8319	0.8549
		179.947	0.7321	0.7338	0.7390	0.7475	0.7592	0.7739	0.7914	0.8111	0.8327	0.8556
73	Ta	180.948	0.7333	0.7351	0.7402	0.7487	0.7604	0.7750	0.7924	0.8121	0.8336	0.8563
		179.947	0.7321	0.7338	0.7390	0.7475	0.7592	0.7739	0.7914	0.8111	0.8327	0.8556
		180.948	0.7333	0.7351	0.7402	0.7487	0.7604	0.7750	0.7924	0.8121	0.8336	0.8563
74	W	183.849	0.7369	0.7386	0.7437	0.7521	0.7637	0.7782	0.7953	0.8147	0.8360	0.8584
		179.947	0.7321	0.7338	0.7390	0.7475	0.7592	0.7739	0.7914	0.8111	0.8327	0.8556
		181.948	0.7346	0.7363	0.7414	0.7499	0.7615	0.7761	0.7934	0.8130	0.8344	0.8571
		182.950	0.7358	0.7375	0.7427	0.7511	0.7627	0.7772	0.7944	0.8139	0.8352	0.8578
		183.951	0.7371	0.7388	0.7439	0.7523	0.7638	0.7783	0.7954	0.8148	0.8361	0.8585
		185.954	0.7395	0.7412	0.7462	0.7546	0.7660	0.7804	0.7974	0.8167	0.8377	0.8599
75	Re	186.207	0.7398	0.7415	0.7465	0.7549	0.7663	0.7807	0.7976	0.8169	0.8379	0.8601
		184.953	0.7383	0.7400	0.7451	0.7534	0.7649	0.7794	0.7964	0.8158	0.8369	0.8592
		186.956	0.7407	0.7424	0.7474	0.7557	0.7671	0.7814	0.7984	0.8175	0.8385	0.8606
76	Os	190.24	0.7445	0.7462	0.7512	0.7594	0.7707	0.7848	0.8015	0.8204	0.8410	0.8629
		183.952	0.7371	0.7388	0.7439	0.7523	0.7638	0.7783	0.7954	0.8148	0.8361	0.8585
		185.954	0.7395	0.7412	0.7462	0.7546	0.7660	0.7804	0.7974	0.8167	0.8377	0.8599
		186.956	0.7407	0.7424	0.7474	0.7557	0.7671	0.7814	0.7984	0.8175	0.8385	0.8606
		187.956	0.7419	0.7436	0.7486	0.7569	0.7682	0.7825	0.7993	0.8184	0.8393	0.8613
		188.958	0.7431	0.7447	0.7497	0.7580	0.7693	0.7835	0.8003	0.8193	0.8401	0.8620
		189.958	0.7442	0.7459	0.7509	0.7591	0.7704	0.7845	0.8012	0.8202	0.8408	0.8627
		191.962	0.7465	0.7482	0.7531	0.7613	0.7725	0.7865	0.8031	0.8219	0.8424	0.8640
77	Ir	192.216	0.7468	0.7485	0.7534	0.7616	0.7727	0.7868	0.8033	0.8221	0.8425	0.8642
		190.961	0.7454	0.7471	0.7520	0.7602	0.7714	0.7855	0.8022	0.8210	0.8416	0.8634
		192.963	0.7477	0.7493	0.7543	0.7624	0.7735	0.7875	0.8040	0.8227	0.8431	0.8647
78	Pt	195.08	0.7500	0.7517	0.7566	0.7646	0.7757	0.7895	0.8059	0.8245	0.8447	0.8660
		189.960	0.7442	0.7459	0.7509	0.7591	0.7704	0.7845	0.8012	0.8202	0.8408	0.8627
		191.961	0.7465	0.7482	0.7531	0.7613	0.7725	0.7865	0.8031	0.8219	0.8424	0.8640
		193.963	0.7488	0.7504	0.7554	0.7634	0.7745	0.7885	0.8049	0.8235	0.8439	0.8653
		194.965	0.7499	0.7516	0.7564	0.7645	0.7756	0.7894	0.8058	0.8244	0.8446	0.8660
		195.965	0.7510	0.7527	0.7575	0.7655	0.7766	0.7904	0.8067	0.8252	0.8453	0.8666
		197.968	0.7532	0.7548	0.7597	0.7676	0.7786	0.7923	0.8085	0.8268	0.8468	0.8679
79	Au	196.967	0.7521	0.7538	0.7586	0.7666	0.7776	0.7913	0.8076	0.8260	0.8460	0.8673
80	Hg	200.588	0.7560	0.7576	0.7624	0.7703	0.7811	0.7947	0.8107	0.8288	0.8486	0.8695
		195.966	0.7510	0.7527	0.7575	0.7656	0.7766	0.7904	0.8067	0.8252	0.8453	0.8666
		197.967	0.7532	0.7548	0.7597	0.7676	0.7786	0.7923	0.8085	0.8268	0.8468	0.8679
		198.968	0.7543	0.7559	0.7607	0.7687	0.7796	0.7932	0.8093	0.8276	0.8475	0.8685
		199.968	0.7554	0.7570	0.7618	0.7697	0.7805	0.7941	0.8102	0.8284	0.8482	0.8691
		200.970	0.7564	0.7580	0.7628	0.7707	0.7815	0.7951	0.8111	0.8292	0.8489	0.8697
		201.971	0.7575	0.7591	0.7638	0.7717	0.7825	0.7960	0.8119	0.8299	0.8496	0.8703
		203.973	0.7595	0.7611	0.7659	0.7736	0.7844	0.7978	0.8136	0.8314	0.8509	0.8715

Table A5.5. Rutherford backscattering spectrometry kinematic factors for N as a projectile (continued).

Atomic no. *(Z)*	**El.**	**Isotopic mass (M_2)** *(amu)*	**180°**	**170°**	**160°**	**150°**	**140°**	**130°**	**120°**	**110°**	**100°**	**90°**
81	Tl	204.383	0.7600	0.7615	0.7663	0.7740	0.7847	0.7981	0.8139	0.8318	0.8512	0.8718
		202.972	0.7585	0.7601	0.7648	0.7727	0.7834	0.7969	0.8127	0.8307	0.8503	0.8709
		204.974	0.7606	0.7621	0.7669	0.7746	0.7853	0.7986	0.8144	0.8322	0.8516	0.8721
82	Pb	207.217	0.7628	0.7644	0.7691	0.7768	0.7873	0.8006	0.8162	0.8339	0.8531	0.8734
		203.973	0.7595	0.7611	0.7659	0.7736	0.7844	0.7978	0.8136	0.8314	0.8509	0.8715
		205.975	0.7616	0.7632	0.7679	0.7756	0.7862	0.7995	0.8152	0.8329	0.8523	0.8727
		206.976	0.7626	0.7642	0.7688	0.7765	0.7871	0.8004	0.8160	0.8337	0.8529	0.8733
		207.977	0.7636	0.7652	0.7698	0.7775	0.7880	0.8012	0.8168	0.8344	0.8536	0.8738
83	Bi	208.980	0.7646	0.7661	0.7708	0.7784	0.7889	0.8021	0.8176	0.8351	0.8542	0.8744
84	Po	208.982	0.7646	0.7661	0.7708	0.7784	0.7889	0.8021	0.8176	0.8351	0.8542	0.8744
85	At	210.000	0.7656	0.7671	0.7718	0.7794	0.7899	0.8030	0.8184	0.8359	0.8549	0.8750
86	Rn	222.018	0.7768	0.7783	0.7827	0.7900	0.8000	0.8126	0.8274	0.8441	0.8622	0.8813
87	Fr	223.000	0.7776	0.7791	0.7835	0.7908	0.8008	0.8133	0.8281	0.8447	0.8628	0.8818
88	Ra	226.025	0.7803	0.7817	0.7861	0.7933	0.8032	0.8156	0.8302	0.8466	0.8645	0.8833
89	Ac	227.000	0.7811	0.7826	0.7869	0.7941	0.8040	0.8163	0.8308	0.8472	0.8650	0.8838
90	Th	232.038	0.7853	0.7867	0.7910	0.7981	0.8078	0.8199	0.8342	0.8503	0.8678	0.8862
91	Pa	231.036	0.7845	0.7859	0.7902	0.7973	0.8071	0.8192	0.8335	0.8497	0.8672	0.8857
92	U	238.018	0.7901	0.7915	0.7957	0.8027	0.8122	0.8240	0.8380	0.8538	0.8709	0.8889
		234.041	0.7869	0.7884	0.7926	0.7997	0.8093	0.8213	0.8355	0.8515	0.8688	0.8871
		235.044	0.7877	0.7892	0.7934	0.8004	0.8100	0.8220	0.8361	0.8520	0.8693	0.8875
		238.040	0.7901	0.7915	0.7957	0.8027	0.8122	0.8240	0.8380	0.8538	0.8709	0.8889
93	Np	237.048	0.7893	0.7908	0.7950	0.8019	0.8115	0.8234	0.8374	0.8532	0.8704	0.8884
94	Pu	244.064	0.7947	0.7961	0.8003	0.8070	0.8164	0.8280	0.8417	0.8571	0.8739	0.8915
95	Am	243.061	0.7940	0.7954	0.7995	0.8063	0.8157	0.8274	0.8411	0.8566	0.8734	0.8911

APPENDIX

6

RUTHERFORD CROSS SECTIONS

Compiled by

A. Dick and J. R. Tesmer

Los Alamos National Laboratory, Los Alamos, New Mexico

CONTENTS

The equation used to calculate the Rutherford cross sections is:

$$\sigma_R(E,\theta)=\left(\frac{Z_1Z_2e^2}{4E}\right)^2\frac{4\left\{\left(M_2^2-M_1^2\sin^2\theta\right)^{1/2}+M_2\cos\theta\right\}^2}{M_2\sin^4\theta\left(M_2^2-M_1^2\sin^2\theta\right)^{1/2}} \qquad \text{(A6.1)}$$

where M_1 and M_2 are the incident and target masses and Z_1 and Z_2 are their respective atomic numbers. Θ is the scattering angle (see Appendix 4). The cross sections calculated in this Appendix use the masses from: *Nuclides and Isotopes*, Fourteenth Edition, Revised 1989, General Electric Company, Nuclear Energy Operations (Walker *et al.*, 1989). Tables A6.1 through A6.9 give the cross sections for 1 MeV protons, deuterons, ^{3}He, ^{4}He, ^{7}Li and ^{12}C, ^{15}N, ^{16}O, and ^{28}Si scattered from a range of target nuclei at various angles. The mass for the scattering element is an average over the isotopic masses.

In some cases, because of electron screening, corrections must be made to the calculated Rutherford cross sections in Tables A6.1 though A6.9. The corrections, called F-Factors, can be found in Table A6.10 and are calculated from the expression given by L'Ecuyer (1979)

$$\sigma_{sc}/\sigma_R = 1-\frac{0.049 Z_1 Z_2{}^{4/3}}{E_{cm}} = F \qquad \text{(A6.2)}$$

where E_{cm}, the center-of-mass kinetic energy in keV, is replaced by the laboratory energy E_{lab}, with negligible error. See Chapter 4, Section 4.2.2.3 and Chapter 12, Section 12.2.1 for more detailed information. Although the F factor is relatively insensitive to angles normally used for backscattering, it is valid only for scattering angles greater than ≈90° (see Hautala and Luomajärvi, 1980). The corrections for forward angles may be much worse! See references in Chapters 4 and 12 for more information.

Table A6.1. Rutherford scattering cross sections (barns) of the elements for 1 MeV ^1H (protons).

Element	Atomic no. (Z_2)	Mass (M_2)	Scattering angle (degrees) 175	170	165	160	150	140	135	120	90	60	30
He	2	4.003	4.564E−03	4.623E−03	4.723E−03	4.868E−03	5.307E−03	5.995E−03	6.459E−03	8.550E−03	2.005E−02	8.223E−02	1.154E+00
Li	3	6.941	1.121E−02	1.135E−02	1.158E−02	1.190E−02	1.290E−02	1.446E−02	1.551E−02	2.023E−02	4.613E−02	1.860E−01	2.597E+00
Be	4	9.012	2.029E−02	2.052E−02	2.093E−02	2.151E−02	2.329E−02	2.606E−02	2.792E−02	3.632E−02	8.237E−02	3.310E−01	4.617E+00
B	5	10.811	3.194E−02	3.231E−02	3.295E−02	3.386E−02	3.663E−02	4.096E−02	4.388E−02	5.700E−02	1.289E−01	5.175E−01	7.215E+00
C	6	12.011	4.615E−02	4.669E−02	4.760E−02	4.891E−02	5.290E−02	5.914E−02	6.334E−02	8.223E−02	1.858E−01	7.453E−01	1.039E+01
N	7	14.007	6.305E−02	6.378E−02	6.502E−02	6.681E−02	7.224E−02	8.073E−02	8.645E−02	1.122E−01	2.532E−01	1.015E+00	1.414E+01
O	8	15.999	8.255E−02	8.350E−02	8.513E−02	8.746E−02	9.456E−02	1.056E−01	1.131E−01	1.467E−01	3.309E−01	1.326E+00	1.847E+01
F	9	18.998	1.047E−01	1.059E−01	1.080E−01	1.109E−01	1.199E−01	1.339E−01	1.434E−01	1.859E−01	4.190E−01	1.678E+00	2.338E+01
Ne	10	20.180	1.294E−01	1.309E−01	1.334E−01	1.370E−01	1.481E−01	1.655E−01	1.771E−01	2.296E−01	5.174E−01	2.072E+00	2.886E+01
Na	11	22.990	1.567E−01	1.585E−01	1.616E−01	1.660E−01	1.794E−01	2.004E−01	2.145E−01	2.780E−01	6.262E−01	2.507E+00	3.492E+01
Mg	12	24.305	1.866E−01	1.887E−01	1.924E−01	1.976E−01	2.136E−01	2.385E−01	2.553E−01	3.309E−01	7.453E−01	2.983E+00	4.156E+01
Al	13	26.982	2.191E−01	2.216E−01	2.259E−01	2.321E−01	2.508E−01	2.801E−01	2.998E−01	3.885E−01	8.749E−01	3.501E+00	4.878E+01
Si	14	28.086	2.542E−01	2.571E−01	2.621E−01	2.692E−01	2.909E−01	3.249E−01	3.478E−01	4.506E−01	1.015E+00	4.061E+00	5.657E+01
P	15	30.974	2.919E−01	2.953E−01	3.010E−01	3.092E−01	3.341E−01	3.731E−01	3.994E−01	5.174E−01	1.165E+00	4.662E+00	6.494E+01
S	16	32.066	3.322E−01	3.360E−01	3.425E−01	3.518E−01	3.802E−01	4.246E−01	4.544E−01	5.888E−01	1.326E+00	5.304E+00	7.388E+01
Cl	17	35.453	3.751E−01	3.794E−01	3.868E−01	3.973E−01	4.294E−01	4.794E−01	5.131E−01	6.648E−01	1.497E+00	5.988E+00	8.341E+01
Ar	18	39.948	4.207E−01	4.255E−01	4.338E−01	4.456E−01	4.815E−01	5.376E−01	5.754E−01	7.454E−01	1.678E+00	6.713E+00	9.351E+01
K	19	39.098	4.687E−01	4.741E−01	4.833E−01	4.964E−01	5.365E−01	5.990E−01	6.411E−01	8.305E−01	1.870E+00	7.480E+00	1.042E+02
Ca	20	40.078	5.194E−01	5.253E−01	5.355E−01	5.501E−01	5.944E−01	6.637E−01	7.104E−01	9.203E−01	2.072E+00	8.288E+00	1.154E+02
Sc	21	44.959	5.727E−01	5.793E−01	5.905E−01	6.066E−01	6.555E−01	7.319E−01	7.834E−01	1.015E+00	2.284E+00	9.138E+00	1.273E+02
Ti	22	47.880	6.287E−01	6.359E−01	6.482E−01	6.659E−01	7.195E−01	8.034E−01	8.598E−01	1.114E+00	2.507E+00	1.003E+01	1.397E+02
V	23	50.942	6.872E−01	6.951E−01	7.085E−01	7.278E−01	7.865E−01	8.781E−01	9.398E−01	1.217E+00	2.740E+00	1.096E+01	1.527E+02
Cr	24	51.996	7.483E−01	7.569E−01	7.715E−01	7.925E−01	8.564E−01	9.562E−01	1.023E+00	1.326E+00	2.983E+00	1.194E+01	1.662E+02
Mn	25	54.938	8.120E−01	8.213E−01	8.372E−01	8.600E−01	9.293E−01	1.038E+00	1.110E+00	1.438E+00	3.237E+00	1.295E+01	1.804E+02
Fe	26	55.847	8.783E−01	8.884E−01	9.055E−01	9.302E−01	1.005E+00	1.122E+00	1.201E+00	1.556E+00	3.501E+00	1.401E+01	1.951E+02
Co	27	58.933	9.472E−01	9.581E−01	9.766E−01	1.003E+00	1.084E+00	1.210E+00	1.295E+00	1.678E+00	3.776E+00	1.511E+01	2.104E+02
Ni	28	58.690	1.019E+00	1.030E+00	1.050E+00	1.079E+00	1.166E+00	1.302E+00	1.393E+00	1.804E+00	4.061E+00	1.625E+01	2.263E+02
Cu	29	63.546	1.093E+00	1.105E+00	1.127E+00	1.157E+00	1.251E+00	1.396E+00	1.494E+00	1.936E+00	4.356E+00	1.743E+01	2.427E+02
Zn	30	65.390	1.169E+00	1.183E+00	1.206E+00	1.239E+00	1.338E+00	1.494E+00	1.599E+00	2.072E+00	4.662E+00	1.865E+01	2.598E+02
Ga	31	69.723	1.249E+00	1.263E+00	1.288E+00	1.323E+00	1.429E+00	1.596E+00	1.708E+00	2.212E+00	4.978E+00	1.991E+01	2.774E+02
Ge	32	72.610	1.331E+00	1.346E+00	1.372E+00	1.409E+00	1.523E+00	1.700E+00	1.820E+00	2.357E+00	5.304E+00	2.122E+01	2.955E+02
As	33	74.922	1.415E+00	1.432E+00	1.459E+00	1.499E+00	1.620E+00	1.808E+00	1.935E+00	2.507E+00	5.641E+00	2.257E+01	3.143E+02
Se	34	78.960	1.502E+00	1.520E+00	1.549E+00	1.591E+00	1.719E+00	1.920E+00	2.054E+00	2.661E+00	5.988E+00	2.395E+01	3.336E+02
Br	35	79.904	1.592E+00	1.610E+00	1.641E+00	1.686E+00	1.822E+00	2.034E+00	2.177E+00	2.820E+00	6.345E+00	2.538E+01	3.536E+02
Kr	36	83.800	1.684E+00	1.704E+00	1.737E+00	1.784E+00	1.928E+00	2.152E+00	2.303E+00	2.983E+00	6.713E+00	2.685E+01	3.740E+02
Rb	37	85.468	1.779E+00	1.800E+00	1.834E+00	1.884E+00	2.036E+00	2.273E+00	2.433E+00	3.151E+00	7.091E+00	2.837E+01	3.951E+02
Sr	38	87.620	1.877E+00	1.898E+00	1.935E+00	1.988E+00	2.148E+00	2.398E+00	2.566E+00	3.324E+00	7.480E+00	2.992E+01	4.168E+02
Y	39	88.906	1.977E+00	2.000E+00	2.038E+00	2.094E+00	2.262E+00	2.526E+00	2.703E+00	3.501E+00	7.879E+00	3.152E+01	4.390E+02
Zr	40	91.224	2.080E+00	2.103E+00	2.144E+00	2.203E+00	2.380E+00	2.657E+00	2.844E+00	3.683E+00	8.288E+00	3.315E+01	4.618E+02
Nb	41	92.906	2.185E+00	2.210E+00	2.253E+00	2.314E+00	2.500E+00	2.792E+00	2.988E+00	3.870E+00	8.708E+00	3.483E+01	4.852E+02
Mo	42	95.940	2.293E+00	2.319E+00	2.364E+00	2.428E+00	2.624E+00	2.929E+00	3.135E+00	4.061E+00	9.138E+00	3.655E+01	5.091E+02
Tc	43	98.000	2.403E+00	2.431E+00	2.478E+00	2.545E+00	2.750E+00	3.071E+00	3.286E+00	4.257E+00	9.578E+00	3.831E+01	5.336E+02
Ru	44	101.070	2.516E+00	2.545E+00	2.594E+00	2.665E+00	2.880E+00	3.215E+00	3.441E+00	4.457E+00	1.003E+01	4.012E+01	5.588E+02
Rh	45	102.906	2.632E+00	2.662E+00	2.714E+00	2.788E+00	3.012E+00	3.363E+00	3.599E+00	4.662E+00	1.049E+01	4.196E+01	5.844E+02
Pd	46	106.420	2.750E+00	2.782E+00	2.836E+00	2.913E+00	3.148E+00	3.514E+00	3.761E+00	4.871E+00	1.096E+01	4.385E+01	6.107E+02

Table A6.1. Rutherford scattering cross sections (barns) of the elements for 1 MeV ^{1}H (protons) (continued).

Element	Atomic no. (Z_2)	Mass (M_2)	Scattering angle (degrees)										
			175	170	165	160	150	140	135	120	90	60	30
Ag	47	107.868	2.871E+00	2.904E+00	2.960E+00	3.041E+00	3.286E+00	3.669E+00	3.926E+00	5.086E+00	1.144E+01	4.577E+01	6.375E+02
Cd	48	122.411	2.995E+00	3.029E+00	3.088E+00	3.172E+00	3.427E+00	3.826E+00	4.095E+00	5.304E+00	1.194E+01	4.774E+01	6.650E+02
In	49	114.820	3.121E+00	3.157E+00	3.218E+00	3.305E+00	3.572E+00	3.987E+00	4.268E+00	5.528E+00	1.244E+01	4.975E+01	6.930E+02
Sn	50	118.710	3.250E+00	3.287E+00	3.350E+00	3.442E+00	3.719E+00	4.152E+00	4.444E+00	5.756E+00	1.295E+01	5.180E+01	7.215E+02
Sb	51	121.750	3.381E+00	3.420E+00	3.486E+00	3.581E+00	3.869E+00	4.320E+00	4.623E+00	5.988E+00	1.347E+01	5.390E+01	7.507E+02
Te	52	127.600	3.515E+00	3.555E+00	3.624E+00	3.723E+00	4.022E+00	4.491E+00	4.806E+00	6.225E+00	1.401E+01	5.603E+01	7.804E+02
I	53	126.904	3.651E+00	3.693E+00	3.765E+00	3.867E+00	4.179E+00	4.665E+00	4.993E+00	6.467E+00	1.455E+01	5.821E+01	8.107E+02
Xe	54	131.290	3.790E+00	3.834E+00	3.908E+00	4.015E+00	4.338E+00	4.843E+00	5.183E+00	6.713E+00	1.511E+01	6.042E+01	8.416E+02
Cs	55	132.905	3.932E+00	3.977E+00	4.054E+00	4.165E+00	4.500E+00	5.024E+00	5.377E+00	6.964E+00	1.567E+01	6.268E+01	8.731E+02
Ba	56	137.327	4.076E+00	4.123E+00	4.203E+00	4.317E+00	4.665E+00	5.208E+00	5.574E+00	7.220E+00	1.625E+01	6.498E+01	9.051E+02
La	57	138.906	4.223E+00	4.272E+00	4.354E+00	4.473E+00	4.833E+00	5.396E+00	5.775E+00	7.480E+00	1.683E+01	6.732E+01	9.377E+02
Ce	58	140.115	4.373E+00	4.423E+00	4.509E+00	4.631E+00	5.004E+00	5.587E+00	5.980E+00	7.745E+00	1.743E+01	6.971E+01	9.709E+02
Pr	59	140.908	4.525E+00	4.577E+00	4.665E+00	4.792E+00	5.178E+00	5.781E+00	6.187E+00	8.014E+00	1.803E+01	7.213E+01	1.005E+03
Nd	60	144.240	4.680E+00	4.734E+00	4.825E+00	4.956E+00	5.355E+00	5.979E+00	6.399E+00	8.288E+00	1.865E+01	7.460E+01	1.039E+03
Pm	61	145.000	4.837E+00	4.893E+00	4.987E+00	5.123E+00	5.535E+00	6.180E+00	6.614E+00	8.567E+00	1.928E+01	7.710E+01	1.074E+03
Sm	62	150.360	4.997E+00	5.054E+00	5.152E+00	5.292E+00	5.718E+00	6.384E+00	6.833E+00	8.850E+00	1.991E+01	7.965E+01	1.109E+03
Eu	63	151.965	5.159E+00	5.219E+00	5.320E+00	5.464E+00	5.904E+00	6.592E+00	7.055E+00	9.138E+00	2.056E+01	8.224E+01	1.146E+03
Gd	64	157.250	5.325E+00	5.386E+00	5.490E+00	5.639E+00	6.093E+00	6.803E+00	7.281E+00	9.430E+00	2.122E+01	8.488E+01	1.182E+03
Tb	65	158.925	5.492E+00	5.555E+00	5.663E+00	5.817E+00	6.285E+00	7.017E+00	7.510E+00	9.727E+00	2.189E+01	8.755E+01	1.219E+03
Dy	66	162.500	5.663E+00	5.728E+00	5.838E+00	5.997E+00	6.480E+00	7.235E+00	7.743E+00	1.003E+01	2.257E+01	9.026E+01	1.257E+03
Ho	67	164.930	5.835E+00	5.903E+00	6.017E+00	6.180E+00	6.678E+00	7.456E+00	7.979E+00	1.034E+01	2.325E+01	9.302E+01	1.296E+03
Er	68	167.260	6.011E+00	6.080E+00	6.197E+00	6.366E+00	6.879E+00	7.680E+00	8.219E+00	1.065E+01	2.395E+01	9.582E+01	1.335E+03
Tm	69	168.934	6.189E+00	6.260E+00	6.381E+00	6.555E+00	7.083E+00	7.907E+00	8.463E+00	1.096E+01	2.466E+01	9.866E+01	1.374E+03
Yb	70	173.040	6.370E+00	6.443E+00	6.567E+00	6.746E+00	7.290E+00	8.138E+00	8.710E+00	1.128E+01	2.538E+01	1.015E+02	1.414E+03
Lu	71	174.967	6.553E+00	6.628E+00	6.756E+00	6.940E+00	7.499E+00	8.372E+00	8.961E+00	1.161E+01	2.611E+01	1.045E+02	1.455E+03
Hf	72	178.490	6.739E+00	6.817E+00	6.948E+00	7.137E+00	7.712E+00	8.610E+00	9.215E+00	1.194E+01	2.685E+01	1.074E+02	1.496E+03
Ta	73	180.948	6.927E+00	7.007E+00	7.142E+00	7.337E+00	7.928E+00	8.851E+00	9.473E+00	1.227E+01	2.761E+01	1.104E+02	1.538E+03
W	74	183.850	7.119E+00	7.201E+00	7.340E+00	7.539E+00	8.146E+00	9.095E+00	9.734E+00	1.261E+01	2.837E+01	1.135E+02	1.580E+03
Re	75	186.207	7.312E+00	7.396E+00	7.539E+00	7.745E+00	8.368E+00	9.342E+00	9.999E+00	1.295E+01	2.914E+01	1.166E+02	1.623E+03
Os	76	190.200	7.509E+00	7.595E+00	7.742E+00	7.952E+00	8.593E+00	9.593E+00	1.027E+01	1.330E+01	2.992E+01	1.197E+02	1.667E+03
Ir	77	192.220	7.708E+00	7.796E+00	7.947E+00	8.163E+00	8.820E+00	9.847E+00	1.054E+01	1.365E+01	3.071E+01	1.229E+02	1.711E+03
Pt	78	195.080	7.909E+00	8.000E+00	8.154E+00	8.377E+00	9.051E+00	1.010E+01	1.081E+01	1.401E+01	3.152E+01	1.261E+02	1.756E+03
Au	79	196.967	8.113E+00	8.207E+00	8.365E+00	8.593E+00	9.285E+00	1.037E+01	1.109E+01	1.437E+01	3.233E+01	1.293E+02	1.801E+03
Hg	80	200.590	8.320E+00	8.416E+00	8.578E+00	8.812E+00	9.521E+00	1.063E+01	1.138E+01	1.473E+01	3.315E+01	1.326E+02	1.847E+03
Tl	81	204.383	8.529E+00	8.627E+00	8.794E+00	9.033E+00	9.761E+00	1.090E+01	1.166E+01	1.511E+01	3.399E+01	1.360E+02	1.894E+03
Pb	82	207.200	8.741E+00	8.842E+00	9.012E+00	9.258E+00	1.000E+01	1.117E+01	1.195E+01	1.548E+01	3.483E+01	1.393E+02	1.941E+03
Bi	83	208.980	8.956E+00	9.059E+00	9.233E+00	9.485E+00	1.025E+01	1.144E+01	1.225E+01	1.586E+01	3.569E+01	1.428E+02	1.988E+03
Po	84	209.000	9.173E+00	9.278E+00	9.457E+00	9.715E+00	1.050E+01	1.172E+01	1.254E+01	1.625E+01	3.655E+01	1.462E+02	2.036E+03
At	85	210.000	9.392E+00	9.500E+00	9.684E+00	9.948E+00	1.075E+01	1.200E+01	1.284E+01	1.663E+01	3.743E+01	1.497E+02	2.085E+03
Rn	86	222.000	9.615E+00	9.725E+00	9.913E+00	1.018E+01	1.100E+01	1.228E+01	1.315E+01	1.703E+01	3.831E+01	1.533E+02	2.135E+03
Fr	87	223.000	9.840E+00	9.953E+00	1.014E+01	1.042E+01	1.126E+01	1.257E+01	1.345E+01	1.743E+01	3.921E+01	1.568E+02	2.185E+03
Ra	88	226.025	1.007E+01	1.018E+01	1.038E+01	1.066E+01	1.152E+01	1.286E+01	1.377E+01	1.783E+01	4.012E+01	1.605E+02	2.235E+03
Ac	89	227.028	1.030E+01	1.042E+01	1.062E+01	1.091E+01	1.178E+01	1.316E+01	1.408E+01	1.824E+01	4.103E+01	1.641E+02	2.286E+03
Th	90	232.038	1.053E+01	1.065E+01	1.086E+01	1.115E+01	1.205E+01	1.345E+01	1.440E+01	1.865E+01	4.196E+01	1.678E+02	2.338E+03
Pa	91	231.036	1.077E+01	1.089E+01	1.110E+01	1.140E+01	1.232E+01	1.375E+01	1.472E+01	1.907E+01	4.290E+01	1.716E+02	2.390E+03
U	92	238.029	1.100E+01	1.113E+01	1.134E+01	1.165E+01	1.259E+01	1.406E+01	1.505E+01	1.949E+01	4.385E+01	1.754E+02	2.443E+03
Np	93	237.048	1.124E+01	1.137E+01	1.159E+01	1.191E+01	1.287E+01	1.437E+01	1.537E+01	1.991E+01	4.480E+01	1.792E+02	2.496E+03
Pu	94	244.000	1.149E+01	1.162E+01	1.184E+01	1.217E+01	1.315E+01	1.468E+01	1.571E+01	2.034E+01	4.577E+01	1.831E+02	2.550E+03

Table A6.2. Rutherford scattering cross sections (barns) of the elements for 1 MeV ^2H (deuterons).

Element	Atomic no. (Z_2)	Mass (M_2)	Scattering angle (degrees) 175	170	165	160	150	140	135	120	90	60	30
He	2	4.003	2.906E−03	2.956E−03	3.042E−03	3.167E−03	3.551E−03	4.165E−03	4.588E−03	6.550E−03	1.791E−02	8.023E−02	1.152E+00
Li	3	6.941	9.819E−03	9.951E−03	1.018E−02	1.050E−02	1.148E−02	1.302E−02	1.406E−02	1.875E−02	4.462E−02	1.845E−01	2.596E+00
Be	4	9.012	1.878E−02	1.902E−02	1.942E−02	2.000E−02	2.177E−02	2.453E−02	2.639E−02	3.476E−02	8.079E−02	3.295E−01	4.616E+00
B	5	10.811	3.029E−02	3.066E−02	3.130E−02	3.221E−02	3.497E−02	3.929E−02	4.220E−02	5.531E−02	1.272E−01	5.158E−01	7.213E+00
C	6	12.011	4.422E−02	4.475E−02	4.566E−02	4.698E−02	5.096E−02	5.719E−02	6.138E−02	8.026E−02	1.839E−01	7.434E−01	1.039E+01
N	7	14.007	6.110E−02	6.184E−02	6.308E−02	6.487E−02	7.029E−02	7.877E−02	8.449E−02	1.102E−01	2.512E−01	1.013E+00	1.414E+01
O	8	15.999	8.060E−02	8.155E−02	8.318E−02	8.551E−02	9.260E−02	1.037E−01	1.111E−01	1.447E−01	3.289E−01	1.324E+00	1.847E+01
F	9	18.998	1.030E−01	1.042E−01	1.062E−01	1.092E−01	1.182E−01	1.322E−01	1.416E−01	1.841E−01	4.172E−01	1.676E+00	2.338E+01
Ne	10	20.180	1.274E−01	1.289E−01	1.315E−01	1.351E−01	1.462E−01	1.635E−01	1.752E−01	2.277E−01	5.155E−01	2.070E+00	2.886E+01
Na	11	22.990	1.549E−01	1.567E−01	1.598E−01	1.642E−01	1.776E−01	1.986E−01	2.127E−01	2.762E−01	6.244E−01	2.505E+00	3.492E+01
Mg	12	24.305	1.847E−01	1.868E−01	1.905E−01	1.957E−01	2.117E−01	2.366E−01	2.534E−01	3.290E−01	7.434E−01	2.981E+00	4.156E+01
Al	13	26.982	2.173E−01	2.198E−01	2.241E−01	2.303E−01	2.490E−01	2.783E−01	2.980E−01	3.867E−01	8.730E−01	3.500E+00	4.877E+01
Si	14	28.086	2.522E−01	2.551E−01	2.601E−01	2.673E−01	2.890E−01	3.229E−01	3.458E−01	4.487E−01	1.013E+00	4.059E+00	5.657E+01
P	15	30.974	2.901E−01	2.934E−01	2.991E−01	3.073E−01	3.323E−01	3.713E−01	3.975E−01	5.156E−01	1.163E+00	4.660E+00	6.494E+01
S	16	32.066	3.302E−01	3.340E−01	3.405E−01	3.499E−01	3.782E−01	4.226E−01	4.525E−01	5.868E−01	1.324E+00	5.302E+00	7.388E+01
Cl	17	35.453	3.733E−01	3.776E−01	3.850E−01	3.955E−01	4.275E−01	4.776E−01	5.113E−01	6.630E−01	1.495E+00	5.986E+00	8.341E+01
Ar	18	39.948	4.191E−01	4.239E−01	4.322E−01	4.440E−01	4.799E−01	5.360E−01	5.738E−01	7.438E−01	1.676E+00	6.712E+00	9.351E+01
K	19	39.098	4.668E−01	4.722E−01	4.814E−01	4.946E−01	5.346E−01	5.971E−01	6.392E−01	8.287E−01	1.868E+00	7.478E+00	1.042E+02
Ca	20	40.078	5.174E−01	5.234E−01	5.335E−01	5.481E−01	5.925E−01	6.618E−01	7.084E−01	9.183E−01	2.070E+00	8.286E+00	1.154E+02
Sc	21	44.959	5.710E−01	5.776E−01	5.888E−01	6.049E−01	6.538E−01	7.302E−01	7.816E−01	1.013E+00	2.282E+00	9.136E+00	1.273E+02
Ti	22	47.880	6.270E−01	6.342E−01	6.465E−01	6.642E−01	7.179E−01	8.017E−01	8.582E−01	1.112E+00	2.505E+00	1.003E+01	1.397E+02
V	23	50.942	6.856E−01	6.935E−01	7.069E−01	7.262E−01	7.849E−01	8.765E−01	9.382E−01	1.216E+00	2.738E+00	1.096E+01	1.527E+02
Cr	24	51.996	7.466E−01	7.552E−01	7.698E−01	7.908E−01	8.547E−01	9.545E−01	1.022E+00	1.324E+00	2.982E+00	1.193E+01	1.662E+02
Mn	25	54.938	8.104E−01	8.197E−01	8.356E−01	8.584E−01	9.277E−01	1.036E+00	1.109E+00	1.437E+00	3.236E+00	1.295E+01	1.804E+02
Fe	26	55.847	8.766E−01	8.867E−01	9.038E−01	9.285E−01	1.003E+00	1.121E+00	1.199E+00	1.554E+00	3.500E+00	1.401E+01	1.951E+02
Co	27	58.933	9.455E−01	9.564E−01	9.749E−01	1.002E+00	1.082E+00	1.209E+00	1.294E+00	1.676E+00	3.774E+00	1.510E+01	2.104E+02
Ni	28	58.690	1.017E+00	1.029E+00	1.048E+00	1.077E+00	1.164E+00	1.300E+00	1.391E+00	1.803E+00	4.059E+00	1.624E+01	2.263E+02
Cu	29	63.546	1.091E+00	1.104E+00	1.125E+00	1.156E+00	1.249E+00	1.395E+00	1.493E+00	1.934E+00	4.355E+00	1.742E+01	2.427E+02
Zn	30	65.390	1.168E+00	1.181E+00	1.204E+00	1.237E+00	1.337E+00	1.493E+00	1.598E+00	2.070E+00	4.660E+00	1.865E+01	2.598E+02
Ga	31	69.723	1.247E+00	1.262E+00	1.286E+00	1.321E+00	1.428E+00	1.594E+00	1.706E+00	2.211E+00	4.976E+00	1.991E+01	2.774E+02
Ge	32	72.610	1.329E+00	1.345E+00	1.371E+00	1.408E+00	1.521E+00	1.699E+00	1.818E+00	2.356E+00	5.303E+00	2.122E+01	2.955E+02
As	33	74.922	1.414E+00	1.430E+00	1.458E+00	1.497E+00	1.618E+00	1.807E+00	1.934E+00	2.505E+00	5.639E+00	2.256E+01	3.143E+02
Se	34	78.960	1.501E+00	1.518E+00	1.548E+00	1.590E+00	1.718E+00	1.918E+00	2.053E+00	2.660E+00	5.987E+00	2.395E+01	3.336E+02
Br	35	79.904	1.591E+00	1.609E+00	1.640E+00	1.685E+00	1.820E+00	2.033E+00	2.176E+00	2.818E+00	6.344E+00	2.538E+01	3.536E+02
Kr	36	83.800	1.683E+00	1.702E+00	1.735E+00	1.783E+00	1.926E+00	2.151E+00	2.302E+00	2.982E+00	6.712E+00	2.685E+01	3.740E+02
Rb	37	85.468	1.778E+00	1.798E+00	1.833E+00	1.883E+00	2.035E+00	2.272E+00	2.432E+00	3.150E+00	7.090E+00	2.837E+01	3.951E+02
Sr	38	87.620	1.875E+00	1.897E+00	1.934E+00	1.986E+00	2.146E+00	2.396E+00	2.565E+00	3.323E+00	7.479E+00	2.992E+01	4.168E+02
Y	39	88.906	1.975E+00	1.998E+00	2.037E+00	2.092E+00	2.261E+00	2.524E+00	2.702E+00	3.500E+00	7.877E+00	3.152E+01	4.390E+02
Zr	40	91.224	2.078E+00	2.102E+00	2.143E+00	2.201E+00	2.378E+00	2.656E+00	2.842E+00	3.682E+00	8.287E+00	3.315E+01	4.618E+02
Nb	41	92.906	2.183E+00	2.208E+00	2.251E+00	2.312E+00	2.499E+00	2.790E+00	2.986E+00	3.868E+00	8.706E+00	3.483E+01	4.852E+02
Mo	42	95.940	2.291E+00	2.318E+00	2.362E+00	2.427E+00	2.622E+00	2.928E+00	3.134E+00	4.059E+00	9.136E+00	3.655E+01	5.091E+02
Tc	43	98.000	2.402E+00	2.429E+00	2.476E+00	2.544E+00	2.749E+00	3.069E+00	3.285E+00	4.255E+00	9.577E+00	3.831E+01	5.336E+02
Ru	44	101.070	2.515E+00	2.544E+00	2.593E+00	2.664E+00	2.878E+00	3.214E+00	3.440E+00	4.455E+00	1.003E+01	4.012E+01	5.588E+02
Rh	45	102.906	2.631E+00	2.661E+00	2.712E+00	2.786E+00	3.011E+00	3.361E+00	3.598E+00	4.660E+00	1.049E+01	4.196E+01	5.844E+02
Pd	46	106.420	2.749E+00	2.781E+00	2.834E+00	2.912E+00	3.146E+00	3.513E+00	3.760E+00	4.870E+00	1.096E+01	4.384E+01	6.107E+02

Table A6.2. Rutherford scattering cross sections (barns) of the elements for 1 MeV ^{2}H (deuterons) (continued).

Element	Atomic no. (Z_2)	Mass (M_2)	Scattering angle (degrees)										
			175	170	165	160	150	140	135	120	90	60	30
Ag	47	107.868	2.870E+00	2.903E+00	2.959E+00	3.040E+00	3.284E+00	3.667E+00	3.925E+00	5.084E+00	1.144E+01	4.577E+01	6.375E+02
Cd	48	122.411	2.994E+00	3.028E+00	3.087E+00	3.171E+00	3.426E+00	3.825E+00	4.094E+00	5.303E+00	1.193E+01	4.774E+01	6.650E+02
In	49	114.820	3.119E+00	3.155E+00	3.216E+00	3.304E+00	3.570E+00	3.986E+00	4.266E+00	5.526E+00	1.244E+01	4.975E+01	6.930E+02
Sn	50	118.710	3.248E+00	3.286E+00	3.349E+00	3.440E+00	3.717E+00	4.151E+00	4.442E+00	5.754E+00	1.295E+01	5.180E+01	7.215E+02
Sb	51	121.750	3.380E+00	3.418E+00	3.484E+00	3.579E+00	3.868E+00	4.318E+00	4.622E+00	5.987E+00	1.347E+01	5.390E+01	7.507E+02
Te	52	127.600	3.514E+00	3.554E+00	3.623E+00	3.721E+00	4.021E+00	4.490E+00	4.805E+00	6.224E+00	1.401E+01	5.603E+01	7.804E+02
I	53	126.904	3.650E+00	3.692E+00	3.763E+00	3.866E+00	4.177E+00	4.664E+00	4.992E+00	6.466E+00	1.455E+01	5.821E+01	8.107E+02
Xe	54	131.290	3.789E+00	3.833E+00	3.907E+00	4.013E+00	4.336E+00	4.842E+00	5.182E+00	6.712E+00	1.510E+01	6.042E+01	8.416E+02
Cs	55	132.905	3.931E+00	3.976E+00	4.053E+00	4.163E+00	4.499E+00	5.023E+00	5.376E+00	6.963E+00	1.567E+01	6.268E+01	8.731E+02
Ba	56	137.327	4.075E+00	4.122E+00	4.202E+00	4.316E+00	4.664E+00	5.207E+00	5.573E+00	7.219E+00	1.624E+01	6.498E+01	9.051E+02
La	57	138.906	4.222E+00	4.271E+00	4.353E+00	4.472E+00	4.832E+00	5.395E+00	5.774E+00	7.479E+00	1.683E+01	6.732E+01	9.377E+02
Ce	58	140.115	4.372E+00	4.422E+00	4.507E+00	4.630E+00	5.003E+00	5.586E+00	5.978E+00	7.743E+00	1.743E+01	6.971E+01	9.709E+02
Pr	59	140.908	4.524E+00	4.576E+00	4.664E+00	4.791E+00	5.177E+00	5.780E+00	6.186E+00	8.013E+00	1.803E+01	7.213E+01	1.005E+03
Nd	60	144.240	4.678E+00	4.732E+00	4.824E+00	4.955E+00	5.354E+00	5.978E+00	6.398E+00	8.287E+00	1.865E+01	7.460E+01	1.039E+03
Pm	61	145.000	4.836E+00	4.891E+00	4.986E+00	5.122E+00	5.534E+00	6.179E+00	6.613E+00	8.565E+00	1.927E+01	7.710E+01	1.074E+03
Sm	62	150.360	4.996E+00	5.053E+00	5.151E+00	5.291E+00	5.717E+00	6.383E+00	6.831E+00	8.849E+00	1.991E+01	7.965E+01	1.109E+03
Eu	63	151.965	5.158E+00	5.217E+00	5.318E+00	5.463E+00	5.903E+00	6.591E+00	7.054E+00	9.136E+00	2.056E+01	8.224E+01	1.146E+03
Gd	64	157.250	5.323E+00	5.385E+00	5.488E+00	5.638E+00	6.092E+00	6.802E+00	7.279E+00	9.429E+00	2.122E+01	8.487E+01	1.182E+03
Tb	65	158.925	5.491E+00	5.554E+00	5.661E+00	5.816E+00	6.284E+00	7.016E+00	7.509E+00	9.726E+00	2.189E+01	8.755E+01	1.219E+03
Dy	66	162.500	5.661E+00	5.726E+00	5.837E+00	5.996E+00	6.479E+00	7.233E+00	7.742E+00	1.003E+01	2.256E+01	9.026E+01	1.257E+03
Ho	67	164.930	5.834E+00	5.901E+00	6.015E+00	6.179E+00	6.677E+00	7.454E+00	7.978E+00	1.033E+01	2.325E+01	9.302E+01	1.296E+03
Er	68	167.260	6.010E+00	6.079E+00	6.196E+00	6.365E+00	6.878E+00	7.679E+00	8.218E+00	1.064E+01	2.395E+01	9.581E+01	1.335E+03
Tm	69	168.934	6.188E+00	6.259E+00	6.380E+00	6.554E+00	7.081E+00	7.906E+00	8.462E+00	1.096E+01	2.466E+01	9.865E+01	1.374E+03
Yb	70	173.040	6.368E+00	6.442E+00	6.566E+00	6.745E+00	7.288E+00	8.137E+00	8.709E+00	1.128E+01	2.538E+01	1.015E+02	1.414E+03
Lu	71	174.967	6.552E+00	6.627E+00	6.755E+00	6.939E+00	7.498E+00	8.371E+00	8.959E+00	1.160E+01	2.611E+01	1.045E+02	1.455E+03
Hf	72	178.490	6.738E+00	6.815E+00	6.947E+00	7.136E+00	7.711E+00	8.609E+00	9.214E+00	1.193E+01	2.685E+01	1.074E+02	1.496E+03
Ta	73	180.948	6.926E+00	7.006E+00	7.141E+00	7.336E+00	7.926E+00	8.850E+00	9.471E+00	1.227E+01	2.760E+01	1.104E+02	1.538E+03
W	74	183.850	7.117E+00	7.199E+00	7.338E+00	7.538E+00	8.145E+00	9.094E+00	9.733E+00	1.261E+01	2.837E+01	1.135E+02	1.580E+03
Re	75	186.207	7.311E+00	7.395E+00	7.538E+00	7.743E+00	8.367E+00	9.341E+00	9.997E+00	1.295E+01	2.914E+01	1.166E+02	1.623E+03
Os	76	190.200	7.507E+00	7.594E+00	7.740E+00	7.951E+00	8.592E+00	9.592E+00	1.027E+01	1.330E+01	2.992E+01	1.197E+02	1.667E+03
Ir	77	192.220	7.706E+00	7.795E+00	7.945E+00	8.162E+00	8.819E+00	9.846E+00	1.054E+01	1.365E+01	3.071E+01	1.229E+02	1.711E+03
Pt	78	195.080	7.908E+00	7.999E+00	8.153E+00	8.375E+00	9.050E+00	1.010E+01	1.081E+01	1.401E+01	3.152E+01	1.261E+02	1.756E+03
Au	79	196.967	8.112E+00	8.205E+00	8.364E+00	8.591E+00	9.283E+00	1.036E+01	1.109E+01	1.437E+01	3.233E+01	1.293E+02	1.801E+03
Hg	80	200.590	8.319E+00	8.414E+00	8.577E+00	8.810E+00	9.520E+00	1.063E+01	1.138E+01	1.473E+01	3.315E+01	1.326E+02	1.847E+03
Tl	81	204.383	8.528E+00	8.626E+00	8.793E+00	9.032E+00	9.759E+00	1.090E+01	1.166E+01	1.510E+01	3.399E+01	1.360E+02	1.894E+03
Pb	82	207.200	8.740E+00	8.840E+00	9.011E+00	9.257E+00	1.000E+01	1.117E+01	1.195E+01	1.548E+01	3.483E+01	1.393E+02	1.941E+03
Bi	83	208.980	8.954E+00	9.057E+00	9.232E+00	9.484E+00	1.025E+01	1.144E+01	1.224E+01	1.586E+01	3.569E+01	1.427E+02	1.988E+03
Po	84	209.000	9.171E+00	9.277E+00	9.456E+00	9.714E+00	1.050E+01	1.172E+01	1.254E+01	1.624E+01	3.655E+01	1.462E+02	2.036E+03
At	85	210.000	9.391E+00	9.499E+00	9.683E+00	9.946E+00	1.075E+01	1.200E+01	1.284E+01	1.663E+01	3.743E+01	1.497E+02	2.085E+03
Rn	86	222.000	9.614E+00	9.724E+00	9.912E+00	1.018E+01	1.100E+01	1.228E+01	1.315E+01	1.703E+01	3.831E+01	1.533E+02	2.135E+03
Fr	87	223.000	9.838E+00	9.952E+00	1.014E+01	1.042E+01	1.126E+01	1.257E+01	1.345E+01	1.743E+01	3.921E+01	1.568E+02	2.185E+03
Ra	88	226.025	1.007E+01	1.018E+01	1.038E+01	1.066E+01	1.152E+01	1.286E+01	1.376E+01	1.783E+01	4.012E+01	1.605E+02	2.235E+03
Ac	89	227.028	1.030E+01	1.041E+01	1.062E+01	1.090E+01	1.178E+01	1.315E+01	1.408E+01	1.824E+01	4.103E+01	1.641E+02	2.286E+03
Th	90	232.038	1.053E+01	1.065E+01	1.086E+01	1.115E+01	1.205E+01	1.345E+01	1.440E+01	1.865E+01	4.196E+01	1.678E+02	2.338E+03
Pa	91	231.036	1.076E+01	1.089E+01	1.110E+01	1.140E+01	1.232E+01	1.375E+01	1.472E+01	1.906E+01	4.290E+01	1.716E+02	2.390E+03
U	92	238.029	1.100E+01	1.113E+01	1.134E+01	1.165E+01	1.259E+01	1.406E+01	1.504E+01	1.949E+01	4.385E+01	1.754E+02	2.443E+03
Np	93	237.048	1.124E+01	1.137E+01	1.159E+01	1.191E+01	1.287E+01	1.436E+01	1.537E+01	1.991E+01	4.480E+01	1.792E+02	2.496E+03
Pu	94	244.000	1.149E+01	1.162E+01	1.184E+01	1.216E+01	1.314E+01	1.467E+01	1.571E+01	2.034E+01	4.577E+01	1.831E+02	2.550E+03

Table A6.3. Rutherford scattering cross sections (barns) of the elements for 1 MeV ^{3}He.

Element	Atomic no. (Z_2)	Mass (M_2)	Scattering angle (degrees)										
			175	170	165	160	150	140	135	120	90	60	30
Li	3	6.941	3.084E−02	3.133E−02	3.216E−02	3.336E−02	3.705E−02	4.289E−02	4.687E−02	6.509E−02	1.680E−01	7.282E−01	1.037E+01
Be	4	9.012	6.567E−02	6.659E−02	6.816E−02	7.042E−02	7.732E−02	8.816E−02	9.549E−02	1.287E−01	3.124E−01	1.307E+00	1.845E+01
B	5	10.811	1.106E−01	1.121E−01	1.146E−01	1.182E−01	1.291E−01	1.462E−01	1.578E−01	2.100E−01	4.975E−01	2.052E+00	2.884E+01
C	6	12.011	1.644E−01	1.666E−01	1.702E−01	1.754E−01	1.912E−01	2.160E−01	2.327E−01	3.079E−01	7.221E−01	2.960E+00	4.154E+01
N	7	14.007	2.318E−01	2.347E−01	2.397E−01	2.468E−01	2.684E−01	3.022E−01	3.250E−01	4.277E−01	9.915E−01	4.038E+00	5.655E+01
O	8	15.999	3.097E−01	3.135E−01	3.200E−01	3.293E−01	3.576E−01	4.018E−01	4.316E−01	5.658E−01	1.302E+00	5.281E+00	7.386E+01
F	9	18.998	4.003E−01	4.052E−01	4.134E−01	4.252E−01	4.610E−01	5.171E−01	5.549E−01	7.248E−01	1.657E+00	6.693E+00	9.349E+01
Ne	10	20.180	4.971E−01	5.031E−01	5.132E−01	5.278E−01	5.721E−01	6.413E−01	6.880E−01	8.978E−01	2.049E+00	8.265E+00	1.154E+02
Na	11	22.990	6.078E−01	6.151E−01	6.273E−01	6.450E−01	6.986E−01	7.824E−01	8.388E−01	1.093E+00	2.486E+00	1.001E+01	1.397E+02
Mg	12	24.305	7.260E−01	7.346E−01	7.492E−01	7.703E−01	8.341E−01	9.338E−01	1.001E+00	1.303E+00	2.961E+00	1.191E+01	1.662E+02
Al	13	26.982	8.571E−01	8.672E−01	8.843E−01	9.090E−01	9.839E−01	1.101E+00	1.180E+00	1.535E+00	3.480E+00	1.399E+01	1.951E+02
Si	14	28.086	9.959E−01	1.008E+00	1.028E+00	1.056E+00	1.143E+00	1.279E+00	1.370E+00	1.782E+00	4.038E+00	1.622E+01	2.263E+02
P	15	30.974	1.148E+00	1.161E+00	1.184E+00	1.217E+00	1.317E+00	1.473E+00	1.578E+00	2.050E+00	4.640E+00	1.863E+01	2.597E+02
S	16	32.066	1.308E+00	1.323E+00	1.349E+00	1.387E+00	1.500E+00	1.677E+00	1.797E+00	2.334E+00	5.281E+00	2.120E+01	2.955E+02
Cl	17	35.453	1.481E+00	1.499E+00	1.528E+00	1.570E+00	1.698E+00	1.898E+00	2.033E+00	2.640E+00	5.967E+00	2.393E+01	3.336E+02
Ar	18	39.948	1.666E+00	1.685E+00	1.718E+00	1.765E+00	1.909E+00	2.133E+00	2.285E+00	2.965E+00	6.695E+00	2.684E+01	3.740E+02
K	19	39.098	1.855E+00	1.877E+00	1.913E+00	1.966E+00	2.126E+00	2.376E+00	2.545E+00	3.302E+00	7.458E+00	2.990E+01	4.167E+02
Ca	20	40.078	2.057E+00	2.081E+00	2.121E+00	2.180E+00	2.357E+00	2.634E+00	2.821E+00	3.660E+00	8.265E+00	3.313E+01	4.618E+02
Sc	21	44.959	2.273E+00	2.299E+00	2.344E+00	2.408E+00	2.604E+00	2.909E+00	3.115E+00	4.041E+00	9.118E+00	3.653E+01	5.091E+02
Ti	22	47.880	2.497E+00	2.526E+00	2.575E+00	2.646E+00	2.860E+00	3.196E+00	3.422E+00	4.438E+00	1.001E+01	4.010E+01	5.587E+02
V	23	50.942	2.732E+00	2.763E+00	2.817E+00	2.894E+00	3.129E+00	3.495E+00	3.742E+00	4.853E+00	1.094E+01	4.383E+01	6.107E+02
Cr	24	51.996	2.975E+00	3.010E+00	3.068E+00	3.152E+00	3.408E+00	3.807E+00	4.076E+00	5.285E+00	1.192E+01	4.772E+01	6.649E+02
Mn	25	54.938	3.231E+00	3.268E+00	3.331E+00	3.423E+00	3.700E+00	4.133E+00	4.425E+00	5.736E+00	1.293E+01	5.178E+01	7.215E+02
Fe	26	55.847	3.495E+00	3.535E+00	3.604E+00	3.703E+00	4.002E+00	4.471E+00	4.786E+00	6.205E+00	1.399E+01	5.601E+01	7.804E+02
Co	27	58.933	3.771E+00	3.815E+00	3.889E+00	3.995E+00	4.318E+00	4.824E+00	5.164E+00	6.694E+00	1.509E+01	6.040E+01	8.416E+02
Ni	28	58.690	4.056E+00	4.102E+00	4.182E+00	4.296E+00	4.644E+00	5.187E+00	5.553E+00	7.199E+00	1.622E+01	6.496E+01	9.051E+02
Cu	29	63.546	4.354E+00	4.404E+00	4.489E+00	4.612E+00	4.985E+00	5.568E+00	5.960E+00	7.726E+00	1.741E+01	6.969E+01	9.709E+02
Zn	30	65.390	4.660E+00	4.714E+00	4.806E+00	4.937E+00	5.336E+00	5.960E+00	6.380E+00	8.269E+00	1.863E+01	7.458E+01	1.039E+03
Ga	31	69.723	4.979E+00	5.036E+00	5.134E+00	5.274E+00	5.700E+00	6.366E+00	6.815E+00	8.832E+00	1.989E+01	7.964E+01	1.109E+03
Ge	32	72.610	5.307E+00	5.368E+00	5.472E+00	5.621E+00	6.076E+00	6.785E+00	7.263E+00	9.412E+00	2.120E+01	8.486E+01	1.182E+03
As	33	74.922	5.645E+00	5.710E+00	5.820E+00	5.979E+00	6.462E+00	7.217E+00	7.725E+00	1.001E+01	2.255E+01	9.025E+01	1.257E+03
Se	34	78.960	5.994E+00	6.063E+00	6.180E+00	6.349E+00	6.862E+00	7.663E+00	8.202E+00	1.063E+01	2.394E+01	9.580E+01	1.335E+03
Br	35	79.904	6.352E+00	6.425E+00	6.550E+00	6.729E+00	7.272E+00	8.121E+00	8.692E+00	1.126E+01	2.537E+01	1.015E+02	1.414E+03
Kr	36	83.800	6.722E+00	6.800E+00	6.931E+00	7.120E+00	7.695E+00	8.593E+00	9.198E+00	1.192E+01	2.684E+01	1.074E+02	1.496E+03
Rb	37	85.468	7.101E+00	7.183E+00	7.322E+00	7.522E+00	8.129E+00	9.078E+00	9.717E+00	1.259E+01	2.835E+01	1.135E+02	1.580E+03
Sr	38	87.620	7.491E+00	7.578E+00	7.724E+00	7.935E+00	8.575E+00	9.576E+00	1.025E+01	1.328E+01	2.990E+01	1.197E+02	1.667E+03
Y	39	88.906	7.891E+00	7.982E+00	8.137E+00	8.359E+00	9.033E+00	1.009E+01	1.080E+01	1.399E+01	3.150E+01	1.261E+02	1.756E+03
Zr	40	91.224	8.302E+00	8.398E+00	8.560E+00	8.794E+00	9.503E+00	1.061E+01	1.136E+01	1.472E+01	3.314E+01	1.326E+02	1.847E+03
Nb	41	92.906	8.723E+00	8.824E+00	8.994E+00	9.240E+00	9.985E+00	1.115E+01	1.193E+01	1.546E+01	3.481E+01	1.393E+02	1.941E+03
Mo	42	95.940	9.155E+00	9.261E+00	9.440E+00	9.697E+00	1.048E+01	1.170E+01	1.252E+01	1.623E+01	3.653E+01	1.462E+02	2.036E+03
Tc	43	98.000	9.597E+00	9.708E+00	9.895E+00	1.017E+01	1.099E+01	1.227E+01	1.313E+01	1.701E+01	3.830E+01	1.532E+02	2.135E+03
Ru	44	101.070	1.005E+01	1.017E+01	1.036E+01	1.064E+01	1.150E+01	1.284E+01	1.375E+01	1.781E+01	4.010E+01	1.605E+02	2.235E+03
Rh	45	102.906	1.051E+01	1.063E+01	1.084E+01	1.113E+01	1.203E+01	1.344E+01	1.438E+01	1.863E+01	4.194E+01	1.678E+02	2.338E+03
Pd	46	106.420	1.099E+01	1.111E+01	1.133E+01	1.164E+01	1.257E+01	1.404E+01	1.503E+01	1.947E+01	4.383E+01	1.754E+02	2.443E+03
Ag	47	107.868	1.147E+01	1.160E+01	1.183E+01	1.215E+01	1.313E+01	1.466E+01	1.569E+01	2.033E+01	4.576E+01	1.831E+02	2.550E+03
Cd	48	122.411	1.197E+01	1.210E+01	1.234E+01	1.267E+01	1.370E+01	1.529E+01	1.637E+01	2.120E+01	4.773E+01	1.910E+02	2.660E+03

Table A6.3. Rutherford scattering cross sections (barns) of the elements for 1 MeV ^{3}He (continued).

Element	Atomic no. (Z_2)	Mass (M_2)	Scattering angle (degrees) 175	170	165	160	150	140	135	120	90	60	30
In	49	114.820	1.247E+01	1.261E+01	1.286E+01	1.321E+01	1.427E+01	1.593E+01	1.706E+01	2.210E+01	4.974E+01	1.990E+02	2.772E+03
Sn	50	118.710	1.298E+01	1.313E+01	1.339E+01	1.375E+01	1.486E+01	1.659E+01	1.776E+01	2.301E+01	5.179E+01	2.072E+02	2.886E+03
Sb	51	121.750	1.351E+01	1.366E+01	1.393E+01	1.431E+01	1.546E+01	1.726E+01	1.848E+01	2.394E+01	5.388E+01	2.156E+02	3.003E+03
Te	52	127.600	1.405E+01	1.421E+01	1.448E+01	1.488E+01	1.608E+01	1.795E+01	1.921E+01	2.489E+01	5.602E+01	2.241E+02	3.122E+03
I	53	126.904	1.459E+01	1.476E+01	1.504E+01	1.545E+01	1.670E+01	1.865E+01	1.996E+01	2.585E+01	5.819E+01	2.328E+02	3.243E+03
Xe	54	131.290	1.515E+01	1.532E+01	1.562E+01	1.604E+01	1.734E+01	1.936E+01	2.072E+01	2.684E+01	6.041E+01	2.417E+02	3.366E+03
Cs	55	132.905	1.571E+01	1.590E+01	1.620E+01	1.664E+01	1.799E+01	2.008E+01	2.149E+01	2.784E+01	6.267E+01	2.507E+02	3.492E+03
Ba	56	137.327	1.629E+01	1.648E+01	1.680E+01	1.726E+01	1.865E+01	2.082E+01	2.228E+01	2.887E+01	6.497E+01	2.599E+02	3.620E+03
La	57	138.906	1.688E+01	1.707E+01	1.740E+01	1.788E+01	1.932E+01	2.157E+01	2.309E+01	2.991E+01	6.731E+01	2.693E+02	3.751E+03
Ce	58	140.115	1.748E+01	1.768E+01	1.802E+01	1.851E+01	2.000E+01	2.233E+01	2.390E+01	3.096E+01	6.969E+01	2.788E+02	3.884E+03
Pr	59	140.908	1.809E+01	1.829E+01	1.865E+01	1.916E+01	2.070E+01	2.311E+01	2.474E+01	3.204E+01	7.212E+01	2.885E+02	4.019E+03
Nd	60	144.240	1.870E+01	1.892E+01	1.929E+01	1.981E+01	2.141E+01	2.390E+01	2.558E+01	3.314E+01	7.458E+01	2.984E+02	4.156E+03
Pm	61	145.000	1.933E+01	1.956E+01	1.993E+01	2.048E+01	2.213E+01	2.471E+01	2.644E+01	3.425E+01	7.709E+01	3.084E+02	4.296E+03
Sm	62	150.360	1.997E+01	2.020E+01	2.059E+01	2.115E+01	2.286E+01	2.552E+01	2.732E+01	3.539E+01	7.964E+01	3.186E+02	4.438E+03
Eu	63	151.965	2.062E+01	2.086E+01	2.126E+01	2.184E+01	2.360E+01	2.635E+01	2.821E+01	3.654E+01	8.223E+01	3.290E+02	4.582E+03
Gd	64	157.250	2.128E+01	2.153E+01	2.195E+01	2.254E+01	2.436E+01	2.720E+01	2.911E+01	3.771E+01	8.486E+01	3.395E+02	4.729E+03
Tb	65	158.925	2.195E+01	2.221E+01	2.264E+01	2.325E+01	2.513E+01	2.805E+01	3.003E+01	3.889E+01	8.753E+01	3.502E+02	4.878E+03
Dy	66	162.500	2.264E+01	2.290E+01	2.334E+01	2.398E+01	2.591E+01	2.893E+01	3.096E+01	4.010E+01	9.025E+01	3.610E+02	5.029E+03
Ho	67	164.930	2.333E+01	2.360E+01	2.405E+01	2.471E+01	2.670E+01	2.981E+01	3.190E+01	4.133E+01	9.300E+01	3.721E+02	5.182E+03
Er	68	167.260	2.403E+01	2.431E+01	2.478E+01	2.545E+01	2.750E+01	3.071E+01	3.286E+01	4.257E+01	9.580E+01	3.833E+02	5.338E+03
Tm	69	168.934	2.474E+01	2.503E+01	2.551E+01	2.621E+01	2.832E+01	3.162E+01	3.384E+01	4.383E+01	9.864E+01	3.946E+02	5.496E+03
Yb	70	173.040	2.547E+01	2.576E+01	2.626E+01	2.697E+01	2.914E+01	3.254E+01	3.483E+01	4.511E+01	1.015E+02	4.061E+02	5.657E+03
Lu	71	174.967	2.620E+01	2.650E+01	2.701E+01	2.775E+01	2.998E+01	3.348E+01	3.583E+01	4.641E+01	1.044E+02	4.178E+02	5.820E+03
Hf	72	178.490	2.694E+01	2.725E+01	2.778E+01	2.854E+01	3.083E+01	3.443E+01	3.685E+01	4.773E+01	1.074E+02	4.297E+02	5.985E+03
Ta	73	180.948	2.770E+01	2.802E+01	2.856E+01	2.933E+01	3.170E+01	3.539E+01	3.788E+01	4.906E+01	1.104E+02	4.417E+02	6.152E+03
W	74	183.850	2.846E+01	2.879E+01	2.934E+01	3.014E+01	3.257E+01	3.637E+01	3.892E+01	5.042E+01	1.135E+02	4.539E+02	6.322E+03
Re	75	186.207	2.924E+01	2.957E+01	3.014E+01	3.096E+01	3.346E+01	3.736E+01	3.998E+01	5.179E+01	1.165E+02	4.662E+02	6.494E+03
Os	76	190.200	3.002E+01	3.037E+01	3.095E+01	3.180E+01	3.436E+01	3.836E+01	4.106E+01	5.318E+01	1.197E+02	4.787E+02	6.668E+03
Ir	77	192.220	3.082E+01	3.117E+01	3.177E+01	3.264E+01	3.527E+01	3.938E+01	4.214E+01	5.459E+01	1.228E+02	4.914E+02	6.845E+03
Pt	78	195.080	3.162E+01	3.199E+01	3.260E+01	3.349E+01	3.619E+01	4.041E+01	4.325E+01	5.602E+01	1.261E+02	5.043E+02	7.024E+03
Au	79	196.967	3.244E+01	3.281E+01	3.345E+01	3.436E+01	3.712E+01	4.145E+01	4.436E+01	5.746E+01	1.293E+02	5.173E+02	7.205E+03
Hg	80	200.590	3.327E+01	3.365E+01	3.430E+01	3.523E+01	3.807E+01	4.251E+01	4.549E+01	5.893E+01	1.326E+02	5.305E+02	7.389E+03
Tl	81	204.383	3.410E+01	3.450E+01	3.516E+01	3.612E+01	3.903E+01	4.358E+01	4.664E+01	6.041E+01	1.359E+02	5.438E+02	7.574E+03
Pb	82	207.200	3.495E+01	3.535E+01	3.604E+01	3.702E+01	4.000E+01	4.466E+01	4.780E+01	6.191E+01	1.393E+02	5.573E+02	7.763E+03
Bi	83	208.980	3.581E+01	3.622E+01	3.692E+01	3.793E+01	4.098E+01	4.575E+01	4.897E+01	6.343E+01	1.427E+02	5.710E+02	7.953E+03
Po	84	209.000	3.668E+01	3.710E+01	3.782E+01	3.885E+01	4.197E+01	4.686E+01	5.016E+01	6.497E+01	1.462E+02	5.848E+02	8.146E+03
At	85	210.000	3.756E+01	3.799E+01	3.872E+01	3.978E+01	4.298E+01	4.799E+01	5.136E+01	6.652E+01	1.497E+02	5.988E+02	8.341E+03
Rn	86	222.000	3.845E+01	3.889E+01	3.964E+01	4.072E+01	4.400E+01	4.912E+01	5.258E+01	6.810E+01	1.532E+02	6.130E+02	8.538E+03
Fr	87	223.000	3.935E+01	3.980E+01	4.057E+01	4.167E+01	4.503E+01	5.027E+01	5.381E+01	6.969E+01	1.568E+02	6.274E+02	8.738E+03
Ra	88	226.025	4.026E+01	4.072E+01	4.151E+01	4.264E+01	4.607E+01	5.144E+01	5.505E+01	7.130E+01	1.605E+02	6.419E+02	8.940E+03
Ac	89	227.028	4.118E+01	4.165E+01	4.245E+01	4.361E+01	4.712E+01	5.261E+01	5.631E+01	7.293E+01	1.641E+02	6.565E+02	9.144E+03
Th	90	232.038	4.211E+01	4.259E+01	4.341E+01	4.460E+01	4.819E+01	5.380E+01	5.758E+01	7.458E+01	1.678E+02	6.714E+02	9.351E+03
Pa	91	231.036	4.305E+01	4.354E+01	4.438E+01	4.559E+01	4.927E+01	5.500E+01	5.887E+01	7.625E+01	1.716E+02	6.864E+02	9.560E+03
U	92	238.029	4.400E+01	4.451E+01	4.537E+01	4.660E+01	5.035E+01	5.622E+01	6.017E+01	7.794E+01	1.754E+02	7.015E+02	9.771E+03
Np	93	237.048	4.496E+01	4.548E+01	4.636E+01	4.762E+01	5.146E+01	5.745E+01	6.148E+01	7.964E+01	1.792E+02	7.169E+02	9.985E+03
Pu	94	244.000	4.593E+01	4.646E+01	4.736E+01	4.865E+01	5.257E+01	5.869E+01	6.281E+01	8.136E+01	1.831E+02	7.324E+02	1.020E+04

Table A6.4. Rutherford scattering cross sections (barns) of the elements for 1 MeV ^{4}He.

Element	Atomic no. (Z_2)	Mass (M_2)	Scattering angle (degrees)										
			175	170	165	160	150	140	135	120	90	60	30
Li	3	6.941	2.090E−02	2.130E−02	2.199E−02	2.299E−02	2.609E−02	3.114E−02	3.466E−02	5.137E−02	1.524E−01	7.145E−01	1.036E+01
Be	4	9.012	5.370E−02	5.456E−02	5.603E−02	5.815E−02	6.465E−02	7.497E−02	8.201E−02	1.143E−01	2.971E−01	1.293E+00	1.844E+01
B	5	10.811	9.691E−02	9.833E−02	1.007E−01	1.042E−01	1.148E−01	1.316E−01	1.429E−01	1.945E−01	4.812E−01	2.036E+00	2.883E+01
C	6	12.011	1.481E−01	1.501E−01	1.537E−01	1.588E−01	1.743E−01	1.987E−01	2.152E−01	2.899E−01	7.033E−01	2.942E+00	4.152E+01
N	7	14.007	2.150E−01	2.179E−01	2.228E−01	2.298E−01	2.512E−01	2.848E−01	3.074E−01	4.096E−01	9.730E−01	4.020E+00	5.653E+01
O	8	15.999	2.926E−01	2.964E−01	3.028E−01	3.121E−01	3.402E−01	3.842E−01	4.139E−01	5.478E−01	1.284E+00	5.263E+00	7.384E+01
F	9	18.998	3.848E−01	3.896E−01	3.978E−01	4.095E−01	4.453E−01	5.012E−01	5.389E−01	7.086E−01	1.641E+00	6.676E+00	9.347E+01
Ne	10	20.180	4.800E−01	4.860E−01	4.961E−01	5.107E−01	5.548E−01	6.239E−01	6.705E−01	8.801E−01	2.031E+00	8.248E+00	1.154E+02
Na	11	22.990	5.918E−01	5.990E−01	6.113E−01	6.289E−01	6.824E−01	7.661E−01	8.225E−01	1.076E+00	2.469E+00	9.991E+00	1.397E+02
Mg	12	24.305	7.089E−01	7.175E−01	7.321E−01	7.531E−01	8.168E−01	9.165E−01	9.836E−01	1.286E+00	2.943E+00	1.190E+01	1.662E+02
Al	13	26.982	8.407E−01	8.508E−01	8.679E−01	8.926E−01	9.674E−01	1.084E+00	1.163E+00	1.518E+00	3.463E+00	1.397E+01	1.951E+02
Si	14	28.086	9.784E−01	9.901E−01	1.010E+00	1.039E+00	1.125E+00	1.261E+00	1.353E+00	1.764E+00	4.020E+00	1.620E+01	2.262E+02
P	15	30.974	1.131E+00	1.145E+00	1.168E+00	1.201E+00	1.300E+00	1.456E+00	1.561E+00	2.033E+00	4.623E+00	1.861E+01	2.597E+02
S	16	32.066	1.290E+00	1.306E+00	1.332E+00	1.369E+00	1.482E+00	1.660E+00	1.779E+00	2.316E+00	5.263E+00	2.118E+01	2.955E+02
Cl	17	35.453	1.465E+00	1.482E+00	1.512E+00	1.554E+00	1.682E+00	1.882E+00	2.017E+00	2.623E+00	5.950E+00	2.392E+01	3.336E+02
Ar	18	39.948	1.651E+00	1.671E+00	1.704E+00	1.751E+00	1.895E+00	2.119E+00	2.270E+00	2.950E+00	6.680E+00	2.682E+01	3.740E+02
K	19	39.098	1.838E+00	1.860E+00	1.897E+00	1.949E+00	2.109E+00	2.359E+00	2.528E+00	3.285E+00	7.441E+00	2.988E+01	4.167E+02
Ca	20	40.078	2.039E+00	2.063E+00	2.103E+00	2.162E+00	2.339E+00	2.616E+00	2.803E+00	3.642E+00	8.247E+00	3.311E+01	4.617E+02
Sc	21	44.959	2.257E+00	2.284E+00	2.328E+00	2.393E+00	2.588E+00	2.894E+00	3.100E+00	4.025E+00	9.102E+00	3.652E+01	5.091E+02
Ti	22	47.880	2.482E+00	2.511E+00	2.560E+00	2.631E+00	2.845E+00	3.181E+00	3.406E+00	4.422E+00	9.994E+00	4.008E+01	5.587E+02
V	23	50.942	2.717E+00	2.749E+00	2.803E+00	2.880E+00	3.114E+00	3.481E+00	3.728E+00	4.838E+00	1.093E+01	4.381E+01	6.107E+02
Cr	24	51.996	2.960E+00	2.995E+00	3.053E+00	3.137E+00	3.392E+00	3.792E+00	4.060E+00	5.269E+00	1.190E+01	4.771E+01	6.649E+02
Mn	25	54.938	3.216E+00	3.253E+00	3.317E+00	3.408E+00	3.685E+00	4.118E+00	4.410E+00	5.722E+00	1.292E+01	5.177E+01	7.215E+02
Fe	26	55.847	3.479E+00	3.520E+00	3.589E+00	3.687E+00	3.987E+00	4.455E+00	4.771E+00	6.190E+00	1.397E+01	5.600E+01	7.804E+02
Co	27	58.933	3.756E+00	3.800E+00	3.874E+00	3.980E+00	4.303E+00	4.809E+00	5.149E+00	6.679E+00	1.507E+01	6.039E+01	8.416E+02
Ni	28	58.690	4.039E+00	4.086E+00	4.166E+00	4.280E+00	4.628E+00	5.171E+00	5.537E+00	7.183E+00	1.621E+01	6.495E+01	9.051E+02
Cu	29	63.546	4.339E+00	4.389E+00	4.475E+00	4.597E+00	4.970E+00	5.553E+00	5.945E+00	7.711E+00	1.739E+01	6.967E+01	9.709E+02
Zn	30	65.390	4.645E+00	4.699E+00	4.791E+00	4.922E+00	5.321E+00	5.945E+00	6.365E+00	8.254E+00	1.861E+01	7.456E+01	1.039E+03
Ga	31	69.723	4.965E+00	5.022E+00	5.120E+00	5.260E+00	5.686E+00	6.352E+00	6.800E+00	8.818E+00	1.988E+01	7.962E+01	1.109E+03
Ge	32	72.610	5.293E+00	5.354E+00	5.458E+00	5.608E+00	6.062E+00	6.771E+00	7.249E+00	9.398E+00	2.119E+01	8.484E+01	1.182E+03
As	33	74.922	5.631E+00	5.696E+00	5.807E+00	5.966E+00	6.448E+00	7.203E+00	7.711E+00	9.997E+00	2.253E+01	9.023E+01	1.257E+03
Se	34	78.960	5.981E+00	6.050E+00	6.167E+00	6.336E+00	6.849E+00	7.650E+00	8.189E+00	1.062E+01	2.392E+01	9.579E+01	1.335E+03
Br	35	79.904	6.338E+00	6.412E+00	6.536E+00	6.715E+00	7.258E+00	8.107E+00	8.679E+00	1.125E+01	2.535E+01	1.015E+02	1.414E+03
Kr	36	83.800	6.709E+00	6.786E+00	6.918E+00	7.107E+00	7.682E+00	8.580E+00	9.185E+00	1.191E+01	2.682E+01	1.074E+02	1.496E+03
Rb	37	85.468	7.088E+00	7.170E+00	7.309E+00	7.509E+00	8.116E+00	9.064E+00	9.703E+00	1.258E+01	2.834E+01	1.134E+02	1.580E+03
Sr	38	87.620	7.478E+00	7.564E+00	7.711E+00	7.922E+00	8.562E+00	9.563E+00	1.024E+01	1.327E+01	2.989E+01	1.197E+02	1.667E+03
Y	39	88.906	7.878E+00	7.969E+00	8.123E+00	8.345E+00	9.020E+00	1.007E+01	1.078E+01	1.398E+01	3.149E+01	1.260E+02	1.756E+03
Zr	40	91.224	8.288E+00	8.384E+00	8.547E+00	8.780E+00	9.490E+00	1.060E+01	1.134E+01	1.470E+01	3.312E+01	1.326E+02	1.847E+03
Nb	41	92.906	8.709E+00	8.810E+00	8.980E+00	9.226E+00	9.971E+00	1.114E+01	1.192E+01	1.545E+01	3.480E+01	1.393E+02	1.941E+03
Mo	42	95.940	9.141E+00	9.247E+00	9.426E+00	9.684E+00	1.047E+01	1.169E+01	1.251E+01	1.621E+01	3.652E+01	1.462E+02	2.036E+03
Tc	43	98.000	9.583E+00	9.694E+00	9.882E+00	1.015E+01	1.097E+01	1.225E+01	1.312E+01	1.700E+01	3.828E+01	1.532E+02	2.135E+03
Ru	44	101.070	1.004E+01	1.015E+01	1.035E+01	1.063E+01	1.149E+01	1.283E+01	1.373E+01	1.780E+01	4.009E+01	1.604E+02	2.235E+03
Rh	45	102.906	1.050E+01	1.062E+01	1.083E+01	1.112E+01	1.202E+01	1.342E+01	1.437E+01	1.862E+01	4.193E+01	1.678E+02	2.338E+03
Pd	46	106.420	1.097E+01	1.110E+01	1.131E+01	1.162E+01	1.256E+01	1.403E+01	1.501E+01	1.946E+01	4.382E+01	1.754E+02	2.443E+03
Ag	47	107.868	1.146E+01	1.159E+01	1.181E+01	1.213E+01	1.311E+01	1.464E+01	1.568E+01	2.031E+01	4.574E+01	1.831E+02	2.550E+03
Cd	48	122.411	1.196E+01	1.209E+01	1.233E+01	1.266E+01	1.369E+01	1.528E+01	1.636E+01	2.119E+01	4.772E+01	1.909E+02	2.660E+03

Table A6.4. Rutherford scattering cross sections (barns) of the elements for 1 MeV ^{4}He (continued).

Element	Atomic no. (Z_2)	Mass (M_2)	Scattering angle (degrees)										
			175	170	165	160	150	140	135	120	90	60	30
In	49	114.820	1.246E+01	1.260E+01	1.284E+01	1.319E+01	1.426E+01	1.592E+01	1.704E+01	2.208E+01	4.972E+01	1.990E+02	2.772E+03
Sn	50	118.710	1.297E+01	1.312E+01	1.337E+01	1.374E+01	1.485E+01	1.658E+01	1.775E+01	2.299E+01	5.177E+01	2.072E+02	2.886E+03
Sb	51	121.750	1.350E+01	1.365E+01	1.392E+01	1.430E+01	1.545E+01	1.725E+01	1.847E+01	2.393E+01	5.387E+01	2.156E+02	3.003E+03
Te	52	127.600	1.403E+01	1.420E+01	1.447E+01	1.486E+01	1.606E+01	1.794E+01	1.920E+01	2.488E+01	5.600E+01	2.241E+02	3.122E+03
I	53	126.904	1.458E+01	1.475E+01	1.503E+01	1.544E+01	1.669E+01	1.863E+01	1.994E+01	2.584E+01	5.818E+01	2.328E+02	3.243E+03
Xe	54	131.290	1.514E+01	1.531E+01	1.561E+01	1.603E+01	1.732E+01	1.935E+01	2.071E+01	2.683E+01	6.040E+01	2.417E+02	3.366E+03
Cs	55	132.905	1.570E+01	1.588E+01	1.619E+01	1.663E+01	1.797E+01	2.007E+01	2.148E+01	2.783E+01	6.265E+01	2.507E+02	3.492E+03
Ba	56	137.327	1.628E+01	1.647E+01	1.679E+01	1.724E+01	1.863E+01	2.081E+01	2.227E+01	2.885E+01	6.496E+01	2.599E+02	3.620E+03
La	57	138.906	1.687E+01	1.706E+01	1.739E+01	1.787E+01	1.931E+01	2.156E+01	2.307E+01	2.989E+01	6.730E+01	2.693E+02	3.751E+03
Ce	58	140.115	1.746E+01	1.767E+01	1.801E+01	1.850E+01	1.999E+01	2.232E+01	2.389E+01	3.095E+01	6.968E+01	2.788E+02	3.884E+03
Pr	59	140.908	1.807E+01	1.828E+01	1.863E+01	1.914E+01	2.069E+01	2.310E+01	2.472E+01	3.203E+01	7.210E+01	2.885E+02	4.019E+03
Nd	60	144.240	1.869E+01	1.891E+01	1.927E+01	1.980E+01	2.139E+01	2.389E+01	2.557E+01	3.313E+01	7.457E+01	2.984E+02	4.156E+03
Pm	61	145.000	1.932E+01	1.954E+01	1.992E+01	2.046E+01	2.211E+01	2.469E+01	2.643E+01	3.424E+01	7.708E+01	3.084E+02	4.296E+03
Sm	62	150.360	1.996E+01	2.019E+01	2.058E+01	2.114E+01	2.285E+01	2.551E+01	2.730E+01	3.537E+01	7.963E+01	3.186E+02	4.438E+03
Eu	63	151.965	2.061E+01	2.085E+01	2.125E+01	2.183E+01	2.359E+01	2.634E+01	2.819E+01	3.652E+01	8.222E+01	3.289E+02	4.582E+03
Gd	64	157.250	2.127E+01	2.152E+01	2.193E+01	2.253E+01	2.435E+01	2.719E+01	2.910E+01	3.770E+01	8.485E+01	3.395E+02	4.729E+03
Tb	65	158.925	2.194E+01	2.220E+01	2.262E+01	2.324E+01	2.512E+01	2.804E+01	3.001E+01	3.888E+01	8.752E+01	3.502E+02	4.878E+03
Dy	66	162.500	2.262E+01	2.289E+01	2.333E+01	2.396E+01	2.590E+01	2.891E+01	3.095E+01	4.009E+01	9.024E+01	3.610E+02	5.029E+03
Ho	67	164.930	2.332E+01	2.358E+01	2.404E+01	2.470E+01	2.669E+01	2.980E+01	3.189E+01	4.131E+01	9.299E+01	3.720E+02	5.182E+03
Er	68	167.260	2.402E+01	2.429E+01	2.476E+01	2.544E+01	2.749E+01	3.069E+01	3.285E+01	4.256E+01	9.579E+01	3.832E+02	5.338E+03
Tm	69	168.934	2.473E+01	2.502E+01	2.550E+01	2.619E+01	2.830E+01	3.160E+01	3.383E+01	4.382E+01	9.863E+01	3.946E+02	5.496E+03
Yb	70	173.040	2.545E+01	2.575E+01	2.624E+01	2.696E+01	2.913E+01	3.253E+01	3.481E+01	4.510E+01	1.015E+02	4.061E+02	5.657E+03
Lu	71	174.967	2.619E+01	2.649E+01	2.700E+01	2.774E+01	2.997E+01	3.346E+01	3.582E+01	4.640E+01	1.044E+02	4.178E+02	5.820E+03
Hf	72	178.490	2.693E+01	2.724E+01	2.777E+01	2.852E+01	3.082E+01	3.441E+01	3.683E+01	4.772E+01	1.074E+02	4.297E+02	5.985E+03
Ta	73	180.948	2.768E+01	2.800E+01	2.854E+01	2.932E+01	3.169E+01	3.538E+01	3.786E+01	4.905E+01	1.104E+02	4.417E+02	6.152E+03
W	74	183.850	2.845E+01	2.878E+01	2.933E+01	3.013E+01	3.256E+01	3.635E+01	3.891E+01	5.040E+01	1.134E+02	4.539E+02	6.322E+03
Re	75	186.207	2.922E+01	2.956E+01	3.013E+01	3.095E+01	3.345E+01	3.734E+01	3.997E+01	5.178E+01	1.165E+02	4.662E+02	6.494E+03
Os	76	190.200	3.001E+01	3.036E+01	3.094E+01	3.179E+01	3.435E+01	3.835E+01	4.104E+01	5.317E+01	1.197E+02	4.787E+02	6.668E+03
Ir	77	192.220	3.081E+01	3.116E+01	3.176E+01	3.263E+01	3.526E+01	3.936E+01	4.213E+01	5.458E+01	1.228E+02	4.914E+02	6.845E+03
Pt	78	195.080	3.161E+01	3.198E+01	3.259E+01	3.348E+01	3.618E+01	4.039E+01	4.323E+01	5.600E+01	1.260E+02	5.043E+02	7.024E+03
Au	79	196.967	3.243E+01	3.280E+01	3.343E+01	3.435E+01	3.711E+01	4.144E+01	4.435E+01	5.745E+01	1.293E+02	5.173E+02	7.205E+03
Hg	80	200.590	3.325E+01	3.364E+01	3.429E+01	3.522E+01	3.806E+01	4.249E+01	4.548E+01	5.892E+01	1.326E+02	5.304E+02	7.389E+03
Tl	81	204.383	3.409E+01	3.448E+01	3.515E+01	3.611E+01	3.902E+01	4.356E+01	4.663E+01	6.040E+01	1.359E+02	5.438E+02	7.574E+03
Pb	82	207.200	3.494E+01	3.534E+01	3.602E+01	3.701E+01	3.999E+01	4.465E+01	4.779E+01	6.190E+01	1.393E+02	5.573E+02	7.763E+03
Bi	83	208.980	3.580E+01	3.621E+01	3.691E+01	3.792E+01	4.097E+01	4.574E+01	4.896E+01	6.342E+01	1.427E+02	5.710E+02	7.953E+03
Po	84	209.000	3.667E+01	3.709E+01	3.780E+01	3.883E+01	4.196E+01	4.685E+01	5.015E+01	6.496E+01	1.462E+02	5.848E+02	8.146E+03
At	85	210.000	3.754E+01	3.798E+01	3.871E+01	3.976E+01	4.297E+01	4.797E+01	5.135E+01	6.651E+01	1.497E+02	5.988E+02	8.341E+03
Rn	86	222.000	3.844E+01	3.888E+01	3.963E+01	4.071E+01	4.399E+01	4.911E+01	5.256E+01	6.809E+01	1.532E+02	6.130E+02	8.538E+03
Fr	87	223.000	3.933E+01	3.979E+01	4.056E+01	4.166E+01	4.502E+01	5.026E+01	5.379E+01	6.968E+01	1.568E+02	6.273E+02	8.738E+03
Ra	88	226.025	4.024E+01	4.071E+01	4.149E+01	4.263E+01	4.606E+01	5.142E+01	5.504E+01	7.129E+01	1.604E+02	6.418E+02	8.940E+03
Ac	89	227.028	4.117E+01	4.164E+01	4.244E+01	4.360E+01	4.711E+01	5.260E+01	5.630E+01	7.292E+01	1.641E+02	6.565E+02	9.144E+03
Th	90	232.038	4.210E+01	4.258E+01	4.340E+01	4.459E+01	4.818E+01	5.379E+01	5.757E+01	7.457E+01	1.678E+02	6.714E+02	9.351E+03
Pa	91	231.036	4.304E+01	4.353E+01	4.437E+01	4.558E+01	4.925E+01	5.499E+01	5.886E+01	7.624E+01	1.716E+02	6.864E+02	9.560E+03
U	92	238.029	4.399E+01	4.450E+01	4.536E+01	4.659E+01	5.034E+01	5.621E+01	6.016E+01	7.793E+01	1.754E+02	7.015E+02	9.771E+03
Np	93	237.048	4.495E+01	4.547E+01	4.635E+01	4.761E+01	5.144E+01	5.744E+01	6.147E+01	7.963E+01	1.792E+02	7.169E+02	9.985E+03
Pu	94	244.000	4.592E+01	4.645E+01	4.735E+01	4.864E+01	5.256E+01	5.868E+01	6.280E+01	8.135E+01	1.831E+02	7.324E+02	1.020E+04

Table A6.5. Rutherford scattering cross sections (barns) of the elements for 1 MeV ^{7}Li.

Element	Atomic no. (Z_2)	Mass (M_2)	Scattering angle (degrees) 175	170	165	160	150	140	135	120	90	60	30
Be	4	9.012	2.919E−02	2.993E−02	3.123E−02	3.314E−02	3.935E−02	5.021E−02	5.832E−02	1.022E−01	4.682E−01	2.755E+00	4.139E+01
B	5	10.811	9.832E−02	1.004E−01	1.040E−01	1.092E−01	1.258E−01	1.532E−01	1.726E−01	2.681E−01	8.868E−01	4.412E+00	6.473E+01
C	6	12.011	1.833E−01	1.868E−01	1.929E−01	2.018E−01	2.293E−01	2.743E−01	3.056E−01	4.548E−01	1.362E+00	6.423E+00	9.326E+01
N	7	14.007	3.223E−01	3.279E−01	3.374E−01	3.512E−01	3.936E−01	4.616E−01	5.083E−01	7.249E−01	1.977E+00	8.848E+00	1.270E+02
O	8	15.999	4.892E−01	4.970E−01	5.103E−01	5.295E−01	5.882E−01	6.815E−01	7.450E−01	1.036E+00	2.682E+00	1.165E+01	1.660E+02
F	9	18.998	7.076E−01	7.180E−01	7.356E−01	7.610E−01	8.384E−01	9.605E−01	1.043E+00	1.419E+00	3.510E+00	1.485E+01	2.102E+02
Ne	10	20.180	9.051E−01	9.180E−01	9.400E−01	9.717E−01	1.068E+00	1.220E+00	1.323E+00	1.788E+00	4.371E+00	1.837E+01	2.595E+02
Na	11	22.990	1.165E+00	1.181E+00	1.208E+00	1.247E+00	1.365E+00	1.551E+00	1.676E+00	2.243E+00	5.372E+00	2.230E+01	3.140E+02
Mg	12	24.305	1.417E+00	1.436E+00	1.468E+00	1.514E+00	1.656E+00	1.878E+00	2.027E+00	2.703E+00	6.428E+00	2.657E+01	3.738E+02
Al	13	26.982	1.720E+00	1.742E+00	1.780E+00	1.835E+00	2.002E+00	2.263E+00	2.440E+00	3.234E+00	7.608E+00	3.125E+01	4.387E+02
Si	14	28.086	2.017E+00	2.043E+00	2.087E+00	2.151E+00	2.345E+00	2.648E+00	2.853E+00	3.775E+00	8.849E+00	3.627E+01	5.088E+02
P	15	30.974	2.370E+00	2.400E+00	2.451E+00	2.525E+00	2.748E+00	3.097E+00	3.332E+00	4.392E+00	1.022E+01	4.169E+01	5.842E+02
S	16	32.066	2.716E+00	2.751E+00	2.809E+00	2.892E+00	3.146E+00	3.544E+00	3.812E+00	5.018E+00	1.165E+01	4.746E+01	6.647E+02
Cl	17	35.453	3.123E+00	3.161E+00	3.227E+00	3.322E+00	3.609E+00	4.058E+00	4.361E+00	5.724E+00	1.321E+01	5.363E+01	7.504E+02
Ar	18	39.948	3.562E+00	3.605E+00	3.679E+00	3.785E+00	4.108E+00	4.612E+00	4.952E+00	6.480E+00	1.487E+01	6.019E+01	8.414E+02
K	19	39.098	3.957E+00	4.006E+00	4.088E+00	4.206E+00	4.565E+00	5.127E+00	5.506E+00	7.209E+00	1.656E+01	6.705E+01	9.374E+02
Ca	20	40.078	4.399E+00	4.452E+00	4.544E+00	4.675E+00	5.073E+00	5.695E+00	6.115E+00	8.002E+00	1.836E+01	7.431E+01	1.039E+03
Sc	21	44.959	4.913E+00	4.972E+00	5.072E+00	5.217E+00	5.656E+00	6.343E+00	6.806E+00	8.887E+00	2.031E+01	8.199E+01	1.145E+03
Ti	22	47.880	5.423E+00	5.488E+00	5.599E+00	5.758E+00	6.240E+00	6.994E+00	7.502E+00	9.787E+00	2.232E+01	9.002E+01	1.257E+03
V	23	50.942	5.958E+00	6.029E+00	6.150E+00	6.323E+00	6.851E+00	7.675E+00	8.230E+00	1.073E+01	2.443E+01	9.842E+01	1.374E+03
Cr	24	51.996	6.497E+00	6.575E+00	6.706E+00	6.895E+00	7.469E+00	8.367E+00	8.971E+00	1.169E+01	2.661E+01	1.072E+02	1.496E+03
Mn	25	54.938	7.077E+00	7.161E+00	7.304E+00	7.509E+00	8.132E+00	9.106E+00	9.762E+00	1.271E+01	2.890E+01	1.163E+02	1.623E+03
Fe	26	55.847	7.663E+00	7.754E+00	7.908E+00	8.130E+00	8.804E+00	9.857E+00	1.057E+01	1.376E+01	3.127E+01	1.258E+02	1.756E+03
Co	27	58.933	8.290E+00	8.389E+00	8.555E+00	8.794E+00	9.521E+00	1.066E+01	1.142E+01	1.486E+01	3.375E+01	1.357E+02	1.893E+03
Ni	28	58.690	8.914E+00	9.019E+00	9.198E+00	9.456E+00	1.024E+01	1.146E+01	1.228E+01	1.598E+01	3.629E+01	1.460E+02	2.036E+03
Cu	29	63.546	9.602E+00	9.716E+00	9.908E+00	1.018E+01	1.102E+01	1.233E+01	1.322E+01	1.719E+01	3.897E+01	1.566E+02	2.184E+03
Zn	30	65.390	1.029E+01	1.041E+01	1.062E+01	1.091E+01	1.181E+01	1.321E+01	1.416E+01	1.841E+01	4.172E+01	1.676E+02	2.338E+03
Ga	31	69.723	1.102E+01	1.115E+01	1.137E+01	1.168E+01	1.264E+01	1.414E+01	1.515E+01	1.969E+01	4.458E+01	1.790E+02	2.496E+03
Ge	32	72.610	1.176E+01	1.190E+01	1.213E+01	1.247E+01	1.349E+01	1.508E+01	1.616E+01	2.100E+01	4.752E+01	1.907E+02	2.660E+03
As	33	74.922	1.252E+01	1.267E+01	1.292E+01	1.327E+01	1.436E+01	1.606E+01	1.720E+01	2.234E+01	5.055E+01	2.029E+02	2.828E+03
Se	34	78.960	1.331E+01	1.347E+01	1.373E+01	1.411E+01	1.527E+01	1.707E+01	1.828E+01	2.374E+01	5.368E+01	2.154E+02	3.003E+03
Br	35	79.904	1.411E+01	1.428E+01	1.456E+01	1.496E+01	1.618E+01	1.809E+01	1.938E+01	2.516E+01	5.689E+01	2.282E+02	3.182E+03
Kr	36	83.800	1.495E+01	1.513E+01	1.542E+01	1.585E+01	1.714E+01	1.916E+01	2.052E+01	2.664E+01	6.021E+01	2.415E+02	3.366E+03
Rb	37	85.468	1.580E+01	1.599E+01	1.630E+01	1.675E+01	1.812E+01	2.025E+01	2.169E+01	2.815E+01	6.361E+01	2.551E+02	3.556E+03
Sr	38	87.620	1.668E+01	1.687E+01	1.720E+01	1.768E+01	1.912E+01	2.137E+01	2.289E+01	2.971E+01	6.711E+01	2.691E+02	3.751E+03
Y	39	88.906	1.758E+01	1.778E+01	1.813E+01	1.863E+01	2.015E+01	2.252E+01	2.411E+01	3.130E+01	7.069E+01	2.834E+02	3.951E+03
Zr	40	91.224	1.850E+01	1.872E+01	1.908E+01	1.961E+01	2.120E+01	2.370E+01	2.538E+01	3.293E+01	7.438E+01	2.982E+02	4.156E+03
Nb	41	92.906	1.945E+01	1.967E+01	2.006E+01	2.061E+01	2.229E+01	2.491E+01	2.667E+01	3.461E+01	7.815E+01	3.133E+02	4.366E+03
Mo	42	95.940	2.042E+01	2.066E+01	2.106E+01	2.164E+01	2.340E+01	2.615E+01	2.800E+01	3.633E+01	8.202E+01	3.288E+02	4.582E+03
Tc	43	98.000	2.141E+01	2.166E+01	2.208E+01	2.269E+01	2.454E+01	2.742E+01	2.936E+01	3.809E+01	8.599E+01	3.446E+02	4.803E+03
Ru	44	101.070	2.243E+01	2.270E+01	2.314E+01	2.377E+01	2.571E+01	2.872E+01	3.076E+01	3.990E+01	9.005E+01	3.608E+02	5.029E+03
Rh	45	102.906	2.347E+01	2.375E+01	2.421E+01	2.487E+01	2.690E+01	3.005E+01	3.218E+01	4.174E+01	9.419E+01	3.774E+02	5.260E+03
Pd	46	106.420	2.454E+01	2.483E+01	2.531E+01	2.601E+01	2.812E+01	3.142E+01	3.364E+01	4.363E+01	9.844E+01	3.944E+02	5.496E+03
Ag	47	107.868	2.563E+01	2.593E+01	2.643E+01	2.716E+01	2.936E+01	3.280E+01	3.512E+01	4.556E+01	1.028E+02	4.117E+02	5.738E+03
Cd	48	122.411	2.678E+01	2.709E+01	2.762E+01	2.837E+01	3.067E+01	3.427E+01	3.668E+01	4.757E+01	1.072E+02	4.295E+02	5.985E+03

Table A6.5. Rutherford scattering cross sections (barns) of the elements for 1 MeV ^{7}Li (continued).

Element	Atomic no. (Z_2)	Mass (M_2)	Scattering angle (degrees)										
			175	170	165	160	150	140	135	120	90	60	30
In	49	114.820	2.788E+01	2.821E+01	2.876E+01	2.954E+01	3.194E+01	3.568E+01	3.820E+01	4.954E+01	1.117E+02	4.476E+02	6.236E+03
Sn	50	118.710	2.905E+01	2.938E+01	2.996E+01	3.078E+01	3.327E+01	3.717E+01	3.979E+01	5.160E+01	1.164E+02	4.660E+02	6.494E+03
Sb	51	121.750	3.023E+01	3.058E+01	3.118E+01	3.203E+01	3.463E+01	3.868E+01	4.141E+01	5.370E+01	1.211E+02	4.849E+02	6.756E+03
Te	52	127.600	3.145E+01	3.181E+01	3.243E+01	3.332E+01	3.602E+01	4.023E+01	4.307E+01	5.584E+01	1.259E+02	5.041E+02	7.024E+03
I	53	126.904	3.267E+01	3.304E+01	3.369E+01	3.461E+01	3.741E+01	4.179E+01	4.474E+01	5.801E+01	1.308E+02	5.237E+02	7.296E+03
Xe	54	131.290	3.392E+01	3.432E+01	3.498E+01	3.594E+01	3.885E+01	4.340E+01	4.646E+01	6.023E+01	1.358E+02	5.436E+02	7.574E+03
Cs	55	132.905	3.520E+01	3.560E+01	3.630E+01	3.729E+01	4.031E+01	4.502E+01	4.820E+01	6.249E+01	1.408E+02	5.639E+02	7.857E+03
Ba	56	137.327	3.650E+01	3.692E+01	3.764E+01	3.867E+01	4.180E+01	4.669E+01	4.998E+01	6.479E+01	1.460E+02	5.847E+02	8.146E+03
La	57	138.906	3.782E+01	3.826E+01	3.900E+01	4.007E+01	4.331E+01	4.838E+01	5.179E+01	6.713E+01	1.513E+02	6.057E+02	8.439E+03
Ce	58	140.115	3.916E+01	3.962E+01	4.039E+01	4.149E+01	4.485E+01	5.009E+01	5.362E+01	6.951E+01	1.566E+02	6.272E+02	8.738E+03
Pr	59	140.908	4.053E+01	4.100E+01	4.179E+01	4.294E+01	4.641E+01	5.184E+01	5.549E+01	7.193E+01	1.621E+02	6.490E+02	9.042E+03
Nd	60	144.240	4.192E+01	4.241E+01	4.323E+01	4.441E+01	4.800E+01	5.362E+01	5.740E+01	7.440E+01	1.676E+02	6.712E+02	9.351E+03
Pm	61	145.000	4.333E+01	4.384E+01	4.469E+01	4.591E+01	4.962E+01	5.542E+01	5.933E+01	7.690E+01	1.733E+02	6.937E+02	9.665E+03
Sm	62	150.360	4.478E+01	4.530E+01	4.618E+01	4.744E+01	5.128E+01	5.727E+01	6.130E+01	7.946E+01	1.790E+02	7.167E+02	9.985E+03
Eu	63	151.965	4.624E+01	4.678E+01	4.768E+01	4.899E+01	5.295E+01	5.913E+01	6.330E+01	8.205E+01	1.849E+02	7.400E+02	1.031E+04
Gd	64	157.250	4.773E+01	4.829E+01	4.922E+01	5.057E+01	5.465E+01	6.104E+01	6.534E+01	8.469E+01	1.908E+02	7.637E+02	1.064E+04
Tb	65	158.925	4.924E+01	4.981E+01	5.078E+01	5.216E+01	5.638E+01	6.297E+01	6.740E+01	8.736E+01	1.968E+02	7.877E+02	1.097E+04
Dy	66	162.500	5.078E+01	5.136E+01	5.236E+01	5.379E+01	5.814E+01	6.493E+01	6.950E+01	9.007E+01	2.029E+02	8.122E+02	1.131E+04
Ho	67	164.930	5.233E+01	5.294E+01	5.396E+01	5.544E+01	5.992E+01	6.692E+01	7.163E+01	9.283E+01	2.091E+02	8.370E+02	1.166E+04
Er	68	167.260	5.391E+01	5.454E+01	5.559E+01	5.711E+01	6.172E+01	6.893E+01	7.379E+01	9.563E+01	2.154E+02	8.622E+02	1.201E+04
Tm	69	168.934	5.551E+01	5.616E+01	5.724E+01	5.881E+01	6.356E+01	7.098E+01	7.598E+01	9.846E+01	2.218E+02	8.877E+02	1.237E+04
Yb	70	173.040	5.714E+01	5.780E+01	5.892E+01	6.053E+01	6.542E+01	7.306E+01	7.821E+01	1.013E+02	2.283E+02	9.136E+02	1.273E+04
Lu	71	174.967	5.879E+01	5.947E+01	6.062E+01	6.228E+01	6.731E+01	7.517E+01	8.046E+01	1.043E+02	2.348E+02	9.399E+02	1.309E+04
Hf	72	178.490	6.047E+01	6.117E+01	6.235E+01	6.405E+01	6.923E+01	7.731E+01	8.275E+01	1.072E+02	2.415E+02	9.666E+02	1.347E+04
Ta	73	180.948	6.216E+01	6.288E+01	6.410E+01	6.585E+01	7.117E+01	7.947E+01	8.507E+01	1.102E+02	2.483E+02	9.936E+02	1.384E+04
W	74	183.850	6.389E+01	6.462E+01	6.587E+01	6.767E+01	7.314E+01	8.167E+01	8.742E+01	1.133E+02	2.551E+02	1.021E+03	1.422E+04
Re	75	186.207	6.563E+01	6.639E+01	6.767E+01	6.952E+01	7.513E+01	8.390E+01	8.981E+01	1.164E+02	2.621E+02	1.049E+03	1.461E+04
Os	76	190.200	6.740E+01	6.818E+01	6.950E+01	7.139E+01	7.716E+01	8.616E+01	9.223E+01	1.195E+02	2.691E+02	1.077E+03	1.500E+04
Ir	77	192.220	6.919E+01	6.999E+01	7.134E+01	7.329E+01	7.920E+01	8.845E+01	9.467E+01	1.227E+02	2.762E+02	1.106E+03	1.540E+04
Pt	78	195.080	7.100E+01	7.182E+01	7.321E+01	7.521E+01	8.128E+01	9.076E+01	9.715E+01	1.259E+02	2.835E+02	1.134E+03	1.580E+04
Au	79	196.967	7.284E+01	7.368E+01	7.510E+01	7.715E+01	8.338E+01	9.311E+01	9.966E+01	1.291E+02	2.908E+02	1.164E+03	1.621E+04
Hg	80	200.590	7.470E+01	7.556E+01	7.702E+01	7.913E+01	8.551E+01	9.549E+01	1.022E+02	1.324E+02	2.982E+02	1.193E+03	1.662E+04
Tl	81	204.383	7.659E+01	7.747E+01	7.897E+01	8.112E+01	8.767E+01	9.790E+01	1.048E+02	1.358E+02	3.057E+02	1.223E+03	1.704E+04
Pb	82	207.200	7.849E+01	7.940E+01	8.094E+01	8.314E+01	8.985E+01	1.003E+02	1.074E+02	1.392E+02	3.133E+02	1.254E+03	1.747E+04
Bi	83	208.980	8.042E+01	8.135E+01	8.292E+01	8.519E+01	9.206E+01	1.028E+02	1.100E+02	1.426E+02	3.210E+02	1.285E+03	1.789E+04
Po	84	209.000	8.237E+01	8.332E+01	8.493E+01	8.725E+01	9.429E+01	1.053E+02	1.127E+02	1.460E+02	3.288E+02	1.316E+03	1.833E+04
At	85	210.000	8.435E+01	8.532E+01	8.697E+01	8.934E+01	9.655E+01	1.078E+02	1.154E+02	1.495E+02	3.367E+02	1.347E+03	1.877E+04
Rn	86	222.000	8.636E+01	8.736E+01	8.905E+01	9.148E+01	9.886E+01	1.104E+02	1.182E+02	1.531E+02	3.447E+02	1.379E+03	1.921E+04
Fr	87	223.000	8.839E+01	8.940E+01	9.113E+01	9.362E+01	1.012E+02	1.130E+02	1.209E+02	1.567E+02	3.527E+02	1.411E+03	1.966E+04
Ra	88	226.025	9.043E+01	9.148E+01	9.325E+01	9.579E+01	1.035E+02	1.156E+02	1.237E+02	1.603E+02	3.609E+02	1.444E+03	2.012E+04
Ac	89	227.028	9.250E+01	9.357E+01	9.538E+01	9.798E+01	1.059E+02	1.182E+02	1.265E+02	1.640E+02	3.691E+02	1.477E+03	2.057E+04
Th	90	232.038	9.460E+01	9.569E+01	9.754E+01	1.002E+02	1.083E+02	1.209E+02	1.294E+02	1.677E+02	3.775E+02	1.510E+03	2.104E+04
Pa	91	231.036	9.671E+01	9.783E+01	9.972E+01	1.024E+02	1.107E+02	1.236E+02	1.323E+02	1.714E+02	3.859E+02	1.544E+03	2.151E+04
U	92	238.029	9.886E+01	1.000E+02	1.019E+02	1.047E+02	1.132E+02	1.264E+02	1.352E+02	1.752E+02	3.945E+02	1.578E+03	2.199E+04
Np	93	237.048	1.010E+02	1.022E+02	1.042E+02	1.070E+02	1.156E+02	1.291E+02	1.382E+02	1.790E+02	4.031E+02	1.613E+03	2.247E+04
Pu	94	244.000	1.032E+02	1.044E+02	1.064E+02	1.093E+02	1.181E+02	1.319E+02	1.412E+02	1.829E+02	4.118E+02	1.648E+03	2.295E+04

Table A6.6. Rutherford scattering cross sections (barns) of the elements for 1 MeV ^{12}C.

Element	Atomic no. (Z_2)	Mass (M_2)	Scattering angle (degrees)										
			175	170	165	160	150	140	135	120	90	60	30
N	7	14.007	1.632E–01	1.679E–01	1.760E–01	1.882E–01	2.285E–01	3.023E–01	3.599E–01	7.040E–01	4.713E+00	3.320E+01	5.067E+02
O	8	15.999	5.757E–01	5.898E–01	6.143E–01	6.504E–01	7.664E–01	9.668E–01	1.114E+00	1.891E+00	7.894E+00	4.433E+01	6.623E+02
F	9	18.998	1.374E+00	1.402E+00	1.451E+00	1.522E+00	1.746E+00	2.116E+00	2.377E+00	3.646E+00	1.171E+01	5.736E+01	8.390E+02
Ne	10	20.180	1.961E+00	1.999E+00	2.065E+00	2.162E+00	2.462E+00	2.952E+00	3.295E+00	4.935E+00	1.499E+01	7.124E+01	1.036E+03
Na	11	22.990	3.004E+00	3.057E+00	3.148E+00	3.281E+00	3.689E+00	4.346E+00	4.799E+00	6.911E+00	1.925E+01	8.714E+01	1.254E+03
Mg	12	24.305	3.861E+00	3.928E+00	4.040E+00	4.204E+00	4.706E+00	5.511E+00	6.063E+00	8.619E+00	2.335E+01	1.041E+02	1.493E+03
Al	13	26.982	5.098E+00	5.179E+00	5.319E+00	5.520E+00	6.138E+00	7.119E+00	7.788E+00	1.086E+01	2.823E+01	1.229E+02	1.753E+03
Si	14	28.086	6.138E+00	6.235E+00	6.399E+00	6.636E+00	7.361E+00	8.511E+00	9.294E+00	1.287E+01	3.305E+01	1.428E+02	2.033E+03
P	15	30.974	7.615E+00	7.729E+00	7.923E+00	8.202E+00	9.056E+00	1.040E+01	1.132E+01	1.547E+01	3.868E+01	1.647E+02	2.335E+03
S	16	32.066	8.870E+00	9.000E+00	9.222E+00	9.543E+00	1.052E+01	1.206E+01	1.310E+01	1.785E+01	4.427E+01	1.876E+02	2.657E+03
Cl	17	35.453	1.061E+01	1.076E+01	1.102E+01	1.139E+01	1.251E+01	1.426E+01	1.546E+01	2.085E+01	5.072E+01	2.125E+02	3.000E+03
Ar	18	39.948	1.256E+01	1.273E+01	1.302E+01	1.344E+01	1.471E+01	1.670E+01	1.804E+01	2.411E+01	5.763E+01	2.390E+02	3.364E+03
K	19	39.098	1.387E+01	1.406E+01	1.438E+01	1.485E+01	1.626E+01	1.847E+01	1.997E+01	2.673E+01	6.408E+01	2.661E+02	3.748E+03
Ca	20	40.078	1.553E+01	1.574E+01	1.609E+01	1.661E+01	1.818E+01	2.063E+01	2.229E+01	2.979E+01	7.118E+01	2.950E+02	4.153E+03
Sc	21	44.959	1.781E+01	1.805E+01	1.844E+01	1.902E+01	2.076E+01	2.348E+01	2.532E+01	3.361E+01	7.926E+01	3.260E+02	4.579E+03
Ti	22	47.880	1.991E+01	2.016E+01	2.060E+01	2.123E+01	2.314E+01	2.614E+01	2.816E+01	3.727E+01	8.738E+01	3.582E+02	5.026E+03
V	23	50.942	2.210E+01	2.238E+01	2.286E+01	2.355E+01	2.564E+01	2.892E+01	3.114E+01	4.110E+01	9.588E+01	3.919E+02	5.494E+03
Cr	24	51.996	2.417E+01	2.448E+01	2.500E+01	2.575E+01	2.804E+01	3.161E+01	3.402E+01	4.487E+01	1.045E+02	4.268E+02	5.982E+03
Mn	25	54.938	2.654E+01	2.687E+01	2.744E+01	2.825E+01	3.074E+01	3.462E+01	3.723E+01	4.902E+01	1.137E+02	4.634E+02	6.491E+03
Fe	26	55.847	2.879E+01	2.916E+01	2.977E+01	3.065E+01	3.334E+01	3.754E+01	4.037E+01	5.311E+01	1.231E+02	5.014E+02	7.021E+03
Co	27	58.933	3.136E+01	3.175E+01	3.241E+01	3.336E+01	3.626E+01	4.079E+01	4.385E+01	5.760E+01	1.331E+02	5.410E+02	7.572E+03
Ni	28	58.690	3.370E+01	3.412E+01	3.483E+01	3.586E+01	3.897E+01	4.385E+01	4.713E+01	6.192E+01	1.431E+02	5.818E+02	8.143E+03
Cu	29	63.546	3.661E+01	3.706E+01	3.783E+01	3.893E+01	4.228E+01	4.751E+01	5.104E+01	6.690E+01	1.540E+02	6.246E+02	8.735E+03
Zn	30	65.390	3.934E+01	3.983E+01	4.065E+01	4.182E+01	4.541E+01	5.101E+01	5.478E+01	7.177E+01	1.650E+02	6.685E+02	9.348E+03
Ga	31	69.723	4.236E+01	4.288E+01	4.375E+01	4.501E+01	4.884E+01	5.482E+01	5.885E+01	7.699E+01	1.765E+02	7.142E+02	9.982E+03
Ge	32	72.610	4.535E+01	4.590E+01	4.684E+01	4.818E+01	5.226E+01	5.864E+01	6.293E+01	8.226E+01	1.883E+02	7.613E+02	1.064E+04
As	33	74.922	4.839E+01	4.898E+01	4.997E+01	5.140E+01	5.574E+01	6.252E+01	6.709E+01	8.765E+01	2.005E+02	8.098E+02	1.131E+04
Se	34	78.960	5.164E+01	5.226E+01	5.332E+01	5.483E+01	5.944E+01	6.664E+01	7.150E+01	9.332E+01	2.131E+02	8.599E+02	1.201E+04
Br	35	79.904	5.478E+01	5.544E+01	5.656E+01	5.817E+01	6.305E+01	7.068E+01	7.582E+01	9.896E+01	2.259E+02	9.112E+02	1.273E+04
Kr	36	83.800	5.820E+01	5.890E+01	6.008E+01	6.178E+01	6.695E+01	7.503E+01	8.047E+01	1.049E+02	2.392E+02	9.643E+02	1.346E+04
Rb	37	85.468	6.158E+01	6.232E+01	6.357E+01	6.536E+01	7.082E+01	7.935E+01	8.510E+01	1.110E+02	2.528E+02	1.019E+03	1.422E+04
Sr	38	87.620	6.508E+01	6.586E+01	6.717E+01	6.907E+01	7.483E+01	8.383E+01	8.989E+01	1.172E+02	2.668E+02	1.075E+03	1.500E+04
Y	39	88.906	6.862E+01	6.944E+01	7.083E+01	7.283E+01	7.889E+01	8.837E+01	9.476E+01	1.235E+02	2.811E+02	1.132E+03	1.580E+04
Zr	40	91.224	7.232E+01	7.318E+01	7.464E+01	7.675E+01	8.313E+01	9.310E+01	9.982E+01	1.300E+02	2.958E+02	1.191E+03	1.662E+04
Nb	41	92.906	7.608E+01	7.698E+01	7.852E+01	8.073E+01	8.743E+01	9.791E+01	1.050E+02	1.367E+02	3.109E+02	1.251E+03	1.746E+04
Mo	42	95.940	8.000E+01	8.095E+01	8.256E+01	8.488E+01	9.192E+01	1.029E+02	1.103E+02	1.436E+02	3.264E+02	1.313E+03	1.833E+04
Tc	43	98.000	8.397E+01	8.497E+01	8.665E+01	8.908E+01	9.646E+01	1.080E+02	1.157E+02	1.507E+02	3.422E+02	1.377E+03	1.921E+04
Ru	44	101.070	8.808E+01	8.912E+01	9.089E+01	9.343E+01	1.012E+02	1.132E+02	1.214E+02	1.579E+02	3.585E+02	1.442E+03	2.011E+04
Rh	45	102.906	9.222E+01	9.331E+01	9.516E+01	9.782E+01	1.059E+02	1.185E+02	1.270E+02	1.653E+02	3.751E+02	1.508E+03	2.104E+04
Pd	46	106.420	9.654E+01	9.768E+01	9.961E+01	1.024E+02	1.108E+02	1.240E+02	1.329E+02	1.729E+02	3.921E+02	1.576E+03	2.198E+04
Ag	47	107.868	1.009E+02	1.020E+02	1.041E+02	1.070E+02	1.158E+02	1.295E+02	1.388E+02	1.805E+02	4.094E+02	1.645E+03	2.295E+04
Cd	48	122.411	1.058E+02	1.070E+02	1.091E+02	1.121E+02	1.213E+02	1.357E+02	1.454E+02	1.889E+02	4.276E+02	1.717E+03	2.394E+04
In	49	114.820	1.099E+02	1.112E+02	1.134E+02	1.166E+02	1.262E+02	1.411E+02	1.512E+02	1.966E+02	4.453E+02	1.789E+03	2.494E+04
Sn	50	118.710	1.146E+02	1.160E+02	1.183E+02	1.215E+02	1.315E+02	1.471E+02	1.576E+02	2.048E+02	4.638E+02	1.863E+03	2.597E+04
Sb	51	121.750	1.194E+02	1.208E+02	1.232E+02	1.266E+02	1.370E+02	1.532E+02	1.641E+02	2.132E+02	4.827E+02	1.938E+03	2.702E+04

Table A6.6. Rutherford scattering cross sections (barns) of the elements for 1 MeV ^{12}C (continued).

Element	Atomic no. (Z_2)	Mass (M_2)	Scattering angle (degrees)										
			175	170	165	160	150	140	135	120	90	60	30
Te	52	127.600	1.243E+02	1.258E+02	1.283E+02	1.318E+02	1.426E+02	1.595E+02	1.708E+02	2.219E+02	5.020E+02	2.015E+03	2.809E+04
I	53	126.904	1.291E+02	1.306E+02	1.332E+02	1.369E+02	1.481E+02	1.656E+02	1.774E+02	2.305E+02	5.215E+02	2.093E+03	2.918E+04
Xe	54	131.290	1.342E+02	1.358E+02	1.384E+02	1.423E+02	1.539E+02	1.721E+02	1.843E+02	2.394E+02	5.415E+02	2.173E+03	3.030E+04
Cs	55	132.905	1.393E+02	1.409E+02	1.437E+02	1.477E+02	1.597E+02	1.786E+02	1.913E+02	2.484E+02	5.618E+02	2.254E+03	3.143E+04
Ba	56	137.327	1.445E+02	1.462E+02	1.491E+02	1.532E+02	1.657E+02	1.853E+02	1.985E+02	2.577E+02	5.826E+02	2.337E+03	3.258E+04
La	57	138.906	1.498E+02	1.516E+02	1.545E+02	1.588E+02	1.718E+02	1.920E+02	2.057E+02	2.670E+02	6.037E+02	2.421E+03	3.376E+04
Ce	58	140.115	1.551E+02	1.570E+02	1.600E+02	1.645E+02	1.779E+02	1.989E+02	2.130E+02	2.765E+02	6.251E+02	2.507E+03	3.495E+04
Pr	59	140.908	1.606E+02	1.624E+02	1.656E+02	1.702E+02	1.841E+02	2.058E+02	2.204E+02	2.862E+02	6.468E+02	2.594E+03	3.617E+04
Nd	60	144.240	1.662E+02	1.681E+02	1.714E+02	1.761E+02	1.905E+02	2.129E+02	2.281E+02	2.961E+02	6.691E+02	2.683E+03	3.740E+04
Pm	61	145.000	1.718E+02	1.738E+02	1.772E+02	1.821E+02	1.969E+02	2.201E+02	2.358E+02	3.060E+02	6.916E+02	2.773E+03	3.866E+04
Sm	62	150.360	1.776E+02	1.797E+02	1.832E+02	1.883E+02	2.036E+02	2.276E+02	2.437E+02	3.163E+02	7.146E+02	2.865E+03	3.994E+04
Eu	63	151.965	1.835E+02	1.856E+02	1.892E+02	1.944E+02	2.103E+02	2.350E+02	2.517E+02	3.267E+02	7.379E+02	2.958E+03	4.124E+04
Gd	64	157.250	1.895E+02	1.917E+02	1.954E+02	2.008E+02	2.172E+02	2.427E+02	2.599E+02	3.373E+02	7.617E+02	3.053E+03	4.256E+04
Tb	65	158.925	1.955E+02	1.978E+02	2.016E+02	2.072E+02	2.240E+02	2.504E+02	2.681E+02	3.479E+02	7.857E+02	3.150E+03	4.390E+04
Dy	66	162.500	2.017E+02	2.040E+02	2.080E+02	2.137E+02	2.311E+02	2.583E+02	2.765E+02	3.588E+02	8.102E+02	3.247E+03	4.526E+04
Ho	67	164.930	2.079E+02	2.103E+02	2.144E+02	2.203E+02	2.382E+02	2.662E+02	2.851E+02	3.699E+02	8.350E+02	3.346E+03	4.664E+04
Er	68	167.260	2.142E+02	2.167E+02	2.209E+02	2.270E+02	2.454E+02	2.743E+02	2.937E+02	3.810E+02	8.601E+02	3.447E+03	4.804E+04
Tm	69	168.934	2.206E+02	2.232E+02	2.275E+02	2.338E+02	2.528E+02	2.824E+02	3.024E+02	3.924E+02	8.857E+02	3.549E+03	4.947E+04
Yb	70	173.040	2.271E+02	2.298E+02	2.343E+02	2.407E+02	2.602E+02	2.908E+02	3.114E+02	4.039E+02	9.116E+02	3.653E+03	5.091E+04
Lu	71	174.967	2.337E+02	2.364E+02	2.410E+02	2.477E+02	2.678E+02	2.992E+02	3.204E+02	4.156E+02	9.379E+02	3.758E+03	5.237E+04
Hf	72	178.490	2.404E+02	2.432E+02	2.480E+02	2.548E+02	2.755E+02	3.078E+02	3.296E+02	4.275E+02	9.646E+02	3.865E+03	5.386E+04
Ta	73	180.948	2.472E+02	2.501E+02	2.550E+02	2.620E+02	2.832E+02	3.165E+02	3.388E+02	4.395E+02	9.916E+02	3.973E+03	5.537E+04
W	74	183.850	2.541E+02	2.571E+02	2.621E+02	2.693E+02	2.911E+02	3.253E+02	3.483E+02	4.517E+02	1.019E+03	4.083E+03	5.689E+04
Re	75	186.207	2.611E+02	2.641E+02	2.693E+02	2.766E+02	2.991E+02	3.342E+02	3.578E+02	4.641E+02	1.047E+03	4.194E+03	5.844E+04
Os	76	190.200	2.682E+02	2.713E+02	2.766E+02	2.842E+02	3.072E+02	3.432E+02	3.675E+02	4.766E+02	1.075E+03	4.307E+03	6.001E+04
Ir	77	192.220	2.753E+02	2.785E+02	2.839E+02	2.917E+02	3.154E+02	3.524E+02	3.773E+02	4.893E+02	1.104E+03	4.421E+03	6.160E+04
Pt	78	195.080	2.826E+02	2.859E+02	2.914E+02	2.994E+02	3.237E+02	3.616E+02	3.872E+02	5.021E+02	1.132E+03	4.536E+03	6.321E+04
Au	79	196.967	2.899E+02	2.933E+02	2.990E+02	3.072E+02	3.321E+02	3.710E+02	3.972E+02	5.151E+02	1.162E+03	4.653E+03	6.484E+04
Hg	80	200.590	2.974E+02	3.008E+02	3.067E+02	3.151E+02	3.406E+02	3.806E+02	4.074E+02	5.283E+02	1.191E+03	4.772E+03	6.649E+04
Tl	81	204.383	3.050E+02	3.085E+02	3.145E+02	3.231E+02	3.493E+02	3.902E+02	4.178E+02	5.417E+02	1.221E+03	4.892E+03	6.817E+04
Pb	82	207.200	3.126E+02	3.162E+02	3.224E+02	3.312E+02	3.580E+02	4.000E+02	4.282E+02	5.552E+02	1.252E+03	5.014E+03	6.986E+04
Bi	83	208.980	3.203E+02	3.240E+02	3.303E+02	3.394E+02	3.669E+02	4.098E+02	4.387E+02	5.689E+02	1.283E+03	5.137E+03	7.158E+04
Po	84	209.000	3.281E+02	3.319E+02	3.383E+02	3.476E+02	3.757E+02	4.197E+02	4.494E+02	5.827E+02	1.314E+03	5.261E+03	7.331E+04
At	85	210.000	3.359E+02	3.398E+02	3.464E+02	3.559E+02	3.848E+02	4.298E+02	4.602E+02	5.967E+02	1.345E+03	5.387E+03	7.507E+04
Rn	86	222.000	3.441E+02	3.481E+02	3.549E+02	3.646E+02	3.941E+02	4.402E+02	4.713E+02	6.110E+02	1.377E+03	5.515E+03	7.684E+04
Fr	87	223.000	3.522E+02	3.563E+02	3.632E+02	3.731E+02	4.033E+02	4.505E+02	4.823E+02	6.253E+02	1.410E+03	5.644E+03	7.864E+04
Ra	88	226.025	3.604E+02	3.646E+02	3.716E+02	3.818E+02	4.127E+02	4.610E+02	4.935E+02	6.398E+02	1.442E+03	5.775E+03	8.046E+04
Ac	89	227.028	3.687E+02	3.729E+02	3.802E+02	3.906E+02	4.222E+02	4.716E+02	5.048E+02	6.545E+02	1.475E+03	5.907E+03	8.230E+04
Th	90	232.038	3.771E+02	3.814E+02	3.888E+02	3.995E+02	4.318E+02	4.823E+02	5.163E+02	6.694E+02	1.509E+03	6.040E+03	8.416E+04
Pa	91	231.036	3.855E+02	3.899E+02	3.975E+02	4.084E+02	4.414E+02	4.931E+02	5.279E+02	6.843E+02	1.542E+03	6.175E+03	8.604E+04
U	92	238.029	3.941E+02	3.987E+02	4.064E+02	4.175E+02	4.513E+02	5.041E+02	5.396E+02	6.995E+02	1.576E+03	6.312E+03	8.794E+04
Np	93	237.048	4.027E+02	4.074E+02	4.153E+02	4.266E+02	4.612E+02	5.151E+02	5.514E+02	7.148E+02	1.611E+03	6.450E+03	8.986E+04
Pu	94	244.000	4.115E+02	4.163E+02	4.244E+02	4.360E+02	4.713E+02	5.264E+02	5.635E+02	7.304E+02	1.646E+03	6.589E+03	9.181E+04

Table A6.7. Rutherford scattering cross sections (barns) of the elements for 1 MeV ^{15}N.

Element	Atomic no. (Z_2)	Mass (M_2)	Scattering angle (degrees)										
			175	170	165	160	150	140	135	120	90	60	30
O	8	15.999	6.011E–02	6.204E–02	6.543E–02	7.055E–02	8.801E–02	1.219E–01	1.499E–01	3.470E–01	5.651E+00	5.811E+01	9.005E+02
F	9	18.998	7.354E–01	7.545E–01	7.877E–01	8.368E–01	9.964E–01	1.278E+00	1.488E+00	2.643E+00	1.262E+01	7.575E+01	1.141E+03
Ne	10	20.180	1.281E+00	1.312E+00	1.366E+00	1.445E+00	1.700E+00	2.139E+00	2.462E+00	4.149E+00	1.698E+01	9.440E+01	1.409E+03
Na	11	22.990	2.550E+00	2.605E+00	2.698E+00	2.835E+00	3.266E+00	3.983E+00	4.492E+00	6.993E+00	2.328E+01	1.162E+02	1.706E+03
Mg	12	24.305	3.526E+00	3.598E+00	3.721E+00	3.900E+00	4.460E+00	5.382E+00	6.031E+00	9.161E+00	2.876E+01	1.391E+02	2.030E+03
Al	13	26.982	5.151E+00	5.248E+00	5.412E+00	5.650E+00	6.390E+00	7.587E+00	8.418E+00	1.233E+01	3.566E+01	1.649E+02	2.384E+03
Si	14	28.086	6.392E+00	6.508E+00	6.706E+00	6.992E+00	7.878E+00	9.306E+00	1.029E+01	1.491E+01	4.206E+01	1.918E+02	2.765E+03
P	15	30.974	8.413E+00	8.556E+00	8.798E+00	9.149E+00	1.023E+01	1.195E+01	1.314E+01	1.860E+01	4.997E+01	2.217E+02	3.176E+03
S	16	32.066	9.968E+00	1.013E+01	1.041E+01	1.082E+01	1.207E+01	1.406E+01	1.542E+01	2.169E+01	5.743E+01	2.527E+02	3.614E+03
Cl	17	35.453	1.243E+01	1.262E+01	1.295E+01	1.343E+01	1.489E+01	1.720E+01	1.878E+01	2.597E+01	6.647E+01	2.868E+02	4.081E+03
Ar	18	39.948	1.525E+01	1.547E+01	1.585E+01	1.640E+01	1.809E+01	2.074E+01	2.254E+01	3.071E+01	7.623E+01	3.231E+02	4.577E+03
K	19	39.098	1.674E+01	1.699E+01	1.742E+01	1.803E+01	1.990E+01	2.284E+01	2.484E+01	3.393E+01	8.462E+01	3.597E+02	5.099E+03
Ca	20	40.078	1.886E+01	1.914E+01	1.961E+01	2.029E+01	2.237E+01	2.565E+01	2.787E+01	3.796E+01	9.416E+01	3.990E+02	5.650E+03
Sc	21	44.959	2.221E+01	2.252E+01	2.305E+01	2.381E+01	2.614E+01	2.980E+01	3.228E+01	4.348E+01	1.055E+02	4.415E+02	6.231E+03
Ti	22	47.880	2.510E+01	2.544E+01	2.603E+01	2.687E+01	2.944E+01	3.348E+01	3.621E+01	4.854E+01	1.167E+02	4.854E+02	6.839E+03
V	23	50.942	2.813E+01	2.851E+01	2.915E+01	3.008E+01	3.290E+01	3.733E+01	4.032E+01	5.383E+01	1.283E+02	5.313E+02	7.476E+03
Cr	24	51.996	3.086E+01	3.127E+01	3.198E+01	3.299E+01	3.607E+01	4.090E+01	4.416E+01	5.887E+01	1.400E+02	5.787E+02	8.140E+03
Mn	25	54.938	3.412E+01	3.457E+01	3.533E+01	3.644E+01	3.979E+01	4.504E+01	4.858E+01	6.457E+01	1.526E+02	6.287E+02	8.833E+03
Fe	26	55.847	3.709E+01	3.758E+01	3.841E+01	3.960E+01	4.323E+01	4.892E+01	5.275E+01	7.005E+01	1.653E+02	6.802E+02	9.554E+03
Co	27	58.933	4.064E+01	4.116E+01	4.206E+01	4.335E+01	4.727E+01	5.341E+01	5.756E+01	7.623E+01	1.790E+02	7.342E+02	1.030E+04
Ni	28	58.690	4.365E+01	4.422E+01	4.518E+01	4.657E+01	5.079E+01	5.739E+01	6.185E+01	8.192E+01	1.924E+02	7.895E+02	1.108E+04
Cu	29	63.546	4.779E+01	4.840E+01	4.944E+01	5.093E+01	5.547E+01	6.256E+01	6.735E+01	8.891E+01	2.074E+02	8.479E+02	1.189E+04
Zn	30	65.390	5.148E+01	5.213E+01	5.324E+01	5.484E+01	5.970E+01	6.730E+01	7.242E+01	9.551E+01	2.224E+02	9.078E+02	1.272E+04
Ga	31	69.723	5.570E+01	5.640E+01	5.759E+01	5.930E+01	6.449E+01	7.262E+01	7.810E+01	1.028E+02	2.382E+02	9.701E+02	1.359E+04
Ge	32	72.610	5.980E+01	6.055E+01	6.182E+01	6.364E+01	6.918E+01	7.784E+01	8.368E+01	1.100E+02	2.543E+02	1.034E+03	1.448E+04
As	33	74.922	6.394E+01	6.474E+01	6.608E+01	6.802E+01	7.392E+01	8.313E+01	8.934E+01	1.173E+02	2.708E+02	1.100E+03	1.540E+04
Se	34	78.960	6.844E+01	6.928E+01	7.072E+01	7.278E+01	7.904E+01	8.883E+01	9.542E+01	1.251E+02	2.881E+02	1.168E+03	1.634E+04
Br	35	79.904	7.265E+01	7.355E+01	7.506E+01	7.725E+01	8.388E+01	9.426E+01	1.012E+02	1.327E+02	3.054E+02	1.238E+03	1.732E+04
Kr	36	83.800	7.737E+01	7.832E+01	7.993E+01	8.224E+01	8.926E+01	1.002E+02	1.076E+02	1.409E+02	3.237E+02	1.311E+03	1.832E+04
Rb	37	85.468	8.194E+01	8.294E+01	8.464E+01	8.708E+01	9.450E+01	1.061E+02	1.139E+02	1.491E+02	3.421E+02	1.385E+03	1.936E+04
Sr	38	87.620	8.669E+01	8.775E+01	8.954E+01	9.212E+01	9.995E+01	1.122E+02	1.204E+02	1.575E+02	3.611E+02	1.461E+03	2.042E+04
Y	39	88.906	9.147E+01	9.259E+01	9.447E+01	9.719E+01	1.054E+02	1.183E+02	1.270E+02	1.661E+02	3.806E+02	1.539E+03	2.150E+04
Zr	40	91.224	9.651E+01	9.768E+01	9.966E+01	1.025E+02	1.112E+02	1.248E+02	1.339E+02	1.750E+02	4.006E+02	1.619E+03	2.262E+04
Nb	41	92.906	1.016E+02	1.028E+02	1.049E+02	1.079E+02	1.170E+02	1.313E+02	1.409E+02	1.841E+02	4.211E+02	1.701E+03	2.377E+04
Mo	42	95.940	1.070E+02	1.083E+02	1.104E+02	1.136E+02	1.232E+02	1.381E+02	1.482E+02	1.935E+02	4.423E+02	1.786E+03	2.494E+04
Tc	43	98.000	1.124E+02	1.137E+02	1.160E+02	1.193E+02	1.293E+02	1.450E+02	1.556E+02	2.031E+02	4.638E+02	1.872E+03	2.614E+04
Ru	44	101.070	1.180E+02	1.194E+02	1.218E+02	1.253E+02	1.358E+02	1.522E+02	1.632E+02	2.130E+02	4.860E+02	1.960E+03	2.737E+04
Rh	45	102.906	1.236E+02	1.251E+02	1.276E+02	1.312E+02	1.422E+02	1.594E+02	1.709E+02	2.230E+02	5.085E+02	2.051E+03	2.863E+04
Pd	46	106.420	1.295E+02	1.311E+02	1.337E+02	1.375E+02	1.490E+02	1.669E+02	1.790E+02	2.334E+02	5.318E+02	2.143E+03	2.992E+04
Ag	47	107.868	1.353E+02	1.370E+02	1.397E+02	1.437E+02	1.556E+02	1.744E+02	1.870E+02	2.438E+02	5.553E+02	2.237E+03	3.123E+04
Cd	48	122.411	1.424E+02	1.441E+02	1.470E+02	1.511E+02	1.636E+02	1.831E+02	1.963E+02	2.555E+02	5.804E+02	2.335E+03	3.258E+04
In	49	114.820	1.478E+02	1.496E+02	1.525E+02	1.568E+02	1.699E+02	1.902E+02	2.039E+02	2.657E+02	6.042E+02	2.433E+03	3.395E+04
Sn	50	118.710	1.542E+02	1.561E+02	1.592E+02	1.636E+02	1.772E+02	1.984E+02	2.127E+02	2.770E+02	6.295E+02	2.533E+03	3.535E+04
Sb	51	121.750	1.607E+02	1.626E+02	1.659E+02	1.705E+02	1.846E+02	2.067E+02	2.216E+02	2.884E+02	6.552E+02	2.636E+03	3.678E+04
Te	52	127.600	1.675E+02	1.695E+02	1.729E+02	1.777E+02	1.924E+02	2.153E+02	2.308E+02	3.003E+02	6.816E+02	2.741E+03	3.824E+04

Table A6.7. Rutherford scattering cross sections (barns) of the elements for 1 MeV ^{15}N (continued).

Element	Atomic no. (Z_2)	Mass (M_2)	Scattering angle (degrees)										
			175	170	165	160	150	140	135	120	90	60	30
I	53	126.904	1.740E+02	1.761E+02	1.795E+02	1.846E+02	1.998E+02	2.236E+02	2.397E+02	3.119E+02	7.080E+02	2.847E+03	3.972E+04
Xe	54	131.290	1.810E+02	1.831E+02	1.867E+02	1.919E+02	2.078E+02	2.325E+02	2.492E+02	3.241E+02	7.353E+02	2.956E+03	4.123E+04
Cs	55	132.905	1.878E+02	1.901E+02	1.938E+02	1.992E+02	2.156E+02	2.413E+02	2.586E+02	3.364E+02	7.630E+02	3.067E+03	4.278E+04
Ba	56	137.327	1.950E+02	1.973E+02	2.012E+02	2.069E+02	2.239E+02	2.505E+02	2.684E+02	3.490E+02	7.913E+02	3.179E+03	4.434E+04
La	57	138.906	2.022E+02	2.046E+02	2.086E+02	2.144E+02	2.321E+02	2.596E+02	2.782E+02	3.617E+02	8.199E+02	3.294E+03	4.594E+04
Ce	58	140.115	2.094E+02	2.119E+02	2.161E+02	2.221E+02	2.404E+02	2.689E+02	2.881E+02	3.746E+02	8.490E+02	3.411E+03	4.757E+04
Pr	59	140.908	2.168E+02	2.193E+02	2.236E+02	2.299E+02	2.488E+02	2.783E+02	2.982E+02	3.877E+02	8.786E+02	3.529E+03	4.922E+04
Nd	60	144.240	2.244E+02	2.271E+02	2.315E+02	2.380E+02	2.575E+02	2.881E+02	3.086E+02	4.012E+02	9.089E+02	3.650E+03	5.091E+04
Pm	61	145.000	2.320E+02	2.347E+02	2.394E+02	2.460E+02	2.662E+02	2.978E+02	3.191E+02	4.147E+02	9.395E+02	3.773E+03	5.262E+04
Sm	62	150.360	2.400E+02	2.429E+02	2.476E+02	2.545E+02	2.754E+02	3.080E+02	3.300E+02	4.288E+02	9.709E+02	3.898E+03	5.436E+04
Eu	63	151.965	2.479E+02	2.509E+02	2.558E+02	2.629E+02	2.844E+02	3.181E+02	3.408E+02	4.429E+02	1.003E+03	4.025E+03	5.613E+04
Gd	64	157.250	2.562E+02	2.592E+02	2.643E+02	2.716E+02	2.939E+02	3.286E+02	3.521E+02	4.574E+02	1.035E+03	4.154E+03	5.792E+04
Tb	65	158.925	2.644E+02	2.675E+02	2.727E+02	2.803E+02	3.032E+02	3.391E+02	3.632E+02	4.719E+02	1.068E+03	4.285E+03	5.975E+04
Dy	66	162.500	2.728E+02	2.760E+02	2.814E+02	2.892E+02	3.129E+02	3.498E+02	3.747E+02	4.867E+02	1.101E+03	4.418E+03	6.160E+04
Ho	67	164.930	2.813E+02	2.846E+02	2.901E+02	2.982E+02	3.225E+02	3.606E+02	3.863E+02	5.017E+02	1.135E+03	4.553E+03	6.348E+04
Er	68	167.260	2.899E+02	2.932E+02	2.990E+02	3.073E+02	3.324E+02	3.716E+02	3.981E+02	5.169E+02	1.169E+03	4.690E+03	6.539E+04
Tm	69	168.934	2.985E+02	3.020E+02	3.079E+02	3.165E+02	3.423E+02	3.827E+02	4.099E+02	5.324E+02	1.204E+03	4.829E+03	6.733E+04
Yb	70	173.040	3.075E+02	3.111E+02	3.172E+02	3.259E+02	3.525E+02	3.941E+02	4.221E+02	5.481E+02	1.239E+03	4.971E+03	6.929E+04
Lu	71	174.967	3.164E+02	3.201E+02	3.264E+02	3.354E+02	3.628E+02	4.056E+02	4.344E+02	5.640E+02	1.275E+03	5.114E+03	7.129E+04
Hf	72	178.490	3.256E+02	3.294E+02	3.358E+02	3.451E+02	3.733E+02	4.173E+02	4.469E+02	5.802E+02	1.311E+03	5.259E+03	7.331E+04
Ta	73	180.948	3.348E+02	3.387E+02	3.454E+02	3.549E+02	3.838E+02	4.291E+02	4.595E+02	5.966E+02	1.348E+03	5.406E+03	7.536E+04
W	74	183.850	3.442E+02	3.482E+02	3.550E+02	3.648E+02	3.946E+02	4.411E+02	4.724E+02	6.132E+02	1.385E+03	5.555E+03	7.744E+04
Re	75	186.207	3.537E+02	3.578E+02	3.648E+02	3.749E+02	4.054E+02	4.532E+02	4.853E+02	6.300E+02	1.423E+03	5.707E+03	7.954E+04
Os	76	190.200	3.634E+02	3.676E+02	3.748E+02	3.851E+02	4.165E+02	4.655E+02	4.986E+02	6.471E+02	1.462E+03	5.860E+03	8.168E+04
Ir	77	192.220	3.731E+02	3.775E+02	3.848E+02	3.954E+02	4.276E+02	4.780E+02	5.119E+02	6.643E+02	1.500E+03	6.015E+03	8.384E+04
Pt	78	195.080	3.830E+02	3.875E+02	3.950E+02	4.059E+02	4.390E+02	4.906E+02	5.254E+02	6.818E+02	1.540E+03	6.173E+03	8.604E+04
Au	79	196.967	3.930E+02	3.976E+02	4.053E+02	4.165E+02	4.504E+02	5.033E+02	5.390E+02	6.995E+02	1.580E+03	6.332E+03	8.826E+04
Hg	80	200.590	4.032E+02	4.079E+02	4.158E+02	4.273E+02	4.620E+02	5.163E+02	5.529E+02	7.175E+02	1.620E+03	6.494E+03	9.051E+04
Tl	81	204.383	4.135E+02	4.183E+02	4.264E+02	4.382E+02	4.738E+02	5.295E+02	5.670E+02	7.357E+02	1.661E+03	6.657E+03	9.278E+04
Pb	82	207.200	4.239E+02	4.288E+02	4.372E+02	4.492E+02	4.857E+02	5.428E+02	5.812E+02	7.541E+02	1.702E+03	6.823E+03	9.509E+04
Bi	83	208.980	4.344E+02	4.394E+02	4.480E+02	4.603E+02	4.977E+02	5.562E+02	5.956E+02	7.727E+02	1.744E+03	6.990E+03	9.742E+04
Po	84	209.000	4.449E+02	4.501E+02	4.588E+02	4.714E+02	5.098E+02	5.697E+02	6.100E+02	7.914E+02	1.786E+03	7.160E+03	9.978E+04
At	85	210.000	4.556E+02	4.609E+02	4.699E+02	4.828E+02	5.220E+02	5.833E+02	6.247E+02	8.104E+02	1.829E+03	7.331E+03	1.022E+05
Rn	86	222.000	4.669E+02	4.723E+02	4.815E+02	4.947E+02	5.349E+02	5.977E+02	6.399E+02	8.301E+02	1.873E+03	7.505E+03	1.046E+05
Fr	87	223.000	4.778E+02	4.834E+02	4.928E+02	5.063E+02	5.474E+02	6.117E+02	6.549E+02	8.496E+02	1.917E+03	7.681E+03	1.070E+05
Ra	88	226.025	4.890E+02	4.947E+02	5.043E+02	5.181E+02	5.602E+02	6.259E+02	6.702E+02	8.693E+02	1.961E+03	7.859E+03	1.095E+05
Ac	89	227.028	5.002E+02	5.060E+02	5.159E+02	5.300E+02	5.731E+02	6.403E+02	6.856E+02	8.892E+02	2.006E+03	8.038E+03	1.120E+05
Th	90	232.038	5.117E+02	5.176E+02	5.277E+02	5.422E+02	5.862E+02	6.549E+02	7.012E+02	9.095E+02	2.052E+03	8.220E+03	1.145E+05
Pa	91	231.036	5.231E+02	5.292E+02	5.395E+02	5.543E+02	5.993E+02	6.695E+02	7.169E+02	9.298E+02	2.098E+03	8.404E+03	1.171E+05
U	92	238.029	5.349E+02	5.411E+02	5.516E+02	5.668E+02	6.128E+02	6.846E+02	7.330E+02	9.506E+02	2.144E+03	8.590E+03	1.197E+05
Np	93	237.048	5.466E+02	5.529E+02	5.637E+02	5.791E+02	6.261E+02	6.995E+02	7.490E+02	9.714E+02	2.191E+03	8.777E+03	1.223E+05
Pu	94	244.000	5.586E+02	5.651E+02	5.761E+02	5.919E+02	6.399E+02	7.149E+02	7.654E+02	9.926E+02	2.239E+03	8.967E+03	1.250E+05

Table A6.8. Rutherford scattering cross sections (barns) of the elements for 1 MeV ^{16}O.

Element	Atomic no. (Z_2)	Mass (M_2)	Scattering angle (degrees)										
			175	170	165	160	150	140	135	120	90	60	30
F	9	18.998	5.745E–01	5.907E–01	6.187E–01	6.605E–01	7.985E–01	1.049E+00	1.242E+00	2.372E+00	1.449E+01	9.786E+01	1.489E+03
Ne	10	20.180	1.155E+00	1.186E+00	1.238E+00	1.315E+00	1.568E+00	2.013E+00	2.347E+00	4.184E+00	2.021E+01	1.221E+02	1.839E+03
Na	11	22.990	2.690E+00	2.751E+00	2.856E+00	3.011E+00	3.502E+00	4.330E+00	4.928E+00	7.941E+00	2.882E+01	1.506E+02	2.227E+03
Mg	12	24.305	3.863E+00	3.947E+00	4.090E+00	4.299E+00	4.958E+00	6.056E+00	6.838E+00	1.069E+01	3.595E+01	1.804E+02	2.651E+03
Al	13	26.982	5.931E+00	6.047E+00	6.247E+00	6.537E+00	7.441E+00	8.919E+00	9.953E+00	1.489E+01	4.512E+01	2.141E+02	3.113E+03
Si	14	28.086	7.463E+00	7.605E+00	7.848E+00	8.200E+00	9.295E+00	1.107E+01	1.231E+01	1.817E+01	5.342E+01	2.492E+02	3.611E+03
P	15	30.974	1.009E+01	1.027E+01	1.057E+01	1.101E+01	1.237E+01	1.456E+01	1.606E+01	2.307E+01	6.388E+01	2.883E+02	4.147E+03
S	16	32.066	1.204E+01	1.225E+01	1.260E+01	1.312E+01	1.470E+01	1.723E+01	1.896E+01	2.702E+01	7.356E+01	3.288E+02	4.719E+03
Cl	17	35.453	1.528E+01	1.552E+01	1.595E+01	1.655E+01	1.842E+01	2.139E+01	2.342E+01	3.272E+01	8.551E+01	3.734E+02	5.329E+03
Ar	18	39.948	1.903E+01	1.932E+01	1.981E+01	2.052E+01	2.269E+01	2.612E+01	2.845E+01	3.905E+01	9.843E+01	4.210E+02	5.977E+03
K	19	39.098	2.085E+01	2.117E+01	2.172E+01	2.250E+01	2.491E+01	2.871E+01	3.129E+01	4.308E+01	1.092E+02	4.686E+02	6.659E+03
Ca	20	40.078	2.355E+01	2.391E+01	2.452E+01	2.539E+01	2.807E+01	3.231E+01	3.518E+01	4.828E+01	1.216E+02	5.198E+02	7.379E+03
Sc	21	44.959	2.802E+01	2.842E+01	2.911E+01	3.010E+01	3.312E+01	3.786E+01	4.108E+01	5.566E+01	1.366E+02	5.755E+02	8.137E+03
Ti	22	47.880	3.181E+01	3.226E+01	3.302E+01	3.411E+01	3.745E+01	4.270E+01	4.625E+01	6.231E+01	1.512E+02	6.329E+02	8.931E+03
V	23	50.942	3.579E+01	3.628E+01	3.712E+01	3.833E+01	4.200E+01	4.776E+01	5.165E+01	6.925E+01	1.665E+02	6.929E+02	9.763E+03
Cr	24	51.996	3.931E+01	3.985E+01	4.076E+01	4.208E+01	4.608E+01	5.237E+01	5.661E+01	7.579E+01	1.817E+02	7.548E+02	1.063E+04
Mn	25	54.938	4.359E+01	4.417E+01	4.517E+01	4.660E+01	5.097E+01	5.780E+01	6.242E+01	8.327E+01	1.982E+02	8.200E+02	1.154E+04
Fe	26	55.847	4.743E+01	4.806E+01	4.914E+01	5.069E+01	5.541E+01	6.282E+01	6.782E+01	9.037E+01	2.147E+02	8.873E+02	1.248E+04
Co	27	58.933	5.208E+01	5.276E+01	5.393E+01	5.561E+01	6.072E+01	6.872E+01	7.412E+01	9.848E+01	2.326E+02	9.578E+02	1.346E+04
Ni	28	58.690	5.593E+01	5.667E+01	5.793E+01	5.973E+01	6.522E+01	7.383E+01	7.964E+01	1.058E+02	2.501E+02	1.030E+03	1.447E+04
Cu	29	63.546	6.142E+01	6.221E+01	6.356E+01	6.551E+01	7.142E+01	8.068E+01	8.692E+01	1.151E+02	2.699E+02	1.106E+03	1.553E+04
Zn	30	65.390	6.622E+01	6.707E+01	6.852E+01	7.060E+01	7.694E+01	8.685E+01	9.353E+01	1.237E+02	2.893E+02	1.185E+03	1.662E+04
Ga	31	69.723	7.179E+01	7.270E+01	7.425E+01	7.648E+01	8.326E+01	9.386E+01	1.010E+02	1.332E+02	3.101E+02	1.266E+03	1.774E+04
Ge	32	72.610	7.716E+01	7.813E+01	7.979E+01	8.216E+01	8.939E+01	1.007E+02	1.083E+02	1.426E+02	3.312E+02	1.350E+03	1.891E+04
As	33	74.922	8.257E+01	8.360E+01	8.536E+01	8.789E+01	9.558E+01	1.076E+02	1.157E+02	1.522E+02	3.527E+02	1.436E+03	2.011E+04
Se	34	78.960	8.848E+01	8.958E+01	9.145E+01	9.414E+01	1.023E+02	1.151E+02	1.237E+02	1.625E+02	3.753E+02	1.525E+03	2.135E+04
Br	35	79.904	9.395E+01	9.511E+01	9.710E+01	9.995E+01	1.086E+02	1.221E+02	1.313E+02	1.724E+02	3.979E+02	1.616E+03	2.262E+04
Kr	36	83.800	1.001E+02	1.014E+02	1.035E+02	1.065E+02	1.157E+02	1.300E+02	1.397E+02	1.831E+02	4.218E+02	1.711E+03	2.393E+04
Rb	37	85.468	1.061E+02	1.074E+02	1.096E+02	1.128E+02	1.225E+02	1.376E+02	1.478E+02	1.938E+02	4.459E+02	1.808E+03	2.528E+04
Sr	38	87.620	1.123E+02	1.137E+02	1.160E+02	1.194E+02	1.296E+02	1.456E+02	1.563E+02	2.048E+02	4.707E+02	1.907E+03	2.666E+04
Y	39	88.906	1.185E+02	1.200E+02	1.224E+02	1.260E+02	1.367E+02	1.536E+02	1.649E+02	2.159E+02	4.961E+02	2.009E+03	2.809E+04
Zr	40	91.224	1.251E+02	1.266E+02	1.292E+02	1.329E+02	1.443E+02	1.620E+02	1.739E+02	2.276E+02	5.223E+02	2.114E+03	2.955E+04
Nb	41	92.906	1.317E+02	1.333E+02	1.361E+02	1.400E+02	1.519E+02	1.705E+02	1.830E+02	2.394E+02	5.490E+02	2.221E+03	3.104E+04
Mo	42	95.940	1.388E+02	1.404E+02	1.433E+02	1.474E+02	1.599E+02	1.794E+02	1.926E+02	2.518E+02	5.767E+02	2.331E+03	3.258E+04
Tc	43	98.000	1.458E+02	1.476E+02	1.505E+02	1.549E+02	1.680E+02	1.884E+02	2.022E+02	2.643E+02	6.048E+02	2.444E+03	3.415E+04
Ru	44	101.070	1.531E+02	1.550E+02	1.581E+02	1.626E+02	1.764E+02	1.978E+02	2.122E+02	2.772E+02	6.338E+02	2.559E+03	3.575E+04
Rh	45	102.906	1.605E+02	1.624E+02	1.657E+02	1.704E+02	1.848E+02	2.072E+02	2.223E+02	2.903E+02	6.632E+02	2.677E+03	3.740E+04
Pd	46	106.420	1.682E+02	1.702E+02	1.737E+02	1.786E+02	1.936E+02	2.170E+02	2.328E+02	3.039E+02	6.936E+02	2.798E+03	3.908E+04
Ag	47	107.868	1.758E+02	1.779E+02	1.815E+02	1.867E+02	2.023E+02	2.268E+02	2.433E+02	3.174E+02	7.243E+02	2.921E+03	4.080E+04
Cd	48	122.411	1.852E+02	1.874E+02	1.912E+02	1.966E+02	2.129E+02	2.384E+02	2.556E+02	3.330E+02	7.573E+02	3.049E+03	4.255E+04
In	49	114.820	1.921E+02	1.944E+02	1.983E+02	2.039E+02	2.209E+02	2.475E+02	2.655E+02	3.461E+02	7.883E+02	3.176E+03	4.434E+04
Sn	50	118.710	2.006E+02	2.029E+02	2.070E+02	2.128E+02	2.306E+02	2.583E+02	2.769E+02	3.609E+02	8.213E+02	3.308E+03	4.617E+04
Sb	51	121.750	2.090E+02	2.115E+02	2.157E+02	2.218E+02	2.403E+02	2.691E+02	2.885E+02	3.758E+02	8.549E+02	3.442E+03	4.804E+04
Te	52	127.600	2.180E+02	2.206E+02	2.250E+02	2.313E+02	2.505E+02	2.804E+02	3.006E+02	3.914E+02	8.894E+02	3.579E+03	4.994E+04
I	53	126.904	2.264E+02	2.291E+02	2.336E+02	2.402E+02	2.601E+02	2.912E+02	3.122E+02	4.065E+02	9.239E+02	3.718E+03	5.188E+04

Table A6.8. Rutherford scattering cross sections (barns) of the elements for 1 MeV ^{16}O (continued).

Element	Atomic no. (Z_2)	Mass (M_2)	Scattering angle (degrees)										
			175	170	165	160	150	140	135	120	90	60	30
Xe	54	131.290	2.355E+02	2.383E+02	2.430E+02	2.498E+02	2.705E+02	3.028E+02	3.246E+02	4.225E+02	9.596E+02	3.860E+03	5.386E+04
Cs	55	132.905	2.445E+02	2.474E+02	2.523E+02	2.593E+02	2.808E+02	3.143E+02	3.369E+02	4.385E+02	9.956E+02	4.004E+03	5.587E+04
Ba	56	137.327	2.539E+02	2.569E+02	2.620E+02	2.693E+02	2.916E+02	3.263E+02	3.497E+02	4.550E+02	1.033E+03	4.152E+03	5.792E+04
La	57	138.906	2.632E+02	2.663E+02	2.716E+02	2.792E+02	3.022E+02	3.383E+02	3.625E+02	4.716E+02	1.070E+03	4.302E+03	6.001E+04
Ce	58	140.115	2.727E+02	2.759E+02	2.814E+02	2.892E+02	3.131E+02	3.503E+02	3.755E+02	4.884E+02	1.108E+03	4.454E+03	6.213E+04
Pr	59	140.908	2.822E+02	2.856E+02	2.912E+02	2.994E+02	3.240E+02	3.626E+02	3.886E+02	5.055E+02	1.147E+03	4.609E+03	6.429E+04
Nd	60	144.240	2.922E+02	2.957E+02	3.015E+02	3.099E+02	3.355E+02	3.754E+02	4.022E+02	5.231E+02	1.186E+03	4.767E+03	6.649E+04
Pm	61	145.000	3.021E+02	3.057E+02	3.117E+02	3.204E+02	3.468E+02	3.881E+02	4.158E+02	5.408E+02	1.226E+03	4.927E+03	6.872E+04
Sm	62	150.360	3.127E+02	3.163E+02	3.226E+02	3.316E+02	3.588E+02	4.014E+02	4.301E+02	5.592E+02	1.267E+03	5.091E+03	7.100E+04
Eu	63	151.965	3.230E+02	3.268E+02	3.332E+02	3.425E+02	3.706E+02	4.146E+02	4.443E+02	5.776E+02	1.309E+03	5.256E+03	7.331E+04
Gd	64	157.250	3.338E+02	3.377E+02	3.444E+02	3.539E+02	3.830E+02	4.284E+02	4.590E+02	5.965E+02	1.351E+03	5.425E+03	7.565E+04
Tb	65	158.925	3.445E+02	3.485E+02	3.554E+02	3.652E+02	3.952E+02	4.420E+02	4.736E+02	6.155E+02	1.394E+03	5.596E+03	7.803E+04
Dy	66	162.500	3.555E+02	3.596E+02	3.667E+02	3.769E+02	4.078E+02	4.561E+02	4.886E+02	6.349E+02	1.437E+03	5.770E+03	8.045E+04
Ho	67	164.930	3.665E+02	3.708E+02	3.781E+02	3.886E+02	4.204E+02	4.702E+02	5.037E+02	6.545E+02	1.481E+03	5.946E+03	8.291E+04
Er	68	167.260	3.778E+02	3.822E+02	3.897E+02	4.005E+02	4.333E+02	4.845E+02	5.191E+02	6.743E+02	1.526E+03	6.125E+03	8.540E+04
Tm	69	168.934	3.891E+02	3.936E+02	4.014E+02	4.125E+02	4.463E+02	4.990E+02	5.346E+02	6.945E+02	1.571E+03	6.307E+03	8.793E+04
Yb	70	173.040	4.008E+02	4.055E+02	4.134E+02	4.249E+02	4.596E+02	5.139E+02	5.505E+02	7.151E+02	1.618E+03	6.491E+03	9.050E+04
Lu	71	174.967	4.125E+02	4.173E+02	4.255E+02	4.373E+02	4.730E+02	5.289E+02	5.665E+02	7.358E+02	1.664E+03	6.678E+03	9.311E+04
Hf	72	178.490	4.244E+02	4.294E+02	4.378E+02	4.499E+02	4.867E+02	5.442E+02	5.829E+02	7.570E+02	1.712E+03	6.868E+03	9.575E+04
Ta	73	180.948	4.365E+02	4.416E+02	4.503E+02	4.627E+02	5.005E+02	5.596E+02	5.994E+02	7.783E+02	1.760E+03	7.060E+03	9.843E+04
W	74	183.850	4.488E+02	4.540E+02	4.629E+02	4.757E+02	5.145E+02	5.752E+02	6.161E+02	8.000E+02	1.809E+03	7.255E+03	1.011E+05
Re	75	186.207	4.612E+02	4.665E+02	4.757E+02	4.888E+02	5.287E+02	5.911E+02	6.331E+02	8.220E+02	1.858E+03	7.453E+03	1.039E+05
Os	76	190.200	4.738E+02	4.794E+02	4.887E+02	5.022E+02	5.432E+02	6.072E+02	6.504E+02	8.443E+02	1.908E+03	7.653E+03	1.067E+05
Ir	77	192.220	4.865E+02	4.922E+02	5.018E+02	5.157E+02	5.577E+02	6.235E+02	6.677E+02	8.669E+02	1.959E+03	7.856E+03	1.095E+05
Pt	78	195.080	4.994E+02	5.053E+02	5.152E+02	5.294E+02	5.725E+02	6.400E+02	6.854E+02	8.897E+02	2.010E+03	8.062E+03	1.124E+05
Au	79	196.967	5.125E+02	5.184E+02	5.286E+02	5.432E+02	5.874E+02	6.566E+02	7.032E+02	9.128E+02	2.062E+03	8.270E+03	1.153E+05
Hg	80	200.590	5.258E+02	5.319E+02	5.423E+02	5.572E+02	6.026E+02	6.736E+02	7.214E+02	9.363E+02	2.115E+03	8.481E+03	1.182E+05
Tl	81	204.383	5.393E+02	5.455E+02	5.562E+02	5.715E+02	6.181E+02	6.908E+02	7.398E+02	9.601E+02	2.169E+03	8.694E+03	1.212E+05
Pb	82	207.200	5.528E+02	5.593E+02	5.702E+02	5.859E+02	6.336E+02	7.081E+02	7.583E+02	9.842E+02	2.223E+03	8.911E+03	1.242E+05
Bi	83	208.980	5.665E+02	5.731E+02	5.843E+02	6.004E+02	6.493E+02	7.256E+02	7.771E+02	1.008E+03	2.277E+03	9.129E+03	1.272E+05
Po	84	209.000	5.802E+02	5.870E+02	5.985E+02	6.149E+02	6.650E+02	7.432E+02	7.959E+02	1.033E+03	2.333E+03	9.351E+03	1.303E+05
At	85	210.000	5.942E+02	6.011E+02	6.129E+02	6.297E+02	6.810E+02	7.611E+02	8.150E+02	1.058E+03	2.388E+03	9.575E+03	1.334E+05
Rn	86	222.000	6.090E+02	6.161E+02	6.281E+02	6.454E+02	6.979E+02	7.798E+02	8.351E+02	1.083E+03	2.446E+03	9.802E+03	1.366E+05
Fr	87	223.000	6.233E+02	6.306E+02	6.429E+02	6.605E+02	7.142E+02	7.981E+02	8.547E+02	1.109E+03	2.503E+03	1.003E+04	1.398E+05
Ra	88	226.025	6.379E+02	6.453E+02	6.579E+02	6.760E+02	7.309E+02	8.168E+02	8.746E+02	1.135E+03	2.561E+03	1.026E+04	1.430E+05
Ac	89	227.028	6.525E+02	6.601E+02	6.730E+02	6.915E+02	7.477E+02	8.355E+02	8.946E+02	1.161E+03	2.620E+03	1.050E+04	1.463E+05
Th	90	232.038	6.676E+02	6.753E+02	6.885E+02	7.074E+02	7.649E+02	8.547E+02	9.151E+02	1.187E+03	2.679E+03	1.074E+04	1.496E+05
Pa	91	231.036	6.824E+02	6.904E+02	7.038E+02	7.232E+02	7.819E+02	8.737E+02	9.355E+02	1.214E+03	2.739E+03	1.098E+04	1.530E+05
U	92	238.029	6.979E+02	7.060E+02	7.198E+02	7.395E+02	7.996E+02	8.934E+02	9.566E+02	1.241E+03	2.800E+03	1.122E+04	1.563E+05
Np	93	237.048	7.131E+02	7.214E+02	7.354E+02	7.556E+02	8.170E+02	9.129E+02	9.775E+02	1.268E+03	2.861E+03	1.146E+04	1.598E+05
Pu	94	244.000	7.289E+02	7.374E+02	7.517E+02	7.724E+02	8.350E+02	9.330E+02	9.990E+02	1.296E+03	2.923E+03	1.171E+04	1.632E+05

Table A6.9. Rutherford scattering cross sections (barns) of the elements for 1 MeV ^{28}Si.

Element	Atomic no. (Z_2)	Mass (M_2)	Scattering angle (degrees) 175	170	165	160	150	140	135	120	90	60	30
P	15	30.974	1.956E+00	2.016E+00	2.121E+00	2.278E+00	2.809E+00	3.810E+00	4.618E+00	9.861E+00	9.804E+01	8.221E+02	1.267E+04
S	16	32.066	3.741E+00	3.851E+00	4.042E+00	4.328E+00	5.282E+00	7.045E+00	8.435E+00	1.696E+01	1.270E+02	9.412E+02	1.441E+04
Cl	17	35.453	1.053E+01	1.081E+01	1.128E+01	1.198E+01	1.427E+01	1.829E+01	2.131E+01	3.781E+01	1.802E+02	1.081E+03	1.628E+04
Ar	18	39.948	2.151E+01	2.201E+01	2.285E+01	2.410E+01	2.806E+01	3.475E+01	3.959E+01	6.405E+01	2.348E+02	1.234E+03	1.826E+04
K	19	39.098	2.199E+01	2.250E+01	2.339E+01	2.470E+01	2.885E+01	3.593E+01	4.107E+01	6.737E+01	2.561E+02	1.371E+03	2.034E+04
Ca	20	40.078	2.689E+01	2.751E+01	2.856E+01	3.011E+01	3.504E+01	4.337E+01	4.938E+01	7.973E+01	2.908E+02	1.524E+03	2.255E+04
Sc	21	44.959	4.232E+01	4.319E+01	4.467E+01	4.684E+01	5.362E+01	6.480E+01	7.267E+01	1.108E+02	3.505E+02	1.703E+03	2.487E+04
Ti	22	47.880	5.363E+01	5.467E+01	5.645E+01	5.904E+01	6.711E+01	8.026E+01	8.943E+01	1.331E+02	3.988E+02	1.880E+03	2.731E+04
V	23	50.942	6.590E+01	6.711E+01	6.919E+01	7.221E+01	8.157E+01	9.670E+01	1.072E+02	1.565E+02	4.489E+02	2.066E+03	2.985E+04
Cr	24	51.996	7.425E+01	7.560E+01	7.791E+01	8.126E+01	9.162E+01	1.083E+02	1.199E+02	1.740E+02	4.930E+02	2.253E+03	3.251E+04
Mn	25	54.938	8.754E+01	8.907E+01	9.168E+01	9.546E+01	1.071E+02	1.259E+02	1.387E+02	1.986E+02	5.461E+02	2.455E+03	3.528E+04
Fe	26	55.847	9.683E+01	9.850E+01	1.014E+02	1.055E+02	1.182E+02	1.387E+02	1.527E+02	2.178E+02	5.940E+02	2.658E+03	3.816E+04
Co	27	58.933	1.117E+02	1.135E+02	1.167E+02	1.213E+02	1.354E+02	1.580E+02	1.734E+02	2.445E+02	6.515E+02	2.876E+03	4.116E+04
Ni	28	58.690	1.195E+02	1.215E+02	1.249E+02	1.298E+02	1.450E+02	1.692E+02	1.858E+02	2.622E+02	6.998E+02	3.093E+03	4.427E+04
Cu	29	63.546	1.395E+02	1.417E+02	1.455E+02	1.510E+02	1.678E+02	1.944E+02	2.126E+02	2.959E+02	7.667E+02	3.332E+03	4.750E+04
Zn	30	65.390	1.533E+02	1.557E+02	1.598E+02	1.657E+02	1.838E+02	2.126E+02	2.321E+02	3.216E+02	8.260E+02	3.571E+03	5.083E+04
Ga	31	69.723	1.726E+02	1.752E+02	1.797E+02	1.861E+02	2.058E+02	2.369E+02	2.581E+02	3.544E+02	8.938E+02	3.824E+03	5.429E+04
Ge	32	72.610	1.894E+02	1.922E+02	1.970E+02	2.040E+02	2.251E+02	2.585E+02	2.812E+02	3.843E+02	9.594E+02	4.081E+03	5.785E+04
As	33	74.922	2.057E+02	2.087E+02	2.139E+02	2.213E+02	2.440E+02	2.796E+02	3.038E+02	4.137E+02	1.026E+03	4.345E+03	6.153E+04
Se	34	78.960	2.255E+02	2.287E+02	2.342E+02	2.422E+02	2.664E+02	3.045E+02	3.304E+02	4.474E+02	1.098E+03	4.621E+03	6.532E+04
Br	35	79.904	2.405E+02	2.440E+02	2.498E+02	2.583E+02	2.840E+02	3.245E+02	3.519E+02	4.760E+02	1.165E+03	4.898E+03	6.922E+04
Kr	36	83.800	2.609E+02	2.646E+02	2.708E+02	2.798E+02	3.072E+02	3.502E+02	3.793E+02	5.110E+02	1.240E+03	5.190E+03	7.324E+04
Rb	37	85.468	2.783E+02	2.822E+02	2.888E+02	2.983E+02	3.273E+02	3.728E+02	4.036E+02	5.428E+02	1.313E+03	5.485E+03	7.737E+04
Sr	38	87.620	2.970E+02	3.011E+02	3.081E+02	3.181E+02	3.488E+02	3.969E+02	4.294E+02	5.764E+02	1.389E+03	5.790E+03	8.161E+04
Y	39	88.906	3.148E+02	3.192E+02	3.266E+02	3.372E+02	3.695E+02	4.202E+02	4.545E+02	6.095E+02	1.466E+03	6.101E+03	8.597E+04
Zr	40	91.224	3.348E+02	3.394E+02	3.472E+02	3.584E+02	3.925E+02	4.459E+02	4.820E+02	6.452E+02	1.546E+03	6.421E+03	9.044E+04
Nb	41	92.906	3.544E+02	3.592E+02	3.674E+02	3.792E+02	4.150E+02	4.713E+02	5.092E+02	6.808E+02	1.628E+03	6.749E+03	9.502E+04
Mo	42	95.940	3.765E+02	3.816E+02	3.902E+02	4.026E+02	4.403E+02	4.994E+02	5.393E+02	7.195E+02	1.713E+03	7.088E+03	9.971E+04
Tc	43	98.000	3.977E+02	4.030E+02	4.121E+02	4.251E+02	4.647E+02	5.267E+02	5.685E+02	7.575E+02	1.799E+03	7.433E+03	1.045E+05
Ru	44	101.070	4.209E+02	4.264E+02	4.359E+02	4.496E+02	4.911E+02	5.561E+02	6.000E+02	7.980E+02	1.889E+03	7.787E+03	1.094E+05
Rh	45	102.906	4.428E+02	4.486E+02	4.586E+02	4.728E+02	5.163E+02	5.844E+02	6.303E+02	8.375E+02	1.979E+03	8.148E+03	1.145E+05
Pd	46	106.420	4.675E+02	4.736E+02	4.840E+02	4.989E+02	5.444E+02	6.156E+02	6.637E+02	8.803E+02	2.073E+03	8.519E+03	1.196E+05
Ag	47	107.868	4.899E+02	4.963E+02	5.072E+02	5.228E+02	5.703E+02	6.447E+02	6.949E+02	9.211E+02	2.166E+03	8.896E+03	1.249E+05
Cd	48	122.411	5.276E+02	5.343E+02	5.456E+02	5.620E+02	6.118E+02	6.896E+02	7.421E+02	9.784E+02	2.277E+03	9.296E+03	1.303E+05
In	49	114.820	5.416E+02	5.485E+02	5.604E+02	5.774E+02	6.291E+02	7.101E+02	7.647E+02	1.011E+03	2.364E+03	9.679E+03	1.357E+05
Sn	50	118.710	5.685E+02	5.757E+02	5.881E+02	6.058E+02	6.597E+02	7.441E+02	8.011E+02	1.057E+03	2.467E+03	1.008E+04	1.414E+05
Sb	51	121.750	5.949E+02	6.024E+02	6.153E+02	6.337E+02	6.899E+02	7.777E+02	8.370E+02	1.104E+03	2.570E+03	1.049E+04	1.471E+05
Te	52	127.600	6.246E+02	6.325E+02	6.458E+02	6.650E+02	7.235E+02	8.149E+02	8.765E+02	1.154E+03	2.679E+03	1.092E+04	1.529E+05
I	53	126.904	6.481E+02	6.563E+02	6.702E+02	6.902E+02	7.509E+02	8.458E+02	9.098E+02	1.198E+03	2.782E+03	1.134E+04	1.588E+05
Xe	54	131.290	6.773E+02	6.858E+02	7.002E+02	7.210E+02	7.840E+02	8.827E+02	9.492E+02	1.248E+03	2.893E+03	1.178E+04	1.649E+05
Cs	55	132.905	7.042E+02	7.131E+02	7.280E+02	7.496E+02	8.150E+02	9.173E+02	9.863E+02	1.297E+03	3.003E+03	1.222E+04	1.711E+05
Ba	56	137.327	7.344E+02	7.435E+02	7.590E+02	7.814E+02	8.492E+02	9.554E+02	1.027E+03	1.349E+03	3.117E+03	1.267E+04	1.773E+05
La	57	138.906	7.623E+02	7.718E+02	7.879E+02	8.110E+02	8.813E+02	9.913E+02	1.065E+03	1.399E+03	3.231E+03	1.313E+04	1.837E+05
Ce	58	140.115	7.904E+02	8.002E+02	8.169E+02	8.409E+02	9.137E+02	1.028E+03	1.104E+03	1.450E+03	3.347E+03	1.359E+04	1.902E+05
Pr	59	140.908	8.187E+02	8.288E+02	8.461E+02	8.709E+02	9.463E+02	1.064E+03	1.144E+03	1.501E+03	3.464E+03	1.407E+04	1.968E+05

Table A6.9. Rutherford scattering cross sections (barns) of the elements for 1 MeV ^{28}Si (continued).

Element	Atomic no. (Z_2)	Mass (M_2)	Scattering angle (degrees)										
			175	170	165	160	150	140	135	120	90	60	30
Nd	60	144.240	8.498E+02	8.603E+02	8.782E+02	9.038E+02	9.818E+02	1.104E+03	1.186E+03	1.556E+03	3.586E+03	1.455E+04	2.036E+05
Pm	61	145.000	8.791E+02	8.900E+02	9.084E+02	9.349E+02	1.016E+03	1.142E+03	1.226E+03	1.609E+03	3.707E+03	1.504E+04	2.104E+05
Sm	62	150.360	9.131E+02	9.243E+02	9.434E+02	9.708E+02	1.054E+03	1.184E+03	1.272E+03	1.667E+03	3.835E+03	1.554E+04	2.174E+05
Eu	63	151.965	9.442E+02	9.558E+02	9.755E+02	1.004E+03	1.090E+03	1.224E+03	1.315E+03	1.723E+03	3.961E+03	1.605E+04	2.245E+05
Gd	64	157.250	9.789E+02	9.909E+02	1.011E+03	1.040E+03	1.129E+03	1.268E+03	1.362E+03	1.782E+03	4.093E+03	1.657E+04	2.316E+05
Tb	65	158.925	1.011E+03	1.023E+03	1.044E+03	1.075E+03	1.166E+03	1.309E+03	1.406E+03	1.840E+03	4.223E+03	1.709E+04	2.389E+05
Dy	66	162.500	1.045E+03	1.058E+03	1.080E+03	1.111E+03	1.205E+03	1.353E+03	1.452E+03	1.900E+03	4.357E+03	1.763E+04	2.463E+05
Ho	67	164.930	1.079E+03	1.092E+03	1.115E+03	1.147E+03	1.244E+03	1.396E+03	1.499E+03	1.960E+03	4.492E+03	1.817E+04	2.539E+05
Er	68	167.260	1.113E+03	1.127E+03	1.150E+03	1.183E+03	1.283E+03	1.440E+03	1.546E+03	2.021E+03	4.629E+03	1.871E+04	2.615E+05
Tm	69	168.934	1.148E+03	1.162E+03	1.185E+03	1.219E+03	1.323E+03	1.484E+03	1.593E+03	2.082E+03	4.767E+03	1.927E+04	2.693E+05
Yb	70	173.040	1.184E+03	1.199E+03	1.223E+03	1.258E+03	1.364E+03	1.530E+03	1.642E+03	2.146E+03	4.910E+03	1.984E+04	2.771E+05
Lu	71	174.967	1.220E+03	1.235E+03	1.260E+03	1.296E+03	1.405E+03	1.576E+03	1.691E+03	2.209E+03	5.053E+03	2.041E+04	2.851E+05
Hf	72	178.490	1.257E+03	1.272E+03	1.298E+03	1.335E+03	1.447E+03	1.623E+03	1.742E+03	2.275E+03	5.199E+03	2.099E+04	2.932E+05
Ta	73	180.948	1.294E+03	1.310E+03	1.336E+03	1.374E+03	1.490E+03	1.670E+03	1.792E+03	2.340E+03	5.346E+03	2.158E+04	3.014E+05
W	74	183.850	1.332E+03	1.348E+03	1.375E+03	1.414E+03	1.533E+03	1.719E+03	1.844E+03	2.407E+03	5.495E+03	2.218E+04	3.097E+05
Re	75	186.207	1.370E+03	1.386E+03	1.414E+03	1.454E+03	1.576E+03	1.767E+03	1.896E+03	2.474E+03	5.647E+03	2.278E+04	3.181E+05
Os	76	190.200	1.409E+03	1.426E+03	1.455E+03	1.496E+03	1.621E+03	1.817E+03	1.949E+03	2.543E+03	5.801E+03	2.340E+04	3.267E+05
Ir	77	192.220	1.448E+03	1.465E+03	1.494E+03	1.537E+03	1.666E+03	1.867E+03	2.002E+03	2.612E+03	5.956E+03	2.402E+04	3.353E+05
Pt	78	195.080	1.487E+03	1.505E+03	1.535E+03	1.579E+03	1.711E+03	1.917E+03	2.056E+03	2.682E+03	6.114E+03	2.465E+04	3.441E+05
Au	79	196.967	1.527E+03	1.545E+03	1.576E+03	1.621E+03	1.756E+03	1.968E+03	2.111E+03	2.752E+03	6.273E+03	2.528E+04	3.530E+05
Hg	80	200.590	1.568E+03	1.587E+03	1.619E+03	1.664E+03	1.803E+03	2.021E+03	2.167E+03	2.825E+03	6.435E+03	2.593E+04	3.620E+05
Tl	81	204.383	1.610E+03	1.629E+03	1.662E+03	1.709E+03	1.851E+03	2.074E+03	2.224E+03	2.898E+03	6.599E+03	2.658E+04	3.711E+05
Pb	82	207.200	1.652E+03	1.671E+03	1.705E+03	1.753E+03	1.899E+03	2.127E+03	2.281E+03	2.972E+03	6.765E+03	2.725E+04	3.803E+05
Bi	83	208.980	1.693E+03	1.713E+03	1.748E+03	1.797E+03	1.946E+03	2.180E+03	2.338E+03	3.046E+03	6.932E+03	2.792E+04	3.896E+05
Po	84	209.000	1.734E+03	1.755E+03	1.790E+03	1.840E+03	1.994E+03	2.233E+03	2.394E+03	3.120E+03	7.100E+03	2.859E+04	3.991E+05
At	85	210.000	1.776E+03	1.798E+03	1.834E+03	1.885E+03	2.042E+03	2.287E+03	2.452E+03	3.195E+03	7.271E+03	2.928E+04	4.086E+05
Rn	86	222.000	1.825E+03	1.847E+03	1.884E+03	1.937E+03	2.097E+03	2.348E+03	2.517E+03	3.278E+03	7.450E+03	2.998E+04	4.183E+05
Fr	87	223.000	1.869E+03	1.891E+03	1.928E+03	1.983E+03	2.147E+03	2.404E+03	2.577E+03	3.355E+03	7.625E+03	3.068E+04	4.281E+05
Ra	88	226.025	1.913E+03	1.936E+03	1.975E+03	2.030E+03	2.198E+03	2.461E+03	2.638E+03	3.434E+03	7.802E+03	3.139E+04	4.380E+05
Ac	89	227.028	1.958E+03	1.981E+03	2.020E+03	2.077E+03	2.249E+03	2.518E+03	2.699E+03	3.513E+03	7.981E+03	3.211E+04	4.480E+05
Th	90	232.038	2.005E+03	2.028E+03	2.069E+03	2.127E+03	2.302E+03	2.577E+03	2.762E+03	3.595E+03	8.164E+03	3.284E+04	4.581E+05
Pa	91	231.036	2.049E+03	2.073E+03	2.114E+03	2.174E+03	2.353E+03	2.634E+03	2.824E+03	3.675E+03	8.346E+03	3.357E+04	4.684E+05
U	92	238.029	2.098E+03	2.123E+03	2.165E+03	2.225E+03	2.409E+03	2.696E+03	2.890E+03	3.760E+03	8.534E+03	3.432E+04	4.787E+05
Np	93	237.048	2.143E+03	2.168E+03	2.211E+03	2.273E+03	2.461E+03	2.755E+03	2.952E+03	3.842E+03	8.720E+03	3.507E+04	4.892E+05
Pu	94	244.000	2.193E+03	2.219E+03	2.263E+03	2.326E+03	2.518E+03	2.818E+03	3.020E+03	3.928E+03	8.913E+03	3.583E+04	4.998E+05

Table A6.10. *F*-Factors. E is the lab energy of the incident ion divided by its atomic number. Z_1 and Z_2 are the incident and target atom's atomic numbers. The factors are only valid for scattering angles greater than 90°.

E/Z_1	**Z_2**										
(MeV)	**8**	**14**	**20**	**29**	**32**	**39**	**47**	**56**	**73**	**79**	**92**
0.100	0.992	0.983	0.973	0.956	0.950	0.935	0.917	0.895	0.851	0.834	0.796
0.200	0.996	0.992	0.987	0.978	0.975	0.968	0.958	0.948	0.925	0.917	0.898
0.300	0.997	0.994	0.991	0.985	0.983	0.978	0.972	0.965	0.950	0.945	0.932
0.400	0.998	0.996	0.993	0.989	0.988	0.984	0.979	0.974	0.963	0.958	0.949
0.500	0.998	0.997	0.995	0.991	0.990	0.987	0.983	0.979	0.970	0.967	0.959
0.600	0.999	0.997	0.996	0.993	0.992	0.989	0.986	0.983	0.975	0.972	0.966
0.700	0.999	0.998	0.996	0.994	0.993	0.991	0.988	0.985	0.979	0.976	0.971
0.800	0.999	0.998	0.997	0.995	0.994	0.992	0.990	0.987	0.981	0.979	0.975
0.900	0.999	0.998	0.997	0.995	0.994	0.993	0.991	0.988	0.983	0.982	0.977
1.000	0.999	0.998	0.997	0.996	0.995	0.994	0.992	0.990	0.985	0.983	0.980
1.100	0.999	0.998	0.998	0.996	0.995	0.994	0.992	0.990	0.986	0.985	0.981
1.200	0.999	0.999	0.998	0.996	0.996	0.995	0.993	0.991	0.988	0.986	0.983
1.300	0.999	0.999	0.998	0.997	0.996	0.995	0.994	0.992	0.989	0.987	0.984
1.400	0.999	0.999	0.998	0.997	0.996	0.995	0.994	0.993	0.989	0.988	0.985
1.500	0.999	0.999	0.998	0.997	0.997	0.996	0.994	0.993	0.990	0.989	0.986
1.600	1.000	0.999	0.998	0.997	0.997	0.996	0.995	0.993	0.991	0.990	0.987
1.700	1.000	0.999	0.998	0.997	0.997	0.996	0.995	0.994	0.991	0.990	0.988
1.800	1.000	0.999	0.999	0.998	0.997	0.996	0.995	0.994	0.992	0.991	0.989
1.900	1.000	0.999	0.999	0.998	0.997	0.997	0.996	0.994	0.992	0.991	0.989
2.000	1.000	0.999	0.999	0.998	0.998	0.997	0.996	0.995	0.993	0.992	0.990
2.100	1.000	0.999	0.999	0.998	0.998	0.997	0.996	0.995	0.993	0.992	0.990
2.200	1.000	0.999	0.999	0.998	0.998	0.997	0.996	0.995	0.993	0.992	0.991
2.300	1.000	0.999	0.999	0.998	0.998	0.997	0.996	0.995	0.994	0.993	0.991
2.400	1.000	0.999	0.999	0.998	0.998	0.997	0.997	0.996	0.994	0.993	0.992
2.500	1.000	0.999	0.999	0.998	0.998	0.997	0.997	0.996	0.994	0.993	0.992
2.600	1.000	0.999	0.999	0.998	0.998	0.998	0.997	0.996	0.994	0.994	0.992
2.700	1.000	0.999	0.999	0.998	0.998	0.998	0.997	0.996	0.994	0.994	0.992
2.800	1.000	0.999	0.999	0.998	0.998	0.998	0.997	0.996	0.995	0.994	0.993
2.900	1.000	0.999	0.999	0.998	0.998	0.998	0.997	0.996	0.995	0.994	0.993
3.000	1.000	0.999	0.999	0.999	0.998	0.998	0.997	0.997	0.995	0.994	0.993

REFERENCES

Hautala, M., and Luomajärvi, M. (1980), *Radiat. Eff.* **45**, 159.

L'Ecuyer, J., Davies, J.A., and Matsunami, N. (1979), *Nucl. Instrum. Methods* **160**, 337.

Walker, F. W., Parrington, J. R., and Feiner, F., Revised. (1989), *Nuclides and Isotopes, Chart of the Nuclides,* Fourteenth Edition, General Electric Company, Nuclear Energy Operations.

APPENDIX

7

NON-RUTHERFORD ELASTIC BACKSCATTERING CROSS SECTIONS

Compiled by

R. P. Cox, J. A. Leavitt, and L. C. McIntyre, Jr.

University of Arizona, Tucson, Arizona

This appendix contains non-Rutherford cross section information for elastic backscattering of ^{1}H and ^{4}He analysis beam ions by target elements with $Z \leq 20$. For ^{1}H projectiles, laboratory energies 0.5-3.0 MeV are emphasized; for ^{4}He projectiles, laboratory energies 1-10 MeV are emphasized. No data for non-Rutherford backscattering of heavier ions are included; for beam energies below 10 MeV, the heavier ion scattering is usually Rutherford (except for electron shell corrections). Further, no useful cross section data were found in the literature for the following beam target pairs: ^{1}H-K, ^{4}He-P, and ^{4}He-S.

The non-Rutherford cross section information in this appendix is presented in graphical form only, as plots of the ratio of the measured cross section to the Rutherford cross section, σ/σ_R, vs. the incident projectile laboratory energy, E_{Lab}, at a specific, large laboratory backscattering angle, θ_{Lab}. In the few instances where the original references give the measured cross section data in tabular form, the data are plotted as *points* on the graphs. In the many instances where only graphical data are presented in the original references, the data were digitized by the procedure described below and used to produce continuous *curves* of σ/σ_R vs. E_{Lab}.

The procedure used to digitize the cross section plots appearing in the original references was as follows. The appropriate figure in the original paper was xerographically enlarged. A smooth curve was drawn by hand through the relevant experimental points. This smooth curve was digitized at irregular intervals with a SUMMA SKETCH II (Suma Graphics) digitizing tablet; the result was stored in a PC ASCII disk file. These data were converted to a σ/σ_R vs. E_{Lab} file, which was also stored on the disk. A plotting program, GRAFTOOL (3-D Visions), was used to produce the final plots of these data (with the digitized points connected with straight lines) showing appropriate regions of interest with appropriate vertical scales.

The absolute accuracy of the σ/σ_R ratios produced by this procedure is difficult to estimate and will vary from plot to plot. The accuracy of the digitizing procedure itself was tested by digitizing plots of reported tabular data for scattering of ^{4}He by O (Leavitt *et al.*, 1990) and comparing the result of the digitizing procedure with the original tabular data. The digitized data agreed with the tabular data to within less than 5%. It should be noted that the plots digitized in this instance are larger and clearer than those normally encountered in the literature. So this test may indicate the lower limit of the uncertainty associated with the use of the digitized curves.

No systematic effort has been made to judge the reliability of the data contained in the original papers themselves. However, the occurance of several well-known instances of discrepancies between results produced by different investigators does not inspire confidence in the reliability of the quoted uncertainties. An example is the ^{1}H-^{16}O cross section for 0.6 MeV < E_{Lab} < 2.5 MeV with $\theta_{Lab} \approx 164°$-$170°$. The values reported by Laubenstein *et al.* (1951), are a factor 2.9 larger than the values reported by Luomajarvi *et al.* (1985) and Chow *et al.* (1975). Another example is the ^{4}He-^{11}B cross section for 2 MeV < E_{Lab} < 2.4 MeV with $\theta_{Lab} \approx 151°$-$170°$. The values reported by Ramirez *et al.* (1972) are a factor of 1.5 larger than the values reported by McIntyre *et al.* (1991).

It should be noted that all available data on non-Rutherford cross sections for ^{1}H and ^{4}He projectiles are not included in the plots in this appendix. The original papers frequently display σ_{CM} vs. E_{Lab} plots for several values of θ_{CM}. The σ/σ_R vs. E_{Lab} curves in this appendix usually display information for one large backscattering angle (150° < θ_{Lab} < 170°) only. Further, several original sources frequently contain essentially the same data; data from only one of the sources is usually included here. For more complete lists of references on non-Rutherford cross sections, see Leavitt and McIntyre (1991) for ^{4}He projectiles, and Rauhala (1991) for ^{1}H projectiles. Further, on-line data bases, such as NNDC and SIGBASE (see References for details) should be checked for the most recent cross section information.

The curves contained in this appendix are intended to serve as a general indication of the energy dependence of the cross section. They may be used to locate flat regions in σ/σ_R for general backscattering work or to locate resonances that may be used for depth profiling. Once the colliding pair and energy region of interest have been chosen, all relevant information in the more complete reference lists should be gathered. This additional information may also be of use in judging the accuracy of the σ/σ_R ratio.

As has been indicated, values of σ/σ_R obtained from the curves should be used with caution in actual calculations since these values may be uncertain by at least 5%. If higher accuracy is desired, or if the analyst needs σ/σ_R vs. E_{Lab} data for θ_{Lab} values not given here (or in the more extensive references), the analyst may wish to actually *measure* σ/σ_R vs. E_{Lab} at the θ_{Lab} of interest. Results of attempts to obtain σ/σ_R values (for θ_{Lab} at values not reported) by interpolation are questionable, at best. For example, results (John *et al.*, 1969) from ^{4}He-^{16}O cross section measurements for E_{Lab} near 8.8 Mev, namely: at $\theta_{Lab} = 165.6°, \sigma/\sigma_R \doteq 28$, while at $\theta_{Lab} = 145.7°, \sigma/\sigma_R \doteq 7$, give an indication of the strong dependence of σ/σ_R on θ_{Lab}. Accurate measurement of the cross sections may require construction of a sample for the purpose. The usual sample consists of a stable thin film containing the target element (*x*) of interest deposited on a light substrate (such as C), and covered by a thin film of heavy metal (*m*). Except for small corrections due to electron shell and film thickness effects, the cross section ratio at energy *E* may be calculated from

$$(\sigma/\sigma_R)_E \doteq \frac{(A_x/A_m)_E}{(A_x/A_m)_{E°}} ,$$

where $(A_x/A_m)_E$ and $(A_x/A_m)_{E°}$ are measured integrated peak count ratios at analysis beam energies *E* and *E*°, respectively. Here E° is sufficiently low energy so that both cross sections (for x and m) are Rutherford; it is assumed that the cross section for m is also Rutherford at energy E. This comparison procedure effectively eliminates the dependence of the result on the accuracy of charge and solid angle measurements. It is capable of producing results with uncertainties of ±2%. For a discussion of the corrections and application of this procedure to the measurement of the ^{4}He-C cross sections, see Leavitt *et al.* (1989).

The non-Rutherford cross section plots for ^{1}H backscattering are followed by those for ^{4}He backscattering; the References conclude this appendix. A target element superscript on the plots indicates that the target was isotopically pure; lack of the superscript indicates the target contained isotopes in the naturally occuring abundances.

We acknowledge the assistance of M.D. Ashbaugh, J.D. Frank, and Z. Lin during the initial stages of this project.

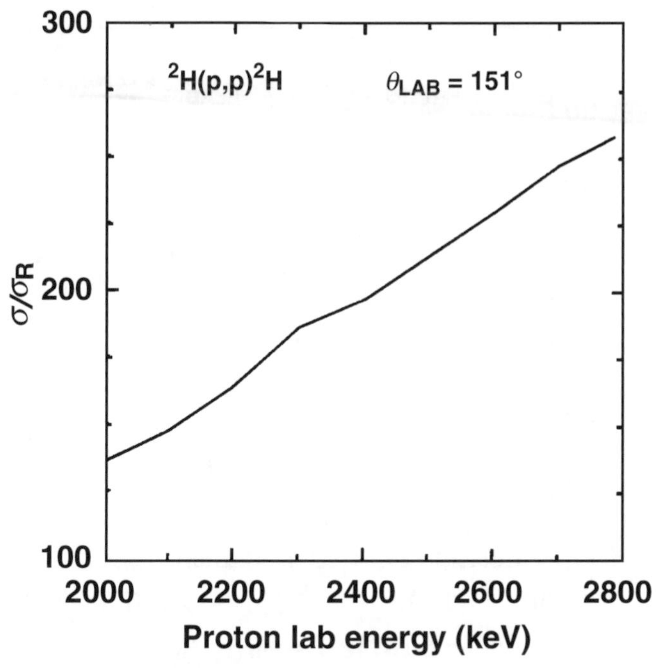

FIG. A7.1. Langley (1976).

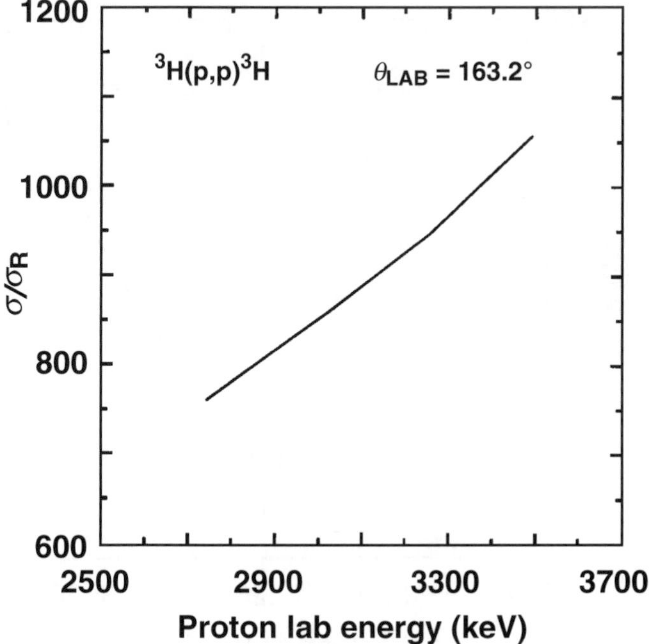

FIG. A7.2. Langley (1976).

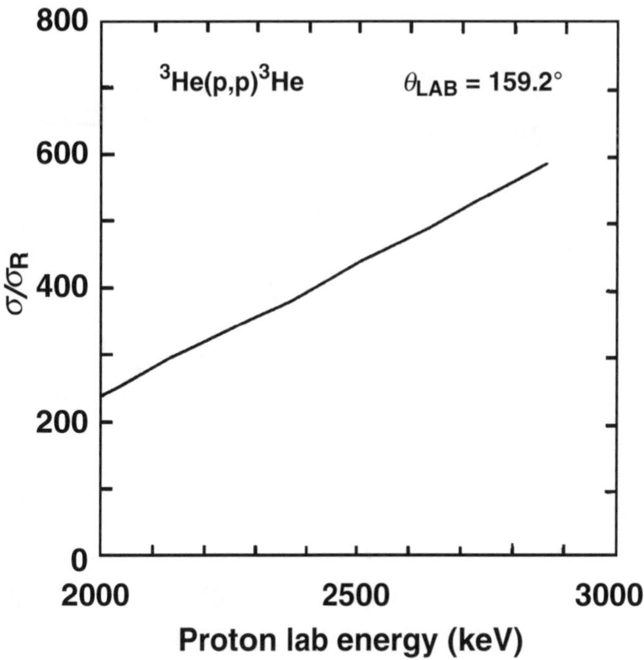

FIG. A7.3. Langley (1976).

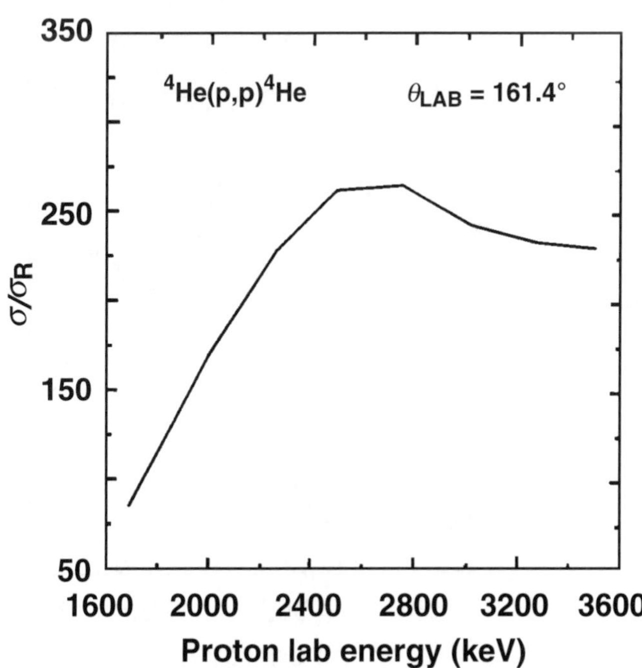

FIG. A7.4. Langley (1976).

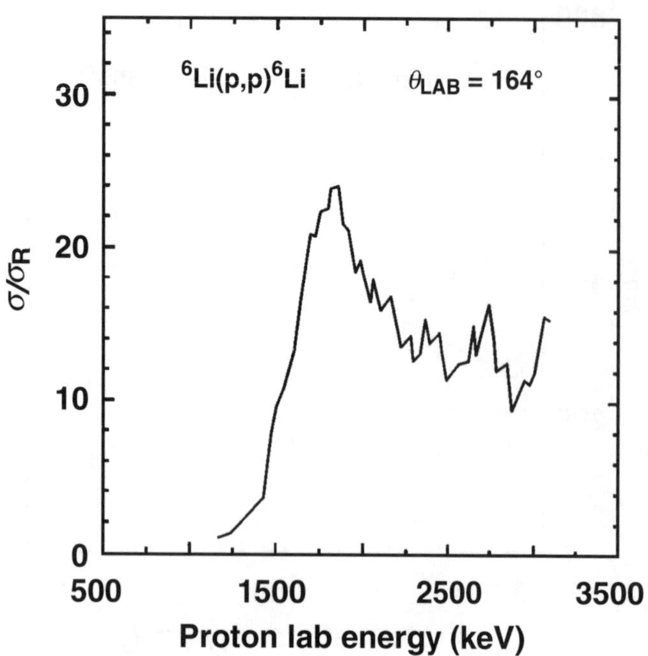

FIG. A7.5. Bashkin and Richards (1951).

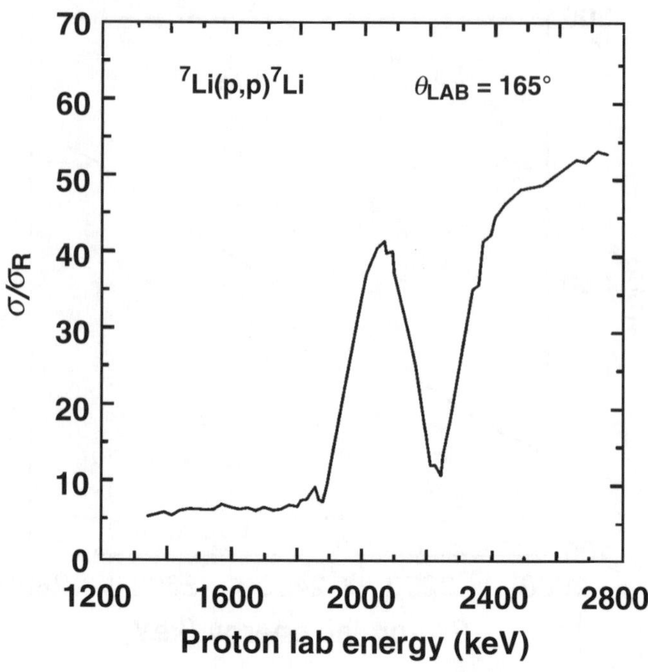

FIG. A7.7. Malmberg (1956).

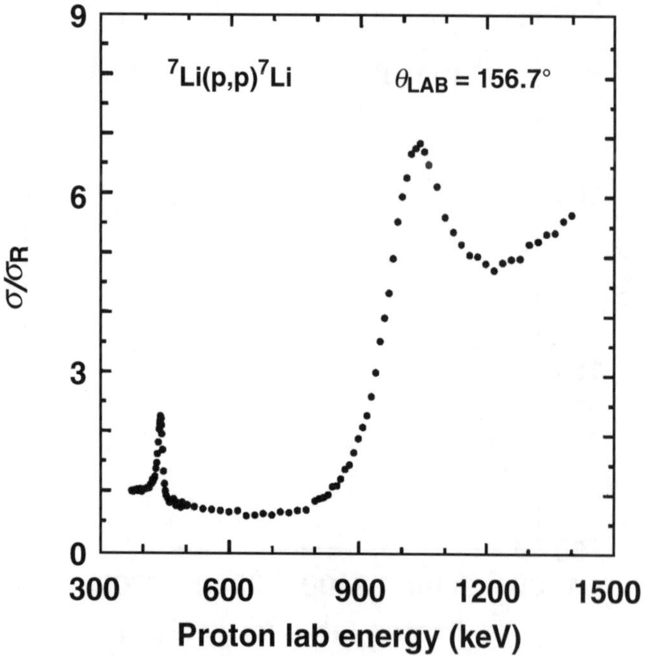

FIG. A7.6. Warters *et al.* (1953).

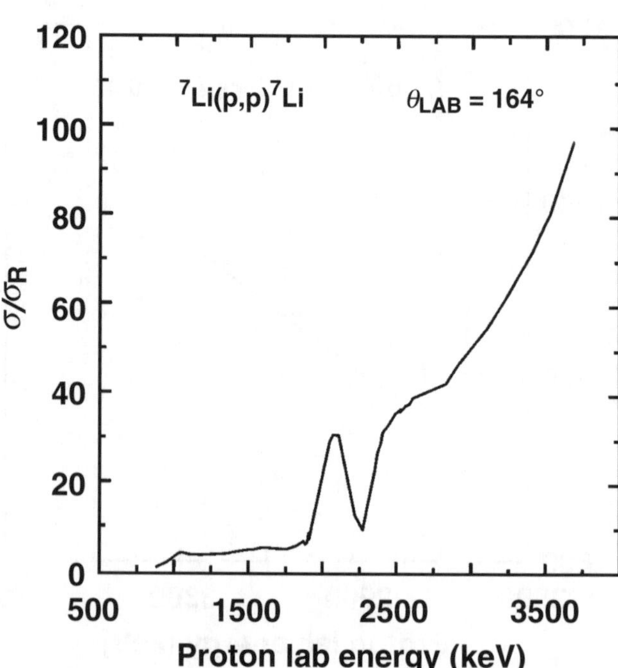

FIG. A7.8. Bashkin and Richards (1951).

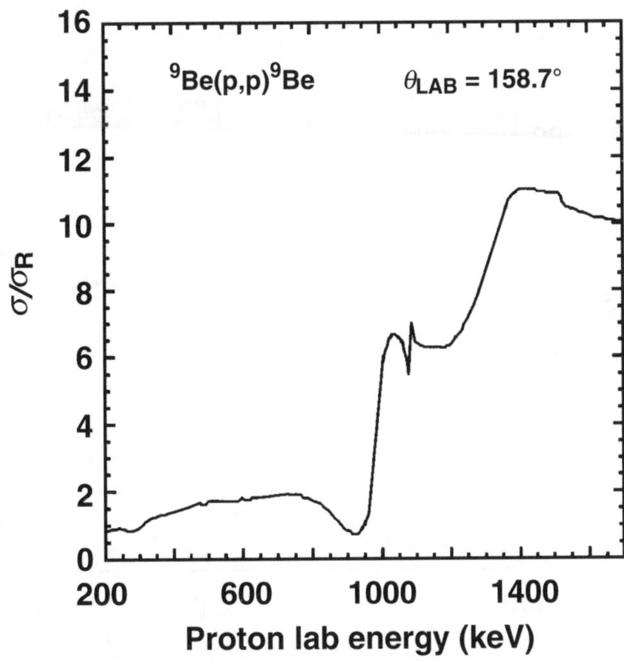

FIG. A7.9. Mozer (1956).

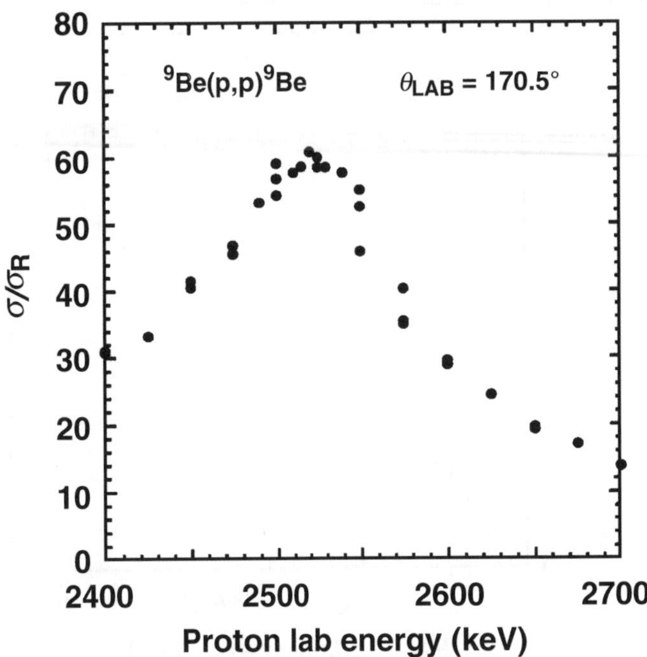

FIG. A7.11. Leavitt *et al.* (1994).

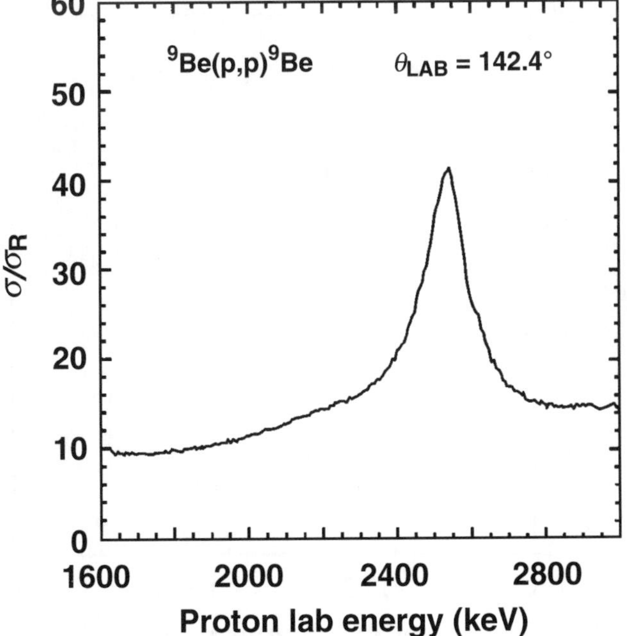

FIG. A7.10. Mozer (1956).

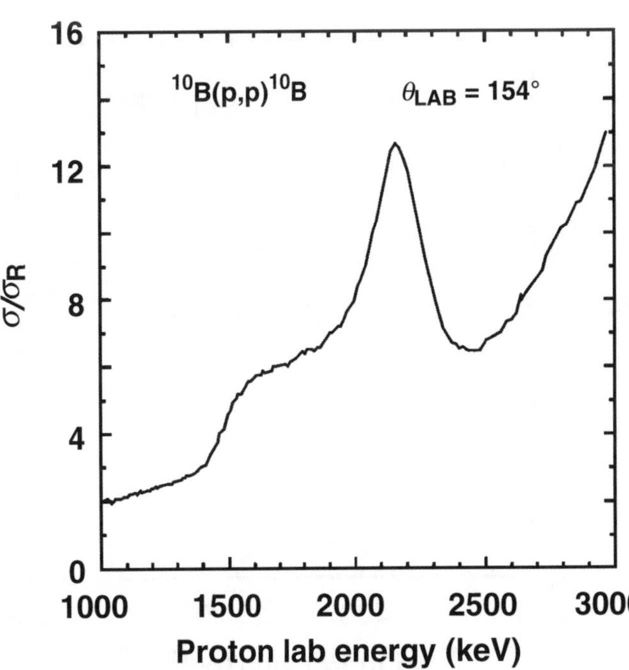

FIG. A7.12. Overley and Whaling (1962).

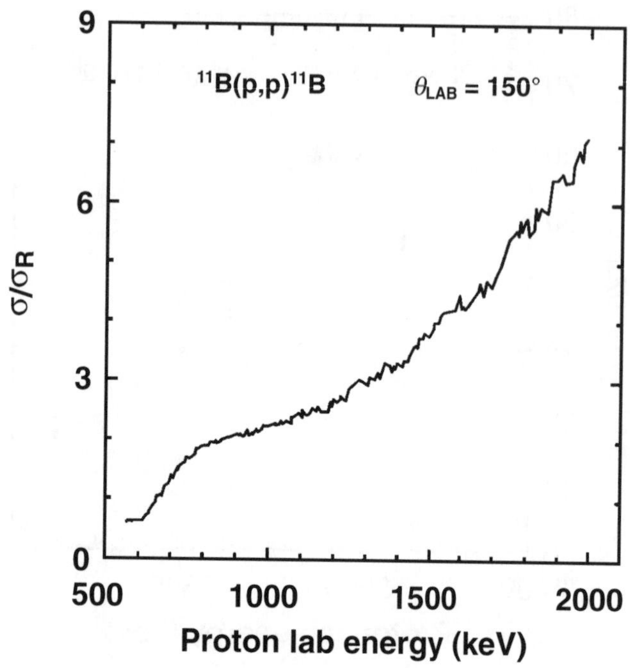

FIG. A7.13. Tautfest and Rubin (1956).

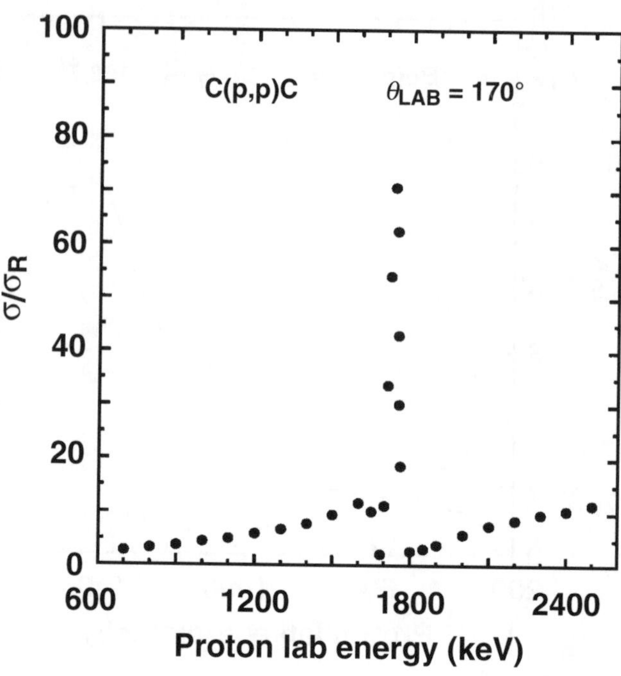

FIG. A7.15. Rauhala (1985).

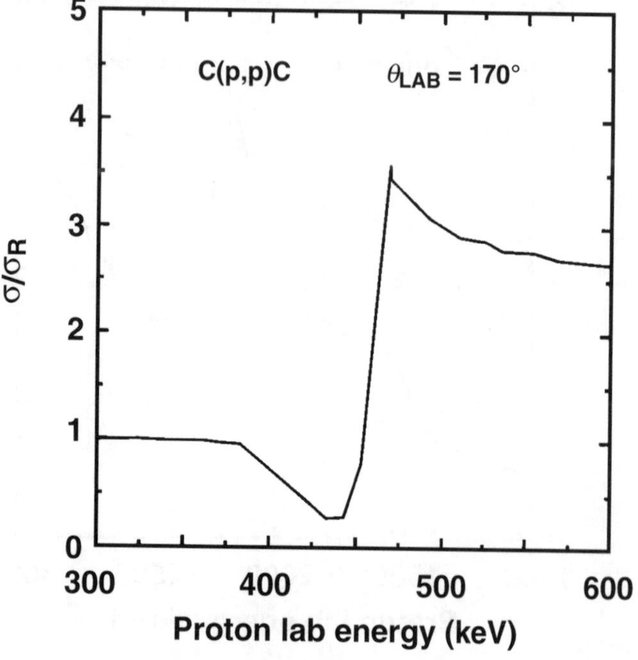

FIG. A7.14. Liu *et al.* (1993).

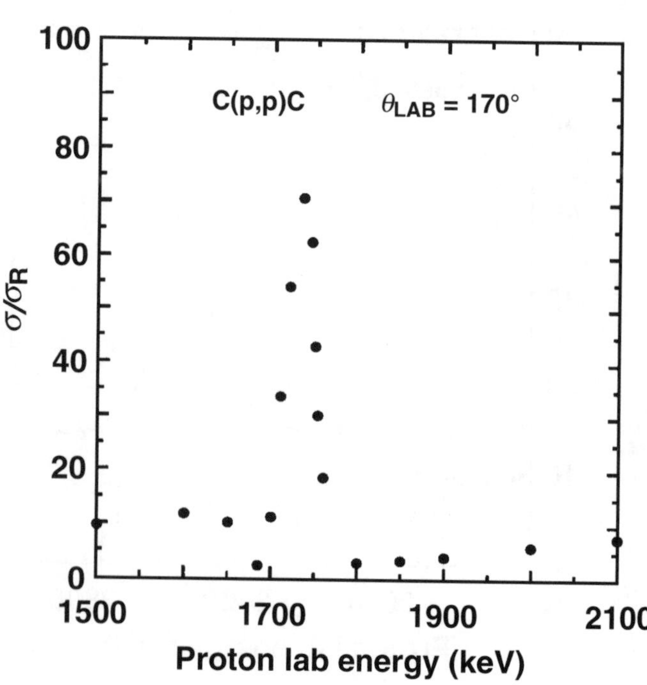

FIG. A7.16. Rauhala (1985).

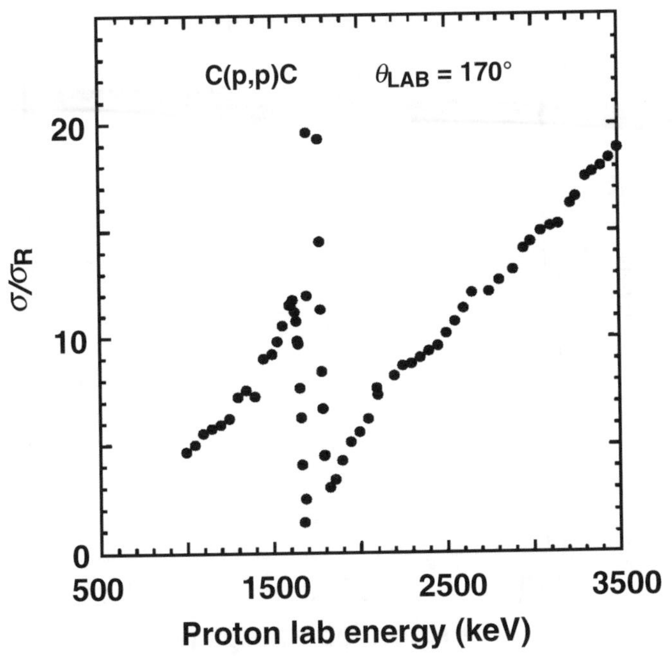

FIG. A7.17. Amirikas *et al.* (1993).

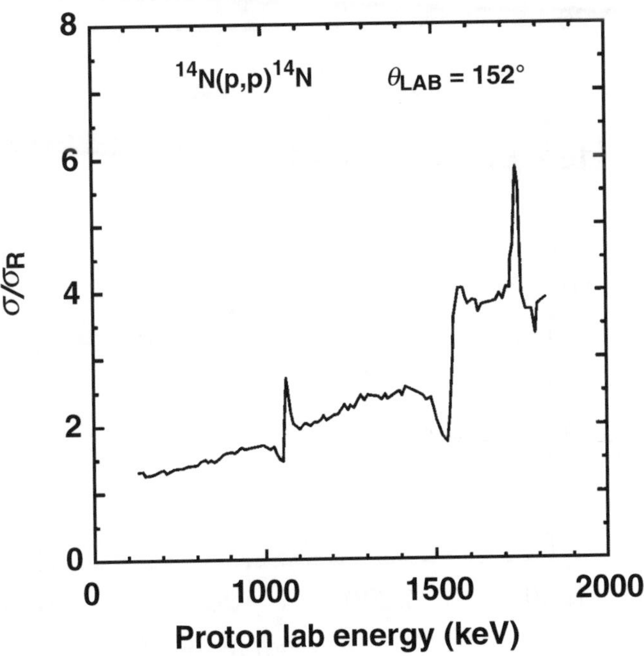

FIG. A7.19. Hagedorn *et al.* (1957).

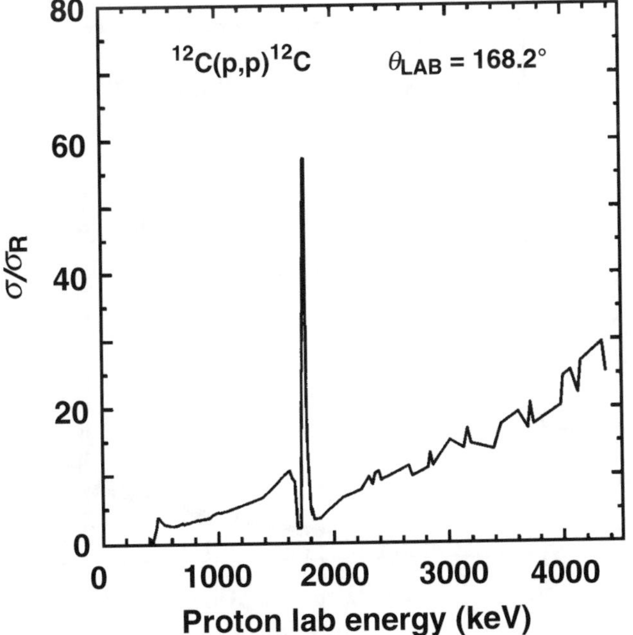

FIG. A7.18. Jackson *et al.* (1953).

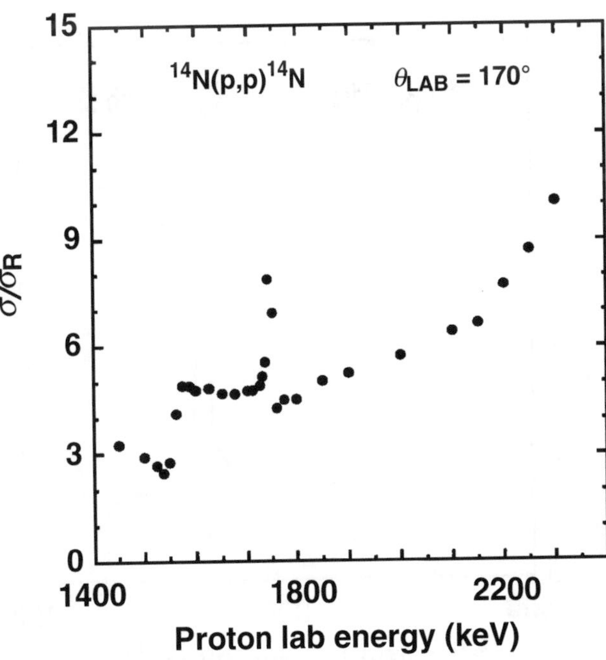

FIG. A7.20. Rauhala (1985).

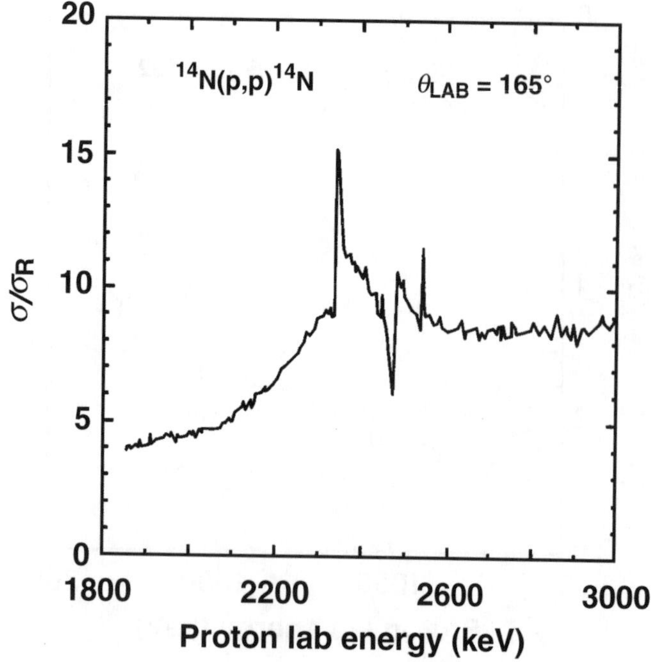

FIG. A7.21. Lambert and Durand (1967).

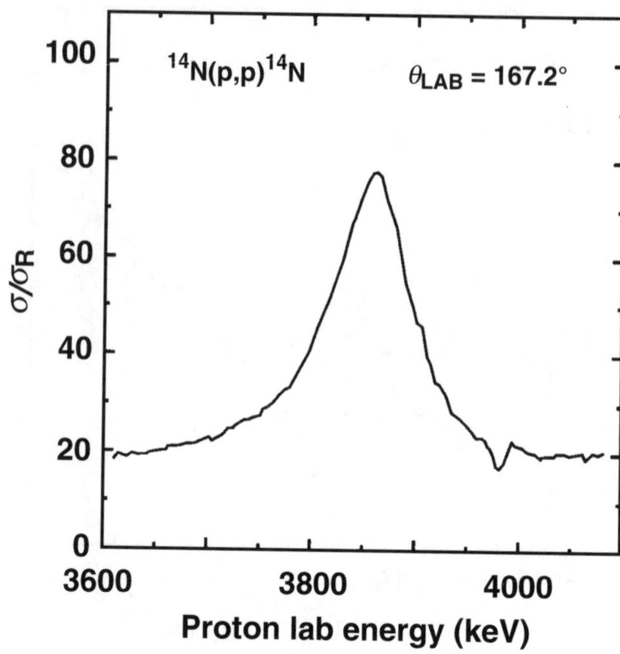

FIG. A7.23. Olness *et al.* (1958a).

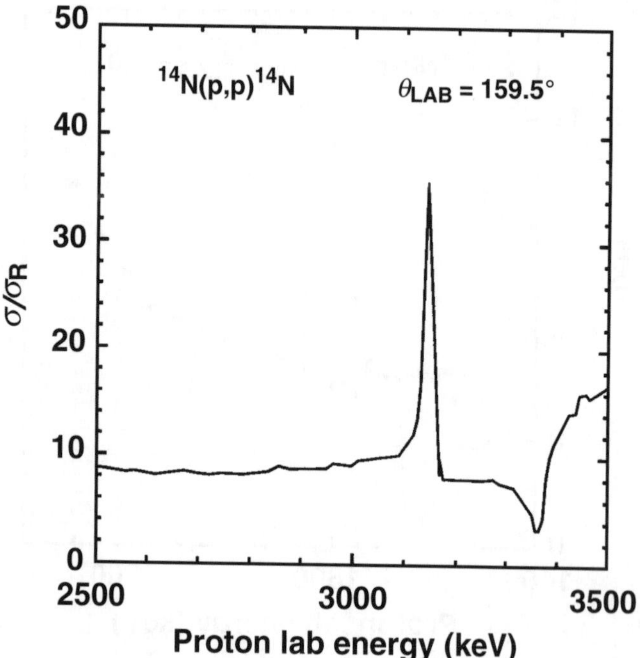

FIG. A7.22. Bashkin *et al.* (1959).

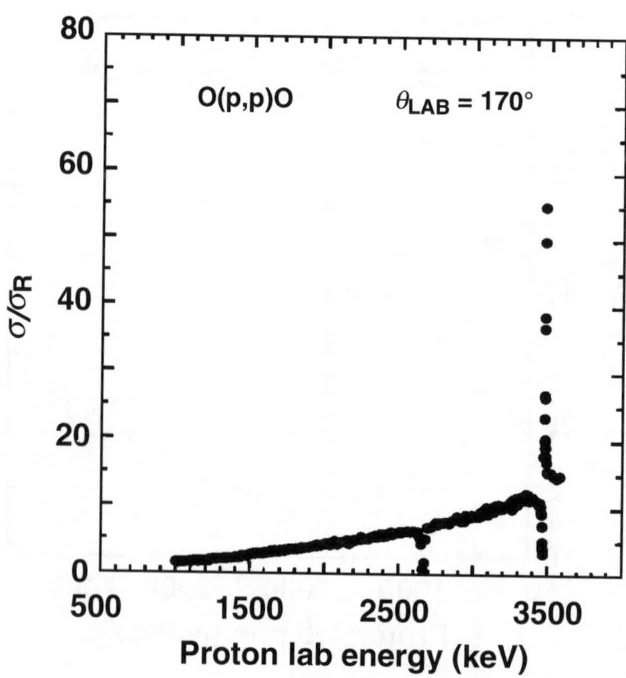

FIG. A7.24. Amirikas *et al.* (1993).

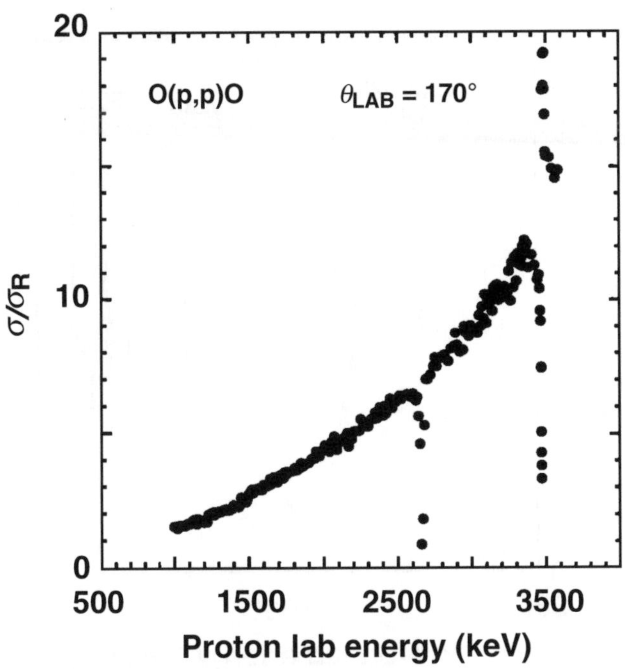

FIG. A7.25. Amirikas *et al.* (1993).

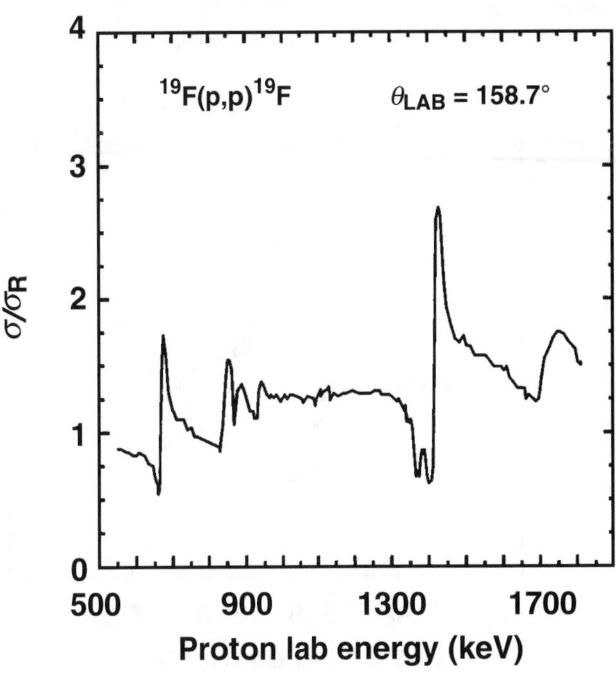

FIG. A7.27. Webb *et al.* (1955).

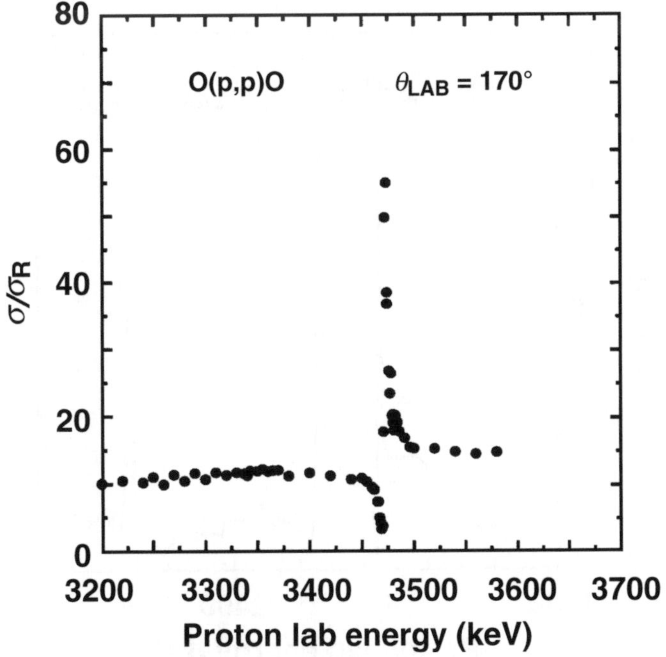

FIG. A7.26. Amirikas *et al.* (1993).

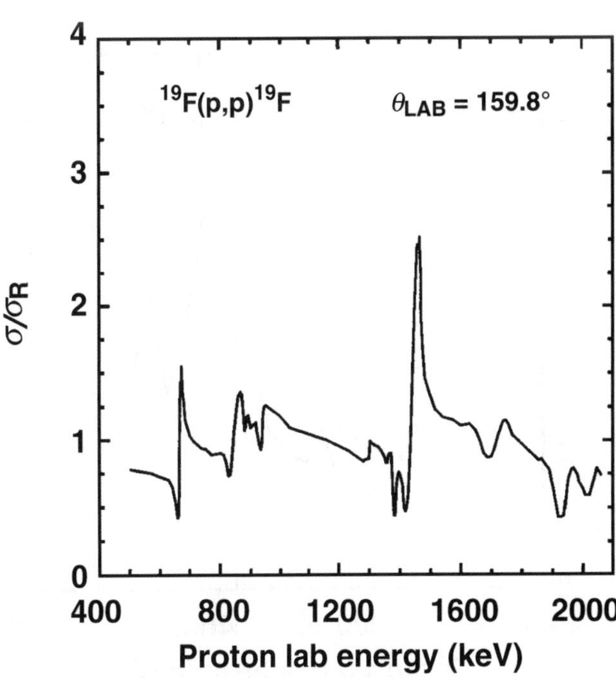

FIG. A7.28. Dearnaly (1956).

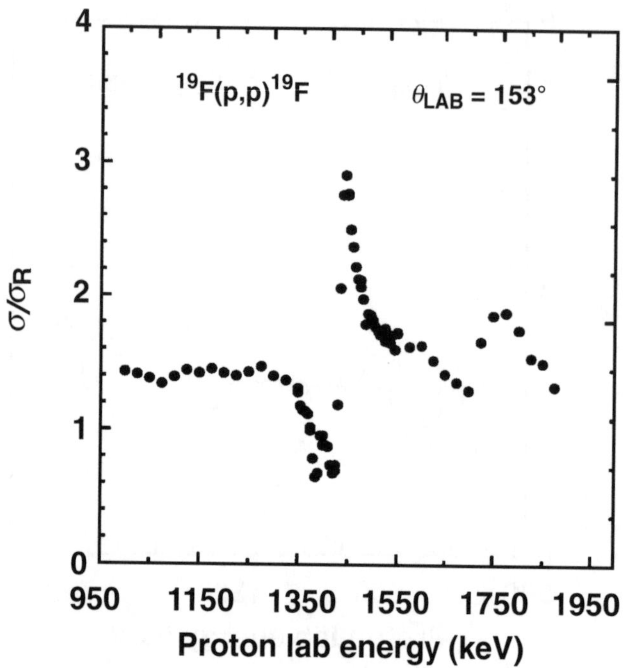

FIG. A7.29. Knox and Harmon (1989).

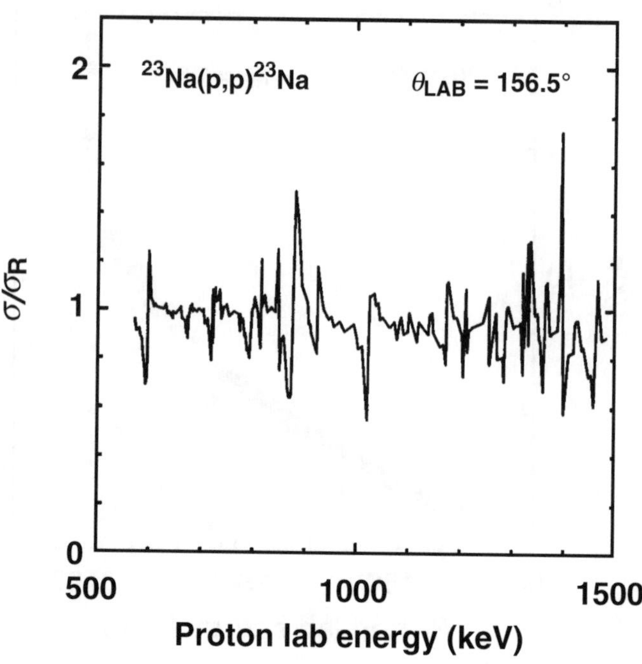

FIG. A7.31. Baumann *et al.* (1956).

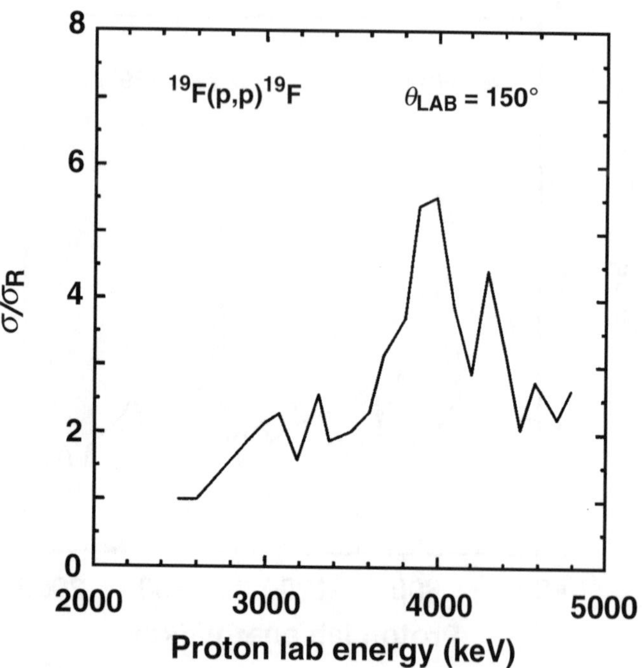

FIG. A7.30. Bogdanovic *et al.* (1993).

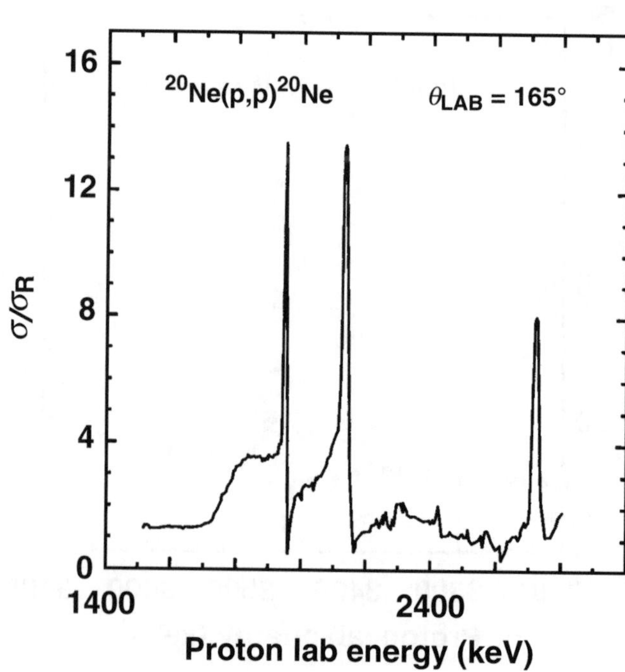

FIG. A7.32. Lambert *et al.* (1971).

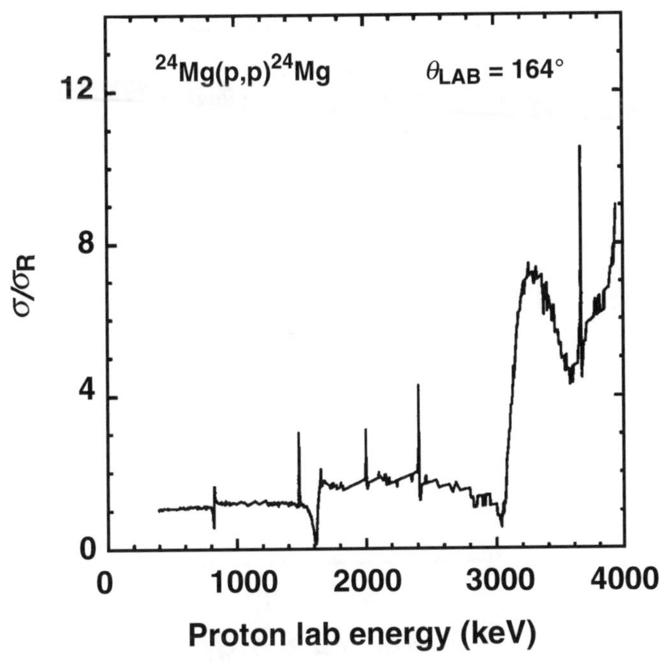

FIG. A7.33. Mooring *et al.* (1951).

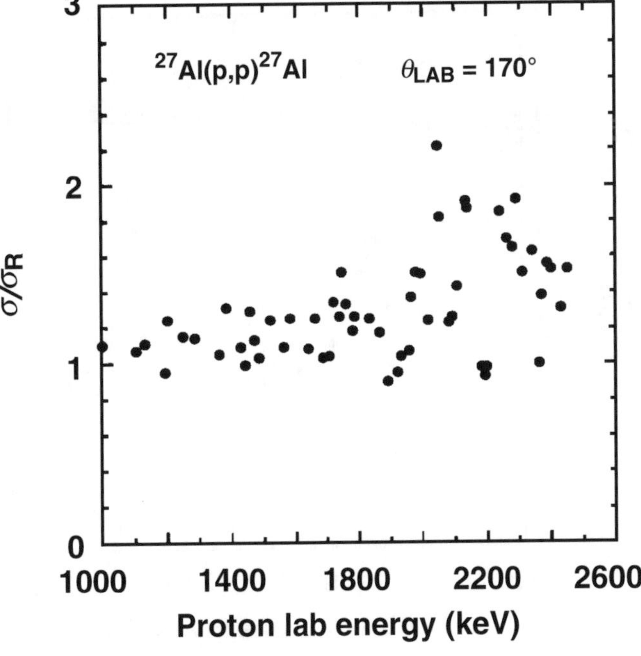

FIG. A7.35. Rauhala (1989).

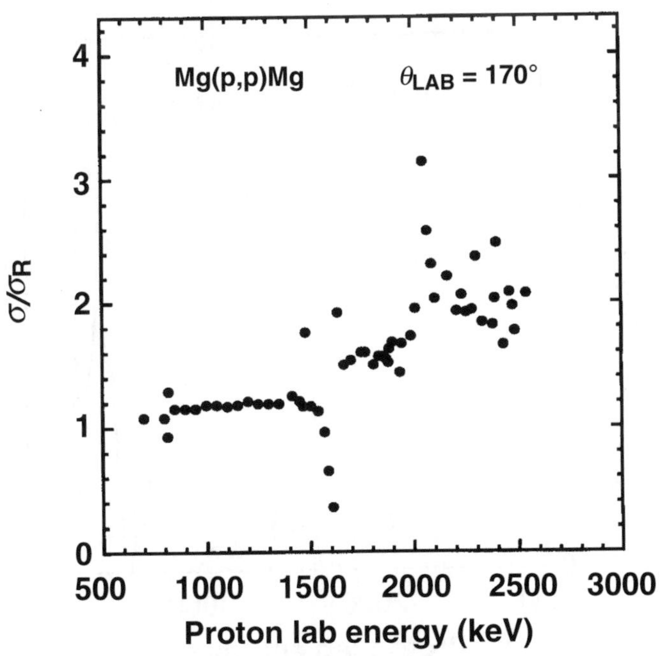

FIG. A7.34. Rauhala and Loumajarvi (1988).

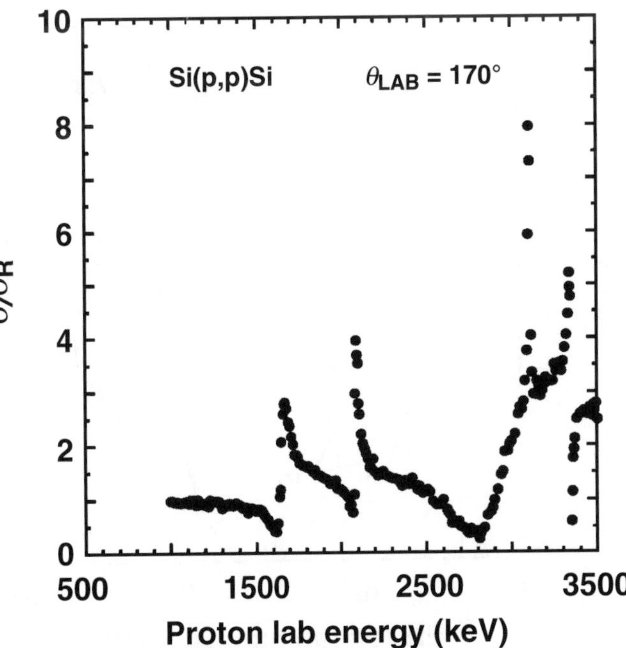

FIG. A7.36. Amirikas *et al.* (1993).

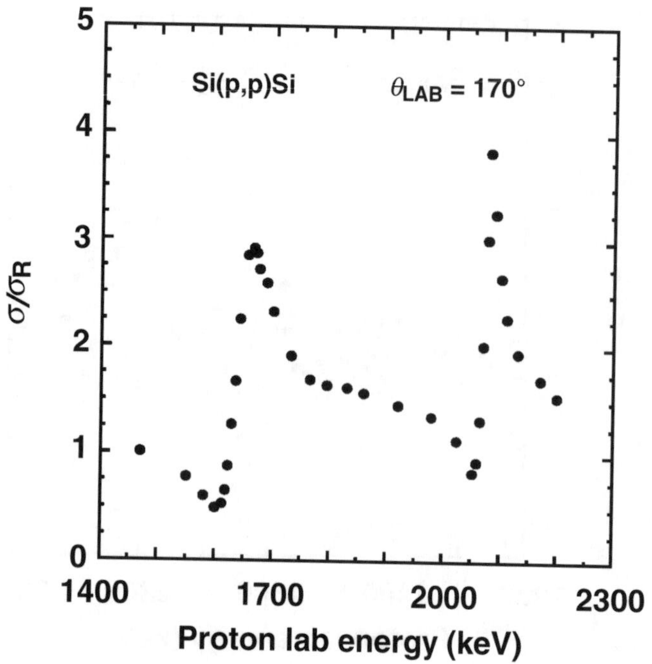

FIG. A7.37. Rauhala (1985).

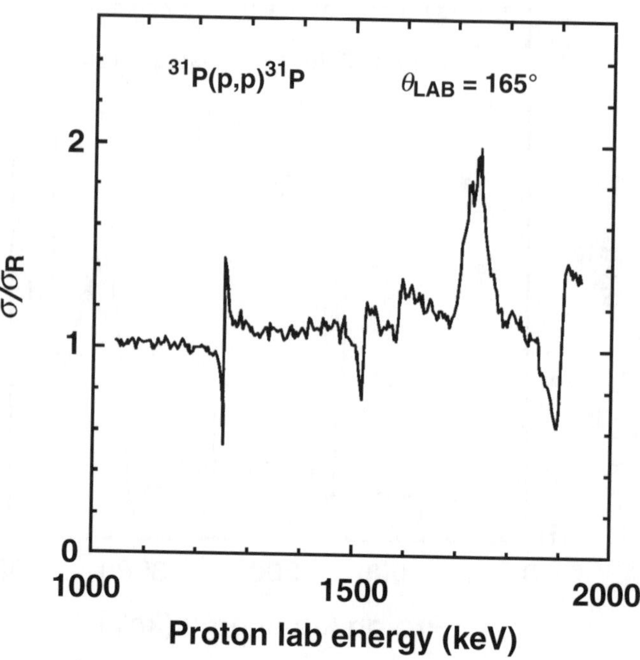

FIG. A7.39. Cohen-Ganouna *et al*. (1963).

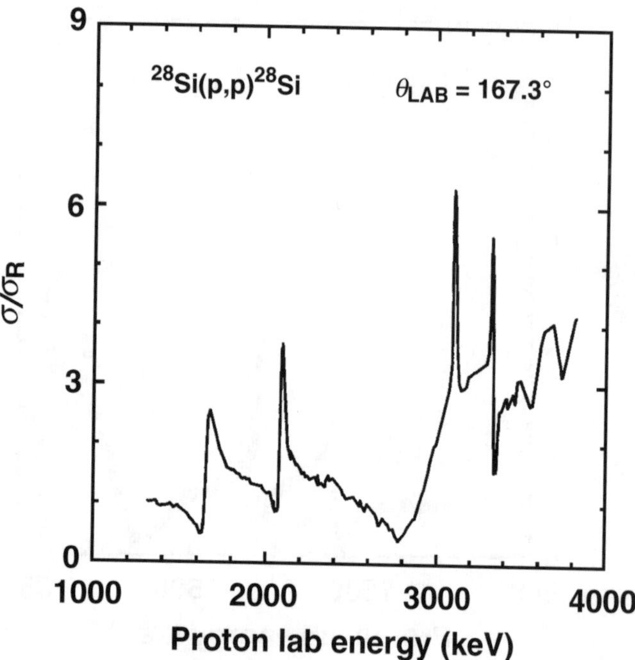

FIG. A7.38. Vorona *et al*. (1959).

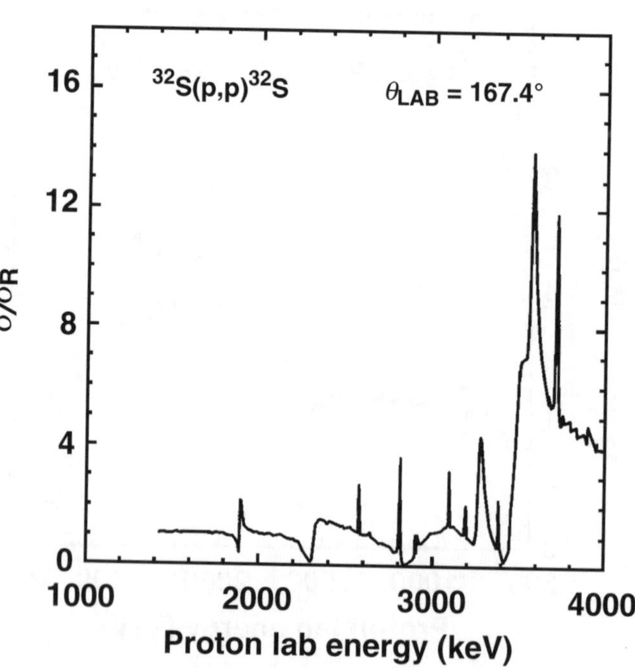

FIG. A7.40. Olness *et al*. (1958b).

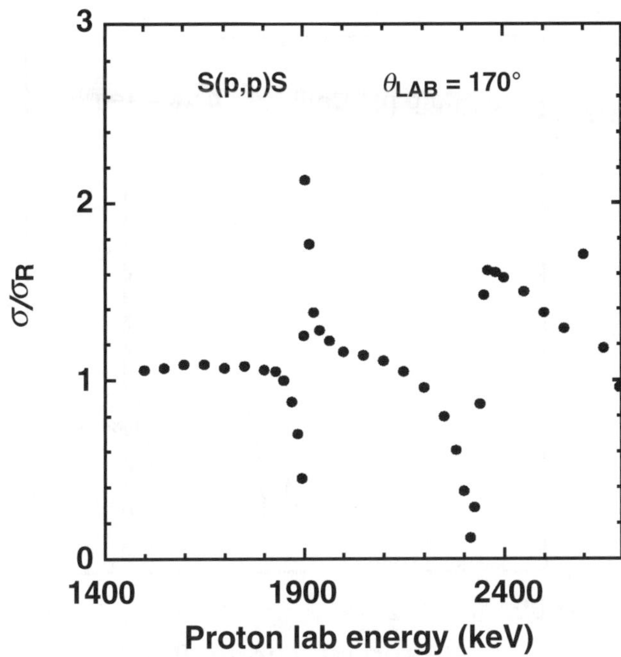

FIG. A7.41. Rauhala and Loumajarvi (1988).

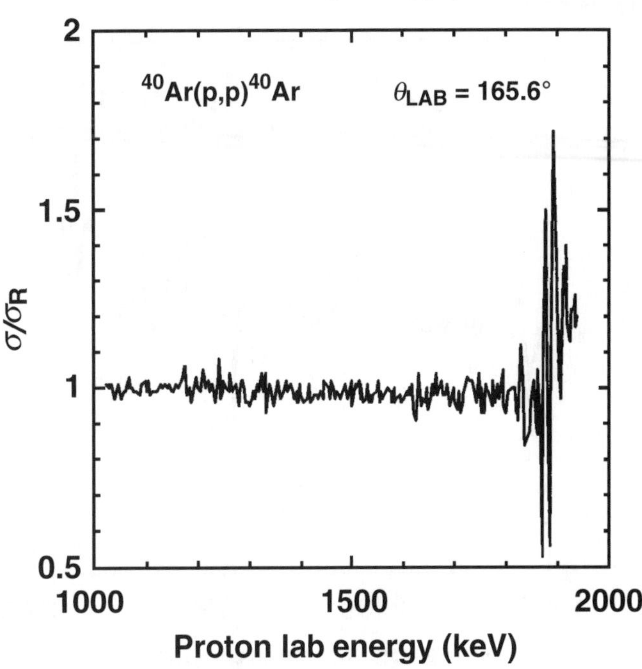

FIG. A7.43. Cohen-Ganouna *et al.* (1963).

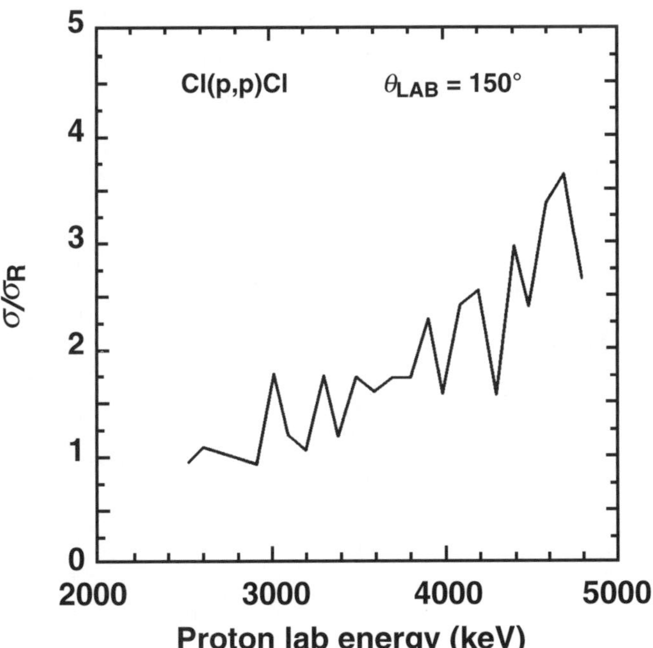

FIG. A7.42. Bogdanovic *et al.* (1993).

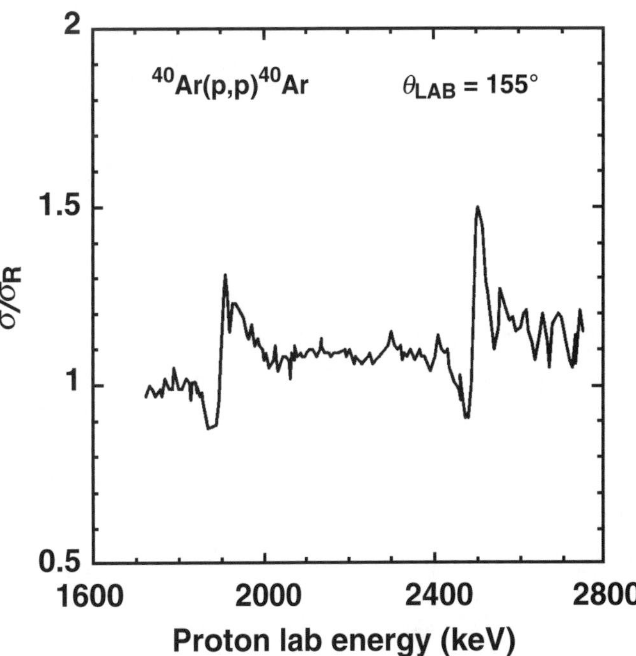

FIG. A7.44. Frier *et al.* (1958).

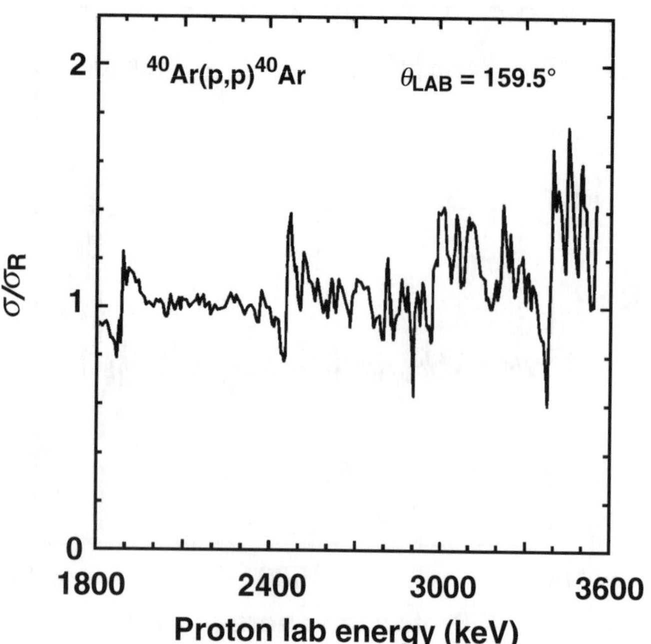

FIG. A7.45. Barnhard and Kim (1961).

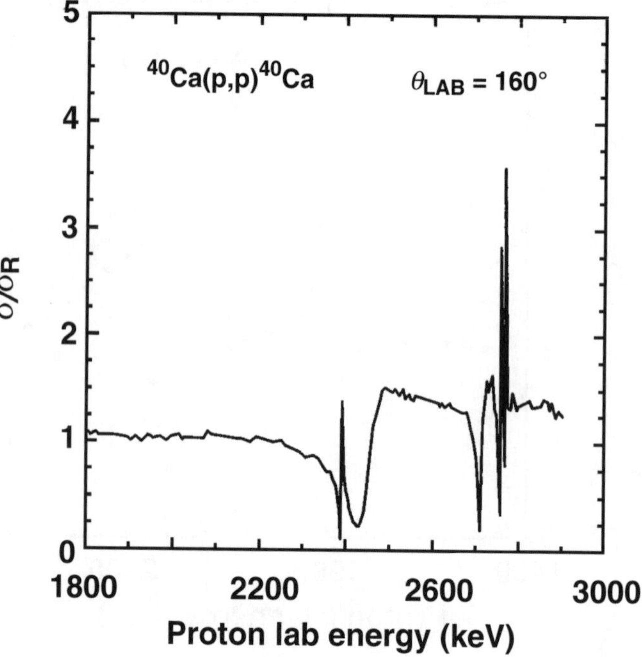

FIG. A7.46. Wilson *et al.* (1974).

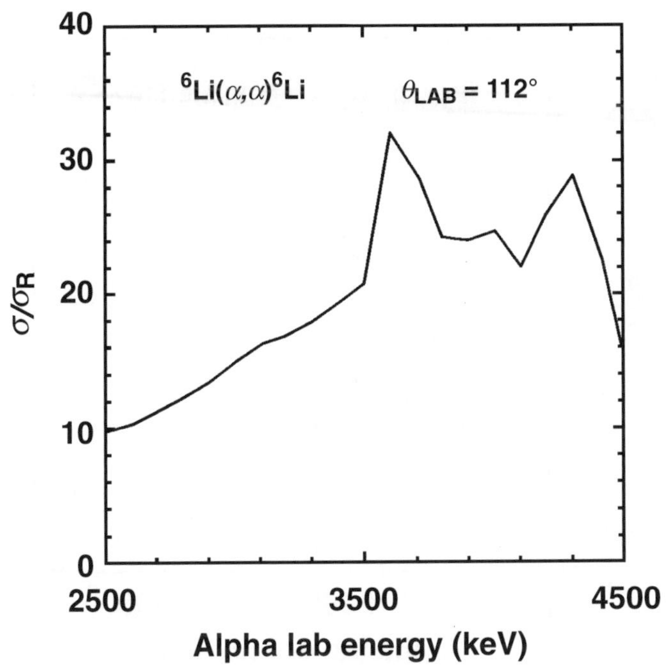

FIG. A7.47. Bolhen *et al.* (1972).

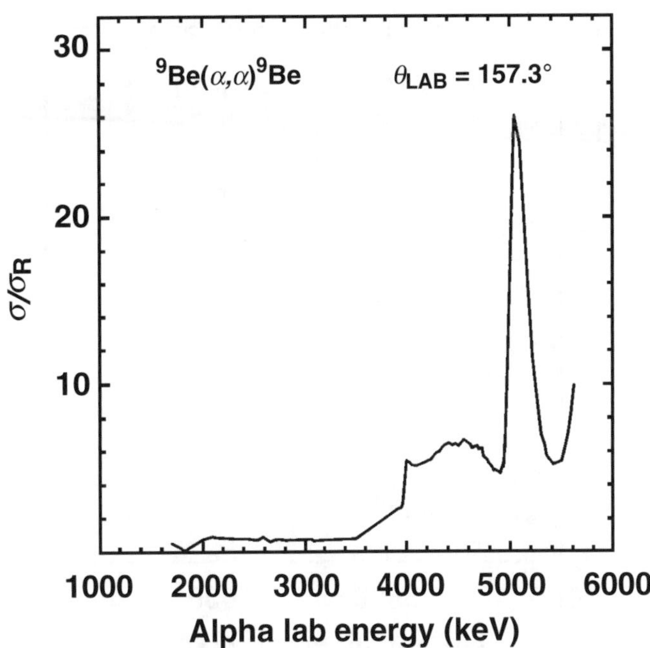

FIG. A7.49. Goss *et al.* (1973).

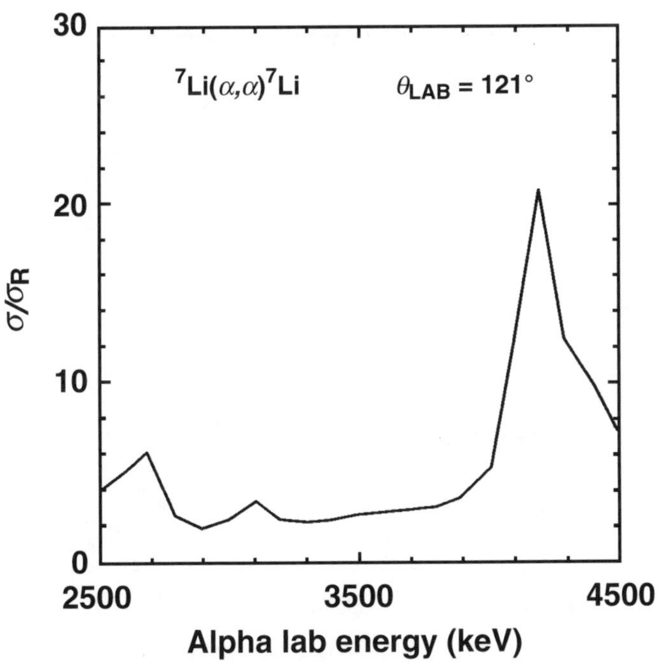

FIG. A7.48. Bolhen *et al.* (1972).

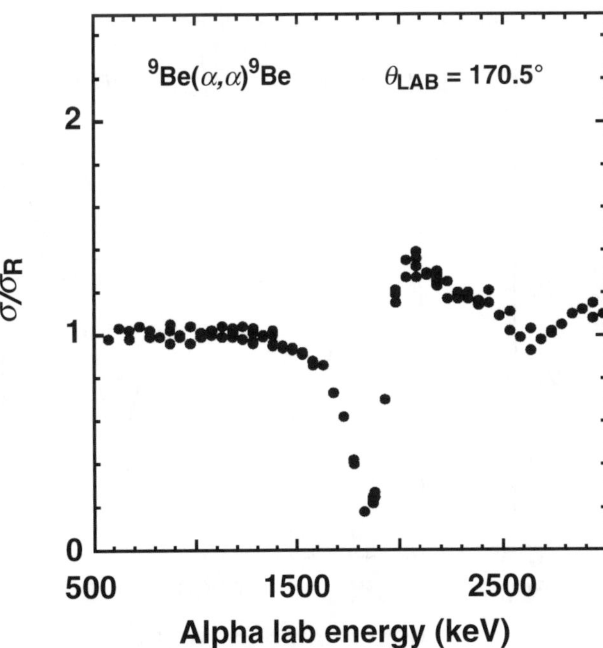

FIG. A7.50. Leavitt *et al.* (1994).

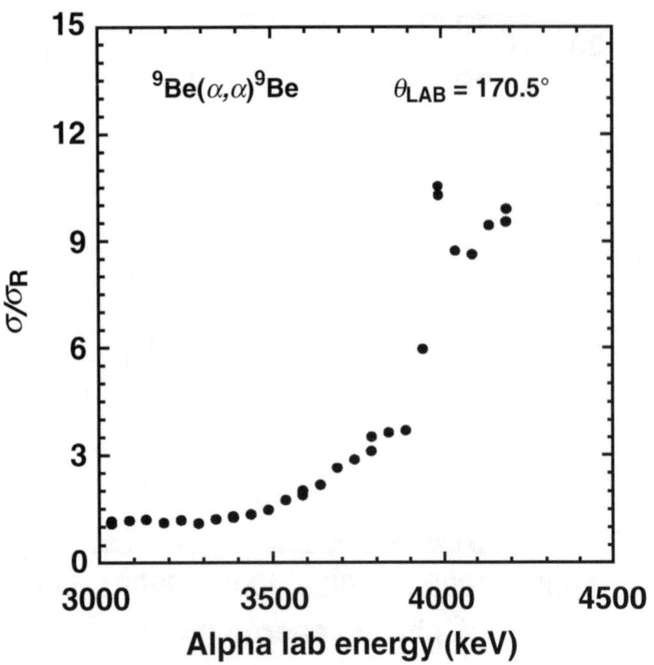

FIG. A7.51. Leavitt *et al.* (1994).

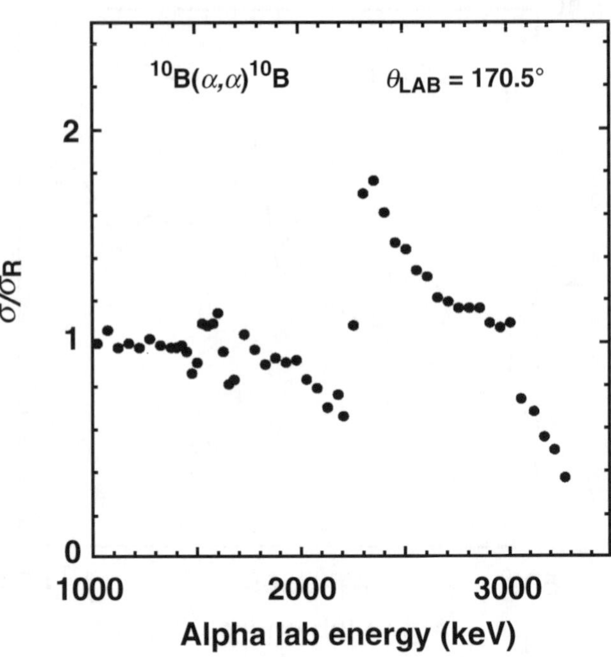

FIG. A7.53. McIntyre *et al.* (1992).

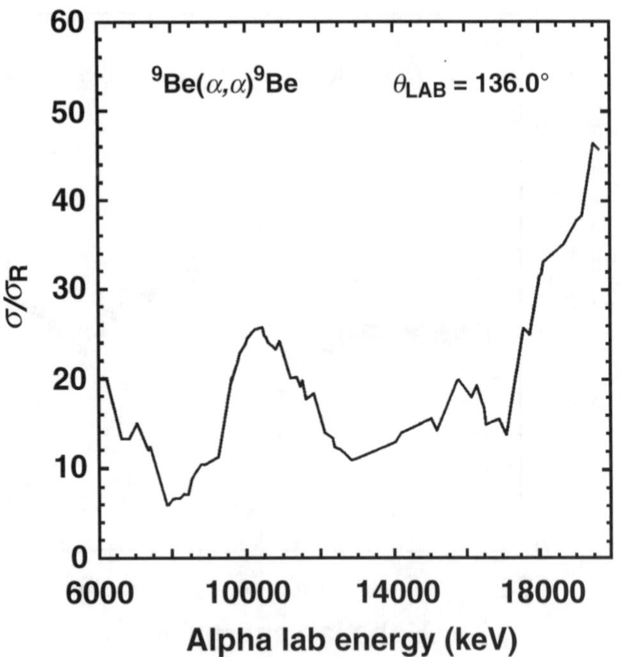

FIG. A7.52. Taylor *et al.* (1965).

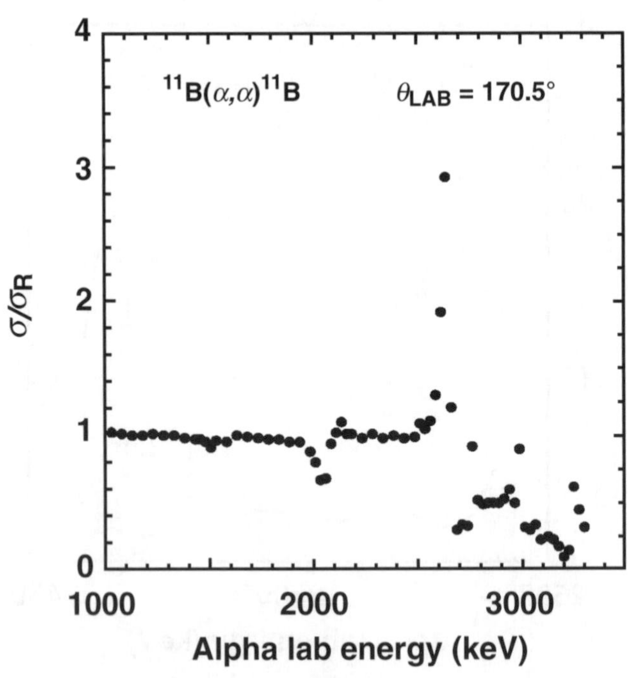

FIG. A7.54. McIntyre *et al.* (1992).

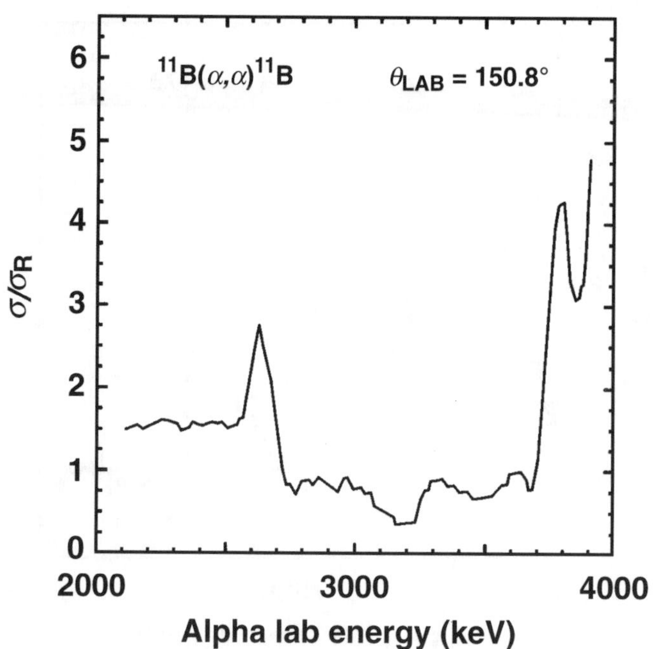

FIG. A7.55. Ramirez *et al.* (1972).

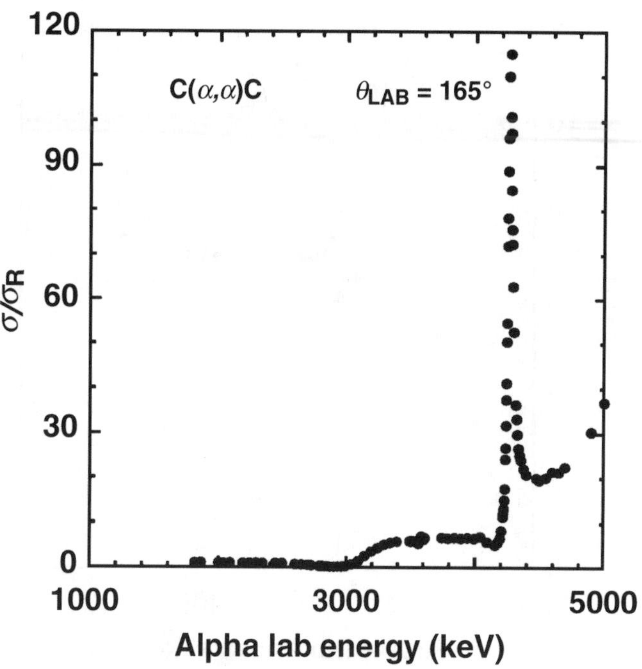

FIG. A7.57. Feng *et al.* (1994a).

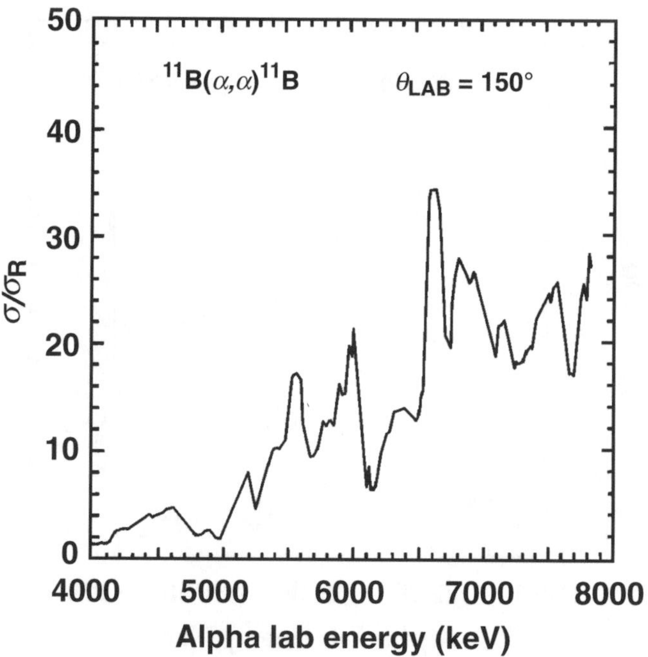

FIG. A7.56. Ott *et al.* (1972).

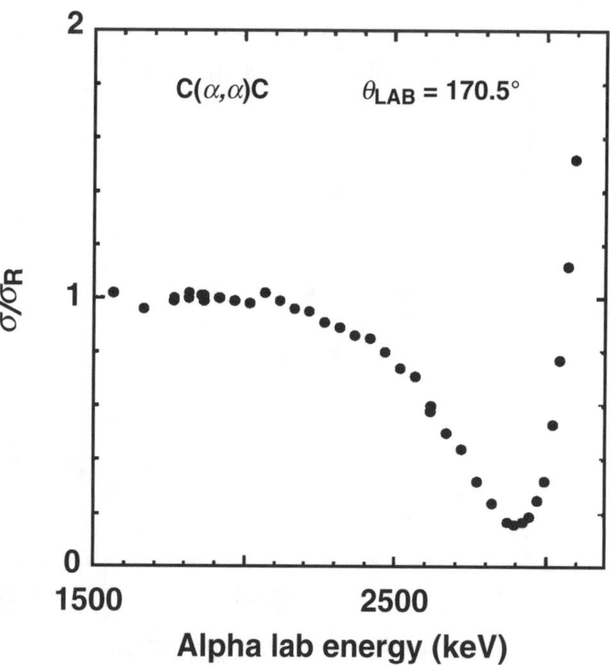

FIG. A7.58. Leavitt *et al.* (1989).

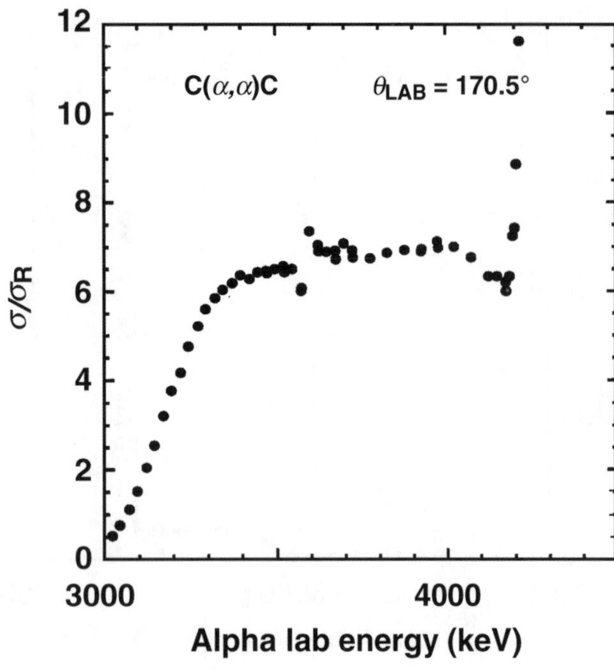

FIG. A7.59. Leavitt *et al.* (1989).

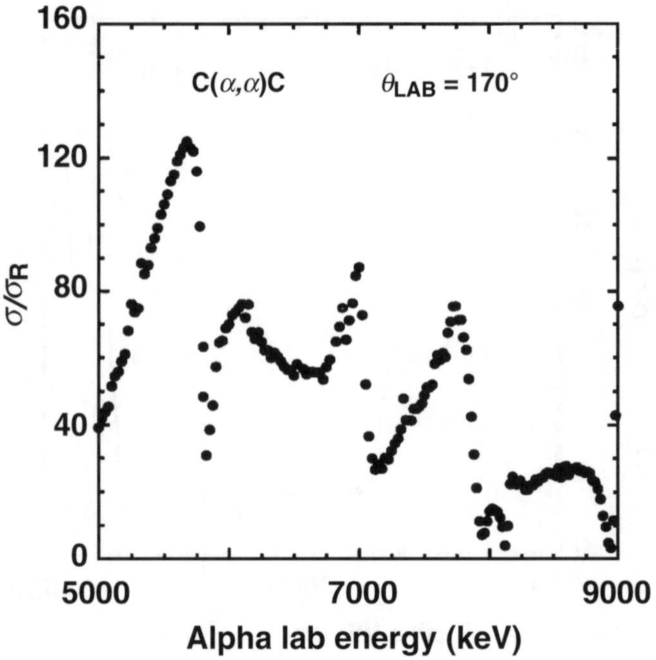

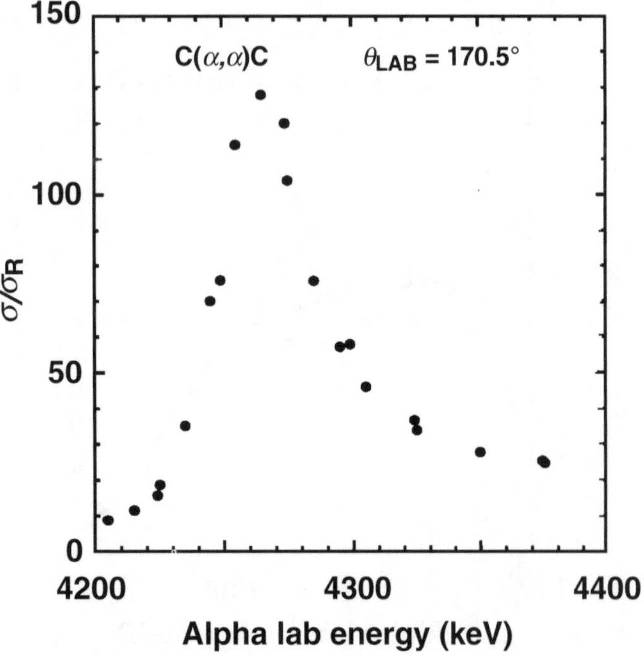

FIG. A7.60. Leavitt *et al.* (1989).

FIG. A7.61. Cheng *et al.* (1994b).

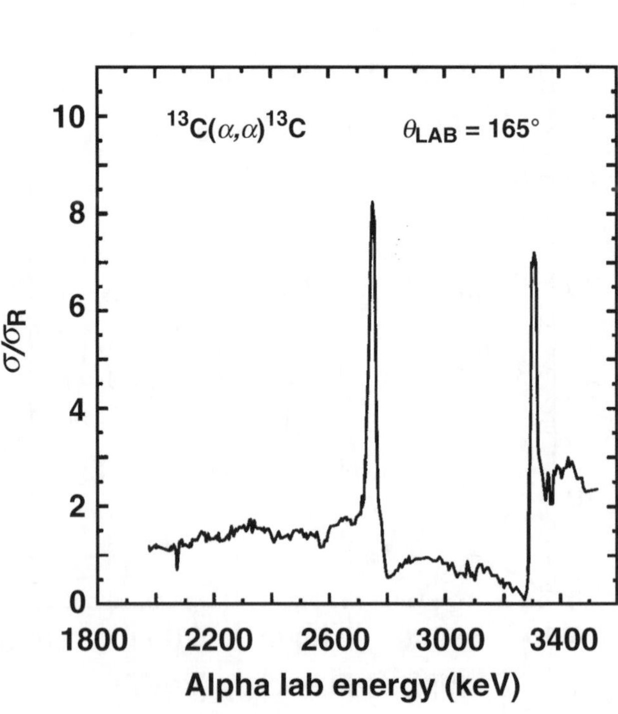

FIG. A7.62. Barnes *et al.* (1965).

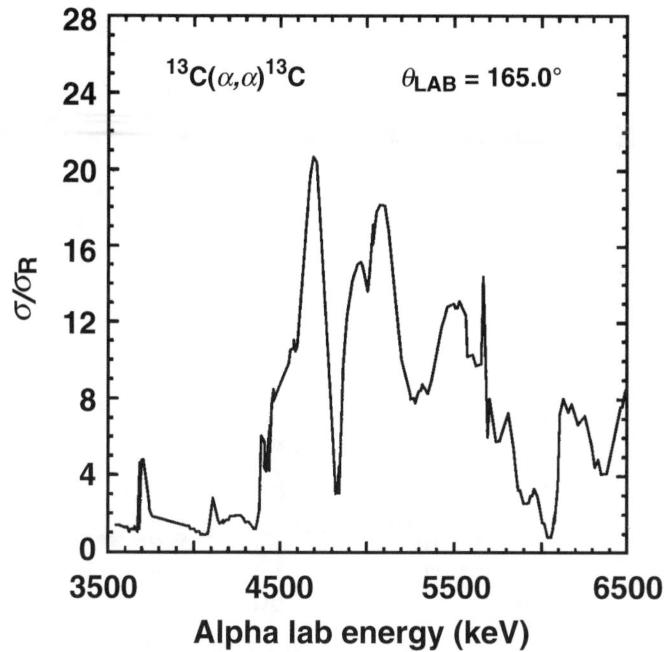

FIG. A7.63. Kerr *et al.* (1968).

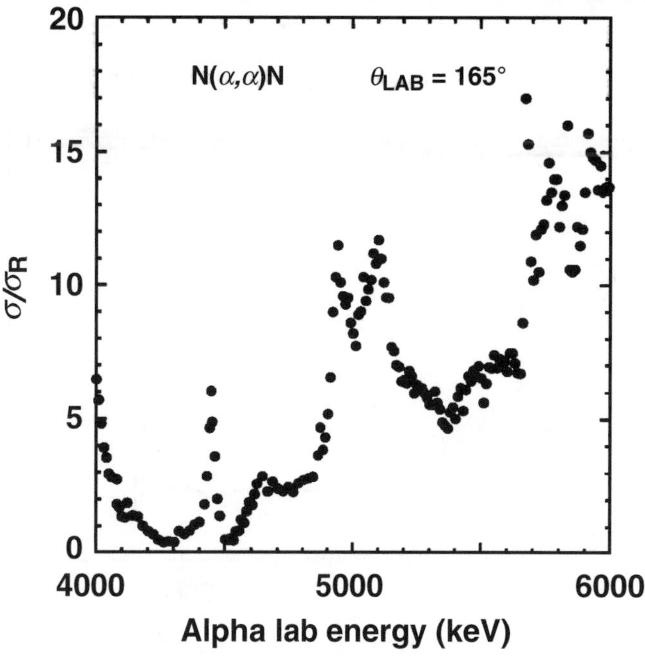

FIG. A7.65. Feng *et al.* (1994b).

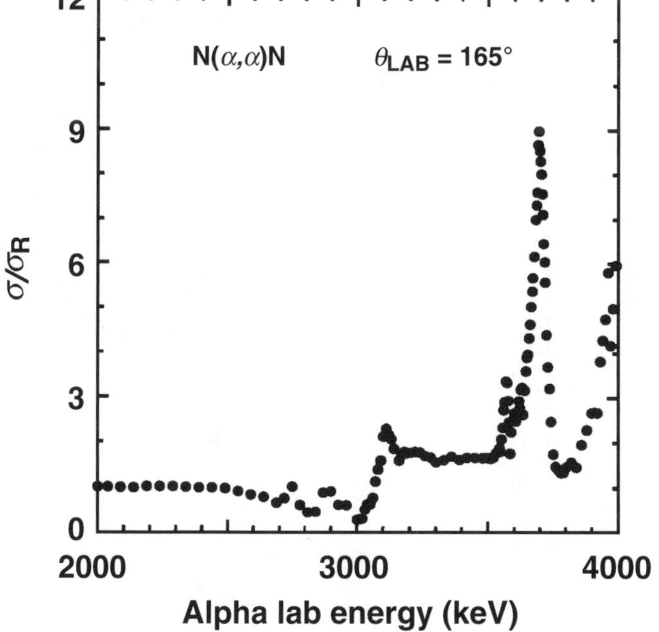

FIG. A7.64. Feng *et al.* (1994b).

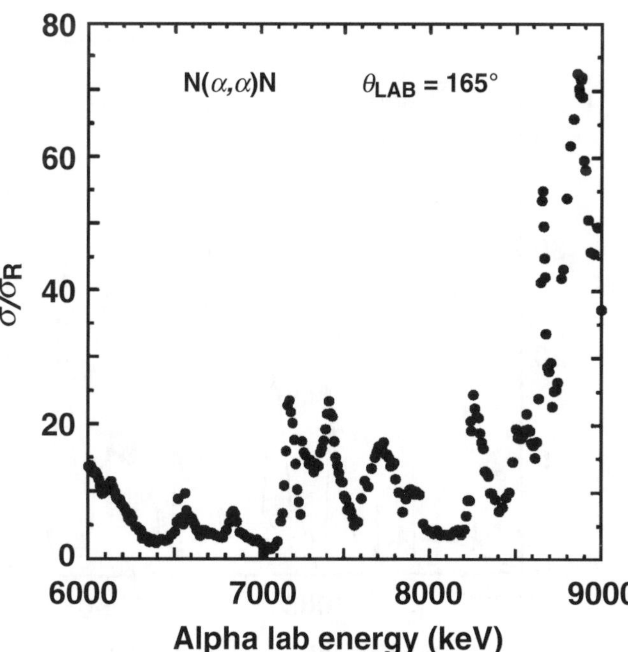

FIG. A7.66. Feng *et al.* (1994b).

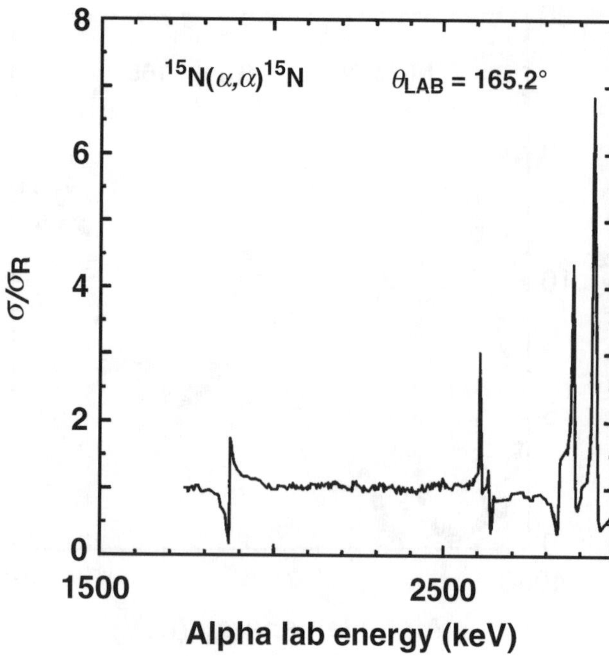

FIG. A7.67. Smotrich *et al.* (1961).

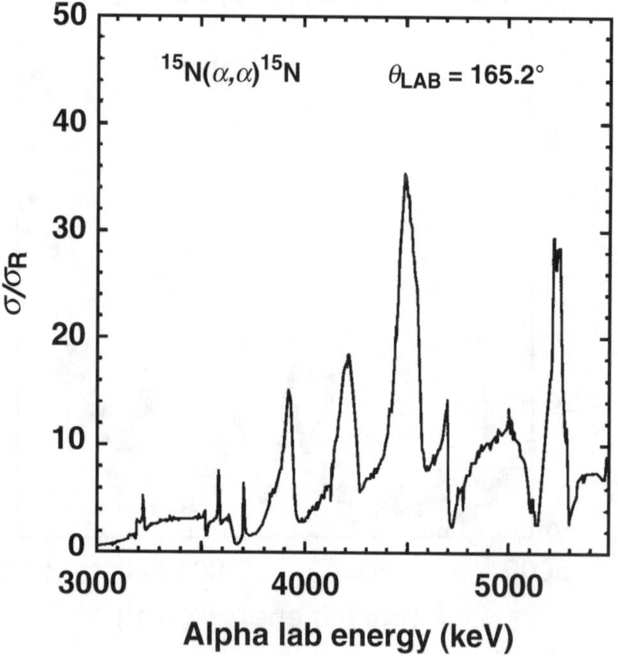

FIG. A7.68. Smotrich *et al.* (1961).

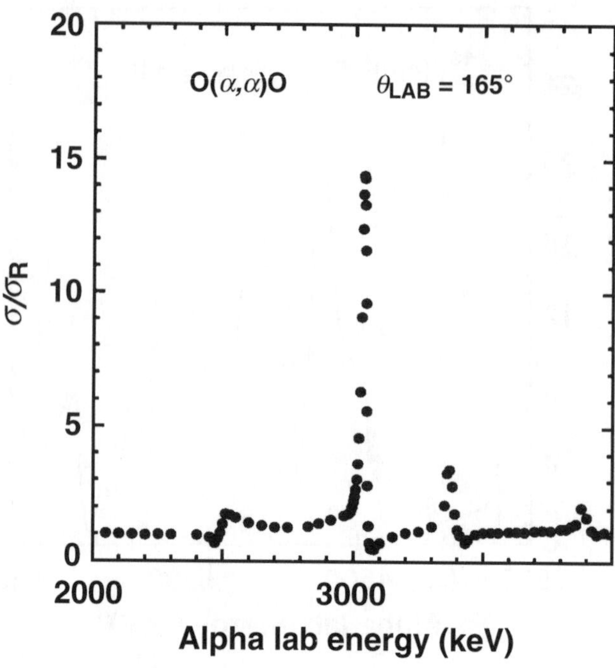

FIG. A7.69. Feng *et al.* (1994c).

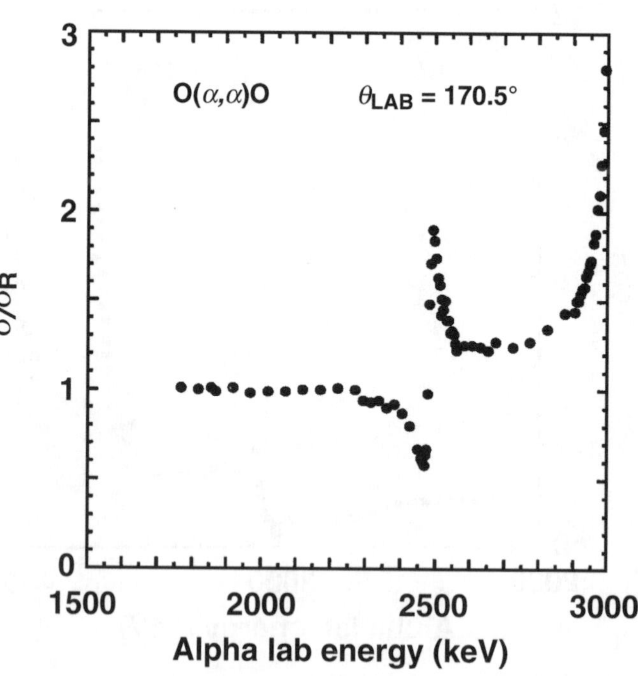

FIG. A7.70. Leavitt *et al.* (1990).

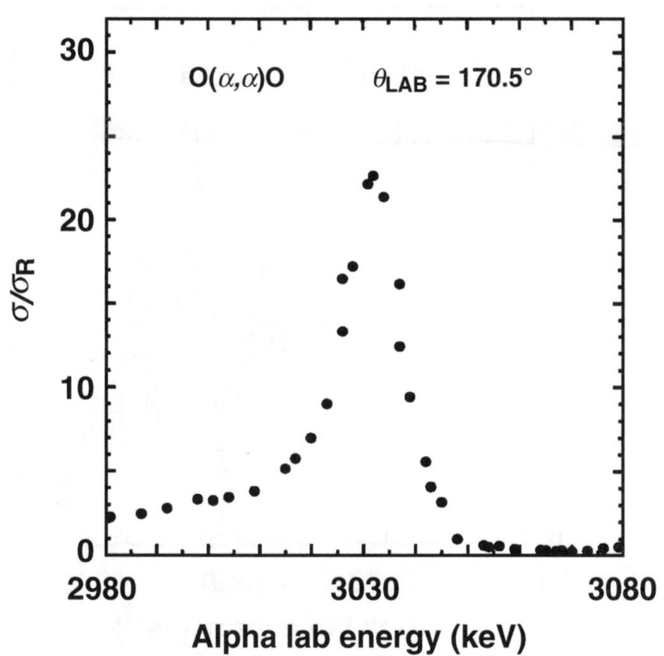

FIG. A7.71. Leavitt *et al.* (1990).

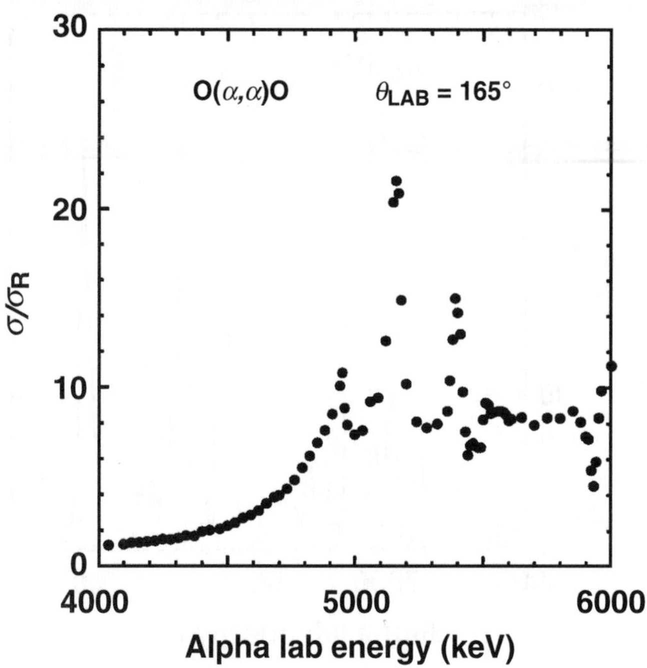

FIG. A7.73. Feng *et al.* (1994c).

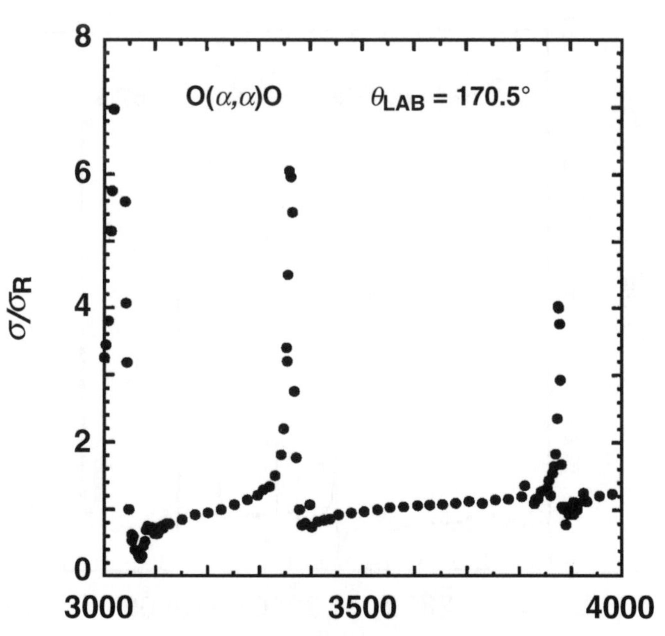

FIG. A7.72. Leavitt *et al.* (1990).

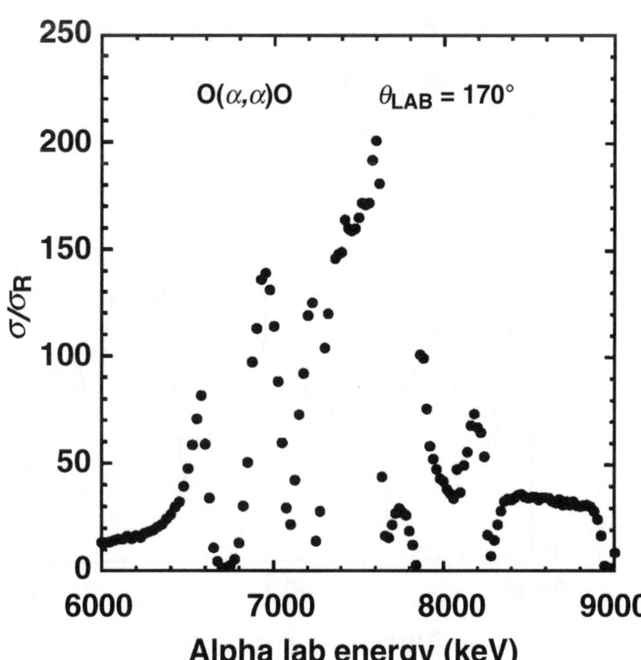

FIG. A7.74 Cheng *et al.* (1993).

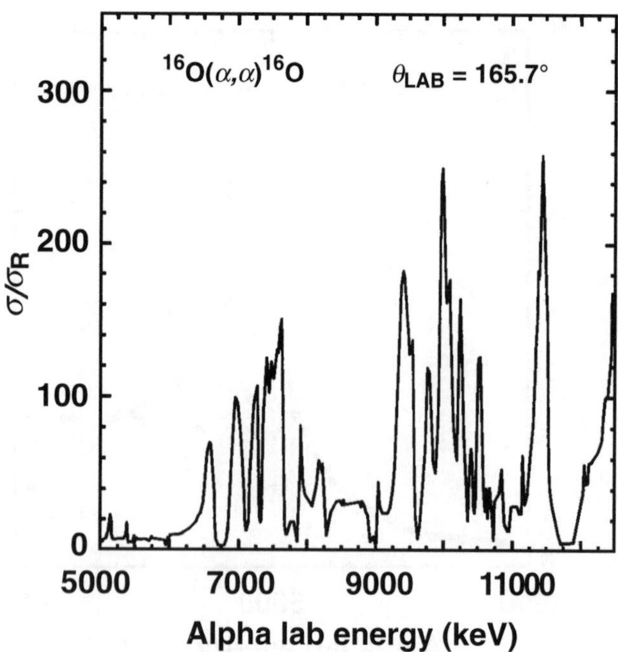

FIG. A7.75. John *et al.* (1969).

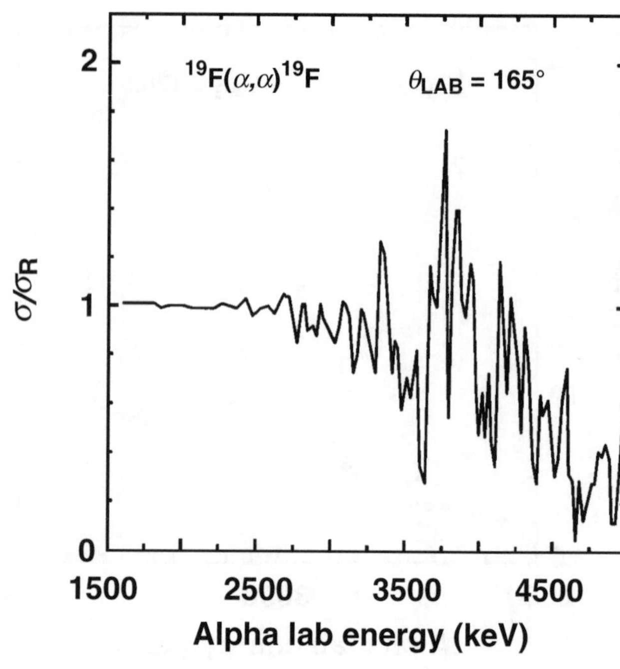

FIG. A7.77. Cheng *et al.* (1994a).

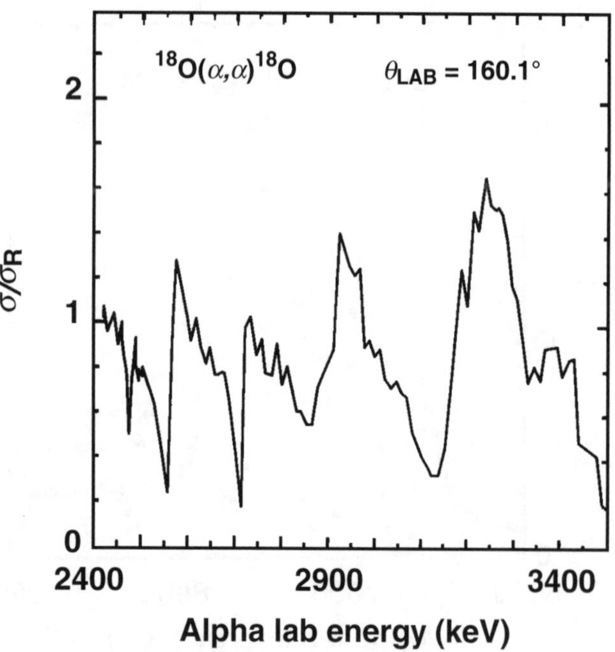

FIG. A7.76. Powers *et al.* (1964).

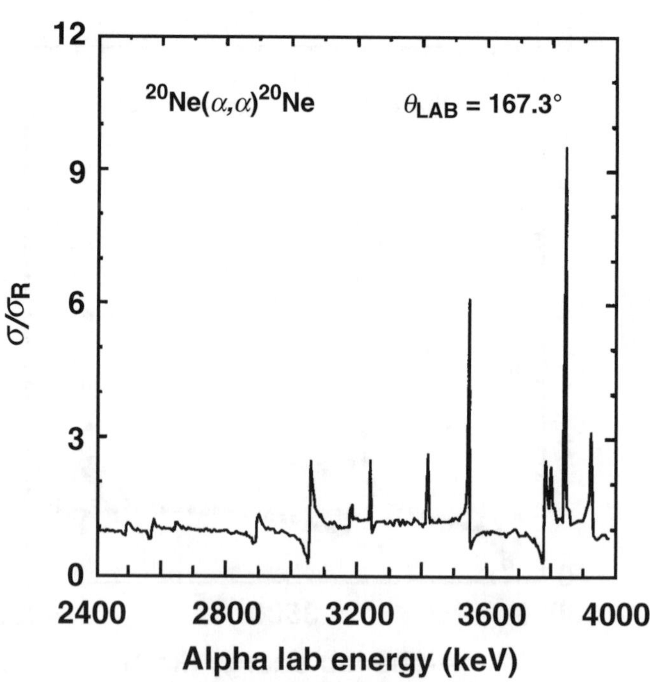

FIG. A7.78. Goldberg *et al.* (1954).

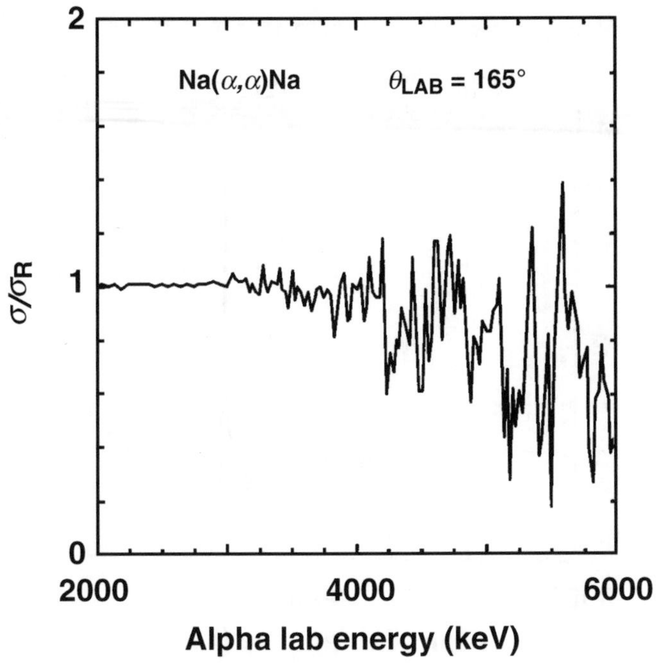

FIG. A7.79. Cheng *et al.* (1991).

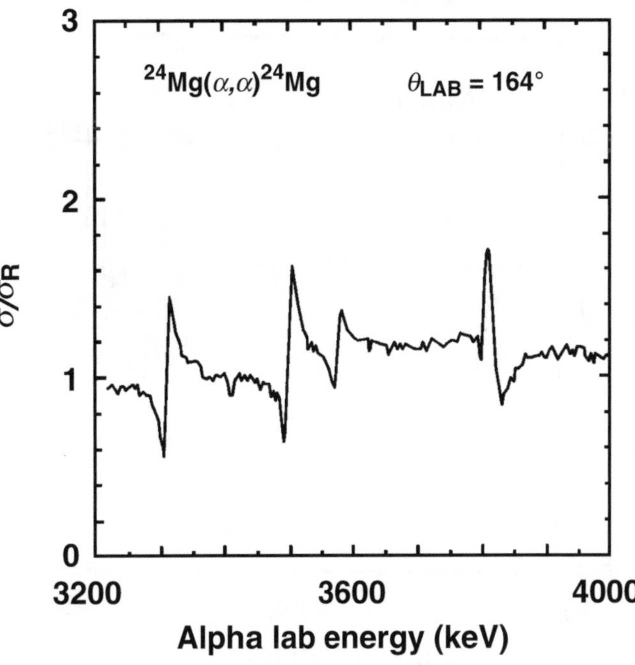

FIG. A7.81. Kaufmann *et al.* (1952).

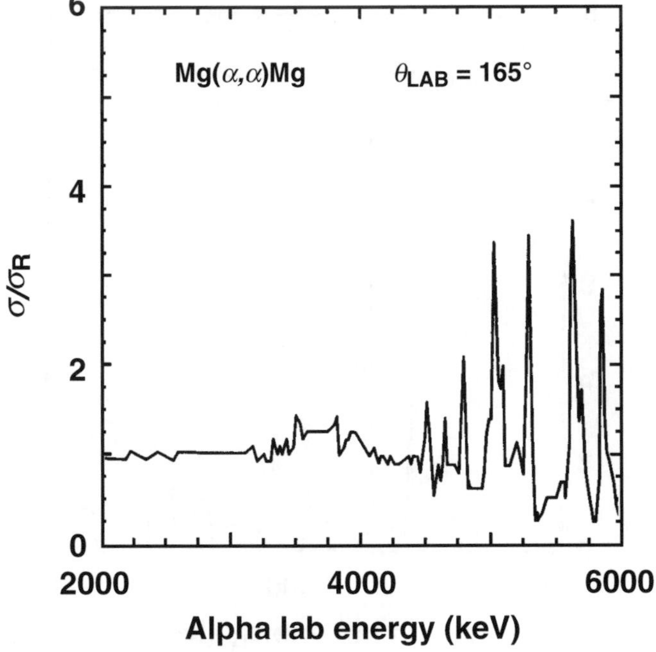

FIG. A7.80. Cheng *et al.* (1994a).

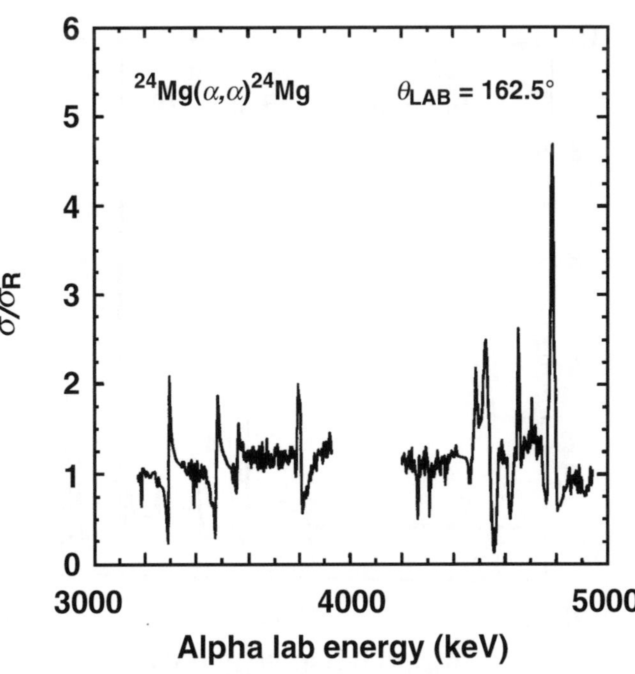

FIG. A7.82. Cseh *et al.* (1982).

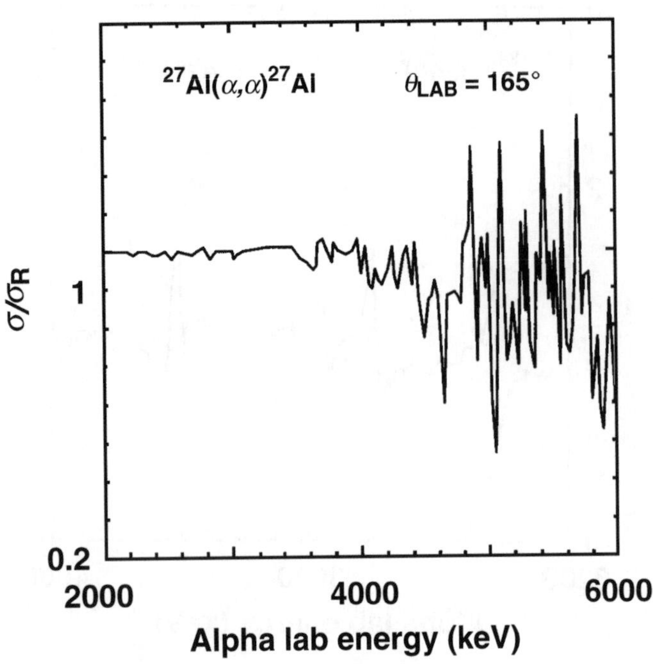

FIG. A7.83. Cheng *et al.* (1991).

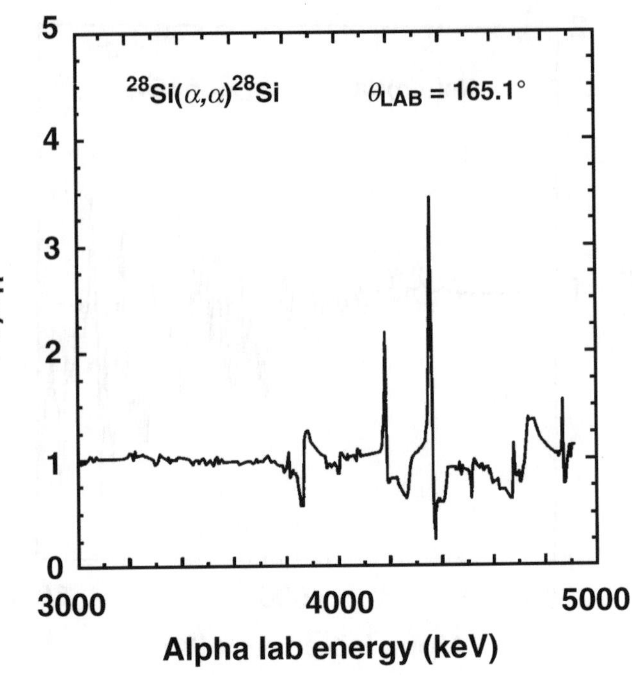

FIG. A7.85. Leung *et al.* (1972).

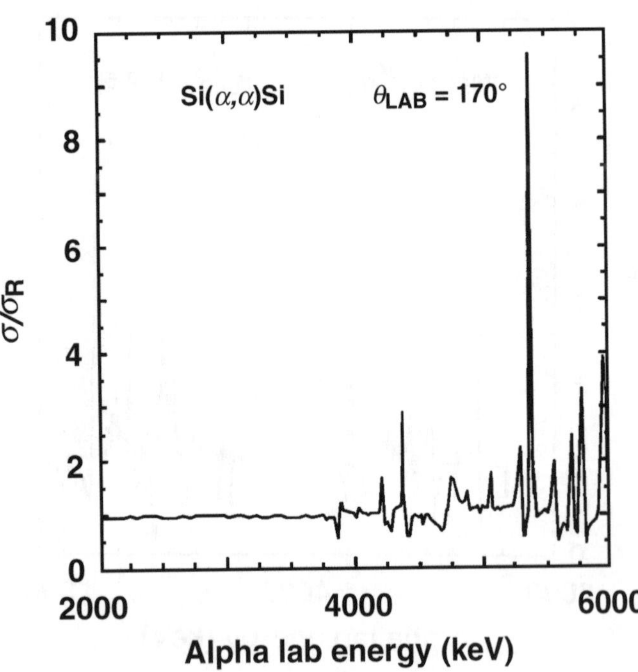

FIG. A7.84. Cheng *et al.* (1994a).

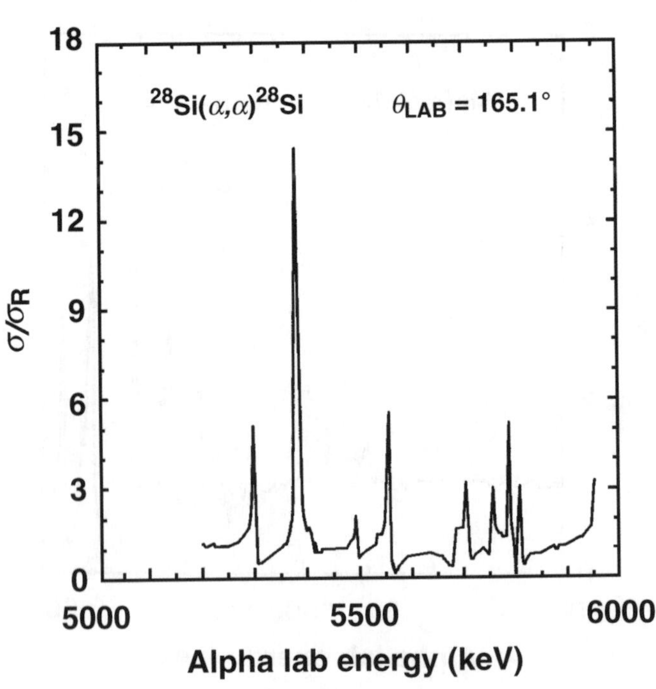

FIG. A7.86. Leung *et al.* (1972).

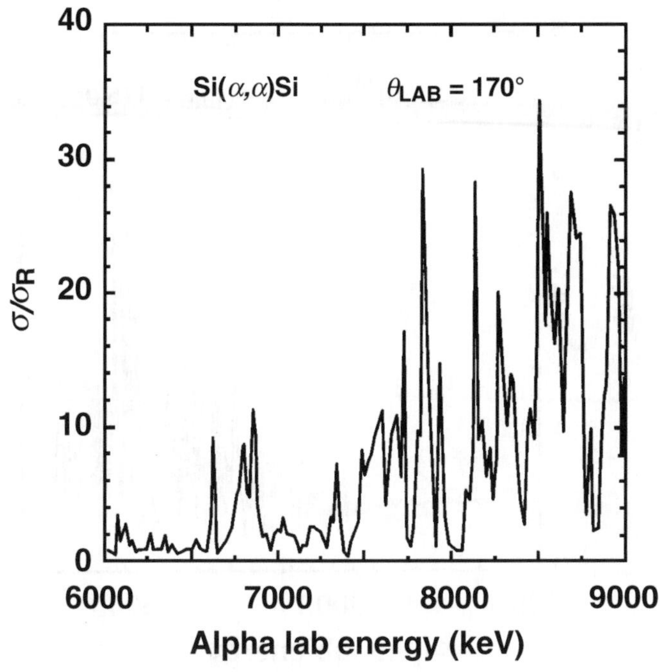

FIG. A7.87. Cheng *et al.* (1994a).

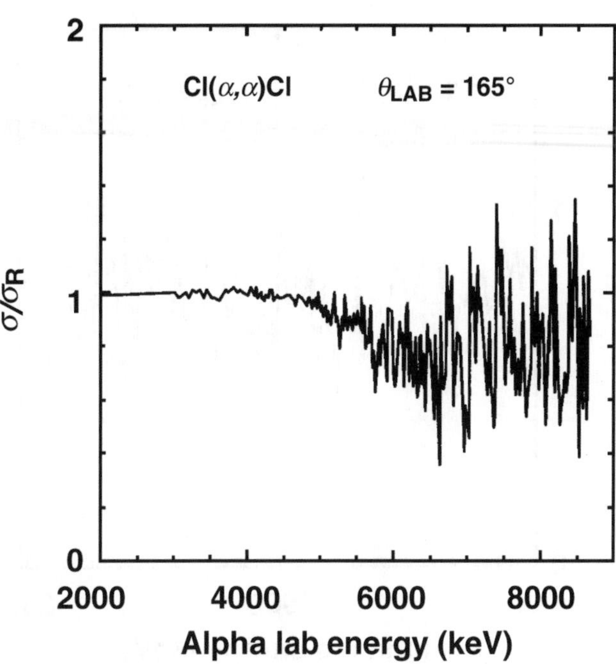

FIG. A7.89. Cheng *et al.* (1994a).

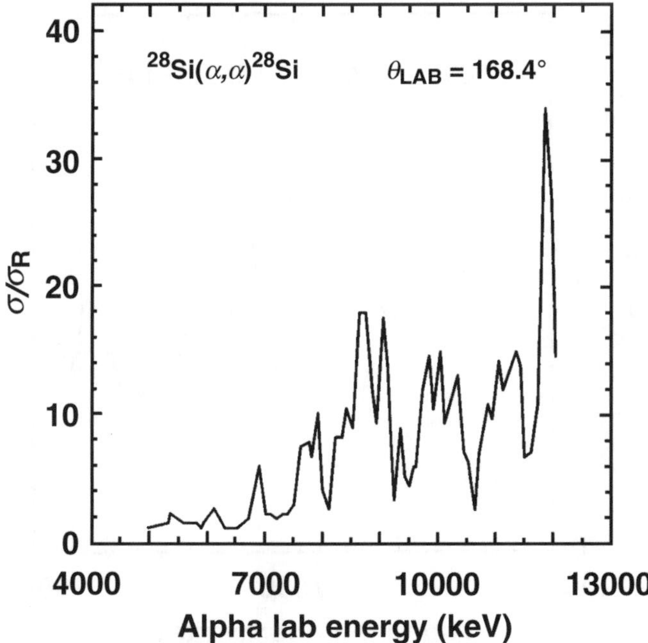

FIG. A7.88. Lawrie *et al.* (1986).

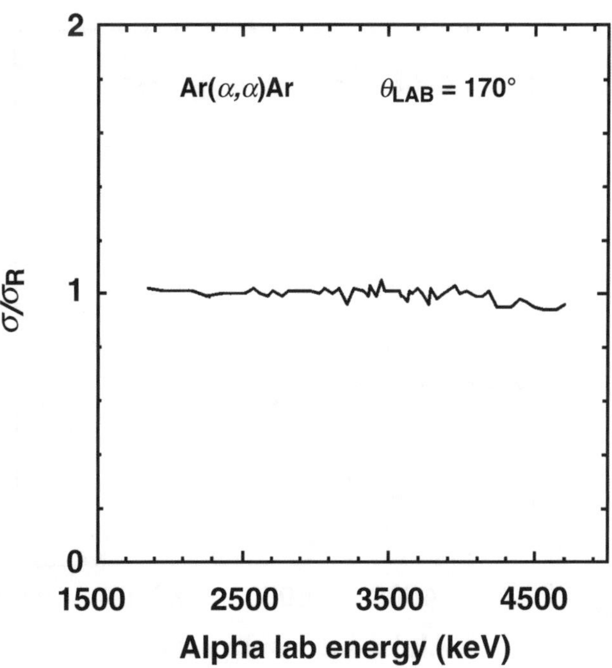

FIG. A7.90. Leavitt *et al.* (1986).

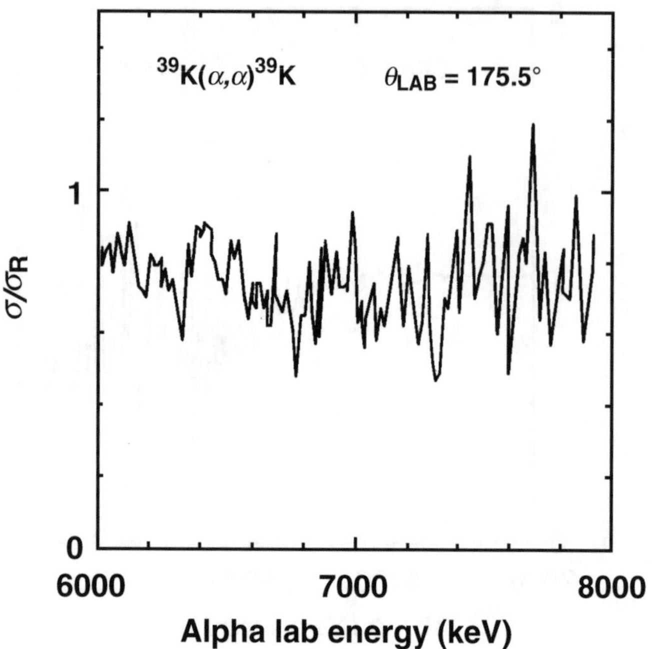

FIG. A7.91. Frekers *et al.* (1983).

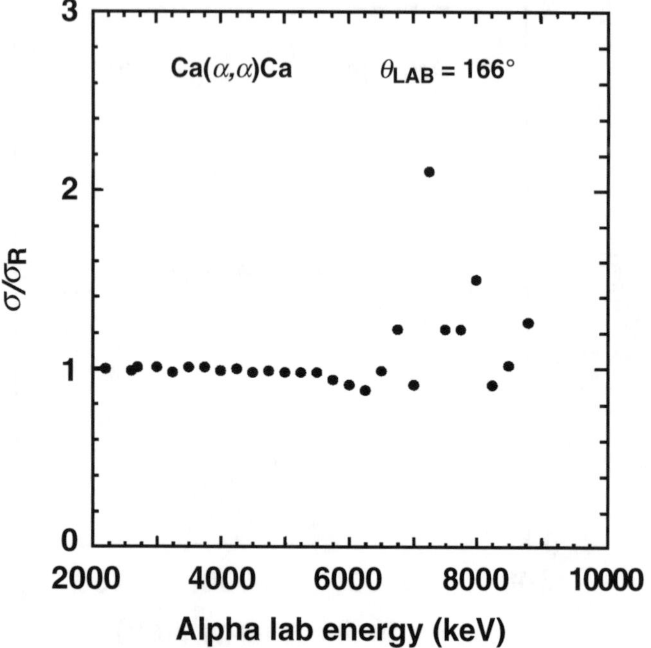

FIG. A7.92. Hubbard *et al.* (1990).

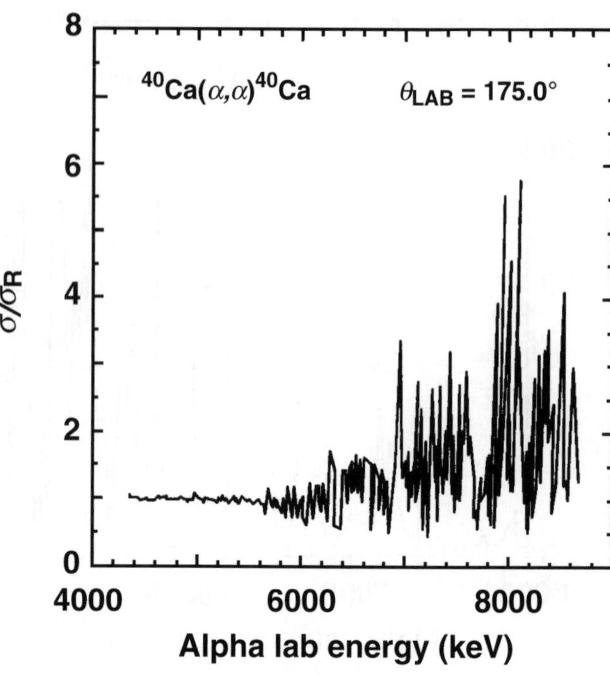

FIG. A7.93. Frekers *et al.* (1983).

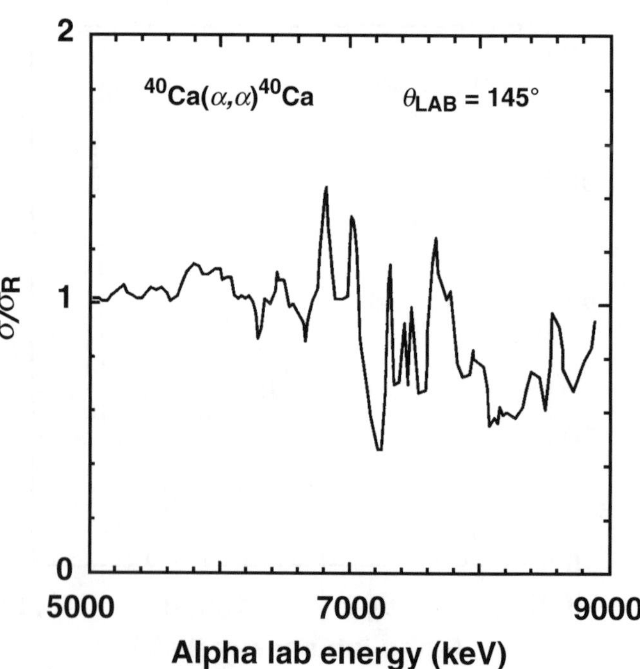

FIG. A7.94. Sellschop *et al.* (1987).

REFERENCES

Amirikas, R., Jamieson, D.N., and Dooley, S.P. (1993), *Nucl. Instrum. Methods* **B77**, 110.

Barnhard, A.C.L., and Kim, C.C. (1961), *Nucl. Phys.* **28**, 428.

Barnes, B.K., Belote, T.A., and Risser, J.R. (1965), *Phys. Rev.* **140**, B616.

Bashkin, S., and Richards, H.T., (1951), *Phys. Rev.* **91**, 917.

Bashkin, S., Carlson, R.R., and Douglas, R.A. (1959), *Phys. Rev.* **114**, 1552.

Baumann, N.P., Prosser, F.W., Read, G.W., and Krone, R.W. (1956), *Phys. Rev.* **104**, 376.

Bogdoanović, I., Fazinić, S., Jakšić, M., Tadić, T., Valković, O., and Valković, V. (1993), *Nucl. Instrum. Methods* **B79**, 524.

Bohlen, H., Marquardt, N., and Von Oertzen, W. (1972), *Nucl. Phys.* **A179**, 504.

Cheng, H.-S., Lee, X.-Y., and Yang, F. (1991), *Nucl. Instrum. Methods* **B56/57**, 749.

Cheng, H.-S., Shen, H., Tang, J., and Yang, F. (1993), *Nucl. Instrum. Methods* **B83**, 449.

Cheng, H.-S., Shen, H., Yang, F., and Tang, J. (1994a), *Nucl. Instrum. Methods* **B85**, 47.

Cheng, H.-S., Shen, H., Tang, J., and Yang, F. (1994b), *Acta Physica Sinica* **43**, 1569.

Chow, H.C., Griffiths, G.M., and Hall, T.H. (1975), *Can. J. Phys.* **53**, 1672.

Cseh, C., Koltay, E., Máté, Z., Somorjai, E., and Zolnai, L. (1982), *Nucl. Phys.* **A385**, 43.

Cohen-Ganouna, J., Lambert, M., and Schmouker, J. (1963), *Nucl. Phys.* **40**, 67.

Dearneley, G. (1956), *Philos. Mag.* **ser. 8,1**, 821.

Feng, Y., Zhou, Z., Zhou, Y., and Zhou, G. (1994a), *Nucl. Instrum. Methods* **B86**, 225.

Feng, Y., Zhou, Z., Zhou, G., and Yang, F. (1994b), *Nucl. Instrum. Methods* **B94**, 11.

Feng, Y. (1994c), private communication.

Frekers, D., Santo, R., and Langke, K. (1983), *Nucl. Phys.* **A394**, 189.

Frier, G.D., Famularo, K.F., Zipoy, D.M., and Leigh, J. (1958), *Phys. Rev.* **110**, 446.

Goldberg, E., Haberli, W., Galonsky, A.L., and Douglas, R.A. (1954), *Phys. Rev.* **93**, 799.

Goss, J.D., Blatt, S.L., Parsignault, D.R., Porterfield, C.D., and Riffle, F.L. (1973), *Phys. Rev.* **C7**, 1837.

Hagedorn, F.B., Mozer, F.S., Webb, T.S., Fowler, W.A., and Lauritsen, C.C. (1957), *Phys. Rev.* **105**, 219.

Hubbard, K.M., Martin, J.A., Muenchhausen, R.E., Tesmer, J.R., and Nastasi, M. (1990), in *High Energy and Heavy Ion Beams in Materials Analysis*, J.R. Tesmer, C.J. Maggiore, M. Nastasi, J.C. Barbour and J.W. Mayer (eds.), Materials Research Society, Pittsburgh, Pennsylvania, p. 129.

Jackson, H.L., Galonsky, A.I., Eppling, F.J., Hill, R.W., Goldberg, E., and Cameron, J.R. (1953), *Phys. Rev.* **89**, 365.

John, J., Aldridge, J.P., and Davis, R.H. (1969), *Phys. Rev.* **181**, 1455.

Kaufman, S., Goldberg, E., Koester, L.J., and Mooring, F.P. (1952), *Phys. Rev.* **88**, 673.

Kerr, G.W., Morris, J.M., and Risser, J.R. (1968), *Nucl. Phys.* **A110**, 637.

Knox, J.M., and Harmon, J.F. (1989), *Nucl. Instrum. Methods* **B44**, 40.

Lambert, M., and Durand, M. (1967), *Phys. Lett.* **24B**, 287.

Lambert, M., Midy, P., Drain, D., Amiel, M., Beaumevielle, H., Dauchy, A., and Meynadier, C. (1972), *J. Phys.* **33**, 155.

Langley, R.A. (1976), in *Proc. Int. Conf. on Radiation Effect and Tritium Technology for Fusion Reactors*, Vol. IV, J.S. Walson and F.W. Wiffen (eds.), Conf. 750989, US Dept. of Commerce, Springfield, Virginia, p. 158.

Laubenstein, R.A., Laubenstein, M.J.W., Koestler, L.J., and Mobley, R.C. (1951), *Phys. Rev.* **84**, 12.

Lawrie, J.J., Cowley, A.A., Whittal, D.M., Mills, S.J., and McMurray, W.R. (1986), *Z. Phys.* **A325**, 175.

Leavitt, J.A., Stoss, P., Cooper, D.B., Seerveld, J.L., McIntyre Jr., L.C., Davis, R.E., Gutierrez, S., and Reith, T.M. (1986), *Nucl. Instrum. Methods* **B15**, 296.

Leavitt, J.A., McIntyre Jr., L.C., Stoss, P., Oder, J.G., Ashbaugh, M.D., Dezfouly-Arjomandy, B., Yang, Z.M., and Lin, Z. (1989), *Nucl. Instrum. Methods* **B40/41**, 776.

Leavitt, J.A., McIntyre Jr., L.C., Ashbaugh, M.D., Oder, J.G., Lin, Z., and Dezfouly-Arjomandy, B. (1990), *Nucl. Instrum. Methods* **B44**, 260.

Leavitt, J.A., and McIntyre Jr., L.C. (1991), *Nucl. Instrum. Methods* **B56/57**, 734.

Leavitt, J.A., McIntyre Jr., L.C., Champlin, R.S., Stoner Jr., J.O., Lin, Z., Ashbaugh, M.D., Cox, R.P., and Frank, J.D. (1994), *Nucl. Instrum. Methods* **B85**, 37.

Leung, M.K. (1972), Ph.D. dissertation, Univ. of Kentucky.

Liu, Z., Li, B. Duan, D., and He, H. (1993), *Nucl. Instrum. Methods* **B74**, 439.

Malmberg, P.R. (1956), *Phys. Rev.* **101**, 114.

McIntyre Jr., L.C., Leavitt, J.A., Ashbaugh, M.D., Lin, Z., and Stoner Jr., J.O. (1992), *Nucl. Instrum. Methods* **B64**, 457.

Mooring, F.P., Koestler, L.J. Jr., Goldberg, E., Saxon, D., and Kaufmann, S.G. (1951), *Phys. Rev.* **84**, 703.

Mozer, F.S. (1956), *Phys. Rev.* **104**, 1386.

NNDC (National Nuclear Data Center) Building 197D, Brookhaven National Laboratory; Upton, New York 11973, USA. Several on line data bases are available. Telephone: (516) 282-2901; FAX: (516) 282-2806; Internet: nndc@bnl.gov. WWW address: http:/ necsØ1.dne.bnl.gov/html/nndc.html.

Olness, J.W., Vorona, J., and Lewis, H.W. (1958a), *Phys. Rev.* **112**, 475.

Olness, J.W., Haeberli, W., and Lewis, H.W. (1958b), *Phys. Rev.* **112**, 1702.

Ott, W.R., and Weller, H.R. (1972), *Nucl. Phys.* **A198**, 505.

Overlay, J.C., and Whaling, W. (1962), *Phys. Rev.* **128**, 315.

Powers, D., Blair, J.K., Ford, J.L.C. Jr., and Willard, H.B. (1964), *Phys. Rev.* **134**, B1237.

Ramirez, J.J., Blue, R.A., and Weller, H.R. (1972), *Phys. Rev.* **C5**, 17.

Rauhala, E. (1985), *Nucl. Instrum. Methods* **B12**, 447.

Rauhala, E., and Luomajärvi, M. (1988), *Nucl. Instrum. Methods* **B33**, 628.

Rauhala, E. (1989), *Nucl. Instrum. Methods* **B40/41**, 790.

Rauhala, E. (1991), "Ion Backscattering Spectrometry," in *Elemental Analysis by Particle Accelerators*, Z. Alfassi and M. Peisach (eds.), CRC-Press.

Sellschop, J.P.F., Zucchiati, A., Mirman, L., Gering, M.Z.I., and DiSalvo, E. (1987), *J. Phys.* **G13**, 1129.

SIGBASE. A service supported on a trial basis by I.C. Vickridge, Institute of Geological and Nuclear Sciences Ltd, P.O. Box 31312, Lower Hutt; New Zealand and G. Vizkelethy, Department of Physics, Idaho State University, Campus Box 8106, Pocatello, ID 83209-8106, USA. Internet: srlnicv@lhn.gns.cri.nz; vizkel@physics.isu.edu; Anonymous ftp to lhn.gns.cri.nz or physics.isu.edu. WWW address: http://ibaserver.physics.isu.edu/sigmabase.

Smotrich, H., Jones, K.W., McDermott, L.C., and Benenson, R.E. (1961), *Phys. Rev.* **122**, 232.

Tautfest, G.W., and Rubin, S. (1956), *Phys. Rev.* **103**, 196.

Taylor, R.B., Fletcher, N.P., and Davis, R.H. (1965), *Nucl. Phys.* **65**, 318.

Vorona, J., Olness, J.W., Haeberli, W., and Lewis, H.W. (1959), *Phys. Rev.* **116**, 1563.

Warters, W.D., Fowler, W.A., and Lauritsen, C.C. (1953), *Phys. Rev.* **91**, 917.

Webb, T.S., Hagedorn, F.B., Fowler, W.A., and Lauritsen, C.C. (1955), *Phys. Rev.* **99**, 138.

Wilson, W.M. Jr., Moses, J.D., and Bilpuch, E.G. (1974), *Nucl. Phys.* **A227**, 277.

APPENDIX

8

ACTUAL COULOMB BARRIERS

Compiled by

M. Bozoian

Los Alamos National Laboratory, Los Alamos, New Mexico

The following tables and figures present optical model calculations of E_{nr}^{cm}, the center-of-mass energy, for a 4% deviation from Rutherford backscattering (RBS) for various optical model parameter sets. Note that the ^1H results are based on both experiments and calculations, and that the ^4He results are based solely on experiments. The ^4He table and figures also include 2% and 6% deviations from RBS.

The tables are self-explanatory. The accompanying figures are plots of the tabular data plus least squares linear fits to the data. Each least squares' straight line equation is included within its figure.

NOTE WELL: An experimentalist's prescription

To find E_{nr}^{lab}, the *laboratory* kinetic energy for departures from RBS, in terms of E_{nr}^{cm}; the projectile mass, M_1; the target mass, M_2; and the laboratory scattering angle, θ_l; the following formula, valid within the stated inequalities, can be used:

For $0 < M_1/M_2 < 1$ and $150° < \theta_{lab} < 180°$,

$$E_{nr}^{lab} = E_{nr}^{cm}(M_1+M_2)/M_2 \ .$$

Thus one can read-off E_{nr}^{cm} from a figure, or extrapolate the figure's straight line fit at either end, providing the above inequalities are true, then use the above formula to find E_{nr}^{lab}. Note that if $M_1/M_2 = 1$, then $\theta_{lab} \le 90°$ for $\theta_{cm} \le 180°$; and if $M_1/M_2 > 1$, then $\theta_{lab} < 90°$. In both these cases there is no RBS. See Appendix 4 and Chap. 3 for discussions on center-of-mass to laboratory relationships.

Finally, because cross section ratios are usually prefered in ion beam analysis (i.e., the desired cross section is measured with respect to a known Rutherford cross section), there is no need to worry about transforming cross sections from the center-of-mass to the laboratory, or vice versa, since the transforming factors will cancel in the cross section ratios.

Table A8.1. ^{1}H-ion energy at which the scattering cross section deviates by 4% from its Rutherford value (Bozoian *et al.*, 1990).

Target atomic number (Z_2)	Target	E_{nr}^{cm} (data, calculation)
6	^{12}C	0.40, 0.54
7	^{14}N	0.64, 0.63
8	^{16}O	0.64, 0.81
10	^{20}Ne	1.10, 1.28
12	^{24}Mg	0.78, 1.63
14	^{28}Si	1.44, 1.63
15	^{31}P	1.21, 1.69
16	^{32}S	1.70, 1.81
18	^{40}Ar	1.78, 2.16
24	^{52}Cr	2.65, 2.83
26	^{56}Fe	2.46, 3.04
29	^{63}Cu	2.95, 3.43

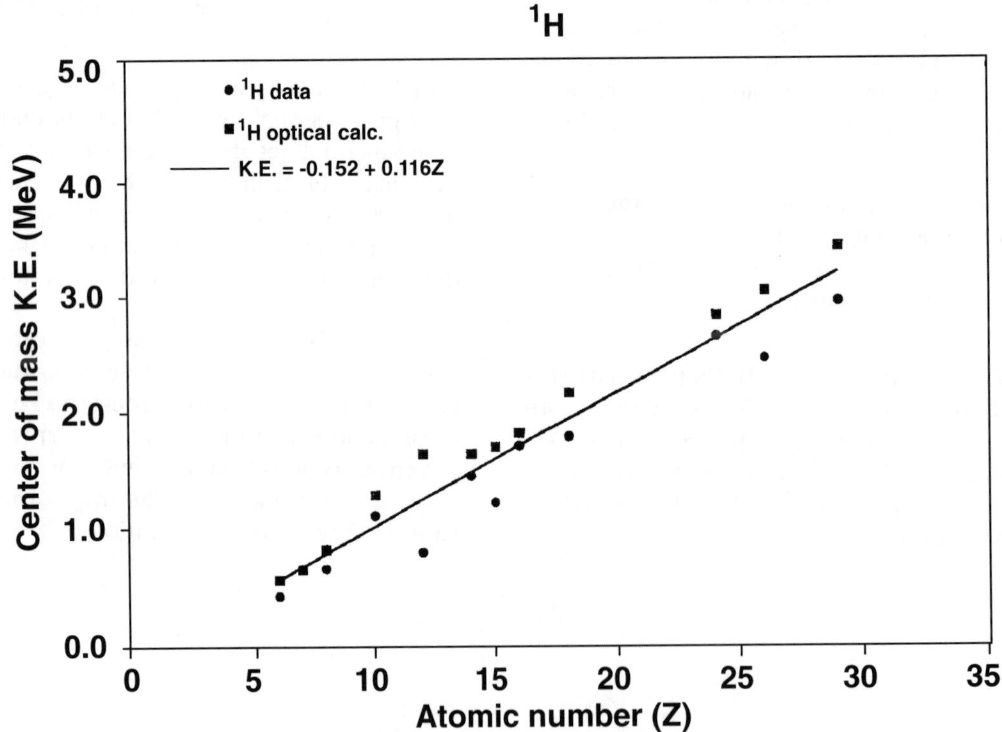

FIG. A8.1. The center-of-mass ^{1}H-ion energy at which the scattering cross section deviates by 4% from its Rutherford value vs. atomic number for several elements. The solid line is a linear least-squares fit to the data.

Table A8.2. ^{2}H-ion energy at which the scattering cross section deviates by 4% from its Rutherford value (Bozoian, 1991a).

Target atomic number (Z_2)	Target	E_{nr}^{cm} (calculation)
6	^{12}C	1.01
7	^{14}N	0.96
8	^{16}O	1.03
12	^{24}Mg	1.40
14	^{28}Si	1.60
16	^{32}S	1.81
24	^{52}Cr	2.63
26	^{56}Fe	2.83
29	^{63}Cu	3.12

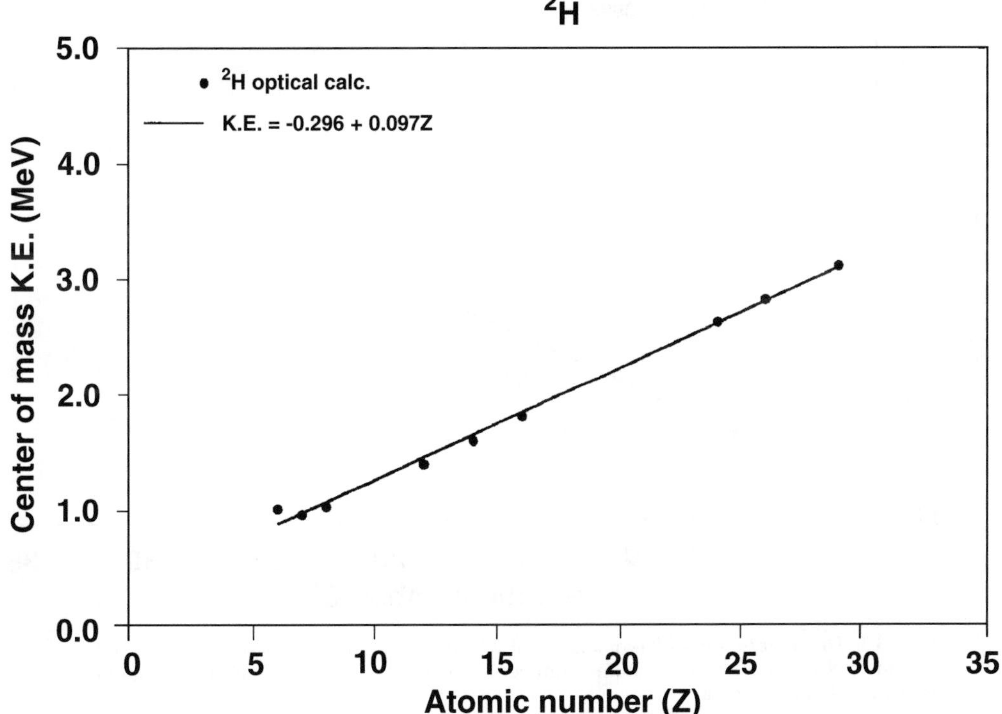

FIG. A8.2. The center-of-mass ^{2}H-ion energy at which the scattering cross section deviates by 4% from its Rutherford value vs. atomic number for several elements. The solid line is a linear least-squares fit to the data.

Table A8.3. ^{3}H-ion energy at which the scattering cross section deviates by 4% from its Rutherford value (Bozoian, 1991a).

Target atomic number (Z_2)	Target	E_{nr}^{cm} (calculation)
6	^{12}C	0.77
8	^{16}O	0.92
20	^{40}Ca	2.15
32	^{74}Ge	3.33

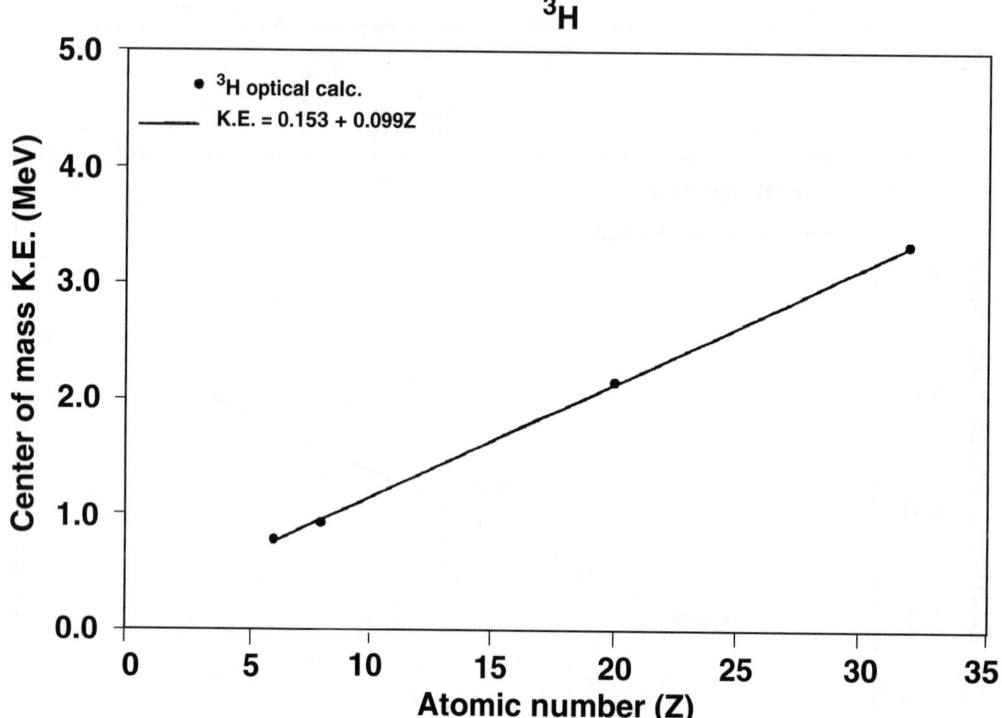

FIG. A8.3. The center-of-mass ^{3}H-ion energy at which the scattering cross section deviates by 4% from its Rutherford value vs. atomic number for several elements. The solid line is a linear least-squares fit to the data.

Table A8.4. ^{3}He-ion energy at which the scattering cross section deviates by 4% from its Rutherford value (Bozoian, 1991a).

Target atomic number (Z_2)	Target	E_{nr}^{cm} (calculation)
6	^{12}C	1.41
7	^{14}N	1.64
8	^{16}O	1.87
9	^{19}F	2.08
12	^{24}Mg	2.78
13	^{27}Al	3.00
14	^{28}Si	3.24
20	^{40}Ca	4.58
22	^{48}Ti	4.96
29	^{63}Cu	6.44
32	^{72}Ge	7.03

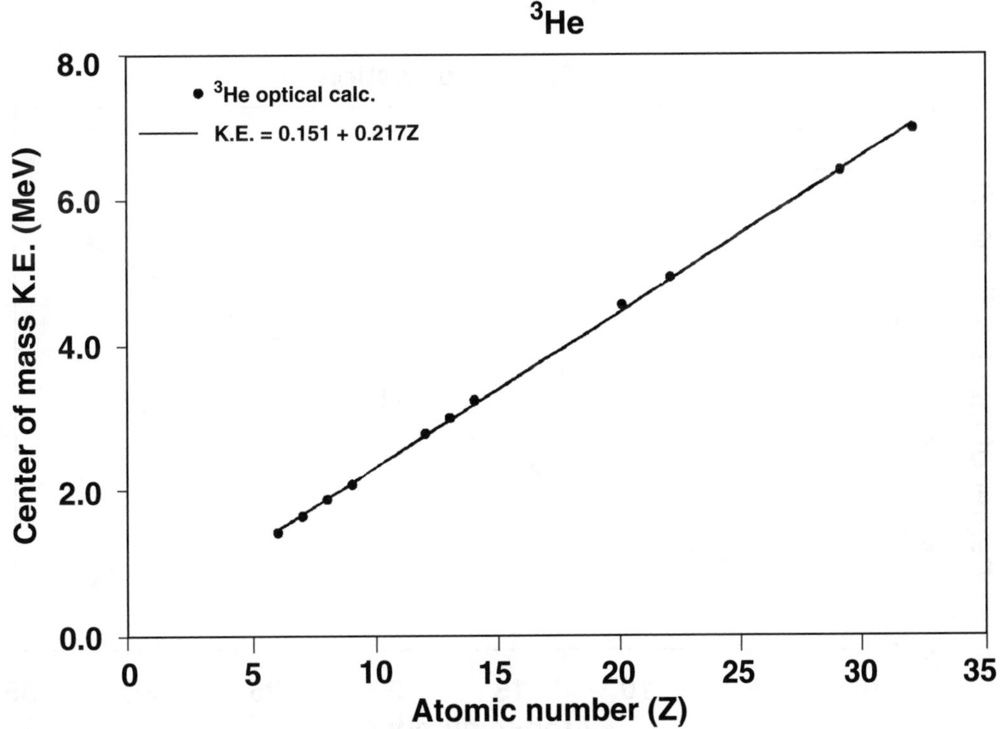

FIG. A8.4. The center-of-mass ^{3}He-ion energy at which the scattering cross section deviates by 4% from its Rutherford value vs. atomic number for several elements. The solid line is a linear least-squares fit to the data.

Table A8.5. ^{4}He-ion energy at which the scattering cross section deviates by 2, 4 and 6% from its Rutherford value. Values are measurements and their associated error (Bozoian *et al.*, 1990).

Target atomic number (Z_2)	Target	E_{nr}^{cm} 2% (data, error)	E_{nr}^{cm} 4% (data, error)	E_{nr}^{cm} 6% (data, error)
6	^{12}C	1.61, 0.04	1.64, 0.02	1.67, 0.02
7	^{14}N	1.93, 0.06	1.96, 0.06	1.98, 0.06
8	^{16}O	1.89, 0.06	1.90, 0.06	1.92, 0.06
9	^{19}F	1.67, 0.06	1.67, 0.06	1.67, 0.06
12	^{24}Mg	2.74, 0.06	2.77, 0.06	2.79, 0.06
13	^{27}Al	3.10, 0.10	3.10, 0.10	3.10, 0.10
14	^{28}Si	3.32, 0.05	3.33, 0.05	3.35, 0.05
20	^{40}Ca	4.70, 0.30	5.10, 0.20	5.30, 0.20
22	^{48}Ti	5.10, 0.20	5.40, 0.20	5.70, 0.20
29	^{63}Cu	6.80, 0.20	7.20, 0.20	7.40, 0.20
32	^{72}Ge	7.50, 0.20	7.80, 0.10	8.00, 0.10

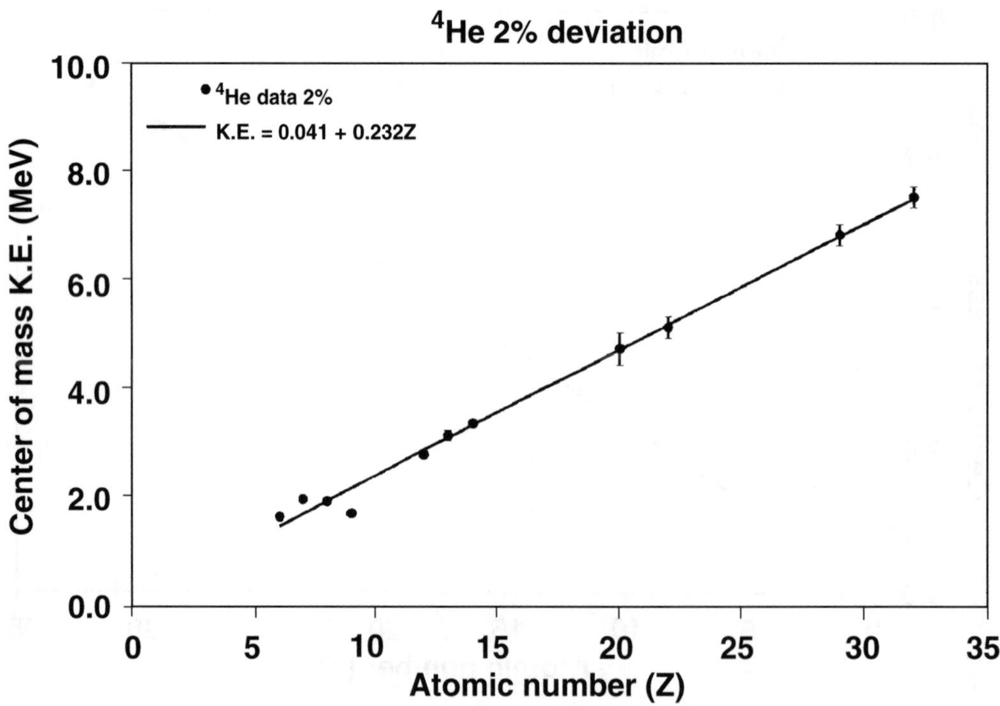

FIG. A8.5(a). The center-of-mass ^{4}He-ion energy at which the scattering cross section deviates by 2% from its Rutherford value vs. atomic number for several elements. The solid line is a linear least-squares fit to the data.

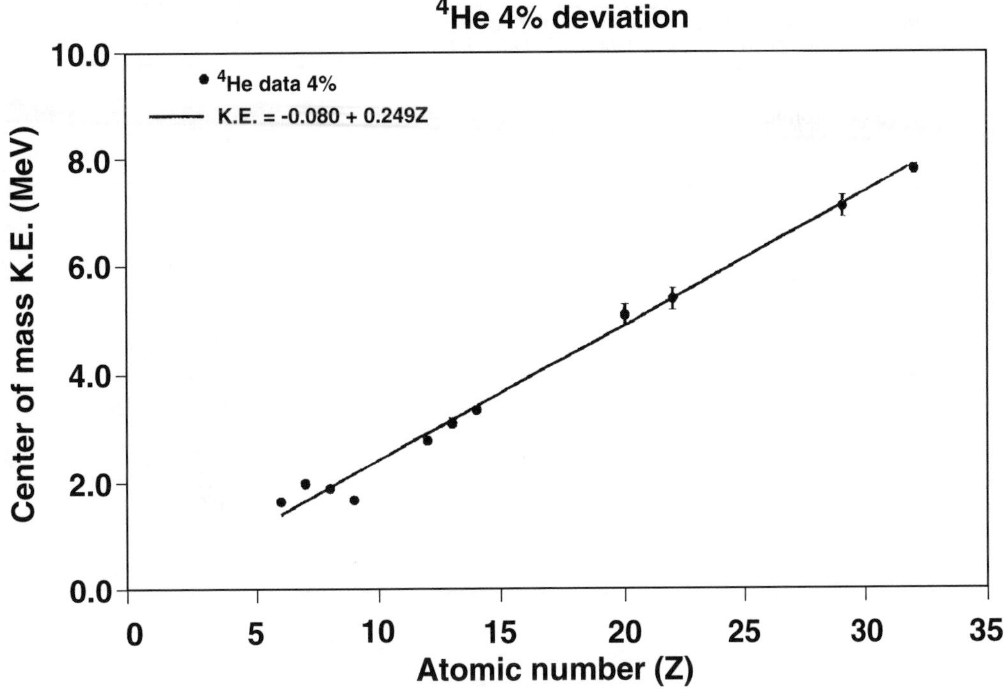

FIG. A8.5(b). The center-of-mass ^{4}He-ion energy at which the scattering cross section deviates by 4% from its Rutherford value vs. atomic number for several elements. The solid line is a linear least-squares fit to the data.

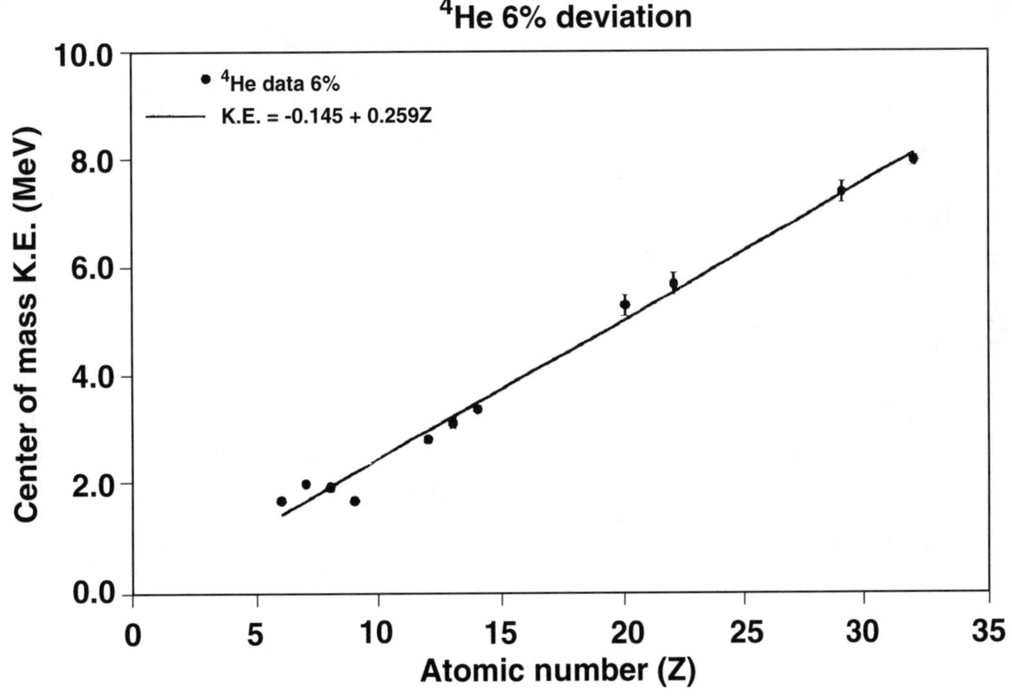

FIG. A8.5(c). The center-of-mass ^{4}He-ion energy at which the scattering cross section deviates by 6% from its Rutherford value vs. atomic number for several elements. The solid line is a linear least-squares fit to the data.

Table A8.6. ^{6}Li-ion energy at which the scattering cross section deviates by 4% from its Rutherford value (Bozoian, 1991b; Hubbard *et al.*, 1991).

Target atomic number (Z_2)	Target	E_{nr}^{cm} (calculation)
4	^{9}Be	1.39
5	^{10}B	2.08
6	^{12}C	2.20, 2.20, 2.23
8	^{16}O	2.93, 2.94
12	^{24}Mg	4.32, 4.37
13	^{27}Al	4.69
14	^{28}Si	5.12, 5.16, 5.19
19	^{39}K	6.19
20	^{40}Ca	7.05, 7.05, 7.13, 7.22
20	^{48}Ca	6.93
27	^{59}Co	9.23
28	^{58}Ni	9.82
29	^{63}Cu	10.09
30	^{64}Zn	10.35
32	^{72}Ge	10.74
40	^{90}Zr	13.34, 13.46

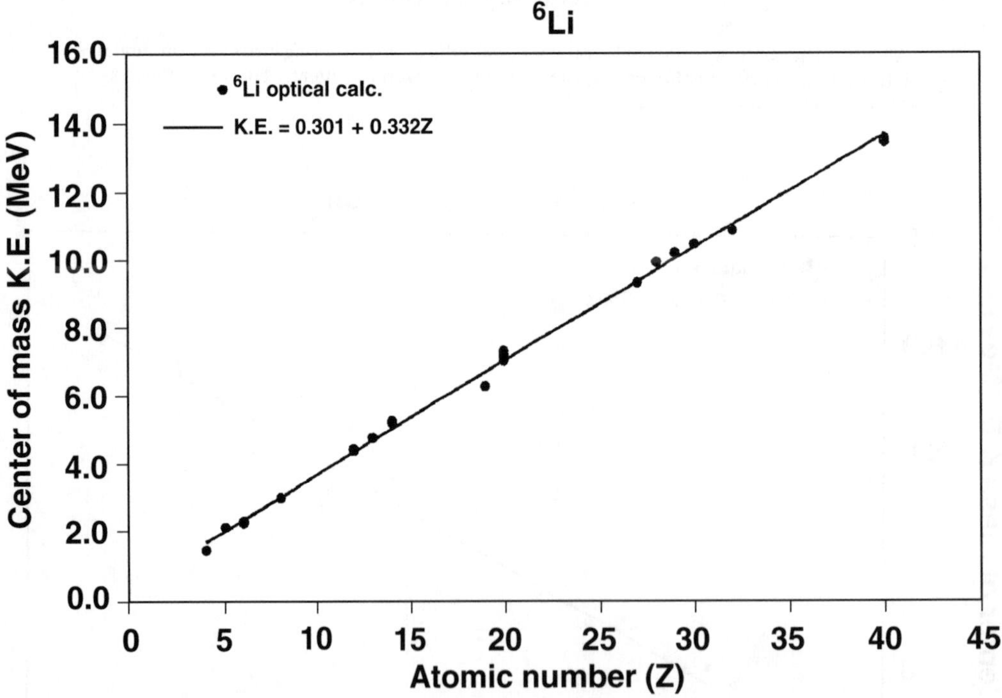

FIG. A8.6. The center-of-mass ^{6}Li-ion energy at which the scattering cross section deviates by 4% from its Rutherford value vs. atomic number for several elements. The solid line is a linear least-squares fit to the data.

Table A8.7. ^{7}Li-ion energy at which the scattering cross section deviates by 4% from its Rutherford value (Bozoian, 1991b, Hubbard *et al.*, 1991).

Target atomic number (Z_2)	Target	E_{nr}^{cm} (calculated)
6	^{12}C	2.26
6	^{13}C	2.23
7	^{15}N	2.60
8	^{16}O	2.99
12	^{24}Mg	4.12, 4.35
12	^{25}Mg	4.36, 4.42
12	^{26}Mg	4.23
13	^{27}Al	4.77
20	^{40}Ca	6.85, 7.12, 7.15
20	^{48}Ca	6.89, 6.90
28	^{62}Ni	9.60
30	^{64}Zn	10.37
32	^{72}Ge	10.86
40	^{90}Zr	13.51

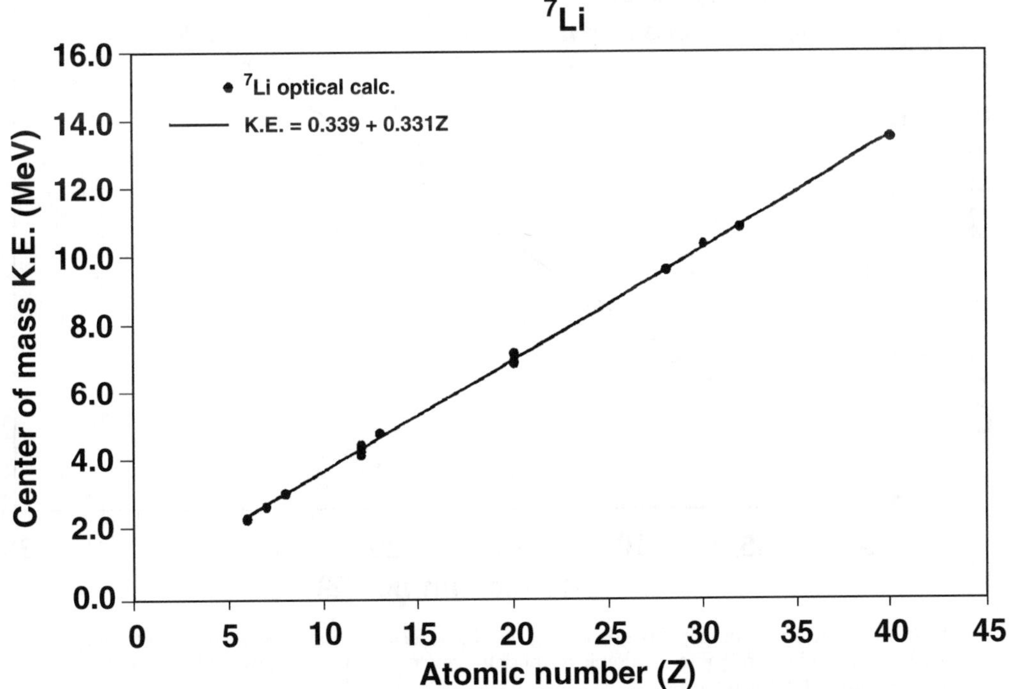

FIG. A8.7. The center-of-mass ^{7}Li-ion energy at which the scattering cross section deviates by 4% from its Rutherford value vs. atomic number for several elements. The solid line is a linear least-squares fit to the data.

Table A8.8. ^{9}Be-ion energy at which the scattering cross section deviates by 4% from its Rutherford value (Bozoian, 1993).

Target atomic number (Z_2)	Target	E_{nr}^{cm} (calculated)
5	^{10}B	2.44
6	^{12}C	2.89, 2.93, 2.95, 3.07
6	^{13}C	2.88, 2.93
7	^{14}N	3.54
8	^{16}O	3.81, 3.98, 4.02, 4.14
12	^{26}Mg	5.64
14	^{28}Si	6.59, 6.64, 6.67, 6.79, 6.84, 7.11
20	^{40}Ca	9.20, 9.23, 9.30, 9.42, 9.64
28	^{58}Ni	12.48
28	^{60}Ni	12.27

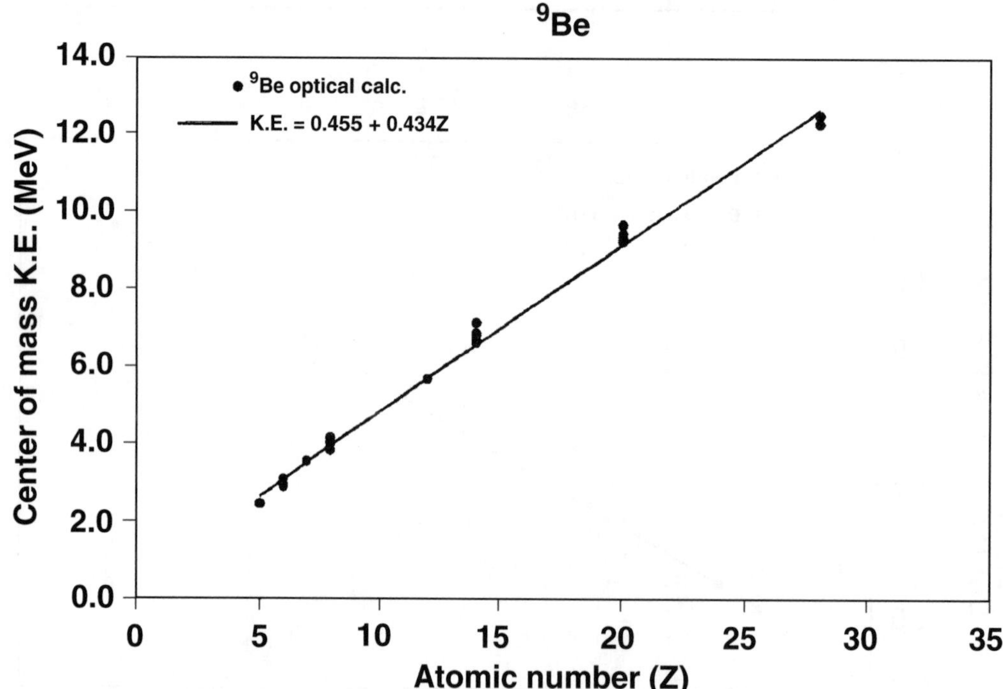

FIG. A8.8. The center-of-mass ^{9}Be-ion energy at which the scattering cross section deviates by 4% from its Rutherford value vs. atomic number for several elements. The solid line is a linear least-squares fit to the data.

Table A8.9. ^{10}B-ion energy at which the scattering cross section deviates by 4% from its Rutherford value (Bozoian, 1993).

Target atomic number (Z_2)	Target	E_{nr}^{cm} (calculated)
6	^{12}C	3.87, 3.98, 3.99
6	^{13}C	3.86
7	^{14}N	4.56, 4.71
8	^{16}O	4.86, 5.09, 5.17,5.25, 5.25, 5.39
10	^{20}Ne	6.28
12	^{24}Mg	7.16
12	^{25}Mg	7.12
13	^{27}Al	7.93
14	^{28}Si	8.76
19	^{39}K	10.68
20	^{40}Ca	12.30
28	^{60}Ni	15.38

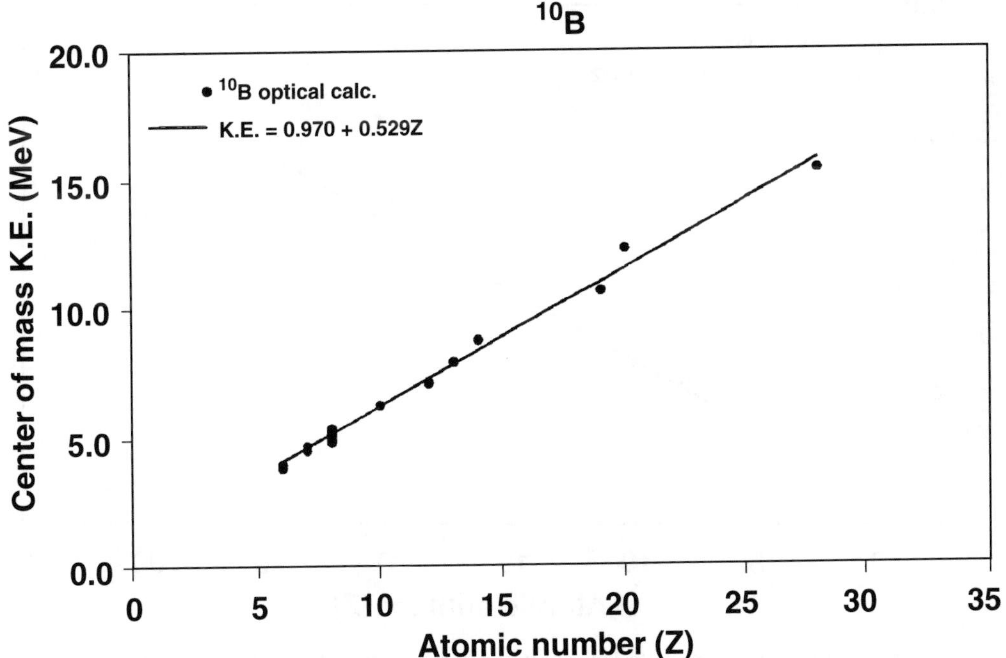

FIG. A8.9. The center-of-mass ^{10}B-ion energy at which the scattering cross section deviates by 4% from its Rutherford value vs. atomic number for several elements. The solid line is a linear least-squares fit to the data.

Table A8.10. ^{11}B-ion energy at which the scattering cross section deviates by 4% from its Rutherford value (Bozoian, 1993).

Target atomic number (Z_2)	Target	E_{nr}^{cm} (calculated)
6	^{12}C	3.70, 3.88, 3.93,3.97, 4.05
7	^{14}N	4.15
8	^{16}O	4.96, 4.97, 4.98, 5.32, 5.40
10	^{20}Ne	6.17
12	^{24}Mg	7.53
13	^{27}Al	7.92, 7.98
14	^{28}Si	8.52
20	^{40}Ca	12.34, 12.35
27	^{59}Co	15.10
28	^{60}Ni	15.33

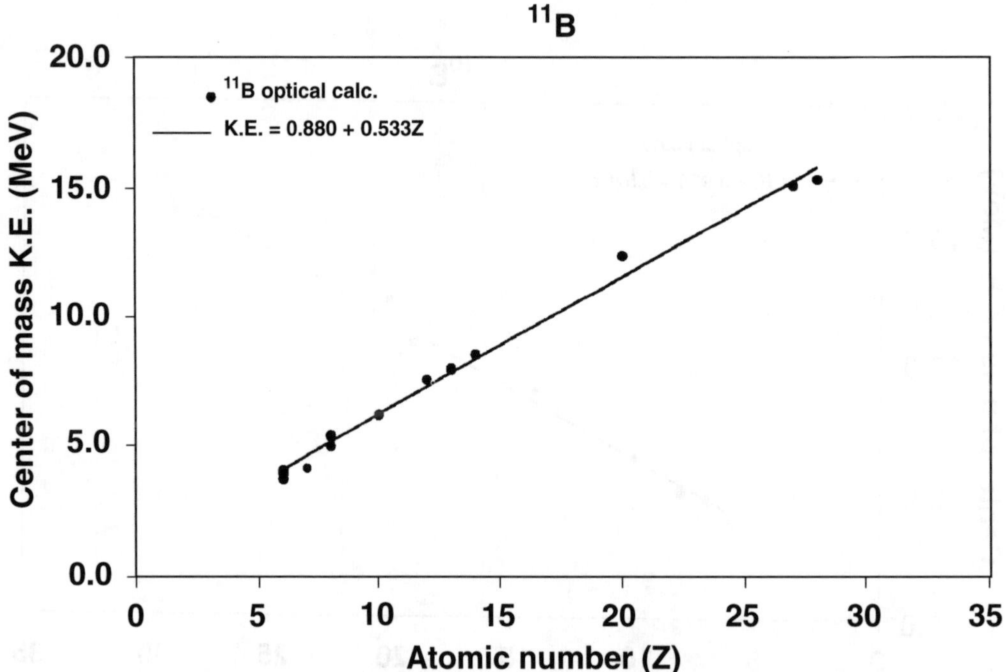

FIG. A8.10. The center-of-mass ^{11}B-ion energy at which the scattering cross section deviates by 4% from its Rutherford value vs. atomic number for several elements. The solid line is a linear least-squares fit to the data.

Table A8.11. ^{12}C-ion energy at which the scattering cross section deviates by 4% from its Rutherford value (Bozoian, 1993).

Target atomic number (Z_2)	Target	E_{nr}^{cm} (calculated)
8	^{16}O	6.81
10	^{20}Ne	7.75
14	^{28}Si	10.82, 11.51
20	^{40}Ca	14.87

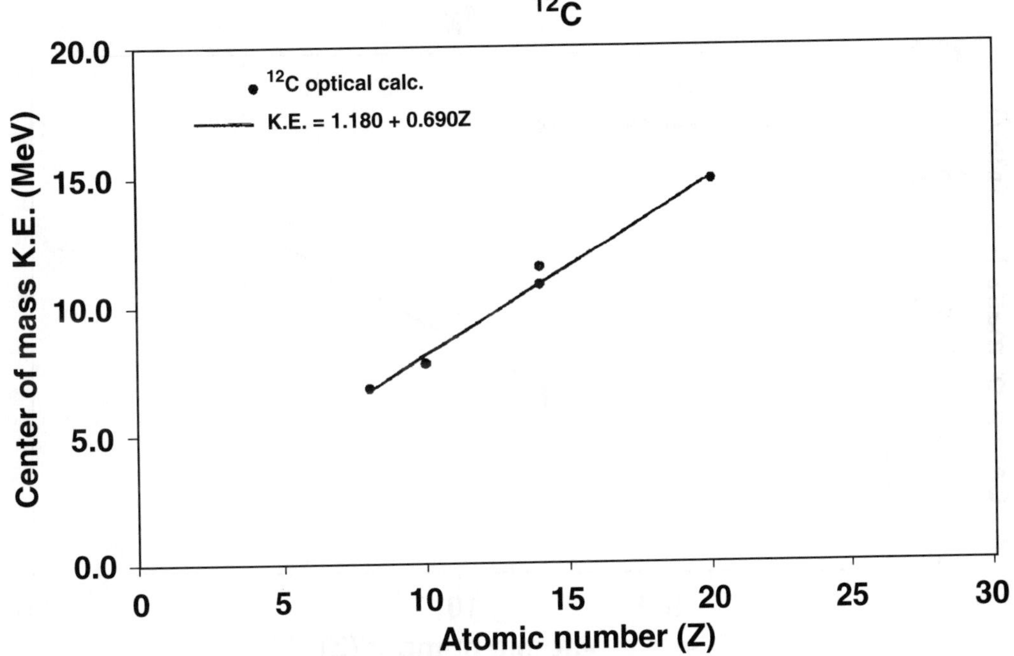

FIG. A8.11. The center-of-mass ^{12}C-ion energy at which the scattering cross section deviates by 4% from its Rutherford value vs. atomic number for several elements. The solid line is a linear least-squares fit to the data.

Table A8.12. ^{14}N-ion energy at which the scattering cross section deviates by 4% from its Rutherford value (Bozoian, 1993).

Target atomic number (Z_2)	Target	E_{nr}^{cm} (calculated)
8	^{16}O	6.46, 6.78, 7.55, 7.79
12	^{24}Mg	10.36
13	^{27}Al	11.45
14	^{28}Si	12.28, 12.60, 13.52

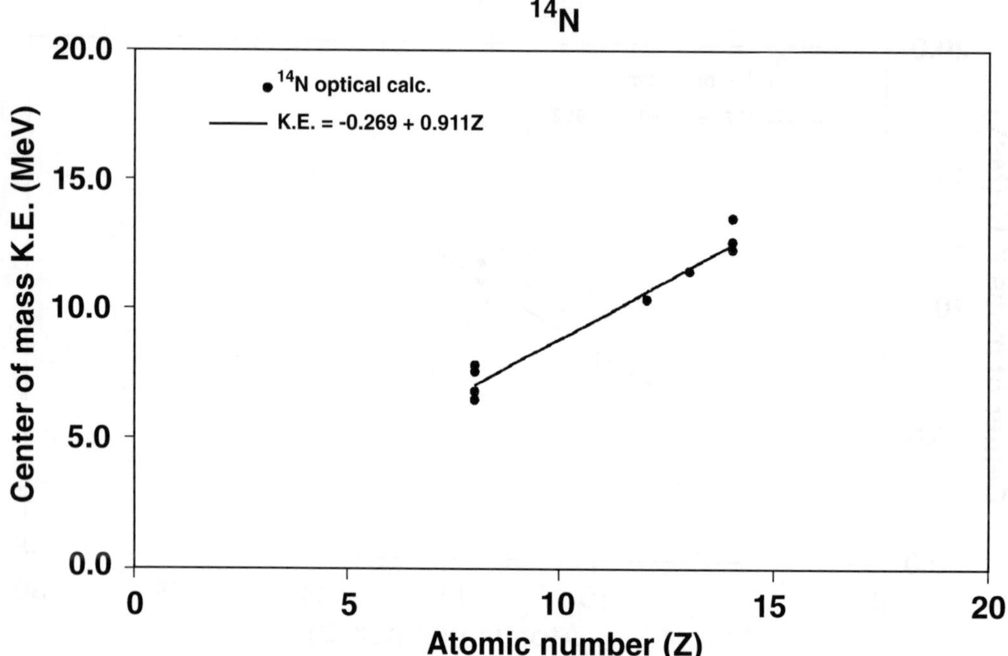

FIG. A8.12. The center-of-mass ^{14}N-ion energy at which the scattering cross section deviates by 4% from its Rutherford value vs. atomic number for several elements. The solid line is a linear least-squares fit to the data.

Table A8.13. ^{16}O-ion energy at which the scattering cross section deviates by 4% from its Rutherford value (Bozoian, 1993).

Target atomic number (Z_2)	Target	E_{nr}^{cm} (calculated)
10	^{20}Ne	10.17
12	^{24}Mg	12.92, 13.27
12	^{26}Mg	13.14
14	^{28}Si	14.35, 14.77, 15.20
14	^{30}Si	14.88
20	^{40}Ca	20.40, 20.74, 20.79
20	^{48}Ca	20.49

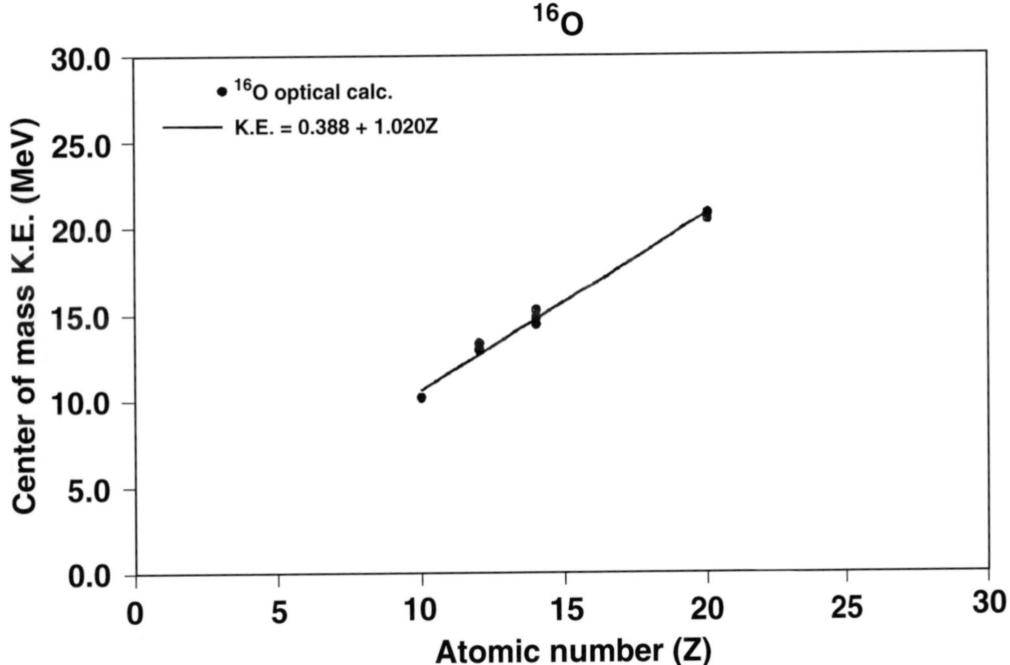

FIG. A8.13. The center-of-mass ^{16}O-ion energy at which the scattering cross section deviates by 4% from its Rutherford value vs. atomic number for several elements. The solid line is a linear least-squares fit to the data.

REFERENCES

Bozoian, M. (1991a), *Nucl. Instrum. Methods* **B58**, 127.

Bozoian, M. (1991b), *Nucl. Instrum. Methods* **B56/57**, 740.

Bozoian, M. (1993), *Nucl. Instrum. Methods* **B82**, 602.

Bozoian, M., Hubbard, K.M., and Nastasi, M. (1990), *Nucl. Instrum. Methods* **B51**, 311.

Hubbard, K.M., Tesmer, J.R., Nastasi, M., and Bozoian, M. (1991), *Nucl. Instrum. Methods* **B58**, 121.

APPENDIX

9

ELASTIC RECOIL DETECTION DATA

Compiled by

J. C. Barbour

Sandia National Laboratories, Albuquerque, New Mexico

CONTENTS

Table A9.1. Elastic recoil detection (forward scattering) kinematic factors for He, C, Si, Cl, and Au, projectiles.

This table gives the ERD kinematic factors K_{M2}, defined by Eq. (5.2), for He, C, Si, Cl, and Au as projectiles of mass M_1 (where M_1=4.0026, 12.0000, 27.9679, 34.9689, and 196.9666 amu respectively) and for the target atoms with isotopic masses (M_2). The kinematic factors are given as a function of forward scattering angle from 20° to 40° measured in the laboratory frame of reference.

	M_1=	He			^{12}C			^{28}Si			^{35}Cl			^{197}Au		
El	angle= M_2	20°	30°	40°	20°	30°	40°	20°	30°	40°	20°	30°	40°	20°	30°	40°
H	1.008	0.5675	0.4821	0.3772	0.2525	0.2144	0.1678	0.1185	0.1007	0.0788	0.0962	0.0817	0.0639	0.0179	0.0152	0.0119
	2.014	0.7866	0.6681	0.5227	0.4347	0.3692	0.2889	0.2213	0.1879	0.1471	0.1819	0.1545	0.1209	0.0354	0.0301	0.0235
He	3.016	0.8656	0.7352	0.5752	0.5669	0.4815	0.3768	0.3103	0.2635	0.2062	0.2582	0.2193	0.1716	0.0525	0.0446	0.0349
	4.003	0.8830	0.7500	0.5868	0.6625	0.5627	0.4403	0.3867	0.3285	0.2570	0.3255	0.2765	0.2163	0.0689	0.0586	0.0458
Li	6.015				0.7856	0.6672	0.5221	0.5144	0.4369	0.3419	0.4423	0.3757	0.2939	0.1016	0.0863	0.0675
	7.016				0.8224	0.6985	0.5465	0.5662	0.4809	0.3763	0.4916	0.4175	0.3267	0.1173	0.0996	0.0780
Be	9.012				0.8652	0.7348	0.5750	0.6509	0.5528	0.4326	0.5755	0.4888	0.3824	0.1478	0.1255	0.0982
B	10.013				0.8758	0.7439	0.5820	0.6856	0.5823	0.4556	0.6112	0.5191	0.4062	0.1626	0.1381	0.1081
	11.009				0.8814	0.7486	0.5857	0.7158	0.6079	0.4757	0.6432	0.5463	0.4275	0.1771	0.1504	0.1177
C	12.000				0.8830	0.7500	0.5868	0.7420	0.6302	0.4931	0.6719	0.5706	0.4465	0.1912	0.1624	0.1271
	13.003							0.7651	0.6499	0.5085	0.6979	0.5928	0.4638	0.2052	0.1743	0.1364
N	14.003							0.7852	0.6669	0.5218	0.7212	0.6125	0.4793	0.2189	0.1859	0.1455
	15.000							0.8025	0.6816	0.5333	0.7420	0.6302	0.4931	0.2323	0.1973	0.1544
O	15.995							0.8175	0.6943	0.5433	0.7606	0.6460	0.5055	0.2454	0.2084	0.1631
	16.999							0.8304	0.7053	0.5519	0.7774	0.6603	0.5167	0.2583	0.2194	0.1717
	17.999							0.8414	0.7147	0.5592	0.7924	0.6730	0.5266	0.2710	0.2302	0.1801
F	18.998							0.8508	0.7226	0.5654	0.8057	0.6843	0.5354	0.2834	0.2407	0.1883

Table A9.2. Rutherford forward-scattering cross sections for C, Si, Cl, and Au projectiles.

This table gives the forward-scattering cross sections used in Elastic Recoil Detection (ERD) of elements with mass M_2. The given cross sections are defined by the Rutherford model (Rutherford, 1911) and expressed in the laboratory frame of reference (Marion and Young, 1968) according to Eq. (5.4). The cross sections are given in barns/steradian for the projectiles C, Si, Cl, and Au of mass M_1 (where M_1=12.0000, 27.9679, 34.9689, and 196.9666 amu respectively), and as a function of forward scattering angle from 20° to 40° measured in the laboratory frame of reference. The energy of the projectiles C, Si, and Cl given in this table correspond to velocities of approximately 1 MeV/amu. These energies are accessible with many tandem accelerators and the cross sections for these energies often obey the Rutherford formula. (Note, 12 MeV C may not obey the Rutherford formula for the lightest elements, and nuclear reactions may also occur which give additional peaks in the spectrum.) The energy of the Au projectile is 10 MeV, which is a useful energy for gaining high depth resolution (such as in time-of-flight ERD), but the actual cross section for this energy is non-Rutherford. A standard should be used to determine the actual cross section for a given target atom, and a useful approach is to express the actual cross section as a ratio to the Rutherford cross section (e.g. $\sigma/\sigma_{Rutherford}$).

	M_1=	12 MeV C			28 MeV Si			35 MeV Cl			10 MeV Au		
	angle=												
El	M_2	20°	30°	40°	20°	30°	40°	20°	30°	40°	20°	30°	40°
H	1.008	0.2601	0.3323	0.4802	1.2916	1.6501	2.3841	1.8778	2.3990	3.4662	15042.7	19217.3	27766.6
	2.014	0.0756	0.0966	0.1395	0.3462	0.4423	0.6391	0.4968	0.6347	0.9171	3804.67	4860.52	7022.83
He	3.016	0.1548	0.1978	0.2858	0.6596	0.8426	1.2175	0.9349	1.1944	1.7258	6855.49	8757.99	12654.2
	4.003	0.0998	0.1275	0.1843	0.3987	0.5094	0.7360	0.5588	0.7138	1.0314	3930.90	5021.77	7255.83
Li	6.015	0.1261	0.1610	0.2327	0.4488	0.5733	0.8284	0.6157	0.7865	1.1364	3995.10	5103.79	7374.34
	7.016	0.1032	0.1319	0.1906	0.3496	0.4466	0.6453	0.4749	0.6067	0.8766	2965.56	3788.54	5473.97
Be	9.012	0.1358	0.1735	0.2507	0.4209	0.5377	0.7769	0.5615	0.7173	1.0365	3258.07	4162.23	6013.90
B	10.013	0.1887	0.2410	0.3483	0.5619	0.7179	1.0373	0.7435	0.9498	1.3723	4164.20	5319.83	7686.48
	11.009	0.1705	0.2178	0.3148	0.4895	0.6254	0.9036	0.6425	0.8208	1.1860	3477.79	4442.93	6419.47
C	12.000	0.2249	0.2873	0.4151	0.6239	0.7970	1.1516	0.8127	1.0382	1.5001	4255.50	5436.46	7855.00
	13.003				0.5583	0.7133	1.0306	0.7220	0.9224	1.3327	3658.98	4674.40	6753.92
N	14.003				0.6877	0.8785	1.2693	0.8831	1.1282	1.6301	4335.56	5538.73	8002.77
	15.000				0.6281	0.8024	1.1594	0.8013	1.0236	1.4790	3814.17	4872.65	7040.37
O	15.995				0.7553	0.9649	1.3941	0.9574	1.2231	1.7673	4422.58	5649.91	8163.41
	16.999				0.6996	0.8937	1.2913	0.8814	1.1260	1.6269	3952.51	5049.39	7295.73
	17.999				0.6520	0.8330	1.2036	0.8167	1.0434	1.5075	3558.52	4546.06	6568.48
F	18.998				0.7733	0.9879	1.4273	0.9631	1.2304	1.7778	4080.14	5212.43	7531.31

Table A9.3a. Elastic recoil detection—mylar foil thickness
Projectiles: He, Si, Cl, and Au.

This table is an aid for determining the thickness of mylar ($C_{10}H_8O_4$) needed to stop various projectiles at different energies from entering the detector. The projectile ions and energies (in MeV) shown in the left column are those which are used most frequently in ERD analyses. The projectiles are assumed to be incident normal to the mylar foil. The electronic stopping, nuclear stopping, range and straggling of possible projectiles used in ERD were calculated using the RSTOP computer program from TRIM-91 (Biersack and Ziegler, 1991; see also Ziegler *et al.*, 1985). A density of 1.397 g/cm^3 was used for mylar and a Bragg correction of 2.05% was used for stopping in mylar. This table shows that a 12-μm-thick mylar foil may be used for stopping several different scattered projectile beams: 2.8 MeV He, 24-28 MeV Si, and 30-36 MeV Cl. This table also shows that the thickness of mylar needed to stop a Au beam varies from 2 to 5 μm. However, a mylar foil may not be necessary if the Au beam is kinematically prohibited from scattering at the detector angle. For example, if the substrate mass is less than 98 amu (Tc), then the critical angle [$\sin^{-1}(M_{substrate}/M_1)$] for scattering Au is less than 30° and the use of a forward scattering angle of 30° or greater would negate the need for a range foil.

Ion E *(MeV)*	S_e *(keV/μm)*	S_n *(keV/μm)*	Range in foil *(μm)*	ΔR_p *(μm)*
2.8 He	171.7	0.1	12.32	0.4115
8 Si	2642	12	5.24	0.2099
12 Si	2816	8	6.70	0.2372
16 Si	2925	7	8.09	0.2588
20 Si	2938	5	9.44	0.2858
24 Si	2894	5	10.81	0.3182
28 Si	2842	4	12.21	0.3485
24 Cl	3631	8	8.98	0.2613
30 Cl	3613	7	10.63	0.2957
36 Cl	3573	6	12.30	0.3407
8 Au	3241	953	2.13	0.1156
12 Au	3848	750	3.04	0.1483
16 Au	4349	625	3.87	0.1715
20 Au	4782	540	4.65	0.1930

Table A9.3b. Elastic recoil detection—aluminum foil thickness
Projectiles: He, Si, Cl, and Au.

This table is an aid for determining the thickness of aluminum needed to stop projectiles at different energies from entering the detector. The projectile ions and energies (in MeV) are shown in the left column. The electronic stopping, nuclear stopping, range, and straggling of possible projectiles used in ERD were calculated as in Table A9.3a. The projectiles were assumed to be incident normal to the Al foil, and a density of 2.698 g/cm^3 was used for Al.

Ion E *(MeV)*	S_e *(keV/μm)*	S_n *(keV/μm)*	Range in foil *(μm)*	ΔR_p *(μm)*
2.8 He	227.9	0.2	10.01	0.3885
8 Si	2989	18	4.27	0.2822
12 Si	3445	13	5.51	0.3034
16 Si	3770	10	6.61	0.3182
20 Si	3910	8	7.64	0.3343
24 Si	3926	7	8.66	0.3525
28 Si	3906	6	9.68	0.3697
24 Cl	4806	13	7.30	0.2904
30 Cl	4902	11	8.53	0.3112
36 Cl	4919	9	9.75	0.3375
8 Au	3321	1512	1.72	0.1708
12 Au	4067	1195	2.50	0.2173
16 Au	4696	1000	3.22	0.2515
20 Au	5251	865	3.89	0.2799

Table A9.4a. Elastic recoil detection—recoil stopping and range in mylar foil   Recoils: H, He, Li, Be, B, C, N, O, and F.

The tables in the previous section gave values for stopping the scattered projectiles in mylar (and Al) foils. However, the choice of foil thickness also depends on the ability of the recoiled atom to get through the range foil with sufficient energy to be detected. These tables give the electronic stopping (S_e) and range in mylar of possible recoiled atoms using the RSTOP computer program from TRIM-91 (Biersack and Ziegler, 1991; see also Ziegler *et al.*, 1985). The recoiled atom species and energy in MeV are given beginning in the far left column. The recoil energy ranges correspond to the possible energies for atoms coming from the surface or in depth of a sample analyzed with the incident projectiles given in Tables A9.3a and A9.3b.

^{1}H *(MeV)*	S_e *(keV/μm)*	Range in foil *(μm)*
0.10	107.9	1.10
0.20	92.81	2.08
0.30	75.29	3.27
0.40	62.82	4.73
0.50	54.20	6.44
0.60	47.99	8.40
0.70	43.30	10.59
0.80	39.62	13.00
0.90	36.63	15.62
1.0	34.13	18.44
1.1	32.01	21.46
1.2	30.18	24.67
1.3	28.58	28.06
1.4	27.17	31.64
1.5	25.90	35.41
1.6	24.77	39.35
1.7	23.74	43.46
1.8	22.80	47.75
2.0	21.15	56.84
2.2	19.75	66.61
2.4	18.54	77.04
2.6	17.48	88.13
2.8	16.55	99.87
3.0	15.72	112.25

^{4}He *(MeV)*	S_e *(keV/μm)*	Range in foil *(μm)*
0.50	304.5	2.35
0.55	307.3	2.52
0.60	308.6	2.68
0.65	308.6	2.84
0.70	307.4	3.00
0.80	302.8	3.33
0.90	295.9	3.66
1.0	287.8	4.00
1.1	279.1	4.36
1.2	270.2	4.72
1.3	261.4	5.09
1.4	252.9	5.48
1.5	244.7	5.88
1.6	236.9	6.30
1.7	229.5	6.73
1.8	222.6	7.17
2.0	209.9	8.09
2.2	198.6	9.07
2.4	188.6	10.10
2.6	179.7	11.19
2.8	171.7	12.32
3.0	164.4	13.51
3.3	154.9	15.39
3.6	146.5	17.38
4.0	136.9	20.20
4.5	126.7	24.00
5.0	118.2	28.08
5.5	110.9	32.44
6.0	104.5	37.09
6.5	98.87	42.00
7.0	93.89	47.19
8.0	85.42	58.35

^{7}Li *(MeV)*	S_e *(keV/μm)*	Range in foil *(μm)*
1.0	450.2	3.48
1.1	455.7	3.70
1.2	458.6	3.92
1.3	459.4	4.13
1.4	458.4	4.35
1.5	456.2	4.57
1.6	453.0	4.79
1.7	450.8	5.01
1.8	447.8	5.23
2.0	440.1	5.68
2.2	431.0	6.14
2.6	411.5	7.08
2.8	401.7	7.58
3.0	391.9	8.08
3.3	378.0	8.86
3.6	365.3	9.66
4.0	349.8	10.78
4.5	332.5	12.24
5.0	317.3	13.78
5.5	303.7	15.39
6.0	291.4	17.07
6.5	280.1	18.82
7.0	269.3	20.64
8.0	250.7	24.48
9.0	235.0	28.60
10.0	221.4	32.98
11.0	209.4	37.62
12.0	198.8	42.52
13.0	189.3	47.67
14.0	180.7	53.07
15.0	172.9	58.73

Table A9.4a. (continued).

^{9}Be *(MeV)*	S_e *(keV/μm)*	Range in foil *(μm)*	^{11}B *(MeV)*	S_e *(keV/μm)*	Range in foil *(μm)*	^{12}C *(MeV)*	S_e *(keV/μm)*	Range in foil *(μm)*
1.0	628.8	2.55	1.0	742.7	2.24	1.0	887.7	1.85
1.1	646.1	2.70	1.1	773.1	2.37	1.1	927.0	1.96
1.2	659.6	2.86	1.2	799.3	2.50	1.2	961.8	2.06
1.3	669.8	3.01	1.3	821.7	2.62	1.3	992.3	2.17
1.4	677.3	3.15	1.4	840.5	2.74	1.4	1019	2.26
1.5	682.3	3.30	1.5	856.3	2.86	1.5	1042	2.36
1.6	685.4	3.44	1.6	869.3	2.97	1.6	1061	2.46
1.7	686.8	3.59	1.7	879.8	3.09	1.7	1078	2.55
1.8	686.7	3.74	1.8	888.1	3.20	1.8	1092	2.64
2.0	683.4	4.03	2.0	899.1	3.42	2.0	1113	2.82
2.2	677.1	4.32	2.2	904.3	3.64	2.2	1127	3.00
2.4	668.8	4.62	2.4	905.1	3.86	2.4	1134	3.17
2.6	659.4	4.92	2.6	902.7	4.08	2.6	1137	3.35
2.8	649.3	5.22	2.8	897.9	4.31	2.8	1137	3.52
3.0	641.3	5.53	3.0	891.5	4.53	3.0	1134	3.70
3.3	629.2	6.00	3.3	880.0	4.87	3.3	1126	3.96
3.6	616.5	6.48	3.6	867.1	5.21	3.6	1115	4.23
4.0	599.6	7.14	4.0	853.0	5.67	4.0	1099	4.59
4.5	579.1	7.99	4.5	836.5	6.26	4.5	1077	5.05
5.0	559.0	8.86	5.0	818.9	6.87	5.0	1059	5.52
5.5	540.6	9.77	5.5	801.3	7.48	5.5	1045	5.99
6.0	523.6	10.71	6.0	784.1	8.11	6.0	1029	6.47
6.5	507.9	11.68	6.5	766.8	8.76	6.5	1013	6.96
7.0	493.3	12.68	7.0	749.8	9.41	7.0	997.8	7.46
8.0	467.0	14.76	8.0	718.5	10.77	8.0	965.9	8.47
9.0	443.8	16.95	9.0	690.2	12.19	9.0	934.8	9.53
10.0	422.2	19.26	10.0	664.4	13.67	10.0	905.9	10.61
11.0	402.6	21.68	11.0	640.7	15.20	11.0	879.0	11.73
12.0	385.1	24.22	12.0	618.8	16.79	12.0	853.8	12.88
13.0	369.4	26.87	13.0	597.7	18.43	13.0	830.0	14.07
14.0	355.0	29.63	14.0	577.5	20.13	14.0	807.5	15.29
15.0	341.9	32.50	15.0	559.0	21.89	15.0	785.1	16.54
16.0	329.8	35.47	16.0	541.9	23.70	16.0	763.6	17.83
17.0	318.6	38.55	17.0	526.0	25.57	17.0	743.5	19.16
18.0	308.1	41.74	18.0	511.1	27.50	18.0	724.6	20.52
			20.0	484.0	31.52	20.0	690.1	23.35

Table A9.4a. (continued).

^{14}N *(MeV)*	S_e *(keV/μm)*	Range in foil *(μm)*	^{16}O *(MeV)*	S_e *(keV/μm)*	Range in foil *(μm)*	^{19}F *(MeV)*	S_e *(keV/μm)*	Range in foil *(μm)*
1.0	972.0	1.66	1.0	984.3	1.64	1.0	941.3	1.66
1.1	1020	1.76	1.1	1040	1.73	1.1	1000	1.76
1.2	1064	1.85	1.2	1092	1.83	1.2	1056	1.86
1.3	1103	1.94	1.3	1140	1.91	1.3	1110	1.95
1.4	1139	2.03	1.4	1184	2.00	1.4	1160	2.04
1.5	1171	2.12	1.5	1225	2.08	1.5	1208	2.12
1.6	1200	2.20	1.6	1263	2.16	1.6	1253	2.20
1.7	1225	2.28	1.7	1298	2.24	1.7	1295	2.28
1.8	1248	2.36	1.8	1329	2.31	1.8	1335	2.35
2.0	1285	2.52	2.0	1384	2.46	2.0	1407	2.50
2.2	1313	2.67	2.2	1430	2.60	2.2	1469	2.63
2.4	1334	2.82	2.4	1466	2.74	2.4	1523	2.77
2.6	1348	2.97	2.6	1495	2.87	2.6	1570	2.89
2.8	1358	3.12	2.8	1518	3.00	2.8	1609	3.02
3.0	1363	3.27	3.0	1535	3.13	3.0	1642	3.14
3.3	1365	3.48	3.3	1553	3.33	3.3	1682	3.32
3.6	1362	3.70	3.6	1564	3.52	3.6	1711	3.50
4.0	1353	4.00	4.0	1569	3.77	4.0	1739	3.73
4.5	1337	4.37	4.5	1566	4.09	4.5	1758	4.01
5.0	1319	4.74	5.0	1557	4.41	5.0	1766	4.29
5.5	1300	5.12	5.5	1544	4.73	5.5	1767	4.57
6.0	1281	5.51	6.0	1530	5.05	6.0	1763	4.86
6.5	1270	5.90	6.5	1515	5.38	6.5	1756	5.14
7.0	1259	6.30	7.0	1501	5.71	7.0	1746	5.42
8.0	1234	7.10	8.0	1487	6.38	8.0	1732	6.00
9.0	1208	7.91	9.0	1469	7.06	9.0	1728	6.57
10.0	1181	8.75	10.0	1449	7.74	10.0	1719	7.15
11.0	1153	9.61	11.0	1428	8.43	11.0	1707	7.74
12.0	1127	10.48	12.0	1403	9.14	12.0	1693	8.32
13.0	1102	11.38	13.0	1379	9.86	13.0	1677	8.91
14.0	1078	12.29	14.0	1356	10.59	14.0	1658	9.51
15.0	1056	13.23	15.0	1333	11.33	15.0	1637	10.12
16.0	1034	14.19	16.0	1311	12.08	16.0	1616	10.73
17.0	1012	15.16	17.0	1289	12.85	17.0	1596	11.35
18.0	990.7	16.16	18.0	1268	13.63	18.0	1576	11.98
20.0	949.6	18.22	20.0	1228	15.24	20.0	1537	13.27
22.0	912.4	20.37	22.0	1186	16.89	22.0	1499	14.58
			24.0	1147	18.60	24.0	1462	15.93

Table A9.4b. Elastic recoil detection—recoil stopping and range in aluminum foil   Recoils: H, He, Li, Be, B, C, N, O, and F.

Table A9.3b gave values for stopping the scattered projectiles in Al foils. However, the choice of foil thickness also depends on the ability of the recoiled atom to get through the range foil with sufficient energy to be detected. These tables give the electronic stopping (S_e) and range in Al of possible recoiled atoms using the RSTOP computer code from TRIM-91 (Biersack and Ziegler, 1991; see also Ziegler *et al.*, 1985). The recoiled atom species and energy in MeV are given beginning in the far left column. The recoil energy ranges correspond to the possible energies for atoms coming from the surface or in depth of a sample analyzed with the incident projectiles given in Table A9.3b.

^{1}H *(MeV)*	S_e *(keV/μm)*	Range in foil *(μm)*
0.10	123.2	0.836
0.20	101.0	1.72
0.30	86.13	2.78
0.40	75.94	4.00
0.50	68.33	5.38
0.60	62.33	6.90
0.70	57.44	8.56
0.80	53.36	10.35
0.90	49.88	12.28
1.0	46.89	14.33
1.1	44.27	16.51
1.2	41.96	18.82
1.3	39.91	21.24
1.4	38.08	23.79
1.5	36.42	26.46
1.6	34.92	29.25
1.7	33.55	32.15
1.8	32.29	35.17
2.0	30.07	41.56
2.2	28.17	48.39
2.4	26.52	55.67
2.6	25.07	63.39
2.8	23.79	71.53
3.0	22.64	80.11

^{4}He *(MeV)*	S_e *(keV/μm)*	Range in foil *(μm)*
0.50	336.7	1.84
0.55	336.6	1.99
0.60	335.9	2.14
0.65	334.8	2.28
0.70	333.3	2.43
0.80	329.6	2.73
0.90	325.1	3.03
1.0	320.2	3.34
1.1	315.0	3.65
1.2	309.5	3.97
1.3	303.9	4.30
1.4	298.2	4.63
1.5	292.5	4.96
1.6	286.8	5.31
1.7	281.2	5.66
1.8	275.6	6.02
2.0	264.9	6.75
2.2	254.8	7.52
2.4	245.2	8.32
2.6	236.3	9.15
2.8	227.9	10.01
3.0	220.1	10.90
3.3	209.4	12.29
3.6	199.7	13.75
4.0	188.2	15.81
4.5	175.7	18.55
5.0	164.8	21.49
5.5	155.4	24.60
6.0	147.0	27.90
6.5	139.6	31.39
7.0	133.0	35.05
8.0	121.6	42.90

^{7}Li *(MeV)*	S_e *(keV/μm)*	Range in foil *(μm)*
1.0	490.0	2.73
1.1	493.5	2.93
1.2	495.8	3.13
1.3	497.1	3.33
1.4	497.7	3.53
1.5	497.8	3.73
1.6	497.4	3.92
1.7	498.4	4.12
1.8	499.0	4.32
2.0	498.8	4.72
2.2	497.0	5.12
2.4	494.0	5.52
2.6	490.2	5.93
2.8	485.7	6.33
3.0	480.3	6.75
3.3	471.7	7.37
3.6	463.0	8.01
4.0	451.1	8.89
4.5	436.4	10.01
5.0	422.1	11.17
5.5	408.3	12.37
6.0	395.2	13.61
6.5	382.6	14.90
7.0	370.0	16.22
8.0	347.6	19.01
9.0	327.9	21.96
10.0	310.6	25.09
11.0	295.1	28.39
12.0	281.1	31.86
13.0	268.5	35.49
14.0	257.0	39.29
15.0	246.5	43.26

Table A9.4b. (continued).

^{9}Be *(MeV)*	S_e *(keV/μm)*	Range in foil *(μm)*	^{11}B *(MeV)*	S_e *(keV/μm)*	Range in foil (μm)	^{12}C *(MeV)*	S_e *(keV/μm)*	Range in foil *(μm)*
1.0	704.8	1.95	1.0	858.9	1.69	1.0	1044	1.38
1.1	715.0	2.09	1.1	879.8	1.80	1.1	1072	1.47
1.2	723.0	2.22	1.2	897.5	1.91	1.2	1096	1.56
1.3	729.3	2.36	1.3	912.4	2.02	1.3	1116	1.65
1.4	734.3	2.50	1.4	925.1	2.13	1.4	1134	1.74
1.5	738.3	2.63	1.5	935.9	2.23	1.5	1150	1.82
1.6	741.4	2.76	1.6	945.3	2.34	1.6	1163	1.91
1.7	743.9	2.90	1.7	953.5	2.44	1.7	1175	1.99
1.8	745.9	3.03	1.8	960.6	2.54	1.8	1186	2.08
2.0	748.4	3.30	2.0	972.4	2.75	2.0	1204	2.24
2.2	749.6	3.56	2.2	981.7	2.95	2.2	1219	2.41
2.4	749.6	3.83	2.4	989.0	3.15	2.4	1232	2.57
2.6	748.8	4.09	2.6	994.8	3.35	2.6	1242	2.73
2.8	747.2	4.36	2.8	999.2	3.55	2.8	1251	2.89
3.0	747.5	4.62	3.0	1003	3.75	3.0	1259	3.05
3.3	747.4	5.02	3.3	1006	4.05	3.3	1268	3.28
3.6	745.2	5.42	3.6	1007	4.34	3.6	1275	3.52
4.0	739.9	5.96	4.0	1011	4.74	4.0	1281	3.83
4.5	730.3	6.64	4.5	1015	5.23	4.5	1285	4.21
5.0	717.7	7.32	5.0	1015	5.72	5.0	1290	4.60
5.5	704.3	8.03	5.5	1010	6.21	5.5	1296	4.99
6.0	690.5	8.74	6.0	1003	6.71	6.0	1297	5.37
6.5	676.6	9.47	6.5	993.9	7.20	6.5	1295	5.75
7.0	662.7	10.21	7.0	982.5	7.71	7.0	1291	6.14
8.0	635.6	11.75	8.0	957.8	8.74	8.0	1274	6.92
9.0	609.8	13.35	9.0	932.0	9.79	9.0	1251	7.71
10.0	584.4	15.03	10.0	905.9	10.88	10.0	1225	8.51
11.0	560.4	16.77	11.0	880.3	12.00	11.0	1199	9.33
12.0	538.6	18.59	12.0	855.4	13.15	12.0	1173	10.18
13.0	518.5	20.48	13.0	830.4	14.33	13.0	1147	11.04
14.0	500.0	22.44	14.0	805.7	15.55	14.0	1121	11.92
15.0	482.9	24.47	15.0	782.6	16.81	15.0	1094	12.82
16.0	466.9	26.57	16.0	760.9	18.10	16.0	1068	13.74
17.0	452.0	28.75	17.0	740.5	19.43	17.0	1043	14.69
18.0	438.1	30.99	18.0	721.3	20.80	18.0	1019	15.66
			20.0	685.8	23.64	20.0	974.7	17.66

Table A9.4b. (continued).

^{14}N (MeV)	S_e (keV/μm)	Range in foil (μm)	^{16}O (MeV)	S_e (keV/μm)	Range in foil (μm)	^{19}F (MeV)	S_e (keV/μm)	Range in foil (μm)
1.0	1176	1.22	1.0	1217	1.20	1.0	1194	1.21
1.1	1213	1.31	1.1	1264	1.28	1.1	1250	1.29
1.2	1246	1.39	1.2	1307	1.35	1.2	1301	1.37
1.3	1274	1.46	1.3	1345	1.43	1.3	1348	1.44
1.4	1299	1.54	1.4	1379	1.50	1.4	1391	1.51
1.5	1321	1.61	1.5	1410	1.57	1.5	1430	1.58
1.6	1340	1.69	1.6	1438	1.64	1.6	1467	1.64
1.7	1358	1.76	1.7	1463	1.70	1.7	1500	1.71
1.8	1373	1.83	1.8	1486	1.77	1.8	1531	1.77
2.0	1401	1.97	2.0	1527	1.90	2.0	1587	1.90
2.2	1423	2.11	2.2	1561	2.03	2.2	1635	2.02
2.4	1443	2.25	2.4	1591	2.15	2.4	1677	2.14
2.6	1460	2.39	2.6	1617	2.27	2.6	1714	2.25
2.8	1474	2.52	2.8	1641	2.40	2.8	1747	2.37
3.0	1487	2.66	3.0	1662	2.52	3.0	1777	2.48
3.3	1504	2.86	3.3	1689	2.69	3.3	1817	2.64
3.6	1518	3.05	3.6	1714	2.87	3.6	1852	2.81
4.0	1534	3.31	4.0	1741	3.10	4.0	1893	3.02
4.5	1548	3.64	4.5	1770	3.38	4.5	1937	3.28
5.0	1558	3.96	5.0	1792	3.66	5.0	1974	3.53
5.5	1565	4.27	5.5	1811	3.93	5.5	2006	3.78
6.0	1568	4.59	6.0	1825	4.21	6.0	2033	4.02
6.5	1579	4.91	6.5	1836	4.48	6.5	2056	4.27
7.0	1586	5.22	7.0	1845	4.75	7.0	2075	4.51
8.0	1590	5.85	8.0	1874	5.28	8.0	2112	4.98
9.0	1585	6.48	9.0	1888	5.81	9.0	2156	5.45
10.0	1570	7.11	10.0	1893	6.34	10.0	2186	5.91
11.0	1551	7.75	11.0	1890	6.87	11.0	2205	6.36
12.0	1529	8.40	12.0	1877	7.40	12.0	2215	6.81
13.0	1506	9.05	13.0	1860	7.93	13.0	2219	7.26
14.0	1482	9.72	14.0	1842	8.47	14.0	2213	7.71
15.0	1457	10.40	15.0	1822	9.01	15.0	2202	8.16
16.0	1433	11.09	16.0	1801	9.56	16.0	2189	8.62
17.0	1409	11.79	17.0	1779	10.12	17.0	2173	9.07
18.0	1383	12.51	18.0	1756	10.69	18.0	2156	9.54
20.0	1332	13.98	20.0	1711	11.84	20.0	2119	10.47
22.0	1286	15.50	22.0	1661	13.02	22.0	2080	11.42
			24.0	1613	14.24	24.0	2039	12.39

Table A9.5. Elastic recoil detection—projectile range in samples   Projectiles: He, Si, Cl, and Au   Samples: C, Si, Fe, Ge, Sn, Pb.

The depth of analysis in ERD depends on the range of the incident projectile ion, the stopping of the scattered recoil atom in the sample, and the stopping of the recoil atom in the range foil. The following tables can be used to estimate the range of a projectile incident upon a sample at 75° from the surface normal. The sample species are shown as titles above each set of three columns containing the electronic stopping power (S_e), the range (R_p), and the straggling. The density used for each sample is as follows: ρ(C)=2.26 g/cm^3, ρ(Si)=2.32 g/cm^3, ρ(Fe)=7.86 g/cm^3, ρ(Ge)=5.32 g/cm^3, ρ(Sn)=7.28 g/cm^3, and ρ(Pb)=1.13 g/cm^3. A Bragg correction of 6.27% was used for calculating the stopping in C. The projectiles and energies (MeV) correspond to those given in Tables A9.3a and A9.3b. The stopping power is for the incident projectile in the sample, and the range in this table (R_p) is the projected range in the sample measured normal to the sample surface. These calculations were determined using the RSTOP computer code from TRIM-91 (Biersack and Ziegler, 1991; see also Ziegler *et al.*, 1985). Note that for the Au projectile, both electronic and nuclear stopping must be considered.

	Carbon			Silicon			Iron		
Ion E *(MeV)*	S_e *(eV/nm)*	R_p *(μm)*	ΔR_p *(nm)*	S_e *(eV/nm)*	R_p *(μm)*	ΔR_p *(nm)*	S_e *(eV/nm)*	R_p *(μm)*	ΔR_p *(nm)*
2.8 He	260.4	2.11	72	202.5	2.75	110	510.5	1.18	61
8 Si	3927	0.89	40	2908	1.12	79	6764	0.53	61
12 Si	4213	1.15	44	3180	1.46	84	7778	0.68	63
16 Si	4412	1.39	48	3392	1.78	89	8466	0.80	65
20 Si	4460	1.62	52	3470	2.08	93	8750	0.92	66
24 Si	4410	1.85	57	3460	2.37	99	8776	1.04	68
28 Si	4343	2.09	62	3429	2.67	104	8733	1.15	69
24 Cl	5506	1.54	47	4275	1.98	81	10760	0.88	56
30 Cl	5507	1.82	53	4321	2.34	87	10960	1.02	58
36 Cl	5461	2.10	60	4316	2.70	95	11000	1.16	60

	Germanium			Tin			Lead		
Ion E *(MeV)*	S_e *(eV/nm)*	R_p *(μm)*	ΔR_p *(nm)*	S_e *(eV/nm)*	R_p *(μm)*	ΔR_p *(nm)*	S_e *(eV/nm)*	R_p *(μm)*	ΔR_p *(nm)*
2.8 He	282.6	2.21	123	313.8	1.88	127	338.9	1.74	162
8 Si	3476	0.98	119	4125	0.81	132	4313	0.73	162
12 Si	4111	1.25	125	4740	1.04	137	5029	0.94	170
16 Si	4591	1.49	128	5174	1.24	141	5534	1.13	175
20 Si	4856	1.70	132	5376	1.44	144	5806	1.31	179
24 Si	4966	1.91	135	5423	1.63	147	5923	1.49	183
28 Si	5023	2.12	137	5428	1.82	150	5999	1.66	186
24 Cl	5946	1.64	113	6604	1.38	123	7117	1.27	157
30 Cl	6201	1.89	116	6770	1.62	127	7394	1.49	161
36 Cl	6346	2.14	121	6844	1.84	131	7582	1.69	165

	Carbon				Silicon			
Ion E *(MeV)*	S_e *(eV/nm)*	S_n *(eV/nm)*	R_p *(μm)*	ΔR_p *(nm)*	S_e *(eV/nm)*	S_n *(eV/nm)*	R_p *(μm)*	ΔR_p *(nm)*
8 Au	5182	1423	0.35	21	3491	1364	0.46	46
12 Au	6033	1119	0.50	27	4276	1081	0.66	57
16 Au	6721	932.7	0.63	31	4937	905	0.84	66
20 Au	7307	805.2	0.77	35	5520	783.9	1.01	72

Table A9.5. (continued).

Ion E (MeV)	Iron S_e (eV/nm)	Iron S_n (eV/nm)	Iron R_p (μm)	Iron ΔR_p (nm)	Germanium S_e (eV/nm)	Germanium S_n (eV/nm)	Germanium R_p (μm)	Germanium ΔR_p (nm)
8 Au	7088	3836	0.19	28	3483	2301	0.33	60
12 Au	8681	3058	0.27	37	4266	1837	0.50	79
16 Au	10020	2570	0.36	43	4926	1546	0.65	94
20 Au	11210	2232	0.43	48	5508	1343	0.80	106

Ion E (MeV)	Tin S_e (eV/nm)	Tin S_n (eV/nm)	Tin R_p (μm)	Tin ΔR_p (nm)	Lead S_e (eV/nm)	Lead S_n (eV/nm)	Lead R_p (μm)	Lead ΔR_p (nm)
8 Au	4325	2652	0.27	64	4283	3215	0.23	75
12 Au	5297	2137	0.40	84	5246	2629	0.34	100
16 Au	6116	1809	0.53	99	6057	2245	0.45	121
20 Au	6838	1578	0.65	112	6772	1972	0.56	138

Table A9.6a. Elastic recoil detection—practical depths of analysis with 12-μm mylar foil Projectiles: 2.8 MeV He, 28 MeV Si, 35 MeV Cl, and 12 MeV Au Recoils: H, C, and N samples: C, Si, Fe, Ge

The following tables present a slab analysis to determine the practical depth of analysis for several projectile target combinations. The projectiles are incident at 75° from the surface normal and the recoiled atoms are detected at a scattering angle of 30°. A 12-μm mylar foil was used as the range foil in front of the detector. The slab analysis is given in sets of three columns. The first column gives the sample type and the depth (ΣX^n) of slab n in the sample, measured normal to the surface. The second column gives the projectile species and energy (keV) as it is incident upon each slab. The third column gives the recoil atom species and its detected energy (keV) corresponding to a depth ΣX^n. A practical limit for the minimum detectable energy in a surface barrier detector is 50 keV. The density used for each sample is as follows: ρ(C)=2.26 g/cm^3, ρ(Si)=2.32 g/cm^3, ρ(Fe)=7.86 g/cm^3, and ρ(Ge)=5.32 g/cm^3. These calculations were determined using the computer program SERDAP from Barbour (1994) and employing stopping powers from Ziegler (1980). The recoil atom stopping powers in the mylar foil were experimentally determined to be 5-8% less than those given by Ziegler (1980) and were scaled accordingly for these calculations. The stopping powers given by Ziegler *et al.* (1985) do match the experimentally determined stopping power more closely. Note that for the Au projectile, both electronic and nuclear stopping must be considered.

Carbon ΣX^n (nm)	Carbon 2.8 He (E_0^n) (keV)	Carbon H recoil (E_d) (keV)	Silicon ΣX^n (nm)	Silicon 2.8 He (E_0^n) (keV)	Silicon H recoil (E_d) (keV)	Iron ΣX^n (nm)	Iron 2.8 He (E_0^n) (keV)	Iron H recoil (E_d) (keV)
0	2800	949	0	2800	949	0	2800	949
80	2727	888	105	2726	886	42	2727	887
160	2653	824	210	2650	820	84	2653	822
240	2578	757	315	2574	752	126	2578	754
320	2502	687	420	2498	680	168	2503	683
400	2425	613	525	2420	604	210	2426	609
480	2348	534	630	2341	523	252	2349	529
560	2269	448	735	2262	435	294	2271	443
640	2188	353	840	2181	337	336	2191	349
720	2107	245	945	2099	226	378	2111	241
800	2025	112	1050	2016	86	420	2030	110
880	1941	–62	1155	1932	–71	462	1947	–61

Table A9.6a. (continued).

Carbon			Carbon			Carbon		
ΣX^n *(nm)*	28 Si (E_0^n) *(keV)*	H recoil (E_d) *(keV)*	ΣX^n *(nm)*	28 Si (E_0^n) *(keV)*	C recoil (E_d) *(keV)*	ΣX^n *(nm)*	28 Si (E_0^n) *(keV)*	N recoil (E_d) *(keV)*
0	28000	2604	0	28000	8293	0	28000	6159
100	26380	2422	45	27271	7449	30	27514	551
200	24749	2237	90	26540	6590	60	27027	4881
300	23109	2048	135	25806	5716	90	26539	4239
400	21462	1854	180	25071	4829	120	26050	3597
500	19808	1655	225	24334	3931	150	25560	2958
600	18151	1450	270	23595	3027	180	25070	2327
700	16492	1236	315	22855	2125	210	24578	1711
800	14836	1011	360	22113	1249	240	24086	1128
900	13186	768	405	21370	479	270	23593	612
1000	11550	492	450	20626	56	300	23100	232
1100	9934	129	495	19880	10	330	22605	56

Silicon			Silicon			Silicon		
ΣX^n *(nm)*	28 Si (E_0^n) *(keV)*	H recoil (E_d) *(keV)*	ΣX^n *(nm)*	28 Si (E_0^n) *(keV)*	C recoil (E_d) *(keV)*	ΣX^n *(nm)*	28 Si (E_0^n) *(keV)*	N recoil (E_d) *(keV)*
0	28000	2604	0	28000	8293	0	28000	6159
130	26331	2416	52	27332	7515	37	27525	5532
260	24657	2225	104	26664	6727	74	27049	4904
390	22980	2031	156	25995	5927	111	26573	4274
520	21303	1833	208	25325	5118	148	26097	3647
650	19627	1630	260	24655	4301	185	25621	3022
780	17956	1421	312	23984	3479	222	25144	2406
910	16292	1205	364	23313	2656	259	24667	1804
1040	14641	977	416	22642	1843	296	24190	1231
1170	13007	733	468	21971	1065	333	23713	713
1300	11395	456	520	21300	403	370	23235	308
1430	9814	82	572	20630	53	407	22758	85

Iron			Iron			Iron		
ΣX^n *(nm)*	28 Si (E_0^n) *(keV)*	H recoil (E_d) *(keV)*	ΣX^n *(nm)*	28 Si (E_0^n) *(keV)*	C recoil (E_d) *(keV)*	ΣX^n *(nm)*	28 Si (E_0^n) *(keV)*	N recoil (E_d) *(keV)*
0	28000	2604	0	28000	8293	0	28000	6159
55	26250	2406	21	27332	7511	16	27491	5485
110	24499	2206	42	26664	6719	32	26982	4809
165	22749	2002	63	25995	5918	48	26473	4134
220	21001	1794	84	25327	5108	64	25963	3462
275	19257	1581	105	24658	4291	80	25454	2796
330	17519	1362	126	23990	3471	96	24945	2143
385	15792	1134	147	23322	2651	112	24435	1512
440	14077	892	168	22654	1842	128	23926	927
495	12381	628	189	21986	1069	144	23417	439
550	10711	315	210	21319	410	160	22908	131
605	9077	−60	231	20653	56	176	22399	33

Table A9.6a. (continued).

Germanium		
ΣX^n *(nm)*	**28 Si (E_o^n)** *(keV)*	**H recoil (E_d)** *(keV)*
0	28000	2604
90	26285	2410
180	24574	2214
270	22869	2016
360	21173	1814
450	19491	1609
540	17827	1400
630	16186	1184
720	14576	960
810	13003	723
900	11476	460
990	10004	124

Germanium		
ΣX^n *(nm)*	**28 Si (E_o^n)** *(keV)*	**C recoil (E_d)** *(keV)*
0	28000	8293
35	27333	7511
70	26667	6720
105	26001	5919
140	25336	5111
175	24671	4298
210	24008	3481
245	23346	2667
280	22685	1864
315	22026	1098
350	21369	439
385	20715	67

Germanium		
ΣX^n *(nm)*	**28 Si (E_o^n)** *(keV)*	**N recoil (E_d)** *(keV)*
0	28000	6159
27	27486	5476
54	26971	4793
81	26458	4110
108	25944	3431
135	25431	2759
162	24918	2101
189	24406	1468
216	23895	885
243	23384	406
270	22875	116
297	22366	30

Carbon		
ΣX^n *(nm)*	**35 Cl (E_o^n)** *(keV)*	**H recoil (E_d)** *(keV)*
0	35000	2621
100	32969	2435
200	30932	2247
300	28890	2054
400	26846	1858
500	24802	1657
600	22761	1449
700	20728	1234
800	18707	1008
900	16704	764
1000	14726	488
1100	12783	124

Carbon		
ΣX^n *(nm)*	**35 Cl (E_o^n)** *(keV)*	**C recoil (E_d)** *(keV)*
0	35000	11232
50	33985	10246
100	32967	9240
150	31949	8216
200	30929	7171
250	29908	6106
300	28886	5023
350	27864	3925
400	26842	2820
450	25820	1728
500	24799	717
550	23778	81

Carbon		
ΣX^n *(nm)*	**35 Cl (E_o^n)** *(keV)*	**N recoil (E_d)** *(keV)*
0	35000	9684
40	34188	8772
80	33374	7848
120	32560	6914
160	31744	5971
200	30928	5022
240	30112	4070
280	29294	3123
320	28477	2191
360	27659	1303
400	26841	538
440	26024	97

Silicon		
ΣX^n *(nm)*	**35 Cl (E_o^n)** *(keV)*	**H recoil (E_d)** *(keV)*
0	35000	2820
130	32908	2639
260	30817	2457
390	28729	2273
520	26649	2089
650	24578	1903
780	22521	1716
910	20483	1528
1040	18468	1338
1170	16484	1145
1300	14536	948
1430	12634	746

Silicon		
ΣX^n *(nm)*	**35 Cl (E_o^n)** *(keV)*	**C recoil (E_d)** *(keV)*
0	35000	19304
63	33986	18500
126	32972	17689
189	31959	16871
252	30946	16047
315	29934	15215
378	28924	14376
441	27915	13529
504	26909	12675
567	25904	11813
630	24903	10944
693	23905	10066

Silicon		
ΣX^n *(nm)*	**35 Cl (E_o^n)** *(keV)*	**N recoil (E_d)** *(keV)*
0	35000	20549
51	34179	19805
102	33358	19057
153	32538	18303
204	31718	17544
255	30898	16780
306	30079	16011
357	29261	15236
408	28444	14456
459	27628	13670
510	26814	12880
561	26001	12084

Table A9.6a. (continued).

Iron			Iron			Iron		
ΣX^n *(nm)*	35 Cl (E_o^n) *(keV)*	H recoil (E_d) *(keV)*	ΣX^n *(nm)*	35 Cl (E_o^n) *(keV)*	C recoil (E_d) *(keV)*	ΣX^n *(nm)*	35 Cl (E_o^n) *(keV)*	N recoil (E_d) *(keV)*
0	35000	2645	0	35000	11232	0	35000	9684
54	32847	2449	26	33963	10211	20	34203	8777
108	30700	2250	52	32928	9174	40	33406	7861
162	28560	2048	78	31894	8121	60	32610	6939
216	26431	1844	104	30861	7051	80	31815	6012
270	24316	1636	130	29831	5966	100	31021	5081
324	22217	1423	156	28803	4867	120	30228	4151
378	20138	1203	182	27777	3760	140	29436	3229
432	18084	973	208	26754	2654	160	28646	2323
486	16061	727	234	25735	1571	180	27858	1455
540	14075	450	260	24719	595	200	27071	686
594	12139	74	286	23706	52	220	26285	168

Germanium			Germanium			Germanium		
ΣX^n *(nm)*	35 Cl (E_o^n) *(keV)*	H recoil (E_d) *(keV)*	ΣX^n *(nm)*	35 Cl (E_o^n) *(keV)*	C recoil (E_d) *(keV)*	ΣX^n *(nm)*	35 Cl (E_o^n) *(keV)*	N recoil (E_d) *(keV)*
0	35000	2645	0	35000	11232	0	35000	9684
92	32803	2444	43	33973	10219	35	34164	8731
184	30618	2242	86	32949	9192	70	33330	7770
276	28449	2037	129	31927	8148	105	32498	6802
368	26301	1831	172	30909	7091	140	31667	5829
460	24179	1621	215	29894	6019	175	30839	4855
552	22090	1408	258	28883	4936	210	30013	3884
644	20041	1190	301	27877	3846	245	29190	2925
736	18040	964	344	26876	2759	280	28370	1992
828	16096	725	387	25881	1694	315	27553	1120
920	14221	463	430	24892	718	350	26739	408
1012	12424	130	473	23910	94	385	25930	63

Table A9.6b. Elastic recoil detection—practical depths of analysis with 10-μm aluminum foil Projectiles: 2.8 MeV He, 28 MeV Si, 35 MeV Cl, and 12 MeV Au Recoils: H, C, and N Samples: C, Si, Fe, Ge

The following tables present a slab analysis similar to that given in Table A9.6a, but in this table a 10-μm Al foil was used as the range foil in front of the detector.

Carbon			Silicon			Iron		
ΣX^n *(nm)*	2.8 He (E_o^n) *(keV)*	H recoil (E_d) *(keV)*	ΣX^n *(nm)*	2.8 He (E_o^n) *(keV)*	H recoil (E_d) *(keV)*	ΣX^n *(nm)*	2.8 He (E_o^n) *(keV)*	H recoil (E_d) *(keV)*
0	2800	910	0	2800	910	0	2800	910
80	2727	848	105	2726	846	42	2727	847
160	2653	784	210	2650	780	84	2653	782
240	2578	717	315	2574	711	126	2578	714
320	2502	646	420	2498	639	168	2503	642
400	2425	572	525	2420	562	210	2426	568
480	2348	492	630	2341	481	252	2349	488
560	2269	407	735	2262	394	294	2271	403
640	2188	314	840	2181	299	336	2191	310
720	2107	211	945	2099	193	378	2111	207
800	2025	91	1050	2016	69	420	2030	89
880	1941	–59	1155	1932	–81	462	1947	–58

Table A9.6b. (continued).

Carbon			Carbon			Carbon		
ΣX^n *(nm)*	28 Si (E_0^n) *(keV)*	H recoil (E_d) *(keV)*	ΣX^n *(nm)*	28 Si (E_0^n) *(keV)*	C recoil (E_d) *(keV)*	ΣX^n *(nm)*	28 Si (E_0^n) *(keV)*	N recoil (E_d) *(keV)*
0	28000	2581	0	28000	7409	0	28000	5160
100	26380	2398	45	27271	6579	30	27514	4563
200	24749	2212	90	26540	5742	60	27027	3972
300	23109	2022	135	25806	4900	90	26539	3391
400	21462	1827	180	25071	4059	120	26050	2824
500	19808	1627	225	24334	3228	150	25560	2277
600	18151	1420	270	23595	2418	180	25070	1755
700	16492	1204	315	22855	1649	210	24578	1269
800	14836	977	360	22113	951	240	24086	829
900	13186	732	405	21370	370	270	23593	449
1000	11550	457	450	20626	12	300	23100	157
1100	9934	113	495	19880	–13	330	22605	–3

Silicon			Silicon			Silicon		
ΣX^n *(nm)*	28 Si (E_0^n) *(keV)*	H recoil (E_d) *(keV)*	ΣX^n *(nm)*	28 Si (E_0^n) *(keV)*	C recoil (E_d) *(keV)*	ΣX^n *(nm)*	28 Si (E_0^n) *(keV)*	N recoil (E_d) *(keV)*
0	28000	2581	0	28000	7409	0	28000	5160
130	26331	2392	52	27332	6644	37	27525	4573
260	24657	2200	104	26664	5874	74	27049	3993
390	22980	2005	156	25995	5102	111	26573	3423
520	21303	1805	208	25325	4331	148	26097	2867
650	19627	1601	260	24655	3567	185	25621	2331
780	17956	1390	312	23984	2818	222	25144	1820
910	16292	1172	364	23313	2096	259	24667	1341
1040	14641	943	416	22642	1418	296	24190	905
1170	13007	697	468	21971	810	333	23713	524
1300	11395	421	520	21300	311	370	23235	219
1430	9814	73	572	20630	8	407	22758	27

Iron			Iron			Iron		
ΣX^n *(nm)*	28 Si (E_0^n) *(keV)*	H recoil (E_d) *(keV)*	ΣX^n *(nm)*	28 Si (E_0^n) *(keV)*	C recoil (E_d) *(keV)*	ΣX^n *(nm)*	28 Si (E_0^n) *(keV)*	N recoil (E_d) *(keV)*
0	28000	2581	0	28000	7409	0	28000	5160
55	26250	2382	21	27332	6640	16	27491	4529
110	24499	2181	42	26664	5867	32	26982	3907
165	22749	1976	63	25995	5093	48	26473	3298
220	21001	1766	84	25327	4322	64	25963	2707
275	19257	1552	105	24658	3558	80	25454	2141
330	17519	1331	126	23990	2811	96	24945	1607
385	15792	1101	147	23322	2091	112	24435	1116
440	14077	858	168	22654	1417	128	23926	681
495	12381	592	189	21986	813	144	23417	320
550	10711	284	210	21319	316	160	22908	70
605	9077	–33	231	20653	12	176	22399	–25

Table A9.6b. (continued).

Germanium			Germanium			Germanium		
ΣX^n (nm)	28 Si (E_0^n) (keV)	H recoil (E_d) (keV)	ΣX^n (nm)	28 Si (E_0^n) (keV)	C recoil (E_d) (keV)	ΣX^n (nm)	28 Si (E_0^n) (keV)	N recoil (E_d) (keV)
0	28000	2581	0	28000	7409	0	28000	5160
90	26285	2387	35	27333	6640	27	27486	4521
180	24574	2189	70	26667	5868	54	26971	3892
270	22869	1990	105	26001	5095	81	26458	3276
360	21173	1787	140	25336	4325	108	25944	2680
450	19491	1580	175	24671	3564	135	25431	2110
540	17827	1369	210	24008	2820	162	24918	1574
630	16186	1152	245	23346	2105	189	24406	1083
720	14576	926	280	22685	1435	216	23895	650
810	13003	687	315	22026	835	243	23384	295
900	11476	425	350	21369	339	270	22875	56
990	10005	109	385	20715	24	297	22366	–26

Carbon			Carbon			Carbon		
ΣX^n (nm)	35 Cl (E_0^n) (keV)	H recoil (E_d) (keV)	ΣX^n (nm)	35 Cl (E_0^n) (keV)	C recoil (E_d) (keV)	ΣX^n (nm)	35 Cl (E_0^n) (keV)	N recoil (E_d) (keV)
0	35000	2623	0	35000	10331	0	35000	8566
100	32969	2437	50	33985	9346	40	34188	7672
200	30932	2249	100	32967	8346	80	33374	6775
300	28890	2057	150	31949	7332	120	32560	5877
400	26846	1862	200	30929	6307	160	31744	4984
500	24802	1662	250	29908	5274	200	30928	4102
600	22761	1456	300	28886	4241	240	30112	3241
700	20728	1243	350	27864	3221	280	29294	2416
800	18707	1020	400	26842	2237	320	28477	1646
900	16704	783	450	25820	1326	360	27659	959
1000	14726	521	500	24799	548	400	26841	394
1100	12783	207	550	23778	38	440	26024	38

Silicon			Silicon			Silicon		
ΣX^n (nm)	35 Cl (E_0^n) (keV)	H recoil (E_d) (keV)	ΣX^n (nm)	35 Cl (E_0^n) (keV)	C recoil (E_d) (keV)	ΣX^n (nm)	35 Cl (E_0^n) (keV)	N recoil (E_d) (keV)
0	35000	2595	0	35000	10331	0	35000	8566
130	32908	2402	63	33986	9340	51	34179	7656
260	30817	2207	126	32972	8336	102	33358	6744
390	28729	2008	189	31959	7321	153	32538	5834
520	26649	1805	252	30946	6297	204	31718	4931
650	24578	1598	315	29934	5268	255	30898	4042
780	22521	1386	378	28924	4242	306	30079	3178
910	20483	1166	441	27915	3233	357	29261	2353
1040	18468	935	504	26909	2260	408	28444	1588
1170	16484	688	567	25904	1361	459	27628	909
1300	14536	410	630	24903	591	510	26814	360
1430	12634	58	693	23905	65	561	26001	25

Table A9.6b. (continued).

Iron			Iron			Iron		
ΣX^n *(nm)*	35 Cl (E_0^n) *(keV)*	H recoil (E_d) *(keV)*	ΣX^n *(nm)*	35 Cl (E_0^n) *(keV)*	C recoil (E_d) *(keV)*	ΣX^n *(nm)*	35 Cl (E_0^n) *(keV)*	N recoil (E_d) *(keV)*
0	35000	2623	0	35000	10331	0	35000	8566
54	32847	2425	26	33963	9312	20	34203	7677
108	30700	2225	52	32928	8281	40	33406	6788
162	28560	2022	78	31894	7239	60	32610	5901
216	26431	1817	104	30861	6190	80	31815	5021
270	24316	1607	130	29831	5140	100	31021	4156
324	22217	1392	156	28803	4095	120	30228	3313
378	20138	1171	182	27777	3072	140	29436	2506
432	18084	939	208	26754	2094	160	28646	1752
486	16061	691	234	25735	1201	180	27858	1074
540	14075	415	260	24719	457	200	27071	504
594	12139	67	286	23706	8	220	26285	103

Germanium			Germanium			Germanium		
ΣX^n *(nm)*	35 Cl (E_0^n) *(keV)*	H recoil (E_d) *(keV)*	ΣX^n *(nm)*	35 Cl (E_0^n) *(keV)*	C recoil (E_d) *(keV)*	ΣX^n *(nm)*	35 Cl (E_0^n) *(keV)*	N recoil (E_d) *(keV)*
0	35000	2623	0	35000	10331	0	35000	8566
92	32803	2421	43	33973	9320	35	34164	7632
184	30618	2217	86	32949	8298	70	33330	6699
276	28449	2011	129	31927	7266	105	32498	5770
368	26301	1803	172	30909	6229	140	31667	4850
460	24179	1592	215	29894	5190	175	30839	3949
552	22090	1377	258	28883	4159	210	30013	3076
644	20041	1157	301	27877	3150	245	29190	2249
736	18040	930	344	26876	2184	280	28370	1488
828	16096	690	387	25881	1298	315	27553	823
920	14221	427	430	24892	549	350	26739	297
1012	12424	113	473	23910	51	385	25930	4

REFERENCES

Biersack, J.P., and Ziegler, J.F. (1991), TRIM-91 computer program, IBM, Yorktown Heights, New York.

Barbour, J.C. (1994), SERDAP computer program, Sandia National Laboratories, Dept. 1111, Albuquerque, New Mexico.

Marion, J.B., and Young, F.C. (1968), *Nuclear Reaction Analysis*, North-Holland Publishing Co., Amsterdam, p. 142.

Rutherford, E. (1911), *Philos. Mag.* **21**, 669.

Ziegler, J.F. (1980), *Handbook of Stopping Cross-Sections for Energetic Ions in all Elements*, Pergamon Press, New York.

Ziegler, J.F., Biersack, J.P., and Littmark, U. (1985), *The Stopping and Range of Ions in Solids*, Pergamon Press, New York.

APPENDIX

10

DEUTERIUM-INDUCED NUCLEAR REACTION PARAMETERS

Compiled by

G. Vizkelethy

Idaho State University, Pocatello, Idaho

This table gives the energy of protons and alphas from (d,p) and (d,α) reactions before (E_{out}) and after passing through (E_{mylar}) an absorber foil of Mylar which stops the deuteron beam. E_d is the energy of deuteron, X_{mylar} is the thickness of the absorber foil, E^{α}_{min} and E^{p}_{min} are the minimum energies with which the α particles and protons can penetrate the absorber foils. (Source: G. Amsel, personal communication.)

a. $E_d = 0.9$ MeV $\Theta = 150°$ $X_{mylar} = 12$ μm

$E^{\alpha}_{min} = 2.9$ MeV $E^{p}_{min} = 0.9$ MeV

Element	Reaction		Q	E_{out}	E_{mylar}
	(d,α)	(d,p)	(MeV)	(MeV)	(MeV)
D		D(d,p)T	4.033	2.36	2.1
^{6}Li		^{6}Li(d,p_0)^{7}Li	5.027	4.36	4.2
		^{6}Li(d,p_1)^{7}Li	4.549	3.97	3.8
		^{6}Li(d,p_2)^{7}Li	0.397	0.67	0
	^{6}Li(d,α)^{4}He		22.36	9.61	8.6
^{7}Li		^{7}Li(d,p)^{8}Li	−0.192	0.29	0
	^{7}Li(d,α)^{5}He		14.163	6.82	5.48
^{9}Be		^{9}Be(d,p)^{10}B	4.585	4.38	4.2
	^{9}Be(d,α_0)^{7}Li		7.152	4.10	2.08
	^{9}Be(d,α_1)^{7}Li		6.674	3.83	1.68
	^{9}Be(d,α_2)^{7}Li		2.522	1.50	0

Element	Reaction (d,α)	Reaction (d,p)	Q (MeV)	E_{out} (MeV)	E_{mylar} (MeV)
^{10}B		$^{10}B(d,p_0)^{11}B$	9.237	8.58	8.48
		$^{10}B(d,p_1)^{11}B$	7.107	6.69	6.56
		$^{10}B(d,p_2)^{11}B$	4.777	4.64	4.50
		$^{10}B(d,p_3)^{11}B$	4.207	4.14	4.00
		$^{10}B(d,p_4)^{11}B$	2.477	2.63	2.40
		$^{10}B(d,p_5)^{11}B$	2.427	2.59	2.35
		$^{10}B(d,p_6)^{11}B$	1.937	2.17	1.90
		$^{10}B(d,p_7)^{11}B$	1.247	1.54	1.23
		$^{10}B(d,p_8)^{11}B$	0.667	1.09	0.58
	$^{10}B(d,\alpha_0)^8Be$		17.819	11.04	10.30
	$^{10}B(d,\alpha_1)^8Be$		14.919	9.22	8.20
^{11}B		$^{11}B(d,p_0)^{12}B$	1.138	1.52	1.16
		$^{11}B(d,p_1)^{12}B$	0.188	0.72	0
	$^{11}B(d,\alpha_0)^9Be$		8.022	5.22	3.60
	$^{11}B(d,\alpha_1)^9Be$		6.272	4.10	2.10
	$^{11}B(d,\alpha_2)^9Be$		5.592	3.67	1.44
	$^{11}B(d,\alpha_3)^9Be$		4.982	3.29	0.76
	$^{11}B(d,\alpha_4)^9Be$		3.282	2.22	0
	$^{11}B(d,\alpha_5)^9Be$		1.262	1.00	0
^{13}B	$^{13}B(d,\alpha)^{11}Be$		5.167	3.73	1.50
^{12}C		$^{12}C(d,p)^{13}C$	2.719	2.95	2.72
^{13}C		$^{13}C(d,p)^{14}C$	5.947	5.89	5.77
	$^{13}C(d,\alpha_0)^{11}B$		5.167	3.73	1.30
	$^{13}C(d,\alpha_1)^{11}B$		3.037	2.30	0
	$^{13}C(d,\alpha_2)^{11}B$		0.707	0.78	0
	$^{13}C(d,\alpha_3)^{11}B$		0.137	0.44	0
^{14}N		$^{14}N(d,p_0)^{15}N$	8.615	8.39	8.25
		$^{14}N(d,p_1)^{15}N$	3.335	3.58	3.40
		$^{14}N(d,p_2)^{15}N$	3.305	3.56	3.37
		$^{14}N(d,p_3)^{15}N$	2.285	2.64	2.40
		$^{14}N(d,p_4)^{15}N$	1.455	1.99	1.60
		$^{14}N(d,p_5)^{15}N$	1.305	1.76	1.47
		$^{14}N(d,p_6)^{15}N$	1.045	1.53	1.17
	$^{14}N(d,\alpha_0)^{12}C$		13.579	9.84	9.05
	$^{14}N(d,\alpha_1)^{12}C$		9.146	6.67	5.35
	$^{14}N(d,\alpha_2)^{12}C$		5.923	4.40	2.44
	$^{14}N(d,\alpha_3)^{12}C$		3.949	3.02	0.20
	$^{14}N(d,\alpha_4)^{12}C$		3.479	2.70	0
	$^{14}N(d,\alpha_5)^{12}C$		2.739	2.19	0
	$^{14}N(d,\alpha_6)^{12}C$		2.479	2.10	0
^{15}N		$^{15}N(d,p)^{16}N$	0.267	0.83	0.08
	$^{15}N(d,\alpha)^{13}C$		7.683	5.80	4.35

Element	Reaction (d,α)	Reaction (d,p)	Q *(MeV)*	E_{out} *(MeV)*	E_{mylar} *(MeV)*
^{16}O		$^{16}O(d,p_o)^{17}O$	1.919	2.36	2.12
		$^{16}O(d,p_1)^{17}O$	1.048	1.58	1.23
	$^{16}O(d,\alpha_0)^{14}N$		3.116	2.61	0
	$^{16}O(d,\alpha_1)^{14}N$		0.804	0.97	0
^{17}O		$^{17}O(d,p_0)^{18}O$	5.842	5.99	5.90
		$^{17}O(d,p_1)^{18}O$	3.860	4.17	4.00
	$^{17}O(d,\alpha_0)^{15}N$		9.812	7.68	6.50
	$^{17}O(d,\alpha_1)^{15}N$		4.532	3.72	1.30
^{18}O		$^{18}O(d,p_0)^{19}O$	1.731	2.24	1.96
		$^{18}O(d,p_1)^{19}O$	1.635	2.15	1.87
		$^{18}O(d,p_2)^{19}O$	0.262	0.90	0.27
	$^{18}O(d,\alpha_0)^{16}N$		4.237	3.58	1.10
	$^{18}O(d,\alpha_1)^{16}N$		4.117	3.49	0.95
	$^{18}O(d,\alpha_2)^{16}N$		3.942	3.36	0.77
	$^{18}O(d,\alpha_3)^{16}N$		3.845	3.29	0.52
	$^{18}O(d,\alpha_4)^{16}N$		0.707	0.97	0
	$^{18}O(d,\alpha_5)^{16}N$		0.257	0.65	0
^{19}F		$^{19}F(d,p_0)^{20}F$	4.379	4.92	4.70
		$^{19}F(d,p_1)^{20}F$	3.729	4.31	4.03
		$^{19}F(d,p_2)^{20}F$	3.549	4.15	3.90
		$^{19}F(d,p_3)^{20}F$	3.389	4.00	3.74
		$^{19}F(d,p_4)^{20}F$	3.319	3.94	3.68
		$^{19}F(d,p_5)^{20}F$	3.069	3.71	3.44
		$^{19}F(d,p_6)^{20}F$	2.409	3.10	2.80
		$^{19}F(d,p_7)^{20}F$	2.329	3.02	2.70
		$^{19}F(d,p_8)^{20}F$	2.179	2.89	2.56
		$^{19}F(d,p_9)^{20}F$	1.509	2.27	1.86
		$^{19}F(d,p_{10})^{20}F$	1.409	2.18	1.79
		$^{19}F(d,p_{11})^{20}F$	0.889	1.71	1.16
		$^{19}F(d,p_{12})^{20}F$	0.849	1.67	1.11
		$^{19}F(d,p_{13})^{20}F$	0.789	1.62	1.05
		$^{19}F(d,p_{14})^{20}F$	0.699	1.54	0.93
		$^{19}F(d,p_{15})^{20}F$	0.419	1.28	0.56
		$^{19}F(d,p_{16})^{20}F$	0.299	1.17	0.35
	$^{19}F(d,\alpha_0)^{17}O$		10.038	8.25	6.50
	$^{19}F(d,\alpha_1)^{17}O$		9.167	7.57	5.60
	$^{19}F(d,\alpha_2)^{17}O$		6.980	5.89	3.36

b. $E_d = 1.2$ MeV $\Theta = 150°$ $X_{mylar} = 19\ \mu m$

$E^{\alpha}_{min} = 4.0$ MeV $E^{p}_{min} = 1.06$ MeV

Element	Reaction (d,α)	(d,p)	Q (MeV)	E_{out} (MeV)	E_{mylar} (MeV)
D		D(d,p)T	4.033	2.306	1.87
^{6}Li		$^{6}Li(d,p_0)^{7}Li$	5.027	4.44	4.15
		$^{6}Li(d,p_1)^{7}Li$	4.549	4.05	3.75
		$^{6}Li(d,p_2)^{7}Li$	0.397	0.80	0
	$^{6}Li(d,\alpha)^{4}He$		22.36	9.87	7.75
^{7}Li		$^{7}Li(d,p)^{8}Li$	–0.192	0.43	0
	$^{7}Li(d,\alpha)^{5}He$		14.163	6.72	4.48
^{9}Be		$^{9}Be(d,p)^{10}B$	4.585	4.522	4.27
	$^{9}Be(d,\alpha_0)^{7}Li$		7.152	4.108	0.04
	$^{9}Be(d,\alpha_1)^{7}Li$		6.674	3.837	0
	$^{9}Be(d,\alpha_2)^{7}Li$		2.522	1.54	0
^{10}B		$^{10}B(d,p_0)^{11}B$	9.237	8.71	8.57
		$^{10}B(d,p_1)^{11}B$	7.107	6.83	6.67
		$^{10}B(d,p_2)^{11}B$	4.777	4.79	4.58
		$^{10}B(d,p_3)^{11}B$	4.207	4.29	4.04
		$^{10}B(d,p_4)^{11}B$	2.477	2.80	2.46
		$^{10}B(d,p_5)^{11}B$	2.427	2.75	2.40
		$^{10}B(d,p_6)^{11}B$	1.937	2.33	1.97
		$^{10}B(d,p_7)^{11}B$	1.247	1.75	1.215
		$^{10}B(d,p_8)^{11}B$	0.667	1.26	0.51
	$^{10}B(d,\alpha_0)^{8}Be$		17.819	11.00	9.6
	$^{10}B(d,\alpha_1)^{8}Be$		14.919	9.19	7.56
^{11}B		$^{11}B(d,p_0)^{12}B$	1.138	1.70	1.15
		$^{11}B(d,p_1)^{12}B$	0.188	0.90	0
	$^{11}B(d,\alpha_0)^{9}Be$		8.022	5.25	2.48
	$^{11}B(d,\alpha_1)^{9}Be$		6.272	4.15	0.4
	$^{11}B(d,\alpha_2)^{9}Be$		5.592	3.72	0
	$^{11}B(d,\alpha_3)^{9}Be$		4.982	3.34	0
	$^{11}B(d,\alpha_4)^{9}Be$		3.282	2.29	0
	$^{11}B(d,\alpha_5)^{9}Be$		1.262	1.09	0
^{13}B	$^{13}B(d,\alpha)^{11}Be$		5.167	3.81	0
^{12}C		$^{12}C(d,p)^{13}C$	2.719	3.13	2.80
^{13}C		$^{13}C(d,p)^{14}C$	5.947	6.07	5.90
	$^{13}C(d,\alpha_0)^{11}B$		5.167	3.81	0
	$^{13}C(d,\alpha_1)^{11}B$		3.037	2.39	0
	$^{13}C(d,\alpha_2)^{11}B$		0.707	0.90	0
	$^{13}C(d,\alpha_3)^{11}B$		0.137	0.55	0

Deuterium-Induced Nuclear Reaction Parameters

Element	Reaction (d,α)	Reaction (d,p)	Q (MeV)	E_{out} (MeV)	E_{mylar} (MeV)
^{14}N		$^{14}N(d,p_0)^{15}N$	8.615	8.56	8.41
		$^{14}N(d,p_1)^{15}N$	3.335	3.78	3.50
		$^{14}N(d,p_2)^{15}N$	3.305	3.75	3.47
		$^{14}N(d,p_3)^{15}N$	2.285	2.83	2.48
		$^{14}N(d,p_4)^{15}N$	1.455	2.10	1.64
		$^{14}N(d,p_5)^{15}N$	1.305	1.96	1.48
		$^{14}N(d,p_6)^{15}N$	1.045	1.73	1.20
	$^{14}N(d,\alpha_0)^{12}C$		13.579	9.88	8.34
	$^{14}N(d,\alpha_1)^{12}C$		9.146	6.74	4.52
	$^{14}N(d,\alpha_2)^{12}C$		5.923	4.48	1.15
	$^{14}N(d,\alpha_3)^{12}C$		3.949	3.12	0
	$^{14}N(d,\alpha_4)^{12}C$		3.479	2.80	0
	$^{14}N(d,\alpha_5)^{12}C$		2.739	2.30	0
	$^{14}N(d,\alpha_6)^{12}C$		2.479	2.12	0
^{15}N		$^{15}N(d,p)^{16}N$	0.267	1.03	0
	$^{15}N(d,\alpha)^{13}C$		7.683	5.88	3.36
^{16}O		$^{16}O(d,p_o)^{17}O$	1.919	2.57	2.20
		$^{16}O(d,p_1)^{17}O$	1.048	1.79	1.27
	$^{16}O(d,\alpha_0)^{14}N$		3.116	2.73	0
	$^{16}O(d,\alpha_1)^{14}N$		0.804	1.11	0
^{17}O		$^{17}O(d,p_0)^{18}O$	5.842	6.19	6.00
		$^{17}O(d,p_1)^{18}O$	3.860	4.37	4.17
	$^{17}O(d,\alpha_0)^{15}N$		9.812	7.78	5.90
	$^{17}O(d,\alpha_1)^{15}N$		4.532	3.85	0
^{18}O		$^{18}O(d,p_0)^{19}O$	1.731	2.45	2.1
		$^{18}O(d,p_1)^{19}O$	1.635	2.37	1.95
		$^{18}O(d,p_2)^{19}O$	0.262	1.13	0.17
	$^{18}O(d,\alpha_0)^{16}N$		4.237	3.72	0
	$^{18}O(d,\alpha_1)^{16}N$		4.117	3.68	0
	$^{18}O(d,\alpha_2)^{16}N$		3.942	3.49	0
	$^{18}O(d,\alpha_3)^{16}N$		3.845	3.42	0
	$^{18}O(d,\alpha_4)^{16}N$		0.707	1.12	0
	$^{18}O(d,\alpha_5)^{16}N$		0.257	0.81	0

Element	Reaction (d,α)	Reaction (d,p)	Q (MeV)	E_{out} (MeV)	E_{mylar} (MeV)
^{19}F		$^{19}F(d,p_0)^{20}F$	4.379	5.13	4.82
		$^{19}F(d,p_1)^{20}F$	3.729	4.53	4.20
		$^{19}F(d,p_2)^{20}F$	3.549	4.37	4.00
		$^{19}F(d,p_3)^{20}F$	3.389	4.22	3.85
		$^{19}F(d,p_4)^{20}F$	3.319	4.16	3.80
		$^{19}F(d,p_5)^{20}F$	3.069	3.93	3.55
		$^{19}F(d,p_6)^{20}F$	2.409	3.32	2.87
		$^{19}F(d,p_7)^{20}F$	2.329	3.25	2.82
		$^{19}F(d,p_8)^{20}F$	2.179	3.11	2.70
		$^{19}F(d,p_9)^{20}F$	1.509	2.50	1.93
		$^{19}F(d,p_{10})^{20}F$	1.409	2.41	1.83
		$^{19}F(d,p_{11})^{20}F$	0.889	1.93	1.30
		$^{19}F(d,p_{12})^{20}F$	0.849	1.90	1.20
		$^{19}F(d,p_{13})^{20}F$	0.789	1.84	1.10
		$^{19}F(d,p_{14})^{20}F$	0.699	1.76	0.94
		$^{19}F(d,p_{15})^{20}F$	0.419	1.51	0.57
		$^{19}F(d,p_{16})^{20}F$	0.299	1.40	0.34
	$^{19}F(d,\alpha_0)^{17}O$		10.038	8.37	5.85
	$^{19}F(d,\alpha_1)^{17}O$		9.167	7.70	4.97
	$^{19}F(d,\alpha_2)^{17}O$		6.980	6.02	2.36

APPENDIX

11

Particle-Particle Nuclear Reaction Cross Sections

Compiled by

L. Foster

Los Alamos National Laboratory, Los Alamos, New Mexico

G. Vizkelethy

Idaho State University, Pocatello, Idaho

M. Lee, J. R. Tesmer, and M. Nastasi

Los Alamos National Laboratory, Los Alamos, New Mexico

The data presented in this appendix are a partial listing of the most relevant reactions discussed in Section 5 of Chapter 6. The data were digitized from the original references given in the figure captions. All values of cross section, scattering angle, and ion energy are presented in laboratory units, unless otherwise noted. The original data were assumed to be in center-of-mass units unless explicitly stated to be in laboratory units. Center-of-mass values of cross section, scattering angle, and ion energy were converted into laboratory units using the conversion formulas given in Appendix 4. The reader should note that the absolute utility of the cross sections presented in this appendix is uncertain at best. This is compounded by the fact that in many instances, the users' scattering geometry is not identical to those of the present data. It is recommended that the cross section data as a function of incident ion energy be used only as a guide in determining the appropriate energy range to perform nuclear reaction analysis. The most reliable analysis will require mapping the cross section as a function of ion energy for the scattering geometry employed and the careful use of well calibrated standards. The use of standards in particle-particle nuclear reaction analysis is discussed in detail in Chapter 6.

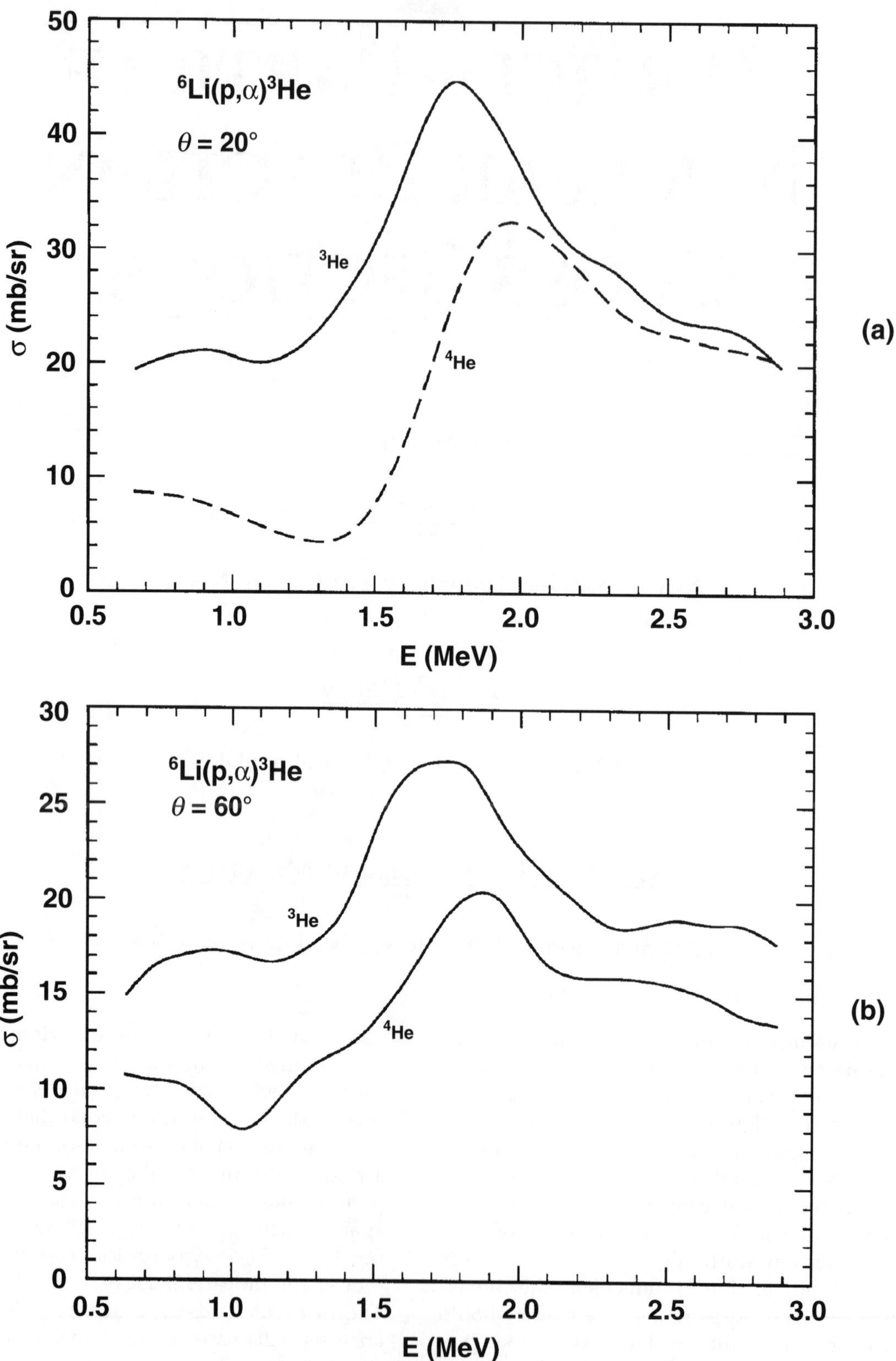

FIG. A11.1. The ^{6}Li(p,α)^{3}He and ^{6}Li(p,^{3}He)^{4}He reactions between 0.5 and 3 MeV at (a) θ = 20° and (b) θ = 60°. From Marion, J.B., Weber, G., and Mozer, F.S. (1956), *Phys. Rev.* **104**, 1402.

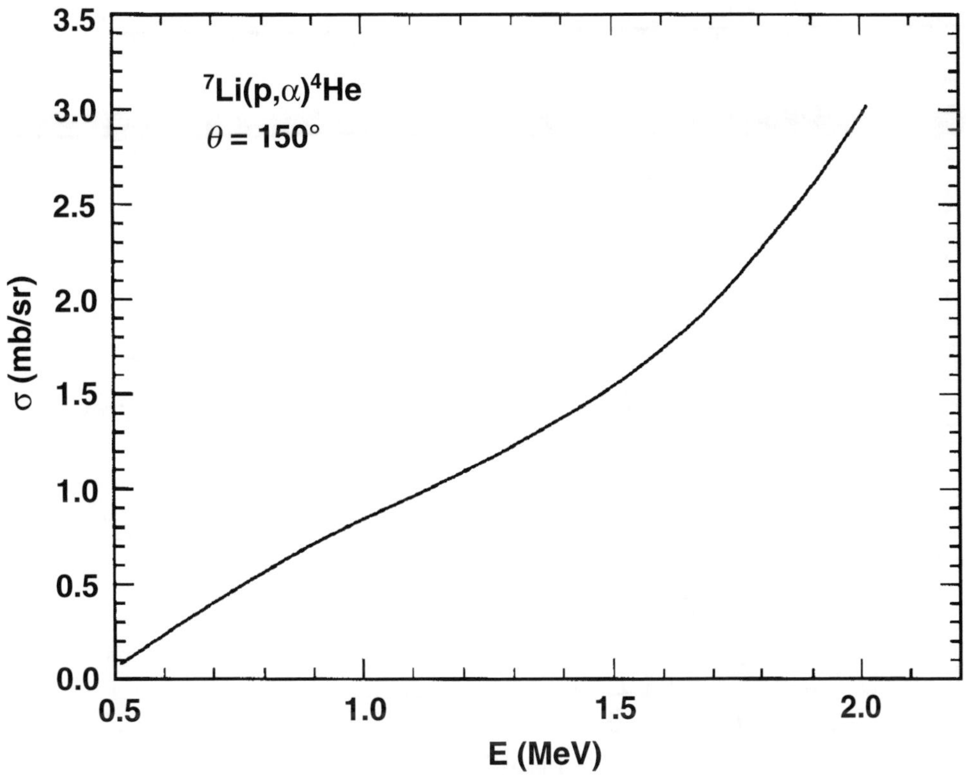

FIG. A11.2. The $^7Li(p,\alpha)^4He$ reaction at $\theta = 150°$ in the energy range of 0.5 - 2 MeV. From Maurel, B., Amsel, G., and Dieumegard, D (1977), in *Ion Beam Handbook*, Mayer, J.W., and Rimini, E. (eds.), Academic Press, New York.

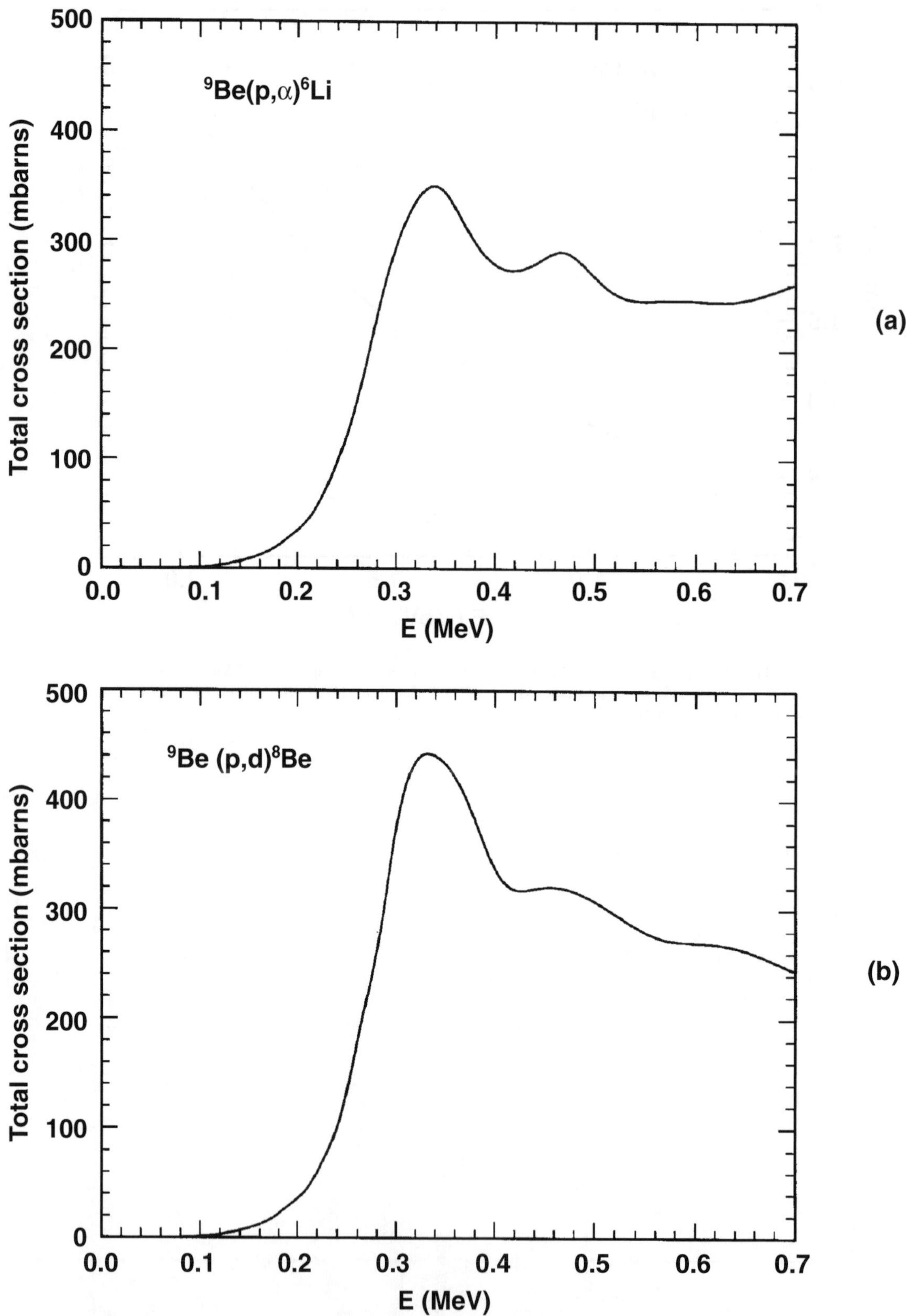

FIG. A11.3. The (a) $^9Be(p,\alpha)^6Li$ and (b) $^9Be(p,d)^8Be$ reactions at $\theta = 90°$ between 0.1 and 0.7 MeV. From Sierk, A.J., and Tombrello, T.A. (1973), *Nucl. Phys.* A**210**, 341.

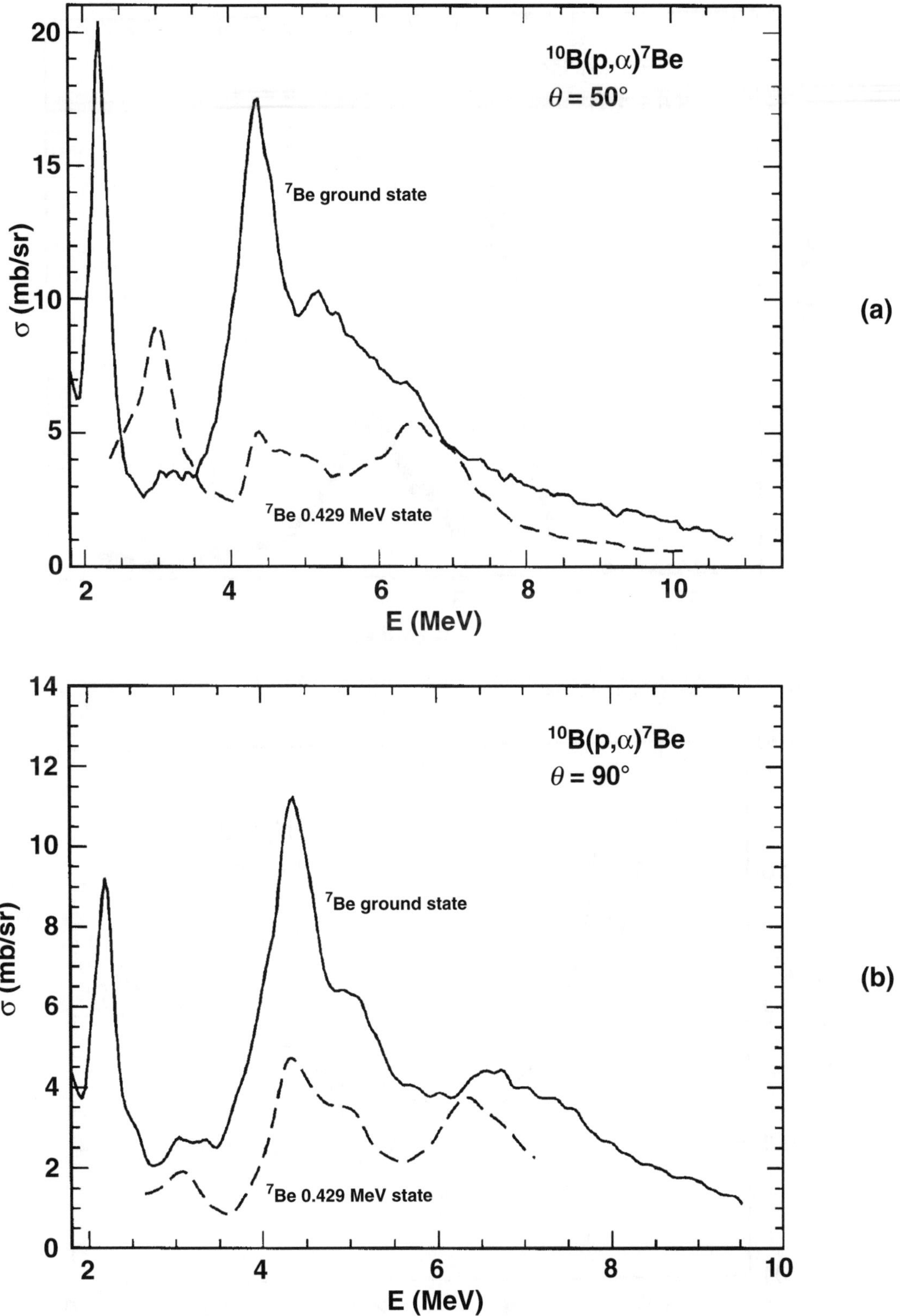

FIG. A11.4. The $^{10}B(p,\alpha)^{7}Be$ reaction between 2 and 11 MeV at (a) θ = 50° and (b) θ = 90°. From Jenkin, J.G., Earwaker, L.G., and Titterton, E.W. (1964), *Nucl. Phys.* **50**, 517.

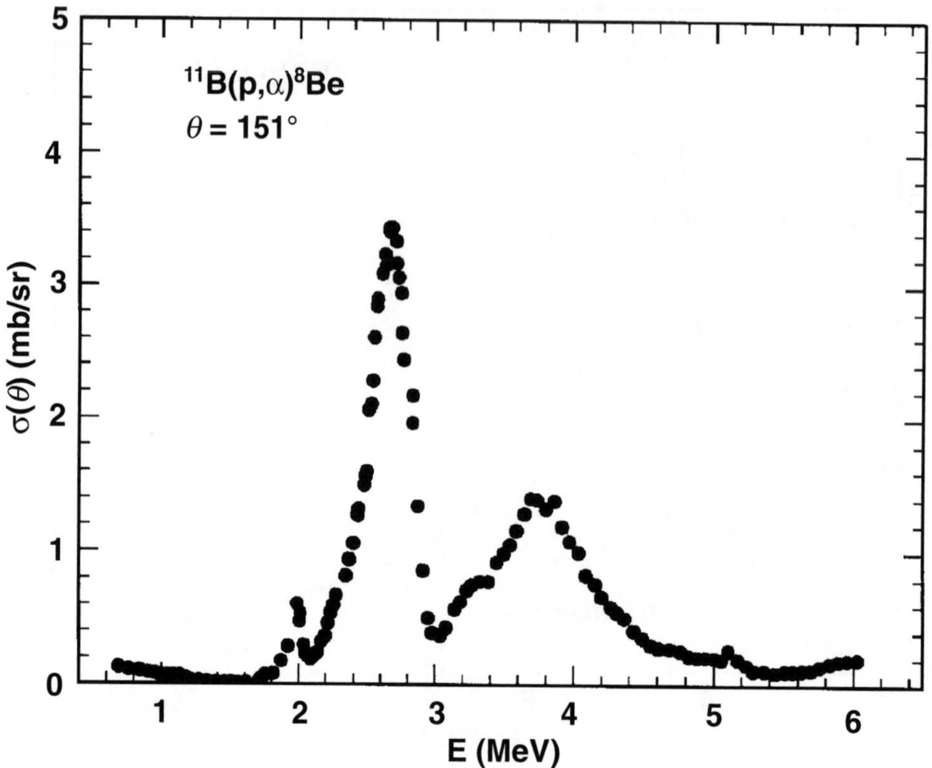

FIG. A11.5. The $^{11}B(p,\alpha)^{8}Be$ reaction at θ = 151° between 0.8 - 6 MeV. From Symons, G.D., and Treacy, P.B. (1963), *Nucl. Phys.* **46**, 93.

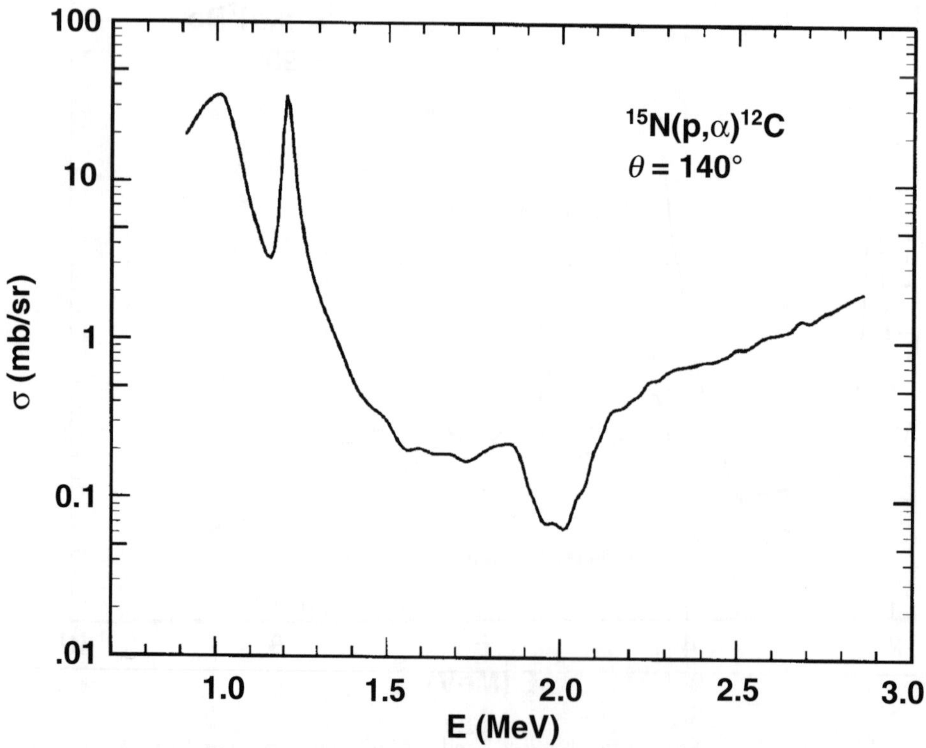

FIG. A11.6. The $^{15}N(p,\alpha)^{12}C$ reaction at θ = 140° between 0.9 - 3 MeV. From Hagendorn, F. B., and Marion, J.B. (1957), *Phys. Rev.* **108**, 1015.

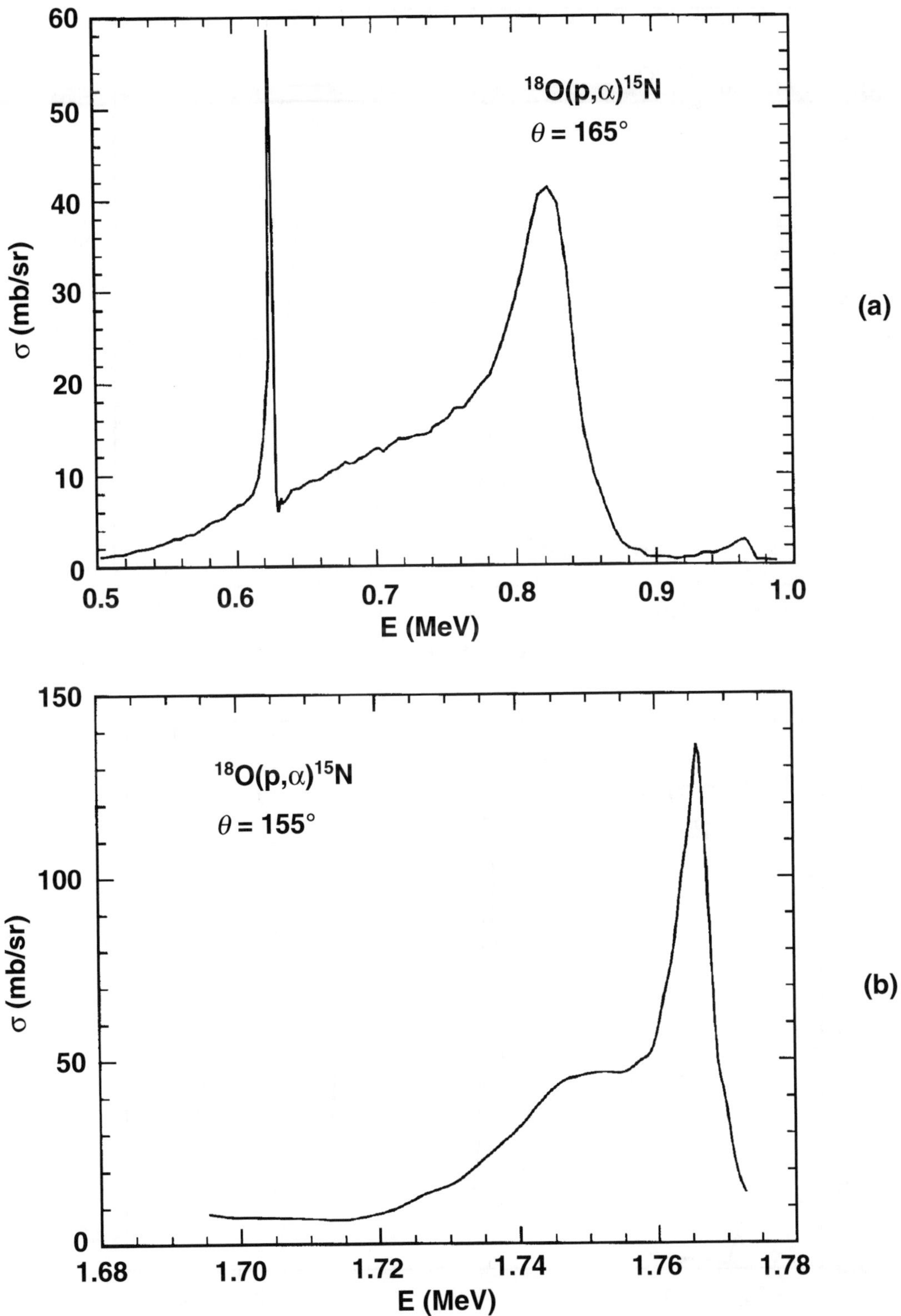

FIG. A11.7. The $^{18}O(p,\alpha)^{15}N$ reaction at (a) θ = 165° between 0.5 - 1 MeV [from Amsel, G., and Samuel, D. (1967), *Anal. Chem.*, **39**, 1689] and (b) at θ = 155° between 1.7 - 1.775 MeV. From Alkemade, P.F.A., Stap, C.A.M., Habraken, F.H.P.M., and van der Weg, W.F. (1988), *Nucl. Instrum. Methods* **B35**, 135.

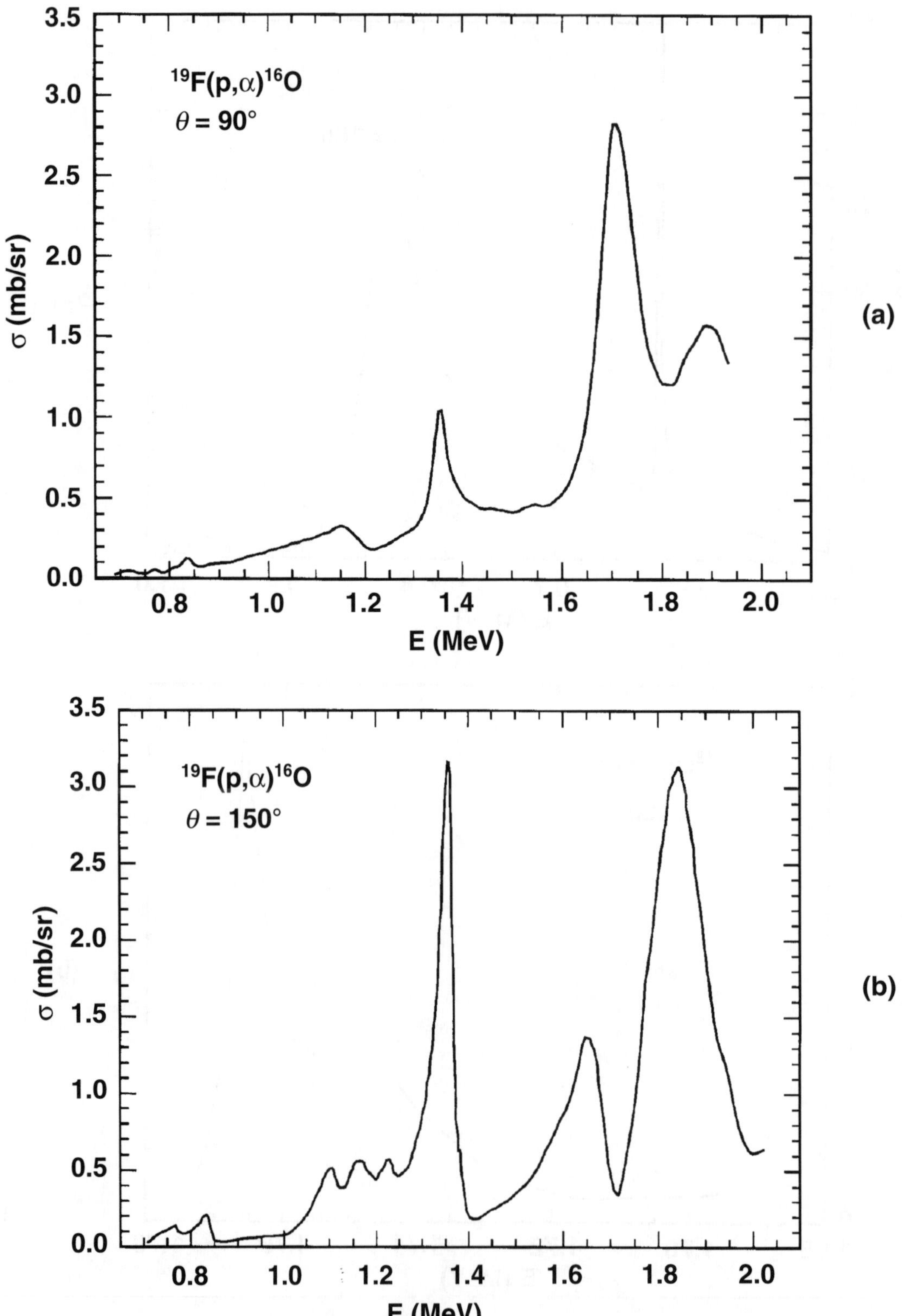

FIG. A11.8. The $^{19}F(p,\alpha)^{16}O$ reaction between 0.5 - 2 MeV at (a) $\theta = 90°$ and (b) $\theta = 150°$. From Dieumegard, D., Dubreuil, D., and Amsel, G. (1979), *Nucl. Instrum. Methods* **166**, 431.

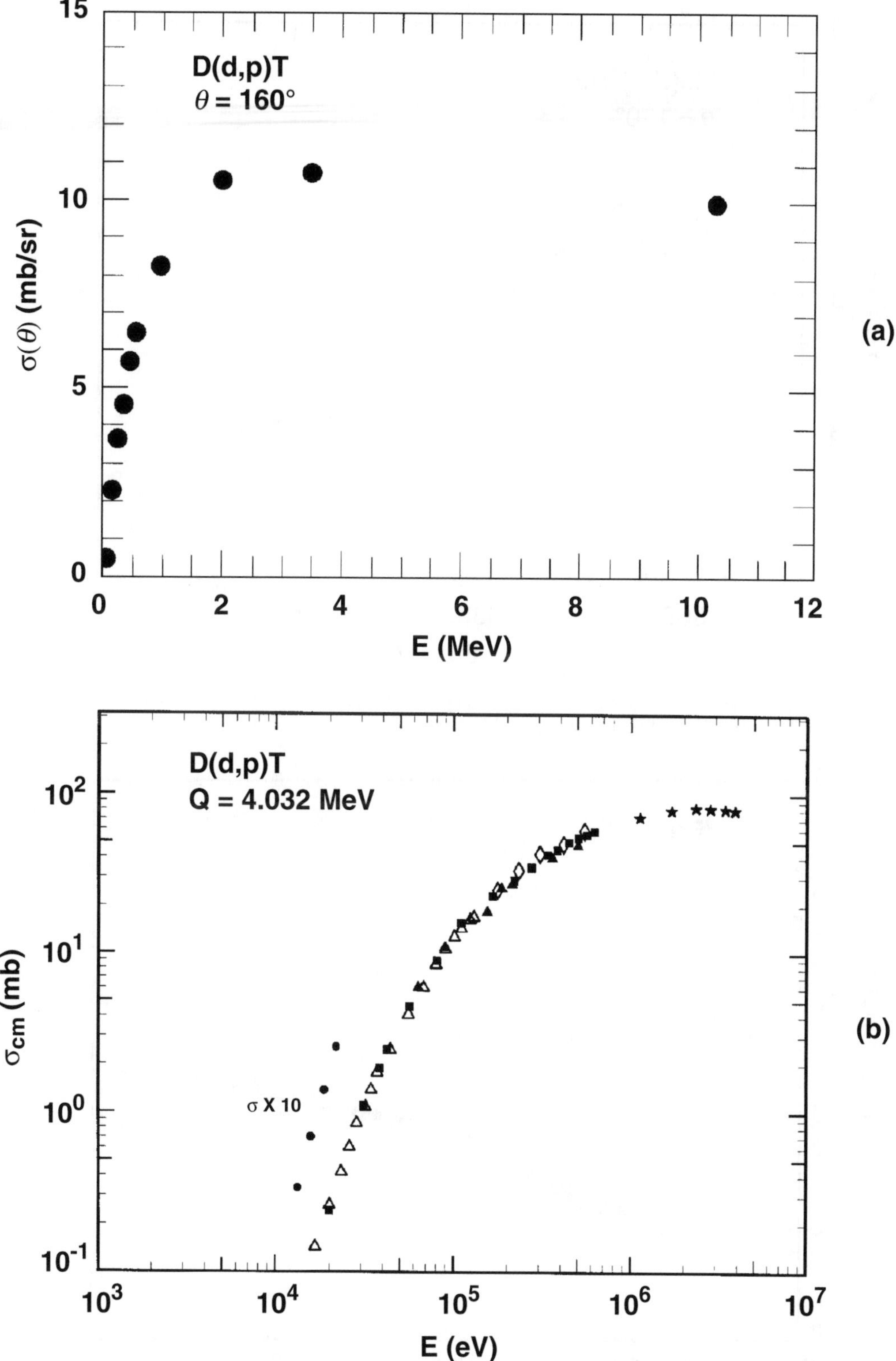

FIG. A11.9. The D(d,p)T reaction, which is always present when a deuteron beam is used. The reaction occurs between the implanted deuterium and the beam (a) the differential cross section at $\theta = 150°$, and (b) the total cross section. From Jarmie, N., and Seagrove, J. (1957), Los Alamos Scientific Laboratory Report, LA-2014.

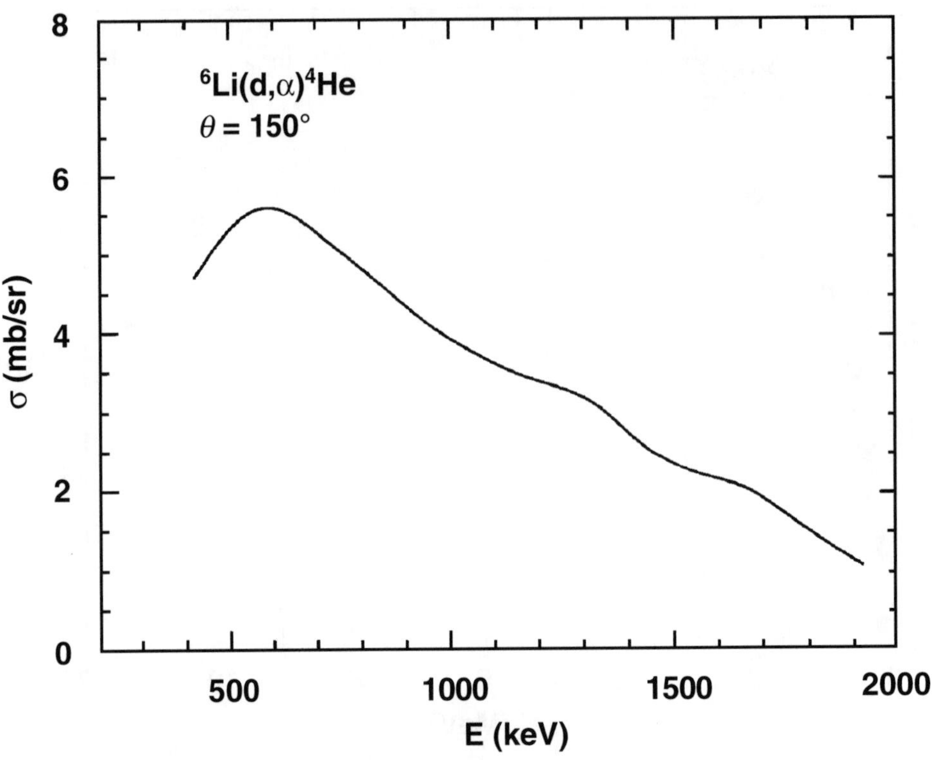

FIG. A11.10. The ^{6}Li(d,α)^{4}He reaction measured at θ = 150° between 0.5 - 2 MeV. From Maurel, B., Amsel, G., and Dieumegard, D. (1981), *Nucl. Instrum. Methods* **191**, 349.

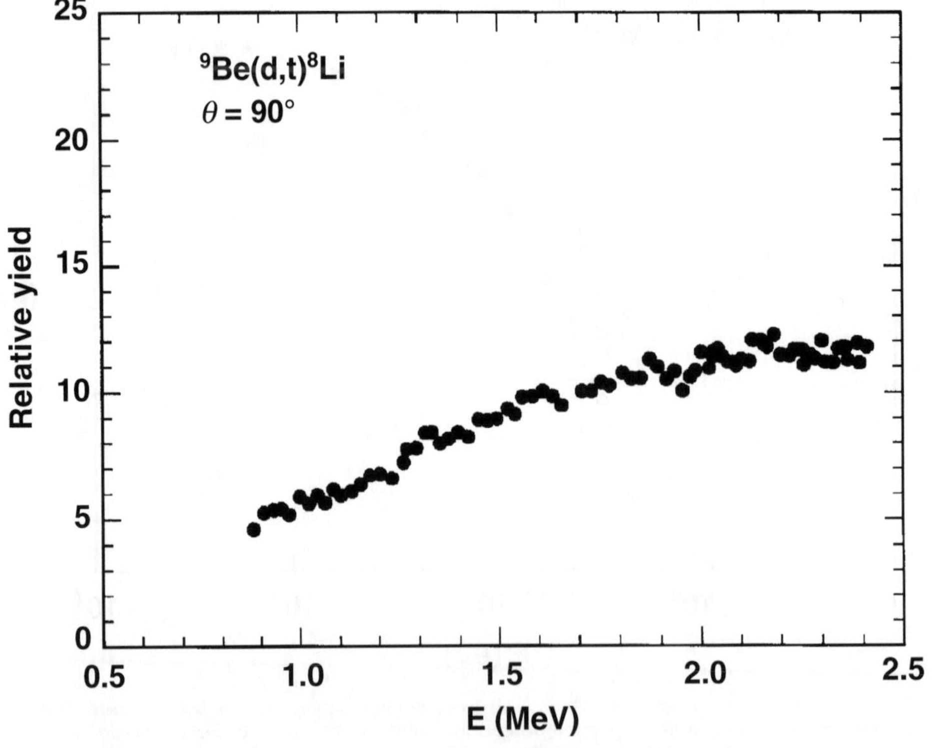

FIG. A11.11. The ^{9}Be(d,t)^{8}Be reaction measured at θ = 90° from 1.5 - 2.5 MeV. From Biggerstaff, J.A., Hood, R.F., Scott, H., and McEllistrem, M.T. (1962), *Nucl. Phys.* **36**, 631.

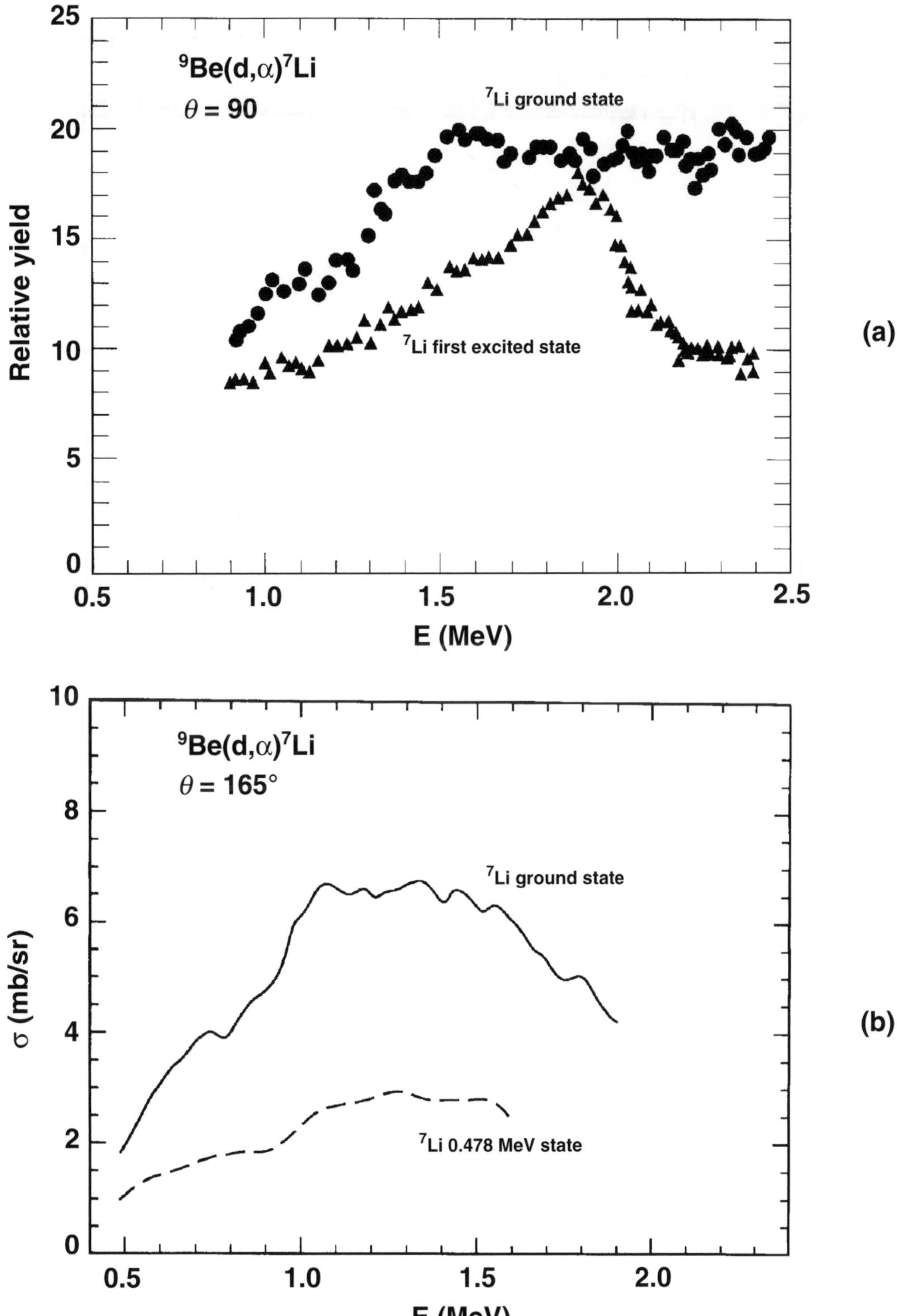

FIG. A11.12. The $^{9}Be(d,\alpha)^{7}Li$ reaction. The relative yield curves at $\theta = 90°$ for the (a) α_o (ground state) and α_1 reactions (first excited state). The cross section (b) for $\theta = 165°$ showing ^{7}Li in the ground state and the 0.478 MeV state. From Biggerstaff, J.A., Hood, R.F., Scott, H., and McEllistrem, M.T. (1962), *Nucl. Phys.* **36**, 631.

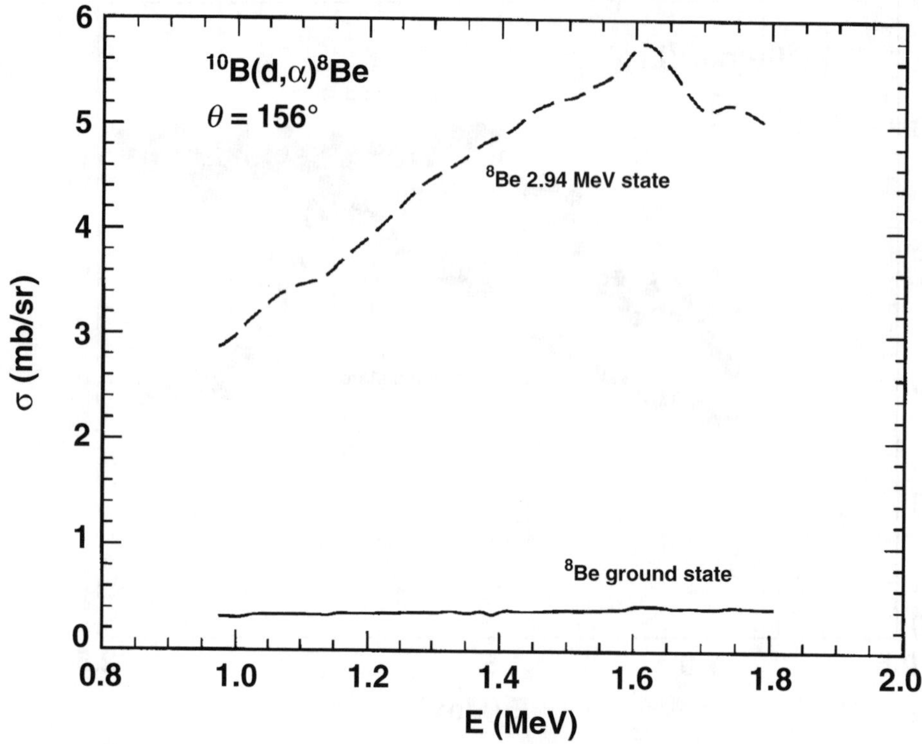

FIG. A11.13. The $^{10}B(d,\alpha)^{8}Be$ reaction at θ = 156° between 1 - 1.8 MeV for both α_0 and α_1. From Purser, K.H., and Wildenthal, B.H. (1963), *Nucl. Phys.* **44**, 22.

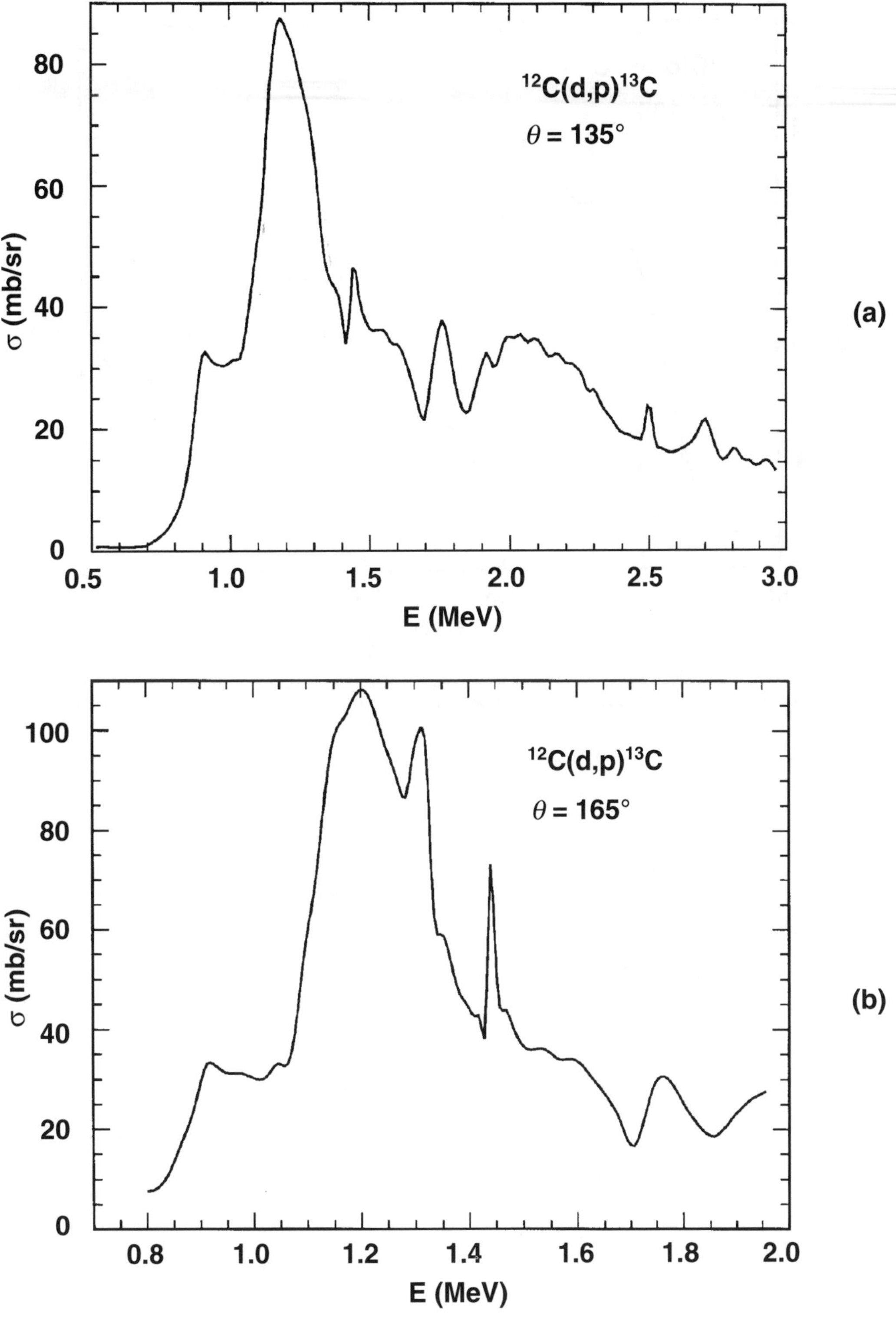

FIG. A11.14. The $^{12}C(d,p)^{13}C$ reaction from 1.8 MeV to 3.5 MeV at (a) θ = 135° and (b) θ = 165°. From McEllistrem, M.T., Jones, K.W., Chiba, R., Douglas, R.A., Herring, D.F., and Silverstein, E.A. (1956), *Phys. Rev.* **104**,1008.

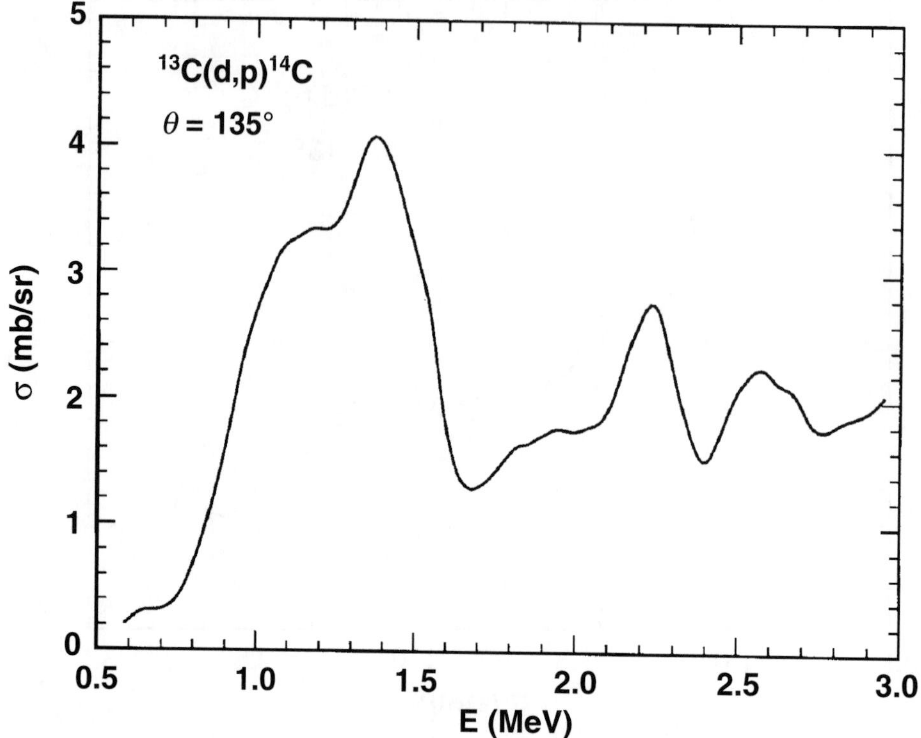

FIG. A11.15. The $^{13}C(d,p)^{14}C$ reaction measured at $\theta = 135°$ in the energy range of 0.6 - 3 MeV. From Marion, J.B., and Weber, G. (1956), *Phys. Rev.* **103**, 167.

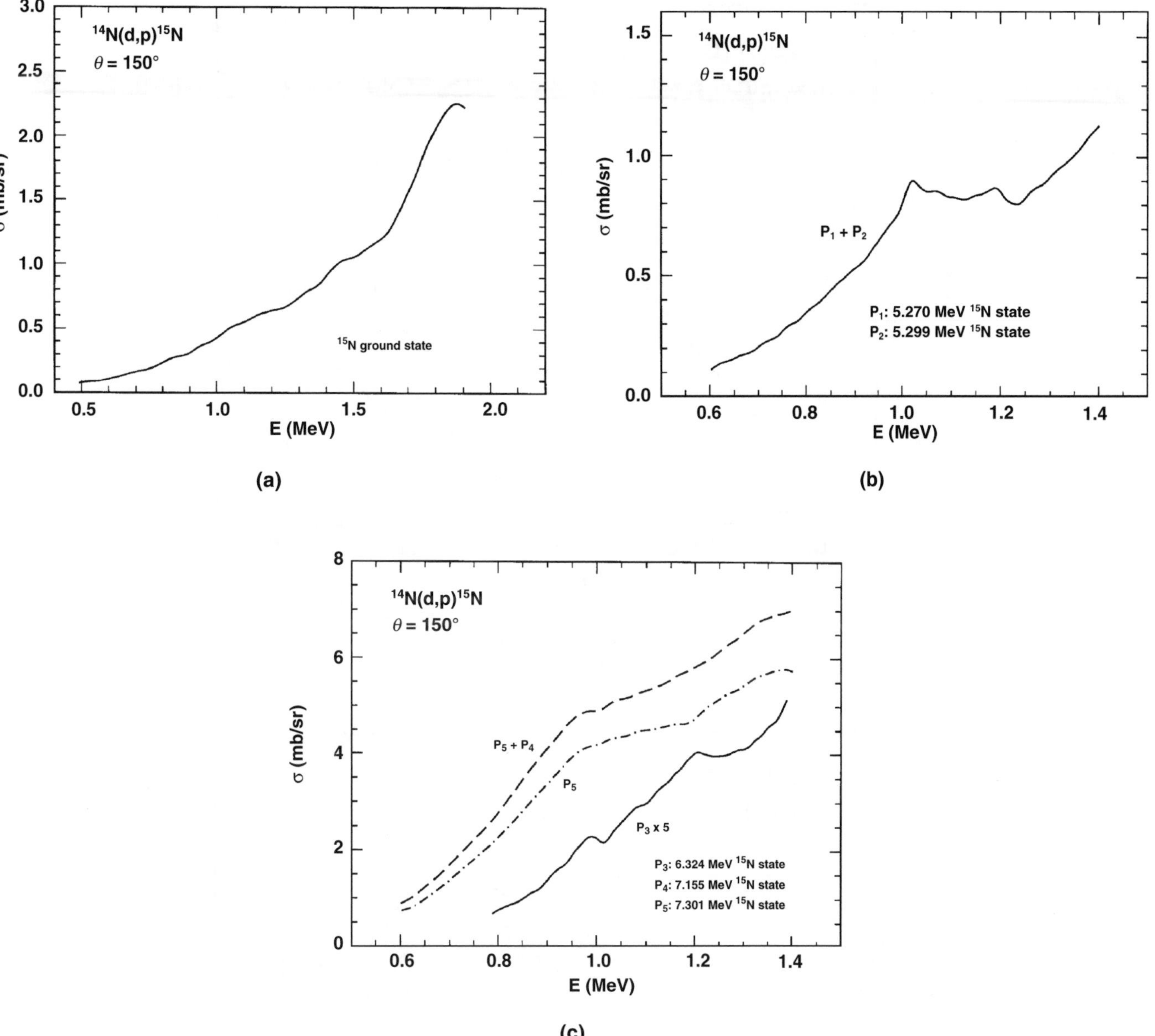

FIG. A11.16. The $^{14}N(d,p)^{15}N$ reaction measured at $\theta = 150°$ from 0.6 to 1.4 MeV for (a) the ground state, p_0 [from Simpson, J.C.B., and Earwaker, L.G. (1984), *Vacuum*, **34**, 899], (b) the p_1 and p_2 states, and (c) the p_3, p_4, and p_5 states. From Amsel, G., and David, D. (1969), *Rev. Phys. Appl.* **4**, 383.

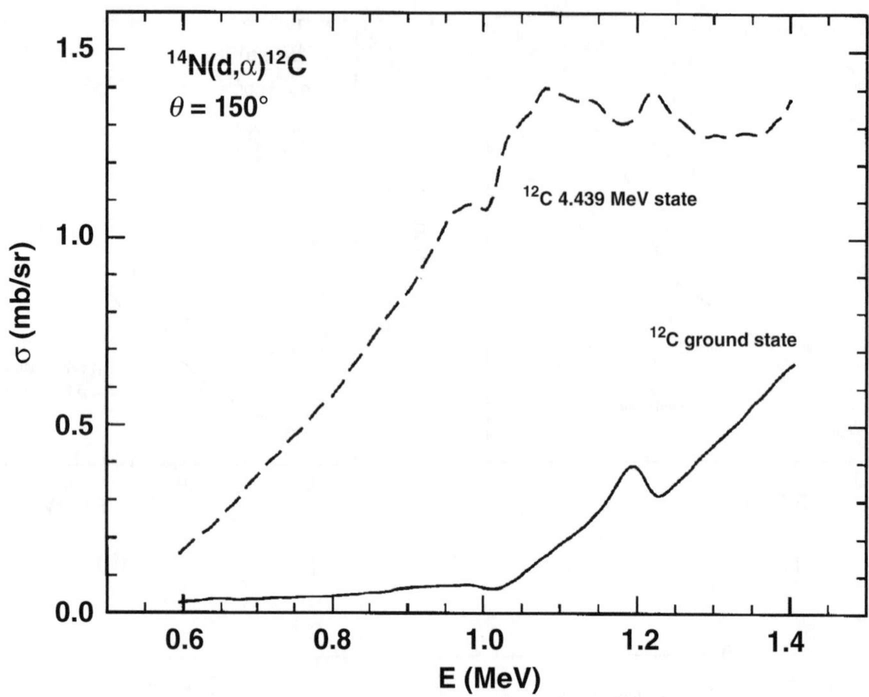

FIG. A11.17. The $^{14}N(d,\alpha)^{12}C$ reaction measured at θ = 150° from 0.6 to 1.4 MeV (α_0 and α_1). From Amsel, G., and David, D. (1969), *Rev. Phys. Appl.* **4**, 383.

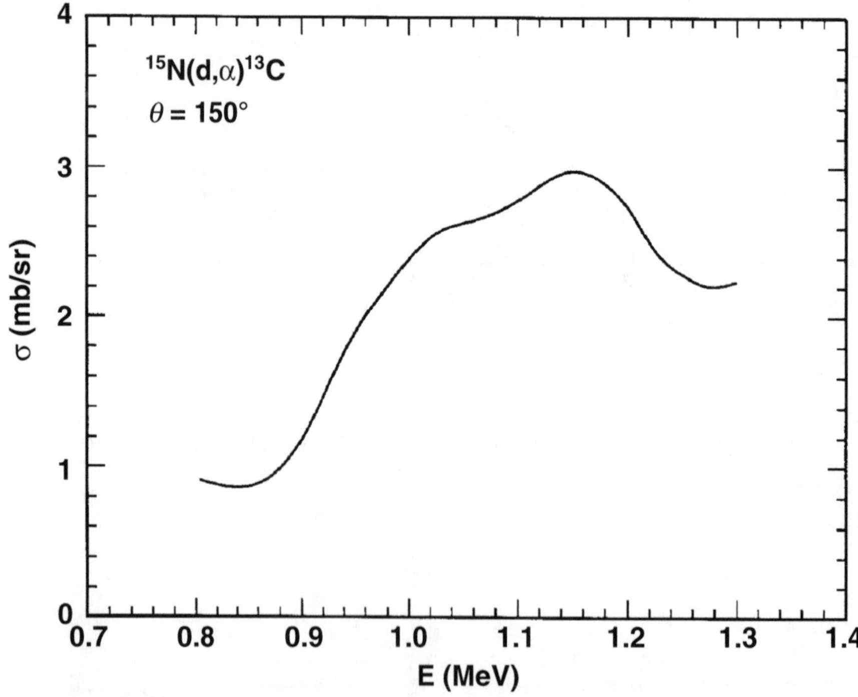

FIG. A11.18. The $^{15}N(d,\alpha)^{13}C$ measured at θ = 150° in the 0.8 -1.3 MeV range. From Sawicki, J.A., Davies, J.A., and Jackman, T.E. (1986), *Nucl. Instrum. Methods* **B15**, 530

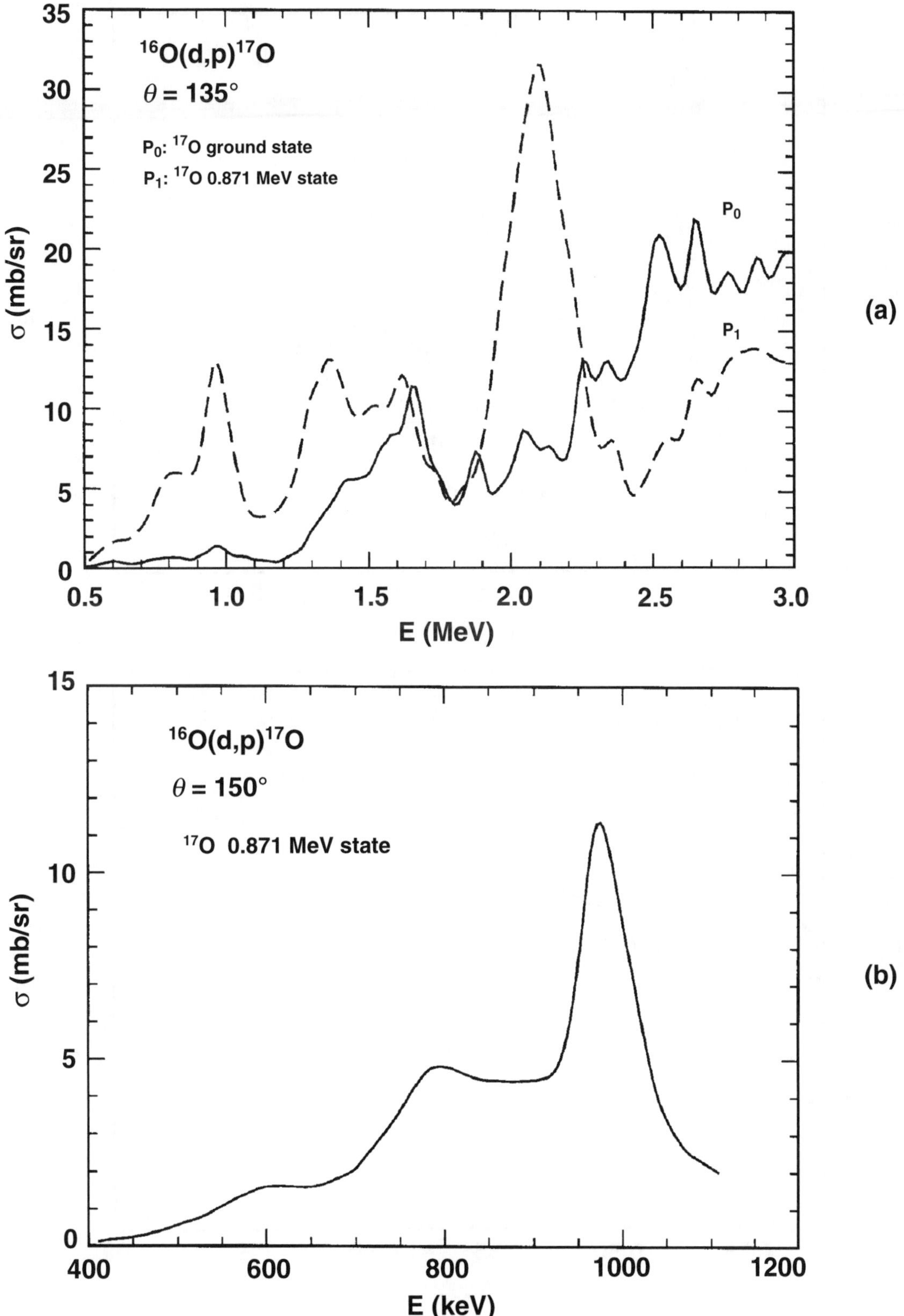

FIG. A11.19. The $^{16}O(d,p)^{17}O$ reaction. Measured at (a) θ = 135° from 0.2 MeV to 3 MeV [from Jarjis, R.A. (1979), Internal Report, University of Manchester] and at (b) θ = 150°, between 0.4 - 1.1 MeV [from Amsel, G., and Samuel, D. (1967), *Anal. Chem.* **39**, 1689].

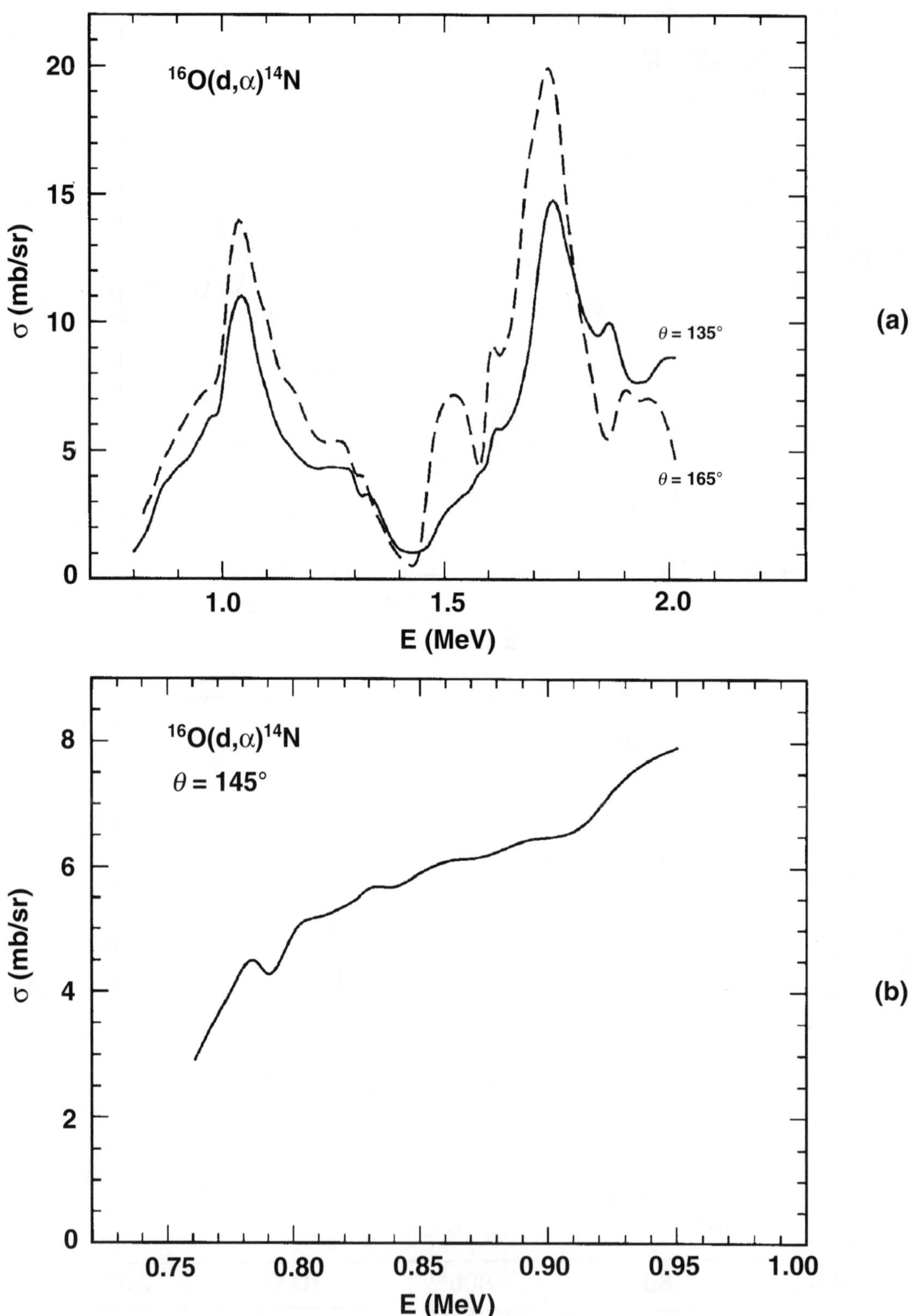

FIG. A11.20. The $^{16}O(d,\alpha)^{14}N$ reaction in the 0.8 - 2 MeV energy range at (a) θ = 135° and 165° [from Amsel, G. (1964), Thesis, *Ann. Phys.* **9**, 297] and (b) θ = 145° between 0.75 MeV and 0.95 MeV [from Turos, A., Wielunski, L., and Batcz, A. (1973), *Nucl. Instrum. Methods* **111**, 605].

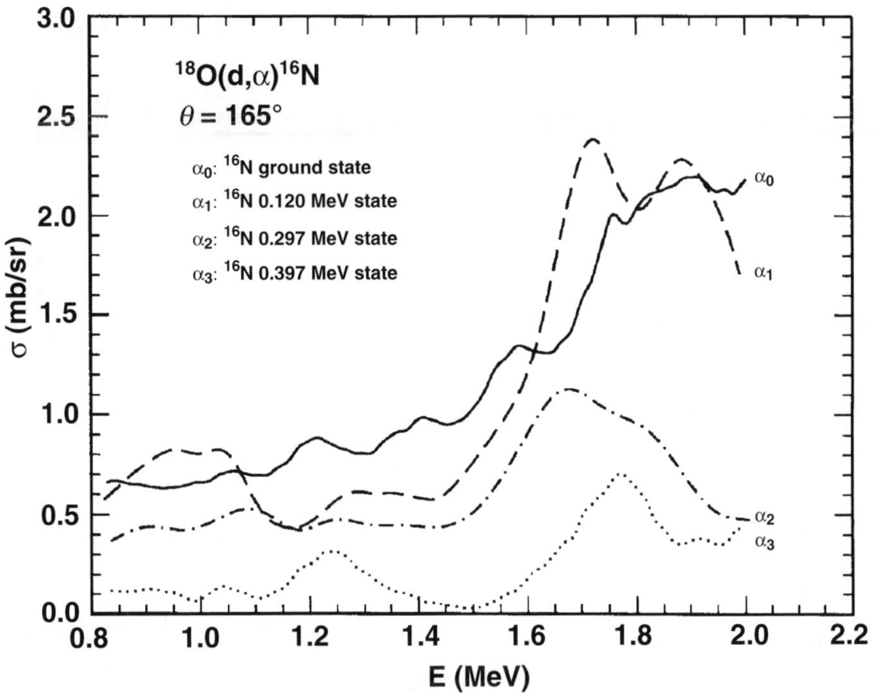

FIG. A11.21. The $^{16}O(d,\alpha)^{14}N$ reaction at different energy states measured at θ = 165° from 0.7 MeV to 2.2 MeV . From Seiler, F., Jones, C.H., Anzick, W.J., Herring, D.F., and Jones, K.W. (1963), *Nucl. Phys.* **45**, 647.

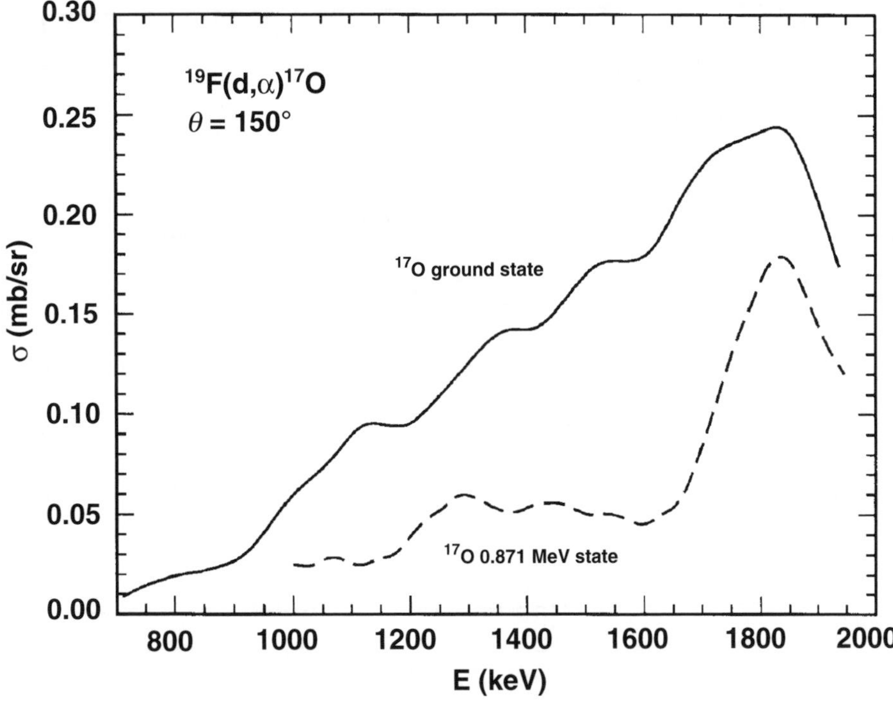

FIG. A11.22. The $^{19}F(d,\alpha)^{17}$ reaction measured at θ = 150° between 0.8 MeV and 2 MeV for both α_0 and α_1. From Maurel, B., Amsel, G., and Dieumegard, D. (1981), *Nucl. Instrum. Methods* **191**, 349.

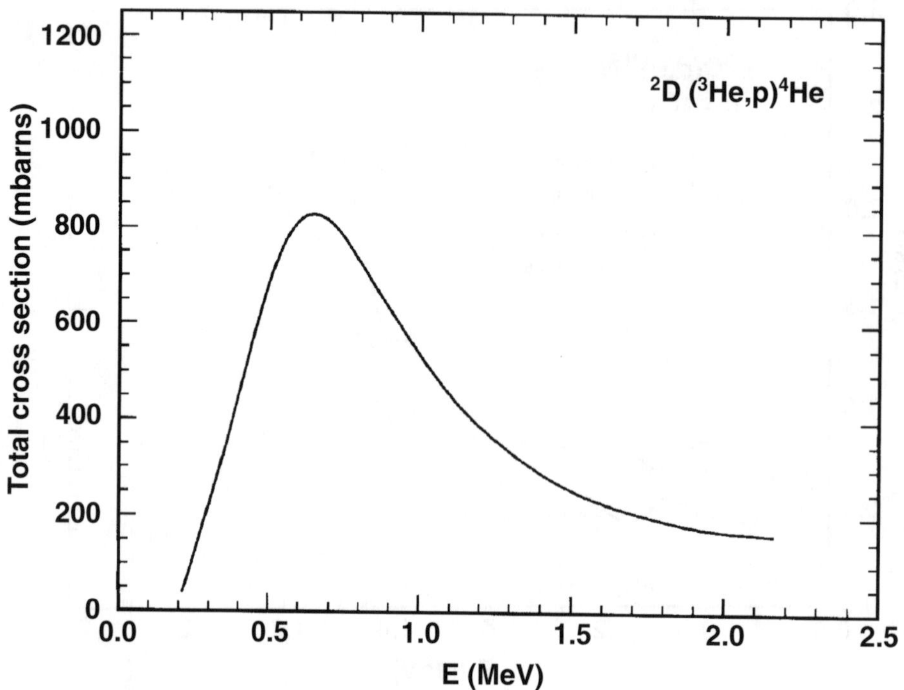

FIG. A11.23. The total cross section for the D(^{3}He,p)^{4}He reaction. From Möller, W., and F.Besenbacher (1980), *Nucl. Instrum. Methods* **168**, 111.

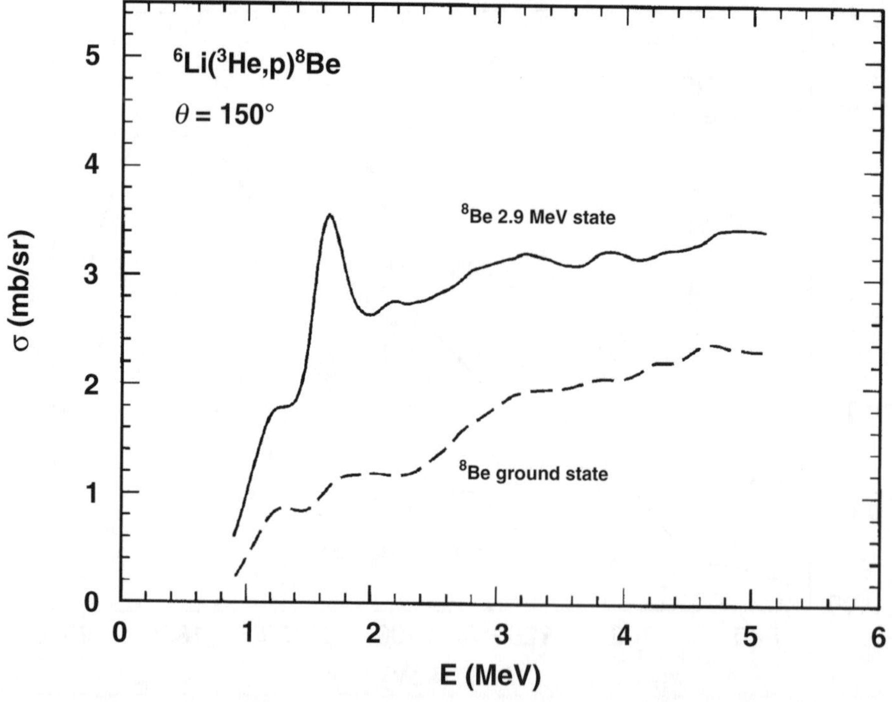

FIG. A11.24. The ^{6}Li(^{3}He,p)^{8}Be reaction for p_0 and p_1 measured at θ = 150° between 1 MeV and 5 MeV. From Schiffer, J.P., Bonner, T.W., Davis, R.H., and Prosser, Jr. F.W. (1956), *Phys. Rev.* **104**, 1064.

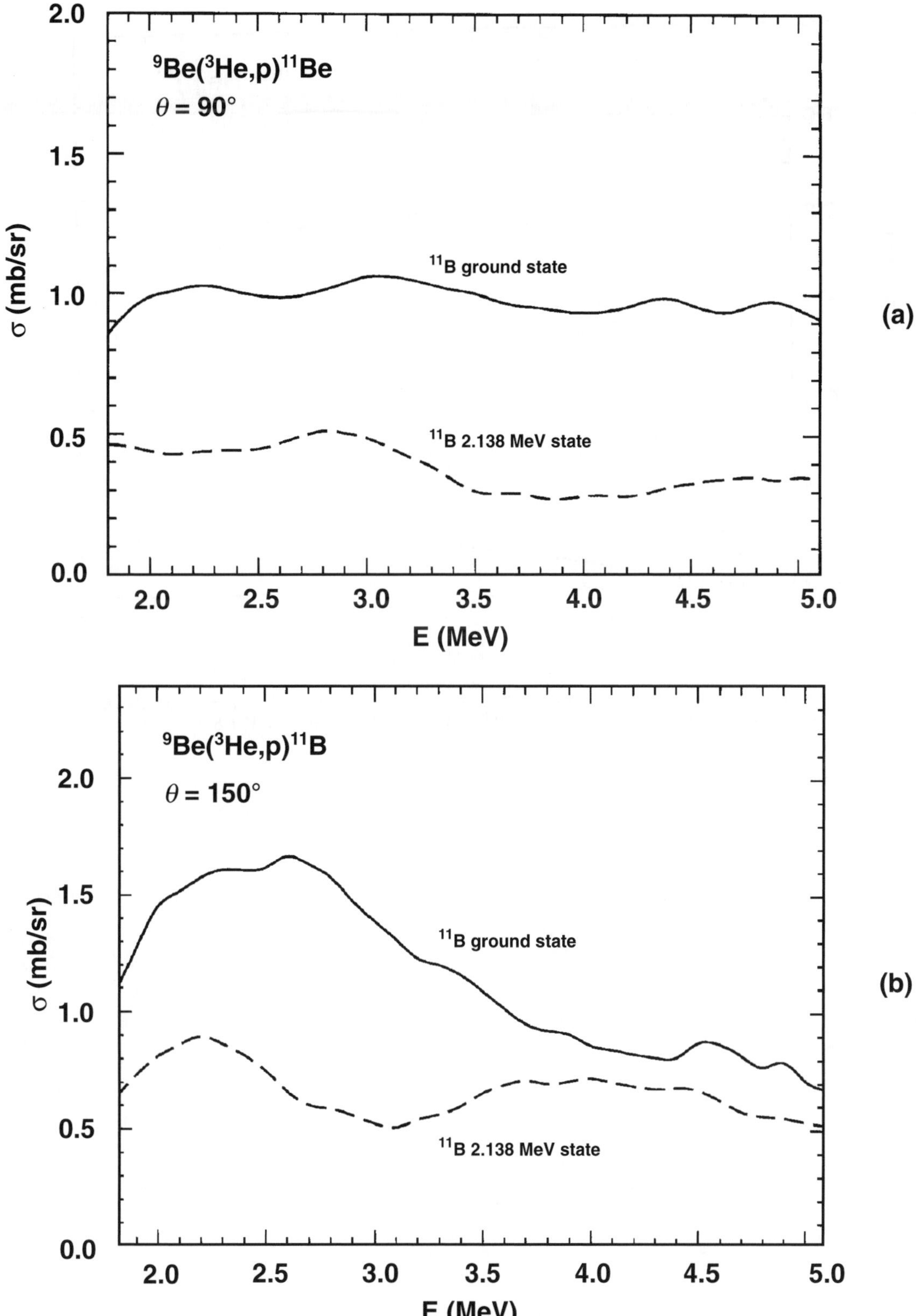

FIG. A11.25. The $^{9}Be(^{3}He,p)^{11}B$ reaction measured for p_0 and p_1 between 1.8 MeV and 5.1 MeV at (a) θ = 90° and (b) θ = 150°. From Wolicki, E.A., Holmgren, H.D., Johnston, R.L., and Geer Illsley, E. (1959), *Phys. Rev.* **116**, 1585.

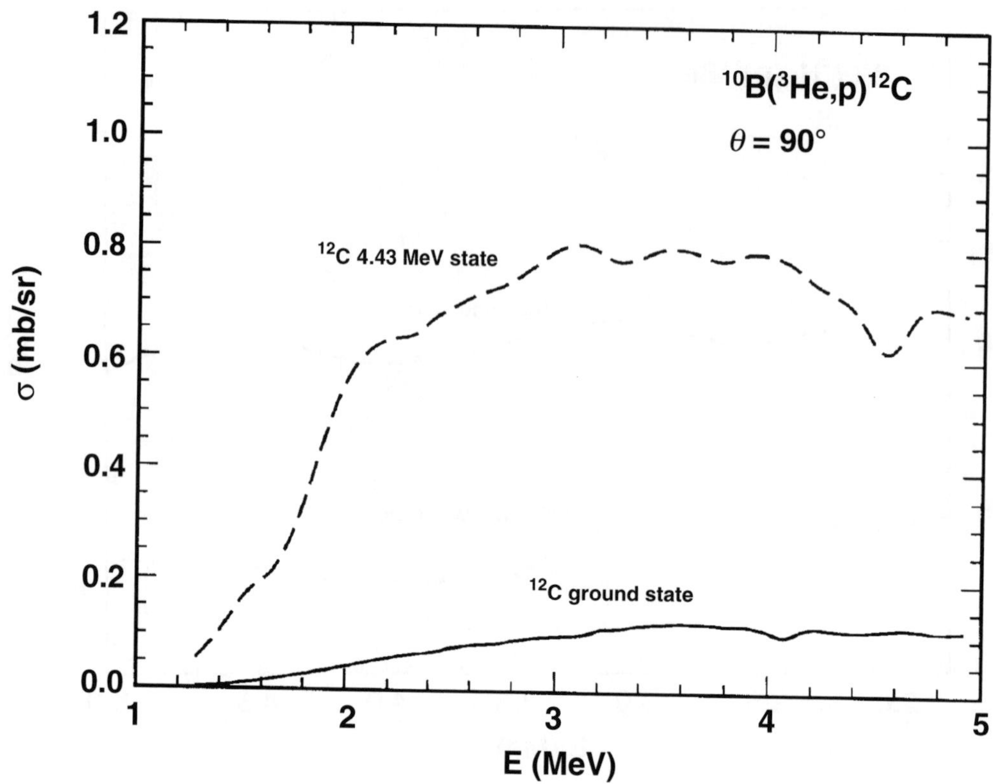

FIG. A11.26. The $^{10}B(^{3}He,p)^{15}N$ reaction measured at θ = 90° for p_0 and p_1 from 0.5 MeV to 5 MeV. From Schiffer, J.P., Bonner, T.W., Davis, R.H., and Prosser, Jr. F.W. (1956), *Phys. Rev* **104**, 1064.

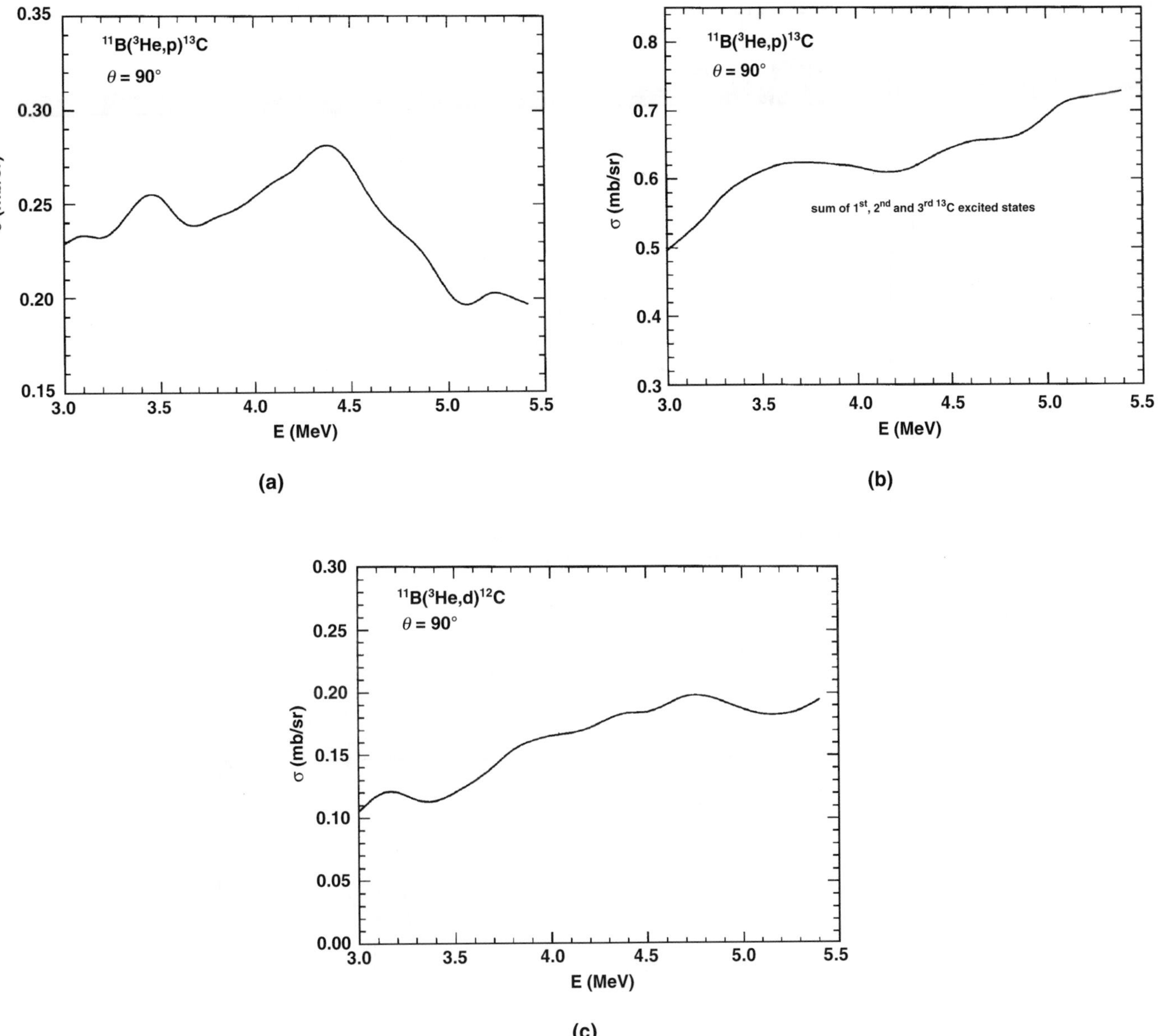

FIG. A11.27. The $^{11}B(^3He,p)^{13}C$ reaction measures at θ = 90° between 3 MeV and 5.5 MeV for (a) p_0 , and (b) p_1+p_2+p_3. The $^{11}B(^3He,d)^{12}C$ reaction (c) for d_0, measures at θ = 90° between 3 MeV and 5.5 MeV. From Holmgren, H.D., Wolicki, E.A., and Johnston, R.L. (1959), *Phys. Rev.* **114**, 1281.

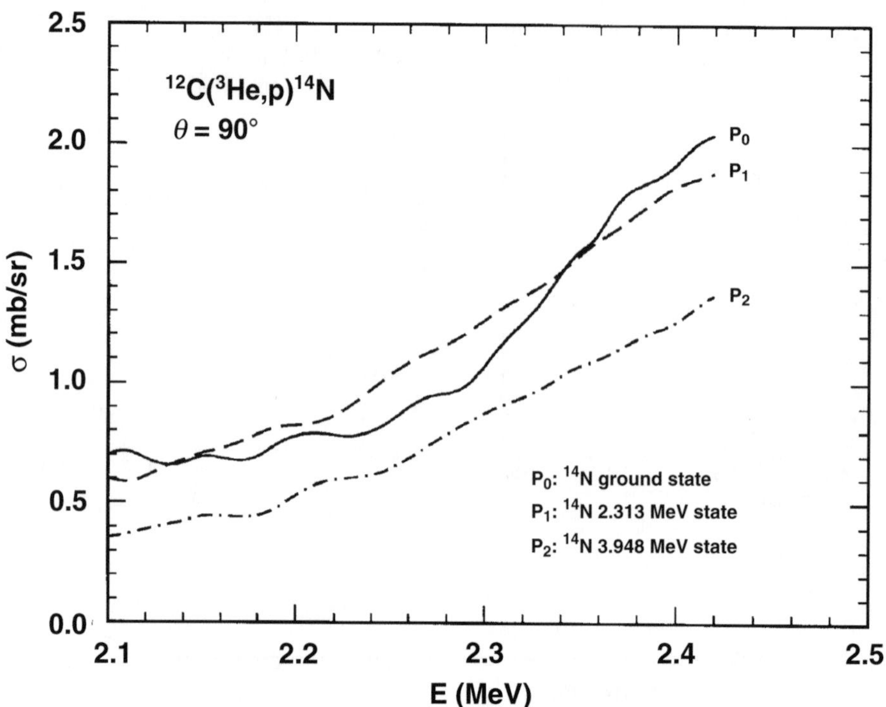

FIG. A11.28. The $^{12}C(^3He,p)^{14}N$ reaction for p_0 - p_2 measured at $\theta = 90°$ from 2.1 - 2.4 MeV. From Tong, S.Y., Lennard, W.N., Alkemada, P.F.A., and Mitchell, I.V. (1990), *Nucl. Instrum. Methods* **B45**, 91.

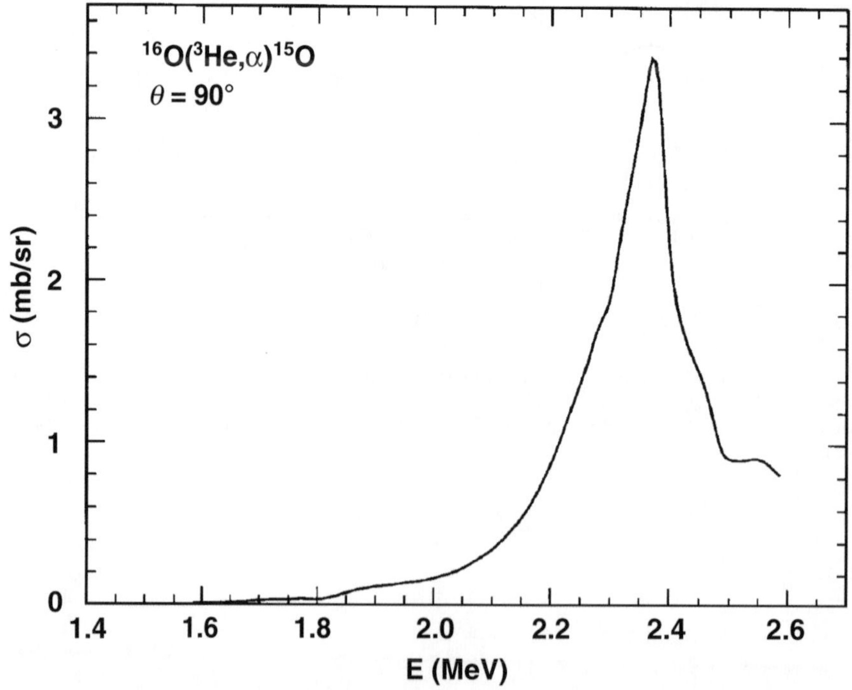

FIG. A11.29. The $^{16}O(^3He,\alpha)^{15}O$ reaction measured at $\theta = 90°$ in the energy range between 1.6 and 2.6 MeV. From Abel, F., Amsel, G., d'Artemare, E., Ortega, C., Siejka, J., and Vízkelethy, G. (1990), *Nucl. Instrum. Methods* **B45**, 10.

APPENDIX

12

PARTICLE-GAMMA DATA

Compiled by

J-P. Hirvonen

Technical Research Centre of Finland, Espoo, Finland

R. Lappalainen

University of Helsinki, Helsinki, Finland

CONTENTS

The (p,γ) data (Table A12.1) have been taken from the work of Butler (1959). The range of bombarding energies has been limited to 3 MeV. It has not been feasible to follow a consistent simple policy with respect to the value of the proton bombarding energy used in the table because of the large number of factors entering into such a choice. In some cases, the value of a single experimenter has been used, but in general, a weighted average of the different measurements has been made. For some of the resonances, the (p,p) bombarding energy for the same resonance has been used with the (p,γ) values to obtain the average.

Columns two through six give respectively the reaction, the energies of emitted gamma rays, cross section, resonance width, and references to the original literature. In order to conserve space, each entry has been limited to one line. This limitation made it necessary to restrict the number of gamma-ray energies and references to three each per entry. For the gamma-ray energies, the three chosen are the ones that are of the greatest aid in identifying the resonance. Therefore, they are as a rule the three having highest energy, but where an exceptionally intense low-energy gamma ray occurs, it has been substituted for one of the highest-energy three.

Where the identification of the gamma-ray transition in the compound nucleus is definite, the gamma-ray energy listed was obtained by taking the difference between the excitation energies of the two states involved in the transition instead of using the directly measured value.

The cross section given is the total cross section in millibarns (mb) at the resonance peak. Where more than one primary gamma ray is emitted, the tabulated value of the cross section is the sum of all such individual primary gamma-ray cross sections. For those resonances which are too narrow for such cross-section measurements, the integrated cross section, $\int \sigma dE$, has been tabulated where this measurement has been made. In these instances the abbreviation "eV-b" for "electron-volt barn" has been inserted in the cross-section column.

The width column gives the measured full width at half maximum of the resonance in the laboratory system of coordinates. In most cases, this is the observed overall experimental width. But for certain narrow well-known resonances, the tabulated width is the actual intrinsic resonance width; that is, it is a processed number that results after the factors such as beam width and Doppler broadening have been removed from the actual experimental number.

Table A12.1. Known (p,γ) resonances vs. bombarding energy (Butler, 1959).

Proton energy *(keV)*	Reaction	Gamma-ray energy *(MeV)*	Cross section *(mb)*	Width *(keV)*	References (author and year)
163	$B^{11}(p,\gamma)C^{12}$	16.11, 11.68, 4.43	0.157	7	Cr 56, Hu 53, Hu 53a
224	$F^{19}(p,\alpha\gamma)O^{16}$	7.12, 6.92, 6.13	>0.2	1	Hu 53a, Hu 52, Ch 50
226	$Mg^{24}(p,\gamma)Al^{25}$	2.06, 1.56, 0.95		1 ?	Ag 56, Va 56, Hu 54
226	$Al^{27}(p,\gamma)Si^{28}$				Hu 53a
251	$Na^{23}(p,\gamma)Mg^{24}$			0.3	Ba 56, Ha 55
261	$C^{14}(p,\gamma)N^{15}$				He 58
278	$N^{14}(p,\gamma)O^{15}$	6.82, 6.14, 1.47		1.6	Ov 56, Du 51, Ta 46
294	$Al^{27}(p,\gamma)Si^{28}$			<1	Hu 53a
295	$Mg^{26}(p,\gamma)Al^{27}$				Va 56b, Ca 53, Ta 46
308	$Na^{23}(p,\gamma)Mg^{24}$	10.6, 7.8, 6.7		0.8	Ba 56, Gr 55, Ha 55
317	$Mg^{25}(p,\gamma)Al^{26}$	6.19, 4.86, 0.82		12	Hu 55, Hu 54, Kl 54
326	$Al^{27}(p,\gamma)Si^{28}$	7.6, 7.2, 6.2		<1	Ca 53, Hu 53a
326	$Si^{29}(p,\gamma)P^{30}$	5.88, 5.17			Kl 55, Mi 55, Ta 46
330	$Be^{9}(p,\gamma)B^{10}$	6.9, 6.2, 5.2		160	Cl 56, Wi 56, Ca 55
339	$Mg^{26}(p,\gamma)Al^{27}$	7.74, 5.85, 5.61			Va 56b, Hu 55, Ca 53
340	$F^{19}(p,\alpha\gamma)O^{16}$	7.12, 6.92, 6.13	160	3	Bo 58, Bu 56, Ch 50
355	$P^{31}(p,\gamma)S^{32}$				Ta 46
356	$C^{14}(p,\gamma)N^{15}$	10.5, 7.1, 5.4			He 58, Ba 55
360	$N^{15}(p,\gamma)O^{16}$	12.43, 6.37	0.007	94	Sc 52

Appendix 12

Table A12.1. Known (p,γ) resonances vs. bombarding energy (Butler, 1959) (continued).

Proton energy *(keV)*	Reaction	Gamma-ray energy *(MeV)*	Cross section *(mb)*	Width *(keV)*	References (author and year)
360	$N^{15}(p,\alpha\gamma)C^{12}$	4.43	0.03	94	He 58
374	$Na^{23}(p,\gamma)Mg^{24}$	6.26	2		Ba 56, Ha 55
392	$Mg^{25}(p,\gamma)Al^{26}$	6.26, 4.6 ?, 3.5 ?	4	8	Hu 55, Hu 54, Kl 54
405	$Al^{27}(p,\gamma)Si^{28}$	7.3, 5.1, 2.8			Ru 54, Ca 53, Hu 53a
414	$Si^{29}(p,\gamma)P^{30}$	5.25, 0.70			Br 56, Mi 55, Ta 46
418	$Mg^{24}(p,\gamma)Al^{25}$	2.70, 2.25, 0.89		1	Ag 56, Va 56c, Hu 55
429	$N^{15}(p,\alpha\gamma)C^{12}$	4.43	300	0.9	Sc 52
429	$N^{15}(p,\gamma)O^{16}$	6.46	0.001	0.9	He 58
437	$Mg^{25}(p,\gamma)Al^{26}$	6.72, 6.30, 4.66 ?			Hu 55, Kl 54, Hu 53a
439	$Al^{27}(p,\gamma)Si^{28}$				Hu 53a
440	$P^{31}(p,\gamma)S^{32}$			34	Ke 56, Gr 51, Ta 46
441	$Li^{7}(p,\gamma)Be^{8}$	17.64, 14.74, 12.24	6	12	Bu 56, Hu 52, Fo 49
444	$Na^{23}(p,\gamma)Mg^{24}$			0.8	Ba 56, Ha 55
448	$C^{13}(p,\gamma)N^{14}$				He 58
454	$Mg^{26}(p,\gamma)Al^{27}$	7.85, 7.68, 5.71			Va 56b, Hu 55
457	$C^{12}(p,\gamma)N^{13}$	2.36	0.127	39.5	Hu 53a, Se 51, Fo 49
473	$Mg^{25}(p,\gamma)Al^{26}$				Br 56a
484	$F^{19}(p,\alpha\gamma)O^{16}$	7.12, 6.92, 6.13	>32	0.9	Bo 58, Bu 56, Ch 50
496	$Mg^{25}(p,\gamma)Al^{26}$	6.36, 4.24, 4.21 ?		5	Hu 55, Hu 54, Kl 54
500	$Si^{30}(p,\gamma)P^{31}$	7.75, 6.48, 4.62			Br 58, Br 56a
504	$Al^{27}(p,\gamma)Si^{28}$	12.07		<0.20	An 59
506	$Al^{27}(p,\gamma)Si^{28}$	10.29		<0.17	An 59
511	$Na^{23}(p,\gamma)Mg^{24}$	10.8, 8.0, 6.9		0.8	Ba 56, Gr 55, Ha 55
513	$Mg^{25}(p,\gamma)Al^{26}$			3	Hu 55, Hu 54, Kl 54
530	$Mg^{25}(p,\gamma)Al^{26}$			3	Hu 55, Hu 54, Kl 54
532	$C^{14}(p,\gamma)N^{15}$	10.7, 5.3			He 58, Ba 55
540	$P^{31}(p,\gamma)S^{32}$				Ke 56, Gr 51, Ta 46
550	$C^{13}(p,\gamma)N^{14}$	8.06, 4.11	1.44	32.5	Br 57, Wo 53, Se 52
580	$Mg^{25}(p,\gamma)Al^{26}$	6.85 ?, 6.43, 4.28			Kl 54, Ta 54
594	$Na^{23}(p,\gamma)Mg^{24}$	10.9, 8.0, 7.0		2	Ba 56, Pr 56, Gr 55
594	$S^{32}(p,\gamma)Cl^{33}$	2.86, 2.05, 0.806			Va 56a
597	$F^{19}(p,\alpha\gamma)O^{16}$	7.12, 6.92, 6.13	7.1	30	Hu 55a, Ch 50, Bo 48
607	$Mg^{25}(p,\gamma)Al^{26}$	6.88 ?, 6.46, 4.34			Kl 54, Ta 54
612	$Al^{27}(p,\gamma)Si^{28}$			<1	An 59, Br 47, Ta 46
625	$Si^{30}(p,\gamma)P^{31}$	7.87			Br 58, Br 56a, Ts 56
630	$O^{18}(p,\gamma)F^{19}$	8.5		2.6	Bu 55
632	$Al^{27}(p,\gamma)Si^{28}$	10.41, 7.59, 1.77		<0.06	An 59, Ru 54, Br 47
636	$Ne^{22}(p,\gamma)Na^{23}$	9.40			Th 58, Br 47a
640	$C^{14}(p,\gamma)N^{15}$	10.8, 5.3			He 58, Ba 55
648	$P^{31}(p,\gamma)S^{32}$			17	Ke 56, Fr 51, Gr 51
650	$Ca^{40}(p,\gamma)Sc^{41}$				Bu 58
654	$Al^{27}(p,\gamma)Si^{28}$	10.43, 7.61		<0.06	An 59, Br 47, Ta 46
660 ?	$Ne^{22}(p,\gamma)Na^{23}$				Th 58, Br 47a
661	$Mg^{26}(p,\gamma)Al^{27}$	7.88, 6.68, 5.9			Va 56b, Ru 54a
667	$Mg^{25}(p,\gamma)Al^{26}$				Kl 54, Ta 54
672	$F^{19}(p,\gamma)Ne^{20}$	11.88, 1.63	0.5	6.0	Fa 55, Hu 55, Si 54

Table A12.1. Known (p,γ) resonances vs. bombarding energy (Butler, 1959) (continued).

Proton energy *(keV)*	Reaction	Gamma-ray energy *(MeV)*	Cross section *(mb)*	Width *(keV)*	References (author and year)
672	$F^{19}(p,\alpha\gamma)O^{16}$	7.12, 6.92, 6.13	57	6.0	Hu 55a, Ch 50, Bo 48
675	$B^{11}(p,\gamma)C^{12}$	12.15, 4.43	0.050	322	Hu 53
675	$Na^{23}(p,\gamma)Mg^{24}$	11.0, 8.1, 7.1		≤1	Ba 56, Pr 56, Fl 54
675	$Mg^{25}(p,\gamma)Al^{26}$	6.55, 5.21, 3.30			Gr 56, Ka 55, Kl 54
675	$Si^{30}(p,\gamma)P^{31}$	7.92, 6.65, 1.27			Br 58, Br 56a
678	$Al^{27}(p,\gamma)Si^{28}$	10.45, 7.63		<1	An 59, Ru 54, Br 47
693	$Si^{29}(p,\gamma)P^{30}$	6.26, 4.29, 3.51			Br 56, Kl 55, Mi 55
700	$N^{14}(p,\gamma)O^{15}$	8.0		100	Du 51
703 ?	Si(p,γ)P				Se 55
710	$N^{15}(p,\gamma)O^{16}$	6.72		40	Ha 57a
717 ?	Si(p,γ)P				Ts 56
720	$Mg^{25}(p,\gamma)Al^{26}$	6.59, 4.93, 2.46			Gr 56, Ka 55, Kl 54
720	$Mg^{26}(p,\gamma)Al^{27}$	6.74, 5.96, 5.28			Va 56b, Ru 54a
725	$Ni^{60}(p,\gamma)Cu^{61}$	≤5.52	0.01 eV-b	<1	Bu 57
730	$Si^{29}(p,\gamma)P^{30}$	3.33			Br 56, Kl 55, Mi 55
731	$Al^{27}(p,\gamma)Si^{28}$			<0.16	An 59, Br 47
736	$Al^{27}(p,\gamma)Si^{28}$			<0.09	An 59, Br 47
740	$Na^{23}(p,\gamma)Mg^{24}$	11		<3	Ba 56, Pr 56
741	$Al^{27}(p,\gamma)Si^{28}$			<1	An 59, Br 47
744	$Na^{23}(p,\gamma)Mg^{24}$	8		<3	Ba 56, Pr 56
759	$Al^{27}(p,\gamma)Si^{28}$			<0.06	An 59, Br 47
760	$Si^{30}(p,\gamma)P^{31}$	6.71, 4.57, 1.27			Br 58, Br 56a, Ts 56
765	$Ne^{21}(p,\gamma)Na^{22}$				Br 47a
766	$Al^{27}(p,\gamma)Si^{28}$			<0.08	An 59, Br 47
773	$Al^{27}(p,\gamma)Si^{28}$	12.33		0.009	An 59, Sm 58, Br 47
775	$Si^{30}(p,\gamma)P^{31}$	8.00, 6.73, 1.27			Br 58, Br 56a, Ts 56
777	$Mg^{25}(p,\gamma)Al^{26}$	6.65 ?, 4.99, 3.90			Gr 56, Ta 54
780	$F^{19}(p,\alpha\gamma)O^{16}$			7.6	Hu 55, Hu 55a
800 ?	Si(p,γ)P				Ts 56
813	$Mg^{26}(p,\gamma)Al^{27}$				Ru 54
816	$P^{31}(p,\gamma)S^{32}$	7.39			Ke 56, Pa 55, Gr 51
820	$Mg^{25}(p,\gamma)Al^{26}$	7.69, 5.04, 4.56			Gr 56, Ta 54
825	$Mg^{24}(p,\gamma)Al^{25}$	3.09, 2.64, 2.14		1.5	Ag 56, Li 56, Gr 55a
825	$P^{31}(p,\gamma)S^{32}$	9.64			Pa 55
828	$Ne^{22}(p,\gamma)Na^{23}$				Br 47a
835	$F^{19}(p,\alpha\gamma)O^{16}$	7.12, 6.92, 6.13	19	6.5	Hu 55a, Ch 50, Bo 48
840	$Mg^{26}(p,\gamma)Al^{27}$				Ru 54
840	$Si^{30}(p,\gamma)P^{31}$	6.82, 4.80, 1.27			Br 58, Br 56a, Ts 56
849	$O^{18}(p,\gamma)F^{19}$	8.8		40	Bu 55
854	$Ne^{22}(p,\gamma)Na^{23}$	9.61, 9.17, 5.70			Th 58, Br 47a
855	$Cl^{35}(p,\gamma)Ar^{36}$	7.2, 5.1, 4.3		≤5	Be 57, To 57, Br 51
855	$Ni^{58}(p,\gamma)Cu^{59}$	≤4.26	0.007 eV-b	<1	Bu 57
872	$F^{19}(p,\alpha\gamma)O^{16}$	7.12, 6.92, 6.13	540	4.5	Bo 58, Bu 56, Ch 50
877	$Na^{23}(p,\gamma)Mg^{24}$	11		8	Ba 56, Pr 56
883 ?	$K^{39}(p,\gamma)Ca^{40}$	9 ?			De 57
884	$Al^{27}(p,\gamma)Si^{28}$			<1	An 59, Br 47

Table A12.1. Known (p,γ) resonances vs. bombarding energy (Butler, 1959) (continued).

Proton energy *(keV)*	Reaction	Gamma-ray energy *(MeV)*	Cross section *(mb)*	Width *(keV)*	References (author and year)
888	$Cl^{35}(p,\gamma)Ar^{36}$				Br 51
890	$Mg^{25}(p,\gamma)Al^{26}$				Ka 55, Ta 54
892	$P^{31}(p,\gamma)S^{32}$			9	Ke 56, Fr 51, Gr 51
895 ?	Si(p,γ)P				Ts 56
895	$Ni^{60}(p,\gamma)Cu^{61}$	≤5.69	0.01 eV-b	<1	Bu 57
898	$N^{15}(p,\alpha\gamma)C^{12}$	4.43	800	2.2	Sc 52
900	$Ar^{40}(p,\gamma)K^{41}$				Br 48
901	$Ne^{22}(p,\gamma)Na^{23}$	9.66 ?, 9.22 ?			Th 58, Br 47a
902	$F^{19}(p,\alpha\gamma)O^{16}$	7.12, 6.92, 6.13	23	5.1	Hu 55a, Ch 50, Bo 48
916	$Si^{29}(p,\gamma)P^{30}$	5.74, 4.48			Br 56a, Ts 56, Mi 55
922	$Al^{27}(p,\gamma)Si^{28}$			<0.19	An 59, Br 47
925 ?	$K^{39}(p,\gamma)Ca^{40}$	9 ?			De 57
933	$Ne^{22}(p,\gamma)Na^{23}$	9.69 ? 9.25 ?			Th 58
935	$F^{19}(p,\alpha\gamma)O^{16}$	7.12, 6.92, 6.13	180	8.6	Hu 55a, Ch 50, Bo 48
936	$Al^{27}(p,\gamma)Si^{28}$			0.34	An 59, Br 47
940	$Mg^{25}(p,\gamma)Al^{26}$	6.99, 5.15			Gr 56, Ka 55, Ta 54
943 ?	$Ne^{22}(p,\gamma)Na^{23}$				Th 58, Br 47a
944 ?	Si(p,γ)P				Ts 56, Se 55
947	$Ni^{58}(p,\gamma)Cu^{59}$	≤4.35	0.14 eV-b	<1	Bu 57
954	$Mg^{26}(p,\gamma)Al^{27}$				Ru 54
955	$Si^{30}(p,\gamma)P^{31}$	8.19, 6.92, 1.27			Br 58, Br 56a
956	$Si^{29}(p,\gamma)P^{30}$	6.49, 504, 4.52			Br 56a, Ts 56, Mi 55
960	$Mg^{25}(p,\gamma)Al^{26}$	5.17, 4.70, 3.57 ?			Gr 56, Ka 55
980 ?	$F^{19}(p,\gamma)Ne^{20}$				Fa 55
980	$Si^{30}(p,\gamma)P^{31}$	4.96, 4.84, 1.27			Br 58, Br 56a, Ts 56
980	$K^{39}(p,\gamma)Ca^{40}$	9 ?			De 57
982	$Ne^{22}(p,\gamma)Na^{23}$				Br 47a
989	$Na^{23}(p,\gamma)Mg^{24}$	9		<1	Ba 56, Pr 56
990	$Mg^{25}(p,\gamma)Al^{26}$				Ka 55, Ta 54
991	$Be^{9}(p,\gamma)B^{10}$	7.5, 6.8, 5.8		89	Mo 56, Hu 55, Ho 53
992	$Mg^{26}(p,\gamma)Al^{27}$				Ru 54
992	$Al^{27}(p,\gamma)Si^{28}$	10.78, 7.93, 1.77		0.05	Bo 58, Bu 56, Bu 56a
995	$Si^{30}(p,\gamma)P^{31}$	6.98, 6.02, 1.27			Br 58, Br 56a
1000	$Si^{30}(p,\gamma)P^{31}$	8.25, 6.98, 5.12			Br 58, Br 56a
1001	$Al^{27}(p,\gamma)Si^{28}$			<1	An 59, Br 47
1002	$Ne^{22}(p,\gamma)Na^{23}$				Br 47a
1006	$Ge^{74}(p,\gamma)As^{75}$			<2.5	Ch 57
1010	$Ni^{58}(p,\gamma)Cu^{59}$	≤4.41	0.007 eV-b	<1	Bu 57
1011	$Na^{23}(p,\gamma)Mg^{24}$			≤0.5	Ba 56, St 54
1015	$Mg^{26}(p,\gamma)Al^{27}$				Ru 54
1022	$Na^{23}(p,\gamma)Mg^{24}$	9		6.6	Ba 56, Pr 56, St 54
1024	$Al^{27}(p,\gamma)Si^{28}$			<0.24	An 59, Br 47
1029	$Ni^{60}(p,\gamma)Cu^{61}$	<5.82	0.02 eV-b	<1	Bu 57
1030	$Li^{7}(p,\gamma)Be^{8}$	18.15, 15.25, 0.478		168	Kr 54, Fo 49
1040	$N^{15}(p,\gamma)O^{16}$	13.09	1	130	Ba 57, Ha 57a, Wi 57a
1040	$N^{15}(p,\alpha\gamma)C^{12}$	4.43	15	130	Co 53, Wi 53, Sc 52

Table A12.1. Known (p,γ) resonances vs. bombarding energy (Butler, 1959) (continued).

Proton energy *(keV)*	Reaction	Gamma-ray energy *(MeV)*	Cross section *(mb)*	Width *(keV)*	References (author and year)
1046	$Mg^{25}(p,\gamma)Al^{26}$				Ka 55, Ta 54
1050	$P^{31}(p,\gamma)S^{32}$			<5	Ke 56, Cl 55, Gr 51
1050	$Ar^{40}(p,\gamma)K^{41}$				Br 48
1056	$Mg^{26}(p,\gamma)Al^{27}$				Ru 54
1059	$N^{14}(p,\gamma)O^{15}$	8.34, 5.27, 3.04		4	Ha 57, Li 53, Du 51
1066	$Ni^{60}(p,\gamma)Cu^{61}$	≤5.86	0.05 eV-b	<1	Bu 57
1068	$P^{31}(p,\gamma)S^{32}$			6	Ke 56, Gr 51
1070	$Ne^{22}(p,\gamma)Na^{23}$				Br 47a
1070	$Cl^{37}(p,\gamma)Ar^{38}$	9.1, 7.5, 6.3		≤5	To 57
1078	$Ni^{60}(p,\gamma)Cu^{61}$	≤5.87	0.03 eV-b	<1	Bu 57
1080	$Ar^{40}(p,\gamma)K^{41}$				Br 48
1084	$Be^{9}(p,\gamma)B^{10}$	6.9, 5.4, 0.7		3.8	Hu 55a, Ho 53, Ha 52
1086	$Mg^{25}(p,\gamma)Al^{26}$				Ka 55, Ta 54
1087	$Na^{23}(p,\gamma)Mg^{24}$			1.1	Ba 56, Pr 56, St 54
1088	$Ne^{22}(p,\gamma)Na^{23}$				Br 47a
1089	$Al^{27}(p,\gamma)Si^{28}$			<0.11	An 59, Br 47
1090	$F^{19}(p,\gamma)Ne^{20}$	12.28, 8.84, 1.63	>0.05	0.7	Fa 55, Hu 55, Si 54
1090	$F^{19}(p,\alpha\gamma)O^{16}$	7.12, 6.92, 6.13	>13	0.7	Hu 55a, Ch 50, Bo 48
1090	$Cl^{37}(p,\gamma)Ar^{38}$				Br 51
1096	$Al^{27}(p,\gamma)Si^{28}$			<1	An 59, Br 47
1094	$Ge^{74}(p,\gamma)As^{75}$			9.5	Ch 57
1100	$Ar^{40}(p,\gamma)K^{41}$				Br 48
1100	$Ni^{58}(p,\gamma)Cu^{59}$	<4.50	0.05 eV-b	<1	Bu 57
1101	$P^{31}(p,\gamma)S^{32}$				Ke 56, Gr 51
1102	$Cl^{35}(p,\gamma)Ar^{36}$				Br 51
1105	$Mg^{25}(p,\gamma)Al^{26}$				Ka 55, Ta 54
1106	$Ne^{22}(p,\gamma)Na^{23}$				Br 47a
1117	$Al^{27}(p,\gamma)Si^{28}$			0.80	An 59, Br 47
1117	$P^{31}(p,\gamma)S^{32}$	9.92		5	Ke 56, Pa 55, Gr 51
1120	$K^{39}(p,\gamma)Ca^{40}$	9.5, 6.1, 3.8		≤5	Be 57, De 57, To 57
1123 ?	$F^{19}(p,\alpha\gamma)O^{16}$			22	Hu 55, Hu 55a
1132	$Ni^{60}(p,\gamma)Cu^{61}$	<5.92	0.04 eV-b	<1	Bu 57
1135	$Cl^{37}(p,\gamma)Ar^{38}$	9.1, 7.5, 6.3		<5	To 57
1140	$F^{19}(p,\alpha\gamma)O^{16}$	7.12, 6.92, 6.13	15	2.5	Hu 55a, Ch 50, Bo 48
1146	$B^{10}(p,\gamma)C^{11}$	9.7, 5.5 ?, 4.2 ?	0.0055	450	Hu 57, Ch 56a
1146	$P^{31}(p,\gamma)S^{32}$	7.71			Ke 56, Pa 55, Gr 51
1160	$C^{13}(p,\gamma)N^{14}$	8.62, 4.67, 2.39	0.56	6	Wi 57, Mi 54, Wo 53
1163	$C^{14}(p,\gamma)N^{15}$	11.30		12	Ba 55
1165	$Ne^{20}(p,\gamma)Na^{21}$	<4			Br 47a
1166	$Na^{23}(p,\gamma)Mg^{24}$			1.2	Ba 56, Pr 56, St 54
1167	$Ni^{60}(p,\gamma)Cu^{61}$	≤5.96	0.15 eV-b	<1	Bu 57
1167	$Ge^{74}(p,\gamma)As^{75}$			4.5	Ch 57
1169	$O^{18}(p,\gamma)F^{19}$	6.3		1	Bu 55
1171	$Al^{27}(p,\gamma)Si^{28}$			0.25	An 59, Br 47
1172	$Mg^{26}(p,\gamma)Al^{27}$				Ru 54
1176	$Na^{23}(p,\gamma)Mg^{24}$			2.5	Ba 56, Pr 56, St 54

Table A12.1. Known (p,γ) resonances vs. bombarding energy (Butler, 1959) (continued).

Proton energy *(keV)*	Reaction	Gamma-ray energy *(MeV)*	Cross section *(mb)*	Width *(keV)*	References (author and year)
1180	$B^{10}(p,\gamma)C^{11}$	9.4	0.0075	570	Ha 55a, Da 54
1182	$Al^{27}(p,\gamma)Si^{28}$			0.71	An 59, Br 47
1185	$Mg^{25}(p,\gamma)Al^{26}$				Ka 55, Ta 54
1189	$F^{19}(p,\alpha\gamma)O^{16}$	7.12, 6.92, 6.13	19	110	Hu 55a, Ch 50, Bo 48
1197	$Ni^{60}(p,\gamma)Cu^{61}$	≤5.99	0.13 eV-b	<1	Bu 57
1198	$Al^{27}(p,\gamma)Si^{28}$			6.3	An 59, Br 47
1200	$Mg^{24}(p,\gamma)Al^{25}$	3.44, 1.83, 1.61		<10	Li 56
1209	$N1^{60}(p,\gamma)Cu^{61}$	≤6.00	0.14 eV-b	<1	Bu 57
1210	$N^{15}(p,\alpha\gamma)C^{12}$	4.43	425	22.5	Ba 57, Sc 52
1212	$Al^{27}(p,\gamma)Si^{28}$			<0.21	An 59, Br 47
1213	$Na^{23}(p,\gamma)Mg^{24}$			0.4	Ba 56
1213	$Ge^{74}(p,\gamma)As^{75}$			<2.5	Ch 57
1227	$Ni^{58}(p,\gamma)Cu^{59}$	≤4.63	0.045 eV-b	<1	Bu 57
1235	$Ar^{40}(p,\gamma)K^{41}$				Br 48
1239	$Ni^{60}(p,\gamma)Cu^{61}$	≤6.03	0.13 eV-b		Bu 57
1247	$Ni^{60}(p,\gamma)Cu^{61}$	≤6.04	0.1 eV-b	<1	Bu 57
1248	$P^{31}(p,\gamma)S^{32}$	10.05, 7.80		9	Ke 56. Pa 55, Gr 51
1250	$C^{13}(p,\gamma)N^{14}$	8.71	0.062	500	Br 57, Mi 54, Wo 53
1255	$Mg^{26}(p,\gamma)Al^{27}$				Ru 54
1257	$Ge^{74}(p,\gamma)As^{75}$			<2.5	Ch 57
1258	$Cl^{35}(p,\gamma)Ar^{36}$				Br 51
1261	$Al^{27}(p,\gamma)Si^{28}$			<0.20	An 59, Br 47
1262	$Ne^{22}(p,\gamma)Na^{23}$				Br 47a
1273	$Na^{23}(p,\gamma)Mg^{24}$				Ba 56, Pr 56
1274	$Al^{27}(p,\gamma)Si^{28}$			<1	An 59, Br 47
1278	$Ne^{22}(p,\gamma)Na^{23}$				Br 47a
1283	$F^{19}(p,\alpha\gamma)O^{16}$	7.12, 6.92, 6.13	29	19	Hu 55a, Ch 50, Bo 48
1295	$Mg^{26}(p,\gamma)Al^{27}$				Ru 54
1300	$K^{39}(p,\gamma)Ca^{40}$	9.6, 6.3, 3.8		≤5	Be 57, To 57
1308	$Ni^{58}(p,\gamma)Cu^{59}$	≤4.71	0.11 eV-b	<1	Bu 57
1312	$C^{14}(p,\gamma)N^{15}$	11.43		43	Fe 56, Ba 55
1313	$Ni^{60}(p,\gamma)Cu^{61}$	≤6.10	0.21 eV-b	<1	Bu 57
1315	$Al^{27}(p,\gamma)Si^{28}$			<0.16	An 59, Br 47
1316	$Ni^{58}(p,\gamma)Cu^{59}$	<4.71	0.08 eV-b	<1	Bu 57
1319	$Ni^{60}(p,\gamma)Cu^{61}$	<6.11	0.25 eV-b	<1	Bu 57
1321	$Na^{23}(p,\gamma)Mg^{24}$	11		2.1	Ba 56, Pr 56, St 54
1322 ?	$F^{19}(p,\gamma)Ne^{20}$	12.50, 1.63	0.081	4.0	Fa 55, Si 54
1322	$Ne^{22}(p,\gamma)Na^{23}$				Br 47a
1323	$Ni^{60}(p,\gamma)Cu^{61}$	≤6.11	0.29 eV-b	<1	Bu 57
1327	$Al^{27}(p,\gamma)Si^{28}$			<0.16	An 59, Br 47
1331	$Ni^{60}(p,\gamma)Cu^{61}$	≤6.12	0.06 eV-b	<1	Bu 57
1332	$Ge^{74}(p,\gamma)As^{75}$			5.0	Ch 57
1338	$K^{39}(p,\gamma)Ca^{40}$	5.91, 5.74, 3.8		≤5	To 57
1343	$Ni^{60}(p,\gamma)Cu^{61}$	≤6.13	0.45 eV-b	<1	Bu 57
1347	$Ni^{60}(p,\gamma)Cu^{61}$	≤6.14	0.40 eV-b	<1	Bu 57
1348	$F^{19}(p,\gamma)Ne^{20}$		0.1	5.6	Fa 55

Table A12.1. Known (p,γ) resonances vs. bombarding energy (Butler, 1959) (continued).

Proton energy *(keV)*	Reaction	Gamma-ray energy *(MeV)*	Cross section *(mb)*	Width *(keV)*	References (author and year)
1348	$F^{19}(p,\alpha\gamma)O^{16}$	7.12. 6.92, 6.13	89	5.6	Hu 55a, Ch 50, Bo 48
1350	$Ne^{22}(p,\gamma)Na^{23}$				Br 47a
1362	$Al^{27}(p,\gamma)Si^{28}$			<0.12	An 59, Br 47
1370 ?	$S^{34}(p,\gamma)Cl^{35}$				Ha 51
1371	$Ni^{60}(p,\gamma)Cu^{61}$	≤6.16	0.15 eV-b	<1	Bu 57
1375	$F^{19}(p,\alpha\gamma)O^{16}$	7.12, 6.92, 6.13	300	11	Hu 55a, Ch 50, Bo 48
1375	$Ne^{22}(p,\gamma)Na^{23}$				Br 47a
1376	$Ni^{58}(p,\gamma)Cu^{59}$	4.77, 4.28, 3.86	0.19 eV-b	<1	Bu 57
1380	$Al^{27}(p,\gamma)Si^{28}$			0.70	An 59, Br 47
1387	$Al^{27}(p,\gamma)Si^{28}$			0.29	An 59, Br 47
1381	$Ni^{60}(p,\gamma)Cu^{61}$	≤6.17	0.2 eV-b	<1	Bu 57
1386	$Ne^{22}(p,\gamma)Na^{23}$				Br 47a
1388	$B^{11}(p,\gamma)C^{12}$	17.23, 12.80	0.053	1270	De 57a, Hu 53
1395	$P^{31}(p,\gamma)S^{32}$			15	Ke 56
1398	$Na^{23}(p,\gamma)Mg^{24}$	8		0.5	Ba 56, Pr 56, St 54
1399	$O^{18}(p,\gamma)F^{19}$	9.3		<15	Bu 55
1408	$P^{31}(p,\gamma)S^{32}$			15	Ke 56, Gr 51
1415	$Ni^{60}(p,\gamma)Cu^{61}$	≤6.20	0.35 eV-b	<1	Bu 57
1419	$Na^{23}(p,\gamma)Mg^{24}$	9		≤0.3	Ba 56, Pr 56, St 54
1422	$Ge^{74}(p,\gamma)As^{75}$			<2.5	Ch 57
1424	$Ni^{58}(p,\gamma)Cu^{59}$	4.82, 4.33	1.7 eV-b	≤0.05	Bo 58, Bu 57
1425	$Mg^{26}(p,\gamma)Al^{27}$				Ru 54
1431 ?	$F^{19}(p,\gamma)Ne^{20}$	12.60, 1.63	0.19	15.7	Fa 55, Si 54
1431	$Ni^{60}(p,\gamma)Cu^{61}$	≤6.22	0.18 eV-b	<1	Bu 57
1433	$Ne^{22}(p,\gamma)Na^{23}$				Br 47a
1443	$P^{31}(p,\gamma)S^{32}$			12	Ke 56, Gr 51
1451	$Ni^{60}(p,\gamma)Cu^{61}$	6.24	0.75 eV-b	<1	Bu 57
1461	$Ni^{60}(p,\gamma)Cu^{61}$	≤6.25	0.14 eV-b	<1	Bu 57
1465	$Mg^{26}(p,\gamma)Al^{27}$				Ru 54
1465	$Ni^{60}(p,\gamma)Cu^{61}$	≤6.25	0.11 eV-b	<1	Bu 57
1470	$C^{13}(p,\gamma)N^{14}$	5.83, 5.10, 3.07	0.074	20	Mi 54, Cl 53, Wo 53
1482	$P^{31}(p,\gamma)S^{32}$			6	Ke 53, Gr 51
1483	$Ni^{60}(p,\gamma)Cu^{61}$	<6.27	0.14 eV-b	<1	Bu 57
1484	$Cl^{35}(p,\gamma)Ar^{36}$	9.9		≤5	To 57, Br 51
1490	$Mg^{24}(p,\gamma)Al^{25}$	3.72, 1.91		0.3	Li 56, Hu 55, Ko 52
1491	$Ni^{60}(p,\gamma)Cu^{61}$	≤6.28	0.14 eV-b	<1	Bu 57
1492	$Ne^{22}(p,\gamma)Na^{23}$				Br 47a
1500	$C^{14}(p,\gamma)N^{15}$	11.61		520	Fe 56, Ba 55
1500	$Al^{27}(p,\gamma)Si^{28}$				Pl 40
1502	$Ne^{22}(p,\gamma)Na^{23}$				Br 47a
1510	$Cl^{35}(p,\gamma)Ar^{36}$	9.9		≤5	Be 57, To 57, Br 51
1515	$Ni^{60}(p,\gamma)Cu^{61}$	≤6.30	0.4 eV-b	<1	Bu 57
1519	$Ni^{60}(p,\gamma)Cu^{61}$	≤6.30	0.7 eV-b	<1	Bu 57
1520 ?	Si(p,γ)P			9.0	Ts 56, Cl 55
1522	$Ni^{58}(p,\gamma)Cu^{59}$	≤4.92	0.012 eV-b	<1	Bu 57
1527	$P^{31}(p,\gamma)S^{32}$			14	Ke 56, Gr 51

Table A12.1. Known (p,γ) resonances vs. bombarding energy (Butler, 1959) (continued).

Proton energy *(keV)*	Reaction	Gamma-ray energy *(MeV)*	Cross section *(mb)*	Width *(keV)*	References (author and year)
1529	$Ni^{60}(p,\gamma)Cu^{61}$	≤6.31	0.06 eV-b	<1	Bu 57
1530	$Ge^{74}(p,\gamma)As^{75}$			9.0	Ch 57
1533	$Cl^{37}(p,\gamma)Ar^{38}$	9.5		≤5	Be 57, To 57, Br 51
1538	$Ni^{60}(p,\gamma)Cu^{61}$	6.32	0.35 eV-b	<1	Bu 57
1540	$Ni^{58}(p,\gamma)Cu^{59}$	<4.93	0.020 eV-b	<1	Bu 57
1544	$N^{14}(p,\gamma)O^{15}$	8.8 ?		34	Fe 58, Ha 57, Du 51
1550	$C^{13}(p,\gamma)N^{14}$	8.99	0.037	7	Mi 54, Wo 53, Se 52
1559	$Ge^{74}(p,\gamma)As^{75}$			6.5	Ch 57
1566	$K^{39}(p,\gamma)Ca^{40}$	9.9, 6.6, 6.1		≤5	Be 57, To 57
1566	$Ni^{60}(p,\gamma)Cu^{61}$	≤6.35	0.22 eV-b	<1	Bu 57
1570	$Al^{27}(p,\gamma)Si^{28}$				Pl 40
1571	$P^{31}(p,\gamma)S^{32}$			7	Ke 56, Gr 51
1577	$Ni^{60}(p,\gamma)Cu^{61}$	≤6.36	0.35 eV-b	<1	Bu 57
1580	$Cl^{35}(p,\gamma)Ar^{36}$	10		≤5	To 57
1582	$Ni^{58}(p,\gamma)Cu^{59}$	≤4.98	0.066 eV-b	<1	Bu 57
1588	$Ni^{60}(p,\gamma)Cu^{61}$	6.37, 5.90	0.9 eV-b	<1	Bu 57
1598	$P^{31}(p,\gamma)S^{32}$			5	Ke 56, Gr 51
1599	$Ni^{60}(p,\gamma)Cu^{61}$	6.38, 5.00	2.3 eV-b	<1	Bu 57
1605	$Ni^{60}(p,\gamma)Cu^{61}$	6.39, 5.01	2.0 eV-b	<1	Bu 57
1607	$F^{19}(p,\alpha\gamma)O^{16}$			6.0	Hu 55, Hu 55a
1610 ?	$S^{34}(p,\gamma)Cl^{35}$				Ha 51
1618 ?	Si(p,γ)P				Ts 56
1620	$Mg^{24}(p,\gamma)Al^{25}$	3.40, 2.90, 1.34		36	Li 56, Hu 55, Ko 52
1620	$Ni^{60}(p,\gamma)Cu^{61}$	6.40, 5.02	1.8 eV-b	<1	Bu 57
1635 ?	Si(p,γ)P				Ts 56
1635	Cl(p,γ)Ar				Br 51
1639	$Ni^{60}(p,\gamma)Cu^{61}$	≤6.42	0.14 eV-b	<1	Bu 57
1640	$N^{15}(p,\alpha\gamma)C^{12}$	4.43	340	68	Ba 57, Ha 57b, Kr 54a
1640	$Al^{27}(p,\gamma)Si^{28}$				Pl 40
1643	$Ni^{60}(p,\gamma)Cu^{61}$	≤6.43	0.35 eV-b	<1	Bu 57
1643	$Ge^{74}(p,\gamma)As^{75}$			~15	Ch 57
1645	Cl(p,γ)Ar				Br 51
1649	$Ni^{60}(p,\gamma)Cu^{61}$	≤6.43	0.29 eV-b	<1	Bu 57
1650	$Si^{28}(p,\gamma)P^{29}$	4.30		50	Ne 57, Vo 57, Ts 56
1653	$Ni^{58}(p,\gamma)Cu^{59}$	≤5.05	0.045 eV-b	<1	Bu 57
1656	$Ni^{60}(p,\gamma)Cu^{61}$	6.44, 5.97	1.0 eV-b	<1	Bu 57
1659	$Al^{27}(p,\gamma)Si^{28}$				Pl 40
1660	$Mg^{24}(p,\gamma)Al^{25}$	3.88, 3.43, 2.93		0.1	Li 56, Ko 52, Mo 51
1660	Cl(p,γ)Ar				Br 51
1663 ?	Si(p,γ)P				Ts 56
1663	$Ni^{58}(p,\gamma)Cu^{59}$	5.06, 4.15, 3.28	0.16 eV-b	<1	Bu 57
1665	$Ge^{74}(p,\gamma)As^{75}$			~15	Ch 57
1669	$Ni^{60}(p,\gamma)Cu^{61}$	≤6.45	0.4 eV-b	<1	Bu 57
1670	Cl(p,γ)Ar				Br 51
1674	$Ni^{60}(p,\gamma)Cu^{61}$	5.50, 5.08	1.0 eV-b	<1	Bu 57
1679	$Ni^{60}(p,\gamma)Cu^{61}$	≤6.46	0.5 eV-b	<1	Bu 57

Table A12.1. Known (p,γ) resonances vs. bombarding energy (Butler, 1959) (continued).

Proton energy *(keV)*	Reaction	Gamma-ray energy *(MeV)*	Cross section *(mb)*	Width *(keV)*	References (author and year)
1680 ?	Si(p,γ)P				Ts 56
1680	Cl(p,γ)Ar				Br 51
1685	O^{18}(p,γ)F^{19}	9.6		15	Bu 55
1690	S^{34}(p,γ)Cl^{35}				Ha 51
1690	Ge^{74}(p,γ)As^{75}			~30	Ch 57
1691	F^{19}(p,αγ)O^{16}	7.12, 6.92, 6.13		35	Hu 55, Hu 55a, Wi 52
1694	Ni^{60}(p,γ)Cu^{61}	6.48, 5.52	1.0 eV-b	<1	Bu 57
1698	C^{12}(p,γ)N^{13}	3.51, 2.37, 1.14	0.035	70	Hu 53b, Se 51
1698	Ni^{60}(p,γ)Cu^{61}	≤6.48	0.3 eV-b	<1	Bu 57
1699 ?	Si(p,γ)P				Ts 56
1700	Al^{27}(p,γ)Si^{28}				Pl 40
1710	Cl(p,γ)Ar				Br 51
1711	Ni^{60}(p,γ)Cu^{61}	≤6.49	0.23 eV-b	<1	Bu 57
1716	Ni^{58}(p,γ)Cu^{59}	5.11, 4.20	0.35 eV-b	<1	Bu 57
1721	Ni^{60}(p,γ)Cu^{61}	≤6.50	0.11 eV-b	<1	Bu 57
1725	Cl^{37}(p,γ)Ar^{38}	5.2, 3		≤5	To 57, Br 51
1726	Al^{27}(p,γ)Si^{28}				Pl 40
1734	Ni^{60}(p,γ)Cu^{61}	6.52	0.7 eV-b	<1	Bu 57
1739	Ni^{60}(p,γ)Cu^{61}	≤6.52	0.3 eV-b	<1	Bu 57
1742	N^{14}(p,γ)O^{15}	9.0 ?		5	Ha 57, Du 51
1748	C^{13}(p,γ)N^{14}	9.17, 6.43, 2.74	340	0.075	Bo 58, Ma 56, Wo 53
1755	Cl(p,γ)Ar				Br 51
1757	Ni^{60}(p,γ)Cu^{61}	≤6.54	0.5 eV-b	<1	Bu 57
1764	Ni^{60}(p,γ)Cu^{61}	≤6.55	0.6 eV-b	<1	Bu 57
1765	Cl(p,γ)Ar				Br 51
1769	O^{18}(p,γ)F^{19}	9.6		4	Bu 55
1770	Ni^{60}(p,γ)Cu^{61}	≤6.55	0.75 eV-b	<1	Bu 57
1774 ?	Si(p,γ)P				Ts 56
1781	Al^{27}(p,γ)Si^{28}				Pl 40
1783	Ni^{60}(p,γ)Cu^{61}	≤6.56	0.55 eV-b	<1	Bu 57
1797	Ni^{60}(p,γ)Cu^{61}	≤6.57	0.45 eV-b	<1	Bu 57
1800 ?	S^{34}(p,γ)Cl^{35}				Ha 51
1805	Ge^{74}(p,γ)As^{75}			20	Ch 57
1810 ?	Si(p,γ)P				Ts 56
1807	N^{14}(p,γ)O^{15}	9.0 ?		5	Ha 57, Du 51
1833	Mg^{24}(p,γ)Al^{25}	4.05, 2.43, 1.62			Li 59
1833	Ni^{58}(p,γ)Cu^{59}	≤5.22	0.063 eV-b	<1	Bu 57
1844	Ni^{58}(p,γ)Cu^{59}	5.23	2.1 eV-b	≤0.1	Bo 58, Bu 57
1849 ?	Si(p,γ)P				Ts 56
1860 ?	S^{34}(p,γ)Cl^{35}				Ha 51
1870	Ca^{40}(p,γ)Sc^{41}				Bu 58
1879 ?	Si(p,γ)P				Ts 56
1890	Al^{27}(p,γ)Si^{28}				Pl 40
1892	P^{31}(p,γ)S^{32}	10.68		24	Ke 56, Cl 55, Pa 55
1906	Ge^{74}(p,γ)As^{75}			~15	Ch 57
1916	P^{31}(p,γ)S^{32}				Ke 56

Table A12.1. Known (p,γ) resonances vs. bombarding energy (Butler, 1959) (continued).

Proton energy *(keV)*	Reaction	Gamma-ray energy *(MeV)*	Cross section *(mb)*	Width *(keV)*	References (author and year)
1926	$Ge^{74}(p,\gamma)As^{75}$			~15	Ch 57
1931	$O^{18}(p,\gamma)F^{19}$	9.8		1.5	Bu 55
1940	$Al^{27}(p,\gamma)Si^{28}$				Pl 40
1945	$F^{19}(p,\alpha\gamma)O^{16}$	6-7		40	Hu 55, Hu 55a, Wi 52
1972	$Ge^{74}(p,\gamma)As^{75}$			35	Ch 57
1979	$N^{15}(p,\alpha\gamma)C^{12}$	4.43	35	23	Ba 57, Ha 57b
1985	$P^{31}(p,\gamma)S^{32}$	10.77			Ke 56, Pa 55
2000 ?	$Li^{7}(p,\gamma)Be^{8}$	19.0 ?, 16.1 ?			Pr 54
2000	$C^{13}(p,\gamma)N^{14}$	5.10, 4.80		~20	Zi 58, Zi 57
2010	$Mg^{24}(p,\gamma)Al^{25}$	3.77, 3.27		0.15	Li 56, Ko 52, Mo 51
2025	$C^{14}(p,\gamma)N^{15}$			18	Sa 56d, Ba 55
2026	$F^{19}(p,\alpha\gamma)O^{16}$	6-7		120	Hu 55, Hu 55a, Wi 52
2026	$Al^{27}(p,\gamma)Si^{28}$				Pl 40
2027	$P^{31}(p,\gamma)S^{32}$	10.81			Pa 55, Cl 55
2074	$Ge^{74}(p,\gamma)As^{75}$			13.5	Ch 57
2079	$C^{14}(p,\gamma)N^{15}$			55	Sa 56, Ba 55, Ro 51
2083	$Al^{27}(p,\gamma)Si^{28}$				Pl 40
2090	$Si^{28}(p,\gamma)P^{29}$	4.74		12	Ne 57, Vo 57
2120	$C^{13}(p,\gamma)N^{14}$	5.10, 4.39	0.20	45	Zi 58, Zi 57, Wo 53
2120	$P^{31}(p,\gamma)S^{32}$	10.90		5	Pa 55
2130	$Li^{7}(p,\gamma)Be^{8}$	19.12, 16.21		400	Ne 57a, Sw 55
2135	$Ne^{20}(p,\gamma)Na^{21}$				Ha 55b, Co 54, Va 53
2161	$Ge^{74}(p,\gamma)As^{75}$			~15	Ch 57
2180	$Al^{27}(p,\gamma)Si^{28}$				Pl 40
2200	$Al^{27}(p,\gamma)Si^{28}$				Pl 40
2210	$Ge^{74}(p,\gamma)As^{75}$			40	Ch 57
2212	$Al^{27}(p,\gamma)Si^{28}$				Pl 40
2282	$Al^{27}(p,\gamma)Si^{28}$				Pl 40
2295	$Ge^{74}(p,\gamma)As^{75}$			27	Ch 57
2315	$F^{19}(p,\alpha\gamma)O^{16}$	6-7		85	Wi 52
2320	$P^{31}(p,\gamma)S^{32}$	11.09		8	Pa 55
2340	$P^{31}(p,\gamma)S^{32}$	11.11		8	Pa 55
2342	$Ge^{74}(p,\gamma)As^{75}$			15	Ch 57
2344	$Al^{27}(p,\gamma)Si^{28}$				Pl 40
2350	$N^{14}(p,\gamma)O^{15}$	9.5 ?		14	Bo 57, Fe 56a, Du 51
2400	$Mg^{24}(p,\gamma)Al^{25}$	3.65		0.3	Li 56, Ko 52, Mo 51
2440	$Ge^{74}(p,\gamma)As^{75}$			11	Ch 57
2480	$N^{14}(p,\gamma)O^{15}$	9.7 ?		11	Bo 57, Fe 56a, Du 51
2510	$F^{19}(p,\alpha\gamma)O^{16}$	6-7		30	Wi 52
2520 ?	$Si(p,\gamma)P$				Ts 56
2528	$Ge^{74}(p,\gamma)As^{75}$			15	Ch 57
2542	$Al^{27}(p,\gamma)Si^{28}$				Pl 40
2543 ?	$Si(p,\gamma)P$				Ts 56
2553 ?	$Si(p,\gamma)P$				Ts 56
2558 ?	$Si(p,\gamma)P$				Ts 56
2564	$Be^{9}(p,\gamma)B^{10}$	8.1, 0.7		39	Ma 54a, Ma 53

Table A12.1. Known (p,γ) resonances vs. bombarding energy (Butler, 1959) (continued).

Proton energy *(keV)*	Reaction	Gamma-ray energy *(MeV)*	Cross section *(mb)*	Width *(keV)*	References (author and year)
2564	$Be^9(p,\alpha\gamma)Li^6$	3.56		39	Hu 55, Ha 52
2570 ?	Si(p,γ)P				Ts 56
2575	$N^{14}(p,\gamma)O^{15}$	9.8 ?		1000	Fe 58, Bo 57, Du 51
2575 ?	Si(p,γ)P				Ts 56
2593	$Ge^{74}(p,\gamma)As^{75}$			44	Ch 57
2630	$B^{11}(p,\gamma)C^{12}$	13.94, 4.43, 2.14		300	Ba 55, Ho 55, Ba 54
2630	$F^{19}(p,\alpha\gamma)O^{16}$	6-7		90	Wi 52
2664	$Ge^{74}(p,\gamma)As^{75}$			10	Ch 57
2800	$F^{19}(p,\alpha\gamma)O^{16}$	6-7		60	Wi 52
3000	$N^{15}(p,\alpha\gamma)C^{12}$	4.43	750	45	Ba 57

Table A12.2. (p,γ) and (p,αγ) resonances by element.

Element	Reaction	E_p *(keV)*	σ *(mb)* or S *(eV)*	Resonance width *(keV)*	Eγ *(MeV)*	Relative intensity
Li	$^7Li(p,\gamma)^8Be$	441.4	6	10	17.65	63
					14.75	37
	$^7Li(p,\gamma)^8Be$	1030		168	18.15	40
					15.25	60
	$^7Li(p,\gamma)^8Be$	2060		310	16.15	100
Be	$^9Be(p,\gamma)^{10}B$	319	4.3 eV	120	6.15	21
					5.15	55
					4.75	11
					1.02	58
					0.72	84
	$^9Be(p,\gamma)^{10}B$	992	84 eV	72	7.5	~100
	$^9Be(p,\gamma)^{10}B$	1083	8.5 eV	2.65	6.85	85
					4.45	4.5
					3.01	9.5
					2.4	15
					0.72	96
B	$^{11}B(p,\gamma)^{12}C$	163	0.157	5.2	16.11	3.5
					11.68	96.5
					4.43	96.5
	$^{11}B(p,\gamma)^{12}C$	675	0.050	322	12.15	~100
					4.43	~100
	$^{10}B(p,\gamma)^{11}C$	1146	0.0055	450	9.7	
	$^{10}B(p,\gamma)^{11}C$	1180	0.0075	570	9.4	
	$^{11}B(p,\gamma)^{12}C$	1390	0.053	1270	17.23	66
					12.80	34
					4.43	34

Table A12.2. (p,γ) and (p,αγ) resonances by element (continued).

Element	Reaction	E_p *(keV)*	σ *(mb)* or S *(eV)*	Resonance width *(keV)*	Eγ *(MeV)*	Relative intensity
C	$^{12}C(p,\gamma)^{13}N$	457	0.127	35	2.36	100
	$^{13}C(p,\gamma)^{14}N$	551	1.44	25	8.06	80
	$^{13}C(p,\gamma)^{14}N$	1152	0.56	4.1	8.62	23
					4.67	24
					2.42	40
					2.31	62
	$^{13}C(p,\gamma)^{14}N$	1320	0.062	440	8.71	~90
	$^{13}C(p,\gamma)^{14}N$	1462	0.074	17	5.83	18
					5.10	53
					3.08	84
	$^{13}C(p,\gamma)^{14}N$	1540	0.037	9	8.98	~100
	$^{12}C(p,\gamma)^{13}N$	1689	0.035	67	3.51	95
					2.36	5
					1.14	5
	$^{13}C(p,\gamma)^{14}N$	1748	340	0.075	9.17	86
					6.45	6
					2.73	9
N	$^{14}N(p,\gamma)^{15}O$	278		1.06	6.79	23
					6.18	58
					1.38	58
	$^{15}N(p,\gamma)^{16}O$	335	0.007	110	12.44	~100
	$^{15}N(p,\alpha\gamma)^{12}C$	335	0.03	110	4.43	100
	$^{15}N(p,\alpha\gamma)^{12}C$	429	300	0.12	4.43	100
	$^{15}N(p,\gamma)^{16}O$	429	0.001	0.12	6.40	60
					6.13	60
	$^{15}N(p,\gamma)^{16}O$	710		40	6.72	
	$^{15}N(p,\alpha\gamma)^{12}C$	897	800	1.7	4.43	100
	$^{15}N(p,\alpha\gamma)^{12}C$	1028	15	140	4.43	100
	$^{15}N(p,\gamma)^{16}O$	1028	1	140	13.09	~100
	$^{14}N(p,\gamma)^{15}O$	1058	0.37	3.9	8.28	53
					5.24	42
					3.04	42
	$^{15}N(p,\alpha\gamma)^{12}C$	1210	425	22.5	4.43	100
	$^{14}N(p,\gamma)^{15}O$	1550	0.09eV	34	6.18	36
					5.18	64
	$^{15}N(p,\alpha\gamma)^{12}C$	1640	340	68	4.43	100
	$^{14}N(p,\gamma)^{15}O$	1742	0.16 eV	3.5	8.92	50
			0.06 eV	8	5.18	39
	$^{14}N(p,\gamma)^{15}O$	1806	0.52	4.2	8.98	94
	$^{15}N(p,\alpha\gamma)^{12}C$	1979	35	23	4.43	100
O	$^{18}O(p,\gamma)^{19}F$	630	0.38	2.0	8.39	42
		841		48	8.68	30
		1167		0.05	6.32	47
					2.58	47
					0.197	>50
					0.110	>35

Table A12.2. (p,γ) and (p,$\alpha\gamma$) resonances by element (continued).

Element	Reaction	E_p (keV)	σ (mb) or S (eV)	Resonance width (keV)	$E\gamma$ (MeV)		Relative intensity
O	$^{18}O(p,\gamma)^{19}F$	1398		3.6	9.32		30
		1684		8	8.24		32
		1768		3.8	9.67		22
		1928		0.3	9.62		41
					γ_1	γ_2	γ_3
					6.13	6.72	7.12
F	$^{19}F(p,\alpha\gamma)^{16}O$	224.0	0.2	1	100		
		340.5	102	2.4	96.5	0.5	3
		483.6	32	0.9	79	1	20
		594	7	30	100	<0.3	<0.3
		668	57	6	81	0.3	19
		832	19	6.5			
		872.1	661	4.5	68	24	8
		902	23	5.1	>90	<5	<5
		935	180	8.6	76	3	21
		1088	13	0.7			
		1136	15	2.5			
		1189	19	110			
		1283	29	19	74	8	18
		1348	89	5.6	55	14	31
		1371	300	11	87	8	5
		1692		35			
Ne	$^{20}Ne(p,\gamma)^{21}Na$	1169	1.6 eV	0.016	3.544		92
					1.828		6
					0.332		8
	$^{22}Ne(p,\gamma)^{23}Na$	640	4.9 eV	0.065	9.406		73
					0.440		12
		851	12 eV	0.006	9.606		27
					9.166		42
					5.693		21
					0.440		>47
Na	$^{23}Na(p,\gamma)^{24}Mg$	309	1.5 eV	<0.036	10.618		28
					7.748		46
					4.239		35
					1.369		>60
	$^{23}Na(p,\gamma)^{24}Mg$	512	1.1 eV	<0.046	10.813		71
					7.943		9
					1.369		>80
		677	8 eV	<0.070	10.970		11
					8.100		45
					7.104		21
					4.239		35
					1.369		>52

Table A12.2. (p,γ) and (p,αγ) resonances by element (continued).

Element	Reaction	E_p *(keV)*	σ *(mb)* or S *(eV)*	Resonance width *(keV)*	Eγ *(MeV)*	Relative intensity
	$^{23}Na(p,\alpha\gamma)^{20}Ne$	1011	55 eV	<0.1	1.634	100
		1164	160 eV	1.2	1.634	100
Mg	$^{24}Mg(p,\gamma)^{25}Al$	419	0.066 eV	<0.044	2.674	26
					2.222	43
					0.452	54
	$^{26}Mg(p,\gamma)^{27}Al$	454	0.9 eV	<0.081	7.864	48
					7.695	16
					0.843	>62
		823	0.98 eV	1.3	2.610	77
					0.452	83
	$^{26}Mg(p,\gamma)^{27}Al$	1548	3 eV	0.020	9.761	20
					8.748	30
					7.552	27
					1.013	>40
Al	$^{27}Al(p,\gamma)^{28}Si$	406	0.15 eV	<0.042	7.360	72
					2.835	>72
					1.779	>80
		632	5.3 eV	0.016	10.416	74
					7.581	15
					1.779	>94
		992	31 eV	0.10	10.763	76
					4.744	10
					1.779	>93
Si	$^{30}Si(p,\gamma)^{31}P$	620	3.9 eV	0.068	7.897	93
P	$^{31}P(p,\gamma)^{32}S$	541	1 eV	<<1	7.159	80
					2.230	~100
		811	1.0 eV	<0.42	7.420	57
					4.956	38
					2.230	>84
	$^{31}P(p,\gamma)^{32}S$	1251	11 eV	1.4	7.846	32
					3.852	42
					2.230	>90
S	$^{34}S(p,\gamma)^{35}Cl$	929	1 eV	0.014	7.274	72
					6.055	20
					1.219	24
		1212	9.7 eV	0.011	4.385	97
					3.163	87
Cl	$^{37}Cl(p,\gamma)^{38}Ar$	766	4.3 eV	<1	8.820	44
					2.168	>60

Table A12.2. (p,γ) and (p,αγ) resonances by element (continued).

Element	Reaction	E_p *(keV)*	σ *(mb)* or S *(eV)*	Resonance width *(keV)*	Eγ *(MeV)*	Relative intensity
Ar	$^{40}Ar(p,\gamma)^{41}K$	1086			8.86	
Ca	$^{40}Ca(p,\gamma)^{41}Sc$	1843	0.260		2.883	
Ti	$^{48}Ti(p,\gamma)^{49}V$	1007			7.582 [a]	
Cr	$^{52}Cr(p,\gamma)^{53}Mn$	1005			2.5-4.7	[b]
Ni	$^{58}Ni(p,\gamma)^{59}Cu$	1424		<0.045	4.82	27
					4.33	50

a. This gamma-ray is specific for this resonance. A resonance at 1013 keV decays with Eγ=7.650 MeV. (C.R. Gossett, *Nucl. Instr. and Meth.* **168**, 217 (1980)).

b. A minor amount of yield from other nearby resonances. (C.R. Gossett, *Nucl. Instr. and Meth.* **168**, 217 (1980)).

This table is based on the similar table for Z=3-9 by Golicheff *et al.* (1972), *J. Radioanal. Chem.* **12**, 233. New data and some corrections are based on the following nuclear data compilations and references therein.

A=5-10 F. Ajzenberg-Selove. (1988), *Nucl. Phys.* **A490**,1.
A=11-12 F. Ajzenberg-Selove. (1990), *Nucl. Phys.* **A506**,1.
A=13-15 F. Ajzenberg-Selove. (1991), *Nucl. Phys.* **A523**,1.
A=16-17 F. Ajzenberg-Selove. (1986), *Nucl. Phys.* **A460**,1.
A=18-20 F. Ajzenberg-Selove. (1987), *Nucl. Phys.* **A475**,1.
A=21-44 P.M. Endt and C. van der Leun. (1978), *Nucl. Phys.* **A310**,1.

Table A12.3. Some useful (α,γ) and (α,pγ) resonances for profiling.

Element	Reaction	E_α *(keV)*	σ *(mb)* or S *(eV)*	Resonance width *(keV)*	Eγ *(MeV)*	Relative intensity
Li	$^{7}Li(\alpha,\gamma)^{11}B$	814	1.2 eV	0.0018	6.74	9
					4.75	87
					4.44	91
		953	6.9 eV	4	9.27	18
					4.83	70
					4.44	73
					2.53	12
B	$^{10}B(\alpha,p\gamma)^{13}C$	1507	6 eV	25	3.85	54
					3.68	42
		1645	8 eV	22	3.85	59
					3.68	39
N	$^{14}N(\alpha,\gamma)^{18}F$	1531	2.6 eV	0.6	4.52	54
					2.47	32
					1.08	62
		1618	1.4	< 0.8	4.59	52
					2.54	29
					1.08	59
Ne	$^{20}Ne(\alpha,\gamma)^{24}Mg$	1929	3 eV		9.56	85

See references for nuclear data compilations in Table A12.2.

Table A12.4. Standard samples for resonance reactions. For most of the metals, high purity foils or sheets are available, and evaporated films can be easily made. These metals include Be, Mg, Al, Ti, V, Cr, Mn, Fe, Co, Ni, Cu, Zn, Y, Zr, Nb, Mo, Ru, Pd, Ag, Cd, In, Sn, Sb, Hf, Ta, W, Pt, Au, and Pb. This table list includes typical standards for non-metallic light elements and metals which can not be used in elemental form.

Element	Stable isotopes	Standards
H	^{1}H 99.985 ^{2}H 0.015	hydrogenated Ta[1] polypropyle $(C_3H_6)_n$, polyester, (Mylar) $(C_{10}H_8O_4)_n$, Kapton® $(C_{22}H_{10}O_5N_2)_n$
He	^{4}He ~100	implanted sample (Ta)
Li	^{6}Li 7.5, ^{7}Li 92.5	LiF[2]
B	^{10}B 19.9, ^{11}B 80.1	B[2]
C	^{12}C 98.9, ^{13}C 1.10	high purity graphite or evaporated film
N	^{14}N 99.63, ^{15}N 0.37	TiN, TaN, NbN[3], N_2O, NO[4], Si_3N_4 ,AlN[5]
O	^{16}O 99.76, ^{17}O 0.0374 ^{18}O 0.2039	Al_2O_3, MgO, SiO_2[5], Ta_2O_5[6]
F	^{19}F 100	CaF_2, LiF[2]
Ne	^{20}Ne 90.48, ^{21}Ne 0.27 ^{22}Ne 9.25	implanted isotopic sample (Ta)
Na	^{23}Na 100	fresh cleaved NaCl (hygroscopic !)
Si	^{28}Si 92.2, ^{29}Si 4.67 ^{30}Si 3.10	Si (channeling possible at certain angles !)
P	^{31}P 100	InP[2]
S	^{32}S 95.02, ^{33}S 0.75 ^{34}S 4.21, ^{36}S 0.02	ZnS, PbS[2]
Cl	^{35}Cl 75.77, ^{37}Cl 24.23	fresh cleaved NaCl, $PbCl_2$ (powder pellet)
Ar	^{36}Ar 0.34, ^{38}Ar 0.06 ^{40}Ar 99.60	implanted isotopic sample (Ta)
K	^{39}K 93.26, ^{40}K 0.012 ^{41}K 6.73	KI, KCl[2]
Ca	^{40}Ca 96.94, ^{42}Ca 0.65 ^{43}Ca 0.135, ^{44}Ca 2.09	CaO, $CaCO_3$[2]

1. B. Hjörvarsson, J. Ryden, T. Ericsson and E. Karlsson. (1989), *Nucl. Instr. Meth.* **B42**, 257.
2. Powder pellet or evaporated film.
3. Powder pellet or deposited film (CVD, sputtering,thermal nitriding, etc.).
4. Frozen gas (J.A. Davies *et al.* (1983), *Nucl. Instr. Meth.* **218**, 141.
5. Sintered ceramic or powder pellet.
6. Films by anodic or thermal oxidation (^{16}O, ^{17}O and ^{18}O) (G. Amsel *et al.* (1978), *Nucl. Instr. Meth.* **149**, 713.

Table A12.5a. Summary of sensitivities (atomic fraction) for the profiling of light elements with resonances leading to gamma-ray emission.

Greater than 1 %	**Between 1 and 0.1 %**	**Better than 0.1 %**
O	Be, B, C, N, ^{18}O, Ne, Na, Mg, P, S, Cl, Ar, Ti, Cr, Ni	^{13}C, ^{15}N, F, ^{22}Ne, Al

Table A12.5b. Summary of sensitivities (atomic fraction) obtainable with 5 MeV $^4He^+$ beam (Giles and Peisach, 1979).

Greater than 1 %	**Between 1 and 0.1 %**	**Better than 0.1 %**
Sc, Cr, Cu, Zn, Rb, Zr, Er, Hf, Hg	O, Mg, Si, Cl, K, Ti, Fe, Br, Mo, Ru, Pd, Ag, Cd, Ta, W, Re, Ir, Pt, Au	Li, B, N, F, Na, Al, P, V, Mn, Rh

Table A12.5c. Summary of sensitivities (atomic fraction) obtainable with protons (E_p < 9 MeV) (based on the data given in Tables A12.6a-A12.6d).

Greater than 1 %	**Between 1 and 0.1 %**	**Better than 0.1 %**
Pd, Sm, Gd, Hf, W, Au, Pb	S, K, Sc, Ti, Cr, Co, Cu, Ge, Y, Zr, Mo, Ru, Ag, Sn, I, Ta, Pt	Li, Be, B, C, N, O, F, Na, Mg, Al, Si, P, Cl, Ca, V, Mn, Fe, Ni, Zn, Nb, Cd, In, Sb

Table A12.6a. Absolute thick target gamma-ray yields for light elements (Z < 21) with 1.0-4.2 MeV protons. The yield values are absolute yields for elemental samples at the angle of 55° with respect to the beam direction and per solid angle (sr) and total accumulated charge (μC) (Anttila *et al.*, 1981; Kiss *et al.*,1984).

		Yield ***(N_γ/μC sr)***						
Element	**E_γ *(keV)***	**1.0 MeV**	**1.7 MeV**	**2.4 MeV**	**3.1 MeV**	**3.8 MeV**	**4.2 MeV**	**Reaction**
Li	429	-	-	-	$9.2 \cdot 10^6$	$2.6 \cdot 10^7$	$3.3 \cdot 10^7$	$^7Li(p, n_1\gamma)^7Be$
	478	$6.5 \cdot 10^5$	$8.6 \cdot 10^6$	$2.6 \cdot 10^7$	$5.6 \cdot 10^7$	$8.9 \cdot 10^7$	$1.1 \cdot 10^8$	$^7Li(p, p_1\gamma)^7Li$
Be	415	$0.4 \cdot 10^3$	$0.6 \cdot 10^3$	$0.6 \cdot 10^3$	-	-	-	$^9Be(p, \gamma_{3\text{-}2})^{10}B$
	718	$1.2 \cdot 10^3$	$3.5 \cdot 10^3$	$5.3 \cdot 10^3$	-	-	-	$^9Be\ (p, \gamma_1)^{10}B$
	1022	$0.4 \cdot 10^3$	$0.8 \cdot 10^3$	$1.3 \cdot 10^3$	-	-	-	$^9Be(p, \gamma_{2\text{-}1})^{10}B$
	3562	-	$0.1 \cdot 10^3$	$2.5 \cdot 10^4$	$2.5 \cdot 10^6$	$5.1 \cdot 10^6$	$6.2 \cdot 10^6$	$^9Be(p, \alpha_1\gamma)^6Li$
	7477	$3.3 \cdot 10^3$	$3.0 \cdot 10^3$	-	-	-	-	$^9Be\ (p,\gamma)^{10}B$
B	429	$1.4 \cdot 10^4$	$9.1 \cdot 10^5$	$3.5 \cdot 10^6$	$7.2 \cdot 10^6$	$8.3 \cdot 10^6$	$8.3 \cdot 10^6$	$^{10}B(p, \alpha_1\gamma)^7Be$
	718	-	$0.4 \cdot 10^4$	$1.2 \cdot 10^5$	$1.3 \cdot 10^6$	$2.6 \cdot 10^6$	$4.7 \cdot 10^6$	$^{10}B(p, p_1\gamma)^{10}B$
	2125	-	-	-	$4.8 \cdot 10^6$	$1.3 \cdot 10^7$	$1.5 \cdot 10^7$	$^{11}B(p, p_1\gamma)^{11}B$
C	1635	6.0	9.0	$2.6 \cdot 10$	$3.5 \cdot 10$	-	-	$^{13}C(p, \gamma_{2\text{-}1})^{14}N$
	2313	$1.7 \cdot 10$	$2.6 \cdot 10$	$3.0 \cdot 10$	$1.4 \cdot 10^2$	-	-	$^{13}C(p, \gamma_1)^{14}N$
	2366	$2.6 \cdot 10^2$	$2.2 \cdot 10^2$	$2.3 \cdot 10^2$	$2.7 \cdot 10^2$	-	-	$^{12}C(p, \gamma_1)^{13}N$
	3089	-	-	-	-	$6.2 \cdot 10^3$	$4.1 \cdot 10^4$	$^{13}C(p, p_1\gamma)^{13}C$
	3511	-	$2.3 \cdot 10^2$	$2.1 \cdot 10^2$	-	-	-	$^{12}C(p, \gamma_2)^{13}N$
	8062	$2.5 \cdot 10$	$2.5 \cdot 10$	-	-	-	-	$^{13}C(p,\gamma)^{14}N$
N	2313	-	-	-	-	$5.5 \cdot 10^4$		$^{14}N(p, p_1\gamma)^{14}N$
	4439	$1.2 \cdot 10^2$	$7.7 \cdot 10^2$	$5.0 \cdot 10^3$	$5.0 \cdot 10^4$	$6.0 \cdot 10^4$		$^{15}N(p, \alpha_1\gamma)^{12}C$
	6793	-	-	$4.0 \cdot 10^2$	-	-		$^{14}N(p,\gamma)^{15}O$

Table A12.6a. Absolute thick target gamma-ray yields for light elements (Z < 21) with 1.0-4.2 MeV protons. The yield values are absolute yields for elemental samples at the angle of 55° with respect to the beam direction and per solid angle (sr) and total accumulated charge (μC) (Anttila *et al.*, 1981; Kiss *et al.*,1984) (continued).

Element	E_γ *(keV)*	Yield $(N_\gamma/\mu C\ sr)$						Reaction
		1.0 MeV	1.7 MeV	2.4 MeV	3.1 MeV	3.8 MeV	4.2 MeV	
O	495	$2.5 \cdot 10$	$2.5 \cdot 10^2$	$9.5 \cdot 10^2$	$7.3 \cdot 10^2$	$1.1 \cdot 10^3$	$2.2 \cdot 10^3$	$^{16}O(p, \gamma_1)^{17}F$
	871	-	-	$3.5 \cdot 10$	$1.3 \cdot 10^3$	$3.8 \cdot 10^3$	$7.1 \cdot 10^3$	$^{17}O(p, p_1\gamma)^{17}O$
	1982	-	-	-	$2.1 \cdot 10^3$	$1.5 \cdot 10^4$	$2.7 \cdot 10^4$	$^{18}O(p, p_1\gamma)^{18}O$
F	110	$0.2 \cdot 10^5$	$1.2 \cdot 10^5$	$3.5 \cdot 10^5$	$7.2 \cdot 10^6$	$1.1 \cdot 10^7$		$^{19}F(p, p_1\gamma)^{19}F$
	197	$0.4 \cdot 10^5$	$2.3 \cdot 10^5$	$2.9 \cdot 10^6$	$2.0 \cdot 10^7$	$3.7 \cdot 10^7$		$^{19}F(p, p_2\gamma)^{19}F$
	1236	-	-	$1.5 \cdot 10^5$	$3.0 \cdot 10^6$	$5.4 \cdot 10^6$		$^{19}F(p, p\gamma_{3\text{-}1})^{19}F$
	1349	-	-	-	$1.3 \cdot 10^6$	$2.1 \cdot 10^6$		$^{19}F(p, p\gamma_{4\text{-}1})^{19}F$
	1357	-	-	$1.0 \cdot 10^5$	$1.4 \cdot 10^6$	$4.2 \cdot 10^6$		$^{19}F(p, p\gamma_{5\text{-}2})^{19}F$
	1459	-	-	-	$9.4 \cdot 10^5$	$3.9 \cdot 10^6$		$^{19}F(p, p_4\gamma)^{19}F$
	6129	$5.3 \cdot 10^5$	$2.6 \cdot 10^6$	$6.0 \cdot 10^6$	$6.7 \cdot 10^7$	$9.5 \cdot 10^7$		$^{19}F(p, p_2\gamma)^{16}O$
	6917	-	-	$1.8 \cdot 10^4$	$2.3 \cdot 10^5$	$2.7 \cdot 10^5$		$^{19}F(p, \alpha_3\gamma)^{16}O$
	7114	-	-	$5.4 \cdot 10^3$	$4.4 \cdot 10^4$	$4.6 \cdot 10^4$		$^{19}F(p, \alpha_4\gamma)^{16}O$
Na	440	$2.2 \cdot 10^3$	$8.3 \cdot 10^5$	$3.4 \cdot 10^6$	$9.6 \cdot 10^6$	$1.6 \cdot 10^7$	$1.5 \cdot 10^7$	$^{23}Na(p, p_1\gamma)^{23}Na$
	1369	$0.5 \cdot 10^3$	$4.6 \cdot 10^3$	-	-	-	-	$^{23}Na(p, \gamma_1)^{24}Mg$
	1634							$^{23}Na(p, \alpha_1\gamma)^{20}Ne$
	and	$0.4 \cdot 10^3$	$2.3 \cdot 10^5$	$1.5 \cdot 10^6$	$9.9 \cdot 10^6$	$1.9 \cdot 10^7$	$1.9 \cdot 10^7$	
	1636							$^{23}Na(p, p\gamma_{2\text{-}1})^{23}Na$
	1951	-	-	-	-	$1.8 \cdot 10^5$	$2.9 \cdot 10^5$	$^{23}Na(p, p\gamma_{3\text{-}1})^{23}Na$
	2391	-	-	-	-	$2.1 \cdot 10^5$	$5.0 \cdot 10^5$	$^{23}Na(p, p_3\gamma)^{23}Na$
	4238	$0.3 \cdot 10^3$	$1.2 \cdot 10^3$	-	-	-	-	$^{23}Na(p, \gamma_2)^{24}Mg$
Mg	390	-	-	$2.4 \cdot 10^4$	$6.2 \cdot 10^4$	$2.5 \cdot 10^5$	$2.8 \cdot 10^5$	$^{25}Mg(p, p\gamma_{2\text{-}1})^{25}Mg$
	417	$0.1 \cdot 10^3$	$0.3 \cdot 10^3$	-	-	-	-	$^{25}Mg(p, \gamma_1)^{26}Al$
	452	$0.2 \cdot 10^3$	$0.3 \cdot 10^3$	-	-	-	-	$^{24}Mg(p, \gamma_1)^{25}Al$
	585	-	$5.1 \cdot 10^3$	$6.7 \cdot 10^4$	$2.2 \cdot 10^5$	$8.6 \cdot 10^5$	$9.9 \cdot 10^5$	$^{25}Mg(p, p_1\gamma)^{25}Mg$
	844	$0.1 \cdot 10^3$	$0.2 \cdot 10^3$	-	-	-	-	$^{26}Mg(p, \gamma_1)^{27}Al$
	975	-	-	$2.2 \cdot 10^4$	$1.2 \cdot 10^5$	$4.6 \cdot 10^5$	$5.2 \cdot 10^5$	$^{25}Mg(p, p_2\gamma)^{25}Mg$
	990	-	-	-	-	$6.2 \cdot 10^4$	$8.6 \cdot 10^4$	$^{25}Mg(p, p\gamma_{4\text{-}2})^{25}Mg$
	1014	$0.1 \cdot 10^3$	$0.3 \cdot 10^3$	-	-	-	-	$^{26}Mg(p, \gamma_2)^{27}Al$
	1369	-	-	$1.5 \cdot 10^5$	$9.3 \cdot 10^5$	-	-	$^{24}Mg(p, p_1\gamma)^{24}Mg$
	and	-	-	-	-	$5.1 \cdot 10^6$	$8.5 \cdot 10^6$	
	1380							$^{25}Mg(p, p\gamma_{4\text{-}1})^{25}Mg$
	1612	-	-	-	$7.6 \cdot 10^4$	$9.0 \cdot 10^5$	$1.2 \cdot 10^6$	$^{25}Mg(p, p_3\gamma)^{25}Mg$
	1809	-	-	-	$6.1 \cdot 10^3$	$7.9 \cdot 10^5$	$1.3 \cdot 10^6$	$^{26}Mg(p, p_1\gamma)^{26}Mg$
	1965	-	-	-	-	$4.1 \cdot 10^4$	$5.7 \cdot 10^4$	$^{25}Mg(p, p_4\gamma)^{25}Mg$
Al	171	-	-	$1.0 \cdot 10^4$	-	-	-	$^{27}Al(p, p\gamma_{2\text{-}1})^{27}Al$
	844	-	$8.4 \cdot 10^3$	$1.5 \cdot 10^5$	$2.3 \cdot 10^6$	$5.5 \cdot 10^6$	$5.1 \cdot 10^6$	$^{27}Al(p, p_1\gamma)^{27}Al$
	1014	-	$1.1 \cdot 10^3$	$3.3 \cdot 10^5$	$4.6 \cdot 10^6$	$1.1 \cdot 10^7$	$1.2 \cdot 10^7$	$^{27}Al(p, p_2\gamma)^{27}Al$
	1369	-	$2.0 \cdot 10^3$	$3.5 \cdot 10^4$	$5.7 \cdot 10^5$	$2.4 \cdot 10^6$	$3.7 \cdot 10^6$	$^{27}Al(p, \alpha_1\gamma)^{24}Mg$
	1720	-	-	-	-	-	$1.6 \cdot 10^5$	$^{27}Al(p, p\gamma_{4\text{-}2})^{27}Al$
	1779	$1.1 \cdot 10^3$	$5.0 \cdot 10^3$	$8.4 \cdot 10^3$	$1.0 \cdot 10^4$	-	-	$^{27}Al(p, p_3\gamma)^{28}Si$
	2211	-	-	-	$1.1 \cdot 10^4$	$8.1 \cdot 10^5$	$2.5 \cdot 10^6$	$^{27}Al(p, p_3\gamma)^{27}Al$
	2839	$0.2 \cdot 10^3$	$0.5 \cdot 10^3$	$0.9 \cdot 10^3$	-	-	-	$^{27}Al(p, \gamma_{2\text{-}1})^{28}Si$
	2734	-	-	-	-	-	$3.2 \cdot 10^4$	$^{27}Al(p, p_4\gamma)^{27}Al$

Table A12.6a. Absolute thick target gamma-ray yields for light elements (Z < 21) with 1.0-4.2 MeV protons. The yield values are absolute yields for elemental samples at the angle of 55° with respect to the beam direction and per solid angle (sr) and total accumulated charge (μC) (Anttila *et al.*, 1981; Kiss *et al.*,1984) (continued).

Element	E_γ *(keV)*	Yield *($N_\gamma/\mu C$ sr)*						Reaction
		1.0 MeV	**1.7 MeV**	**2.4 MeV**	**3.1 MeV**	**3.8 MeV**	**4.2 MeV**	
	2981	-	-	-	-	-	$3.6 \cdot 10^{4}$	$^{27}Al(p, p_5\gamma)^{27}Al$
	3004	-	-	-	-	-	$7.7 \cdot 10^{4}$	$^{27}Al(p, p_6\gamma)^{27}Al$
Si	677	4.0	$2.1 \cdot 10$	$3.1 \cdot 10$	-	-		$^{29}Si(p, \gamma_1)^{30}P$
	709	3.0	$1.9 \cdot 10$	$4.4 \cdot 10$	-	-		$^{29}Si(p, \gamma_2)^{30}P$
	755	-	-	-	$1.0 \cdot 10^{5}$	$6.3 \cdot 10^{5}$		$^{29}Si(p, p\gamma_{2\text{-}1})^{29}Si$
	1266	$1.2 \cdot 10$	$9.6 \cdot 10$	$1.9 \cdot 10^{2}$	-	-		$^{30}Si(p, \gamma_1)^{31}P$
	1273	-	-	$1.8 \cdot 10^{3}$	$1.2 \cdot 10^{5}$	$7.6 \cdot 10^{5}$		$^{29}Si(p, p_1\gamma)^{29}Si$
	1384	-	$5.2 \cdot 10$	$9.1 \cdot 10$	-	-		$^{28}Si(p, \gamma_1)^{29}P$
	1779	-	-	$2.3 \cdot 10^{2}$	$1.2 \cdot 10^{6}$	$7.2 \cdot 10^{6}$		$^{28}Si(p, p_1\gamma)^{28}Si$
	2028	-	-	-	$2.8 \cdot 10^{3}$	$5.3 \cdot 10^{4}$		$^{29}Si(p, p_2\gamma)^{29}Si$
	2233							$^{30}Si(p, \gamma_2)^{31}P$
	and	-	-	$1.3 \cdot 10^{2}$ [b]	$4.7 \cdot 10^{3}$	$3.4 \cdot 10^{4}$		
	2235							$^{30}Si(p, p_1\gamma)^{30}Si$
	4343	-	$1.6 \cdot 10^{2}$	$1.8 \cdot 10^{2}$	-	-		$^{28}Si(p,\gamma)^{29}P$
P	1266	-	$7.2 \cdot 10^{2}$	$3.8 \cdot 10^{4}$	$1.6 \cdot 10^{6}$	$5.2 \cdot 10^{6}$	$2.3 \cdot 10^{7}$	$^{31}P(p, p_1\gamma)^{31}P$
	1779	-	-	$2.0 \cdot 10^{3}$	$2.1 \cdot 10^{5}$	$6.5 \cdot 10^{5}$	$1.6 \cdot 10^{6}$	$^{31}P(p, \alpha_1\gamma)^{28}Si$
	2230							$^{31}P(p, \gamma_1)^{32}S$
	and	$1.3 \cdot 10^{2}$	$1.7 \cdot 10^{3}$	$3.5 \cdot 10^{3}$	$1.2 \cdot 10^{4}$	$4.0 \cdot 10^{5}$	$9.5 \cdot 10^{5}$	
	2233							$^{31}P(p, p_2\gamma)^{31}P$
S	811	$2.1 \cdot 10$	$3.4 \cdot 10$	$4.5 \cdot 10$	$1.2 \cdot 10^{2}$	-	-	$^{32}S(p, \gamma_1)^{33}Cl$
	841	-	9.1	$1.4 \cdot 10^{2}$	$8.7 \cdot 10^{3}$	$1.7 \cdot 10^{4}$	$2.9 \cdot 10^{4}$	$^{33}S(p, p_1\gamma)^{33}S$
	1219	$1.0 \cdot 10$	$3.7 \cdot 10$	$6.5 \cdot 10$	$4.3 \cdot 10^{3}$	$4.6 \cdot 10^{4}$	$2.8 \cdot 10^{5}$	$^{34}S(p, \gamma_1)^{35}Cl$
	2035	$2.0 \cdot 10$	$2.3 \cdot 10$	-	-	-	-	$^{32}S(p, \gamma_{2\text{-}1})^{33}Cl$
	2127	-	-	-	-	$2.1 \cdot 10^{4}$	$4.8 \cdot 10^{4}$	$^{34}S(p, p_1\gamma)^{34}S$
	2230	-	-	-	$5.3 \cdot 10^{3}$	$1.5 \cdot 10^{5}$	$8.9 \cdot 10^{5}$	$^{32}S(p, p_1\gamma)^{32}S$
Cl	670	$0.1 \cdot 10^{2}$	$2.1 \cdot 10^{2}$	$5.2 \cdot 10^{2}$	$3.8 \cdot 10^{2}$	-		$^{37}Cl(p, \gamma_{3\text{-}2})^{38}Ar$
	1219	-	-	$3.5 \cdot 10^{3}$	$2.2 \cdot 10^{5}$	$1.5 \cdot 10^{6}$		$^{35}Cl(p, p_1\gamma)^{35}Cl$
	1410	-	-	-	-	$2.1 \cdot 10^{5}$		$^{37}Cl(p, n_1)^{37}Ar$
	1611	-	-	-	-	$1.9 \cdot 10^{5}$		$^{37}Cl(p, n_2\gamma)^{37}Ar$
	1642	$0.3 \cdot 10^{2}$	$4.9 \cdot 10^{2}$	$1.0 \cdot 10^{3}$	-	-		$^{37}Cl(p, \gamma_{2\text{-}1})^{38}Ar$
	1727	-	-	-	$3.8 \cdot 10^{3}$	$6.9 \cdot 10^{4}$		$^{37}Cl(p, p_1\gamma)^{37}Cl$
	1763	-	-	-	$5.2 \cdot 10^{4}$	$6.8 \cdot 10^{5}$		$^{35}Cl(p, p_2\gamma)^{35}Cl$
	1970	$2.2 \cdot 10^{2}$	$4.5 \cdot 10^{2}$	$1.1 \cdot 10^{3}$	$3.3 \cdot 10^{3}$	-		$^{35}Cl(p, \gamma_1)^{36}Ar$
	2127	-	-	$1.2 \cdot 10^{3}$	$3.1 \cdot 10^{4}$	$1.8 \cdot 10^{5}$		$^{37}Cl(p, \alpha_1\gamma)^{38}Ar$
	2168	$1.6 \cdot 10^{2}$	$1.6 \cdot 10^{3}$	$2.9 \cdot 10^{3}$	$3.9 \cdot 10^{3}$	$5.3 \cdot 10^{3}$		$^{37}Cl(p, \gamma_1)^{38}Ar$
	2208	$1.2 \cdot 10^{2}$	$2.1 \cdot 10^{2}$	$4.5 \cdot 10^{2}$	-	-		$^{35}Cl(p, \gamma_{2\text{-}1})^{36}Ar$
K	313	4.0	$5.6 \cdot 10$	-	-	-	-	$^{41}K(p, \gamma_{2\text{-}1})^{42}Ca$
	755	-	$6.1 \cdot 10$	-	-	-	-	$^{39}K(p, \gamma_{2\text{-}1})^{40}Ca$
	899	7.0	$1.1 \cdot 10$	$2.5 \cdot 10^{2}$	-	-	-	$^{41}K(p, \gamma_{3\text{-}1})^{42}Ca$
	980		-	$5.2 \cdot 10^{2}$	$1.1 \cdot 10^{4}$	$5.5 \cdot 10^{4}$	$9.3 \cdot 10^{4}$	$^{41}K(p, p_1\gamma)^{41}K$
	1294		-	-	-	$3.0 \cdot 10^{4}$	$1.2 \cdot 10^{5}$	$^{41}K(p, p_2\gamma)^{41}K$
	1525	$4.3 \cdot 10$	$5.4 \cdot 10^{2}$	$1.3 \cdot 10^{3}$	-	-	-	$^{41}K(p, \gamma_1)^{42}Ca$
	1943	-	-	-	-	$8.0 \cdot 10^{4}$	$1.4 \cdot 10^{5}$	$^{41}K(p, n_1\gamma)^{41}Ca$

Table A12.6a. Absolute thick target gamma-ray yields for light elements (Z < 21) with 1.0-4.2 MeV protons. The yield values are absolute yields for elemental samples at the angle of 55° with respect to the beam direction and per solid angle (sr) and total accumulated charge (μC) (Anttila *et al.*, 1981; Kiss *et al.*,1984) (continued).

		Yield ($N_\gamma/\mu C\ sr$)						
Element	E_γ (*keV*)	**1.0 MeV**	**1.7 MeV**	**2.4 MeV**	**3.1 MeV**	**3.8 MeV**	**4.2 MeV**	**Reaction**
	2010	-	-	-	-	$3.5 \cdot 10^4$	$1.2 \cdot 10^5$	$^{41}K(p, n_2\gamma)^{41}Ca$
	2168	-	$2.0 \cdot 10^2$	$2.6 \cdot 10^3$	$2.5 \cdot 10^4$	$9.5 \cdot 10^4$	$1.6 \cdot 10^5$	$^{41}K(p, \alpha_1\gamma)^{38}Ar$
	2522	-	-	-	-	-	$8.2 \cdot 10^4$	$^{39}K(p, n_1\gamma)^{41}Ca$
	2576	-	-	-	-	-	$1.4 \cdot 10^4$	$^{41}K(p, n_1\gamma)^{41}Ca$
	2604	-	-	-	-	-	$9.7 \cdot 10^3$	$^{41}K(p, n_1\gamma)^{41}Ca$
Ca	364	-	$4.3 \cdot 10$	$2.4 \cdot 10^2$	-	-		$^{44}Ca(p, \gamma_{2\text{-}1})^{45}Sc$
	371	-	-	-	$3.7 \cdot 10^3$	$1.6 \cdot 10^4$		$^{48}Ca(p, n\gamma_{2\text{-}1})^{48}Sc$
	373	-	$3.2 \cdot 10$	$8.7 \cdot 10^2$	-	-		$^{43}Ca(p, p_1\gamma)^{43}Ca$
	520	-	-	-	$7.3 \cdot 10^2$	$3.6 \cdot 10^3$		$^{48}Ca(p, n\gamma_{3\text{-}2})^{48}Sc$
	531	-	$1.5 \cdot 10$	$6.9 \cdot 10$	-	-		$^{44}Ca(p, p\gamma_{3\text{-}1})^{44}Ca$
	543	-	$1.2 \cdot 10$	$4.6 \cdot 10$	-	-		$^{44}Ca(p, \gamma_3)^{45}Sc$
	720	-	$2.9 \cdot 10$	$1.0 \cdot 10^2$	-	-		$^{44}Ca(p, \gamma_4)^{45}Sc$
	779	-	-	-	$1.2 \cdot 10^3$	$6.5 \cdot 10^2$		$^{48}Ca(p, n\gamma_{4\text{-}2})^{48}Sc$
	1157	-	-	-	$6.9 \cdot 10^3$	$8.6 \cdot 10^3$		$^{44}Ca(p, p_1\gamma)^{44}Ca$
	1525	-	-	-	$5.9 \cdot 10^2$	$7.7 \cdot 10^3$		$^{42}Ca(p, p_1\gamma)^{42}Ca$
Sc	364	-	$1.4 \cdot 10^3$	$3.6 \cdot 10^4$	$1.3 \cdot 10^5$	$3.6 \cdot 10^5$	$5.1 \cdot 10^5$	$^{45}Sc(p, p\gamma_{2\text{-}1})^{45}Sc$
	431	-	-		$1.8 \cdot 10^4$	$5.8 \cdot 10^4$	-	$^{45}Sc(p, p\gamma_{6\text{-}3})^{45}Sc$
	531	-	-	$2.5 \cdot 10^4$	$1.1 \cdot 10^5$	$3.4 \cdot 10^5$	$3.6 \cdot 10^5$	$^{45}Sc(p, p\gamma_{3\text{-}1})^{45}Sc$
	543	-	-	-	$9.8 \cdot 10^4$	$2.5 \cdot 10^5$	$2.3 \cdot 10^5$	$^{45}Sc(p, p_3\gamma)^{45}Sc$
	720	-	-	$2.5 \cdot 10^4$	$2.5 \cdot 10^5$	$7.2 \cdot 10^5$	$6.0 \cdot 10^5$	$^{45}Sc(p, p_4\gamma)^{45}Sc$
	889	$0.6 \cdot 10^3$	$9.8 \cdot 10^3$	$6.4 \cdot 10^4$	$1.9 \cdot 10^5$	$3.4 \cdot 10^5$	$2.8 \cdot 10^5$	$^{45}Sc(p, \gamma_1)^{46}Ti$
	962	-	-	-	$1.9 \cdot 10^5$	$3.7 \cdot 10^5$	$2.6 \cdot 10^5$	$^{45}Sc(p, p\gamma_{6\text{-}1})^{45}Sc$
	974	-	-	-	$9.3 \cdot 10^4$	$3.3 \cdot 10^5$	$2.9 \cdot 10^5$	$^{45}Sc(p, p_6\gamma)^{45}Sc$
	1049	$0.1 \cdot 10^3$	$1.7 \cdot 10^3$	-	$2.9 \cdot 10^4$	$3.6 \cdot 10^4$	-	$^{45}Sc(p, \gamma_{5\text{-}2})^{46}Ti$
	1121	$0.4 \cdot 10^3$	$6.1 \cdot 10^3$	$4.2 \cdot 10^4$	$1.2 \cdot 10^5$	$1.5 \cdot 10^5$	-	$^{45}Sc(p, \gamma_{2\text{-}1})^{46}Ti$
	1236	-	-	-	$1.0 \cdot 10^5$	$4.7 \cdot 10^5$	$4.0 \cdot 10^5$	$^{45}Sc(p, p_8\gamma)^{45}Sc$
	1409	-	-	-	-	$2.1 \cdot 10^5$	$2.3 \cdot 10^5$	$^{45}Sc(p, p_{10}\gamma)^{45}Sc$
	1662	-	-	-	-	$4.9 \cdot 10^4$	$7.5 \cdot 10^4$	$^{45}Sc(p, p_{13}\gamma)^{45}Sc$

Table A12.6b. Absolute thick target gamma-ray yields for heavy elements (Z > 30) with 1.7 and 2.4 MeV protons. The yield values are absolute yields for elemental samples at the angle of 55° with respect to the beam direction and per solid angle (sr) and total accumulated charge (μC) (Räisänen and Hänninen, 1983).

Element	E_γ *(keV)*[1]	Yield E_p $(\mu C\ sr)^{-1}$ 1.7 MeV	2.4 MeV	Reaction
Ga	175	320	1.2×10^4	$^{69}Ga(p,\gamma)^{70}Ge$, $^{71}Ga(p, n\gamma)^{71}Ge$
	318	160	1.5×10^3	$^{69}Ga(p, p'\gamma)^{69}Ga$
	327		3.0×10^3	$^{71}Ga(p, n\gamma)^{71}Ge$
	391		1.7×10^3	$^{71}Ga(p, n\gamma)^{71}Ge$, $^{71}Ga(p, p'\gamma)^{71}Ga$
	500		4.3×10^3	$^{71}Ga(p, n\gamma)^{71}Ge$
	668	80	3.0×10^3	$^{69}Ga(p,\gamma)^{70}Ge$
	1040	450	1.2×10^4	$^{69}Ga(p,\gamma)^{70}Ge$
	1708	670	2.0×10^3	$^{69}Ga(p,\gamma)^{70}Ge$
Ge	147	70	1.6×10^3	$^{70}Ge(p,\gamma)^{71}As$
	199	50	2.1×10^3	$^{74}Ge(p,\gamma)^{75}As$
	254	45	1.1×10^3	$^{72}Ge(p,\gamma)^{73}As$, $^{73}Ge(p, n\gamma)^{73}As$
	265	70	2.3×10^3	$^{74}Ge(p,\gamma)^{75}As$
	280	40	2.0×10^3	$^{74}Ge(p,\gamma)^{75}As$
	361	40	1.9×10^3	$^{72}Ge(p,\gamma)^{73}As$, $^{73}Ge(p, n\gamma)^{73}As$
	563	60	1.4×10^3	$^{76}Ge(p, p'\gamma)^{76}Ge$
	596	1.9×10^3	5.6×10^3	$^{74}Ge(p, p'\gamma)^{74}Ge$, $^{74}Ge(p,\gamma)^{75}As$
As	112	30	4.8×10^3	$^{75}As(p, n\gamma)^{75}Se$
	199	1.3×10^3	4.3×10^3	$^{75}As(p, p'\gamma)^{75}As$
	280	1.9×10^3	8.8×10^3	$^{75}As(p, p'\gamma)^{75}As$
	287		7.4×10^3	$^{75}As(p, n\gamma)^{75}Se$
	559	220	1.2×10^3	$^{75}As(p,\gamma)^{76}Se$
	573	95	2.3×10^3	$^{75}As(p, p'\gamma)^{75}Se$
Se	239	1.6×10^3	4.1×10^3	$^{77}Se(p, p'\gamma)^{77}Se$
	276	80	2.6×10^3	$^{80}Se(p,\gamma)^{81}Br$
	559	130	2.1×10^3	$^{76}Se(p, p'\gamma)^{76}Se$
	614	120	2.4×10^3	$^{78}Se(p, p'\gamma)^{78}Se$
	666	80	2.7×10^3	$^{80}Se(p, p'\gamma)^{80}Se$
Br	190		1.3×10^3	$^{81}Br(p, n\gamma)^{81}Kr$
	217	1.4×10^3	6.2×10^3	$^{79}Br(p, p'\gamma)^{79}Br$
	276	1.2×10^3	6.2×10^3	$^{81}Br(p, p'\gamma)^{81}Br$
	306	310	2.1×10^3	$^{79}Br(p, p'\gamma)^{79}Br$
	523	75	1.6×10^3	$^{79}Br(p, p'\gamma)^{79}Br$
	617	55	3.9×10^3	$^{79}Br(p,\gamma)^{80}Kr$
Rb	130	130	780	$^{85}Rb(p, p'\gamma)^{85}Rb$
	151	35	1.6×10^3	$^{85}Rb(p, p'\gamma)^{85}Rb$
	232		5.0×10^3	$^{85}Rb(p, n\gamma)^{85}Sr$
	485	15	350	$^{87}Rb(p, n\gamma)^{87}Sr$
	1077	40	790	$^{85}Rb(p,\gamma)^{86}Sr$
	1229		470	$^{87}Rb(p, n\gamma)^{87}Sr$
	1854		150	$^{85}Rb(p,\gamma)^{86}Sr$

1. Values from Table of Isotopes, (1978), C.M. Lederer and V.S. Shirley (eds.), Wiley, New York.

Table A12.6b. Absolute thick target gamma-ray yields for heavy elements (Z > 30) with 1.7 and 2.4 MeV protons. The yield values are absolute yields for elemental samples at the angle of 55° with respect to the beam direction and per solid angle (sr) and total accumulated charge (μC) (Räisänen and Hänninen, 1983) (continued).

Element	E_γ *(keV)*[1]	Yield E_p *(μC sr)*$^{-1}$ 1.7 MeV	2.4 MeV	Reaction
Sr	232		50	$^{87}Sr(p,\gamma)^{88}Y$
	793	1	20	$^{86}Sr(p,\gamma)^{87}Y$
	909	4	110	$^{88}Sr(p,\gamma)^{89}Y$
	1313	2	50	$^{88}Sr(p,\gamma)^{89}Y$
	1507	5	220	$^{88}Sr(p,\gamma)^{89}Y$
	1745		35	$^{88}Sr(p,\gamma)^{89}Y$
Y	420	7	60	$^{89}Y(p,\gamma)^{90}Zr$
	562	9	100	$^{89}Y(p,\gamma)^{90}Zr$
	891		35	$^{89}Y(p,\gamma)^{90}Zr$
	2186	40	650	$^{89}Y(p,\gamma)^{90}Zr$
	2319	9	80	$^{89}Y(p,\gamma)^{90}Zr$
Zr	656		60	$^{92}Zr(p,\gamma)^{93}Nb$
	744		20	$^{92}Zr(p,\gamma)^{93}Nb$
	919		10	$^{94}Zr(p, p'\gamma)^{94}Zr$
	1082		50	$^{90}Zr(p,\gamma)^{91}Nb$
	1208		75	$^{90}Zr(p,\gamma)^{91}Nb$
Nb	449		15	$^{93}Nb(Zr(p,\gamma)^{94}Mo$
	703	2	230	$^{93}Nb(p,\gamma)^{94}Mo$
	742		50	$^{93}Nb(p,\gamma)^{94}Mo$, $^{93}Nb(p, p'\gamma)^{93}Nb$
	850		60	$^{93}Nb(p,\gamma)^{94}Mo$
	871	3	270	$^{93}Nb(p,\gamma)^{94}Mo$
Mo	204	210	940	$^{95}Mo(p, p'\gamma)^{95}Mo$
	481	2	60	$^{97}Mo(p, p'\gamma)^{97}Mo$
	535	20	930	$^{100}Mo(p, p'\gamma)^{100}Mo$
	778		110	$^{96}Mo(p, p'\gamma)^{96}Mo$
	787		150	$^{98}Mo(p, p'\gamma)^{98}Mo$
Ru	90	380	1.1×10^3	$^{99}Ru(p, p'\gamma)^{99}Ru$
	127	310	1.3×10^3	$^{101}Ru(p, p'\gamma)^{101}Ru$
	358	610	7.4×10^3	$^{104}Ru(p, p'\gamma)^{104}Ru$
	475	130	3.8×10^3	$^{102}Ru(p, p'\gamma)^{102}Ru$
	540	15	700	$^{100}Ru(p, p'\gamma)^{100}Ru$
Rh	295	1.6×10^3	1.6×10^4	$^{103}Rh(p, p'\gamma)^{103}Rh$
	357	880	1.2×10^4	$^{103}Rh(p, p'\gamma)^{103}Rh$
Pd	280	20	170	$^{105}Pd(p, p'\gamma)^{105}Pd$
	374	250	4.0×10^3	$^{110}Pd(p, p'\gamma)^{110}Pd$
	434	210	4.8×10^3	$^{108}Pd(p, p'\gamma)^{108}Pd$
	442		980	$^{105}Pd(p, p'\gamma)^{105}Pd$
	556	7	480	$^{102}Pd(p, p'\gamma)^{102}Pd$, $^{104}Pd(p, p'\gamma)^{104}Pd$

1. Values from Table of Isotopes, (1978), C.M. Lederer and V.S. Shirley (eds.), Wiley, New York.

Table A12.6b. Absolute thick target gamma-ray yields for heavy elements (Z > 30) with 1.7 and 2.4 MeV protons. The yield values are absolute yields for elemental samples at the angle of 55° with respect to the beam direction and per solid angle (sr) and total accumulated charge (μC) (Räisänen and Hänninen, 1983) (continued).

Element	E_γ *(keV)*[1]	Yield E_p *(μC sr)$^{-1}$* 1.7 MeV	2.4 MeV	Reaction
Ag	311	810	7.3×10^3	$^{109}Ag(p, p'\gamma)^{109}Ag$
	325	610	6.0×10^3	$^{107}Ag(p, p'\gamma)^{107}Ag$
	415	220	4.2×10^3	$^{109}Ag(p, p'\gamma)^{109}Ag$
	423	180	3.6×10^3	$^{107}Ag(p, p'\gamma)^{107}Ag$
Cd	298	85	770	$^{113}Cd(p, p'\gamma)^{113}Cd$
	342	45	480	$^{111}Cd(p, p'\gamma)^{111}Cd$
	558	15	980	$^{114}Cd(p, p'\gamma)^{114}Cd$
	617	5	450	$^{112}Cd(p, p'\gamma)^{112}Cd$
	658		140	$^{110}Cd(p, p'\gamma)^{110}Cd$
In	115		60	$^{115}In(p, n\gamma)^{115}Sn$
	497		30	$^{115}In(p, n\gamma)^{115}Sn$
Sn				
Sb	160	40	320	$^{123}Sb(p, p'\gamma)^{123}Sb$
	382	1	50	$^{123}Sb(p, p'\gamma)^{123}Sb$
	508		30	$^{121}Sb(p, p'\gamma)^{121}Sb$
	542	0.3	25	$^{123}Sb(p, p'\gamma)^{123}Sb$
	573	0.6	55	$^{121}Sb(p, p'\gamma)^{121}Sb$
Te	666		570	$^{126}Te(p, p'\gamma)^{126}Te$
I	145	30	240	$^{127}I(p, p'\gamma)^{127}I$
	172	10	220	$^{127}I(p, p'\gamma)^{127}I$
	203	340	2.2×10^3	$^{127}I(p, p'\gamma)^{127}I$
	375	10	210	$^{127}I(p, p'\gamma)^{127}I$
	418	4	120	$^{127}I(p, p'\gamma)^{127}I$
	629		130	$^{127}I(p, p'\gamma)^{127}I$
Cs	161	160	620	$^{133}Cs(p, p'\gamma)^{133}Cs$
	303		410	$^{133}Cs(p, p'\gamma)^{133}Cs$
	384		170	$^{133}Cs(p, p'\gamma)^{133}Cs$
Ba	221		30	$^{135}Ba(o, p'\gamma)^{135}Ba$
	279	9	75	$^{137}Ba(p, p'\gamma)^{137}Ba$
	481	4	55	$^{135}Ba(p, p'\gamma)^{135}Ba$
	605		10	$^{134}Ba(p, p'\gamma)^{134}Ba$
La				
Ce				
Pr	145	7	55	$^{141}Pr(p, p'\gamma)^{141}Pr$
Nd	130	400	4.4×10^3	$^{150}Nd(p, p'\gamma)^{148}Nd$
	302	60	1.0×10^3	$^{148}Nd(p, p'\gamma)^{148}Nd$
	454	7	400	$^{146}Nd(p, p'\gamma)^{146}Nd$
	696		25	$^{144}Nd(p, p'\gamma)^{144}Nd$

1. Values from Table of Isotopes, (1978), C.M. Lederer and V.S. Shirley (eds.), Wiley, New York.

Table A12.6b. Absolute thick target gamma-ray yields for heavy elements (Z > 30) with 1.7 and 2.4 MeV protons. The yield values are absolute yields for elemental samples at the angle of 55° with respect to the beam direction and per solid angle (sr) and total accumulated charge (μC) (Räisänen and Hänninen, 1983) (continued).

Element	E_γ (keV)[1]	Yield E_p ($\mu C\ sr$)$^{-1}$ 1.7 MeV	2.4 MeV	Reaction
Sm	82	2.4×10^3	1.5×10^4	^{154}Sm(p, p′γ)^{154}Sm
	122	3.2×10^3	3.4×10^4	^{152}Sm(p, p′γ)^{152}Sm, ^{147}Sm(p,p′γ)^{147}Sm
	197	20	220	^{147}Sm(p, p′γ)^{147}Sm
	334	40	990	^{150}Sm(p, p′γ)^{150}Sm
	550	6	70	^{148}Sm(p, p′γ)^{148}Sm
Eu	83	1.1×10^4	3.0×10^4	^{153}Eu(p, p′γ)^{153}Eu
	111	490	3.0×10^3	^{151}Eu(p, p′γ)^{151}Eu
	193	1.1×10^3	6.2×10^3	^{153}Eu(p, p′γ)^{153}Eu
	197	230	1.6×10^3	^{151}Eu(p, p′γ)^{151}Eu
	286	25	400	^{151}Eu(p, p′γ)^{151}Eu
	307	170	2.7×10^3	^{151}Eu(p, p′γ)^{151}Eu
Gd	80	3.9×10^3	9.0×10^3	^{158}Gd(p, p′γ)^{158}Gd
	89	6.8×10^3	1.7×10^4	^{156}Gd(p, p′γ)^{156}Gd
	123	620	2.2×10^3	^{154}Gd(p, p′γ)^{154}Gd
	131	110	440	^{157}Gd(p, p′γ)^{157}Gd
	146	130	560	^{155}Gd(p, p′γ)^{155}Gd
	344		20	^{152}Gd(p, p′γ)^{152}Gd
Tb				
Dy	81	3.3×10^3	8.5×10^3	^{162}Dy(p, p′γ)^{162}Dy
	167	580	3.3×10^3	^{163}Dy(p, p′γ)^{163}Dy, ^{164}Dy(p, p′γ)^{164}Dy
Ho	95	4.0×10^4	1.2×10^3	^{165}Ho(p, p′γ)^{165}Ho
	115	2.3×10^3	1.6×10^4	^{165}Ho(p, p′γ)^{165}Ho
	210	110	760	^{165}Ho(p, p′γ)^{165}Ho
Er				
Tm	110	1.4×10^4	1.8×10^5	^{169}Tm(p, p′γ)^{169}Tm
	118	1.4×10^3	1.4×10^4	^{169}Tm(p, p′γ)^{169}Tm
Yb	181	20	150	^{172}Yb(p, p′γ)^{172}Yb, ^{173}Yb(p, p′γ)^{173}Yb
Lu	114	3.8×10^3	1.6×10^4	^{175}Lu(p, p′γ)^{175}Lu
	138	180	2.4×10^3	^{175}Lu(p, p′γ)^{175}Lu
	185	90	680	^{176}Lu(p, p′γ)^{176}Lu
	251	120	1.6×10^3	^{175}Lu(p, p′γ)^{175}Lu
Hf	93	1.2×10^4	4.2×10^4	^{178}Hf(p, p′γ)^{178}Hf, ^{180}Hf(p, p′γ)^{180}Hf
	113	2.5×10^3	9.0×10^3	^{177}Hf(p, p′γ)^{177}Hf
	123	1.2×10^3	4.9×10^3	^{179}Hf(p, p′γ)^{179}Hf
	146	40	280	^{179}Hf(p, p′γ)^{179}Hf
	250	110	930	^{177}Hf(p, p′γ)^{177}Hf
Ta	136	3.1×10^4	1.6×10^5	^{181}Ta(p, p′γ)^{181}Ta
	165	80	1.5×10^3	^{181}Ta(p, p′γ)^{181}Ta
	302	55	1.1×10^3	^{181}Ta(p, p′γ)^{181}Ta

1. Values from Table of Isotopes, (1978), C.M. Lederer and V.S. Shirley (eds.), Wiley, New York.

Table A12.6b. Absolute thick target gamma-ray yields for heavy elements (Z > 30) with 1.7 and 2.4 MeV protons. The yield values are absolute yields for elemental samples at the angle of 55° with respect to the beam direction and per solid angle (sr) and total accumulated charge (μC) (Räisänen and Hänninen, 1983) (continued).

Element	E_γ (keV)[1]	Yield E_p ($\mu C\ sr)^{-1}$ 1.7 MeV	2.4 MeV	Reaction
W	100	5.0×10^3	1.5×10^4	$^{182}W(p, p'\gamma)^{182}W$
	111	6.1×10^3	2.1×10^4	$^{184}W(p, p'\gamma)^{184}W$
	123	5.4×10^3	2.2×10^4	$^{186}W(p, p'\gamma)^{186}W$
	292	4	60	$^{183}W(p, p'\gamma)^{183}W$
Pt	211	110	1.2×10^3	$^{195}Pt(p, p'\gamma)^{195}Pt$
	239	40	430	$^{195}Pt(p, p'\gamma)^{195}Pt$
	329	60	1.9×10^3	$^{194}Pt(p, p'\gamma)^{194}Pt$
	356	20	930	$^{196}Pt(p, p'\gamma)^{196}Pt$
	407	2	110	$^{198}Pt(p, p'\gamma)^{198}Pt$
Au	191	20	240	$^{197}Au(p, p'\gamma)^{197}Au$
	269	1	23	$^{197}Au(p, p'\gamma)^{197}Au$
	279	100	1.3×10^3	$^{197}Au(p, p'\gamma)^{197}Au$
	548		60	$^{197}Au(p, p'\gamma)^{197}Au$
Hg	158	70	460	$^{199}Hg(p, p'\gamma)^{199}Hg$
	208	15	160	$^{199}Hg(p, p'\gamma)^{199}Hg$
	368	7	310	$^{200}Hg(p, p'\gamma)^{200}Hg$
	412		110	$^{198}Hg(p, p'\gamma)^{198}Hg$
Tl	204	65	670	$^{205}Tl(p, p'\gamma)^{205}Tl$
	279	9	230	$^{203}Tl(p, p'\gamma)^{203}Tl$
	416		17	$^{205}Tl(p, p'\gamma)^{205}Tl$
Pb				

1. Values from Table of Isotopes, (1978), C.M. Lederer and V.S. Shirley (eds.), Wiley, New York.

Appendix 12

Table A12.6c. Absolute thick target gamma-ray yields for most elements by 7 and 9 MeV protons.

Isotope	E_γ *(keV)*[a]	Reaction[b]	Absolute γ-ray yield[c] $(\mu C\ sr)^{-1}$ 7 MeV	9 MeV
^{7}Li[d]	429	2	4.33 (7)	2.23 (8)
	478	1	3.59 (8)	2.09 (9)
^{9}Be	3562	3	1.95 (8)	1.86 (8)
^{10}B	718	1	1.16 (8)	1.54 (8)
^{11}B	2000	2	2.25 (7)	9.49 (7)
	2125	1	2.79 (8)	4.27 (8)
^{12}C	4439	1	7.53 (8)	2.89 (9)
^{14}N[e]	1635	1	1.26 (7)	3.12 (7)
	2313	1	1.82 (7)	4.24 (7)
	5106	1	2.21 (6)	2.47 (7)
^{16}O[f]	6129	1	1.15 (7)	1.80 (9)
	6919	1		8.29 (7)
^{19}F[d]	1236	1; 2	3.44 (7)	3.00 (8)
	1346; 1357	1; 1	1.35 (8)	7.70 (8)
	6129	3	1.94 (8)	9.43 (8)
^{23}Na[g]	440	1	7.31 (8)	6.67 (8)
	1633	1; 3	4.75 (8)	4.60 (8)
	3915	1	1.51 (7)	3.24 (7)
^{24}Mg[h]	1369	1	7.28 (8)	9.28 (8)
^{26}Mg	1809	1	7.81 (7)	8.58 (7)
^{24}Mg	2754	1	1.12 (7)	7.38 (7)
	4239	1	5.20 (7)	9.25 (7)
^{27}Al	844	1	2.06 (8)	3.78 (8)
	1014	1	4.97 (8)	8.89 (8)
	1369	3	3.92 (8)	7.62 (8)
	1720	1	1.33 (8)	2.38 (8)
	2210	1	4.17 (8)	8.00 (8)
	2981	1	7.32 (7)	1.01 (8)
	3004	1	2.04 (8)	5.44 (8)
^{28}Si[i]	1779	1	2.08 (8)	4.68 (8)
^{29}Si; ^{30}Si	2028	1; 4	4.87 (6)	1.08 (7)
^{30}Si	2235	1; 4	4.30 (6)	1.15 (7)
^{28}Si	2839	1	4.87 (6)	2.16 (7)
	3200	1	2.78 (6)	2.91 (7)
^{31}P[j]	1266	1	9.74 (7)	1.74 (8)
	1779	3	5.22 (7)	8.10 (7)
	2230; 2235	4; 1	7.01 (7)	1.33 (8)
^{32}S[k]	2230	1	6.17 (7)	1.48 (8)
	4282	1	5.99 (6)	3.66 (7)
^{35}Cl[l]	1219	1	5.67 (7)	8.55 (8)
	1763	1	9.51 (7)	1.62 (8)
	2230	3	5.13 (7)	1.06 (8)
	3163	1	2.27 (7)	5.71 (7)
^{39}K[m]	1970	3	1.37 (7)	3.12 (7)
^{39}K; ^{41}K	2523	1; 2	1.57 (7)	2.46 (7)
^{39}K	2814	1	3.98 (7)	1.03 (8)
	3019	1	1.80 (7)	2.92 (7)
	3598	1	6.99 (6)	1.51 (7)
^{40}Ca[n]	755	1	5.68 (5)	2.55 (7)
	3736	1	7.40 (7)	2.65 (8)
	3904	1	8.59 (7)	2.84 (8)
^{47}Ti; ^{48}Ti	308	4; 2	3.88 (7)	1.46 (8)
^{46}Ti	889	1	9.57 (6)	2.57 (7)
^{48}Ti	983	1	6.15 (7)	1.27 (8)
	1312	1	4.34 (6)	1.72 (7)
	1437	1	6.97 (6)	1.71 (7)
^{51}V	319	1	1.46 (8)	2.46 (8)
	749	2	3.58 (8)	5.38 (8)
	808	2	1.47 (8)	2.20 (8)
	1148	2	8.21 (7)	1.23 (8)
	1164	2	3.29 (8)	4.85 (8)
	1480	2	1.01 (8)	1.72 (8)
^{52}Cr	936	1	5.05 (6)	3.95 (7)
	1334	1	2.68 (6)	4.62 (7)
	1434	1	1.10 (8)	3.26 (8)
	1531	1	5.06 (6)	3.75 (7)
	1728	1	2.66 (6)	2.39 (7)
^{55}Mn	411	2	1.04 (8)	1.26 (8)

a. Table of Isotopes, (1978), C.M. Lederer and V.S. Shirley (eds.), Wiley, New York. The γ-energies which can lead to an ambiguous analysis are: 288 (Cd), 292 (W), 297 (Gd); 307 (Cd), 308 (Ti); 319 (V), 321 (I), 326 (Gd), 328 (Pt, Hf); 336 (Mo), 339 (Co), 365 (Ag, Ta); 375 (I), 382 (Sb); 411 (Mn, I), 429 (Li), 434 (Pd); 583 (Au), 588 (Y); 700 (Sn), 702 (Ag); 804 (Mn), 808 (V, Zn), 833 (Ru), 834 (Ge), 844 (Al), 847 (Fe,Co); 931 (Mn), 934 (In), 936 (Cr); 1039 (Zn), 1040 (Ge); 1230 (Sn), 1236 (F), 1237 (In), 1238 (Fe); 1312 (Ti), 1314 (Mn), 1315 (Zn); 1327 (Cu), 1332 (Cu, Co, Ni), 1334 (Cr), 1338 (Co); 1369 (Mg, Al); 1408 (Mn, Fe), 1434 (Cr), 1437 (Ti); 1477 (Nb), 1480 (V); 1507 (Y), 1510 (Mo); 1627 (Y), 1632 (Na), 1635 (N); 1779 (P, Si); 1809 (Mg), 1811 (Fe); 2230 (S, Cl), 2235 (P, S); 6126 (O, F).

b. 1: (p,p'), 2: (p,n), 3: (p,α), 4: (p, γ).

c. (n) means $\times 10^n$.

d. LiF target.

e. TaN target.

f. MgO, SiO_2, and Sm_2O_3 targets.

g. NaCl target.

h. MgO target.

i. Si and SiO_2 targets.

j. InP target.

k. PbS target.

l. NaCl and $PbCl_2$ targets.

m. KI target.

n. CaO target.

Table A12.6c. Absolute thick target gamma-ray yields for most elements by 7 and 9 MeV protons (continued).

Isotope	E_γ *(keV)*[a]	Reaction[b]	Absolute γ-ray yield[c] *(μC sr)⁻¹* 7 MeV	9 MeV
	804	2	6.83 (7)	1.01 (8)
	931	2	3.45 (8)	4.73 (8)
	1314	2	2.11 (8)	3.09 (8)
	1408	2	7.39 (7)	9.90 (7)
^{56}Fe	847	1	6.01 (8)	7.61 (8)
	1238	1	3.67 (7)	1.01 (8)
^{54}Fe	1408	1	2.94 (7)	4.64 (7)
^{56}Fe	1811	1	4.49 (7)	7.13 (7)
	2113	1	2.71 (7)	4.29 (7)
^{59}Co	339	2	1.01 (8)	1.88 (8)
	847	3	1.02 (7)	2.52 (7)
	878	2	1.13 (7)	1.93 (7)
	998	2	2.40 (7)	4.77 (7)
	1332; 1338	4; 2	1.83 (7)	3.36 (7)
	1428	2	1.90 (7)	3.80 (7)
^{60}Ni	826	1	3.19 (7)	8.22 (7)
^{58}Ni	1005	1	2.26 (7)	1.61 (8)
	1321	1	2.59 (7)	1.18 (8)
^{60}Ni	1332	1	1.21 (8)	3.29 (8)
^{58}Ni	1454; 1448	1	3.12 (8)	8.77 (8)
	1584	1	1.02 (7)	5.02 (7)
^{63}Cu	962	1	1.37 (7)	3.39 (7)
	1065	2	5.09 (6)	1.54 (7)
	1327	1	5.78 (6)	1.54 (7)
	1332	3	4.51 (6)	1.39 (7)
^{64}Zn	808	1	5.15 (7)	1.66 (8)
	992	1	2.52 (8)	7.60 (8)
^{66}Zn	1039	1	9.73 (7)	2.13 (8)
^{64}Zn	1315	1	1.17 (7)	7.50 (7)
	1799	1	2.21 (7)	7.27 (7)
^{72}Ge	834	1	3.63 (7)	7.84 (7)
^{70}Ge	1040	1	7.22 (7)	2.13 (8)
	1708	1	9.44 (6)	2.87 (7)
^{89}Y	588	2	1.35 (8)	3.25 (8)
	770	2	1.54 (7)	4.41 (7)
	863	2	2.06 (7)	8.64 (7)
	1155	2	1.39 (7)	4.25 (7)
	1507	1	1.81 (7)	4.07 (7)
	1627	2	1.71 (7)	5.95 (7)
	1833	2	1.42 (7)	5.08 (7)
^{90}Zr	1761	1	1.27 (6)	1.53 (7)
	2186	1		4.18 (7)
^{93}Nb	685; 687	2; 1	3.29 (7)	1.01 (8)
	1363	2	3.18 (7)	9.30 (7)
	1477	2	1.11 (8)	2.84 (8)
	1520	2	1.97 (7)	4.48 (7)
^{95}Mo	336	2	1.13 (7)	5.28 (7)
	627	2	1.21 (7)	3.34 (7)
^{92}Mo	1510	1	6.58 (6)	5.41 (7)
^{104}Ru	358	1	1.70 (7)	4.01 (7)
^{96}Ru	833	1	4.75 (6)	1.94 (7)
^{108}Pd	434	1	2.41 (6)	8.99 (6)
^{107}Ag	365	2	2.13 (7)	9.00 (7)
^{109}Ag	614	2	3.28 (7)	5.07 (7)
	624	2	2.75 (7)	4.69 (7)
^{107}Ag; ^{109}Ag	702	2: 1	9.20 (6)	4.44 (7)
^{114}Cd	210	2	1.04 (8)	3.07 (8)
	288	2	6.94 (7)	1.84 (8)
	307	2	3.26 (7)	1.71 (8)
^{114}Cd; ^{111}Cd	536, 537	2	2.91 (7)	1.43 (8)
^{111}Cd	1102	2	5.81 (6)	1.55 (7)
^{113}In; ^{115}In[o]	497	2	1.05 (8)	4.28 (8)
^{115}In	934	1	2.43 (7)	9.50 (7)
	1078	1	3.27 (7)	1.49 (8)
	1237	2	1.64 (6)	7.18 (6)
^{119}Sn	700	2	1.63 (7)	5.36 (7)
^{118}Sn	1230	1	3.29 (6)	3.02 (7)
^{116}Sn	1294	1	1.95 (7)	1.11 (8)
^{121}Sb	212	2	1.84 (8)	5.84 (8)
	230	2	6.56 (7)	1.76 (8)
	244	2	4.55 (7)	1.36 (8)
^{123}Sb	281	2	3.52 (7)	1.13 (8)
	382	1	3.32 (7)	1.10 (8)
^{127}I	321	2	2.77 (7)	9.60 (7)
	375	1	3.42 (7)	1.15 (8)
	411	2	857 (6)	1.15(8)
^{147}Sm[p]	777	2	7.37 (5)	2.05 (6)
^{156}Gd; ^{157}Gd	297	1; 2	4.87 (6)	3.36 (7)
	326	2	5.22 (5)	7.78 (6)
	598	2	3.55 (5)	2.70 (6)
^{179}Hf; ^{178}Hf	325; 328;	2; 1; 1	1.80 (6)	9.29 (6)
^{180}Hf	328			
^{181}Ta[q]	365	2	7.02 (6)	3.10 (7)
	661	2	9.46 (5)	3.04 (6)
^{183}W	292	1	5.14 (5)	6.00 (6)
^{195}Pt	262	2	4.33 (5)	4.69 (7)
^{194}Pt	328	1	1.88 (7)	3.44 (7)
^{197}Au	548	1	1.64 (6)	4.03 (6)
	583	1	5.48 (5)	2.56 (6)
^{207}Pb; ^{208}Pb	570; 571	1; 2	1.37 (5)	6.62 (6)
^{208}Pb	970	2	1.21 (5)	2.56 (6)

a. Table of Isotopes, (1978), C.M. Lederer and V.S. Shirley (eds.), Wiley, New York. The γ-energies which can lead to an ambiguous analysis are: 288 (Cd), 292 (W), 297 (Gd); 307 (Cd), 308 (Ti); 319 (V), 321 (I), 326 (Gd), 328 (Pt, Hf); 336 (Mo), 339 (Co), 365 (Ag, Ta); 375 (I), 382 (Sb); 411 (Mn, I), 429 (Li), 434 (Pd); 583 (Au), 588 (Y); 700 (Sn), 702 (Ag); 804 (Mn), 808 (V, Zn), 833 (Ru), 834 (Ge), 844 (Al), 847 (Fe,Co); 931 (Mn), 934 (In), 936 (Cr); 1039 (Zn), 1040 (Ge); 1230 (Sn), 1236 (F), 1237 (In), 1238 (Fe); 1312 (Ti), 1314 (Mn), 1315 (Zn); 1327 (Cu), 1332 (Cu, Co, Ni), 1334 (Cr), 1338 (Co); 1369 (Mg, Al); 1408 (Mn, Fe), 1434 (Cr), 1437 (Ti); 1477 (Nb), 1480 (V); 1507 (Y), 1510 (Mo); 1627 (Y), 1632 (Na), 1635 (N); 1779 (P, Si); 1809 (Mg), 1811 (Fe); 2230 (S, Cl), 2235 (P, S); 6126 (O, F).

b. 1: (p,p′), 2: (p,n), 3: (p,α), 4: (p, γ).

c. (n) means $\times 10^n$.

o. In and InP targets.

p. Sm_2O_3 target.

q. Ta and TaN targets.

Table A12.6d. Absolute neutron yields by 7 and 9 MeV protons. The yield values are absolute yields for elemental samples at the angle of 55° with respect to the beam direction and per solid angle (sr) and total accumulated charge (μC) (Räisänen *et al.*, 1987).

Target	Relative neutron yields[a]	
	7 MeV	**9 MeV**
LiF	8.8 (4)	1.6 (5)
Be	2.7(5)	4.9 (5)
B	4.7 (4)	8.9 (4)
C	8.2 (2)	2.7 (3)
TaN	3.5 (2)	5.0 (3)
MgO	1.5 (3)	6.6 (3)
SiO_2	4.3 (2)	1.9 (3)
Sm_2O_3	9.7 (2)	1.0 (4)
NaCl	1.5 (4)	4.0 (4)
Al	2.8 (3)	1.8 (4)
Si	5.8 (2)	3.4 (3)
InP	5.3 (3)	2.9 (4)
PbS	1.4 (2)	1.7 (3)
$PbCl_2$	6.7 (3)	1.6 (4)
KI	4.0 (3)	2.0 (4)
CA	7.5 (2)	2.1 (3)
Ti	3.0 (4)	9.6 (4)
V	7.7 (4)	1.2 (5)
Fe	7.6 (3)	4.4 (4)
Co	4.9 (4)	1.1 (5)
Ni	2.0 (3)	1.1 (4)
Cu	3.2 (4)	8.4 (4)
Zn	1.1 (4)	4.4 (4)
Ge	2.6 (4)	8.8 (4)
Y	1.2 (4)	5.5 (4)
Zr	9.4 (3)	4.0 (4)
Nb	1.5 (4)	5.7 (4)
Mo	1.3 (4)	5.2 (4)
Ru	1.1 (4)	5.1 (4)
Pd	9.0 (3)	4.1 (4)
Ag	8.6 (3)	3.9 (4)
Cd	7.1 (3)	3.7 (4)
In	7.0 (3)[b]	3.5 (4)
Sn	5.6 (3)[b]	3.1 (4)
Sb	5.4 (3)[b]	3.1 (4)
Gd	1.0 (3)[b]	1.2 (4)
Yb	5.2 (2)	6.1 (3)
Hf	5.4 (2)	7.3 (3)
Ta	3.5 (2)	4.5 (3)[c]
W	8.6 (2)	5.7 (3)[c]
Pt	2.0 (2)	5.3 (3)[c]
Au	1.7 (2)	2.6 (3)[c]
Pb	1.0 (2)	4.0 (3)[c]

a. Yield obtained in the geometry (used in the original article) per μC.

b. The values given in Elwyn *et al.* (1966) are: 7.0 (3) for In, 6.0 (3) for Sn, 5.6 (3) for Sb, 1.1 (3) for Gd.

c. The values given in Elwyn *et al.* (1966) are: 4.5 (3) for Ta, 3.7 (3) for W, 2.6 (3) for Pt, 2.3 (3) for Au, 1.8 (3) for Pb.

Table A12.7a. Absolute thick target gamma-ray yields for light elements with 2.4 MeV $^4He^+$ beam. The yield values are absolute yields for elemental samples at the angle of 55° with respect to the beam direction and per solid angle (sr) and total accumulated charge (μC) (Lappalainen *et al.*, 1983).

Element	E_γ *(keV)*	Absolute γ-yield $\gamma/(\mu C \cdot Sr)$	Detection limit[b]
Li	478	9.6×10^5	0.45 ppm
Be	4439	1.6×10^6	0.42 ppm
B	170	2.5×10^4	
	3088	3.1×10^3	
	3684	4.9×10^4	
	3854	3.1×10^4	10 ppm
C	a		
N	937	5.9	
	1041	5.6	
	1080	13	
	2125	4.5	
	2471	4.4	
	2542	7.8	
	4525	5.9	1.1%
O	351	15	
	1634	16	2.0%
F	110	9.2×10^2	
	197	1.6×10^4	22 ppm
	1275	6.2×10^3	
Na	440	1.7×10^3	260 ppm
	1130	15	
	1809	610	
Mg	1273	8.8	
	1779	15	1.3%
	2839	0.6	
Al	2235	15	3.2%
Si	2230	0.3	65%

a. No separate γ-ray was observable.

b. The detection limits are based on the volcanic stone measurement.

Table A12.7b. Alpha-induced prompt gamma-rays generated by 5 MeV α particles. The notation x(m,n) means reaction with light ion product x and gamma-ray transition between states m and n in the residual nucleus. $-m_e$ and $-2m_e$ mean single and double escape peaks, respectively. The states are counted starting from 0 for the ground state. c/mC and bg/mC mean net gamma-ray and background counts/accumulated charge. Sensitivity values corresponds to the net count in the peak equal to three times the standard deviation of the peak background (Giles and Peisach, 1979).

Element	**Assignment**	E_γ ***(keV)***	**Relative intensity *(%)***	**c/mC**	**bg/mC**	**‰**
Lithium	^{7}Li α(1,0)	478	100.0	72000	6300	0.03
Boron	^{10}B p(3,2)	170	100.0	4700	3600	1.2
	^{10}B p(2,1)	598	not possible to integrate			
	^{10}B α(1,0)	718	66.0	3100	1800	1.3
	^{10}B p(3,1)	768	0.8	38	710	66
	^{11}B $-2m_e$	1291	3.6	170	830	16
	^{11}B $-m_e$	1802	0.4	18	280	88
	^{11}B n(1,0)	2313	25.5	1200	570	1.9
	^{10}B $-2m_e$	2662	18.7	880	1100	
	^{10}B $-2m_e$	2832	8.3	390	570	5.9
	^{10}B p(1,0)	3086	4.9	230	890	12
	^{10}B $-m_e$	3173	12.3	580	1200	5.6
	^{10}B $-m_e$	3343	4.5	210	470	9.9
	^{10}B p(2,0)	3684	27.7	1300	210	1.1
	^{10}B p(3,0)	3854	9.8	460	18	0.87
Nitrogen	^{14}N pl(1,0)	871	100.0	1500	40	0.04
Oxygen	^{18}O n(1,0)	351	100.0	120	12	12
	^{18}O n(2,1)	1395	5.7	6.8	3.3	11
	^{17}O n(1,0)	1634	1.8	2.2	1.6	24
	^{18}O n(3,1)	2438	1.0	1.2	0.94	34
	^{17}O n(2,1)	2614	0.1	0.15	0.4	170
Fluorine	^{19}F n(2,1)	74	2.5	430	3500	7.9
	^{19}F α(1,0)	110	56.5	9600	3700	0.37
	^{19}F α(2,0)	197	100.0	17000	2500	0.17
	^{19}F n(1,0)	583	21.2	3600	660	0.42
	^{19}F n(4,3)	637	0.2	32	500	40
	^{19}F n(3,0)	891	5.2	890	550	1.5
	^{19}F n(3,1)	1236	1.4	240	180	3.3
	^{19}F p(1,0)	1275	18.8	3200	280	0.30
	^{19}F n(1,0)	1280			not resolved	
	^{19}F α(4,1)	1349	1.3	220	170	3.4
	^{19}F α(5,2)	1357	0.3	50	82	11
	^{19}F n(6,1)	1369	0.4	63	100	9.3
	^{19}F n(7,1)	1400	0.1	21	74	23
	^{19}F α(4,0)	1459	0.3	47	150	15
	^{19}F n(4,0)	1528	1.1	180	85	3.0
	^{19}F $-m_e$	1570	0.1	9.6	56	45
	^{19}F p(2,1)	2081	1.4	240	41	1.5
	^{19}F $-2m_e$	2160	0.1	16	35	21
	^{19}F $-m_e$	2671	0.04	6.7	18	37
	^{19}F p(3,1)	3182	0.2	34	7	4.5

Table A12.7b. Alpha-induced prompt gamma-rays generated by 5 MeV α particles. The notation x(m,n) means reaction with light ion product x and gamma-ray transition between states m and n in the residual nucleus. $-m_e$ and $-2m_e$ mean single and double escape peaks, respectively. The states are counted starting from 0 for the ground state. c/mC and bg/mC mean net gamma-ray and background counts/accumulated charge. Sensitivity values corresponds to the net count in the peak equal to three times the standard deviation of the peak background (Giles and Peisach, 1979) (continued).

Element	Assignment	E_γ *(keV)*	Relative intensity *(%)*	c/mC	bg/mC	‰
Sodium	^{23}Na n(2,0)	417	66.2	860	320	0.54
	^{23}Na α(1,0)	440	100.0	1300	280	0.32
	^{23}Na $-2m_e$	787	7.7	100	440	5.4
	^{23}Na n(3,1)	830	2.0	26	150	12
	^{23}Na p(4,2)	1003	8.5	110	170	3.0
	^{23}Na p(2,1)	1130	51.5	670	240	0.59
	^{23}Na $-m_e$	1298	2.8	36	220	11
	^{23}Na p(7,2)	1412	2.6	34	190	11
	^{23}Na p(3,1)	1780	2.8	37	74	6.1
	^{23}Na p(1,0)	1809	100.0	1300	94	0.19
	^{23}Na p(8,2)	1897	0.1	1.5	16	71
	^{23}Na $-2m_e$	1916	0.1	1.8	17	60
	^{23}Na p(4,1)	2132	2.3	30	31	4.9
	^{23}Na $-m_e$	2427	0.4	4.9	12	19
	^{23}Na p(5,1)	2511	2.3	30	15	3.3
	^{23}Na p(6,1)	2524	1.5	19	9.0	4.0
	^{23}Na p(7,1)	2541	1.2	16	9.7	5.1
	^{23}Na p(2,0)	2938	2.1	27	5.5	2.3
	^{23}Na p(9,1)	3092	0.2	2.3	1.5	14
Magnesium	^{25}Mg α(1,0)	585	58.6	170	660	5.6
	^{26}Mg n(2,1)	755	9.0	26	520	32
	^{24}Mg p(1,0)	844	93.1	270	950	4.3
	^{24}Mg p(2,0)	1015	55.2	160	650	5.9
	^{26}Mg n(1,0)	1273	100.0	290	280	2.1
	^{25}Mg n(1,0)	1779	58.6	170	86	2.0
	^{26}Mg n(4,1)	1794	5.5	16	56	18
	^{26}Mg n(2,0)	2028	10.7	31	27	6.2
	^{25}Mg $-m_e$	2328	1.4	4.2	14	32
	^{26}Mg n(3,0)	2426	12.4	36	20	4.6
	^{25}Mg n(2,1)	2839	8.3	24	16	6.2
Aluminum	^{27}Al n(1,0)	677	4.0	56	220	25
	^{27}Al n(2,0)	709	27.1	380	340	4.6
	^{27}Al α(1,0)	844	6.5	91	270	17
	^{27}Al α(2,0)	1015	5.5	77	280	21
	^{27}Al $-2m_e$	1214	15.7	220	300	7.4
	^{27}Al p(2,1)	1263	36.4	510	330	3.4
	^{27}Al p(5,2)	1311	4.5	63	220	22
	^{27}Al p(6,2)	1332	0.9	13	200	100
	^{27}Al n(3,0)	1454	3.1	43	210	32
	^{27}Al n(3,1)	1534	4.9	69	240	21
	^{27}Al n(4,1)	1552	2.1	30	240	50
	^{27}Al $-m_e$	1725	7.9	110	410	10
	^{27}Al p(1,0)	2236	100.0	1400	170	0.87

Table A12.7b. Alpha-induced prompt gamma-rays generated by 5 MeV α particles. The notation x(m,n) means reaction with light ion product x and gamma-ray transition between states m and n in the residual nucleus. $-m_e$ and $-2m_e$ mean single and double escape peaks, respectively. The states are counted starting from 0 for the ground state. c/mC and bg/mC mean net gamma-ray and background counts/accumulated charge. Sensitivity values corresponds to the net count in the peak equal to three times the standard deviation of the peak background (Giles and Peisach, 1979) (continued).

Element	**Assignment**	E_γ ***(keV)***	**Relative intensity *(%)***	**c/mC**	**bg/mC**	**‰**
	^{27}Al $-2m_e$	2476	6.9	96	110	10
	^{27}Al p(5,1)	2574	0.4	6.2	63	120
	^{27}Al p(6,1)	2595	7.1	100	94	9.2
	^{27}Al $-2m_e$	2748	1.0	14	98	68
	^{27}Al $-m_e$	2987	4.2	59	94	16
	^{27}Al $-m_e$	3259		obscured by Compton edge		
	^{27}Al p(2,0)	3498	11.4	160	39	3.8
	^{27}Al p(3,0)	3770	2.9	40	26	12
Silicon	^{29}Si p(1,0)	78	100.0	97	140	12
	(^{28}Si+^{30}Si) $-2m_e$	1213	5.6	5.4	8.1	50
	^{28}Si p(1,0)	1266	7.6	7.4	4.3	26
	^{29}Si α(1,0)	1273	3.2	3.1	3.9	60
	(^{28}Si+^{30}Si) $-m_e$	1724	2.8	2.7	4.5	75
	^{28}Si α(1,0)	1779	4.3	4.2	6.1	56
	^{30}Si α(1,0)	2235	29.9	29	1.9	4.6
	^{28}Si p(2,0)				not resolved	
Phosphorus	^{31}P $-2m_e$	1105	12.1	5.7	9.6	2.3
	^{31}P p(2,1)	1176	15.7	7.4	7.9	1.6
	^{31}P α(1,0)	1266	3.4	1.6	5.1	6.1
	^{31}P $-m_e$	1616	4.3	2	7	5.6
	^{31}P p(1,0)	2127	100.0	47	2.6	0.15
	^{31}P $-2m_e$	2282	1.1	0.52	0.77	7.1
	^{31}P $-m_e$	2793	1.0	0.49	0.54	6.4
	^{31}P p(2,0)	3304	2.1	1.0	1.4	4.8
Sulphur		-				
Chlorine	^{35}Cl $-2m_e$	1146	11.6	3.7	5.0	30
	^{35}Cl p(2,1)	1210	3.4	1.1	3.1	77
	^{37}Cl p(1,0)	1461	1.8	0.56	3.6	170
	^{35}Cl p(3,1)	1643	11.3	3.6	6.5	35
	^{35}Cl $-m_e$	1657	6.9	2.2	4.5	49
	^{35}Cl p(1,0)	2168	100.0	32	1.5	1.9
Potassium	^{39}K p(2,1)	313	20.0	2.8	8.6	17
	^{39}K p(3,1)	899	11.4	1.6	3.9	19
	^{39}K p(1,0)	1524	100.0	14	2.4	1.7
Calcium		-				
Scandium	^{55}Sc α(2,1)	364	100.0	43	650	31
Titanium	^{47}Ti α(1,0)	159	100.0	390	91	2.3
	^{48}Ti n(1,0)	749	4.1	16	9.9	19
	^{46}Ti α(1,0)	889	2.8	11	3.6	4.4
	^{48}Ti α(1,0)	983	10.3	40	3.4	4.4
Vanadium	^{51}V n(1,0)	54	0.8	5.1	110	190
	^{51}V n(2,0)	157	5.2	32	130	34
	^{51}V n(3,2)	207	1.6	10	43	61
	^{51}V α(1,0)	320	100.0	620	36	0.91

Table A12.7b. Alpha-induced prompt gamma-rays generated by 5 MeV α particles. The notation x(m,n) means reaction with light ion product x and gamma-ray transition between states m and n in the residual nucleus. $-m_e$ and $-2m_e$ mean single and double escape peaks, respectively. The states are counted starting from 0 for the ground state. c/mC and bg/mC mean net gamma-ray and background counts/accumulated charge. Sensitivity values corresponds to the net count in the peak equal to three times the standard deviation of the peak background (Giles and Peisach, 1979) (continued).

Element	Assignment	E_γ *(keV)*	Relative intensity *(%)*	c/mC	bg/mC	‰
	^{51}V α(2,1)	609	0.1	0.75	1.9	170
	^{51}V α(2,0)	929	0.4	2.5	1.3	43
Chromium	^{53}Cr α(1,0)	564	97.6	8.2	13	42
	^{50}Cr α(1,0)	783	100.0	8.4	5.1	26
	^{54}Cr α(1,0)	835	41.7	3.5	4.0	54
Manganese	^{55}Mn α(1,0)	126	100.0	3700	610	0.63
	^{55}Mn α(2,1)	858	0.1	3.6	1.6	34
Iron	^{57}Fe α(2,1)	122	100.0	70	44	9.0
	^{57}Fe α(2,0)	137	12.3	8.6	34	64
	^{57}Fe α(3,2)	230	2.3	1.6	18	250
	^{57}Fe α(3,1)	353		obscured by ^{18}O		
	^{57}Fe α(3,0)	367	3.6	2.5	10	120
	^{58}Fe α(1,0)	811	0.8	0.54	1.1	180
	^{56}Fe α(1,0)	847	84.3	59	1.7	2.1
Cobalt		-				
Nickel		-				
Copper	^{63}Cu α(1,0)	670	100.0	9.6	2.4	15
	^{65}Cu α(1,0)	771	18.8	1.8	0.70	46
	^{63}Cu α(2,0)	962	24.0	2.3	1.1	43
	^{65}Cu α(2,0)	1116	5.8	0.56	1.4	200
Zinc	^{67}Zn α(1,0)	93	17.2	6.2	40	97
	^{67}Zn α(2,0)	185	100.0	36	7.1	11
	^{64}Zn α(1,0)	992	17.2	6.2	0.85	14
	^{66}Zn α(1,0)	1039	5.6	2.0	0.71	41
	^{68}Zn α(1,0)	1077	2.5	0.90	0.63	84
Gallium		-				
Bromine	^{79}Br α(2,0)	217	100.0	130	43	1.6
	^{79}Br α(3,0)	261	9.2	12	22	13
	^{81}Br α(1,0)	276	84.6	110	26	1.4
	^{79}Br α(4,0)	306	25.4	33	17	4.0
	^{79}Br α(5,0)	397	1.9	2.5	7.0	33
	^{79}Br α(6,0)	523	12.3	16	5.2	4.6
	^{81}Br α(3,0)	538	3.3	4.3	6.9	19
	^{79}Br α(7,0)	606	0.7	0.93	5.0	76
Rubidium	^{85}Rb α(1,0)	151	100.0	82	2400	46
Strontium		-				
Yttrium		-				
Zirconium	^{96}Zr α(1,0)	1594	100.0	0.26	0.54	270
Niobium		-				
Molybdenum	^{95}Mo α(1,0)	204	100.0	57	30	9.0
	^{97}Mo α(1,0)	481	3.3	1.9	2.7	83
	^{100}Mo α(1,0)	536	40.4	23	2.1	6.0
	^{96}Mo α(1,0)	778	2.6	1.5	0.50	44
	^{95}Mo α(2,0)	786	2.6	1.5	0.60	50

Table A12.7b. Alpha-induced prompt gamma-rays generated by 5 MeV a particles. The notation x(m,n) means reaction with light ion product x and gamma-ray transition between states m and n in the residual nucleus. $-m_e$ and $-2m_e$ mean single and double escape peaks, respectively. The states are counted starting from 0 for the ground state. c/mC and bg/mC mean net gamma-ray and background counts/accumulated charge. Sensitivity values corresponds to the net count in the peak equal to three times the standard deviation of the peak background (Giles and Peisach, 1979) (continued).

Element	Assignment	E_γ *(keV)*	Relative intensity *(%)*	c/mC	bg/mC	‰
	^{98}Mo α(2,0)	787		not resolved		
	^{94}Mo α(1,0)	871		obscured by ^{14}N		
Ruthenium	^{99}Ru α(1,0)	89	40.4	38	330	45
	^{101}Ru α(1,0)	127	51.1	48	250	31
	^{104}Ru α(1,0)	358	100.0	94	57	7.6
	^{102}Ru α(1,0)	475	26.6	25	73	33
	^{100}Ru α(1,0)	540	3.9	3.7	39	160
Rhodium	^{103}Rh α(3,0)	295	100.0	1000	130	1.1
	^{103}Rh α(4,0)	358	69.0	690	31	0.77
Palladium	^{105}Pd α(1,0)	281	5.9	8.8	31	60
	^{110}Pd α(1,0)	374	100.0	150	33	3.6
	^{108}Pd α(1,0)	434	86.7	130	40	4.5
	^{110}Pd α(2,1)	440	8.7	13	14	28
	^{104}Pd α(1,0)	556	7.3	11	3.9	17
	^{102}Pd α(1,0)	557				
Silver	^{107}Ag α(2,1)	98	5.3	20	150	58
	^{109}Ag α(3,2)	103	4.7	18	180	70
	^{109}Ag α(2,0)	312	100.0	380	69	2.1
	^{107}Ag α(1,0)	325	81.6	310	41	2.0
	^{109}Ag α(3,0)	415	36.8	140	16	2.7
	^{109}Ag α(2,0)	423	31.6	120	11	2.6
Cadmium	^{113}Cd α(2,0)	299	100.0	25	14	14
	^{111}Cd α(2,0)	342	68.0	17	9.5	17
	^{114}Cd α(1,0)	558	80.0	20	2.9	8.1
	^{113}Cd α(6,0)	584	13.2	3.3	1.7	38
	^{112}Cd α(1,0)	617	32.8	8.2	2.2	17
	^{110}Cd α(1,0)	658	8.8	2.2	1.0	43
Indium		-				
Tin		-				
Tellurium		-				
Barium		-				
Lanthanum		-				
Cerium		-				
Praseodymium		-				
Neodymium		-				
Erbium	^{167}Er α(2,0)	178	100.0	16	140	61
	^{170}Er α(2,1)	182	22.5	3.6	100	230
	^{166}Er α(2,1)	184			not resolved	
	^{168}Er α(2,1)	184			not resolved	
Hafnium	^{178}Hf α(1,0)	93	100.0	120	190	11
	^{180}Hf α(1,0)	93			not resolved	
	^{177}Hf α(1,0)	113	34.2			
	^{179}Hf α(1,0)	123	23.3	41	120	25

Table A12.7b. Alpha-induced prompt gamma-rays generated by 5 MeV α particles. The notation x(m,n) means reaction with light ion product x and gamma-ray transition between states m and n in the residual nucleus. $-m_e$ and $-2m_e$ mean single and double escape peaks, respectively. The states are counted starting from 0 for the ground state. c/mC and bg/mC mean net gamma-ray and background counts/accumulated charge. Sensitivity values corresponds to the net count in the peak equal to three times the standard deviation of the peak background (Giles and Peisach, 1979) (continued).

Element	Assignment	E_γ *(keV)*	Relative intensity *(%)*	c/mC	bg/mC	‰
Tantalum	^{181}Ta α(1,0)	136	100.0	1200	200	1.1
	^{181}Ta α(2,1)	165	5.2	62	78	14
	^{181}Ta α(2,0)	301	3.3	39	16	9.7
Tungsten	^{183}W α(2,0)	99			not resolved	
	^{182}W α(1,0)	100	27.5	110	210	12
	^{184}W α(1,0)	111	65.0	260	150	4.5
	^{186}W α(1,0)	122	100.0	400	100	2.4
Rhenium	^{185}Re α(1,0)	125	59.1	650	420	3.0
	^{187}Re α(1,0)	134	100.0	1100	210	1.2
	^{185}Re α(2,1)	159	4.1	45	63	17
	^{187}Re α(3,1)	167	5.3	58	53	12
	^{185}Re α(2,0)	285	0.5	5.7	15	65
	^{187}Re α(3,0)	301	0.6	6.4	14	55
Iridium	^{193}Ir α(1,0)	73	100.0	510	260	3.0
	^{193}Ir α(3,1)	107	1.9	9.9	97	94
	^{191}Ir α(2,0)	129	56.9	290	150	3.9
	^{193}Ir α(2,0)	139	88.2	450	89	2.0
	^{191}Ir α(4,0)	179	1.3	6.4	32	84
	^{193}Ir α(3,0)	180			not resolved	
	^{191}Ir α(5,2)	214	2.5	13	24	36
	^{193}Ir α(4,2)	219	3.7	19	19	21
	^{191}Ir α(5,0)	343	2.7	14	5.7	16
	^{193}Ir α(4,0)	358	0.6	2.9	4.3	67
	^{193}Ir α(5,0)	362	3.1	16	4.6	13
Platinum	^{195}Pt α(1,0)	99	5.9	4.1	72	200
	^{195}Pt α(2,0)	130	4.6	3.2	59	230
	^{195}Pt α(5,1)	140	9.6	6.7	54	100
	^{195}Pt α(4,0)	211	92.9	65	35	8.6
	^{195}Pt α(5,0)	239	32.9	23	18	17
	^{192}Pt α(1,0)	317	4.0	2.8	9.6	100
	^{194}Pt α(1,0)	329	100.0	70	10	4.3
	^{196}Pt α(1,0)	356	42.9	30	9.0	9.5
	^{198}Pt α(1,0)	407	4.3	3.0	2.6	52
Gold	^{197}Au α(1,0)	77	100.0	190	140	5.7
	^{197}Au α(2,1)	192	7.9	15	36	39
	^{197}Au α(2,0)	269	1.1	2.0	6.8	120
	^{197}Au α(3,0)	279	36.8	70	15	5.0
	^{197}Au α(6,0)	548	0.6	1.2	1.5	97
Mercury	^{199}Hg α(1,0)	158	100.0	16	65	42
	^{199}Hg α(2,0)	208	16.9	2.7	28	160
	^{200}Hg α(1,0)	368	21.9	3.5	16	94
	^{198}Hg α(1,0)	412	9.4	1.5	15	220
	^{202}Hg α(1,0)	439	11.9	1.9	7.7	120
Lead		-				
Bismuth		-				

Table A12.8. Typical gamma-ray background peaks. 4n and 4n+2 refer to the natural radioactive decay chains (Giles and Peisach, 1979).

E_γ *(keV)*	**Assignment**	**Origin or natural decay chain**
57	Ta Kα X-rays	Chamber lining
61		
66	Ta Kβ X-rays	Chamber lining
74	^{208}Bi; ^{212}Pb	^{209}Bi(n, 2n) ^{208}Bi; 4n unresolved Kα X-ray
86	^{208}Bi; ^{212}Pb	^{209}Bi(n, 2n) ^{208}Bi; 4n unresolved Kβ X-ray
91		
110	^{19}F n,n′(1,0)	Neutron bombardment of detector
129	^{228}Ac	4n
136	^{181}Ta, α,α'(1,0)	Collimators and chamber lining
186	^{226}Ra	(4n + 2)
197	^{19}F n,n′(2,0)	Neutron bombardment of detector
239	^{212}Pb; ^{214}Pb	4n; (4n + 2)
277	^{208}Tl; ^{228}Ac	4n; 4n
285	^{214}Bi	(4n + 2)
296	^{210}Tl; ^{214}Pb	(4n + 2); (4n + 2)
301	^{181}Ta α,α' (2,0)	Collimators and chamber lining
322		
328	^{228}Ac	4n
339	^{228}Ac	4n
351	^{18}O α,n(1,0)	Oxygen on target surface
352	^{214}Pb	(4n + 2)
417	^{23}Na α,n(2,0)	Sodium contamination of target
440	^{23}Na α,α'(1,0)	Sodium contamination of target
463	^{228}Ac	4n
478	^{7}Be	Cosmic-ray produced
511	β^+; ^{208}Tl	Various; 4n
563	^{76}Ge n,n′(1,0)	Neutron bombardment of detector
583	^{208}Tl	4n
596	^{74}Ge n,n′(1,0)	Neutron bombardment of detector
604		
608	^{74}Ge n,n′(1,0)	Neutron bombardment of detector
609	^{214}Bi	(4n + 2)
666	^{214}Bi	(4n + 2)
691	^{72}Ge n,n′(1,0)	Neutron bombardment of detector
718		
727	^{212}Bi	4n
769	^{214}Bi	(4n + 2)
795	^{210}Tl; ^{228}Ac	(4n + 2); 4n
806		
834	^{72}Ge n,n′(2,0)	Neutron bombardment of detector
835	^{54}Mn	^{54}Fe(n, p)^{54}Mn by fast neutrons
844	^{27}Mg	^{27}Al(n, p)^{27}Mg by fast neutrons
	^{27}Al α,α'(1,0)	Scattered-alpha excitation of chamber
861	^{208}Tl	4n
871	^{14}N α,p(1,0)	Nitrogen on target surface
885	^{73}Ge n,α(1,0)	Neutron bombardment of detector
894	^{72}Ge n,n′(4,2)	Neutron bombardment of detector
911	^{228}Ac	4n
935	^{214}Bi	(4n + 2)
969	^{228}Ac	4n
1015	^{27}Al α,α'(2,0)	Scattered-alpha excitation of chamber
1120	^{214}Bi	(4n + 2)
1155	^{214}Bi	(4n + 2)
1214	(2236-2m_e)	^{27}Al
1223		
1238	^{214}Bi	(4n + 2)
1275		
1369	^{24}Na	^{27}Al(n, α)^{24}Na by fast neutrons
1378	^{214}Bi	(4n + 2)
1406		
1408	^{214}Bi	(4n + 2)
1461	^{40}K	0.012 atom% of natural K
1464	^{72}Ge n,n′(3,0)	Neutron bombardment of detector
1509	^{214}Bi	(4n + 2)
1588	^{228}Ac	4n
1592	(2614-2m_e)	^{208}Bi; ^{208}Tl
1693	(2204-m_e)	^{214}Bi
1725	(2236-m_e)	^{27}Al
1732	(2754-2m_e), ^{214}Bi	^{24}Na; (4n + 2)
1756		
1764	^{214}Bi	(4n + 2)
1809	^{23}Na α,p(1,0)	Sodium contamination of targets
1850	^{214}Bi	(4n + 2)
2103	(2614-m_e)	^{208}Bi; ^{208}Tl
2204	^{214}Bi	(4n + 2)
2236	^{27}Al, α,p(1,0)	Scattered-alpha excitation of chamber
2243	(2754-m_e)	^{24}Na
2614	^{208}Bi; ^{208}Tl	^{209}Bi(n, 2n)^{208}Bi by fast neutrons; 4n
2754	^{24}Na	^{27}Al(n,α)^{24}Na by fast neutrons

REFERENCES

Anttila, A., Hänninen, R., and Räisänen, J. (1981), *J. Radioanal. Chem.* **62**, 441.

Butler, J.W. (1959), U.S. Naval Research Laboratory, NRL Report 5282.

Elwyn, A.J., Marinov, A., and Schiffer, J.P. (1966), *Phys. Rev.* **146**, 957.

Giles, I., and Peisach, M. (1979), *J. Radioanal. Chem.* **50**, 307.

Kiss, A.Z., Koltay, E., Nyako, B., Somorjai, E., Anttila, A., and Räisänen J. (1985), *J. Radioanal. Chem.* **89**, 123.

Lappalainen, R., Anttila, A., and Räisänen, J. (1983), *Nucl. Instr. Meth.* **212**, 441.

Räisänen, J., and Hänninen, R. (1983), *Nucl. Instr. Meth.* **205**, 259.

Räisänen, J., Witting, T., and Keinonen, J. (1987), *Nucl. Instr. Meth.* **B28**, 199.

Ag 56 Ager-Hanssen, H., Lönsjö, O.M., and Nordhagen, R. (1956), *Phys. Rev.* **101**, 1779.

An 59 Andersen, S.L., Bö, H., Holtebekk, T., Lönsjö, O., and Tangen, R. (1959), *Nucl. Phys.* **9**, 509.

Ba 55 Bartholomew, G.A., Brown, F., Gove, H.E., Litherland, A.E., and Paul, E.B. (1955), *Can. J. Phys.* **33**, 441.

Ba 56 Baumann, N.P., Prosser, F.W., Jr., Read, W.G., and Krone, R.W. (1956), *Phys. Rev.* **104**, 376.

Ba 57 Bashkin, S., and Carlson, R.R. (1957), *Phys. Rev.* **106**, 261.

Be 57 Berenbaum, R., and Matthews, J.H. (1957), *Proc. Phys. Soc.* **70A**, 445.

Bo 48 Bonner, T.W., and Evans, J.E. (1948), *Phys. Rev.* **73**, 666.

Bo 57 Bolmgren, C.R., Freier, G.D., Likely, J.G., and Famularo, K.F. (1957), *Phys. Rev.* **105**, 210.

Bo 58 Bondelid, R.O., and Kennedy, C.A. (1958), "A Two-Meter Positive Ion Beam Electrostatic Analyzer," NRL Report 5083.

Br 47 Broström, K.J., Huus, T., and Tangen, R. (1947), *Phys. Rev.* **71**, 661.

Br 47a Broström, K.J., Huus, T., and Koch, J. (1947), *Nature* **160**, 498 (L).

Br 48 Broström, K.J., Huus, T., and Koch, J. (1948), *Nature* **162**, 695 (L).

Br 51 Broström, K.J., Madsen, B.S., and Madsen, C.B. (1951), *Phys. Rev.* **83**, 1265 (L).

Br 56 Broude, C., Green, L.L., Singh, J.J., and Willmott, J.C. (1956), *Phys. Rev.* **101**, 1052.

Br 56a Broude, C., Green, L.L., Willmott, J.C., and Singh, J.J. (1956), *Physica* **22**, 1139 (A).

Br 57 Broude, C., Green, L.L., Singh, J.J., and Willmott, J.C. (1957), *Phil. Mag.* **2**, 499.

Bu 55 Butler, J.W., and Holmgren, H.D. (1955), *Phys. Rev.* **99**, 1649 (A).

Bu 56 Bumiller, F., Staub, H.H., and Weaver, H.E. (1956), *Helv. Phys. Acta* **29**, 83.

Bu 56a Bumiller, F., Müller, J., and Staub, H.H. (1956), *Helv. Phys. Acta* **29**, 234.

Bu 57 Butler, J.W., and Gossett, C.R. (1957), *Phys. Rev.* **108**, 1473.

Bu 58 Butler, J.W. (1958), verbal report, Am. Phys. Soc. Meeting, Washington.

Ca 53 Casson, H. (1953), *Phys. Rev.* **89**, 809.

Ca 55 Carlson, R.R., and Nelson, E.B. (1955), *Phys. Rev.* **98**, 1310.

Ch 50 Chao, C.Y., Tollestrup, A.V., Fowler, W.A., and Lauritsen, C.C. (1950), *Phys. Rev.* **79**, 108.

Ch 56 Chick, D.R., Evans, W.W., Hancock, D.A., Hunt, S.E., and Pope, R.A. (1956), *Proc. Phys. Soc.* (London) **69A**, 624.

Ch 56a Chadwick, G.B., Alexander, T.K., and Warren, J.B. (1956), *Can. J. Phys.* **34**, 381.

Ch 57 Chick, D.R., and Hunt, S.E. (1957), *Nature* **180**, 88.

Cl 55 Clarke, R.L., Almqvist, E., and Paul, E.B. (1955), *Phys. Rev.* **99**, 654 (A).

Cl 56 Clegg, A.B. (1956), *Phil. Mag.* **1**, 1116.

Co 53 Cohen, A.V., and French, A.P. (1953), *Phil. Mag.* **44**, 1259.

Co 54 Cox, M.C., Van Loef, J.J., and Lind, D.A. (1954), *Phys. Rev.* **93**, 925 (A).

Cr 56 Craig, D.S., Cross, W.G., and Jarvis, R.G. (1956), *Phys. Rev.* **103**, 1414.

Cr 56a Cronin, J.W. (1956), *Phys. Rev.* **101**, 298.

Da 54 Day, R.B., and Huus, T. (1954), *Phys. Rev.* **95**, 1003.

De 57 DeVeiga Simão, J., and Sellschop, J.P.F. (1957), *Phys. Rev.* **106**, 98.

De 57a Dearnaley, G., Dissanaike, G.A., French, A.P., and Jones, G.L. (1957), *Phys. Rev.* **108**, 743.

Du 51 Duncan, D.B., and Perry, J.E. (1951), *Phys. Rev.* **82**, 809.

Fa 55 Farney, G.K., Givin, H.H., Kern, B.D., and Hahn, T.M. (1955), *Phys. Rev.* **97**, 720.

Fe 56 Ferguson, A.J., and Gove, H.E. (1956), *Bull. Am. Phys Soc.* **1**, 180.

Fe 56a Ferguson, A.J., Clarke, R.L., Gove, H.E., and Sample, J.T. (1956), Chalk River Report, PD-261.

Fe 58 Ferguson, A.J. (1958), *Bull. Am. Phys. Soc.* **3**, 26.

Fl 54 Flack, F.C., Rutherglen, J.G., and Grant, P.J. (1954), *Proc. Phys. Soc.* **67A**, 973.

Fo 49 Fowler, W.A., and Lauritsen, C.C. (1949), *Phys. Rev.* **76**, 314.

Fr 51 Freeman, J.M., and Seed, J. (1951), *Proc. Phys. Soc.* (London) **64A**, 313 (L).

Gr 51 Grove, C.R., and Cooper, J.N. (1951), *Phys. Rev.* **82**, 505.

Gr 55 Grant, P.J., Rutherglen, J.G., Flack, F.C., and Hutchinson, G.W. (1955), *Proc. Phys. Soc.* (London) **68A**, 369.

Gr 55a Green, L.L., Singh, J.J., and Willmott, J.C. (1955), *Phil. Mag.* **46**, 982.

Gr 56 Green, L.L., Singh, J.J., and Willmott, J.C. (1956), *Proc. Phys. Soc.* (London) **69A**, 335.

Ha 51 Hanscome, T.D., and Malich, C.W. (1951), *Phys. Rev.* **82**, 304 (A).

Ha 52 Hahn, T.M., Snyder, C.W., Willard, H.B., Bair, J.K., Klema, E.D., Kington, J.D., and Green, F.P. (1952), *Phys. Rev.* **85**, 934.

Ha 55 Hancock, D.A., and Verdaguer, F. (1955), *Proc. Phys. Soc.* (London) **68A**, 1080.

Ha 55a Hahn, T.M., Jr., Kern, B.D., and Farney, G.K. (1955), *Phys. Rev.* **98**, 1183 (A).

Ha 55b Haeberli, W. (1955), *Phys. Rev.* **99**, 640 (A).

Ha 57 Hagedorn, F.B., Mozer, F.S., Webb, T.S., Fowler, W.A., and Lauritsen, C.C. (1957), *Phys. Rev.* **105**, 219.

Ha 57a Hagedorn, F.B. (1957), *Phys. Rev.* **108**, 735.

Ha 57b Hagedorn, F.B., and Marion, J.B. (1957), *Phys. Rev.* **108**, 1015.

He 58 Hebbard, D., private communication to F. Ajzenberg-Selove.

Ho 53 Hornyak, W.F., and Coor, T. (1953), *Phys. Rev.* **92**, 675.

Ho 55 Holland, R.E., Inglis, D.R., Malm, R.E., and Mooring, F.P. (1955), *Phys. Rev.* **99**, 92.

Hu 52 Hunt, S.E. (1952), *Proc. Phys. Soc.* (London) **65A**, 982.

Hu 53 Huus, T., and Day, R.B. (1953), *Phys. Rev.* **91**, 599.

Hu 53a Hunt, S.E., and Jones, W.M. (1953), *Phys. Rev.* **89**, 1283.

Hu 54 Hunt, S.E., Jones, W.M., Churchill, J.L.W., and Hancock, D.A. (1954), *Proc. Phys. Soc.* (London) **67A**, 443.

Hu 55 Hunt, S.E., and Hancock, D.A. (1955), *Phys. Rev.* **97**, 567 (L).

Hu 55a Hunt, S.E., and Firth, K. (1955), *Phys. Rev.* **99**, 786.

Hu 57 Hunt, S.E., Pope, R.A., and Evans, W.W. (1957), *Phys. Rev.* **106**, 1012.

Ka 55 Kavanagh, R.W., Mills, W.R., and Sherr, R. (1955), *Phys. Rev.* **97**, 248 (L).

Ke 56 Kern, B.D., and Cochran, L.W. (1956), *Phys. Rev.* **104**, 711.

Kl 54 Kluyver, J.C., van der Leun, C., and Endt, P.M. (1954), *Physica* **20**, 1287.

Kl 55 Kluyver, J.C., and Endt, P.M. (1955), Conference on Electromagnetically Enriched Isotopes, Harwell.

Ko 52 Koester, L.J., Jr. (1952), *Phys. Rev.* **85**, 643.

Kr 54 Kraus, A.A., Jr. (1954), *Phys. Rev.* **93**, 1308.

Kr 54a Kraus, A.A., Jr. (1954), *Phys. Rev.* **94**, 975.

Li 53 Li, C.W. (1953), *Phys. Rev.* **92**, 1084 (A).

Li 56 Litherland, A.E., Paul, E.B., Bartholomew, G.A., and Gove, H.E. (1956), *Phys. Rev.* **102**, 208.

Li 59 Litherland, A.E., Gove, H.E., and Ferguson, A.J. (1959), *Phys. Rev.* **114**, 1312.

Ma 53 Mackin, R.J., Jr. (1953), Ph.D. Thesis, California Institute of Technology.

Ma 54 Mackin, R.J., Jr., Mims, W.B., and Mills, W.R. (1954), *Phys. Rev.* **93**, 950 (A).

Ma 54a Mackin, R.J., Jr. (1954), *Phys. Rev.* **94**, 648.

Ma 56 Marion, J.B., and Hagedorn, F.B. (1956), *Phys. Rev.* **104**, 1028.

Mi 54 Milne, E.A. (1954), *Phys. Rev.* **93**, 762.

Mi 55 Milani, S., Cooper, J.N., and Harris, J.C. (1955), *Phys. Rev.* **99**, 645 (A), plus verbal report to Nuclear Data Group.

Mo 51 Mooring, F.P., Koester, L.J., Jr., Goldberg, E., Saxon, D., and Kaufmann, S.G. (1951), *Phys. Rev.* **84**, 703.

Mo 56 Mozer, F.S. (1956), *Phys. Rev.* **104**, 1386.

Na 53 Nathans, R., and Halpern, J. (1953), *Phys. Rev.* **92**, 207.

Ne 57 Newton, J.O., private communication to P.M. Endt.

Ne 57a Newson, H.W., Williamson, R.M., Jones, K.W., Gibbons, J.H., and Marshak, H. (1957), *Phys. Rev.* **108**, 1294.

Ov 56 Overley, J.C., Pixley, R.E., and Whaling, W. (1956), *Bull. Am. Phys. Soc.* **1**, 387, and verbal report.

Pa 55 Paul, E.B., Gove, H.E., Litherland, A.E., and Bartholomew, G.A. (1955), *Phys. Rev.* **99**, 1339.

Pl 40 Plain, G.P., Herb, R.G., Hudson, C.M., and Warren, R.E. (1940), *Phys. Rev.* **57**, 187.

Pr 54 Price, P.C. (1954), *Proc. Phys. Soc.* (London) **67A**, 849.

Pr 56 Prosser, F.W., Jr., Baumann, N.P., Brice, D.K., Read, W.G., and Krone, R.W. (1956), *Phys. Rev.* **104**, 369.

Ro 51 Roseborough, W.D., McCue, J.J.G., Preston, W.M., and Goodman, C. (1951), *Phys. Rev.* **83**, 1133.

Ru 54 Rutherglen, J.G., Grant, P.S., Flack, F.C., and Deuchars, W.M. (1954), *Proc. Phys. Soc.* (London) **67A**, 101.

Ru 54a Russell, L.N., Taylor, W.E., and Cooper, J.N. (1954), *Phys. Rev.* **95**, 99.

Sa 56 Sanders, R.M. (1956), *Phys. Rev.* **104**, 1434.

Sc 52 Schardt, A., Fowler, W.A., and Lauritsen, C.C. (1952), *Phys. Rev.* **86**, 527.

Se 51 Seagrave, J.D. (1951), *Phys. Rev.* **84**, 1219.

Se 52 Seagrave, J.D. (1952), *Phys. Rev.* **85**, 197.

Se 55 Seiler, M.R., Cooper, J.N., and Harris, J.C. (1955), *Phys. Rev.* **99**, 340 (A).

Si 54 Sinclair, R.M. (1954), *Phys. Rev.* **93**, 1082.

Sm 58 Smith, P.B., and Endt, P.M. (1958), *Phys. Rev.* **110**, 397.

St 54 Stelson, P.H., and Preston, W.M. (1954), *Phys. Rev.* **95**, 974.

Sw 55 Swann, C.P., Rothman, M.A., Porter, W.C., and Mandeville, C.E. (1955), *Phys. Rev.* **98**, 1183 (A).

Ta 46 Tangen, R. (1946), *Kgl. Norske Videnskabers Selskabs Skrifter*, No. 1 pp. 1-91.

Ta 54 Taylor, W.E., Russell, L.N., and Cooper, J.N. (1954), *Phys. Rev.* **93**, 1056.

Th 58 Thornton, D.E.J., Meads, R.E., and Collie, C.H. (1958), *Phys. Rev.* **109**, 480.

To 57 Towle, J.H., Berenbaum, R., and Matthews, J.H. (1957), *Proc. Phys. Soc.* (London) **70A**, 84.

Ts 56 Tsytko, S.P., and Antuf'ev, Iu.P. (1956), Zhur. Eksptl. i Teoret. Fiz. **30**, 1171 (1956); *Soviet Phys.* **JETP 3,** 993.

Va 53 Van Loef, J.J., private communication to P.M. Endt.

Va 56 Van Patter, D.M., Swann, C.P., Porter, W.C., and Mandeville, C.E. (1956), *Phys. Rev.* **103**, 656.

Va 56a van der Leun, C., and Endt, P.M. (1956), *Physica* **22**, 1234 (L).

Va 56b van der Leun, C., and Endt, P.M., Kluyver, J.C., and Vrenken, L.E. (1956), *Physica* **22**, 1223.

Va 56c Varma, J., and Jack, W. (1956), *Physica* **22**, 1139 (A).

Vo 57 Vorona, J., Olness, J.W., Haeberli, W., and Lewis, H.W. (1957), *Bull. Am. Phys. Soc.* **2**, 34.

Wi 52 Willard, H.B., Bair, J.K., Kington, J.D., Hahn, T.M., Snyder, C.W., and Green, F.P. (1952), *Phys. Rev.* **85**, 849.

Wi 53 Wilkinson, D.H. (1953), *Phys. Rev.* **90**, 721.

Wi 56 Wilkinson, D.H., Toppel, B.J., and Alburger, D.E. (1956), *Phys. Rev.* **101**, 673.

Wi 57 Wilkinson, D.H., and Bloom, S.D. (1957), *Phil. Mag.* **2**, 63.

Wi 57a Wilkinson, D.H., and Bloom, S.D. (1957), *Phys. Rev.* **105**, 183.

Wo 53 Woodbury, H.H., Day, R.B., and Tollestrup, A.V. (1953), *Phys. Rev.* **92**, 1199.

Zi 57 Zipoy, D.M., Freier, G., and Famularo, K. (1957), *Phys. Rev.* **106**, 93.

Zi 58 Zipoy, D.M. (1958), *Phys. Rev.* **110**, 995 (L).

APPENDIX 13

HYDROGEN NUCLEAR REACTION DATA

Compiled by

W. A. Lanford

State University of New York, Albany, New York

CONTENTS

Appendix 13

Table A13.1. Nuclear reaction cross section characteristics[1, 2].

Reaction	Units	$^{7}Li + H$	$^{15}N + H$	$^{15}H + H$	$^{19}F + H$	$^{19}F + H$
Resonance energy	MeV	3.07	6.385	13.35	6.418	16.44
Cross section (σ) at resonance	mbarn	4.8	1650	1050	88	440
Resonance width (Γ)	keV	81	1.8	25.4	44	86
σ·Γ	mbarn × keV	389	2970	26700	3870	37800
Relative yield		0.13	1.000	9.0	1.3	12.7
dE/dx in Si	MeV/micron	0.442	1.45	1.35	1.94	1.94
(Γ)/dE/dx	microns	0.183	0.0012	0.0188	0.0226	0.0443
Energy of next resonance	MeV	7.11	13.35	18.0	9.1	17.6
Gamma-ray energy	MeV	17.7,14.7	4.43	4.43	6.13,6.98,7.12	6.13,6.98,7.12

1. The ^{7}Li results are from Trocellier and Engelmann (1986), see Chapter 8.
2. The ^{15}N and ^{19}F results are from Xiong *et al.,* (1987), see Chapter 8.

Table A13.2. dE/dx for ^{7}Li, ^{15}N and ^{19}F at their resonance energies.

Element	Atomic number	Density	^{7}Li at 3.07 MeV *(MeV/mg/cm^2)*	^{15}N at 6.385 MeV *(MeV/mg/cm^2)*	^{15}N at 13.35 MeV *(MeV/mg/cm^2)*	^{19}F at 6.418 MeV *(MeV/mg/cm^2)*	^{19}F at 16.44 MeV *(MeV/mg/cm^2)*
H	1		8.715E+00	2.838E+01	2.166E+01	4.145E+01	3.222E+01
He	2		3.508E+00	1.148E+01	9.233E+00	1.606E+01	1.370E+01
Li	3	0.54	2.606E+00	8.514E+00	7.399E+00	1.166E+01	1.087E+01
Be	4	1.82	2.330E+00	7.693E+00	7.055E+00	1.022E+01	1.032E+01
B	5	2.47	2.528E+00	8.307E+00	7.227E+00	1.143E+01	1.062E+01
C	6	3.52	2.574E+00	8.453E+00	7.397E+00	1.160E+01	1.086E+01
N	7		2.607E+00	8.537E+00	7.112E+00	1.192E+01	1.048E+01
O	8		2.425E+00	7.965E+00	6.923E+00	1.087E+01	1.016E+01
F	9		2.096E+00	6.915E+00	6.266E+00	9.046E+00	9.178E+00
Ne	10		2.043E+00	6.747E+00	6.198E+00	8.799E+00	9.077E+00
Na	11	1.01	2.117E+00	6.987E+00	6.442E+00	9.204E+00	9.422E+00
Mg	12	1.74	1.990E+00	6.535E+00	6.022E+00	8.682E+00	8.795E+00
Al	13	2.70	1.775E+00	5.858E+00	5.556E+00	7.633E+00	8.099E+00
Si	14	2.33	1.896E+00	6.242E+00	5.677E+00	8.421E+00	8.297E+00
P	15	1.82	1.843E+00	6.055E+00	5.458E+00	8.288E+00	7.976E+00
S	16	2.07	1.779E+00	5.820E+00	4.924E+00	8.231E+00	7.218E+00
Cl	17		1.883E+00	6.162E+00	5.222E+00	8.699E+00	7.655E+00
Ar	18		1.649E+00	5.416E+00	4.834E+00	7.464E+00	7.059E+00
K	19	0.91	1.875E+00	6.133E+00	5.456E+00	8.222E+00	8.010E+00
Ca	20	1.53	1.823E+00	5.978E+00	5.354E+00	7.942E+00	7.839E+00
Sc	21	2.99	1.713E+00	5.613E+00	4.872E+00	7.655E+00	7.148E+00
Ti	22	4.51	1.571E+00	5.178E+00	4.737E+00	6.909E+00	6.921E+00
V	23	6.09	1.599E+00	5.263E+00	4.566E+00	7.195E+00	6.698E+00
Cr	24	7.19	1.456E+00	4.825E+00	4.477E+00	6.330E+00	6.538E+00
Mn	25	7.47	1.374E+00	4.550E+00	4.264E+00	5.936E+00	6.221E+00
Fe	26	7.87	1.373E+00	4.531E+00	4.261E+00	5.926E+00	6.211E+00
Co	27	8.9	1.283E+00	4.243E+00	4.037E+00	5.510E+00	5.877E+00
Ni	28	8.91	1.250E+00	4.160E+00	4.116E+00	5.277E+00	5.971E+00
Cu	29	8.93	1.137E+00	3.776E+00	3.814E+00	4.778E+00	5.516E+00
Zn	30	7.13	1.145E+00	3.797E+00	3.783E+00	4.801E+00	5.481E+00
Ga	31	5.91	1.092E+00	3.606E+00	3.537E+00	4.651E+00	5.119E+00
Ge	32	5.32	1.071E+00	3.547E+00	3.593E+00	4.536E+00	5.184E+00
As	33	5.77	1.058E+00	3.501E+00	3.417E+00	4.588E+00	4.945E+00
Se	34	4.81	1.003E+00	3.321E+00	3.319E+00	4.311E+00	4.794E+00
Br	35		1.041E+00	3.422E+00	3.464E+00	4.500E+00	4.988E+00
Kr	36		1.048E+00	3.434E+00	3.340E+00	4.572E+00	4.828E+00
Rb	37	1.63	1.153E+00	3.783E+00	3.570E+00	4.952E+00	5.203E+00
Sr	38	2.58	1.146E+00	3.749E+00	3.431E+00	5.089E+00	4.986E+00
Y	39	4.48	1.118E+00	3.669E+00	3.404E+00	4.934E+00	4.946E+00
Zr	40	6.51	1.126E+00	3.704E+00	3.405E+00	5.031E+00	4.952E+00
Nb	41	8.58	1.158E+00	3.828E+00	3.601E+00	5.009E+00	5.248E+00
No	42	10.22	1.037E+00	3.440E+00	3.249E+00	4.502E+00	4.730E+00
Tc	43	11.5	1.044E+00	3.455E+00	3.302E+00	4.527E+00	4.799E+00
Ru	44	12.36	1.036E+00	3.431E+00	3.271E+00	4.458E+00	4.759E+00
Rh	45	12.42	1.006E+00	3.332E+00	3.211E+00	4.307E+00	4.667E+00
Pd	46	12.00	9.752E-01	3.244E+00	3.169E+00	4.155E+00	4.600E+00
Ag	47	10.5	9.602E-01	3.187E+00	3.032E+00	4.139E+00	4.406E+00
Cd	48	8.65	9.329E-01	3.082E+00	2.998E+00	3.993E+00	4.350E+00
In	49	7.29	9.298E-01	3.066E+00	2.984E+00	3.968E+00	4.329E+00
Sn	50	5.76	9.053E-01	2.985E+00	2.855E+00	3.907E+00	4.147E+00
Sb	51	6.69	8.942E-01	2.955E+00	2.866E+00	3.839E+00	4.160E+00
Te	52	6.25	8.879E-01	2.932E+00	2.782E+00	3.850E+00	4.040E+00
I	53		8.977E-01	2.947E+00	2.859E+00	3.920E+00	4.137E+00
Xe	54		8.820E-01	2.894E+00	2.825E+00	3.832E+00	4.086E+00
Cs	55	1.99	9.601E-01	3.151E+00	2.994E+00	4.111E+00	4.359E+00
Ba	56	3.59	9.256E-01	3.037E+00	2.891E+00	4.028E+00	4.199E+00
La	57	6.17	9.380E-01	3.083E+00	2.866E+00	4.071E+00	4.180E+00
Ce	58	6.77	8.289E-01	2.729E+00	2.732E+00	3.611E+00	3.937E+00
Pr	59	6.78	8.666E-01	2.859E+00	2.828E+00	3.680E+00	4.097E+00
Nd	60	7.00	8.349E-01	2.753E+00	2.731E+00	3.552E+00	3.953E+00

Table A13.2. dE/dx for ^{7}Li, ^{15}N and ^{19}F at their resonance energies (continued).

Element	Atomic number	Density	^{7}Li at 3.07 MeV *(MeV/mg/cm^2)*	^{15}N at 6.385 MeV *(MeV/mg/cm^2)*	^{15}N at 13.35 MeV *(MeV/mg/cm^2)*	^{19}F at 6.418 MeV *(MeV/mg/cm^2)*	^{19}F at 16.44 MeV *(MeV/mg/cm^2)*
Pm	61		8.362E-01	2.755E+00	2.709E+00	3.587E+00	3.920E+00
Sm	62	7.54	8.612E-01	2.844E+00	2.809E+00	3.623E+00	4.074E+00
Eu	63	5.25	7.640E-01	2.523E+00	2.613E+00	3.192E+00	3.760E+00
Gd	64	7.89	7.704E-01	2.541E+00	2.541E+00	3.281E+00	3.670E+00
Tb	65	8.27	7.389E-01	2.441E+00	2.503E+00	3.106E+00	3.608E+00
Dy	66	8.53	7.273E-01	2.405E+00	2.477E+00	3.041E+00	3.571E+00
Ho	67	8.80	6.587E-01	2.181E+00	2.308E+00	2.731E+00	3.315E+00
Er	68	9.04	7.032E-01	2.327E+00	2.413E+00	2.916E+00	3.478E+00
Tm	69	9.32	6.543E-01	2.165E+00	2.276E+00	2.733E+00	3.269E+00
Yb	70	6.97	6.562E-01	2.164E+00	2.207E+00	2.774E+00	3.174E+00
Lu	71	9.84	6.336E-01	2.094E+00	2.184E+00	2.670E+00	3.137E+00
Hf	72	13.2	6.221E-01	2.060E+00	2.157E+00	2.617E+00	3.098E+00
Ta	73	16.66	6.286E-01	2.092E+00	2.191E+00	2.635E+00	3.151E+00
W	74	19.25	6.258E-01	2.083E+00	2.083E+00	2.652E+00	3.013E+00
Re	75	21.03	6.200E-01	2.062E+00	2.137E+00	2.583E+00	3.078E+00
Os	76	22.58	6.085E-01	2.025E+00	2.101E+00	2.532E+00	3.027E+00
Ir	77	22.55	5.951E-01	1.981E+00	2.082E+00	2.481E+00	2.994E+00
Pt	78	21.47	5.751E-01	1.917E+00	2.015E+00	2.433E+00	2.890E+00
Au	79	19.28	6.070E-01	2.022E+00	2.043E+00	2.568E+00	2.944E+00
Hg	80	14.26	5.900E-01	1.949E+00	2.000E+00	2.504E+00	2.872E+00
Tl	81	11.87	5.901E-01	1.948E+00	2.017E+00	2.496E+00	2.898E+00
Pb	82	11.34	6.185E-01	2.042E+00	2.022E+00	2.649E+00	2.915E+00
Bi	83	9.8	6.470E-01	2.134E+00	2.077E+00	2.794E+00	3.001E+00
Po	84	9.24	6.363E-01	2.097E+00	2.090E+00	2.763E+00	3.009E+00
At	85		6.462E-01	2.122E+00	2.113E+00	2.787E+00	3.046E+00
Rn	86		6.182E-01	2.030E+00	2.025E+00	2.669E+00	2.917E+00
Fr	87		6.369E-01	2.088E+00	2.021E+00	2.766E+00	2.919E+00
Ra	88	5	6.498E-01	2.129E+00	2.040E+00	2.849E+00	2.946E+00
Ac	89		6.751E-01	2.217E+00	2.110E+00	2.962E+00	3.055E+00
Th	90	11.2	6.432E-01	2.117E+00	2.035E+00	2.804E+00	2.942E+00
Pa	91	15.4	6.568E-01	2.207E+00	2.090E+00	2.933E+00	3.028E+00
U	92	18.7	5.880E-01	1.948E+00	1.962E+00	2.521E+00	2.828E+00

Table A13.3. Hydrogen bearing materials.

Common name	Chemical name	Composition
Celluloid	Cellulose acetate	$(C_9H_{13}O_7)_n$
Kapton	Polyamide film	$(C_{22}H_{10}N_2O_4)_n$
Kel-F	Chlorotrifluoro-ethylene polymer	$(CF_2CHCl)_n$
Lucite (Plexiglas)	methyl methacrylate	$(C_5H_8O_2)_n$
Mica	Muscovite	$(K_2O)3(Al_2O_3)6(SiO_2)2(H_2O)$
Mylar	Polyester film	$(C_{10}H_8O_4)_n$
Nylon	polyamides	$(C_{12}H_{22}N_2O_2)_n$
Polythylene		$(CH_2)_n$
Polystyrene		$(CH)_n$

APPENDIX 14

ACTIVATION ANALYSIS DATA

Compiled by

G. Blondiaux and J. L. Debrun

CNRS-CERI, Orleans, France

C. J. Maggiore

Los Alamos National Laboratory, Los Alamos, New Mexico

All references for this appendix appear in the bibliography of Chapter 9, "Charged Particle Activation Analysis."

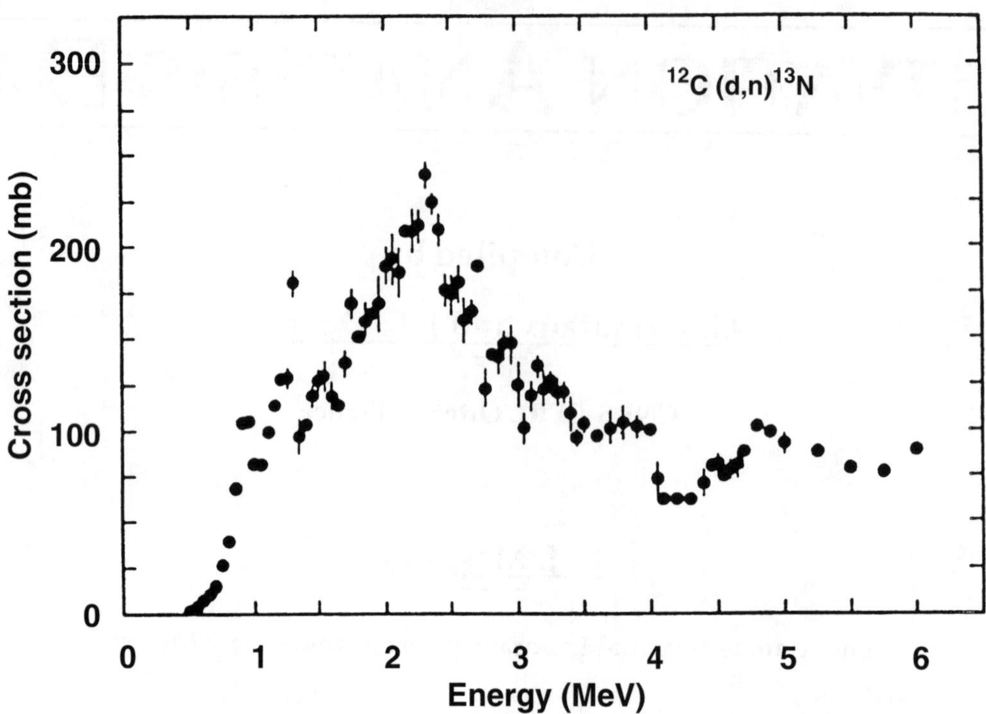

FIG. A14.1. Cross-section for the $^{12}C(d, n)^{13}N$ reaction, from Michelmann *et al.* (1990).

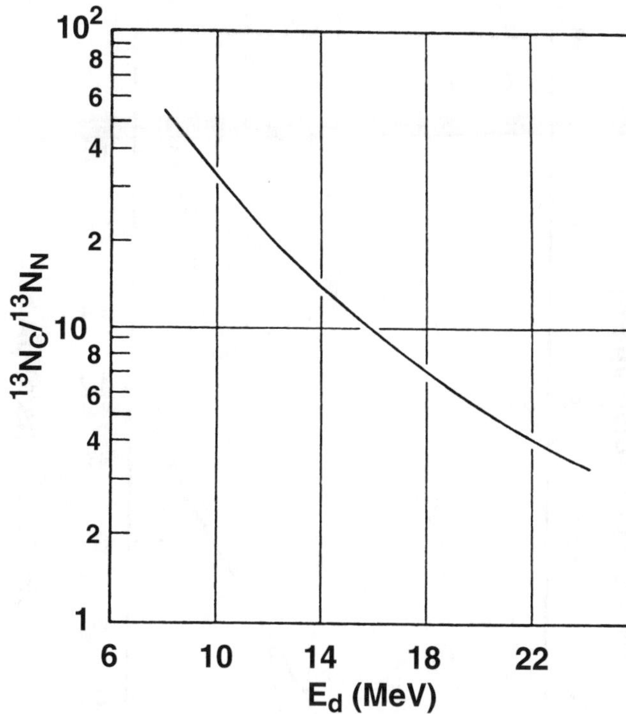

FIG. A14.2. Ratio of the ^{13}N activity from the $^{12}C(d, n)^{13}N$ reaction, to that from the $^{14}N(d, t)^{13}N$ + $^{14}N(d, dn)^{13}N$ reactions, for equal C and N concentrations, versus the irradiation energy. Engelmann and Marschal (1971).

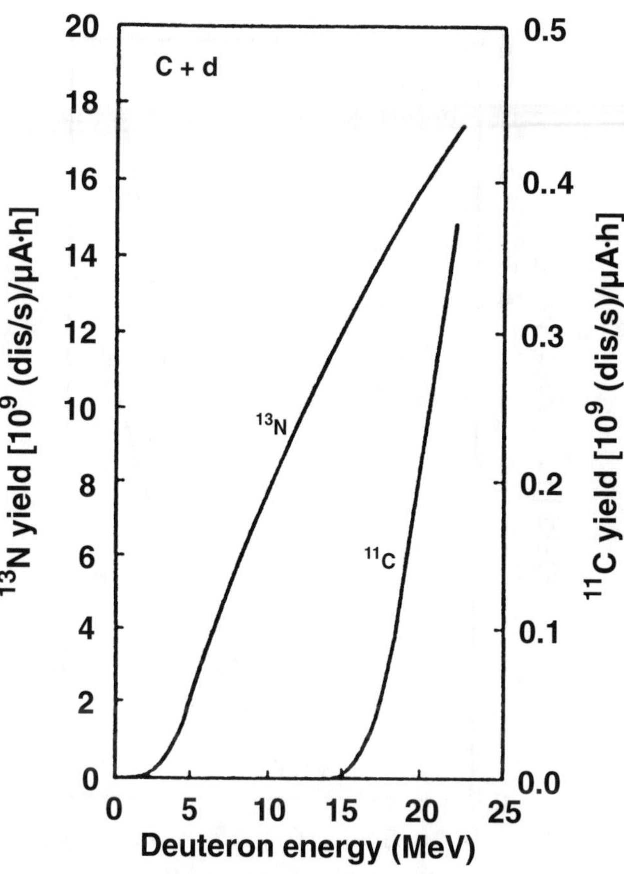

FIG. A14.3. Thick target yield for the $^{12}C(d, n)^{13}N$ reaction, from Krasnov *et al.* (1970). Also shown $^{12}C(d, t)^{11}C$. Target is graphite.

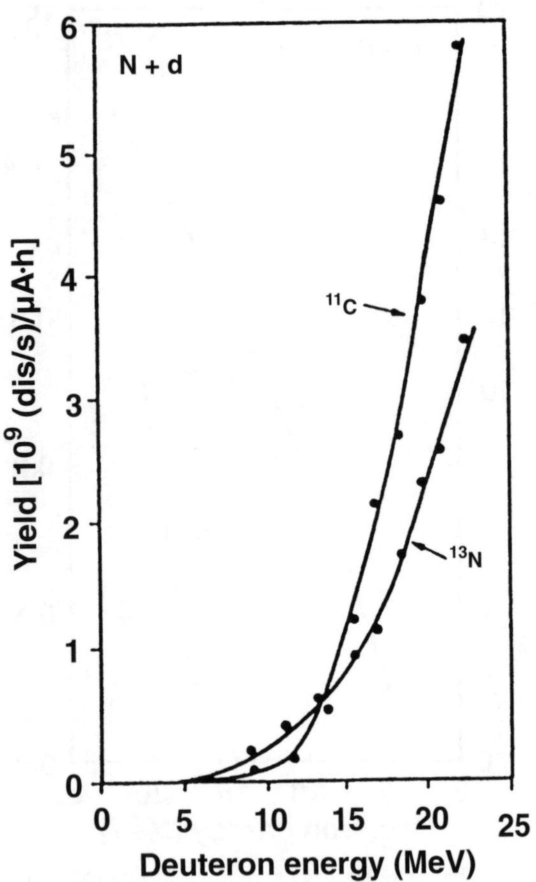

FIG. A14.4. Thick target yield for the production of ^{13}N by the $^{14}N(d, t)^{13}N$ + $^{14}N(d,dn)^{13}N$ reactions from Krasnov *et al.* (1970). Also shown $^{14}N(d, \alpha n)^{11}C$. Target is AlN.

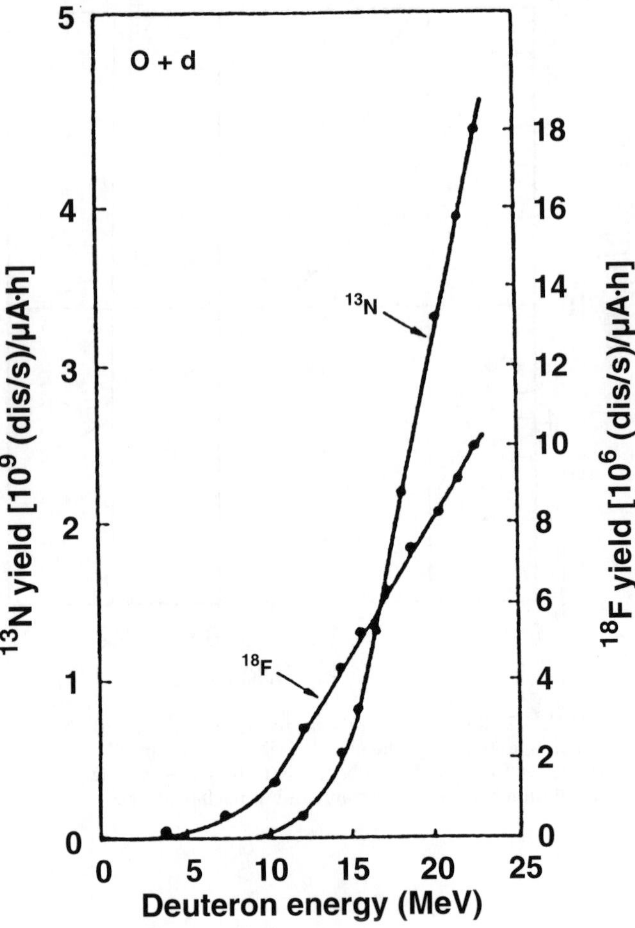

FIG. A14.5. Thick target yield for the production of ^{13}N by the $^{16}O(d, \alpha n)^{13}N$ reaction, from Krasnov *et al.* (1970). Also shown $^{17}O(d, n)^{18}F$. Target is Al_2O_3.

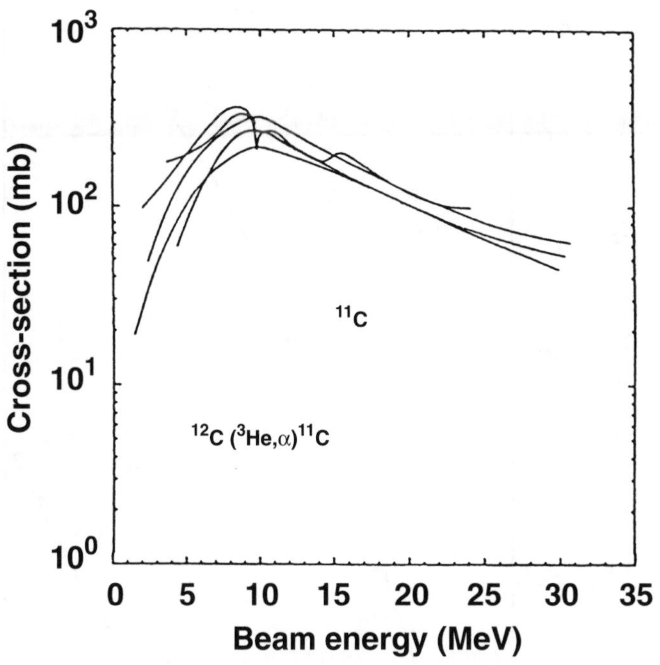

FIG. A14.6. Cross-sections for the $^{12}C(^{3}He, \alpha)^{11}C$ reaction, compiled by Lamb (1969).

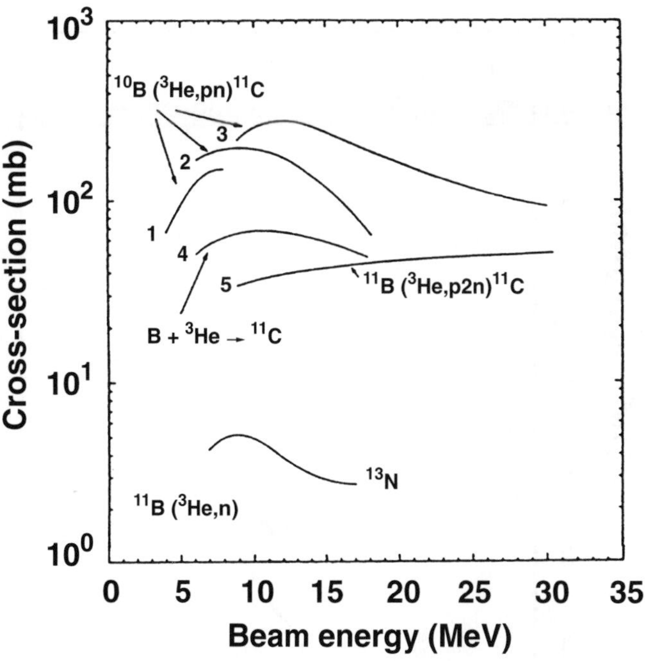

FIG. A14.7. Cross-sections for the production of ^{11}C by irradiation of boron, compiled by Lamb (1969).

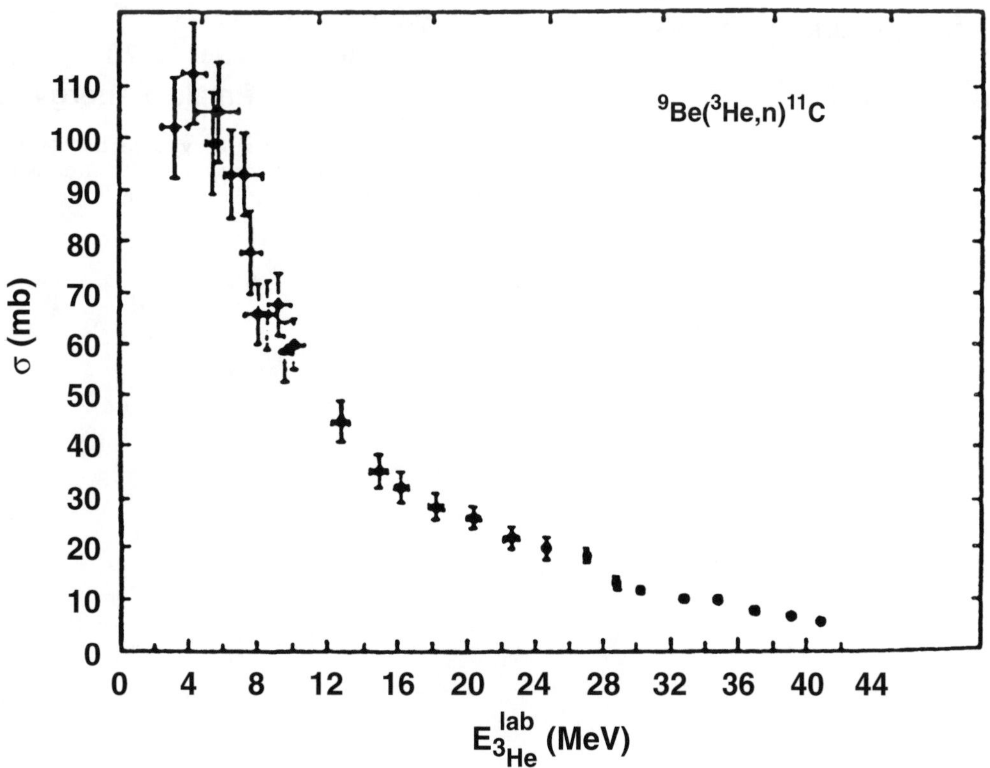

FIG. A14.8. Cross-section for the $^{9}Be(^{3}He, n)^{11}C$ reaction, from Anders (1981).

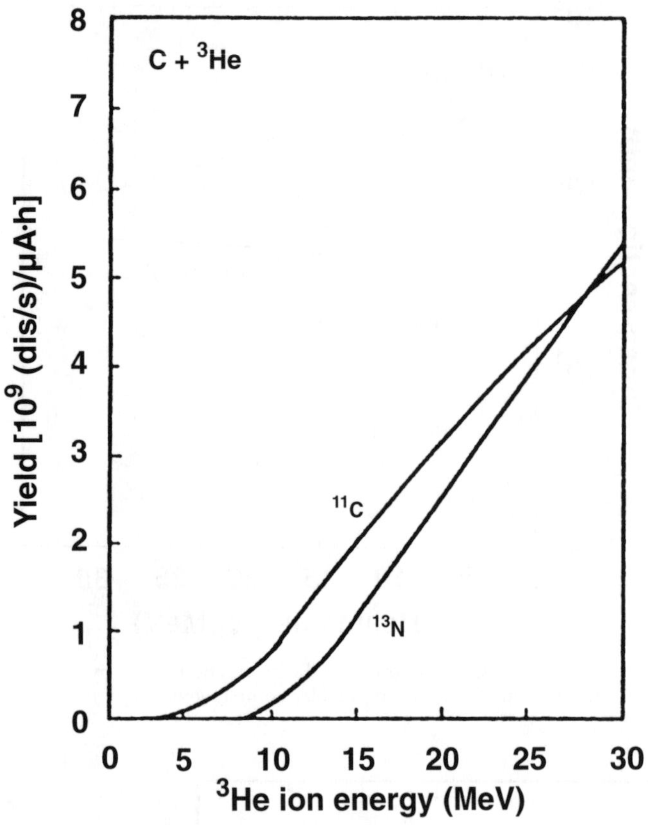

FIG. A14.9. Thick target yield for the production of ^{11}C by the $^{12}C(^{3}He, \alpha)^{11}C$ reaction, from Krasnov *et al.* (1970). Also shown $^{12}C(^{3}He, d)^{13}N$. Target is graphite.

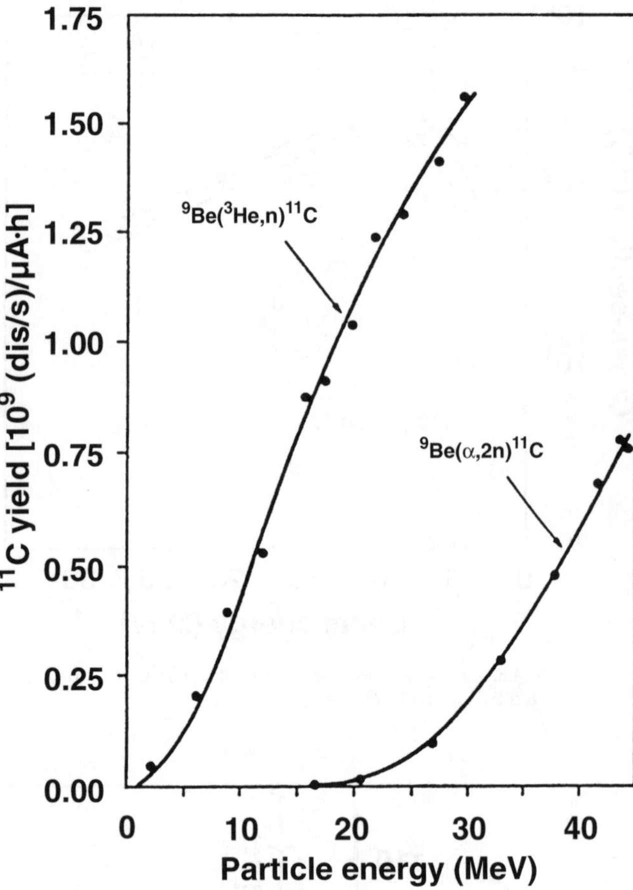

FIG. A14.10. Thick target yield for the $^{9}Be(^{3}He, n)^{11}C$ reaction, from Krasnov *et al.* (1970). Also shown $^{9}Be(\alpha, 2n)^{11}C$. Target is beryllium.

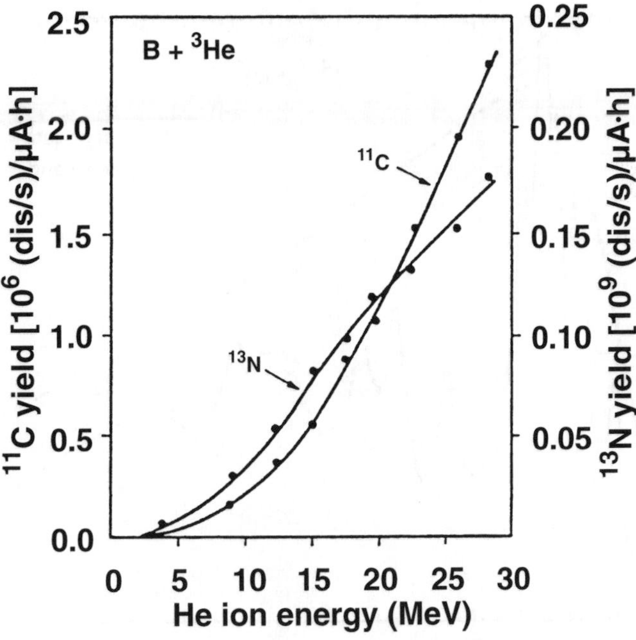

FIG. A14.11. Thick target yield for the production of ^{11}C by ^{3}He irradiation of boron, from Krasnov *et al.* (1970). Also shown, production of ^{13}N. Target is boron.

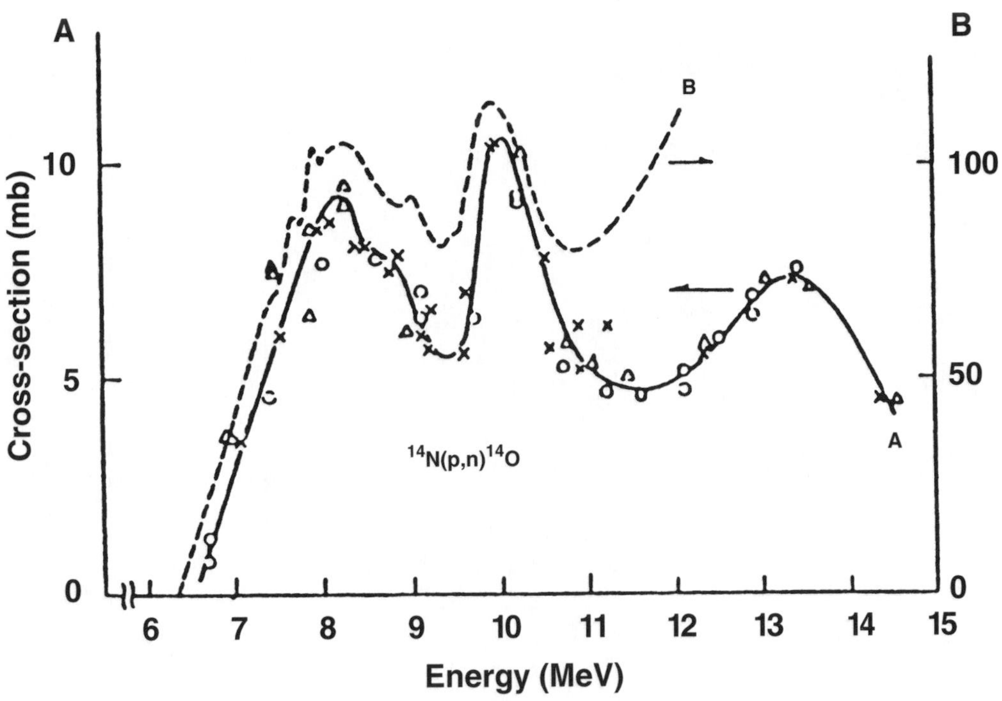

FIG. A14.12. Cross-section for the $^{14}N(p, n)^{14}O$ reaction, from Nozaki and Iwamoto (1981) and Kuan and Risser (1964).

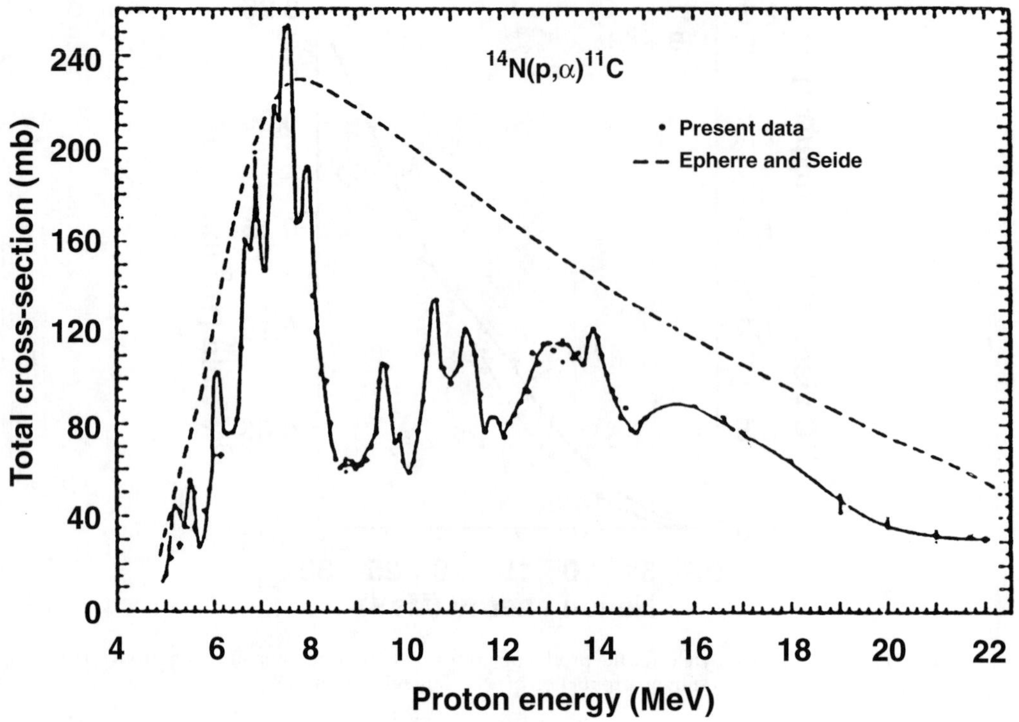

FIG. A14.13. Cross-section for the ^{14}N(p, α)^{11}C reaction, from Jacobs *et al.* (1974).

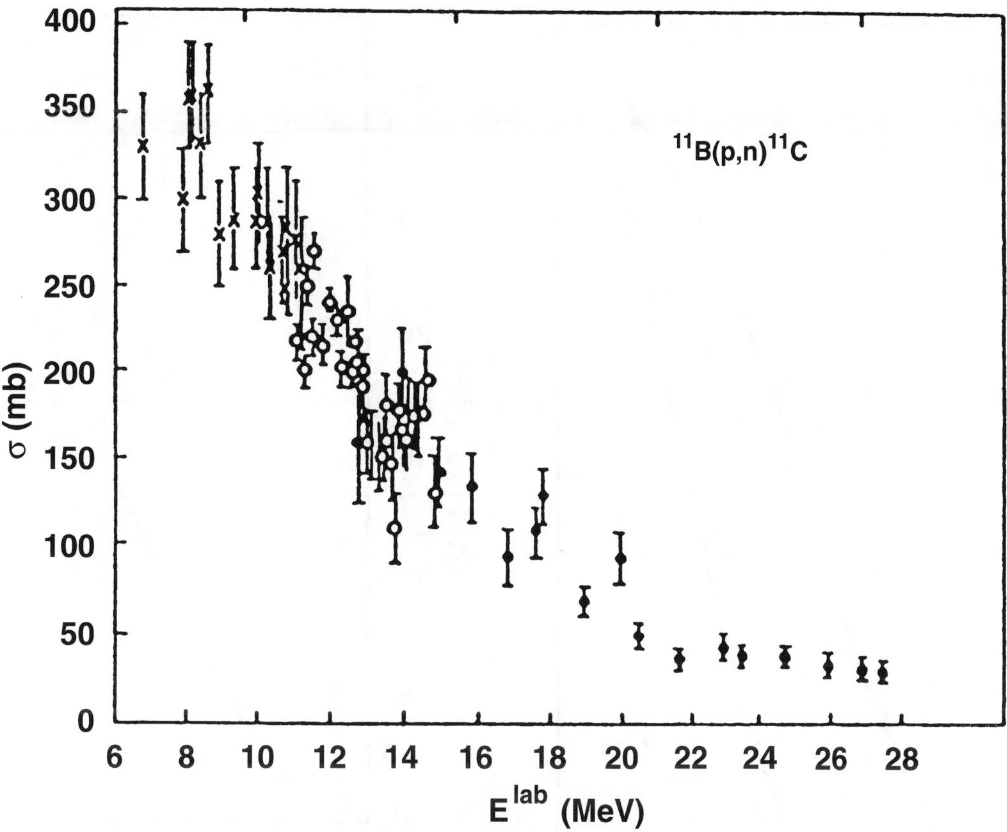

FIG. A14.14. Cross-section for the $^{11}B(p, n)^{11}C$ reaction, from Anders (1981).

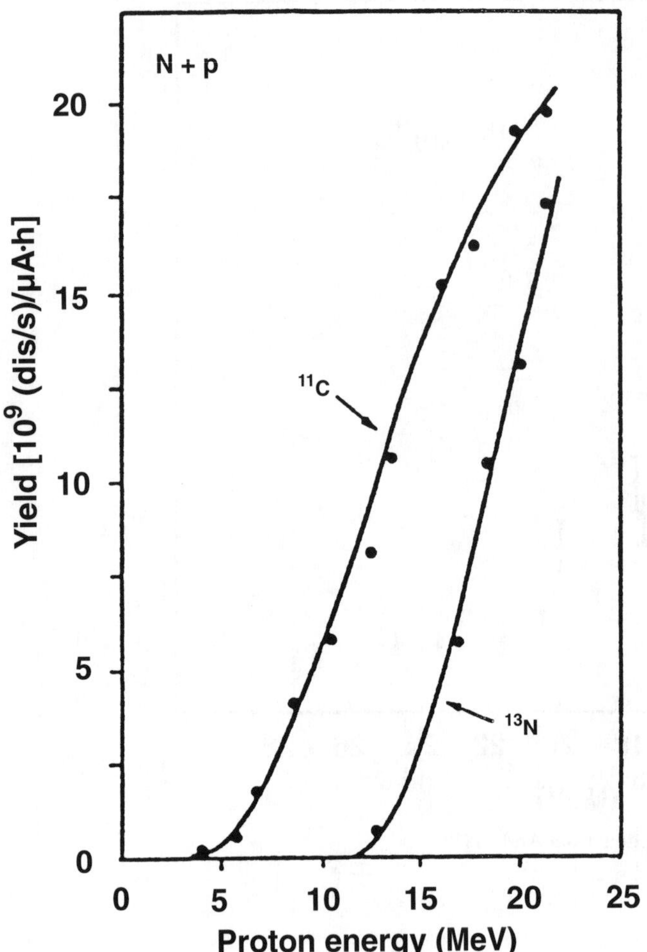

FIG. A14.15. Thick target yield for the $^{14}N(p, \alpha)^{11}C$ reaction. Target is AlN.

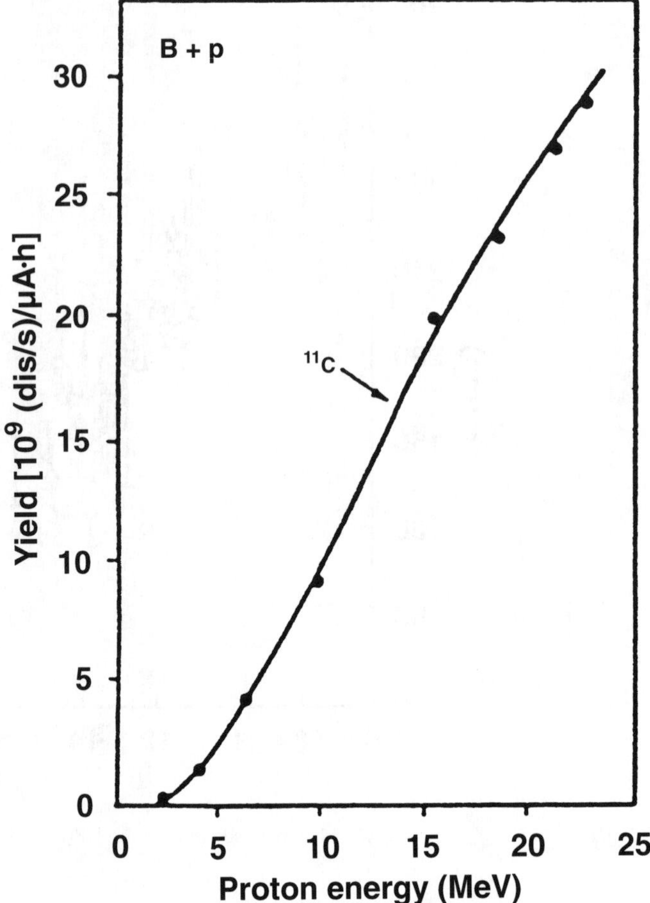

FIG. A14.16. Thick target yield for the $^{11}B(p, n)^{11}C$ reaction.

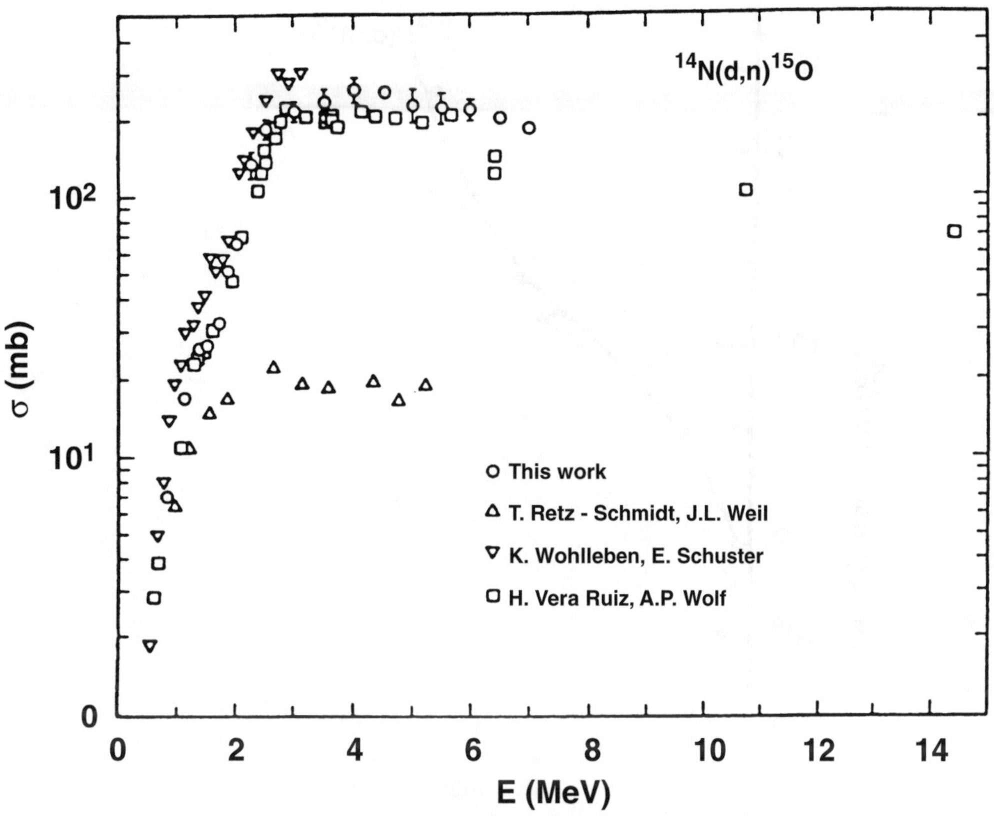

FIG. A14.17. Cross-section for the $^{14}N(d, n)^{15}O$ reaction, from Kohl *et al.* (1990).

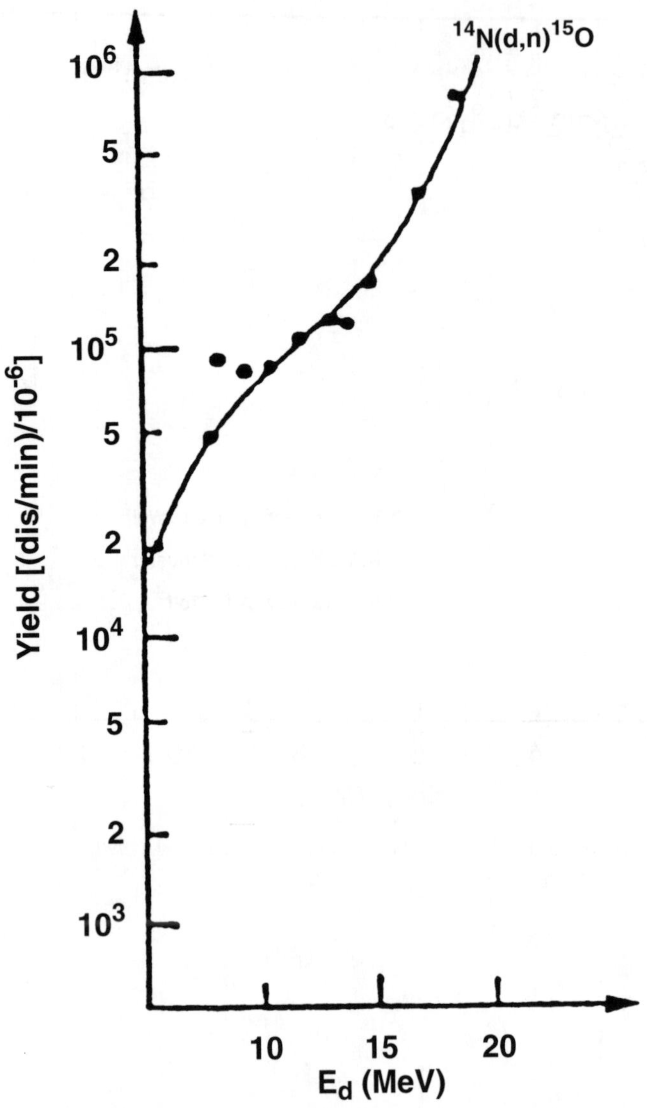

FIG. A14.18. Thick target yield for the $^{14}N(d, n)^{15}O$ reaction, from Engelmann (1971a). Target is AlN.

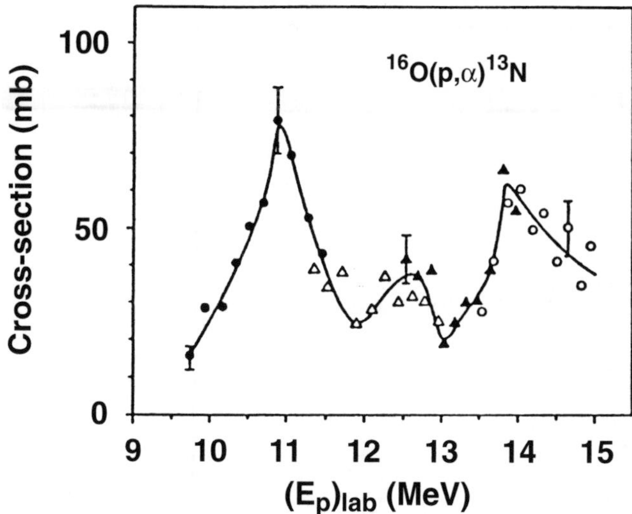

FIG. A14.19. Cross-section for the $^{16}O(p, \alpha)^{13}N$ reaction, from Furukawa and Tanaka (1960).

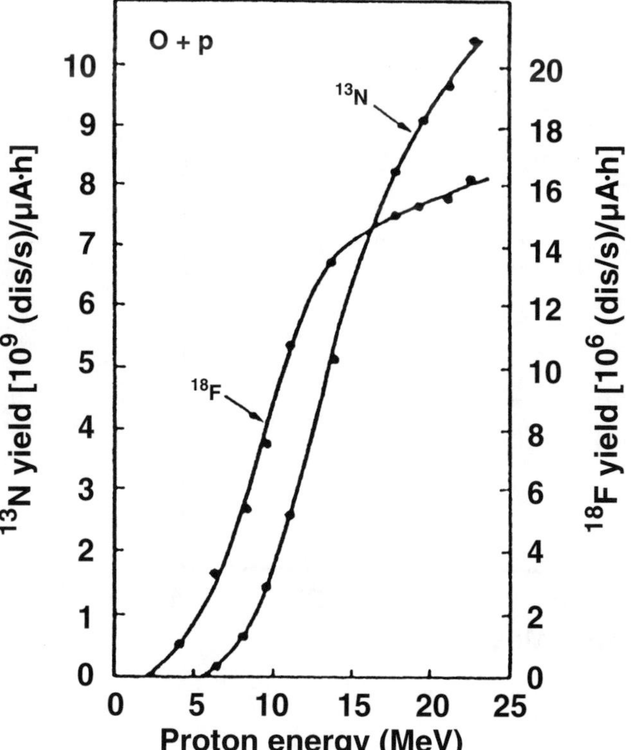

FIG. A14.20. Thick target yield for the $^{16}O(p, \alpha)^{13}N$, from Krasnov *et al.* (1970). Also shown $^{18}O(p, n)^{18}F$. Target is Al_2O_3.

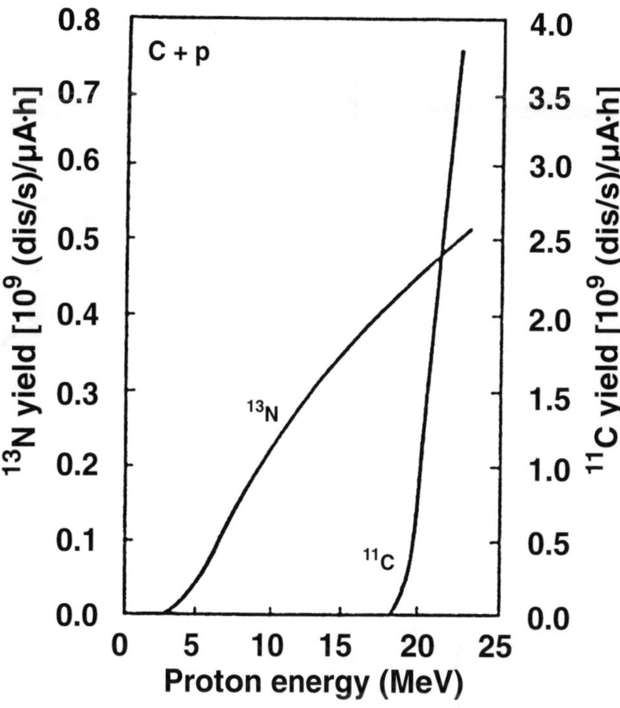

FIG. A14.21. Thick target yield for the production of ^{13}N by irradiation of carbon with protons, from Krasnov *et al.* (1970). Also shown $^{12}C(p, pn)^{11}C$. Target is graphite.

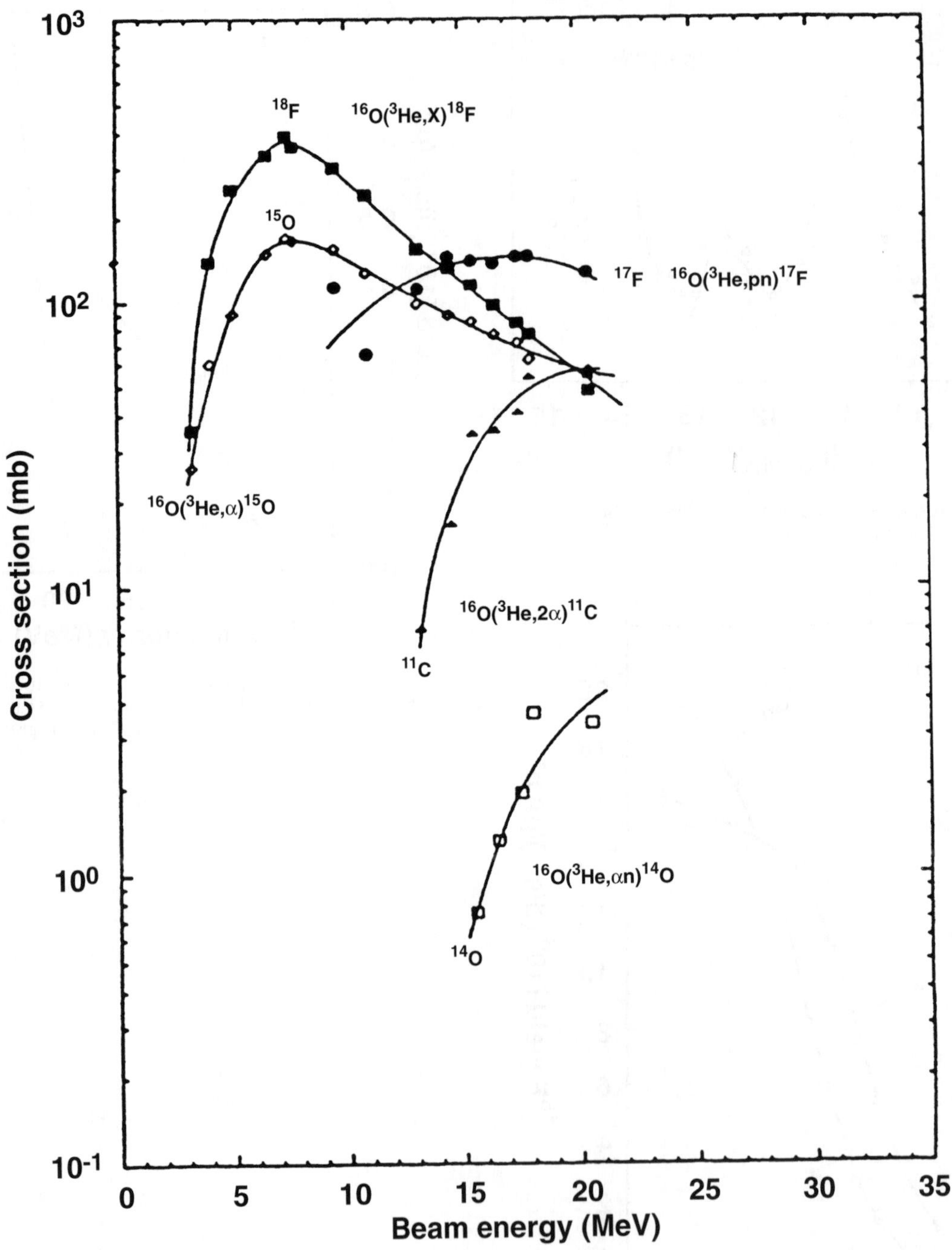

FIG. A14.22. Cross-section for the production of ^{18}F by irradiation of oxygen with ^{3}He, from Lamb (1969). Also shown production of ^{11}C, ^{14}O, ^{15}O, and ^{17}F.

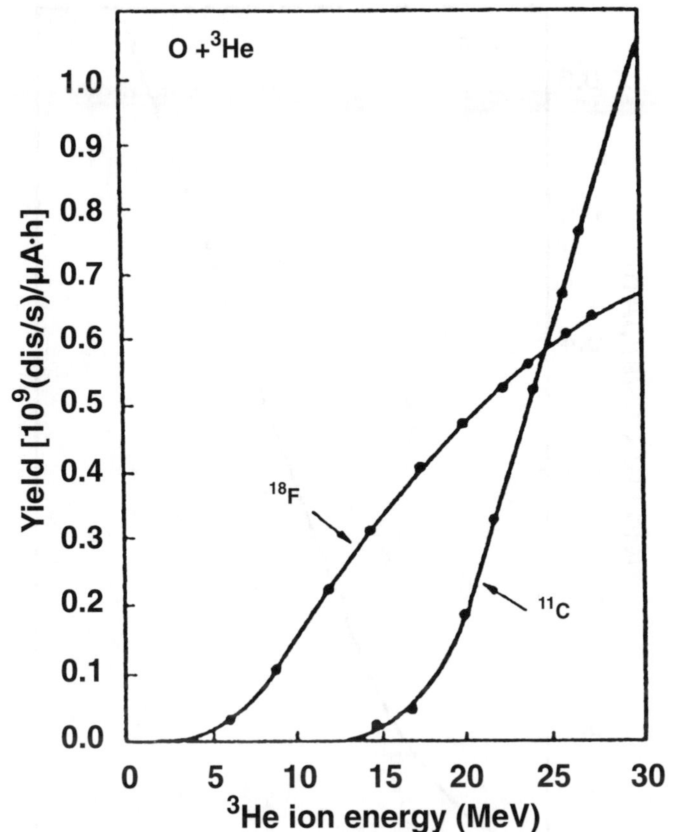

FIG. A14.23. Thick target yield for the production of ^{18}F by irradiation of oxygen with ^{3}He, from Krasnov *et al.* (1970). Also shown $^{16}O(^{3}He, 2\alpha)^{11}C$. Target is Al_2O_3.

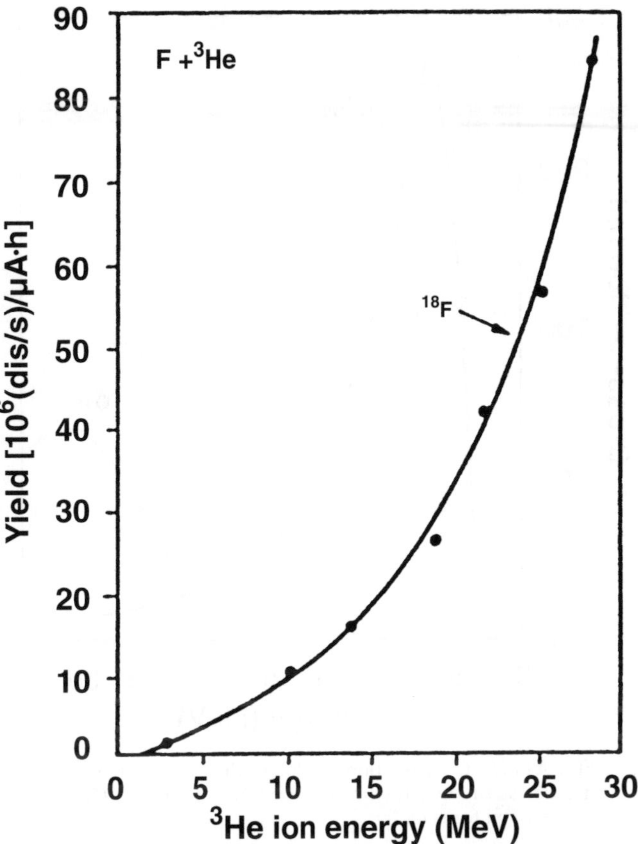

FIG. A14.24. Thick target yield for the production of $^{19}F(^{3}He, \alpha)^{18}F$ reaction, from Krasnov *et al.* (1970). Target is Teflon.

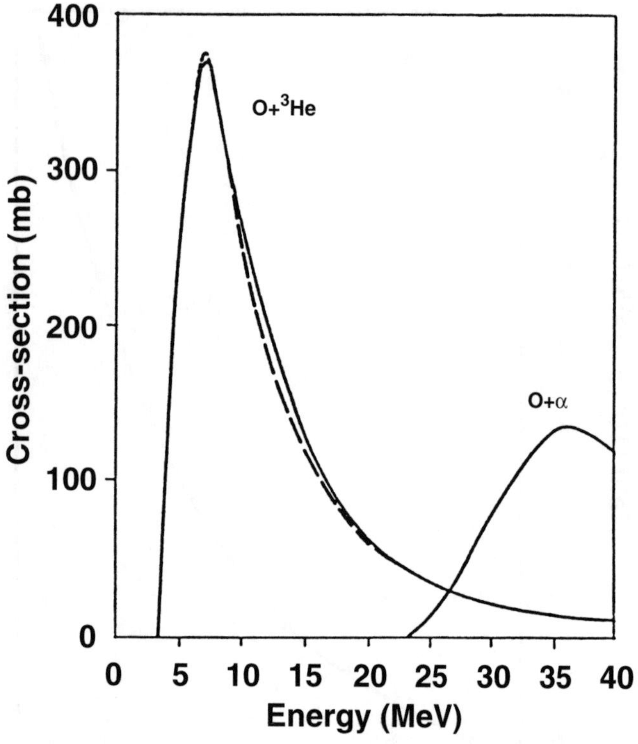

FIG. A14.25. Cross-section for the production of ^{18}F by irradiation of oxygen with ^{4}He, from Nozaki *et al.* (1974a). Also shown for ^{3}He.

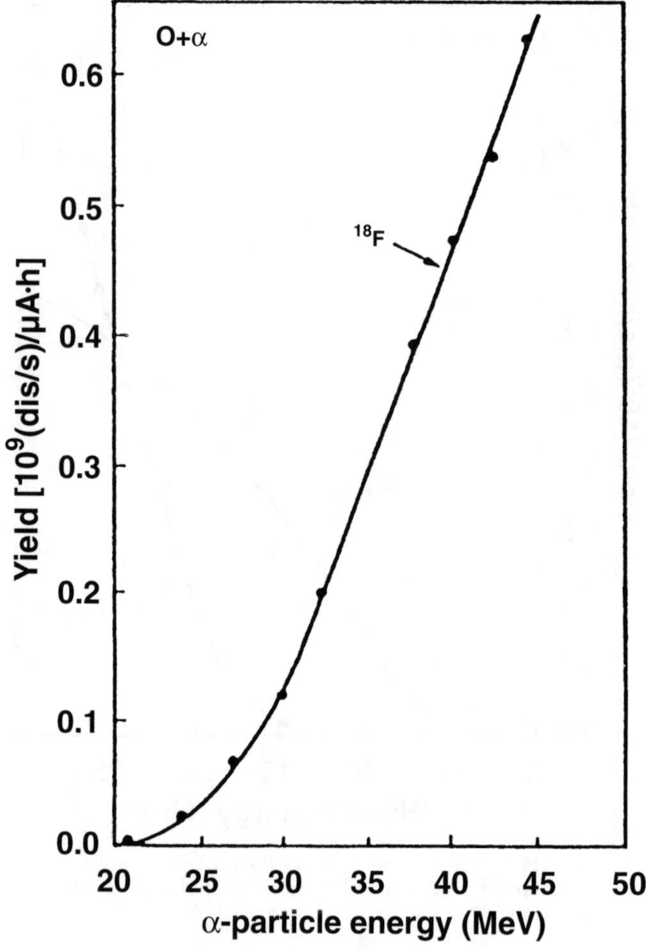

FIG. A14.26. Thick target yield for the production of ^{18}F by irradiation of oxygen with ^{4}He.

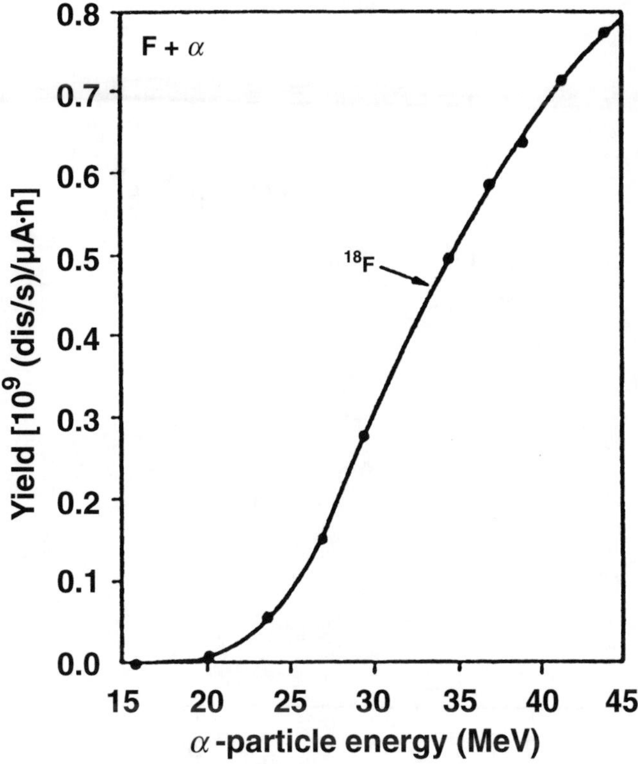

FIG. A14.27. Thick target yield for the $^{19}F(\alpha, \alpha n)^{18}F$ reaction, from Krasnov *et al.* (1970). Target is Teflon.

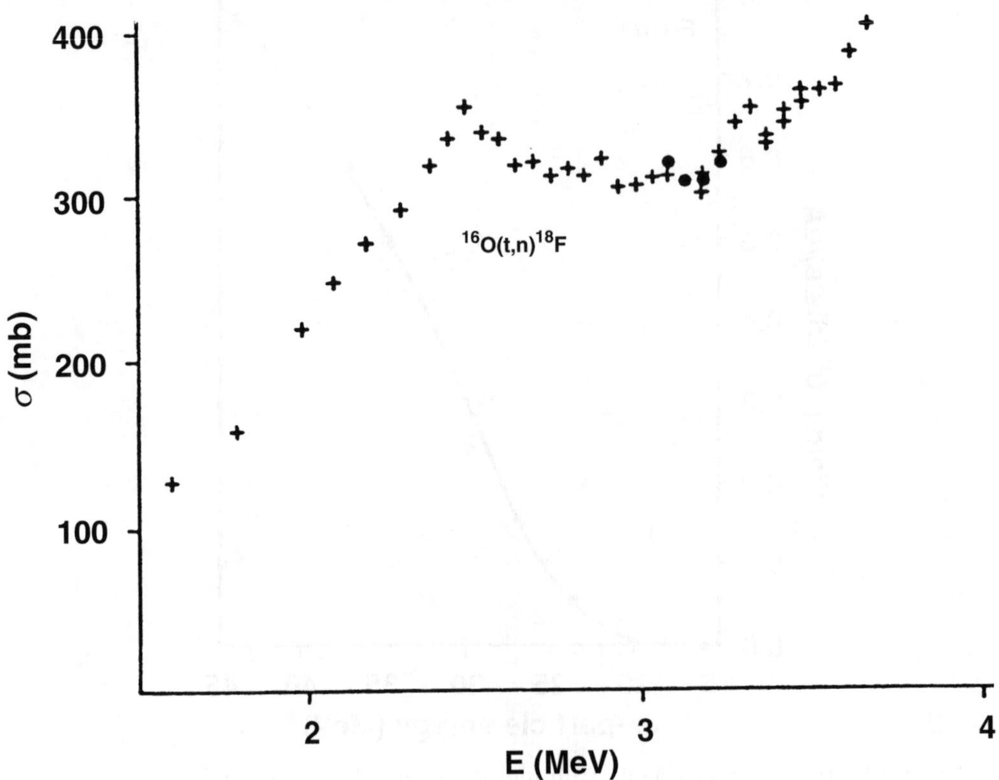

FIG. A14.28. Cross-section for the $^{16}O(t, n)^{18}F$ reaction, from Revel *et al.* (1977).

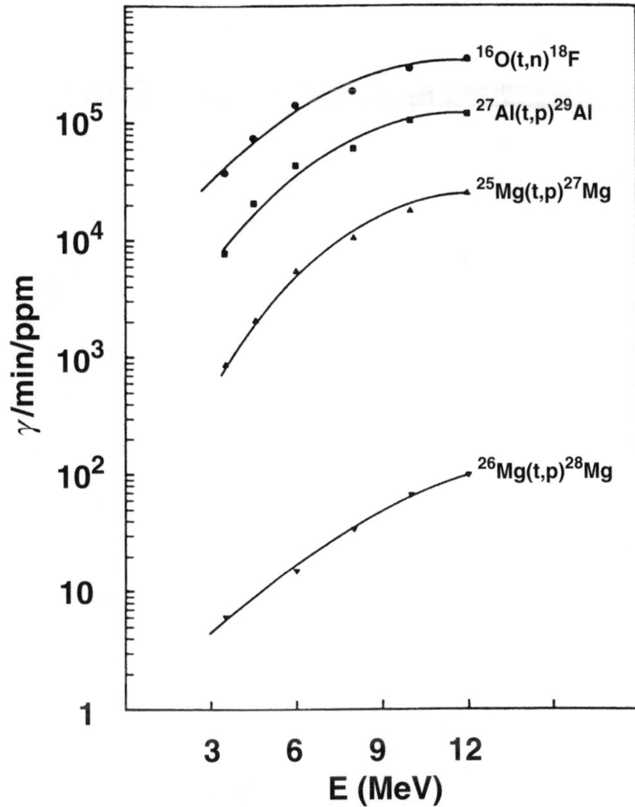

FIG. A14.29. Thick target yield for the $^{16}O(t, n)^{18}F$ reaction, from Bordes *et al.* (1987). Also shown yields for production of ^{27}Mg, ^{28}Mg, ^{29}Al by triton irradiation of Mg and Al. Target is Al_2O_3. Yield expressed in γ/min/ppm at the end of a 1 hour irradiation at 1 μA.

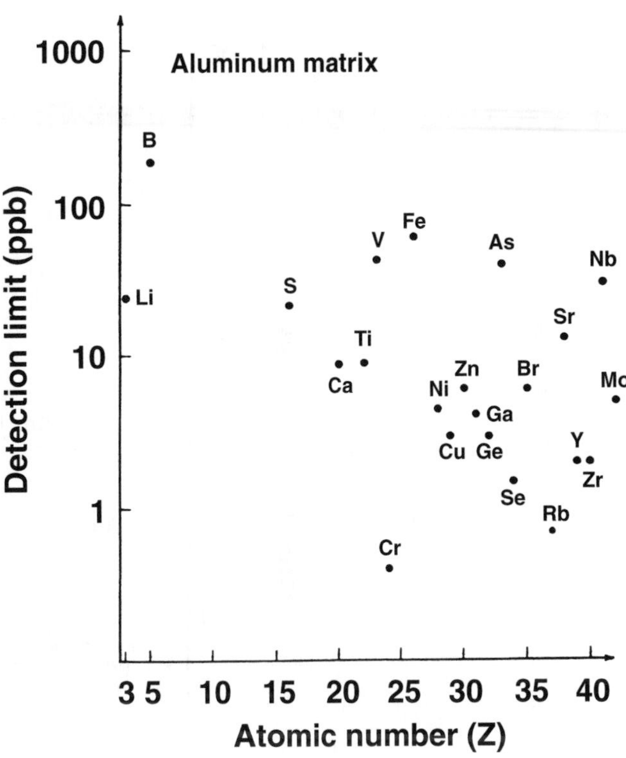

FIG. A14.30. Calculated best detection limits for 22 elements from Z = 3 to Z = 42 in an aluminum matrix. Standard irradiation conditions : 1 hour, 1 microampere, 10 MeV protons. From Debrun *et al.* (1976).

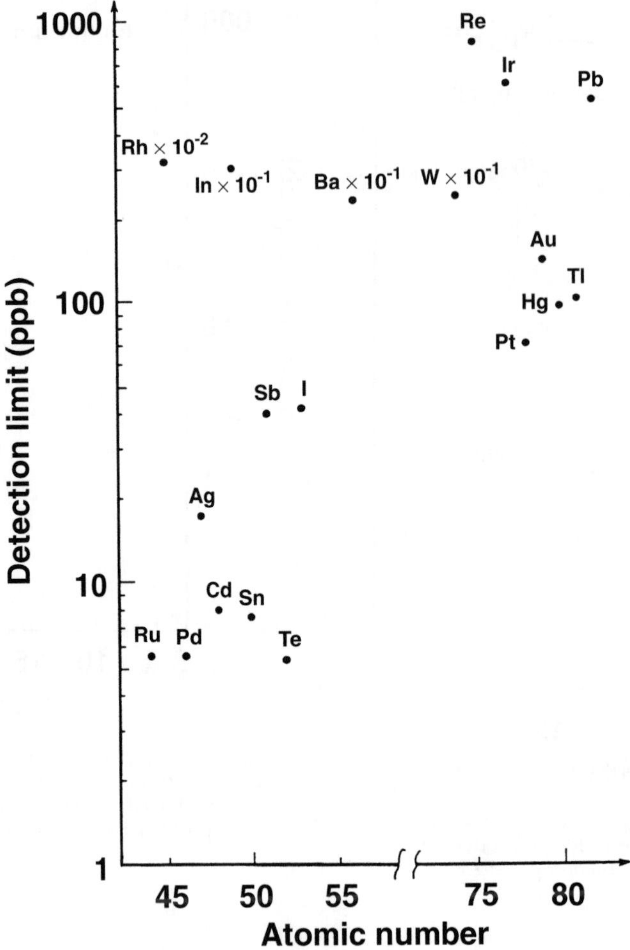

FIG. A14.31. Calculated best detection limits for 19 elements from Z = 44 to Z = 82, in an aluminum matrix. Standard irradiation conditions : 1 hour, 1 microampere, 10 MeV protons. From Barrandon *et al.* (1976).

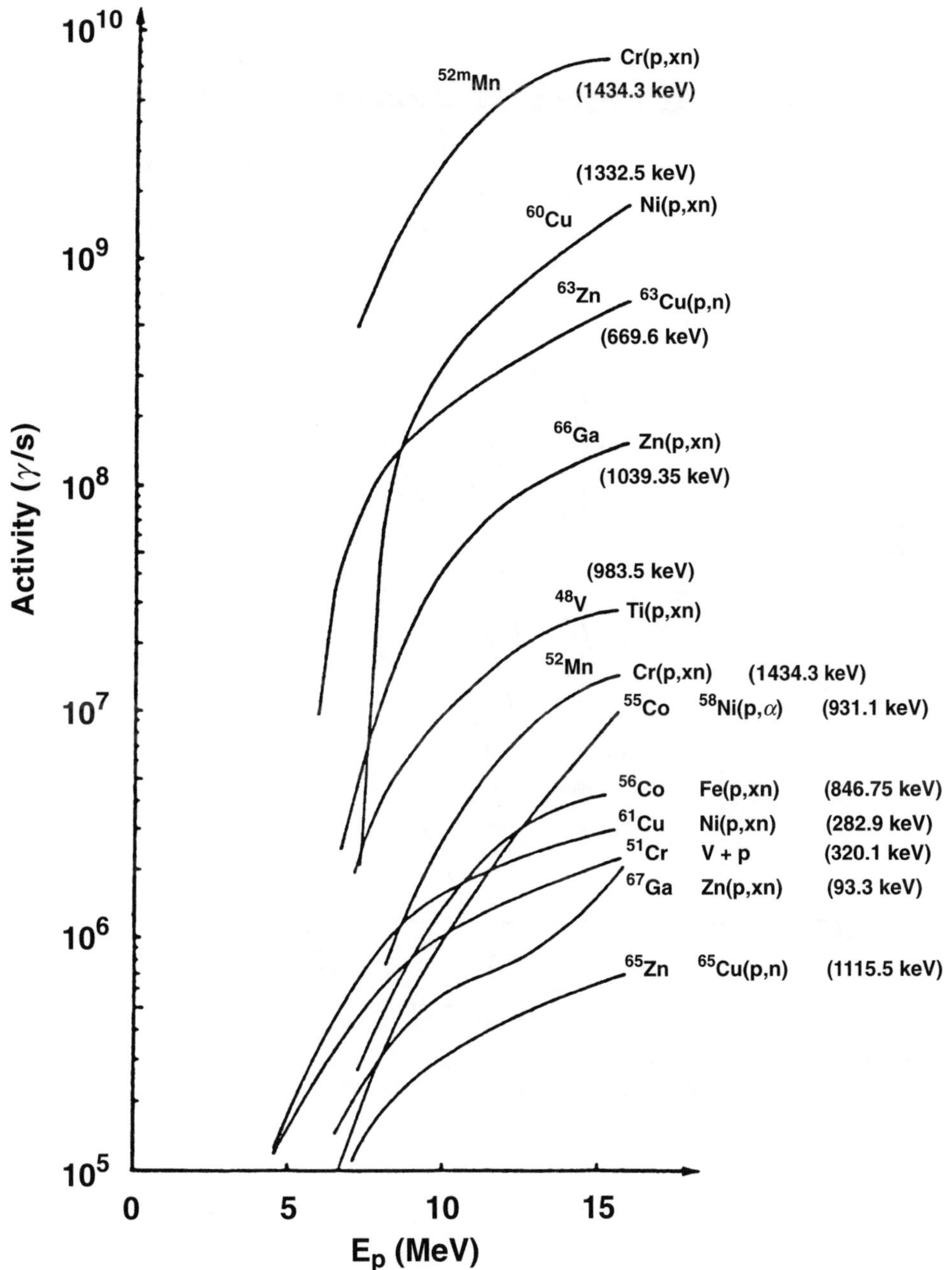

FIG. A14.32. Thick target yields for the proton activation of Ti, V, Cr, Fe, Ni, Cu, and Zn. Expressed in gamma-rays/second at the end of an irradiation of 1 hour at 1 microampere. From Albert (1987).

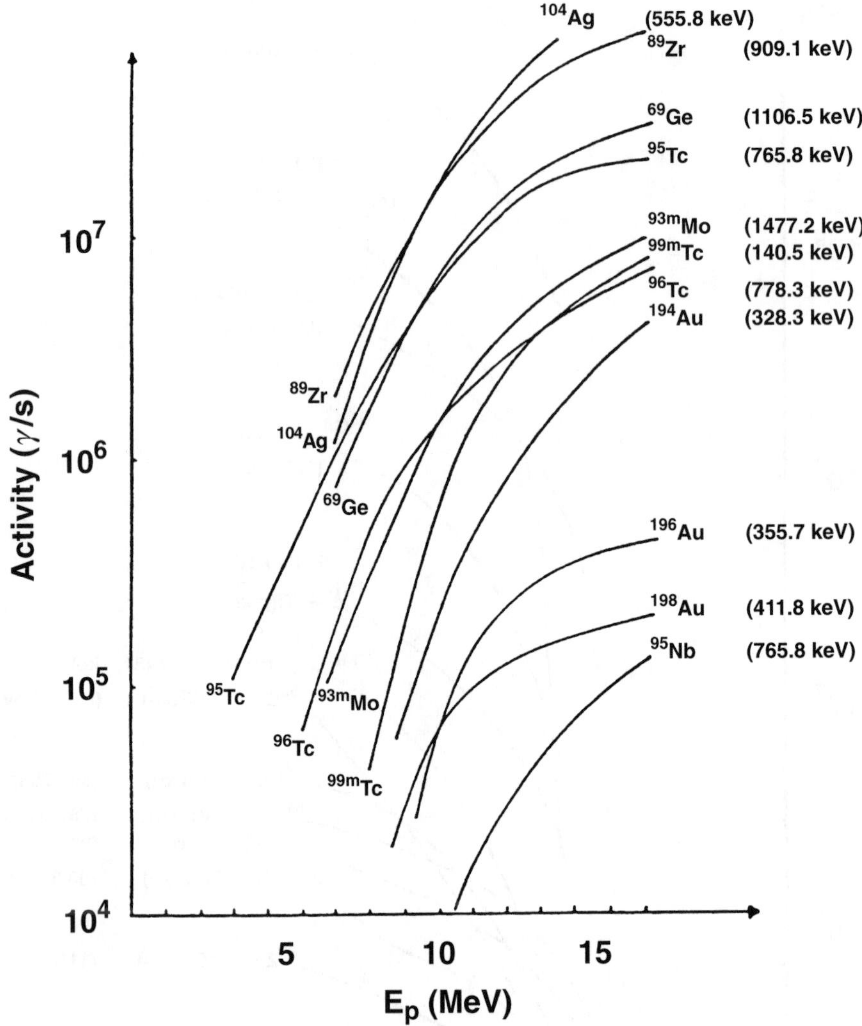

FIG. A14.33. Thick target yields for the proton activation of Y, Zr, Nb, Mo, and Pd. Expressed in gamma-rays/second at the end of an irradiation of 1 hour at 1 microampere. From Albert (1987).

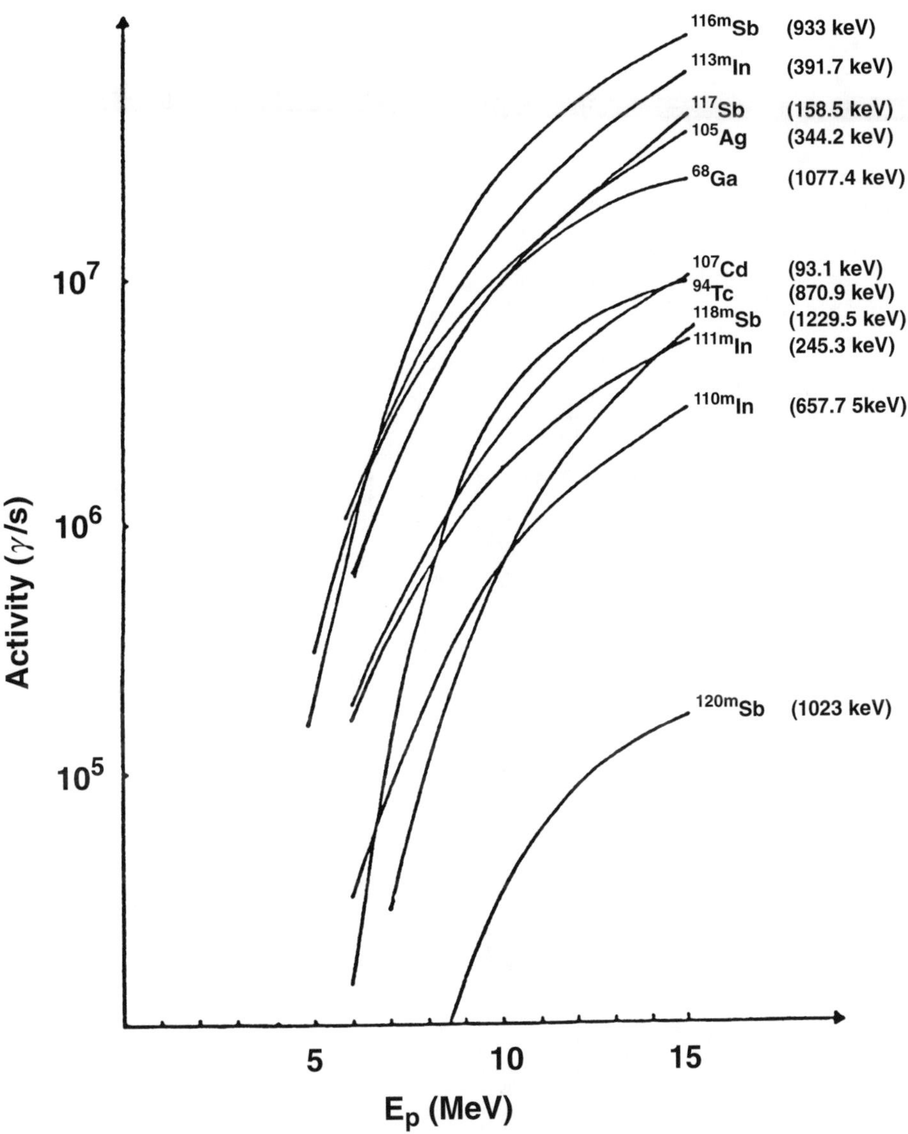

FIG. A14.34. Thick target yields for the proton activation of Mo, Pd, Ag, Cd, and Sn. Expressed in gamma-rays/second at the end of an irradiation of 1 hour at 1 microampere. From Albert (1987).

APPENDIX

15

CHANNELING DATA

Compiled by

M. L. Swanson

University of North Carolina, Chapel Hill, North Carolina

CONTENTS

NOTATION FOR CHANNELING

a = Thomas-Fermi screening length

a_o = Bohr radius = 5.292×10^{-11} m

d = atomic spacing along axial direction

d_o = lattice constant

d_p = interplanar spacing

e^2 = electronic charge squared = 1.44×10^{-13} cm MeV

E = ion energy

dE/dx = energy loss along the ion beam path

F_{RS} = square root of the Molière string potential [Eq.(10.7)]

F_{PS} = square root of the planar potential (Fig. A15.3)

H = number of counts per unit energy loss in an RBS spectrum [Eq.(10.34)]

M_1, M_2 = atomic masses of ions and target atoms

N = atomic density per unit volume

N_d = defect concentration per unit volume

T = crystal temperature

u_1 = one dimensional vibrational amplitude

u_2 = two dimensional vibrational amplitude (= 1.414 u_1)

x = distance along ion beam path

Z_1, Z_2 = atomic numbers of ions and target atoms

χ = normalized yield = (aligned yield)/(random yield)

χ_h = normalized yield from host atoms

χ_s = normalized yield for solute atoms

χ_v = normalized yield for undamaged crystal

$\chi_h^{<uvw>}$, $\chi_s^{<uvw>}$ = normalized yields for host and solute atoms for alignment along <uvw> axial channels.

$\chi_h^{\{hkl\}}$, $\chi_s^{\{hkl\}}$ = normalized yields for host and solute atoms for alignment along {hkl} planar channels.

χ_{hD} = dechanneled fraction of ions

ϕ = crystal azimuthal angle

ϕ_D = the Debye function

λ_p = dechanneling length caused by dislocations [Eq. (10.33)]

θ = crystal tilt angle

θ_D = Debye temperature (K)

σ_d = cross section for dechanneling by defects

σ_{th} = cross section for dechanneling by thermal vibrations

ψ_1 = characteristic angle for channeling

$\psi_{1/2}$ = half-width of the channeling dip

EQUATIONS FOR CHANNELING

(A15.1) $a = 0.04685(Z_1^{1/2} + Z_2^{1/2})^{-2/3} \cong 0.04685\ Z_2^{-1/3}$ (units of nm) (see Table A15.3)

(A15.2) $u_1 = 12.1\ ((\phi_D(x'')/x'' + 1/4)(M_2\theta_D)^{-1})^{1/2}$ (10^{-8} cm)

where $x'' = \theta_D/T$ and the Debye function ϕ_D is given in Fig. A15.1

(A15.3) $\psi_1 = (2Z_1Z_2e^2/Ed)^{1/2}$ (radians)

$= 0.307\ (Z_1Z_2/Ed)^{1/2}$ (degrees)

(where d is in units of 10^{-8} cm and the ion energy E is in MeV)

(A15.4) $\psi_{1/2} = 0.8\ F_{RS}(x')\psi_1$ for **axial channels**, if ψ_1 (rad) < a/d; i.e., for MeV light ions, or very high energy heavy ions (Barrett, 1971), where $x' = 1.2\ u_1/a$, and $F_{RS}\ (x') = (f_M(x'))^{1/2}$ is the square root of the Molière string potential [see Eq. (10.7) and Fig. A15.2]. Experimental and calculated values are compared in Table A15.5.

(A15.5) $\psi_{1/2} = 7.57\ (a\psi_1/d)^{1/2}$ (degrees) for **axial channels**, if ψ_1 (rad) > a/d; i.e., for heavy ions in the keV to MeV range; e.g. for 50 keV As on <110> Si, $\psi_{1/2} = 4.85°$. Thermal vibrations are neglected.

(A15.6) $\psi_{1/2} = 0.72\ F_{PS}(x',y')\psi_a$ for **planar channels** (Barrett, 1971), where $x' = 1.6u_1/a$, $y' = d_p/a$, F_{PS} = square root of adimensional planar potential using Molière's screening function (see Fig. A15.3), and
$\psi_a = 0.545\ (Z_1Z_2Nd_pa/E)^{1/2}$ (degrees).

(A15.7) $\chi_h^{<uvw>} = Nd\pi(2u_1^2 + a^2)$ [**axial channel** approximation sometimes attributed to Lindhard; see Eq. (10.9)].

(A15.8) $\chi_h^{<uvw>} = 18.8\ Ndu_1^2(1 + \xi^{-2})^{1/2}$ for **axial channels**. This is Barrett's (1971) Monte Carlo result, where $\xi = 126\ u_1/(\psi_{1/2}d)$, $\psi_{1/2}$ given in degrees. See Fig. A15.4 for fitting with somewhat different parameters.

At high energy where $\psi^{1/2} << u_1/d$, Eq. (A15.8) becomes Eq. (A15.9):

(A15.9) $\chi_h^{<uvw>} = 18.8\ Ndu_1^2$

Example: For 0.5 MeV He at 293K on Ge <110>, $\chi_{Ge}^{<110>} = 0.020$ from Eq. (A15.7), 0.0268 from Eq. (A15.8) and 0.0240 from Eq. (A15.9).

(A15.10) $\chi_h^{\{hkl\}} = 2a/d_p$, for **planar channels**, as estimated after Lindhard (1965); this approximation is independent of energy and temperature; e.g. for He on {110} Ge : d_p =0.200 nm from Tables A15.1 and A15.3, a = 0.0148 nm from Table A15.3, and thus $\chi_{Ge}^{\{110\}} = 0.148$.

(A15.11) $\chi_h = P(\theta_c)$, for the normalized yield of a crystal covered by an amorphous overlayer, where the reduced critical angle θ_c is given by

(A15.12) $\theta_c = aE\psi_{1/2}\ /\ (2Z_1Z_2e^2) = 1.5 \times 10^2 a\ (E/(Z_1Z_2d))^{1/2}F_{RS}$, with a and d in units of 10^{-8} cm and E in MeV. Values of $P(\theta_c)$ are shown in Fig. A15.5 for values of the reduced thickness m given by $m = \pi a^2Nt$. Values of m from 0.2 to 20 correspond to Si thicknesses of 34 to 3400 × 10^{-8} cm (Mayer and Rimini, 1977; Lugujjo and Mayer, 1973).

Appendix 15

Table A15.1. Values by which the lattice constant must be multiplied to compute the interatomic spacings d in axial directions and the interplanar spacings d_p for planar directions in the simplest (monatomic) cubic structures (Gemmell, 1974).

Structure	Atoms per unit cell	Axis <100>	Axis <110>	Axis <111>	Plane {100}	Plane {110}	Plane {111}
f. c. c.	4	1	$1/\sqrt{2}$	$\sqrt{3}$	1/2	$1/2\sqrt{2}$	$1/\sqrt{3}$
b. c. c.	2	1	$\sqrt{2}$	$\sqrt{3}/2$	1/2	$1/\sqrt{2}$	$1/2\sqrt{3}$
diamond cubic	8	1	$1/\sqrt{2}$	$\sqrt{3}/4$, $3\sqrt{3}/4$	1/4	$1/2\sqrt{2}$	$1/4\sqrt{3}$, $\sqrt{3}/4$

Table A15.2. Values by which the lattice constant must be multiplied to compute the interatomic spacings d in axial directions and the interplanar spacings d_p for planar directions in the most common simple diatomic compounds (atoms labeled "A" and "B") having cubic structures* (Gemmell, 1974).

Structure	Atoms per unit cell	Axis <100>	Axis <110>	Axis <111>	Plane {100}	Plane {110}	Plane {111}
Rocksalt (like NaCl)	4A + 4B	ABAB ... 1/2	pure $1/\sqrt{2}$	ABAB ... $\sqrt{3}/2$	mixed 1/2	mixed $1/2\sqrt{2}$	pure ABAB .. $1/2\sqrt{3}$
Fluorite (like CaF_2)	4A + 8B	pure 1 (A), 1/2(B)	pure $1/\sqrt{2}$	BAB..BAB $\sqrt{3}/4$, $\sqrt{3}/2$	pure ABAB 1/4	mixed $1/2\sqrt{2}$	pure BAB..BAB.. $1/4\sqrt{3}$, $\sqrt{3}/4$
Zinc Blende (like ZnS)	4A + 4B	pure 1	pure $1/\sqrt{2}$	AB...AB $\sqrt{3}/4$, $3\sqrt{3}/4$	pure ABAB 1/4	mixed $1/2\sqrt{2}$	pure AB..AB.. $1/4\sqrt{3}$, $\sqrt{3}/4$

*For the axial case, the term "pure" indicates that each row contains only one atomic species. For rows containing both species, the ordering in the row is given. For the planar case, the term "pure" indicates that each sheet of atoms contains only one atomic species and the way in which the sheets are ordered is shown. The term "mixed" refers to cases where each planar sheet of atoms contains both atomic species.

Table A15.3. Crystal parameters at room temperature (Gemmell, 1974). (θ_D values are approximate - from specific heat data).

Z_2	M_2	Element	Structure	Screening length $a=0.04685Z_2^{-1/3}$ (nm)	Debye temp.(θ_D) (K)	Vibrational amplitude u_1(293K) (nm)	Lattice constant d_0 (nm)
6	12.01	C	diamond cubic	0.0258	2000	0.004	0.3567
13	26.98	Al	f. c. c.	0.0199	390	0.0105	0.4050
14	28.09	Si	diamond cubic	0.0194	543	0.0075	0.5431
23	50.94	V	b. c. c.	0.0165	360	0.0082	0.3024
24	52.00	Cr	b. c. c.	0.0162	485	0.0061	0.2884
26	55.85	Fe	b. c. c.	0.0158	420	0.0068	0.2867
28	58.71	Ni	f. c. c.	0.0154	425	0.0065	0.3524
29	63.54	Cu	f. c. c.	0.0152	315	0.0084	0.3615
32	72.59	Ge	diamond cubic	0.0148	290	0.0085	0.5657
41	92.91	Nb	b. c. c.	0.0136	275	0.0079	0.3300
42	95.94	Mo	b. c. c.	0.0135	380	0.0057	0.3147
45	102.91	Rh	f. c. c.	0.0132	340	0.0061	0.3803
46	106.40	Pd	f. c. c.	0.0131	275	0.0074	0.3890
47	107.87	Ag	f. c. c.	0.0130	215	0.0093	0.4086
73	180.95	Ta	b. c. c.	0.0112	245	0.0064	0.3306
74	183.85	W	b. c. c.	0.0112	310	0.0050	0.3165
78	195.09	Pt	f. c. c.	0.0110	225	0.0066	0.3923
79	196.97	Au	f. c. c.	0.0109	170	0.0087	0.4078
82	207.19	Pb	f. c. c.	0.0108	88	0.0164	0.4951

Table A15.4. Commonly studied diatomic compounds having cubic lattice structures at room temperature (Gemmell, 1974).

Compound	**Structure**	**Lattice constant *(nm)***
AlSb	ZnS	0.6135
BaF_2	CaF_2	0.6200
CaF_2	CaF_2	0.5463
CsI	CsCl	0.4567
GaP	ZnS	0.5451
GaAs	ZnS	0.5654
GaSb	ZnS	0.6118
InAs	ZnS	0.6036
InSb	ZnS	0.6478
KCl	NaCl	0.6293
KBr	NaCl	0.6600
KI	NaCl	0.7066
LiF	NaCl	0.4017
MgO	NaCl	0.4211
NaF	NaCl	0.4620
NaCl	NaCl	0.5640
NaI	NaCl	0.6473
PbS	NaCl	0.5936
RbBr	NaCl	0.6854
SrF_2	CaF_2	0.5800
ThO_2	CaF_2	0.5600
UC	NaCl	0.4959
UO_2	CaF_2	0.5468

Table A15.5. Comparison of calculated and measured* values of $\psi_{1/2}$ for axial channeling, using Eq. (A15.4) (Barrett, 1971).

Target	Direction	Ion	Energy *(MeV)*	$\psi_{1/2}$ (degrees) Calculated	$\psi_{1/2}$ (degrees) Measured
C (diamond)	<011>	H	1.0	0.53	0.54
		He	1.0	0.75	0.75
Al	<011>	H	0.4	0.84	0.90
	<011>		1.4	0.45	0.42
Si	<011>	H	0.25	1.03	1.02
			0.5	0.73	0.68
			1.0	0.51	0.53
			2.0	0.36	0.36
			3.0	0.30	0.26
		He	0.5	1.03	1.10
			1.0	0.73	0.75
			2.0	0.51	0.55
Ge	<011>	He	1.0	0.93	0.95
W	<111>	H	3.0	0.83	0.85
		He	6.0	0.83	0.80
	<001>	H	2.0	0.95	1.00
		He	2.0	1.34	1.39
		C	10.0	0.98	1.10
		O	10.0	1.13	1.23
		Cl	10.0	1.60	1.82
Au	<011>	Cl	20.0	0.84	1.10
		I	60.0	0.80	1.00

*Estimated error for measurements about ± 0.07°

Table A15.6. Comparison of calculated and measured* values of $\psi_{1/2}$ for planar channeling (Barrett, 1971).

Target	Plane	Ion	Energy *(MeV)*	$\psi_{1/2}$ (degrees) Calculated	$\psi_{1/2}$ (degrees) Measured
Si	{100}	H	3.0	0.07	0.07
	{110}	H	3.0	0.10	0.09
Ge	{110}	He	1.0	0.31	0.30
	{100}	He	1.0	0.23	0.18
	{211}	He	1.0	0.18	0.20
W	{100}	H	2.0	0.25	0.22
	{110}	H	2.0	0.32	0.26
	{100}	He	2.0	0.35	0.27
	{110}	He	2.0	0.45	0.38
	{100}	C	10.0	0.25	0.20
	{110}	C	10.0	0.32	0.28
	{110}	O	10.0	0.36	0.33
	{110}	Cl	10.0	0.50	0.42
Au	{100}	Cl	20.0	0.27	0.31
	{110}	Cl	20.0	0.21	0.24
	{111}	Cl	20.0	0.30	0.32

*Estimated error for measurements about ± 0.03°

Table A15.7. Measured critical angles in diamond type lattice for monatomic and diatomic compounds (S.T. Picraux, 1969).

Crystal	Z_1	E *(MeV)*	$\psi_{1/2}$ axial		$\psi_{1/2}$ planar				
			<111>	<110>	{110}	{111}	{001}	{112}	{113}
Si	H^+	0.25		1.02	0.32	0.33	0.25	0.20	0.16
α=1.12$_5$		0.50		0.68	0.24	0.26	0.17	0.16	0.13
		1.00		0.53	0.16	0.20	0.12	0.11	0.10
		2.00		0.36	0.12	0.13	0.08	0.08	0.08
	He^+	0.50	0.98	1.10	0.30	0.32	0.23	0.19	0.16
		1.00	0.69	0.75	0.22	0.26	0.18	0.13	0.16
		2.00	0.46	0.55	0.17	0.16	0.13	0.09	0.09
Ge	He^+	0.50	1.13		0.40				
		1.00	0.80	0.95	0.30	0.23	0.18	0.20	
α=1.00		1.90	0.54		0.23				
C (diamond)	H^+	1.00	0.46	0.54	0.16				
α=1.44	He^+	1.00	0.58	0.75					
GaAs	He^+	0.50	1.07		0.33				
α=1.03		1.00	0.81		0.24				
		1.90	0.48		0.14				
GaP	He^+	0.50	1.03	0.38					
α=1.14		1.00	0.74	0.99	0.25	0.26	0.18	0.18	
		1.90	0.59		0.18				
GaSb	He^+	0.50	1.16						
α=0.94		1.00	0.88	1.10	0.28	0.25	0.17		

Estimated error for axial measurements ±0.06°
Estimated error for planar measurments ±0.03°
Calculated values are given in Tables A15.5 and A15.6

Table A15.8. Angles between planes in cubic crystals.

HKL	hkl	Angles between HKL and hkl planes					
100	100	0.00	90.00				
	110	45.00	90.00				
	111	54.75					
	210	26.56	63.43	90.00			
	211	35.26	65.90				
	221	48.19	70.53				
	310	18.43	71.56	90.00			
	311	25.24	72.45				
	320	33.69	56.31	90.00			
	321	36.70	57.69	74.50			
	322	43.31	60.98				
	331	46.51	76.74				
	332	50.24	64.76				
	410	14.04	75.96	90.00			
	411	19.47	76.37				
110	110	0.00	60.00	90.00			
	111	35.26	90.00				
	210	18.43	50.77	71.56			
	211	30.00	54.74	73.22	90.00		
	221	19.47	45.00	76.37	90.00		
	310	26.56	47.87	63.43	77.08		
	311	31.48	64.76	90.00			
	320	11.31	53.96	66.91	78.69		
	321	19.11	40.89	55.46	67.79	79.11	
	322	30.96	46.69	80.12	90.00		
	331	13.26	49.54	71.07	90.00		
	332	25.24	41.08	81.33	90.00		
	410	80.96	46.69	59.04	80.12		
	411	33.56	60.00	70.53	90.00		
111	111	0.00	70.53				
	210	39.23	75.04				
	211	19.47	61.87	90.00			
	221	15.79	54.74	78.90			
	310	43.09	68.58				
	311	29.50	58.52	79.98			
	320	36.81	80.76				
	321	22.21	51.89	72.02	90.00		
	322	11.42	65.16	81.95			
	331	22.00	48.53	82.39			
	332	10.02	60.50	75.75			
	410	45.56	65.16				
	411	35.26	57.02	74.21			
210	210	0.00	36.87	53.13	66.42	78.46	90.00
	211	24.09	43.09	56.79	79.48	90.00	
	221	26.56	41.81	53.40	63.43	72.65	90.00
	310	8.13	31.95	45.00	64.90	73.57	81.87
	311	19.29	47.61	66.14	82.25		
	320	7.12	29.74	41.91	60.25	68.15	75.64
		82.87					
	321	17.02	33.21	53.30	61.44	68.99	83.14
		90.00					

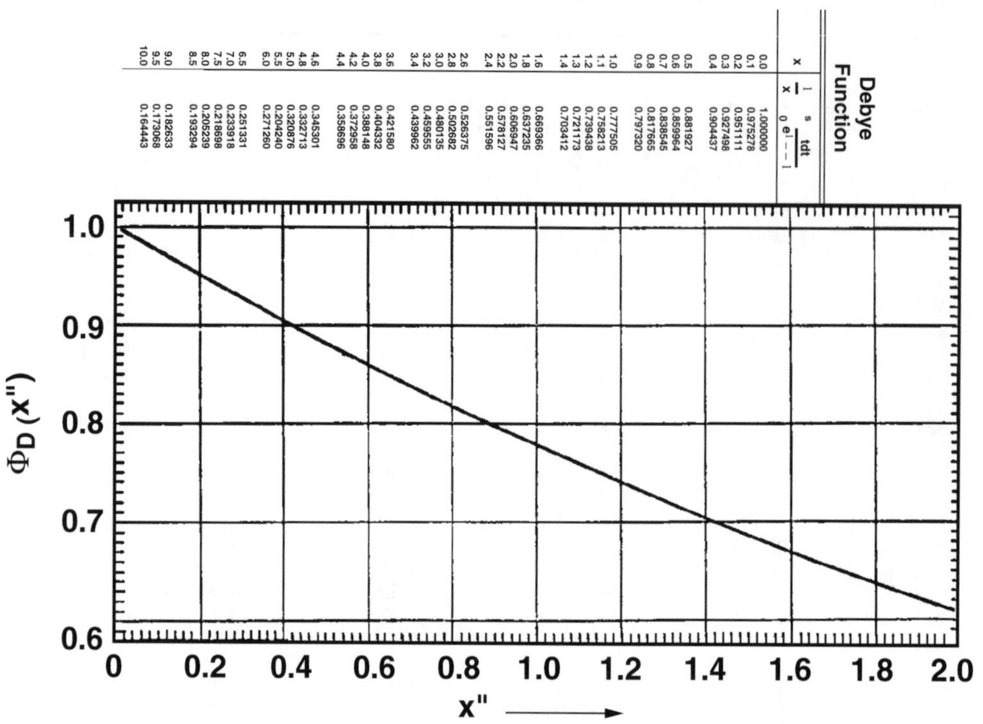

Debye Function

x	$\frac{1}{x}\int_0^s \frac{tdt}{e^t - 1}$
0.0	1.000000
0.1	0.975278
0.2	0.951111
0.3	0.927498
0.4	0.904437
0.5	0.881927
0.6	0.859964
0.7	0.838545
0.8	0.817665
0.9	0.797320
1.0	0.777505
1.1	0.758213
1.2	0.739438
1.3	0.721173
1.4	0.703412
1.6	0.669366
1.8	0.637235
2.0	0.606947
2.2	0.578127
2.4	0.551596
2.6	0.526375
2.8	0.502682
3.0	0.480135
3.2	0.459555
3.4	0.439962
3.6	0.421580
3.8	0.404332
4.0	0.388148
4.2	0.372958
4.4	0.358696
4.6	0.345301
4.8	0.332713
5.0	0.320876
5.5	0.204240
6.0	0.271260
6.5	0.251331
7.0	0.233918
7.5	0.218698
8.0	0.205239
8.5	0.193294
9.0	0.182633
9.5	0.173068
10.0	0.164443

FIG. A15.1. The Debye function $\phi_D(x'')$, where $x'' = \theta_D/T$ (see Eq. A15.2). From Gemmell (1974).

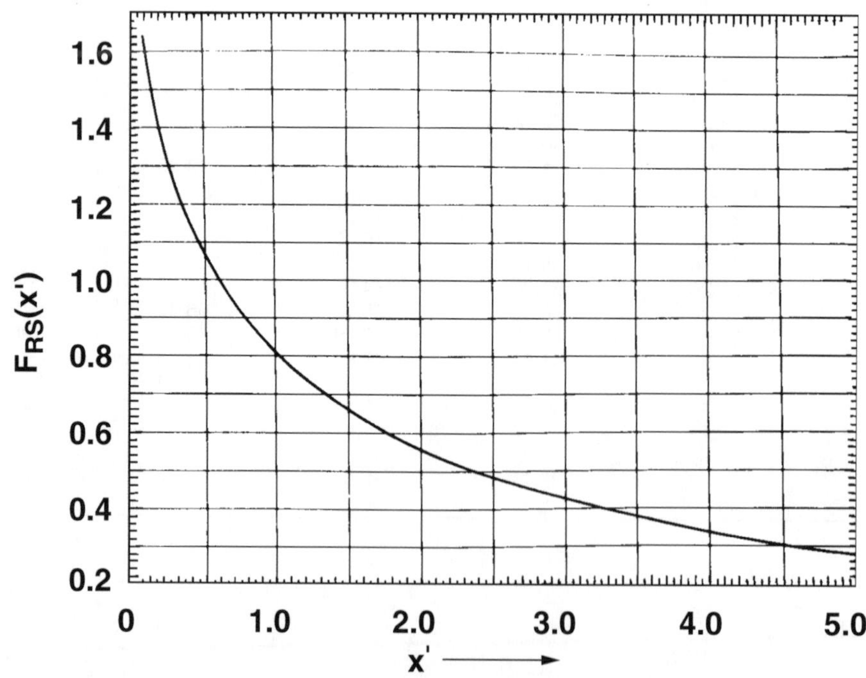

FIG. A15.2. A plot of $F_{RS}(x')$ versus x', where $F_{RS}(x')$ is the square root of the Molière string potential f_M and $x' = 1.2u_1/a$ (see Eqs. (10.7) and A15.4). From Mayer and Rimini (1977).

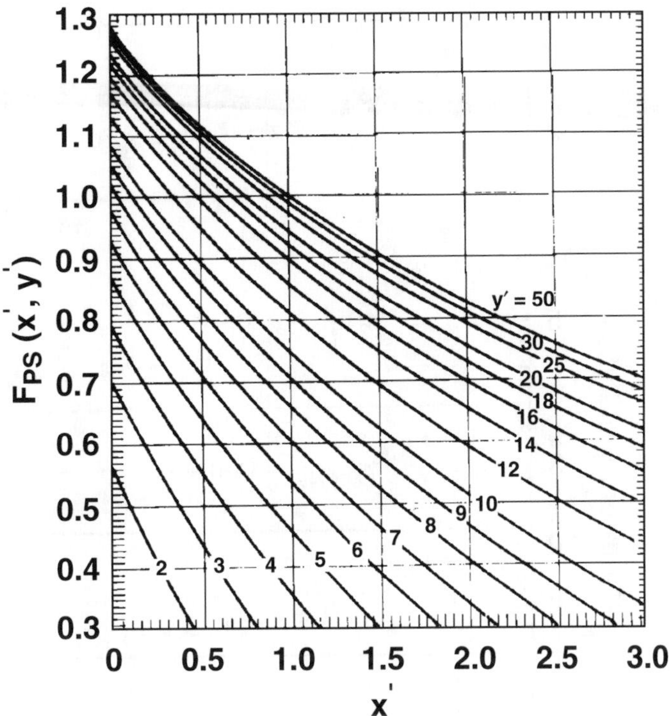

FIG. A15.3. A plot of $F_{PS}(x',y')$ versus x', for different values of y', where $F_{PS}(x')$ is the square root of the adimensional planar potential, $x' = 1.6u_1/a$ and $y' = d_p/a$ (see Eq. A15.6). From Barrett (1971).

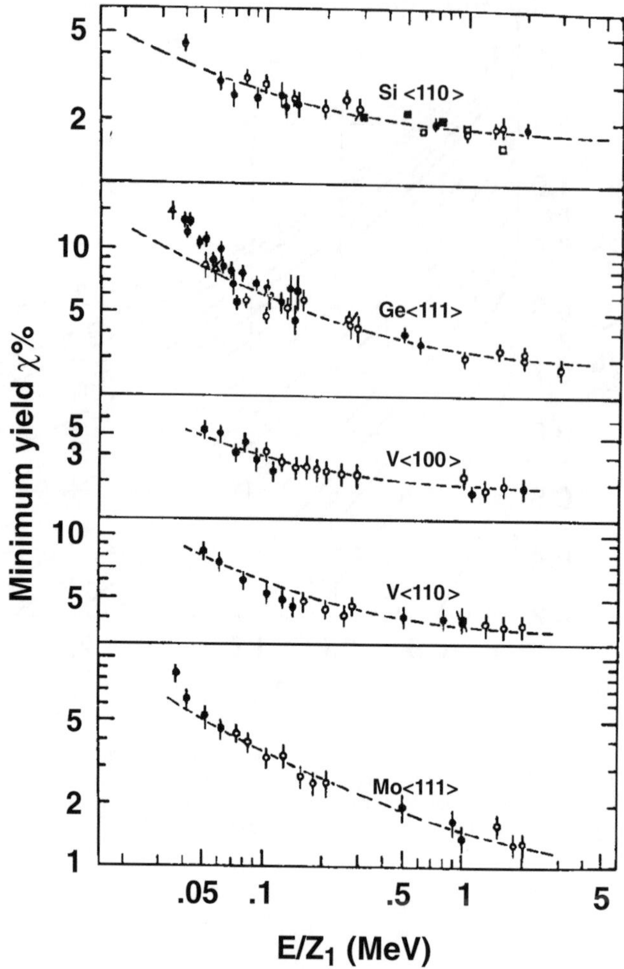

FIG. A15.4. Axial minimum yields χ for several ion-target combinations are plotted as a function of ratio E/Z_1, where E and Z_1 are the energy and the atomic number of the incident ion, respectively.

The dashed lines represent a fit of experimental data using the formula (Barrett, 1971):

$$\chi = 2\pi N d c u_1^2 \left(1 + \frac{\Psi_{1/2}^2 d^2}{k^2 u_1^2}\right)^{1/2}$$

where the constants c and k have the following values

		c	k
Si	<110>	2.7	1.2
Ge	<111>	2.8	1.0
V	<100>	2.6	1.2
V	<110>	2.6	1.1
Mo	<111>	2.8	0.8

Good averaged values for c and k, are 2.7 and 1.1 respectively. All experimental data, shown in Figure A15.4, are well fitted by the following formula [the constants are somewhat different from Eq. (A15.8)]:

$$\chi = 17 N d u_1^2 \left(1 + \xi^{-2}\right)^{1/2}$$

$$\xi = 63 u_1 / \left(\Psi_{1/2}\, d\right)$$

where N is the atomic density (atoms/(nm)^3), d is the atomic spacing along an axial direction (nm), u_1 is the one dimensional rms vibration amplitude (nm), and $\psi_{1/2}$ is the axial half angle (degrees). From Mayer and Rimini (1977): *Catania Handbook*.

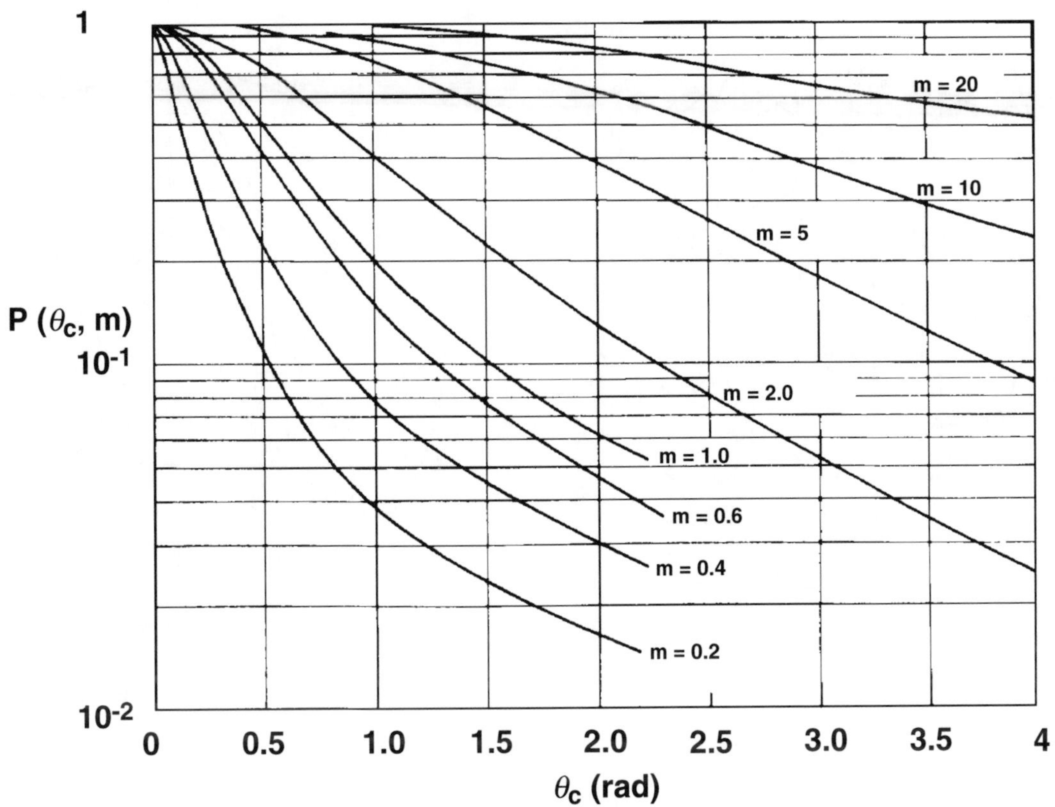

FIG. A15.5. Normalized integrated distribution P (using the Meyer treatment) versus reduced angle θ_c for various m values (see Eqs. A15.11 and A15.12). From Lugujjo and Mayer (1973). The function P is used as a measure of the normalized yield χ for crystals covered with an amorphous layer.

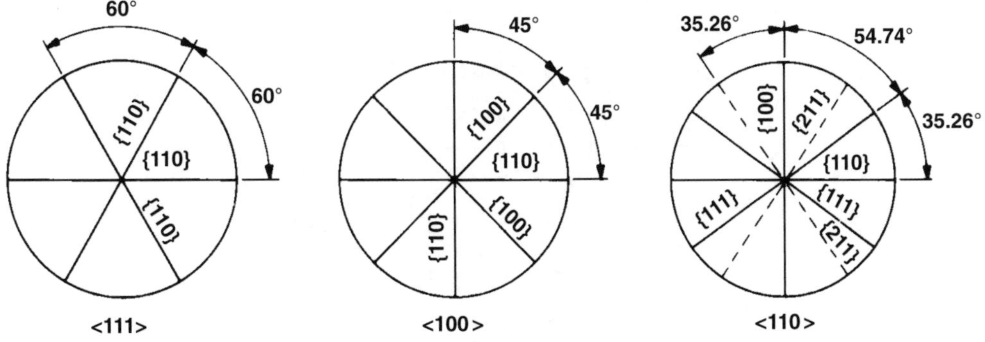

FIG. A15.6. Low index planes around the <111>, <100>, and <110> axes in a cubic structure. To assist in orienting crystals, the major axes and planes are listed below in order of increasing spacing along the rows and planes; thus the channeling dips progress from strong to weak from left to right; e.g., <110> channels give the lowest minimum yields for the fcc and diamond structures.

Structure	Axes <uvw>	Planes {hkl}
f.c.c.	110, 100, 111	111, 100, 110
b.c.c.	111, 100, 110	110, 100, 211
diamond	110, 111, 100	110, 111, 100

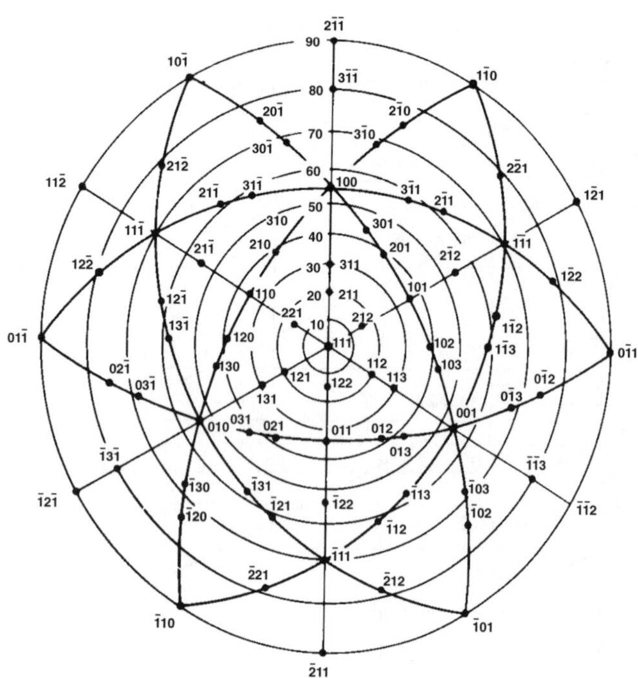

FIG. A15.7. Standard <111> stereographic projection for a cubic crystal.

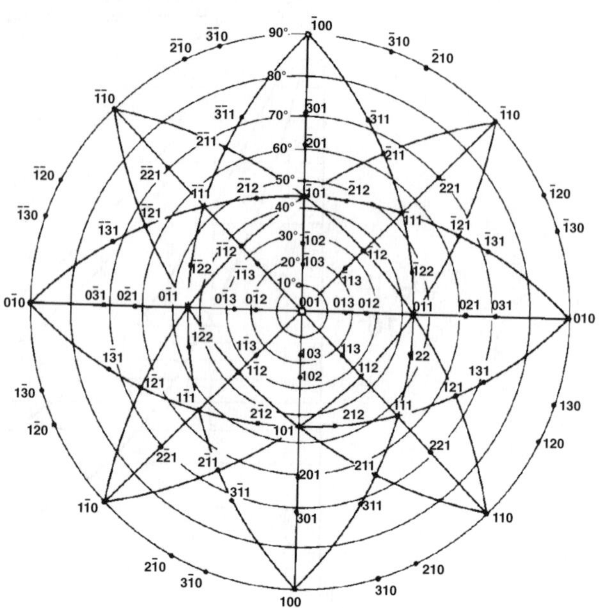

FIG. A15.8. Standard <100> stereographic projection for a cubic crystal.

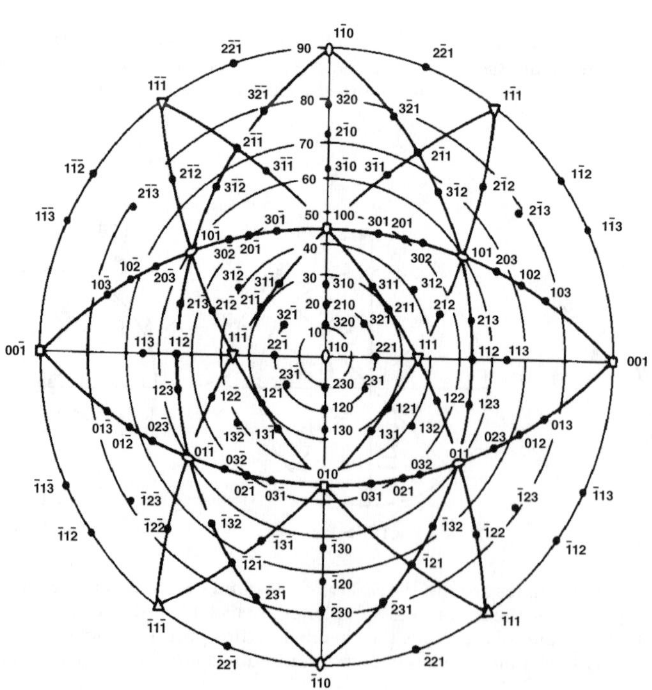

FIG. A15.9. Standard <110> stereographic projection for a cubic crystal.

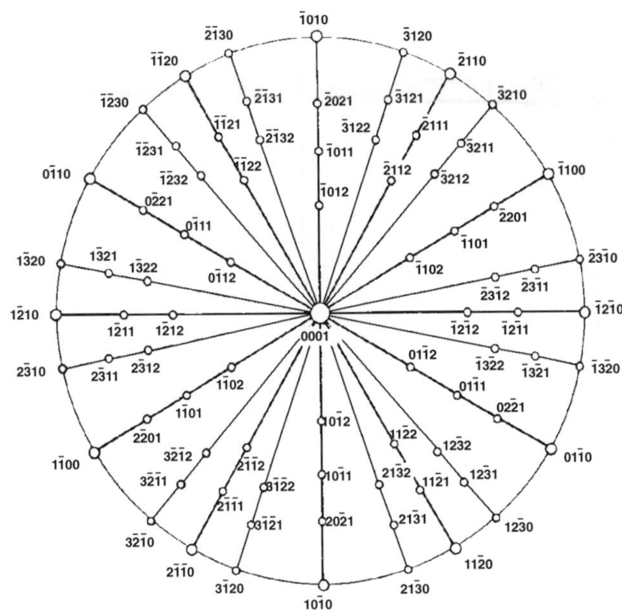

FIG. A15.10. Standard <0001> stereographic projection for a close packed hexagonal crystal.

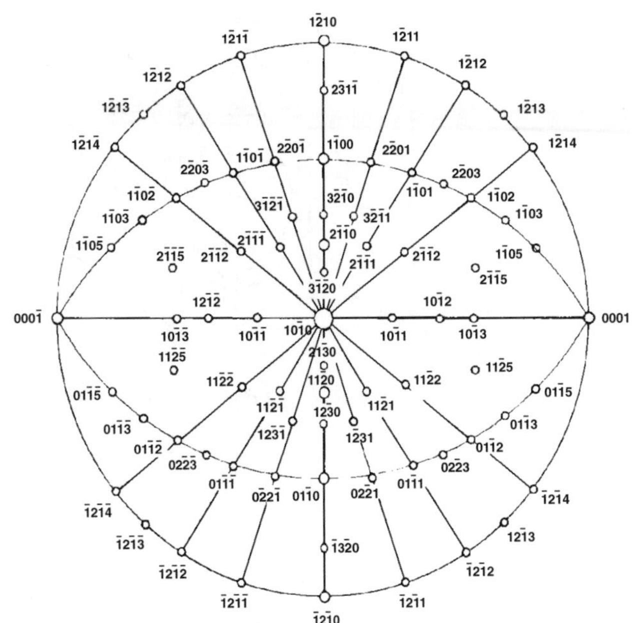

FIG. A15.11. Standard < $10\bar{1}0$ > stereographic projection for a close packed hexagonal crystal.

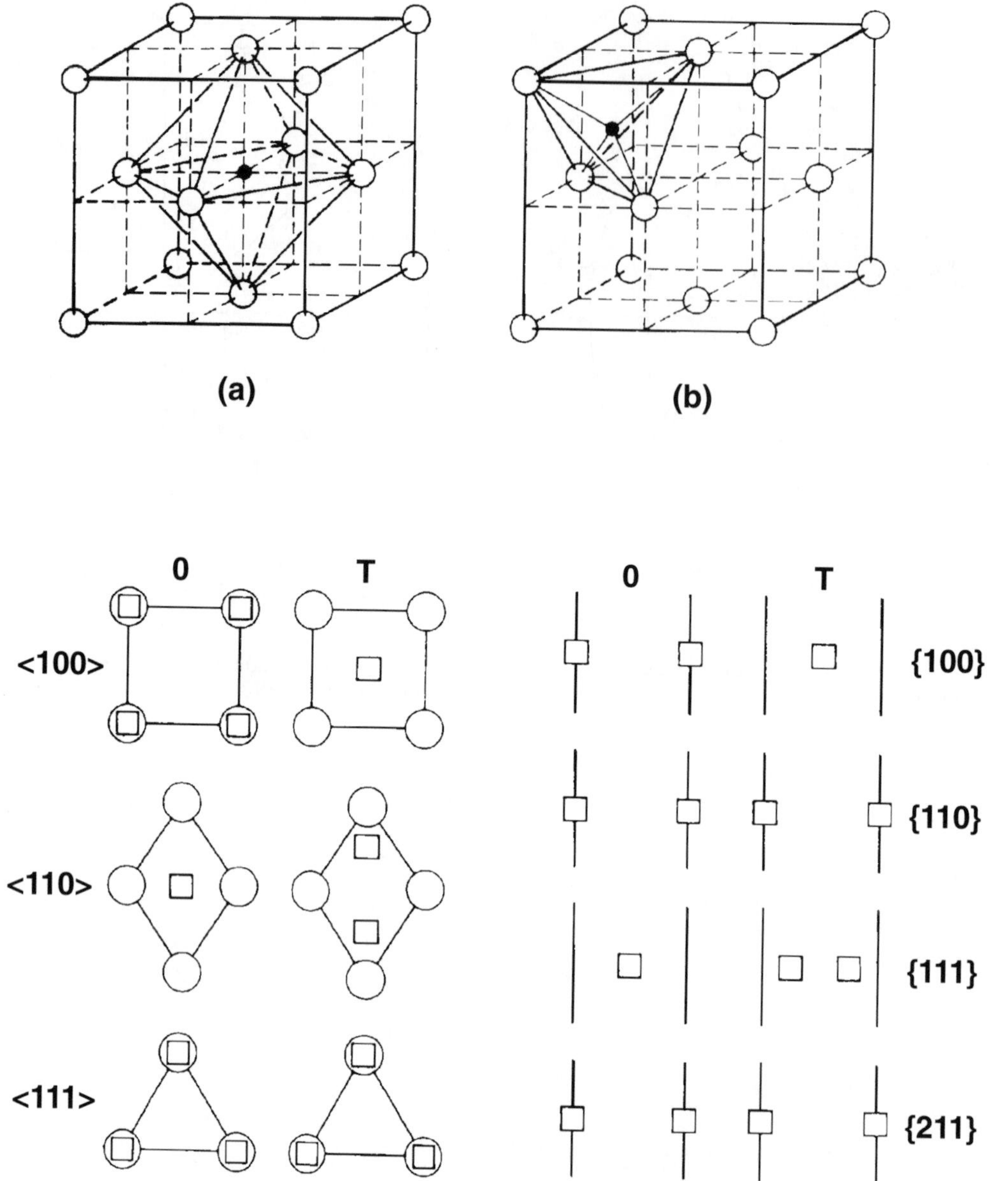

FIG. A15.12. Top part: octahedral (a) and tetrahedral (b) interstitial sites in a f.c.c crystal. Bottom part: projections of octahedral and tetrahedral interstitial sites into the major axial and planar channels of a f.c.c. structure (from Carstanjen, 1980).

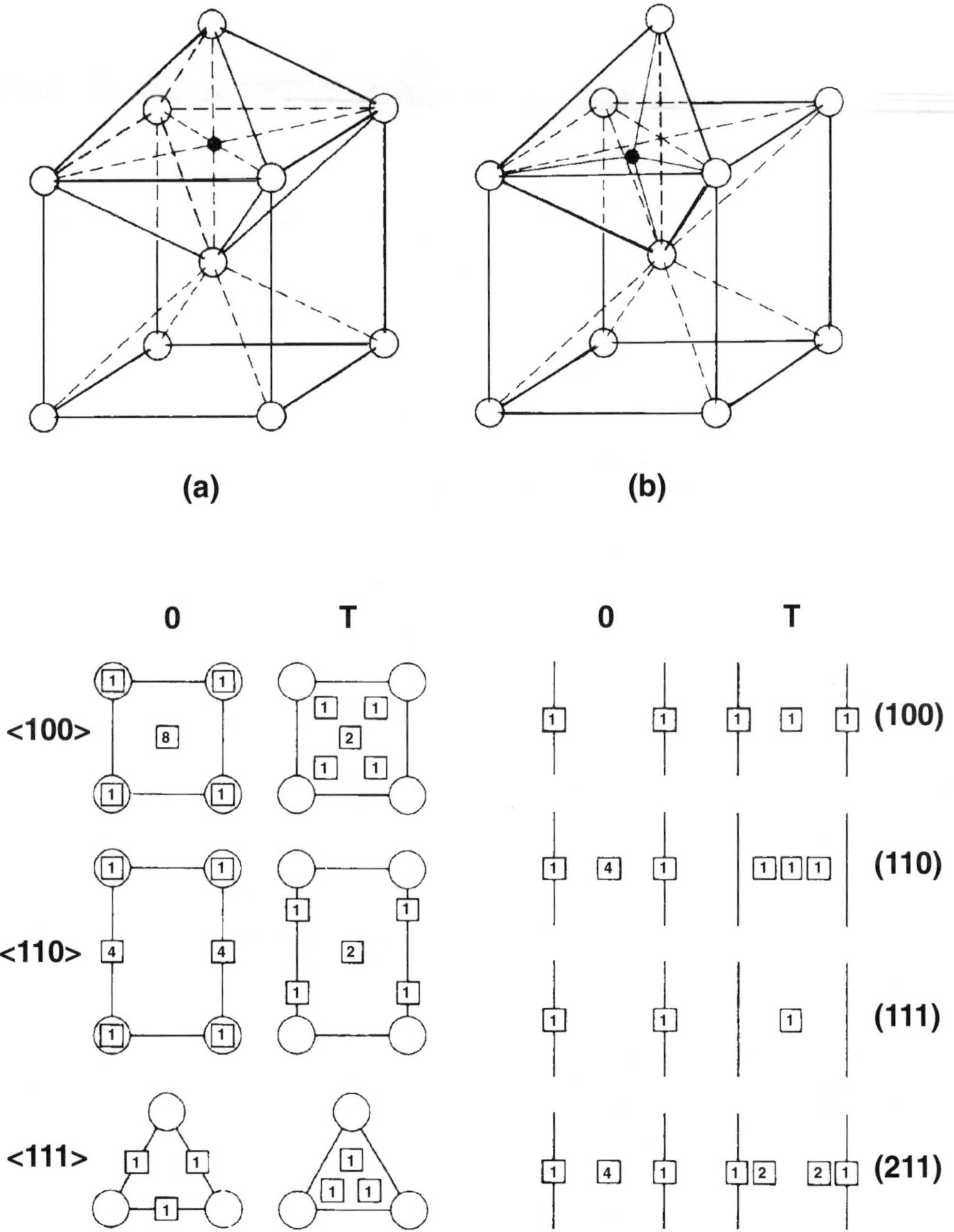

FIG. A15.13. Top part: octahedral (a) and tetrahedral (b) interstitial sites in a b.c.c crystal. Bottom part: projections of octahedral and tetrahedral interstitial sites into the major axial and planar channels of a b.c.c. structure (from Carstanjen, 1980).

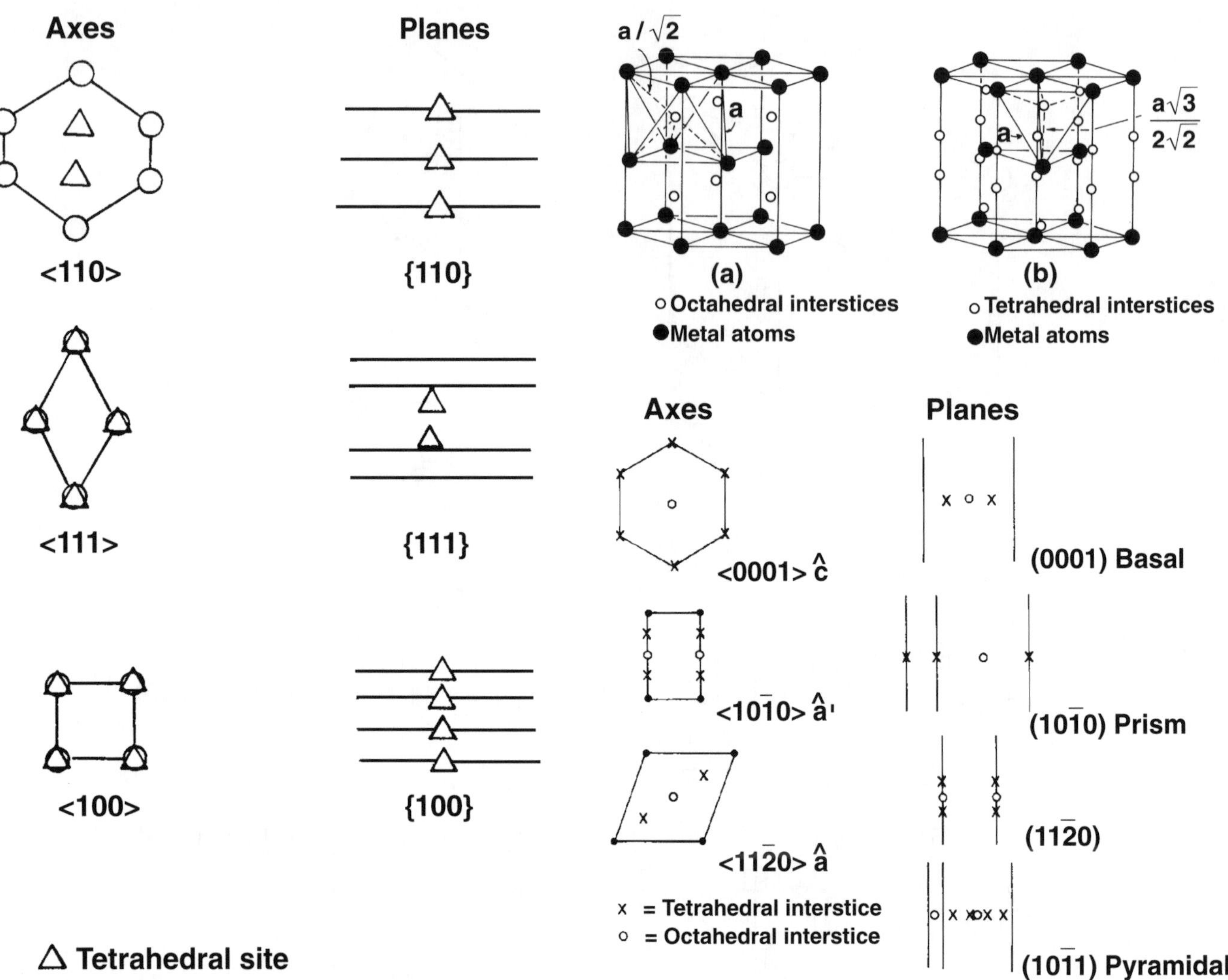

FIG. A15.14. Projection of tetrahedral interstitial sites into the major axial and planar channels of a diamond crystal (from Mayer and Rimini, 1977). See also Morita (1976) for other interstitial positions.

FIG. A15.15. Top: Octahedral (a) and tetrahedral (b) interstitial sites in a hexagonal close-packed crystal. Bottom: Projections of tetrahedral and octahedral interstitial sites into the major axial and planar channels of a hexagonal close packed structure (from Mayer and Rimini, 1977 and Howe).

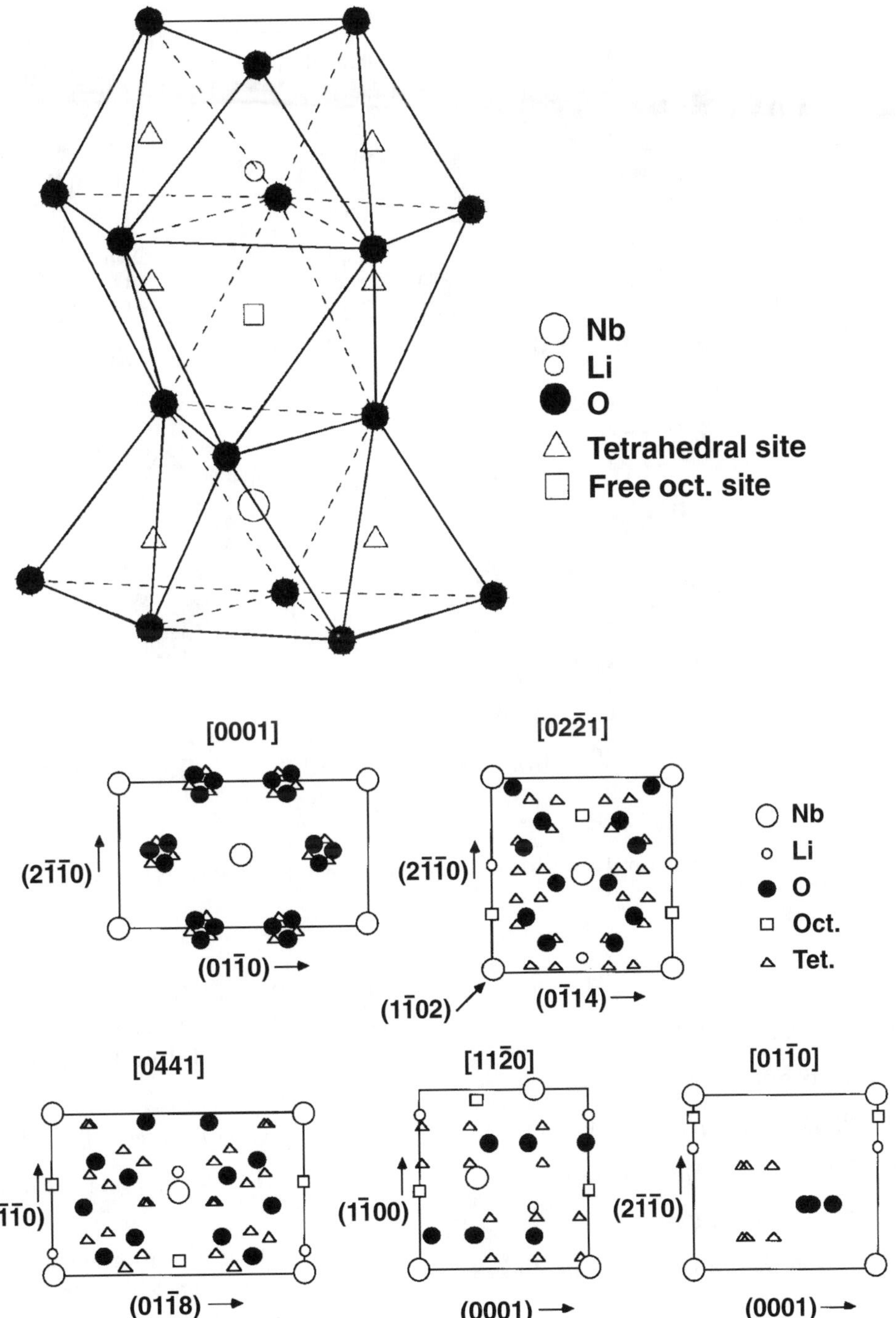

FIG. A15.16. Top part: structure of $LiNbO_3$ below the ferroelectric Curie temperature, 1480 K. Bottom part: projection of Li, Nb and O sites, and tetrahedral and octahedral interstitial sites into the major axial and planar channels of trigonal $LiNbO_3$ (from Soares).

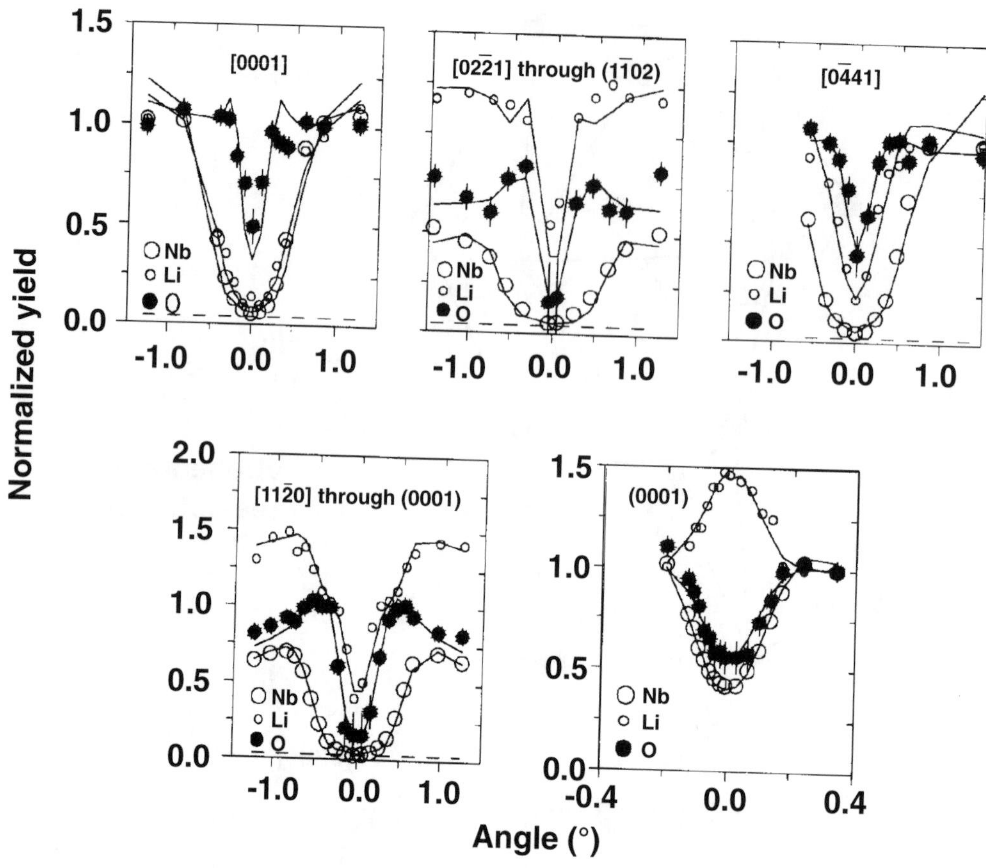

FIG. A15.17. Experimental and calculated (solid lines) yields for 1.6 MeV $^{1}H^{+}$ ions on a $LiNbO_3$ crystal (from Boerma).

REFERENCES

Barrett, J.H. (1971), *Phys. Rev.* **B3**, 1527.

Boerma, D.O. Private communication.

Carstanjen, H.-D. (1980), *Phys. Stat. Sol.* **A59**, 11.

Gemmell, D.S. (1974), *Rev. Mod. Phys.* **46**, 129.

Howe, L.M. Private communication.

Lindhard, J. (1965), *Mat.-Fys. Medd.: K. Dan. Vidensk. Selsk.* **34**, No. 14.

Lugujjo, E. and Mayer, J.W. (1973), *Phys. Rev.* **B7**, 1782.

Mayer, J.W., and Rimini, E. (eds.). (1977), *Ion Beam Handbook for Material Analysis*, Academic Press, New York. This is the old Ion Beam Handbook, ("Catania Handbook").

Morita, K. (1976), *Radiat. Effects* **28**, 65.

Picraux, S.T. (1969), Ph.D. Thesis, California Inst. of Technology.

Soares, J.C. Private communication [see also Rebouta, L., Soares, J.C., daSilva, M.F., Sanz-Garcia, J.A., Dieguez, E. and Agullo-Lopez, F. (1990), *Nucl. Instr. Meth.* **B50**, 428; and (1992), *J. Mater. Res.* **7**, 130].

APPENDIX

16

THIN-FILM MATERIALS AND PREPARATION

Compiled by

R. A. Weller

Vanderbilt University, Nashville, Tennessee

CONTENTS

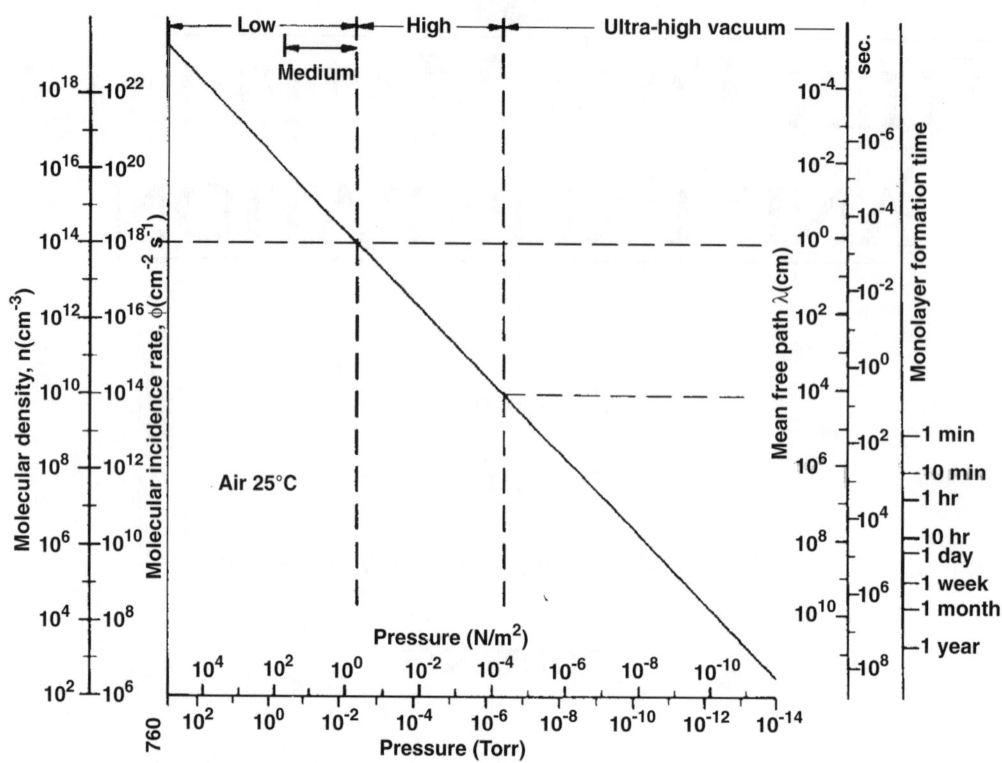

FIG. A16.1. Relationship between different measures of vacuum. From Roth, A., 1982, *Vacuum Technology*, 2nd Edition, North Holland, Amsterdam. Used with permission.

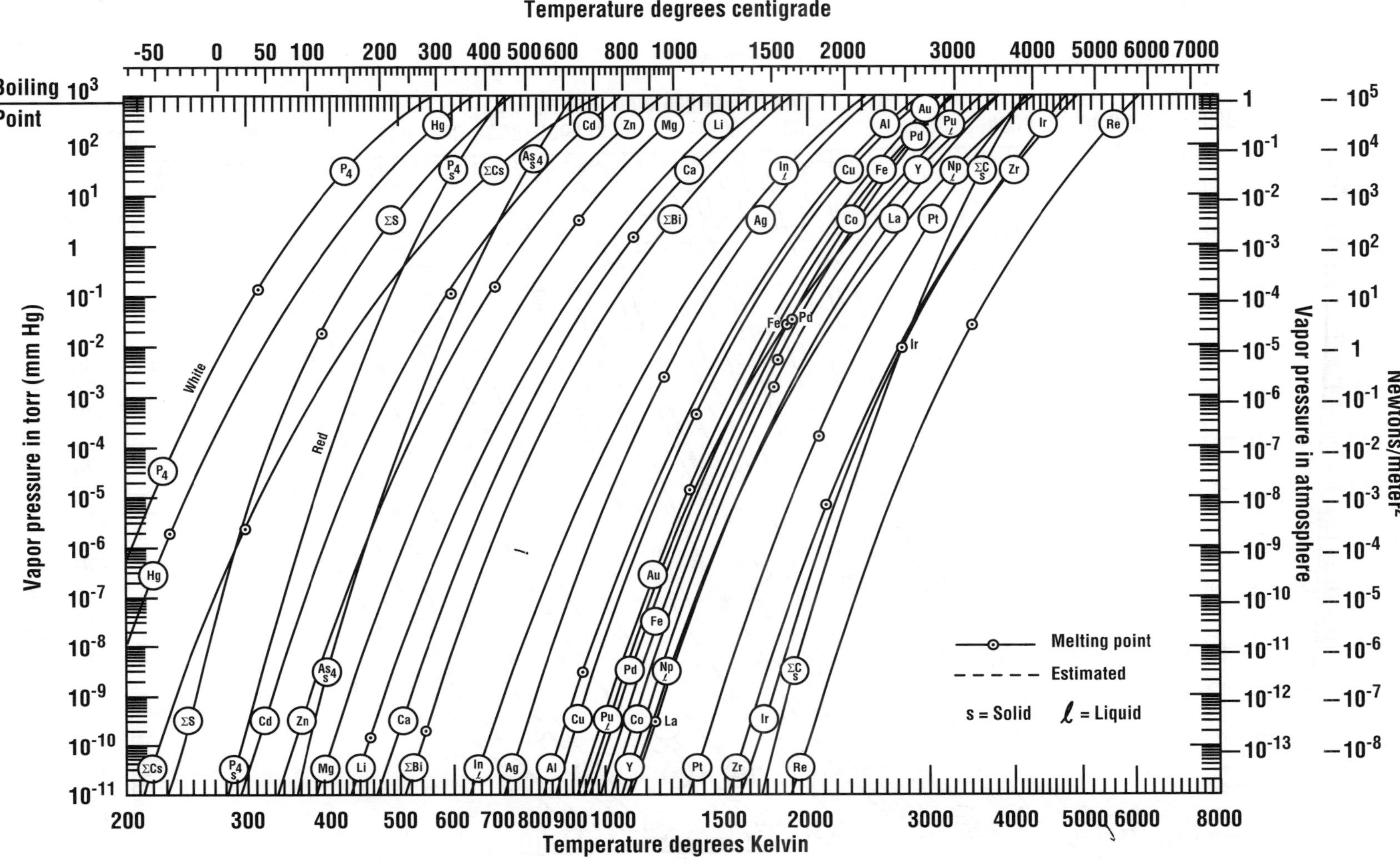

FIG. A16.2. Vapor pressures of the elements. From R. E. Honig and D. A. Kramer, 1969, *RCA Review* **30**, 285. Used with permission.

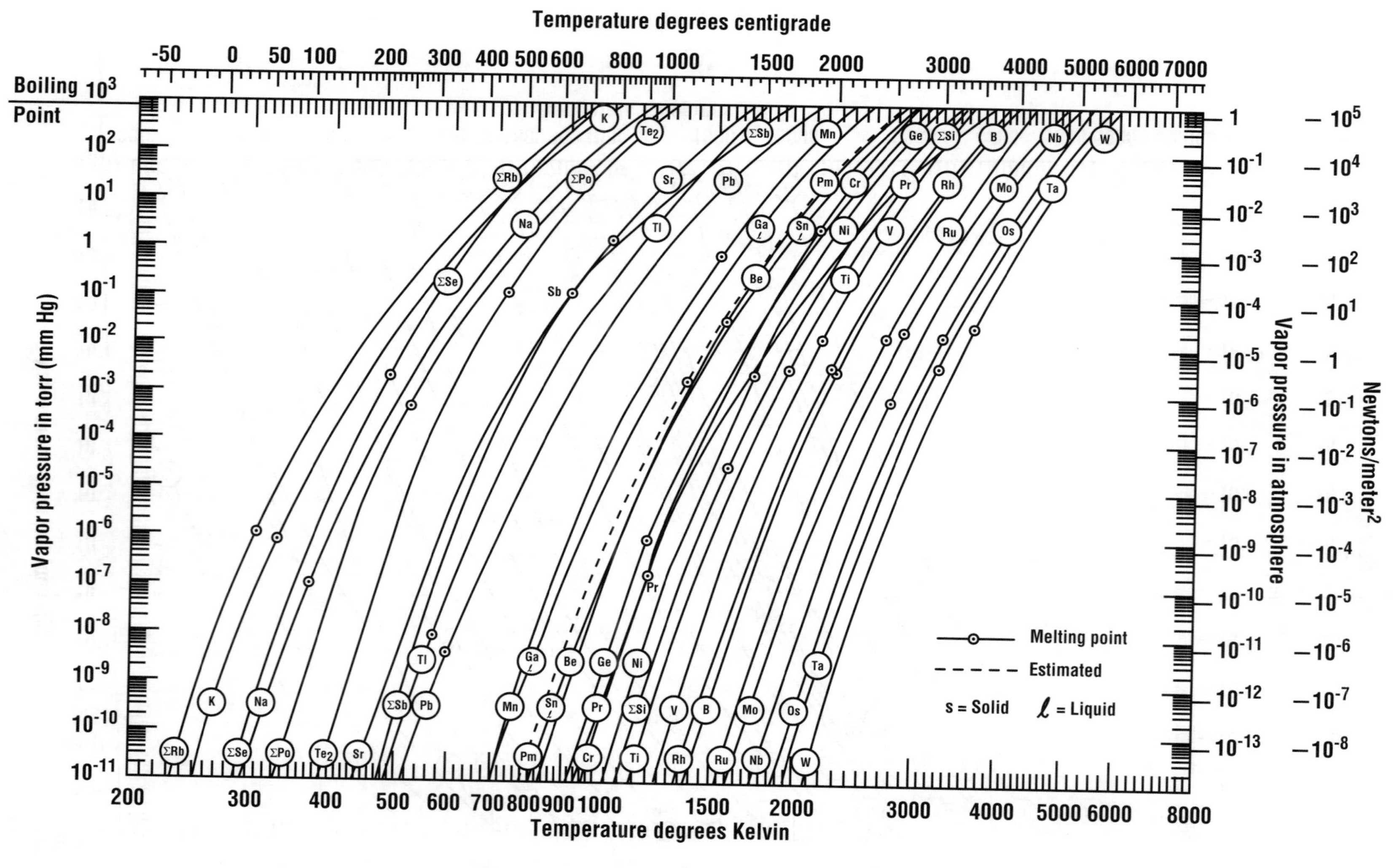

FIG. A16.2. (continued).

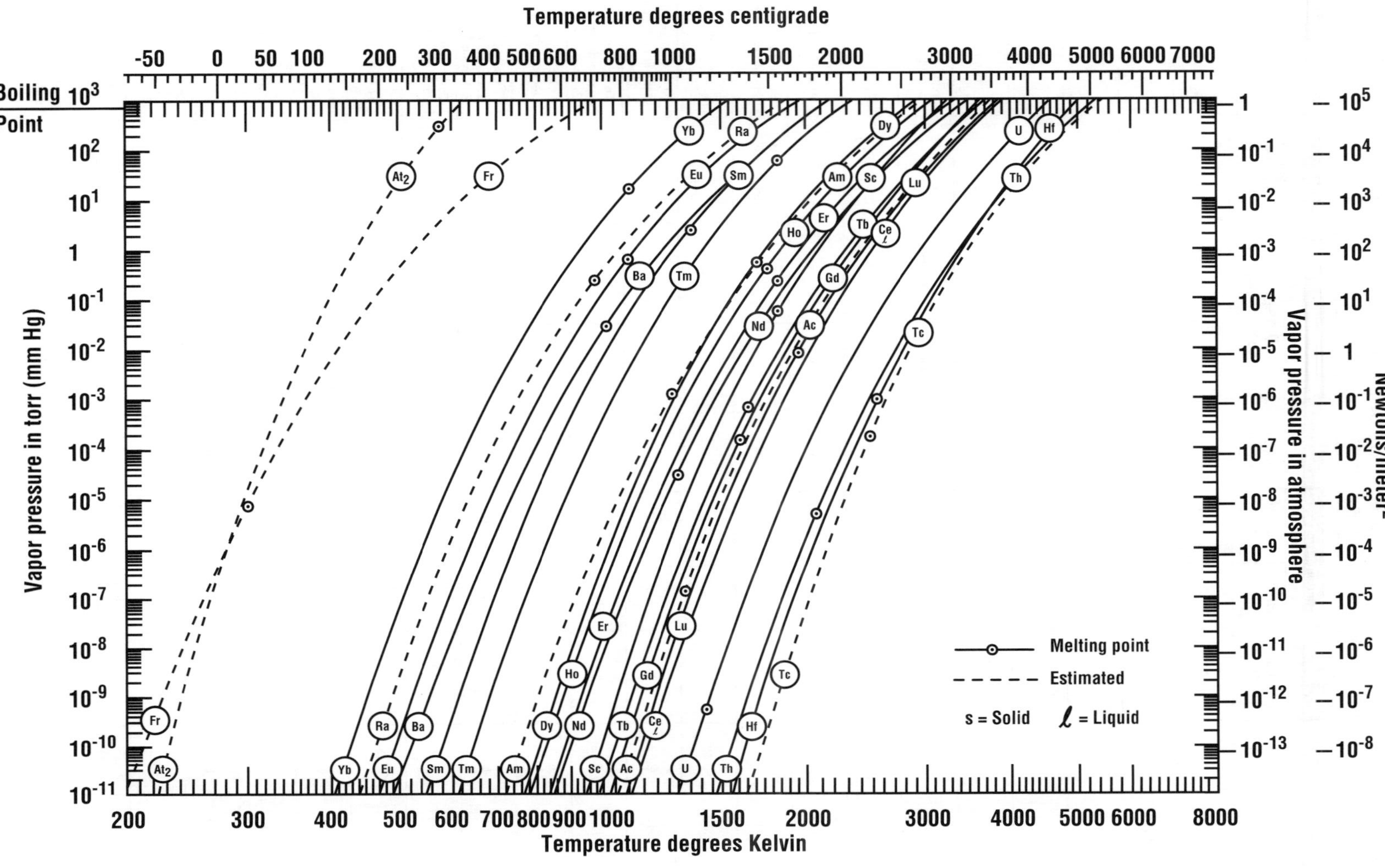

FIG. A16.2. (continued).

Appendix 16

Table A16.1. Lebow thin film databook. Compiled by Edward Graper, Lebow Corporation. Copyright 1990. Used by permission. All reproduction for any purpose without the written permission of Lebow Company is prohibited.

Name	Symbol	Melting point (°C)	Bulk density (g/cm^3)	Acoustic impedance ratio (z)	Temperature (°C) vap. press.		Evaporation techniques			Index of refraction @ microns	Remarks
					(10^{-8} torr)	(10^{-4} torr)	Electron beam	Resistance Source	Resistance Material		
Aluminum	Al	660	2.70	1.05	677	1010	Xlnt.	Coil Bar	W TiB_2-BN	1.0 @ .85 33 @ 12	Alloys and wets tungsten; stranded superior.
Aluminum Antimonide	AlSb	1080	4.3	–	–	–	–	–	–	3.62	–
Aluminum Arsenide	AlAs	1600	3.7	–	–	1300	–	–	–	–	–
Aluminum Bromide	$AlBr_3$	97	2.64	–	–	50	–	Boat	Mo	–	–
Aluminum Carbide	Al_4C_3	2100	2.36	–	–	800	Fair	–	–	2.75 @ .70	Sublimes
Aluminum 2% Copper	Al2%Cu	640	2.82	–	–	–	–	Bar	TiB_2-BN	–	Wire feed and flash, difficult from dual - sources.
Aluminum Fluoride	AlF_3	2191	2.88	–	410	700	Poor	Boat	Mo	1.38 @ .55	Sublimes.
Aluminum Nitride	AlN	2200	3.26	–	–	1750	Fair	–	–	–	Decomposes. Reactive evap Al in 10^{-3} N_2 with glow discharge.
Aluminum Oxide (α) (alumina)	Al_2O_3	2045	3.96	.36	–	1550	Xlnt.	–	–	1.59 @ .60 1.56 @ 1.6	Sapphire xlnt. in EB, forms smooth hard films.
Aluminum Phosphide	AlP	2000	2.42	–	–	–	–	–	–	–	–
Aluminum 2% Silicon	Al2%Si	640	2.69	–	–	1010	–	Bar	TiB_2-BN	–	Wire feed and flash, difficult from dual source.
Antimony	Sb	631	6.62	.59	278	428	Fair	Boat	Mo	3.4 @ 1.0 5.1 @ 11	Toxic. Film structure is rate dependent Use Mo E.B. liner.
Antimony Telluride	Sb_2Te_3	629	6.50	–	–	600	–	Boat	Mo	–	Toxic. Decomposes over 750°C.
Antimony Oxide	Sb_2O_3	656	5.82	–	–	300	Good	Boat	Mo	2.10 @ .50	Toxic, sublimes. Decomposes on W. Use low rate. Z. Physik 165,202 (1961).
Antimony Selenide	Sb_2Se_3	629	6.50	1.87	–	–	–	Crucible	C	–	Toxic. Stoichiometry variable.
Antimony Sulfide	Sb_2S_3	550	4.12	–	–	200	Good	Boat	Mo	3.01 @ .55	Toxic. No decomposition.
Arsenic	As	814	5.73	–	107	210	Poor	Crucible Boat	C Mo	–	Toxic. Sublimes rapidly at low temperature.
Arsenic Selenide	As_2Se_3	360	4.75	–	–	–	–	Boat	Mo	3.03 @ .82 2.9 @ 92	Toxic. JVST 10, 748 (1973).
Arsenic Sulfide	AS_2S_3	300	3.43	–	–	400	Fair	Boat	Mo	2.69 @ .56 2.84 @ 88	Toxic. JVST 10, 748 (1973).
Arsenic Telluride	AS_2Te_3	362	5.0	–	–	–	–	Flash	W	–	Toxic. JVST 10, 748 (1975).
Barium	Ba	725	3.51	–	545	735	Fair	Boat	Mo	.85 @ .50	Toxic. Wets w/o alloying, reacts with ceramics.
Barium Chloride	$BaCl_2$	961	3.86	–	–	650	Good	Boat	Mo	1.74 @ .58	Use gentle preheat to outgas.

Table A16.1. (continued).

Name	Symbol	Melting point (°C)	Bulk density (g/cm^3)	Acoustic impedance ratio (z)	Temperature (°C) vap. press. (10^{-8} torr)	(10^{-4} torr)	Evaporation techniques Electron beam	Resistance Source	Resistance Material	Index of refraction @ microns	Remarks
Barium Fluoride	BaF_2	1280	4.89	.90	–	480	Good	Boat	Mo	1.51 @ .27 1.40 @ 10.3	Sublimes. Density rate dependent.
Barium Oxide	BaO	1923	5.72	–	–	1300	Poor	Boat	Pt	1.98 @ .59	Decomposes slightly.
Barium Sulfide	BaS	1200	4.25	–	–	1100	–	Boat	Mo	2.16 @ .59	Sublimes.
Barium Titanate	$BaTiO_3$	1620	5.85	.32	decomposes ...		–	–	–	2.4 @ .8	Decomposes, yields free Ba; sputter or co-evaporate.
Beryllium	Be	1283	1.85	.55	710	1000	Xlnt.	Boat	W Ta	2.5 @ .5	Metal powder and oxides very toxic. Wets W/Mo/Ta.
Beryllium Chloride	$BeCl_2$	440	1.90	–	–	150	–	Boat	Mo	–	Very toxic.
Beryllium Fluoride	BeF_2	800	1.99	–	–	480	Good	Boat	Mo	1.33 @ .59	Very toxic, sublimes.
Beryllium Oxide	BeO	2575	3.01	–	–	1900	Good	Boat	Ta	1.82 @ .19 1.72 @ .55	Powders very toxic. No decomposition from EB guns.
Bismuth	Bi	271	9.80	.81	330	520	Xlnt.	Boat	Mo	.82 @ .35 4.5 @ 1.0	Vapors are toxic. High resistivity.
Bismuth Fluoride	BiF_3	727	5.32	–	–	300	–	Crucible	C	1.74 @ 1.0 1.64 @ 10	Toxic, sublimes. App. Opt. 18,105 (1979).
Bismuth Oxide	Bi_2O_3	811	8.9	–	–	1390	Poor	Boat	W	2.48 @ .58	Vapors are toxic. JVST 12, 63 (1975).
Bismuth Selenide	Bi_2Se_3	710	7.66	–	–	650	Good	–	–	–	Toxic. Sputter or co-evaporate.
Bismuth Telluride	Bi_2Te_3	585	6.82	–	–	600	–	Boat	Mo	–	Toxic. Sputter or co-evaporate.
Bismuth Titanate	$Bi_2Ti_2O_7$	–	–	–	decomposes ...		–	–	–	–	Toxic. Decomposes. Sputter or co-evaporate in 10^{-2} O_2.
Bismuth Sulfide	Bi_2S_3	685	7.39	–	–	–	–	–	–	1.5	Toxic.
Boron	B	2100	2.34	.45	1278	1797	Poor	Crucible	C	–	Material explodes with rapid cooling. Forms carbide with container.
Boron Carbide	B_4C	2350	2.52	–	2500	2650	Xlnt.	–	–	–	Similar to chromium.
Boron Nitride	BN	2300	2.25	–	–	1600	Poor	–	–	–	Sputtering pref. Decomposes. JVST A5(4),2696 (1987).
Boron Oxide	B_2O_3	460	2.46	–	–	1400	Good	Boat	Mo	1.46	–
Boron Sulfide	B_2S_3	310	1.55	–	–	800	–	–	–	–	–
Cadmium	Cd	321	8.65	.6	64	180	Fair	Boat	Mo	1.13 @ .6	Poisons vacuum systems, low sticking coefficient. Use Mo E.B. liner.
Cadmium Antimonide	CdSb	456	6.92	–	–	–	–	Boat	Mo	–	–

Inficon Z-Ratio° = acoustic impedance ratio, z.
Z-Ratio° Leybold Inficon

Maxtec Inc. Acoustic Impedance (A. I.) = 8.83 ÷ z

Table A16.1. (continued).

Name	Symbol	Melting point (°C)	Bulk density (g/cm³)	Acoustic impedance ratio (z)	Temperature (°C) vap. press. (10⁻⁸ torr)	Temperature (°C) vap. press. (10⁻⁴ torr)	Evaporation techniques: Electron beam	Resistance: Source	Resistance: Material	Index of refraction @ microns	Remarks
Cadmium Arsenide	Cd_3As_2	721	6.21	–	–	–	–	–	–	–	Toxic.
Cadmium Bromide	$CdBr_2$	567	5.19	–	–	300	–	Boat	Mo	–	Sublimes.
Cadmium Chloride	$CdCl_2$	960	4.05	–	–	400	–	–	–	–	Sublimes.
Cadmium Fluoride	CdF_2	1100	6.64	–	–	600	–	–	–	1.56 @ .58	–
Cadmium Iodide	CdI_2	387	5.67	–	–	250	–	–	–	–	–
Cadmium Oxide	CdO	1430	8.15	–	–	530	–	–	–	2.49 @ .67	Disproportionates.
Cadmium Selenide	CdSe	1351	5.79	–	–	580	Good	Box	Mo	2.4 @ .58	Toxic, sublimes.
Cadmium Silicide	$CdSiO_2$	–	–	–	–	600	–	–	–	1.69	–
Cadmium Sulfide	CdS	1750	4.82	1.02	–	550	Good	Box Crucible	Mo Quartz	2.43 @ .67 2.31 @ 1.4 2.27 @ 7.0	Sublimes. Sticking coeff. affected by sub temp. Comp. variable JVST 12,188 (1975)
Cadmium Telluride	CdTe	1041	6.20	.98	–	450	–	Box Boat	Mo Mo	2.68 @ 4.0 2.51 @ 32	Toxic. Stoichiometry depends on substrate temp. JVST 8,412 (1971).
Calcium	Ca	845	1.55	2.36	272	459	Poor	Boat	Mo	.29 @ .58	Flammable, sublimes. Corrodes in air. Optic 18,59 (1961).
Calcium Fluoride	CaF_2	1360	3.18	.85	–	1100	Xlnt.	Boat	Mo	1.47 @ .24 1.32 @ 9.4	Rate control important. Use gentle preheat to Outgas.
Calcium Oxide	CaO	2580	3.35- 3.38	–	–	1700	–	Boat	W	1.84 @ .59	Forms volatile oxides with W and Mo.
Calcium Silicate	$CaO•SiO_2$	1540	2.90	–	–	–	Good	–	–	1.61	–
Calcium Sulfide	CaS	subl.	2.5	–	–	1100	–	Box	Mo	2.14 @ .59	Decomposes.
Calcium Titanate	$CaTiO_3$	1975	4.10	–	1490	1690	Poor	–	–	2.34 @ .59	Disproportionates except in sputtering.
Calcium Tungstate	$CaWO_4$	1620	6.06	–	–	–	Good	Boat	W	1.92 @ .59	–
Carbon (Diamond)	C	3727	3.52	.22	–	–	–	–	–	2.94 @ .19 2.42 @ .66	Deposit by CVD.
Carbon (Graphite)	C	subl.	2.26	4.33	1657	2137	Fair	–	–	1.47	Sublimes. EB preferred, Arc evaporat. Poor film adhesion.
Cerium	Ce	795	6.67	.86	970	1380	Good	Boat	W	1.91 @ .59	Films oxidize easily.
Cerium (III) Fluoride	CeF_3	1460	6.16	–	–	900	Good	Box Boat	Mo W	1.63 @ .55 1.55 @ 12	Use gentle preheat to outgas.
Cerium (IV) Oxide	CeO_2	2600	7.13	–	1890	2310	Good	Boat	W	2.18 @ .55	Sublimes. Use 250°C sub. temperature. Reacts with W. J Opt Soc Am 48,324 (1958).

Table A16.1. (continued).

Name	Symbol	Melting point (°C)	Bulk density (g/cm³)	Acoustic impedance ratio (z)	Temperature (°C) vap. press. (10⁻⁸ torr)	Temperature (°C) vap. press. (10⁻⁴ torr)	Evaporation techniques: Electron beam	Evaporation techniques: Resistance Source	Evaporation techniques: Resistance Material	Index of refraction @ microns	Remarks
Cerium Oxide	Ce_2O_3	1691	6.89	.41	–	–	Fair	Boat	.02 W	2.18 @ .58	Alloys with source. J.Opt.Soc.Am 48,324 (1958).
Cesium	Cs	29	1.89	–	–16	30	–	Boat	Mo	–	Flammable.
Cesium Bromide	CsBr	636	4.44	1.72	–	400	–	Boat	Mo	1.79 @ .36 1.56 @ 39	–
Cesium Chloride	CsCl	646	3.99	–	–	500	–	Boat	Mo	1.64	Hygroscopic.
Cesium Fluoride	CsF	682	4.11	–	–	500	–	Boat	Mo	–	–
Cesium Hydroxide	CsOH	272	3.67	–	–	550	–	Boat	Pt	–	–
Cesium Iodide	CsI	621	4.51	2.95	–	500	–	Boat	W	1.99 @ .23 1.62 @ 53	–
Chiolote	$Na_5Al_3F_{14}$	–	2.9	–	–	800	–	Boat	Mo	1.33	–
Chromium	Cr	1875	7.19	.31	837	1157	Good	plated rod or basket	W	.83 @ .13 3.19 @ .63	Sublimes. Films very adherent. High rates possible.
Chromium Boride	CrB	2760	6.17	–	–	–	–	–	–	–	–
Chromium Bromide	$CrBr_2$	842	4.36	–	–	550	–	Boat	Mo	–	–
Chromium Carbide	Cr_3C_2	1890	6.68	–	–	2000	Fair	Boat	W	–	–
Chromium Chloride	$CrCl_2$	824	2.88	–	–	550	–	Box	Mo	–	Sublimes easily.
Chromium Oxide	Cr_2O_3	2435	5.21	–	–	2000	Good	Box	Mo	2.5 @ .59	Disproportionates to lower oxides, reoxidizes @ 600°C in air.
Chromium Silicide	Cr_3Si_2	1710	6.51	–	–	–	–	–	–	–	–
Chromium-Silicon Monoxide	Cr-SiO	influenced by composition					Good	Bar	W	–	Flash.
Cobalt	Co	1495	8.92	.34	850	1200	Xlnt.	Boat Coil	.02 W W	1.10 @ .23 5.65 @ 2.2	Alloys with refractory metals.
Cobalt Bromide	$CoBr_2$	678	4.91	–	–	400		Box	Mo	–	Sublimes.
Cobalt Chloride	$CoCl_2$	724	3.37	–	–	472		Box	Mo	1.51 @ .63	Sublimes.
Cobalt Oxide	CoO	1935	6.45	–	–	–		–	–	–	Sputtering preferred.
Copper	Cu	1083	8.94	.43	727	1017	Xlnt.	Boat Coil	Mo W	.87 @ .45 15.5 @ 12	Films do not adhere well. Use intermediate Cr layer, O_2 free Cu req'd.
Copper Chloride	CuCl	431	4.19	–	–	580	–	–	–	1.93	–

Inficon Z-Ratio° = acoustic impedance ratio, z.
Z-Ratio° Leybold Inficon

Maxtec Inc. Acoustic Impedance (A. I.) = 8.83 ÷ z

Table A16.1. (continued).

Name	Symbol	Melting point (°C)	Bulk density (g/cm³)	Acoustic impedance ratio (z)	Temperature (°C) vap. press. (10^{-8} torr)	Temperature (°C) vap. press. (10^{-4} torr)	Evaporation techniques: Electron beam	Evaporation techniques: Resistance Source	Evaporation techniques: Resistance Material	Index of refraction @ microns	Remarks
Copper Oxide	Cu_2O	1235	6.0	–	–	600	Good	Box	Ta	2.70 @ .59	Sublimes. Evaporate in 10^{-2} to 10^{-4} of O_2; J.Electro-chem. Soc. 110,119 (1967).
Copper (I) Sulfide	Cu_2S	1100	5.6	.68	–	–	–	Boat	Mo	–	–
Copper (II) Sulfide	CuS	1113	6.75	.82	–	500	–	–	–	1.45	Sublimes.
Cryolite	Na_3AlF_6	1000	2.9	–	1020	1480	Xlnt.	Boat	Mo	2.34 @ .63	Large chunks reduce spitt-ing. Little decomposition. App.Opt.15, 1969 (1976).
Dysprosium	Dy	1407	8.54	.60	625	900	Good	Boat	Ta	–	Flammable.
Dysprosium Fluoride	DyF_3	1360	–	–	–	800	Good	Box	Ta	–	Sublimes.
Dysprosium Oxide	Dy_2O_3	2340	7.81	–	–	1400	–	Boat	W	–	Loses oxygen.
Erbium	Er	1461	9.09	.74	680	980	Good	Boat	Ta	–	Sublimes.
Erbium Fluoride	ErF_3	1350	7.81	–	–	750	–	Boat	Mo	–	JVST A3(6),2320.
Erbium Oxide	Er_2O_3	2400	8.64	–	–	1600	–	Boat	W	–	Loses oxygen.
Europium	Eu	826	5.26	1.62	280	480	Fair	Boat	Ta	–	Flammable, sublimes. Low tantalum solubility.
Europium Fluoride	EuF_2	1390	6.5	–	–	950	–	Box	Mo	–	–
Europium Oxide	Eu_2O_3	2056	7.42	–	–	1600	Good	Boat	W	–	Loses oxygen; films clear and hard.
Europium Sulfide	EuS	–	5.75	–	–	–	Good	–	–	–	–
Gadolinium	Gd	1312	7.89	.67	760	1175	Xlnt.	Boat	Ta	–	High Ta solubility. Flammable.
Gadolinium Oxide	Gd_2O_3	2310	7.41	–	–	–	Fair	Box	W	1.8 @ .55	Loses oxygen.
Gallium	Ga	30	5.91	.59	619	907	Good	–	–	–	Alloys with refractory metals. Use EB gun.
Gallium Antimonide	GaSb	710	5.6	–	–	–	Fair	Boat	W Ta	3.80 @ 2.2	Flash evaporate.
Gallium Arsenide	GaAs	1238	5.31	1.59	–	–	Good	Boat	W	3.34 @ .78 2.12 @ 23	Flash evaporate.
Gallium Nitride	GaN	800	6.1	–	–	200	–	–	–	–	Sublimes. Evaporate Ga in $10^{-3}N_2$.
Gallium Oxide(β)	Ga_2O_3	1900	5.88	–	–	–	–	Boat	W	–	Loses oxygen.
Gallium Phosphide	GaP	1348	4.1	–	520	920	–	Boat	W	3 @ 2.15	Decomposes. Vapor mostly P.

Table A16.1. (continued).

Name	Symbol	Melting point (°C)	Bulk density (g/cm^3)	Acoustic impedance ratio (z)	Temperature (°C) vap. press.		Evaporation techniques			Index of refraction @ microns	Remarks
					(10^{-8} torr)	(10^{-4} torr)	Electron beam	Resistance			
								Source	Material		
Germanium	Ge	937	5.32	.51	812	1167	Xlnt.	Boat Crucible	.02-.04 W C	4.20 @ 2.1 4.00 @ 20	–
Germanium Nitride	Ge_3N_2	450	5.25	–	–	650	–	–	–	–	Sublimes. Sputtering preferred.
Germanium Oxide	GeO_2	1086	6.24	–	–	625	Good	Box	Mo	1.61 @ .59	Similar to SiO, film predominately GeO.
Germanium Telluride	GeTe	725	6.20	–	–	381	–	Box	Mo	–	–
Glass, Schott 8329	–	–	2.20	–	–	–	Xlnt.	–	–	1.47	Evaporable alkali glass. Melt in air before evaporating.
Gold	Au	1962	19.32	.39	807	1132	Xlnt.	Boat Coil	W W	1.50 @ .13 32 @ 12	Films soft, not very adherent.JVST 12,704 (1975).
Hafnium	Hf	2222	13.09	.34	2160	3090	Good	–	–	–	–
Hafnium Boride	HfB_2	3250	10.5	–	–	–	–	–	–	–	–
Hafnium Carbide	HfC	3890	12.20	–	–	2600	–	–	–	–	Sublimes.
Hafnium Nitride	HfN	3305	13.8	–	–	–	–	–	–	–	–
Hafnium Oxide	HfO_2	2811	9.69	–	–	2500	Fair	Boat	W	2.08 @ .48	Film HfO. App. Opt. Apr. 1977.
Hafnium Silicide	$HfSi_2$	1680	8.02	–	–	–	–	–	–	–	–
Holnium	Ho	1461	8.80	.58	650	950	Good	Boat	W	–	Sublimes.
Holnium Fluoride	HoF_3	1143	7.64	–	–	800	–	–	–	–	–
Holnium Oxide	Ho_2O_3	2360	8.36	–	–	–	–	Boat	W	–	Loses oxygen. App. Opt. 16,439
Inconel	Ni/Cr/Fe	1425	8.5	.33	–	–	Good	Boat Coil	.02 W W	–	Use fine wire pre-wrapped on W. Low rate required for smooth films.
Indium	In	157	7.31	.84	487	742	Xlnt.	Boat	Mo	–	Wets W and Cu; use Mo liner in guns.
Indium Antimonide	InSb	535	5.76	.77	500	400	–	Boat	W	1.00 @ .55 4.0 @ 7.9 3.8 @ 22	Toxic. Decomposes; sputter preferred; or co-evaporate from 2 sources; flash.
Indium Arsenide	InAs	943	5.7	–	780	970	–	Boat	W	3.3 @ 10	Toxic. Sputtering preferred; or co-evaporate frome 2 sources; flash.
Indium Oxide	In_2O_3	1565	7.18	–	–	1200	Good	Boat	W	–	Sublimes. Film In_2O; transparent conductor. JVST 12,99 (1975).
Indium Phosphide	InP	1071	4.9	–	580	730	–	Boat	W	3.3 @ 8.8	Deposits P rich. Flash evaporate.
Indium Selenide	In_2Se_3	890	5.67	–	–	–	–	–	–	–	Sputter, co-evaporate or flash.

Inficon Z-Ratio° = acoustic impedance ratio, z.
Z-Ratio° Leybold Inficon

Maxtec Inc. Acoustic Impedance (A. I.) = 8.83 ÷ z

Table A16.1. (continued).

Name	Symbol	Melting point (°C)	Bulk density (g/cm³)	Acoustic impedance ratio (z)	Temperature (°C) vap. press. (10^{-8} torr)	Temperature (°C) vap. press. (10^{-4} torr)	Evaporation techniques: Electron beam	Evaporation techniques: Resistance Source	Evaporation techniques: Resistance Material	Index of refraction @ microns	Remarks
Indium (III) Sulfide	In_2S_3	1050	4.45	–	–	850	–	–	–	–	Sublimes. Film In_2S.
Indium (I) Sulfide	In_2S	653	5.87	–	–	650	–	–	–	–	–
Indium Telluride	In_2Te_3	667	5.8	–	–	–	–	–	–	–	Sputter, co-evaporate, or flash.
Iridium	Ir	2454	22.45	.13	1850	2380	Fair	–	–	–	–
Iron	Fe	1536	7.87	.35	858	1180	Good	Boat	W	2.0 @ .58	Attacks W. Films hard, smooth. Use gentle preheat to outgas.
Iron Bromide	$FeBr_2$	684	4.64	–	–	561	–	–	–	–	–
Iron Chloride	$FeCl_2$	674	3.16	–	–	300	–	–	–	1.57 @ .59	Sublimes.
Iron Iodide	FeI_2	592	5.31	–	–	400	–	–	–	–	–
Iron (II) Oxide	FeO	1420	5.7	–	–	–	Poor	–	–	2.32 @ .59	Decomposes; sputtering preferred.
Iron (III) Oxide	Fe_2O_3	1538	5.18	–	–	–	Good	Boat	W	3.0	Disproportionates to Fe_3O_4 at 1530°C.
Iron Sulfide	FeS	1195	4.74	–	–	–	–	–	–	–	Decomposes.
Kanthal	FeCrAl	1500	7.1	–	–	1150	–	Boat Coil	.02 W W	1.74 @ .58	JVST 7, 739 (1980).
Lanthanium	La	920	6.1	.93	990	1368	Xlnt.	Boat	W	–	Films will burn in air if scraped.
Lanthanium Boride	LaB_6	2210	2.61	–	–	–	Good	–	–	–	Toxic.
Lanthanium Fluoride	LaF_3	1491	5.99	–	–	900	Good	Boat	Mo	1.40 @ .30 1.20 @ 8.8	Sublimes. No decomposition. Heat substrate over 300°C.
Lanthanium Oxide	La_2O_3	2315	6.51	–	–	1400	Good	Boat	W	1.95 @ .55 1.89 @ 2.0	Loses oxygen.
Lead	Pb	327	11.34	1.10	342	497	Xlnt.	Boat	Mo	2.6 @ .58	Toxic. Use Mo liner in E.B. gun.
Lead Bromide	$PbBr_2$	373	6.68	–	–	300	–	Boat	Mo	–	Toxic.
Lead Chloride	$PbCl_2$	5015.85	–	–	325		–	Boat	Mo	2.3 @ .55 2.0 @ 10	Toxic. Little decomposition.
Lead Fluoride	PbF_2	855	8.24	–	–	400	–	Boat	Mo	1.92 @ .30 1.60 @ 11	Toxic, sublimes. Z.Physic 159,117 (1959).
Lead Iodide	PbI_2	502	6.16	–	–	320	–	Crucible	Quartz	2.7	Toxic. J. Opt. Soc. 65,914.
Lead Oxide	PbO	888	9.53	–	–	550	–	Boat	Pt Mo	2.51 @ .59	Toxic. No decomposition. J.Opt.Soc.Am. 52,161 (1962).

Table A16.1. (continued).

Name	Symbol	Melting point (°C)	Bulk density (g/cm³)	Acoustic impedance ratio (z)	Temperature (°C) vap. press.		Evaporation techniques			Index of refraction @ microns	Remarks
					(10^{-8} torr)	(10^{-4} torr)	Electron beam	Resistance Source	Resistance Material		
Lead Stannate	$PbSnO_3$	1115	8.1	–	670	905	Poor	Boat	Pt Mo	–	Toxic. Disproportionates.
Lead Selenide	PbSe	1065	8.10	–	–	500	–	Box	Mo	3.5 @ 1.0	Toxic, sublimes.
Lead Sulfide	PbS	1114	7.5	.56	–	550	–	Box	Mo	3.9 @ .5	Toxic, sublimes. Little decomposition.
Lead Telluride	PbTe	917	8.16	–	780	1050	–	Boat	Ta	5.6 @ 5 3.4 @ 30	Vapors toxic. Deposits Te rich. Sputter or co-evaporate.
Lead Titanate	$PbTiO_3$	–	7.52	–	–	–	–	Boat	Ta	–	Toxic.
Lithium	Li	180	0.53	5.95	227	407	Good	Boat	Mo	–	Metal reacts rapidly in air.
Lithium Bromide	LiBr	550	3.46	–	–	500	–	Boat	Mo	1.78 @ .59	–
Lithium Chloride	LiCl	614	2.07	–	–	400	–	Boat	Mo	1.66 @ .59	Use gently preheat for outgas.
Lithium Fluoride	LiF	841	2.59	.78	878	1180	Good	Boat	Ta	1.44 @ .19 1.36 @ 3.5	Rate control important. Use preheat for outgas. App. Opt.11,2245 (1972).
Lithium Iodide	LiI	450	3.49	–	–	400	–	Boat	Mo	1.96 @ .59	–
Lithium Oxide	Li_2O	1427	2.01	–	–	850	–	Boat	W	1.64 @ .59	–
Lutetium	Lu	1652	9.84	.48	–	1300	Xlnt.	Boat	Ta	–	Ta impurity a problem.
Lutetium Oxide	Lu_2O_3	2487	9.41	–	–	1400	–	Boat	W	–	Decomposes.
Magnesium	Mg	650	1.74	1.61	180	327	Good	Coil Boat	W Mo	.52 @ .40	Flammable, sublimes. Extremely high rates possible.
Magnesium Aluminate	$MgAl_2O_4$	2135	3.6	–	–	–	Good	–	–	–	Natural spinel.
Magnesium Bromide	$MgBr_2$	700	3.72	–	–	450	–	Boat	Mo	–	Decomposes.
Magnesium Chloride	$MgCl_2$	714	2.32	–	–	400	–	Boat	Mo	1.6	Decomposes.
Magnesium Fluoride	MgF_2	1248	3.0	.68	–	–	Xlnt.	Boat	Mo	1.52 @ .20 1.36 @ 2.0	Rate control & sub. heat required for optical films. App. Opt.11, 2245 (1972).
Magnesium Iodide	MgI_2	700	4.24	–	–	200	–	Boat	Mo	–	–
Magnesium Oxide	MgO	2800	3.58	.38	–	1300	Good	–	–	1.77 @ .36 1.63 @ 5.1	W produces volatile oxides. App. Opt.11, 2243 (1972).
Manganese	Mn	1241	7.39	.43	508	648	Good	Basket Boat	W Mo	2.59 @ .59	Flammable, sublimes.
Manganese Bromide	$MnBr_2$	695	4.38	–	–	500	–	Boat	Mo	–	–

Inficon Z-Ratio° = acoustic impedance ratio, z.
Z-Ratio° Leybold Inficon

Maxtec Inc. Acoustic Impedance (A. I.) = 8.83 ÷ z

Table A16.1. (continued).

Name	Symbol	Melting point (°C)	Bulk density (g/cm³)	Acoustic impedance ratio (z)	Temperature (°C) vap. press.		Evaporation techniques			Index of refraction @ microns	Remarks
					(10^{-8} torr)	(10^{-4} torr)	Electron beam	Resistance Source	Resistance Material		
Manganese Chloride	$MnCl_2$	650	2.98	–	–	450	–	Boat	Mo	–	–
Manganese Oxide	Mn_3O_4	1705	4.86	–	–	–	–	Boat	W	1.73	–
Manganese Sulfide	MnS	1615	3.58	.94	–	1300	–	Boat	Mo	2.7	Decomposes.
Mercury	Hg	–39	13.55	.74	–68	–6	–	–	–	–	Toxic.
Mercury Sulfide	HgS	583	8.10	–	–	250	–	–	–	–	Toxic, decomposes.
Molybdenum	Mo	2610	10.22	.27	1592	2117	Xlnt.	–	–	3.65 @ .59	Films smooth, hard. Careful degas req'd.
Molybdenum Boride	Mo_2B_3	2200	7.48	–	–	–	Poor	–	–	–	–
Molybdenum Carbide	Mo_2C	2687	9.18	–	–	–	Fair	–	–	–	Evaporation of $Mo(CO)_6$ yields Mo_6C.
Molybdenum Silicide	$MoSi_2$	2050	6.31	–	–	–	–	Boat	Mo	1.9	Slight O_2 loss.
Molybdenum Sulfide	MoS_2	1185	4.80	–	–	50	–	Boat	W	–	Decomposes.
Molybdenum Oxide	MoO_3	795	4.69	–	–	900	–	–	–	–	–
Neodymium	Nd	1024	7.00	.84	731	1062	Xlnt.	Boat	Ta	.89 @ .39 .30 @ .88	Flammable. Low Ta solubility.
Neodymium Fluoride	NdF_3	1410	6.51	–	–	900	Good	Boat	Mo	1.61 @ .55 1.58 @ 2.0	Very little decomposition.
Neodymium Oxide	Nd_2O_3	1900	7.24	–	–	1400	Good	Boat	W	2.0 @ .55 1.95 @ 2.0	Loses O_2, films clear, EB preferred. Hygroscopic. n varies with substrate temp.
Nichrome IV	Ni/Cr	1395	8.50	–	847	1217	Xlnt.	Coil Boat	W .02-.04 W	3.74 @ 8.8 10.2 @ 12.5	Alloys with refractory metals.
Nickel	Ni	1453	8.91	.33	927	1262	Fair	Coil Boat	W .02-.04 W	3.74 @ 8.8 10.2 @ 12.5	Alloys with refractory metals. Forms smooth adherent films.
Nickel Bromide	$NiBr_2$	963	4.64	–	–	362	–	Boat	Mo	–	Sublimes.
Nickel Chloride	$NiCl_2$	1001	3.55	–	–	444	–	Boat	Mo	–	Sublimes.
Nickel Oxide	NiO	1990	6.69	–	–	1480	–	–	–	2.18 @ .48	Dissociates upon heating.
Niobium (Columbium)	Nb	2468	8.57	.47	1728	2287	Xlnt.	Coil	W	1.80 @ .58	Attacks W source.
Niobium Boride	NbB_2	3050	6.97	–	–	–	–	–	–	–	–
Niobium Carbide	NbC	3500	7.82	–	–	–	Fair	–	–	–	–

Table A16.1. (continued).

Name	Symbol	Melting point (°C)	Bulk density (g/cm^3)	Acoustic impedance ratio (z)	Temperature (°C) vap. press. (10^{-8} torr)	Temperature (°C) vap. press. (10^{-4} torr)	Evaporation techniques: Electron beam	Evaporation techniques: Resistance Source	Evaporation techniques: Resistance Material	Index of refraction @ microns	Remarks
Niobium Nitride	NbN	2573	8.4	–	–	–	–	–	–	–	Reactive, evaporate Nb in 10^{-3} N_2.
Niobium Oxide	NbO	–	7.30	–	–	1100	–	Boat	W	–	–
Niobium Oxide (V)	Nb_2O_5	1520	4.47	–	–	–	–	Boat	.02 W	2.3	–
Niobium Telluride	$NbTe_5$	–	7.6	–	–	–	–	–	–	–	Composition variable.
Niobium-Tin	Nb_3Sn	–	–	–	–	–	Xlnt.	–	–	–	Co-evaporate from 2 sources.
Osmium	Os	3045	22.6	.13	2170	2760	Fair	–	–	–	Toxic.
Palladium	Pd	1552	12.02	.38	842	1192	Xlnt.	Boat	.02 W	1.5 @ .30 2.3 @ .54	Alloys with refractory metals; rapid evaporation suggested. Spits in EB.
Palladium Oxide	PdO	870	8.70	–	–	575	–	–	–	–	Decomposes.
Parylene-N	C_6H_8	300-400	1.1	–	–	–	–	–	–	–	Vapor depositable plastic. (Union Carbide).
Permalloy	Ni/Fe	1395	8.7	–	947	1307	Good	Boat	.02 W	–	Film low in Ni content. Use 84% Ni source. JVST 7(6),573 (1970).
Phosphorus	P	44.2	1.82	–	327	402	–	Boat	Mo	–	Metal reacts violently in air.
Platinum	Pt	1769	21.45	.24	1292	1747	Xlnt.	–	–	3.42 @ 1.0	Alloys, E.B. req'd. Films soft. Poor adhesion.
Platinum Oxide	PtO_2	450	10.2	–	–	–	–	–	–	–	–
Plutonium	Pu	635	19	–	–	–	–	Boat	W	–	Toxic, radioactive.
Polonium	Po	254	9.4	–	117	244	–	Boat	Mo	–	Radioactive.
Potassium	K	64	0.86	–	23	125	–	Boat	Mo	.74 @ .25	Metal reacts violently in air. Use gentle preheat to outgas.
Potassium Bromide	KBr	731	2.79	–	–	450	–	Boat	Mo	1.56 @ .48 1.47 @ 24	Use gentle preheat to outgas.
Potassium Chloride	KCl	776	2.51	2.05	–	510	Good	Boat	Mo	1.72 @ .20 1.25 @ 24	Melt in air to outgas.
Potassium Fluoride	KF	846	2.48	–	–	500	Poor	Boat	Mo	1.35 @ 1.4	Melt in air to outgas.
Potassium Hydroxide	KOH	360	2.04	–	–	400	–	Boat	Mo	–	Melt in air to outgas. Hygroscopic.
Potassium Iodide	KI	686	3.13	2.0	–	500	–	Boat	Mo	1.92 @ .27 1.56 @ 28	Melt in air to outgas.
Praseo-dymium	Pr	936	6.77	–	800	1150	Good	Boat	Ta	–	Flammable.

Inficon Z-Ratio° = acoustic impedance ratio, z.
Z-Ratio° Leybold Inficon

Maxtec Inc. Acoustic Impedance (A. I.) = 8.83 ÷ z

Table A16.1. (continued).

Name	Symbol	Melting point (°C)	Bulk density (g/cm^3)	Acoustic impedance ratio (z)	Temperature (°C) vap. press. (10^{-8} torr)	Temperature (°C) vap. press. (10^{-4} torr)	Evaporation techniques Electron beam	Evaporation techniques Resistance Source	Resistance Material	Index of refraction @ microns	Remarks
Praseodymium Chloride	$PrCl_3$	786	4.02	–	–	500	–	Boat	Mo	1.86	–
Praseodymium Oxide	Pr_2O_3	2125	6.88	–	–	1400	Good	Boat	W	1.92 @ .27 1.83 @ 2.0	Loses oxygen.
Radium	Ra	700	5.0	–	246	416	–	Boat	Mo	–	–
Rhenium	Re	3180	21.04	.14	1928	2571	Poor	–	–	3.18 @ .59	Fine wire will self-evaporate.
Rhenium Oxide	Re_2O_7	297	6.10	–	–	100	–	Boat	Mo	–	–
Rhodium	Rh	1966	12.41	.24	1272	1707	Good	Coil	W	1.62 @ .55	EB gun preferred.
Rubidium	Rb	38	1.53	2.54	–3	111	–	Boat	Mo	1.03 @ .25	–
Rubidium Chloride	RbCl	715	2.80	–	–	550	–	Boat	Mo	1.49	–
Rubidium Iodide	RbI	641	3.59	–	–	400	–	Boat	Mo	1.68 @ .58	–
Ruthenium	Ru	2500	12.45	.20	1780	2260	Poor	Boat	W	–	Spits violently in EB. Requires long degas.
Samarium	Sm	1072	7.54	.91	373	573	Good	Boat	Ta	–	–
Samarium Oxide	Sm_2O_3	2350	8.35	–	–	–	Good	Boat	Ir	1.97	Loses O_2. Films smooth, clear.
Samarium Sulfide	Sm_2S_3	1900	5.72	–	–	–	Good	–	–	–	A.IP Conf.Proc. on Mag.& Mag. Mat.B, 5,860 (1971).
Scandium	Sc	1539	2.99	.91	714	1002	Xlnt.	Boat	W	–	Alloys with Ta. Flammable.
Scandium Fluoride	ScF_3	1550	2.50	–	–	1400	Good	Boat	.02 W	–	–
Scandium Oxide	Sc_2O_3	2300	3.86	–	–	400	Fair	Boat	Ta	1.88 @ .55	Loses O_2.
Selenium	Se	217	4.79	.87	89	170	Xlnt.	Boat	Mo	1.88 @ .24 2.43 @ 2.36	Very toxic. Poisons vacuum systems. JVST 9, 387 (1972) JVST 12, 573 & 807 (1975)
Silicon	Si	1410	2.33	.88	992	1337	Fair	Boat	.04 W	3.49 @ 1.4 3.42 @ 32	Alloys with W; Some SiO produced above $4x10^{-6}$ Torr. App.Opt. 15,2348 (1976).
Silicon Boride	SiB_4	1870	2.47	–	–	–	Poor	–	–	–	–
Silicon Carbide	SiC	2700	3.22	–	–	1000	–	–	–	2.62 @ .69 6.86 @ 13	Sputtering preferred.
Silicon Dioxide	SiO_2	1610 –1710	2.20 –2.70	1.00 influenced by composition	–	1025	Xlnt.	–	–	1.47 @ .30 1.45 @ 2.8	Quartz xlnt. in EB.
Silicon (II) Monoxide	SiO	1702	2.1	.50	–	850	Poor	Box	Ta	2.10 @ .10 1.67 @ 6 2.75 @ 11	Sublimes. Baffle box source best.

Table A16.1. (continued).

Name	Symbol	Melting point (°C)	Bulk density (g/cm³)	Acoustic impedance ratio (z)	Temperature (°C) vap. press. (10^{-8} torr)	(10^{-4} torr)	Evaporation techniques: Electron beam	Resistance: Source	Resistance: Material	Index of refraction @ microns	Remarks
Silicon Nitride	Si_3N_4	1900	3.44	–	–	800	–	Boat	Mo	2.0 @ .12 2.05 @ 4	Sublimes.
Silicon Selenide	SiSe	–	–	–	–	550	–	Boat	Mo	–	Toxic.
Silicon Sulfide	SiS	subl.	1.85	–	–	450	–	Boat	Mo	–	–
Silicon Telluride	$SiTe_2$	–	4.39	–	–	550	–	Boat	Mo	–	Toxic.
Silver	Ag	961	10.49	.50	847	1105	Xlnt.	Coil Boat	W Mo	1.2 @ .30 14.5 @ 12	Evaporates well from any source.
Silver Bromide	AgBr	431	6.49	1.18	–	380	–	Boat	Mo	2.28 @ .58	–
Silver Chloride	AgCl	455	5.56	1.41	–	520	–	Boat	Mo	2.13 @ .43 1.91 @ 19	–
Silver Iodide	AgI	558	6.01	–	–	500	–	Boat	Ta	2.02 @ .59	–
Sodium	Na	97	0.97	4.8	74	192	–	Boat	Mo	.03 @ .59	Metal reacts violently in air.
Sodium Bromide	NaBr	755	3.20	–	–	400	–	Boat	Mo	2.12 @ .21 1.64 @ .59	Use gentle preheat to outgas.
Sodium Chloride	NaCl	801	2.16	1.57	–	530	Good	Boat	Mo	1.79 @ .20 1.20 @ 27	Little decomposition. Use gentle preheat to outgas. Hygroscopic.
Sodium Cyanide	NaCN	563	–	–	–	550	–	Boat	Mo	1.45 @ .59	Toxic. Use gentle preheat to outgas.
Sodium Fluoride	NaF	988	2.56	–	–	700	Good	Boat	Mo	1.39 @ .19 1.25 @ 23	Use gentle preheat to outgas.
Sodium Hydroxide	NaOH	318	2.13	–	–	470	–	Boat	Pt	1.36	Melt in air to outgas. Deliquescent.
Sodium Iodide	NaI	651	3.67	–	–	700	–	Boat	Mo	1.80 @ .49 1.76 @ .66	–
Spinel	MgO_3• $5Al_2O_3$	–	8.0	–	–	–	Good	Boat	Ta	1.72 @ .66	–
Strontium	Sr	769	2.6	–	239	403	Poor	Boat	Mo	.61 @ .58	Toxic. Wets but does not alloy with refractory metal May react violently in air.
Strontium Fluoride	SrF_2	1450	4.24	–	–	1000	–	Boat	Ta	1.44 @ 1.4	–
Strontium Oxide	SrO	2461	4.9	–	–	1500	–	Boat	Ta	1.88 @ .58	Sublimes. Reacts with Mo and W.
Strontium Sulfide	SrS	>2000	3.70	–	–	–	–	–	–	2.11 @ .59	Decomposes.
Sulfur	S	115	2.07	2.29	13	57	Poor	Boat	Mo	–	Toxic. Poisons vacuum system.
Supermalloy	Ni/Fe/Mo	1410	8.9	–	–	–	Good	–	–	–	Sputtering preferred; or co-evap. from 2 sources: Permalloy and Mo.

Inficon Z-Ratio° = acoustic impedance ratio, z.
Z-Ratio° Leybold Inficon

Maxtec Inc. Acoustic Impedance (A. I.) = 8.83 ÷ z

Table A16.1. (continued).

Name	Symbol	Melting point (°C)	Bulk density (g/cm^3)	Acoustic impedance ratio (z)	Temperature (°C) vap. press. (10^{-8} torr)	(10^{-4} torr)	Evaporation techniques Electron beam	Resistance Source	Material	Index of refraction @ microns	Remarks
Tantalum	Ta	2996	16.6	.26	1960	2590	Xlnt.	–	–	2.05 @ .58	Forms good films.
Tantalum Boride	TaB_2	3000	11.15	–	–	–	–	–	–	–	–
Tantalum Carbide	TaC	3880	14.65	–	–	2500	–	–	–	–	JVST 12, 811 (1975).
Tantalum Nitride	TaN	3360	16.30	–	–	–	–	–	–	–	Reactive; evaporate Ta in 10^{-3} N_2.
Tantalum Oxide	Ta_2O_5	1800	8.74	.30	1550	1920	Good	Boat	Ta	2.28 @ .40 2.0 @ 1.5	Slight decomposition; evap. in 10^{-3} Torr of O_2. App. Opt. 19, 1737 (1980).
Tantalum Sulfide	TaS_2	1300	–	–	–	–	–	–	–	–	–
Technetium	Tc	2200	11.5	–	1570	2090	–	–	–	–	–
Teflon	PTFE	330	2.9	–	–	–	–	Box	Mo	–	Baffled source. Film structure doubtful.
Tellurium	Te	452	6.25	.53	157	277	Poor	Boat	Mo	4.9 @ 6.0	Toxic. Wets w/o alloying.
Terbium	Tb	1357	8.27	.64	800	1150	Xlnt.	Boat	Ta	–	–
Terbium Fluoride	TbF_3	1176	–	–	800	–	–	Boat	Mo	–	–
Terbium Oxide	Tb_2O_3	2387	7.87	–	–	1300	–	Boat	Ta	–	Partially decomposes.
Terbium Oxide	Tb_4O_7	–	–	–	–	–	–	Boat	Ta	–	Films TbO.
Thallium	Tl	301	11.89	1.58	280	480	Poor	Boat	Mo	–	Wets freely, very toxic.
Thallium Bromide	TlBr	480	7.56	1.77	–	200	–	Boat	Mo	2.65 @ .44 2.32 @ 24	Toxic, sublimes.
Thalium Chloride	TlCl	430	7.00	1.21	–	150	–	Boat	Mo	2.20 @ .75 2.6 @ 12	Toxic, sublimes.
Thallium Iodide (B)	TlI	440	7.09	–	–	200	–	Boat	Mo	2.78 @ .75	Toxic, sublimes.
Thallium Oxide	Tl_2O_3	717	9.65	–	–	350	–	Boat	Mo	–	Toxic. Goes to Tl_2 at 850°C.
Thorium	Th	1875	11.7	.54	1430	1925	Xlnt.	Boat	W	–	Toxic, radioactive.
Thorium Bromide	$ThBr_4$	–	5.67	–	–	–	–	Boat	Mo	2.47 @ 5	Radioactive, sublimes.
Thorium Carbide	ThC_2	2773	8.96	–	–	2300	–	–	–	–	Radioactive.
Thorium Fluoride	ThF_4	900	6.32	.74	–	750	Fair	Boat	Ta	1.52 @ .40 1.25 @ 12	Radioactive. Heat substrate to above 150°C. JVST 12,919 (1975).

Table A16.1. (continued).

Name	Symbol	Melting point (°C)	Bulk density (g/cm^3)	Acoustic impedance ratio (z)	Temperature (°C) vap. press. (10^{-8} torr)	Temperature (°C) vap. press. (10^{-4} torr)	Evaporation techniques: Electron beam	Evaporation techniques: Resistance Source	Evaporation techniques: Resistance Material	Index of refraction @ microns	Remarks
Thorium Oxide	ThO_2	3050	9.86	–	–	2100	Good	Boat	.02 W	1.8 @ .55 1.75 @ 2.0	Radioactive.
Thorium Oxyfluoride	$ThOF_2$	900	9.1	–	–	–	–	Boat	Mo	1.52	Radioactive. Films often ThF_4.
Thorium Sulfide	ThS_2	1925	6.80	–	–	–	–	–	–	–	Radioactive. Sputtering preferred; or co-evaporate from 2 sources.
Thulium	Tm	1545	9.32	.52	461	680	Good	Boat	Ta	–	Sublimes.
Thulium Oxide	Tm_2O_3	–	8.90	–	–	1500	–	Boat	Ir	–	Decomposes.
Tin	Sn	232	7.29	.74	682	997	Xlnt.	Coil Boat	W Mo	1.48 @ .59	Wets Mo; use Ta liner in EB guns.
Tin Oxide	SnO_2	1131	6.99	–	–	1000	Xlnt.	Boat	W	2.08 @ .58	Films from W oxygen deficient, oxidize in air.
Tin Selenide	SnSe	861	6.18	–	–	400	Good	Boat	Mo	–	JVST 12, 110 (1975).
Tin Sulfide	SnS	882	5.22	–	–	450	–	Boat	Mo	–	–
Tin Telluride	SnTe	780	6.44	–	–	450	–	Boat	Mo	–	–
Titanium	Ti	1668	4.50	.63	1067	1453	Xlnt.	Boat	Ta	2.04 @ .45	Alloys with refractory metals; evolves gas on first heating.
Titanium Boride	TiB_2	2900	4.50	–	–	–	Poor	–	–	–	–
Titanium Carbide	TiC	3140	4.93	–	–	2300	–	–	–	–	JVST 12, 851 (1975).
Titanium Oxide (IV) (rutile)	TiO_2	1840	4.26	.40	–	1300	Fair	Boat	.02 W	2.55 @ .38 2.30 @ 1.0	Evaporate in 10^{-4} of O_2 onto 350°C substrates. App.Opt.15,2986 (1976).
Titanium (II) Oxide	TiO	1700	4.95	–	–	1500	Good	Boat	.02 W	2.2	Use gentle preheat to outgas. Films TiO_2 if evaporated like TiO_2.
Titanium Nitride	TiN	2930	5.43	–	–	–	Good	Boat	Mo	–	Sputter preferred. Decomposes with thermal evaporation.
Titanium Silicide	$TiSi_2$	1540	4.39	–	–	–	–	Boat	W	–	–
Tungsten	W	3387	19.3	.16	2117	2757	Good	–	–	2.76 @ .58	Forms volatile oxides. Films hard and adherent.
Tungsten Boride	WB_2	2900	12.75	–	–	–	Poor	–	–	–	–
Tungsten Carbide	W_2C	267	15.6	.18	1480	2120	Xlnt.	–	–	–	–
Tungsten Oxide	WO_3	1473	7.16	–	–	980	Good	Boat	W	2.0 @ .5 2.0 @ 2.0	Sublimes Preheat to outgas. W reduces oxide slightly. App. OPT 28, 1497.
Tungsten Selenide	WSe_2	2150	9.0	–	–	–	–	–	–	–	–

Inficon Z-Ratio° = acoustic impedance ratio, z.
Z-Ratio° Leybold Inficon

Maxtec Inc. Acoustic Impedance (A. I.) = 8.83 ÷ z

Table A16.1. (continued).

Name	Symbol	Melting point (°C)	Bulk density (g/cm³)	Acoustic impedance ratio (z)	Temperature (°C) vap. press. (10^{-8} torr)	Temperature (°C) vap. press. (10^{-4} torr)	Evaporation techniques: Electron beam	Evaporation techniques: Resistance Source	Evaporation techniques: Resistance Material	Index of refraction @ microns	Remarks
Tungsten Silicide	WSi_2	2165	9.4	–	–	–	–	–	–	–	–
Tungsten Sulfide	WS_2	1250	7.51	–	–	–	–	–	–	–	–
Tungsten Telluride	WTe_3	–	9.49	–	–	–	–	–	–	–	–
Uranium	U	1132	19.07	.24	1132	1582	Good	Boat	.02 W	–	Films oxidize.
Uranium Carbide	UC_2	2260	11.28	–	–	2100	–	–	–	–	Decomposes.
Uranium Fluoride	UF_4	1000	–	–	–	300	–	Boat	Mo	–	–
Uranium (IV) Oxide	UO_2	2500	10.9	–	–	–	–	Boat	W	–	Ta causes decomposition.
Uranium Oxide	U_3O_3	dec	8.30	–	–	–	–	Boat	W	–	Decomposes at 1300°C to UO_2.
Uranium Phosphide	UP_2	–	8.57	–	–	1200	–	Boat	Ta	–	Decomposes.
Uranium Sulfide	U_2S_3	–	–	–	–	1400	–	Boat	W	–	Slight decomposition.
Vanadium	V	1890	6.11	.53	1890	1547	Xlnt.	Boat	.02 W	3.03 @ .58	Wets Mo. EB evaporated films preferred.
Vanadium Boride	VB_2	2400	5.10	–	–	–	–	–	–	–	–
Vanadium Carbide	VC	2810	5.77	–	–	1800	–	–	–	–	–
Vanadium Nitride	VN	2320	6.13	–	–	–	–	–	–	–	–
Vanadium (IV) Oxide	VO_2	1967	4.34	–	–	575	–	Boat	W	2.51 @ .63 2.76 @ 3.4	Sublimes. Deposit V metal @ $7x10^3$ O_2 JVST A2(2),301 (1984) & A7(3),1310 (1989).
Vanadium (V) Oxide	V_2O_5	690	3.36	–	–	500	–	–	–	–	–
Vanadium Silicide	VSi_2	1700	4.42	–	–	–	–	–	–	–	–
Ytterbium	Yb	824	6.98	1.27	520	690	Good	Boat	Ta	–	Sublimes.
Ytterbium Fluoride	YbF_3	1161	8.19	–	–	780	–	Boat	Mo	1.48 @ 2.2 1.32 @ 14	–
Ytterbium Oxide	Yb_2O_3	2227	9.17	–	–	1500	–	–	–	–	Sublimes. Loses oxygen.
Yttrium	Y	1509	4.47	.82	830	1157	Xlnt.	Boat	W	–	High Ta solubility.
Yttrium Aluminum Oxide	$Y_3Al_5O_{12}$	1990	–	–	–	–	Good	–	–	–	Films not ferroelectric.

Table A16.1. (continued).

Name	Symbol	Melting point (°C)	Bulk density (g/cm³)	Acoustic impedance ratio (z)	Temperature (°C) vap. press.		Evaporation techniques			Index of refraction @ microns	Remarks
					(10^{-8} torr)	(10^{-4} torr)	Electron beam	Resistance Source	Resistance Material		
Yttrium Fluoride	YF_3	1152	4.01	–	–	–	–	–	–	1.46 @ 2.5 1.42 @ 10	–
Yttrium Oxide	Y_2O_3	2410	5.01	–	–	2000	Good	Boat	.02 W	1.79 @ 1	Sublimes. Loses oxygen, films smooth and clear.
Zinc	Zn	419	7.14	.50	127	250	Xlnt.	Boat Coil	Mo W	2.62 @ .69	Evaporates well under wide range of conditions. Use Mo E.B. liner.
Zinc Antimonide	Zn_3Sb_2	570	6.33	–	–	–	–	–	–	–	–
Zinc Bromide	$ZnBr_2$	391	4.99	–	–	300	–	Boat	Mo	1.58 @ .58	Decomposes.
Zinc Fluoride	ZnF_2	872	4.95	–	–	800	–	Boat	Ta	–	–
Zinc Nitride	Zn_3N_2	–	6.22	–	–	–	–	Boat	Mo	–	Decomposes.
Zinc Oxide	ZnO	1975	5.60	.55	–	1800	Fair	–	–	2.02 @ .59	Anneal in air at 450°C to reoxidize. JVST 12,879 (1975).
Zinc Selenide	ZnSe	1526	5.42	.72	–	660	–	Boat	Mo	2.61 @ .40 2.43 @ 18	Toxic. Use gentle preheat to outgas. Sublimes well. Z.Angew.Phys.19,392 (1965).
Zinc Sulfide	ZnS	1700	4.10	.77	–	800	Good	Boat	Mo	2.35 @ .55 2.60 @ 4.0 2.13 @ 13	Sublimes Gentle preheat rqd Sticking coeff varies with sub temp. JVST 6,433 (1969)
Zinc Telluride	ZnTe	1240	6.34	–	–	600	–	Boat	Mo	3.56 @ .59 2.80 @ 8.0	Toxic, sublimes. Use gentle preheat to outgas.
Zircon	$ZrSiO_4$	2550	4.56	–	–	–	–	–	–	1.96 @ .59	–
Zirconium	Zr	1851	6.39	.58	1748	1987	Xlnt.	Boat	.02 W	–	Flammable. Alloys with W. Films oxidize readily.
Zirconium Bromide	ZrB_2	3000	6.09	–	–	–	Good	–	–	–	–
Zirconium Carbide	ZrC	3540	6.73	–	–	2500	–	–	–	–	–
Zirconium Nitride	ZrN	2980	7.09	–	–	–	–	–	–	–	Reactively evaporate in 10^{-3} N_2 atmosphere.
Zirconium Oxide	ZrO_2	2715	5.49	–	–	2200	Good	Boat	W	2.05 @ .50 2.0 @ 2	Films oxygen deficient, clear and hard.
Zirconium Silicide	$ZrSi_2$	1790	4.88	–	–	–	–	–	–	–	–

Inficon Z-Ratio° = acoustic impedance ratio, z.
Z-Ratio° Leybold Inficon

Maxtec Inc. Acoustic Impedance (A. I.) = 8.83 ÷ z

APPENDIX

17

ACCELERATOR ENERGY CALIBRATION AND STABILITY

Compiled by

J. R. Tesmer

Los Alamos National Laboratory, Los Alamos, New Mexico

CONTENTS

A17.1 INTRODUCTION

The precision needed in energy calibration for ion beam materials analysis is generally not as high as that needed in low-energy nuclear physics (± a few keV). In fact, it has been stated by the noted materials scientist, B. Manfred Ullrich, in his Second Law (Ullrich, 1988), "No one cares about the beam energy!" This observation is substantiated by the large number of relative measurements made, and by the slow energy variation of stopping powers and Coulomb cross sections. For measurements in the Coulomb regime, energies within several 10s of keV of the actual value are usually acceptable. What is expected, however, is that the energy be at the same value for each sample measured. This is not always easy to accomplish. Some accelerator systems cannot maintain day-to-day calibrations because of their compact size and lack of high resolution energy analyzing systems.

In practice, both absolute energy and beam energy spread can be important for many measurements. These are particularly important for many resonance reactions. Most of today's accelerators have adequate energy spread, i.e., certainly the beam energy spread is much less than the detector resolution and in most cases approaches that of the narrower resonances. Hence the problems are (1) calibration of the beam energy, (2) day-to-day consistency of that energy, and (3) measurement of the energy spread.

A17.2 ABSOLUTE ENERGY CALIBRATION

For absolute energy calibration, the reactions in Table A17.1 are well known and extensively used. There are, however, other reactions which are not listed which may be useful in specific cases. The reader is referred to the most recent references or to J.W. Butler's section (4.9.5) in Mayer and Rimini (1977) for guidance to other reactions. Where possible the references refer to compilations of data which combine several measurements to produce a best value. For very accurate energy calibrations, it is usually not enough to calibrate the energy at just one point. The non-linearity in the analyzing magnet as well as the variability of other accelerator components may make it necessary to calibrate over a range of energies, and, possibly over a range of ions that are used (see Overley *et al.*, 1968). Hysteresis effects in the analyzing magnet make the procedure for setting of its field very critical if great precision is needed. There are other effects as well; hence, the article by J.B. Marion (1966) is a good starting point for those interested in the details of energy calibration.

A17.3 ENERGY CALIBRATION STANDARDS

A17.3.1 Gamma resonances

Either NaI(Tl) or BGO scintillation detectors are the detectors of choice for high-energy gamma rays which are produced by the (p, γ) or (α, γ) reactions (see Chap. 11 for a discussion of these detectors). Normally the detectors used are as large as you can afford and placed very close to the target. Generally, a thick target is much easier to find and is sufficient for most reactions. The gamma yield is measured as a function of energy (the resulting curve can be fitted by a step function). The energy of the resonance corresponds to the mid-point on the yield curve halfway between the 12% and 88% height of the net yield. The energy difference between these points is the energy spread of the beam assuming that the resonance width is much narrower than the energy spread and that the resonance and energy spread of the beam are Gaussian. In many cases the wider resonances have Lorentzian (Breit-Wigner) shapes (see Chap. 3, Sec. 3.3.7). The convolution of Gaussians with Lorentzians is discussed in Chapters 7 and 8.

A17.3.2 Neutron thresholds

For neutron threshold reactions a neutron detector such as a Long Counter (a BF_3 proportional counter surrounded by a neutron moderator) or a plastic scintillator is used. However, as the energies of the thresholds increase there is an increase in backgrounds so that at the higher energies it is common to measure the positrons produced by the reactions in the targets (or, rather, their annihilation radiation) instead of the neutrons (for example, see Overley *et al.*, 1968). The net neutron yield (neutrons minus background) to the 2/3 power is plotted vs. energy and a linear fit to the net yield is extrapolated to zero yield, which is the threshold energy. It is usually the practice to stay within several keV of the threshold to obtain the best straight-line fit (see Overley *et al.*, 1968; Roush *et al.*, 1970).

Table A17.1. Energy calibrating reactions.

Energy (keV)	Detected radiation [energy (MeV)]	Width (keV)	Reaction	Reference
Protons				
340.46 ± 0.04	γ (7.12, 6.92, 6.13), α	2.34 ± 0.04	$^{19}F(p,\alpha\gamma)^{16}O$	Aj, 1983; Uhrmacher *et al.*, 1985; Marion, 1966
872.11 ± 0.20	γ (7.12, 6.92, 6.13)	4.7 ± 0.2	$^{19}F(p,\alpha\gamma)^{16}O$	Marion, 1966
991.88 ± 0.04	γ (10.78, 7.93, 1.77)	0.1 ± 0.02	$^{27}Al(p,\gamma)^{28}Si$	Roush *et al.*, 1970; Marion, 1966
1747.6 ± 0.9	γ (9.17, 6.43, 2.74)	0.077 ± 0.012	$^{13}C(p,\gamma)^{14}N$	Marion, 1966
1880.42 ± 0.05	n, threshold	—	$^{7}Li(p,n)^{7}Be$	Barker *et al.*, 1984
3235.7 ± 0.07	n, threshold	—	$^{13}C(p,n)^{13}N$	Marion, 1966
4234.3 ± 0.80	n, threshold	—	$^{19}F(p,n)^{19}Ne$	Marion, 1966
5803.3 ± 0.26	n, threshold	—	$^{27}Al(p,n)^{27}Si$	Naylor and White, 1977
14230.75 ± 0.02	p	1.2	$^{12}C(p,p)^{12}C$	Huenges *et al.*, 1973
Deuterons				
1829.2 ± 0.6	n, threshold	—	$^{15}O(d,n)^{17}F$	Bondelid *et al.*, 1960
^{3}He				
1437.9 ± 0.6	n, threshold	—	$^{12}C(^{3}He,n)^{14}O$	Roush *et al.*, 1970
Alphas				
2435.1 ± 0.3	γ (12.1, 10.3)	<0.25	$^{24}Mg(\alpha,\gamma)^{28}Si$	Endt and Van der Leun, 1978; Maas *et al.*, 1978
2865.8 ± 0.3	γ (12.4)	<1	$^{24}Mg(\alpha,\gamma)^{28}Si$	Endt and Van der Leun, 1978; Maas *et al.*, 1978
3035.9 ± 2.3	α	8.1 ± 0.3	$^{16}O(\alpha,\alpha)$	MacArthur *et al.*, 1980; Leavitt *et al.*, 1990
3198.3 ± 0.3	γ (12.7, 10.9)	0.76 ± 0.17	$^{24}Mg(\alpha,\gamma)^{28}Si$	Endt and Van der Leun, 1978; Maas *et al.*, 1978
3363 ± 5	α	5	$^{16}O(\alpha,\alpha)$	Leavitt *et al.*, 1990
3576 ± 4	α	<4	$^{14}N(\alpha,\alpha)$	Herring, 1958
3877 ± 5	α	2	$^{16}O(\alpha,\alpha)$	Leavitt *et al.*, 1990
4265 ± 5	α	27 ± 3	$^{12}C(\alpha,\alpha)$	Kettner *et al.*, 1982; Aj, 1986; Leavitt *et al.*, 1989
5058 ± 3	α	0.11 ± .02	$^{16}O(\alpha,\alpha)$	Häusser *et al.*, 1972; Aj, 1983

A17.3.3 Elastic backscattering

Commonly, backscattering is used more than other analysis methods and the resonances for the (p,p) and (α,α) reactions listed in Table A17.1 can be easily used for quick, fairly accurate energy calibrations. Thin films of carbon and oxygen can be readily found on most samples.

A17.3.4 Geometry

Care must be taken when detecting either neutrons or gammas to place the detector at an angle where the particles of interest will be emitted. For neutron-threshold reactions the detector should be at 0° scattering angle and as close as possible to the target. For gammas, 0° scattering angle will also work if the solid angle of the detector approaches 2π. For smaller solid angles, let the experimenter beware, the distribution of the emitted gamma must be taken into account. Angles which will always work (but may not be optimum for a given reaction) are 55° and 125° (see Chap. 7, Sec. 7.1.2.3). A great deal of time can be wasted by placing your detector at the wrong angle! Check the literature to determine the exact angular distribution of the gamma rays for the chosen reaction if sensitivity is a problem.

The detector angle for the resonances in the particle-in particle-out reactions vary, but backscattering angles are generally used.

A17.3.5 Other calibration methods

Another method is to place a long-lived alpha source in the chamber and have the alphas from this source appear in the spectrum along with your data. An extensive list of alpha sources is given in Table 11.9 of Chap. 11. Determining the positions of this peak along with two other known points will not only give the detector system gain but, in theory, also the energy of the beam if the scattering geometry is known. In fact one of the popular analysis programs, RUMP (Doolittle, 1985; Doolittle, 1986; Doolittle, 1990) has this procedure built in. Care must be taken if this method is used to accurately calibrate the beam energy because of several problems relating to sources and detector windows. See Chap. 12 and Lennard *et al.* (1990). Corrections for the dead layer, the pulse-height defect, and the non-linearity of the surface barrier detector must be taken into account as well as surface contamination and aging of the source for accurate beam energy determination. In some cases an exothermic reaction has been used to replace the alpha source (Scott and Paine, 1983).

Another method which uses a backscattering or "cross-over" technique relies on kinematics and the determination of the angle where particles scattered from different elements or different reactions in the target have the same energy. Normally, this is applied to beam energies above a few MeV (see Birattari *et al.*, 1992).

A17.4 CONSISTENCY

The problem of achieving day-to-day consistency can be solved in several ways. For backscattering analysis, one very useful reaction for energy calibration as well as a consistency check is the $^{16}O(\alpha,\alpha)$ (3.036 MeV). This has the lowest energy of any easily obtained alpha reactions, and, therefore, is useful for day-to-day checks of the beam energy. The resonance energy is sufficiently well known and has a large cross section. A thin oxide layer can be found in almost any scattering chamber for use in quickly calibrating the beam energy. This resonance is not Gaussian and the peak height of the resonance should be used to obtain the calibration energy, i.e., for thick targets the energy that produces the largest height (not integrated area) determines the resonance energy. This is one of the preferred reactions for day-to-day calibrations at normal RBS energies.

Another very useful technique to assure consistency is to prepare a calibration standard such as a thin layer (<1 nm) of gold on silicon dioxide. One then checks to make sure the particles scattered from the gold and the Si (edge) appear in the same channels as in previous measurements before new data is taken. This method should be used with caution since the effect of the detector-system gain shift or zero-energy offset can be overcome by a change in beam energy.

If depth profiling with (p,γ) resonances is the analysis technique, then consistency is easily obtained by observing at what energy the resonance of interest is first generated in the sample.

REFERENCES

Ajzenberg-Selove, F. (1983), *Nucl. Phys.* **A392** , 1.

Ajzenberg-Selove, F. (1986), *Nucl. Phys.* **A460**, 1.

Biratarri, C., Castiglioni, M., and Silari, M. (1992), *Nucl. Instrum. Methods* **A320,** 413.

Bondelid, R.O., Butler, J.W., and Kennedy, C.A. (1960), *Phys. Rev.* **120**, 889.

Doolittle, L.R. (1985), *Nucl. Instrum. Methods* **B9,** 344.

Doolittle, L.R. (1986), *Nucl. Instrum. Methods* **B15**, 227.

Doolittle, L.R. (1990), in *High Energy and Heavy Ion Beams in Materials Analysis,* J.R. Tesmer, C.J. Maggiore, M. Nastasi, J.C. Barbour, and J.W. Mayer, (eds.), Materials Research Society, Pittsburgh, Pennsylvania, p. 175.

Endt, P.M., and Van der Leun, C. (1978), *Nucl. Phys.* **A310**, 1.

Häusser, O., Ferguson, A.J., McDonald, A.B., Szöghy, I.M., Alexander, T.K., and Disdier, D.L. (1972), *Nucl. Phys.* **A179**, 465.

Herring, D.F. (1958), *Phys. Rev.* **112**, 1217.

Huenges, E., Rösler, H., and Vonach, H. (1973), *Phys. Lett.* **B46**, 361.

Kettner, K.U., Becker, H.W., Buchmann, L., Gorres, J., Kräwinkel, H., Rolfs, C. Schmalbrock, P., Trautvetter, H.P., and Vlieks, A. (1982), *Z. Phys.* **A308**, 73.

Lennard, W.N., Tong, S.Y., Massoumi, G.R., and Wong, L. (1990), *Nucl. Instrum. Methods* **B45,** 281.

Leavitt, J.A., McIntyre, Jr., L.C., Stoss, P., Oder, J.G., Ashbaugh, M.D., Dezfouly-Arjomandy, B., Yang, Z.M., and Lin, Z. (1989), *Nucl. Instrum. Methods* **B40/41** 776.

Leavitt, J.A., McIntyre, Jr., Asbaugh, M.D., Oder, J.G., Lin, Z., and Dezfouly-Arjomandy, B. (1990), *Nucl. Instrum. Methods* **B44,** 260.

Marion, J.B. (1966), *Rev. Mod. Phys.* **38**, 660.

Maas, J.W., Somorai, E., Graber, H.D., Van den Wijngaart, C.A., Van der Leun, C., and Endt, P.M. (1978), *Nucl. Phys.* **A301**, 213.

Naylor, H., and White, R.E. (1977), *Nucl. Instrum. Methods* **144,** 331.

Overley, J.C., Parker, P.D., and Bromley, D.A. (1968), *Nucl. Instrum. Methods* **68**, 61.

Roush, M.L., West, L.A., and Marion, J.B. (1970), *Nucl. Phys.* **A147**, 235.

Scott, D.M., and Paine, B.M. (1983), *Nucl. Instrum. Methods* **218,** 154.

Uhrmacher, M., Pampus, K., Bergmeister, F.J., Purschke, D., and Lieb, K.P., (1985), *Nucl. Instrum. Methods* **B9**, 234.

APPENDIX 18

RADIATION HAZARDS OF (α,n) REACTIONS

S. N. Basu[1], M. Nastasi, and J. R. Tesmer

Los Alamos National Laboratory, Los Alamos, New Mexico

CONTENTS

Reprinted from *High Energy and Heavy Ion Beams in Materials Analysis*, (1990), J.R. Tesmer, C.J. Maggiore, M. Nastasi, J.C. Barbour, and J.W. Mayer (eds.), Materials Research Society, Pittsburgh, Pennsylvania, p. 279.

1. Current address: Boston University, Boston, Massachusetts.

A18.1 INTRODUCTION

The analysis of materials using α irradiation is an established technique. However, the irradiation of low Z elements by energetic α particles can lead to the production of neutrons (Bair and del Campo, 1979; West and Sherwood, 1982; Wilson *et al.*, 1988). These neutrons can ionize body tissues causing health hazards (Cember, 1976). The maximum allowable radiation dosage recommended by the International Commission on Radiological Protection (ICRP) is 5 rem per year (ICRP, 1987). In this paper, it is assumed that an individual would not prefer to be exposed to more than a tenth of this maximum dose. This paper presents the neutron radiation dosage associated with (α,n) reactions that occur when the low Z elements and some compounds are irradiated by α particles having energies up to 10 MeV. The number of hours of radiation exposure a week required to exceed 0.5 rem per year are also presented.

A18.2 CALCULATIONS

The α-induced thick target neutron yields have been measured experimentally (Bair and del Campo, 1979; West and Sherwood, 1982) and calculated theoretically (Wilson *et al.*, 1988). The thick target neutron yields, *Y*, for the lighter atomic number elements are listed in column 3 of Table A18.1 as a function of the energy of the α particles. The neutron yield data for 10 MeV α particles has been obtained by extrapolation of data presented by West and Sherwood (1982), who reported neutron yields for α particles having energies up to 9.9 MeV.

Assuming that the α particles are singly charged and the neutron generation is isotropic in space, the neutron flux, *n*, for a given beam current, *i*, at a distance, *d*, from the target can be calculated as

$$n = \frac{iY}{e4\pi d^2} \qquad \text{(A18.1)}$$

where *e* is the electronic charge. Assuming d = 10 ft ≈ 300 cm, then

$$\frac{n}{i} = 5.52 \times 10^3 Y \text{ neutrons/cm}^2\text{-sec-nA} . \qquad \text{(A18.2)}$$

This flux can be converted to α radiation dose using the appropriate Neutron Flux to Dose Rate Conversion Factor. It should be noted that for a given energy of α particles, neutrons of various energies can be emitted from a target. However, the conversion factors for 1 MeV and 20 MeV neutrons differ by less than a factor of 2 (ANSI, 1977). Since, the conversion factor is not very sensitive to the neutron energy, the value of the conversion factor for 10 MeV neutrons (2.45×10^{-7} neutrons/cm^2 = 1 rem) has been used for all the calculations. See Chapter 13 for the actual values. The neutron radiation dose in rem/nA-hour, calculated at a distance of 10 feet from the target, is listed in column 4 of Table A18.1.

The maximum allowable ionizing radiation dose is 5 rem/year. It is assumed that an individual would prefer not to be exposed to more than a tenth of this maximum dose. Assuming a beam current of 100 nA, a calculation has been made for the number of hours a week a person has to be exposed to exceed 0.5 rem/year. The results of this calculation is listed in column 5 of Table A18.1. The "-" in this column indicates that the number of hours per week exceeds 40 and the radiation dosage associated such experiments is not very significant. The cases where the number of hours per week is less than 4 (length of a typical ion beam experimental session) have been printed in bold letters, indicating that extra caution is required.

It should be noted that the radiation dosage is proportional to the beam current and inversely proportional to the square of the distance. Thus, the values presented can be adjusted for any beam current and distance. Also, if the α particles are doubly charged, the radiation dosage is cut by half.

A18.3 RADIATION YIELD FROM COMPOUNDS

Up to now, all the data presented have been for elemental targets. The neutron yield of compounds can be calculated if the individual yields of its constitutive elements are known (West, 1979). The thick target yield of a compound, *Y*, is given as

$$Y = \frac{\sum_j a_j A_j Y_j C_{j1}}{\sum_j a_j A_j C_{j1}} \qquad \text{(A18.3)}$$

Table A18.1. Neutron radiation dose for pure element targets.

Target	MeV	Neutrons/α	Rem/na-hr	Number of hours/week[1]	Reference
Li	4.5	2.80e-8	2.27e-8	-	Bair and del Campo, 1979
	5.0	6.29e-7	5.10e-7	-	Bair and del Campo, 1979
	5.5	2.15e-6	1.41e-7	-	Bair and del Campo, 1979
	6.0	4.87e-6	3.95e-6	25	Bair and del Campo, 1979
	6.5	1.04e-5	8.44e-6	11	Bair and del Campo, 1979
Be	2.0	3.16e-6	2.56e-6	39	Wilson *et al.*, 1988
	2.5	7.73e-6	6.27e-6	16	Wilson *et al.*, 1988
	3.0	9.79e-6	7.94e-6	12	Bair and del Campo, 1979
	3.5	1.28e-5	1.04e-5	10	Bair and del Campo, 1979
	4.0	2.00e-5	1.62e-5	6	Bair and del Campo, 1979
	4.5	**3.33e-5**	**2.70e-5**	**4**	Bair and del Campo, 1979
	5.0	**4.94e-5**	**4.01e-5**	**2**	Bair and del Campo, 1979
	5.5	**7.18e-5**	**5.80e-5**	**1.5**	Bair and del Campo, 1979
	6.0	**1.10e-4**	**8.92e-4**	**1**	West and Sherwood, 1982
	6.5	**1.39e-4**	**1.13e-4**	**1**	West and Sherwood, 1982
	7.0	**1.69e-4**	**1.37e-4**	**0.5**	West and Sherwood, 1982
	7.5	**2.03e-4**	**1.64e-4**	**0.5**	West and Sherwood, 1982
	8.0	**2.42e-4**	**1.96e-4**	**0.5**	West and Sherwood, 1982
	8.5	**2.82e-4**	**2.29e-4**	**<0.5**	West and Sherwood, 1982
	9.0	**3.29e-4**	**2.67e-4**	**<0.5**	West and Sherwood, 1982
	9.5	**3.80e-4**	**3.08e-4**	**<0.5**	West and Sherwood, 1982
	10.0	**4.32e-4***	**3.50e-4**	**<0.5**	West and Sherwood, 1982
B	3.5	3.15e-6	2.55e-6	39	Bair and del Campo, 1979
	4.0	6.24e-6	5.06e-6	20	Bair and del Campo, 1979
	4.5	1.06e-5	8.60e-6	12	Bair and del Campo, 1979
	5.0	1.56e-5	1.26e-5	8	Bair and del Campo, 1979
	5.5	2.06e-5	1.67e-5	6	Bair and del Campo, 1979
	6.0	2.54e-5	2.06e-5	5	Bair and del Campo, 1979
	6.5	**2.99e-5**	**2.43e-5**	**4**	Bair and del Campo, 1979
C	3.0	2.40e-8	1.95e-8	-	Bair and del Campo, 1979
	3.5	4.00e-8	3.24e-8	-	Bair and del Campo, 1979
	4.0	4.20e-8	3.41e-8	-	Bair and del Campo, 1979
	4.5	4.70e-8	3.81e-8	-	Bair and del Campo, 1979
	5.0	6.30e-8	5.11e-8	-	Bair and del Campo, 1979
	5.5	1.10e-7	8.92e-8	-	Bair and del Campo, 1979
	6.0	1.70e-8	1.38e-8	-	Bair and del Campo, 1979
	6.5	2.52e-7	2.04e-7	-	Bair and del Campo, 1979
	7.0	3.43e-8	2.78e-7	-	Bair and del Campo, 1979
	7.5	4.55e-7	3.69e-7	-	West and Sherwood, 1982
	8.0	6.10e-7	4.95e-7	-	West and Sherwood, 1982
	8.5	7.40e-7	6.00e-7	-	West and Sherwood, 1982
	9.0	8.71e-7	7.06e-7	-	West and Sherwood, 1982
	9.5	9.92e-7	8.50e-7	-	West and Sherwood, 1982
	10.0	1.14e-6*	9.25e-7	-	West and Sherwood, 1982
N	6.5	4.86e-7	3.94e-7	-	Wilson *et al.*, 1988
O	3.0	3.00e-9	2.43e-9	-	Wilson *et al.*, 1988
	3.5	6.00e-9	4.87e-9	-	Bair and del Campo, 1979
	4.0	1.40e-8	1.14e-8	-	Bair and del Campo, 1979
	4.5	2.60e-8	2.11e-8	-	Bair and del Campo, 1979
	5.0	4.50e-8	3.65e-8	-	Bair and del Campo, 1979
	5.5	6.80e-8	5.52e-8	-	Bair and del Campo, 1979
	6.0	9.20e-8	7.46e-8	-	Bair and del Campo, 1979
	6.5	1.32e-7	1.07e-7	-	Bair and del Campo, 1979

1. Values represent the number of hours of exposure each week for a year, under the stated conditions (see text), which result in a dose equivalent of 0.5 Rem. Bold-face numbers (4) indicate extra caution is required.

* Neutrons/α data at 10 MeV are an extrapolation of data presented in West and Sherwood (1982).

Table A18.1. Neutron radiation dose for pure element targets (continued).

Target	MeV	Neutrons/α	Rem/na-hr	Number of hours/week[1]	Reference
F	3.5	2.16e-7	1.75e-7	-	Wilson *et al.*, 1988
	4.0	8.79e-7	7.13e-7	-	Bair and del Campo, 1979
	4.5	2.16e-6	1.75e-6	-	Bair and del Campo, 1979
	5.0	4.39e-6	3.56e-6	28	Bair and del Campo, 1979
	5.5	7.75e-6	6.29e-6	16	Bair and del Campo, 1979
	6.0	1.23e-5	9.98e-6	10	Bair and del Campo, 1979
	6.5	1.79e-5	1.45e-5	7	Bair and del Campo, 1979
Ne	3.0	2.55e-7	2.07e-7	-	Wilson *et al.*, 1988
	3.5	5.20e-7	4.20e-7	-	Wilson *et al.*, 1988
	4.0	9.07e-7	7.36e-7	-	Wilson *et al.*, 1988
	4.5	1.42e-6	1.15e-6	-	Wilson *et al.*, 1988
	5.0	2.03e-6	1.65e-6	-	Wilson *et al.*, 1988
	5.5	2.74e-6	2.22e-6	-	Wilson *et al.*, 1988
	6.0	3.58e-6	2.90e-6	34	Wilson *et al.*, 1988
	6.5	4.55e-6	3.69e-6	27	Wilson *et al.*, 1988
Na	4.0	4.30e-8	3.48e-8	-	Wilson *et al.*, 1988
	4.5	4.34e-7	3.50e-7	-	Wilson *et al.*, 1988
	5.0	1.46e-6	1.18e-6	-	Wilson *et al.*, 1988
	5.5	2.53e-6	2.05e-6	-	Wilson *et al.*, 1988
	6.0	6.53e-6	5.30e-6	19	Wilson *et al.*, 1988
	6.5	1.11e-5	9.00e-6	11	Wilson *et al.*, 1988
Mg	4.0	7.70e-8	6.24e-8	-	Bair and del Campo, 1979
	4.5	2.63e-7	2.13e-7	-	Bair and del Campo, 1979
	5.0	6.44e-7	5.22e-7	-	Bair and del Campo, 1979
	5.5	1.26e-6	1.02e-6	-	Bair and del Campo, 1979
	6.0	2.14e-6	1.74e-6	-	Bair and del Campo, 1979
	6.5	3.25e-6	2.64e-6	38	Bair and del Campo, 1979
	7.0	5.01e-6	4.06e-6	25	West and Sherwood, 1982
	7.5	6.92e-6	5.61e-6	18	West and Sherwood, 1982
	8.0	9.15e-6	7.42e-6	13	West and Sherwood, 1982
	8.5	1.16e-5	9.41e-6	11	West and Sherwood, 1982
	9.0	1.45e-5	1.18e-5	8	West and Sherwood, 1982
	9.5	1.75e-5	1.42e-5	7	West and Sherwood, 1982
	10.0	2.09e-5*	1.69e-5	6	West and Sherwood, 1982
Al	4.0	1.69e-8	1.37e-8	-	Bair and del Campo, 1979
	4.5	8.02e-8	6.51e-8	-	Bair and del Campo, 1979
	5.0	2.64e-7	2.14e-7	-	Bair and del Campo, 1979
	5.5	6.97e-7	5.65e-7	-	Bair and del Campo, 1979
	6.0	1.44e-6	1.17e-6	-	Bair and del Campo, 1979
	6.5	2.78e-6	2.25e-6	-	Bair and del Campo, 1979
	7.0	5.00e-6	4.06e-6	25	West and Sherwood, 1982
	7.5	7.72e-6	6.26e-6	16	West and Sherwood, 1982
	8.0	1.10e-5	8.92e-6	11	West and Sherwood, 1982
	8.5	1.50e-5	1.22e-5	8	West and Sherwood, 1982
	9.0	1.96e-5	1.59e-5	6	West and Sherwood, 1982
	9.5	2.50e-5	2.03e-5	5	West and Sherwood, 1982
	10.0	**3.13e-5***	**2.54e-5**	**4**	West and Sherwood, 1982

1. Values represent the number of hours of exposure each week for a year, under the stated conditions (see text), which result in a dose equivalent of 0.5 Rem. Bold-face numbers (4) indicate extra caution is required.

* Neutrons/α data at 10 MeV are an extrapolation of data presented in West and Sherwood (1982).

Table A18.1. Neutron radiation dose for pure element targets (continued).

Target	MeV	Neutrons/α	Rem/na-hr	Number of hours/week[1]	Reference
Si	4.0	3.97e-9	3.22e-9	-	West and Sherwood, 1982
	4.5	1.60e-8	1.30e-8	-	Bair and del Campo, 1979
	5.0.	5.20e-8	4.22e-8	-	Bair and del Campo, 1979
	5.5	1.14e-7	9.25e-8	-	Bair and del Campo, 1979
	6.0	2.31e-7	1.87e-7	-	Bair and del Campo, 1979
	6.5	3.85e-7	3.12e-7	-	Bair and del Campo, 1979
	7.0	6.51e-7	5.28e-7	-	West and Sherwood, 1982
	7.5	9.44e-7	7.66e-7	-	West and Sherwood, 1982
	8.0	1.34e-6	1.09e-6	-	West and Sherwood, 1982
	8.5	1.81e-6	1.47e-6	-	West and Sherwood, 1982
	9.0	2.37e-6	1.92e-6	-	West and Sherwood, 1982
	9.5	2.95e-6	2.39e-6	-	West and Sherwood, 1982
	10.0	3.69e-6*	2.99e-6	33	West and Sherwood, 1982
Fe	5.0	2.05e-10	1.66e-10	-	West and Sherwood, 1982
	5.5	3.35e-10	2.72e-10	-	West and Sherwood, 1982
	6.0	4.22e-9	3.42e-9	-	West and Sherwood, 1982
	6.5	2.07e-8	1.68e-8	-	West and Sherwood, 1982
	7.0	9.00e-8	7.30e-8	-	West and Sherwood, 1982
	7.5	3.10e-7	2.51e-7	-	West and Sherwood, 1982
	8.0	7.63e-7	6.19e-7	-	West and Sherwood, 1982
	8.5	1.61e-6	1.31e-6	-	West and Sherwood, 1982
	9.0	3.14e-6	2.55e-6	39	West and Sherwood, 1982
	9.5	5.72e-6	4.64e-6	22	West and Sherwood, 1982
	10.0	9.19e-6*	7.45e-6	13	West and Sherwood, 1982

1. Values represent the number of hours of exposure each week for a year, under the stated conditions (see text), which result in a dose equivalent of 0.5 Rem. Bold-face numbers (≤4) indicate extra caution is required.

* Neutrons/α data at 10 MeV are an extrapolation of data presented in West and Sherwood (1982).

where Y_j is the thick target yield of the elemental material of atomic weight A_j, which constitutes a_j mole fraction of the compound. The symbol C_{j1} is a constant, defined as

$$C_{j1} = \frac{\left(\frac{1}{\rho}\frac{dE}{dx}\right)_{j,E}}{\left(\frac{1}{\rho}\frac{dE}{dx}\right)_{1,E}} \tag{A18.4}$$

where subscript *1* refers to some chosen reference element among the constituents. The term

$$\left(\frac{1}{\rho}\frac{dE}{dx}\right)_{j,E}$$

is the stopping cross section for α particles of energy E in element j, and can be calculated using the program TRIM (Ziegler *et al.*, 1985). The radiation yields of some compounds are listed in Table A18.2.

Table A18.2. Neutron radiation dose for compound targets.

Target	MeV	Neutrons/α	Rem/na-hr	Number of hours/week[1]	Reference
BN	4.5	4.23e-6	3.43e-6	29	West and Sherwood, 1982
	5.0	6.26e-6	5.08e-6	20	West and Sherwood, 1982
	5.5	9.00e-6	7.30e-6	14	West and Sherwood, 1982
	6.0	1.18e-5	9.57e-6	10	West and Sherwood, 1982
	6.5	1.42e-5	1.15e-5	9	West and Sherwood, 1982
316SS	5.0	7.03e-10	5.70e-10	-	West and Sherwood, 1982
	5.5	1.85e-9	1.50e-9	-	West and Sherwood, 1982
	6.0	8.19e-9	6.64e-9	-	West and Sherwood, 1982
	6.5	3.18e-8	2.58e-8	-	West and Sherwood, 1982
	7.0	1.17e-7	9.49e-8	-	West and Sherwood, 1982
	7.5	3.51e-7	2.85e-7	-	West and Sherwood, 1982
	8.0	8.09e-7	6.56e-7	-	West and Sherwood, 1982
	8.5	1.71e-6	1.39e-6	-	West and Sherwood, 1982
	9.0	3.29e-6	2.67e-6	37	West and Sherwood, 1982
	9.5	5.72e-6	4.64e-6	22	West and Sherwood, 1982
	10.0	9.08e-6*	7.36e-6	14	West and Sherwood, 1982

1. Bold-face number indicates extra caution is required.

* Neutrons/α data at 10 MeV are an extrapolation of data presented in West and Sherwood (1982).

ACKNOWLEDGEMENT

The authors would like to thank Larry Andrews and Richard Olsher for providing the flux-to-dose-rate conversion factors, and M. Bozoian for providing the paper on calculated thick target neutron yields.

REFERENCES

ANSI *(American National Standard, ANSI/ANS-6.1.1)* (1977), "Neutron and gamma-ray flux-to dose-rate factors", 4.

Bair, J.K., and del Campo, J.G. (1979), *Nucl. Sci. Eng.*, 71, 18.

Cember, H. (1976), *Introduction to Health Physics*, Pergamon Press.

ICRP (International Commission on Radiological Protection). (1987), "Radiation - a fact of life", IAEA/PI/A9E 85-00740.

West, D., and Sherwood, A.C. (1982), *Ann. Nucl. Energy*, **9**, 551.

West, D. (1979), *Ann. Nucl. Energy* **6**, 549.

Wilson, W.B., Bozoian, M., and Perry, R.T. (1988), "Calculated Alpha-induced Thick Target Yields and Spectra, with Comparison to Measured Data", *Proc. Int. Conf. on Nuclear Data for Science and Technology,* May 30-June 3, Mito, Japan, p. 1193.

Ziegler, J.F., Biersack, J.P., and Littmark, U. (1985), "The Stopping and Range of lons in Solids", *The Stopping and Ranges of Ions in Matter, Vol. 1,* J.F. Ziegler (ed.), Pergamon, New York.

INDEX

D

E

M

N

P

Q

R

S

T

U

V